WITHDRAWN

Practical Handbook of
SOIL, VADOSE ZONE, and GROUND-WATER CONTAMINATION
Assessment, Prevention, and Remediation

J. Russell Boulding

LEWIS PUBLISHERS
Boca Raton   Ann Arbor   London   Tokyo

Library of Congress Cataloging-in-Publication Data

Boulding, Russell
    Practical handbook of soil, vadose zone and ground water contamination: assessment, prevention and remediation / J. Russell Boulding.
        p. cm.
    Includes bibliographical references and index.
    ISBN 1-56670-051-5
    1. Soil pollution. 2. Soil remediation. 3. Groundwater pollution. 4. Groundwater-Quality-Management. 5. Zone of aeration. I. Title
    TD978.B684   1995
    628.5'5—dc20

                                                                                                                                       94-37622
                                                                                                                                         CIP

    This book contains information obtained from authentic and highly regarded sources. Reprinted material is quoted with permission, and source are indicated. A wide variety of references are listed. Reasonable efforts have been made to publish reliable data and information, but the author and the publisher cannot assume responsibility for the validity of all materials or the consequences of their use.

    Except as noted below, neither this book nor any part may be reproduced or transmitted in any form or by any means, electronic or mechanical, including photocopying, microfilming, and recording, or by any information storage and retrieval system, without prior permission in writing from the publisher. CRC Press, Inc.'s consent does not extent to copying for general distribution, for promotion, for creating new works, or for resale. Specific permission must be obtained in writing from CRC Press for such copying. Permission is hereby granted for copying of worksheets and checklists in Appendix D, provided they are only used by the purchaser of the this book.

    Direct all inquiries to CRC Press, Inc., 2000 Corporate Blvd., N.W., Boca Raton, Florida 33431.

© 1995 by CRC Press, Inc.
Lewis Publishers is an imprint of CRC Press

No claim to original U.S. Government works
International Standard Book Number 1-56670-051-5
Library of Congress Card Number 94-37622
Printed in the United State of America 1 2 3 4 5 6 7 8 9 0
Printed on acid-free paper

Preface

My goal in writing this book has been to create a single, convenient, reference source to be used: (1) as a starting point for obtaining answers to questions that any environmental manager or professional might have about contaminated soil or ground-water, or (2) as a textbook for a college course that focuses on technical aspects of soil and ground-water contamination assessment and management.

Environmental professionals, managers, and regulators without specialized training in soils, geology, geomorphology, vadose-zone and ground-water hydrology, soil and ground-water chemistry and microbiology will find that the four chapters in Part I (Basic Concepts) provide a comprehensive, yet accessible introduction to those topics as they relate to soil and ground water contamination.

Project planning and field personnel involved in investigating contaminated sites will find that the six chapters in Part II (Assessment and Monitoring) provide up-to-date and comprehensive information on methods for characterizing contaminated sites and analyzing site data.

Decision-makers, project planners and other environmental professionals will find that the four chapters in Part III (Prevention and Remediation) provide systematic and practical information for identifying and implementing methods for (1) preventing or minimizing contamination and (2) cleaning up sites that are already contaminated.

Note About Reference Citations in This Handbook

In order to minimize duplication of references in this handbook, references are cited in four ways:

- Major references are grouped into 23 topics which follow index tables at the end of each chapter. A master list of these index tables and bibliographies can be found at the end of the table of contents.

- Citations in the text are given as footnotes, unless they are contained in a reference table, in which case the reference table number is given. For example, Sara (1994/T9-10) indicates that the full citation can be found in Table 9-10.

- Sources for figures and tables in Chapters 1 through 14 are given in Appendix F, which also serves as a master list with the page on which figures and tables can be found. Full citations for references cited in Appendices A through E appear at the end of each appendix.

- ASTM standard methods are cited by the ASTM designation number in the text. Full citations for ASTM test methods can be found in Appendix A (Table A-14).

Where government documents are cited, the place where documents can be obtained is given, or, if the document is available from the National Technical Information Service, the NTIS acquisition number is given. U.S. EPA's Center for Environmental Research Information

(CERI) in Cincinnati is the best place to contact for obtaining EPA documents (ORD Publications, PO Box 19963, Cincinnati, OH 45268-0963; 513/569-7562). If CERI does not have the document, it will refer the request to the appropriate EPA office, or provide the NTIS acquisition number. Table 5-3 identifies phone numbers for other EPA hotlines and information sources. Documents available from NTIS can be obtained by calling 800/553-6847 or writing to: National Technical Information Service, Springfield, VA 22161.

Note About Ground Water Hyphenation Conventions

There are few words in the environmental literature that have greater editorial inconsistency than ground water. The preferred ASTM usage of two words for *ground water* (to parallel *surface water*, which is never written as a single word) is used in this handbook, except when reference citations use a single word, *groundwater*, in the title, in which case they are cited as written. Given the inconsistency in hyphenation practices for the terms vadose zone and ground water in the literature, l have chosen as a matter of personal preference to use the following conventions: (1) *neither* vadose zone or ground water is hyphenated when it serve as a modifier in chapter, section, figure and table titles because I think it looks better that way; (2) vadose zone is *not* hyphenated in normal text when it serves as a modifier because that seems to be the most common usage in the vadose zone literature; and (3) ground water *is* hyphenated when it serves as a modifier because that seems to be the more common usage in the ground-water literature. I think it was Ralph Waldo Emerson who said: "Foolish consistency is the hobgoblin of little minds." I have tried to consistently follow the above inconsistent conventions, but am not overly concerned if I have not entirely succeeded.

ABBREVIATIONS AND ACRONYMS

AGWSE	Association of Ground Water Scientists and Engineers (NWWA/NGWA)
API	American Petroleum Institute
CERI	Center for Environmental Research Information (U.S. EPA)
DNAPL	Dense nonaqueous phase liquid
EPA	U.S. Environmental Protection Agency
HEW	U.S. Department of Health, Education and Welfare
HMCRI	Hazardous Materials Control Research Institute
IGWMC	International Ground Water Modeling Center
LNAPL	Light nonaqueous phase liquid
NAPL	Nonaqueous phase liquid
NWWA/NGWA	National Water Well Association (named changed to National Ground Water Association in 1992)
PAH	Polynuclear aromatic hydrocarbon
PNA	Polynuclear aromatic compound
WHPA	Wellhead protection area

J. Russell Boulding first began working in the environmental field in 1973 when he helped set up the Environmental Defense Fund's Denver Office, and has been a free-lance environmental consultant since 1977 when he established Boulding Soil-Water Consulting in Bloomington, Indiana. He has a B.A. in Geology (1970) from Antioch College, Yellow Springs, Ohio, and an M.S. in Water Resources Management (1975) from the University of Wisconsin/Madison. From 1975 to 1977 he was a soil scientist with the Indiana Department of Natural Resources and mapped soils in southern Indiana on a cooperative program with the U.S. Soil Conservation Service. Since 1984 he has been Senior Environmental Scientist with Eastern Research Group, Inc. in Lexington, Massachusetts.

Mr. Boulding is the author of more than 100 books, chapters, articles and consultant reports in the areas of soil and ground-water contamination assessment, geochemical fate assessment of hazardous wastes, mined land reclamation, and natural resource management and regulatory policy. From 1978 to 1980 he served as a member of the Environmental Subcommittee of the Committee on Surface Mining and Reclamation (COSMAR) of the National Academy of Sciences (NAS) and as a consultant to the NAS Committee on Soil as a Resource in Relation to Surface Mining for Coal. Mr. Boulding is an ARCPACS Certified Professional Soil Classifier and his professional memberships include the Soil Science Society of America, International Society of Soil Science, Association of Ground Water Scientists and Engineers, and the International Association of Hydrogeologists. Since 1992 he has been a member of the American Society for Testing and Material's Committee D18 (Soil and Rock) and active in subcommittees D18.01 (Surface and Subsurface Characterization), D18.07 (Identification and Classification of Soils) and D18.21 (Ground Water and Vadose Zone Investigations). In 1993 he became chair of D18.01's Section on Site Characterization for Environmental Purposes.

Acknowledgments

This book represents the synthesis and, in a sense, the culmination of work done during the past six years on a series of technology transfer documents written wholly or in part by the author for U.S. EPA's Center for Environmental Research Information. At the project management level, Heidi Schultz (Eastern Research Group), and Carol Grove and Sue Schock (U.S. EPA/CERI) require special recognition.

The starting point for this book was a 700 page manuscript that was an expanded and extensively revised version of EPA's 1987 **Ground Water Handbook** (EPA/625/6-87/016) which was written by Michael Barcelona, Joseph Keely, Wayne Pettyjohn and Allen Wehrmann. Chapter 7 of that document drew heavily upon another EPA report, **Introduction to Ground Water Tracers** (EPA/600/2-85/022) by S.N. Davis, D.J. Campbell, H.W. Bentley, and T.J. Flynn, and Chapter 8 on Joseph Keely's monograph **The Use of Models in Managing Ground-Water Protection Programs** (EPA/600/8-87/003). As it turned out, the author's revised versions of Chapters 7 and 8 are the only ones that were published in EPA's second edition of the **Ground Water Handbook, Volume I: Ground Water and Contamination** (EPA/625/6-90-16a) and **Volume II:Methodology** (EPA/625/6-90-16a), both of which are available from the Center for Environmental Information. These, with third-generation revisions and updating, appear in this handbook as Chapters 8 (Soil and Ground-Water Tracers) and 10 (Use of Models and Computers in Contaminant Investigations). To the extent that material originally written by the above-mentioned individuals can be found in this handbook, they deserve credit. Maureen Casey and Leslie Sparrow, HydroQual, Inc., New Jersey, made significant contributions to chapters on monitoring well design and construction, ground-water sampling, and ground-water restoration in the manuscript mentioned at the beginning of this paragraph, some of which has probably made its way into this handbook.

Other individuals who require special recognition for textual contributions to this handbook are Ron Sims and Judy Sims, whose chapters in EPA's **Site Characterization for Subsurface Remediation** on basic approaches to soil and ground-water remediation and on remediation techniques for contaminated soils formed the starting point for Chapters 12 and 13 in this Handbook. I was technical editor of that document, and co-author with Michael Barcelona of chapters on basic statistical and analytical concepts and geochemical sampling of subsurface solids and ground-water which provided the basis for much of Chapter 5 of this handbook.

Anyone bothering to read this acknowledgement page may wonder if I can take credit for any of the contents of this handbook. I am the original author of Chapter 3 (Soil and Ground Water Chemistry and Biology), Chapter 4 (Sources and Behavior of Subsurface Contaminants, Chapter 6 (Geophysical and Remote Sensing Techniques), Chapter 11 (Prevention and Minimization of Contamination), all reference tables and compilations and all the Appendices. All other chapters represent second to fourth generation edits, revisions, or updates on my part.

In various incarnations, the material in this handbook has benefitted from the technical review and suggestions from many individuals: Gina Bochicchio, Fred Cornell, Larry Eccles, Lorne Everett, Malcolm Field, Pete Haeni, Jan Hendrickx, Paul Heigold, Beverly Herzog, Dave Kaminski, Jack Keeton, Scott Keys, Eric Koglin, Robert Powell, Robert Puls, Duncan McNeill, Gary Olhoeft, Charlie Riggs, Ron Schalla, Ron Sims, Jim Ursic, Paul van der Heijde, Mark Vendl, and John Williams. Special thanks are due to my colleagues on ASTM's Environmental Site Characterization Task Group (Joe Downey, Ed Gutentag, and Mario Fernandez) and to Gareth Davies, and Mark Kram for review of individual or multiple chapters.

Part I: Basic Concepts

Chapter 1 Geology, Soils, and Geomorphology ... 1

1.1 Geologic Materials ... 2
1.2 Geologic Processes ... 12
1.3 Stratigraphy and Structure ... 14
1.4 Basic Soil Concepts ... 20
1.5 Geomorphology and Ground Water ... 28
1.6 Geologic Settings of Ground Water Occurrence and Quality ... 32
1.7 Guide to Major References ... 38

Chapter 2 Ground Water and Vadose Zone Hydrology ... 59

2.1 Ground Water in the Hydrologic Cycle ... 60
2.2 Ground Water-Atmospheric Relationships ... 61
2.3 Ground Water-Surface Water Relationships ... 65
2.4 Water in the Vadose Zone ... 72
2.5 Water in the Saturated Zone (1): Fundamental Concepts ... 76
2.6 Water in the Saturated Zone (2): Water Storage and Flow ... 86
2.7 Guide to Major References ... 94

Chapter 3 Soil and Ground Water Geochemistry and Microbiology ... 117

3.1 Key Characteristics of Chemical Systems ... 118
3.2 Subsurface Distribution Processes ... 123
3.3 Subsurface Transformation Processes ... 129
3.4 Subsurface Environmental Parameters ... 137
3.5 Microbial Ecology of the Subsurface ... 148
3.6 Guide to Major References ... 156

Chapter 4 Sources and Behavior of Subsurface Contaminants ... 169

4.1 Definitions of Contamination ... 170
4.2 Extent and Sources of Subsurface Contamination ... 171
4.3 General Mechanisms of Ground-Water Contamination ... 179
4.4 Contaminant Transport Processes ... 182
4.5 Contaminant Retardation ... 192
4.6 Contaminant Plume Behavior ... 198
4.7 Guide to Major References ... 202

Part II: Assessment and Monitoring

Chapter 5 Planning Field Investigations ... 237

5.1 Overview of Investigation Planning Process ... 238
5.2 Existing Information Sources ... 246
5.3 Developing a Sampling and Monitoring Plan ... 253
5.4 Data Measurement and Reliability ... 269
5.5 Analytical and QA/QC Concepts ... 276
5.6 Statistical Techniques ... 281
5.7 Guide to Major References ... 284

Page

Chapter 6 Geophysical and Remote Sensing Techniques 303

6.1 Overview of Remote Sensing and Geophysical Techniques 304
6.2 Airborne Remote Sensing 308
6.3 Surface Geophysical Methods 312
6.4 Borehole Geophysics 322
6.5 Selection of Geophysical Methods 331
6.6 Guide to Major References 335

Chapter 7 Characterization of Vadose Zone and Ground Water Hydrology 371

7.1 Measurement of Vadose Zone Hydrologic Parameters 372
7.2 Preparation and Use of Potentiometric Maps 377
7.3 Field and Laboratory Measurement of Aquifer Parameters 388
7.4 Estimation of Subsurface Hydrologic Parameters 394
7.5 Special Considerations in Hydrogeologic Mapping 398
7.6 Guide to Major References 413

Chapter 8 Soil and Ground-Water Tracers 439

8.1 Types and Uses of Tracer Tests 440
8.2 Tracer Selection 443
8.3 Types of Tracers 448
8.4 Tracer Tests in Karst and Fractured Rock 467
8.5 Tracer Tests in Porous Media 469
8.6 Guide to Major Reference 478

Chapter 9 Field Sampling and Monitoring of Contaminants 493

9.1 Types of Monitoring Installations 494
9.2 Drilling Methods for Sampling and Well Installation 499
9.3 Conventional Monitoring Well Installations 504
9.4 Sampling Subsurface Solids and the Vadose Zone 514
9.5 Sampling Ground Water 520
9.6 Field Screening and Analytical Methods 525
9.7 Guide to Major References 528

Chapter 10 Use of Models and Computers in Contaminant Investigations 561

10.1 Uses of Models and Computers 562
10.2 Mathematical Approaches to Modeling 565
10.3 Classification of Ground Water Computer Codes 573
10.4 General Code Selection Considerations 581
10.5 Other Geoenvironmental Computer Applications 589
10.6 Guide to Major Information Sources 599

Part III: Prevention and Remediation

Chapter 11 Prevention and Minimization of Contamination 623

11.1 General Approaches 624
11.2 Delineation of Wellhead Protection Areas 626
11.3 Identification of Potential Contaminant Sources 641
11.4 Assessing The Risk From Potential Contaminants 644
11.5 Wellhead Protection Area Management 652
11.6 Guide to Major References 661

Chapter 12 Remediation Planning 695

12.1 Conceptual Approach to Soil and Ground Water Remediation 696
12.2 Methodology 699
12.3 Selection of Treatment Methods 710
12.4 Measurement and Interpretation of Treatment Effectiveness 715
12.5 Guide to Major References 716

Chapter 13 Remediation of Contaminated Soils 731

13.1 General Approaches to Soil Remediation 731
13.2 Soil Vacuum Extraction (SVE) 732
13.3 Bioremediation 745
13.4 Other Treatment Approaches 750
13.5 Prepared Bed Reactors 754
13.6 Guide to Major References 756

Chapter 14 Remediation of Contaminated Ground Water 761

14.1 Overview 762
14.2 Source Control 766
14.3 Containment: Ground Water Barriers and Flow Control 769
14.4 Ground Water Collection 772
14.5 Ground Water Pump-and-Treat Methods 779
14.6 In Situ Treatment 790
14.7 Guide to Major References 795

Appendix A Summary Information on Major Subsurface Characterization and Monitoring Techniques and Index to ASTM Methods 805
Appendix B Manufacturers and Distributors of Field Characterization and Monitoring Equipment 835
Appendix C Tables and Figures for Estimation of Aquifer Parameters 855
Appendix D Worksheets and Checklists for Ground-Water and Wellhead Protection 877
Appendix E PC-Based Geoenvironmental Software 911
Appendix F Master List of Figures and Tables with Credits 927

Key Word Index 948

List of Reference Index Tables (see Appendix F for list of figures and other tables)

		Page
1-3	Index to Major References on Geology, Soils and Geomorphology	40
1-4	Index to Major References on Karst Geology, Geomorphology and Hydrology	54
2-4	Index to Major References on Surface and Vadose Zone Hydrology, Hydrogeology and Hydraulics	97
2-5	Index to Major References on Regional Hydrology and Ground Water in the United States	110
3-12	Index to Major References on Soil and Ground Water Geochemistry and Microbiology	157
4-4	Index to Major References on Types and Sources of Contamination in Soil and Ground Water	206
4-5	Index to Major References on Contaminant Chemical Characteristics and Behavior in the Subsurface	219
5-13	Index to Major References on Existing Environmental Information and Data Management	288
5-14	Index to Major References on Statistics and Sampling Design	295
6-8	Index to Major References on Remote Sensing and Surface Geophysics	338
6-9	Index to Major References on Borehole Geophysics	357
7-5	Index to Major References on Hydraulic Conductivity and Water Balance Test Methods	415
7-6	Index to Source References on Pump Text Analytical Solutions and Methods for Characterizing Anisotropic and Fractured Rock Aquifers	426
8-5	Index to Major References on Tracer Methods	480
9-10	Index to Major Reference Sources on Subsurface Sampling and Monitoring Methods	529
9-11	Index to Major References on Field and Laboratory Analytical Methods	549
10-6	Index to Major References on Ground Water and Vadose Zone Flow and Contaminant Transport Modeling	602
10-7	Index to Major References on Geographic Information Systems (GIS)	616
11-10	Index to Major References on Ground-Water Vulnerability Mapping and Chemical Hazard and Risk Assessment	661
11-11	Index to Major References on Pollution Prevention and Soil and Ground Water Protection Management	678
12-7	Index to Major References on Soil and Ground Water Remediation Planning	718
13-9	Index to Major References on Soil Treatment and Remediation	757
14-9	Index to Major References on Ground Water Treatment and Remediation	796

PART I:

BASIC CONCEPTS

CHAPTER 1

GEOLOGY, SOILS, AND GEOMORPHOLOGY

1.1 Geologic Materials 2

    1.1.1 Mineralogy 3
    1.1.2 Texture and Fabric 7
    1.1.3 Rocks 7
    1.1.4 Unconsolidated Materials 11

1.2 Geologic Processes 12

    1.2.1 Mechanical Weathering 12
    1.2.2 Chemical Weathering 12
    1.2.3 Erosion and Deposition 12

1.3 Stratigraphy and Structure 14

    1.3.1 Stratigraphic Relationships 15
    1.3.2 Age and Relationship of Stratigraphic Units ... 15
    1.3.3 Folds and Fractures 16
    1.3.4 Geologic Maps and Cross Sections 18

1.4 Basic Soil Concepts 20

    1.4.1 Factors of Soil Formation 20
    1.4.2 Soil Classification 22
    1.4.3 Soil Physical Properties 23
    1.4.4 Soil Chemical Properties 28

1.5 Geomorphology and Ground Water 28

    1.5.1 Hydrogeomorphology 29
    1.5.2 Karst Geomorphology and Hydrology 30

1.6 Geologic Settings of Ground Water Occurrence and Quality 32

    1.6.1 Ground Water in Igneous and Metamorphic Rocks 32
    1.6.2 Ground Water in Sedimentary Rocks 35
    1.6.3 Ground Water in Unconsolidated Sediments 36
    1.6.4 Regional Relationships in Ground-Water Quality 36

1.7 Guide to Major References* 38

*Appendix F contains citations for table and figure sources.

Geology, the study of the earth, includes the investigation of earth materials, the processes that act on these materials, the products that are formed, the history of the earth, and the origin and development of life forms. There are several subfields of geology. *Physical geology* deals with all aspects of the earth and includes most earth science specialties. *Historical geology* is the study of the origin of the earth, continents and ocean basins, and life forms. *Economic geology* is an applied approach focusing on the search and exploitation of mineral resources, such as metallic ores, fuels, and water. *Structural geology* deals with the various structures of the earth and the forces that produce them. *Geophysics* is the examination of the physical properties of the earth and includes the study of earthquakes and methods to evaluate the subsurface.

All of the geology subfields are used to some extent in the study of ground water. Probably the most difficult concept to comprehend by individuals with little or no geological training is the complexity of the subsurface, which is hidden from view and, at least presently, cannot be adequately sampled. A guiding principle in geologic and hydrogeologic studies is that the present is the key to the past. The processes occurring today are the same processes that have occurred throughout geologic history, although their magnitude may vary with time. Furthermore, an understanding of present processes, and how they have acted in the past, can be used as a guide to predict the future.

Soil science, also called *pedology*, and *geomorphology*, the study of surface landforms, are disciplines related to geology but focusing on the earth's surface. Geology, soil science and geomorphology are intimately related and concerned with many of the same earth processes such as weathering, erosion and deposition. Nevertheless, each of these disciplines has a distinct perspective that is usually helpful and often essential in the study of ground-water contamination.

This chapter provides a brief description of fundamental concepts in geology, soil science and geomorphology as they relate to ground-water contamination. Sections 1.1 (Geologic Materials), 1.2 (Geologic Processes) and 1.3 (Stratigraphy and Structure) focus on geology. Section 1.4 examines basic soil concepts and Section 1.5 geomorphic concepts with special emphasis on karst geomorphic settings because of their distinctive hydrogeologic characteristics, which include very rapid movement of contaminants in the subsurface.

1.1 Geologic Materials

Geologic materials result from constant changes at the earth's surface and subsurface. Over long periods of time these changes affect the location, quality, and movement of ground water. Rocks that rise as mountains over millions of years will be gradually eroded and transported by wind, water or gravity to low-lying areas. Variations in these transport processes alter particles physically and chemically, giving rise to deposits of unique texture and composition. Grain-size variations and degree of sorting will cause differences in permeability and ground-water velocity, while changes in mineral composition can lead to variations in water quality.

1.1.1 Mineralogy

Minerals are the basic building blocks of rocks. Most minerals contain two or more elements, but of all the elements known, only eight account for nearly 98% of the rocks and minerals:

Oxygen 46%
Silicon 27.7%
Aluminum 8.1%
Iron 5.0%
Calcium 3.6%
Sodium 2.8%
Potassium 2.6%
Magnesium 2.1%

A general understanding of mineralogy is important to the study of ground water because it is the mineral composition of rocks that, to a large extent, controls the quality of water that a rock contains under natural conditions and the chemical reactions between rock and contaminants or naturally occurring substances.

The most common rock-forming minerals can be divided into four broad groups: (1) oxides, carbonates, and sulfates, (2) silicates, (3) clay minerals, and (4) common ores. Organic matter, from which oil shales, coal and petroleum deposits form, is another important material that is discussed in Section 3.4.4 because of its geochemical importance in soil and ground-water chemistry.

Oxides, Carbonates, and Sulfates. *Quartz* (SiO_2), one of the most common minerals, is hard and resistant to both chemical and mechanical weathering. In sedimentary rocks, quartz occurs as sand-size grains (sandstone) or finer silt- and clay-sized grains. It may also appear as a silica cement. Because of the low solubility of silica, it generally appears in ground water in concentrations less than 25 mg/L.

Limonite is actually a group name for the hydrated ferric oxide minerals ($Fe_2O \cdot 3H_2O$), that occur so commonly in many types of rocks. Limonite is generally rusty or blackish with a dull, earthy luster. It is a common weathering product of other iron minerals. Because limonite and other iron-bearing minerals are nearly universal, dissolved iron is a very common constituent in water, causing staining of clothing and plumbing fixtures.

The major carbonate minerals are *calcite* ($CaCO_3$), the major component of limestone, and *dolomite* ($CaMg(CO_3)_2$). Dolomite is also the name of carbonate sediments enriched with this magnesium carbonate. *Gypsum*, a hydrated calcium sulfate ($CaSO_4 \cdot 2H_2O$), occurs as a sedimentary evaporite deposit and as crystals in shale and some clay deposits. Quite soluble, it is the major source of sulfate in ground water.

Silicates. The most common rock-forming silicate minerals include the feldspars, micas, pyroxenes, amphiboles, and olivine. Feldspars, the most abundant minerals on earth, are alumino-silicates of potassium or sodium and calcium. Most of the minerals in this group are white, gray, or pink. Upon weathering they turn to clay and release the remaining chemical elements to water. Muscovite and biotite mica are platy alumino-silicate minerals that are common and easily recognized in igneous, metamorphic, and sedimentary rocks. Pyroxenes, a group of silicates of calcium, magnesium, and iron, as well as amphiboles, which are complex hydrated silicates of calcium, magnesium, iron, and aluminum, are common in most igneous and metamorphic rocks. They appear as small, dark crystals. Olivine, a magnesium-iron silicate, is generally green or yellow and is common in certain igneous and metamorphic rocks. None of the rock-forming silicate minerals have a major impact on water quality in most situations.

Clay Minerals. Next to organic matter (see Section 3.4.4), clay minerals are the most chemically active materials in soil and unconsolidated geologic materials. Both consolidated materials composed of clay minerals (shales) and clayey unconsolidated materials tend to have low permeabilities and consequently water movement is very slow. Because of their geochemical significance in the study of ground-water contamination, clay minerals are described in more detail here than the other major mineral groups. Two broad groups of clay minerals are recognized: silicate clays and hydrous oxide clays.

Silicate clays form from the weathering of primary silicate minerals such as feldspars and olivine. They have a sheet-like lattice structure with either silicon (Si) in coordination with four oxygen atoms (silica tetrahedra) or aluminum (Al) in coordination with six oxygen atoms (alumina octahedra). The strong sorptive capacity of clay derives from the negative charges created at the edges of these crystalline sheets where oxygen atoms (O^{-2}) have extra electrons that are not bonded to the cations in the crystalline structure. The negative charge can be further increased when ions with a lower valence substitute for ions with a higher valence in the sheet structure (for example Al^{+3} substitutes for Si^{+4} in tetrahedral sheets, and Mg^{+2} substitutes for Al^{+3} in octahedral sheets).

Silicate clays are classified according to different stacking arrangements of the tetrahedral (silica) and octahedral (alumina) lattice layers and their tendency to expand in water. The stacking type strongly affects certain properties of clays, including (1) surface area, (2) the tendency to swell during hydration, and (3) *cation exchange capacity* (CEC), which is a quantitative measure of the ability of a mineral surface to adsorb ions. CEC is the sum of exchangeable cations that a material can adsorb at a specific pH. Standard International (SI) units for CEC are centimoles per kilogram, but it is also commonly reported as milliequivalents per 100 grams (meq), where 1 meq is defined as 1 mg of hydrogen or the amount of any other ion that will combine with or displace 1 mg hydrogen. These units are interchangeable (1 cmole/kg = 1 meq/100 g).

Table 1-1 summarizes some properties of different silicate clay minerals. The montmorillonite group of silicate clays is most sensitive to swelling and has a high CEC. This type of clay has these characteristics because the 2:1 lattice structure (two octahedral sheets separated by a tetrahedral sheet) forms sheets that are loosely connected by exchangeable cations. The exchange sites between 2:1 lattice layers can be easily hydrated (i.e., adsorb water molecules) under certain conditions. Because the water molecules have a greater diameter than the cations that hold the sheets together, hydration pushes the layers farther apart. Vermiculite has stronger negative charges on its inner surfaces than montmorillonite because of the substitution of lower-valence magnesium ions for aluminum. This factor results in an even higher CEC than that found in montmorillonite, but it also has the effect of bonding the 2:1 sheets more strongly. Consequently, vermiculite clays are less susceptible to swelling.

In Table 1-1, the clays are listed in sequence from most reactive (montmorillonite and vermiculite) to least reactive (kaolinite). The 1:1 lattice structure in kaolinite creates strong bonds between the paired sheets, resulting in a low surface area and CEC. Illite and chlorite have intermediate surface areas, CEC, and sensitivities to swelling.

Clay minerals in sedimentary formations are usually mixtures of different groups. In addition, *mixed-layer* clay minerals can form. These minerals have properties and compositions that are intermediate between two well-defined clay types (i.e., chlorite-illite, illite-montmorillonite, etc.). Where soils have a high clay content, clay mineralogy is a criterion for classification of soils in the USDA soil taxonomy (Section 1.4.2). Such soils are identified by the dominant clay (hallosyite, illite, kaolinite, montmorillonite, etc.) or as having mixed mineralogy.

Hydrous oxide clays are less well understood than silicate clays. These clays are oxides of iron, magnesium, and aluminum that are associated with water molecules, although the exact mechanism by which the water molecules are held together is somewhat uncertain. Because of the lower overall valence of the cations in hydrous oxide clays compared to silicate clays, CEC is lower in hydrous oxide clays. However, hydrous oxides of magnesium (Mn) and iron (Fe) can furnish the principal control on the fixation of cobalt (Co), nickel (Ni), copper (Cu), and zinc (Zn) heavy metals in soils and freshwater sediments.[1]

Ores. The three most common ore minerals are galena, sphalerite, and pyrite. Galena, a lead sulfide (PbS), is heavy, brittle, and breaks into cubes. Sphalerite is a zinc sulfide (ZnS) mineral that is brownish, yellowish, or black. It ordinarily occurs with galena and is a major zinc ore. The iron sulfide pyrite (FeS), which is also called fool's gold, is common in all types of rocks. Weathering of this mineral leads to acid-mine drainage, a major surface and ground-water quality problem in certain coal mining areas of the Midwest and Appalachia and in metal sulfide mining regions.

---

[1] Jenne, E.A. 1968. Controls on Mn, Fe, Co, Ni, Cu, and Zn Concentrations in Soils and Water: the Significant Role of Hydrous Mn and Fe Oxides. In: Adsorption From Aqueous Solution, ACS Adv. in Chem. Ser. 79, pp. 337-387.

Table 1-1 Important Characteristics of Silicate Clay Minerals

Property	Type of Clay[a]				
	Montmorillonite (Smectite)[b]	Vermiculite	Illite	Chlorite	Kaolinite
Lattice type[c]	2:1	2:1	2:1	2:2	1:1
Expanding?	Yes	Slightly	No	No	No
Specific surface area (m^2/g)	700-800	700-800	65-120	25-40	7-30
External surface area	High	High	Medium	Medium	Low
Internal surface area	Very high	High	Medium	Medium	None
Swelling capacity	High	Med-High	Medium	Low	Low
Cation exchange capacity (meq/100 g)	80-150	100-150+	10-40[e]	10-40[e]	3-15[e]
Other similar clays	Beidellite Nontronite Saponite Bentonite[d]				Halloysite Anauxite Dickite Nacrite

[a] Clays are arranged from most reactive (montmorillonite) to least reactive (kaolinite).

[b] The term smectite is now used to refer to the montmorillonite group of clays (Soil Science Society of America, 1987/T1-3).

[c] Tetrahedral:octahedral layers.

[d] Bentonite is a clay formed from weathering of volcanic ash and is made up mostly of montmorillonite and beidellite.

[e] Upper range occurs with smaller particle size.

Source: Boulding (1990).

1.1.2 Texture and Fabric

The term *texture* has different meanings in geology and soil science. In soil science it is simply the relative proportions of clay-, silt-, and sand-sized particles in soil or unconsolidated material. In geology, describing the texture also involves characterization of grain size, but also grain shape, degree of crystallization, and contact relationships of grains. The term *fabric* applies to the total of all physical features of a rock or soil that can be observed macroscopically and microscopically. In solid rock this includes texture, porosity, orientation of mineral grains, cleavage, joints and fractures, all of which may influence water transmitting characteristics. Soil fabric analysis involves the study of the distinctive physical features resulting from soil-forming processes, which also strongly influence the location and rate of water movement in soil.

A variety of scales are available for the classification of unconsolidated materials based on particle-size distribution. In geology, the *Wentworth-Udden* scale is most widely used: boulder (>256 mm), cobble (64-256 mm), pebble (4-64 mm), granule or gravel (2-4 mm), sand (1/16-2 mm), silt (1/256-1/16 mm), and clay (<1/256 mm). The U.S. Department of Agriculture (USDA) soil textural classification system is most widely used by soil scientists; engineers usually use the ASTM version (ASTM D2488/Table A-14) of the unified soil classification system (USCS), and less commonly the AASTHO (American Association of State Highway Officials) soil classification system. Figure 1-1 compares the grain-size limits for the ASTM, AASTHO, USDA, Federal Aviation Administration, and the U.S. Army Corps of Engineers/Bureau of Reclamation. Figure 1-2 shows the twelve USDA soil texture classes based on relative percentages of silt, sand and clay. The hydrologic properties of soils are strongly related to particle-size distribution, and the USDA system is the most useful system for estimating a number of these properties (see, for example, Figures C-2, C-8, and C-9).

For several reasons, the term *clay* may be confusing. First, the various particle-size distribution schemes define clay differently. The Wentworth-Udden scale for clay (<.0039 mm) lies between the AASTHO (<.005) and the USDA (<.002) definitions. The ASTM and FAA systems define clay as <0.004 mm, which includes silt-sized particles in the USDA system (see Figure 1-1). Secondly, clay minerals are usually clay-sized particles, but may be silt-sized. Similarly, non-clay minerals may be clay particles if they are small enough.

1.1.3 Rocks

Three major types of rock make up the earth. *Igneous rocks* have solidified from molten material either within the earth (intrusive) or on or near the surface (extrusive). *Metamorphic rocks* originally were igneous or sedimentary rocks that have subsequently been modified by temperature, pressure, and/or chemically active fluids. *Sedimentary rocks* result from the weathering of preexisting rocks, erosion, and deposition. While geologists have developed elaborate systems of nomenclature and classification of rocks, only basic rock descriptions having the most value in hydrogeologic studies will be presented here.

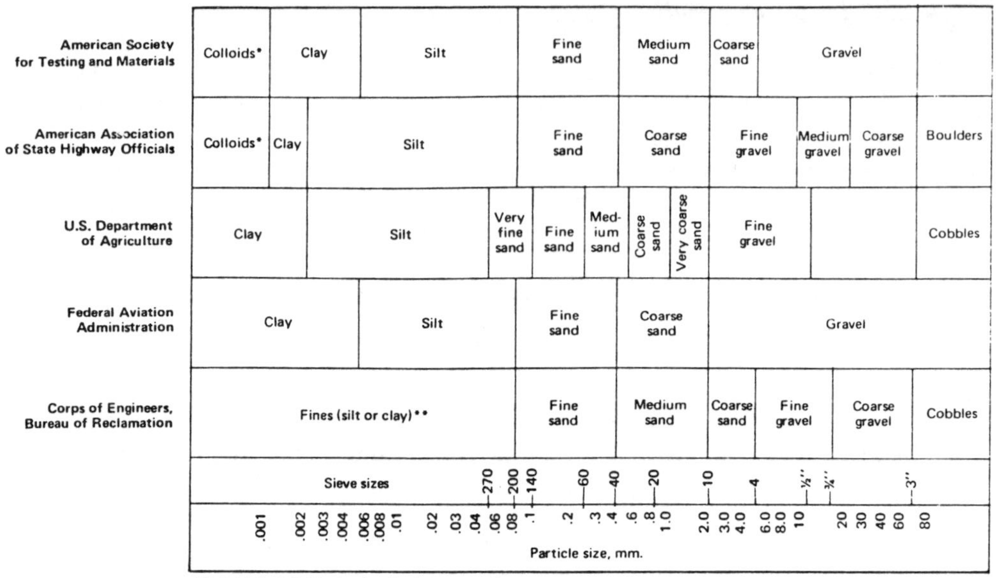
American Society for Testing and Materials
Colloids*
Clay
Silt
Fine sand
Medium sand
Coarse sand
Gravel
American Association of State Highway Officials
Colloids*
Clay
Silt
Fine sand
Coarse sand
Fine gravel
Medium gravel
Coarse gravel
Boulders
U.S. Department of Agriculture
Clay
Silt
Very fine sand
Fine sand
Medium sand
Coarse sand
Very coarse sand
Fine gravel
Cobbles
Federal Aviation Administration
Clay
Silt
Fine sand
Coarse sand
Gravel
Corps of Engineers, Bureau of Reclamation
Fines (silt or clay)**
Fine sand
Medium sand
Coarse sand
Fine gravel
Coarse gravel
Cobbles
Sieve sizes
270 200 140 60 40 20 10 4 ½" ¾" 3"
.001 .002 .003 .004 .006 .008 .01 .02 .03 .04 .06 .08 .1 .2 .3 .4 .6 .8 1.0 2.0 3.0 4.0 6.0 8.0 10 20 30 40 60 80
Particle size, mm.

*Colloids included in clay fraction in test reports.
**The LL and PI of "Silt" plot below the "A" line on the plasticity chart, Table 4, and the LL and PI for "Clay" plot above the "A" line.

Figure 1-1 Particle-size limits of different U.S. textural classification systems (Mercer and Spalding, 1991b, after Portland Cement Association, 1973).

Igneous Rocks. Igneous rocks are classified on the basis of their composition and grain size. *Basic* igneous rocks consist mostly feldspar and a variety of dark, iron- and magnesium-rich minerals (olivine, pyroxenes and hornblendes). *Acidic* igneous rocks are lighter in color, with feldspar, quartz and micas being the dominant minerals. If the parent molten material cools slowly deep below the surface, minerals will have an opportunity to grow and the rock will be coarse grained (*gabbro*, if cooled from basic magma, and *granites*, if cooled from acidic magma). Magma that cools rapidly, such as that derived from volcanic activity, is so fine grained that individual minerals generally cannot be seen even with a hand lens. *Basalt* is the fine-grained equivalent of gabbro, and *andesite* is the fine-grained equivalent of granite. In some cases, the molten material initially cools slowly, allowing large mineral crystals to grow. If the cooling rate increases, rapid cooling crystallizes the remaining melt into a fine-grained matrix. This texture, consisting of large crystals in a fine-grained matrix, is called *porphyritic*.

Intrusive igneous rocks can only be seen where they have been exposed by erosion. They are *concordant* if they generally parallel the bedding of the enclosing rocks and *discordant* if they cut across the bedding. The largest discordant igneous masses are called batholiths and occur in the eroded centers of many ancient mountains. Their dimensions are in the range of tens of miles. Batholiths usually consist largely of granite, which is surrounded by metamorphic rocks. Other discordant igneous rocks include *dikes* ranging in thickness from a few inches to

Geology, Soils, and Geomorphology

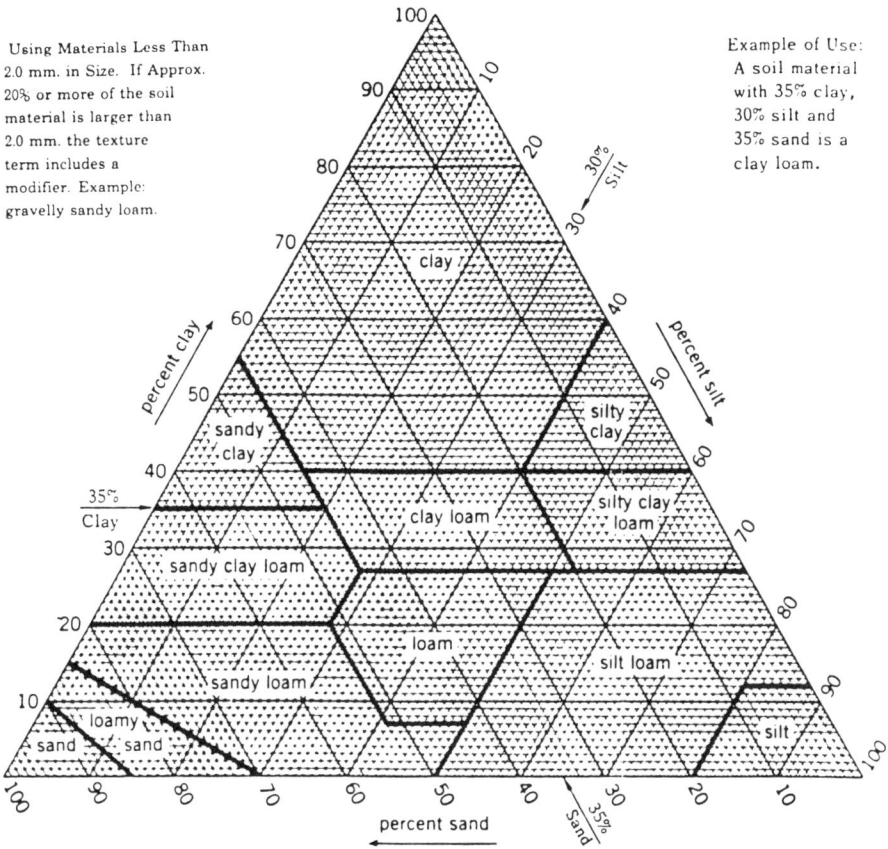

Figure 1-2 Guide for USDA soil textural classification (SCS, 1971).

thousands of feet. Many are several miles long. *Sills* are concordant bodies that have invaded sedimentary rocks along bedding planes. They are relatively thin. Both sills and dikes, when intruded into existing rocks, tend to cool quite rapidly and are fine grained. *Pegmatite* dikes, on the other hand, form in the last stages of a magma's cooling, and can form centimeter- to meter-length crystals.

Extrusive rocks include lava flows or other types associated with volcanic activity, such as the consolidated ash called *tuff*. These are fine grained or even glassy. Some extrusive rocks, like pumice, have sufficiently high porosity resulting from gas bubbles in the magma during cooling that they can float on water.

Igneous rocks are typically dense and have very low porosity and permeability. Most, however, are fractured to some degree and can store and transmit a modest amount of water. These fractures form preferential flow paths for contaminants, resulting in relatively rapid movement, even though the overall hydraulic conductivity is low. Some lava flows are notable exceptions because they contain large diameter tubes or a permeable zone at the top of the flow where gas bubbles migrated to the surface before the rock solidified. These rocks are called scoria.

Metamorphic Rocks. Metamorphism is a process that changes preexisting rocks into new forms because of increases in temperature, pressure, and chemically active fluids. Metamorphism may affect igneous, sedimentary, or other metamorphic rocks. The changes brought about include the formation of new minerals, increase in grain size, and modification of rock structure or texture, all of which depend on the original rock's composition and the intensity of the metamorphism.

Some of the most obvious changes are in texture, which serves as a means of classifying metamorphic rocks into two broad groups: foliated and nonfoliated rocks. *Foliated* metamorphic rocks typify regions that have undergone severe deformation, such as mountain ranges. Shale, which consists mainly of silt and clay, is transformed into slate by the change of clay to mica. Mica, a platy mineral, grows with its long axis perpendicular to the principle direction of stress, forming a preferred orientation. This orientation, as in the development of cleavage in slate, may differ greatly from the original bedding.

With increasing degrees of metamorphism, the grains of mica grow larger so that the rock has a distinct foliation texture, characteristic of the metamorphic rock *schist*. At even higher grades of metamorphism, the mica may be transformed to a much coarser grained feldspar, producing the strongly banded texture of *gneiss*.

Nonfoliated rocks include the hornfels and another group formed from rocks that consist mainly of a single mineral. The *hornfels* occur around an intrusive body and were changed by "baking" during intrusion. The second group includes marble and quartzite, as well as several other forms. *Marble* is metamorphosed limestone and *quartzite* is metamorphosed quartz sandstone.

There are many different types of metamorphic rocks, but from a hydrogeologic viewpoint they neither store nor transmit much water and are of only minor importance as aquifers. Their primary permeability is notably small, if it exists at all, and fluids are forced to migrate through secondary openings, such as faults, joints, or other types of fractures. As with igneous rocks, the concentration of flow in fractures means that contaminants may move relatively rapidly, especially in response to ground water pumping.

Sedimentary Rocks. Sedimentary rocks are deposited either in a body of water or on the land by running water, wind, and glaciers. Sediments are first derived by the weathering and erosion of preexisting rocks, and each depositional agent leaves a characteristic stamp on the material it deposits (Section 1.2.3). The change from a loose, unconsolidated sediment to a rock is the process of *lithification*. Unconsolidated sediments are discussed in Section 1.1.4. The most common sedimentary rocks are shale, siltstone, sandstone, and limestone. Although sedimentary rocks appear to be the dominant type, in reality they make up but a small percentage of the earth. They are most readily evident, however, because they form a thin crust over much of the earth's surface. Sedimentary rocks and the unconsolidated materials that serve as precursors to sedimentary rocks are the primary sources of ground water.

Most sedimentary rocks are deposited in a sequence of layers or strata. Each layer or stratum is separated by a bedding plane, which reflects variations in sediment supply or short-term erosion. Bedding planes commonly represent changes in grain size. *Stratigraphic correlation* is the process of matching strata between wells or outcrops (Section 1.3).

Sedimentary rocks are classified on the basis of texture (grain size and shape) and composition. *Clastic* rocks consist of particles of broken or worn material and include shale, siltstone, sandstone, and conglomerate. These rocks were lithified by compaction, in the case of shale, and by cementation. The most common cements are clay, calcite, quartz, and limonite. The last three, carried by ground water, precipitate in the unconsolidated material under specific geochemical conditions.

The *organic* or *chemical* sedimentary rocks consist of strata formed from or by organisms and by chemical precipitates from sea water or other solutions. Most have a crystalline texture. Some consist of well-preserved organic remains, such as reef deposits and coal seams. Chemical sediments include limestones, dolomites, and evaporites such as halite (sodium chloride), gypsum, and anhydrite (anhydrous calcium sulfate).

The major features of marine sedimentary rocks are their widespread occurrence and generally uniform thickness and composition. If not disturbed by some type of earth movement, they are stratified and horizontal. Furthermore, each lithologic type is unique relative to adjacent units. The bedding planes or contacts that divide them represent distinct differences in texture or composition. From a hydrologic perspective, differences in texture from one rock type to another produce boundaries that strongly influence ground-water flow. Ground water tends to flow parallel to these boundaries, that is, within particular geologic formations rather than across them.

1.1.4 Unconsolidated Materials

Unconsolidated materials may result from the in situ weathering of rock or, more commonly, from erosion of weathered material with subsequent deposition at another location. The major characteristics of unconsolidated material are *sorting*, *rounding*, and *stratification*. A sediment is well sorted if the grains are nearly all the same size. Wind is the most effective sorting agent, followed by water. Glacial till is unsorted and consists of a wide mixture of material that ranges from large boulders to clay. Section 1.2.3 describes further the characteristics of waterborne, windborne, and glacial deposits.

While being transported, sedimentary material loses its sharp, angular configuration and develops some degree of rounding. The amount of rounding depends on the original shape, composition, transporting medium, and distance traveled.

Sorting and rounding are important features of both consolidated and unconsolidated material because they are key to controlling permeability and porosity. The greater the degree of sorting and rounding, the higher will be the water-transmitting and storage properties (see Table 7-2). This is why a sand deposit, in contrast to glacial till, can be such a productive aquifer.

1.2 Geologic Processes

Generally speaking, a rock is stable only in the environment in which it was formed. Once removed from that environment, it begins to change, rapidly in some cases, but more often slowly, by weathering. The two major processes of weathering are mechanical and chemical, and they usually proceed in concert.

1.2.1 Mechanical Weathering

Mechanical weathering is the physical breakdown of rocks and minerals. Fracturing results when water in a crack turns to ice, or from thermal expansion and contraction resulting from daily and seasonal temperature fluctuations. Abrasion occurs during transport by water, ice, or wind. Gravity causes rocks to fall and shatter. Weathering detritus ranges in size from boulders to silt. Mechanical weathering alone only reduces the size of the rock; its chemical composition does not change. Quartz, for example, is very resistant to chemical weathering. However, it does mechanically weather to quartz sand.

1.2.2 Chemical Weathering

Chemical weathering is an actual change in composition as minerals are modified from one type to another. Many, if not most, of the changes are accompanied by a volumetric increase or decrease, which in itself further promotes additional chemical weathering. The rate depends on temperature, surface area, and available water. The major reactions that occur during chemical weathering are oxidation, hydrolysis, and carbonation. *Oxidation* is a reaction with oxygen (air) to form an oxide, *hydrolysis* is a reaction with water, and *carbonation* is a reaction with CO_2 to form a carbonate. Section 3.3 discusses these chemical processes further. Some of the feldspars weather to clay and release calcium, sodium, silica, and many other elements that are transported in water. The iron-bearing minerals leach iron and magnesium weathering products.

1.2.3 Erosion and Deposition

Once a rock begins to weather, material is transported or eroded and deposited. The major agents involved in this part of the rock cycle are running water, wind, and glacial ice.

Waterborne Deposits. Sediment reaching a stream by gravity or surface erosion is carried to a temporary or permanent site of deposition. During transportation some sorting occurs and the finer silt and clay are carried farther downstream. The streams, constantly filling, eroding, and widening their channels, deposit material that provides clues to much of the history of the region. Alluvial deposits are distinct but highly variable in grain size, composition, and thickness. Where they consist of glacially derived sand and gravel, called *outwash*, they form some of the most productive water-bearing units in the world.

Windborne Deposits. Wind-laid or *eolian* deposits are relatively rare in the geologic record. The massively cross-bedded sandstone of the Navajo Sandstone in Utah's Zion National Park and surrounding areas is a classic example in the United States. Other deposits are more or less local and are represented by dunes formed along beaches of large water bodies or streams. Their major characteristic is the high degree of sorting. Dunes, being relatively free of silt and clay, are very permeable and porous, unless the openings have been filled by cement. They allow rapid infiltration of water and, if the topographic and geologic conditions are such that the water does not rapidly drain, can form major water-bearing units.

Another wind-deposited sediment is *loess*, which consists largely of silt- and clay-sized particles. It lacks bedding but is typified by vertical jointing. Silt is transported by wind from deserts, flood plains, and glacial deposits. Loess weathers to a fertile soil and is very porous. It is common along the major rivers in the glaciated parts of the United States and in China, parts of Europe, and adjacent to deserts and deposits of glacial outwash.

Glacial Deposits. Glaciers erode, transport, and deposit sediments that range from clay to huge boulders. They rework the land surface over which they flow and bury former river systems. The areas covered by glaciers during the last Ice Age in the United States are shown in Figure 1-3, but the deposits extend far beyond the former margins of the ice. The two major types of glaciers include valley or mountain glaciers and the far more extensive continental glaciers. The deposits they leave are similar, differing for the most part only in scale.

As a glacier passes slowly over the land surface it incorporates material from the underlying rocks into the ice mass. This material is transported and deposited elsewhere when the ice melts. During this process, glaciers modify the land surface, both through erosion and deposition. The debris associated with glacial activity is collectively termed *glacial drift*. Unstratified drift, usually deposited directly by the ice, is *glacial till*, a heterogeneous mixture of boulders, gravel, sand, silt, and clay. Till often has low porosity due to the extreme pressure exerted on it by the glacial overburden, with ground-water flow concentrated in fractures, as with igneous rocks (Section 1.1.3). Glacial debris reworked by streams and in lakes is stratified drift. Stream-laid deposits are called *glacial outwash*. Although stratified drift may range widely in grain size, the sorting far surpasses that of glacial till. Glacial lake or *lacustrine* clays are particularly well sorted.

Glacial geologists usually map on the basis of landforms resulting from glacial action, such as moraines, outwash, drumlins, etc. The various kinds of moraines and associated landforms are composed largely of unstratified drift with incorporated layers of sand and gravel. Stratified drift is found along existing or former stream valleys or lakes that were either in the glacier or extended downgradient from it. Meltwater

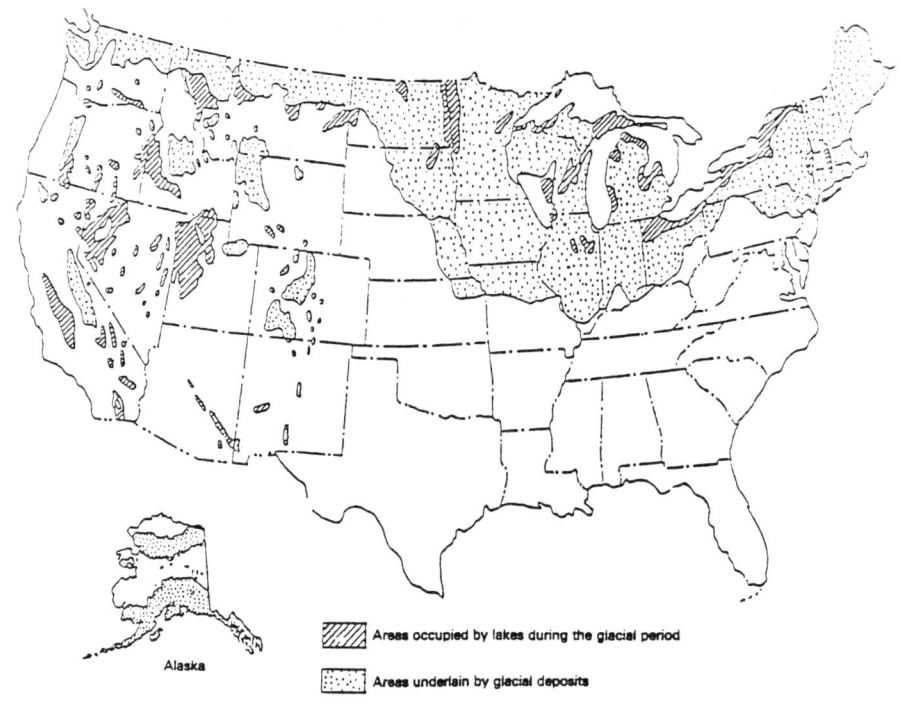

Figure 1-3 Areal extent of glacial deposits in the United States (Heath, 1984).

stream deposits are mixtures of sand and gravel. In some places, they have coalesced into extensive outwash plains.

Glaciers advanced and retreated many times, reworking, overriding, and incorporating sediments from previous advances into the ice, subsequently redepositing them elsewhere. There was a constant inversion of topography as buried ice melted, causing adjacent, waterlogged till to slump into the low areas. During advances, the ice might have overridden older outwash layers so that upon melting, these sand and gravel deposits were covered by a younger layer of till. Regardless of the cause, the final effect is a complex history and stratigraphy. When working with glacial till deposits, it is nearly always impossible to predict the lateral extent or thickness of a particular lithology in the subsurface. Surficial stratified drift is more uniform than till in thickness, extent, and texture.

1.3 Stratigraphy and Structure

Stratigraphy and structural geology strongly influence the occurrence and behavior of ground water, although structural geology is less important where sediments are flat-lying and undisturbed by folding or faulting.

1.3.1 Stratigraphic Relationships

A general principle of geology is that the youngest unit is on the top in any sequence of sedimentary rocks that has not been disturbed by folding or faulting. A second general principle is that sedimentary rocks are deposited in a horizontal or nearly horizontal position. The fact that rocks are found overturned, displaced vertically or laterally, and squeezed into open or tight folds clearly indicates that the earth's crust is a dynamic system.

An *unconformity* is a break in the geologic record. It is caused by a cessation in deposition that is followed by erosion and subsequent deposition. When periods of erosion are protracted and over large geographic areas, significant portions of the geologic record can be destroyed.

If a sequence of strata is horizontal but the contact between two rock groups in the sequence represents an erosional surface, that surface is said to be a *disconformity*. Where a sequence of strata has been tilted and eroded and then younger, horizontal rocks are deposited over them, the contact is an *angular unconformity*. A *nonconformity* occurs where eroded igneous or metamorphic rocks are overlain by sedimentary rocks. A *paraconformity* exists where underlying horizontal sedimentary layers have been eroded and then covered by sediments deposited above them. This type of unconformity is difficult to identify because there is no obvious inconsistency in the strata type or orientation.

1.3.2 Age and Relationship of Stratigraphic Units

Dating of rock units deals with the relation between the emplacement or disturbance of rocks and time. The geologic time scale was developed to provide a standard classification system (Table 1-2) and is based on a sequence of rocks that were deposited during a particular time interval. The divisions are commonly based on some type of unconformity. In considering geologic time, three types of units are defined: rock units, time-rock units, and time units.

A *rock unit* refers to some particular lithology or type of rock. These may be further divided into geologic formations that are of sufficient size and uniformity to be mapped in the field. The Pierre Shale, for example, is a widespread and, in places, thick geologic formation that extends over much of the Northern Great Plains. Formations can also be divided into smaller units called members. Formations have a geographic name that may be coupled with a term that describes the major rock type. Two or more formations comprise a group.

Time-rock units refer to the rock that was deposited during a certain period of time. These units are divided into system, series, and stage. *Time units* refer to the time during which a sequence of rocks was deposited. The time-rock term "system" has the equivalent time term, "period." That is, during the Cretaceous Period, for example, rocks of the Cretaceous System were deposited, consisting of many groups and formations. Time units are named in such a way that the eras reflect the complexity of life forms that existed, such as the Mesozoic or "middle life." System or

2period nomenclature is largely based on the geographic location in which the rocks were first described, such as Jurassic, which relates to the Jura Mountains of Europe.

The terms used by geologists to describe rocks relative to geologic time are useful for ground-water investigations in providing a general overview of the regional geology of an area. The terms alone have no significance as far as water-bearing properties are concerned.

1.3.3 Folds and Fractures

Folds and fractures are the major types of structural features described by geologists. Structural features are usually mapped using a combination of surface outcrop observations, subsurface geophysical measurements, and borehole observations.

Table 1-2 Geologic Time Scale

Era	Period	Epoch	Millions of Years Ago
Cenozoic	Quaternary	Recent	0-0.1
		Pleistocene	0.1-1
	Tertiary	Pliocene	1-5
		Miocene	5-24
		Oligocene	24-38
		Eocene	38-58
		Paleocene	58-63
Mesozoic	Cretaceous		63-138
	Jurassic		138-205
	Triassic		205-240
Paleozoic	Permian		240-290
	Pennsylvanian		290-330
	Mississippian		330-360
	Devonian		360-410
	Silurian		410-435
	Ordovician		435-500
	Cambrian		500-570
Precambrian			570-4550

Source: Update of U.S. EPA (1987a).

Geology, Soils, and Geomorphology 17

Folds. Rocks folded by compressional forces are common in and adjacent to former or existing mountain ranges. The folds range from a few inches to 50 miles or so across. *Anticlines* are rocks folded upward into an arch; their counterpart, *synclines*, are folded downward like a valley (Figure 1-4). A *monocline* is a flecture in which the rocks are horizontal, or nearly so, on either side of the flecture.

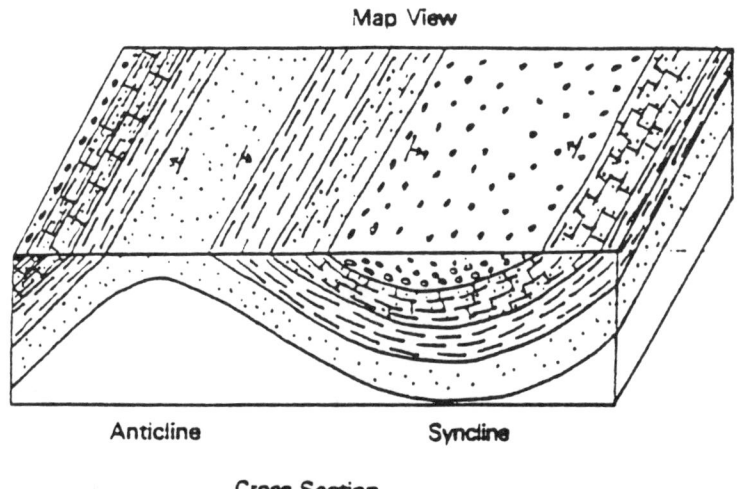

The arrow indicates the direction of dip. In an anticline, the rocks dip away from the crest and in a syncline they dip toward the center.

Figure 1-4 Block diagram of an anticline and syncline (U.S. EPA, 1987a).

Although many rocks have been folded into various structures, this may not be reflected by the topographic features. As uplift proceeds, erosion removes weathering products from the rising mass, and deposits them elsewhere. The final topography is related to the erodibility of the rocks, with resistant strata such as sandstone forming ridges, and the less resistant material such as shale forming valleys. Consequently, the geologic structure of an area may bear little resemblance to its topography.

The structure of an area can be determined from field studies or a geologic map, if one exists. Various types of folds and their dimensions appear as unusual patterns on geologic maps. An anticline, for example, will be depicted as a series of rock units in which the oldest is in the middle, while a syncline is represented by the youngest rock in the center (Figure 1-4). More or less equidimensional anticlines and synclines are termed domes and basins, respectively.

The inclination of the top of a fold is the plunge. Folds may be symmetrical, asymmetrical, overturned, or recumbent in relation to the fold axis. The inclination of the rocks is indicated by dip and strike symbols. The strike is perpendicular to the dip and dip is commonly recorded as the number of degrees with respect to the horizontal plane. The dip may range from less than a degree to vertical.

Fractures. Fractures in rocks are either joints or faults. A *joint* is a fracture along which no movement has taken place; a *fault* implies movement. Movement along faults is as little as a few inches to tens of miles. Probably all consolidated rocks and a good share of the unconsolidated deposits contain joints. Joints exert a major control on ground-water movement and chemical quantity. Characteristically, joints are open and serve as major conduits or pipes. Water can move through them quickly, perhaps carrying contaminants, and, being open, the filtration effect is lost. The outbreak of many waterborne diseases can be traced to ground-water supplies containing infectious agents that have been transmitted through fractures to wells and springs.

Faults are most common in the deformed rocks of mountain ranges, suggesting either lengthening or shortening of the crust. Movement along a fault may be horizontal, vertical, or a combination of both. The most common types of faults are called normal, reverse, and lateral (Figure 1-5). A *normal* fault, which indicates stretching of the crust, is one in which the upper or hanging wall has moved down relative to the lower or foot wall.

Death Valley in California, the Red Sea, and the large lake basins in the east African highlands, among many others, lie in a *graben*, which is a block bounded by normal faults (Figure 1-5). A *reverse* or *thrust* fault implies compression and shortening of the crust. It is distinguished by the fact that the hanging wall has moved up relative to the foot wall. A *lateral* fault is one in which the movement has been largely horizontal. The San Andreas Fault, extending some 600 miles from San Francisco Bay to the Gulf of California, is the most notable lateral fault in the United States.

1.3.4 Geologic Maps and Cross Sections

Geologists use a number of techniques to graphically represent surface and subsurface conditions. Some of the more important methods that may have value in ground-water investigations are described here briefly:

- *Surface geologic maps* depict the geographic extent of formations and their structure at the earth's surface. The map view portion of Figure 1-4 represents such a map.

- *Subsurface geologic maps* show the areal location of buried rock and other geologic units. These include (1) *structure contour* maps that show the elevation of a particular rock unit such as bedrock below glacial deposits or the top of a single rock formation; (2) *isopach maps* that show variations in thickness of a unit and are based largely or entirely on well logs.

Geology, Soils, and Geomorphology 19

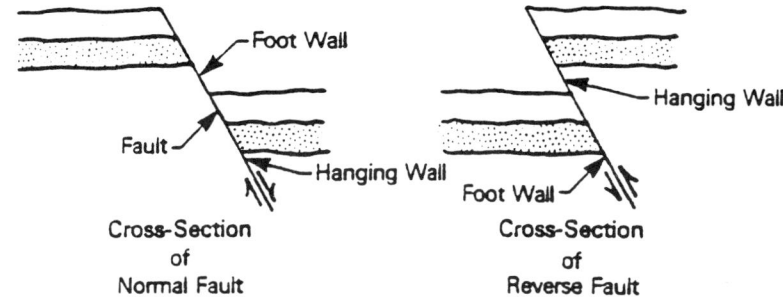

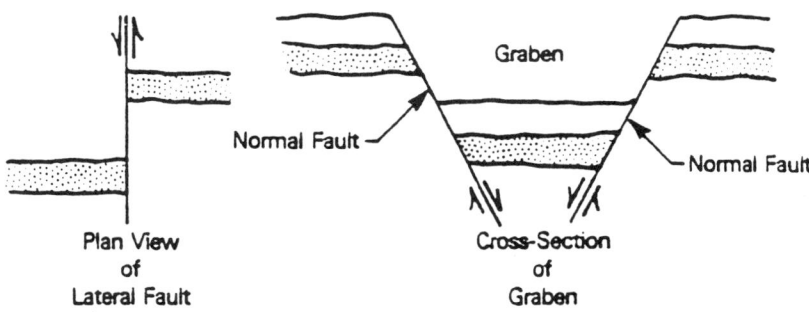

Figure 1-5 Cross sections of normal and reverse faults and a graben; plan view of a lateral fault (U.S. EPA, 1987a).

- *Cross sections* may take several forms: (1) *geologic cross sections* illustrate the subsurface distribution of rock units between points of control, such as outcrops or well bores; (2) *columnar sections* describe the vertical distribution of rock units, their lithology, and thickness; and (3) graphical representation of borehole logs without interpolation as is required for geologic cross sections.

- *Three-dimensional* representations of geologic data can be accomplished by *block diagrams* (Figure 1-4) and *panel diagrams* (also called *fence diagrams*) in which cross sections are combined to create a three-dimensional image (Figure 1-6).

Whatever the graphic techniques, they are not exact because the features they attempt to show are complex, nearly always hidden from view, and difficult to sample. Nevertheless, graphic representations are valuable, if not essential, to subsurface studies.

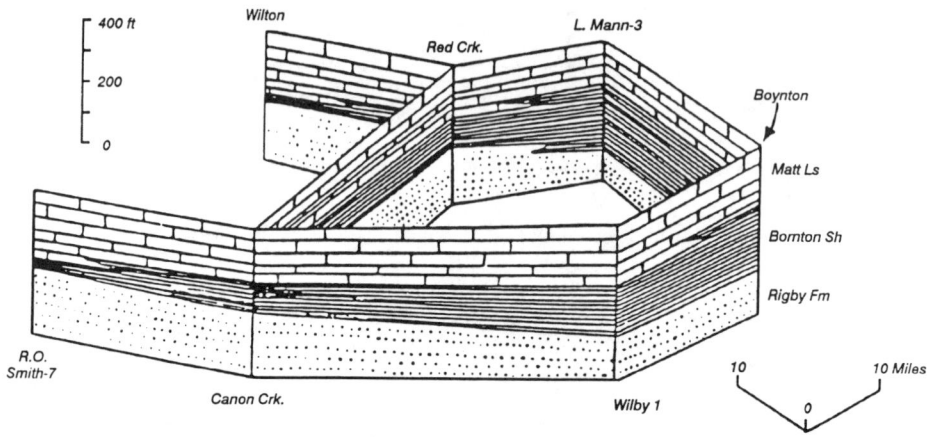

Figure 1-6 Sample fence diagram construction (Mercer and Spalding, 1991b, after Compton, 1962).

1.4 Basic Soil Concepts

Although the term soil is often loosely used to refer to any unconsolidated material, soil scientists distinguish it from other unconsolidated geologic materials by observable features that result from soil-forming processes, such as accumulation of organic matter, formation of soil structure and leaching. Soil forms at the land surface in geologic materials and consequently the study of soil is at the interface between geology (particularly glacial and quaternary geology) and geomorphology.

1.4.1 Factors of Soil Formation

The soil at a particular location is the result of the interaction of five factors: (1) parent material, (2) topography, (3) climate, (4) biota, and (5) time:

- *Parent material*, in the form of unweathered consolidated or unconsolidated geologic material, provides the initial physical and chemical framework for soil formation.

- *Topography* affects soil formation by its influence on erosion and wetness. For example, soils on steep slopes tend to be thin and poorly developed because the rate of erosion tends to counterbalance the effects of weathering by climate and biota.

- *Climate* influences soil formation primarily by the amount of precipitation. Soils can be broadly classified based on the relationship between precipitation and evapotranspiration. In humid climates, where precipitation exceeds evapotranspiration, soluble constituents leach from the soil. Where evapotranspiration exceeds precipitation, as in arid and semi-arid climates, salts tend to accumulate in the soil profile.

- *Biota* affect soil formation primarily through the process of organic matter formation. Vegetation is the major biological factor. For example, in the same parent material, prairie grassland will form an entirely different type of soil than a forest.

- The length of *time* that parent material is subjected to the weathering processes of climate and biological activity strongly influences soil type. A young soil in fresh geologic materials will look very different from a soil in the same material where weathering processes have operated for tens or hundreds of thousands of years.

The interaction of the above factors of soil formation result in the formation of a soil *profile*, the description of which forms the basis for classifying a soil. Specific soil-forming processes that influence soil profile development include (1) organic matter accumulation; (2) weathering of minerals to clays; (3) the depletion of clay and other sesquioxide minerals from upper horizons, called *eluviation*, with subsequent enrichment in lower horizon, called *illuviation*; (4) leaching or accumulation of soluble salts; and (5) the formation of *soil structure* by the aggregation of soil particles into larger units called *peds*, and (6) the formation of slowly permeable layers, such as *fragipans* in humid climates and *duripans* in arid climates.

Perhaps the most distinctive feature of a soil profile are its major horizons:

- The *O horizon*, if present, is a layer of partially decomposed organic material.

- The *A horizon* is a mineral horizon characterized by maximum accumulation of organic matter lying at or near the ground surface. It usually has a distinctly darker color than lower horizons.

- The *E horizon*, whose main feature is the loss of silicate clay, iron or aluminum, is typically found between the A and B horizon (note that soil textbooks published before 1981 call this the B1 horizon). It may also occur within a B horizon above a fragipan.

- The *B horizon* is the zone of most active weathering, is often enriched in clays, and has a well-defined soil structure. In humid climate soluble cations, such as calcium, are often depleted, whereas in dryer climates calcium carbonate and other soluble salts often accumulate in this horizon. Soil formed in recent geologic materials typically are missing a B horizon or it is observable only by a slightly redder color compared to the C horizon.

- The *C horizon* is unconsolidated material that has experienced little or no weathering. In arid zones minerals may precipitate in the C horizon to form cemented petrocalcic layers (also called *caliche*) or duripans (cementation by silica).

- The *R horizon* is solid rock.

Depending on the interaction between the five factors of soil formation at a site, the transport of contaminants in the subsurface can be increased or decreased relative to unweathered materials with similar physical and chemical composition (Sections 1.4.3 and 1.4.4). Many soil properties that affect potential for contaminant transport in the subsurface can be evaluated using soil profile descriptions, prepared using USDA soil description procedures. Section 1.7 identifies major references on these methods.

1.4.2 Soil Classification

The dominant system for classifying soils in the United States is the USDA soil taxonomy. This system went through seven "approximations" before being formalized in 1975 with the publication of **Soil Taxonomy** (Agricultural Handbook 436). It is still an evolving system, which is updated biannually by the Soil Conservation Service's soil survey staff in **Keys to Soil Taxonomy** (the fifth edition was published in 1992). Although the nomenclature may seem intimidating to the uninitiated, the USDA soil classification system is a very useful tool for assessing potential for transport of contaminants in the subsurface. The agricultural origins of the system resulted in a strong emphasis on features affecting soil water and nutrient status. These are the same soil properties that are most significant when evaluating soil and ground water contamination. The following is a brief overview of the system.

The USDA soil taxonomy is a hierarchical classification system with six levels:

- *Orders* and *suborders* form the highest level. The original system had 10 orders (see legend to Figure 1-7) and 47 suborders that are differentiated by the presence or absence of diagnostic horizons and the kind and degree of dominant soil-forming processes. In 1990 an 11th order called andosols was added for soils formed in volcanic ashes.

- *Great groups* and *subgroups* are the next level. There are about 185 great groups and about 970 subgroups. Great groups are differentiated based on the whole assemblage of soils horizons and moisture and temperature regimes. Subgroups are defined based on significant subordinate soil processes that result in intergrades or transitional forms to other orders, suborders, or great groups.

- *Families* are a category within a subgroup that have similar physical and chemical properties including (1) particle-size distribution, (2) mineralogy, (3) temperature regime, and (4) rooting depth. About 4,500 families are currently recognized in the United States.

- *Series* is the lowest category in the system and is defined by a more limited range of characteristics than the family grouping. Over 10,500 soil series have been recognized in the United States. In soil mapping a soil series is further subdivided into *map units* reflecting occurrence on different slope classes and sometimes degrees of erosion. However, soil map units are not considered to be a level in the classification system.

Figure 1-7 shows the patterns of soil orders and suborders in the United States. The legend describes their salient characteristics and the origin of major word roots. Familiarity with this system provides a powerful tool for interpreting soil conditions at any site where a soil survey conducted by the U.S. Soil Conservation Service is available.

1.4.3 Soil Physical Properties

Soil physical properties such as texture (see Section 1.1.2), structure, and pore-size distribution are the major determinants in water movement in soil and consequently of major concern in ground-water contamination studies. Depending on the specific soil, water movement may be enhanced or retarded compared to unweathered geologic materials. Organic matter enhances water-holding capacity and infiltration. The formation of soil structure also enhances permeability, particularly in clayey soils. Buried soil horizons form zones of preferential lateral movement of contaminants in the subsurface, which may be overlooked by environmental professionals who are not trained to recognize such horizons. On the other hand, the formation of restrictive layer such as fragipans may substantially reduce infiltration compared to unweathered materials.

The study of soil water is primarily the domain of the soil physicist. Physical properties affecting movement of water in soil are discussed further in Sections 2.2.2 (Infiltration) and Section 2.4 (Water in the Soil and Vadose Zone). Soil micromorphology and fabric analysis are methods for studying other soil physical properties. These methods typically involve the preparation of thin sections and examination of pores and other ordered features through a microscope.

Micromorphological and general fabric analysis of soil is used infrequently in the study of ground-water contamination, more because of unfamiliarity with the methods than their lack of value. For example, Collins and McGown (1981) used micromorphologic and fabric analysis of layered alluvial soils and glacial soils to

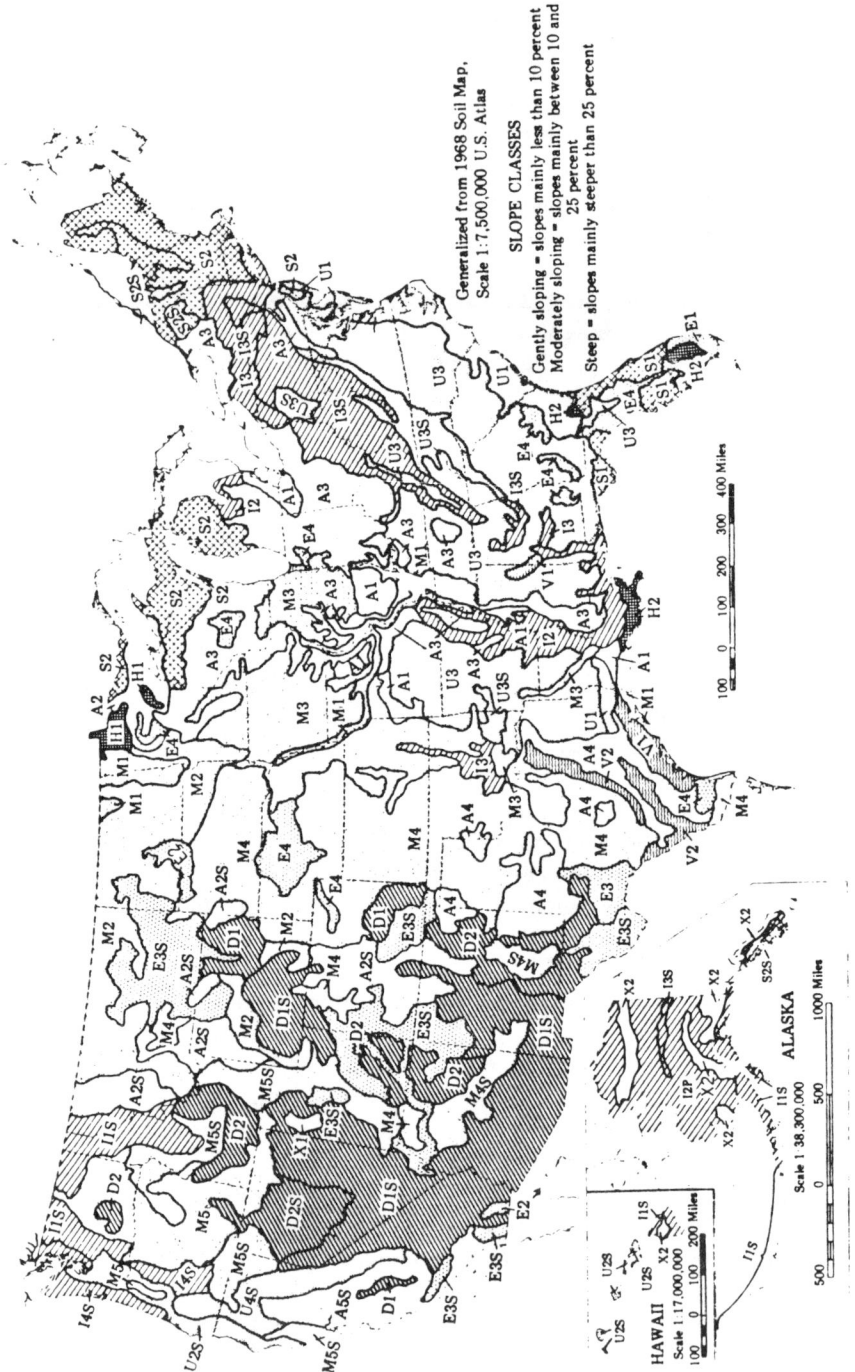

Figure 1-7 Patterns of soil orders and suborders of the United States (Birkeland, 1984).

Geology, Soils, and Geomorphology

Figure 1-7 Legend

LEGEND

Only the dominant orders and suborders are shown. Each delineation has many inclusions of other kinds of soil. General definitions for the orders and suborders follow. For complete definitions, see Soil Survey Staff.[12] Approximate equivalents in the modified 1938 soil classification system are indicated for each suborder.

ALFISOLS ... Soils with gray to brown surface horizons, medium to high base supply, and subsurface horizons of clay accumulation; usually moist but may be dry during warm season

A1 AQUALFS (seasonally saturated with water) gently sloping; general crops if drained, pasture and woodland if undrained (Some Low-Humic Gley soils and Planosols)

A2 BORALFS (cool or cold) gently sloping; mostly woodland, pasture, and some small grain (Gray Wooded soils)

A2S BORALFS steep; mostly woodland

A3 UDALFS (temperate, or warm, and moist) gently or moderately sloping; mostly farmed, corn, soybeans, small grain, and pasture (Gray-Brown Podzolic soils)

A4 USTALFS (warm and intermittently dry for long periods) gently or moderately sloping; range, small grain, and irrigated crops (Some Reddish Chestnut and Red-Yellow Podzolic soils)

A5S XERALFS (warm and continuously dry in summer for long periods, moist in winter) gently sloping to steep; mostly range, small grain, and irrigated crops (Noncalcic Brown soils)

ARIDISOLS ... Soils with pedogenic horizons, low in organic matter, and dry more than 6 months of the year in all horizons

D1 ARGIDS (with horizon of clay accumulation) gently or moderately sloping; mostly range, some irrigated crops (Some Desert, Reddish Desert, Reddish Brown, and Brown soils and associated Solonetz soils)

D1S ARGIDS gently sloping to steep

D2 ORTHIDS (without horizon of clay accumulation) gently or moderately sloping; mostly range and some irrigated crops (Some Desert, Reddish Desert, Sierozem, and Brown soils, and some Calcisols and Solonchak soils)

D2S ORTHIDS gently sloping to steep

 ENTISOLS ... Soils without pedogenic horizons

E1 AQUENTS (seasonally saturated with water) gently sloping; some grazing

E2 ORTHENTS (loamy or clayey textures) deep to hard rock; gently to moderately sloping; range or irrigated farming (Regosols)

E3 ORTHENTS shallow to hard rock; gently to moderately sloping; mostly range (Lithosols)

E3S ORTHENTS shallow to rock; steep; mostly range

E4 PSAMMENTS (sand or loamy sand textures) gently to moderately sloping; mostly range in dry climates, woodland or cropland in humid climates (Regosols)

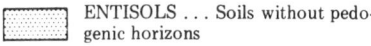 HISTOSOLS ... Organic soils

H1 FIBRISTS (fibrous or woody peats, largely undecomposed) mostly wooded or idle (Peats)

H2 SAPRISTS (decomposed mucks) truck crops if drained, idle if undrained (Mucks)

 INCEPTISOLS ... Soils that are usually moist, with pedogenic horizons of alteration of parent materials but not of accumulation

I1S ANDEPTS (with amorphous clay or vitric volcanic ash and pumice) gently sloping to steep; mostly woodland; in Hawaii mostly sugar cane, pineapple, and range (Ando soils, some Tundra soils)

I2 AQUEPTS (seasonally saturated with water) gently sloping; if drained, mostly row crops, corn, soybeans, and cotton; if undrained, mostly woodland or pasture (Some Low-Humic Gley soils and Alluvial soils)

(continued)

Figure 1-7 Legend (continued)

I2P AQUEPTS (with continuous or sporadic permafrost) gently sloping to steep; woodland or idle (Tundra soils)

I3 OCHREPTS (with thin or light-colored surface horizons and little organic matter) gently to moderately sloping; mostly pasture, small grain, and hay. (Sols Bruns Acides and some Alluvial soils)

I3S OCHREPTS gently sloping to steep; woodland, pasture, small grains

I4S UMBREPTS (with thick dark-colored surface horizons rich in organic matter) moderately sloping to steep; mostly woodland (Some Regosols)

MOLLISOLS... Soils with nearly black, organic-rich surface horizons and high base supply

M1 AQUOLLS (seasonally saturated with water) gently sloping; mostly drained and farmed (Humic Gley soils)

M2 BOROLLS (cool or cold) gently or moderately sloping, some steep slopes in Utah; mostly small grain in North Central States, range and woodland in Western States (Some Chernozems)

M3 UDOLLS (temperate or warm, and moist) gently or moderately sloping; mostly corn, soybeans, and small grains (Some Brunizems)

M4 USTOLLS (intermittently dry for long periods during summer) gently to moderately sloping; mostly wheat and range in western part, wheat and corn or sorghum in eastern part, some irrigated crops (Chestnut soils and some Chernozems and Brown soils)

M4S USTOLLS moderately sloping to steep; mostly range or woodland

M5 XEROLLS (continuously dry in summer for long periods, moist in winter) gently to moderately sloping; mostly wheat, range, and irrigated crops (Some Brunizems, Chestnut, and Brown soils)

M5S XEROLLS moderately sloping to steep; mostly range

SPODOSOLS... Soils with accumulations of amorphous materials in subsurface horizons

S1 AQUODS (seasonally saturated with water) gently sloping; mostly range or woodland; where drained in Florida, citrus and special crops (Ground-Water Podzols)

S2 ORTHODS (with subsurface accumulations of iron, aluminum, and organic matter) gently to moderately sloping; woodland, posture, small grains, special crops (Podzols, Brown Podzolic soils)

S2S ORTHODS steep; mostly woodland

ULTISOLS... Soils that are usually moist, with horizon of clay accumulation and a low base supply

U1 AQUULTS (seasonally saturated with water) gently sloping; woodland and pasture if undrained, feed and truck crops if drained (Some Low-Humic Gley soils)

U2S HUMULTS (with high or very high organic matter content) moderately sloping to steep; woodland and pasture if steep, sugar cane and pineapple in Hawaii, truck and seed crops in Western States (Some Reddish-Brown Lateritic soils)

U3 UDULTS (with low organic-matter content; temperate or warm, and moist) gently to moderately sloping; woodland, pasture, feed crops, tobacco, and cotton (Red-Yellow Podzolic soils, some Reddish-Brown Lateritic soils)

U3S UDULTS moderately sloping to steep, woodland, pasture

U4S XERULTS (with low to moderate organic-matter content, continuously dry for long periods in summer) range and woodland (Some Reddish-Brown Lateritic soils)

VERTISOLS... Soils with high content of swelling clays and wide deep cracks at some season

V1 UDERTS (cracks open for only short periods, less than 3 months in a year) gently sloping; cotton, corn, pasture, and some rice (Some Grumusols)

V2 USTERTS (cracks open and close twice a year and remain open more than 3 months); general crops, range, and some irrigated crops (Some Grumusols)

Geology, Soils, and Geomorphology

Figure 1-7 Legend (continued)

☐ AREAS with little soil . . .

X1 Salt flats X2

X2 Rock land (plus ice fields in Alaska)

NOMENCLATURE

The nomenclature is systematic. Names of soil orders end in *sol* (L. *solum*, soil), e.g., ALFISOL, and contain a formative element used as the final syllable in names of taxa in suborders, great groups, and subgroups.

Names of suborders consist of two syllables, e.g., AQUALF. Formative elements in the legend for this map and their connotations are as follows:

and	— Modified from Ando soils; soils from vitreous parent materials
aqu	— L. *aqua*, water; soils that are wet for long periods
arg	— Modified from L. *argilla*, clay; soils with a horizon of clay accumulation
bor	— Gr. *boreas*, northern; cool
fibr	— L. *fibra*, fiber; least decomposed
hum	— L. *humus*, earth; presence of organic matter
ochr	— Gr. base of ochros, pale; soils with little organic matter
orth	— Gr. *orthos*, true; the common or typical
psamm	— Gr. *psammos*, sand; sandy soils
sapr	— Gr. *sapros*, rotten; most decomposed
ud	— L. *udus*, humid; of humid climates
umbr	— L. *umbra*, shade; dark colors reflecting much organic matter
ust	— L. *ustus*, burnt; of dry climates with summer rains
xer	— Gr. *xeros*, dry; of dry climates with winter rains

evaluate discontinuities for engineering purposes.[2] Paglai et al. (1981) used micromorphological methods to evaluate the effect of sewage sludges applied to soil on pore size and density.[3]

1.4.4 Soil Chemical Properties

Minerals in the soil are the chemical signature of the bedrock from which they originated. Rainfall and temperature are two significant factors that dictate the rate and extent to which mineral solids in the soil react with water. As water passes through soil horizons, it dissolves the chemical remnants of the parent material. In arid and semiarid climates dissolved constituents often precipitate in a lower horizon when plants transpire soil water to the atmosphere. In humid climates, soil water that is not taken up by plant roots carries the dissolved minerals to the ground water. The more water that flows through the soil, the more solids react with the undersaturated solvent.

Organic matter and clay content are major parameters of importance in studying the transport and fate of contaminants in soil. The geochemical properties of clay have been described in Section 1.1.1, and the importance of organic matter in adsorption of organic chemicals in Section 3.4.4. Chapter 3 covers basic concepts related to soil chemistry.

1.5 Geomorphology and Ground Water

Geomorphology is the study of the evolution of surface landforms. Careful observation of surface features at a site (landforms, streams and stream patterns, locations of springs, seeps, and lakes, as well as vegetation) may reveal considerable information about both geology and ground water. Landforms are controlled by the geology and many hills are capped by resistant strata, such as sandstone, while valleys are usually carved into soft, less resistant material, such as shale. Likewise, many changes in topographic slope are related to differences in rock type. These, in turn, provide a general impression of the types of rocks present, their areal extent, and composition. Rock exposures in stream channels and road cuts are very useful also when attempting to understand the local geology. Large scale fracture systems can be mapped as linear features on aerial photographs. Lineations on aerial photographs may also serve as indicators of changes in lithology and geologic structure. At a smaller scale, joint and fracture systems, their directional trends, density, and size can

---

[2] Collins, K. and A. McGown. 1981. Micromorphological Studies in Soil Engineering. In: Soil Micromorphology, Vol. 1 Techniques and Applications, P. Bullock and C.P. Murphy (eds.), Academic Publishers, New York, pp. 195-217.

[3] Paglai, M., M. La Marca, and G. Lucamante. 1981. Micromorphological Investigation of the Effect of Sewage Sludges Applied to Soil. In: Soil Micromorphology, Vol. 1 Techniques and Applications, P. Bullock and C.P. Murphy (eds.), Academic Publishers, New York, pp. 219-225.

Geology, Soils, and Geomorphology 29

all be measured on rock outcrops. Fluid movement through joints and other fractures may be the controlling factor for migration of contaminants.

Geomorphology is a logical starting point for site investigations, because it allows preliminary interpretations of subsurface conditions without the cost of drilling or other subsurface investigation methods. For example, Hatheway and Bliss (1980) used surficial geologic maps to develop geomorphic units of similar engineering and hydrogeologic properties as a starting point for evaluating siting options for hazardous waste facilities.[4]

1.5.1 Hydrogeomorphology

A lot can be inferred about subsurface flow of water from examination of a topographic map, because slope steepness and shape strongly influences how much precipitation enters the ground. Figure 1-8 illustrates a number of geomorphic and hillslope components. Refer to this figure for help in visualizing the following common relationships between surface runoff or infiltration (the entry of water into the soil) and geomorphic and hillslope features:

- *Headslopes* concentrate surface runoff; *noseslopes* disperse surface runoff.

- Infiltration is usually highest on *footslopes* and *toeslopes*, followed by interfluvs/hill summits, and lowest on shoulders and backslopes. At all topographic situations, infiltration is highest in dry soils and slows as the soil gets wetter.

- Surface runoff is usually greatest on steep surfaces such as *headslopes/sideslopes* and *shoulders/backslopes*, lowest on flat surfaces (broad interfluvs and alluvial fill). At all topographic positions surface runoff is at a maximum when the soil is saturated.

- A *concave* sideslope will concentrate water in the soil more than a *convex* sideslope (this is not explicitly illustrated in Figure 1-8, but the principle is the same as the headslope/hillslope relationship).

- Alluvial fill will usually have more ground water than interfluvs.

The above relationships can be useful in developing a preliminary conceptual model of how water is flowing in the subsurface. Very subtle changes in surface topography may affect the distribution of water between the surface and ground. These are often evident in vegetation, with relative greenness marking differences in the availability of ground water. Vegetation can sometimes also be used to map

---

[4] Hatheway, A.W. and Z.F. Bliss. 1980. Geomorphology as an Aid to Hazardous Waste Facility Siting, Northeast United States. In: Applied Geomorphology, R.G. Craig and J.L. Craft (eds.), George Allen & Unwin, Boston, pp. 55-71.

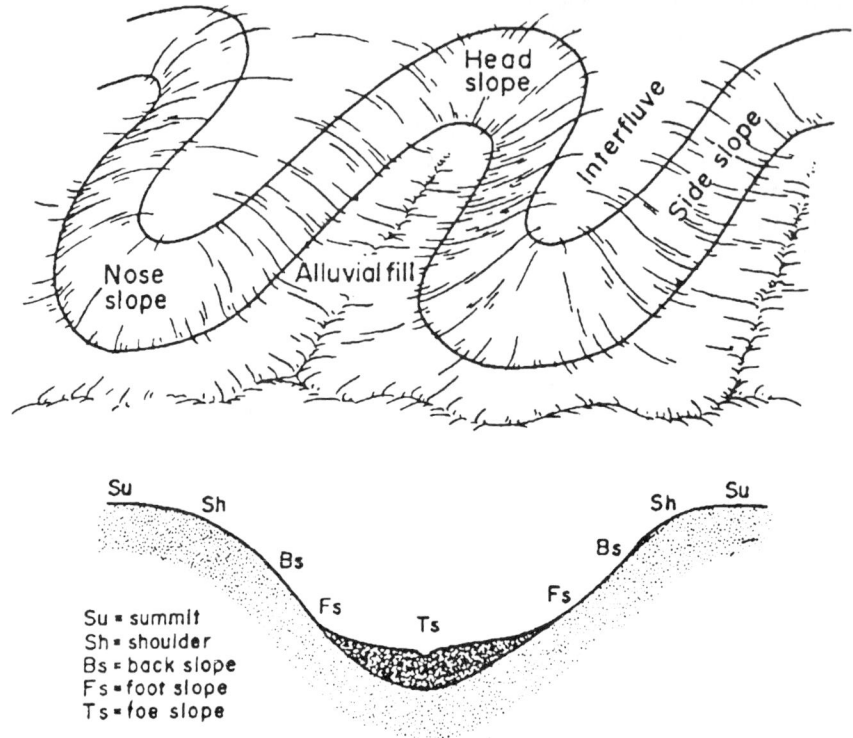

Figure 1-8 Geomorphic and hillslope components (Mausbach and Nielsen, 1991, after Ruhe and Walker, 1968). Reprinted with permission.

certain rock types. For example, cader trees can be an indicator of limestone bedrock. Springs and seeps are zones of ground-water discharge. They develop in the vicinity of strata of low permeability that are overlain by a unit of greater permeability.

Stream patterns also are related to the geology, especially geologic structure and fracture or joint systems. Regional stream patterns provide an idea of the relative difference in discharge from one stream to another. Surface streams can also provide useful information on basin permeability, shallow ground-water quality, and local sites where the ground water is contaminated (Section 2.3).

1.5.2 Karst Geomorphology and Hydrology

The term *karst*, named after the Dinaric Karst limestone region of the former Yugoslavia, refers to a distinctive set of geomorphic landforms resulting from the development of extensive subsurface solution channels and caves in carbonate rocks. These channels form where circulating ground water has dissolved carbonates along fractures and bedding planes (see Figure 1-9). Karst terrane (also spelled terrain in the karst literature) is usually characterized by sinkholes and general absence of perennial surface streams. Springs are abundant where impermeable rock below the cavernous limestone crops out at the surface.

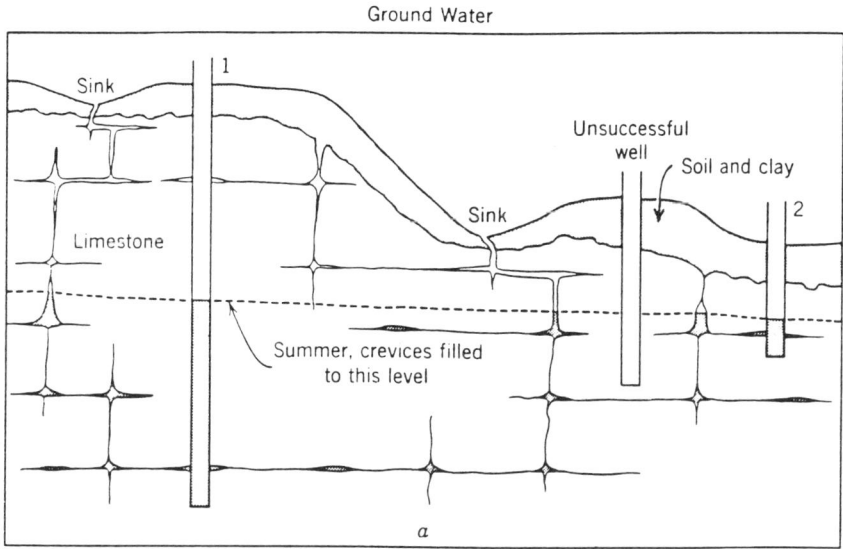

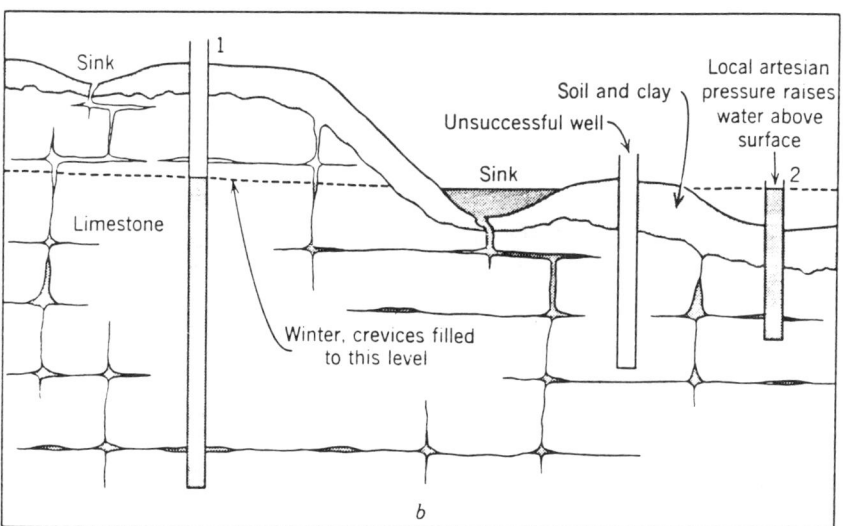

Figure 1-9 Diagram of a karst aquifer showing seasonal artesian conditions (Walker, 1956).

Conduit flow, which does not obey Darcy's law (Section 2.5.4), is a salient characteristic of karst aquifers. As the term implies, flow in solution channels is rapid, more like flow in a pipe or open channel. This feature of karst aquifers make them characteristically idiosyncratic in behavior, and surface water entering such a system may reappear at unexpected locations and at different locations depending on whether low- or high-flow conditions exist. As shown in Figure 1-9, large fluctuations can result in seasonal artesian conditions. Ground-water tracing experiments are the only way that karst ground-water flow patterns can be accurately characterized (Section 8.4).

Karst areas are troublesome water sources even though they can provide large quantities of water to wells and springs. Rapid infiltration rates limited filtering action to retard contaminants, pollutants move rapidly once they reach a karst conduit and are less attenuated by adsorption compared to porous aquifers (Field, 1989).[5] Consequently, karst terrane is generally unsuitable for the disposal of polluting wastes.

Figure 1-10 shows the distribution of karst areas in the United States. Near-surface karst areas are shaded, and other areas with carbonate or sulfate rocks near the surface are stippled. Karst areas in this figure are divided into four major regions: A = Atlantic and Gulf Coastal Plain; B = east-central region of Paleozoic and other old rock; C = Great Plains; and D = western mountain region. The distinctive geomorphic and hydrogeologic features of karst terrane have resulted in a scientific literature that is probably out of proportion to its actual distribution on the face of the earth (see next section).

1.6 Geologic Settings of Ground Water Occurrence and Quality

The occurrence of ground water is intimately related to its geologic setting. Heath (1982)[6] describes 12 major ground-water regions in the continental United States based on geologic setting: (1) Western mountain ranges, (2) alluvial basins, (3) Columbia lava plateau, (4) Colorado plateau and Wyoming, (5) high plains, (6) nonglaciated central region, (7) glaciated central region (Figure 1-3), (8) Piedmont Blue Ridge region, (9) northeast and superior uplands, (10) Atlantic and Gulf coastal plain, (11) southeast coastal plain, and (12) alluvial valleys. Figure 1-11 shows the boundaries of the first eleven regions. The alluvial valleys region consists mainly of the floodplains of the Mississippi, Missouri and Ohio Rivers. Aller et al. (1987/T11-10) have further subdivided Heath's major regions into 85 subregions for purposes of evaluating ground-water pollution potential. Section 11.2.3 discusses ground-water vulnerability mapping further.

1.6.1 Ground Water in Igneous and Metamorphic Rocks

Nearly all of the porosity and permeability of igneous and metamorphic rocks are the result of secondary openings such as fractures and faults and the dissolution of certain minerals. A few notable exceptions include large lava tunnels present in

---

[5] Field, M.S. 1989. The Vulnerability of Karst Aquifers to Chemical Contamination. In: Recent Advances in Ground-Water Hydrology, J.E. Moore et al. (eds.), American Institute of Hydrology, Minneapolis, MN, pp. 130-142.

[6] Heath, R.C. 1982. Classification of Ground-Water Systems of the United States. Ground Water 20(4):393-401. Additional information on Heath's ground water regions can be found in U.S. Geological Professional Paper 2242 published in 1984.

Geology, Soils, and Geomorphology 33

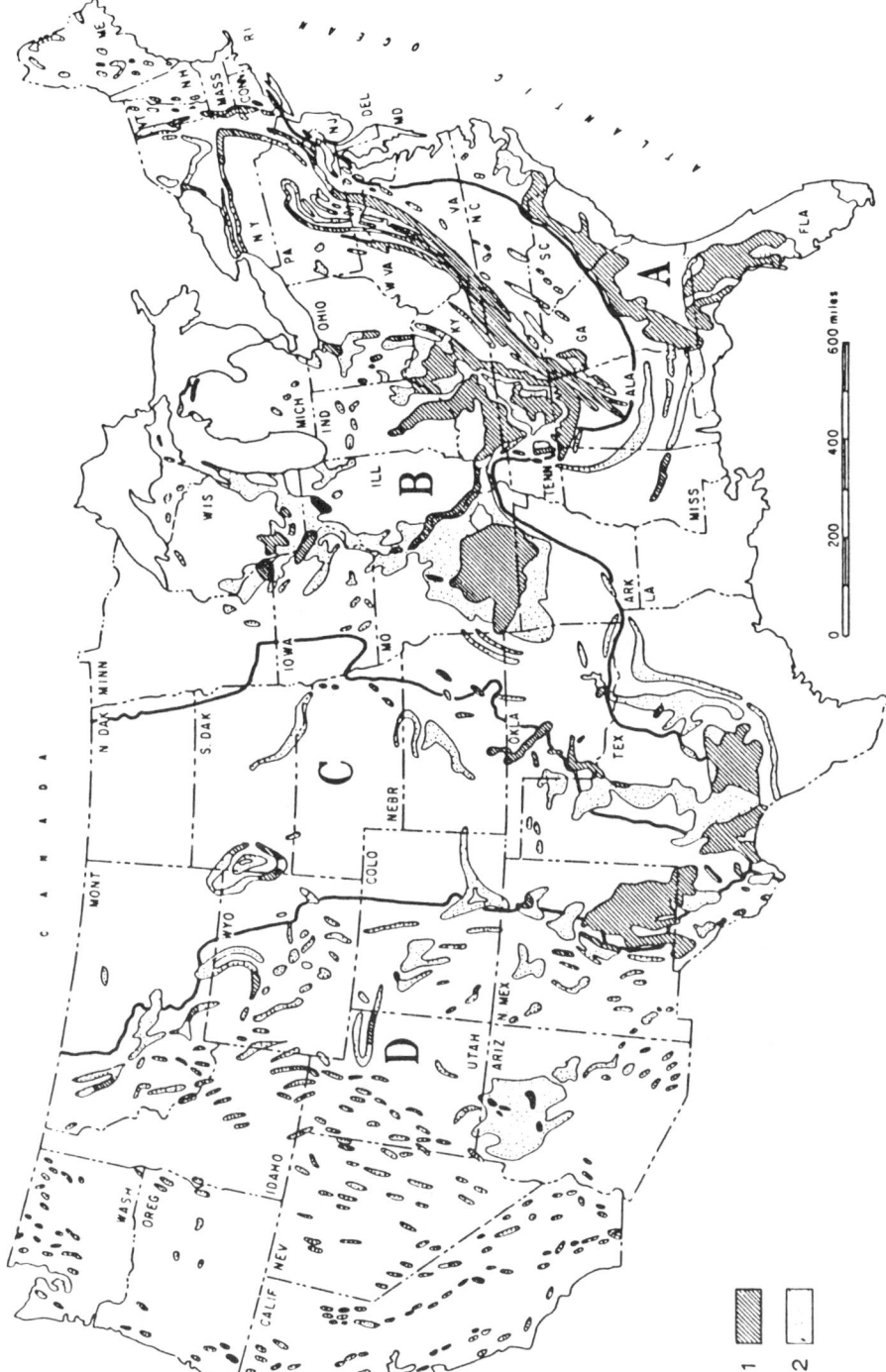

Figure 1-10 Distribution of karst areas in relation to carbonate and sulfate rocks in the United States (Davies and LeGrand, 1972; reprinted by permission). 1 = karst areas; 2 = carbonate and sulfate rocks at or near the surface.

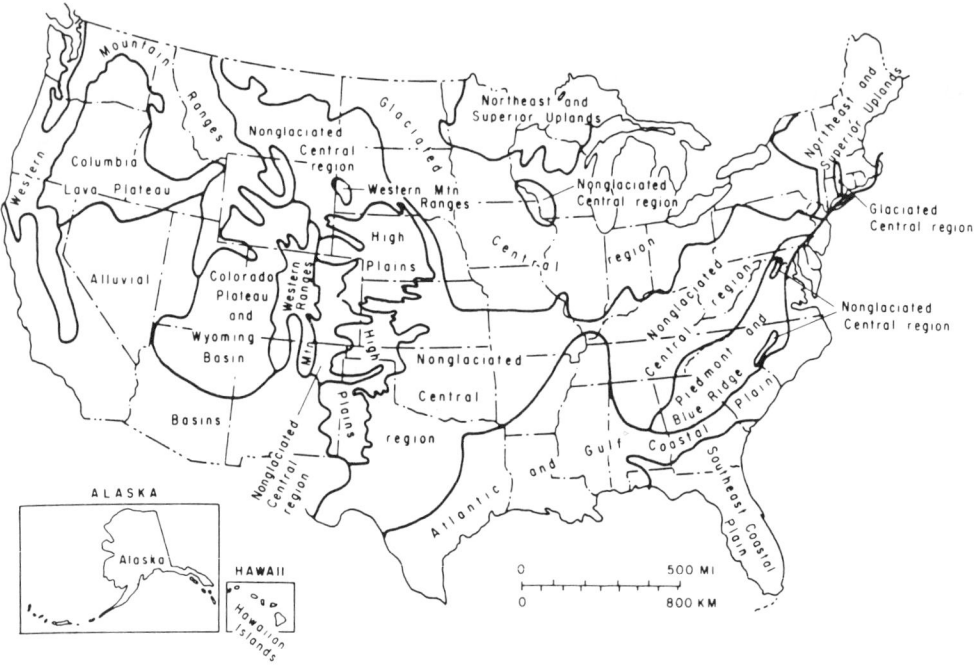

Figure 1-11 Major ground-water regions of the United States (Heath, 1984).

some flows, interflow or coarse sedimentary layers between individual lava flows, and deposits of selected pyroclastic materials.

Because the openings in igneous and metamorphic rocks are quite small volumetrically, rocks of this type are poor suppliers of ground water. The supplies that are available commonly drain rapidly after a period of recharge by infiltration of precipitation. In addition they are subject to contamination from the surface where these rocks outcrop.

Evaluating water and contaminant movement in fractured rocks is difficult, because the actual direction of movement may not be in the direction of decreasing head, but rather in some different though related direction. The problem is further compounded by the difficulty in locating the fractures. Because of these characteristics, evaluating water availability, direction of movement, and velocity is exceedingly difficult.

Unless some special circumstance exists, such as where rocks crop out at the surface, water obtained from igneous and metamorphic rocks is nearly always of excellent chemical quality. Dissolved solids are present in crystalline rocks; however, they are commonly in concentrations of less than 100 mg/L. In the case of water from metamorphosed carbonate rocks, moderate to high concentrations of hardness may be found.

1.6.2 Ground Water in Sedimentary Rocks

Usable supplies of ground water can be obtained from all types of sedimentary rocks, but the fine-grained strata such as shale and siltstone may only provide a few gallons per day and even this can be highly mineralized. Even though fine-grained rocks may have relatively high porosities, their primary permeability is usually low. On the other hand, shale is likely to contain a great number of joints that are both closely spaced and extend to considerable depths. Therefore, in this case, rather than being impermeable, they can be quite transmissive. This is of considerable importance in waste disposal schemes because of the potential for flow through fractures. In addition, leachate formed as water infiltrates through waste might be small in quantity but highly mineralized. Because of the low bulk permeability, it would be difficult to pump out the contaminated water or even to properly locate monitoring wells.

From another perspective, fine-grained sedimentary rocks, owing to their high porosity, can store huge quantities of water. Some of this water can be released to adjacent aquifers when a head difference is developed due to pumping. On a regional scale, fine-grained confining units provide a great deal of water to aquifer systems. The porosity, however, decreases with depth because of compaction brought about by the weight of overlying sediments.

The porosity of sandstones ranges from less than 1% to a maximum of about 30%. This is a function of sorting, grain shape, and cementation. Cementation can vary both in space and time, and, on outcrops, cementation can differ greatly from that in the subsurface.

As is the case in igneous and metamorphic rocks, fractures also play an important role in the movement of fluids through sandstones. Transmissivities may be as much as two orders of magnitude greater in a fractured rock than in an unfractured part of the same geologic formation.

Sandstone units that were deposited in a marine or near-marine environment can be very widespread, covering tens of thousands of square miles, such as the St. Peter Sandstone of Cambrian age. Those representing ancient alluvial channel fills, deltas, and related environments of deposition are more likely to be discontinuous and erratic in thickness. Individual units are exceedingly difficult to trace in the subsurface. Regional ground-water flow and storage may be strongly influenced by the geologic structure.

Carbonate rocks are formed in many different environments and the original porosity and permeability are modified rapidly after burial. Some special carbonate rocks, such as coquina and some breccias, which tend to have a coarse texture, may remain very porous and permeable, but these are exceptions. When calcite changes to dolomite ($CaMg(CO_3)_2$), the resulting 13% reduction in volume creates considerable pore space. High yielding aquifers develop from fractures and other secondary openings in carbonate formations (see discussion of karst in Section 1.5.2).

1.6.3 Ground Water in Unconsolidated Sediments

Unconsolidated sediments accumulate in many different environments, all of which leave their mark on the characteristics of the deposit. Some are thick and areally extensive, as the alluvial fill in the Basin and Range Province; others are exceedingly long and narrow, such as the alluvial deposits along streams and rivers; and others may cover only a few hundred square feet, for example, some glacial forms. In addition to serving as major aquifers, unconsolidated sediments are also important as sources of raw materials for construction.

Closely related to sorting, the porosities of unconsolidated materials range from less than 1 to more than 90%, the latter representing the porosity of uncompacted mud. Permeabilities also range widely. Cementing of some type and degree is probably universal, but not obvious, with silt and clay being the predominant form.

Most unconsolidated sediments owe their emplacement and texture (sorting, grain size, etc.) to running water. Water as an agent of transport varies in both volume and velocity, which are climate dependent, and this variation leaves an imprint on the sediments. Stream-related, unconsolidated material varies in extent, thickness, and grain size. The water-bearing properties of glacial drift are highly variable, but stratified drift is more uniform and better sorted than glacial till. Some knowledge of the stratigraphy of the most common depositional environments is essential for adequate characterization.

1.6.4 Regional Relationships in Ground-Water Quality

As water infiltrates in a recharge area, the mineral content is relatively low. The quality changes, however, along the flow path and dissolved solids as well as other constituents increase with increasing distances traveled in the ground. The water eventually flows into a stream or body of surface water and, due to the different lengths of flow paths and rock solubility, even streams and small lakes in close proximity may differ greatly in both flow and quality.

The availability of ground-water supplies and their chemical quality are closely related to precipitation. As a general rule, the least mineralized water, both in streams and underground, occurs in areas of the greatest amount of rainfall. Inland, precipitation decreases, water supplies diminish, and quality deteriorates. Because water-bearing rocks exert a strong influence on ground-water quality, however, the solubility of the rocks may override the role of precipitation.

Where precipitation exceeds 40 inches per year, shallow ground water usually contains less than 500 mg/L and commonly less than 250 mg/L of dissolved solids. Where precipitation ranges between 20 and 40 inches, dissolved solids may range between 400 and 1,000 mg/L, and in drier regions dissolved solids commonly exceed 1,000 mg/L.

The dissolved solids concentration of ground water increases toward the interior of the continent. The increase is closely related to precipitation and the solubility of the aquifer framework. The least mineralized ground water is found in a broad belt that extends southward from the New England States along the Atlantic Coast to Florida, and then continues to parallel much of the Gulf Coast. Similarly, along the Pacific Coast from Washington to central California the mineral content is also very low. Throughout this belt, dissolved solids concentrations are generally less than 250 mg/L and commonly less than 100 mg/L (Figure 1-12).

The Appalachian region consists of a sequence of strata that range from nearly horizontal to complexly folded and faulted. Likewise, ground-water quality in this region is also highly variable, being generally harder and containing more dissolved minerals than water along the coastal belt. Much of the difference in quality, however, is related to the abundance of carbonate aquifers, which provide waters rich in calcium and magnesium.

Westward from the Appalachian Mountains to about the position of the 20-inch precipitation line (eastern North Dakota to Texas), dissolved solids in ground water progressively increase. They are generally less than 1,000 mg/L and are most commonly in the 250- to 750-mg/L range. The water is moderately to very hard, and in some areas concentrations of sulfate and chloride are excessive.

From the 20-inch precipitation line westward to the northern Rocky Mountains, dissolved solids are in the 500- to 1,500-mg/L range. Much of the water from glacial drift and bedrock formations is very hard and contains significant concentrations of calcium sulfate. Other bedrock formations may contain soft sodium bicarbonate, sodium sulfate, or sodium chloride water.

Throughout much of the Rocky Mountains, ground-water quality is variable, although the dissolved solids concentrations commonly range between 250 and 750 mg/L. Stretching southward from Washington to southern California, Arizona, and New Mexico is a vast desert region. Here the difference in ground-water quality is wide and dissolved solids generally exceed 750 mg/L. In the central parts of some desert basins, the ground water is highly mineralized, but along the mountain flanks the mineral content may be quite low.

Extremely hard water is found over much of the interior lowlands, Great Plains, Colorado Plateau, and Great Basin. Isolated areas of high hardness are present in northwestern New York, eastern North Carolina, the southern tip of Florida, northern Ohio, and parts of southern California. In general, the hardness is of the carbonate type.

On a regional level, chloride does not appear to be a significant problem, although it is troublesome locally due largely to industrial activities, the intrusion of seawater caused by overpumping coastal aquifers, or interaquifer leakage related to pressure declines brought about by withdrawals.

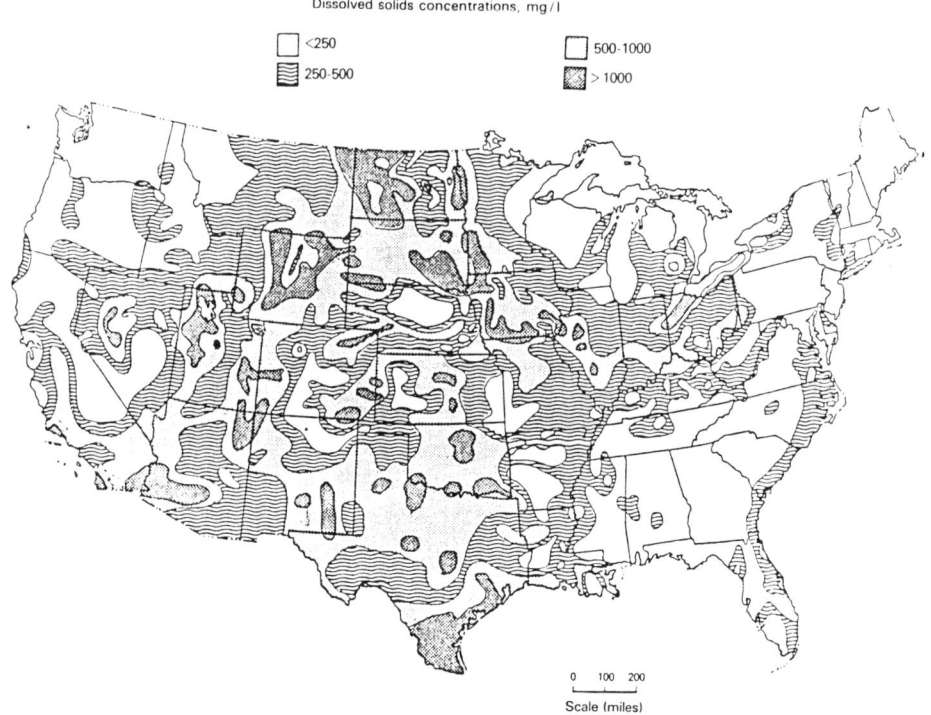

Figure 1-12 Dissolved solids concentrations in ground water used for drinking in the United States (U.S. EPA, 1987a, after Pettyjohn et al., 1979).

In many locations, sulfate levels exceed the federal recommended limit of 250 mg/L; regionally, sulfate may be a problem only in the Great Plains, eastern Colorado Plateau, Ohio, and Indiana. Iron problems are ubiquitous; concentrations exceeding only 0.3 mg/L will cause staining of clothing and fixtures. Fluoride is abnormally high in several areas, particularly parts of western Texas, Iowa, Illinois, Indiana, Ohio, New Mexico, Wyoming, Utah, Nevada, Kansas, New Hampshire, Arizona, Colorado, North and South Dakota, and Louisiana.

1.7 Guide to Major References

Table 1-3 identifies major text references in five major areas: (1) geology, (2) soils, (3) geomorphology, (4) interfaces between geology, soils and geomorphology, and (5) engineering applications. The last category, soils and geologic engineering, have not been discussed in any detail in this chapter, but are an essential element in planning and design for remediation of contaminated soils and ground water.

Almost any text on physical geology, stratigraphy, and structural geology, general soils, and geomorphology will provide more in-depth coverage of topics covered in this chapter, but generally will not emphasize principles as they relate to soil and groundwater contamination. This chapter has emphasized the value of USDA Soil

Conservation Service (SCS) soil description and survey methods for contaminant investigations, and more needs to be said about key reference sources in this area.

For years the standard reference for soil horizon nomenclature was the 1962 supplement to the **Soil Survey Manual** (Soil Survey Staff, 1951, 1962) and elaborated further in **Soil Taxonomy** (Soil Survey Staff, 1975).[7] In 1981 conventions for describing soil horizons and subordinate distinctions within master horizons were significantly changed as part of a comprehensive revision of the 1951 **Soil Survey Manual**, which has recently become available (Soil Survey Staff, 1994a). The best up-to-date source of official SCS horizon designations and naming conventions can be found in the **Keys to Soil Taxonomy** (Soil Survey Staff, 1994b), which is updated biannually.

None of the above documents are specifically oriented toward use of SCS soil description at contaminated sites. In 1988, the Indiana Department of Environmental Management (IDEM) developed draft guidelines for describing unconsolidated deposits at hazardous waste sites, which adapted SCS soil description methods for this purpose.[8] In 1991 U.S. EPA's Center for Environmental Research Information published **Description and Sampling of Contaminated Soils: A Field Pocket Guide** (Boulding, 1991), which presented a detailed, field-oriented adaptation of SCS soil description procedures for use with EPA's Environmental Sampling Expert Systems (Cameron, 1991). A second edition of the 1991 guide was published two years later (Boulding, 1993), which incorporated new SCS procedures for describing soil wetness conditions. These were adopted in 1992 and represent a significant improvement in the ability to characterize soil hydrology based on soil morphology.

Table 1-4 provides an index of major references on karst in the following categories: (1) hydrology and ground water, (2) karst tracing, (3) geomorphology and geology, (4) geochemistry, (5) engineering applications, (6) environmental applications, and (7) major symposia. Ford and Williams (1989) is probably the best single text that covers both karst geomorphology and hydrology with a strong U.S. focus. The triennial conferences on hydrogeology, ecology, monitoring, and management of ground water in karst terranes (NWWA, 1986, 1989; AGWSE, 1991) are very useful references for information on recent developments in karst studies, again with a strong U.S. focus.

---

[7] Citations are to references included in Table 1-3.

[8] Indiana Department of Environmental Management. 1988. Requirements for Describing Unconsolidated Deposits (draft revised 11/18/88). IDEM, Indianapolis, IN.

Table 1-3 Index to Major References on Geology, Soils and Geomorphology

Topic	References
Geology	
Terminology	Allaby and Allaby (1990), Bates and Jackson (1984), Michel and Fairbridge (1992), SCS (1977), Weller (1960a), Whitten and Brooks (1972)
Physical Geology	Birkeland and Larson (1989), Dercourt and Pacquet (1985), Flint and Skinner (1977), Foster (1983), Gilluly et al. (1975), Mears (1977), Hamblin (1978), Press and Siever (1986), Sawkins et al. (1978), Strahler (1976), Tarbuck and Lutgens (1984), Verhoogen et al. (1979)
Stratigraphy	Blatt et al. (1980), Bouma (1969), Folk (1968), Garrels and Mackenzie (1971), Krumbein and Sloss (1963), Matthews (1984), Pettijohn (1975), Trask (1950), Weller (1960b)
Structural Geology	Billings (1972), Hills (1972), Ragan (1973), Ramsay and Huber (1983, 1987), Spencer (1977)
Field Geology	Bishop (1960), Compton (1962, 1985), Dietrich et al. (1990), Kempton (1981), Lahee (1961), LeRoy et al. (1987), Low (1957); Field Rock Description: Fry (1984), Thorpe and Brown (1985), Tucker (1982)
Soils	
General	Brady (1974), Courtney and Trudgill (1984), Fairbridge and Finkl (1979), Fitzpatrick (1980, 1986), Foth (1971), Harpstead and Hole (1980), Hausenbuiller (1972), Jenny (1980), Stefferud (1957); Bedrock Soils: Cremeens et al. (1994); Forest Soils: Armson (1979), Burns (1959), Lutz and Chandler (1947), Pritchett (1979), Valentine (1986), Wilde (1958); Soil and Vegetation: Trudgill (1988); Terminology: ASAE (1967), Soil Science Society of America (1987)
Classification and Mapping	Amundson et al. (1994), Bailey (1987—bibliography), Buol et al. (1989), Butler (1980), Fanning and Fanning (1989), Finkl (1982), Forest Service (1961, 1963), McRae (1988), Milne et al. (1991), Olson (1981, 1984), Soil Survey Staff (1951, 1962, 1975, 1992, 1994a, 1994b), Webster (1977); Contaminated Sites: Boulding (1991, 1994), Cameron (1991)
Micromorphology and Fabric Analysis	Brewer (1976), Brewer and Sleeman (1988), Bullock and Murphy (1983), Bullock et al. (1985), Douglas (1990), Douglas and Thompson (1985), Fitzpatrick (1984, 1993), Miedema and Mermut (1990), Stoops and Eswarin (1986), Thompson et al. (1993)

Geology, Soils, and Geomorphology 41

Table 1-3 (Continued)

Topic	References
Geomorphology	
General	Chorley et al. (1984), Dury (1960), Lobeck (1939), Pitty (1971), Ritter (1986), Ruhe (1975), Selby (1985), Sparks (1986), Thornbury (1969); Terminology: SCS (1977)
Geomorphic Regions	Austin (1972), Fenneman (1931, 1938), Fenneman et al. (1946), Hunt (1967), Snead (1980), Thornbury (1965), USDA (1981)
Specific Topics	Environmental: Cook and Doornkamp (1990); Coasts: Trenhaile (1987); Deserts: Cook et al. (1993); Fluvial: Leopold et al. (1964), Heede (1992), Morisawa (1985), Richards (1982); Ground Water: Higgings and Coates (1990); Hillslope: Selby (1993); Phytogeomorphology: Howard and Mitchell (1985)
Interfaces Between Geology, Soils, and Geomorphology	
Soils/Geomorphology	Birkeland (1984), Cruikshank (1972), Daniels and Hammer (1992), Foth and Shafer (1980), Gerrard (1981, 1992), Hole and Campbell (1981), Richards et al. (1985)
Glacial and Quaternary Soils/Geology/ Geomorphology	Bell and Walker (1992), Catt (1986, 1988), Embleton and King (1968), Eyles (1983), Flint (1971), Ruhe (1965)
Engineering Applications	
Soil	Asphalt Institute (1969), Bureau of Reclamation (1969, 1974, 1990), Droshevska (1962), Hough (1969), Kezdi (1980), Portland Cement Association (1992), SCS (1990), Sowers (1979), Terghazi and Peck (1967); Unified Soil Classification System: Howard (1986)
Soil Engineering Properties	Bell (1992), Bowles (1978, 1984), Lamb and Whitman (1969), Means and Parcher (1963), Mitchell (1976), Obert and Duvall (1967), Spengler and Handy (1982), Taylor (1948), U.S. Navy Facilities Command (1971), Yong and Wartentin (1975); Foundation Engineering: Bowles (1982), Leonards (1962), Peck et al. (1974)
Engineering Geology/ Rock Mechanics	Attewell and Farmer (1976), Bell (1992), Bureau of Reclamation (1988, 1989), Dennen and Moore (1986), Heley and McIver (1971), Holtz and Kovacs (1981), Hunt (1972), Institution of Civil Engineers (1976), Legget and Hatheway (1988), Rahn (1986), Stagg and Zienkiewicz (1968); Terminology: International Society for Rock Mechanics (1972)

Table 1-3 References (Appendix F contains references for figure and table sources)

Allaby, A. and M. Allaby. 1990. The Concise Oxford Dictionary of Earth Sciences. Oxford University Press, Oxford, UK, 432 pp.

American Society of Agricultural Engineers (ASAE). 1967. Glossary of Soil and Water Terms. ASAE, St. Joseph, MI, 62 pp.

Amundson, R., J. Harden, and M. Singer (eds.). 1994. Factors of Soil Formation: A Fiftieth Anniversary Retrospective. SSSA Sp. Pub. No. 33, Soil Science Society of America, Madison, WI, 160 pp. [8 contributed chapters]

Armson, K.A. 1979. Forest Soils, Properties and Processes. University of Toronto Press, Toronto, 390 pp.

Asphalt Institute. 1969. Soil Manual. Manual Series No. 10, College Park MD, 265 pp.

Attewell, P.B. and I.W. Farmer. 1976. Principles of Engineering Geology. John Wiley & Sons, NY, 1045 pp.

Austin, M.E. 1972. Land Resource Regions and Major Land Resource Areas of the United States (Exclusive of Alaska and Hawaii). Agricultural Handbook 296, Soil Conservation Service, U.S. Department of Agriculture, Washington, DC, 82 pages. [First issued 1965, slightly revised 1972; superseded by USDA (1981)]

Bailey, G.D. (compiler). 1987. Bibliography of Soil Taxonomy, 1960-1979. CAD International, Tucson, AZ, 194 pp. [World literature on USDA soil taxonomy]

Bates, R. and J. Jackson (eds.). 1984. Dictionary of Geological Terms, 3rd ed. AGI, Washington, DC, 788 pp. [Supersedes Weller (1960)]

Bell, F.G. 1992. Engineering Properties of Soils and Rocks, 3rd ed. Butterworth-Heinemann, Boston, MA, 345 pp.

Bell, M. and M.J.C. Walker. 1992. Late Quaternary Environmental Change. Longman Scientific and Technical, Harlow, UK, 273 pp.

Billings, M.P. 1972. Structural Geology, 3rd ed. Prentice-Hall, Englewood Cliffs, NJ, 606 pp. [2nd edition 1972]

Birkeland, P.W. 1984. Soils and Geomorphology. Oxford University Press, New York, NY, 372 pp. [Revision of Pedology, Weathering, and Geomorphological Research published in 1973]

Birkeland, P.W. and E.E. Larson. 1989. Putnam's Geology, 5th ed. Oxford University Press, New York, NY. [19 chapters]

Bishop, M.S. 1960. Subsurface Mapping. Wiley, New York, NY, 198 pp.

(Table 1-3 Basic Geology, Soils and Geomorphology References)

Blatt, H., G. Middleton, and R. Murray. 1980. Origin of Sedimentary Rocks, 2nd ed. Prentice-Hall, Englewood Cliffs, NJ, 782 pp. [Text on methods in the study of sedimentary rock, sedimentary processes and characteristics of different sediments; 1st edition 1972]

Boulding, J.R. 1991. Description and Sampling of Contaminated Soils: A Field Pocket Guide. EPA/625/2-91/002, 122 pp. Available from CERI.*

Boulding, J.R. 1994. Description and Sampling of Contaminated Soils: A Field Guide, Revised and Expanded 2nd Edition. Lewis Publishers, Chelsea, MI, 220 pp. [Boulding (1991) is recommended for use in the field; the second edition corrects minor errors in the first edition and includes new SCS procedures for field description of redoximorphic soil features]

Bouma, A.H. 1969. Methods for the Study of Sedimentary Structures. Wiley-Interscience, New York, NY, 458 pp. [Text focussing on method for the sampling and study of sedimentary structures]

Bowles, J.E. 1978. Engineering Properties of Soils and Their Measurement. McGraw-Hill, New York, NY. [1st edition 1968, 2nd edition 1968]

Bowles, J.E. 1982. Foundation Analysis and Design, 3rd ed. McGraw-Hill, New York, NY, 816 pp.

Bowles, J.E. 1984. Physical and Geotechnical Properties of Soils, 2nd ed. McGraw-Hill, New York, NY, 578 pp.

Brady, N.C. 1974. The Nature and Property of Soils, 8th ed. MacMillan, New York, NY, 639 pp. [Comprehensive text with emphasis on physical and chemical properties significant for plant growth]

Brewer, R. 1976. Fabric and Mineral Analysis of Soils, 2nd ed. Krieger Publ. Co., Melbourne, FL, 470 pp. [1st edition published in 1964 by John Wiley & Sons]

Brewer, R. and J.R. Sleeman. 1988. Soil Structure and Fabric. CSIRO, Division of Soils, Adelaide, Australia.

Bullock, P. and C.P. Murphy (eds.). 1983. Soil Micromorphology, Vol. 1 Techniques and Applications, Vol. 2 Soil Genesis. AB Academic Publishers, Berkhamstead, UK, 705 pp. [Proceedings of 1981 international workshop; volume 1 contains 27 papers on soil micromorphological techniques and applications]

Bullock, P., N. Fedoroff, A. Jongerius, G. Stoops, and T. Tursina. 1985. Handbook for Soil Thin Section Description. Waine Research Publications, Wolverhampton, UK.

Buol, S.W., F.D. Hole, and R.J. McCracken. 1989. Soil Genesis and Classification, 3rd ed. Iowa State University Press, Ames, IA, 446 pp. [Introductory text focussing on the USDA soil taxonomy; 1st edition 1973, 2nd edition 1980]

Bureau of Reclamation. 1969. Soil as an Engineering Material. Research Report No. 17, U.S. Department of the Interior, Bureau of Reclamation, Denver, CO, 45 pp.

(Table 1-3 Basic Geology, Soils and Geomorphology References)

Bureau of Reclamation. 1974. Earth Manual, 2nd ed. U.S. Department of the Interior, Bureau of Reclamation, Denver, CO. [First 3 chapters reprinted in 1990, 326 pp.; remaining chapters superseded by 1990 3rd edition]

Bureau of Reclamation. 1988. Engineering Geology Office Manual. U.S. Department of the Interior, Bureau of Reclamation, Denver, CO, 58 pp.

Bureau of Reclamation. 1989. Engineering Geology Field Manual. U.S. Department of the Interior, Bureau of Reclamation, Denver, CO, 598 pp.

Bureau of Reclamation. 1990. Earth Manual, 3rd ed, Part 2. U.S. Department of the Interior, Bureau of Reclamation, Denver, CO, 1270 pp. [Part 1 consists of a 1990 reprint of the first 3 chapters of the 1974 2nd edition]

Burns, P.Y. 1959. Southern Forest Soils. Louisiana State University Press, Baton Rouge, LA, 132 pp.

Butler, B.E. 1980. Soil Classification for Soil Survey. Oxford University Press, New York, NY, 129 pp. [Advanced text (British)]

Cameron, R.E. 1991. Guide To Site and Soil Description for Hazardous Waste Sites, Vol. 1, Metals. EPA 600/4-91-029 (NTIS PB92-146158).

Catt, J.A. 1986. Soils and Quaternary Geology: A Handbook for Field Scientists. Oxford University Press, New York, NY, 267 pp. [British focus, but includes review of quaternary geology of the midwestern U.S.]

Catt, J.A. 1988. Quaternary Geology for Scientists and Engineers. Halstead Press, New York, NY, 340 pp.

Chorley, R.J., S.A. Schumm and D.E. Sugden. 1984. Geomorphology. Methuen, New York, NY, 605 pp. [Four parts: (1) introduction, (2) geological geomorphology, (3) geomorphic processes and landforms, and (4) climatic geomorphology]

Cook, R., A. Warren, and A. Goudie. 1993. Desert Geomorphology. UCL Press, London, UK, 526 pp.

Cooke, R.U. and J.C. Doornkamp. 1990. Geomorphology in Environmental Management, 2nd ed. Oxford University Press, New York, NY, 434 pp. [1st edition 1974]

Compton, R.R. 1962. Manual of Field Geology. John Wiley & Sons, New York, NY, 378 pp.

Compton, R.R. 1985. Geology in the Field. John Wiley & Sons, New York, NY, 398 pp.

Courtney, F.M. and S.T. Trudgill. 1984. The Soil: An Introduction to Soil Study. Edward Arnold Publishers, London, UK, 123 pp. [Focusses on soil properties, management, and soil classification]

(Table 1-3 Basic Geology, Soils and Geomorphology References)

Geology, Soils, and Geomorphology

Cremeens, D.L., R.B. Brown, and J.H. Huddleston (eds.). 1994. Whole Regolith Pedology. SSSA Sp. Pub. No. 34, Soil Science Society of America, Madison, WI, 136 pp. [7 contributed chapters]

Cruikshank, J.G. 1972. Soil Geography. Halstead Press Division, John Wiley & Sons, New York, NY, 256 pp.

Daniels, R.B. and R.D. Hammer. 1992. Soil Geomorphology. John Wiley & Sons, New York, NY, 256 pp.

Dennen, W.H. and B.R. Moore. 1986. Geology and Engineering. W.C. Brown, Dubuque, IA, 378 pp.

Dercourt, J. and J. Pacquet. 1985. Geology: Principles and Methods. Gulf Publishing, Houston, TX, 384 pp. [Four major parts: (1) minerals to rocks, (2) the earth's interior, oceans, and continents, (3) global dynamics, and (4) applied geology]

Dietrich, R.V., J.V. Dutro, Jr., and R.M. Foose (Compilers). 1990. AGI Data Sheets for Geology in Field, Laboratory, and Office, 3rd Edition. American Geological Institute, Washington, DC, 294 pp. [2nd edition dated 1982]

Douglas, L.A. (ed.). 1990. Soil Micromorphology: A Basic and Applied Science. Elsevier, New York, NY, 734 pp. [Over 70 papers presented at the 8th International Working Meeting of Soil Micromorphology in San Antonio, TX]

Douglas, L.A. and M.L. Thompson (eds.). 1985. Soil Micromorphology and Soil Classification. SSSA Sp. Pub. No. 15, Soil Science Society of America, Madison, WI, 216 pp. [10 contributed chapters]

Droshevska, L. 1962. Review of Recent USSR Publications in Selected Fields of Engineering Soil Science. In: Reviews in Engineering Geology, T. Fluhr and R.F. Legget (eds.), Geological Society of America, Boulder, CO, 1:197-256.

Dury, G.H. 1960. Map Interpretation, 2nd ed. Pitman, London, UK, 209 pp. [1st edition, 1957]

Embleton, C. and C.A.M. King. 1968. Glacial and Periglacial Geomorphology. Edward Arnold Publishers, London, 608 pp. [Advanced text]

Eyles, N. (ed.). 1983. Glacial Geology: An Introduction for Engineers and Earth Scientists. Pergamon Press, New York, NY, 409 pp. [Emphasizes site investigation procedures]

Fairbridge, R.W. and C.W. Finkl, Jr. (eds.). 1979. The Encyclopedia of Soil Science, Part 1: Physics, Chemistry, Biology, Fertility, and Technology. Dowden, Hutchinson & Ross, Stroudsburg, PA, 646 pp.

Fanning, D.S. and M.C.B. Fanning. 1989. Soil: Morphology, Genesis and Classification. John Wiley & Sons, New York, NY.

(Table 1-3 Basic Geology, Soils and Geomorphology References)

Fenneman, N.M. 1931. Physiography of the Western United States. McGraw-Hill, New York, NY, 534 pp.

Fenneman, N.M. 1938. Physiography of the Eastern United States. McGraw-Hill, New York, NY, 714 pp.

Fenneman, N.M. et al. 1946. Physiographic Division of the United States. U.S. Geological Survey Map Scale 1:7,000,000.

Finkl, Jr., C.W. (ed.). 1982. Soil Classification. Benchmark Papers in Soil Science, Vol. 1, Hutchinson Ross Publishing Co., 391 pp. [27 contributed chapters focussing on classification systems used in different countries]

Fitzpatrick, E.A. 1980. Soils, Their Formation, Classification and Distribution. Longman, New York, NY, 353 pp.

Fitzpatrick, E.A. 1984. Micromorphology of Soils. Chapman and Hall, New York, NY, 433 pp. [Focusses on use of thin sections for study of soil mineralogy and fabric]

Fitzpatrick, E.A. 1986. An Introduction to Soil Science, 2nd ed. John Wiley & Sons, New York, NY, 255 pp. [Text focussing on factors of soil formation, processes in the soil system and properties of soils]

Fitzpatrick, E.A. 1993. Soil Microscopy and Micromorphology. John Wiley & Sons, New York, NY, 304 pp.

Flint, R.F. 1971. Glacial and Quaternary Geology. John Wiley & Sons, New York, NY, 892 pp.

Flint, R.F. and B.J. Skinner. 1977. Physical Geology, 2nd ed. John Wiley & Sons, New York, NY, 679 pp. [Six major parts: (1) basics, (2) external processes on land, (3) external processes in the seas, (4) internal processes, (5) the planets, and (6) man and earth]

Folk, R.L. 1968. Petrography of Sedimentary Rocks. Hemphills, Austin, TX, 170 pp.

Forest Service. 1961. Handbook of Soils. U.S. Department of Agriculture, Forest Service, Washington, DC.

Forest Service. 1963. Handbook of Soils, Amendment No. 1: Physical Properties of Soils. U.S. Department of Agriculture, Forest Service, Washington, DC.

Foster, R.J. 1983. Physical Geology, 4th ed. Charles E. Merrill, Columbus, OH. [18 chapters]

Foth, H.D. and J.W. Schafer. 1980. Soil Geography and Land Use. John Wiley & Sons, New York, NY, 484 pp.

Foth, H.D. et al. 1971. Laboratory Manual for Introductory Soil Sciences, 2nd ed. William C. Brown Co., Dubuque, IA.

(Table 1-3 Basic Geology, Soils and Geomorphology References)

Fry, N. 1984. The Field Description of Metamorphic Rocks. John Wiley & Sons, New York, NY, 110 pp.

Garrels, R.M. and F.T. Mackenzie. 1971. Evolution of Sedimentary Rocks. N.W. Norton & Co., New York, NY, 397 pp.

Gerrard, A.J. 1981. Soils and Landforms: An Integration of Geomorphology and Pedology. George Allen and Unwin, Boston, MA, 219 pp.

Gerrard, J. 1992. Soil Geomorphology. Chapman & Hall, New York, NY, 279 pp.

Gilluly, J., A.C. Waters, and A.O. Woodford. 1975. Principles of Geology, 4th ed. W.H. Freeman, San Francisco, CA, 527 pp. [19 chapters]

Hamblin, W.K. 1978. The Earth's Dynamic Systems. Burgess Publishing Co., Minneapolis, MN, 459 pp.

Harpstead, M.I. and F.D. Hole. 1980. Soil Science Simplified. Iowa State University Press, Ames, IA.

Hausenbuiller, R.L. 1972. Soil Science—Principles and Practices. W.C. Brown Co., Dubuque, IA, 504 pp.

Heede, B.H. 1992. Stream Dynamics: An Overview for Land Managers. General Technical Report RM-72. USDA Forest Service Rocky Mountain Forest and Range Experiment Station, Fort Collins, CO, 26 pp. [Revision of 1980 publication with same title]

Heley, W. and B.N. McIver. 1971. Engineering Properties of Clay Shales. In: Development and Classification Indices of Clay Shales. Soil and Pavement Laboratory Technical Report TR-S-71-6. USCE Water Experiment Stations.

Higgings, C.G. and D.R. Coates (eds.). 1990. Groundwater Geomorphology: The Role of Subsurface Water in Earth-Surface Processes and Landforms. Geological Society of America Special Paper 252, 368 pp.

Hills, E.S. 1972. Elements of Structural Geology, 2nd ed. John Wiley & Sons, New York, NY, 502 pp. [Text emphasizing methods of fabric analysis in the study of geological structures; 1st edition 1963]

Hole, F.D. and J.B. Campbell. 1985. Soil Landscape Analysis. Rowman & Allanheld Publishers, Totowa, NJ, 196 pp.

Holtz, R.D. and W.D. Kovacs. 1981. An Introduction to Geotechnical Engineering. Prentice-Hall, Englewood Cliffs, NJ, 733 pp.

Hough, B.K. 1969. Basic Soils Engineering, 2nd ed. The Ronald Press Co., NY, 513 pp.

(Table 1-3 Basic Geology, Soils and Geomorphology References)

Howard, A.K. 1986. *Unified Soil Classification System Geotechnical Branch Training Manuals, 2nd ed.*: Laboratory Classification of Soils (No. 4, 102 pp.), Visual Classification of Soils (No. 5, 106 pp.), Soil Classification Handbook (No. 6, 81 pp.). Bureau of Reclamation, Denver, CO.

Howard, J.A. and C.W. Mitchell. 1985. Phytogeomorphology. John Wiley & Sons, New York, NY, 222 pp.

Hunt, C.B. 1967. Physiography of the United States. W.H. Freeman and Co., San Francisco, CA, 480 pp.

Hunt, C.B. 1972. Geology of Soils: Their Evolution, Classification and Uses. W.H. Freeman, San Francisco, 344 pp. [Emphasis on engineering aspects]

Institution of Civil Engineers. 1976. Manual of Applied Geology for Engineers. Institution of Civil Engineers, London, 378 pp.

International Society for Rock Mechanics. 1972. Final Document on Terminology, English Versions. Committee on Terminology, Symbols and Graphic Representation, 19 pp.

Jenny, H. 1980. The Soil Resource: Origin and Behavior. Springer-Verlag, New York, NY, 377 pp. [Text focussing on soil-forming processes and their effect on soil genesis]

Kempton, J.P. 1981. Three-Dimensional Geologic Mapping for Environmental Studies in Illinois. Illinois State Geological Survey Environmental Geology Note 100, 43 pp.

Kezdi, A. 1980. Soil Testing: Handbook of Soil Mechanics, Vol. 2. Elsevier, New York, NY, 258 pp.

Krumbein, W.C. and L.L. Sloss. 1963. Stratigraphy and Sedimentation. W.H. Freeman, San Francisco, CA, 660 pp.

Lahee, F.H. 1961. Field Geology, 6th ed. McGraw-Hill, New York, NY, 926 pp.

Lambe, W.T. and R.V Whitman. 1969. Soil Mechanics. John Wiley & Sons, New York, NY, 553 pp.

Legget, R.F. and A.W. Hatheway. 1988. Geology and Engineering, 3rd ed. McGraw-Hill, New York, NY, 613 pp. [2nd edition 1962 by Legget]

Leonards, G.S. (ed.). 1962. Foundation Engineering. McGraw-Hill, New York, NY, 1138 pp.

LeRoy, L.W., D.O. LeRoy, S.D. Schwochow, and J.W. Raese (eds.). 1987. Subsurface Geology, 5th ed. Colorado School of Mines, Golden, CO. [1st edition: LeRoy and Cran (1947), 2nd edition: LeRoy (1951), 3rd and 4th editions: Huan and LeRoy (1958, 1977)]

Lobeck, A.K. 1939. Geomorphology: An Introduction to the Study of Landscapes. McGraw-Hill, New York, NY, 731 pp.

Low, J.W. 1957. Geologic Field Methods. Harper, New York, NY, 489 pp.

(Table 1-3 Basic Geology, Soils and Geomorphology References)

Geology, Soils, and Geomorphology

Leopold, L.B., M.G. Wolman, and J.P. Miller. 1964. Fluvial Processes in Geomorphology. W.H. Freeman, San Francisco, CA, 522 pp.

Lutz, H.J. and R.F. Chandler. 1947. Forest Soils. John Wiley & Sons, New York, NY, 514 pp.

Matthews, R.K. 1984. Dynamic Stratigraphy, 2nd ed. Prentice-Hall, Englewood Cliffs, NJ, 489 pp. [1st edition 1974]

McRae, S.G. 1988. Practical Pedology: Studying Soils in the Field. Halstead Press, New York, NY, 253 pp. [Introductory text (British)]

Means, R.E. and J.V. Parcher. 1963. Physical Properties of Soils. Charles E. Merril, Columbus OH, 464 pp.

Mears, Jr., B. 1977. The Changing Earth: Introduction to Geology, 2nd ed. Van Nostrand, New York, NY, 593 pp. [17 chapters covering physical and historical geology]

Michel, J.-P. and R.W. Fairbridge. 1992. Dictionary of Earth Sciences. John Wiley & Sons, New York, NY, 300 pp.

Miedema, R. and A.R. Mermut. 1990. Soil Micromorphology: An Annotated Bibliography 1968-1986. CAB International, Tucson, AZ, 250 pp.

Milne, J.D.G., B. Clayden, P.L. Singleton, and A.D. Wilson. 1991. Soil Description Handbook. DSIR Land Resources, Lower Hutt, New Zealand, 133 pp.

Mitchell, J.K. 1976. Fundamentals of Soil Behavior. John Wiley & Sons, New York, NY, 422 pp.

Morisawa, M. 1985. Rivers: Form and Processes. Longman, New York, NY, 222 pp.

Obert, L. and W.I. Duvall. 1967. Rock Mechanics and the Design of Structures in Rocks. John Wiley & Sons, New York, NY, 650 pp.

Olson, G.W. 1981. Soils and the Environment: A Guide to Soil Surveys and Their Applications. Chapman and Hall, New York, NY, 178 pp.

Olson, G.W. 1984. Field Guide to Soils and the Environment: Applications of Soil Surveys. Chapman and Hall, New York, NY, 219 pp.

Peck, R.B., W.E. Hanson, and T.H. Thornburn. 1974. Foundation Engineering, 3rd ed. John Wiley & Sons, New York, NY, 514 pp.

Pettijohn, F.J. 1975. Sedimentary Rocks, 3rd ed. Harper and Row, New York, NY, 628 pp.

Pitty, A.F. 1971. Introduction to Geomorphology. Methuen, New York, NY, 526 pp. [Five sections: (1) definitions, nature and basic postulates, (2) landforms and structure, (3) physical, chemical and biological basis of geomorphological processes, (4) inter-relationships between processes and landforms, and (5) landforms and time]

(Table 1-3 Basic Geology, Soils and Geomorphology References)

Portland Cement Association. 1992. PCA Soil Primer, Revised ed. Engineering Bulletin EB007.045. Portland Cement Association, Skokie, IL, 40 pp. [1st edition published 1973]

Press, F. and R. Siever. 1986. Earth, 4th ed. W.H. Freeman, San Francisco, CA, 656 pp.

Pritchett, W.L. 1979. Properties and Management of Forest Soils. John Wiley & Sons, New York, NY, 500 pp.

Ragan, D.N. 1973. Structural Geology: An Introduction to Geometrical Techniques, 2nd ed. John Wiley & Sons, New York, NY.

Rahn, P. 1986. Engineering Geology. Elsevier, New York, NY, 589 pp.

Ramsay, J.G. and M. Huber. 1983. The Techniques of Modern Structural Geology, Vol. 1: Strain Analysis. Academic Press, New York, NY, 302 pp.

Ramsay, J.G. and M. Huber. 1987. The Techniques of Modern Structural Geology, Vol. 2: Folds and Fractures. Academic Press, New York, NY, 695 pp.

Richards, K. 1982. Rivers: Form and Process in Alluvial Channels. Methuen, New York, NY, 358 pp.

Richards, K.S., R.R. Arnett, and S. Ellis (eds.). 1985. Geomorphology and Soils. George Allen and Unwin, Boston, MA, 441 pp.

Ritter, D.F. 1986. Process Geomorphology. Wm. C. Brown Publishers, Dubuque, IA, 579 pp.

Ruhe, R.V. 1975. Geomorphology: Geomorphic Processes and Surficial Geology. Houghton Mifflin, Boston, 246 pp. [Introductory text with 11 chapters]

Sawkins, F.J., C.G. Chase, D.G. Darby, and G. Rapp, Jr. 1978. The Evolving Earth, A Text in Physical Geology, 2nd ed. MacMillan, New York, NY, 558 pp.

Schmidt, K.-H. and J. de Ploey (eds.). 1992. Functional Geomorphology: Landform Analysis and Models. Catena Supplement 23, Catena Verlag, Lawrence, KS.

Selby, M.J. 1985. Earth's Changing Surface: An Introduction to Geomorphology. Oxford University Press, New York, NY, 607 pp. [Three parts: (1) nature and structure of earth's major physical features, (2) processes of weathering, erosion and deposition, and (3) major bioclimatic zones of the earth]

Selby, M.J. 1993. Hillslope Materials and Processes, 2nd ed. Oxford University Press, New York, NY, 332 pp.

Snead, R.E. 1980. World Atlas of Geomorphic Features. Robert E. Krieger Publishing Co., Melbourne, FL, 301 pp.

Soil Conservation Service (SCS). 1977. Glossary of Selected Geologic and Geomorphic Terms. U.S. Department of Agriculture, Soil Conservation Service Western Technical Service Center, Portland, OR, 24 pp.

(Table 1-3 Basic Geology, Soils and Geomorphology References)

Geology, Soils, and Geomorphology 51

Soil Conservation Service (SCS). 1990. Elementary Soil Engineering. In: Engineering Field Manual, SCS, U.S. Department of Agriculture, Washington, DC, Chapter 4.

Soil Science Society of America. 1987. Glossary of Soil Science Terms. SSSA, Madison, WI, 44 pp.

Soil Survey Staff. 1951. Soil Survey Manual. U.S. Dept of Agric. Agricultural Handbook No. 18. Supplement issued in 1962 replaced pages 173-188 (Identification and Nomenclature of Soil Horizons). [Superseded by Soil Survey Staff (1994)]

Soil Survey Staff. 1975. Soil Taxonomy: A Basic System of Soil Classification for Making and Interpreting Soil Surveys. U.S. Dept. of Agric. Agricultural Handbook No. 436, 754 pp. [See Soil Survey Staff (1994) for latest revisions]

Soil Survey Staff. 1994a. Examination and Description of Soils. In: Soil Survey Manual (new edition). U.S. Dept. of Agric. Agricultural Handbook No. 18. Soil Conservation Service, Washington, DC, Chapter 3. [Note that this supersedes the 1951 Handbook by the same title, and the 1962 supplement. U.S. Government Printing Office Stock No. 001-000-04611-0]

Soil Survey Staff. 1994b. Keys to Soil Taxonomy, 6th ed. U.S. Government Printing Office, Washington, DC, Stock No. 001-000-04612-8. [Updated every 2 years]

Sowers, G.F. 1979. Introductory Soil Mechanics and Foundations: Geotechnical Engineering, 4th ed. Macmillan, New York, NY, 621 pp. [2nd edition 1961, 3rd edition 1970]

Sparks, B.W. 1986. Geomorphology, 3rd ed. Longman, New York, NY, 561 pp. [Introductory text]

Spencer, E.W. 1977. Introduction to the Structure of the Earth, 2nd ed. McGraw-Hill, New York, NY, 640 pp. [Structural geology and tectonics]

Spengler, M.G. and R.L. Handy. 1982. Soil Engineering, 4th ed. Harper & Row, New York, NY, 819 pp. [3rd edition 1973]

Stagg, K.C. and O.C. Zienkiewicz. 1968. Rock Mechanics in Engineering Practice. John Wiley & Sons, NY, 492 pp.

Stefferud, A. (ed.). 1957. Soil: The 1957 Yearbook of Agriculture. U.S. Department of Agriculture, Washington, DC, 784 pp.

Stoops, S. and H. Eswarin (eds.). 1986. Soil Micromorphology. Van Nostrand Reinhold, New York, NY, 345 pp.

Strahler, A.N. 1976. Principles of Earth Science. Harper & Row, New York, NY, 434 pp. [25 chapters]

Tarbuck, E.J. and F.K. Lutgens. 1984. The Earth, An Introduction to Physical Geology. Charles E. Merrill Publishing, Columbus, OH, 594 pp.

(Table 1-3 Basic Geology, Soils and Geomorphology References)

Taylor, D.W. 1948. Fundamentals of Soil Mechanics. John Wiley & Sons, New York, NY, 700 pp.

Terzaghi, K. and R.B. Peck. 1967. Soil Mechanics in Engineering Practice, 2nd ed. John Wiley & Sons, New York, NY, 729 pp. [1st edition 1948]

Thompson, M.L., A.R. Mermut, W.D. Nettleton, L.D. Darrell, and S. Pawluk (compilers). 1993. A Reference Slide Collection for Soil Micromorphology. Soil Science Society of America, Madison, WI, 135 pp. + 115 slides.

Thornbury, W.D. 1965. Regional Geomorphology of the United States. John Wiley & Sons, New York, NY, 609 pp. [Chapters on 27 geomorphic regions in the U.S.]

Thornbury, W.D. 1969. Principles of Geomorphology, 2nd ed. John Wiley & Sons, New York, NY, 618 pp. [Introductory text with 22 chapters]

Thorpe, R. and G. Brown. 1985. The Field Description of Igneous Rocks. John Wiley & Sons, New York, NY, 154 pp.

Trask, P.D. (ed.). 1950. Applied Sedimentation. John Wiley & Sons, New York, NY, 707 pp.

Trenhaile, A.S. 1987. The Geomorphology of Rock Coasts. Oxford University Press, New York, NY, 394 pp.

Trudgill, S.T. 1988. Soil and Vegetation Systems, 2nd ed. Oxford University Press, New York, NY, 211 pp. [1st edition 1977]

Tucker, M.E. 1982. The Field Description of Sedimentary Rocks. John Wiley & Sons, New York, NY, 112 pp.

U.S. Department of Agriculture (USDA). 1981. Land Resource Regions and Major Land Resource Areas of the United States. Agricultural Handbook 296, Soil Conservation Service, U.S. Department of Agriculture, Washington, DC, 156 pp. [Supersedes Austin 1972]

U.S. Navy Facilities Command. 1971. Design Manual - Soil Mechanics, Foundations and Earth Structures. NAV-FAC DM-&, U.S. Government Printing Office, Washington, DC.

Valentine, K.W.G. 1986. Soil Resource Surveys for Forestry. Oxford University Press, New York, NY, 200 pp.

Verhoogen, J., F.J. Turner, C.E. Weiss, C. Wahrhaftig, and W.S. Frye. 1979. The Earth: An Introduction to Physical Geology. Holt, Rhinehart and Wilson, New York, NY, 748 pp.

Webster, R. 1977. Quantitative and Numerical Methods in Soil Classification and Survey. Oxford University Press, New York, NY, 269 pp. [British focus]

Weller, J.M. (ed.). 1960a. Glossary of Geology and Related Sciences with Supplement, 2nd edition. American Geological Institute, Washington, DC, 325 pp., Supplement 72 pp. [Superseded by Bates and Jackson (1984)]

(Table 1-3 Basic Geology, Soils and Geomorphology References)

Weller, J.M. 1960b. Stratigraphic Principles and Practice. Harper & Brothers, New York, NY, 725 pp.

Whitten, D.G.A. and J.R.V. Brooks. 1972. The Penguin Dictionary of Geology. Penguin Books, Baltimore, MD, 514 pp.

Wilde, S.A. 1958. Forest Soils. Ronald Press, New York, NY.

Yong, R.N. and B.P. Wartentin. 1975. Soil Properties and Behavior. Elsevier, New York, NY, 449 pp.

* See Preface for information on how to obtain documents from CERI (U.S. EPA Center for Environmental Research Information) and NTIS.

(Table 1-3 Basic Geology, Soils and Geomorphology References)

Table 1-4 Index to Major References on Karst Geology, Geomorphology and Hydrology

Topic	References
Glossary	Monroe (1970)
Hydrology/Ground Water	Bibliographies: LaMoreaux (1986), LaMoreaux et al. (1970, 1989, 1993), Warren and Moore (1975); Texts: Back et al. (1990), Bögli (1980), Bonacci (1987), Burger and Dubertret (1975), Ford and Williams (1989), LaMoreaux (1986), LaMoreaux et al. (1975, 1984), Milanović (1981), Stringfield et al. (1974), White (1988); Review Papers: Field (1989), Kresic (1993), LeGrand and Stringfield (1973); Case Histories: Burger and Dubertret (1984), White and White (1989); Proceedings: AGWSE (1991), Beck and Wilson (1987), Doaxin (1988), Günay and Johnson (1986), IASH (1967), Rauch and Werner (1974), Tolson and Doyle (1977), Yevjevich (1976)
Karst Tracing	Aley and Fletcher (1976), Back and Zoetl (1975), Bögli (1980), Brown (1972), Ford and Williams (1989), Gospodaric and Habic (1976), Gunn (1982), Jones (1984), LaMoreaux (1984, 1989), Milanović (1981), Mull et al. (1988), Quinlan (1986, 1989), Sweeting (1973), SUWT (1966, 1970, 1976, 1981, 1986), Thrailkill et al. (1983)
Geomorphology/Geology	Dreybodt (1988), Ford and Williams (1989), Herak and Stringfield (1972), Jakucs (1977), Jennings (1985), Rauch and Werner (1974), Sweeting (1973), Trudgill (1985), White (1988)
Geochemistry	Dreybodt (1988)
Engineering Aspects	Davies et al. (1976), James (1992); Proceedings: Beck (1984, 1989), Beck and Wilson (1987)
Environmental Aspects	AGWSE (1991), Beck (1984, 1990), Beck and Wilson (1987), Doaxin (1988), NWWA (1986, 1988)
Conference Proceedings	AGWSE (1991), Beck (1984, 1990), Beck and Wilson (1987), Doaxin (1988), Günay and Johnson (1986), IASH (1967), NWWA (1986, 1988), Rauch and Werner (1974), Tolson and Doyle (1977), Yevjevich (1976)

Table 1-4 References (Appendix F contains references for figure and table sources)

Aley, T. and M.W. Fletcher. 1976. The Water Tracer's Cookbook. Missouri Speleology 16(3):1-32.

Association of Ground Water Scientists and Engineers (AGWSE). 1991. Proceedings of the Third Conference on Hydrogeology, Ecology, Monitoring, and Management of Ground Water in Karst Terranes (Nashville, TN). Ground Water Management, Book 10. National Ground Water Association, Dublin, OH, 793 pp.

Back, W. and J. Zoetl. 1975. Application of Geochemical Principles, Isotopic Methodology, and Artificial Tracers to Karst Hydrology. In: Hydrogeology of Karstic Terrains, A. Burger and L. Dubertret (eds.), Int. Ass. Hydrogeologists, Paris, pp. 105-121.

Back, W., J.S. Herman, and H. Paloc (eds.). 1990. Hydrogeology of Selected Karst Regions. International Contributions to Hydrogeology Vol. 13, Int. Ass. of Hydrogeologists, Verlag Heinz Heise, Hannover, Germany, 494 pp. [31 chapters]

Beck, B.F. (ed.). 1984. Sinkholes: Their Geology, Engineering and Environmental Impact. Balkema, Accord, MA, 429 pp. [[Proc. 1st Multidisciplinary Conference on Sinkholes and Environmental Impacts of Karst (Orlando, FL); over 60 papers]

Beck, B.F. (ed.). 1989. Engineering and Environmental Impacts of Sinkholes and Karst. Balkema, Brookfield, VT, 384 pp. [Proc. 3rd Multidisciplinary Conference (Petersburg Beach, FL); 46 papers]

Beck, B.F. and W.L. Wilson (eds.). 1987. Karst Hydrogeology: Engineering and Environmental Applications. Balkema, Accord, MA, 467 pp. [Proc. 2nd Multidisciplinary Conference on Sinkholes and Environmental Impacts of Karst (Orlando, FL); over 60 papers]

Bögli, A. 1980. Karst Hydrology and Physical Speleology. Springer-Verlag, New York, NY, 284 pp. [Text focusing on karst hydrology and the development and classification of underground cavities]

Bonacci, O. 1987. Karst Hydrology with Special Reference to the Dinaric Karst. Springer-Verlag, New York, NY, 184 pp. [Text on karst hydrology focusing on the Dinaric karst of Yugoslavia; includes chapters on tracing]

Brown, M.C. 1972. Karst Hydrology of the Lower Maligne Basin, Jasper, Alberta. Cave Studies No. 13. Cave Research Associates, Castro Valley, CA. [Chapter III reviews tracer methods]

Burger, A. and L. Dubertret (eds.). 1975. Hydrogeology of Karstic Terrains. International Union of Geological Sciences, Series B, Number 3. Int. Ass. Hydrogeologists, Paris, 190 pp. [Eleven contributed chapters on the hydrogeology of karst terrains with a multi-lingual glossary of specific terms]

Burger, A. and L. Dubertret (eds.). 1984. Hydrogeology of Karstic Terrains: Case Histories. International Contributions to Hydrogeology, Vol. 1, Int. Ass. of Hydrogeologists, Paris, 264 pp. [61 case histories]

(Table 1-4 Major Karst References)

Daoxian, Y. (ed.). 1988. Karst Hydrogeology and Karst Environment Protection: Proc. 21st Congress of the IAH (Guilin, China), 2 volumes. Int. Ass. Sci. Hydrology Publ. No. 176, 1261 pp. [Vol. 1 contains 119 papers and abstracts; Vol. 2 contains 143 papers and abstracts]

Davies. W.E., J.H. Simpson, G.C. Olmacher, W.S. Kirk, and E.G. Newton. 1976. Map Showing Engineering Aspects of Karst in the United States. U.S. Geological Survey Open File Map 76-623.

Dreybodt, W. 1988. Processes in Karst Systems: Physics, Chemistry and Geology. Springer-Verlag, New York, NY.

Field, M.S. 1989. The Vulnerability of Karst Aquifers to Chemical Contamination. In: Recent Advances in Ground-Water Hydrology, American Institute of Hydrology, Minneapolis, MN, pp. 130-142.

Ford, D.C. and P.W. Williams. 1989. Karst Geomorphology and Hydrology. Unwin Hyman, Winchester, MA, 601 pp.

Gospordaric, R. and P. Habic (eds.). 1976. Underground Water Tracing: Investigations in Slovenia 1972-1975. Institute Karst Research, Ljubljana, Yugoslavia.

Günay, G. and A.I. Johnson (eds.). 1986. Karst Water Resources. Int. Ass. Sci. Hydrology Pub. No. 161, 642 pp. [Symposium proceedings with 45 papers]

Gunn, J. 1982. Water Tracing in Ireland: A Review with Special References to the Cuillcagh Karst. Irish Geography 15:94-106.

Herak, M. and V.T. Stringfield (eds.). 1972. Karst: Important Karst Regions of the Northern Hemisphere. Elsevier, New York, NY, 551 pp. [15 contributed chapters on major karst regions of the northern hemisphere]

International Association of Scientific Hydrology (IASH). 1967. Hydrology of Fractured Rocks (Proc. of 1965 Dubrovnik Symposium), 2 Vols. IASH Publ. No. 73.

Jakucs, L. 1977. Morphogenetics of Karst Regions: Variants of Karst Evolution. Adam Hilger, Bristol, UK, 283 pp.

James, A.N. 1992. Soluble Materials in Civil Engineering. Ellis Horwood, U.K. [Dam construction in karst]

Jennings, J.N. 1985. Karst Geomorphology, 2nd ed. Basil Blackwell, New York, NY, 293 pp.

Jones, W.K. 1984. Dye Tracers in Karst Areas. National Speleological Society Bulletin 36:3-9.

Kresic, N.A. 1993. Review and Selected Bibliography on Quantitative Definition of Karst Hydrogeological Systems. In: Annotated Bibliography of Karst Terranes, Volume 5 with Three Review Articles, P.E. LaMoreaux, F.A. Assaad, and A. McCarley (eds.), International Contributions to Hydrogeology, Vol. 14, International Association of Hydrogeologists, Verlag Heinz Heise, Hannover, West Germany, pp. 51-87.

(Table 1-4 Major Karst References)

LaMoreaux, P.E. (ed.). 1986. Hydrology of Limestone Terranes. Int. Ass. Hydrogeologists, Verlag Heinz Heise, Hannover, West Germany. [Includes an annotated bibliography for the literature published since 1975; see White and Moore (1976) for bibliography to 1975]

LaMoreaux, P.E., D. Raymond, and T.J. Joiner. 1970. Hydrology of Limestone Terranes: Annotated Bibliography of Carbonate Rocks. Geological Survey of Alabama Bulletin 94A, 242 pp.

LaMoreaux, P.E., H.E. LeGrand, V.T. Stringfield, and J.S. Tolson. 1975. Hydrology of Limestone Terranes: Progress of Knowledge About Hydrology of Carbonate Terranes. Geological Survey of Alabama Bulletin 94E, pp. 1-30.

LaMoreaux, P.E., B.M. Wilson, and B.A. Mermon (eds.). 1984. Guide to the Hydrology of Carbonate Rocks. UNESCO, Studies and Reports in Hydrology No. 41.

LaMoreaux, P.E., E. Prohic, J. Zoetl, J.M. Tanner, and B.N. Roche (eds.). 1989. Hydrology of Limestone Terranes: Annotated Bibliography of Carbonate Rocks, Volume 4. International Association of Hydrogeologists Int. Cont. to Hydrogeology Volume 10. Verlag Heinz Heise GmbH., Hannover, West Germany, 267 pp.

LaMoreaux, P.E., F.A. Assaad, and A. McCarley (ed.). 1993. Annotated Bibliography of Karst Terranes, Volume 5 with Three Review Articles. International Contributions to Hydrogeology, Vol. 14, International Association of Hydrogeologists, Verlag Heinz Heise, Hannover, West Germany, 425 pp.

LeGrand, H.E. and V.T. Stringfield. 1973. Karst Hydrology--A Review. J. Hydrology 20(2):97-120.

Milanović, P.T. 1981. Karst Hydrogeology. Water Resources Publications, Littleton, CO, 444 pp. [May also be cited with 1979 date]

Monroe, W.H. (compiler). 1970. A Glossary of Karst Terminology. U.S. Geological Survey Water Supply Paper 1899-K, 26 pp.

Mull, D.S., T.D. Lieberman, J.L. Smoot, and L.H. Woosely, Jr. 1988. Application of Dye-Tracing Techniques for Determining Solute-Transport Characteristics of Ground Water in Karst Terranes. EPA 904/6-88-001, Region 4, Atlanta, GA.

National Water Well Association (NWWA). 1986. Proceedings 1st Conference on Environmental Problems in Karst Terranes and Their Solutions. NWWA, Dublin, OH.

National Water Well Association (NWWA). 1988. Proceedings 2nd Conference on Environmental Problems in Karst Terranes and Their Solutions. NWWA, Dublin, OH. [22 papers]

Quinlan, J.F. 1986. Discussion of "Ground Water Tracers" by Davis et al. (1985) with Emphasis on Dye Tracing, Especially in Karst Terranes. Ground Water 24(2):253-259 and 24(3):396-397 (References).

Quinlan, J.F. 1989. Ground-Water Monitoring in Karst Terranes: Recommended Protocols and Implicit Assumptions. EPA 600/X-89/050, EMSL, Las Vegas, NV.

(Table 1-4 Major Karst References)

Rauch, H.W. and E. Werner (eds.). 1974. Proceeding of the Fourth Conference on Karst Geology and Hydrology. West Virginia Geological and Economic Survey, Morgantown, WV. [32 papers]

Stringfield, V.T., P.E. LaMoreaux, and H.E. LeGrand. 1974. Karst and Paleohydrology of Carbonate Rock Terranes in Semiarid and Arid Regions with a Comparison to Humid Karst of Alabama. Geological Survey of Alabama Bulletin 105, 106 pp.

Sweeting, M.M. 1973. Karst Landforms. Columbia University Press, New York, NY, 362 pp. [Includes chapter on tracing]

Symposium on Underground Water Tracing (SUWT). 1966. 1st SUWT (Graz, Austria). Published in: Steirisches Beitraege zur Hydrogeologie Jg. 1966/67.

Symposium on Underground Water Tracing (SUWT). 1970. 2nd SUWT (Freiburg/Br., West Germany). Published in: Steirisches Beitraege zur Hydrogeologie 22(1970):5-165, and Geologisches Jahrbuch, Reihe C. 2(1972):1-382.

Symposium on Underground Water Tracing (SUWT). 1976. 3rd SUWT (Ljubljana-Bled, Yugoslavia). Published by Ljubljana Institute for Karst Research: Volume 1 (1976), 213 pp., Volume 2 (1977), 182 pp. See also Gospodaric and Habic (1976).

Symposium on Underground Water Tracing (SUWT). 1981. 4th SUWT (Bern, Switzerland). Published in: Steirisches Beitraege zur Hydrogeologie 32(1980):5-100; 33(1981):1-264; and Beitraege zur Geologie der Schweiz--Hydrologie 28 pt.1(1982):1-236; 28 pt.2(1982):1-213.

Symposium on Underground Water Tracing (SUWT). 1986. 5th SUWT (Athens, Greece). Published by Institute of Geology and Mineral Exploration, Athens.

Thrailkill, J. et al. 1983. Studies in Dye-Tracing Techniques and Karst Hydrogeology. Univ. of Kentucky, Water Resources Research Center Research Report No. 140.

Tolson, J.S. and F.L. Doyle (eds.). 1977. Karst Hydrogeology: Memoirs of the 12th Int. Congress, Int. Ass. Hydrogeologists. University of Alabama, Huntsville, AL, 578 pp. [60 papers]

Trudgill, S.T. 1985. Limestone Geomorphology. Longman, New York, NY, 196 pp.

Warren, W.M. and J.D. Moore. 1975. Hydrology of Limestone Terranes: Annotated Bibliography of Carbonate Rocks. Geological Survey of Alabama Bulletin 94E, pp. 31-163.

White, W.B. 1988. Geomorphology and Hydrology of Karst Terrains. Oxford University Press, New York, NY, 454 pp.

White, W.B. and E.L. White (eds.). 1989. Karst Hydrology: Concepts from the Mammoth Cave Area. Van Nostrand Reinhold, New York, NY, 343 pp. [12 contributed papers]

Yevjevich, V. (ed.) 1976. Karst Hydrology and Water Resources, Vol. 1 Karst Hydrology, Vol. 2 Karst Water Resources. Water Resources Publications, Fort Collins, CO, 873 pp. [Symposium proceedings with 38 papers]

(Table 1-4 Major Karst References)

CHAPTER 2

GROUND WATER AND VADOSE ZONE HYDROLOGY

2.1 Ground Water in the Hydrologic Cycle 60

2.2 Ground Water-Atmospheric Relationships 61

    2.2.1 Precipitation 61
    2.2.2 Infiltration 63
    2.2.3 Evapotranspiration 64
    2.2.4 Distribution of Precipitation in the Hydrologic Cycle 65

2.3 Ground Water-Surface Water Relationships 65

    2.3.1 Characteristics of Surface Water Flow 65
    2.3.2 Drainage Basins 66
    2.3.3 Stream Types 70
    2.3.4 Surface Water Quality 71

2.4 Water in the Vadose Zone 72

    2.4.1 Soil-Water Energy Concepts 73
    2.4.2 Subdivisions of the Vadose Zone 74
    2.4.3 Saturated vs. Unsaturated Flow 76

2.5 Water in the Saturated Zone (1): Fundamental Concepts 76

    2.5.1 Hydraulic Head and Gradients 76
    2.5.2 Unconfined and Confined Aquifers 78
    2.5.3 Heterogeneity and Anisotropy 79
    2.5.4 Porous Media vs. Fracture/Conduit Flow 80
    2.5.5 Ground Water Fluctuations 82
    2.5.6 Ground Water Divides and Other Aquifer Boundaries 85

2.6 Water in the Saturated Zone (2): Water Storage and Flow 86

    2.6.1 Aquifer Storage Properties 86
    2.6.2 Water Transmitting Properties 88
    2.6.3 Darcy's Law 88
    2.6.4 Flow Between Aquifers 90
    2.6.5 Interstitial Velocity and Time of Travel 90
    2.6.6 Ground Water Pumping Concepts 92

2.7 Guide to Major References* 94

*Appendix F contains citations for table and figure sources.

Hydrogeology is the study of ground water--its origin, occurrence, movement, and quality. Modern hydrogeology in the past century has developed along three more or less separate lines:[1] (1) elaboration of the relation between geology and ground-water occurrences (Section 1.6), (2) development of mathematical equations to describe the movement of water through rocks and unconsolidated sediments (this chapter), and (3) the study of the chemistry of ground water (Chapter 3). Another line of study, movement of water in the *vadose*, or unsaturated zone, has been mainly studied by agronomists and soil physicists. Only in the last 10 years or so has the importance of the vadose zone in the study of the movement and fate of contaminants in the subsurface been recognized; this part of the ground-water system is still often overlooked.

2.1 Ground Water in the Hydrologic Cycle

The hydrologic cycle involves the continual movement of water between the atmosphere, surface water, and the ground (Figure 2-1). The ground-water system must be understood in relation to both surface water and moisture in the atmosphere. Most additions (recharge) to ground water come from the atmosphere in the form of precipitation, but surface water in streams, rivers, ponds, lakes, and artificial impoundments will move into the ground-water system wherever the hydraulic head of the water surface is higher than the water table (Section 2.5.1). Most water entering the ground as precipitation returns to the atmosphere by evapotranspiration before reaching the saturated zone. Most water that reaches the saturated zone eventually returns to the surface again by flowing to a point of discharge at the ground surface. Typically, these points of surface discharge are rivers, lakes, or the ocean; locally, they may also take the form of springs or soil seeps. Soil, geology, and climate will in large measure determine the amounts and rates of flow among the atmospheric, surface, and ground-water systems.

Ground water is the most difficult part of the hydrologic cycle to study because it is hidden from view and occurs in a complex environment of soil and geologic materials. The movement of water in the atmosphere and surface water can be directly observed, and boundary conditions (air-ground, air-surface water, and surface water-ground) are readily defined. Inferences concerning the movement of ground water rely largely on indirect observations supplemented by a limited number of direct observations (monitoring wells). Even data from direct observations may have large margins of error as a result of variability in the materials through which the ground water is flowing.

Hydrogeology is not an exact science, but the fundamental principles of ground-water flow are well enough understood that a reasonably good characterization of a particular system is possible. In fact, at the site-specific level, the ground-water system is more predictable than either atmospheric or surface water systems that have large stochastic (random) elements. Provided that the subsurface system is adequately

---

[1] Davis, S.N. and R.J.M. DeWiest. 1966. Hydrogeology. John Wiley & Sons, New York, 463 pp.

Ground Water and Vadose Zone Hydrology 61

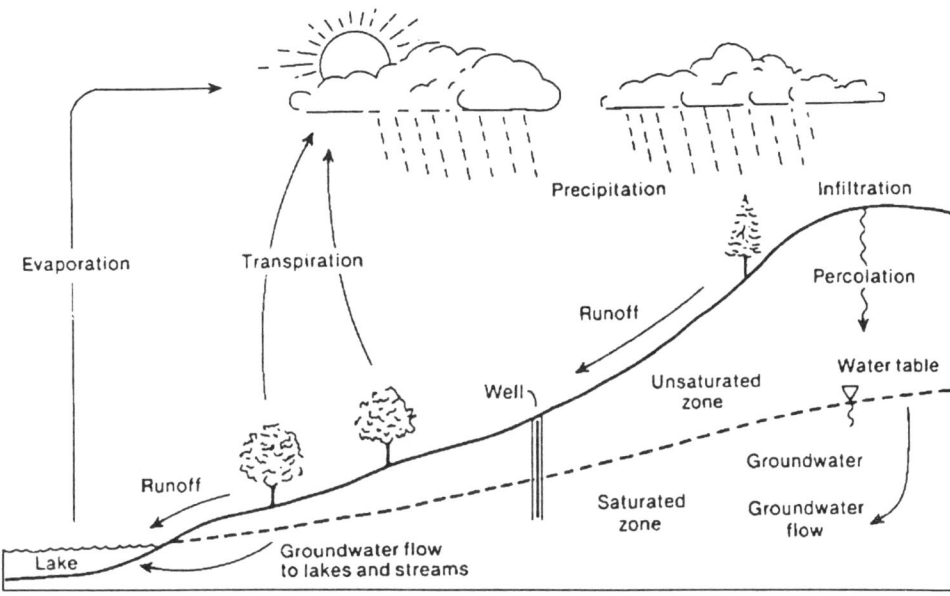

Figure 2-1 The hydrologic cycle (Muldoon and Payton, 1993).

characterized, greater confidence can be placed in what direction and how far a contaminant will travel in a specified time period than in how much rain will fall or how much water will flow through a cross section of a stream during the same period of time.

2.2 Ground Water-Atmospheric Relationships

Precipitation, infiltration, and evapotranspiration are the key elements governing the flow of water between the atmosphere and the ground.

2.2.1 Precipitation

Precipitation (rain, snow, sleet, etc.) is usually the starting point of any mass balance analysis of the flow of water in the hydrologic cycle. Properties of precipitation that affect how much reaches the ground surface include:

- *Amount.* The total amount of precipitation falling on the surface is the first parameter required in any water budget calculation. There is likely to be more ground water, and it will tend to be nearer the surface, in an area of high precipitation than in one with low precipitation. Figure 2-2 shows average annual precipitation in the continental United States.

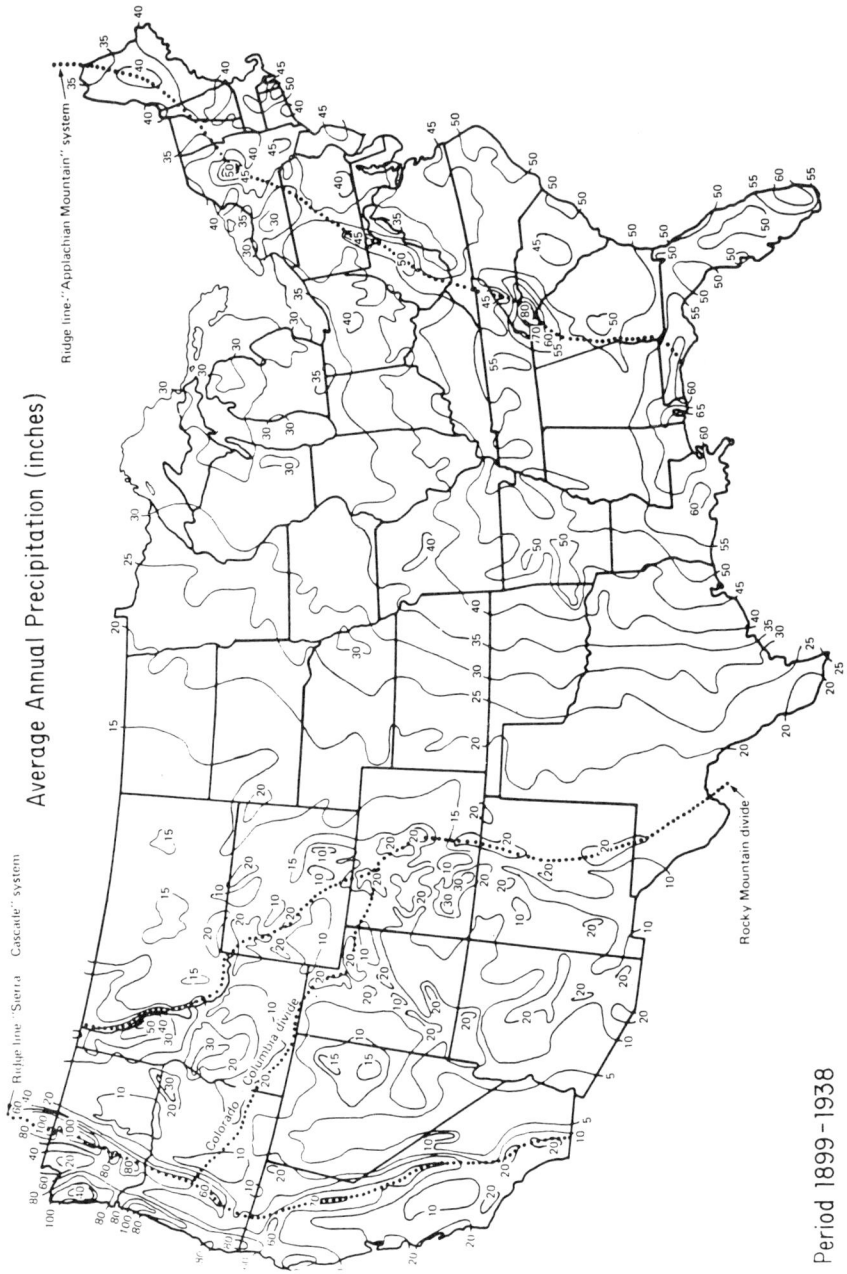

Figure 2-2 Mean annual precipitation (Viessman et al., 1972, after USDA Soil Conservation Service).

- *Form.* Whether precipitation reaches the ground as rain or snow will influence how much is likely to reach the ground-water system. In humid areas rain will more readily enter the soil than snow, although snow and ice may be significant sources of ground-water recharge in the spring once the ground has thawed. In arid, semi-arid, and alpine areas snow-melt may be the dominant form of recharge to ground water.

- *Seasonal Distribution.* Equal amounts of precipitation during different seasons will result in different amounts of ground-water recharge. When the ground is frozen, ground-water recharge is low because snow returns to the atmosphere through evaporation or runs off the surface when it melts. In spring and early summer, when the ground is saturated or has a high moisture content, recharge will be the greatest. In late summer and fall, when soils tend to be dryer, recharge will be less because precipitation goes to replenish soil moisture. Conversely, when plants are dormant (i.e., after freeze in the fall, or in early spring before plants start their normal growth) recharge will be greater.

- *Intensity and Duration.* Precipitation events can be characterized by intensity (inches per hour) and duration (length of time over which the precipitation falls). *Convectional* precipitation (thunderstorms with high intensity and short duration) is common during the summer and results in less ground-water recharge than *cyclonic* precipitation (low intensity-long duration events caused by large low pressure systems that cross the United States from the northwest of the Gulf of Mexico). *Orographic* precipitation (caused by topographic barriers that force moisture laden air to rise and cool) also tend to be of low intensity and long duration.

2.2.2 Infiltration

Not all precipitation reaches the ground; some is *intercepted* by buildings and trees and evaporates. The amount of water reaching the ground that then enters the soil is determined by the *infiltration capacity*. The infiltration capacity of a given soil is controlled by several factors:

- *Antecedent precipitation and soil-moisture conditions.* Soil moisture fluctuates seasonally; it is usually high during winter and spring and low during the summer and fall. If the soil is dry, wetting the top of it will create a strong capillary potential just under the surface, supplementing gravity. When wetted, some clayey soils swell, closing pore spaces, thereby reducing the infiltration capacity shortly after a rain starts.

- *Compaction of the soil due to rain.* The impact of raindrops on the soil surface during an intense rainfall is more likely to reduce infiltration than a gentle rain.

- *Inwash of fine material* into soil openings reduces infiltration capacity. This is especially important if the soil is dry.

- *Compaction of the soil* due to animals, roads, trails, urban development, etc. reduces infiltration.

- Certain *microstructures* in the soil will promote infiltration, such as channels created by the interface between soil structural units, openings caused by burrowing animals, insects, decaying roots and other vegetative matter, frost heaving, desiccation cracks, and other macropores.

- *Vegetative cover* tends to increase infiltration because it promotes populations of burrowing organisms and retards surface runoff, erosion, and compaction by raindrops.

- *Decreasing temperature*, which increases water viscosity, reduces infiltration.

- *Entrapped air* in the unsaturated zone tends to reduce infiltration.

- *Surface gradient*. Flat topography favors infiltration because any surface water will move more slowly than it will on sloping topography.

Infiltration capacity is usually greater at the start of a rain that follows a dry period, but it decreases rapidly. As the duration of rainfall increases, infiltration is determined by the saturated hydraulic conductivity of the soil, and becomes nearly constant. In fine-grained soil infiltration may be lower than the hydraulic conductivity immediately below the surface as a result of clogging by particles. If precipitation is greater than the infiltration capacity, surface runoff occurs. If it is less than the infiltration capacity, all the moisture enters the subsurface. The amount of infiltrating water that actually enters the ground-water system as recharge will depend on the amount of evapotranspiration.

2.2.3 Evapotranspiration

Water moves from the ground to the atmosphere by direct evaporation, and by plant transpiration. During the growing season, water intercepted by vegetation before reaching the ground, which then evaporates, typically amounts to 10 to 20% of total precipitation in humid areas. Water removal from the soil by transpiration occurs to whatever depth plant roots are able to penetrate. Depending on the type of vegetation and soil conditions, this depth is typically 3 to 4 feet for grains and pasture grasses (although under favorable conditions alfalfa roots will penetrate as deep as 10 feet). Deciduous trees have deeper roots and consequently remove soil moisture from greater depths (6 to 12 feet or more); desert plants may extend roots tens of feet below the surface to obtain moisture. Obviously, these rooting depths will be shallower where soil or geologic conditions limit rooting.

The relationship between evapotranspiration and precipitation determines whether soluble salts in soil will leach (precipitation > evapotranspiration) or accumulate (precipitation < evapotranspiration). This has important implications for contaminant transport, because a leaching environment will tend to move soluble contaminants into the saturated zone (generally in the humid East, Midwest, and coastal areas of the Pacific Northwest), whereas contaminants will tend to stay in the soil in a salt-accumulating environment (the semi-arid plains and deserts of the West).

2.2.4 Distribution of Precipitation in the Hydrologic Cycle

Very little precipitation that reaches the ground surface actually reaches the saturated zone. The mass balance for precipitation at a site in southwestern Indiana might look something like this:

$$P = E + T + G + S \qquad (2\text{-}1)$$

where:

P = Precipitation = 41 inches
E = Evaporation (interception) = 4 inches
T = Transpiration = 24 inches
G = Ground-water recharge = 1 inch
S = Surface runoff = 12 inches

This example shows that about 10% of the precipitation is intercepted and returns to the atmosphere by evaporation, about 30% runs off the surface to enter streams, and about 60% of the precipitation enters the soil. However, of the water entering the soil by infiltration, the amount actually reaching the water table represents only about 2% of total precipitation, while transpiration returns the rest to the atmosphere (58% of total precipitation).

2.3 Ground Water-Surface Water Relationships

Surface and near-surface ground water are intimately connected. Ground water that reaches the surface becomes surface water and vice versa. The direction(s) of flow between these two systems must be understood because contaminated ground water may contaminate surface water (ground water flows to the surface) or contaminated surface water may contaminate ground water (surface water flows to the ground).

2.3.1 Characteristics of Surface Water Flow

Channel storage refers to all of the water contained at any instant within the permanent stream channel. Runoff includes all of the water in a stream channel flowing past a cross section; this water may consist of precipitation that falls directly into the channel, surface runoff, ground-water runoff (also called *base flow*), and effluent. Streamflow, runoff, discharge, and yield of drainage basin are all nearly synonymous terms. Bank storage (Section 2.3.3) will also move into channel storage after flow in a stream channel drop from high-flow to intermediate and low-flow conditions.

Rates of flow are generally reported as cubic feet per second (cfs); millions of gallons per day (mgd); acre-feet per day, month, or year; cfs per square mile of drainage basin (cfs/mi^2); or inches depth on drainage basin per day, month, or year. In the United States, the most common unit of measurement for rate of flow is cubic feet per second (cfs).

Surface water discharge (Q) is determined by measuring the cross-sectional area of the channel (A), in square feet, and the average velocity of the water (v), in feet per second, so that:

$$Q = Av \qquad (2\text{-}2)$$

The cross-sectional area generally shows little change with time, so velocity is the main variable that must be measured to calculate streamflow. Typically, streamflow is measured by developing rating curves, in which accurate discharge measurements are made at low, intermediate, and high flows, and plotted against a gage that measures the height of the stream at each discharge measurement. Once the rating curve has been fitted to the initial measurements, the flow at any gage height can be estimated (provided that the channel geometry remains stable).

Surface water flow generally shows short-term fluctuations in response to individual precipitation events, and longer-term seasonal fluctuations. *Stream hydrographs* (plots of discharge as a function of time) are a useful way of viewing these fluctuations. Figure 2-3 shows a hydrograph in which individual peaks represent streamflow response to specific precipitation events. Seasonal fluctuations are also evident in this hydrograph, with lowest flow occurring in the fall. Another useful way to describe streamflow is the *flow-duration curve*, which shows the percentage of time discharge for the period of record equalled or exceeded various rates of discharge. *Base flow* (labeled as ground-water runoff in Figure 2-3) is the contribution of ground-water flow to a stream after all other contributions to streamflow have been subtracted out. Figure 2-3 shows that base flow fluctuates seasonally with a maximum in the spring and a minimum in the fall.

2.3.2 Drainage Basins

The drainage basin or *watershed* is the basic geographic unit for studying surface water. The watershed is the area that contributes water to a particular channel or set of channels. When the ground surface is saturated or precipitation exceeds the infiltration rate, surface runoff moves downhill until it reaches a stream. When several streams come together, the flows from their separate watersheds combine to form a larger watershed. Surface drainage basins are readily defined by topographic maps on which *drainage divides* mark the boundaries between watersheds.

Surface drainage patterns can be classified according to type of pattern and density. Figure 2-4a illustrates six basic drainage patterns and Figure 2-4b shows coarse, medium and fine densities. Stream segments can also be classified according to the number of tributaries. *First-order* streams have no tributaries and tend to be located in the upper reaches of a watershed. *Second-order* streams have as tributaries

Ground Water and Vadose Zone Hydrology 67

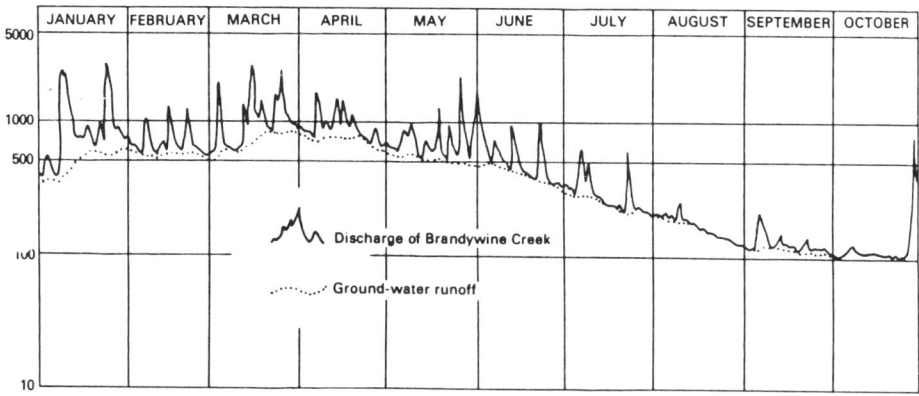

Figure 2-3 Hydrograph of Brandywine Creek, Chadd's Ford, Pennsylvania, 1952-1953 (U.S. EPA, 1987a, after Olmsted and Hely, 1962).

only first-order channels, *third-order* streams receive as tributaries only first- and second-order channels, and so on. Figure 2-4b shows stream orders for the coarse drainage density. Preliminary interpretation concerning geology and hydrogeology can be made by examining drainage patterns on aerial photographs or topographic maps. Table 2-1 summarize preliminary geologic and hydrogeologic interpretations that can be made based on drainage patterns illustrated in Figure 2-4a.

Surface and ground-water drainage basins commonly coincide, but a perfect match should never be assumed. Situations where near-surface ground-water and surface water watersheds may not coincide include:

- Karst limestone terranne where subsurface drainage may have developed independently of surface drainage.

- Sedimentary formations where the regional dip is in the opposite direction of surface water flow.

- Areas where pumping of ground water has created disturbed normal ground-water flow patterns.

- Areas where underground mining has altered subsurface flow patterns.

- Heterogeneities in the subsurface near drainage divides, which may result in ground-water divides that differ from the topographic divide.

68
I: Basic Concepts

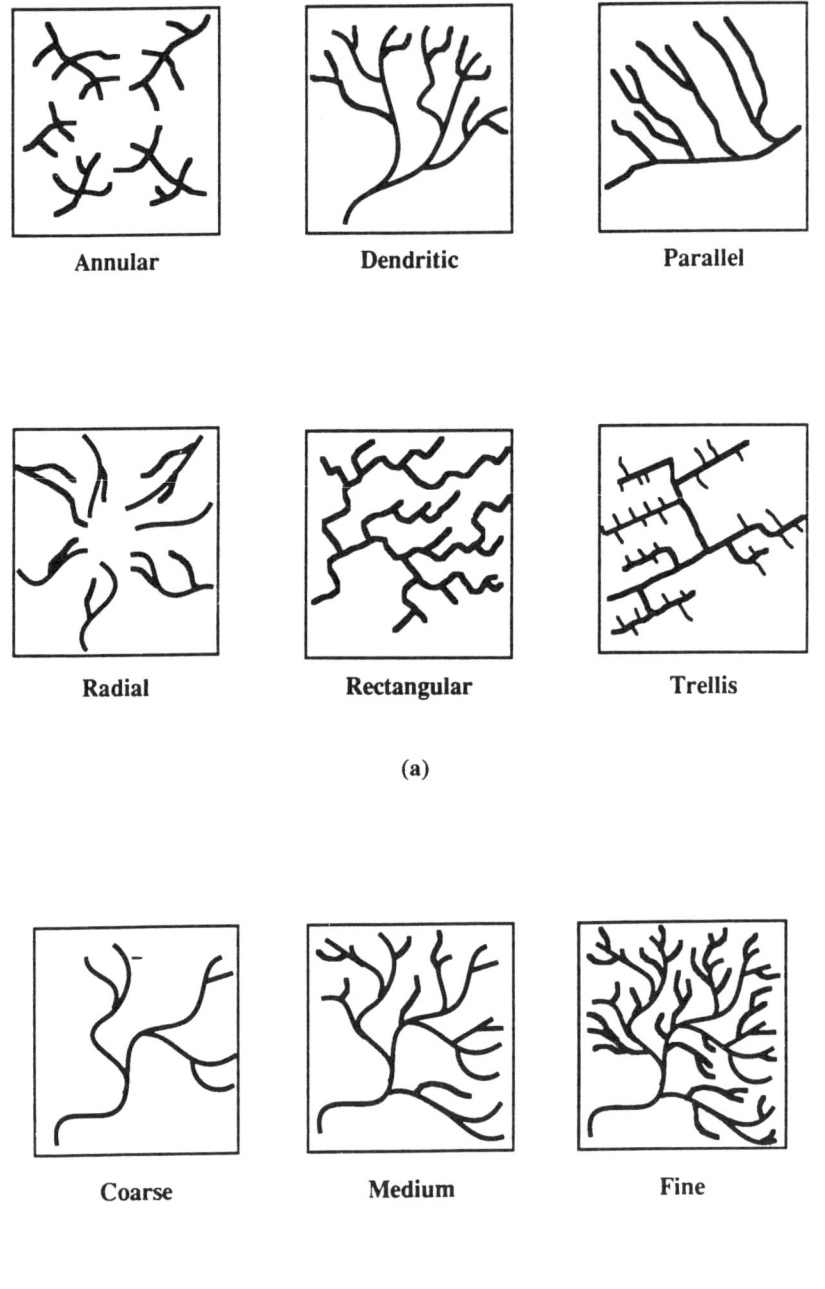

Figure 2-4 Drainage patterns: (a) six basic patterns; (b) drainage density variations (Kolm, 1993, after Way, 1973).

Table 2-1 Drainage Network Patterns as Indicators of Geologic/Hydrogeologic Systems

Drainage Pattern	Geologic/Geomorphic Indicator	Hydrogeologic Indicator
Annular	Structural and lithologic control, dome (alternating tilted stratified units; volcanic or sedimentary units) or caldera features; low or high slope gradients.	Local flow systems prevail, sedimentary or volcanic hydrology, local and regional recharge areas; fracture and matrix flow.
Dendritic	Lithologic control, flat-lying volcanic or sedimentary units; low slope gradient.	Extensive regional and local flow systems possible; controlled by degree of topographic/geologic continuity. Uniform lithology; matrix flow dominates.
Parallel/Subparallel	Structural and/or lithologic control; surficial units; slope control.	Local flow systems prevail in surficial materials, matrix flow dominates.
Radial	Structural and lithologic control, dome structures (intrusive), volcanic structures, erosional remnants, volcanic, crystalline or sedimentary units; low and high slope gradients.	Edges of regional flow systems; local flow systems prevail; local and regional recharge areas; fracture flow dominates if structural, mixed fracture and matrix flow dominate if volcanic, matrix flow dominates if erosional features.
Rectangular	Structural control, uniform crystalline units (granite, carbonate and volcanic rocks).	Regional (karst) and local (karst or igneous/metamorphic/sedimentary units) flow systems possible; fault and fracture zones control flow systems and location of discharge areas.
Trellis	Structural and lithologic control, alternating tilted stratified units; volcanic or sedimentary units; low and high sloe gradients.	Local flow systems prevail; alternating aquifers and confining units; matrix and fracture flow possible.

Source: Adapted from Kolm (1993).

2.3.3 Stream Types

From a hydrogeologic point of view, there are three major stream types: ephemeral, intermittent, and perennial. Stream type is determined by the relation between the water table and the stream channel.

An *ephemeral* stream owes its entire flow to surface runoff. It may have no well-defined channel and the water table consistently remains below the bottom of the channel (Figure 2-5, A-A') Water leaks from the channel into the ground, recharging the underlying strata.

Intermittent streams flow only part of the year, generally from spring to midsummer, as well as during wet periods. During dry weather, these streams flow only because of ground water that discharges into them when the water table rises above the base of the channel (Figure 2-5, B-B'). Eventually sufficient ground water discharges throughout the basin to lower the water table below the channel, which then becomes dry. This reflects a decrease in the quantity of ground water in storage. During late summer or fall, a wet period may temporarily raise the water table enough for ground water to again discharge into the stream. Thus, during part of the year the floodplain materials are full to overflowing, causing the discharge to increase in a downstream direction; at other times, water will leak into the ground, reducing the discharge.

Perennial streams flow year-round. Typically, the water table is always above the stream bottom, hence ground water is discharged to the surface, and streamflow increases downstream (Figure 2-5, C-C'). A stream in which the discharge increases downstream is called a *gaining stream*. When the discharge of a stream decreases downstream due to leakage, it is called a *losing stream*. In a losing stream, the water table is below the bottom of the stream, but the amount discharged to the subsurface is not enough to eliminate surface flow during periods of low flow. During wet periods, surface flow in perennial streams comes from a mixture of surface runoff and ground-water inflow. During dry periods, the flow of perennial streams comes primarily from ground-water discharge, and is called the *base flow*. Figure 2-3 shows the ground-water base flow component of Brandywine Creek. In this figure most of the flow in the months of August through October is ground-water base flow.

Normally, ground water flows into a gaining stream. However, short-term reversals in the direction of flow may occur in response to precipitation events. When the crest of runoff from a precipitation event passes a particular stream cross section, the stream level may rise higher than the water table, resulting in stream flow into the unsaturated portion of the stream bank. This occurrence blocks the ground water that would normally flow into the stream and causes the water table in the floodplain to rise. Once the stream begins to fall, the water that was recently added to the ground water will begin to flow back into the stream, rapidly at first and then more slowly as the water-table gradient declines. This temporary storage of water in the near vicinity of the stream channel is called *bank storage* (Figure 2-6).

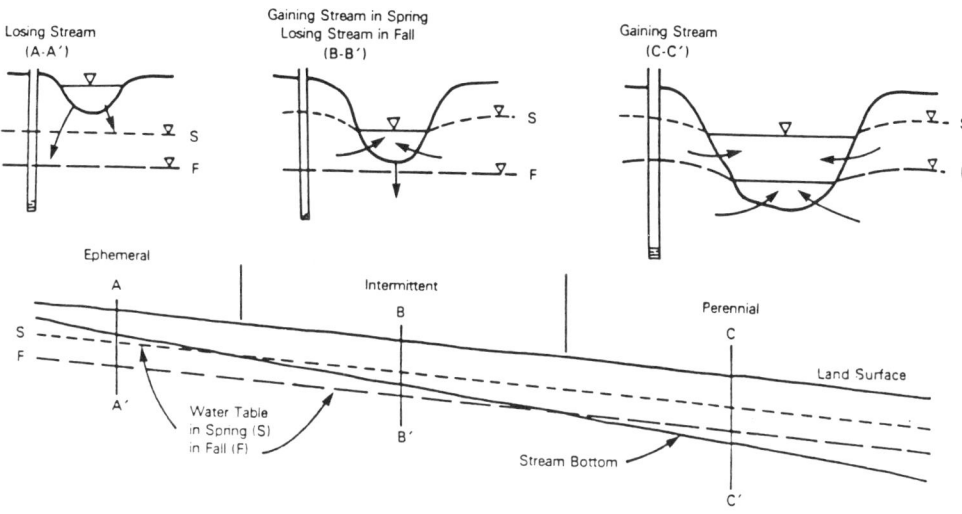

Figure 2-5 Relation between water table and stream type (U.S. EPA, 1987a).

Nearly all water courses have headwater regions characterized by ephemeral streams. Farther downbasin, intermittent streams predominate and, even farther, the water courses are perennial. Some streams fed by springs or glacial melt water are perennial throughout their entire length. The natural gradation from one stream type to another may be interrupted by either natural or manmade causes. Irrigation may provide enough recharge to raise the water table sufficiently to increase ground-water runoff, while pumping from wells may have the opposite effect.

Streams flowing through saturated permeable deposits, such as sand and gravel, are normally gaining streams, but streams flowing through karst regions, or areas of extensive underground mining, may be losing in one reach and gaining in another. High dry-weather flow may reflect the discharge of water from mine workings.

2.3.4 Surface Water Quality

Under *base flow* conditions (i.e., stream flow comes entirely from ground-water discharge into the stream channel), the chemical quality of water in a perennial stream reflects the quality of ground water in the zone of active circulation within the basin, provided the stream is not contaminated by some surface source. During wet weather, the chemical quality of water in a stream varies largely because of the mixing of dilute surface runoff with the more highly mineralized ground-water runoff. The sediment load, reflecting erosion in the basin and stream channel, also affects the quality of the

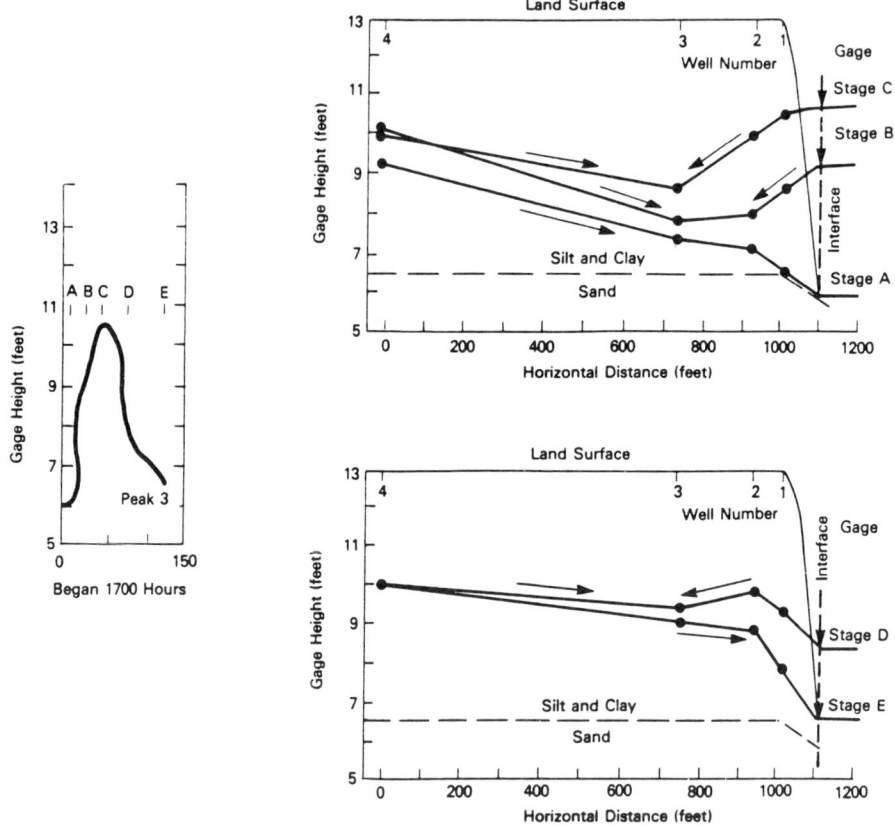

Figure 2-6 Movement of water into and out of bank storage along a stream in Indiana (Daniels et al., 1970).

stream. The loading of a stream with either sediment or dissolved constituents is commonly reported in units of tons per day: short tons per day = discharge x concentration x .0027.[2]

2.4 Water in the Vadose Zone

Historically, hydrogeologists have focused their attention on the *phreatic* (saturated) zone without paying much attention to water moving through the *vadose* (unsaturated) zone. The vadose cannot be ignored in the study of contaminant hydrogeology because it may be a significant reservoir for the capture, storage, and release of contaminants. Unsaturated fluid flow is complicated by matric and osmotic

---

[2] U.S. Environmental Protection Agency (EPA). 1987a. Handbook: Ground Water. EPA/625/6-87/016, 212 pp.

Ground Water and Vadose Zone Hydrology

energy potentials (Section 2.4.1). Although the term "unsaturated zone" is often used loosely to refer to the vadose zone, part or all of this zone may be intermittently saturated and may contain several important subdivisions (Section 2.4.2).

2.4.1 Soil-Water Energy Concepts

The retention and movement of water in the subsurface are energy-related phenomena. *Free energy* is the general term used to characterize the energy status of water. Water will tend to move or change from a higher to a lower free energy level, with the critical factor being differences in energy levels from one contiguous site to another.

Free energy of soil-water is influenced by three major types of energy potentials:

- *Matric potential* (P_m) is the attraction of water to solids in the subsurface. Matric potential arises from both adsorption of water onto solids and capillary action in soil pores. The forces causing this energy potential *reduce* the free energy of water, and are sometimes called *matric suction* (see Figure 2-7). Generally, the smaller the particle and pore size, the greater the matric potential.

- *Osmotic potential* (P_o) results from dissolved constituents in subsurface water. The attraction of solute ions to water molecules reduces the free energy of water. Consequently, pure water will move across a semipermeable membrane to the side with a higher solute concentration. This is sometimes called *osmotic suction*. The higher the solute concentration differential across a membrane, the greater the osmotic suction. The negative pressure potential shown in Figure 2-7 represents the combined matric and osmotic suction.

- *Gravitational potential* (P_g) is the attraction of the force of gravity toward the earth's center. $P_g = Gh$, where G is the acceleration of gravity and h is height above a reference elevation (usually chosen below the lowest point at which this potential will be measured so that the gravitational potential will always be positive).

Total soil-water potential is the sum of the contributions of the various forces acting on soil water:

$$P_t = P_g + P_m + P_o + \ldots \quad (2\text{-}3)$$

where the gravitational, matric, and osmotic potentials are as defined above and other less significant potentials are indicated by dots. Since gravity is a positive potential and matric and osmotic potentials are negative, water will only move through the soil profile if $P_g > P_m + P_o$. As discussed below, matric and osmotic potentials are significant forces affecting the movement of water in the unsaturated zone. In the saturated zone, gravitational potential is the dominant force.

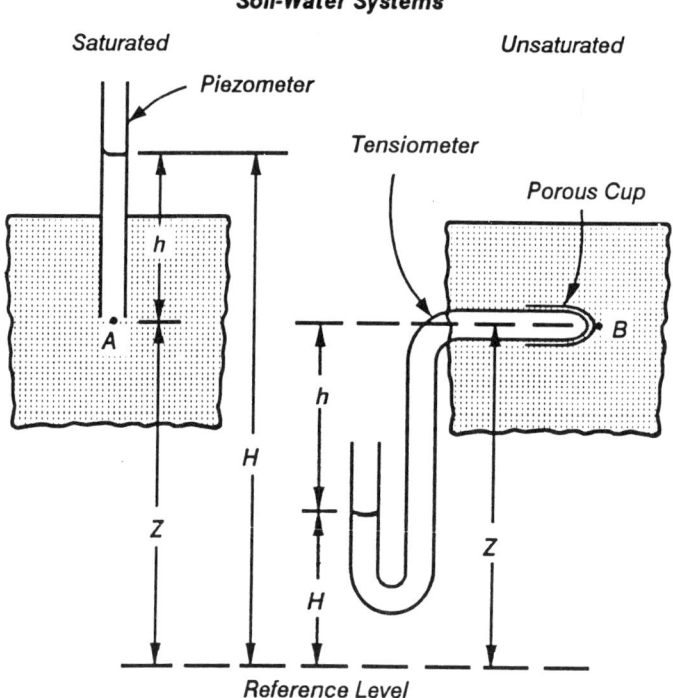

Figure 2-7 Diagram of the relationship between hydraulic head, H, pressure head, h and gravitational head, Z (Mercer and Spalding, 1991c).

2.4.2 Subdivisions of the Vadose Zone

The vadose zone has three major subdivisions (see Figure 2-8):

- *Soil-water* or *root zone*. This zone lies between the ground surface and the maximum depth to which roots penetrate. It is characterized by large fluctuations in the quantity and quality of moisture in response to transpiration and evaporation. The rooting zone is commonly called the *soil-water zone* in the hydrogeologic literature, but this term is not entirely accurate, since roots can penetrate well below the weathered zone of a soil profile. Local differences in gravitational, osmotic (created across root membranes), and matric potential will determine whether water entering the soil attaches to soil particles, is transpired by plants, or moves through the rooting zone.

- *Intermediate vadose*. This zone contains a residual moisture content determined by the matric potential. In coarse-grained material (sand and gravel), the amount of water held by matric potential is low; in fine-grained materials, particularly clays, the amount of water held may be very high. Since this zone also contains a significant amount of air in pore spaces, gravitational water

Ground Water and Vadose Zone Hydrology

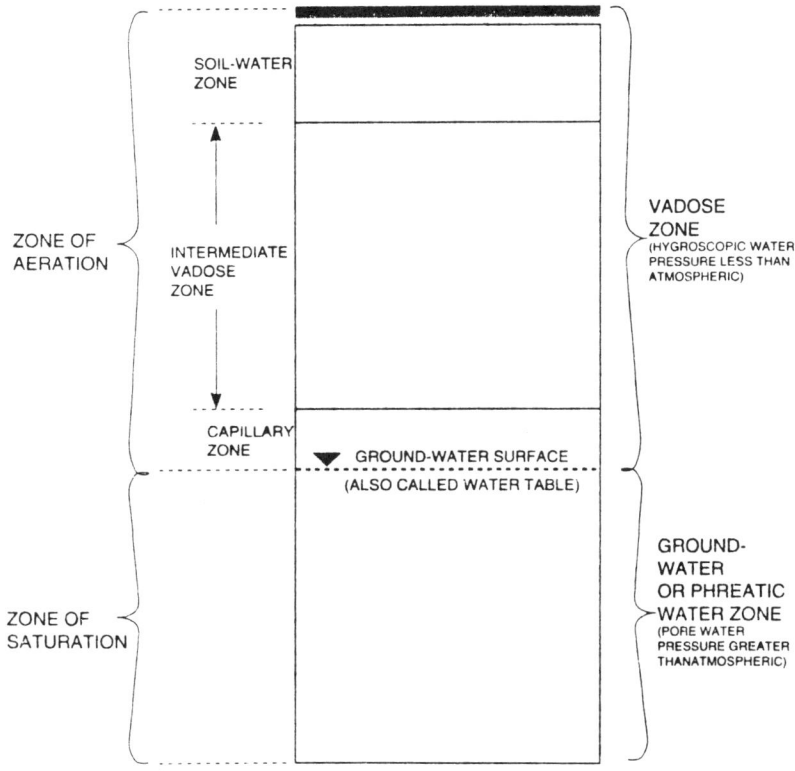

DISTRIBUTION OF FLUID PRESSURES IN THE GROUND
WITH RESPECT TO THE GROUND WATER SURFACE

Figure 2-8 Classification of subsurface water (Sara, 1994). Reprinted with permission.

reaching this zone moves relatively slowly to the saturated zone by unsaturated flow until it reaches the capillary fringe.

- *Capillary fringe.* This zone marks the final transition between the vadose zone and the saturated zone. Capillarity is due to (a) the attractive forces of water solids, and (b) the surface tension of water. For water, the height of capillary rise in a tube can be calculated by the expression $h = 0.15/r$, where r is the radius of the tube. In other words, the smaller the diameter, the greater the capillary rise. Depending on the texture of the soil or geologic materials, this fringe may range from around 1 foot (sand) to 3 feet or more (loam).

Contaminants entering the vadose zone will tend to move more slowly than in the saturated zone because of the processes described above that retard the rate of water movement. For example, in the rooting zone, contaminants may be removed from the soil and incorporated into plant tissue, or they may remain in the soil after water is removed by transpiration until more water enters the soil and moves them farther down the soil profile.

2.4.3 Saturated vs. Unsaturated Flow

Saturated flow will occur in the vadose zone when the amount of water moving into the ground is sufficient to fill all the pore space in a soil. This is usually a temporary state because as soon as water stops flowing into the ground, gravitational water will flow relatively quickly into channels and the larger pore spaces in the soil. Usually gravitational water (where $P_g > P_m + P_o$) drains from the soil within 1 to 3 days. At this time, the soil is holding its *field capacity* of water, with approximately one half of the pore space filled with air and one half with water. A sandy soil has a low field capacity of about 1 inch per foot of soil in the plant root zone that is reached quickly; clay-rich soils are characterized by high field capacities (3.5 to 4 inches/foot) that are reached slowly.

As noted above, water can flow by gravity through the soil even when it is not saturated. However, *rates* of unsaturated flow are substantially slower than rates of saturated flow. Figure 2-9 shows that *hydraulic conductivity* (the rate of flow in cm/day) at saturation (at or near zero matric suction) is about six orders of magnitude higher than at a matric suction of 10 bars. Saturated flow takes place at or near zero suction, while most of the unsaturated flow occurs at a suction of 0.1 bar or above. Fifteen bars is the *wilting point* at which plants are no longer able to effectively remove moisture from the soil.

Figure 2-9 also illustrates that the *saturated* hydraulic conductivity of coarse-textured sandy loam is almost an order of magnitude higher than that of fine-grained clay soil. The relationship reverses under unsaturated conditions, such that the *unsaturated* hydraulic conductivity is an order of magnitude *lower* than that of the clay soil. These variations, plus variations in texture and pore-size distribution in the vadose zone, make moisture status and movement difficult to measure.

2.5 Water in the Saturated Zone (1): Fundamental Concepts

2.5.1 Hydraulic Head and Gradients

The water level in a well, usually expressed as feet above sea level, is the total head (ht), which consists of elevation head (z) and pressure head (hp)--see Figure 2-7.

$$ht = z + hp \tag{2-4}$$

In an unconfined aquifer, pressure head (hp) equals zero at the water table surface because it marks the transition from negative pressure head in the vadose zone to a pressure head that may be either negative or positive in the saturated zone. Serious inaccuracies in defining ground-water flow paths may result from measuring water levels in monitoring wells without considering the pressure potential component (Section 7.2).

In a ground water *recharge* zone, the pressure head *decreases* with increasing depth (i.e., hp in equation 2-4 is negative); in a *discharge* zone, the pressure head

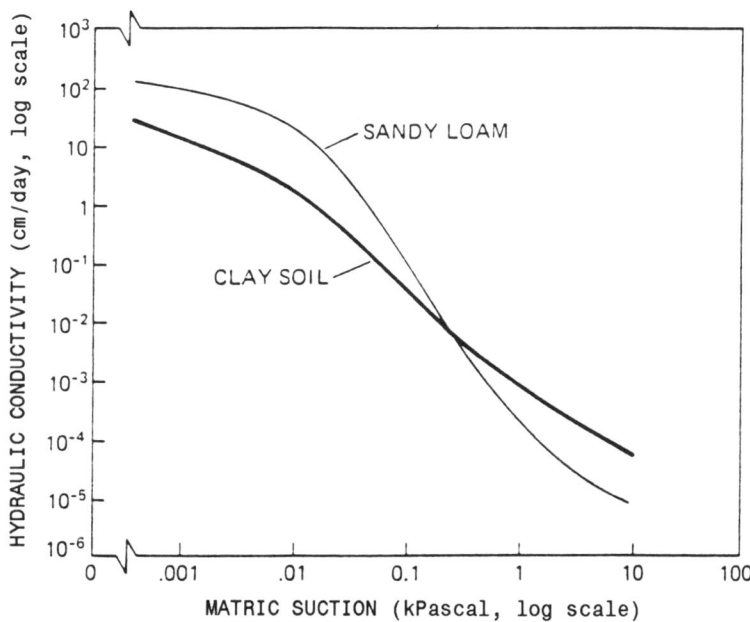

Figure 2-9 Generalized relationship between matric suction and hydraulic conductivity for a sandy loam and a clay soil (adapted from Brady, 1974).

increases with depth. This is illustrated in Figure 2-10. In the figure, the water level in piezometer c is the same as well b since it lies along the same equipotential line. The water level in well b is lower than the water table surface at that well location. This is because the well is cased to a depth where it is actually measuring the pressure potential of the water table at well c. Conversely, wells d and e in the discharge area are measuring the pressure potential of the water table upslope from the actual discharge area. Wells d and e will flow like artesian wells even though there is no confining layer.

Typically, wells are not installed at different depths in the same location to allow determination of whether the area is in a recharge or discharge zone. Topography is a simple indicator, with discharge in topographically low areas and recharge in topographically high areas.

The hydraulic gradient (I or i) is measured as the change in water level per unit of distance along the direction of maximum head decrease. It is determined by measuring the water level in several wells that measure the true unconfined water table or the same confined aquifer. The hydraulic gradient is the driving force that causes ground water to move in the direction of decreasing total head, and is generally expressed in consistent units such as feet per foot. For example, if the difference in water level in two wells 1,000 feet apart is 8 feet, the gradient is 8/1,000 or 0.008. The

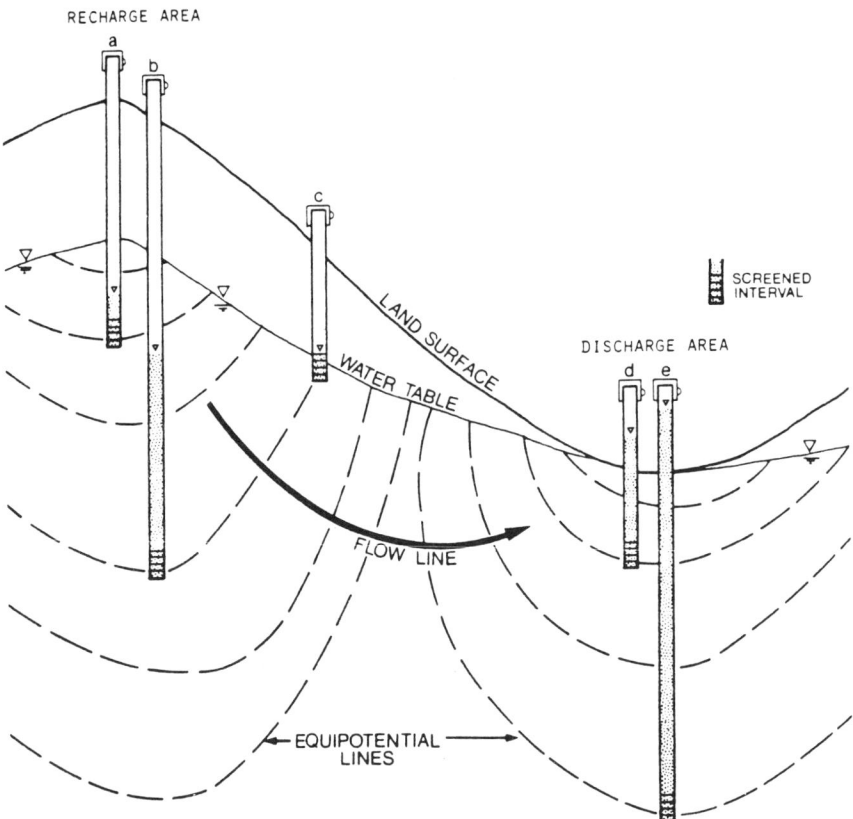

Figure 2-10 Cross-sectional diagram showing the water level as measured by piezometers at various depths (Mills et al., 1985).

direction of ground water movement and the hydraulic gradient can be determined with information from three wells (Section 7.2.2).

2.5.2 Unconfined and Confined Aquifers

Aquifers are broadly classified as *unconfined*, where the top of the saturated zone is at atmospheric pressure, and *confined*, where a slowly permeable geologic layer prevents upward flow when the hydraulic head is above the level of the confining layer, causing pressure head at the top of the aquifer to exceed atmospheric pressure. Confining layers or units may also be called *aquitards* or *aquicludes* in the older literature. Confined aquifers are classified as either *semiconfined* or *leaky* and *confined*, depending on how permeable the confining layer is. Because no confining unit is completely impermeable, Kreitler and Senger (1991) have suggested using the term *highly confined* for confining units with very low permeability.[3] Section 7.5.3 identifies methods for characterizing presence and degree of confinement. Aquifer classification is especially important in selecting methods for interpreting pump test data and serves as an indicator of the vulnerability to ground water contamination.

In humid and semi-arid regions in particular, the water table in an unconfined aquifer generally conforms to the surface topography, although it usually has greater depth under hills than under valleys (Figure 2-10). The hydraulic gradient (Section 2.5.1) slopes away from divides and topographically high areas toward adjacent low areas, such as streams and rivers. The high areas serve as ground-water recharge areas, while the low areas are ground-water discharge zones. In general, the water table lies at depths ranging from 0 to about 20 feet in humid and semi-arid regions, but often lies hundreds to thousands of feet deep in some desert environments. Generally, surface streams and waterbodies such as swamps, ponds, lakes, and flooded excavations (abandoned gravel pits, highway borrow pits, etc.) can be considered surface expressions of the water table.

Unconfined water tables may be either *perched* or *regional*. Perched water tables rest on impermeable strata, below which unsaturated flow occurs (see Figure 2-11, upper right corner). In regional aquifers, all water moves by saturated flow until it reaches a point of surface discharge (Figure 2-11, Aquifer C). Aquifers A and B in Figure 2-11 exhibit characteristics of both perched and regional water tables. Most of their water is part of the regional water, although it may travel part-way by unsaturated flow before reaching Aquifer C. Some water, however, reaches the surface as springs, a common situation with perched aquifers.

2.5.3 Heterogeneity and Anisotropy

Aquifers in which the hydraulic conductivity or other properties are nearly uniform are called *homogeneous*; those in which properties are variable are *heterogeneous* or nonhomogeneous. If hydraulic conductivity at a given point in an aquifer differs in the vertical or horizontal directions, it is *anisotropic*. If hydraulic conductivity is uniform in all directions, which is rare, the aquifer is *isotropic*. Figure 2-12a illustrates four possible combinations of these characteristics. The distinctions between these terms may not seem obvious at first, but a careful examination of this figure should provide a clearer understanding.

Figure 2-12b illustrates three different types of aquifer heterogeneity: (1) varying thickness (A, common with fluvial and deltaic sediments), (2) layering with different hydraulic conductivities (B, common with sediments deposited under water bodies), and (3) lateral changes in hydraulic conductivity (C, common with glaciofluvial deposits and areas of complex geologic structure). Because both unconsolidated and consolidated sedimentary strata are typically deposited in horizontal units (example B

---

[3] Kreitler, C.W. and R.K. Senger. 1991. Wellhead Protection Strategies for Confined-Aquifer Settings. EPA/570/9-91-008, 168 pp.

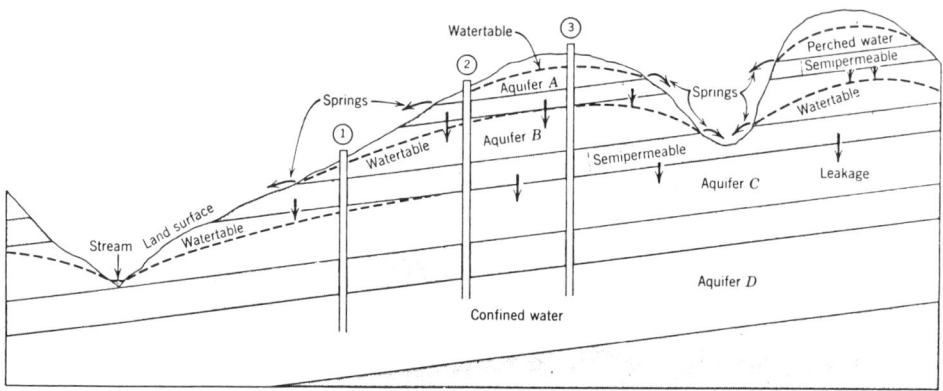

Figure 2-11 Confined, unconfined, and perched water in a simple stratigraphic section of sandstone and shale (Davis and DeWiest, 1966). © 1966 John Wiley & Sons. Reprinted by permission.

in the figure), hydraulic conductivity is generally greater horizontally than vertically by at least an order of magnitude. The third example (C in the figure) is most likely to occur as a result of faulting or other tectonic activity. Failure to consider heterogeneity and anisotropy can lead to significant underestimation of time of travel of contaminants and incorrect delineation of the direction of ground-water flow.

2.5.4 Porous Media vs. Fracture/Conduit Flow

Ground water flows in the interconnected pore spaces between solid particles in an aquifer. Most ground-water flow equations assume that the water is flowing through material where the pore sizes are small enough that water flows without turbulence. This is generally true in aquifers where *primary* or *matrix porosity* has not been altered by geologic or soil-forming processes that create secondary openings, often called *secondary porosity*. Secondary openings are classified as *fractures*, which develop as a result of deformation and stress release by geologic processes, and *solution* openings, which are formed from the enlargement of fractures by dissolution of soluble minerals such as carbonate in limestone.

Flow in fractures is most significant in crystalline rocks (granites, various metamorphic rocks) because primary porosity of these rocks is very low. Many consolidated sedimentary aquifers are fractured to varying degrees. Aquifers where fracture flow is significant tend to be anisotropic. Ground-water flow directions in these aquifers may depart significantly from the directions indicated by potentiometric surface maps. Analysis of aquifer test data in fractured rocks requires special care because most analytical solutions assume porous-media flow (Section 7.5.4). However, fractures are typically narrow enough to prevent turbulent flow, making adaptation of ground-water flow equations possible. Fracture flow is a major contributor to macro-scale hydrodynamic dispersion, causing contaminants to move much more quickly in an aquifer than would be predicted by flow calculations based on primary porosity.

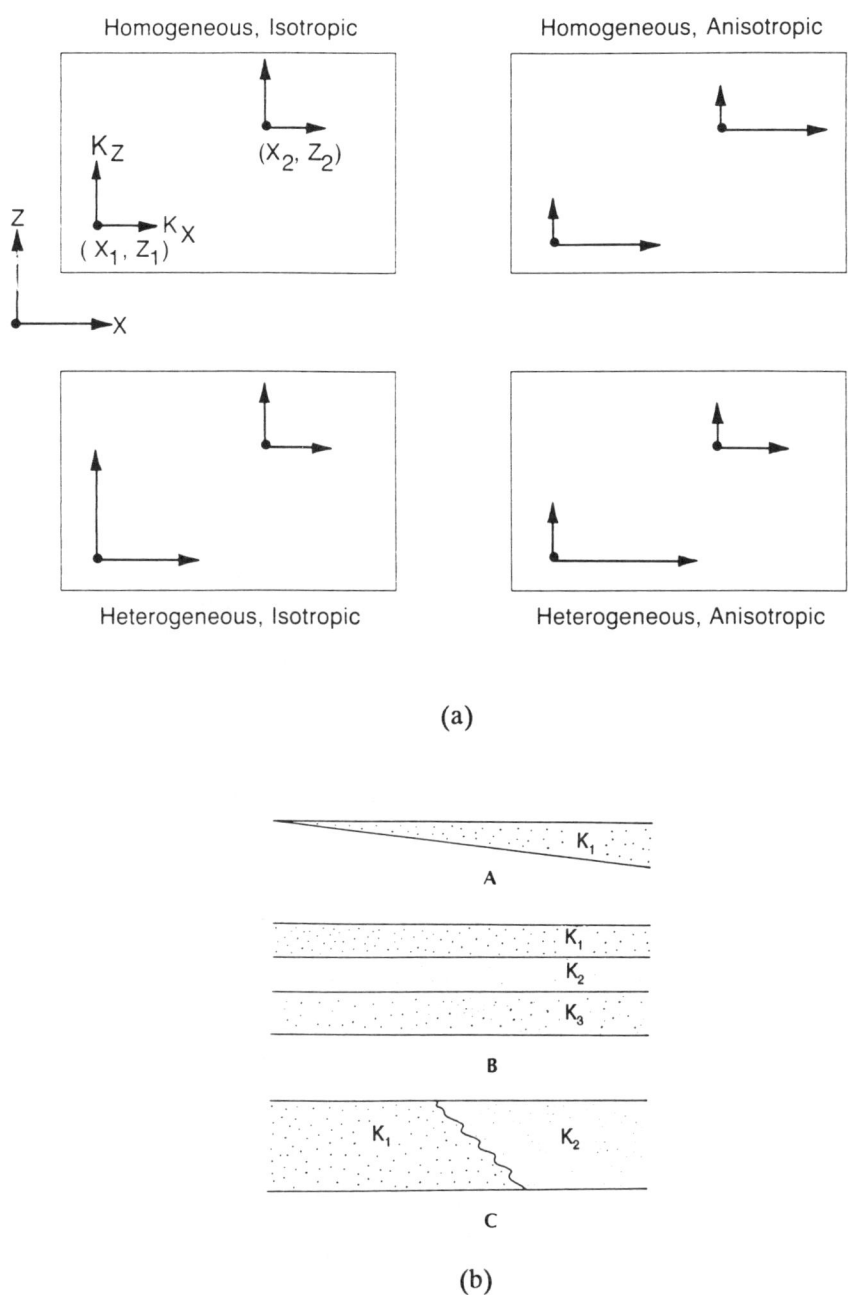

Figure 2-12 Heterogeneity and anisotropy: (a) four possible combinations (Freeze and Cherry, 1979-see Appendix F for full credit); (b) three types of aquifer heterogeneity (adapted from Fetter, 1994).

Flow in cavernous limestones and dolomites is called *conduit* flow. The subsurface channels can be large and continuous enough that the system is more like a series of interconnected pipes than a porous material. As with crystalline rocks, primary porosity of limestones tends to be very low, so that most ground-water flow is concentrated in fractures and solution channels. Aquifers where conduit flow dominates are called *karst* aquifers. However, unlike fracture-rock aquifers, ground-water flow in karst aquifers is often rapid enough that Darcy's Law (Section 2.6.3) is not valid (Section 7.5.4). The irregular shape of solution channels in these aquifers makes the use of conventional methods for analyzing pump test data and modeling ground-water flow essentially useless. Figure 1-9 illustrates the wide fluctuation in ground-water levels that can occur in a karst aquifer. Table 1-4 identifies major references where more information can be obtained about karst geomorphology and hydrology.

2.5.5 Ground Water Fluctuations

Ground-water levels fluctuate throughout the year in response to natural changes in recharge and discharge (or storage), changes in pressure, and artificial stresses. Fluctuations brought about by changes in pressure are limited to confined aquifers. Most of these changes are short-term and are caused by loading, such as by a passing train compressing the aquifer, or by an increase in discharge from an overlying stream. Others are related to changes in barometric pressure, tides, and earthquakes. Languth and Treskatis (1989) describe an unusual situation where an aquifer test in a semiconfined aquifer system temporarily increased water levels in observation wells tapping the overlying confining bed instead of resulting in the usual immediate lowering.[4] None of these fluctuations reflect a change in the volume of water in storage. Table 2-2 summarizes 13 mechanisms that lead to fluctuations in ground water levels.

Water level fluctuations in confined aquifers can be characterized by the *barometric efficiency*, the ratio of change in head to change in atmospheric pressure. This ratio usually falls in the range of 0.20 to 0.75. The possibility of using barometric efficiency to estimate the storage properties of confined aquifers was first suggested by Jacob (1940).[5] Barometric efficiency can be used to estimate a range of aquifer properties, including storage coefficient, transmissivity, and bulk elastic properties.[6]

Fluctuations that involve changes in storage are generally more long lived. Most ground-water recharge takes place during the spring and causes the water level to rise.

---

[4] Languth, H.R. and C. Treskatis. 1989. Reverse Water Level Fluctuations in Semiconfined Aquifer Systems--"Rhade Effect". J. Hydrology 109:79-93.

[5] Jacob, C.E. 1940. On the Flow of Water in an Elastic Artesian Aquifer. Trans. Am. Geophys. Union 21:574-586.

[6] Ritzi, R.W., S. Sorooshian, and P.A. Hsieh. 1991. The Estimation of Fluid Flow Properties from the Response of Water Levels in Wells to the Combined Atmospheric and Earth Tide Forces. Water Resources Research 27(5):883-893.

Ground Water and Vadose Zone Hydrology

Following this period of a month or two, the water level declines in response to natural discharge, largely to streams. Although the major period of recharge occurs in the spring, minor events can happen any time it rains. A number of human activities cause long-term fluctuations in ground-water levels. Ground-water pumpage reduces ground-water levels; activities such as agricultural irrigation, artificial recharge, leakages from ponds, lagoons and landfills tend to cause localized increases in ground-water levels. Deep well injection into confined aquifers also causes elevation in the potentiometric surface.

Table 2-2 Summary of Mechanisms that Lead to Fluctuations in Ground Water Levels

	Unconfined	Confined	Natural	Man-Induced	Short-term	Diurnal	Seasonal	Long-term	Climatic
Ground-water recharge	x		x				x		x
Air entrapment during recharge	x		x		x				x
Evapotranspiration	x		x			x			x
Stream bank storage effects	x		x				x		x
Tidal effects near ocean	x	x	x			x			
Atmospheric pressure effects	x	x	x		x				x
Confined aquifer external loading		x		x	x				
Earthquakes		x	x		x				
Ground-water pumpage	x	x		x				x	
Deep-well injection		x		x				x	
Artificial recharge/ leakage	x			x				x	
Agriculture irrigation/ drainage	x			x			x		x
Geotechnical drainage	x			x				x	

Source: Adapted from Freeze and Cherry (1979).

Evapotranspiration effects on a surficial or shallow aquifer are both seasonal and daily. Plants, serving as minute pumps, remove water from the capillary fringe or even from beneath the water table during hours of daylight in the growing season. This results in a diurnal fluctuation in the water table and stream flow.

Table 2-3 summarizes typical natural conditions affecting ground-water fluctuations in response to (1) freezing, (2) moisture regime, (3) surface drainage and degree of slope, and (4) thickness of the zone of aeration. All these factors need to be considered in compiling data on water levels in wells when preparing potentiometric surface maps (Section 7.2).

Table 2-3 Factors and Natural Conditions Affecting Natural Ground Water Fluctuations

Factor/Zone	Ground Water Conditions and General Characteristics of Water Level Fluctuations
Soil Freezing	
1. Permafrost areas.	Two summer water level rises.
2. Uniform freezing in the soil zone at the land surface.	Marked water level rise in the spring, followed by water level recession until autumn. A second smaller water level rise in autumn, followed by gradual decline until spring. Aquifer may change from water table conditions during summer to confining conditions when the soil is frozen.
3. Sporadic freezing of the zone of aeration.	Water level rises mainly in the winter.
4. Complete absence of soil freezing.	Water level rises during rainy season.
Soil Moisture Regime	
1. Region of high moisture.	The amount of precipitation is higher than evapotranspiration. Water levels affected rapidly by small rains and small temperature variations. Small amplitude of water fluctuations.
2. Region of moderate moisture.	As water table is at greater depth than in zone 1, amplitudes of water level fluctuations are more distinct and greater than in zones 1 and 3.
3. Region of small moisture.	Evapotranspiration is a dominant factor in water level fluctuations.
Surface Drainage and Degree of Slope	
1. Well-developed drainage (generally mountainous topography).	High runoff and low infiltration to ground water. Water level fluctuation amplitude may be high.

Table 2-3 (cont.)

Factor/Zone	Ground Water Conditions and General Characteristics of Water Level Fluctuations
Surface Drainage and Degree of Slope (cont.)	
2. Moderately developed drainage (generally uplands).	Moderate runoff and infiltration to ground water. Water level fluctuation amplitudes are lower than in zone 1 but higher than in zone 3.
3. Poorly developed drainage (generally plains and valley bottoms).	Low runoff and high infiltration to ground water. Water table at shallow depth. High evapotranspiration.
Thickness of Zone of Aeration (d)	
1. d is less than 0.5 m.	Water level fluctuations of small amplitude. Evapotranspiration from the water table prevails over spring discharge.
2. d is between 0.5 and 4 m thick.	Water level fluctuations of larger amplitude than in zone 1. Spring discharge prevails over evapotranspiration.
3. d is greater than 4 m.	Water level fluctuations of small amplitude and evapotranspiration might be of limited importance.

Source: Adapted from Brown et al. (1983).

2.5.6 Ground Water Divides and Other Aquifer Boundaries

In surface hydrology, a drainage divide forms the boundary between two watersheds. Ground-water drainage basins are similar to surface watersheds, except that they are defined by contours of equal hydraulic head (equipotential lines) rather than topographic contours. In unconfined, homogenous, isotropic aquifers, these contours generally follow the surface topography, albeit with a more subdued gradient (Figures 2-10 and 2-11). However, topography is only one of many factors that influence the location of ground-water divides and the flow of water within a basin. An understanding of the *boundary conditions* in an aquifer is essential when mapping aquifers and when using ground-water computer models (Chapter 10).

Figure 2-11 illustrates several ground-water divides. Infiltrating water entering the aquifer at the recharge area flows to a discharge point determined by where the water table intersects the ground surface. Note that the topographic divide for Aquifer A does not quite coincide with the ground-water divide due to the dip of the sediments.

Figure 2-13 illustrates more than 40 boundary conditions that may define the edges of a ground-water drainage area. These boundary conditions are classified as: (1) *barrier boundaries*, created by geologic or other materials of contrasting (lower) permeability compared to the aquifer, (2) *permeable recharge* boundaries, and (3) *permeable discharge* boundaries. Figure 2-13 further classifies boundary conditions according to whether they represent head conditions or flow conditions. It also shows the number of dimensions required to represent the condition: (1) points (one-dimensional), (2) lines (two-dimensional), and (3) areas (three-dimensional). These distinctions become important when analytical and numerical ground-water models are selected and used (Chapter 10). Where streams form boundaries to an aquifer, it is important to determine whether they are *gaining* or *losing* streams (Section 2.3.3) to develop correct interpretations of the ground-water flow system.

2.6 Water in the Saturated Zone (2): Water Storage and Flow

Key questions that most ground-water studies must answer are (1) how much water is available from an aquifer, and (2) how fast is water moving through the aquifer? The answer to these questions requires measurement or estimation of aquifer storage properties (porosity, and specific yield or storativity) and aquifer transmitting properties (hydraulic conductivity and transmissivity).

2.6.1 Aquifer Storage Properties: Porosity and Specific Yield/Storativity

Porosity, expressed as a percentage or decimal fraction, is the ratio between the openings in the rock and the total rock volume. It defines the amount of water a saturated rock volume can store. If a unit volume of saturated rock drains by gravity, not all of the water it contains will be released. The volume drained is the *specific yield*, a percentage, and the volume retained is the *specific retention*. Therefore, porosity is equal to specific yield plus specific retention. Knowing any two of these terms allows calculation of the third.[7] Tables C-1 and C-2 show some typical values for porosity and specific yield for various soil and geologic materials. Actual values for specific materials can differ substantially from the numbers shown here.

Another important term is *storativity* (S), which describes the quantity of water that an aquifer will release from storage or take into storage per unit of its surface area per unit change in head. In unconfined aquifers, the storativity is, for all practical purposes, equal to the *specific yield*. The storativity of confined aquifers is substantially smaller than specific yield (Table C-2), because the water released from storage when the head declines comes from the expansion of water and compression of the aquifer, both of which are very small. For confined aquifers, storativity generally ranges between 0.005 and 0.00005, with leaky confined aquifers falling in the high end of this range.[8] The small storativity of confined aquifers means that a large pressure change throughout a wide area is needed to obtain a sufficient supply from a well. This is not the case with unconfined aquifers, because the water derived is not related to

---

[7] This includes only interconnected pores through which water can flow. Isolated pores, whether air- or water-filled, can be considered part of the solid volume of a rock for purposes of ground-water flow analysis.

[8] 0.0001 to 0.00001 may also be cited in the literature as a typical range.

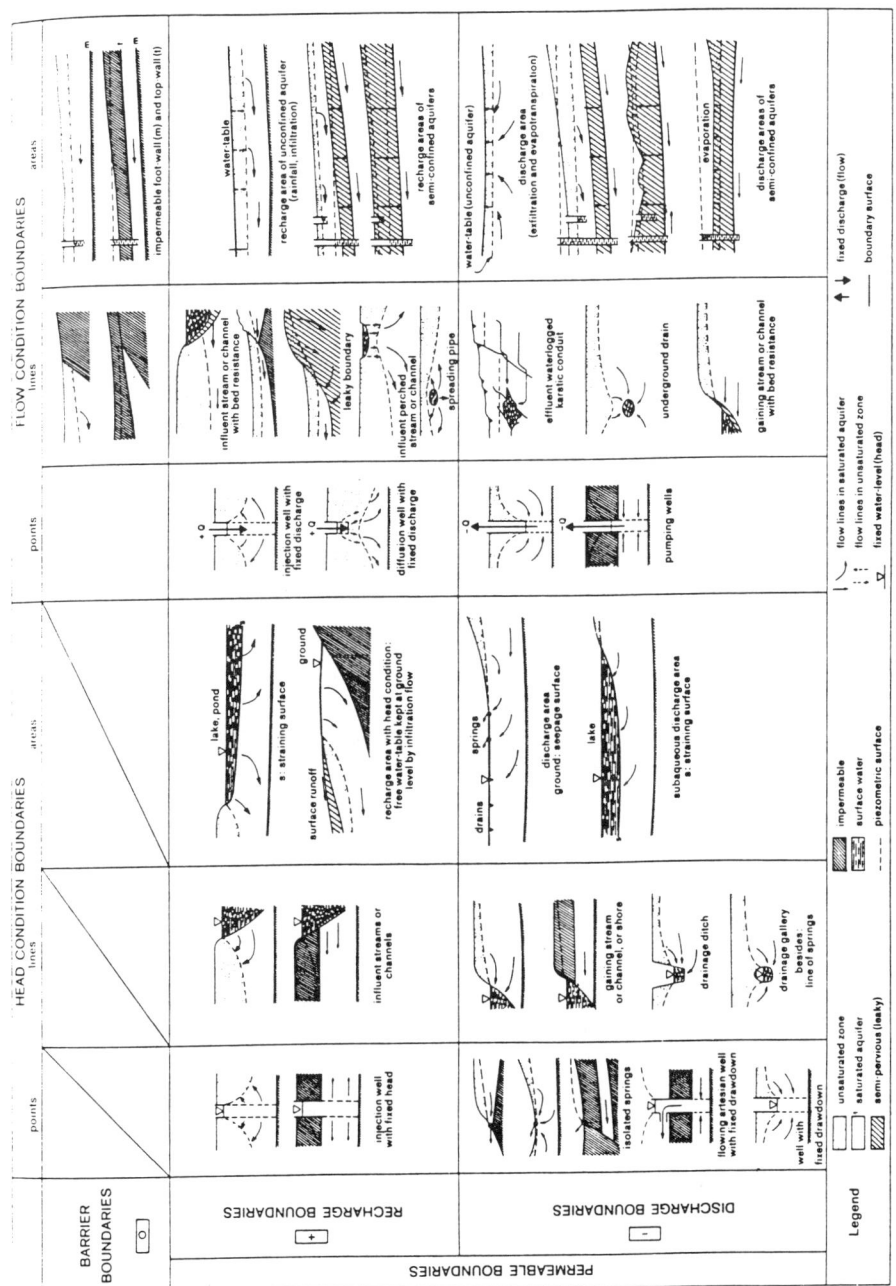

Figure 2-13 Types of aquifer boundary conditions (Struckmeier et al., 1986, after Castany and Margat, 1977).

expansion and compression, but instead comes from gravity drainage and dewatering of the aquifer.

2.6.2 Water-Transmitting Properties: Hydraulic Conductivity and Transmissivity

The terms *permeability* (P) and *hydraulic conductivity* (K) are often used interchangeably to refer to the ease with which water moves through soil or an aquifer under saturated conditions. Hydrogeologists draw a distinction between *intrinsic* permeability (k—a property of the porous medium alone that is independent of the nature of the liquid or potential field) and *hydraulic conductivity* (K—a function of both the medium and the fluid flowing through it). A precise definition of hydraulic conductivity is:

> The quantity of water that will flow through a unit cross-sectional area of a porous material per unit of time under a hydraulic gradient of 1.0 (measured at right angles to the direction of flow) at a specified temperature.[9]

ASTM defines hydraulic conductivity measured from field aquifer tests as follows:

> The volume of water at the existing kinematic viscosity that will move in a unit time under unit hydraulic gradient through a unit area measured at right angles to the direction of flow (ASTM 4043/Table A-14).

The terms hydraulic conductivity and permeability in this handbook refer to saturated hydraulic conductivity unless otherwise specified. Soil permeability rates are typically reported in units of inches/hour based on percolation tests. Hydraulic conductivity may be reported in a variety of units: μm/second, cm/second, m/second, ft/day, and gpd/ft^2 (gallons per day per square foot). Currently, centimeters per second is probably the most commonly used unit. Hydraulic conductivity values range widely from one rock type to another and even within the same rock. Table C-3 and Figures C-3 through C-15 identify ranges of hydraulic conductivity for a variety of soil and geologic materials.

Transmissivity (T), a term derived from hydraulic conductivity, describes the capacity of an aquifer to transmit water. Transmissivity is equal to the product of the aquifer's saturated thickness (b) and the hydraulic conductivity (K). It is commonly measured in units of gpd/ft of aquifer thickness:

$$T = Kb \qquad (2\text{-}5)$$

2.6.3 Darcy's Law

Darcy's Law, expressed in many different forms, allows calculation of the quantity of water flowing through a defined area of an aquifer, provided that the

---

[9] From glossary in Nielsen, D.M. (ed.). 1991. Practical Handbook of Ground Water Monitoring. Lewis Publishers, Chelsea, MI, 717 pp.

Ground Water and Vadose Zone Hydrology 89

hydraulic conductivity and the hydraulic gradient are known. One means of expressing Darcy's Law is:

Q = KiA (2-6)

where:

Q = quantity of flow per unit of time, in gpd
K = hydraulic conductivity, in gpd/ft^2
i = hydraulic gradient
A = cross-sectional area through which the flow moves, in ft^2

Darcy's Law assumes that flow is *laminar*, which means that the water will follow distinct flow lines rather than mix with other flow lines. Most ground-water flow in porous media is laminar. The equation does not work for *turbulent* flow, as in the case of the unusually high velocity that might be found in fractures or solution openings or in granular aquifers adjacent to some pumping wells.

Figure 2-14 shows an example of the use of Darcy's Law. In this case, a sand aquifer about 30 feet thick lies within the flood plain of a river about 1 mile wide. The aquifer is covered by a confining unit of glacial till, the bottom of which is about

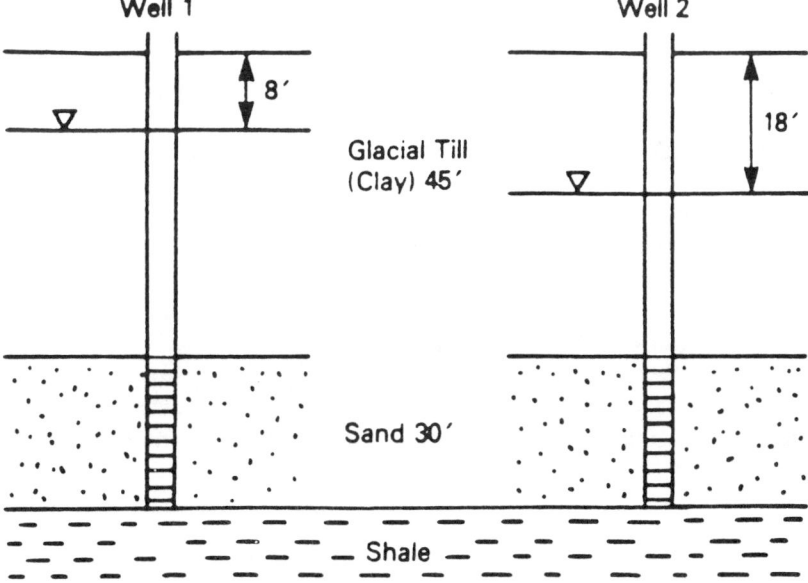

Figure 2-14 Using Darcy's Law to estimate underflow in an aquifer (U.S. EPA, 1987a).

45 feet below the land surface. The difference in water level in two wells 1 mile apart is 10 feet, and the hydraulic conductivity of the sand is 500 gpd/ft². Therefore, the quantity of underflow moving through the cross section in Figure 2-14 is:

$$Q = KiA = 500 \text{ gpd/ft}^2 \times (10 \text{ ft}/5280 \text{ ft}) \times (5280 \times 30) = 150,000 \text{ gpd}$$

2.6.4 Flow Between Aquifers

To determine the flow from one aquifer to another via a confining unit, a slightly modified form of Darcy's Law can be used:

$$Q_l = (p/m)A\Delta h \tag{2-7}$$

where:

Q_l = quantity of leakage, in gpd
p = vertical hydraulic conductivity of the confining unit, in gpd/ft²
m = thickness of the confining unit, in ft
A = cross-sectional area, in ft²
Δh = difference in head between the two wells

Figure 2-15 illustrates two aquifers separated by a layer of silt. The silty confining unit is 10 feet thick and has a vertical hydraulic conductivity of 2 gpd/ft². The difference in water level between wells tapping the upper and lower aquifers is 15 feet. Assuming these hydrogeologic conditions exist in an area of 1 square mile, the daily quantity leaking from the shallower aquifer to the deeper one within the area is:

$$Q_l = (2 \text{ gpd/ft}^2/10 \text{ ft}) \times 5280^2 \times 15 \text{ ft} = 83,635,200 \text{ gpd}$$

This calculation clearly shows that the quantity of leakage, either upward or downward, can be highly significant even if the hydraulic conductivity of the confining unit is small.

2.6.5 Interstitial Velocity and Time of Travel

The time it takes ground water to travel a specified distance is particularly important in contamination studies. Time of travel can be estimated using the form of Darcy's Law that describes average linear velocity:

$$\bar{v} = Ki/n \tag{2-8}$$

where:

$\bar{v}$ = average interstitial (linear) velocity
K = horizontal hydraulic conductivity
i = horizontal hydraulic gradient
n = porosity

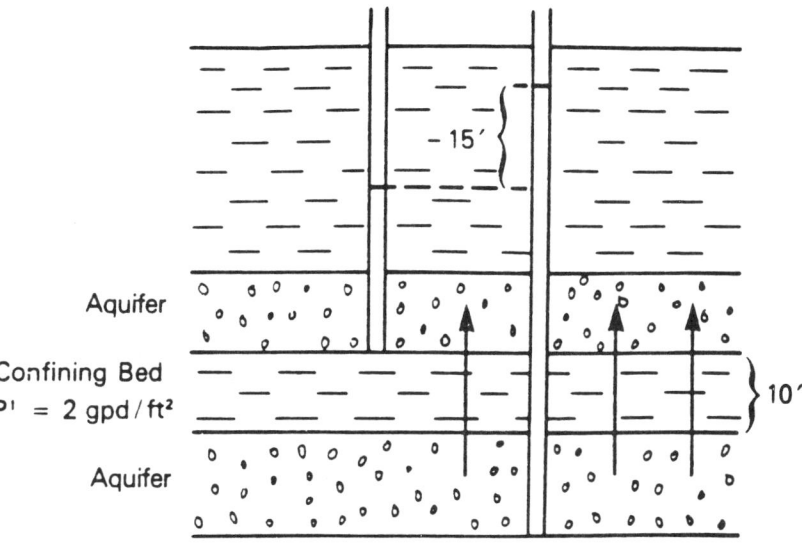

Figure 2-15 Using Darcy's Law to calculate the quantity of leakage from one aquifer to another (U.S. EPA, 1987a).

This equation is most easily used when a potentiometric map of the aquifer is available for measuring hydraulic gradients. For preliminary calculations, K and n can be estimated (see Appendix C). Once average velocity is known, the time of travel over a given distance can be easily calculated:

$$t = d/\bar{v} = dn/Ki \qquad (2\text{-}9)$$

where:

t = time of travel
d = distance

Or the distance to time of travel contours is calculated as follows:

$$d = \bar{v}t = tKi/n \qquad (2\text{-}10)$$

Where terms are the same as described in Equations 2-8 and 2-9. This equation is most applicable to the following situations:

- To calculate time of travel in a highly confined aquifer with a nearly flat potentiometric surface (gradient of <0.0005 to 0.001).

- To calculate time of travel in an unconfined aquifer with a nearly flat water table and when drawdown is small compared to the aquifer thickness or screened interval (<10 %).

- To calculate time of travel of a contaminant from a point source to a downgradient point of interest, *if* the equipotential lines are approximately equally spaced between the two points (i.e., the aquifer is homogeneous).

Somewhat more complex methods are required for wells with steep gradients in the cone of depression and wells in areas where there is a sloping regional water table.

The following example involves a spill of a conservative substance such as chloride. The liquid waste infiltrates through the unsaturated zone and quickly reaches an unconfined aquifer that consists of sand and gravel with a hydraulic conductivity of 2,000 gpd/ft^2 and an effective porosity of 0.20. The water level in a well at the spill lies at an altitude of 1,525 feet and, at a well a mile directly downgradient, is at 1,515 feet (Figure 2-16). The velocity of the water and the contaminant and the time it will take for the chloride to contaminate the second well can be determined by the following equations:

v = (2,000 gpd/ft^2) x (10 ft/5,280 ft)/7.48 x .20 = 2.5 ft/d
Time = 5,280 ft/2.5 ft/d = 2,112 days or 5.8 yr

This velocity value is crude at best and can only be used as an estimate. For example, the equation does not consider hydrodynamic dispersion (see Section 4.4.2). Velocity of most chemical species is further affected by reactions with the geologic framework, particularly with certain clays, soil-organic matter, and certain hydroxides (Section 4.5). Only conservative substances such as the chloride ion will move unaffected by retardation.

In addition, not only the water below the water table is moving, but also fluids within the capillary fringe. There the velocity diminishes rapidly upward from the water table. Movement in the capillary fringe is important where the contaminant is gasoline or another substance less dense than water.

2.6.6 Ground-Water Pumping Concepts

Cone of Depression. When a well is pumped, the water level declines to provide a gradient that drives water toward the discharge point. The gradient becomes steeper closer to the well, because the flow is converging from all directions and the area through which the water flows gets smaller. This results in a cone of depression around the well (Figure 2-17). The cone of depression around a well tapping an unconfined aquifer is relatively small compared to that around a well in a confined system because more water is released per foot of drawdown. In a confined aquifer the release of pressure with pumping moves very quickly through the aquifer, whereas

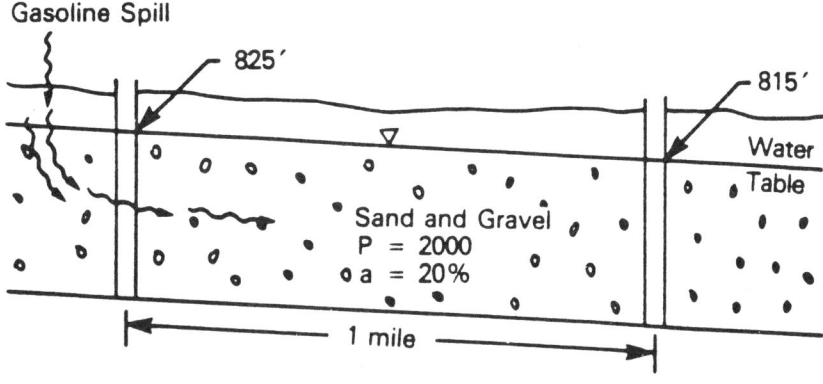

Figure 2-16 Ground water velocity calculations (U.S. EPA, 1987a).

water moves more slowly. The cone of depression of unconfined aquifers may be tens to a few hundred feet in diameter, while in confined aquifers it may extend outward for miles.

Cones of depression from several pumping wells may overlap and, since their drawdown effects are additive, the water level decline throughout the area of influence is greater than that from a single cone. In ground-water studies, particularly of contamination problems, evaluation of the cone or cones of depression can be critical; they represent an increase in the hydraulic gradient, which controls ground-water velocity and direction of flow. On the other hand, properly spaced and pumped wells provide a mechanism to control the migration of leachate plumes (Section 14.3.5). Discharging and recharging well schemes are commonly used to restore contaminated aquifers.

Specific Capacity. The *static* water level in a well is the level prior to any pumping. The *drawdown* refers to the decline of the water level in a pumping well and is the difference between the static level and the level to which the water drops during pumping (Figure 2-17). The discharge rate of the well divided by the drawdown is the *specific capacity*. The specific capacity indicates how much water the well will produce per foot of drawdown and should not be confused with *specific yield* (Section 2.6.1). It can be calculated by the following equation:

Specific capacity = Q/s (2-11)

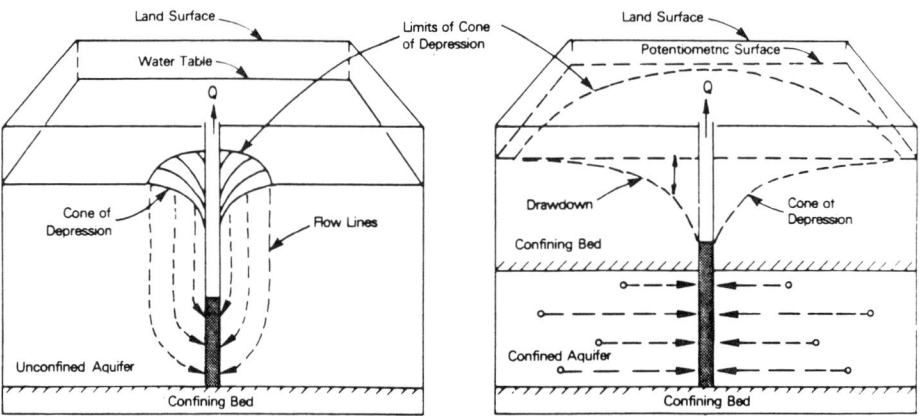

The cone of depression surrounding a pumping well in an unconfined aquifer is relatively small compared to that in a confined system.

Figure 2-17 Cones of depression in unconfined and confined aquifers (Heath, 1983).

where:

Q = discharge rate, in gpm
s = drawdown, in ft

If a well produces 100 gpm and the drawdown is 8 feet, the well will produce 12.5 gpm for each foot of available drawdown. Specific capacity is a useful measurement because it is easy to obtain and multiplying specific capacity by 2,000 gives a crude estimate of transmissivity (gpd/ft) for confined aquifers.

2.7 Guide to Major References

Table 2-4 identifies major references in the following areas: (1) water resources and hydrology (which tend to focus on surface hydrology), (2) vadose zone hydrology, (3) hydrogeology, and (4) hydraulics. Most texts on hydrogeology will provide some coverage of surface and vadose zone hydrology and the hydraulics of ground-water flow, along with investigation methods, occurrence in different geologic settings, and ground-water chemistry. Heath (1983) serves as an excellent introductory text on hydrogeology. Freeze and Cherry (1979) is probably the most commonly cited intermediate hydrogeology text, but many other good texts are also available.

The hydraulics texts in Table 2-4 are divided into four major categories: (1) ground-water hydraulics, (2) porous media flow (which includes both saturated and unsaturated flow), (3) drainage and seepage, and (4) engineering hydraulics. Bear (1979) is probably the most widely cited text on ground-water hydraulics. Hydraulic principles are used in the design and analysis of aquifer pump tests. Major references

Ground Water and Vadose Zone Hydrology 95

focussing on pump test analysis are indexed in Table 7-5, and Table 7-6 provides an index to the primary literature on analytical solutions for aquifer tests.

Texts on drainage and seepage may be especially useful for design of hydrodynamic controls for ground-water protection or remediation (Section 14.4). Earlier editions and the current third edition of Cedergren (1989) is probably the most commonly used text in this category by the ground-water community. Individuals with a focus on engineering or agricultural applications may prefer other texts. Texts on engineering hydraulics tend to focus of flow in pipes and open channels.

Table 2-5 identifies major references on the characteristics of ground-water occurrence in the United States in the following categories: (1) national overviews, (2) regional summary appraisals prepared by the U.S. Geological Survey in the 1970s, (3) other regional assessments, and (4) state-specific reports. The Water Resources Division of USGS publishes foldouts for each state listing water-resource publications and maps by USGS and cooperating agencies. These can be obtained from the USGS Water Resources Division District Office in a state, or the principal state water resource agency.

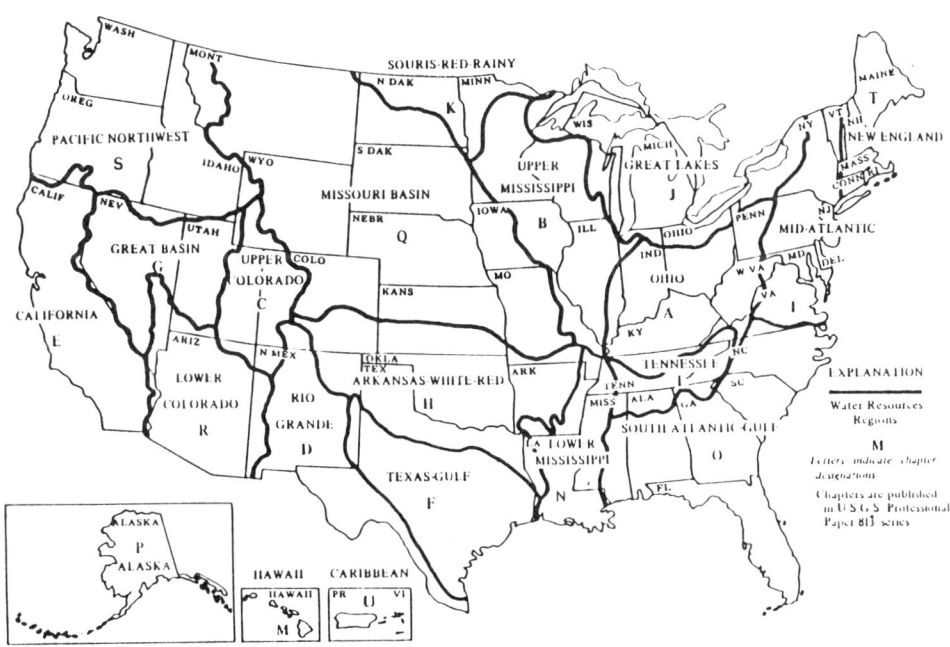

Figure 2-18 Geographic index to U.S. Geological Survey Professional Paper 813 ground-water reports.

Figure 2-18 provides a geographic index to the 21 summary appraisals of ground-water resources that were published as USGS Professional Paper 813. The boundaries on this map are those established by the U.S. Water Resources Council for water-resources regions in the United States, and generally do not follow major hydrogeologic boundaries. These papers have also been reprinted as a single volume (Todd, 1983). Figure 2-19 shows the boundaries of major regional aquifers for which studies are underway or completed under the U.S. Geological Survey's regional aquifer system analysis (RASA) program. Sun and Weeks (1991) provide a comprehensive bibliography of more than 800 USGS publications prepared under this program. Table 2-5 also includes a number of publications by the American Water Resources Association, prepared in cooperation with the U.S. Geological Survey's RASA program.

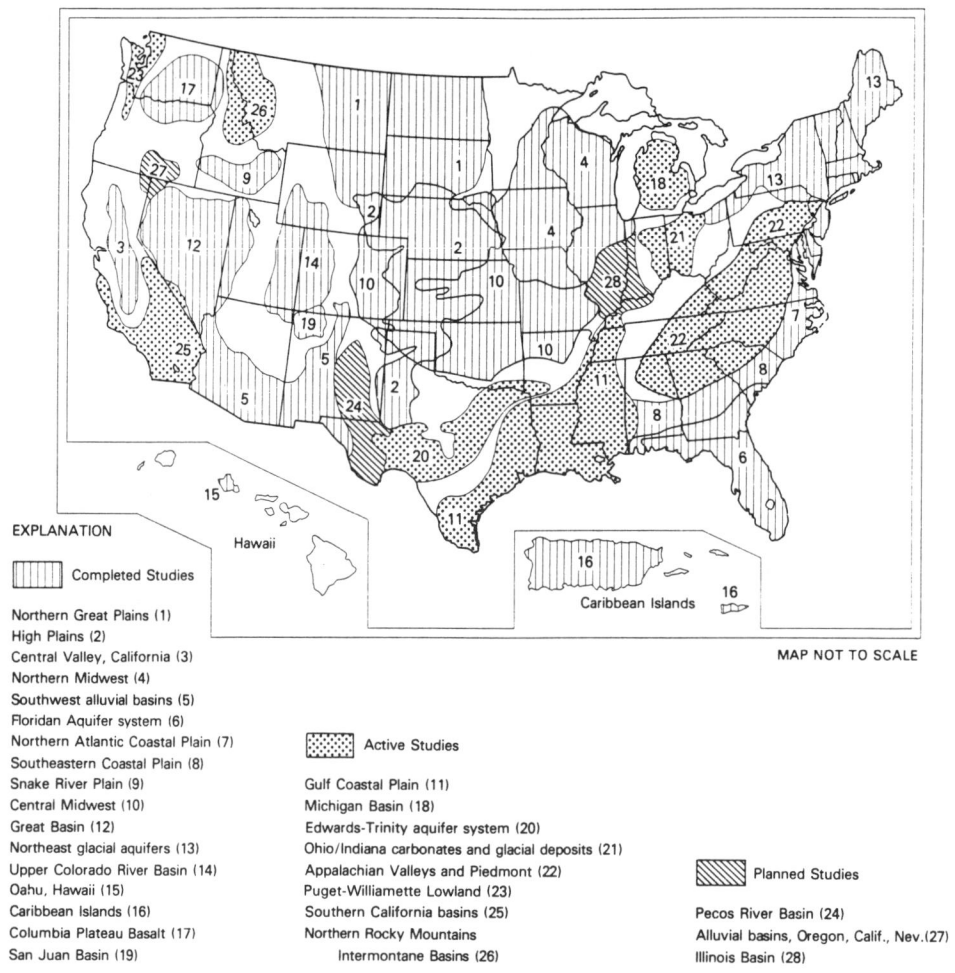

Figure 2-19 Geographic index to U.S. Geological Survey regional aquifer system studies (Sun and Weeks, 1991).

Table 2-4 Index to Major References on Surface and Vadose Zone Hydrology, Hydrogeology and Hydraulics

Topic	References
Water Resources/ Hydrology	Bras (1990), Bowen (1982), Branson et al. (1981), Chow (1964), Chow et al. (1988), Downing and Wilkinson (1992), Dunne and Leopold (1978), Gray (1973), Grigg (1985), Kazmann (1988), Leopold and Langbein (1960), Linsley et al. (1949), Maidment (1993), Meinzer (1942), Newsom (1994), Rhoda (1985), Shaw (1988), Tebutt (1973), Todd (1970), van der Leeden et al. (1990) Viessman et al. (1977), Watson and Burnett (1993), Willis and Yeh (1987), Wisler and Brater (1959); Engineering: ASCE (1952), Butler (1957), Linsley et al. (1958), Linsley and Franzini (1972), Skeat (1969), Shaw (1989), Wilson (1974)
Vadose Zone Hydrology	Guymon (1994); Soil Physics: Baver et al. (1972), Campbell (1985), Childs (1969), Ghildyal (1987), Hanks (1992), Hanks and Ashcroft (1980), Hillel (1980a, 1980b, 1982), Jury et al. (1991), Kirkham and Powers (1972), Kohnke (1968), Koorevaar et al. (1983), Marshall and Holmes (1988); Soil-Water Relations: Hillel (1971), Marshall (1960), Miyazaki (1993), Nielsen et al. (1972), Rode (1965); Plant-Water Relations: Slayter (1967); Edited Volumes: Hillel and Elrick (1990), Kienitz et al. (1991), Rijtema and Wassink (1969)
Hydrogeology*	Bibliography/Glossary: Lohman et al. (1972), Pfannkuch (1969), van der Leeden et al. (1991); Introductory: AWWA (1989), Baldwin and McGuiness (1963), Barton et al. (1985), Heath (1980, 1983), Heath and Trainer (1981), Mills et al. (1985), Pinnaker (1983), Smith (1982), Rau (1970), Redwine et al. (1991), U.S. EPA (1985, 1990); Intermediate-Advanced: Bouwer (1978), Bowen (1980), Cooley et al. (1972), Custodio and Llama (1975), Davis and DeWiest (1966), Driscoll (1986), Fetter (1994), Freeze and Cherry (1979), Gelher (1993), Johnson (1966), Kovács et al. (1981), Matthess (1982), McWhorter and Sunada (1981), Raghunath (1982), Todd (1980), Tolman (1937); Investigations: Brassington (1988), Brown et al. (1983), Erdélyi and Gálfi (1988), Hamill and Bell (1986), Mandel and Shifton (1981), Moss (1990), U.S. Geological Survey (1980), Walton (1970); Ground-Water Engineering: De Marsily (1986), Hunt (1983), Kashef (1986), Rethati (1984), Walton (1991); Edited Volumes: Back and Freeze (1982), Back and Stephenson (1979), Back et al. (1988), IAH (1985), IAHS (1967), Jones and Laenen (1992), Moore et al. (1989, 1991), Sakura (1993), Saleem (1976), Simpson and Sharp (1990), Zaporozec (1990)

Table 2-4 (cont.)

Topic	References
Chemical/Contaminant Hydrogeology	See Table 4-5
Hydraulics**	<u>Fluid Mechanics</u>: Dodge and Thompson (1937), Streeter and Wylie (1979), White (1974); <u>Ground-Water Flow</u>: Bear (1979), Bennett (1976), Bureau of Reclamation (1960, 1981), Campbell and Lehr (1973), Chapman (1981), Daly (1984-flow lines), DeWiest (1965), Edelman (1983), Freeze and Witherspoon (1967), Glover (1964, 1974), Halek and Svec (1979), Hantush (1964), Hubbert (1940, 1969), Hunt (1983), Jacob (1950), Lohman (1972), De Marsily (1986), McWhorter and Sunada (1981), Peterson et al. (1952), Randkivi and Callender (1976), Rosenshein and Bennett (1984), Strack (1989), Verruijt (1982), Zijl and Nawalany (1993); <u>Porous Media Flow</u>: Bear (1972), Bear and Corapcioglu (1987), Bird et al. (1960), Brooks and Corey (1964), Case (1993-unsaturated), Collins (1961), Corapcioglu (1991), Corey (1977-heterogenous fluids), Cushman and Hall (1991), Dagan (1989), DeWiest (1966), Dullien (1979), Elrick (1972), Greenkorn (1983), IAHR (1972), Milne-Thompson (1968), Muskat (1937), Scheidegger (1960); <u>Drainage/Seepage</u>: Bear et al. (1968), Bureau of Reclamation (1968), Cedergren (1989), Craig and Gray (1985), Harr (1977), Luthin (1978), Marino and Luthin (1982), Powers (1992), Rushton and Redshaw (1979); <u>Engineering Hydraulics</u>: Colt Industries (1974), Hauser (1991), Lencastre (1987), Rouse (1950), Simon (1981)

* See Table 1-4 for index of major references on karst geology, geomorphology and hydrology.

** References listed under hydrogeology will also cover hydraulics.

Table 2-4 References (Appendix F contains references for figure and table sources)

American Society of Civil Engineers (ASCE). 1952. Hydrology Handbook. Manual of Engineering Practice No. 28. ASCE, New York, NY.

American Water Works Association (AWWA). 1989. Ground Water. Manual MP21. Denver, CO, 160 pp.

Back, W. and R.A. Freeze. 1982. Physical Hydrogeology. Benchmark Papers in Geology No. 72, Hutchinson Ross, Stroudsburg, PA, 431 pp.

Back, W. and D.A. Stephenson (eds.). 1979. Contemporary Hydrogeology. Elsevier, New York, NY.

Back, W., J.S. Rosenshein, and P.R. Seaber (eds.). 1988. Hydrogeology. Geological Society of America, Boulder, CO, 524 pp.

Baldwin, H.L. and C.L. McGuiness. 1963. A Primer on Ground Water. U.S. Geological Survey, Washington, DC, 25 pp.

Barton, Jr., A.R. et al. 1985. Groundwater Manual for the Electric Utility Industry, Vol. 1: Geological Formations and Groundwater Aquifers, 1st ed. EPRI CS-3901. Electric Power Research Institute, Palo Alto, CA. [See also Redwine et al. (1991)]

Baver, L.D., W.H. Gardner and W.R. Gardner. 1972. Soil Physics, 4th ed. John Wiley & Sons, New York, NY, 498 pp. [Introductory text with 14 chapters]

Bear, J. 1972. Dynamics of Flow in Porous Media. Elsevier, New York, NY, 764 pp. [Reissued in paperback in 1988 by Dover Publications, Mineola, NY]

Bear, J. 1979. Hydraulics of Groundwater. McGraw-Hill, New York, NY, 567 pp.

Bear, J., D. Zazlavsky, and S. Irmay. 1968. Physical Principles of Water Percolation and Seepage. Arid Zone Research, Vol. 29, UNESCO, Paris, 465 pp.

Bear, J. and M.Y. Corapcioglu (eds.). 1987. Advances in Transport Phenomena in Porous Media. NATO Advanced Studies Institutes Series E, Vol. 128. Martinus Nijhoff Publishers, Dordrecht, The Netherlands.

Bedinger, M.S. and J.E. Reed. 1988. Practical Guide to Aquifer Test Analysis. USGS/EPA Interagency Agreement DW 14932431-01-1. U.S. Environmental Protection Agency Environmental Monitoring Systems Laboratory, Las Vegas, NV

Bennett, G.D. 1976. Introduction to Ground-Water Hydraulics: A Programmed Text for Self-Instruction. U.S. Geological Survey Techniques of Water Resources Investigations TWRI 3-B2, 172 pp.

Bird, R.B., W.E. Stewart, and E.N. Lightfoot. 1960. Transport Phenomena. John Wiley & Sons, New York, NY.

(Table 2-4 Basic Hydrology References)

Bouwer, H. 1978. Groundwater Hydrology. McGraw-Hill, New York, NY, 480 pp. [General text covering ground-water hydraulics, quality, and management]

Bowen, R. 1980. Ground Water. John Wiley & Sons, New York, NY, 227 pp. [General text with 13 chapters]

Bowen, R. 1982. Surface Water. John Wiley & Sons, New York, NY, 289 pp.

Branson, F.A., G.F. Gifford, K.G. Denard, and R.F. Hadley. 1981. Rangeland Hydrology, 2nd ed. Kendall/Hunt, Dubuque, IA.

Bras, R.L. 1990. Hydrology: An Introduction to Hydrologic Science. Addison-Wesley, Reading, MA, 643 pp.

Brassington, R. 1988. Field Hydrogeology. Halsted Press, New York, NY, 175 pp. [Introductory field manual for field techniques in hydrogeologic investigations]

Brooks, R.H. and A.T. Corey. 1964. Hydraulic Properties of Porous Media. Hydrology Paper 3, Colorado State University, Fort Collins, CO, 27 pp.

Brown, R.H., A.A. Konoplyantsev, J. Ineson, and V.S. Kovalensky. 1983. Ground-Water Studies: An International Guide for Research and Practice. Studies and Reports in Hydrology No. 7. UNESCO, Paris. [Chapter 6 covers aquifer tests]

Bureau of Reclamation. 1960. Studies of Ground-Water Movement. Technical Memorandum No. 657. U.S. Department of The Interior, Denver, CO, 180 pp. [Collection of 19 office memoranda on studies of technical problems arising from ground-water movement on Bureau of Reclamation projects]

Bureau of Reclamation. 1978. Drainage Manual. U.S. Department of the Interior, Denver, CO, 286 pp.

Bureau of Reclamation. 1981. Ground Water Manual—A Water Resources Technical Publication, 2nd ed. U.S. Department of the Interior, Bureau of Reclamation, Denver, CO, 480 pp. [1st edition 1977; 7 chapters covering hydraulics and pumping tests]

Butler, S.S. 1957. Engineering Hydrology. Prentice-Hall, Englewood Cliffs, NJ.

Campbell, G.J. 1985. Soil Physics with BASIC. Elsevier, New York, NY.

Campbell, M.D. and J.H. Lehr. 1973. Water Well Technology. McGraw-Hill, New York, NY, 681 pp. [Chapter 10 covers well hydraulics]

Case, C.M. 1993. Physical Principles of Flow in Unsaturated Porous Media. Oxford University Press, New York, NY, 336 pp.

Cedergren, H.R. 1989. Seepage, Drainage, and Flow Nets, 3rd ed. John Wiley & Sons, New York, NY, 465 pp. [1st edition 1967, 2nd edition 1977]

(Table 2-4 Basic Hydrology References)

Chapman, R.E. 1981. Geology and Water—An Introduction to Fluid Mechanics for Geologists. Martinus Nijhoff Publishers, The Hague, The Netherlands, 228 pp.

Childs, E.C. 1969. An Introduction to the Physical Basis of Soil Water Phenomena. Wiley Interscience, New York, NY, 493 pp.

Chow, V.T. (ed.). 1964. Handbook of Applied Hydrology: A Compendium of Water-Resources Technology. McGraw-Hill, New York, NY, 1453 pp.

Chow, V.T., D.R. Maidment, and L.W. Mays (eds.). 1988. Applied Hydrology. McGraw-Hill, New York, NY, 572 pp.

Collins, R.E. 1961. Flow of Fluids in Porous Media. Reinhold Publishing Corp., New York, NY, 275 pp.

Colt Industries. 1974. Hydraulic Handbook. Fairbanks Morse Pump Division, Colt Industries, Kansas City, KS, 246 pp.

Cooley, R.L., J.F. Harsh, and D.L. Levy. 1972. Principles of Ground-Water Hydrology. Hydrologic Engineering Methods for Water Resource Development, Vol. 10. U.S. Army Corps of Engineers Hydrologic Engineering Center, Davis, CA.

Corapcioglu, M.Y. (ed.). 1991. Advances in Porous Media, Vol. 1. Elsevier, New York, NY, 309 pp.

Corey, A.T. 1977. Mechanics of Heterogeneous Fluids in Porous Media. Water Resources Publications, Fort Collins, CO.

Craig, D.J. and I. Gray. 1985. Groundwater Lowering by Horizontal Drains. GCO Publication No. 2/85, Geotechnical Control Office, Hong Kong, 123 pp.

Cushman, J.H. and L. Hall. 1991. Dynamics of Fluids in Hierarchical Porous Media. Academic Press, New York, NY, 528 pp.

Custodio, E. and M.R. Llama. 1975. Hidrologia Subterránea, 2 Vols. Ediciones Omega, Barcelona, 2,359 pp.

Dagan, G. 1989. Flow and Transport in Porous Formations. Springer-Verlag, New York, NY, 465 pp. [Focuses on stochastic modeling of subsurface flow and transport at different scales]

Daly, C.J. 1984. A Procedure for Calculating Ground Water Flow Lines. CRREL Special Report 84-9. U.S. Army Corps of Engineers Cold Regions Research and Engineering Laboratory, Hanover, NH.

Davis, S.N. and R.J.M. DeWiest. 1966. Hydrogeology. John Wiley & Sons, New York, NY, 463 pp. [General text focusing on geologic aspect of ground water; includes chapter on radionuclides in ground water]

De Marsily, G. 1986. Quantitative Hydrogeology: Groundwater Hydrology for Engineers. Academic Press, New York, NY, 440 pp.

(Table 2-4 Basic Hydrology References)

DeWiest, R.J.M. 1965. Geohydrology. John Wiley & Sons, New York, NY, 366 pp.

DeWiest, R.J.M. (ed.). 1969. Flow Through Porous Media. Academic Press, New York, NY, 366 pp. [11 contributed chapters]

Dodge, R.A. and M.J. Thompson. 1937. Fluid Mechanics. McGraw-Hill, New York, NY.

Downing, R.A. and W.B. Wilkinson (eds.). 1992. Applied Groundwater Hydrology: A British Perspective. Oxford University Press, New York, NY, 352 pp. [19 contributed chapters on ground-water management, quality and waste disposal]

Driscoll, F.G. 1986. Groundwater and Wells, 2nd ed. Johnson Division, UOP Inc., St. Paul, MN, 1089 pp. First edition by Johnson, UOP, 1966. [Chapter 9 covers well hydraulics and Chapter 16 discusses collection and analysis of pumping test data]

Dullien, F.A.L. 1979. Porous Media: Fluid Transport and Structure. Academic Press, New York, NY, 396 pp.

Dunne, T. and L.B. Leopold. 1978. Water in Environmental Planning. W.H. Freeman, San Francisco, CA, 818 pp.

Edelman, J.H. 1983. Groundwater Hydraulics of Extensive Aquifers, 2nd ed. ILRI Bulletin No. 13. International Institute for Land Reclamation and Improvement, Wageningen, The Netherlands, 216 pp. [1st edition 1972]

Elrick, D.E. (ed.). 1972. Proceedings of the 2nd Symposium on Fundamentals of Transport Phenomena in Porous Media. Office of Continuing Education, University of Guelph, Guelph, Ontario. [51 papers]

Fetter, Jr., C.W. 1994. Applied Hydrogeology, 3rd ed. Macmillan, New York, NY, 691 pp. [Textbook focusing on ground-water occurrence and flow; 1st edition 1980; 2nd edition 1988 published by Charle E. Merrill Publishing Co.]

Freeze, R.A. and J.A. Cherry. 1979. Groundwater. Prentice-Hall, Englewood Cliffs, NJ, 604 pp. [Comprehensive text covering all aspects of ground-water flow, ground-water contamination and geochemistry]

Freeze, R.A. and P.A. Witherspoon. 1967. Theoretical Analysis of Regional Ground-Water Flow: 3. Quantitative Interpretations. Water Resources Research 4:581-590.

Gelher, L.W. 1993. Stochastic Subsurface Hydrology. Prentice-Hall, Englewood Cliffs, NJ, 390 pp.

Ghildyal, B.P. 1987. Soil Physics: Theory and Practice. John Wiley & Sons, New York, NY. [Five sections: (1) soil solids, (2) soil liquid, (3) soil gas, (4) soil heat, and (5) transport processes]

Glover, R.E. 1964. Ground-Water Movement. Tech. Eng. Monograph No. 31. U.S. Bureau of Reclamation, Denver, CO, 76 pp.

(Table 2-4 Basic Hydrology References)

Glover, R.E. 1974. Transient Ground Water Hydraulics. Water Resources Publications, Fort Collins, CO, 413 pp.

Gray, D.M. (ed.). 1973. Handbook on the Principles of Hydrology (with special emphasis directed to Canadian conditions in the discussions, applications and presentation of data). Water Information Center, Port Washington, NY, 720 pp. [Reprint of 1970 edition published in Canada]

Greenkorn, R.A. 1983. Flow Phenomena in Porous Media: Fundamentals and Applications in Petroleum, Water and Food Production. Marcel Dekker, New York, NY, 550 pp.

Grigg, N.S. 1985. Water Resources Planning. McGraw-Hill, New York, NY, 328 pp.

Guymon, G.L. 1994. Unsaturated Zone Hydrology. Prentice Hall, Englewood Cliffs, NJ, 210 pp.

Halek, V. and J. Svec. 1979. Ground-Water Hydraulics. Developments in Water Science, Vol. 7, Elsevier, New York, NY, 620 pp.

Hammill, L. and F.G. Bell. 1986. Groundwater Resource Development. Butterworth, London, UK, 344 pp.

Hanks, R.J. 1992. Applied Soil Physics: Soil Water and Temperature Applications, 2nd ed. Springer-Verlag, New York, NY, 176 pp.

Hanks, R.J. and G.L. Ashcroft. 1980. Applied Soil Physics. Springer-Verlag, New York, NY, 159 pp.

Hantush, M.S. 1964. Hydraulics of Wells. Advances in Hydroscience 1:181-432.

Harr, M.E. 1977. Ground Water and Seepage. McGraw-Hill, New York, NY, 315 pp.

Hauser, B.A. 1991. Practical Hydraulics Handbook. Lewis Publishers, Chelsea, MI, 347 pp. [Focus on applications for drinking and wastewater operators]

Heath, R.C. 1980. Basic Elements of Ground-Water Hydrology with Reference to Conditions in North Carolina. U.S. Geological Survey Open File Report OFR 80-44, 93 pp.

Heath, R.C. 1983. Basic Ground-Water Hydrology. U.S. Geological Survey Water-Supply Paper 2220, 85 pp. Republished in a 1984 edition by National Water Well Association, Dublin, OH. [Contains one- and two-page synopses of fundamental concepts and terms in hydrogeology; most of this material can also be found in Chapter 2 of U.S. EPA (1985)]

Heath, R.C. and F.W. Trainer. 1981. Introduction to Ground Water Hydrology, 2nd ed. John Wiley & Sons, New York, NY, 284 pp. [Introductory text including laboratory exercises]

Hillel, D. 1971. Soil and Water: Physical Principles and Processes. Academic Press, New York, NY, 288 pp. [Part I has five chapters on physical principles of soil-water relationships and Part II contains six chapters on the field water cycle]

(Table 2-4 Basic Hydrology References)

Hillel, D. 1980a. Fundamentals of Soil Physics. Academic Press, New York, NY, 413 pp. [Intermediate-level text in soil physics]

Hillel, D. 1980b. Applications of Soil Physics. Academic Press, New York, NY, 385 pp.

Hillel, D. 1982. Introduction to Soil Physics. Academic Press, New York, NY, 364 pp. [Introductory text with 17 chapters focusing on solid, liquid and gaseous phases and the field water cycle]

Hillel, D. and D.E. Elrick (eds.). 1990. Scaling in Soil Physics: Principles and Applications. SSSA Sp. Pub. No. 25. Soil Science Society of America, Madison, WI, 122 pp.

Hubbert, M.K. 1940. The Theory of Ground-Water Motion. J. Geology 48:785-944.

Hubbert, M.K. 1969. The Theory of Ground-Water Motion and Related Papers. Hafner Publishing Co., 311 pp.

Hunt, B. 1983. Mathematical Analysis of Groundwater Resources. Butterworth, Boston, 271 pp.

International Association for Hydraulic Research (IAHR). 1972. Fundamentals of Transport Phenomena in Porous Media. Elsevier, New York, NY. [Conference proceedings containing 31 papers]

International Association of Hydrogeologists (IAH). 1985. Hydrogeology of Rocks of Low Permeability, 2 Parts. Vol. XVII, Int. Congr. of IAH Memoires (Tucson, AZ), 850 pp.

International Association of Scientific Hydrology (IASH). 1967. Hydrology of Fractured Rocks (Proc. of 1965 Dubrovnik Symposium), 2 Vols. IASH Publ. No. 73.

Jacob, C.E. 1950. Flow of Ground Water. In: Engineering Hydraulics, H. Rouse (ed.), Wiley & Sons, New York, NY, pp. 321-386.

Johnson, E.E., Inc. 1966. Ground Water and Wells. Johnson Division, UOP, St. Paul, MN, 440 pp. [See Driscoll (1986) for 2nd edition]

Jones, M.E. and A. Laenen (eds.). 1992. Interdisciplinary Approaches in Hydrology and Hydrogeology. American Institute of Hydrology, Minneapolis, MN, 644 pp. [Proc. AIH 1992 Annual Meeting, Portland, OR]

Jury, W.A., W.R. Gardner, and W.H. Gardner. 1991. Soil Physics, 5th ed. John Wiley & Sons, New York, NY, 352 pp.

Kashef, A.I. 1986. Groundwater Engineering. McGraw-Hill, New York, NY, 512 pp.

Kazmann, R.G. 1988. Modern Hydrology, 3rd ed. Harper and Row, New York, NY. Earlier edition 1972, 635 pp. [Comprehensive text covering water resources from physical, environmental, economic and societal perspectives]

(Table 2-4 Basic Hydrology References)

Kienitz, G., P.C.D. Milly, M.Th. Van Genuchten, D. Rosbjerg, and W.J. Shuttleworth (eds.). 1991. Hydrological Interactions Between Atmosphere, Soil and Vegetation. IAHS Publ. No. 204, IAHS Press, Wallingford, Oxfordshire, UK, 494 pp.

Kirkham, D. and W.L. Powers. 1972. Advanced Soil Physics. Wiley-Interscience, New York, NY, 534 pp.

Kohnke, H. 1968. Soil Physics. McGraw-Hill, New York, NY, 224 pp. [Introductory text with 10 chapters]

Koorevaar, P., G. Menlik, and C. Dirksen. 1983. Elements of Soil Physics. Elsevier, New York, NY.

Kovács, G., J. Gálfi, and N. Pataki. 1981. Subterranean Hydrology. Water Resource Publications, Littleton, CO, 988 pp.

Lencastre, A. 1987. Handbook of Hydraulic Engineering. John Wiley & Sons, New York, NY, 540 pp.

Leopold, L.B. and W.B. Langbein. 1960 A Primer on Water. U.S. Government Printing Office 1970-0-398-800, 50 pp.

Linsley, Jr., R.K. and J.B. Franzini. 1972. Water Resources Engineering, 2nd ed. McGraw-Hill, New York, NY, 690 pp.

Linsley, Jr., R.K., M.A. Kohler, and J.L.H. Paulhus. 1949. Applied Hydrology. McGraw-Hill, New York, NY, 689 pp.

Linsley, Jr., R.K. and M.A. Kohler. 1982. Hydrology for Engineers, 3rd ed. McGraw-Hill, New York, NY, 512 pp. [1st edition by Linsley, Kohler and Paulhus published in 1958]

Lohman, S.W. 1972. Ground-Water Hydraulics. U.S. Geological Survey Professional Paper 708. [Covers methods for estimating aquifer parameters]

Lohman, S.W. et al. 1972. Definitions of Selected Ground-Water Terms--Revisions and Conceptual Refinements. U.S. Geological Survey Water-Supply Paper 1988, 21 pp.

Luthin, J.N. 1978. Drainage Engineering. R.E. Krieger Publ. Co., Huntington, NY, 281 pp.

Maidment, D.R. (ed.). 1993. Handbook of Hydrology. McGraw-Hill, New York, NY, 1,000 pp.

Mandel, S. and Z.L. Shifton. 1981. Groundwater Resources: Investigation and Development. Academic Press, New York, NY, 288 pp.

Marino, M.A. and J.N. Luthin. 1982. Seepage and Groundwater. Elsevier, New York, NY, 492 pp.

Marshall, T.J. 1960. Relations Between Water and Soil. Technical Communication No. 50, Commonwealth Bureau of Soil Science, England.

(Table 2-4 Basic Hydrology References)

Marshall, T.J. and J.W. Holmes. 1988. Soil Physics, 2nd ed. Cambridge University Press, New York, NY. [First edition 1979]

Matthess, G. 1982. Properties of Groundwater. John Wiley & Sons, New York, NY, 406 pp. [Text focusing on geochemical aspects of ground water]

McWhorter, D.B. and D.K. Sunada. 1981. Ground-Water Hydrology and Hydraulics. Water Resources Publications, Littleton, CO, 492 pp. [Earlier edition published in 1977]

Meinzer, O.E. (ed.). 1942. Hydrology. McGraw-Hill, New York, NY, 712 pp. [Reprinted by Dover Publications, New York, NY]

Mills, W.B. et al. 1985. Water Quality Assessment: A Screening Procedure for Toxic and Conventional Pollutants, Part II. EPA 600/6-85/002b. [Part 2 covers basic hydrogeologic concepts for assessing water quality impacts of toxic and conventional pollutants]

Milne-Thompson, L.M. 1968. Theoretical Hydrodynamics, 5th ed. Macmillan, New York, NY, 743 pp.

Miyazaki, T. 1993. Water Flow in Soils. Marcel Dekker, New York, NY, 312 pp.

Moore, J.E., A.A. Zaporozec, S.C. Csallany, and T.C. Varney (eds.). 1989. Recent Advances in Ground-Water Hydrology. American Institute of Hydrology, Minneapolis, MN, 602 pp. [Proc. of 1988 Int. Conf. on Ground-Water Hydrology, Tampa, FL]

Moore, J.E., R.A. Kanivetsky, J.S. Rosenshein, C. Zenone, and S.C. Csallany (eds.). 1991. First USA/USSR Joint Conference on Environmental Hydrology and Hydrogeology. American Institute of Hydrology, Minneapolis, MN, 464 pp. [Proc. 1990 Int. Conf., Leningrad, USSR]

Moss, R. 1990. Handbook of Ground Water Development. Wiley Interscience, New York, NY, 493 pp.

Muskat, M. 1937. The Flow of Homogenous Fluids Through Porous Media. McGraw-Hill, New York, NY, 763 pp.

Newson, M. 1994. Hydrology and the River Environment. Oxford University Press, New York, NY, 256 pp.

Nielsen, D.R., R.D. Jackson, J.W. Cary, and D.D. Evans (eds.). 1972. Soil Water. American Society of Agronomy, Madison, WI, 175 pp. [Focuses on theoretical analysis and field investigation methods for the study of soil water]

Peterson, D.F. et al. 1952. Hydraulics of Wells. Agric. Exp. Sta. Bull. 351, Utah State College, Logan, UT.

Pinnaker, E.V. (ed.). 1983. General Hydrogeology. Cambridge University Press, New York, NY, 141 pp. [Translated from Russian]

Pfannkuch, H.O. 1969. Elsevier's Dictionary of Hydrogeology. Elsevier, NY, 168 pp.

(Table 2-4 Basic Hydrology References)

Powers, J.P. 1992. Construction Dewatering: A Guide to Theory and Practice, 2nd ed. Wiley & Sons, Somerset, NJ, 494 pp. [1st edition 1981]

Raghunath, H.M. 1982. Groundwater. John Wiley, Somerset, NJ, 456 pp.

Randkivi, A.J. and R.A. Callander. 1976. Analysis of Groundwater Flow. John Wiley & Sons, New York, NY, 214 pp.

Rau, J. 1970. Ground Water Hydrology for Water Well Drilling Contractors. National Water Well Association, Columbus, OH, 257 pp.

Redwine, J.C. et al. 1991. Groundwater Manual for the Electric Utility Industry, Second Edition, Vol. 1: Geological Formations and Groundwater Aquifers. EPRI GS-7534. Electric Power Research Institute, Palo Alto, CA. [First edition by Barton et al. (1985)]

Rethati, L. 1984. Groundwater in Civil Engineering. Elsevier, New York, NY, 474 pp.

Rhoda, J.C. (ed.). 1985. Facets of Hydrology, 2 Vols. John Wiley & Sons, Chichester, UK, 368 pp.

Rijtema, P.E. and H. Wassink (eds.). 1969. Water in the Unsaturated Zone (Proc. Wageningen Symp.), 2 Vols. IASH-UNESCO Studies and Reports in Hydrology 2. UNESCO, Paris.

Rode, A.A. 1965. Theory of Soil Moisture, Vol. I: Moisture Properties of Soils and Movement of Soil Moisture. Israel Program for Scientific Translations, Jerusalem. [Translated from Russian]

Rosenshein, J. and G. Bennett (eds.). 1984. Groundwater Hydraulics. American Geophysical Union Water Resources Monograph 9, 407 pp.

Rouse, H. (ed.). 1950. Engineering Hydraulics. Wiley and Sons, New York, NY. [Proceedings of the 1949 Hydraulics Conference, University of Iowa, Iowa City; may be cited with a 1949 date]

Rushton, K.R. and S.C. Redshaw. 1979. Seepage and Groundwater Flow. John Wiley & Sons, 339 pp.

Sakura, Y. (ed.). 1993. Selected Papers on Environmental Hydrogeology, Vol. 4. Int. Ass. Hydrogeologists, Verlag Heinz Heise, Hannover, Germany, 245 pp. [19 papers from 29th Int. Geol. Congr., Kyoto, August 24-September 3, 1992]

Saleem, Z.A. (ed.). 1976. Advances in Groundwater Hydrology. American Water Resources Association, Minneapolis, MN, 333 pp.

Scheidegger, A.E. 1974. The Physics of Flow Through Porous Media, 3rd ed. University of Toronto Press, Toronto, Ontario, 353 pp. [1st edition by MacMillan in 1957, 2nd edition 1960]

Shaw, E.M. 1988. Hydrology in Practice, 2nd ed. Van Nostrand Reinhold, New York, NY, 569 pp. [Introductory text focusing on surface hydrology]

(Table 2-4 Basic Hydrology References)

Shaw, E.M. 1989. Engineering Hydrology Techniques in Practice. Halsted Press, New York, NY, 349 pp.

Simon, A.L. 1981. Practical Hydraulics, 2nd ed. John Wiley & Sons, New York, NY, 403 pp. [1st edition 1976]

Simpson, E.S. and J.M. Sharp, Jr. (eds.). 1990. Selected Papers on Hydrogeology, Vol. 1. Int. Ass. Hydrogeologists, Verlag Heinz Heise, Hannover, Germany, 508 pp. [37 papers from 28th Int. Geol. Congr., Washington, DC, July 9-19, 1989]

Skeat, W.O. (ed.). 1969. Manual of British Water Engineering Practice, Vol. II, Engineering Practice, 4th ed. W. Heffer and Sons, Cambridge.

Slayter, R.D. 1967. Plant-Water Relationships. Academic Press, New York, NY, 366 pp.

Smith, S. (ed.). 1982. Ground Water Hydrology for Water Well Contractors. National Water Well Association, Dublin, OH, 288 pp.

Strack, O.D.L. 1989. Ground Water Mechanics. Prentice-Hall, Englewood Cliffs, NJ, 732 pp. [Advanced mathematically oriented text]

Streeter, V.L. and E.B. Wylie. 1979. Fluid Mechanics. McGraw-Hill, New York, NY, 562 pp.

Tebutt, T.H.Y. 1973. Water Science and Technology. Barnes & Noble Books, New York, NY, 240 pp.

Todd, D.K. 1970. See van der Leeden et al. (1990)

Todd, D.K. 1980. Groundwater Hydrology, 2nd ed. John Wiley & Sons, New York, NY, 535 pp. [First edition 1959. Basic text on the fundamentals of ground water hydrology with 14 chapters]

Tolman, C.F. 1937. Ground Water. McGraw-Hill, New York, NY, 593 pp. [Text on groundwater hydrology with 17 chapters]

U.S. Environmental Protection Agency (EPA). 1985. Protection of Public Water Supplies from Ground-Water Contamination. Seminar Publication, EPA/625/4-85/016, 181 pp. Available from CERI.* [Chapter 2 contains most of the material in Heath (1983)]

U.S. Environmental Protection Agency (EPA). 1990. Ground Water Handbook, Vol I: Ground Water and Contamination. EPA/625/6-90/016a. Available from CERI.*

U.S. Geological Survey. 1980. Ground Water. In: National Handbook of Recommended Methods for Water Data Acquisition, Office of Water Data Coordination, Reston, VA, Chapter 2.

van der Leeden, F. 1991. Geraghty & Miller's Groundwater Bibliography, 5th ed. Water Information Center, Plainview, New York, NY, 507 pp.

(Table 2-4 Basic Hydrology References)

van der Leeden, F., F.L. Troise, and D.K. Todd (eds.). 1990. The Water Encyclopedia, 2nd ed. Lewis Publishers, Chelsea, MI, 808 pp. [1st edition edited by Todd published in 1970]

Verruijt, A. 1982. Theory of Ground Water Flow, 2nd ed. Gordon and Breach, New York, NY, 144 pp. [1st edition 1970]

Viessman, Jr., W., T.E. Harbaugh, and J.W. Knapp. 1977. Introduction to Hydrology, 2nd ed. Intext Educational Publishers, New York, NY, 704 pp. [General text on surface and ground water hydrology; 1st edition 1972]

Walton, W.C. 1970. Groundwater Resource Evaluation. McGraw-Hill, New York, NY, 664 pp.

Walton, W.C. 1991. Principles of Groundwater Engineering. Lewis Publishers, Chelsea, MI, 346 pp.

Watson, I. and A. Burnett. 1993. Hydrology: An Environmental Approach. Buchanan Books, Ft. Lauderdale, FL, 702 pp.

White, F.M. 1974. Viscous Fluid Flow. McGraw-Hill, New York, NY, 725 pp.

Willis, R. and W. W-G. Yeh. 1987. Groundwater Systems Planning and Management. Prentice Hall, Englewood Cliffs, NJ.

Wilson, E.M. 1974. Engineering Hydrology, 2nd ed. John Wiley & Sons, New York, NY, 232 pp.

Wisler, C.O. and E.R. Brater. 1959. Hydrology, 2nd ed. John Wiley & Sons, New York, NY.

Zaporozec, A. (ed.). 1990. Minimizing Risk to the Hydrologic Environment. American Institute of Hydrology, Minneapolis, MN, 266 pp. [AIH Spring Meeting, Las Vegas, NV]

Zijl, W. and M. Nawalany. 1993. Natural Groundwater Flow. Lewis Publishers, Chelsea, MI, 321 pp.

* See Preface for information on how to obtain documents from CERI (U.S. EPA Center for Environmental Research Information) and NTIS.

(Table 2-4 Basic Hydrology References)

Table 2-5 Index to Major References on Regional Hydrology and Ground Water in the United States

Topic	References
National	Information Sources/Bibliographies: Giefer (1976), Ralston (1975), Randolph and Deike (1966), Riggs (1962), Sun and Weeks (1991), USGS (1982, 1984, 1988a), U.S. Water Resources Council (1978a), van der Leeden et al. (1990); Hydrologic Atlases: Geraghty et al. (1973), Gerlach (1970), Langbein et al. (1949); Ground Water Occurrence/Supply: Aller et al. (1987), Back et al. (1988), Heath (1982, 1984), McGuiness (1963), Meinzer (1923), Thomas (1952); Ground Water Quality: See Table 4-4
Computer Databases	See Table 5-13
USGS Regional Summary Appraisals*	East: Boyd (1974—Ohio Region), Cederstrom et al. (1979—South Atlantic Gulf), Sinnott (1982—New England), Sinnot and Cushing (1978—Mid Atlantic), Weist (1978—Great Lakes), Zurawski (1978—Tennessee); Central: Baker and Wall (1976—Texas-Gulf), Bedinger and Sniegocki (1976—Arkansas-White-Red), Boyd (1975—Upper Mississippi), Reeder (1978—Souris-Red-Rainy), Taylor (1978—Missouri Basin), Terry and Bryant (1979—Lower Mississippi), West and Broadhurst (1975—Rio Grande); West: Davidson (1979—Lower Colorado), Eakin et al. (1976—Great Basin), Foxworthy (1979—Pacific Northwest), Price and Arnow (1974—Upper Colorado), Thomas and Phoenix (1976—California); Noncontinental U.S.: Gomez-Gomez and Heisel (1980—Caribbean), Takasaki (1978—Hawaii), Zenone and Anderson (1978—Alaska)
Other Regional	Atlantic/Gulf Coastal Plain: Johnston and Bush (1988), USGS (1987, 1988b), Vecchioli and Johnson (1987)**; Basin and Range: Bedinger et al. (1989); Caribbean: Gomez-Gomez et al. (1991)**; Midwest: Swain and Johnson (1989)**; Northeast Glacial Aquifers: Randall and Johnson (1992)**; Southeast: Hotchkiss and Johnson (1992)**; Southwest Alluvial Basins: Anderson and Johnson (1986)**; Western Mountain Area: McLean and Johnson (1988)**; Far West: Prince and Johnson (1992)**, Smith (1988)

Table 2-5 (cont.)

Topic	References
States	Giefer and Todd (1972, 1976); <u>Alabama</u>: Avrett (1968); <u>Alaska</u>: Fuelner et al. (1971), U.S. Water Resource Council (1978c); <u>Colorado</u>: Robson (1987); <u>Hawaii</u>: Mink (1977), Stearns (1946), U.S. Water Resource Council (1978b); <u>Florida</u>: Franks (1981), Hyde (1975), Southeastern Geological Society (1986); <u>Idaho</u>: Graham et al. (1981), Yee and Souza (1987); <u>Indiana</u>: Marie (1976); <u>Massachusetts</u>: Massachusetts Department of Environmental Quality Engineering (1986); <u>Mississippi</u>: Mitchell (1986); <u>South Dakota</u>: Meyer (1986); <u>Washington</u>: Molennar et al. (1980)

* Reports with state names do not strictly follow state boundaries—see Figure 2-18; also published as a single volume (Todd, 1983).

** Published in AWRA/USGS Regional Aquifer System Analysis Program (RASA) Series; see Sun and Weeks (1991) for a bibliography of USGS RASA publications, and Figure 2-19 for region boundaries.

Table 2-5 References (Appendix F contains references for figure and table sources)

Aller, L. T. Bennett, J.H. Lehr, R.J. Petty, and G. Hackett. 1987. DRASTIC: A Standardized System for Evaluating Ground Water Pollution Potential Using Hydrogeologic Settings. EPA/600/2-87/035, NTIS PB87-213914. Also published in NWWA/EPA series, National Water Well Association, Dublin, OH. [An earlier version dated 1985 with the same title (EPA/600/2-85/018) does not have the chapter on application of DRASTIC to maps or the 10 case studies contained in the later report]

Anderson, T.W. and A.I. Johnson (eds.). 1986. Regional Aquifer Systems of the United States: Southwest Alluvial Basins of Arizona. AWRA Monograph #7, American Water Resources Association, Bethesda, MD, 116 pp. [9 papers]

Avrett, J.R. 1968. A Compilation of Ground Water Quality Data in Alabama. Alabama Geological Survey Circular 37, 336 pp.

Back, W., J.S. Rosenshein, and P.R. Seaber. 1988. Hydrogeology, The Geology of North America. Geological Society of America, Boulder, CO, 534 pp.

Baker, Jr., E.T. and J.R. Wall. 1976. Summary Appraisals of the Nation's Ground-Water Resources--Texas-Gulf Region. U.S. Geological Survey Professional Paper 813-F.

Bedinger, M.S. and R.T. Sniegocki. 1976. Summary Appraisals of the Nation's Ground-Water Resources--Arkansas-White-Red Region. U.S. Geological Survey Professional Paper 813-H.

Bedinger, M.S., K.A. Sargen, W.H. Langer, F.B. Sherman, J.E. Reed, and B.T. Brady. Studies of Geology and Hydrology in the Basin and Range Province, Southwestern United States, for Isolation of High-Level Radioactive Waste—Basis of Characterization and Evaluation. U.S. Geological Survey Professional Paper 1370-A.

Boyd, Jr., R.M. 1974. Summary Appraisals of the Nation's Ground-Water Resources--Ohio Region. U.S. Geological Survey Professional Paper 813-A.

Boyd, Jr., R.M. 1975. Summary Appraisals of the Nation's Ground-Water Resources--Upper Mississippi Region. U.S. Geological Survey Professional Paper 813-B.

Cederstrom, D.J., E.H. Boswell, and G.R. Tarver. 1979. Summary Appraisals of the Nation's Ground-Water Resources--South Atlantic-Gulf Region. U.S. Geological Survey Professional Paper 813-O.

Davidson, E.S. 1979. Summary Appraisals of the Nation's Ground-Water Resources--Lower Colorado Region. U.S. Geological Survey Professional Paper 813-R.

Eakin, T.E., D. Price, and J.R. Harrill. 1976. Summary Appraisals of the Nation's Ground-Water Resources--Great Basin Region. U.S. Geological Survey Professional Paper 813-G.

Foxworthy, B.L. 1979. Summary Appraisals of the Nation's Ground-Water Resources--Pacific Northwest Region. U.S. Geological Survey Professional Paper 813-S.

(Table 2-5 Regional Hydrogeology References)

Fuelner, A.J., J.M. Childers, and V.W. Norman. 1971. Water Resources of Alaska. U.S. Geological Survey Open File Report, 60 pp.

Franks, B.J. 1981. Principal Aquifers in Florida. U.S. Geological Survey Water Resources Investigation, Open File Report 82-255.

Geraghty, J.J., D.W. Miller, F. van der Leeden, and F.L. Troise. 1973. Water Atlas of the United States. Water Information Center, Port Washington, New York, 122 plates

Gerlach, A.C. (ed.). 1970. The National Atlas of the United States of America. U.S. Geological Survey, Washington, DC.

Giefer, G.J. 1976. Sources of Information in Water Resources. Water Information Center, Syosett, New York, 290 pp.

Giefer, G.J. and D.K. Todd (eds.). 1972. Water Publications of State Agencies. Water Information Center, Syosett, New York, 319 pp.

Giefer, G.J. and D.K. Todd (eds.). 1976. Water Publications of State Agencies, First Supplement, 1971-1974. Water Information Center, Syosett, New York, 189 pp.

Gomez-Gomez, F. and J.E. Heisel. 1980. Summary Appraisals of the Nation's Ground-Water Resources--Caribbean Region. U.S. Geological Survey Professional Paper 813-U.

Gomez-Gomez, F., V. Quinones-Aponte, and A.I. Johnson (eds.). 1991. Regional Aquifer Systems of the United States: Aquifers of the Caribbean Islands. AWRA Monograph #15, American Water Resources Association, Bethesda, MD, 113 pp. [7 papers]

Graham, W.G. and L.J. Campbell. 1981. Groundwater Resources of Idaho. Idaho Department of Water Resources, Boise, ID, 100 pp.

Heath, R.C. 1982. Classification of Ground-Water Systems of the United States. Ground Water 20(4):393-401.

Heath, R.C. 1984. Ground-Water Regions of the United States. U.S. Geological Survey Water-Supply Paper 2242.

Hotchkiss, W.R. and A.I. Johnson (eds.). 1992. Regional Aquifer Systems of the United States: Aquifers of the Southeastern United States. AWRA Monograph #17, American Water Resources Association, Bethesda, MD.

Hyde, L.W. 1975. Principal Aquifers in Florida. Florida Bureau of Geology Map Series 16, Tallahassee, FL.

Johnston, R.H. and P.W. Bush. 1988. Summary of the Hydrology of the Floridan Aquifer System in Florida, and Parts of Georgia, South Carolina, and Alabama. U.S. Geological Survey Water Supply Paper 1403-A, 24 pp.

Langbein, W.B. et al. 1949. Annual Runoff in the United States. U.S. Geological Survey Circular 52, 14 pp.

(Table 2-5 Regional Hydrogeology References)

Marie, J.R. 1976. Preliminary Evaluation of the Ground Water Data Network in Indiana. U.S. Geological Survey Water Resources Investigations Report 76-24.

Massachusetts Department of Environmental Quality Engineering. 1986. Massachusetts Hydrogeologic Information Matrix. Boston, MA.

McGuiness, C.C. 1963. The Role of Ground Water in the National Water Situation. U.S. Geological Survey Water Supply Paper 1800, 1,121 pp.

McLean, J.S. and A.I. Johnson (eds.). 1988. Regional Aquifer Systems of the United States: Aquifers of the Western Mountain Area. AWRA Monograph, American Water Resources Association, Bethesda, MD, 229 pp. [13 papers]

Meinzer, O.E. 1923. The Occurrence of Ground Water in the United States. U.S. Geological Survey Water Supply Paper 489, 321 pp.

Meyer, M. 1986. A Summary of South Dakota's Ground Water Information Resources Data, Data Management Efforts, and Data Needs. EPA/SEA 2.3.1, 51 pp.

Mink, J.F. 1977. Handbook-Index of Hawaii Groundwater and Resources Data. Extracted from Reports of the Water Resources Research Center, University of Hawaii.

Mitchell, G.F. 1986. Assessment and Compilation of Groundwater Quality Data for Mississippi. Water Resources Research Institute, Mississippi State University, 8 pp.

Molennar, D., P. Grimstad, and K.L. Walters. 1980. Principle Aquifers and Well Yields in Washington. Washington State Department of Ecology and U.S. Geological Survey.

Price, D. and T. Arnow. 1974. Summary Appraisals of the Nation's Ground-Water Resources--Upper Colorado Region. U.S. Geological Survey Professional Paper 813-C.

Prince, K.R. and A.I. Johnson (eds.). 1992. Regional Aquifer Systems of the United States: Aquifers of the Far West. AWRA Monograph #16, American Water Resources Association, Bethesda, MD, 127 pp. [9 papers]

Ralston, V.H. 1975. Water Resources--A Bibliographic Guide to Reference Sources. Institute of Water Resources Report No. 23, University of Connecticut, 123 pp.

Randall, A.D. and A.I. Johnson (eds.). 1988. Regional Aquifer Systems of the United States: The Northeast Glacial Aquifers. AWRA Monograph, American Water Resources Association, Bethesda, MD, 156 pp. [7 papers]

Randolph, J.R. and R.G. Deike. 1966. Bibliography of Hydrology of the United States, 1963. U.S. Geological Survey Water-Supply Paper 1863, 166 pp.

Reeder, H.O. 1978. Summary Appraisals of the Nation's Ground-Water Resources--Souris-Red-Rainy Region. U.S. Geological Survey Professional Paper 813-K.

Riggs, H.C. 1962. Annotated Bibliography on Hydrology and Sedimentation, United States and Canada, 1955-1958. U.S. Geological Survey Water Supply Paper 1546, 236 pp.

(Table 2-5 Regional Hydrogeology References)

Robson, S.G. 1987. Bedrock Aquifers in the Denver Basin, Colorado-A Quantitative Water Resources Appraisal. U.S. Geological Survey Professional Paper 1257, 73 pp.

Sinnott, A. 1982. Summary Appraisals of the Nation's Ground-Water Resources--New England Region. U.S. Geological Survey Professional Paper 813-T.

Sinnott, A. and E.M. Cushing. 1978. Summary Appraisals of the Nation's Ground-Water Resources--Mid-Atlantic Region. U.S. Geological Survey Professional Paper 813-I.

Smith, Z.A. 1988. Groundwater in the West. Academic Press, New York, NY. (Covers 19 states including Alaska and Hawaii)

Southeastern Geological Society. 1986. Hydrogeological Units of Florida. Florida Geological Survey Special Publication No. 28, Tallahassee, FL.

Stearns, H.T. 1946. Geology of the Hawaiian Islands. Hawaii Division of Hydrography Bulletin 8, 106 pp.

Sun, R.J. and J.B. Weeks. 1991. Bibliography of Regional Aquifer-System Analysis Program of the U.S. Geological Survey, 1978-91. U.S. Geological Survey Water-Resources Investigations Report 91-4122, 92 pp. [Bibliography listing 876 published reports]

Swain, L.A. and A.I. Johnson (eds.). 1989. Regional Aquifer Systems of the United States: Aquifers of the Midwestern Area. AWRA Monograph, American Water Resources Association, Bethesda, MD, 252 pp. [13 papers]

Takasaki, K.J. 1978. Summary Appraisals of the Nation's Ground-Water Resources--Hawaii Region. U.S. Geological Survey Professional Paper 813-M.

Taylor, O.J. 1978. Summary Appraisals of the Nation's Ground-Water Resources--Missouri Basin Region. U.S. Geological Survey Professional Paper 813-Q.

Terry, J.E. and C.T. Bryant. 1979. Summary Appraisals of the Nation's Ground-Water Resources--Lower Mississippi Region. U.S. Geological Survey Professional Paper 813-N.

Thomas, H.E. 1952. Ground-Water Regions of the United States, Their Storage Facilities. Interior and Insular Affairs Committee Report, U.S. Congress, 78 pp.

Thomas, H.E. and D.A. Phoenix. 1976. Summary Appraisals of the Nation's Ground-Water Resources--California Region. U.S. Geological Survey Professional Paper 813-E.

Todd, D.K. (compiler). 1983. Ground-Water Resources of the United States. Premier Press, Berkeley, CA. [Publication containing all USGS Summary Groundwater appraisal reports except USGS Professional Paper 813-U (Caribbean Region)]

U.S. Geological Survey (USGS). 1982. Codes for the Identification of Hydrologic Units in the United States and the Caribbean Outlying Areas. U.S. Geological Survey Circular 878-A, 115 pp. [See also USGS (1988a)]

(Table 2-5 Regional Hydrogeology References)

U.S. Geological Survey (USGS). 1984. National Water Summary 1984. U.S. Geological Survey Water Supply Paper 2275.

U.S. Geological Survey (USGS). 1987. Geophysical Well-Log Data Base for the Gulf Coast Aquifer Systems, South Central United States. U.S. Geological Survey Open File Report 87-677, 213 pp.

U.S. Geological Survey (USGS). 1988a. Codes for the Identification of Hydrologic Units in the United States and the Caribbean Outlying Areas. U.S. Geological Survey Circular 878-C, 8 pp. [Current list of codes available from National Water Data Exchange, USGS, 421 National Center, Reston, VA 22092]

U.S. Geological Survey (USGS). 1988b. Geohydrology and Regional Ground Water Flow of the Coastal Lowlands Aquifer System in Parts of Louisiana, Mississippi, Alabama and Florida. U.S. Geological Survey Water-Resources Investigations Report 88-4100.

U.S. Water Resources Council. 1978a. The Nation's Water Resources, 1975-2000: Second National Water Assessment, 4 Vols. U.S. Government Printing Office, Washington, DC.

U.S. Water Resources Council. 1978b. Hawaii Region, The Nation's Water Resources, 1975-2000: Second National Water Assessment, Vol. 4. U.S. Government Printing Office, Washington, DC, 52 pp.

U.S. Water Resources Council. 1978c. Alaska Region, The Nation's Water Resources, 1975-2000: Second National Water Assessment, Vol. 4. U.S. Government Printing Office, Washington, DC, 59 pp.

van der Leeden, F., F.L. Troise, and D.K. Todd. 1990. The Water Encyclopedia, 2nd ed. Lewis Publishers, Chelsea, MI, 808 pp. [First edition by Todd published in 1970]

Vecchioli, J. and A.I. Johnson (eds.). 1987. Regional Aquifer Systems of the United States: Aquifers of the Atlantic and Gulf Coastal Plain. AWRA Monograph #9, American Water Resources Association, Bethesda, MD, 179 pp.

Weist, Jr., W.G. 1978. Summary Appraisals of the Nation's Ground-Water Resources--Great Lakes Region. U.S. Geological Survey Professional Paper 813-J.

West, S.W. and W.L. Broadhurst. 1975. Summary Appraisals of the Nation's Ground-Water Resources--Rio Grande Region. U.S. Geological Survey Professional Paper 813-D.

Zenone, C. and G.S. Anderson. 1978. Summary Appraisals of the Nation's Ground-Water Resources--Alaska. U.S. Geological Survey Professional Paper 813-P.

Zurawski, A. 1978. Summary Appraisals of the Nation's Ground-Water Resources--Tennessee Region. U.S. Geological Survey Professional Paper 813-L.

Yee, J.S.J. and W.R. Souza. 1987. Quality of Ground Water in Idaho. U.S. Geological Survey Water Supply Paper 2272, 53 pp.

(Table 2-5 Regional Hydrogeology References)

CHAPTER 3

SOIL AND GROUND-WATER GEOCHEMISTRY AND MICROBIOLOGY

3.1 Key Characteristics of Chemical Systems 118

    3.1.1 Equilibrium, Thermodynamics, and Kinetics 118
    3.1.2 Heterogeneity and Reversibility 120
    3.1.3 Phases and Speciation 121
    3.1.4 Distribution vs. Transformation Processes 122

3.2 Subsurface Distribution Processes 123

    3.2.1 Acid-Base Equilibria 123
    3.2.2 Sorption 124
    3.2.3 Precipitation and Dissolution 126
    3.2.4 Immiscible Phase Separation 128
    3.2.5 Volatilization 129

3.3 Subsurface Transformation Processes 129

    3.3.1 Complexation 129
    3.3.2 Hydrolysis 131
    3.3.3 Oxidation-Reduction 132
    3.3.4 Biotransformation 135
    3.3.5 Other Transformation Processes 136

3.4 Subsurface Environmental Parameters 137

    3.4.1 pH 137
    3.4.2 Eh and other Redox Indicators 141
    3.4.3 Salinity 144
    3.4.4 Soil and Aquifer Matrix 145
    3.4.5 Temperature and Pressure 147

3.5 Microbial Ecology of the Subsurface 148

    3.5.1 Classification of Microorganisms 148
    3.5.2 Natural Biological Activity in the Subsurface 148
    3.5.3 Aerobic vs. Anaerobic Degradation 149
    3.5.4 Biotransformation of Organic Contaminants 152

3.6 Guide to Major References* 156

*Appendix F contains citations for table and figure sources.

Understanding the basic geology and subsurface flow patterns is only the first step in studying contaminated soils and ground water. The rate and distance of transport of any specific contaminant will be strongly influenced by physical, geochemical and microbiological processes. Such processes may reduce or enhance the mobility of a contaminant. These processes may transform a contaminant to substances which may be less or more toxic than the original contaminant.

The physical and chemical interactions of individual contaminants with existing environmental conditions the soil will affect the amount of contaminant that reaches the ground water, and the concentrations in contaminant plumes. These geochemical and microbiological processes will also largely determine what combination of remediation techniques will be required if cleanup becomes necessary.

The same difficulties inherent in studying ground-water flow (limited direct observation and variability of geologic materials) apply equally to subsurface chemistry and microbiology. In fact, the difficulties are compounded because of the large number of geochemical reactions that are possible and the fact that the outcome of these reactions will vary depending on environmental conditions such as pH, Eh (redox potential), salinity, mineralogy of the solid matrix, temperature, and pressure. Nevertheless, steady advances in the study of the geochemistry and microbiology of soil and ground-water systems in recent years provide a basis for developing a qualitative analysis of the potential behavior of many contaminants in the subsurface.

This chapter begins by discussing key characteristics of chemical systems (Section 3.1) followed by identification of major subsurface chemical distribution processes (Section 3.2), transformation processes (Section 3.3), and environmental parameters that affect chemical processes (Section 3.4). The chapter concludes by describing basic concepts related to microbial ecology of the subsurface (Section 3.5).

3.1 Key Characteristics of Chemical Systems

A chemical system is a mixture of individual chemical components. *Homogeneous* systems consist of components in one phase of matter (solid, liquid, or gas), whereas *heterogeneous* systems contain two or more phases. A chemical system can be described by the interactions that occur within it and by the effects that these processes have on the system's chemical composition and phases. Interactions that change the chemical structure of system components are called chemical reactions. Other interactions, such as processes that alter the solubility of system components, change the system without altering chemical structures.

3.1.1 Equilibrium, Thermodynamics, and Kinetics

The *equilibrium* state implies that as long as no significant changes in major components or environmental factors affect the system, the chemical speciation and phases of the system will tend towards a specific composition. An equilibrium state does not imply that chemical processes cease. Within the time-frames of observation the rates of equilibrium reactions in the forward direction (i.e. towards products) can be related to the rate of the reverse reactions by the equilibrium constant.

In nonequilibrium systems, chemical processes act to alter the chemical composition and/or phase of the system and equilibrium predictions may not apply. Simple systems, such as mixtures of sodium chloride and water, attain solution equilibrium relatively rapidly. More complex systems, particularly those involving solution and solid phases may never reach equilibrium under normal environmental conditions.

Thermodynamically, a chemical system is in equilibrium when its free energy is minimized. Thus *thermodynamic* principles define the stability of substances within the system and whether a reaction can occur. If complex solid-liquid systems are perturbed by the introduction of a contaminant species or phase, chemical reactions will proceed and it may be very difficult to predict the approach to equilibrium. *Exothermic* chemical reactions release energy in the form of heat, and *endothermic* reactions require an input of energy to take place. At equilibrium, reactions may continue, but for every exothermic reaction there will be a compensating endothermic reaction such that the overall distribution of chemical species remains the same. Geochemical distribution-of-species computer codes are based primarily on equilibrium thermodynamic principles (Section 10.3.3).

Thermodynamic calculations can predict whether a chemical reaction is likely to occur under specified conditions, but give no indication of how fast the reaction will occur. *Kinetics* describes the rate of chemical reactions. Some reactions, such as the ionization of a strong acid in water (Section 3.2.1), will occur almost instantaneously. These reactions often occur in one or more concerted steps. Other reactions, such as the hydrolysis of cyanides at low pH (Section 3.3.2), may take tens of thousands of years. Complex, multi-step reactions involving several chemical species and their solution equilibria may have very complicated rate laws. Three empirically derived rate laws can be used to approximate the rates of concerted chemical processes (Bedient et al., 1982):[1]

$$dC_A/dt = -k_0 \qquad \text{Zero-order}$$

$$dC_B/dt = -k_1 C_A \qquad \text{First-order}$$

$$dC_B/dt = -k_2 C_A C_B \qquad \text{Second-order}$$

where:

k_0, k_1, k_2 = rate constants (mol/L-sec,/sec,L/mol-sec, respectively)
C_A, C_B = some reacting species

The rate of zero-order reactions proceeds independently of concentration,

---

[1] Bedient, P.B., N.K. Springer, C.J. Cook, and M.B. Tomson. 1982. Modeling Chemical Reactions and Transport in Groundwater Systems: A Review. In: Modeling the Fate of Chemicals in the Aquatic Environment, K.L. Dickson, A.W. Maki and J. Cairns, Jr. (eds.), Ann Arbor Science, Ann Arbor, MI, pp. 215-246.

whereas the speed of first-order reactions is directly proportional to the concentration of a single reacting species. Higher-order reaction kinetic rates depend on the relative concentration of several reacting species. Fractional kinetic orders are quite common in reactions which occur in solution where several equilibria control the concentrations of speciation of reactive chemical species. Most solute transport computer codes assume first-order kinetics when modeling ground-water systems (Section 10.3.2).

3.1.2 Heterogeneity and Reversibility

Chemical processes can be broadly classified as either homogeneous or heterogeneous and as either reversible or irreversible (Table 3-1). *Homogeneous* reactions in ground water take place in only the aqueous phase. In general, these reactions occur uniformly throughout the phase and, are easier, typically, to study and predict than heterogeneous reactions. *Heterogeneous* reactions tend to occur at the interface between different phases, thus involving more than one phase. An example is sorption. Some reactions, such as precipitation, may result in phase changes. Heterogeneous reactions also tend to occur more actively at some locations in the chemical system than at others. Bacterial decomposition of wastes is an example of a complex heterogeneous process that may be more active in locations with conditions favorable for organisms and less active in other, less favorable locations.

Table 3-1 Characteristics of Chemical Processes that May Be Significant in the Subsurface

Characteristic	Types of Reactions
Homogeneous	Acid-base, hydrolysis, hydration, neutralization, oxidation-reduction, polymerization, thermal degradation
Heterogeneous	Adsorption-desorption, precipitation-dissolution, immiscible phase separation, biodegradation, complexation
Reversible	Acid-base, neutralization, oxidation-reduction (biologically mediated), adsorption-desorption, precipitation-dissolution, complexation
Irreversible	Hydrolysis, oxidation-reduction (inorganic), biomineralization, immiscible phase separation

Source: Boulding (1990).

The *reversibility* of specific chemical reactions is another important characteristic in assessing the fate of contaminants in ground water. Depending on environmental conditions, reversible reactions may proceed in either one or both directions. Acid-base reactions exemplify reversible processes. In aqueous solutions, relatively minor changes in factors such as pH or concentration or reactants can change the direction of these reactions. Irreversible reactions, typified by hydrolysis, have a strong tendency to go in one direction only.

Table 3-1 lists several reversible and irreversible processes that may be significant in ground water. The characteristics of the specific contaminant and the environmental factors present in an aquifer (Section 3.4) strongly influence which processes will occur and whether they will tend to be irreversible. Irreversible reactions are of particular interest in the study of contaminants in ground water. A contaminant that is rendered nontoxic through irreversible reactions may be considered to be permanently transformed to a nonhazardous state.

3.1.3 Phases and Speciation

Chemical reactions may result from interactions among solids, liquids, and gases. The major interactions that occur between contaminants and the subsurface include:

- *Liquid-liquid interactions*. These occur when nonaqueous phase liquids (NAPLs) reach the water table.

- *Liquid-solid interactions*. Water itself can react chemically with solids in the subsurface. Also contaminants dissolved in water may react with solids through sorption or ion exchange.

- *Liquid-gas interactions*. Volatile NAPLs such as benzene and carbon tetrachloride may shift to a vapor phase in the vadose zone. This may also occur with dissolved volatile organic contaminants at the interface between the water table and the vadose zone.

A substance may exist in several forms, or *species*. Five major types occur in ground water:

1. "Free" ions are surrounded only by water molecules and are very mobile in ground water. Acid-base (Section 3.2.1) and dissolution reactions (Section 3.2.3) create free ions.

2. Insoluble species may exist in solid form (such as Ag_2S, $BaSO_4$) or liquid form (such as gasoline). Precipitation reactions (3.2.3) and immiscible phase separation (Section 3.2.4) are important processes affecting this type of speciation.

3. Metal/ligand complexes (such as $Al(OH)^{2+}$, Cu-humate) and organic/ligand complexes tend to be mobile in ground water (see Section 3.3.1).

4. Physically or chemically sorbed species are immobile in ground water but can be remobilized if replaced by other species with a stronger affinity to the solid surface or via transformation reactions (see Section 3.2.2).

5. Species may differ by oxidation state (manganese (II) and (IV); iron (II) and (III); and chromium (III and (VI)). Oxidation state is influenced by the redox potential (see Section 3.3.3). Mobility is affected because oxidation state influences precipitation-dissolution reactions (Section 3.2.3) and also toxicity in the case of heavy metals.

Dissolved species may be either ionic or nonionic. *Ionic* species, possess a net positive or negative charge. *Nonionic* species, are neutral molecules which have no net excess charge. *Cations* are positively charged ions (Na^+, Ca^{2+}) and *anions* are negatively charged (SO_4^{-2}). The ability of a solids with a net neutral charge to dissociate into ionic species is more common with inorganic contaminants than with organic contaminants. Acid-base reactions (Section 3.2.1) and hydrolysis reactions (Section 3.3.2) often determine the distribution between ionic and nonionic species.

Neutral substances may be either nonpolar or polar. In *nonpolar* species there is no overall excess charge. In *polar* species, the chemical structure stabilizes charged poles on the molecule, even though the net charge is zero. Water (H_2O) is a polar molecule with the positive pole on the side of the hydrogen atom and the negative pole on the side of the oxygen atoms. Nonpolar molecules tend to be *hydrophobic* (water avoiding) for reasons discussed in Section 3.2.2.

Many substances may exist as several species in the subsurface depending on geochemical conditions. Identifying active chemical forms of an element and predicting the reactivity in the vadose and the saturated zones is an essential part of evaluating the fate and transport of contaminants.

3.1.4 Distribution vs. Transformation Processes

Geochemical processes in the subsurface can be broadly classified into distribution processes and transformation processes. *Distribution processes* affect the form or state of association of a specific chemical substance in the aqueous or solid at a given time or under specific environmental conditions. Thus, a substance may be associated with a solid phase or in solution (described by the distribution processes). Regardless of the phase, the chemical properties and toxicity of chemical species may remain unaltered. The physical state of a substance, however, influences the transformation and transport processes that can occur. For this reason, distribution processes are important to define during a fate assessment.

Transformation processes alter the chemical structure of a substance. In the subsurface, the transformation processes that may occur are largely determined by the conditions created by distribution processes, reactivity, and the prevalent environmental factors. Transport processes may not need to be considered if transformation processes irreversibly change a hazardous waste to a nontoxic form.

Table 3-2 lists major distribution and transformation processes and also indicates whether processes are biotic (mediated or initiated by organisms in the environment) or abiotic (not involving biological mediation), or can be both. Biotic processes are

Soil and Ground-Water Chemistry and Microbiology

often limited to environmental conditions that favor growth of the organisms that are capable of transforming a particular compound. Abiotic processes, on the other hand, can generally occur under a wide range of environmental conditions. Volatilization, neutralization, sorption, precipitation, and complexation are abiotic processes, though they may be catalyzed by biochemical factors.

3.2 Subsurface Distribution Processes

Phase distribution usually does not affect the toxic properties of the substance. It can, however, affect the mobility of the contaminants in the subsurface. The major distribution processes and the sections in which they are discussed are as follows:

- Acid-base equilibrium (Section 3.2.1)
- Sorption (Section 3.2.2)
- Precipitation-dissolution (Section 3.2.3)
- Immiscible phase separation (Section 3.2.4)
- Volatilization (Section 3.2.5)

3.2.1 Acid-Base Equilibria

Acid-base equilibrium reactions affect pH, a function of the concentration of hydrogen ions in solution, which is a controlling factor in the type and rate of many other chemical reactions (see Section 3.4.1).

Acids ionize in solution to form hydrogen ions and anions according to the general reaction:

HA (neutral) <====> H$^+$ (cation) + A$^-$ (anion)

As the equation shows, the ionization is reversible. The acid anion (acting as a weak base) can recombine with the hydrogen ion to reform neutral HA. Both reactions occur continuously in solution, with the extent of ionization dependent on the strength of the acid. Strong acids, such as HCl, ionize completely in dilute aqueous solution. Thus, a 0.01 molar (10^{-2} molar) solution has a pH of 2. Weak acids, such as acetic acid and other organic acids, do not ionize completely in aqueous solution and form solutions with pH generally ranging from 4 to 6.

In the above example, the acid anion A$^-$ functions as a base when it combines with a hydrogen ion. By definition, any substance that combines with hydrogen ions is a base. Like strong acids, strong bases ionize completely in dilute aqueous solution. Thus, NaOH dissolves in water to form hydroxide ions, which in turn function as a base when they combine with hydrogen ions to form water, as shown by the general equation:

MOH <====> M$^+$ + OH$^-$

Table 3-2 Significance of Chemical Processes in the Subsurface

Process	Detoxication	Mobility	Biotic/Abiotic
PHASE DISTRIBUTION			
Acid-base equilibrium	No	Yes	Both
Adsorption-desorption	No	Yes	Abiotic
Precipitation-dissolution	No	Yes	Abiotic
Immiscible phase separation	No	Yes	Both
Volatilization	No	Yes	Abiotic
TRANSFORMATION			
Biodegradation	Yes	Yes	Biotic
Complexation	No	Yes	Abiotic
Hydrolysis	Yes	Yes	Both
Neutralization	Yes	No	Abiotic
Oxidation-reduction	Yes	Yes	Both

Source: Boulding (1990).

Acid-base equilibrium reactions generally occur quickly. When the pH of a solution changes, acids and bases readily attain a new equilibrium between neutral and ionic forms. Weak acids or weak bases have little influence, if any, on the pH of the aqueous solution. Mills et al. (1985/T2-4) describe the procedures for calculating the fraction of a toxic organic acid or base that is in the nonionic, neutral form. This procedure is especially useful for evaluating the volatilization of organics at near-surface conditions (because only electrically neutral species are directly volatile).

3.2.2 Sorption

Sorption is a phase-distribution process in which dissolved metals or toxic organics (solutes) are transferred from the aqueous phase (water) to the solid phase (rock, soil, or particles of organic matter). Sorption is a major mechanism affecting the mobility of heavy metals and toxic organic substances and is thus a major consideration when assessing transport. *Adsorption*, which occurs on the solid-solution interface, such as on clays, is usually fully or partly reversible (desorption). Although adsorption does not directly affect the toxicity of a substance, the substance may be rendered nontoxic by transformation processes, such as hydrolysis, while it is adsorbed. *Absorption* is a process whereby a chemical specie may be sorbed into the solid matrix material, such as with activated charcoal. In some respects it may be considered solid-solution formation. Absorption may occur after surface sorption or adsorption has occurred.

Sorption is a widespread chemical phenomena that has been studied by many different disciplines, resulting in a diverse and often confusing terminology. Thus, the general term *sorption* is often used where the distinction between adsorption and absorption is not clear. In this handbook, the term *sorption* is used to describe any one of a number of intermolecular interactions involved in phase distribution from the aqueous to the solid phase. Under carefully controlled laboratory conditions individual sorption mechanisms can be studied, but in the real world sorption is the result of multiple processes acting on both multiple species and solid surfaces. In fact, the distinction between sorption and precipitation is not easily made at the liquid-solid interface (Sposito, 1984/T3-12).

Sorption and desorption are caused by interactions between molecules in solution and those in the structure of solid surfaces. Many chemical and physical properties of both aqueous and solid phases affect sorption, and the physical chemistry of the process itself is complex. For example, sorption of one ion or neutral specie may result in the desorption of another (see discussion of ion exchange below).

Sorption reactions are often exothermic (Section 3.1.1) and sorption processes can be broadly classified into two groups based on the energies involved: chemical sorption (high energy) and physical sorption (low energy). Sorption can also be described in terms of type of bond and forces involved. Table 3-3 summarizes these parameters and also lists the type of contaminant that is most likely to be involved in the different types of bonds.

Chemical sorption (also called chemisorption) involves the formation of chemical bonds between the sorbate molecule and the sorbent surface. Ion exchange, protonation, and hydrogen bonding are examples of chemical sorption processes (see Table 3-3). These bonds typically involve energies on the order of 10 kcal/mole. In *ion exchange*, a cation at a negatively charged site on a mineral surface is replaced by another cation. The cation exchange capacity measures the sorption capacity of a material (see discussion of clay minerals in Section 1.1.1). Anions such as sulfate and nitrate may also be sorbed but at much lower levels than cations because many mineral surfaces are negatively charged. *Protonation* involves the attachment of a molecule to a previously sorbed hydrogen ion by an acid-base reaction. *Hydrogen bonds* result from the attraction of polar molecules to hydrogen atoms in already sorbed molecules or ionic species.

Table 3-3 Major Intermolecular Interactions Involved in Sorption in the Subsurface

Type of Bond/Attraction	Forces	Sorbate	Energy (kcal/mole)
CHEMICAL BONDS			
Ion exchange	Electrostatic	Metal cations Organic acid/Cation	up to 50
Protonation	Electrostatic	Organic bases	up to 35
Hydrogen	Electrostatic	Polar organic	0.5-15
PHYSICAL SORPTION FORCES			
Van der Waals	Electrostatic	Small molecules	1-2
		Large molecules	11+
Hydrophobic	Entropically driven	Nonpolar organic	1

Source: Boulding (1990).

Physical sorption processes involve physical forces and are associated with lower energies, typically less than 10 kcal/mole. *Van der Waals* attractions, are weak electrostatic forces that operate between all atoms, ions, and molecules. *Hydrophobic bonds*, which involve the "pushing" of nonpolar organic molecules toward solid surfaces by polar water molecules to achieve a more thermodynamically stable liquid structure (Table 3-3).

3.2.3 Precipitation and Dissolution

Precipitation is a phase-distribution process whereby insoluble solids are formed and separate from a solution. Dissolution involves a change from the solid or gaseous phase to the aqueous phase. The solubility of a compound, its tendency to dissolve in water or other solutions, is the main property affecting the precipitation-dissolution process. In the subsurface, precipitation-dissolution reactions are often evaluated by the use of mineral stability diagrams that delineate the pH, Eh (or pe), temperature and pressure conditions under which a particular mineral is stable. Figure 3-1 illustrates such a diagram for the iron in water.

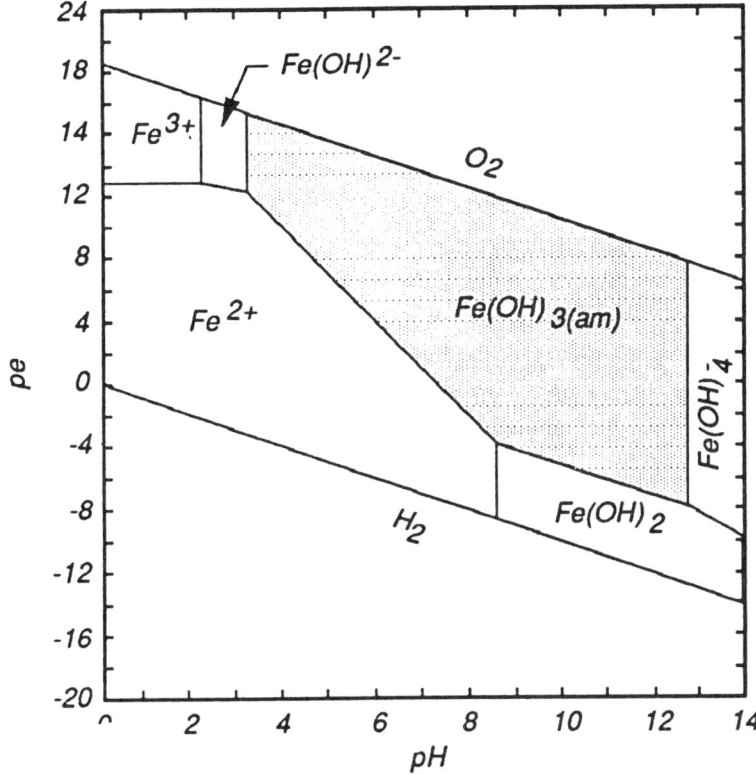

Figure 3-1 pe-pH diagram for the Fe-H$_2$O system (Palmer and Fish, 1991).

Precipitation reduces mobility, and dissolution increases mobility of contaminants. Ionic precipitation-dissolution reactions are often fully reversible. Precipitation can only be considered to effectively immobilize a contaminant if environmental conditions in an aquifer are sufficiently stable to prevent dissolution from occurring at some future time.

The equilibrium concentration of a compound in water is determined by its *equilibrium solubility*, or *solubility product constant*, the maximum amount of a compound that will dissolve in a solution at a specified temperature and pressure. The solubility of toxic organic compounds is generally much lower than that of inorganic salts, a characteristic that is particularly true of nonpolar compounds because of their hydrophobic character. Equilibrium solubility will change with changing environmental parameters such as temperature, pressure, and pH. For example, the solubility of many organic compounds increases when solution temperature is increased.

Precipitation may occur when the concentration of a compound in solution exceeds the equilibrium solubility. For organic chemical wastes, precipitation is not generally a significant phase-distribution process in the subsurface, although in certain circumstances, it may need to be considered. For example, pentachlorophenol tends to precipitate from aqueous solution when the solution has a pH of less than 5 (Choi and Aomine, 1974).[2] Polychlorophenols form insoluble precipitates in water that is high in Mg^{+2} and Ca^{+2} ions (Davis, 1967/T3-12). Also, organic anions react with such elements as Ca^{+2}, Fe^{+2}, and Al^{+3} to form sparingly soluble to nearly insoluble compounds.

Precipitation reactions are most likely to be significant for heavy metals and other inorganic species in ground water. For example, sulfide ions precipitate with metal ions as metal sulfides. Biological activity caused by organic contaminants may initiate redox reactions resulting in the precipitation of iron and manganese oxides (see Section 3.3.3). Mixing of ground waters from different aquifers or injection by artificial recharge may cause precipitation and dissolution reactions. Precipitation reactions can lead to plugging of injection or monitoring wells.

Coprecipitation is a phase-distribution process whereby metal ions precipitate from the aqueous phase even if the equilibrium solubility has not been exceeded. This process occurs when metals are incorporated into the crystal lattice structure of silicon (Si), aluminum (Al), and iron (Fe) oxides when these latter compounds precipitate out of solution. The incorporation of metals in coprecipitated solids varies. Iron oxides and hydroxides may incorporate a number of metal ions, such as chromium (Cr), nickel (Ni), arsenic (As), selenium (Se), cadmium (Cd), and thorium (Th), during precipitation. Coprecipitation with carbonate minerals may be an important mechanism for removal of Co, Pb, Zn, and Cd (Förstner and Wittmann, 1981/T4-5).

Cosolvation is a process by which mixing of several water-miscible organic solvents (such as ethanol, methanol, and acetone) with ground water increases the solubility of hydrophobic organic contaminants. Cosolvents reduce hydrophobicity by increasing the interactions between the solute and the solvent, thereby decreasing physical sorption (Nkedi-Kizza et al., 1985).[3] For hydrophobic contaminants, the presence of biologically derived or anthropogenic compounds in the range of 20% or greater by volume increase their solubility by an order of magnitude or more.

3.2.4 Immiscible Phase Separation

A liquid or a gas that is poorly soluble in water may separate from the water, resulting in an immiscible phase. Nonaqueous phase liquids (NAPLs) that are lighter than water (LNAPLs) will tend to float on the surface of the water table, and those that are denser than water (DNAPLs) will tend to sink to the bottom of aquifers. The behavior of NAPL contaminants in the subsurface is described in more detail in Section 4.4.3 (Density/Viscosity Differences).

---

[2] Choi, J. and S. Aomine. 1974. Mechanisms of Pentachlorophenol Adsorption by Soils. Soil Sci. Plant Nutr. 20(4):371-379.

[3] Nkedi-Kizza, P., P.S.C. Rao, and A.G. Hornsby. 1985. Influence of Organic Cosolvents on Sorption of Hydrophobic Organic Chemicals by Soils. Environ. Sci. Technol. 19:975-979.

3.2.5 Volatilization

Transfer of contaminants from the free-product or aqueous phase to the gaseous phase is called *volatilization*. Volatilization is not a major process in the natural subsurface, but is very significant when contaminants such as chlorinated solvents and petroleum based liquids are involved. Major factors that affect volatilization include (1) the area of contact between the free products and the unsaturated zone, (2) the vapor pressure of the spilled compound, and (3) the rate at which the compound diffuses in the subsurface. Contaminants with vapor pressures above 14 mm Hg at 20°C are generally volatile enough to use soil vapor extraction (Section 13.2.1) as a remediation technique.

3.3 Subsurface Transformation Processes

Transformation processes change the chemical structure of a compound. The evaluation of contaminants in the subsurface must consider both the full range of transformation processes that may occur and the toxicity and mobility of the resulting reaction products. Transformation processes in the subsurface may result in one or more of the following:

- *Detoxification*, an irreversible change in a substance from a toxic to a nontoxic form. This occurs most commonly when an organic substance breaks down into its inorganic constituents with water and carbon dioxide being the main by-products.

- *Transtoxification*, the conversion of one toxic compound to another toxic compound. In the process, toxicity may remain the same, increase, or be reduced. For example tetrachloroethylene is readily degraded under anaerobic conditions to vinyl chloride which is both more mobile and more toxic than the original compound.

- *Toxification*, the conversion of a nontoxic compound to a toxic substance.

The major transformation processes in the subsurface are complexation (Section 3.3.1), hydrolysis (Section 3.3.2), oxidation-reduction (Section 3.3.3) and biotransformation (Section 3.3.4). Two additional transformation processes, polymerization and thermal degradation, occur less frequently and are discussed briefly in Section 3.3.5 along with catalysis, which may facilitate the occurrence of various transformation processes.

3.3.1 Complexation

A complex ion is one which may contain several molecules or ionic species. Metal ions in solution commonly form complexes with water, chloride and other species which are called *ligands*. Depending on the structure and reactivity of a contaminant, complexation by inorganic or organic ligands may increase or decrease its mobility, reactivity and persistence. Because of these effects, complex formation is a very important process for transition series metals (i.e. Fe, Mn, Co, Ni, Zn, Cu), and may also be significant for some organic contaminants. Heavy metals (i.e., Pb, Cd, Hg) are particularly prone to complexation because their atomic structure (specifically the presence of unfilled d orbitals) favors the formation of bonds with polar molecules, such as water or ammonia (NH_3) and anions such as chloride (Cl^- and cyanide (CN^-).

Complex formation is likely in solutions with high ionic strength, because the large number of ions present in solution increases the number of chemical species that can form. Many variables affect the stability of a complex ion relative to ions and metals that can serve as potential ligands to the central metal, the most important of which are the valence (charge) and radius of the central cation. As a rule, the stability of complexes formed with a given ligand increases with cation charge and decreases with cation radius (Langmuir, 1979).[4]

The total solubility of metals is often much higher when they are in the form of organometallic complexes. Naturally occurring chemicals that can partially complex with metal compounds and increase the solubility of the metal include aliphatic acids, aromatic acids, alcohols, aldehydes, ketones, amines, aromatic hydrocarbons, esters, ethers, and phenols.

Chelation is the process of forming complex ions with organic ligands that have more than one functional group available for bonding to the central metal ion in the complex. The complex ion formed by this process is called a *chelate*. The ligands in chelates are classified according to the number of binding sites in the molecule: monodentate (one-site), bidentate (two sites), etc. Metal solubility (i.e., mobility) is often substantially increased when chelation occurs, and metal-chelate compounds are very stable when the metal ion is chelated by a heterocyclic ring of an organic molecule. Many simple organic-metal complexes will tend to dissociate if solutions become more dilute. Often chelated complexes do not (Martell, 1971).[5] Even sorbed metals may be remobilized into solution by organic chelates. For example, the synthetic chelate, nitrilotriacetic acid (NTA) has been used as a detergent-builder alternative to polyphosphate in detergents. NTA has been observed to remobilize sorbed heavy metals in the near-surface environment (Förstner and Wittmann, 1981/T4-5).

Metal ions in solution readily form complex ions by the process of *hydration* (bonding to water molecules). Because of the dipolar nature of water molecules, the oxygen atoms with a slight excess negative charge are attracted to the positively charged metal ion. Hydration tends to increase the complexity of chemical reactions

---

[4] Langmuir, D. 1979. Techniques of Estimating Thermodynamic Properties for Some Aqueous Complexes of Geochemical Interest. In: Chemical Modeling in Aqueous Systems: Speciation, Sorption, Solubility and Kinetics, E.A. Jenne, E.A. (ed.), ACS Symp. Series 93, American Chemical Society, Washington, DC, pp. 353-387.

[5] Martell, A.E. 1971. Principles of Complex Formation. In: Organic Compounds in Aquatic Environments, S.D. Faust and J.V. Hunter (eds.), Marcel Dekker, New York, pp. 239-263.

because hydrated polyvalent metal ions may form multiple associations with other metals to create complex polynuclear ions. Hydration may also reduce mobility of metal ions through physical sorption. However, mobility can also be reduced by dehydration when organic ligands replace water molecules in complex ions. Polynuclear metal ions and large organic complexes can be readily sorbed onto mineral surfaces because of their large molecular weights, which enhance the sorption processes (Section 3.2.2).

3.3.2 Hydrolysis

Hydrolysis occurs when a compound reacts chemically with water, and new chemical species are formed by the reaction. Hydrolysis reactions fall into two major types:

- *Replacement* is the most common hydrolysis reaction. In this reaction, one functional group is replaced by an -OH (hydroxide ion) originating from a water molecule. For example, an hydroxide ion can replace the halide ion in an alkyl halide to form a soluble alcohol, leaving the halide ion in solution.

- *Addition reactions* involve the incorporation of water into the chemical structure of a compound. An example of this type of reaction is that by which alcohols can form by the addition of water to a carbon-carbon double bond.

Replacement and addition reactions are essentially irreversible transformation processes, and may result in intermediate compounds that are subject to further hydrolysis (e.g., nitriles to amides and then on to acids). Whether hydrolysis results in detoxification, transtoxification, or toxification depends on the toxicity of the most stable end-product of any series of hydrolysis reactions.

Hydrolysis reaction rates are commonly reported in terms of half-life, the time it takes for half of the original concentration of the substance to be hydrolyzed. Hydrolysis half-lives of various hazardous organic contaminants range from days to thousands of years. Hydrolysis rates greatly depend on pH and also vary widely for an individual compound under acidic to basic conditions. Hydrogen cyanide illustrates the strong effect that pH can have on hydrolysis rates. Cyanides hydrolyze to amides, which then hydrolyze to acids and ammonia. At a pH of greater than 10, this reaction has a half-life of about 10 years. At a pH of 4, however, the reaction takes over 10,000 years (Scrivner et al., 1986).[6] Furthermore, metal-cyanide complexes do not hydrolyze readily. These complexes can reduce the concentration of free cyanide in

---

[6] Scrivner, N.C., K.E. Bennet, R.A. Pease, A. Kopatsis, S.J. Sanders, D.M. Clark, and M. Rafal. 1986. Chemical Fate of Injected Wastes. In: Proc. Int. Symp. on Subsurface Injection of Liquid Wastes, National Water Well Association, Dublin, OH, pp. 560-609. [A summary of this paper with the same title was published the same year in Ground Water Monitoring Review 6(3):53-58]

solution, but can also increase the time needed for the total cyanide concentration to decrease due to hydrolysis.

Many classes of organic compounds hydrolyze in aqueous solutions, whereas other classes are resistant to hydrolysis. Table 3-4 summarizes organic functional groups that are potentially susceptible to hydrolysis and those that are generally resistant.

Table 3-4 Susceptibility of Organic Functional Groups to Hydrolysis

Potentially Susceptible	Generally Resistant[a]
Alkyl halides	Alkanes
Amides	Alkenes
Amines	Alkynes
Carbamates	Benzenes/Biphenyls
Carboxylic acid esters	Polycyclic aromatic hydrocarbons
Epoxides	Heterocyclic polycyclic aromatic hydrocarbons
Nitriles	
Phosphonic acid esters	Halogenated aromatics/PCBs
Phosphoric acid esters	Dieldrin/Aldrin and related halogenated hydrocarbon pesticides
Sulfonic acid esters	
Sulfuric acid esters	Aromatic nitro compounds
	Aromatic amines
	Alcohols
	Phenols
	Glycols
	Ethers
	Aldehydes
	Ketones
	Carboxylic acids
	Sulfonic acids

[a] Multifunctional organic compounds in these categories may be hydrolytically reactive if they contain other functional group(s) that are hydrolyzable.

Source: Boulding (1990) after Guswa et al. (1984).

3.3.3 Oxidation-Reduction

Along with sorption, oxidation-reduction (redox) processes are probably the most significant transformation reactions affecting contaminants in the subsurface. Abiotic and biotic redox reactions strongly affect the solubility and mobility of heavy metals. Biologically mediated redox reactions are able to transform most organic contaminants.

Oxidation-reduction (redox) reactions involve the loss of electrons and increase in oxidation number (oxidation) by one substance or system, with an associated gain of electrons and decrease in oxidation number (reduction) by another substance or system. Thus, for every oxidation, there must be a reduction. The *oxidation number* of an atom represents the hypothetical charge an atom would have if the ion or molecule were to dissociate. The oxidation number is normally expressed as a roman numeral, as in Fe(II) and Fe(III). In these cases, the oxidation number is the same numerically as the ionic charges of the ions (Fe^{+2} and Fe^{+3}).

Since redox reactions involve the transfer of electrons, the intensity of redox reactions is measured by equilibrium electrode potential differences, termed Eh (redox potential). Highly oxidizing conditions in the environment will have an equilibrium potential Eh of about +800 millivolts (mV); highly reducing conditions, an Eh of about -400 mV. Eh as an environmental factor is discussed in more detail in Section 3.4.2. Eh is difficult to measure accurately, and ground-water systems are often out of equilibrium with respect to redox reactions. Consequently, the Eh of a chemical system indicates the types of redox reactions that may occur, rather than predicts the specific reactions that will occur. Inorganic chemical systems tend to have redox reactions that are irreversible, but many reactions can be reversed with biological mediation. In many ways, therefore, oxidation-reduction equilibria are analogous to acid-base equilibria.

Redox Reactions Involving Simple Hydrocarbons. The simplest example of an oxidation reaction of an organic compound is the transformation of a methane to carbon dioxide and water in the presence of oxygen:

$$CH_4 + 2O_2 ---> CO_2 + 2H_2O$$

This type of reaction is called *aerobic respiration*, and without biological mediation, it is irreversible. In ground water, aerobic respiration depletes dissolved oxygen and, unless a continual supply of fresh oxygen is available, a sequence of reducing reactions begins.

Table 3-5 shows the sequence of reducing reactions involving formaldehyde that will occur after oxygen is depleted in a closed ground-water system (i.e., in which there is no source of oxygen replenishment). The reactions shown in the table are not stoichiometrically correct. Anthropogenic organic chemicals can also be reduced abiotically under anaerobic conditions (Macalady et al. (1986).[7]

Redox Reactions Involving Complex Organic Compounds. Oxidation reactions involving cyclic hydrocarbons and hydrocarbon derivatives are often more complex than those involving simple hydrocarbons. It is not always obvious how to classify these chemical reactions in redox terms. Organic redox reactions most often involve the transfer of electrons via the transfer of functional groups. Oxidation frequently

---

[7] Macalady, D.L., P.G. Tratnyek, and T.J. Grundl. 1986. Abiotic Reduction Reactions of Anthropogenic Organic Chemicals in Anaerobic Systems. J. Contaminant Hydrology 1:1-28.

involves a gain in oxygen and a loss in hydrogen atoms, whereas reduction involves the reverse process. Organic functional groups can be classified in order of increasing oxidation state to facilitate the classification of reactions as either oxidation or reduction. Table 3-6 summarizes relative formal oxidation states of several major functional groups. A functional group is considered to be oxidized if its principal atom is converted to a higher oxidation state. Similarly, reduction is defined by conversion of the principal atom within a group to a lower oxidation state.

Table 3-5 Redox Reactions in a Closed Ground-Water System

Reaction	Equation
1. Aerobic respiration	$CH_2O + O_2 = CO_2 + H_2O$
2. Denitrification	$CH_2O + Nitrate\ (NO_3^-) = Nitrogen + CO_2 + H_2O$
3. Mn(IV) reduction	$CH_2O + MnO_2 = Mn^{2+} + CO_2 + H_2O$
4. Fe(III) reduction	$CH_2O + H^+ + Fe(OH)_3 = Fe^{2+} + CO_2 + H_2O$
5. Sulfate reduction	$CH_2O + Sulfate\ (SO_4^{2-}) + H^+ = HS^- + CO_2 + H_2O$
6. Methane fermentation	$CH_2O + CO_2 = Methane\ (CH_4) + CO_2$
7. Nitrogen fixation	$CH_2O + H_2O + N_2 + H^+ = Ammonia\ (NH_4) + CO_2$

Note: Equations are not stoichiometrically balanced. Reactions will tend to go to completion at equilibrium in sequence from top to bottom.

Source: Boulding (1990) after Champ et al (1979).

Table 3-7 lists some organic compounds according to their susceptibility to oxidation. In general, the importance of chemical redox reactions involving organic compounds in the soil and water environment is not well documented (Valentine, 1986).[8] In anaerobic environments, reduction of chemicals by both biological and nonbiological processes can occur. Reduction of organochlorine compounds (such as DDT and toxaphene), where a chlorine atom is replaced by a hydrogen atom, is an example of this type of reaction.

---

[8] Valentine, R.L. 1986. Nonbiological Transformation. In: Vadose Zone Modeling of Organic Pollutants, S.C. Hern and S.M. Melancon (eds.), Lewis Publishers, Chelsea, MI, pp. 223-243.

Table 3-6 Relative Oxidation States of Organic Functional Groups

Functional Group		Increasing Oxidation State		
Least Oxidized			Most Oxidized	
RH	ROH	RC(O)R	RCOOH	CO_2
	RCl	$(R)_2CCl_2$	$RC(O)NH_2$	CCl_4
	RNH_2		$RCCl_3$	
	C=C	-C≡C-		
-4	-2	0	+2	+4

Source: Boulding (1990) after Valentine (1986).

Table 3-7 Susceptibility of Organic Compounds to Oxidation in Water

Most Susceptible	Least Susceptible
Phenols	Alkenes
Aromatic amines	Haloalkanes
Olefins and dienes (electron-rich)	Alcohols
Alkyl sulfides	Esters
Enamines	Ketones

Source: Boulding (1990) after Mill (1980).

3.3.4 Biotransformation

Biotransformation is the alteration of a compound due to the influence of living organisms. It is one of the most prevalent processes causing the breakdown of organic compounds in the subsurface. *Biodegradation* is a more specific term used to describe the biologically mediated change of a chemical into simpler products, although the simpler daughter products may be as toxic or more toxic than the original compounds. For example, anaerobic biotransformation of tetra- and trichloroethylene yields equally toxic and more persistent dichlorethenes and vinyl chloride (Wood et al., 1985).[9]

Mineralization is the complete conversion of an organic compound to its inorganic constituents (primarily water and carbon dioxide). This generally results in complete detoxification unless one of the products is of environmental concern, such as nitrate or sulfide under certain conditions. *Cometabolism* is the conversion by a microorganism of an organic compound to another compound without the microorganism using the compound as a nutrient for growth (Alexander, 1981).[10] *Cooxidation* is a similar term that is applied to oxidation of nongrowth-producing hydrocarbons in the presence of growth-stimulating hydrocarbons acting as cosubstrates (Perry, 1979).[11]

Almost all of the specific chemical reactions involved in biotransformation can be classified as either oxidation-reduction, hydrolysis, or conjugative. Biotic *oxidation-reduction* reactions are broadly classified as *aerobic* (taking place in the presence of oxygen) and *anaerobic* (taking place without oxygen). See Section 3.5.2 for additional discussion of these types of reactions. *Hydrolysis* reactions involve chemical reactions with water that may be biologically mediated. Goring et al. (1975) identified 26 oxidative, 7 reductive, and 14 hydrolytic transformations of pesticides.[12] *Conjugation* involves the addition of functional groups or a hydrocarbon moiety to an organic molecule or inorganic species. For example, conjugation occurs when microbial processes transform inorganic mercury into dimethyl mercury.

3.3.5 Other Transformation Processes

Polymerization is a process by which large molecules (polymers) are formed by bonding together many smaller molecules. Polymerization of amino acids to peptides catalyzed by clay surfaces has been observed to increase sorption of amino acids 1,000 times that expected for amino acids alone (Degens and Metheja, 1971).[13] Sorption of phenol and benzene as a result of polymerization on smectite clay surfaces has been

---

[9] Wood, P.R., R.F. Lang, and I.L. Payan. 1985. Anaerobic Transformation, Transport, and Removal of Volatile Chlorinated Organics in Ground Water. 1985. In: Ground Water Quality, C.H. Ward, W. Giger, and P.L. McCarty, (eds.), Wiley Interscience, New York, pp. 493-511.

[10] Alexander, M. 1981. Biodegradation of Chemicals of Environmental Concern. Science 211:132-138.

[11] Perry, J.J. 1979. Microbial Cooxidations Involving Hydrocarbons. Microbiol. Rev. 43(1):59-72.

[12] Goring, C.A.I., D.A. Laskowski, J.W. Hamaker, and R.W. Meikle. 1975. Principles of Pesticide Degradation in Soil. In: Environmental Dynamics of Pesticides, R. Haque and V.H. Freed (eds.), Plenum Press, New York, pp. 135-172.

[13] Degens, E.T. and J. Matheja. 1971. Formation of Organic Polymers on Minerals and Vice Versa. In: Organic Compounds in Aquatic Environments, S.D. Faust and J.V. Hunter (eds.), Marcel Dekker, New York, pp. 29-41.

observed in the laboratory (Mortland and Halloran, 1976).[14] Polymerization is probably not a major process affecting contaminants in near-surface ground-water systems.

Thermal degradation is a process by which compounds undergo structural changes in response to heat, leading to the formation of simpler species. For example, many organophosphorus esters isomerize under the influence of heat to break down to component molecules (Crosby, 1973/T4-5). In near-surface ground-water systems, thermal degradation is unlikely to be a significant process, but it may need to be considered in relation to deep-well-injection of hazardous wastes.

The rates of many transformation reactions can increase in the presence of a *catalyst*, which itself remains unchanged in quantity and chemical composition when the reaction has been completed. Although the catalyst itself is not transformed, the catalytic process causes transformation by speeding up reactions that would occur naturally, or by promoting reactions that would not occur in the absence of the catalyst. For example, metal ions catalyze the hydrolysis and oxidation reactions in biochemical systems (Martell, 1971--Footnote 5), and clays catalyze numerous acid-base and redox reactions involving organic compounds (Laszlo, 1987).[15]

3.4 Subsurface Environmental Parameters

The type and outcome of chemical reactions that will occur when contaminants reach the subsurface depend on the chemical characteristics of the contaminant and the environmental conditions that exist in the vadose and saturated zones. Six major environmental parameters of the subsurface must be considered when evaluating the fate of contaminants: (1) pH, (2) Eh, (3) salinity, (4) reservoir matrix, (5) temperature, and (6) pressure. The first four are chemical properties or measures of chemical properties of a system, and provide information on what types of chemical reactions may occur and how the reactions might be expected to proceed. The last two, temperature and pressure, are physical properties of the system that primarily influence the rate of chemical reactions.

3.4.1 pH

The symbol *pH* stands for the negative logarithm of the hydronium ion [H_3O^+] concentration and is a convenient way of expressing the very low concentrations of H_3O^+ that are present in aqueous solutions. In chemical reactions, the term H^+ is often used instead of H_3O^+. Pure water has a pH of 7.0. Solutions of pH less than 7.0 are acidic, and those of pH greater than 7.0 are basic. Acid-base reactions (Section 3.2.1) determine the pH of a solution at equilibrium.

---

[14] Mortland, M.M. and L.J. Halloran. 1976. Polymerization of Aromatic Molecules on Smectite. Soil Sci. Soc. Amer. J. 40:367-370.

[15] Laszlo, P. 1987. Chemical Reactions on Clays. Science 235:1473-1477.

The pH of a system greatly influences what chemical processes will occur in the subsurface environment. Directly or indirectly, pH also affects most of the other environmental factors that are discussed in this chapter. Table 3-8 summarizes the significance and some major effects of changes in pH on chemical processes and environmental factors in the subsurface.

Very small changes in acidity greatly affect chemical reactions and the form of chemical species in solution. Figure 3-2 illustrates how pH influences the distribution of molecular and ionic species of cadmium, mercury, and lead respectively.

Table 3-8 Effects of pH on Subsurface Geochemical Processes and Other Environmental Factors

Process/Factor	pH Effect
DISTRIBUTION PROCESS	
Acid-base	Measures acid-base reactions. Strong acids (bases) will tend to change pH; weak acids (bases) will buffer solutions to minimize pH changes.
Sorption	Strongly influences sorption, because hydrogen ions play an active role in both chemical and physical bonding processes. Mobility of heavy metals is strongly influenced by pH. Adsorption rates of organics may also be pH dependent.
Precipitation-dissolution	Strongly influences precipitation-dissolution reactions. Mixing of solutions with different pH often results in precipitation reactions. See also reservoir matrix below.
TRANSFORMATION PROCESS	
Complexation	Strongly influences positions of equilibria involving complex ions and metal chelate formation.
Hydrolysis	Strongly influences rates of hydrolysis. Hydrolysis of aliphatic and alkylic halides optimum at neutral to basic conditions. Other hydrolysis reactions tend to be faster at either high or low pH.
Oxidation-reduction	Redox systems generally become more reducing with increasing pH.

Table 3-8 (cont.)

Process/Factor	pH Effect
	ENVIRONMENTAL FACTOR
Biotransformation	In combination with Eh, strongly influences the types of bacteria that will be present. High- to medium-pH, low-Eh environments will generally restrict bacterial populations to sulfate reducers and heterotrophic anaerobes. In reducing conditions, pH strongly affects whether methanogenic or sulfate-reducing bacteria predominate.
Eh	Increasing pH generally lowers Eh.
Salinity	pH-induced dissolution increases salinity; pH-induced precipitation decreases salinity.
Reservoir matrix	Acidic solutions tend to dissolve carbonates and clays; highly alkaline solutions tend to dissolve silica and clays. Greater pH generally increases cation-exchange capacity of clays.
Temperature	pH-driven exothermic (heat-releasing) reactions will increase fluid temperature; pH-driven endothermic (heat-consuming) reactions will decrease fluid temperature.
Pressure	Will not influence pressure unless pH-induced reactions result in a significant change in the volume of reaction products.

Source: Boulding (1990).

Buffer capacity is a measure of how much the pH changes when a strong acid or base is added to a solution. A highly buffered solution will show little change in pH with the addition of a strong acid or base. Conversely, the pH of a solution with low buffering capacity will change rapidly if an acid or base is added to that solution. Weak acids or bases buffer a solution, and the higher their concentration in solution, the greater the buffering capacity. *Alkalinity*, which is usually expressed in calcium carbonate equivalents required to neutralize acid to a specified pH, is a measure of the buffering capacity of a solution.

Acid-base equilibrium reactions of buffers act to either add or remove hydrogen ions to or from the solution so as to maintain a nearly constant equilibrium concentration of H^+. For example, carbon dioxide acts as a buffer when it dissolves in water to form carbonic acid, which then dissociates to carbonate and bicarbonate

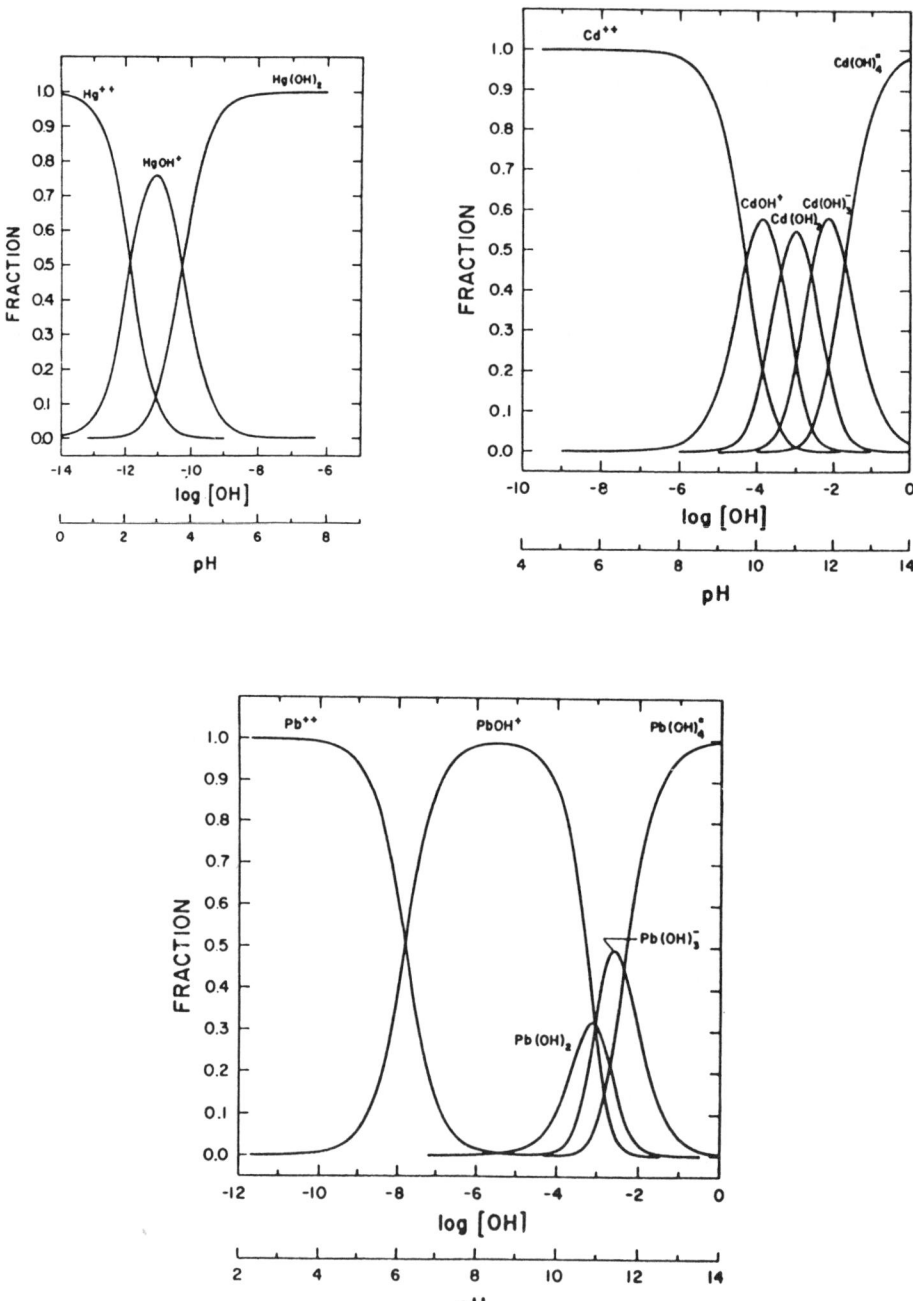

Figure 3-2 Distribution of molecular and ionic species of divalent cadmium, mercury and lead at different pH values (Boulding, 1990, after Hahne and Kroontje, 1973).

ions according to the following formula:

$$CO_2(aq) + H_2O ===> H_2CO_3 <===> HCO_3^- + H^+ <===> CO_3^{-2} + 2H^+$$
(carbonic acid)   (carbonate)   (bicarbonate)

At equilibrium, the concentration of H$^+$ will remain constant. When a strong acid (represented by H$^+$) is introduced into solution, the concentration of H$^+$ is increased. The buffer compensates by reacting with the excess H$^+$ ions, moving the direction of the above reaction to the left. By combining with bicarbonate and carbonate ions to form the nonionic carbonic acid, equilibrium is reestablished at a pH nearly the same as that existing before the acid was added. The buffer capacity in this case is determined by the total concentration of carbonate and bicarbonate ions. When carbonate or bicarbonate ions are no longer available to combine with excess H$^+$ ions, the buffer capacity has been exceeded and pH will decrease rapidly upon addition of further acid.

3.4.2 Eh and other Redox Indicators

The term Eh, which is the *oxidation-reduction potential* (often referred to as *redox potential*), is an expression of the tendency of a reversible redox system to be oxidized or reduced. It is especially significant in its influence on biotransformation processes (see Section 3.5.3). The energy of oxidation (electron-escaping tendency) present in a reversible oxidation-reduction system (in volts (V) or millivolts (mV)) is measured as the potential difference between a standard hydrogen electrode and the system being measured. Large positive values (up to about +800 mV) indicate a strong oxidizing tendency, and large negative values (down to about -500 mV) indicate a strong reducing tendency. Eh values of +200 mV and lower are used as indicators of reducing conditions in near-surface soils and sediments (Ponnamperuma, 1972).[16] Figure 3-3 shows typical Eh-ph ranges of natural surface and subsurface aquatic environments.

As long as dissolved oxygen is present in ground water, aerobic respiration should be possible (Section 3.3.3). Typically the bulk of dissolved oxygen in recharge water is consumed in the soil and unsaturated zone by microbial respiration and the decomposition of organic matter. Dissolved oxygen concentration may be expected to be highest in ground water near recharge zones. In general, redox conditions will become progressively more reducing as water travels into deeper aquifers. Figure 3-4 illustrates this process. Sand and gravel aquifers will tend to have higher levels of dissolved oxygen than silty and clayey materials. Winograd and Robertson (1982) reported dissolved oxygen values from 2 to 8 mg/L from a variety of deep (100 to 1,000 meters) aquifers, so reducing conditions should not always be assumed on the basis of depth or distance from the recharge areas alone.[17]

---

[16] Ponnamperuma, F.N. 1972. The Chemistry of Submerged Soils. Adv. Agron. 24:29-98.

[17] Winograd, I.J. and F.N. Robertson. 1982. Deep Oxygenated Ground Water: Anomaly or Common Occurrence? Science 216:1227-1229.

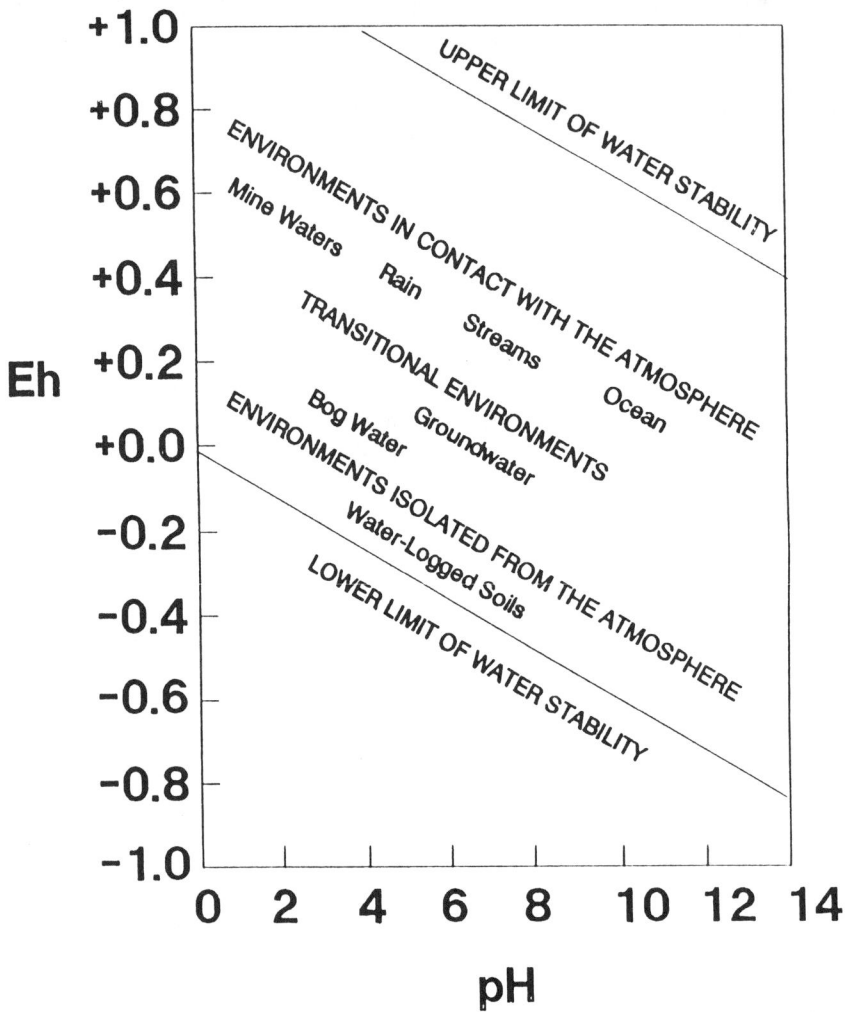

Figure 3-3 pH-Eh diagram showing the ranges of various aquatic environments (Johnson et al., 1989).

The Eh of connate waters (i.e., water entrapped in the interstices of sediment at the time of deposition) typically ranges from 0 to -200 mV (Baas-Becking et al., 1960).[18] Barcelona et al. (1989) provide a good review of the literature on horizontal and vertical gradients in subsurface oxidation-reduction conditions at both site and regional scales.[19]

---

[18] Baas-Becking, L.G.M., I.R. Kaplan, and D. Moore. 1960. Limits of the Natural Environment in Terms of pH and Oxidation-Reduction Potentials. J. Geology 68(3):243-284.

[19] Barcelona, M.J., T.R. Holm, M.R. Schock, and G.K. George. 1989. Spatial and Temporal Gradients in Aquifer Oxidation-Reduction Conditions. Water Resources Res. 25(5):991-1003.

Soil and Ground-Water Chemistry and Microbiology

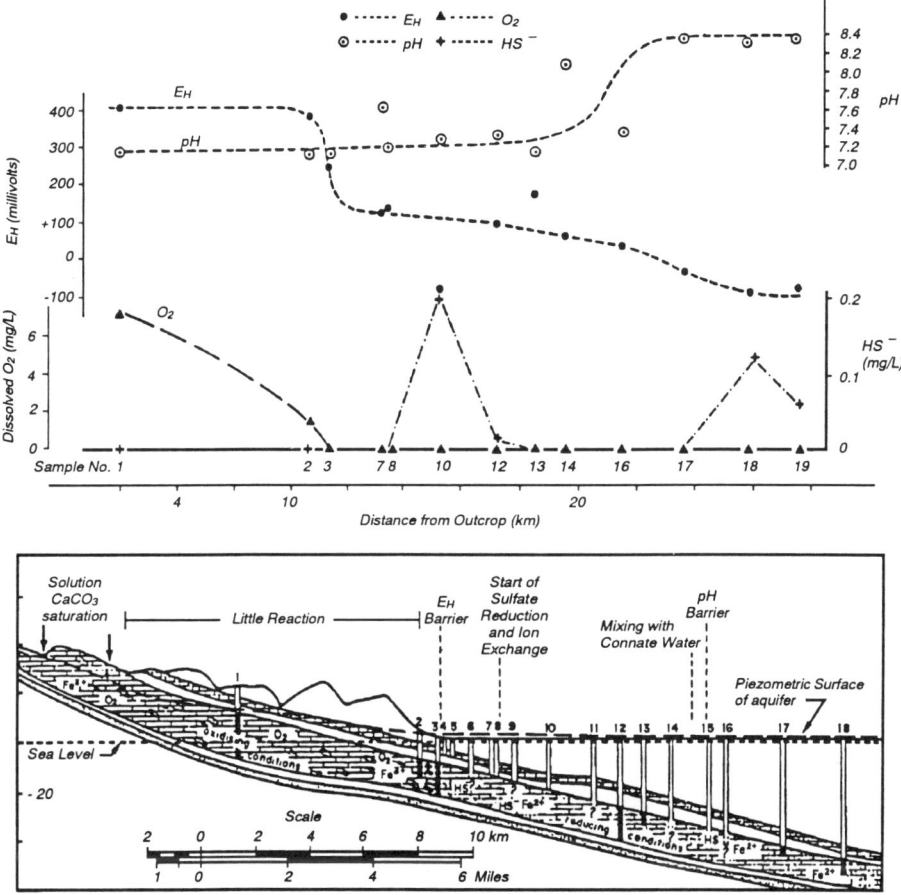

Figure 3-4 Horizontal gradients in uncontaminated oxidation-reduction conditions (Boulding and Barcelona, 1991b, after Champ et al., 1979).

Several measures of organic pollutant loading to waters indicate the redox status of a system: (1) biochemical oxygen demand (BOD), (2) chemical oxygen demand (COD), (3) total organic carbon (TOC), (4) dissolved organic carbon (DOC), and (5) suspended organic carbon (SOC).[20] When high values exist for any of these parameters, oxygen may be rapidly depleted in ground waters and reducing conditions will develop. BOD and COD were designed to measure oxygen consumption during the microbial degradation of municipal sewage. They are, however, only semiquantitative indicators of organic loading because measurement procedures for

---

[20] SOC may also be used as an abbreviation for soil organic carbon. The meaning of the abbreviation should usually be evident from the context of the discussion.

these parameters have no particularly direct geochemical significance. DOC and SOC are the most useful parameters for geochemical interpretation.[21]

Introduction of organic contaminants usually reduces Eh because microbiological activity rapidly depletes available oxygen unless toxic effects inhibit growth (see Section 3.5.3). On the other hand, artificial recharge of ground water by injection of treated wastes tends to create aerobic conditions in the immediate vicinity of injection wells. For example, Ragone et al. (1973) observed a change from reducing to oxidizing conditions when tertiary-treated sewage (reclaimed water) was injected into the Magothy aquifer, Long Island, New York, at a depth of 400 feet. The reclaimed water had 6.6 mg/L dissolved oxygen compared to no detectable dissolved oxygen in the formation water.[22]

3.4.3 Salinity

Salinity is defined as the concentration of total dissolved solids (TDS) in a solution, usually expressed in mg/L. The TDS concentration in water is usually determined from the weight of the dry residue remaining after the volatile portion of the original solution evaporates. Ground water can be classified into four salinity classes:

- Slightly saline (1,000 to 3,000 mg/L)

- Moderately saline (3,000 to 10,000 mg/L)

- Very saline (10,000 to 35,000 mg/L)

- Brine (more than 35,000 mg/L)

Seawater lies in the upper range of very saline water (about 34,000 mg/L); deep sedimentary basins may have salinities as high as 300,000 mg/L (Kreitler, 1986).[23] Most contamination of near-surface aquifers involves slightly to moderately saline waters. Pumping of ground water where denser seawater is present at greater depths may contaminate an aquifer by saltwater intrusion.

---

[21] Malcolm, R.L. and J.A. Leenheer. 1973. The Usefulness of Organic Carbon Parameters in Water Quality Investigations. In: Proc. of the Inst. of Env. Sciences 1973 Annual Meeting, Anaheim, California, April 1-6, pp. 336-340.

[22] Ragone, S.E., J. Vecchioli, and H.F.H. Ku. 1973. Short-Term Effect of Injection of Tertiary-Treated Sewage on Iron Concentration of Water in Magothy Aquifer, Bay Park, New York. In: Symposium on Underground Waste Management and Artificial Recharge, J. Braunstein (ed.), Pub. No. 110, Int. Assoc. of Hydrological Sciences, pp. 273-290.

[23] Kreitler, C.W. 1986. Hydrogeology of Sedimentary Basins as It Relates to Deep-Well Injection of Chemical Wastes. In: Proc. Int. Symp. on Subsurface Injection of Liquid Wastes, National Water Well Association, Dublin, OH, pp. 398-416.

Solutions of substances that are good conductors of electricity are called *electrolytes*. For example, sodium chloride, a major constituent of seawater, is a strong electrolyte. Most salts, as well as strong acids and bases, are strong electrolytes because they remain in solution primarily in ionic (charged) forms. Weak acids and bases are weak electrolytes because they do not fully dissociate into ionic species in solution. Pure water is a nonconductor of electricity. The high concentration of ions in very saline and briny waters creates more complex thermodynamic properties of the fluid and hence makes geochemical modeling more difficult. In evaluating the geochemistry of near-surface aquifers, these kinds of interactions often require special consideration.

The electrical conductivity of solutions is measured as *specific conductance*, usually expressed as microsiemens per centimeter (μS/cm-SI units) or micromhos per centimeter (μmhos/cm) at 25°C. Seawater has a specific conductance of about 50,000 μmhos/cm. Specific conductance shows a high correlation with salinity at low to moderate TDS levels, and is thus a very useful parameter to measure in the field characterization of ground-water systems.

3.4.4 Soil and Aquifer Matrix

The solid matrix of aquifer is typically unconsolidated or lithified sediments (see Sections 1.1.3 and 1.1.4). The reactions that take place when a contaminant enters an aquifer are largely determined by the physical and chemical properties of the aquifer solids. The most important physical properties of subsurface materials in relation to geochemical interactions are *texture* (Section 1.1.2) and *specific surface area*. The most important chemical property is *mineralogy*, which is defined by the types and proportions of minerals present (Section 1.1.1). *Organic matter*, the most geochemically reactive portion of the solid phase in the subsurface, is the focus of this section.

There are two major classes of organic matter: (1) nonhumic substances, which are largely water soluble, of low molecular weight, and susceptible to microbial metabolism, and (2) humic substances, which are largely water insoluble, of high molecular weight, and relatively resistant to further microbial degradation. Table 3-9 shows further subdivisions of these classes and estimated half-lives for decomposition.

Humic substances consist of an operationally-defined class of biogenic, refractory, yellow-black organic substances that occur in all terrestrial and aquatic environments. Although they have been studied by scientists for about 200 years, no fundamental, or even generally accepted, understanding of the nature, origin, and geochemical role of humic substances has been developed (Aiken et al., 1985).[24]

---

[24] Aiken, G.R., D.M. McNight, R.L. Wershaw, and P. MacCarthy. 1985. An Introduction to Humic Substances in Soil, Sediment, and Water. In: Humic Substances in Soil, Sediment, and Water: Geochemistry, Isolation, and Characterization, G.R. Aiken, D.M. McKnight, R.L. Wershaw, and P. MacCarthy (eds.), John Wiley & Sons, New York, pp. 1-9.

Humic substances are classified into three major groups: (1) humic acids (insoluble at <pH 2, soluble at higher pH), (2) fulvic acids (soluble under all pH conditions), and (3) humins (insoluble residue). Humic acids have colloidal properties, and their structure is based primarily on six-membered aromatic and heterocyclic rings, which may include benzene, naphthalene and anthracene (Manaskaya and Drozdova, 1968).

Table 3-9 Half-Lives of Different Forms of Organic Matter

Form	Half-Life (Years)
Nonhumic Substances	
Decomposable plant material	0.17
Living biomass	1.7
Woody plant material	2.3
Humic Substances	
Physically stabilized	50
Chemically stabilized	1,000

Source: Boulding (1990), after Jenkinson and Raynor (1977).

The mobility and toxicity of contaminants in the subsurface may be affected by a number of important chemical properties and characteristics of humic substances:

- High sorption capacity for metals and organic pollutants

- Ability to readily form complexes with heavy metals

- Ability to incorporate organic pollutants with similar structures to the building blocks of humus (such as chlorinated phenols, naphtholic compounds and halogenated anilines) when humus forms

- Ability to solubilize organic compounds that are otherwise water insoluble

- Ability to increase hydrolysis reactions as a catalyst or, conversely, to slow the rate of hydrolysis reactions by sorption

- Ability to affect the rate and pathways of oxidation-reduction reactions

Humic substances dissolved in ground water are important as substrates for microorganisms and may be geochemically significant as complexing agents. The amounts of dissolved organic matter in ground waters are sufficient to support small but diverse populations of microorganisms that may be able to adapt and degrade organic contaminants (see Section 3.5). Complexation of heavy metals by humic substances will generally be limited by the low dissolved organic carbon content of most ground waters.

Except as noted below, most ground waters contain less than 1 mg dissolved organic carbon/L. Thurman (1985),[25] using data primarily from Leenheer et al. (1974),[26] reports the following median concentrations of organic carbon in various types of aquifers: sand and gravel, limestone, and sandstone, 0.7 mg/L; igneous, 0.5 mg/L; oil shales, 3.0 mg/L; organically rich recharge waters, 10.0 mg/L; and petroleum-associated waters, 100.0 mg/L. Dissolved organic carbon represents typically less than 1% of the total organic carbon in an aquifer system. The bulk of the carbon in the aquifer solids remains uncharacterized.

3.4.5 Temperature and Pressure

Temperature and pressure are environmental factors that primarily influence the rate of chemical reactions. *Temperature* is measured in degrees using three main temperature scales: Fahrenheit (°F), Centigrade (°C), and Kelvin (°K). *Pressure* is measured in a variety of units, the most common scales being bars, atmospheres (atm), megapascals (MPa), and pounds per square inch (psi). Both temperature and pressure increase with depth below the earth's surface. Consequently, temperatures and pressures in deep aquifers are significantly higher than those encountered in near-surface aquifers and should receive particular attention when assessing the geochemical fate of deep-well injected wastes.

The rates of most acid-base and dissolution reactions increase as temperature increases. Increased temperature often increases the rate of redox reactions as well. However, the exact effect is difficult to predict because the interactions between competing reactions may offset the effect of increased temperature (Valentine, 1986--Footnote 8).

---

[25] Thurman, E.M. 1985. Humic Substance in Groundwater. In: Humic Substances in Soil, Sediment, and Water: Geochemistry, Isolation, and Characterization, G.R. Aiken, D.M. McKnight, R.L. Wershaw, and P. MacCarthy (eds.), John Wiley & Sons, New York, pp. 87-103.

[26] Leenheer, J.A., R.L. Malcolm, P.W. McKinley, and L.A. Eccles. 1974. Occurrence of Dissolved Organic Carbon in Selected Groundwater Samples in the United States. J. Res. U.S. Geological Survey 2:361-369.

3.5 Microbial Ecology of the Subsurface

3.5.1 Classification of Microorganisms

Microorganisms are by far the most significant group of organisms involved in biotransformation. Numerous biochemical pathways for degradation are present in this group because high rates of reproduction and mutation contribute to a great diversity of species, adapted strains, and enzyme systems.

Microorganisms fall into three main groups: (1) *bacteria* (single-cell organisms), (2) *actinomycetes* (filamentous unicellular organisms with characteristics of both bacteria and fungi), and (3) *fungi* such as mold and yeasts. Bacteria are dominant in the saturated zone; actinomycetes and fungi may be very active in the soil rooting zone.

Bacteria may be classified as either *heterotrophic*, requiring organic matter for growth, or *autotrophic*, capable of growth on inorganic carbon (carbon dioxide). They may also be *aerobic*, requiring oxygen for respiration, or *anaerobic*, not requiring oxygen for respiration.

Anaerobic microorganisms may be either obligately anaerobic or facultatively anaerobic. Oxygen is toxic to *obligate* microorganisms, whereas *facultative* microorganisms can live with or without oxygen. Groups of anaerobic microorganisms that may be important in the degradation of contaminants in the subsurface include:

- *Fermenters* degrade complex organic compounds to simpler organic compounds and may also mineralize organic compounds to hydrogen and carbon dioxide. Fermenters called *acetogens* produce acetate from biomonomers (sugars, amino acids, peptides, and alcohols), which can then be used by sulfate-reducing and methanogenic bacteria. Acetogens are particularly important in this group.

- *Denitrifiers* mineralize an appropriate carbon source in the presence of nitrate to form nitrogen, carbon dioxide, and water (see equation 2, Table 3-5).

- *Sulfate reducers* can mineralize an appropriate carbon source in the presence of sulfate and hydrogen to form hydrogen sulfide, carbon dioxide, and water (see equation 5, Table 3-5).

- *Methanogens* can mineralize an appropriate carbon source (for example, acetate or methylated amines) to form methane and carbon dioxide (see equation 6, Table 3-5). Methanogens may also use carbon dioxide as a growth source in the presence of hydrogen (H_2), producing methane as a byproduct.

Iron- and *manganese-reducing bacteria* are another important group of bacteria that may significantly affect geochemical reactions in ground-water (Ehrlich, 1987;

Ghiorse, 1988; Lovely, 1987).[27] However, the role that these bacteria may have in degrading organic contaminants has not received much attention by researchers.

3.5.2 Natural Biological Activity in the Subsurface

Microorganisms are ubiquitous in the shallow and deep subsurface. Microorganisms have adapted to live in essentially the complete range of environmental conditions that exist on and below the earth's surface. They have been observed at pressures up to 25,000 psi, temperatures up to 100°C, and salt concentrations up to 300,000 mg/L (Kuznetsov et al., 1963/T3-12). In recent years thermophilic bacteria have been isolated from marine and terrestrial geothermal sites whose optimum temperature for growth approaches and in some cases exceeds 100°C (Borman, 1991).[28] Ghiorse and Wilson (1988/T3-12) summarized data on studies of the types, abundance, and activities of microorganisms in subsurface materials in pristine sites. Depths of ground water sampled in these studies ranged from 1.2 to 1,752 meters, and included a wide diversity of geologic materials (unconsolidated sands, gravels, clays, sandstone, and limestone). In all instances, evidence of microbiological activity was found at the maximum depth sampled. Typically, viable counts of microorganisms ranged from 10^3 to 10^6 viable organisms/gram.

Particle size is a major factor affecting microbiological activity in the subsurface. Messineva (1962), in studies on the geological activity of bacteria in the Soviet Union, found that mineralization of organic matter by bacteria occurs most rapidly in sand-silt sediments.[29] In clay and clay-silt sediments, the process of mineralization slows down, despite the fact that the number of bacteria in the clay sediments is considerably greater than in sand-silt sediments. Other major environmental factors that affect subsurface microbial activity include: nutrient availability, redox conditions, nutritional ecology, and surface activity of the solid matrix.

3.5.3 Aerobic vs. Anaerobic Degradation

In aerobic degradation, oxygen serves as an electron acceptor and is converted to water, releasing energy that can be used by microorganisms in the process. More

---

[27] Ehrlich, H.L. 1987. Manganese Oxide Reduction as a Form of Anaerobic Respiration. Geomicrobiology J. 5:423-429.
Ghiorse, W.C. 1988. Microbial Reduction of Manganese and Iron. In: Biology of Anaerobic Microorganisms, A.J.B. Zehnder (ed.), John Wiley & Sons, New York, pp. 305-331.
Lovely, D.R. 1987. Organic Mineralization with the Reduction of Ferric Iron: A Review. Geomicrobiology J. 5:375-399.

[28] Borman, S. 1991. Bacteria that Flourish Above 100°C Could Benefit Industrial Processing. Chemical and Engineering News, November 4, pp. 31-34.

[29] Messineva, M.A. 1962. The Geological Activity of Bacteria and Its Effect on Geochemical Processes. In: Geologic Activity of Microorganisms, S.I. Kuznetsov (ed.), Transactions of the Institute of Microbiology No. IX (trans. from Russian), Consultants Bureau, New York, pp. 6-24.

energy can be obtained from this reaction (125.1 kJ) than from competing potential anaerobic reactions. Consequently, this type of reaction will predominate as long as dissolved oxygen is present in the ground water. The toxicity of oxygen to obligate anaerobes also inhibits anaerobic activity in the presence of oxygen.

Once the oxygen is depleted, anaerobic degradation reactions may follow the equilibrium sequence shown in Table 3-5 using nitrate, Mn(IV), Fe(III), sulfate, nitrate (again, if all the nitrate has not been consumed earlier in the sequence), and carbon dioxide as electron acceptors. The reaction sequence is listed in order of energy released, with nitrate reduction (denitrification) releasing the most (118.8 kJ), and methane fermentation the least (23.2 kJ). Thermodynamics thus favors bacteria that participate in higher energy reactions over lower energy reactions, as long as the higher energy electron acceptors are available. Once higher energy electron acceptors are depleted, biological activity will focus on reactions the next step down in the reaction sequence. The redox reaction sequence in Table 3-5 shows only some of the most common reactions in ground water. Zehnder and Stumm (1988) list a total of 23 redox reactions pertinent in aquatic conditions.[30] Electron acceptors other than the ones already mentioned include NO_2^-, S, CH_2O, CH_3OH, and $HCOO^-$.

Decomposition of organic matter in anaerobic environments often depends on the interaction of metabolically different bacteria. Degradation in this situation is a multistep process in which complex organic compounds are first degraded to short chain acids by facultative bacteria (see Figure 3-5). Acetogenic bacteria degrade the acids to produce hydrogen (H_2) and acetate. H_2, CO_2, and acetate are termed *competitive substrates* because they can be metabolized by both methanogenic and sulfate-reducing bacteria.

Where sulfate is present, sulfate reduction will be favored over methane production. Winfrey and Zeikus (1977) have shown that sulfate inhibits methanogenesis in freshwater sediments by altering normal carbon and electron flow during anaerobic mineralization.[31] They proposed that sulfate reducers assume the role of methanogens in sulfate-containing sediments by metabolizing methanogenic precursors. Sulfate-reducing bacteria will outcompete methanogens for H_2 and acetate. Martens and Berner (1974) found that methanogenesis in marine sediments is not initiated until sulfate is depleted from interstitial water.[32] Where sulfate is absent, methanogenic bacteria will be the end stage of the biodegradation process.

---

[30] Zehnder, A.J.B. and W. Stumm. 1988. Geochemistry and Biogeochemistry of Anaerobic Habitats. In: Biology of Anaerobic Microorganisms, A.J.B. Zehnder (ed.), John Wiley & Sons, New York, pp. 1-38.

[31] Winfrey, M.R. and J.G. Zeikus. 1977. Effect of Sulfates on Carbon and Electron Flow During Microbial Methanogenesis in Freshwater Sediments. Appl. Environ. Microbiol. 33:275-281.

[32] Martens, C.S. and R.A. Berner. 1974. Methane Production in the Interstitial Waters of Sulfate-Depleted Marine Sediments. Science 185:1167-1169.

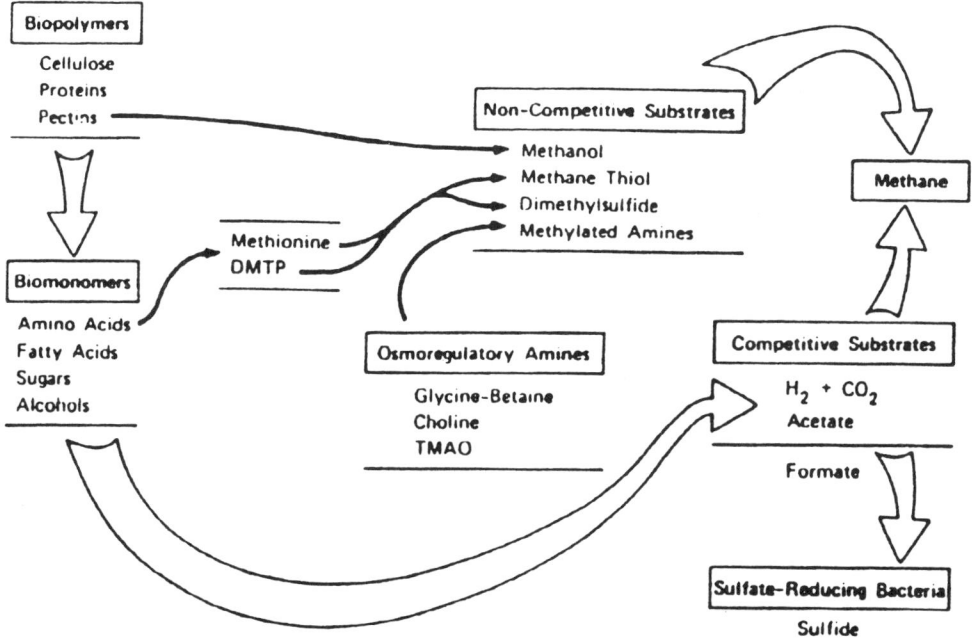

Figure 3-5 Anaerobic food web for aquatic microbial ecosystems (Oremland, 1988).

Bivalent cations may inhibit sulfate-reducing bacteria, although the reason is not understood (Kuznetsov et al., 1963/T3-12). Porter (1946/T3-12) reports the following sequence of relative inhibition of sulfate-reducing bacteria by cations: Na < K < NH_4 < Mg < Ca < Ba < Mn < Fe^{+2} < Zn < Fe^{+3} < Al < Pb < Cu < Hg < Ag.

Methanogenic bacteria may survive in a wide range of environmental conditions, but require more restricted conditions for optimal growth. The optimum temperature range for methane production by species in sediments of Lake Mendota, Wisconsin, measured in vitro was 35° to 42°C, well above the maximum in situ temperature of 23°C (Zeikus and Winfrey, 1976).[33] DiTommaso and Elkan (1973) report that 20 to 50 micrograms of iron per liter are required for methane fermentation.[34] Wolin and Miller (1987) describe in detail the interactions between acetogenic bacteria and

---

[33] Zeikus, J.G. and M. Winfrey. 1976. Temperature Limitation of Methanogenesis in Aquatic Sediments. Appl. Environ. Microbiol. 31:99-107.

[34] DiTommaso, A. and G.H. Elkan. 1973. Role of Bacteria in Decomposition of Injected Liquid Waste at Wilmington, North Carolina. In: Symp. on Underground Waste Management and Artificial Recharge, J. Braunstein (ed.), IASH Pub. No. 110, pp. 585-599.

methanogens involved in the complete bioconversion of organic carbon to methane and carbon dioxide.[35]

The addition of sulfate and other compounds (nitrate, nitrite, acetylene) to sediments has been shown to inhibit methanogenesis (Zeikus, 1977).[36] Cappenberg (1975) has suggested that the inhibition of methanogenesis by the presence of sulfates results from the production of toxic H_2S.[37]

The inhibiting effects of sulfates on methane production would seem to indicate that sulfate-reduction will take place in preference to methanogenesis as long as sulfates are present. However, the ecological significance of sulfate reducer-methanogen interrelationships is not well understood. Cappenberg (1975--Footnote 37) found that lactate metabolism by sulfate-reducing bacteria in the upper sulfate-containing sediment layers at Lake Vechten provides the main energy source for acetate-fermenting methanogens located lower in the sediment.

3.5.4 Biotransformation of Organic Contaminants

As noted in Section 3.5.2, most ground waters support a small but diverse population of microorganisms. In near-surface, pristine ground waters, aerobic bacteria will typically dominate unless unusually high concentrations of dissolved organic carbon (recharge from stagnant swamps, for example) have created more reducing conditions. However, denitrifiers, sulfate-reducers, and methanogens are likely to be present in low numbers. Microbiological activity in ground water is usually characterized by a dominant consortium of species favored by the predominant redox potential, with smaller numbers of other species in a quiescent state or thriving in microenvironments with differing redox potentials from the dominant one.

When a potential energy source in the form of an organic contaminant enters the water, the group most capable of utilizing the substrate at the environmental conditions existing in the aquifer will adapt and increase in population. The population of other indigenous microbes will remain small, or possibly be eliminated if the new environment is unfavorable for their survival. The microbial population will change as the favored electron acceptors are depleted.

Whether an organic contaminant will be biodegraded in the subsurface depends on (1) the existence of microorganisms capable of degrading the contaminant and (2) environmental conditions that favor the growth of these microbes. Since different contaminants may be degraded by different microorganisms, the concept of redox

---

[35] Wolin, M.J. and T.L. Miller. 1987. Bioconversion of Organic Carbon to CH_4 and CO_2. Geomicrobiology J. 5:239-259.

[36] Zeikus, J.G. 1977. The Biology of Methanogenic Bacteria. Bacteriol. Rev. 41:514-541.

[37] Cappenberg, T.E. 1975. A Study of Mixed Continuous Culture of Sulfate-Reducing and Methane-Producing Bacteria. Microbial Ecol. 2:60-72.

zones for biotransformation of micropollutants has been proposed by Bouwer and McCarty (1984).[38] Table 3-10 presents a hypothetical sequence for biotransformation of microcontaminants in treated sewage effluent that is injected into ground water. Any contaminants that are not degraded in their appropriate redox zone will tend to migrate unchanged once they enter subsequent zones (unless other distribution or transformation processes occur).

Table 3-10 Redox Zones for Biotransformation of Organic Micropollutants

Increasing Distance from Pollutant Source ---->

Biological Conditions			
Aerobic heterotrophic respiration	Denitrification	Sulfate respiration	Methanogenesis

Organic Pollutants Transformed			
Chlorinated benzenes Ethylbenzene Styrene Naphthalene	Carbon tetrachloride Bromodichloromethane Dibromochloromethane Bromoform		C_1 and C_2 Halogenated aliphatics (alkenes)

Source: Boulding (1990), after Bouwer and McCarty (1984).

Where contaminants are highly concentrated (as with deep-well injection of hazardous wastes), the redox zone model shown in Table 3-10 does not work because of toxic effects. In this situation, biotransformation is restricted to a relatively narrow moving front where contaminant concentrations are low enough to prevent toxic effects but high enough to allow significant growth of adapted microorganisms (see Figure 3-6). The sequence shown in this figure has been observed at a site in Wilmington, North Carolina, where wastes containing organic acids, formaldehyde and methanol were injected (Leenheer and Malcolm, 1973).[39]

---

[38] Bouwer, E.J. and P.L. McCarty. 1984. Modeling of Trace Organics Biotransformation in the Subsurface. Ground Water 22:433-440.

[39] Leenheer, J.A. and R.L. Malcolm. 1973. Case History of Subsurface Waste Injection of an Industrial Organic Waste. In: Symposium on Underground Waste Management and Artificial Recharge, J. Braunstein (ed.), Pub. No. 110, Int. Assoc. of Hydrological Sciences, pp. 565-584.

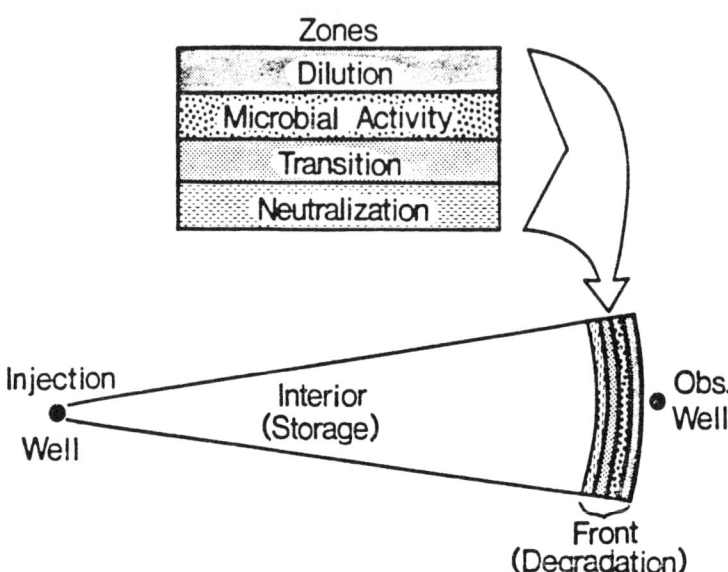

Figure 3-6 Geochemical zones with subsurface injection of concentrated toxic wastes (Boulding, 1990, after Leenheer and Malcolm, 1973).

The zone of microbial activity in Figure 3-6 may also develop redox zones similar to those shown in Table 3-10 but are likely to be reversed in order, since redox potential will be lowest where contaminant concentrations are highest. In fact, this reverse zonation is likely to exist where near-surface ground water is heavily contaminated. For example, Ehrlich et al. (1983) observed evidence of both denitrifying and methanogenic bacteria in phenol-depleted zones of a creosote-contaminated aquifer, and concluded that the denitrifying bacteria contributed to the degradation process.[40] Denitrifiers and iron reducers were the dominant anaerobes in contaminated wells, but methane production was highest in the most contaminated wells closest downgradient from the creosote source. Thus the redox zones shown in Table 3-10 were reversed, with methanogenic conditions strongest close to the contaminant source, grading to denitrifying conditions farther from the site, where contaminant concentrations were lower.

Table 3-11 identifies the susceptibility of 34 types of RCRA-regulated organic compounds to biodegradation by aerobic and anaerobic respiration, fermentation, oxidation and cooxidation.

---

[40] Ehrlich, G.G., E.M. Godsy, D.F. Goerlitz, and M.F. Hult. 1983. Microbial Ecology of a Creosote-Contaminated Aquifer at St. Louis Park, Minnesota. Dev. Ind. Microbiol. 24:235-245.

Table 3-11 Biodegradable RCRA-Regulated Compounds

Substrate Compounds	Respiration Aerobic	Respiration Anaerobic	Fermentation	Oxidation	Co-oxidation
Straight Chain Alkanes	+	+	+	+	+
Branched Alkanes	+	+	+	+	+
Saturated Alkyl Halides		+		+	+
Unsaturated Alkyl Halides		+		+	
Esters, Glycols, Epoxides	+	+	+	+	
Alcohols	+	+		+	
Aldehydes, Ketones	+	+		+	
Carboxylic Acids	+	+		+	
Amides	+	+			
Esters	+	+			
Nitriles	+	+			
Amines	+	+			
Phthalate Esters	+	+		+	
Nitrosamines		+			
Thiols					
Cyclic Alkanes	+		+	+	+
Unhalogenated Aromatics	+	+		+	
Halogenated Aromatics	+	+		+	+
Simple Aromatic Nitro Compounds	+	+			
Aromatic Nitro Compounds with Other Functional Groups	+	+			+
Phenols	+	+	+	+	+
Halogenated Side Chain Aromatics	+		+	+	
Fused Ring Hydroxy Compounds	+				
Nitrophenols		+			
Halophenols	+			+	
Phenols – Dihydrides Polyhydrides	+			+	+
Two & Three Ring Fused Polycyclic Hydrocarbons	+			+	
Biphenyls	+				
Chlorinated Biphenyls	+				
Four Ring Fused Polycyclic Hydrocarbons	+				
Five Ring Fused Polycyclic Hydrocarbons	+				
Fused Polycyclic Hydrocarbons	+				
Organophosphates	+	+			
Pesticides and Herbicides	+	+			

Source: U.S. EPA (1985).

Aerobic biotransformation of contaminants has been widely studied because one of the most common ground-water contaminants, gasoline, is composed of compounds that are readily degraded aerobically. Unpolluted near-surface aquifers typically contain enough oxygen for aerobic conditions to prevail. In general, organic contaminants are more readily degraded aerobically than anaerobically. For example, of about 90 examples of hazardous anthropogenic compounds and microorganisms that can degrade them, compiled by Kobayashi and Rittmann (1982/T4-5), less than a third involved anaerobic degradation processes. Groups most susceptible to aerobic degradation include chlorobenzenes, alkylbenzenes, phenol and alkyl phenols, chlorophenols, aliphatic hydrocarbons, and two- and three-ring polynuclear aromatic hydrocarbons. Hazardous organic compounds that are susceptible to biotransformation under anaerobic conditions include halogenated aliphatic hydrocarbons, brominated methanes, phenol and alkyl phenols, and chlorophenols.

Ten years ago anaerobic degradation of organic contaminants was considered to be relatively unimportant in ground water. Anaerobic biotransformation of organic contaminants has received increasing attention in recent years because of the recognition that (1) certain organic contaminants can be degraded anaerobically that are not amenable to aerobic degradation, and (2) reducing conditions are common in the subsurface, especially where ground water is contaminated.

3.6 Guide to Major References

Table 3-12 provides an index to major references in the following broad categories: (1) environmental chemistry, (2) soil and geochemistry, (3) aqueous geochemistry, and (4) biogeochemistry and subsurface microbial ecology. Sposito (1989) provides a good general introduction to soil chemistry. The geochemistry texts identified in Table 3-12 tend to focus more on the chemistry of rock-forming materials, which is useful for classification of geologic materials, but need to be used in combination with texts of soil mineralogy and weathering for developing geochemical interpretations of contaminated sites.

Stumm and Morgan (1981) is probably the most widely cited text on aqueous geochemistry. Good recent texts include Drever (1989) and Pankow (1991). Paul and Clark (1989) is a good source for more detailed information on soil microbiology and biochemistry. Section 4.7 discusses major references dealing with contaminant hydrogeology and microbiology of contaminated ground water.

Table 3-12 Index to Major References on Soil and Ground-Water Geochemistry and Microbiology*

Topic	References
Environmental Chemistry	Bonati (1993), Hutzinger (various dates), Fortescue (1980), Irgolic and Martell (1985), Manahan (1993, 1994), Thornton (1983); <u>Biogeochemistry</u>: Hallberg (1981), Ingerson (1973), Schlesinger (1991), Zajic (1969)
Soil and Geochemistry	
General	<u>Soil Chemistry</u>: Bear (1964), Bohn et al. (1985), Bollag and Stotzky (1993-soil biochemistry), Bolt (1982), Bolt and Burggenwert (1978), Bonneau and Souchier (1982), Cresser et al. (1993), Fairbridge and Finkl (1979), Greenland and Hayes (1978, 1980), Marshall (1964, 1977), Sparks (1986, 1989), Sposito (1989), Tan (1993, 1994); <u>Geochemistry</u>: Berner (1981), Brookins (1988), Fortescue (1980), Goldschmidt (1954), Hawkes and Webb (1962), Mason (1966), Rankama and Sahama (1950), Richardson and McSween (1989), Wedephol (1974); <u>Geochemical Thermodynamics</u>: Anderson and Crerar (1993), Nordstrom and Munoz (1985)
Soil Fertility and Nutrient Chemistry	Barber (1984), Baver (1968), Donahue et al. (1977), Dowdy et al. (1981), Plaster (1985), Soon (1985), Thompson and Troeh (1978), Tisdale and Nelson (1975), Troeh and Thompson (1993)
Mineralogy/Weathering	<u>Soil Mineralogy</u>: Marshall (1964, 1977), Dinauer (1977), Dixon and Weed (1989); <u>Clay Mineralogy</u>: Eslinger and Pevear (1988), Grim (1968), Theng (1974, 1979); <u>Rock Mineralogy</u>: Deer et al. (1977), Pirrson and Knopf (1958), Garrels and Christ (1965), Gribble and Hall (1992), Iler (1979-silica), Mason and Berry (1968), Reeves and Brosler (1978), Silverman and Ehrlich (1964), Van Johnso and Maxwell (1981); <u>Trace Elements</u>: See Table 4-5; <u>Weathering</u>: Colman and Dethier (1986), Kittrick (1986)
Soil Organic Matter/ Humic Substances	Aiken et al. (1985), Allard et al. (1991), Baker (1991a, 1991b, 1991c), Baver (1968), Beck et al. (1994), Broadbent (1953), Christman and Gjessing (1983), Kononova (1966), Kubát (1992), MacCarthy et al. (1990), Manaskaya and Drozdova (1968), Schnitzer and Khan (1972, 1978), Stevenson (1982), Tate (1987)
Sorption/Ion Exchange/ Complexation	<u>Chelation</u>: Barton and Hemming (1993); <u>Colloids</u>: Hiemenez (1986), Iler (1955), Van Olphen (1963, 1977); <u>Complexation</u>: Dzombak and Morel (1990), Theng (1979), <u>Ion Exchange</u>: Helfferich (1962), Marrinsky (1966); <u>Sorption</u>: Anderson and Rubin (1981); <u>Surface Chemistry</u>: Hiemenez (1986), Jaycock and Parfitt (1981), Sposito (1984), Davis and Hayes (1986), Theng (1974)

Table 3-12 (cont.)

Topic	References
Aqueous Geochemistry	
Hydro(geo)chemistry	Chapelle (1993), Drever (1989), Erikkson (1985), Faust and Aly (1981), Hem (1985), Lloyd and Heathcote (1985), Mazor (1990), Morel (1983), Pagendorf (1978), Pankow (1991), Stumm and Morgan (1981); <u>Symposia/Edited Volumes</u>: Back and Letaile (1981), Erikkson (1984), Faust and Hunter (1967), Ingerson (1973), Rice (1989), Stumm (1987); <u>Oilfield Waters</u>: Collins (1975)
Chemical Equilibria	Brookins (1988), Denbigh (1981), Garrels and Christ (1965), Harned and Owen (1958), Pytkowicz (1983), Sillen and Martell (1964), Smith and Missen (1982), Van Zeggeren and Storey (1970); <u>Electrochemistry</u>: Bard et al. (1985), Bockris and Reddy (1973), Davies (1967), Harned and Owen (1958), Robinson and Stokes (1965); <u>Soil Chemical Equilibria/Solution Chemistry</u>: Elprince (1986), Lindsay (1979), Sposito (1994); <u>Symposia/Edited Volumes</u>: Hem (1971), Bernard (1986), Jenne (1979), Melchior and Bassett (1990), Stumm (1967)
Other Topics	<u>Humic Substances</u>: Aiken et al. (1985), Baker (1991a, 1991b, 1991c), Beck et al. (1994), Christman and Gjessing (1983); <u>Hydrolysis</u>: Baes and Mesmer (1976); <u>Oxidation Potentials</u>: Brookins (1988), Latimer (1952)
Subsurface Microbial Ecology	
General Microbiology and Microbial Ecology	Atlas (1988), Atlas and Bartha (1987), Moat (1988), Pointdexter and Leadbetter (1986), Porter (1946), Rosswall (1973), Zehnder (1988); <u>Environmental/Contaminant Microbiology</u>: Mitchell (1991), see also Biogeochemistry above, and Table 4-5
Subsurface Microbiology, Biology, and Ecology	<u>Soil</u>: Alexander (1971, 1977), Barton and Hemming (1993), Crossley et al. (1991), Edwards et al. (1988), Hattori (1973), Killham (1994), Parr et al. (1981), Paul and Clark (1989), Metting (1993); <u>Ground Water</u>: Chapelle (1993); <u>Papers</u>: Ghiorse and Wilson (1988), Harvey and Widdowson (1993)
Microbiology of Natural Geologic Systems	Ehrlich (1981), Kuznetsov (1962), Kuznetsov et al. (1963), Odum and Singleton (1993), Silverman and Ehrlich (1964), Zajic (1969); <u>Petroleum Microbiology</u>: Atlas (1984), Davis (1967), Litchfield and Clark (1973)

* See Table 4-5 for major references on contaminant chemical behavior

Table 3-12 References (Appendix F contains references for figure and table sources)

Aiken, G.R., D.M. McKnight, R.L. Wershaw, and P. MacCarthy (eds.). 1985. Humic Substances in Soil, Sediment, and Water: Geochemistry, Isolation, and Characterization. John Wiley & Sons, New York, NY.

Alexander, M. 1971. Microbial Ecology. John Wiley & Sons, New York, NY. [Focusses on soil systems]

Alexander, M. 1977. Introduction to Soil Microbiology, 2nd ed. John Wiley & Sons, New York, NY.

Allard, B., H. Borén, and A. Grimvall (eds.). 1991. Humic Substances in the Aquatic and Terrestrial Environment. Springer-Verlag, New York, NY, 514 pp. [Proceedings of 1989 international symposium in Sweden]

Anderson, G.M. and D.A. Crerar. 1993. Thermodynamics in Geochemistry: The Equilibrium Model. Oxford University Press, New York, NY, 608 pp.

Anderson, M.A. and A.J. Rubin (eds.). 1981. Adsorption of Inorganics at Solid-Liquid Interfaces. Ann Arbor Science.

Atlas, R.M. (ed.). 1984. Petroleum Microbiology. Macmillan, New York, NY.

Atlas, R.M. 1988. Microbiology: Fundamentals and Applications, 2nd ed. Macmillan, New York, NY, 807 pp.

Atlas, R.M. and R. Bartha. 1987. Microbial Ecology: Fundamentals and Applications, 2nd ed. Benjamin/Cummings, Menlo Park, CA, 533 pp. [First edition published in 1981]

Back, W. and R. Letaile (eds.). 1981. Symposium on Geochemistry of Groundwater (26th Int. Geol. Cong., Paris, 1980). J. Hydrology 54(1-3). [Special Issue]

Baes, Jr., C.F. and R.E. Mesmer. 1976. The Hydrolysis of Cations. John Wiley & Sons, New York, NY, 489 pp.

Baker, R.A. (ed.). 1991a. Organic Substances in Sediments and Water, Vol. 1: Humics and Soils. Lewis Publishers, Chelsea, MI, 392 pp. [Proceedings of 1990 American Chemical Society symposium in Boston, 20 papers]

Baker, R.A. (ed.). 1991b. Organic Substances in Sediments and Water, Vol. 2: Process and Analytical. Lewis Publishers, Chelsea, MI, 547 pp. [Proceedings of 1990 American Chemical Society symposium in Boston, 22 papers]

Baker, R.A. (ed.). 1991c. Organic Substances in Sediments and Water, Vol. 3: Biological. Lewis Publishers, Chelsea, MI, 332 pp. [Proceedings of 1990 American Chemical Society symposium in Boston, 18 papers]

Barber, S.A. 1984. Soil Nutrient Bioavailability, A Mechanistic Approach. John Wiley & Sons, New York, NY, 398 pp.

(Table 3-12 Basic Geochemistry and Microbiology References)

Bard, A.J., R. Parsons, and J. Jordan. 1985. Standard Potentials in Aqueous Solution. Marcel Dekker, New York, NY, 834 pp.

Barton, L.L. and B.C. Hemming (eds.). 1993. Iron Chelation in Plants and Soil Microorganisms. Academic Press, New York, NY, 512 pp.

Baver, L.D. 1968. The Effect of Organic Matter and Fertility. John Wiley & Sons, New York, NY, 1,017 pp.

Bear, F. (ed.). 1964. Chemistry of the Soil. Reinhold Publishing Corp., New York, NY.

Beck, A.J., K.C. Jones, M.H.B. Hayes, and U. Mingelgrin. 1994. Organic Substances in Soil and Water: Natural Constituents and Their Influence on Contaminant Behavior. Lewis Publishers, Boca Raton, FL, 220 pp.

Bernard, M. et al. (eds.). 1986. The Importance of Chemical Speciation in Environmental Processes. Dahlem Workshop Report, Vol. 33, Springer-Verlag, New York, NY.

Berner, R.A. 1981. A New Geochemical Classification of Sedimentary Environments. J. Sedimentary Petrology 51:359-365.

Bockris, J.O'M. and A.K.N. Reddy. 1973. Modern Electrochemistry: An Introduction to an Interdisciplinary Area. Plenum Press, New York, NY.

Bohn, H.L., B.L. McNeal, and G.A. O'Connor. 1985. Soil Chemistry, 2nd ed. Wiley Interscience, New York, NY. [Introductory text with 11 chapters; first edition 1979]

Bollag, J.-M. and G. Stotzky. 1993. Soil Biochemistry. Marcel Dekker, New York, NY, 418 pp.

Bolt, G.H. (ed.). 1982. Soil Chemistry, Part B: Physico-Chemical Models, 2nd ed. Developments in Soil Science, Vol. 5. Elsevier, New York, NY, 528 pp.

Bolt, G.H. and M.G.M. Bruggenwert (eds.). 1978. Soil Chemistry, Part A: Basic Elements, 2nd ed. Developments in Soil Science, Vol. 5. Elsevier, New York, NY, 282 pp. [Advanced text with 10 chapters]

Bonati, L. et al. (eds.). 1993. Trends in Ecological Physical Chemistry. Elsevier, New York, NY, 380 pp.

Bonneau, M. and B. Souchier (eds.). 1982. Constituents and Properties of Soils. Academic Press, New York, NY.

Broadbent, F.E. 1953. The Soil Organic Fraction. Adv. Agronomy 5:153-183.

Brookins, D.G. 1988. Eh-pH Diagrams for Geochemistry. Springer-Verlag, New York, NY, 176 pp.

Chappelle, F.H. 1993. Ground-Water Microbiology and Geochemistry. John Wiley & Sons, New York, NY, 448 pp.

(Table 3-12 Basic Geochemistry and Microbiology References)

Christman, R.F. and E.T. Gjessing (eds.). 1983. Aquatic and Terrestrial Humic Materials. Ann Arbor Science, Ann Arbor, MI.

Collins, A.G. 1975. Geochemistry of Oilfield Waters. Elsevier, New York, NY, 496 pp.

Colman, S.M. and D.P. Dethier (eds.). 1986. Rates of Chemical Weathering of Rocks and Minerals. Academic Press, New York, NY, 608 pp.

Cresser, M., K. Killham, and A. Edwards. 1993. Soil Chemistry and Its Applications. Cambridge University Press, New York, NY, 203 pp.

Crossley, Jr., D.A. et al. (eds.). 1991. Modern Techniques in Soil Ecology. Elsevier, New York, NY, 512 pp. [40 papers presented at 1989 international workshop held in Athens, GA; also published in Volume 34 of Agriculture, Ecosystems and Environment]

Davies, C.W. 1967. Electrochemistry. Philosophical Library, London, UK, 234 pp.

Davis, J.B. 1967. Petroleum Microbiology. Elsevier, New York, NY.

Davis, J.A. and K.F. Hays (eds.). 1986. Geochemical Processes at Mineral Surfaces. ACS Symp. Series 323, American Chemical Society, Washington, DC.

Deer, W.A., R.A. Howe, and J. Zussman. 1977. An Introduction to the Rock Forming Minerals. Longman Grays Ltd., London, 528 pp.

Denbigh, K. 1981. The Principles of Chemical Equilibrium, 4th ed. Cambridge University Press, Oxford, 494 pp. [1st edition 1955, 2nd edition 1968, 3rd edition 1971]

Dinauer, R.C. (ed.). 1977. Minerals in Soil Environments. Soil Science Society of America, Madison, WI, 948 pp. [Replaced by 2nd edition, see Dixon and Weed, 1989 below]

Dixon, J.B. and S.B. Weed (eds.). 1989. Minerals in Soil Environments, 2nd ed. Soil Science Society of America, Madison, WI, 1,264 pp. [Replaces first edition published in 1977 and edited by R.D. Dinauer]

Donahue, R.L., R.W. Miller, and J.C. Shickluma. 1977. An Introduction to Soils and Plant Growth. Prentice-Hall, Englewood Cliffs, NJ.

Dowdy, R.H., D.E. Baker, J.A. Ryan, and V.V. Volk (eds.). 1981. Chemistry in the Soil Environment. ASA Sp. Pub. No. 40, American Society of Agronomy, Madison, WI, 259 pp. [Focusses on chemistry of elements accumulated by growing plants, their movement, sorption and equilibria in soils]

Drever, J.I. 1989. The Geochemistry of Natural Water, 2nd ed. Prentice-Hall, Englewood Cliffs, NJ.

Dzombak, D.A. and F.M.M. Morel. 1990. Surface Complexation Modelling, Hydrous Ferric Hydroxide. John Wiley & Sons, New York, NY, 393 pp.

(Table 3-12 Basic Geochemistry and Microbiology References)

Edwards, C.A., B.R. Stinner, and S. Rabatin. 1988. Biological Interactions in Soil. Elsevier, New York, NY, 380 pp. [More than 30 papers from workshop on interactions between soil-inhabiting invertebrates and microorganisms in relation to plant growth]

Ehrlich, H.L. 1981. Geomicrobiology. Marcel Dekker, New York, NY. [Focuses on microbiology of natural geological systems]

Elprince, A.M. (ed.). 1986. Chemistry of Soil Solutions. Van Nostrand Reinhold, New York, NY.

Eriksson, E. (ed.). 1984. Hydrochemical Balances of Freshwater Systems. IAHS Pub. No. 150. International Association for Hydrological Sciences, Washington, DC. [Includes 12 papers on water interaction with soil and rock]

Eriksson, E. 1985. Principles and Applications of Hydrochemistry. Chapman and Hall, New York, NY. [Special emphasis on use of environmental isotopes]

Eslinger, E. and D. Pevear. 1988. Clay Minerals for Petroleum Geologists and Engineers. SEPM Short Course Notes No. 22. Society of Economic Paleontologists and Mineralogists, Tulsa, OK.

Fairbridge, R.W. and C.W. Finkl, Jr. (eds.). 1979. The Encyclopedia of Soil Science, Part 1: Physics, Chemistry, Biology, Fertility, and Technology. Dowden, Hutchinson & Ross, Stroudsburg, PA, 646 pp.

Faust, S.D. and O.M. Aly. 1981. Chemistry of Natural Waters. Ann Arbor Science Publishers, Ann Arbor, MI.

Faust, S.D. and S.J. Hunter (eds.). 1967. Principles and Applications of Water Chemistry. John Wiley & Sons, NY.

Fortescue, J.A.C. 1980. Environmental Geochemistry: A Holistic Approach. Springer Verlag, New York, NY, 347 pp.

Garrels, R.M. and C.L. Christ. 1965. Solutions, Minerals and Equilibria. Freeman, Cooper & Co., San Francisco, CA.

Ghiorse, W.C. and J.T. Wilson. 1988. Microbial Ecology of the Terrestrial Subsurface. Adv. Appl. Microbiol. 33:107-172. [Literature review of natural and contaminated systems with more than 160 citations]

Goldschmidt, V.M. 1954. Geochemistry. Clarendon Press, Oxford, England, 730 pp.

Greenland, D.J. and Hayes, M.H.B. (eds.). 1978. The Chemistry of Soil Constituents. Wiley Interscience, New York, NY.

Greenland, D.J. and Hayes, M.H.B. (eds.). 1980. The Chemistry of Soil Processes. John Wiley & Sons, New York, NY, 520 pp.

(Table 3-12 Basic Geochemistry and Microbiology References)

Gribble, C.D. and A.J. Hall. 1992. Optical Mineralogy: Principles & Practice. UCL Press, London, 303 pp.

Grim, R.E. 1968. Clay Mineralogy, 2nd ed. McGraw-Hill, New York, NY, 596 pp.

Hallberg, R. (ed.). 1981. Environmental Biogeochemistry. Ecological Bulletin No. 35, Swedish Natural Science Research Council, Stockholm. [49 papers on biogeochemical activity in air, soil, and water]

Harned, H.S. and B.B. Owen. 1958. The Physical Chemistry of Electrolyte Solutions. Reinhold, New York, NY.

Harvey, R.W. and M.A. Widdowson. 1993. Microbial Distributions, Activities and Movement in the Terrestrial Subsurface: Experimental and Theoretical Studies. In: Interacting Processes in Soil Science, R.J. Wagenet, P. Baveye, and B.A. Stewart (eds.), Lewis Publishers, Boca Raton, FL.

Hattori, T. 1973. Microbial Life in the Soil. Marcel Dekker, New York, NY. [Introductory text]

Hawkes, H. and J. Webb. 1962. Geochemistry in Mineral Exploration. Harper and Row, New York, NY, 415 pp.

Helfferich, F. 1962. Ion Exchange. McGraw-Hill, New York, NY.

Hem, J.D. (ed.). 1971. Non-Equilibrium Concepts in Natural Water Chemistry, ACS Adv. in Chemistry Series 106, American Chemical Society, Washington, DC. [13 papers]

Hem, J.D. 1985. Study and Interpretation of the Chemical Characteristics of Natural Water, 3rd ed. U.S. Geological Survey Water-Supply Paper 2254. [Replaces 1970 second edition, WSP 1473].

Hiemenez, P.C. 1986. Principles of Colloid and Surface Chemistry. Marcel Dekker, New York, NY, 815 pp.

Hutzinger, O. Various Dates. Handbook of Environmental Chemistry. Volume 1: The Natural Environment and the Biogeochemical Cycles; Volume 2: Reactions and Processes. Springer-Verlag, New York, NY. [Editors of individual parts vary. Volume 1 covers the natural environment and biogeochemical cycles (Part A, 1980; Part B, 1982; Part C, 1984; Part D, 1985); Volume 2 covers reactions and processes (Part A, 1980; Part B, 1982; Part C, 1985; Part D, 1988)]

Iler, R.K. 1955. The Colloid Chemistry of Silica and Silicates. Cornell University Press, Ithaca, NY, 324 pp.

Iler, R.K. 1979. The Chemistry of Silica. John Wiley & Sons, New York, NY, 866 pp.

Ingerson, E. (ed.). 1973. Proceeding of the Symposium on Hydrogeochemistry and Biogeochemistry (Tokyo, 1970). Clarke Company, Washington, DC

(Table 3-12 Basic Geochemistry and Microbiology References)

Irgolic, K.J. and A.E. Martell (eds.). 1985. Environmental Inorganic Chemistry. VCH Publishers, Deerfield Beach, FL.

Jaycock, M.J. and G.D. Parfitt. 1981. Chemistry of Interfaces. Ellis Horwood, Chichester, England, 279 pp.

Jenne, E.A. (ed.). 1979. Chemical Modeling in Aqueous Systems: Speciation, Sorption, Solubility, and Kinetics. ACS Symp. Series 93. American Chemical Society, Washington, DC. [38 papers focusing on speciation, sorption, solubility and kinetics]

Killham, K. 1994. Soil Ecology. Cambridge University Press, New York, NY, 250 pp.

Kittrick, J.A. (ed.). 1986. Soil Mineral Weathering. Van Nostrand Reinhold, New York, NY, 269 pp.

Kononova, M.M. 1966. Soil Organic Matter: Its Nature, Its Role in Soil Formation and in Soil Fertility, 2nd ed. Pergamon Press, New York, NY, 574 pp. [Translated by T.Z. Nowakowlski and A.C.D. Newman; first edition published in 1961]

Kubát, J. (ed.). 1992. Humus, its Structure and Role in Agriculture and Environment. Elsevier, New York, NY, 201 pp. [Proc. 10th Int. Symp Humus et Planta, Prague, August, 1991]

Kuznetsov, S.I. (ed.). 1962. Geologic Activity of Microorganisms. Transactions of the Institute of Microbiology No. IX (trans. from Russian), Consultants Bureau, New York, NY. [20 papers on geological activity of organisms, primarily associated with petroleum deposits in the Soviet Union]

Kuznetsov, S.I., M.V. Ivanov, and N.N. Lyalikova. 1963. Introduction to Geological Microbiology. McGraw-Hill, New York, NY. [Text focusing on natural microbiological activity in subsurface geologic systems]

Latimer, W.M. 1952. Oxidation Potentials, 2nd ed. Prentice-Hall, Englewood Cliffs, NJ, 392 pp.

Lindsay, W.L. 1979. Chemical Equilibria in Soils. John Wiley & Sons, New York, NY, 449 pp. [Chemical equilibria for 20 elements in soils]

Litchfield, J.H. and L.C. Clark. 1973. Bacterial Activities in Ground Waters Containing Petroleum Products. API Publ. No. 4211, American Petroleum Institute, Washington, DC.

Lloyd, J.W. and J.A. Heathcote. 1985. Natural Inorganic Hydrochemistry in Relation to Groundwater. Oxford University Press, New York, NY.

MacCarthy, P., C.E. Clapp, R.L. Malcolm, and P.R. Bloom (eds.). 1990. Humic Substances in Soil and Crop Sciences: Selected Readings. Soil Science Society of America and American Society of Agronomy, Madison, WI, 304 pp.

Manahan, S.E. 1993. Fundamentals of Environmental Chemistry. Lewis Publishers, Boca Raton, FL, 864 pp.

(Table 3-12 Basic Geochemistry and Microbiology References)

Manahan, S.E. 1994. Environmental Chemistry, 6th ed. Lewis Publishers, Chelsea, MI, 832 pp.

Manaskaya, S.M. and T.V. Drozdova. 1968. Geochemistry of Organic Substances. Pergamon Press, New York, NY, 347 pp.

Marrinsky, J.A. (ed.). 1966. Ion Exchange, Vol. 1. Marcel Dekker, New York, NY.

Marshall, C.E. 1964. The Physical Chemistry and Mineralogy of Soils, Vol. 1: Soil Materials. John Wiley & Sons, New York, NY.

Marshall, C.E. 1977. The Physical Chemistry and Mineralogy of Soils, Vol. 2: Soils in Place. John Wiley & Sons, New York, NY.

Mason, B. 1966. Principles of Geochemistry, 3rd ed. 329 pp. [2nd edition 1958]

Mason, B. and L.G. Berry. 1968. Elements of Mineralogy. W.H. Freeman, San Francisco, CA, 550 pp.

Mazor, E. 1990. Applied Chemical and Isotopic Ground Water Hydrology. John Wiley & Sons, New York, NY, 256 pp.

Melchior, D.C. and R.L. Bassett. 1990. Chemical Modeling of Aqueous Systems II. ACS Symp. Series 416. American Chemical Society, Washington, DC, 556 pp. [See Jenne (1970) for first symposium]

Metting, Jr., F.B. (ed.). 1993. Soil Microbial Ecology. Marcel Dekker, New York, NY, 646 pp.

Mitchell, R. 1992. Environmental Microbiology. John Wiley & Sons, New York, NY, 434 pp.

Moat, A.G. 1988. Microbial Physiology, 2nd ed. John Wiley & Sons, New York, NY, 597 pp.

Morel, F.M.M. 1983. Principles of Aquatic Chemistry. Wiley Interscience, New York, NY.

Nordstrom, D.K. and J.L. Munoz. 1985. Geochemical Thermodynamics. Benjamin Cummings Publishing Co., Menlo Park, CA, 477 pp.

Odom, J.M. and R. Singleton, Jr. (eds.). 1993. The Sulfate-Reducing Bacteria: Contemporary Perspectives. Springer-Verlag, New York, NY, 280 pp.

Pagendorf, G.K. 1978. Introduction to Natural Water Chemistry. Marcel Dekker, New York, NY.

Pankow, J.F. 1991. Aquatic Chemistry Concepts. Lewis Publishers, Chelsea, MI, 683 pp.

Parr, J.F. (ed.). 1981. Water Potential Relations in Soil Microbiology. SSSA Special Publication 9, Soil Science Society of America, Madison, WI, 151 pp. [10 contributed chapters on water potential applied to soil microbiology and biochemistry, plant pathology and the microbial ecology of soils]

Paul, E.A. and F.E. Clark. 1989. Soil Microbiology and Biochemistry. Academic Press.

Pirrson, L.V. and A. Knopf. 1958. Rock and Rock Minerals, 3rd ed. John Wiley & Sons, New York, NY, 349 pp.

Plaster, E.J. 1985. Soil Fertility. Delmar Publishers, Albany, NY, 1,284 pp.

Poindexter, J.S. and E.R. Leadbetter (eds.). 1986. Bacteria in Nature, Vol. 2: Methods and Special Applications in Bacterial Ecology. Plenum, New York, NY, 385 pp.

Porter, J.R. 1946. Bacterial Chemistry and Physiology. John Wiley & Co., New York, NY.

Pytkowicz, R.M. 1983. Equilibria, Nonequilibria, and Natural Waters, 2 Vols. Wiley-Interscience, New York, NY.

Rankama, K, and T.G. Sahama. 1950. Geochemistry. University of Chicago Press, Chicago, IL, 912 pp.

Reeves, R.D. and R.R. Brosler. 1978. Trace Element Analysis of Geological Minerals. John Wiley & Sons, New York, NY, 421 pp.

Rice, R. (ed.). 1989. Chemistry of Ground Water. Lewis Publishers, Chelsea, MI.

Richardson, S.M. and H.Y. McSween. 1989. Geochemistry Pathways and Processes. Prentice-Hall, Englewood Cliffs, NJ, 488 pp.

Robinson, R.A. and R.H. Stokes. 1965. Electrolyte Solutions, 2nd ed. Butterworths, London, 571 pp.

Rosswall, T. (ed.). 1973. Modern Methods in the Study of Microbial Ecology. Ecological Bulletin No. 17, Swedish Natural Science Research Council, Stockholm. [More than 80 papers and short communications]

Schlesinger, W.H. 1991. Biogeochemistry: An Analysis of Global Change. Academic Press, New York, NY, 443 pp.

Schnitzer, M. and S. Kahn. 1972. Humic Substances in the Environment. Marcel-Dekker, New York, NY, 327 pp.

Schnitzer, M. and S.U. Khan. 1978. Soil Organic Matter. Developments in Soil Science, Vol. 8. Elsevier, New York, NY, 320 pp.

Sillen, L.G. and A.E. Martell. 1964. Stability Constants in Metal-Ion Complexes. Chemical Society Special Publication 17, 754 pp.

Silverman, M.P. and H.L. Ehrlich. 1964. Microbial Formation and Degradation of Minerals. Adv. Appl. Microbiol. 6:153-206. [Literature review with more than 270 citations]

Smith, W.R. and R.W. Missen. 1982. Chemical Reaction Equilibrium Analysis. Wiley-Interscience, New York, NY, 364 pp.

(Table 3-12 Basic Geochemistry and Microbiology References)

Soon, Y.K. (ed.). 1985. Soil Nutrient Availability: Chemistry and Concepts. Van Nostrand Reinhold, NY, 353 pp.

Sparks, D.L. (ed.). 1986. Soil Physical Chemistry. CRC Press, Boca Raton, FL, 320 pp.

Sparks, D.L. 1989. Kinetics of Soil Chemical Processes. Academic Press, New York, NY.

Sposito, G. 1984. The Surface Chemistry of Soils. Oxford University Press, New York, NY, 248 pp.

Sposito, G. 1989. The Chemistry of Soils. Oxford University Press, New York, NY. [Introductory text with 13 chapters]

Sposito, G. 1994. Chemical Equilibria and Kinetics in Soils. Oxford University Press, New York, NY, 256 pp.

Stevenson, F.J. 1982. Humus Chemistry: Genesis, Compositions, Reactions. Wiley Interscience, New York, NY.

Stumm, W. (ed.). 1967. Equilibrium Concepts in Natural Water Systems. ACS Adv. in Chemistry Series 67, American Chemical Society, Washington, DC. [16 papers]

Stumm, W. (ed.). 1987. Aquatic Surface Chemistry. John Wiley and Sons, New York, NY.

Stumm, W. and J.J. Morgan. 1981. Aquatic Chemistry, 2nd ed. Wiley Interscience, New York, NY, 780 pp.

Tan, K.H. 1993. Principles of Soil Chemistry, 2nd ed. Marcel Dekker, New York, NY, 376 pp.

Tan, K.H. 1994. Environmental Soil Science. Marcell Dekker, New York, NY, 320 pp.

Tate, III, R.L. 1987. Soil Organic Matter: Biological and Ecological Effects. Academic Press, New York, NY.

Theng, B.K.G. 1974. The Chemistry of Clay-Organic Reactions. Halstead, New York, NY.

Theng, B.K.G. 1979. Formation and Properties of Clay-Polymer Complexes. Developments in Soil Science, Vol. 9. Elsevier, New York, NY, 362 pp.

Thompson, L.M. and F.R. Troeh. 1978. Soil and Soil Fertility, 4th ed. McGraw-Hill, New York, NY. [See Troeh and Thompson (1993) for 5th edition]

Thornton, I. (ed.). 1983. Applied Environmental Geochemistry. Academic Press, New York, NY, 528 pp. [16 contributed chapters]

Tisdale, S.L. and W.L. Nelson. 1975. Soil Fertility and Fertilizers, 3rd ed. MacMillan, New York, NY, 694 pp.

Troeh, F.R. and L.M. Thompson. 1993. Soils and Soil Fertility, 5th ed. Oxford University Press, New York, NY, 462 pp. [4th edition Thompson and Troeh (1978)]

(Table 3-12 Basic Geochemistry and Microbiology References)

Van Johnso, W.M. and J.A. Maxwell. 1981. Rock and Mineral Analysis, 2nd ed. John Wiley & Sons, New York, NY.

Van Olphen, H. 1963. Clay Colloid Chemistry. John Wiley & Sons, New York, NY, 301 pp.

Van Olphen, H. 1977. An Introduction to Clay Colloid Chemistry. Wiley-Interscience, New York, NY, 318 pp.

Van Zeggeren, F. and S.H. Storey. 1970. The Computation of Chemical Equilibria. Cambridge University Press, 176 p.

Wedephol, K.H. (ed.). 1974. Handbook of Geochemistry. Springer-Verlag, Berlin, 2,311 pp.

Zajic, J.E. 1969. Microbial Biogeochemistry. Academic Press, New York, NY. [Text focusing on the microbial biogeochemistry of natural geological systems]

Zehnder, A.J.B. (ed.) 1988. Biology of Anaerobic Microorganisms. Wiley-Interscience, New York, NY. [14 review papers]

(Table 3-12 Basic Geochemistry and Microbiology References)

CHAPTER 4

SOURCES AND BEHAVIOR OF SUBSURFACE CONTAMINANTS

4.1 Definitions of Contamination 170

4.2 Extent and Sources of Subsurface Contamination 171

    4.2.1 Extent of Contamination 172
    4.2.2 Major Types of Contaminants 173
    4.2.3 Major Sources of Contamination 174

4.3 General Mechanisms of Ground-Water Contamination 179

    4.3.1 Infiltration 179
    4.3.2 Recharge from Surface Water 179
    4.3.3 Direct Migration 180
    4.3.4 Interaquifer Exchange 181

4.4 Contaminant Transport Processes 182

    4.4.1 Advection 183
    4.4.2 Hydrodynamic Dispersion 184
    4.4.3 Density/Viscosity Differences 186
    4.4.4 Osmotic Potential 191
    4.4.5 Facilitated Transport 192

4.5 Contaminant Retardation 192

    4.5.1 Filtration 192
    4.5.2 Sorption 194
    4.5.3 Precipitation 198
    4.5.4 Transformation 198

4.6 Contaminant Plume Behavior 198

    4.6.1 Geologic Influences 199
    4.6.2 pH and Eh 200
    4.6.3 Leachate Composition 200
    4.6.4 Source Characteristics 200
    4.6.5 Interactions of Various Factors on Contaminant Plumes 202

4.7 Guide to Major References* 202

* Appendix F contains citations for table and figure sources.

4.1 Definitions of Contamination

The Safe Drinking Water Act (SDWA) of 1974 defines the term "contaminant" broadly to include "any physical, chemical, biological, or radiological substance or matter in water." This definition draws no distinction between contamination from natural and anthropogenic sources, nor does it distinguish between acceptable and unacceptable levels of contamination. Definitions of contamination from several other sources give some useful additional perspectives on the meaning of the term:

- A *contaminated* freshwater aquifer has been impaired in quality--directly or indirectly by human activity--to such a degree that the U.S. Public Health Service would not recommend it for drinking water (Deutsch, 1963).[1]

- Ground-water *contamination* is the degradation of the natural quality of ground water as a result of human activity (U.S. EPA, 1977/T4-4).

- *Contaminants* are all solutes introduced into the hydrologic environment as a result of human activity regardless of whether or not the concentrations reach levels that cause significant degradation of water quality; *pollution* results when contaminant concentrations reach levels that are considered to be objectionable (Freeze and Cherry, 1979/T2-4).

- Boundaries of *polluted* ground-water zones are the lines at which the concentration of all pollutants have fallen below the maximum permissible concentration for potable water, or where all water properties have taken on the normal values of the environment concerned (Matthess, 1982/T4-5).

- A *substance* is any organic or inorganic chemical, microorganism, radionuclide, or other material, such as sediment. Whether or not a substance is a *contaminant* depends on its association with adverse impacts and on other site-specific factors such as hydrogeology (OTA, 1984/T4-4).

This handbook adopts the middleground taken by U.S. EPA (1977/T4-4). Natural ground waters may be *unsuitable* for a specified use, but the term *contamination* is reserved for degradation of natural water quality as a result of human activity. Whether the degree of contamination results in *unacceptable* degradation of ground water lies in the realm of psychology, sociology, political science, and law. This determination may be site specific and a matter of personal preference (i.e., ground water at a particular location may meet drinking water standards but not be considered potable by an individual due to bad taste), or it may be defined by law and regulation at the county, state, or national level. Even regulatory definitions of acceptable concentrations of specific contaminants may differ from the county to the national level, with the differences generally being based on different judgements as to the degree of risk posed by a specific contaminant.

---

[1] Deutsch, M. 1963. Ground-Water Contamination and Legal Controls in Michigan. U.S. Geological Survey Water-Supply Paper 1691.

A definition of ground-water contamination tied to human activity does not necessarily require that the contaminants themselves be of artificial origin, although this is usually the case (Section 4.2.3). Acid-mine drainage from coal mining and intrusion of saltwater into freshwater aquifers by overpumping are examples of natural contaminants being released due to human activity (see discussion of Categories IV and VI in Section 4.2.3).

4.2 Extent and Sources of Subsurface Contamination

Ground-water contamination is not a new problem. Probably the earliest scientific documentation of disease spread by ground-water contamination is the classic work of Dr. John Snow, who in 1854 linked 500 deaths from cholera in London's Soho District to a single well (Stamp, 1964).[2] The U.S. Public Health Service (Stiles et al., 1927)[3] probably conducted the earliest systematic scientific study of ground-water contamination. This research examined the movement, velocity, and persistence of chemical and bacterial pollutants in ground water at Fort Casell, North Carolina. Fiedler (1936) provides one of the earliest systematic descriptions of the general geologic and hydrologic controls governing the migration of contaminants in various types of aquifers.[4]

Identification of ground-water contamination as a serious environmental problem in the United States is relatively recent. Zanoni (1971/T4-4), in a review of the scientific literature on ground-water pollution and sanitary landfills, identified only one citation on the topic prior to 1950 (Calvert, 1932)[5] and eight citations in the 1950s. That number jumped to 47 in the 1960s. The U.S. Department of Health, Education, and Welfare sponsored the first symposium in the United States devoted to ground-water contamination in 1961 (U.S. Public Health Service, 1961).[6] Deutsch (1963-- Footnote 1), in the first systematic state-level study, identified over 50 cases of actual or suspected ground-water contamination in Michigan. Lindorff and Cartwright (1977/T4-4), in a compilation of 116 case histories of ground-water contamination, listed twice as many references for the period 1970 to 1975 than for the entire period prior to 1970.

---

[2] Stamp, D.L. 1964. The Geography of Life and Death. Cornell University Press, Ithaca, NY, 160 pp.

[3] Stiles, C.W., H.R. Crohurst, and G.E. Thompson. 1927. Experimental Bacterial and Chemical Pollution of Wells via Ground Water, and the Factors Involved. U.S. Public Health Service Hygienic Lab. Bull. 147.

[4] Fiedler, A.G. 1936. Occurrence of Ground Water with Reference to Contamination. J. Am. Water Works Ass. 28(12):1954-1962.

[5] Calvert, C.K. 1932. Contamination of Ground Water by Impounded Garbage Waste. J. Am. Water Works Ass. 24:266-270.

[6] U.S. Public Health Service. 1961. Proceedings of the 1961 Symposium, Ground Water Contamination. U.S. Public Health Service Tech. Rept. W61-5.

The 1980s has seen an explosion in the amount of scientific literature on ground-water contamination. The number of symposia on the topic seems to grow each year. Table 5-12 identifies more than 90 symposia and conference series that focus on or address issues of soil and ground-water contamination. Perhaps 1986 can be pinpointed as the year when the study of ground-water contamination became a discipline in its own right, with the publication of the first issue of the **Journal of Contaminant Hydrology**.

4.2.1 Extent of Contamination

A variety of factors make precise estimates of the extent of ground-water contamination in the United States difficult. These include (1) the great variety and ranges of toxicity of waste materials, (2) variable patterns of waste disposal and accidental release of contaminants in the ground, (3) variable patterns of ground-water use from wells, (4) differing behaviors of each contaminant in the soil, water, and rock environment, (5) the wide range in geologic and hydrologic conditions in different parts of the country, and (6) changes in hydrologic conditions in time (LeGrand, 1965).[7]

In addition, many potentially hazardous contaminants are colorless, odorless, and tasteless, and therefore difficult to detect by passive means. Many of the synthetic organic chemicals require sophisticated and expensive sampling and analytical techniques, further burdening detection efforts. For example, one round of organic compound testing of the 3,400 public water supply wells in Illinois required an estimated 4 to 5 years to complete (Illinois EPA, 1986).[8] Systematic testing of the estimated 500,000 private wells in the state would not only be prohibitive in terms of cost, but extremely time consuming.

An assessment of the extent and severity of contamination is further complicated by the almost exponential growth of the synthetic organic chemistry industry in the United States since the early 1940s. At least 63,000 synthetic organic chemicals are in common industrial and commercial use in the United States and this number continues to grow by approximately 500 to 1,000 new compounds every year (U.S. EPA, 1979).[9] The human health effects of many of these chemicals, particularly over long periods of time at low exposure levels, are not known. It would take many years to properly test all these compounds and then prepare a complete contamination assessment.

---

[7] LeGrand, H.E. 1965. Patterns of Contaminated Zones of Water in the Ground. Water Resources Research 1(1):83-95.

[8] Illinois Environmental Protection Agency. 1986. A Plan for Protecting Illinois Groundwater. Illinois Environmental Protection Agency, Springfield, IL. [Note: this plan has been replaced by the 1987 Illinois Groundwater Protection Act]

[9] U.S. Environmental Protection Agency (EPA). 1979. Environmental Assessment: Short-Term Tests for Carcinogens, Mutagens and Other Genotoxic Agents. EPA/615/9-79/003 (NTIS PB300-611).

Nevertheless, a relatively small percentage of all ground water is estimated to be contaminated. Lehr (1982), using simple assumptions of total ground water and the extent of ground-water contamination, estimated that 0.2% was contaminated.[10] The Office of Technology Assessment (OTA, 1984/T4-4) cited a range of 1 to 2%, and concluded that the extent of contamination is likely to be greater because substances known to contaminate ground water are used throughout society, whereas efforts to detect contamination have focused primarily on public drinking water supplies and point sources, such as landfills and hazardous waste sites. Furthermore, even if only a small percentage of potentially available ground water is contaminated, this percentage may be significant because (1) contamination is often near heavily populated areas, and (2) reliance on ground water is increasing.

4.2.2 Major Types of Contaminants

Over 200 chemical substances have been found in ground water, many of which could have potentially adverse impacts on human health (OTA, 1984/T4-4). This number includes approximately 175 organic chemicals, over 50 inorganic chemicals (metals, nonmetals, and inorganic acids), and radionuclides. Many of these chemicals occur naturally in ground water, especially minerals that dissolve from geologic earth materials contacting the water. Most of these, however, have been introduced to the ground-water system by humans (Section 4.2.3).

Page (1981/T4-4) tested the concentration of 56 toxic substances (9 heavy metals and 47 organic compounds) in over 1,000 ground-water samples and over 600 surface water samples selected to be representative of the entire state of New Jersey. Each substance tested was found in detectable concentrations in one or more samples. Five organic compounds were found in more than 50% of the ground-water samples (1,1,1-trichloroethane--78%, chloroform and carbon tetrachloride--64%, 1,1,2-trichloroethylene--58%, and trans-dichloroethylene--50%). An additional 20 organic compounds were detected in 10 to 50% of the samples. This study concluded that overall, ground water was as polluted as surface water in New Jersey.

The Ground Water Supply Survey (GWSS) conducted by the U.S. EPA provided information on the frequency with which volatile organic compounds (VOCs) were detected in 466 randomly selected public ground-water supply systems (Westrick et al., 1984/T4-4). The survey detected one or more VOCs in 16.8% of the small systems and 28.0% of large systems sampled. Two or more VOCs were found in 6.8% and 13.4% of the samples from small and large systems, respectively. The two VOCs found most often were trichloroethylene (TCE) and tetrachloroethylene (PCE).

---

[10] Lehr, J.H. 1982. How Much Ground Water Have We Really Polluted? Ground Water Monitoring Review 2(1):4.

A reliable determination of the extent and severity of ground-water degradation and associated health risks in the United States is probably not feasible because (1) tens of thousands of sites where a potential exists for contamination are not being monitored, and (2) comprehensive analyses of water quality at hundreds of thousands of wells would be required (Miller, 1985/T4-4).

4.2.3 Major Sources of Contamination

The U.S. Office of Technology Assessment (OTA, 1984/T4-4) has grouped 33 types of ground-water contamination sources into six major categories (Table 4-1) based on the general nature of the contaminating activity. Figure 4-1 depicts a number of these sources.

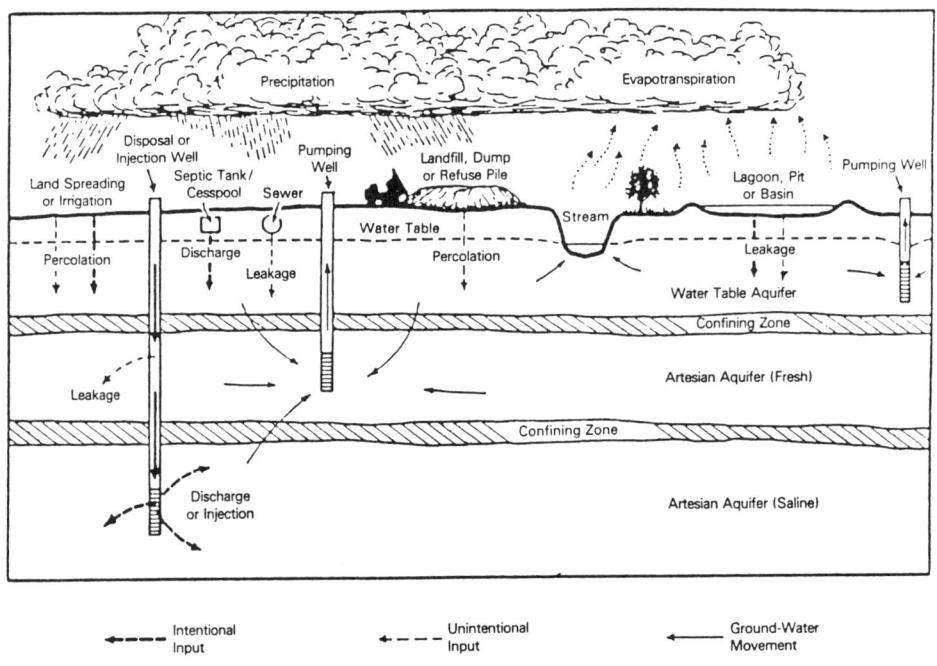

Figure 4-1 Sources of ground-water contamination (U.S. EPA, 1987a, after Geraghty and Miller, 1985).

Category I includes sources that are intentionally designed to discharge substances. Subsurface percolation systems such as septic tanks and cess pools, injection wells, and land application of wastewater or sludges fall within this category. Such systems are designed primarily to use the natural capacity of the soil materials to degrade wastewaters. Septic tanks and cess pools have been estimated to discharge the largest volume of wastewater into the ground and are the most frequently reported

Table 4-1 Sources of Ground Water Contamination

Category I—Sources Designed to Discharge Substances

Subsurface percolation (e.g., septic tanks and cess pools)
Injection wells
    Hazardous waste
    Nonhazardous waste (e.g., brine disposal and drainage)
    Nonwaste (e.g., enhanced recovery, artificial recharge, solution mining, and in situ mining)
Land application
    Wastewater (e.g., spray irrigation)
    Wastewater by-products (e.g., sludge)
    Hazardous waste
    Nonhazardous waste

Category II—Sources Designed to Store, Treat, and/or Dispose of Substances; Discharge through Unplanned Release

Landfills
    Industrial hazardous waste
    Industrial nonhazardous waste
    Municipal sanitary
Open dumps, including illegal dumping (waste)
Residential (or local) disposal (waste)
Surface impoundments
    Hazardous waste
    Nonhazardous waste
Waste tailings
Waste piles
    Hazardous waste
    Nonhazardous waste
Materials stockpiles (nonwaste)
Graveyards
Animal burial
Aboveground storage tanks
    Hazardous waste
    Nonhazardous waste
    Nonwaste
Underground storage tanks
    Hazardous waste
    Nonhazardous waste
    Nonwaste
Containers
    Hazardous waste
    Nonhazardous waste
    Nonwaste
Open burning and detonation sites
Radioactive disposal sites

Category III—Sources Designed to Retain Substances during Transport or Transmission

Pipelines
    Hazardous waste
    Nonhazardous waste
    Nonwaste
Materials transport and transfer operations
    Hazardous waste
    Nonhazardous waste
    Nonwaste

Category IV—Sources Discharging Substances as Consequence of Other Planned Activities

Irrigation practices (e.g., return flow)
Pesticide applications
Fertilizer applications
Animal feeding operations
De-icing salts applications
Urban runoff
Percolation of atmospheric pollutants
Mining and mine drainage
    Surface mine-related
    Underground mine-related

Table 4-1 (cont.)

Category V—Sources Providing Conduit or Inducing Discharge through Altered Flow Patterns	Category VI—Naturally Occurring Sources Whose Discharge Is Created and/or Exacerbated by Human Activity

Production wells
    Oil (and gas) wells
    Geothermal and heat recovery wells
    Water supply wells
Other wells (nonwaste)
    Monitoring wells
    Exploration wells
Construction excavation
	Ground water-surface water interactions
	Natural leaching
	Saltwater intrusion/brackish water upcoming (or intrusion and other poor quality natural water)

Source: OTA (1984).

sources of ground-water contamination (U.S. EPA, 1977). More than 23 million homes in the United States rely on onsite wastewater disposal systems; and the use of septic system cleaners that remove grease and kill roots may result in ground-water contamination by halogenated hydrocarbons and heavy metals, respectively (Noss, 1989).[11]

Injection wells are another potential source of contamination. Injected wastewaters are often placed in unusable zones to be assimilated with poor quality ground water of natural origin. Current regulations prohibit injection of wastes into an underground source of drinking water (USDW) or contamination of a USDW by deep-well injection. Injection of hazardous wastes is regulated under EPA's Underground Injection Control Program. Injection of wastes can contaminate ground water in five major ways (U.S. EPA, 1985/T4-4):

1. Upward migration around the injection well casing through faulty construction
2. Upward migration through improperly plugged or completed wells placed in the injection zone from oil exploration or other reasons
3. Upward migration through faults or fractures in the confining layer
4. Lateral migration of injected wastes
5. Direct injection into a USDW (no longer permitted)

Land application, a popular and inexpensive method of disposing of wastewater and sludge, can pollute ground water in several ways: (1) organic and inorganic

---

[11] Noss, R.R. 1989. Septic System Cleaners: A Significant Threat to Groundwater Quality. Journal of Environmental Health 51(4):201-204.

contaminants in directly applied wastewater can move directly into ground water if the soil's filtration capacity is exceeded, and (2) precipitation infiltrating through land-applied sludges may leach contaminants into the ground-water system. EPA (1983/T4-4) estimated that 40 to 50% of the municipal sludge generated every year is applied to the land.

Category II includes sources designed to store, treat, or dispose of substances, but not to release contaminants to the subsurface. Landfills, open dumps, local residential disposal, surface impoundments, waste tailings and piles, materials stockpiles, graveyards, aboveground and underground storage tanks, containers, open burning sites, and radioactive disposal sites all fall into this broad category. It is important to note here that while a number of sources in this category are considered "waste" sources (e.g., landfills, dumps, impoundments, etc.), many others are "nonwaste" sources such as petroleum and other chemical products. Storage tanks, stockpiles, and a variety of containers with residues of commercial products have been found to contribute contaminants to ground water if not properly designed and maintained.

Category III consists of sources designed to retain substances during transport or transmission. Such sources consist primarily of pipelines and materials transport or transfer operations. Contaminant releases generally occur by accident or neglect; for example, as a result of pipeline breakage or a traffic accident. Again, most substances subject to release from sources within this category are not wastes but raw materials or products to be used for some beneficial purpose.

Category IV includes those sources discharging substances as a consequence of other planned activities. This category contains a number of agriculturally related sources such as irrigation return flows, feedlot operations, and pesticide and fertilizer applications. A number of sources related to urban activities, such as highway desalting, urban runoff, and atmospheric deposition, are included. Surface and underground mine-related drainage also fall within this category. Sources in this category tend to be spread over large areas and are generally more difficult to regulate.

Category V comprises sources providing conduits or inducing discharge through altered flow patterns. Such sources include water, oil, and gas production wells, monitoring wells, exploration holes, and construction excavations. Ground-water contamination from production wells stems from poor installation and operation methods, and incorrect plugging or abandonment procedures. Such practices create opportunities for cross-contamination by vertical migration of contaminants.

Finally, Category VI includes naturally occurring sources whose discharge is induced or intensified by human activity. Ground-water/surface water interactions, described in the previous section, and saltwater intrusion or upconing (ground-water movement upward as a result of pumpage) provide the basis for this category. Withdrawals that are significantly more than recharge can affect ground-water quality. Saltwater intrusion in coastal areas and brine-water upconing from deeper formations in inland areas both can occur when pumpage exceeds an aquifer's natural recharge rate.

Contaminant releases are also referred to as originating from point or nonpoint sources. *Point sources* are those that release contaminants from a discrete geographic location, including leaking underground storage tanks, and ruptured or corroded transfer pipes (the cause of most fuel leaks), septic systems, and injection wells. *Nonpoint* sources of contamination are more extensive in area and diffuse in nature. It is therefore difficult to trace contaminants from nonpoint sources back to their origin. Agricultural activities (i.e., application of pesticides and fertilizers), urban runoff, and atmospheric deposition are potential nonpoint contaminant sources.

Systematic data on specific contaminants in ground water are not readily available for reasons discussed in Section 4.2.1. Table 4-2 shows types and sources of soil contamination from 100 sites in The Netherlands. Gasworks were the largest source of contamination (45%) followed by waste dumps and landfills (26%). The main contaminants identified at these sites were aromatic and halogenated hydrocarbons.

Table 4-2 Classification of Types and Sources of Soil Contamination in The Netherlands Based on a Sample of 100 Cases

Source of Contamination	Type of Contamination	Frequency (%)
Gasworks	Aromatic hydrocarbons, phenols, CN^-	45
Waste dumps and landfills	Halogenated hydrocarbons, alkyl-benzenes; metals like As, Pb, Cd, Ni, CN^-; pesticides	26
Chemicals production and handling sites (including painting industries and tanneries)	Halogenated hydrocarbons, alkyl-benzenes; metals like Pb, Cr, Zn, As	13
Metal plating and cleaning industries	Tri- and tetrachloroethylene, benzene, toluene, Cr, Cd, Zn, CN^-	9
Pesticide manufacturing sites	Pesticides, Hg, As, Cu	4
Automobile service facilities (including gasoline storage tanks)	Hydrocarbons, Pb	3

Source: Zoeteman (1985). © John Wiley & Sons. Reprinted by permission.

Palmer et al. (1988) reviewed data on Superfund sites according to the primary hazardous substances detected (Figure 4-2). Sites contaminated by organics made up the largest group, including 136 sites; 78 sites were contaminated by heavy metals. Individual organic compounds frequently singled out as major contaminants include TCE, polychlorinated biphenyls (PCBs), toluene, and phenol. Arsenic and chromium are the most frequently identified individual heavy metal contaminants.

4.3 General Mechanisms of Ground Water Contamination

Contaminant releases to ground water can occur by design, by accident, or through neglect. Most ground-water contamination incidents involve substances released at or only slightly below the land surface. Consequently, most contaminant releases affect shallow ground water initially. Certain activities, however, such as oil and gas exploration, deep-well waste injection, and pumping of ground water underlain by saltwater, initially tend to affect deeper ground water.

Ground water contamination can occur by infiltration, recharge from surface water, direct migration, and interaquifer exchange. The first and second mechanisms primarily affect surface aquifers; the third and fourth may affect either surface or deep aquifers.

4.3.1 Infiltration

Infiltration is probably the most common ground-water contamination mechanism. A portion of the water that falls to the earth as precipitation slowly infiltrates the soil through pore spaces in the soil matrix. As the water moves downward under the influence of gravity, it dissolves materials with which it comes into contact. Water percolating downward through a contaminated zone can dissolve contaminants, forming leachate that may contain inorganic and organic constituents. The leachate will continue to migrate downward under the influence of gravity until it reaches the saturated zone. In the saturated zone, contaminants in the leachate will spread horizontally in the direction of ground-water flow, and vertically due to gravity (Figure 4-3). This process can occur beneath any surface or near-surface contaminant source exposed to the weather and the effects of infiltrating water. Section 4.4.3 discusses how nonaqueous phase liquids (NAPLs) move into the subsurface.

4.3.2 Recharge from Surface Water

Normally, ground water moves toward or "discharges" to surface water bodies. However, movement of contaminants from surface water to ground water can occur in losing streams (where normal elevation of the water table lies below the stream channel) and during flooding. Flood stages may cause a temporary reversal in the hydraulic gradient, with a flow of contaminants into bank storage, or contaminant entry through improperly cased wells (Figure 4-4a). For example, Schwarzenbach et al.

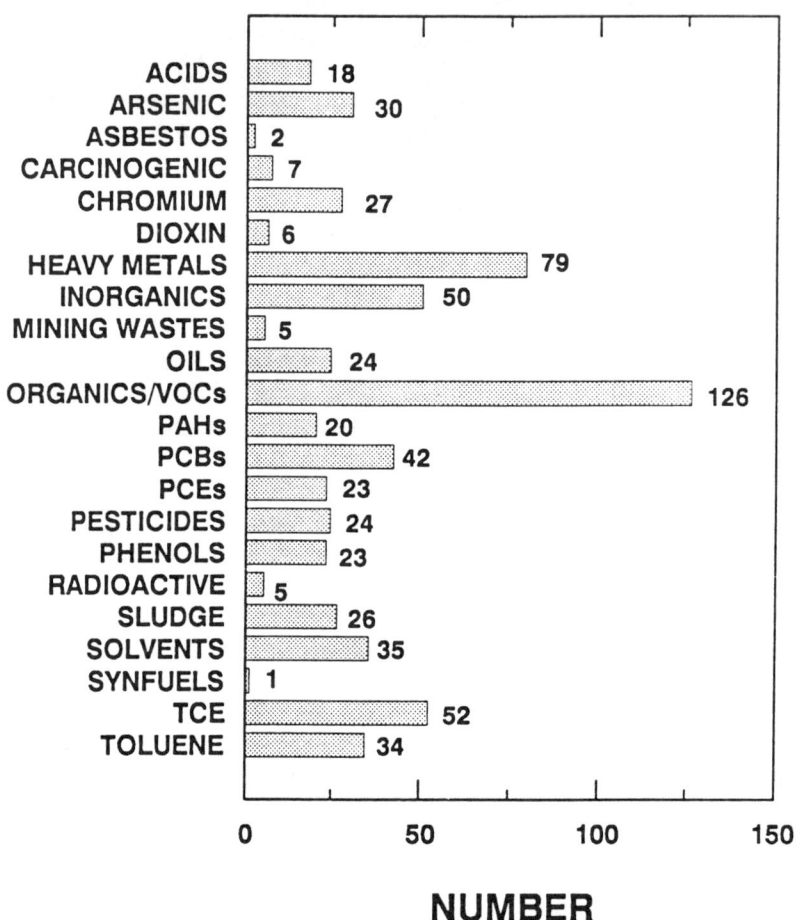

Figure 4-2 Major contaminants at Superfund Sites (Palmer and Fish, 1991).

$(1983)^{12}$ documented movement of organic contaminants in river water into glacial sand and gravel aquifers in the Aare and Glatt valleys in Switzerland. Contaminated surface water can also enter an aquifer if the ground-water level adjacent to a surface water body is lowered by pumping (Figure 4-4b).

4.3.3 Direct Migration

Contaminants can migrate directly into ground water from below-ground sources (e.g., storage tanks, pipelines) that lie within the saturated zone. Much greater

---

[12] Schwarzenbach, R., W. Giger, E. Hoehn, and J. Schneider. 1983. Behavior of Organic Compounds During Infiltration of River Water to Ground Water--Field Studies. Environ. Sci. Technol. 17(8):472-479.

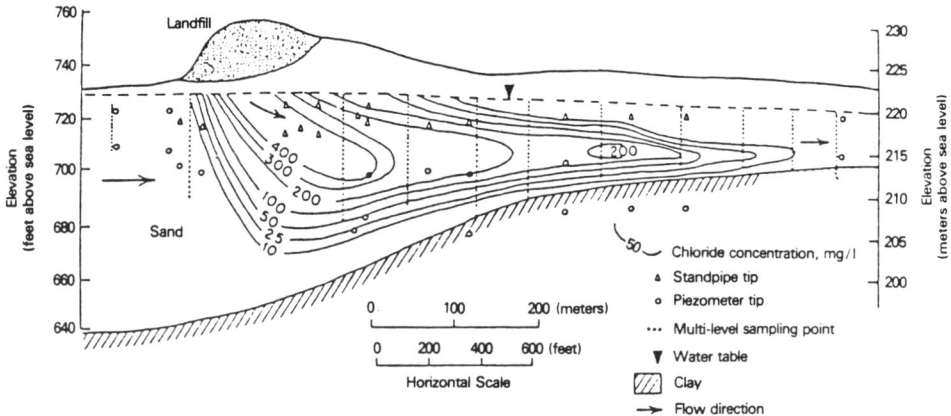

Figure 4-3 Plume of leachate migrating from a sanitary landfill on a sandy aquifer using contours of chloride concentration (U.S. EPA, 1987a, after Freeze and Cherry, 1979).

concentrations of contaminants may occur from these sources because of the continually saturated conditions. Storage sites and landfills excavated to a depth near the water table may also permit direct contact of contaminants with ground water. In addition, contaminants can enter the ground-water system from the surface by vertical leakage through the seals around well casings, through wells abandoned without proper procedures, or as a result of contaminant disposal through deteriorated or improperly constructed wells.

4.3.4 Interaquifer Exchange

Contaminated ground water can mix with uncontaminated ground water through a process known as interaquifer exchange, in which one water-bearing unit communicates hydraulically with another. This occurs most commonly in bedrock aquifers where a well penetrates more than one water-bearing formation to increase its yield. Each water-bearing unit has its own range of head potential. When the well is not being pumped, water moves from the formations with the greatest potential to formations of lesser potential. If the formation with the greater potential contains contaminated or poorer quality water, it may degrade the quality of water in another formation.

In a process similar to direct migration, old and improperly abandoned wells with deteriorated casings or seals may contribute to interaquifer exchange. Vertical movement may be induced by pumping, or may occur under natural gradients. For example, Figure 4-5 depicts an improperly abandoned well with a corroded casing that formerly tapped only a lower uncontaminated aquifer. The corroded casing allows water from an overlying contaminated zone to communicate directly with the lower aquifer. The pumping of a nearby well tapping the lower aquifer creates a downward gradient between the two water-bearing zones. As pumping continues, contaminated water migrates through the lower aquifer to the pumping well. Downward migration

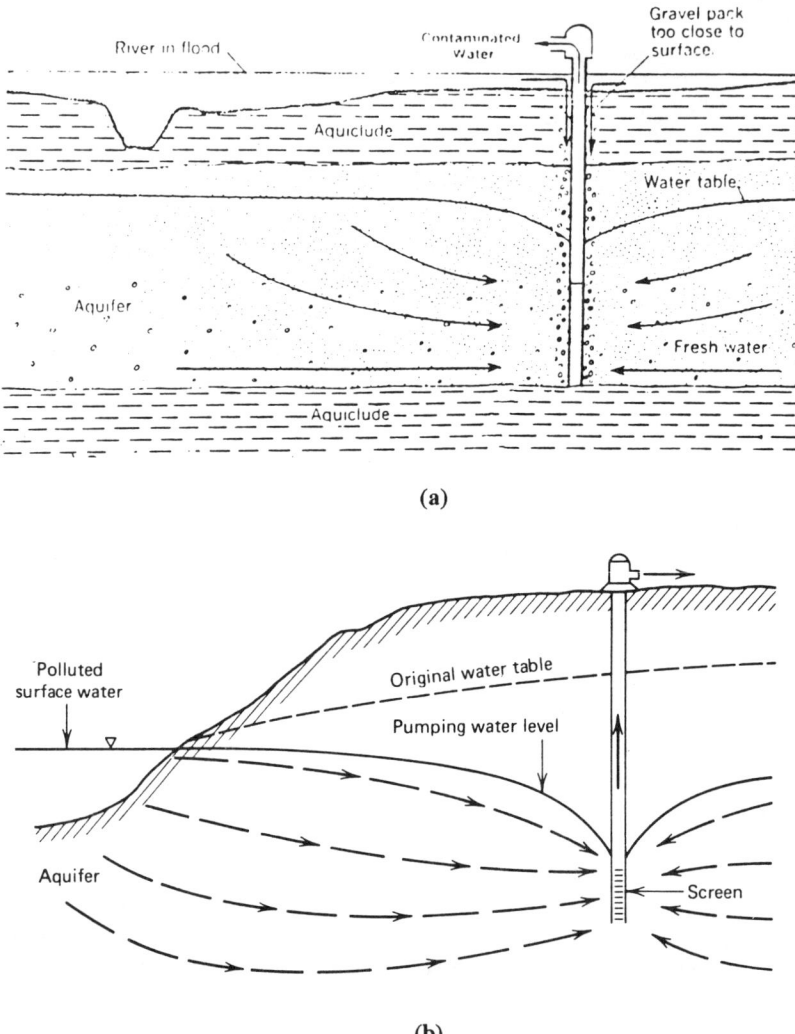

Figure 4-4 Ground-water contamination from surface water: (a) floodwater entering improperly cased well (Deutsch, 1963); (b) induced flow from pumping (Deutsch, 1965-see Appendix F for full credit).

of the contaminant may also occur through the confining layer that separates the upper and lower aquifers. However, the rate of contaminant movement through a confining layer is often much slower than the rate of movement through the direct connection of an abandoned well.

4.4 Contaminant Transport Processes

The extent to which a contaminant moves in ground water depends on its behavior in relation to various processes that encourage transport (Sections 4.4.1 through 4.4.5) and other processes that serve to retard movement (Section 4.5). The

Sources and Behavior of Subsurface Contaminants 183

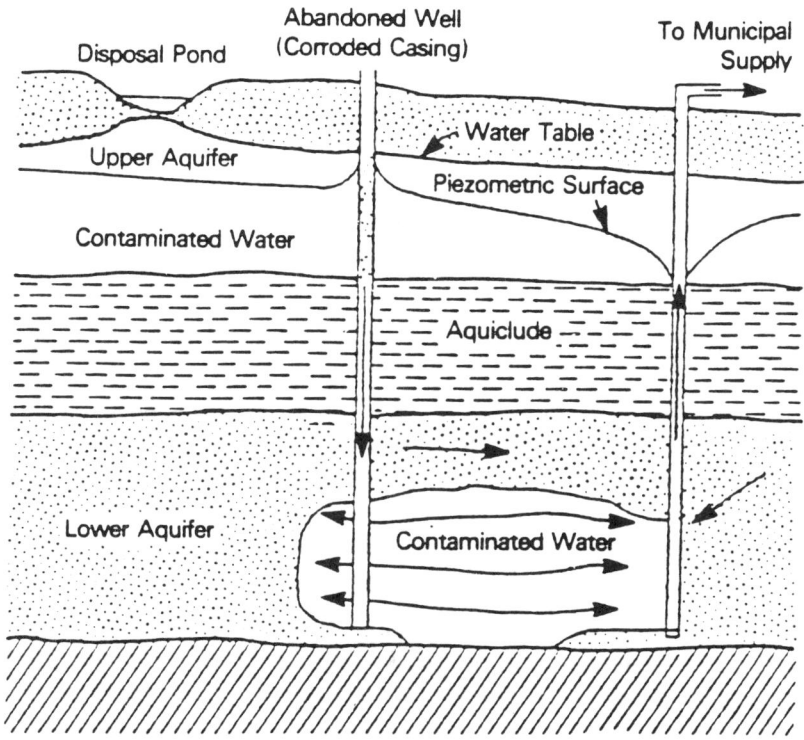

Figure 4-5 Vertical movement of contaminants along an old, abandoned, or improperly constructed well (U.S. EPA, 1977, after Deutsch, 1961).

shape and speed of contaminant plumes are determined by these processes and by factors relating to the aquifer materials and characteristics of the contaminants (Section 4.6). In addition to the discussion in Chapter 3, EPA's Seminar Publication on Transport and Fate of Contaminants in the Subsurface (U.S. EPA, 1989/T4-5), and Part II (Physical and Chemical Processes in the Subsurface) of EPA's Seminar Publication on Site Characterization for Subsurface Remediation (U.S. EPA, 1991/T9-10) provide more detailed treatment of contaminant transport and retardation processes.

In broad terms, three processes govern the extent to which chemical constituents migrate in ground water: (1) advection, movement caused by the flow of ground water; (2) dispersion, movement caused by the irregular mixing of waters during advection; and (3) retardation, principally chemical mechanisms that occur during advection, which tend to slow down the rate of contaminant migration.

4.4.1 Advection

Ground water in its natural state is constantly in motion, although in most cases it is moving very slowly, typically at a rate of inches or feet per day. Ground-water flow, or advection, is calculated using Darcy's Law (Section 2.6.3) and is governed by the hydraulic principles discussed in Chapter 2. Time-of-travel calculations based on advective flow (Section 2.6.5) may *underestimate* the rate of migration of dissolved constituents such as chlorides and nitrates, which are minimally retarded by aquifer solids, as a result of hydrodynamic dispersion (Section 4.4.2). On the other hand, time-of-travel estimates may *overestimate* the rate of migration for contaminants subject to retardation processes.

Figure 4-6a shows the relative concentration of a dissolved constituent emanating from a constant source of contamination versus distance along the flow path. Figure 4-6b shows a similar plot for a discontinuous contaminant source that produced a single slug of dissolved contaminant. Considering advective flow only, no diminution of concentration appears as a straight line moving at the rate of ground-water flow.

Several mechanisms influence the spread of a contaminant in the flow field. Dispersion and density/viscosity differences may accelerate contaminant movement, while various retardation processes slow the rate of movement compared to that predicted by simple advective transport.

4.4.2 Hydrodynamic Dispersion

Hydrodynamic dispersion is the net effect of a variety of microscopic, macroscopic, and regional conditions that influence the spread of a solute concentration front through an aquifer (Anderson, 1984; Schwartz, 1977).[13] Quantifying dispersion may be important in fate assessment, because contaminants can move more rapidly through an aquifer by this process than by simple plug flow (i.e., uniform movement of water through an aquifer with a vertical front). In other words, physical conditions (such as the presence of permeable zones where water can move more quickly) and chemical processes (such as molecular diffusion of dissolved species to ground water with lower concentrations ahead of the contaminant front) result in more rapid contaminant movement than would be predicted by ground-water equations for physical flow, which assume average values for permeability.

Dispersion on the *microscopic* scale is caused by (1) external forces acting on the ground-water fluid, (2) variations in pore geometry, (3) molecular diffusion along concentration gradients, and (4) variations in fluid properties such as density and viscosity. Dispersion at this scale, also called *mechanical dispersion*, is generally less accurate than estimated advective flow, and for this reason is often ignored. Lehr (1988) warns against putting too much effort to quantify dispersion at this scale.[14]

---

[13] Anderson, M.P. 1984. Movement of Contaminants in Groundwater: Groundwater Transport--Advection and Dispersion. In: Groundwater Contamination, National Academy Press, Washington, DC, pp. 37-45.
Schwartz, F.W. 1977. Macroscopic Dispersion in Porous Media: The Controlling Factors: Water Resources Research 13(4):743-752.

[14] Lehr, J.H. 1988. An Irreverent View of Contaminant Dispersion. Ground Water Monitoring Review 8(4):4-6.

Sources and Behavior of Subsurface Contaminants 185

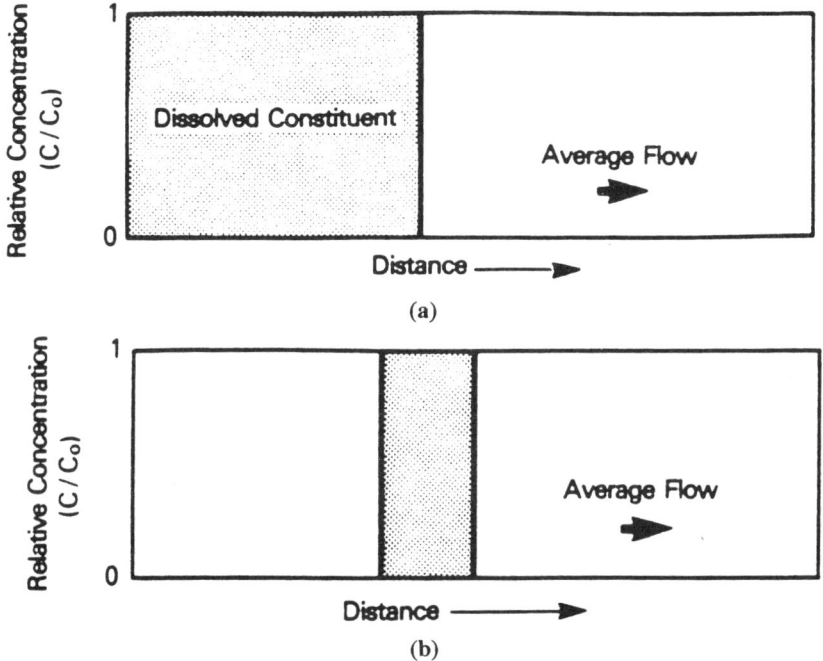

Figure 4-6 Movement of a concentration front by advection only: (a) continuous source; (b) slug (U.S. EPA, 1987a).

Dispersion on the *macroscopic* scale is caused by variations in hydraulic conductivity and porosity, which create irregularities in velocity and consequent additional mixing of the solute. Over large distances, regional variations in hydrogeologic units can affect the amount of dispersion that occurs. Macroscopic dispersion may result in substantially faster travel times of contaminants than predicted by equations for mechanical dispersion. Therefore, it should be the focus of efforts to characterize dispersion (Wheatcraft, 1989).[15] Macroscopic features, such as lenses of higher conductivity in unconsolidated materials, solution channeling and fracturing are the major macroscopic features that may contribute to contaminant dispersion.

Figure 4-7a shows the effect of dispersion as a plot of relative constituent concentration versus distance along a flow path. In the figure, the front of the dissolved constituent distribution is no longer straight, but instead appears "smeared." Some of the dissolved constituent actually moves ahead of what would have been predicted if only advection were considered. Figure 4-7b gives an aerial view of dispersion of a contaminant plume from a continuous source.

In a similar manner, the concentration of a slug of material introduced to a flow field appears as shown in Figure 4-8a, with the peak concentration declining over time

---

[15] Wheatcraft, S.W. 1989. An Alternate View of Contaminant Dispersion. Ground Water Monitoring Review 9(3):11-12.

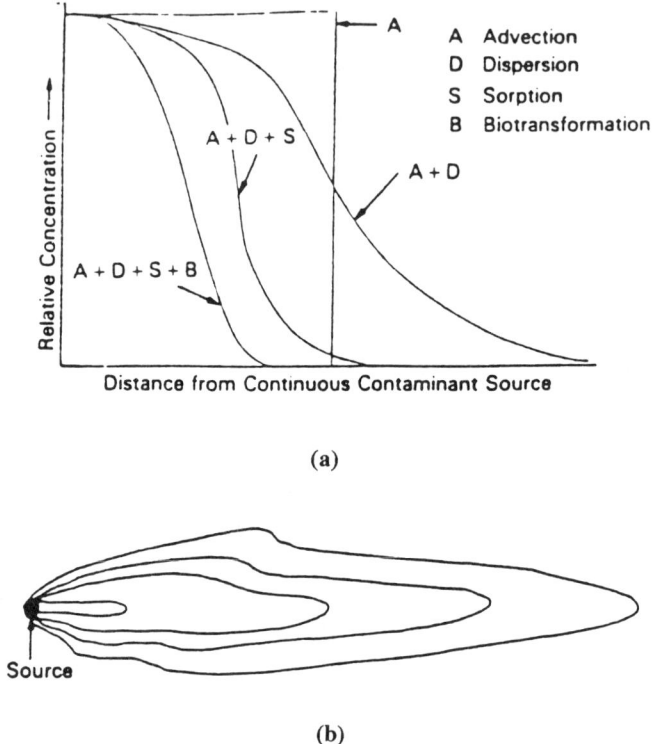

Figure 4-7 Effect of dispersion and retardation on movement of a contaminant front from a continuous source: (a) relative concentrations compared to advection only; (b) plan view of plume (U.S. EPA, 1987a).

and distance. In such a situation, the total mass of dissolved constituent remains the same; however, it occupies a larger volume, effectively reducing the concentration found at any distance along the flow path. An aerial view of intermittent sources affected by dispersion is shown in Figure 4-8b. Dispersion dilutes the concentration of a contaminant, thus reducing peak concentrations encountered in the ground-water system.

4.4.3 Density/Viscosity Differences

Contaminants having a density lower than ground water tend to concentrate in the upper portions of an aquifer, while those having a higher density concentrate in the lower portions. The viscosity (tendency to resist internal flow) of specific contaminants affects their rate of migration through soil and within an aquifer. Density and viscosity effects are significant when nonaqueous phase liquids (NAPLs) are present in the subsurface, and where the salinity of ground waters contrast strongly (freshwater and saltwater).

Nonaqueous Phase Liquids (NAPLs). NAPLs moving through the subsurface can displace both water and air. Water in the presence of NAPLs tends to line the

Sources and Behavior of Subsurface Contaminants

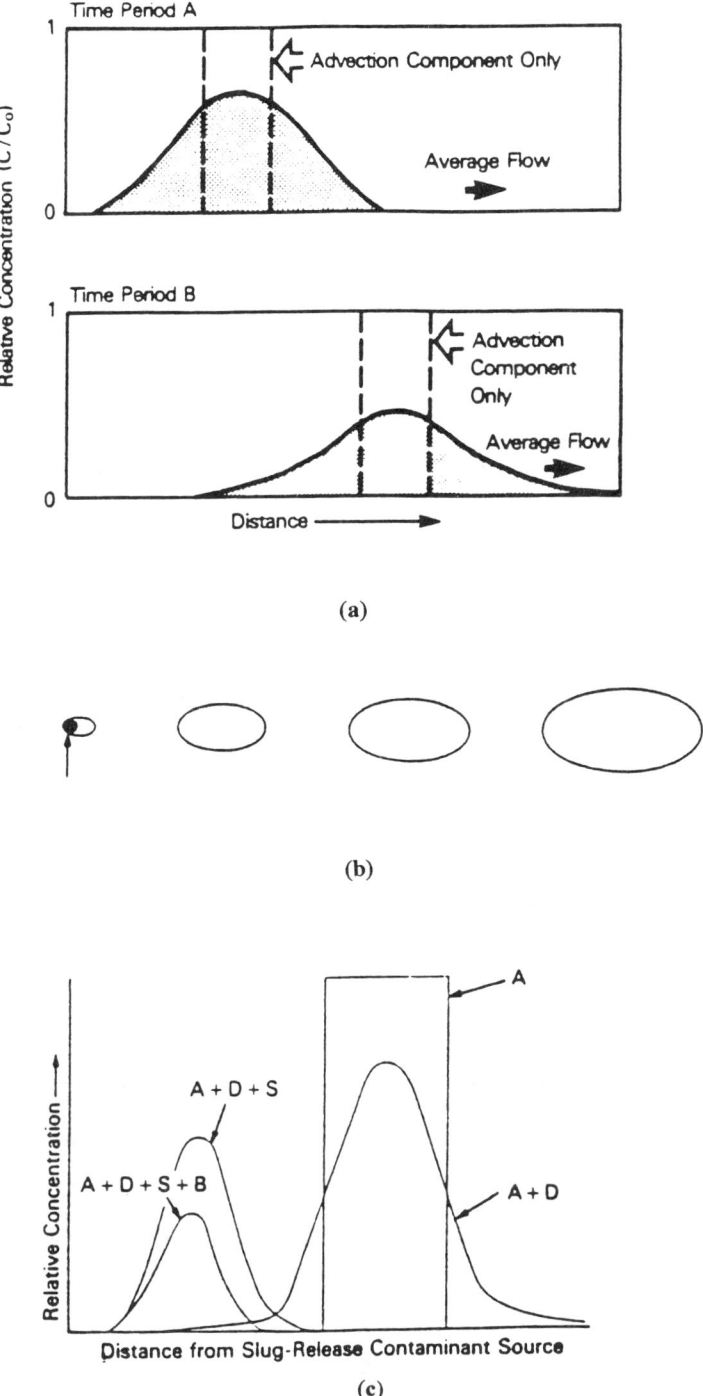

Figure 4-8 Effect of dispersion and retardation on movement of a contaminant slug: (a) dispersion over time; (b) plan view of plume from intermittent point source; (c) sorption and biodegradation (U.S. EPA, 1987a).

edges of the pores and adhere to soil particles, while the NAPL tends to move through the central portions of the pores. In unsaturated soil, when neither water nor the NAPL occupies the entire pore space, the NAPL and water move more slowly through the soil than they would if one or the other filled the pore space. A certain percentage of nonaqueous phase liquids will adhere to soil particles in the vadose zone (called *residual saturation*), with further movement dependent on volatilization or dissolution by percolating ground water.

The behavior and movement of NAPLs that are lighter than water (LNAPL) differ from those that are denser than water (DNAPLs). Figure 4-9 illustrates the movement of a spill of an LNAPL in the subsurface. If the amount of product spilled is small the LNAPL will flow into the unsaturated zone until residual saturation is reached (Figure 4-9A). Infiltrating water will gradually dissolve components of the LNAPL, such as benzene, toluene and xylene, and carry the contaminants to the ground water. Since LNAPLs tend to be volatile, some of the spilled material will also partition into the soil air and move through the vadose zone by molecular diffusion. Parts of the gaseous phase may also be dissolved by percolating water and enter the ground water by that path. The movement of soil gas in the vadose zone does not necessary follow the ground-water gradient, so soil gas measurements must be supplemented to ground-water sampling to delineate contaminant plumes.

A larger spill will result in product reaching the water table. The dissolved components of the infiltrating NAPL precede the product and may change the wetting properties of the water, causing a reduction in the residual water content and a collapse of the capillary fringe and depression of the water table (Figure 4-9B). If the flow from the NAPL source is stopped, it will flow through the vadose zone until residual saturation is reached. The NAPL will then tend to spread laterally along the top of the capillary fringe (Figure 4-9C).[16] Draining of the upper portions of the vadose zone reduces the total head at the interface between the NAPL and the ground water allowing the water table to rebound somewhat. However, product remains in the aquifer at residual saturation and is gradually released by dissolution as ground water moves through this zone.

DNAPLs can be very mobile in the subsurface as a result of their relatively low solubility, high density, and low viscosity. Figure 4-10 illustrates the behavior of spilled DNAPLs in the subsurface. Displacement of water, which has lower density and higher viscosity than DNAPLs, creates an unstable liquid front with viscous fingering resulting. A relatively small spill will result in fingers of DNAPL penetrating the vadose zone until residual saturation is reached. A contaminant plume in ground

---

[16] The term "free product", although it is often used when describing subsurface flow of NAPLs at concentrations above residual saturation, is not strictly correct in this context. Where an LNAPL occupies most of the pore space, the LNAPL will seep from pore space into a borehole cavity and form free product that floats on the surface of the ground water, but free product in the matrix pure fuel will not fill 100 % of the pore space. For example, at the capillary fringe there exists more of a mixing zone where hydrocarbon saturation gets up to a certain level, but does not displace water completely.

Sources and Behavior of Subsurface Contaminants 189

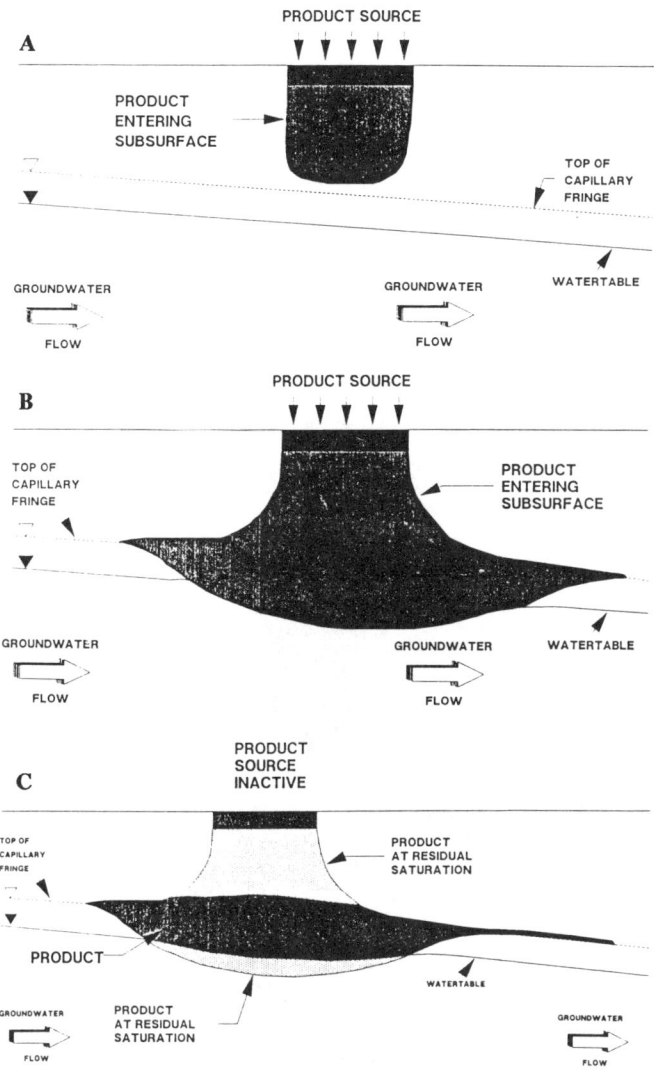

Figure 4-9 Movement of LNAPLs into the subsurface (Palmer and Johnson, 1989b).

water generally develops as a result of dissolution by percolating water and dense vapors that come in contact with the capillary fringe (Figure 4-10A).

A larger spill may result in flow of the DNAPL until it reaches the capillary fringe. If the thickness is sufficient to overcome the capillary force between the water and the aquifer matrix it will flow into the saturated zone and continue until residual saturation is reached (Figure 4-10B). A very large spill can result in penetration to the bottom of the aquifer, with pools of DNAPL forming in depressions. Figure 4-10C illustrates how a DNAPL can actually flow against the direction of ground-water flow when the impermeable base of an aquifer is sloping.

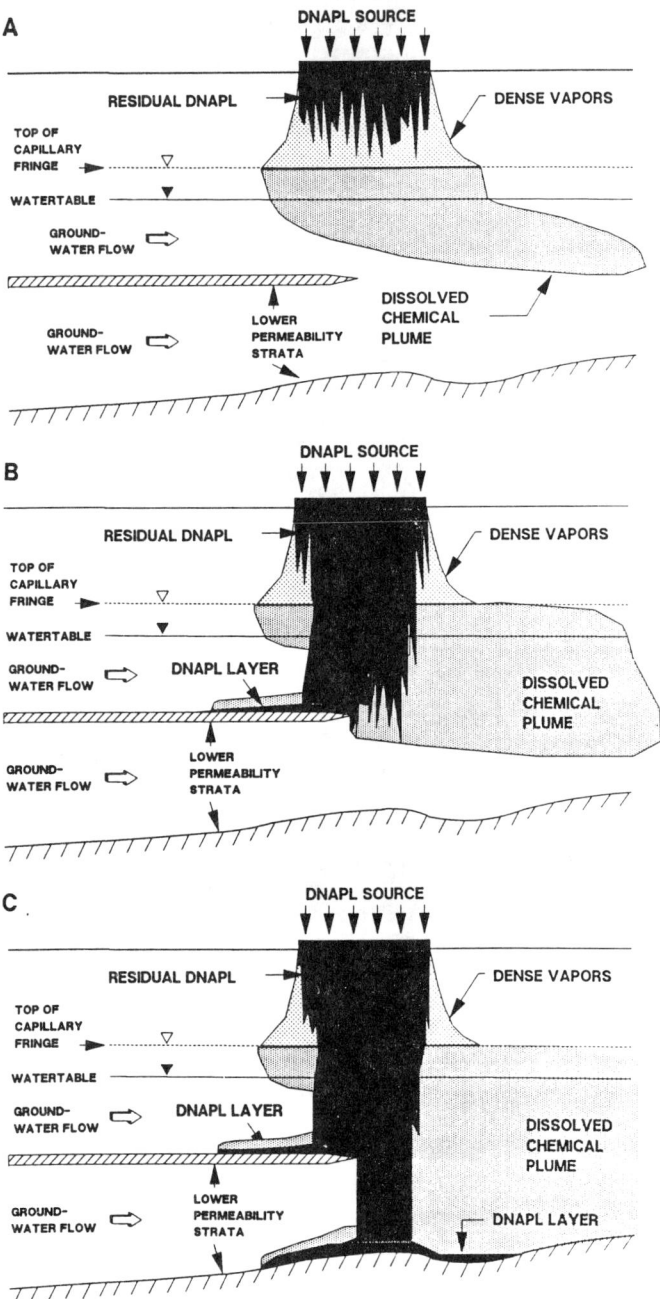

Figure 4-10 Movement of DNAPLs into the subsurface (Palmer and Johnson, 1989b).

Ground Water Density Effects. Freshwater is less dense than water with high concentrations of dissolved solids. For example, in coastal areas, freshwater tends to float on top of seawater where sediments are hydrologically connected with the ocean. Pumping of ground water can disrupt this equilibrium, resulting in contamination of freshwater aquifers by sodium chloride. Consideration of density differences is essential when modeling interactions between freshwater and seawater (Frind, 1982).[17] Density variations in ground water in deep boreholes may also result in errors in estimating flow directions, because they do not necessarily follow hydraulic gradients (Oberlander, 1989).[18]

Temperature Effects. The viscosity of water decreases as temperature increases. For example, Sniegocki (1963)[19] found that viscosity differences resulting from surface water at 66°F injected into ground water at 43°F reduced the specific capacity (gallons per minute per foot of drawdown) of an artificial recharge well in the Grand Prairie Region in Arkansas by 30%. Kaufman and McKenzie (1975)[20], on the other hand, observed that the apparent hydraulic conductivity of an injection zone in the Floridan aquifer receiving hot organic wastes increased about 2.5 times because of temperature differences alone.

4.4.4 Osmotic Potential

Osmotic potential, the energy required to pull water away from ions in solution that are attracted to polar water molecules, is a primary factor affecting solute transport in the vadose zone (see Section 2.4.1). Shales that serve as confining layers can sometimes act as semipermeable membranes, if the ionic concentrations on both sides of the clay layer differ greatly (Hanshaw, 1972).[21] Bredehoeft et al. (1963)[22] have suggested this as a mechanism for the development of brines in deep sedimentary

---

[17] Frind, E.O. 1982. Simulation of Long-Term Transient Density-Dependent Transport in Groundwater. Adv. Water Resources 5(June):73-88.

[18] Oberlander, P.L. 1989. Fluid Density and Gravitational Variations in Deep Boreholes and their Effect on Fluid Potential. Ground Water 27(3):341-350.

[19] Sniegocki, R.T. 1963. Problems in Artificial Recharge through Wells in the Grand Prairie Region, Arkansas. U.S. Geological Survey Water-Supply Paper 1615-F.

[20] Kaufman, M.I and D.J. McKenzie. 1975. Upward Migration of Deep-Well Waste Injection Fluids in Floridan Aquifer, South Florida. J. Res. U.S. Geol. Survey 3:261-271.

[21] Hanshaw, B.B. 1972. Natural-Membrane Phenomena and Subsurface Waste Emplacement. In: Symposium on Underground Waste Management and Environmental Implications, T.D. Cook, (ed.), American Association of Petroleum Geologists Memoir 18, pp. 308-315.

[22] Bredehoeft, J.D., C.R. Blyth, W.A. White, and G.B. Maxey. 1963. Possible Mechanism for Concentration of Brines in Subsurface Formations. Bull. Am. Ass. Petroleum Geologists 47(2):257-269.

basins where concentrations three to six or more times the concentration of seawater occur. In laboratory experiments, Kharaka (1973)[23] found that ion mobility across geologic membranes varied with the material, but that monovalent and divalent cations generally followed the same sequences: Li < Na < NH_3 < K < Rb < Cs and Mg < Ca < Sr < Ba. Osmosis, if it occurs at all, is likely to occur in deep sedimentary basins rather than near-surface aquifers where salinities are higher and strong contrasts in ionic strength between water-bearing units are more likely to exist.

4.4.5 Facilitated Transport

Facilitated transport, in which the mobility of a contaminant is increased relative to "expected" retardation by adsorption to subsurface solids, is a relatively new area of study in the field of contaminant transport. Processes such as chelation (the formation of complex ions with organic ligands) have long been known to increase the mobility of metal ions because the bonding sites that would be available to the substrate become occupied by the chelating agent. More recently, attention has been focused on increased mobility of organic compounds by (1) *cosolvation* (increased solubility of hydrophobic organic contaminants when water-miscible organic solvents, such as ethanol, methanol, and acetone, are present in ground water), and (2) attachment to colloidal particles that are often mobile in the unsaturated and saturated zones of the subsurface (Huling, 1989/T4-5). In addition, when DNAPLs, such as chlorinated solvents, and LNAPLs are both present at a site, the DNAPL can dissolve the LNAPL and carry it beneath the water to deeper aquifer zones.

4.5 Contaminant Retardation

In ground-water contaminant transport, a number of chemical and physical mechanisms retard or slow the movement of constituents in ground water. Four major mechanisms that retard contaminant movement are: (1) filtration, (2) sorption, (3) precipitation, and (4) transformation or degradation.

Figures 4-7a and 4-8c illustrate the movement of a concentration front by advection only (A), advection plus dispersion (A+D), and with the addition of sorption, a partitioning process (A+D+S). The greatest retardation, however, results from the combined effects of advection, dispersion, sorption, and biotransformation (A+D+S+B). The amount of retardation resulting from sorption, other partition processes, and biotransformation depends on physical and chemical properties of the aquifer, including biologic populations, and chemical properties of the contaminant.

4.5.1 Filtration

Filtration is the entrapment of solid particles and large dissolved molecules in the pore spaces of the soil and aquifer media. Figure 4-11 shows three major

---

[23] Kharaka, Y.K. 1973. Retention of Dissolved Constituents of Waste by Geologic Membranes. In: Symposium on Underground Waste Management and Artificial Recharge, J. Braunstein, (ed.), Int. Ass. of Hydrological Sciences Pub. No. 110, pp. 420-435.

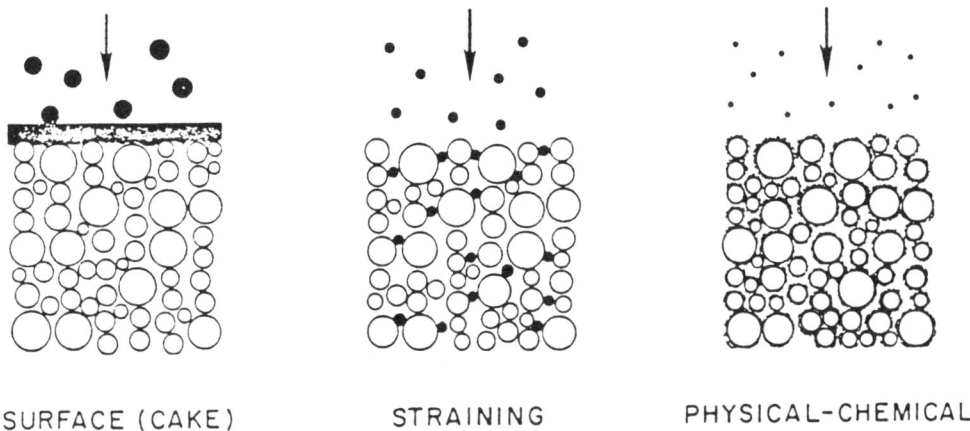

Figure 4-11 The three filtration mechanisms that limit particle migration through porous media (Palmer and Johnson, 1989a, after McDowell-Boyer et al., 1986).

mechanisms of filtration: surface filtration, straining, and physical-chemical interactions. *Surface filtration* results when particles are larger than the pore spaces and form a cake on the surface, at which the pore size becomes too small. Caking may also result from biological activity, as in the clogging mat that develops in septic tank absorption trenches. *Straining* happens when the particles are about the same size as the pore spaces. In this process, particles move through pores until they become lodged at the entrance to a pore that is too small. Filtration resulting from *physical-chemical interactions* with solid surfaces is discussed under partitioning process in the next section.

Filtration limits flow by clogging pore spaces and reducing the hydraulic conductivity of the material. Most dissolved species are retarded by partitioning or transformation, but if the molecular size of a chemical reaction product exceeds the pore size of the soil or aquifer, mechanical filtration occurs. Flocculation of colloidal material resulting from the precipitation of iron and manganese oxides, as well as clogging resulting from microbial activity, may hinder the movement of dissolved constituents. Gas bubble formation may also eventually clog pore spaces, resulting in a filtering effect. For example, a 10% increase in the air content of media voids can cause a 15% decrease in effective porosity, resulting in a 35% decrease in permeability, and about a 50% reduction in dispersion (Orlob and Radhakrishna, 1958).[24]

Filtration may also result in residual contamination that is highly resistant to both mobilization by desorption into air and water and microbial degradation. For

---

[24] Orlob, G.T. and G.N. Radhakrishna. 1958. The Effects of Entrapped Gases on the Hydraulic Characteristics of Porous Media. Trans. Am. Geophysical Union 39(4):648-659.

example, the soil fumigant 1,2-dibromomethane, which is readily biodegraded under aerobic conditions, has been found in agricultural soils up to 19 years after its last known application, due to entrapment in soil micropores (Steinberg et al., 1987).[25]

4.5.2 Sorption

Ion exchange involves the replacement of a cation attached to a negatively charged site on a mineral surface by another cation. The mineralogy and cation exchange capacity of an aquifer gives a general indication of its effectiveness in retarding cationic contaminants. As long as the ionic contaminant has a greater affinity for the solid surface than for existing adsorbed ions, retardation will occur. Once the exchangeable sites are filled, the contaminant will travel unretarded (see A+D+S curve in Figures 4-7a and 4-8c). Precise predictions of retardation by ion exchange are not possible because of interactions among multiple ions. Furthermore, changes in environmental conditions such as pH and Eh (Section 4.6.2) or groundwater solution composition may remobilize contaminants formerly bound to geologic materials. In fact, the release of ions by exchange processes may aggravate a contamination problem. For example, Rovers et al. (1976)[26] observed release of aluminum to solution from soil contaminated by industrial waste.

Most organic contaminants are nonionic and, consequently, partitioning to aquifer solids usually occurs by *physical adsorption* processes such as Van der Waals and hydrophobic bonding (Section 3.2.2).

The *adsorption isotherm* is a measure of changes in the amount of a substance adsorbed at different concentrations at a constant temperature. It is the simplest and most widely used method for predicting physical adsorption. Empirical constants can be calculated from adsorption isotherms, and these constants then can be used to predict the amount of adsorption at concentrations other than those measured. However, this method assumes that temperature and other environmental conditions are the same as those under which the isotherms were measured originally.

There are three major types of adsorption isotherms: (1) the *linear distribution coefficient*, (2) the *Langmuir* adsorption isotherm, and 3) the *Freundlich* adsorption isotherm.

Linear Distribution Coefficient. The simplest type of isotherm is the linear distribution coefficient, K_d (also called the partition coefficient, K_p), which assumes that the amount of contaminant sorbed is directly proportional to the concentration

---

[25] Steinberg, S.M., J.J. Pignatello, and B.L. Sawhney. 1987. Persistence of 1,2-Dibromomethane in Soils: Entrapment in Intraparticle Micropores. Environ. Sci. Technol. 21:1201-1213.

[26] Rovers, F.A., H. Mooij, and G.J. Farquhar. 1976. Contaminant Attenuation - Dispersed Soil Studies. In: Residual Management by Land Disposal, W.H. Fuller (ed.), EPA 600/9-76/015 (NTIS PB256 768), pp. 224-234.

of the compound in solution. The equation for calculating adsorption at different concentrations is:

$$S = K_d C \tag{4-1}$$

where:

S = amount sorbed (micrograms [µg]/g solid)
C = concentration of substance in solution (µ/milliliter [mL])
K_d = distribution coefficient

This equation is widely used to describe adsorption in soil and near-surface aquatic environments. Another widely used linear coefficient is the organic carbon partition coefficient (K_{oc}), which is equal to the distribution coefficient divided by the percentage of organic carbon present in the system ($K_{oc} = K_d/\%$ organic carbon), as proposed by Hamaker and Thompson (1972).[27] Equations are available for predicting the K_{oc} of both polar and nonpolar organic molecules based on molecular topology, provided the organic matter percentage exceeds 0.1% (Sabljić, 1987).[28] Karickhoff (1984)[29] discusses in some detail sorption processes of organic pollutants in relation to K_{oc}.

There are several major problems associated with using the linear distribution coefficient for describing adsorption/desorption reactions in ground-water systems (Reardon, 1981).[30] Some of these problems include:

- The coefficient actually measures multiple processes (reversible and irreversible adsorption, precipitation, and coprecipitation). Consequently, it is a purely empirical number with no theoretical basis on which to predict adsorption under differing environmental conditions, or to give information on the types of bonding mechanisms that are involved in the adsorption.

---

[27] Hamaker, J.W. and J.M. Thompson. 1972. Adsorption. In: Organic Chemicals in the Soil Environment, Vol. I, C.A.I. Goring and J.W. Hamaker (eds.), Marcel Dekker, New York, pp. 49-143.

[28] Sabljić, A. 1987. On the Prediction of Soil Sorption Coefficients of Organic Pollutants from Molecular Structure: Application of Molecular Topology Model. Environ. Sci. Technol. 21(4):358-366.

[29] Karickhoff, S.W. 1984. Organic Pollutant Sorption in Aquatic Systems. J. Hydraulic Engineering 110:707-735.

[30] Reardon, E.J. 1981. Kd's--Can They Be Used to Describe Reversible Ion Sorption Reactions in Contaminant Migration? Ground Water 19(3):279-286.

- Contaminated aquifers may undergo a dynamic chemical evolution in which changing environmental parameters may result in variations of K_d values by several orders of magnitude at different locations and at the same location at different times.

- All methods used to measure the K_d value involve some disturbance of the solid material and, consequently, may not accurately reflect in situ conditions. Furthermore, K_d values taken from the literature may have been developed using solid material that differs significantly in physical and chemical characteristics from the site of interest.

In spite of these shortcomings, K_d values from the published literature can provide a qualitative assessment of a contaminant's mobility, and adsorption batch laboratory tests using simulated contaminated solutes and samples of actual soils or aquifer materials from a site can provide valuable information about contaminant behavior. Use of adsorption batch tests is required for more accurate assessment of contaminant fate and transport because they allow evaluation of sorption as a function of varying subsurface properties (organic carbon, grain size distribution, and mineralogy). This information, in turn, can be used to assign different values of K_d in numerical subsurface flow and transport models.

Langmuir Equation. The Langmuir adsorption equation was originally developed to describe adsorption of gases on homogeneous surfaces and is commonly expressed as follows:

$$C/S = 1/kS_{max} + 1/CS_{max} \tag{4-2}$$

where:

S_{max} = maximum adsorption capacity (μg/g soil)
k = Langmuir coefficient related to adsorption bonding energy (mL/μg)
S = amount adsorbed (μg/g solid)
C = concentration of adsorbed substance in solution (μ/mL)

A plot of C/S versus $1/C$ allows the coefficients k and S_{max} to be calculated. When kC is much less than 1, adsorption will be linear as represented by Equation 4-1. Figure 4-12a shows the graphic form of the Langmuir isotherm and a different form of the basic equation.

The Langmuir model has been used to describe adsorption behavior of some organic compounds at near-surface conditions (Alben et al., 1988).[31] However, the Langmuir model makes three important assumptions:

---

[31] Alben, K.T., E. Shpirt, and J.H. Kaczmarczyk. 1988. Temperature Dependence of Trihalomethane Adsorption on Activated Carbon: Implications for Systems with Seasonal Variations in Temperature and Concentration. Environ. Sci. Technol. 22:406-412.

- The energy of adsorption is the same for all sites and is independent of degree of surface coverage.

- Adsorption occurs only on localized sites with no interaction between adjoining adsorbed molecules.

- The maximum adsorption capacity (S_{max}) represents coverage on only a single layer of molecules.

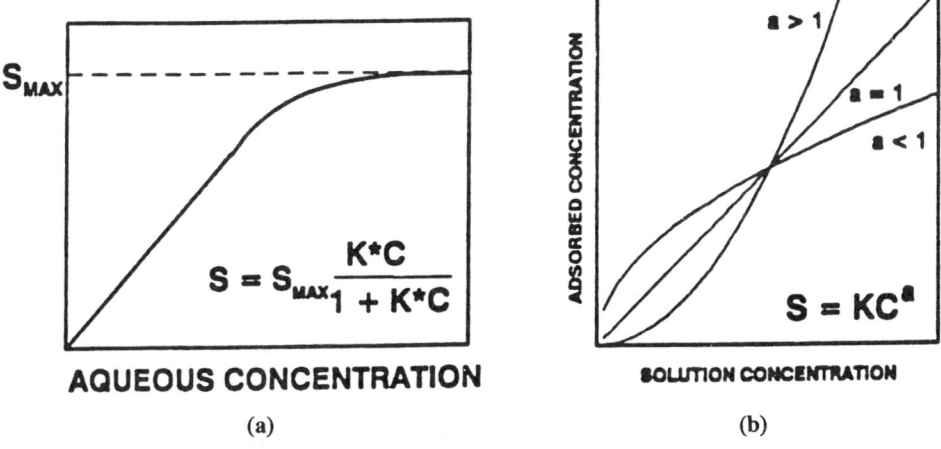

Figure 4-12 Adsorption isotherms: (a) Langmuir; (b) Freundlich (Palmer and Johnson, 1991).

These assumptions do not generally hold true in a complex heterogeneous media such as soil (Rao and Davidson, 1980).[32] For example, Bailey et al. (1968),[33] in a study of adsorption of organic herbicides by montmorillonite, found that none of the compounds conformed to the Langmuir adsorption equation. Of the 23 compounds they tested, however, only a few did not conform well to the Freundlich equation, discussed below.

Freundlich Equation. The Freundlich equation is expressed as:

---

[32] Rao, P.S.C. and J.M. Davidson. 1980. Estimation of Pesticide Retention and Transformation Parameters Required in Nonpoint Source Pollution Models. In: Environmental Impact of Nonpoint Source Pollution, M.R. Overcash and J.M. Davidson (eds.), Ann Arbor Science Publishers, Ann Arbor, MI, pp. 23-67.

[33] Bailey, G.W., J.L. White, and T. Rothberg. 1968. Adsorption of Organic Herbicides by Montmorillonite: Role of pH and Chemical Character of Adsorbate. Soil Sci. Soc. Am. Proc. 32:222-234.

$$S = KC^a \qquad (4\text{-}3)$$

where S and C are as defined in Equation 4-2 and K and a are empirical coefficients. Figure 4-12b shows the shape of the Freundlich isotherm at different values of a.

Taking the natural logarithms of both sides of Equation 4-3:

$$\ln S = \ln K + a \ln C \qquad (4\text{-}4)$$

Thus, a log-log plot of S versus C, provides an easy way to obtain values for K (the intercept) and a (the slope of the line). When values for K and a have been obtained, the amount of sorption at a given concentration can be calculated using Equation 4-3. The log-log plot can also be used for graphic interpolation of adsorption at other concentrations. When $a = 1$, Equation 4-3 simplifies to Equation 4-1 (i.e., adsorption is linear).

4.5.3 Precipitation

Precipitation reactions, in which geochemical reactions in the aquifer result in a contaminant moving from a dissolved form to an insoluble form can be an important retardation process for inorganic contaminants. As with adsorption, precipitation reactions are reversible, so it is possible for a contaminant to remobilize if environmental conditions change in the aquifer. Precipitation-dissolution reactions are largely determined by acid-base equilibria and redox conditions (Section 3.4.2). Section 3.2.3 provides additional discussion of precipitation and dissolution reactions. Geochemical distribution-of-species and reaction progress codes (Section 10.3) may help identify important inorganic precipitation reactions.

4.5.4 Transformation

All processes that transform a contaminant retard migration in that the original contaminant will no longer be present. However, unless the contaminant's reaction products are nontoxic inorganic elements, contamination may still persist. Complexation reactions involving heavy metals may even increase toxicity and mobility (Section 3.3.1). Some organic contaminants may be transformed by hydrolysis in ground water, but they often produce intermediate organic compounds of varying toxicity (Section 3.3.2). Microbiological activity is probably the most important means by which contaminants are transformed in the subsurface. Oxidation-reduction reactions (Section 3.3.3) and biotransformation (Section 3.3.4) are probably the two transformation processes that are most significant for most organic contaminants. Section 3.5.4 discusses biotransformation of organic contaminants in more detail.

4.6 Contaminant Plume Behavior

The physical mechanisms of advection and dispersion, as well as a variety of chemical and microbial reactions, interact to influence the movement of contaminants in ground water. The degree to which these mechanisms influence contaminant

Sources and Behavior of Subsurface Contaminants 199

movement depends on a number of factors, including geologic material properties, pH and Eh, leachate composition, and source characteristics.

4.6.1 Geologic Influences

The rate of ground-water movement is largely dependent on the type of geologic material through which it is moving. More rapid movement can be expected through coarse-textured materials such as sand or gravel than through fine-textured materials such as silt and clay. The physical and chemical composition of the geologic material is equally important. Fine-textured materials with a high clay content tend to impede contaminant migration both by having a low hydraulic conductivity and through ion exchange and physical adsorption. Figure 4-13a illustrates the effect of fracture flow on a chloride plume. Basalt rock has a very low permeability, but fractures allowed movement of the plume a distance of 5 miles downgradient from the source in a period of just 16 years. It is evident that the main orientation of the fractures runs across the direction of ground-water flow because the plume is 6 miles wide. Figure 4-13b shows a chromium plume in a sand and gravel aquifer. This plume moved 4,000 feet downgradient from the source with a maximum width of 1,000 feet after 13 years. Such a plume is indicative of a relatively homogenous isotropic aquifer. The illustration of time-of-travel calculations in Section 2.6.5 showed that ground water in a sand and gravel aquifer might typically take about 6 years to go a mile. Since the chromium plume traveled less than a mile in 13 years, it is apparent that sorption and possibly precipitation have reduced contaminant movement in this aquifer to about half of what would have been expected without retardation.

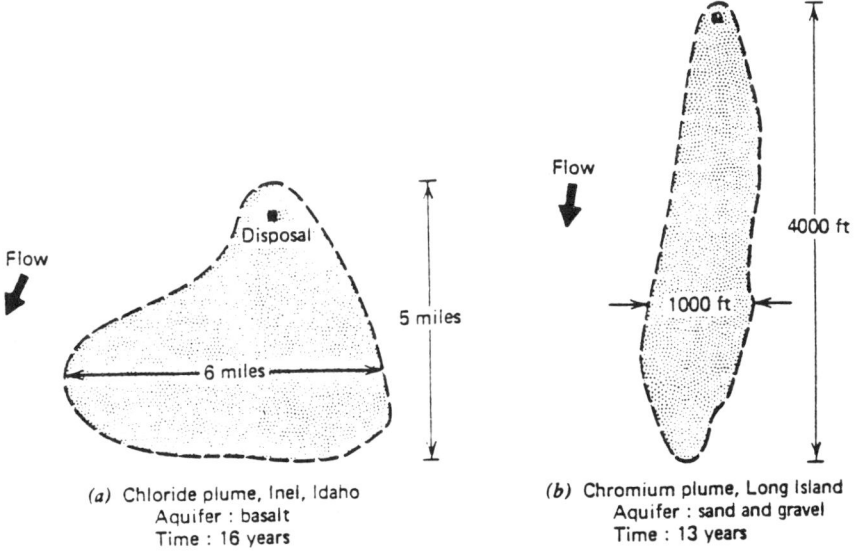

(a) Chloride plume, Inel, Idaho
Aquifer : basalt
Time : 16 years

(b) Chromium plume, Long Island
Aquifer : sand and gravel
Time : 13 years

Figure 4-13 Effect of differences in geology on shapes of contamination plumes (Miller, 1985). © John Wiley & Sons. Reprinted by permission.

4.6.2 pH and Eh

The pH and Eh of the geologic materials and the waste stream strongly influence contaminant mobility. The pH affects the speciation of many dissolved chemical constituents, which in turn determines solubility and reactivity. Ion exchange and hydrolysis reactions are also particularly sensitive to pH. Eh influences many precipitation and dissolution reactions, particularly those involving iron and manganese, and determines in large measure the type of biodegradation that occurs. Sections 3.4.1 and 3.4.2 discuss the significance of pH and Eh in more detail.

4.6.3 Leachate Composition

The influence of all other factors on contaminant migration ultimately depends on the composition of the leachate or contaminants entering the ground-water system. Similar contaminants may behave differently in the same environment due to the influence of other constituents in a complex leachate. Solubility (which affects the mobile concentration), density, chemical structure, and many other properties can affect net contaminant migration. For example, Figure 4-14 illustrates the appearance of two chemicals, benzene and chloride, in a monitoring well. Even though both contaminants may have entered the ground-water system at the same time and in the same concentration, their detection in the monitoring well reveals significantly different migration rates. Chloride has migrated essentially unaffected, while benzene has been retarded significantly.

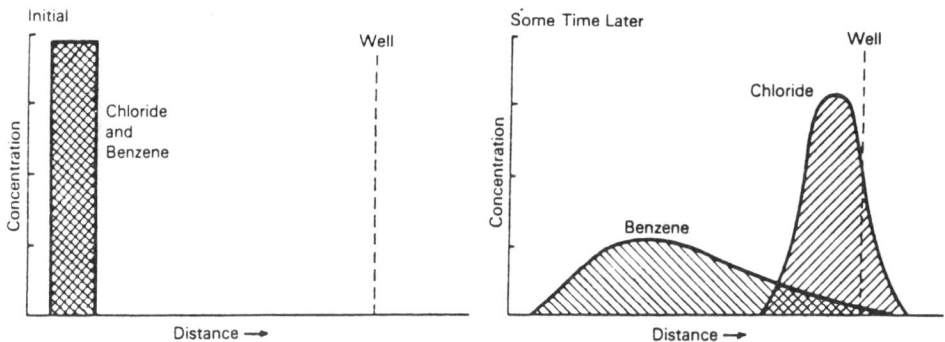

Figure 4-14 Benzene and chloride appearance in a monitoring well (U.S. EPA, 1987a, after Geraghty and Miller, 1985).

4.6.4 Source Characteristics

Source characteristics include the type of source (Section 4.2.2), the source mechanism (Section 4.3), and temporal features. Temporal characteristics include the manner in which a contaminant is released over time and the time elapsed since the contaminant's release.

Figure 4-15 presents the effects caused by changes in the rate of a point waste discharge on plume size and shape. *Plume enlargement* results from an increase in the rate of waste discharge to the ground-water system. Similar effects can be produced if the retardation capacity of the geologic materials is exceeded, or if the water table rises closer to the source, causing an increase in dissolved constituent concentration. Decreases in waste discharge, lowering of the water table, retardation through sorption, and reductions in ground-water flow rate can *diminish* the size of the plume. *Stable* plume configurations suggest that the rate of waste discharge is at a steady state with respect to retardation and transformation processes. A plume will *shrink* in size when contaminants are no longer released to the ground-water system and a mechanism to reduce contaminant concentrations is present. Unfortunately, many contaminants, particularly complex chlorinated hydrocarbons and heavy metals, may persist in ground water for extremely long time periods without appreciable transformation. Lastly, an intermittent or seasonal source can produce a *series* of plumes that are separated by the advection of ground water during periods of no contaminant discharge.

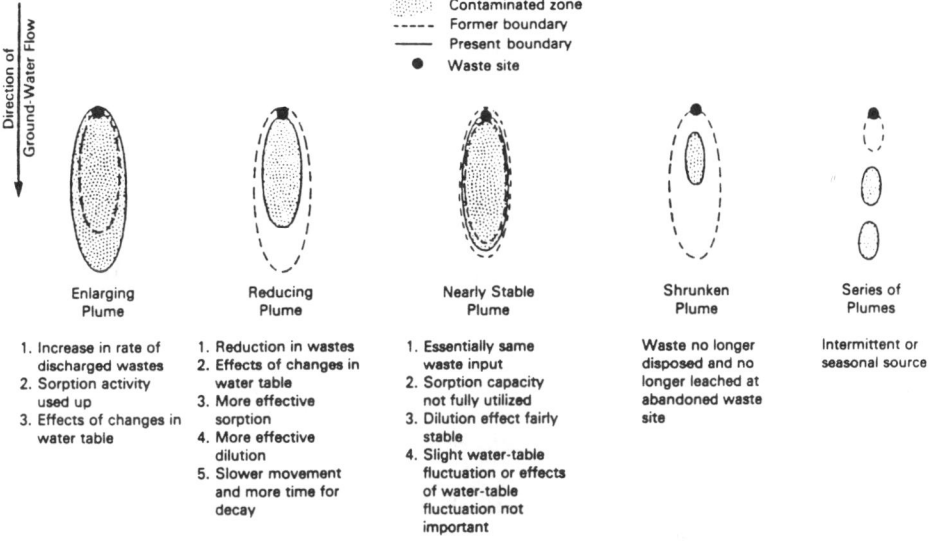

Figure 4-15 Changes in plumes and factors causing the changes (U.S. EPA, 1987a, after U.S. EPA, 1977 and LeGrand, 1965).

When contamination comes from a point source, concentrations are typically highest near the source. Nonpoint sources of contamination, such as from agricultural chemicals, tend to result in relatively low concentrations of contaminants distributed over a relatively large area. In this situation, plumes with distinctive boundaries are less easily defined.

4.6.5 Interactions of Various Factors on Contaminant Plumes

The various factors discussed above can result in varying sizes and shapes of contaminant plumes. Figure 4-16 shows 18 different types of contaminated zones. An *x* marks the contaminant source in each example. Table 4-3 explains the relative importance of dilution, degradation, and sorption in each plume and lists examples of the types of contaminants typically involved. A few minutes spent examining this Figure and the accompanying text in Table 4-3 should give a good feeling for types of interactions between contaminant sources and geology that can be looked for in the field. Contaminant plumes where contamination moves *upgradient* from the direction of ground-water flow is a special type that is not identified in Table 4-3. Figure 4-10 illustrates this for a DNAPL. This can also occur with LNAPLs, such a gasoline, when vapors that have moved in the vadose zone to areas where the saturated zone is upgradient of the source and then transported to the ground water by dissolution or entrainment by percolating water.

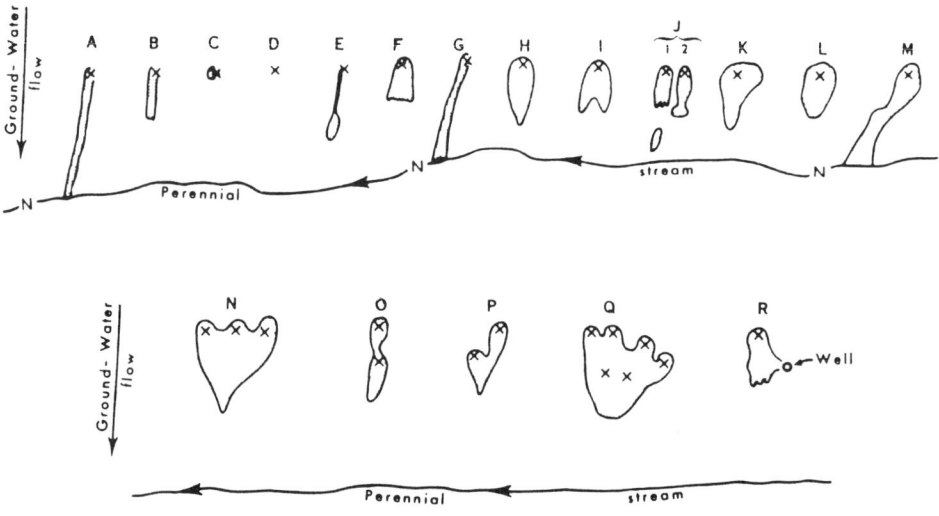

Figure 4-16 Map view of various types of contaminant plumes; see Table 4-3 for interpretations (LeGrand, 1965). © American Geophysical Union.

4.7 Guide to Major References

Table 4-4 provides an index to major references on types and sources of contamination in soil and ground water. The table includes the following major categories: (1) general references on ground-water contamination, (2) sources of information on baseline chemistry that may be useful for evaluating contaminant concentrations, (3) reviews of types of contaminants, (4) national, regional and specific-source contamination assessment, and (5) general and specific references of contamination sources.

Sources and Behavior of Subsurface Contaminants

Table 4-3 Explanation of Contaminant Plumes Shown in Figure 4-16

Site	Contaminant Plume Governed by: Dilution	Decay	Sorption	Liquid Waste Recharge Forming Water-Table Mound	Composite Waste Sites	Examples of Type of Contaminant	Remarks
A	Not appreciable in ground; some in stream	No	No	No	No	Chlorides, nitrates	
B	Not appreciable	Either decay or sorption or both		No	No	--	
C	Improbable	Perhaps	Perhaps	No	No	Sewage, radioactive wastes	Probably small waste release or good attenuation in zone of aeration
D	No plume formed (see remarks)	Either decay or sorption or both		No	No	Sewage, radioactive wastes	Contaminant is completely attenuated in zone of aeration and does not reach zone of saturation
E	Slight near waste site; some at greater distance	Possibly	Possibly	No	No	--	Lack of dispersion near waste site typical of linear openings in rock; contaminated water downgradient disperses into different type of material
F	Yes, suggestive of nearly homogeneous porous materials	Improbable	Improbable	No	No	Chlorides, nitrates	
G	Not appreciable in ground, some near and in stream	Not appreciable	Not appreciable	No	No	Chlorides, nitrates	Irregularities in permeability cause deviation in plume
H	Yes, suggestive of nearly homogeneous porous material	Probably either decay or sorption or both		No	No	Sewage, radioactive wastes	

Table 4-3 (cont.)

Site	Contaminant Plume Governed by: Dilution	Decay	Sorption	Liquid Waste Recharge Forming Water-Table Mound	Composite Waste Sites	Examples of Type of Contaminant	Remarks
I	Yes	Perhaps	Perhaps	No	No	--	Downgradient split in plume may be due to dense impermeable rock or great increase in sorptive materials
J	Slight	Not appreciable	Probably not appreciable	No	No	Chlorides, nitrates	Downgradient plume is due to discharge of contaminant to land surface at a surface seep and reinfiltration of contaminant
K	Yes; suggestive of nearly homogeneous porous materials	Either decay or sorption or both		Yes, forming a water-table mound	No	Sewage, radioactive wastes	Irregularities in plume caused by changes in permeability and/or sorption
L	Yes; suggestive of nearly homogeneous materials	Either decay or sorption or both		Yes, forming a water-table mound	No	Sewage, radioactive wastes	
M	Some in ground and stream	Not appreciable	Not appreciable	Yes, forming a water-table mound	--	Chlorides, nitrates	Deviation in plume due to impermeable zone
N	Yes	Either decay or sorption or both		Yes, forming a water-table mound	No	Sewage, radioactive wastes	Contaminated water from three waste sites at right angles to ground water flow, merging to form a composite plume
O	Yes	Either decay or sorption or both		No	Yes	Sewage, radioactive wastes	Contaminated water from two waste sites parallel to ground water flow, forming a composite plume

Table 4-3 (cont.)

Site	Contaminant Plume Governed by: Dilution	Decay	Sorption	Liquid Waste Recharge Forming Water-Table Mound	Composite Waste Sites	Examples of Type of Contaminant	Remarks
P	Some	Either decay or sorption or both		No	Yes	Sewage, radioactive wastes	Contaminated water from two waste sites at an angle with ground-water flow, forming a composite plume
Q	Some	Either decay or sorption or both		No	Yes	Sewage, radioactive wastes	Large composite plume formed by several waste sites
R	Yes	Either decay or sorption or both		No	No	Sewage, radioactive wastes	Pumping well draws plume toward it; contaminated water is greatly diluted at well

Source: Adapted from LeGrand (1965).

As discussed in Section 4.2.2, the number of potential ground-water contaminants is far too large to provide any detailed discussion of the chemical characteristics of specific contaminants. Table 4-5 provides an index to major references containing more detailed information about specific chemical processes, and chemical characteristics and behavior of contaminants in the subsurface. Generally, only texts, edited volumes, and conference proceedings are indexed in Table 4-5, but some important review papers published in scientific journals are also included. The references include: (1) general chemical references, (2) compilations of chemical fate constants, (3) references on ground water and vadose zone/soil chemistry, (4) references on trace elements and heavy metals, (5) references on toxic and other organic chemicals, and (6) references on microbial ecology and biodegradation.

Table 4-4 Index to Major References on Types and Sources of Contamination in Soil and Ground Water

Topic	References
General	Canter and Knox (1987), Cole (1972-Europe), Guswa et al. (1984), Haimes and Snyder (1986), Meyer (1973), Miller (1980, 1985), Pettyjohn (1972), U.S. Public Health Service (1961), van Duijvenbooden and van Waegeningen (1987), van Duijvenbooden et al. (1981), Ward et al. (1985); <u>Bibliographies/Literature Reviews</u>: Bader (1973), Congressional Research Service (1984), Lindorff and Cartwright (1977), Rima et al. (1971), Summers and Spiegel (1974), Todd and McNulty (1974), U.S. EPA (1972), van der Leeden (1991), Zanoni (1971)
Baseline Chemistry	Durfor and Becker (1964); <u>Soil</u>: Connor and Shacklette (1975), Dragun and Chiasson (1992), Ebens and Shacklette (1982), Shacklette et al. (1971a, 1971b, 1973, 1974); <u>Ground/Surface Water</u>: Clarke (1924), Durum and Haffty (1961), Durum et al. (1971), Ebens and Shacklette (1982), Feth (1981), Feth et al. (1968), Fishman and Hem (1976), Hem (1972), Kopp and Kroner (1968), Ledin et al. (1989), Leenheer et al. (1974), Love (1960), Pettyjohn et al. (1979), Skougstad and Horr (1963), Thurman (1985), White et al. (1963, 1970)
Types of Contaminants	Page (1981), Palmer et al. (1988), Pettyjohn and Hounslow (1983), Zoeteman (1985); <u>National Water Quality Assessments</u>: Francis et al. (1981), NAS (1987), U.S. EPA (1985a), Westrick et al. (1984)
Ground Water Contamination Assessments	<u>U.S.</u>: Ballentine et al. (1972), Lehr (1982), Patrick et al. (1987), Pye and Kelley (1984); <u>Regional Assessment</u>: Fuhriman and Barton (1971-AZ, CA, NV, UT), Miller and Scalf (1974), Miller et al. (1974-northeast), Miller et al. (1977-southeast), Scalf et al. (1973-southcentral), van der Leeden et al. (1975-northwest); <u>Source Assessments</u>: U.S. EPA (1977--waste disposal); U.S. EPA (1978, 1983-surface impoundments), U.S. EPA (1985b-injection of hazardous waste), U.S. EPA (1984-rural water, 1986a, 1986b-underground storage tanks), U.S. EPA (Pesticides: 1986d, 1990c, 1994)
<u>Contamination Sources</u>	
General	Cape Cod Aquifer Management Project (1988), LaSpina and Palmquist (1992), Meyer (1973), Miller (1982), Noake (1988), Shineldecker (1992), U.S. EPA (1977, 1987a, 1990b, 1991b), U.S. Fish and Wildlife Service (1986), U.S. OTA (1984)

Table 4-4 (cont.)

Topic	References
Contamination Sources (cont.)	
State Contaminant Inventory Guidance	Nebraska Department of Environmental Quality (1992), New Hampshire Office of State Planning (1991), North Dakota State Department of Health (1993), Ohio Environmental Protection Agency (1991), Oregon Department of Environmental Quality (1992), RIDEM (1992), Washington State Department of Health (1993)
Commercial/Industrial	Dotson (1991), U.S. EPA (1990a, 1991a, 1992), Ward et al. (1990), WEF (1993), see also Table 11-6 and references on waste minimization in Table 11-11; Underground Storage Tanks: U.S. EPA (1986a, 1986b); Estimating Chemical Releases: PEI Associates (1990), U.S. EPA (1987b, 1988a, 1990e)
Rural/Nonpoint	Ashton and Underwood (1975), Delfino (1977), D'Itri and Wolfson (1987), Nielsen and Lee (1987), Novotny and Chesters (1981), Overcash and Davidson (1980), U.S. EPA (1984, 1990d, 1991b)
Agricultural Chemicals	Bloom and Degler (1969), Fairchild (1987), Gustafson (1993), Hallberg (1986), Irvine and Knights (1974), Jenkins (1979), Rack and Leslie (1993), U.S. EPA (1986c), WEF (1993)
Resource Extraction	Energy Production/Use: Boulding (1992), Dotson (1991), U.S. Army Engineers Waterways Experiment Station (1979), U.S. EPA (1988b); Mineral Extraction: U.S. EPA (1985c)
Septic Systems	California Assembly Office of Research (1985), Canter and Knox (1984, 1985), Cartwright and Sherman (1974), Noss (1989), Scalf et al. (1977), Thomson (1984); see also references on bacterial and viral tracers in Table 8-5.
Waste Disposal Sites	Landfills: Zanoni (1971), see also landfill gases in Table 12-7; Surface Impoundments: Silka and Swearingen (1978), U.S. EPA (1978, 1983)
Wells	Abandoned: Aller (1984), Frischknecht et al. (1983), Gass et al. (1977), Texas Water Commission (1989b); Waste Injection: Rima et al. (1971), U.S. EPA (1985b, 1990)
Other Sources	Accidental Spills: Guswa et al. (1984); Military Sites: Kimball et al. (1993)

Table 4-4 References (Appendix F contains references for figure and table sources)

Aller, L. 1984. Methods for Determining the Location of Abandoned Wells. EPA-600/2-83-123 (NTIS PB84-141530). Also published in NWWA/EPA Series, National Water Well Association, Dublin, OH, 130 pp.

Ashton, P.M. and R.C. Underwood. 1975. Non-Point Sources of Water Pollution. Virginia Water Resources Center, Virginia Tech, Blacksburg, VA.

Bader, J.S. et al. 1973. Selected References—Ground-Water Contamination, United States and Puerto Rico. U.S. Geological Survey, Washington, DC. [834 references indexed according to geographic areas, states, and kinds and sources of contamination]

Ballentine, R.K., S.R. Reznek, and C.W. Hall. 1972. Subsurface Pollution Problems in the United States. EPA TS-00-72-02 (NTIS PB210 293).

Bloom, S.C. and S.E. Degler. 1969. Pesticides and Pollution. Bureau of International Affairs, Washington, DC.

Boulding, J.R. 1992. Disposal of Coal Combustion Waste in Indiana: An Analysis of Technical and Regulatory Issues, Final Report. Prepared for Hoosier Environmental Council, Indianapolis, IN, 104 pp. [Contains comprehensive review of literature on potential for ground-water contamination from coal ash and flue gas desulfurization wastes]

California Assembly Office of Research. 1985. The Leaching Fields: A Nonpoint Threat to Groundwater. California State Assembly, Sacramento.

Canter, L.W. and R.C. Knox. 1984. Evaluation of Septic Tank System Effects on Ground Water Quality. EPA/600/2-84/107 (NTIS PB84-244441), 381 pp.

Canter, L.W. and R.C. Knox. 1985. Septic Tank Systems Effects on Ground Water Quality. Lewis Publishers, Chelsea, MI.

Canter, L.W., R.C. Knox, and D.M. Fairchild. 1987. Ground Water Quality Protection. Lewis Publishers, Chelsea, MI.

Cape Cod Aquifer Management Project. 1988. Guide to Contamination Sources for Wellhead Protection. EPA/901/3-88/004 (NTIS PB89-155790), 320 pp. [Executive Summary, EPA/901/3-88/003 (NTIS PB89-155782), 20 pp]

Cartwright, K. and F.B. Sherman, Jr. 1974. Assessing Potential for Pollution from Septic Systems. Ground Water 12:239-240.

Clarke, F.W. 1924. The Composition of the River and Lake Waters of the United States. U.S. Geological Survey Professional Paper 135, 199 pp.

Cole, J.A. (ed.). 1972. Groundwater Pollution in Europe. Water Information Center, Port Washington, NY. [More than 50 papers and case histories]

Congressional Research Service. 1984. Groundwater Contamination by Toxic Substances: A Digest of Reports. U.S. Library of Congress, Washington, DC.

Connor, J.J. and H.T. Shacklette. 1975. Background Geochemistry of Some Rocks, Soils, Plants, and Vegetables in the Conterminous United States. U.S. Geological Survey Professional Paper 574-F.

Delfino, J.J. 1977. Contamination of Potable Groundwater Supplies in Rural Areas. In: Drinking Water Quality Enhancement Through Source Protection, R.B. Pojacek (ed.), Ann Arbor Science Press, Ann Arbor, MI, pp. 275-295.

D'Itri, F.M. and L.G. Wolfson (eds.). 1987. Rural Groundwater Contamination. Lewis Publishers, Chelsea, MI.

Dotson, G.K. 1991. Migration of Hazardous Substances through Soils: Part II—Determination of the Leachability of Metals from Five Industrial Wastes and their Movement within Soil; Part III—Flue-Gas Desulfurization and Fly-Ash Wastes; Part IV—Development of a Serial Batch Extraction Method and Application to the Accelerated Testing of Seven Industrial Wastes. EPA/600/2091/017 (Part II, incorporating unpublished portions of Part I interim report: NTIS AD-A 158990; Part III: AD-A 182108; Part IV AD-A 191856). [Waste from electroplating, secondary zinc refining, inorganic pigment, zinc-carbon battery, titanium dioxide pigment, nickel-cadmium battery, hydrofluoric acid, water-based paint, white phosphorus, chlorine production, oil re-refining, flue-gas desulfurization, and coal fly ash]

Dragun, J. and A. Chiasson. 1992. Elements in North American Soils. Hazardous Materials Control Research Institute, Greenbelt, MD, 300 pp.

Durfor, C.N. and E. Becker. 1964. Public Water Supplies of the 100 Largest Cities in the United States, 1962. U.S. Geological Survey Water-Supply Paper 1812, 364 pp.

Durum, W.H. and J. Haffty. 1961. Occurrence of Minor Elements in Water. U.S. Geological Survey Circular 445.

Durum, W.H., J.D. Hem, and S.G. Heidel. 1971. Reconnaissance of Selected Minor Elements in Surface Waters of the United States, October 1970. U.S. Geological Survey Circular 643.

Ebens, R.J. and H.T. Shacklette. 1982. Geochemistry of Some Rocks, Mine Spoils, Stream Sediments, Soils, Plants and Waters in the Western Energy Region of the Conterminous United States. U.S. Geological Survey Professional Paper 1238.

Fairchild, D.M. (ed.). 1987. Ground Water Quality and Agricultural Practices. Lewis Publishers, Chelsea, MI.

Feth, J.H. 1981. Chloride in Natural Continental Water-A Review. U.S. Geological Survey Water-Supply Paper 2176, 30 pp.

Feth, J.H. et al. 1965. Preliminary Map of the Conterminous U.S. Showing Depth to and Quality of Shallowest Ground Water Containing More Than 1,000 ppm Dissolved Solids. U.S. Geological Survey Hydrologic Atlas HA-199. [2 sheets with 31 pages of text]

(Table 4-4 Contaminant Source References)

Fishman, M.J. and J.D. Hem. 1976. Lead Content of Water. In: Lead in the Environment, T.G. Lovering (ed.), U.S. Geological Survey Professional Paper 957, pp. 35-41.

Francis, J.D., B.L. Brower, and W.F. Graham. 1981. National Statistical Assessment of Rural Water Conditions. U.S. Environmental Protection Agency, Washington, DC.

Frischknecht, F.C., L. Muth, R. Grette, T. Buckley, and B. Kornegay. 1983. Geophysical Methods for Locating Abandoned Wells. U.S. Geological Survey Open-File Report 83-702, 211 pp. Also published as EPA/600/4-84-065 (NTIS PB84-212711).

Fuhriman, D.K. and J.R. Barton. 1971. Ground Water Pollution in Arizona, California, Nevada and Utah. EPA 16060 ERU 12/71 (NTIS PB211 145).

Gass, T.E., J.H. Lehr, and H.W. Heiss, Jr. 1977. Impact of Abandoned Wells on Ground Water. EPA/600/3-77-095 (NTIS PB272-665)

Gustafson, D. 1993. Pesticides in Drinking Water. Van Nostrand Reinhold, New York, NY, 256 pp.

Guswa, J.H., W.J. Lyman, A.S. Donigian, Jr., T.Y.R. Lo, and E.W. Shanahan. 1984. Groundwater Contamination and Emergency Response Guide. Noyes Publications, Park Ridge, NY.

Haimes, Y.Y. and J.H. Snyder (eds.). 1986. Groundwater Contamination. Engineering Foundation, New York, NY.

Hallberg, G.R. 1986. Overview of Agricultural Chemicals in Ground Water. In: Agricultural Impacts on Ground Water--A Conference, National Water Well Association, Dublin, OH, pp. 1-66.

Hem, J.D. 1972. Chemistry and Occurrence of Cadmium and Zinc in Surface Water and Ground Water. Water Resources Research 8:661-679.

Irvine, D.E.G. and B. Knights. 1974. Pollution and the Use of Chemicals in Agriculture. Butterworth, London.

Jenkins, S.H. (ed.). 1979. The Agricultural Industry and Its Effects on Water Quality. Pergamon Press, New York, NY.

Kimball, D., L. Siegel, and P. Tyler. 1993. Covering the Map: A Survey of Military Pollution Sites in the United States. Physicians for Social Responsibility, Washington, DC.

Kopp, J.F. and R.C. Kroner. 1968. Trace Metals in Water in the United States, October 1, 1962-September 30, 1967. U.S. Department of the Interior, Federal Water Pollution Control Administration, 48 pp.

LaSpina, J. and R. Palmquist. 1992. Catalog of Contaminant Databases: A Listing of Databases of Actual or Potential Contaminant Sources. Washington State Department of Ecology, Olympia, WA.

(Table 4-4 Contaminant Source References)

Ledin, A., C. Pettersson, B. Allard, and M. Aastrup. 1989. Background Concentration Ranges of Heavy Metals in Swedish Groundwaters from Crystalline Rocks: A Review. Water, Air, and Soil Pollution 47:419-426. [Includes Cr, Cu, Zn, Cd, Pb]

Leenheer, J.A., R.L. Malcolm, P.W. McKinley, and L.A. Eccles. 1974. Occurrence of Dissolved Organic Carbon in Selected Groundwater Samples in the United States. J. Res. U.S. Geological Survey 2:361-369.

Lehr, J.H. 1982. How Much Ground Water Have We Really Polluted? Ground Water Monitoring Review 2(1):4.

Lindorff, D.E. and K. Cartwright. 1977. Ground-Water Contamination: Problems and Remedial Actions. Illinois Geological Survey Environmental Geology Note 81, 58 pp. [75 references, 116 ground-water contamination case histories]

Love, S. 1960. Quality of Surface Water of the United States, 1958. U.S. Geological Survey Water Supply Paper 1571, Parts 1-4, 733 pp.

Meyer, C.F. (ed.). 1973. Polluted Groundwater: Some Causes, Effects, Controls, and Monitoring. EPA 600/4-73-001b (NTIS PB232 117).

Miller, D.W. (ed.). 1980. Waste Disposal Effects on Ground Water. Premier Press, Berkeley, CA. [Note: this report is the same as U.S. EPA (1977)]

Miller, D.W. 1982. Groundwater Contamination: A Special Report. Geraghty & Miller, Inc., Syosset, NY.

Miller, D.W. 1985. Chemical Contamination of Ground Water. In: Ground Water Quality, C.H. Ward, W. Giger, and P.L. McCarty, (eds.), Wiley Interscience, New York, NY, pp. 39-52.

Miller, D.W. and M.R. Scalf. 1974. New Priorities for Groundwater Quality Protection. Ground Water 12(6):335-347.

Miller, D.W., F.A. DeLuca, and T.L. Tessier. 1974. Ground Water Contamination in the Northeast States. EPA 660/2-74/056 (NTIS PB235 702).

Miller, J.C., P.S. Hackenberry, and F.A. DeLuca. 1977. Ground-Water Pollution Problems in the Southeastern United States. EPA 600/3-77/012 (NTIS PB268 234).

Nebraska Department of Environmental Quality. 1992. Contaminant Source Inventory. Wellhead Protection Newsletter III, NDEQ, Lincoln, NE, 12 pp.

Nielsen, E.G. and L.K. Lee. 1987. The Magnitude and Costs of Groundwater Contamination from Agricultural Chemicals--A National Perspective. Economic Research Service, U.S. Dept. of Agriculture, Washington, DC, 54 pp.

New Hampshire Office of State Planning. 1991. Developing a Local Inventory of Potential Contamination Sources. Prepared for New Hampshire Department of Environmental Services, Water Supply and Pollution Control Division, Concord, NH, 63 pp.

(Table 4-4 Contaminant Source References)

Noake, K.D. 1988. Guide to Contamination Sources for Wellhead Protection (Draft). Massachusetts Department of Environmental Quality Engineering, Boston, MA.

National Academy of Science (NAS). 1987. National Water Quality Monitoring and Assessment. National Academy of Science Press, Washington, DC.

North Dakota State Department of Health. 1993. North Dakota Wellhead Protection User's Guide. Division of Water Quality, Bismarck, ND.

Noss, R.R. 1989. Septic System Cleaners: A Significant Threat to Groundwater Quality. Journal of Environmental Health 51(4):201-204.

Novotny, V. and G. Chesters. 1981. Handbook of Nonpoint Source Pollution Sources and Management. Van Nostrand Reinhold, New York, NY.

Office of Technology Assessment (OTA). 1984. Protecting the Nation's Groundwater from Contamination, Vols. I and II. OTA-0-233 and OTA-0-276. OTA, Washington, DC. [Chapter 2 of Volume I and Appendix A of Volume II focus on ground-water contamination and its impacts]

Ohio Environmental Protection Agency. 1991. Guidance for Conducting Pollution Source Inventories in Wellhead Protection Areas (Draft). OEPA, Division of Ground Water, Columbus, OH, 17 pp.

Oregon Department of Environmental Quality. 1992. Guidelines for Potential Source of Contamination for Wellhead Protection in Oregon. Oregon Department of Environmental Quality, Portland, OR. [Based on Noake (1988)]

Overcash, M.R. and J.M. Davidson (eds.). 1980. Environmental Impact of Nonpoint Source Pollution. Ann Arbor Science Press, Ann Arbor, MI.

Page, G.W. 1981. Comparison of Groundwater and Surface Water for Patterns and Levels of Contaminations by Toxic Substances. Environ. Sci. Technol. 15:1475-1481.

Palmer, C.D., W. Fish, and J.F. Keely. 1988. Inorganic Contaminants: Recognizing the Problem. In: Proc. 2nd Nat. Outdoor Action Conf. on Aquifer Restoration, Ground Water Monitoring and Geophysical Methods, National Water Well Association, Dublin, OH, pp. 555-579.

Patrick, R., E. Ford, and J. Quarles. 1987. Groundwater Contamination in the United States, 2nd ed. University of Pennsylvania Press, Philadelphia, PA. (First edition, published in 1983, was by Pye, Patrick and Quarles). [Contains special summaries for 19 states: AZ, CA, CT, FL, ID, IL, MA, MT, NE, NJ, NM, ND, OR, PA, RI, SC, TX, VT, and WA]

PEI Associates. 1990. Guidance for Food Processors: Section 313, Emergency Planning and Community Right-to-Know Act. EPA 560/4-90-014. Available from EPCRA Hotline.*

Pettyjohn, W.A. 1972. Water Quality in Stressed Environments. Burgess Pub. Co., Minneapolis, MN, 309 pp.

(Table 4-4 Contaminant Source References)

Pettyjohn, W.A. and A.W. Hounslow. 1983. Organic Compounds and Ground-Water Pollution. Ground Water Monitoring Review 3(4):41-47.

Pettyjohn, W.A., J.R.J. Studlick, R.C. Bain, and J.H. Lehr. 1979. A Groundwater Quality Atlas of the United States. National Demonstration Water Project, 272 pp.

Pye, V.I. and J. Kelley. 1984. The Extent of Groundwater Contamination in the United States. In: Groundwater Contamination, National Academy Press, Washington, DC, pp. 23-33.

Pye, Patrick and Quarles (1983) -- see Patrick et al. (1987)

Rack, K.D. and A.R. Leslie (eds.). 1993. Pesticides in Urban Environments: Fate and Significance. ACS Symp. Series No. 522, American Chemical Society, Washington, DC, 385 pp.

Rhode Island Department of Environmental Management (RIDEM). 1992. Inventory of Potential Sources of Groundwater Contamination in Wellhead Protection Areas: RIDEM Guidance Document. RIDEM, Providence, RI, 38 pp. + appendices.

Rima, D.R., E.B. Chase and B.M. Myers. 1971. Subsurface Waste Disposal by Means of Wells-A Selected Annotated Bibliography. U.S. Geological Survey Water-Supply Paper 2020. [692 references]

Scalf, M.R., J.W. Keeley, and C.J. LaFevers. 1973. Ground Water Pollution in the South Central States. EPA R2-73/268 (NTIS PB222 178).

Scalf, M.R., W.J. Dunlap, and J.F. Kreissl. 1977. Environmental Effects of Septic Tank Systems. EPA/600/3-77-096 (NTIS PB272-702), 43 pp.

Shacklette, H.T. et al. 1971a. Elemental Composition of Surficial Materials in the Conterminous United States. U.S. Geological Survey Professional Paper 574-D. Includes: Al, Ba, Be, Bo, Ca, Ce, Cr, Co, Cu, Ga, Fe, La, Pb, Mg, Mo, Ne, Ni, Nb, P, K, Sc, Na, Sr, Ti, V, Y, Yb, Zn, Zr.

Shacklette, H.T. et al. 1971b. Mercury in the Environment—Surficial Materials of the Conterminous United States. U.S. Geological Survey Circular 644.

Shacklette, H.T. et al. 1973. Lithium in Surficial Materials of the Conterminous United States and Partial Data on Cadmium. U.S. Geological Survey Circular 673.

Shacklette, H.T. et al. 1974. Selenium, Fluorine, and Arsenic in Surficial Materials of the Conterminous United States. U.S. Geological Survey Circular 692.

Shineldecker, C.L. 1992. Handbook of Environmental Contaminants. Lewis Publishers, Chelsea, MI, 371 pp. [Key to contaminants that are likely to be associated with specific types of facilities, processes, and products]

Silka, L.R. and T.L. Swearingen. 1978. Manual for Evaluating Contamination Potential of Surface Impoundments. EPA-570/9-78-003 (NTIS PB85-211433).

(Table 4-4 Contaminant Source References)

Skougstad, M.W. and C.A. Horr. 1963. Occurrence and Distribution of Strontium in Natural Water. U.S. Geological Survey Water-Supply Paper 1496-D, pp. D55-D97.

Summers, W.K. and Z. Spiegel. 1974. Ground Water Pollution: A Bibliography. Ann Arbor Science Publishers, Ann Arbor, MI. [Partially annotated; more than 400 references organized by topic]

Texas Water Commission. 1989b. On Dangerous Ground: The Problem of Abandoned Wells in Texas. Austin, TX.

Thomson, M. et al. 1984. Characterization of Soil Disposal System Leachates. EPA/600/2-84/101 (NTIS PB84-196229).

Thurman, E.M. 1985. Humic Substances in Groundwater. In: Humic Substances in Soil, Sediment, and Water: Geochemistry, Isolation, and Characterization, G.R. Aiken, D.M. McKnight, R.L. Wershaw, and P. MacCarthy (eds.), John Wiley & Sons, New York, NY, pp. 87-103.

Todd, D.K. and D.E.O. McNulty. 1974. Polluted Groundwater: A Review of the Significant Literature. EPA 680/4-74-001 (NTIS PB235 556). Also published in 1976 under same title by Water Information Center, Plainview, NY. [661 references]

U.S. Army Engineers Waterways Experiment Station. 1979. Effects of Flue Gas Cleaning Waste on Groundwater Quality and Soil Characteristics. EPA/600/2-79/164 (NTIS PB80-118656).

U.S. Environmental Protection Agency (EPA). 1972. Subsurface Water Pollution—A Selective Annotated Bibliography, Part I—Subsurface Waste Injection (NTIS PB211 340), Part II—Saline Water Intrusion (NTIS PB211 341), Pt. III—Percolation from Surface Sources (NTIS PB211 342). [Total of 319 references]

U.S. Environmental Protection Agency (EPA). 1977. The Report to Congress, Waste Disposal Practices and Their Effects on Ground Water. EPA/570/9-77/001 (NTIS PB265-081), 512 pp. [Note: this report is the same as Miller (1980)]

U.S. Environmental Protection Agency (EPA). 1978. Surface Impoundments and their Effects on Ground Water Quality in the U.S.--A Preliminary Survey. EPA-570/9-78-005.

U.S. Environmental Protection Agency (EPA). 1979. Environmental Assessment: Short-Term Tests for Carcinogens, Mutagens and Other Genotoxic Agents. EPA/615/9-79/003 (NTIS PB300 611).

U.S. Environmental Protection Agency (EPA). 1983. Surface Impoundment Assessment National Report. EPA 570/9-84-002 (NTIS DE84-901182).

U.S. Environmental Protection Agency. (EPA). 1984. National Statistical Assessment of Rural Water Conditions. Executive Summary (EPA/570/9-84-003--Also included in Technical Summary); Technical Summary (EPA/570/9-84-004; NTIS PB84-213517); Set of four Volumes (EPA/570/9-84-004; NTIS PB84-222322); Vol. I (EPA/570/9-84-004a; NTIS PB84-222330, 424 pp.); Vol. II (EPA/570/9-84-004b; NTIS PB84-222348, 444 pp.); Vol. III

(Table 4-4 Contaminant Source References)

Sources and Behavior of Subsurface Contaminants 215

(EPA/570/9-84-004c; NTIS PB84-222355, 465 pp.); Vol. IV (EPA/570/9-84-004d; NTIS PB84-222363, 316 pp.).

U.S. Environmental Protection Agency (EPA). 1985a. National Water Quality Inventory 1984: National Report to Congress. EPA 440/4-85-029.

U.S. Environmental Protection Agency (EPA). 1985b. Report to Congress on Injection of Hazardous Wastes. EPA 570/9-85/003 (NTIS PB86-203056).

U.S. Environmental Protection Agency (EPA). 1985c. Wastes from the Extraction and Beneficiation of Metallic Ores, Phosphate Rock, Asbestos, Overburden from Uranium Mining, and Oil Shale: Report to Congress. EPA/530-SW-85-033, 290 pp.

U.S. Environmental Protection Agency (EPA). 1986a. Summary of State Reports on Releases from Underground Storage Tanks. EPA 600/M-86/020.

U.S. Environmental Protection Agency (EPA). 1986b. Underground Motor Fuel Storage Tanks: A National Survey: Vol. 1, Technical Report; Vol. 2, Appendices. EPA 560/5-86-013 (NTIS PB86-216512), 556 pp.

U.S. Environmental Protection Agency (EPA). 1986c. Pesticides in Ground Water: Background Document. EPA/440/6-86-002 (NTIS PB88-111976).

U.S. Environmental Protection Agency (EPA). 1986d. National Survey of Pesticides in Drinking Water Wells. U.S. EPA, Washington, DC.

U.S. Environmental Protection Agency (EPA). 1987a. EPA Activities Related to Sources of Ground-Water Contamination. EPA/440/6-87/002 (NTIS PB88-111901), 125 pp.

U.S. Environmental Protection Agency (EPA). 1987b. Estimating Releases and Waste Treatment Efficiencies for the Toxic Chemical Release Inventory Form. EPA/560/4-88-002. Available from EPCRA Hotline.*

U.S. Environmental Protection Agency (EPA). 1988a. *Industry-Specific Guidance Documents for Estimating Releases*: Monofilament Fiber Manufacture (EPA/560/4-88-004a); Printing Operations (EPA/560/4-88-004b); Electrodeposition of Organic Coatings (EPA/560/4-88-004c); Spray Application of Organic Coatings (EPA/560/4-88-004d); Semiconductor Manufacturers (EPA/560/4-88-004e); Formulation of Aqueous Solutions (EPA/560/4-88-004f); Electroplating Operations (EPA/560/4-88-004g); Textile Dyeing (EPA/560/4-88-004h); Presswood & Laminated Wood Products Manufacturing (EPA/560/4-88-004i); Roller, Knife and Gravure Coating Operations (EPA/560/4-88-004j); Paper and Paperboard Production (EPA/560/4-88-004k); Leather Tanning and Finishing (EPA/560/4-88-004l); Wood Preserving (EPA/560/4-88-004p); Rubber Production and Compounding (EPA/560/4-88-004q). Available from EPCRA Hotline.*

U.S. Environmental Protection Agency (EPA). 1988b. Report to Congress: Waste from the Combustion of Coal by Electric Utility Power Plants. EPA/530-SW-88-002.

(Table 4-4 Contaminant Source References)

U.S. Environmental Protection Agency (EPA). 1990a. Does Your Business Produce Hazardous Waste? Many Small Businesses Do. EPA/530/SW-90-027, 5 pp. Available from RCRA Hotline--see Table 5-3. [2- to 4-page business-specific reports (EPA/530/SW-90-027A to S) are also available from RCRA Hotline--see Table 5-3: Vehicle Maintenance (A); Drycleaning and Laundry (B); Furniture/Wood Finishing (C); Equipment Repair (D); Textile Manufacturing (E); Wood Preserving (F); Printing and Allied Industry (G); Chemical Manufacturers (H); Pesticide End-Users (I); Construction (J); Motor Freight Terminals/Railroad Transport (K); Educational/Vocational (L); Laboratories (M); Metal Manufacturing (N); Pulp and Paper Industry (O); Formulators (P); Cleaning and Cosmetics (Q); Leather and Leather Products (R); Uniform Hazardous Waste Manifest Instructions (S)]

U.S. Environmental Protection Agency (EPA). 1990b. Ground Water Handbook, Vol I: Ground Water and Contamination. EPA/625/6-90/016a. Available from CERI.*

U.S. Environmental Protection Agency. (EPA). 1990c. National Survey of Pesticides in Drinking Water: Phase I Report. EPA/570/9-90-014 (NTIS PB91-125765).

U.S. Environmental Protection Agency (EPA). 1990d. An Annotated Bibliography of the Literature Addressing Nonpoint Source Contaminated Ground-Water Discharge to Surface Water. EPA/440/6-90-006.

U.S. Environmental Protection Agency (EPA). 1990e. Toxic Chemical Release Inventory: Clarification and Guidance for the Metal Fabrication Industry (Section 313 Issue Reporting Paper). EPA/560/4-90-012. Available from EPCRA Hotline.*

U.S. Environmental Protection Agency (EPA). 1991a. A Review of Sources of Ground-Water Contamination from Light Industry. EPA/440/6-90-005 (NTIS PB91-145938).

U.S. Environmental Protection Agency (EPA). 1991b. A Review of Methods for Assessing Nonpoint Source Contaminated Ground-Water Discharge to Surface Water. EPA/570/9-91-010 (NTIS PB92-188697).

U.S. Environmental Protection Agency (EPA). 1992. Publications Office of Science and Technology: Catalog. EPA-820-B-92-002. Available from U.S. EPA Office of Water Resource Center (WH-556) 401 M Street, SW, Washington DC 20460; 202/260-7786. [List of titles for over 200 EPA documents used to develop industrial effluent limitations and guidelines along with information on how documents can be obtained].

U.S. Environmental Protection Agency (EPA). 1994. Pesticides in Ground Water Database-A Compilation of Monitoring Studies, 1971-1991: National Summary. EPA/734-12-92-001 (NTIS PB93-163715), 200 pp. Also available from U.S. Government Printing Office S/N 055-000-00413-7. [Separate reports published for each EPA Region: Region 1 (PB93-163723), Region 2 (PB93-163731), Region 3 (PB93-163749), Region 4 (PB93-163756), Region 5 (PB93-163764), Region 6 (PB93-163772), Region 7 (PB93-163780), Region 8 (PB93-163798), Region 9 (PB93-163806), Region 10 (PB93-163814); also available electronically from EPA's Pesticide Information Network: 703/305-7499]

U.S. Fish and Wildlife Service. 1986. Contaminant Issues of Concern—National Wildlife Refuges. Washington, DC.

(Table 4-4 Contaminant Source References)

U.S. Office of Technology Assessment (OTA). 1984. Protecting the Nation's Groundwater from Contamination, 2 Vols. OTA-O-233 and OTA-O-276. Washington, DC.

U.S. Public Health Service. 1961. Proceedings of the 1961 Symposium, Ground Water Contamination. U.S. Public Health Service Tech. Rept. W61-5.

van der Leeden, F., L.A. Cerrillo, and D.W. Miller. 1975. Ground-Water Pollution Problems in the Northwestern United States. EPA 660/3-75/018 (NTIS PB242 860).

van der Leeden, F. 1991. Geraghty & Miller's Groundwater Bibliography, 5th ed. Water Information Center, Plainview, NY. 4th ed. 1987 [Some 5,000 selected references in 32 categories]

van Duijvenbooden, W. and H.G. van Waegeningen (eds.). 1987. Vulnerability of Soil and Groundwater to Pollutants. Nat. Inst. of Public Health and Environmental Hygiene, Noordwijk aan Zee, The Netherlands, Vol. 38.

van Duijvenbooden, W., P. Glasbergen, and H. van Lelyveld. 1981. Quality of Groundwater. Elsevier, New York, NY. [Sections 1 (Effects of diffuse polluting sources, land and precipitation) and 2 (effects of local polluting sources) contain 45 papers]

Ward, C.H., W. Giger, and P.L. McCarty (eds.). 1985. Ground Water Quality. Wiley-Interscience, New York, NY. [Part One contains 8 contributed chapters on sources, types, and quantities of contaminants in ground waters]

Ward, W.D., L.E. Oates, and K.B. McCormack. 1990. Tools for Wellhead Protection: Control and Identification of Light Industrial Sources. Ground Water Management 1:579-593 (Proc. of the 1990 Cluster of Conferences: Ground Water Management and Wellhead Protection).

Washington State Department of Health. 1993. Inventory for Potential Contaminant Sources Within Washington's Wellhead Protection Areas. Washington State Department of Health, Olympia, WA, 25 pp.

Water Environmental Federation (WEF). 1993. *Proceedings Second USA/CIS Joint Conference on Environmental Hydrology and Hydrogeology, 5 Vols.*: Environmental Impact of Industrial Activities (21 papers, 197 pp.); Environmental Impact of Agricultural Practices and Agrichemicals (20 papers, 217 pp.); Global and Regional Issues in Environmental Hydrology (31 papers, 251 pp.); Hydrogeologic Investigations and Monitoring, Groundwater Modeling (28 papers, 367 pp.); Water Management and Protection (24 papers, 239 pp.). WEF, Alexandria, VA. [Conference organized by American Institute of Hydrology]

Westrick, J.J., J.W. Mello, and R.F. Thomas. 1984. The Ground Water Supply Survey. J. Am. Water Works Ass. 76(5):52.

White, D.G., J.D. Hem, and G.A. Waring. 1963. Chemical Composition of Subsurface Waters. In: Data of Geochemistry, 6th ed., U.S. Geological Survey Professional Paper 440-F, 67 pp.

White, D.E., M.E. Hinkle, and I. Barnes. 1970. Mercury Contents of Natural Thermal and Mineral Fluids. In: Mercury in the Environment, U.S. Geological Survey Professional Paper 713, pp. 25-28

Zanoni, A.E. 1971. Ground-Water Pollution and Sanitary Landfills—A Critical Review. In: Proceedings of the National Water Quality Symposium, EPA 1606 ERB 08/71 (NTIS PB214 614), pp. 97-110. [61 references]

Zoeteman, B.C.J. 1985. Overview of Contaminants in Ground Water. In: Ground Water Quality, C.H. Ward, W. Giger, and P.L. McCarty, (eds.), Wiley Interscience, New York, NY, pp. 27-37.

* See Introduction for information on how to obtain documents from CERI (U.S. EPA Center for Environmental Research Information) and NTIS. Table 5-3 gives the phone number for the EPCRA Hotline.

(Table 4-4 Contaminant Source References)

Sources and Behavior of Subsurface Contaminants

Table 4-5 Index to Major References on Contaminant Chemical Characteristics and Behavior in the Subsurface

Topic	References
General Chemical References	ACS (annual), Budavari (1989), Dean (1992), Howard and Neal (1992), Lewis (1992a), Lide (1993), Perry and Chiltin (1973), Verschueren (1983); <u>Hazardous Chemicals</u>: ACGIH (1992), Armour (1991), Government Institutes (annual), Keith (1993), Lewis (1990, 1991, 1992b, 1993, 1994), NIOSH (1990), Occupational Safety Health Services (1990), Patnalk (1992), Shafer (1993), Shineldecker (1992), U.S. Coast Guard (1985), U.S. DOT (1990), U.S. EPA (1985, 1992a); <u>Agrochemicals</u>: Fisher (1991), James and Kidd (1992), Kidd and James (1991), Montgomery (1993), Walker and Keith (1992)
Chemical Fate Data	Boulding (1990), Callahan et al. (1979), Gherini et al. (1988, 1989), Howard (1989, 1990a, 1990b, 1992a, 1992b, 1993), Howard et al. (1991), Kollig (1993), Kollig et al. (1991), Lyman et al. (1990, 1992), Mabey et al. (1982), Montgomery (1991), Montgomery and Welkom (1989), Ney (1990), Rai and Zachara (1984), Rai et al. (1984); <u>Sorption/Partition Coefficients</u>: Ellington et al. (1991), Fochman (1981), Howard and Meylan (1992), Leo et al. (1971), Sablić (1988); <u>Henry's Law Constants</u>: Yaws et al. (1991); <u>Hydrolysis Rate Constants</u>: Ellington et al. (1991), Kollig et al. (1990); <u>Diffusion Coefficients</u>: Boynton and Brattain (1929-gases/vapors), Bruins (1929-liquids); <u>Speciation</u>: NBS (1981)
Natural Baseline Chemistry	See Table 4-4
Contaminant Sources	See Table 4-4
Chemical/Contaminant Hydrogeology	<u>Texts</u>: Back and Freeze (1983), Bedient et al. (1994), Devinny et al. (1990), Domencio and Schwartz (1991), Fetter (1992), Matthess (1982), Mazor (1990), Palmer (1992), Tinsley (1979), Weber (1972); <u>Papers</u>: Back and Baedecker (1989), Brusseau (1993), Mackay et al (1985); <u>Subsurface Transport/Retardation Processes</u>: Gelhar et al. (1985), Guarmaccia et al. (1992-multiphase), Güven et al. (1992a, 1992b), Huling (1989-facilitated transport), Huling and Weaver (1991-DNAPLs), Knox et al. (1993), Luckner and Schestakow (1991), Piwoni and Keeley (1990), U.S. EPA (1989, 1992b), van der Zee and Desouni (1993)

Table 4-5 (cont.)

Topic	References
Vadose Zone/Soil/ Sediment Chemistry	Environmental Science and Engineering (1985), Yaron et al. (1984), Yong et al. (1992); <u>Inorganic Chemicals</u>: Bar-Yosef et al. (1989); <u>Toxic Organic Chemicals</u>: Dragun (1988), Gerstl et al. (1989), Goring and Hamaker (1972), TNO/BMFT (1985, 1989); <u>Contaminated Sediments</u>: Baker (1980), Gambrell et al. (1977), Horowitz (1991), Shear and Watson (1977)
Biodegradation/ Contaminant Microbiology	Alexander (1994), Borchardt et al. (1977), Fochman (1981), Gibson (1984), Kobayashi and Rittman (1982), Mitchell (1971), Ratledge (1993), Rogers (1986), Rogers and Abramowicz (1993), Scow (1982), Sims et al. (1991), Zehnder (1988); <u>Soil</u>: Huang and Schnitzer (1986), Nelson et al. (1983), Ramsey et al. (1972); <u>Ground Water</u>: Bitton and Gerba (1984), Bouwer and McCarty (1984), Ghiorse and Wilson (1988), Maki et al. (1980), Tabak et al. (1981), Wilson and McNabb (1983)
Trace Elements/ Heavy Metals	Beck et al. (1992/T3-12), Bowen (1966), Chappelle and Peterson (1976), Hem (1964), Jarrell et al. (1980-molybdenum), National Research Council Canada (1976, 1978a, 1978b, 1979a, 1979b, 1981, 1982), Purves (1978), Thibodeaux (1979), Thornton (1983), Shaw (1989), Watras and Huckabee (1994-mercury); <u>Sediment</u>: Gambrell et al. (1977), Horowitz (1991); <u>Soil</u>: Allaway (1968, 1991), Aubert and Pinta (1978), Copenhaver and Wilkinson (1979a), Dotson (1991), Fuller (1977), Gibb and Cartwright (1987), Jacob (1989-selenium), Kabata-Pendias and Pendias (1992), Kotaby-Amacher and Gambrell (1988), Lisk (1972), McBride (1989), McLean and Bledsoe (1992), Page (1974), Rai and Zachara (1988), Zachara et al. (1992); <u>Water</u>: Allen et al. (1990, 1993), Salomons and Förstner (1984), Förstner and Wittman (1981), Kramer and Duinker (1984), Krenkl (1975), Moore and Ramamoorthy (1984a), Rai and Zachara (1986), Salbu (1994), Singer (1973), Singh and Subramian (1983); <u>Sewage Sludge/Wastewater</u>: Page (1974), Page et al. (1981); <u>Biogeochemistry</u>: Adriano (1992), Wildung and Drucker (1977)
Inorganic Agricultural Chemicals	<u>Nitrates</u>: Burt et al. (1993), Follet et al (1991)

Table 4-5 (cont.)

Topic	References
Toxic and Other Organic Chemicals	Beck et al. (1994/T3-12), Larson and Weber (1994), Linn et al. (1993), Lyman et al. (1992), NAS (1972), Thibodeaux (1979); Soil: Meikle (1972), Minnich (1993-VOCs), Morril et al. (1982), Nelson et al. (1983), Overcash (1981), Sawhney and Brown (1989); Ground/Surface Water: Borchardt et al. (1977), Faust and Hunter (1971), Gerstl et al. (1989), Moore and Ramamoorthy (1984b), Nielsen (1994); Aliphatic Hydrocarbons: Barbee (1994), Britton (1984), Moore and Ramamoorthy (1984b); Halogenated Hydrocarbons: Sims et al. (1991), Rogers and Abramowicz (1993)Monocyclic Aromatic Hydrocarbons and Halides: Chapman (1972), Gibson and Subramian (1984), Moore and Ramamoorthy (1984b), Reinike (1984); Phthalate Esters: Ribbons (1984), Pierce et al. (1980); Polycyclic Aromatic Hydrocarbons: Moore and Ramamoorthy (1984b), Safe (1984); Pesticides: Cheng (1990), Copenhaver and Wilkinson (1979b), Crosby (1973), Guenzi (1974), Hamaker (1972), Hamaker and Thompson (1972), Haque and Freek (1975), Honeycutt and Schabacker (1994), Kearney and Kaufman (1972), Linn et al. (1993), Moore and Ramamoorthy (1984b), NAS (1972), Ou et al. (1980), Rao and Davidson (1980), Somasundarum and Coats (1991)--see also references for agricultural chemicals in Table 4-4; Explosives: Environmental Science and Engineering (1985)

Table 4-5 References (Appendix F contains references for figure and table sources)

Adriano, D.C. (ed.) 1992. Biogeochemistry of Trace Metals. Lewis Publishers, Boca Raton, FL, 513 pp. [16 contributed chapters]

Allen, E.M. Perdue, and D. Brown (eds.). 1990. Metal Speciation in Groundwater. Lewis Publishers, Chelsea, MI.

Allen, H.E., E.M. Perdue, and D. Brown. 1993. Metals in Groundwater. Lewis Publishers, Chelsea, MI, 300 pp.

Alexander, M. 1994. Biodegradation and Bioremediation. Academic Press, New York, NY, 436 pp.

Allaway, W.H. 1968. Agronomic Controls Over the Environmental Cycling of Trace Elements. Adv. Agron. 20:235-274.

Allaway, B. 1991. Heavy Metals in Soils. John Wiley & Sons, New York, NY, 339 pp.

American Conference of Governmental Industrial Hygienists (ACGIH). 1992. 1992-1993 Threshold Limit Values for Chemical Substances and Physical Agents and Biological Exposure Indices. ACGIH, Technical Information Office, 6500 Glenway Ave., Bldg. D-7, Cincinnati, OH 45211-4438.

American Chemical Society (ACS). Annual. Chemcyclopedia: The Manual of Commercially Available Chemicals. ACS, Washington, DC.

Armour, M.A. 1991. Hazardous Laboratory Chemicals Disposal Guide. CRC Press, Boca Raton, FL, 464 pp.

Aubert, H. and M. Pinta. 1978. Trace Elements in Soils. Elsevier, New York, NY, 396 pp. [Includes chapters on Bo, Cr, Co, Cu, I, Pb, Mn, Mo, Ni, Se, Ti, V, and Zn, and a chapter on 10 other minor elements (Li, Rb, Cs, Ba, Sr, Bi, Ga, Ge, Ag, and Sn)]

Back, W. and M.J. Baedecker. 1989. Chemical Hydrogeology in Natural and Contaminated Environments. J. Hydrology 106:1-28.

Back, W. and R.A. Freeze. 1983. Chemical Hydrogeology. Benchmark Papers in Geology, No. 73, Hutchinson Ross, Stroudsburg, PA, 416 pp.

Baker, R (ed.). 1980. Contaminants and Sediments, 2 Vols. Ann Arbor Science, Ann Arbor, MI, Vol. 1, 558 pp, Vol. 2, 627 pp. [Proceedings of 1979 American Chemical Society symposium in Honolulu]

Barbee, G.C. 1994. Fate of Chlorinated Aliphatic Hydrocarbons in the Vadose Zone and Ground Water. Ground Water Monitoring and Remediation 14(1):129-140.

Bar-Yosef, B., N.J. Barrow, and J. Goldschmid (eds.). 1989. Inorganic Chemicals in the Vadose Zone. Springer-Verlag, New York, NY.

(Table 4-5 Contaminant Chemistry References)

Sources and Behavior of Subsurface Contaminants

Bedient, P.B., H.S. Rifai, and C.J. Newell. 1994. Ground Water Contamination: Transport and Remediation. Prentice-Hall, Englewood Cliffs, NJ.

Bitton, G. and C.P. Gerba (eds.). 1984. Groundwater Pollution Microbiology. Wiley-Interscience, New York, NY. [14 papers covering health and environmental aspects]

Borchardt, J.A., J.K. Cleland, W.J. Redman, and G. Olivier (eds.). 1977. Viruses and Trace Contaminants in Water and Wastewater. Ann Arbor Science, Ann Arbor, MI. [19 seminar papers focusing and health and treatment aspects]

Boulding, J.R. 1990. Assessing the Geochemical Fate of Deep-Well-Injected Hazardous Waste: A Reference Guide. EPA 625/6-89-025a. Available from CERI.* [Appendix B provides an index of more than 90 references that provide data on sorption and/or biodegradation of more than 150 organic compounds]

Bouwer, E.J. and P.L. McCarty. 1984. Modeling of Trace Organics Biotransformation in the Subsurface. Ground Water 22:433-440.

Bowen, H.J.M. 1966. Trace Elements in Biochemistry. Academic Press, London, 241 pp.

Boynton, W.B. and W.H. Brattain. 1929. Interdiffusion of Gases and Vapors. Int. Crit. Tables 5:62-63. [Vapor diffusion coefficients in air]

Britton, L.N. 1984. Microbial Degradation of Aliphatic Hydrocarbons. In: Microbial Degradation of Organic Compounds, D.T. Gibson (ed.), Marcel Dekker, New York, NY, pp. 89-130.

Bruins, R. 1929. Coefficients of diffusion in Liquids. Int. Crit. Tables 5:63-72. [Liquid diffusion coefficient in water]

Brusseau, M.L. 1993. Complex Mixtures and Groundwater Quality. Ground Water Issue Paper EPA/600/S-93/004, 15 pp. Available from CERI.*

Budavari, S. (ed.). 1989. The Merck Index: An Encyclopedia of Chemicals, Drugs, and Biologicals, 11th ed. Merck and Co., Rahway, NJ 07065. [Around 10,000 listings with extensive index and cross-index]

Burt, T.P., A.L. Heathwaite, and S.T. Trudgill (eds.). 1993. Nitrate, Processes, Patterns and Management. John Wiley & Sons, New York, NY, 444 pp.

Callahan, M.A. et al. 1979. Water-Related Environmental Fate of 129 Priority Pollutants, 2 Volumes. EPA 440/4-79/029a-b (NTIS PB80-204373 and PB80-204381).

Chapman, P.J. 1972. An Outline of Reaction Sequences Used for the Bacterial Degradation of Phenolic Compounds. In: Degradation of Synthetic Organic Molecules in the Biosphere. National Academy of Sciences, Washington, DC, pp. 17-53.

Chappelle, W. and K. Peterson (eds.). 1976. Symposium on Molybdenum, 2 Vols. Marcel Dekker, New York, NY.

(Table 4-5 Contaminant Chemistry References)

Cheng, H.H. (ed.). 1990. Pesticides in the Soil Environment: Processes, Impacts and Modeling. Soil Science Society of America, Madison, WI, 554 pp.

Copenhaver, E.D. and B.K. Wilkinson. 1979a. Movement of Hazardous Substances in Soil: A Bibliography, Vol. 1: Selected Metals. EPA 600/9-79-024a (NTIS PB80-113103, 152 pp. [Bibliography with abstracts of articles from 1970 to 1974 on mobility of As, asbestos, Be, Cd, Cr, Cu, cyanide, Pb, Hg, Se, and Zn in soil]

Copenhaver, E.D. and B.K. Wilkinson. 1979b. Movement of Hazardous Substances in Soil: A Bibliography, Vol. 2: Pesticides. EPA 600/9-79-024b (NTIS PB80-113111).

Crosby, D.G. 1973. The Fate of Pesticides in the Environment. Ann. Rev. Plant Physiol. 24:467-492.

Dean, J.A. (ed.). 1992. Lange's Handbook of Chemistry, 14th ed. McGraw-Hill, New York, NY, 1472 pp. [Data on chemical and physical properties of elements, minerals, inorganic compounds, organic compounds, and miscellaneous tables of specific properties; 13th edition published in 1985]

Devinny, J.S., L.R. Everett, J.C.S. Lu, and R.L. Stollar. 1990. Subsurface Migration of Hazardous Wastes. Van Nostrand Reinhold, New York, NY.

Domenico, P. and F. Schwartz. 1991. Physical and Chemical Hydrogeology. John Wiley & Sons, New York, NY, 824 pp.

Dotson, G.K. 1991. Migration of Hazardous Substances through Soils: Part II—Determination of the Leachability of Metals from Five Industrial Wastes and their Movement within Soil; Part III—Flue-Gas Desulfurization and Fly-Ash Wastes; Part IV—Development of a Serial Batch Extraction Method and Application to the Accelerated Testing of Seven Industrial Wastes. EPA/600/2091/017 (Part II, incorporating unpublished portions of Part I interim report: NTIS AD-A 158990; Part III: AD-A 182108; Part IV: AD-A 191856). [Waste from electroplating, secondary zinc refining, inorganic pigment, zinc-carbon battery, titanium dioxide pigment, nickel-cadmium battery, hydrofluoric acid, water-based paint, white phosphorus, chlorine production, oil re-refining, flue-gas desulfurization, and coal fly ash]

Dragun, J. 1988. The Soil Chemistry of Hazardous Materials. Hazardous Materials Control Research Institute, Silver Spring, MD, 458 pp.

Ellington, J.J., C.T. Jafvert, H.P. Kollig, E.J. Weber, and N.L. Wolfe. 1991. Chemical-Specific Parameters for Toxicity Characteristic Contaminants. EPA/600/3-91/004 (NTIS PB91-148361). [Acid, base, and neutral hydrolysis rate constants and partition coefficients for 44 "toxicity characteristic" contaminants]

Environmental Science and Engineering, Inc. 1985. Evaluation of Critical Parameters Affecting Contaminant Migration Through Soils. Report No. AMXTH-TE-CR-85030. U.S. Army Toxic and Hazardous Materials Agency, Aberdeen Proving Ground, MD. [Focus on explosive and propellant (PEP) contaminants]

(Table 4-5 Contaminant Chemistry References)

Faust, S.D. and J.V. Hunter (eds.). 1971. Organic Compounds in Aquatic Environments. Marcel Dekker, New York, NY. [24 papers on the origin, occurrence, and behavior of organic compounds in aquatic environments]

Fisher, N. (ed.). 1991. Farm Chemicals Handbook '91. Meister Publishing Co., Willoughby, OH, 216/942-2000. [Pesticides and Fertilizers]

Fetter, C.W. 1992. Contaminant Hydrogeology. Macmillan, New York, NY, 457 pp.

Fochman, E.G. 1981. Biodegradation and Carbon Adsorption of Carcinogenic and Hazardous Organic Compounds. EPA/600/2-81-032. [Data on 12 polynuclear aromatic compounds]

Follet, R.F., D.R. Keeney, and R.M. Cruse (eds.). 1991. Managing Nitrogen for Groundwater Quality and Farm Profitability. Soil Science Society of America, Madison, WI, 357 pp.

Förstner, U. and G.T.W. Wittmann. 1981. Metal Pollution in the Aquatic Environment, 2nd ed. Springer-Verlag, New York, NY, 486 pp. [First edition published in 1979]

Fuller, W.H. 1977. Movement of Selected Metals, Asbestos and Cyanide in Soils: Applications to Waste Disposal Problems. EPA 600/2-77-020 (NTIS PB 266905). [Review containing over 200 references on the movement of metals in soil]

Gambrell, R., R. Khalid, M. Verloo, and W. Patrick. 1977. Transformations of Heavy Metals and Plant Nutrients in Dredged Sediments as Affected by Oxidation-Reduction Potential and pH. U.S. Report D 77-4, U.S. Army Corps of Engineers Waterways Experiment Station, Vicksburg, MI, 309 pp.

Gelhar, L.W., A. Mantaglou, C. Welty, and K.R. Rohfelt. 1985. A Review of Field Scale Physical Solute Transport Processes in Saturated and Unsaturated Porous Media. EPRI RP-2485-05. Electric Power Research Institute, Palo Alto, CA.

Gerstl, Z., Y. Chen, U. Mingelgrin, and B. Yaron (eds.). 1989. Toxic Organic Chemicals in Porous Media. Springer-Verlag, New York, NY.

Gherini, S.A., K.V. Summers, R.K. Munson, and W.B. Mills. 1988. Chemical Data for Predicting the Fate of Organic Compounds in Water, Vol. 2: Database. EPRI EA-5818. Electric Power Research Institute, Palo Alto, CA. [Data relevant to predicting the release, transport, transformation, and fate of more than 50 organic compounds]

Gherini, S.A., K.V. Summers, R.K. Munson, and W.B. Mills. 1989. Chemical Data for Predicting the Fate of Organic Compounds in Water, Vol. 1: Technical Basis. EPRI EA-5818. Electric Power Research Institute, Palo Alto, CA.

Ghiorse, W.C. and J.T. Wilson. 1988. Microbial Ecology of the Terrestrial Subsurface. Adv. Appl. Microbiol. 33:107-172. [Literature review with more than 160 citations]

Gibb, J.P. and K. Cartwright. 1987. Retention of Zinc, Cadmium, Copper and Lead by Geologic Materials. EPA/600/2-86/108 (NTIS PB88-232819).

(Table 4-5 Contaminant Chemistry References)

Gibson, D.T. (ed.). 1984. Microbial Degradation of Organic Compounds. Marcel Dekker, New York, NY. [16 papers on aerobic and anaerobic degradation of major groups of contaminants]

Gibson, D.T. and V. Subramanian. 1984. Microbial Degradation of Aromatic Hydrocarbons. In: Microbial Degradation of Organic Compounds, D.T. Gibson (ed.), Marcel Dekker, New York, NY, pp. 181-252.

Goring, C.A.I. and J.W. Hamaker. 1972. Organic Chemicals in the Soil Environment, 2 Volumes. Marcel Dekker, New York, NY. [13 chapters]

Government Institutes, Inc. Annual. Book of Lists for Regulated Hazardous Substances, 1993 ed. Government Institutes, Inc., 4 Research Place, Suite 200, Rockville, MD, 20850; 301/921-2355, 345 pp. [Contains 70 regulatory lists of hazardous substances; updated annually]

Guarmaccia, J.F. et al. 1992. Multiphase Chemical Transport in Porous Media. EPA-600/S-92-002, 19 pp.

Guenzi, W.D. (ed.). 1974. Pesticides in Soil and Water. Soil Science Society of America, Madison, WI.

Güven, O., J.H. Dane, W.E. Hill, and J.G. Melville. 1992a. Mixing and Plume Penetration Depth at the Groundwater Table. EPRI TR-100576. Electric Power Research Institute, Palo, Alto, CA.

Güven, O., J.H. Dane, M. Oostrom, and J.S. Hayworth. 1992b. Physical Model Studies of Dense Solute Plumes in Porous Media. EPRI TR-101387. Electric Power Research Institute, Palo, Alto, CA.

Hamaker, J.W. 1972. Decomposition: Quantitative Aspects. In: Organic Chemicals in the Soil Environment, Vol.I, C.A.I. Goring and J.W. Hamaker (eds.), Marcel Dekker, New York, NY, pp. 253-340.

Hamaker, J.W. and J.M. Thompson. 1972. Adsorption. In: Organic Chemicals in the Soil Environment, Vol.I, C.A.I. Goring and J.W. Hamaker (eds.), Marcel Dekker, New York, NY, pp. 49-143.

Haque, R. and W.H. Freek (eds.). 1975. Environmental Dynamics of Pesticides. Plenum Press, New York, NY.

Hem, J.D. 1964. Deposition and Solution of Manganese Oxides. U.S. Geological Survey Water-Supply Paper 1667-B, 42 pp.

Honeycutt, R.C. and D.J. Schabacker (eds.). 1994. Mechanisms of Pesticide Movement into Ground Water. Lewis Publishers, Boca Raton, FL, 224 pp.

Horowitz, A.J. 1991. A Primer on Sediment-Trace Element Chemistry, 2nd ed. Lewis Publishers, Chelsea, MI, 136 pp. [Originally published in 1985 as U.S. Geological Survey Water-Supply Paper 2277.

(Table 4-5 Contaminant Chemistry References)

Howard, P.H. (ed.). 1989. Handbook of Environmental Fate and Exposure Data for Organic Chemicals: Vol. I, Large Production and Priority Pollutants. Lewis Publishers, Chelsea, MI, 600 pp.

Howard, P.H. (ed.). 1990a. Handbook of Environmental Fate and Exposure Data for Organic Chemicals: Vol. II, Solvents. Lewis Publishers, Chelsea, MI, 536 pp.

Howard, P.H. (ed.). 1990b. Handbook of Environmental Fate and Exposure Data for Organic Chemicals: Vol. III, Pesticides. Lewis Publishers, Chelsea, MI, 712 pp.

Howard, P.H. 1992a. PC Environmental Fate Databases: Datalog, Chemfate, Biolog, and Biodeg. Lewis Publishers, Chelsea, MI. [Each database comes with a manual and diskettes: **Datalog** contains 180,000 records for 13,000 chemicals; **Chemfate** contains actual physical property values and rate constants for 1,700 chemicals; **Biolog** contains 40,000 records on microbial toxicity and biodegradation data on about 6,000 chemicals; **Biodeg** contains data on biodegradation studies for about 700 chemicals]

Howard, P.H. 1992b. Database of SMILES Notations. Lewis Publishers, Chelsea, MI, 16 pp. plus disks. [Contains about 20,000 SMILES (Simplified Molecular Input Line Entry System) notations. Can be used with Lewis Publishers Koc program (Howard and Meylan, 1992), Henry's Law Constant Program, Biodegradation Probability Program, and Hydrolysis Rate Program]

Howard, P.H. (ed.). 1993. Handbook of Environmental Fate and Exposure Data for Organic Chemicals: Vol. IV, Solvents 2. Lewis Publishers, Chelsea, MI, 608 pp.

Howard, P.H. and W.M. Meylan. 1992. Soil/Sediment Adsorption Constant Program. Lewis Publishers, Chelsea, MI, 78 pp. plus disk. [Uses SMILES notation structural input to calculate Koc; see also, Howard, 1992]

Howard, P.H. and M.W. Neal. 1992. Dictionary of Chemical Names and Synonyms. Lewis Publishers, Chelsea, MI, 2544 pp. [Basic information on more than 20,000 chemicals]

Howard, P.H., W.F. Jarvis, W.M. Meylan, and E.M. Mikalenko. 1991. Handbook of Environmental Degradation Rates. Lewis Publishers, Chelsea, MI, 700+ pp. [Provides rate constants and half-life ranges for different media for more than 430 organic chemicals; processes include aerobic and anaerobic degradation, direct photolysis, hydrolysis and reaction with various oxidants or free radicals]

Huang, P.M. and M. Schnitzer (eds.). 1986. Interactions of Soil Minerals with Natural Organics and Microbes. SSSA Sp. Pub. No. 17. Soil Science Society of America, Madison, WI, 606 pp. [15 contributed chapters]

Huling, S.G. 1989. Facilitated Transport. Ground Water Issue Paper, EPA/540/4-89/003. Available from CERI.*

Huling, S.G. and J.W. Weaver. 1991. Dense Nonaqueous Phase Liquids. Ground Water Issue Paper, EPA/540/4-91-002, 21 pp. Available from CERI.*

(Table 4-5 Contaminant Chemistry References)

Jacobs, L.W. (ed.). 1989. Selenium in Agriculture and the Environment. SSSA Sp. Pub. No. 23. Soil Science Society of American, Madison, WI, 233 pp. [11 contributed chapters]

James, D.R. and H. Kidd. 1992. Pesticide Index, 2nd ed. Lewis Publishers/Royal Society of Chemistry, Chelsea, MI, 288 pp. [Listing of about 800 active ingredients and 25,000 trades of pesticides containing the ingredients]

Jarrell, W.M., A.L. Page, and A.A. Elseewi. 1980. Molybdenum in the Environment. Residue Rev. 74:1-43.

Kabata-Pendias, A. and H. Pendias. 1992. Trace Elements in Soils and Plants, 2nd ed. CRC Press, Boca Raton, FL, 365 pp. [First edition published in 1984]

Kearney, P.C. and D.D. Kaufman. 1972. Microbial Degradation of Some Chlorinated Pesticides. In: Degradation of Synthetic Organic Molecules in the Biosphere. National Academy of Sciences, Washington, DC, pp. 166-188.

Keith, L.H. (ed.). 1992. IRIS: EPA's Chemical Information Database. Lewis Publishers, Chelsea, MI. [Manual and annual subscription product updated on a quarterly basis; information on acute hazard information and physical and chemical properties on about 500 regulated and unregulated hazardous substances]

Kidd, H. and D.R. James (eds.). 1991. The Agrochemicals Handbook, 3rd ed. Lewis Publishers/Royal Society of Chemistry, Chelsea, MI, 1,500 pp.

Knox, R.C., D.A. Sabatini, and L.W. Canter. 1993. Subsurface Transport and Fate Processes. Lewis Publishers, Chelsea, MI, 430 pp.

Kobayashi, H. and B.E. Rittmann. 1982. Microbial Removal of Hazardous Organic Compounds. Environ. Sci. Technol. 16:170A-183A. [Literature review summarizing about 90 examples of biodegradation of hazardous organic compounds; more than 150 citations]

Kollig, H.P. 1993. Environmental Fate Constants for Organic Chemicals Under Consideration for EPA's Hazardous Waste Identification Projects. EPA/600/R-93/132 (NTIS PB93-221646).

Kollig, H.P., J.J. Ellington, E.J. Wever, and N.L. Wolfe. 1990. Pathway Analysis of Chemical Hydrolysis for 14 RCRA Chemicals. EPA/600/M-89/009, 6 pp. [Acrylonitrile, carbon tetrachloride, chlordane, *bis*(2-chlorethyl)ether, chloroform, 1,2-dichloroethane, heptachlor, lindane, methoxychlor, methylene chloride, 1,1,1-trichloroethane, 1,1,2-trichloroethane, 1,1,1,2-tetrachloroethane, 1,1,2,2-tetrachloroethane]

Kollig, H.P., K.J. Hamrick, and B.E. Kitchens. 1991. FATE, The Environmental Fate Constants Information System Database. EPA/600/3-91/045 (NTIS PB91-216192).

Kotaby-Amacher, J. and R.P. Gambrell. 1988. Factors Affecting Trace Metal Mobility in Subsurface Soils. EPA/600/2-88-036 (NTIS PB88-224829).

Kramer, C.J.M. and J.C. Duinker (eds.). 1984. Complexation of Trace Metals in Natural Waters. Martinus Nijhoff/Dr W. Junk Publishers, Boston. [42 papers]

(Table 4-5 Contaminant Chemistry References)

Krenkl, P. (ed.). 1975. Heavy Metals in the Aquatic Environment. Pergamon Press, New York, NY.

Larson, R.A., and E.J. Weber. 1994. Reaction Mechanisms in Environmental Organic Chemistry. Lewis Publishers, Boca Raton, FL, 448 pp.

Leo, A., C. Hansch, and D. Elkins. 1971. Partition Coefficients and Their Uses. Chemical Reviews 71(6):525-616. [First major literature review on partition coefficients and their uses; compilation of coefficients from more than 500 references]

Lewis, Sr., R.J. 1990. Carcinogenically Active Chemicals: A Reference Guide. Van Nostrand Reinhold, New York, NY, 1,184 pp. [Information on more than 3,400 chemicals]

Lewis, Sr., R.J. 1991. Reproductively Active Chemicals. Van Nostrand Reinhold, New York, NY, 1,184 pp. [Information on about 3,500 chemicals]

Lewis, Sr., R.J. 1992a. Hawley's Condensed Chemical Dictionary, 12th ed. Van Nostrand Reinhold, New York, NY, 1,288 pp. [More than 19,000 entries on chemicals, reactions and processes, state of matter, and compounds. N.I. Sax and R.J. Lewis were authors of 11th edition, published in 1987]

Lewis, Sr., R.J. 1992b. Sax's Dangerous Properties of Industrial Materials, 8th ed., 3 Vols. Van Nostrand Reinhold, NY, 4,300 pp. [Contains some 20,000 chemical entries covering physical and carcinogenic properties, clinical aspects, exposure standards, and regulations. N.I. Sax and R.J. Lewis were authors of 7th edition, published in 1989. Earlier editions: 1963 (2nd), 1968 (3rd), 1975 (4th), 1976 (5th), 1984 (6th)]

Lewis, Sr., R.J. 1993. Hazardous Chemicals Desk Reference, 3rd ed. Van Nostrand Reinhold, New York, NY, 1,752 pp. [Covers more than 6,000 of the most hazardous chemicals; each entry provides the chemical's hazard rating, a toxic and hazard review paragraph, CAS, NIOSH and DOT numbers, description of physical properties, synonyms, and current standards for exposure limits. Lewis was author of 2nd edition, published in 1990; N.I. Sax and R.J. Lewis were authors of 1st edition, published in 1987]

Lewis, G.R. 1994. 1,001 Chemicals in Everyday Products. Van Nostrand Reinhold, New York, NY, 344 pp.

Lide, D.R. 1993. CRC Handbook of Chemistry and Physics, 74th ed. CRC Press, Boca Raton, Fl, 2472 pp. [New edition published annually]

Linn, D.M., T.H. Carkski, M.L. Brusseau, and F.-H. Chang. 1993. Sorption and Degradation of Pesticides and Organic Chemicals in Soil. SSSA Sp. Pub. No. 32, Soil Science Society of America, Madison, WI, 260 pp. [14 contributed chapters]

Lisk, D.J. 1972. Trace Metals in Soils, Plants and Animals. Advances in Agronomy 24:267-325.

Luckner, L. and W.M. Schestakow. 1991. Migration Processes in the Soil and Groundwater Zone. Lewis Publishers, Chelsea, MI, 485 pp.

(Table 4-5 Contaminant Chemistry References)

Lyman, W.J., W.F. Reehl, and D.H. Rosenblatt (eds.). 1990. Handbook of Chemical Property Estimation Methods: Environmental Behavior of Organic Compounds, 2nd ed. American Chemical Society, Washington, DC, 960 pp. [First edition published by McGraw-Hill in 1982]

Lyman, W.J., P.J. Reidy, and B. Levy. 1992. Mobility and Degradation of Organic Contaminants in Subsurface Environments. Lewis Publishers, Chelsea, MI, 416 pp.

Mabey, W.R. et al. 1982. Aquatic Fate Process Data for Organic Priority Pollutants. EPA 440/4-81-014 (NTIS PB87-169090).

Mackay, D.M., P.V. Roberts, and J.A. Cherry. 1985. Transport of Organic Contaminants in Groundwater. Environ. Sci. Technol. 19(5):384-392.

Maki, A.W., K.L. Dickson, and J. Cairns, Jr. (eds.). 1980. Biotransformation and Fate of Chemicals in the Aquatic Environment. American Society for Microbiology, Washington, DC. [19 workshop papers]

Matthess, G. 1982. The Properties of Groundwater. John Wiley & Sons, New York, NY.

Mazor, E. 1990. Applied Chemical and Isotopic Ground Water Hydrology. John Wiley & Sons, New York, NY, 256 pp.

McBride, M.A. 1989. Reactions Controlling Heavy Metal Solubility in Soils. In: Advances in Soil Science, B.A. Stewart (ed.), Springer-Verlag, New York, NY, Vol. 10.

McLean, J.E. and B.E. Bledsoe. 1992. Behavior of Metals in Soils. Ground Water Issue Paper, EPA/540/S-92/018, 25 pp. Available from CERI.*

Meikle, R.W. 1972. Decomposition: Qualitative Relationships. In: Organic Chemicals in the Soil Environment, Vol. I, C.A.I. Goring and J.W. Hamaker (eds.), Marcel Dekker, New York, NY, pp. 145-251. [Reviews qualitative relationships in the biodegradation of 21 groups of organic compounds]

Minnich, M. 1993. Behavior and Determination of Volatile Organic Compounds in Soil: A Literature Review. EPA/600/R-93/140 (NTIS PB94-100153), 104 pp.

Mitchell, R. (ed.). 1971. Water Pollution Microbiology, 2 Vols. Wiley-Interscience, New York, NY. [Volume 1 contains 17 contributed chapters and Volume 2 has 16 chapters focussing primarily on surface water microbiology]

Montgomery, J.H. 1991. Ground Water Chemical Desk Reference, Vol. 2. Lewis Publishers, Chelsea, MI, 944 pp. [Data on 267 additional compounds not included in Montgomery and Welkom (1989)]

Montgomery, J.H. 1993. Agrochemicals Desk Reference: Environmental Data. Lewis Publishers, Chelsea, MI, 672 pp. [Physical/chemical data on 200 compounds including pesticide, herbicides and fungicides, partition coefficients, transformation products, etc.]

(Table 4-5 Contaminant Chemistry References)

Montgomery, J.H. and L.M. Welkom. 1989. Ground Water Chemicals Desk Reference. Lewis Publishers, Chelsea, MI, 640 pp. [Data on 137 organic compounds commonly found in ground water and the unsaturated zone, include: appearance, odor, boiling point, dissociation constant, Henry's law constant, log Koc, Log Kow, melting point, solubility in water and organics, specific density, transformation products, vapor pressure, fire hazard data (lower and upper explosive limits), and health hazards (IDLH, PEL). See also Montgomery (1991)]

Moore, J.W. and S. Ramamoorthy. 1984a. Heavy Metals in Natural Waters: Applied Monitoring and Impact Assessment. Springer-Verlag, New York, NY. [Covers As, Cd, Cr, Cu, Hg, Ni, and Zn]

Moore, J.W. and S. Ramamoorthy. 1984b. Organic Chemicals in Natural Waters: Applied Monitoring and Impact Assessment. Springer-Verlag, New York, NY. [Covers aliphatic hydrocarbons, mono- and polycyclic aromatic hydrocarbons, chlorinated pesticides, petroleum hydrocarbons, phenols, PCBs, and PCDD]

Morril, L.G., B. Mahalum, and S.H. Mohiuddin. 1982. Organic Compounds in Soils: Sorption, Degradation and Persistence. Ann Arbor Science/The Butterworth Group, Woburn, MA, 326 pp.

National Academy of Science (NAS). 1972. Degradation of Synthetic Organic Molecules in the Biosphere. National Academy of Science, Washington, DC. [16 papers focussing mostly on pesticides]

National Bureau of Standards (NBS). 1981. Proceedings of the NBS Workshop on Aqueous Speciation of Dissolved Contaminants. CONF-810588-4, NBS, Gaithersburg, MD.

National Institute for Occupational Safety and Health (NIOSH). 1990. NIOSH Pocket Guide to Chemical Hazards. DHHS (NIOSH) Publication No. 90-117, 245 pp. [Summarizes information from the three-volume **NIOSH/OSHA Occupational Health Guidelines for Chemical Hazards**; data are presented in tables, and the source includes chemical names and synonyms, permissible exposure limits, chemical and physical properties, and other toxicological information]

National Research Council Canada, 1976. Effects of Chromium in the Canadian Environment. NRCC Report No. 15018. Ottawa, Ontario.

National Research Council Canada, 1978a. Effects of Arsenic in the Canadian Environment. NRCC Report No. 15391. Ottawa, Ontario.

National Research Council Canada, 1978b. Effects of Lead in the Environment - 1978: Quantitative Aspects. NRCC Report No. 16736. Ottawa, Ontario.

National Research Council Canada, 1979a. Effects of Mercury in the Canadian Environment. NRCC Report No. 16739. Ottawa, Ontario.

National Research Council Canada, 1979b. Effects of Cadmium in the Canadian Environment. NRCC Report No. 16743. Ottawa, Ontario.

(Table 4-5 Contaminant Chemistry References)

National Research Council Canada, 1981. Effects of Nickel in the Canadian Environment. NRCC Report No. 18568 (Reprint). Ottawa, Ontario.

National Research Council Canada. 1982. Data Sheets on Selected Toxic Elements. NRCC Report No. 19252. Ottawa, Ontario. [Includes: Sb, Ba, Be, Bi, B, Cs, Ga, Ge, In, Mo, Ag, Te, Tl, Sn (inorganic and organic), U, Zr]

Nelson, D.W., D.E. Elrick, and K.K. Tanji. 1983. Chemical Mobility and Reactivity in Soil Systems. SSSA Sp. Pub. No. 11. Soil Science Society of America, Madison, WI, 262 pp. [17 contributed chapters with sections on principles of chemical mobility and reactivity, biological activity and chemical mobility, and environmental impacts of toxic chemical transport]

Ney, Jr., R.E. 1990. Where Did That Chemical Go? A Practical Guide to Chemical Fate and Transport in the Environment. Van Nostrand Reinhold, New York, NY, 200 pp. [Information on more than 100 organic and inorganic chemicals]

Neilsen, A.H. 1994. Organic Chemicals in the Aquatic Environment: Distribution, Persistence and Toxicity. Lewis Publishers, Boca Raton, FL, 448 pp.

Ou, L.T., J.M. Davidson, and P.S.C. Rao. 1980. Rate Constants for Transformation of Pesticides in Soil-Water Systems: A Review of the Available Data Base. U.S. Environmental Protection Agency.

Occupational Safety Health Services. 1990. PESTLINE: Material Safety Data Sheets for Pesticides and Related Chemicals, 2 Vols. Van Nostrand Reinhold, New York, NY, 2100 pp. [Information on about 1,200 pesticides]

Overcash, M.R. (ed.). 1981. Decomposition of Toxic and Nontoxic Organic Compounds in Soil. Ann Arbor Science/The Butterworth Group, Woburn, MA, 375 pp. [43 papers on decomposition of chlorinated organics, agricultural chemical, phenols, aromatic and polynuclear aromatics, urea resins, and surfactants in soil]

Page, A.L. 1974. Fate and Effects of Trace Elements in Sewage Sludge When Applied to Agricultural Lands: A Literature Review Study. EPA/670/2-74-005 (NTIS PB231-171), 107 pp.

Page, A.L., A.C. Chang, G. Sposito, and S. Mattigod. 1981. Trace Elements in Wastewater: Their Effects on Plant Growth and Composition and Their Behavior in Soils. In: Modeling Wastewater Renovation Land Treatment, I.K. Iskander (ed.), Wiley Interscience, New York, NY, pp. 182-222.

Palmer, C.M. 1992. Principles of Contaminant Hydrogeology. Lewis Publishers, Chelsea, MI, 211 pp.

Patnalk, P. 1992. A Comprehensive Guide to the Hazardous Properties of Chemical Substances. Van Nostrand Reinhold, New York, NY, 800 pp. [Information on the 1,000 most commonly encountered hazardous chemicals]

Perry, H.P. and C.H Chiltin (eds.). 1973. Chemical Engineers Handbook. McGraw-Hill, New York, NY.

Pierce, R.C., S.P. Mathur, D.T. Williams, and M.J. Boddington. 1980. Phthalate Esters in the Aquatic Environment. NRCC Report No. 17583. National Research Council of Canada, Ottawa.

Piwoni, M.D. and J.W. Keeley. 1990. Basic Concepts of Contaminant Sorption at Hazardous Waste Sites. Ground Water Issue Paper EPA/540/4-90/053, 7 pp. Available from CERI.*

Purves, D. 1978. Trace-Element Contamination of the Environment. Elsevier, New York, NY.

Rai, D. and J.M. Zachara. 1984. Chemical Attenuation Rates, Coefficients and Constants in Leachate Migration. Vol. 1: A Critical Review. EPRI EA-3356. Electric Power Research Institute, Palo Alto, CA. [Data on 21 elements related to leachate migration: Al, Sb, As, Ba, Be, B, Cd, Cr, Cu, F, Fe, Pb, Mn, Hg, Mo, Ni, Se, Na, S, V, and Zn; see Rai et al. (1984) for annotated bibliography]

Rai, D. and J.M. Zachara. 1986. Geochemical Behavior of Chromium Species. EPRI EA-4544. Electric Power Research Institute, Palo Alto, CA.

Rai, D. and J.M. Zachara. 1988. Chromium Reactions in Geologic Materials. EPRI EA-5741. Electric Power Research Institute, Palo Alto, CA. [Contains laboratory data and equilibrium constants for key reactions needed to predict the geochemical behavior of chromium in soil and ground water]

Rai, D., J.M. Zachara, R.A. Schmidt, and A.P. Schwab. 1984. Chemical Attenuation Rates, Coefficients, and Constants in Leachate Migration, Vol. 2: An Annotated Bibliography. EPRI EA-3356. Electric Power Research Institute, Palo Alto, CA. [See Rai and Zachara (1984) for elements covered]

Ramsey, R.H., C.R. Wetherill, and H.C. Duffer. 1972. Soil Systems for Municipal Effluents-A Workshop and Selected References. EPA-16080-6WF-02172 (NTIS PB217-853), 60 pp.

Rao, P.S.C. and J.M. Davidson. 1980. Estimation of Pesticide Retention and Transformation Parameters Required in Nonpoint Source Pollution Models. In: Environmental Impact of Nonpoint Source Pollution, M.R. Overcash and J.M. Davidson (eds.), Ann Arbor Science Publishers, Ann Arbor, MI, pp. 23-67.

Ratledge, C. (ed.). 1993. Biochemistry of Microbial Degradation. Kluwer Academic Publishers, Hingham, MA, 584 pp.

Reinke, W. 1984. Microbial Degradation of Halogenated Aromatic Compounds. In: Microbial Degradation of Organic Compounds, D.T. Gibson (ed.), Marcel Dekker, New York, NY, pp. 319-360.

Ribbons, D.W., P. Keyser, R.W. Eaton, B.N. Anderson, D.A. Kunz, and B.F. Taylor. 1984. Microbial Degradation of Phthalates. In: Microbial Degradation of Organic Compounds, D.T. Gibson (ed.), Marcel Dekker, New York, NY, pp. 371-398.

Rogers, J.E. 1986. Anaerobic Transformation Processes: A Review of the Microbiological Literature. EPA/600/3-86/042, NTIS PB86-230042. [Review of the microbiological literature on anaerobic transformation processes with more than 200 references]

(Table 4-5 Contaminant Chemistry References)

Rogers, J.E. and D.A. Abramowicz. 1993. Anaerobic Dehalogenation and Its Environmental Implications. EPA/600/R-93/131 (NTIS PB93-217799). [Abstracts of 33 papers presented at 1992 American Society of Microbiology conference]

Sablić, A. 1988. On the Prediction of Soil Sorption Coefficients of Organic Pollutants by Molecular Topology. Environ. Sci. Technol. 21(4):358-366. [Sorption coefficient data for 72 nonpolar and 159 polar and ionic organic compounds]

Safe, S.H. 1984. Microbial Degradation of Polychlorinated Biphenyls. In: Microbial Degradation of Organic Compounds, D.T. Gibson (ed.), Marcel Dekker, New York, NY, pp. 261-370.

Salbu, B. (ed.). 1994. Trace Elements in Natural Waters. Lewis Publishers, Boca Raton, FL, 336 pp.

Salomons, W. and U. Förstner. 1984. Metals in the Hydrocycle. Springer-Verlag, New York, NY, 349 pp.

Sawhney, B.L. and K. Brown (eds.). 1989. Reactions and Movement of Organic Chemicals in Soils. SSSA Sp. Pub. No. 22. American Society of Agronomy, Madison, WI. [18 contributed chapters]

Scow, K.M. 1982. Rate of Biodegradation. In: Handbook of Chemical Property Estimation Methods: Environmental Behavior of Organic Compounds, W.J. Lyman, W.F. Reehl, and D.H. Rosenblatt (eds.), McGraw-Hill, New York, NY, pp. 9-1 to 9-85. [Literature review with more than 170 citations]

Shafer, D. (ed.). 1993. The Book of Chemical Lists. Business & Legal Reports, Inc., Madison, CT, 800/727-5257. [Two loose-leaf volumes: Section I (Master Chemical Cross-Reference), Section II (Environmental Planning and Reporting), Section III (Health and Safety Guidelines), Section IV (State Chemical Lists); updated annually, supplements available for earlier editions]

Shaw, A.J. (ed.). 1989. Heavy Metal Tolerance in Plants: Evolutionary Aspects. CRC Press, Boca Raton, FL, 355 pp.

Shear, H. and A. Watson (eds.). 1977. The Fluvial Transport of Sediment-Associated Nutrients and Contaminants. IJC/LUARG, Windsor, Ontario.

Shineldecker, C.L. 1992. Handbook of Environmental Contaminants. Lewis Publishers, Chelsea, MI, 371 pp. [Key to contaminants that are likely to be associated with specific types of facilities, processes, and products]

Sims, J.L., J.M. Suflita, and H.H. Russell. 1991. Reductive Dehalogenation of Organic Contaminants in Soils and Ground Water. Ground Water Issue Paper EPA/540/4-90/054, 12 pp. Available from CERI.*

Singer, P.C. 1973. Trace Metals and Metal Organic Interactions in Natural Waters. Ann Arbor Science, Ann Arbor, MI. [13 contributed chapters]

(Table 4-5 Contaminant Chemistry References)

Singh, S. and V. Subramian. 1983. Hydrous Fe and Mn Oxides - Scavengers of Heavy Metals in the Aquatic Environment. CRC Critical Reviews in Environmental Control 14:33-90.

Somasundarum, L. and J.R. Coats (eds.). 1991. Pesticide Transformation Products: Fate and Significance in the Environment. ACS Symp. Series No. 459, American Chemical Society, Washington, DC, 320 pp.

Tabak, H.H. et al. 1981. Biodegradability Studies with Organic Priority Pollutant Compounds. J. Water Pollution Control Federation 53(10):1503-1518. [Results of biodegradability studies for 114 organic priority pollutants]

Thibodeaux, L.J. 1979. Chemodynamics: Environmental Movement of Chemicals in Air, Water and Soil. John Wiley & Sons, New York, NY, 501 pp.

Thornton, I. (ed.). 1983. Applied Environmental Geochemistry. Academic Press, New York, NY. [16 contributed chapters with emphasis on heavy metals]

Tinsley, I.J. 1979. Chemical Concepts in Pollutant Behavior. John Wiley & Sons, New York, NY.

TNO/BMFT. 1985. First International Conference on Contaminated Soil. Kluwer Academic Publishers, Hingham, MA.

TNO/BMFT. 1989. Second International Conference on Contaminated Soil. Kluwer Academic Publishers, Hingham, MA.

U.S. Coast Guard. 1985. CHRIS: Chemical Hazard Response Information System: Vol. 1, Condensed Guide to Chemical Hazards (CG-446-1); Vol. 2, Hazardous Substance Data Manual (CG-446-2—3 binders, GPO Stock No. 050-012-00147-2); Vol. 3, Hazard Assessment Handbook (CG-446-3); Vol. 4, Response Methods Handbook (CG-446-4).

U.S. Department of Transportation (DOT). 1990. Emergency Response Guidebook. DOT P5600.5, U.S. DOT, Office of Hazardous Materials Transportation, Washington, DC. [Information on potential hazards of DOT-regulated hazardous chemicals; updated every three years]

U.S. Environmental Protection Agency (EPA). 1985. Chemical, Physical and Biological Properties of Compounds Present at Hazardous Waste Sites. EPA/530/SW-89-010 (NTIS PB88-224829).

U.S. Environmental Protection Agency (EPA). 1989. Transport and Fate of Contaminants in the Subsurface. Seminar Publication EPA/625/4-89/019, 148 pp. Available from CERI.*

U.S. Environmental Protection Agency (EPA). 1992a. Handbook of RCRA Ground-Water Monitoring Constituents: Chemical and Physical Properties (Appendix IX to 40 CFR part 264). EPA/530-R-92-022, 267 pp. Office of Solid Waste, Washington, DC.

U.S. Environmental Protection Agency (EPA). 1992b. Dense Nonaqueous Phase Liquids--A Workshop Summary. EPA/600/R-92/030.

(Table 4-5 Contaminant Chemistry References)

van der Zee, S.E.A.T.M. and G. Desouni. 1993. Transport of Inorganic Solutes in Soil. In: Interacting Processes in Soil Science, R.J. Wagenet, P. Baveye, and B.A. Stewart (eds.), Lewis Publishers, Boca Raton, FL.

Verschueren, K. 1983. Handbook of Environmental Data on Organic Chemicals, 2nd ed. Van Nostrand Reinhold, NY, 1,310 pp. [Data on more than 1,300 organic chemicals]

Walker, M.M. and L.W. Keith. 1992. EPA's Pesticide Fact Sheet Database. Lewis Publishers, Chelsea, MI. [Manual and two 3.5-inch/four 5.25-inch diskettes containing comprehensive source of information on several hundred pesticides and formulations]

Watras, C.J. and J.W. Huckabee (eds.). 1994. Mercury Pollution: Integration and Synthesis. Lewis Publishers, Boca Raton, FL, 768 pp.

Weber, Jr., W.J. 1972. Physicochemical Processes for Water Quality. Wiley Interscience, New York, NY, 640 pp.

Wildung, R. and H. Drucker (eds.). 1977. Biological Implications of Metals in the Environment. CONF-750929, NITS, Springfield, VA.

Wilson, J.T. and J.F. McNabb. 1983. Biological Transformation of Organic Pollutants in Groundwater. Eos (Trans. Am. Geophysical Union) 20:997-1002.

Yaron, B., G. Dagan, and J. Goldschmid (eds.). 1984. Pollutants in Porous Media: The Unsaturated Zone Between Soil Surface and Groundwater. Springer-Verlag, New York, NY.

Yaws, C., H.-C. Yang, and X. Pan. 1991. Henry's Law Constants for 363 Organic Compounds in Water. Chemical Engineering 98(11):179-185.

Yong, R.N., A.M.O. Mohamed, and B.P. Warkentin. 1992. Principles of Contaminant Transport in Soils. Elsevier, New York, NY, 327 pp.

Zachara, J.M. et al. 1992. Aqueous Complexation, Precipitation, and Adsorption Reactions of Cadmium in the Geologic Environment. EPA TR-100751. Electric Power Research Institute, Palo Alto, CA.

Zehnder, A.J.B. (ed.). 1988. Biology of Anaerobic Microorganisms. Wiley-Interscience, New York, NY. [14 papers on the biology of anaerobic microorganisms, including biodegradation of contaminants]

* See Introduction for information on how to obtain documents from CERI (U.S. EPA Center for Environmental Research Information) and NTIS.

CHAPTER 5

PLANNING FIELD INVESTIGATIONS

5.1 Overview of Investigation Planning Process 238

    5.1.1 Investigation Objectives 238
    5.1.2 Investigation Scale 239
    5.1.3 Planning and Conducting Contaminant Investigations . . . 240

5.2 Existing Information Sources 246

    5.2.1 Soil and Geomorphic Data 246
    5.2.2 Geologic and Hydrologic Data 251
    5.2.3 Airphoto Interpretation 252

5.3 Developing a Sampling and Monitoring Plan 253

    5.3.1 Types of Monitoring 253
    5.3.2 Sampling Protocol 253
    5.3.3 Sample Location 255
    5.3.4 Sample Parameter/Analyte Selection 259
    5.3.5 Sampling Frequency, Type and Size 262
    5.3.6 Sample Collection and Handling 264

5.4 Data Measurement and Reliability 269

    5.4.1 Deterministic vs. Random Geochemical Data 269
    5.4.2 Data Representativeness 269
    5.4.3 Measurement Bias, Precision, and Accuracy 271
    5.4.4 Sources of Error 271

5.5 Analytical and QA/QC Concepts 276

    5.5.1 Instrumentation and Analytical Methods 276
    5.5.2 Limit of Detection 276
    5.5.3 Types of QA/QC Samples 279

5.6 Statistical Techniques 281

    5.6.1 Statistical Approaches to Geochemical Variability 281
    5.6.2 Geostatistics 282

5.7 Guide to Major References* 284

* Appendix F contains citations for table and figure sources.

Well-planned field investigations for assessing soil and ground-water quality or contamination reduce costs and yield better results than investigations that are conducted using a cookbook or haphazard approach. This chapter focuses on the things that should be done before initiating a significant level of field work, and reviews some statistical concepts and techniques that are especially important in contaminant investigations. Subsequent chapters cover more specific field methods for local and site investigations:

- Geophysical and remote sensing techniques (Chapter 6)

- Characterization of vadose zone and ground-water hydrology (Chapter 7)

- Soil and ground-water tracers (Chapter 8)

- Field sampling and monitoring (Chapter 9)

5.1 Overview of Investigation Planning Process

5.1.1 Investigation Objectives

A clear definition of objectives is the first step in effective planning. The objectives of an investigations will in large measure define the approach and methods that will be used. Some common objectives include:

- *Collection of Baseline Data.* For many years the U.S. Geological Survey (USGS) has collected both quality and quantity data of both surface and ground water. These data are collected to provide historical documentation of changes in amount and quality of ground water, and baseline data on "natural" levels of chemical constituents in ground water.

- *Pollution Prevention.* Field investigations for the siting of waste disposal facilities aid in the selection of areas that are less vulnerable to the effects of contamination and design of measures to prevent contamination. Field investigations of varying intensity are required to delineate drinking water wellhead protection areas and develop management practices to prevent pollution.

- *Regulatory Compliance.* Field investigations and monitoring are required to demonstrate compliance with regulatory standards dealing with underground storage tanks, and waste disposal, storage, or treatment facilities.

- *Litigation and Regulatory Enforcement.* Soil and ground-water investigations often play an important role in litigation. Results may be used to establish liability under the civil laws of trespass or in response to federal legislation such as RCRA and Superfund. In this situation, special attention must be given to documentation of field and laboratory procedures, and use of stringent quality assurance/quality control

(QA/QC) procedures. At the beginning of the study, legal counsel should provide guidance in developing the work plan.

- *Research.* Soil and ground-water quality investigations for research purposes are as varied as the nature of the research itself. Objectives of these investigations range from model validation to determining the rates and breakdown products of contaminant degradation. Such investigations often require specialized field equipment and technologies to obtain representative samples of subsurface materials for use in column and microcosm studies. Usually more observation points are employed in research studies than in other types of ground-water investigations. As with investigations involving litigation or regulatory enforcement, stringent QA/QC procedures must be followed.

A clear statement and understanding of an investigation program's objectives must precede decisions concerning sampling locations, piezometer or well types, numbers, location, depth, constituents of interest, and methods of sample collection, storage, transport, and analysis. Each decision requires consideration of the data needs and associated costs involved in each phase of a site characterization and monitoring program. In general, a phased approach is best, in which experimental design focusses on the reasoning behind each piece of data that is collected, and use of less expensive field methods guides use of more expensive methods as needed. For example, surface geophysical methods and hand-held or truck-mounted subsurface probing devices can guide location of boreholes and monitoring well installations using larger drill rigs. In the same way, use of less expensive chemical field screening and analytical techniques on a relatively large number of samples can guide selection of a fewer number of samples for chemical analysis in fixed laboratories (Section 9.6).

5.1.2 Investigation Scale

Soil and ground-water quality investigations can be divided into three general types based on scale or geographic scope: regional, local, and site-specific evaluations.

Regional Investigations. Regional evaluations, which may encompass several hundred or even thousands of square miles, are mainly reconnaissance efforts, and are used to obtain an overall evaluation of the ground-water situation. This broad brush type of investigation can be the starting point for two general types of explorations. The purpose of the first type is to locate potential individual sources or sites of ground-water contamination to determine if a problem exists. The second type aims at determining the occurrence and availability of ground water on a more regional scale. Then, if necessary, the exploration would ascertain prevalent hydrologic properties of earth materials, generalized flow directions in both major and minor aquifers, primary sources and rates of recharge and discharge, chemical quality of the aquifers and surface water, and locations and yields of pumping centers. These data can be useful in more detailed studies because they provide information on the geology and flow direction, both of which impact smaller scale studies.

Local Investigations. Local investigations are intermediate in scope and areal extent between regional evaluations and site investigations. For example, an investigation of the hydrogeology of an area encompassing a few tens or several hundred square miles in order to evaluate the effect of oil-field brine production and disposal would fall short of being called regional. At the other end of the scale, local investigations grade into site investigations when a hydrogeologic assessment must extend beyond the specific boundaries of a site. Investigations of this nature usually include only a few square miles. Their purpose is to define in greater detail the geology and hydrology in an area surrounding a specific site or sites of concern. The information obtained is used in designing and carrying out more detailed site investigations.

Site Investigations. The purpose of the site evaluation is to ascertain, with considerable certainty, the extent of soil and ground-water contamination, its local source or sources, and hydraulic and chemical properties that influence contaminant migration. The site investigation is the most detailed, complex, costly, and, from a legal and restoration viewpoint, the most critical of the three types of field studies. A site investigation must address a large number of pertinent parameters affecting contaminant transport and transformation. These include the soils, geology, hydrogeology, geochemical interactions, biotic and abiotic degradation processes, and the rate of contaminant movement through the unsaturated and saturated zones. All of the phenomena that may influence the movement of contaminant plumes, such as pumping wells, local streams, and multi-aquifer interactions, must be characterized.

At the same time, ancillary assessments at the site might include tank inventories, toxicological evaluations, air pollution monitoring, studies of manufacturing and waste handling procedures, as well as many other studies, all of which will eventually interface in the development of a comprehensive site investigation report.

5.1.3 Planning and Conducting Contaminant Investigations

Regardless of the complexity or detail of the investigation, a logical series of steps should be followed. Although each investigation is unique, most investigations include the following steps: (1) definition of objectives, (2) collection of existing data (including literature review), (3) field investigations and sampling, (4) laboratory analysis, (5) data interpretation and conclusions, and (6) report preparation and recommendations. These elements interact iteratively as part of a larger process that is illustrated in Figure 5-1. In contaminant investigations, the term *conceptual site model* is now widely used to refer to the "system behavior hypothesis" in Figure 5-1.

Failure to follow these steps may result in excessive cost and, in the worst case, rejection of the study results as inadequate. Common mistakes include:

- Failure to define precise objectives appropriate for the purpose and scale of the investigation.

Planning Field Investigations 241

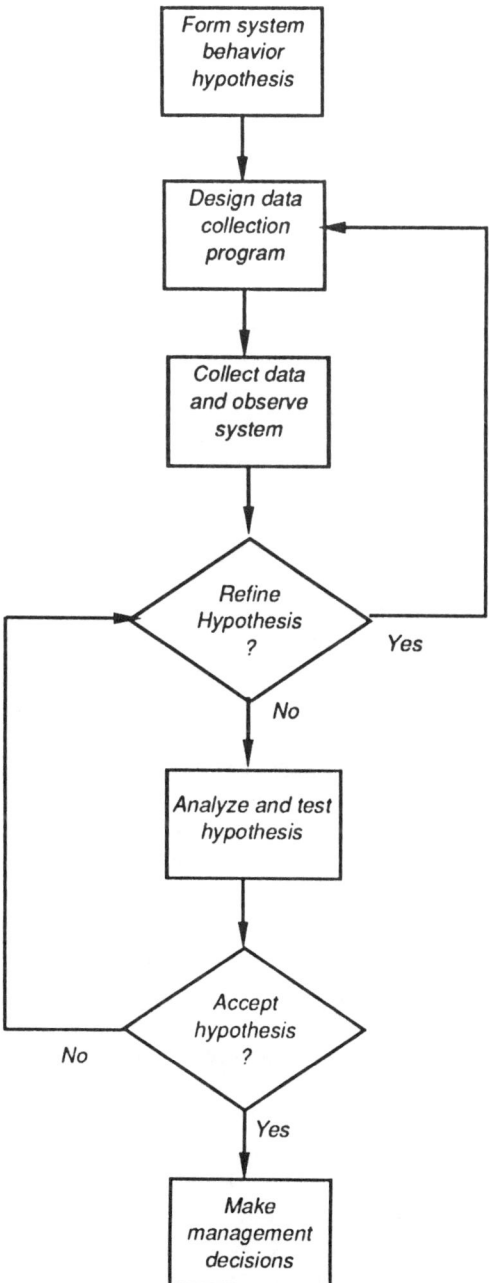

Figure 5-1 Site characterization phases (Mercer and Spalding, 1991a, after Bouwer et al., 1988).

- Failure to review all available data on the site prior to initiating the field investigation. For example, failure to identify offsite sources of ground-water pumping may lead to incorrect interpretations of ground-water flow patterns.

- Failure to obtain necessary basic hydrogeologic data before selecting locations for more intensive water quality sampling. This may result in unnecessary expense in water quality analyses.

- Misinterpretation of the data collected during the investigation.

Developing Study Objectives and a Work Plan. Establishing an investigation's major objectives is essential to a successful and cost-effective project. All interested parties should clearly define and agree upon these points. These objectives should be stated in writing and referred to often during the life of the study. Otherwise, the study may drift from the original objectives, resulting in the costly collection of superfluous information at the expense of other essential information.

The approach, time requirements, and funding can vary greatly between a regional reconnaissance evaluation and a site investigation. The former, which deals with regional properties, may require a relatively short time while the latter, which necessitates minute detail, may demand years. In either case, the time and resource requirements are dictated by the goals, and the success of the work is measured by how directly the investigation pursues these goals.

For example, a stated objective to "measure the water level surface in a given township using existing wells" indicates the limited nature of the study. On the other hand, an objective to "evaluate the degradation rate of tetrachloroethylene at a specific spill site, define the plumes of the parent and degradation contaminants, and predict the location and concentrations of these contaminants in ten years" clearly requires a substantial level of effort. The investigation would require a detailed knowledge of the site's soils, geochemistry, geology, and hydrogeology, along with sophisticated analytical capabilities, predictive models, and the information necessary to drive them.

A *work plan* provides the overall administrative framework for an investigation and should be responsive to the stated objectives, using existing data and information to the fullest extent possible. The field portion of an investigation is guided by a more specific *sampling and monitoring plan* and *health and safety plan*. Data quality assurance and quality control can be addressed in a separate *QA/QC plan* or in the appropriate section of the work plan and sampling and monitoring plan. Section 5.3 addresses more specific aspects of developing a sampling and monitoring plan, including QA/QC. Health and safety issues are addressed further in Section 12.2.6.

Unlike the project goals, which are fixed, the work and sampling/monitoring plans need to be flexible. For example, the position of all test wells, and boring and monitoring points cannot be determined at the start of an investigation. Rather, these locations should be adjusted on the basis of information obtained as each hole is completed. In this way, the usefulness of the data obtained from each drill site can be maximized, and future sampling points located to increase understanding of the ground water and contamination situation at the site under study.

Similarly, the contaminants of interest, appropriate analytical methods, details of sampling techniques, and the required number of samples cannot be accurately estimated at the beginning of a project. These must be refined as data are collected and their results interpreted.

A work plan may need to incorporate procedures or guidelines established by EPA such as the RCRA Ground Water Monitoring Technical Enforcement Guidance (U.S. EPA, 1986 and 1993/T9-10), and provisions for the safety and health of workers at the site (NIOSH/OSHA/USCG/EPA, 1985/T12-7). State regulatory agencies may have even more stringent requirements. Also, in the case of Superfund and RCRA sites, investigators will probably need to coordinate with and use data collected by consultants for potentially responsible parties (PRPs).

Existing Data Collection. The amount of existing data will affect the sophistication of the preliminary conceptual site model. Knowledge of site history, and the location of anthropogenic features such as buried cables and pipelines are also essential for worker health and safety when field investigations begin. When feasible, a preliminary or reconnaissance site visit provides a basis for evaluating the quality of existing data as it applies to the specific area of interest. Section 5.2 addresses existing information sources in some detail.

Field Investigation. The initial field phase of a ground-water investigation is the most intensive and important part of the project, and the data collected during this phase will determine the project's success. Some of the main factors affecting the quality of field data collected include an understanding of the basic geology and hydrogeology of the site (Chapters 1 and 2), well and aquifer testing (Section 7.3), a knowledge of the types of contaminants involved and their behavior in the subsurface (Chapters 3 and 4), the appropriate use of geophysical methods (Chapter 6), the location and construction of monitoring wells (Section 9.3), and the sampling and analytical techniques (Chapter 9). Tracer tests (Chapter 8) may also provide valuable data.

In the late 1980s there was a general recognition that a lot of money was being wasted in contaminant investigations, especially at uncontrolled hazardous waste sites that had been targeted for action under EPA's Superfund program. Tens of thousands of dollars were being spent on expensive chemical analysis of ground-water samples taken from monitoring wells that were placed without a good understanding of the hydrogeology of a site. Figure 5-2 provides a qualitative illustration of the longer-term cost savings that result from increasingly intensive site characterization efforts. Table 5-1 identifies the types of site characterization activities associated with the "conventional", "state-of-the-art", and "state-of-the-science" approaches, along with the advantages and disadvantages. As of the mid-1990s, the conventional approach has moved closer to the state-of-the-art defined in Table 5-1.

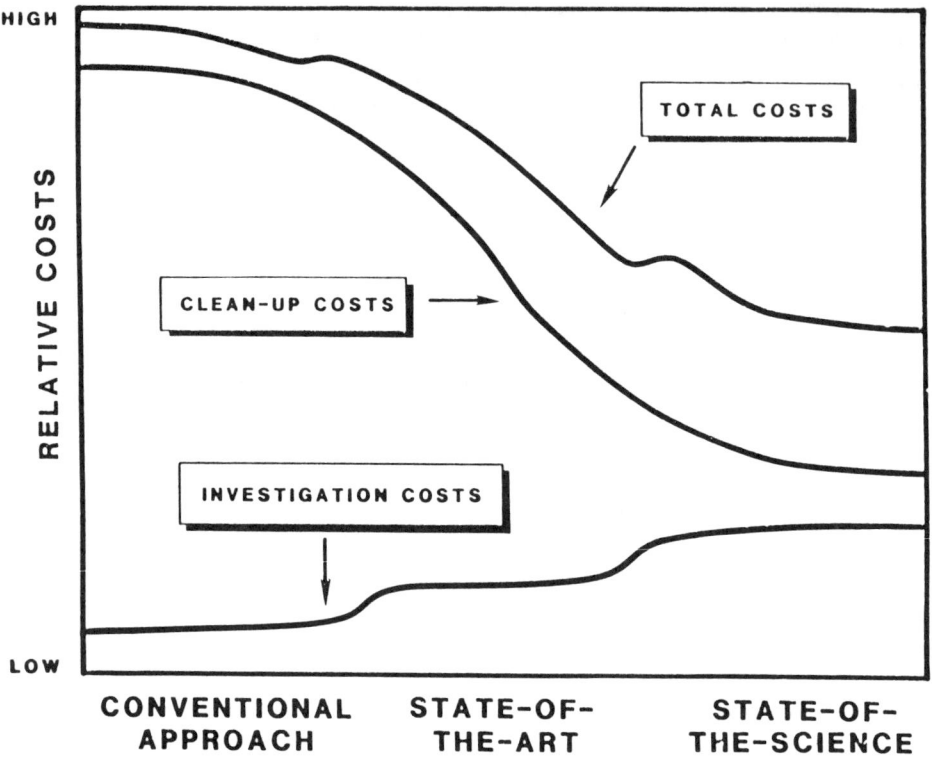

Figure 5-2 General relationship between site characterization costs and cleanup costs as a function of the characterization approach (Keely, 1987).

Data Analysis and Conclusions. Data interpretation should begin with the development of the work plan. Even at a moderate-sized site, a ground-water investigation of limited scope can result in the collection of a great deal of data. The time loss and frustration avoided during data interpretation are directly proportional to the degree to which the study anticipates: (1) the types and amounts of data collected; (2) the calculations required to determine contaminant transformation process rates, support conclusions, and make projections; (3) the correlations required to prove cause and effect, define relationships, and determine reaction coefficients; and (4) the graphic displays needed for reports and presentations.

To the extent possible, the amounts and types of data should be anticipated early in the project and provisions made for the continuous input of collected information as the work progresses. The widespread availability and use of computers greatly facilitates data processing, and software exists for a wide variety of data handling requirements (Section 10.5). The QA/QC program should include the data handling system so that data quality can be monitored.

Table 5-1 Comparison of Approaches to Site Characterization Efforts

Approach	Advantages	Disadvantages
Conventional		
-Install a few dozen shallow monitoring wells -Sample and analyze numerous times for 129+ pollutants -Define geology primarily by driller's log and cuttings -Evaluate hydrology with water level maps only -Possibly obtain soil and core samples (chemical extractions)	-Rapid screening of problem -Moderate costs involved -Field and lab techniques standardized -Data analysis relatively straightforward -Tentative identification of remedial options possible	-True extent of problem often misunderstood -Selected remedial alternative may not be appropriate -Optimization of remedial actions not possible -Cleanup costs unpredictable and excessive -Verification of compliance uncertain and difficult
State-of-the-Art		
-Conduct geophysical surveys (resistivity soundings, etc.) -Install depth-specific piezometers and well clusters -Sample and analyze for 129+ pollutants initially -Analyze selected contaminants in subsequent samples -Define geology by extensive coring/split-spoon samples -Evaluate hydrology with well clusters and geohydraulic tests -Perform limited tests on solids (grain size, clay content)	-Conceptual understanding of problem more complete -Better prospect for optimization of remedial actions -Predictability of remediation effectiveness increased -Cleanup costs lowered; estimate improved -Verification of compliance soundly based, more certain	-Characterization costs somewhat higher -Detailed understanding of problem still difficult -Full optimization of remedial actions not likely -Field tests may create secondary problems -Demand for specialists increased
State-of-the-Science		
-Assume "State-of-the-Art Approach" as starting point -Conduct tracer tests and borehole geophysical surveys -Determine % organic carbon, exchange capacity, etc., of solids -Measure redox potential, pH, dissolved oxygen, etc., of fluids -Evaluate sorption-desorption behavior using select cores -Identify bacteria and assess potential for biotransformation	-Thorough conceptual understanding of problem obtained -Full optimization of remedial actions possible -Predictability of remediation effectiveness maximized -Cleanup costs lowered significantly; estimates reliable -Verification of compliance assured	-Characterization costs significantly higher -Few previous field applications of advanced theories -Field and laboratory techniques not yet standardized -Availability of specialized equipment low -Demand for specialists dramatically increased

Source: Adapted from Keely (1987).

If predictive models are required at some point in the investigation, or later in the development of an aquifer remediation project, they should be formulated or selected from basic flow and transport models as early as possible (Section 10.4). Then requirements for acquisition of the appropriate data can be built into the project plan. Steps should also be taken for model calibration and validation as the project proceeds.

The groundwork for the development of conclusions also should be laid in the early stages of the project by establishing hypotheses pertaining to the project objectives. The project should be designed and modified around their acceptance or rejection. If done correctly, this approach ensures that the project design responds efficiently to the project goals and minimizes the collection of extraneous information.

5.2 Existing Information Sources

Before field work begins in earnest, available information about the area of interest should be obtained and reviewed. This includes first reviewing published maps and reports about soils, geology, and hydrology of the area. The next step is finding and analyzing any unpublished data, such as well drill logs, and hydrologic and water quality data on file at local, state, or federal government offices. EPA's STORET and USGS's NAWDEX databases may have ground-water quality data from the area (see Table 5-13). Finally, examination of aerial photographs provides an opportunity to relate knowledge gained in reviewing published and unpublished information to the specific wellhead area, and helps focus field efforts to collect additional required information.

The above steps do not have to be followed in strict sequential order, but an intensive initial effort to identify and review published and other existing information will generally pay off by (1) avoiding field effort spent in collecting data that is already available, and (2) targeting the location and type of field data collection to yield the greatest benefits. Table 5-2 identifies major potential sources of information, and Table 5-3 provides a list of major EPA information hotlines, information numbers, and electronic databases.

Getting to know one or more individuals in the various state and federal agencies that publish and maintain files of information on soils, geology, and water resources can facilitate the process of determining what is available for the area of interest. The planning and utility departments of local government are also sources of potentially valuable information that may not be available from other sources. Worksheet D-W2 provides a form for listing personal contacts and identifying available maps that can provide a starting point for compiling a hydrogeologic map of an area.

5.2.1 Soil and Geomorphic Data

Soil maps can be used as an aid in geologic interpretation because soil types are often related to both the original rock and the weathering processes from which they were derived. Soil information is necessary also to evaluate the potential for movement of organic and inorganic compounds through the unsaturated zone.

Planning Field Investigations

Table 5-2 Sources and Types of Existing Data for Soil and Ground-Water Quality Investigations

Sources	Types/Comments
Cross-Cutting Information Sources	
Remote Sensing Imagery	Drainage patterns, land use, vegetation stress, historical land development, and geologic structure; see ASTM D5518 (Table A-14)
Computer Databases	Wide variety of reference data and bibliographies; see Table 5-13
U.S. EPA and State Environmental Agency	
RCRA permits and applications	EPA identification numbers
Waste generators and transporters	Generator annual reports; may require special clearance for reviewer
NPDES permits and applications	Liquid waste types; treatment processes; production information
Uncontrolled waste disposal sites	
Spills of oil and hazardous materials	
Water supplies	Nearest water supply
Enforcement actions	Problem history; previous findings
Surveillance reports	Plans, concerns, and past problems
Other Federal Agencies	
U.S. Geological Survey	Technical geologic and hydrologic reports, maps, aerial photographs, and water monitoring data
U.S. Department of Agriculture (ASCS, SCS)	Soil maps, types, physical characteristics, depths association, and uses
U.S. Fish and Wildlife Service	Endangered species
National Oceanic and Atmospheric Administration (NOAA)	Climatic data
National Ocean Survey	Tidal data; historic, recent, and projected
Land Management Agencies (BIA, BLM, Forest Service)	Air photos, vegetation/land use maps, hydrologic data
Occupational Safety and Health Administration (OSHA)	Processes, hazards, protective equipment needs
Other State Agencies	
Geological Survey	Technical geologic and hydrologic reports, state geologic maps, and monitoring data
Water Division/Survey	Well locations and logs
Regional Planning Agencies	
CWA-Section 208 Agency	
Other	Areawide solid waste/wastewater treatment management

Table 5-2 (cont.)

Sources	Types/Comments
County Offices	
Health Department	Problems, complaints, analytical results
Planning and zoning	Land use restrictions
Assessor	Plant maps and land owners
Surveyor	Survey bench mark locations
City Offices	
Chamber of Commerce	Information and local industries including number of employees, principal products, and facility addresses
Clerk	
Engineer	Foundation and inspection reports; survey benchmark locations
Fire department	History of fires and/or explosions at facility
Law enforcement	Complaints and violations of local ordinances
Water and sewer	Location of buried mains and lines
Company Files and Records	Confidential records require special handling and storage
Contractors	
Building	Local soils, geology, and shallow water levels
Soil/foundation, water well drillers	Local soils, geology, hydrogeology, water levels, regulations, and equipment availability
Utility Companies	
Gas	Location of buried lines
Electric	
Water	
Petroleum or natural gas pipelines	

Source: Adapted from Sisk (1981).

Section 7.4.1 discusses the use of soil survey data in the estimation of aquifer parameters. Soil surveys published by the Soil Conservation Service (SCS) of the U.S. Department of Agriculture are typically at a scale of 1:15,840 or 1:20,000 and mapped on an airphoto base. Simplified geomorphic maps can be readily developed from a soil map by grouping soil map units into larger geomorphic units (floodplains, terraces, uplands, etc.). Nonfloodplain soils are differentiated on the basis of slope with letter designations in the map symbol. This allows development of geomorphic units based on slope range. Slope range, combined with the infiltration characteristics of the soil, allow qualitative interpretations of infiltration-runoff characteristics of an area (see, for example, index surface runoff classes in Table 2-1 of Boulding, 1994/Table 9-10).

Planning Field Investigations

Table 5-3 Environmental Hotlines, Clearinghouses, and Electronic Databases (EPA-sponsored unless otherwise indicated)

General

Center for Environmental Research Information, Research Information Unit: 513/569-7562. [Technology transfer publications]

Pollution Prevention Information Clearinghouse (PPIC): 703/821-4800.

Public Information Center: 703/821-4800.

Small Business Ombudsman Clearinghouse/Hotline: 800/368-5888; 703/557-1938.

Water

National Small Flows Clearinghouse: 800/624-8301.

National Drinking Water Clearinghouse: 800/624-8301.

Safe Drinking Water Hotline: 800/426-4791; 202/382-5533.

Nonpoint Source (NPS) Information Exchange: 202/260-7085.

Wetlands Protection Hotline: 800/832-7828.

Clean Lakes Clearinghouse: 202/260-7111.

Hazardous, Toxic, and Solid Waste

Solid Waste Information Clearinghouse and Hotline (SWICH): 800/67-SWICH. [Library and electronic bulletin board]

RCRA/Superfund Industry Assistance Hotline: 800/424-9346; 703/920-9810. Also called RCRA/CERCLA Hotline. [Answers to regulatory and technical questions on RCRA, CERCLA, UST programs; some documents available]

Emergency Planning and Community Right-To-Know Act (EPCRA) Hotline: 800/535-0202. Sometimes referred to as SARA Title III Hotline. [Documents available]

Alternative Treatment Technology Information Center (ATTIC): 301/231-5250. Center includes RCRA/Superfund Hotline, Cleanup Information (CLU-IN) Bulletin Board, and Hazardous Waste Superfund Collection (library).

Hazardous Waste Ombudsman Program: 202/260-9361.

National Response Center: 800/424-8802; 202/267-2675. [U.S. Coast Guard; for notification of accidental hazardous releases]

Table 5-3 (cont.)

Hazardous, Toxic, and Solid Waste (cont.)

National Hazardous Materials Information Exchange (HMIX): 800/752-6367; 800/367-9592 (Illinois residents). [Sponsored by Federal Emergency Management Agency (FEMA) and U.S. Department of Transportation (DOT)]

CERCLIS (Comprehensive Environmental Response, Compensation and Liability Information System) Helpline: 202/252-0056. [Assistance for users of CERCLIS database]

Records of Decision System (RODS) Hotline: 202/245-3770. [Assistance for users of RODS database]

Toxic Substances Control Act (TSCA) Assistance Information Service: 202/554-1404.

Air and Radiation

Air Risk Information Support Center (AIR RISC) Hotline: 919/541-0888. [Technical assistance and information relating to health, exposure and risk assessments for toxic air pollutants]

BACT/LAER Clearinghouse: 919/541-5432. Best Available Control Technology at Lowest Achievable Emission Rate.

Control Technology Center (CTC) Hotline: 919/541-0800.

National Air Toxics Information Clearinghouse (NATICH): 919/541-0850.

Emission Factors Clearinghouse: 919/541-5477.

Electronic Databases

NSFC WTIE-BBS (National Small Flows Clearinghouse Wastewater Treatment Information Exchange Bulletin Board System, Morgantown, WV, 800/624-8301; access to NSFC's online databases, including EPA's EFIN, the Environmental Financing Information Network, electronic mail, conferences, bulletins and feature articles.

NDWC BBS (National Drinking Water Clearinghouse Bulletin Board System, Morgantown, WV, voice support 800/624-8301; access to articles in On Tap, sharing by users of questions and solutions to specific water problems).

ORD BBS (EPA Office of Research and Development's Bulletin Board System, Cincinnati, OH, voice support 513/569-7272; provides access to ORD's expert systems software, methods standardization/QA news, and information on technical aspects of ORD research programs).

IRIS (Intergrated Risk Information System, voice support 513/569-7574, provides access to a database containing summaries of health risk and regulatory information on more than 500 chemicals).

Source: U.S. EPA (1991) and other sources.

Planning Field Investigations

5.2.2 Geologic and Hydrologic Data

Geologic reports, maps, and cross sections provide details of the surface and subsurface, including the areal extent, thickness, composition, and structure of rock units. Geology and hydrogeologic settings are key to any ground-water investigation. These sources of information should be supplemented, if possible, by the original field *logs* of wells and test holes. These logs provide first-hand information on types and characteristics of rocks in the subsurface, their thickness, and areal extent. They also may describe drilling conditions that allow inference of relative permeability values, describe well construction details, and report water-level measurements.

The Hydrologic Atlas (HA) and Water Resource Investigation (WRI) series of the U.S. Geological Survey are some of the best sources of hydrogeologic information. These maps are based on the interpretation of all available geologic information from soil profiles, test wells, rock outcrops, observation wells, seismic surveys, and other means of subsurface observation. The location of aquifers on these maps is estimated by examining surficial geology, depth to bedrock, and depth to the water table. A hydrologic atlas contains information about ground-water availability, well locations, ground-water quality, surficial deposits influencing transmissivity, basin boundaries, flow characteristics of surface water, and other hydrologic factors. The U.S. Geological Survey has established a National Water Data Exchange (NAWDEX), which maintains a national database called WATSTORE, and U.S. EPA maintains a ground-water monitoring database called STORET (see Table 5-13). The USGS Water Resource Division District Office in each state can assist in obtaining data available from NAWDEX.

A water table or potentiometric surface map, if available, is another valuable source of hydrogeologic information (Section 7.2). Such maps may be available from the state water resource agency or geological survey. SCS-published soil surveys usually give summary data on monthly distribution, averages, and ranges of temperature and precipitation. The National Oceanic and Atmospheric Administration (NOAA) is the primary source for other climatological data, which may be required to evaluate recharge of unconfined aquifers. Hatch (1988/T5-13) provides a guide to types of data from NOAA's National Climate Data Center.

Geologic information is available from many sources. The U.S. Geological Survey and state geological surveys are the primary source for surficial and bedrock geologic maps. Important surface hydrologic features include drainage basins (watersheds), surface water bodies, wetlands, and flood zones. Wetlands can be identified on topographic maps; however, more detailed wetland maps may be available from the state wetlands regulatory agency or regional office of the U.S. Army Corps of Engineers. Flood mapping for every state has been prepared by the Federal Emergency Management Agency (FEMA). Two types of flood mapping are available: Flood Insurance Rate Maps (FIRM) and Flood Boundary and Floodway Maps. These maps delineate the areas adjacent to surface waters that would be under water in 100-year and 500-year floods. Historic flood data may also be available from community and state libraries.

If published information sources are lacking or scarce, a review of well logs, both public and private, and test boring logs becomes the primary method for developing preliminary hydrogeologic interpretations for an area. Well records provide geological data (although the quality of descriptions prepared by water well drillers may be problematic). Records of well discharge and water level fluctuations may provide a basis for evaluating an aquifer's hydraulic conductivity, transmissivity, and storativity.

5.2.3 Airphoto Interpretation

Aerial photographs and *satellite imagery* are useful tools, both for office and field study. The latter should be examined first to detect trends of lineaments, which may indicate the presence of faults or joints. These may reflect zones of high permeability that exert a strong influence on fluid movement from the land surface or through the subsurface. Satellite imagery also can be used to indicate the presence of shallow ground water, where subtle tonal changes and differences in vegetation are caused by the higher moisture content. Rock types may also be evident.

Aerial photographs provide a relatively inexpensive way to directly observe natural and artificial features on the land surface. Much information can be obtained from stereopairs of black-and-white airphotos, which provide a three-dimensional image of the surface when viewed with a stereoscope. Patterns of vegetation, variations in gray tones in soil and rock, drainage patterns, and linear features allow preliminary interpretations of geology, soils, and hydrogeology. All airphoto interpretations should be field checked and revised where "ground truthing" indicates features that were missed or incorrectly delineated. Section 6.2 provides additional information on remote sensing techniques.

Black-and-white airphotos are available from various federal agencies for almost any location in the United States. These are the cheapest type of airphoto to obtain. The nearest county office of the Soil Conservation Service or Agricultural Stabilization and Conservation Service (they will often be in the same building) is the best starting place to determine what is available. Many of these offices have airphoto coverage that extends back to the 1930s. When photographs for multiple years are available, all should be examined, because significant features that are obscured in one set may be evident in another. Also, sequential examination of airphotos taken at different times provides valuable information on changes in land use.

Airphotos often reveal linear features, called fracture traces, that indicate zones of relatively higher permeability in the subsurface. Fracture-trace analysis using airphotos can provide preliminary information on possible preferential movement of contaminants. Fetter (1980, pp. 406-411/T2-5) provides a good introduction to fracture-trace analysis. Parizek (1976) provides a good review of the North American literature on fracture trace and lineament analysis.[1]

---

[1] Parizek, R.R. 1976. On the Nature and Significance of Fracture Traces and Lineaments in Carbonate and Other Terranes. In: Karst Hydrology and Water Resources, V. Yevjevich (ed.), Water Resources Publications, Fort Collins, CO, Vol. 1, pp. 3-1 to 3-62.

5.3 Developing a Sampling and Monitoring Plan

Soil and ground-water sampling is conducted for a variety of reasons ranging from detection or assessment of the extent of a contaminant release to evaluation of trends in regional water quality. Initial sampling of soil and ground water should focus on locating the source and distribution of contaminants, if present, and characterizing geochemical variability. This information helps guide selection of locations for ongoing monitoring and identifies potential chemical problems that may affect selection and design of systems for soil and ground-water treatment. Soil-solute and ground-water sampling over a period of time provides information on the rate and magnitude of changes in ground-water quality.

Reliable sampling of the subsurface is inherently more difficult than either air or surface water sampling because of the inevitable disturbance caused by well drilling or pumping and the inaccessibility of the sampling zone. Consequently, monitoring well design and installation (Section 9.3) and sampling activities should be conducted so as to minimize disturbance of the subsurface geochemical and hydrogeologic conditions. A trained field team following a well-developed sampling protocol and communication between field and laboratory personnel are essential for coordinating sample handling and analysis.

5.3.1 Types of Monitoring

A complete sampling program for subsurface site characterization includes several types of monitoring, each with its own goal. The goal of *detection monitoring* is generally to determine the presence of contaminated conditions. For example, vadose zone monitoring (Section 9.1.1) allows early detection of subsurface contaminants before they enter the ground-water system. *Assessment monitoring* seeks to identify the extent and magnitude of contamination. For example, soil gas sampling allows relatively rapid initial delineation of areas contaminated by gasoline and other volatile nonaqueous phase liquids. If assessment monitoring results indicate a degree of contamination requiring remediation, *evaluation monitoring* is used to provide data necessary to design the remediation system. *Performance monitoring* is designed to evaluate the success of remediation efforts. Each stage of monitoring often requires the placement of additional monitoring wells and piezometers for water level measurements. Other types of monitoring include litigation monitoring in response to legal actions at contaminated sites and research monitoring aimed at specific scientific objectives.

5.3.2 Sampling Protocol

A sampling protocol ensures the collection of data of known quality. Uncertainty, hydrogeologic variability, and quality-assurance decision-making need to be addressed from the initial design stage. Most initial effort and fiscal resources should be spent on characterizing basic site geology and hydrology, and sampling protocols should leave room for evolutionary development of the network design. For example, sampling experiments can be used to determine spatial correlation for solid samples. A large number of surface samples or split-spoon samples can be collected but it may only be necessary to analyze a certain percentage (20 to 50%) to achieve adequate spatial coverage. If the initial sample groups indicate sufficient sampling resolution, the other samples need not be analyzed. If necessary, additional samples can be analyzed until geostatistical analysis indicates an adequate sampling intensity has been achieved.

Reliable protocols and the optimization of sampling procedures require particular attention to the effect of (1) sample location (Section 5.3.3), (2) sample frequency (Section 5.3.5), (3) sampling mechanism (Sections 9.4 and 9.5), (4) operator error, and (5) sample collection and handling procedures (Section 5.3.6) on both the sensitivity and reliability of chemical constituent monitoring results. Protocols should include documentation of all field activities, and deviations from the established protocol are necessary for a complete record of sampling activities, and useful if data collected for a particular purpose ends up being used for other objectives.

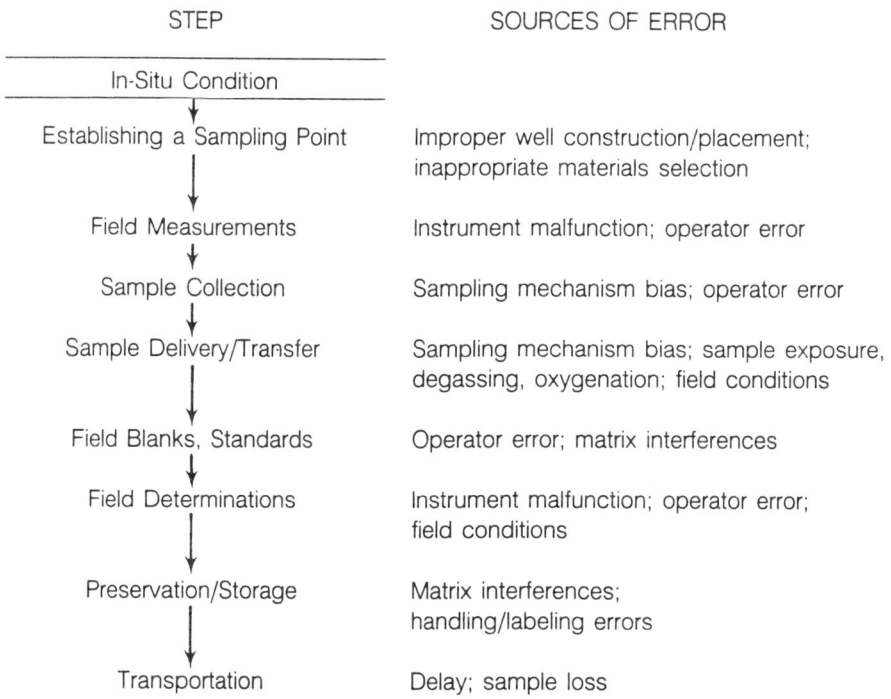

Figure 5-3 Steps in ground-water sampling and sources of error (Barcelona et al., 1985).

Each step within the protocol affects the quality and completeness of the information being collected. Figure 5-3 lists the steps and corresponding sources of potential error in each phase of field collection of ground-water samples after a well has been properly purged of stagnant water. Figure 5-4 identifies potential sources of error once a soil or ground-water sample reaches the laboratory. Section 5.4 discusses sources of error further. Sections 9.4 and 9.5 address in more detail vadose zone and ground-water sampling methods, respectively.

Planning Field Investigations 255

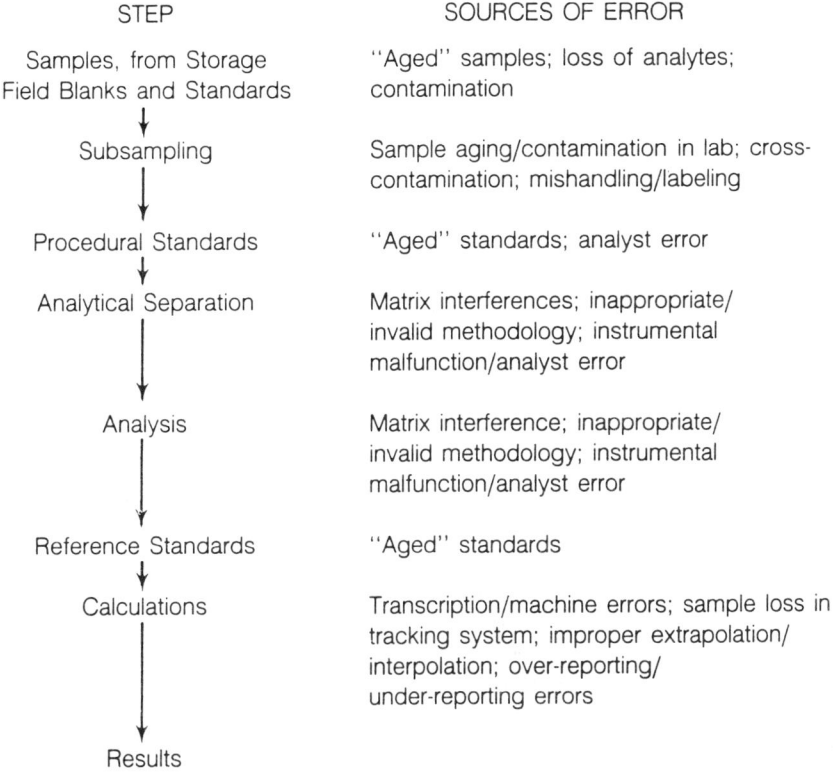

Figure 5-4 Steps in water sample analysis and sources of error (Barcelona et al., 1985).

Sampling protocols at contaminated sites or any site where there is a possibility for analyte cross-contamination should specify solutions and procedures for decontamination of portable sampling equipment after sampling at each location is completed. ASTM D5088 (Table A-14) provides guidance on decontamination procedures at nonradioactive waste sites. The appropriate state or federal agency should be consulted to identify required or preferred procedures. Mickam et al. (1989) provide a good review of decontamination procedures in different EPA and state regulatory programs.[2]

5.3.3 Sample Location

Sample location and frequency (Section 5.3.5) are among the most critical aspects of sampling because sample collection at the wrong location and time can give entirely erroneous results even when executed carefully. Initial selection of locations for sampling must be based on a good preliminary characterization of the geology and

---

[2] Mickam, J.T., R. Bellandi, and E.C. Tifft, Jr. 1989. Equipment Decontamination Procedures for Ground Water and Vadose Zone Monitoring Programs: Status and Prospects. Ground Water Monitoring Review 9(2):100-121.

hydrogeology of the site, including good vertical and horizontal resolution of hydrogeologic conditions. This may require spending more of the available financial resources on hydrogeologic characterization than on chemical sampling and analyses. Additional sample locations should be added as understanding of the site evolves.

Table 5-4 summarizes major types of sampling designs and when they should be used for characterizing subsurface geochemistry. In general, haphazard water-quality or solid sampling is not an appropriate approach to designing sampling for subsurface geochemical characterization, even though professional judgment alone is probably the most frequently used method for siting ground-water monitoring wells. Figure 5-5 illustrates some two-dimensional probability sampling designs for spatial characterization. The trends or patterns that commonly exist in subsurface contamination mean that simple random sampling will not give as accurate an estimate of population characteristics as stratified random and grid sampling designs.

Hydrogeologic characterization, initially using surface geophysical techniques followed by piezometers and preliminary well tests to estimate the distribution of hydrogeologic parameters, should come before the location and installation of monitor wells. Good vertical resolution is essential in sampling to characterize distribution of oxidized and reduced species, contaminants, and microbiota. Achieving this resolution requires more discrete well completions with short screens. In most cases, 5-ft to 1.5-m well screens should give adequate vertical resolution.

The spatial distribution of contamination is a major concern with sampling solids. The intensity and number of samples depends on the nonsampling variance, which is the variability of concentration that is unrelated to sampling procedures. Spatial structure determines the distance between samples that have essentially the same concentration, called the range of correlation, to avoid oversampling (Section 5.6.2).

There are two broad designs for soil sampling: (1) grids in which samples are taken from a matrix of squares or quadrants at a site, and (2) transects in which samples are taken at specified intervals along a line. Grids presume an aerial or dispersed source of some kind, and transects presume a preferential source. For example, Starks et al. (1986) established sampling transects where the length was proportional to the frequency with which wind blew in a particular direction to characterize metal contamination from a smelter near Palmerton, Pennsylvania.[3] Flatman (1986/T5-14) describes use of geostatistics to determine sampling intensity. Grids can be used to estimate short-range correlation. Transects along the path of ground-water or contaminant movement provide the best way to look at long-range correlation. The combination of the two strategies coupled with the initial analysis of selected solid samples at alternate grid or transect locations can be quite effective.

---

[3] Starks, T.H., K.W. Brown, and N.J. Fisher. 1986. Preliminary Monitoring Design for Metal Pollution in Palmerton, Pennsylvania. In: Quality Control in Remedial Site Investigation: Hazardous and Industrial Solid Waste Testing, 5th Vol., C.L. Perket (ed.), ASTM STP 925, American Society for Testing and Materials, Philadelphia, PA, pp. 57-66.

Table 5-4 Summary of Sampling Designs and Conditions for their Use

Type of Sampling Design	Conditions When the Sampling Design is Useful
Haphazard sampling	A very homogeneous population over time and space is essential if unbiased estimates of population parameters are needed. This method of selection is not recommended due to difficulty in verifying this assumption.
Judgment sampling	The target population should be clearly defined, homogeneous, and completely assessable so that sample selection bias is not a problem. Or specific environmental samples are selected for their unique value and interest rather than for making inferences to a wider population.
Probability sampling	
Simple random	The simplest random sampling design. Other designs below will frequently give more accurate estimates of means if the population contains trends or patterns of contamination.
Stratified random	Useful when a heterogeneous population can be broken down into parts that are internally homogeneous.
Multistage	Needed when measurements are made on subsamples or aliquots of the field sample.
Cluster	Useful when population units cluster together (schools of fish, clumps of plants, etc.) and every unit in randomly selected cluster can be measured. Soil and ground-water contamination rarely, if ever, exhibit this characteristic.
Systematic	Usually the method of choice when estimating trends or patterns of contamination over space. Also useful for estimating the mean when trends and patterns in concentrations are not present or they are known a priori or when strictly random methods are impractical.
Double	Useful when there is a strong linear relationship between the variable of interest and a less expensive or more easily measured variable.
Search Sampling	Useful when historical information, site knowledge, or prior samples indicate where the object of the search may be found.

Source: Boulding and Barcelona (1991a), after Gilbert (1987).

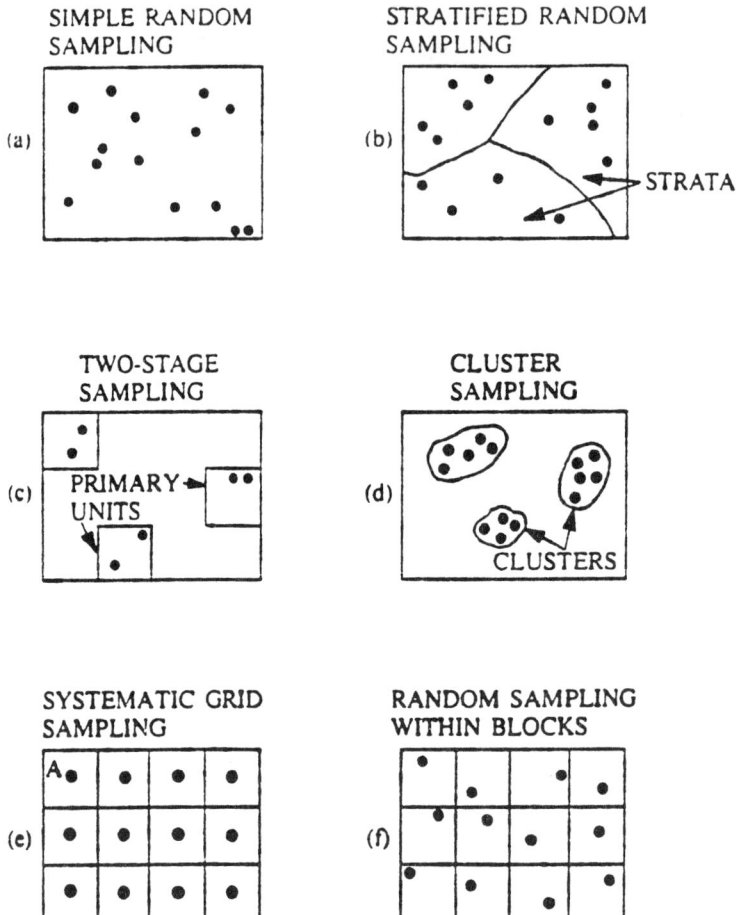

Figure 5-5 Some two-dimensional probability sampling designs for sampling over space (Boulding and Barcelona, 1991a, after Gilbert, 1987).

The combined strategy also can avoid the potential collection of redundant information. Using geostatistical analysis techniques of successive analytical subsets minimizes the number of samples actually analyzed. Transects could be both parallel and perpendicular to the axis of ground-water movement, along with some random samples from a grid, as shown in Figure 5-5(f). Analysis of samples from four equally spaced locations on a transect or grid within the area of influence is a good starting point to estimate the distance of short-range correlation. For soils, at least 5% of sampling points should be duplicated to help determine the sampling variability so it can be analyzed with geostatistical techniques. At least 5% of the samples should be split as well.

Preliminary efforts that can help guide the location of initial wells for ground-water sampling include (1) surface geophysical techniques for mapping the extent of contaminant plumes; (2) soil gas sampling techniques; (3) drive-point ground-water samplers; and (4) selective sampling of drive-point piezometers for simple constituents such as pH, conductance, and possibly iron or dissolved oxygen concentrations.

Soil gas monitoring (Section 9.4.2), drive-point ground-water samplers such as Hydropunch® and BAT samplers (Section 9.5.2), and sequential sampling of drive-point piezometers probably give the best pictures of short-range variability in three dimensions. Sampling from monitoring wells usually gives some sort of integrated value depending on the relative width or thickness of the hydrogeologic formation of interest and the length of the screen. Disadvantages of soil gas concentrations include (1) lack of the ability to directly calibrate, because all values are relative and difficult to reproduce, (2) decontamination, and (3) short circuiting of air from the surface, which can distort results.

Locating both piezometers and monitoring wells spatially and vertically to ensure that they are monitoring the ground-water flow regime of concern is one of the most important components in a ground-water quality monitoring program. The placement and number of sampling points will depend on the complexity of the hydrologic setting, diversity of source characteristics, and the degree of spatial and temporal detail needed to meet the program's goals. Piezometer and well placement should also be viewed as an evolutionary activity that may expand or contract as the needs of the program dictate.

In most monitoring situations, the goal is to determine the effect some surface or near-surface activity has had on nearby ground-water quality. Most dissolved constituents will descend vertically through the unsaturated zone beneath the area of activity and then, upon reaching the saturated zone, move horizontally in the direction of ground-water flow. Therefore, source monitoring points are normally completed downgradient in the first permeable water-bearing unit encountered. Upgradient points would be constructed to provide background or ambient water quality information.

In karst limestone terranes ground-water monitoring wells are unlikely to intercept conduits that serve as the path for contaminant transport (Section 2.5.4). In this situation, monitoring of springs that have been identified as discharge points for the area of interest by dye tracing is the preferred method for monitoring ground-water quality. Section 7.5.4 discusses special considerations in hydrogeologic characterization of fractured rock and karst aquifers, and Section 8.4 addresses tracer tests in karst.

5.3.4 Sample Parameter/Analyte Selection

Soil Parameter Selection. Halocarbons, chlorinated hydrocarbon solvents (e.g., tetra- and trichloroethylene), and fuel constituents (e.g., toluene, benzene, ethyl benzene, and xylenes) are amenable to preliminary delineation by soil gas methods (see Table 9-7). Soil gas samples for carbon dioxide, methane, oxygen, and nitrogen can provide additional insights into subsurface chemistry, particularly microbiological activity. In addition to examining chemical constituents, solid samples should be analyzed for grain-size distribution and other parameters related to hydrologic properties (Section 7.4). Boulding (1994/T9-10) includes a checklist of more than 50 potential soil physical, hydrologic and chemical parameters for evaluating contaminated sites.

Ground Water Parameter Selection. Tables 5-5 and 5-6 identify chemical constituents of interest for various types of ground-water monitoring activities. In hazardous waste site investigations, regulations will generally specify the contaminants to be tested for. Focusing on priority pollutants alone, however, may not provide a complete geochemical picture of contamination. The source of contamination may involve a large number of individual contaminants that are not classified as hazardous. Also, determination of redox-sensitive constituents (dissolved oxygen and dissolved iron), pH, and conductance, may provide valuable insight into subsurface contaminant geochemistry. Highly mineralized ground water, commonly encountered in formations being evaluated for deep-well injection of wastes, may require more complete analyses for natural inorganic and organic constituents.

Table 5-5 Chemical Constituents of Interest in Ground-Water Monitoring

Type of Analyte	Where Done L = Lab F = Field FF = Field Filtered	Water quality	Drinking water	Contamination	Possible source impacts	Geochemical evaluation of data
Geochemical						
pH, Eh	F	X	X	X	X	X
Conductivity	F	X	X	X	X	X
Temperature	F	X	X	X	X	X
Dissolved oxygen	F	X				X
Alkalinity	F(FF)			X	X	X
Ca^{+2}, Mg^{+2}	L(FF)					X
Na^+, K^+	L(FF)	X	X			
Cl^-, SO_4^{-2}, PO_4^{-2}	F(FF)	X	X		X	X
Silicate	L(FF)					X
Water quality						
Trace Metals (Fe, Mn, Cr, Cd, Pb, Cu)	L(FF)	X	X	X	X	X
NO_3^-, NH_4^+	L(FF)	X	X	X	X	X
F^-	L	X	X		X	
TOC	L	X	X	X	X	
TOX	L	X	X	X	X	
TDS	L(FF)	X	X	X	X	X
Organic compounds	L	X	X	X	X	

Source: Modified from Barcelona et al. (1989).

Table 5-6 Recommended Analytical Parameters for Detective Monitoring

Type of Parameter	Where Measured F = Field L = Lab	Analytes	
		Required by regulation	Suggested for Completeness
Well purging	F	pH, conductivity	Temperature Redox potential
Contamination indicators	F L L	pH, conductivity TOC (total organic carbon) TOX (total organic halogen)	
Water quality[a]	L L	Cl^-, Fe, Mn, Na^+, SO_4^{-2} Phenols	Alkalinity (F) or acidity (F) Ca^{+2}, Mg^{+2}, K^+, NO_3^-, PO_4^{-2}, NH_4^+, silicate
Drinking water suitability[b]	L L L L	As, Ba, Cd, Cr, F, Pb, NO_3^-, Se, Ag Endrin, lindane, methoxychlor, toxaphene, 2,4-D, 2,4,5-TP (Silvex) Radium, gross alpha/beta Coliform bacteria	

[a] All parameters required to be determined quarterly for the first year of network operations (RCRA Part 265.92).

[b] These parameters are excluded from the annual reporting requirements of RCRA after the first year.

Source: Barcelona et al. (1985).

Iron, an inexpensive constituent to determine analytically, can be used as an indicator of redox conditions and potential mobility for heavy metals. Dissolved gases are excellent indicators of redox conditions and microbial activity. For example, Leenheer and Malcolm (1973) analyzed for H_2, N_2, CH_4, CO_2, and H_2S in serial samples from a well through which a plume of deep-well injected wastes passed. They used changes in the relative percentages of the different gases as indicators of changing microbial activity.[4]

---

[4] Leenheer, J.A. and R.L. Malcolm. 1973. Case History of Subsurface Waste Injection of an Industrial Organic Waste. In: Symposium on Underground Waste Management and Artificial Recharge, J. Braunstein (ed.), Int. Ass. of Hydrological Sciences Pub. No. 110, pp. 565-579.

Battista and Connelly (1989) found that inorganic parameters such as chemical oxygen demand, specific conductance, chloride, alkalinity, and hardness were reasonably good indicators for predicting VOC contamination from landfills. When the inorganic parameters were detected above background levels in monitoring wells, VOCs were also usually present.[5] Out of 49 ground-water samples at landfill sites in Wisconsin, VOCs and elevated inorganic parameters were detected at about the same frequency in 20 (41%), elevated inorganic parameters without VOCs were detected in 11 (22%), and VOCs without elevated inorganic parameters were detected in 3 wells (6%). The remaining 15 wells in the study showed neither VOCs nor elevated inorganic parameters.

When organic contaminants are susceptible to biotransformation (Section 3.5.4), the list of analytes should include known daughter products. For example, trichloroethylene (TCE) yields a number of biotransformation products under anaerobic conditions (1,2-dichloroethene, 1,1-dichloroethene, and vinyl chloride), which are also toxic. In fact, vinyl chloride is more toxic than TCE and is resistant to degradation under anaerobic conditions.

Calcium carbonate and iron/manganese concentrations are especially important parameters if remediation involves air stripping. Air-stripping towers are particularly susceptible to fouling by calcium carbonate and metal oxide precipitates.

5.3.5 Sampling Frequency, Type and Size

Ground-Water Sampling Frequency. Table 5-7 shows estimated ranges of sampling frequency in months necessary to maintain information loss at less than 10% for selected types of chemical parameters. For many chemical constituents, quarterly sampling is adequate for characterizing short-term (i.e., monthly to 1 or 2 years) changes over time. For some reactive constituents such as iron and other redox-sensitive constituents, bimonthly sampling may be required.

With intermittent sources of contamination, it is especially important that the frequency of sampling not allow a contaminant to be missed. Barcelona et al. (1985/T9-10) describe a procedure for estimating sample frequency to detect contaminant plumes based on the type of plume (slug, intermittent, or continuous) and hydrogeologic parameters of gradient, hydraulic conductivity, effective porosity, and distance along the flow path. Figure 5-6 shows a nomograph that can be used when these parameters are known. When the contaminant plume is a slug source or intermittent, sampling frequency should probably be more frequent to ensure that the plume is not missed. One advantage to the slow movement of ground water is that if there are questions about a sample, resampling a week later will yield roughly the same ground water.

---

[5] Battista, J.R. and J.P. Connelly. 1989. VOC Contamination at Selected Wisconsin Landfills--Sampling Results and Policy Implications. WDNR PUBL-SW-094 89. Wisconsin Department of Natural Resources, Madison, WI.

Table 5-7 Estimated Ranges of Sampling Frequency (in Months) to Maintain Information Loss at <10% for Selected Types of Chemical Parameters

Type of Parameter	Pristine background conditions	Contaminated	
		Upgradient	Downgradient
Water quality			
Trace constituents (<1.0 mg/L)	2 to 7	1 to 2	2 to 10
Major constituents	2 to 7	2 to 38	2 to 10
Geochemical			
Trace constituents (<1.0 mg/L)	1 to 2	<2	1 to 5
Major constituents	1 to 2	7 to 14	1 to 5
Contaminant indicator			
TOC	2	3	3
TOX	6 to 7	24	7
Conductivity	6 to 7	24	7
pH	2	2	1

Source: Boulding and Barcelona (1991c).

Precise estimation of optimum sampling frequency is probably impractical for most investigations. For example, Bell and DeLong (1988) found that tetrachloroethylene at concentrations of 200 to 300 μg/L exhibited variations of a factor of one or two over the course of a year. Their work points out that data collection may be required for 4 years or more in order to estimate the optimal sampling frequency to determine seasonal variability. Therefore, it is important to select sampling frequency on the basis of an initial period of monitoring in the context of the duration of the program.[6]

Soil Sample Type and Size. Soil sampling must take into account fractures in earth materials and the fact that the subsurface is heterogeneous (at scales ranging from centimeters to meters). If the soil has obvious fractures and channels in the subsurface, sampling should sample both affected and apparently nonfractured areas for comparison. Soil sample quantities of less than 100 g tend to be unrepresentative even of the areas where the sample is taken. In the laboratory, the sample can be mixed and subsampled prior to analysis.

---

[6] Bell, H.F. and H.P. DeLong. 1988. Data Characteristics of Ground Water Monitoring's Catch 22. ACS Abstracts 28(2):20-24.

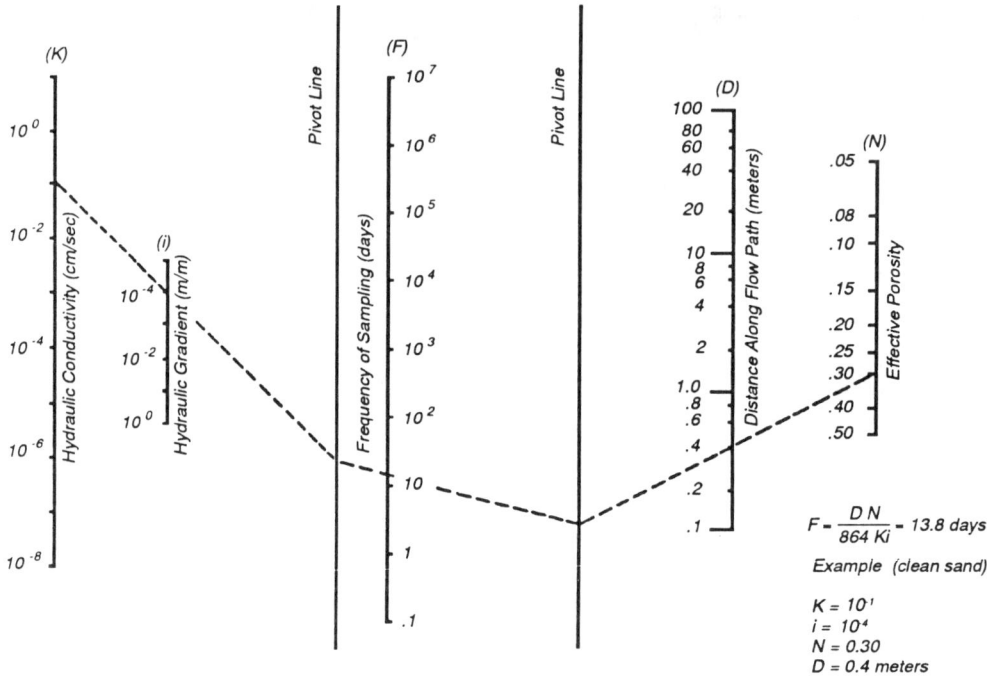

Figure 5-6 Sampling frequency nomograph (Barcelona et al., 1985).

Compositing samples is often beneficial for soil investigations where constituents of interest are not volatile or redox sensitive. For example, Williams et al. (1989) compared the results of one 500-g sample, twenty 25-g composite samples, and ten 50-g composite samples and found that a single 25-g composite sample was the most accurate and precise technique for determining radium concentrations in contaminated surface soil.[7] Initial soil samples of 100-g are about the best size for such composite analyses.

5.3.6 Sample Collection and Handling

Container Selection. Sample containers should be free of contaminants prior to use and made of a material that will not alter the sample. Multiple sample containers will usually be required for each sampling point, with the number depending on the parameters to be analyzed. Each container must be clearly labeled indicating the soil or well sample number and location. Table 5-8 lists the volume of sample required and the recommended container type for the analyses in a detective monitoring program. Table A-1 in Boulding (1994) provides similar information for soil sample containers.

---

[7] Williams, L.R., R.W. Leggett, M.L. Espegren, and C.A. Little. 1989. Optimization of Sampling for the Determination of Mean Radium-226 Concentration in Surface Soil. Environmental Monitoring and Assessment 12:83-96.

Table 5-8 Recommended Sample Handling and Preservation Procedures for a Detective Monitoring Program

Parameters (Type)	Volume Required(mL) 1 Sample[a]	Containers (Material)	Preservation Method	Maximum Holding Period
Well purging				
pH (grab)	50	T,S,P,G	None; field det.	<1 hr[b]
Ω^{-1} (grab)	100	T,S,P,G	None; field det.	<1 hr[b]
T (grab)	1,000	T,S,P,G	None; field det.	None
Eh (grab)	1,000	T,S,P,G	None; field det.	None
Contamination indicators				
pH, Ω^{-1} (grab)	As above	As above	As above	As above
TOC	40	G,T	Dark, 4°C	24 hr[d]
TOX	500	G,T	Dark, 4°C	5 days
Water quality				
Dissolved gases (O_2, CH_4, CO_2)	10 mL minimum	G,S	Dark, 4°C	<24 hr
Alkalinity/ acidity	100	T,G,P	4°C/None	<6 hr[b] <24 hr
	Filtered under pressure with appropriate media			
(Fe, Mn, Na^+, K^+, Ca^{+2}, Mg^{+2})	All filtered 1,000 mL[f]	T,P	Field acidified to pH <2 with HNO_3	6 months[c]
(PO_4^-, Cl^-, Silicate)	@50	(T,P,G glass only)	4°C	24 hr/ 7 days[e]; 7 days
NO_3^-	100	T,P,G	4°C	24 hr[d]
SO_4^-	50	T,P,G	4°C	7 days[e]
OH_4^+	400	T,P,G	4°C/H_2SO_4 to pH <2	24 hr/ 7 days
Phenols	500	T,G	4°C/H_3PO_4 to pH <4	24 hr

Table 5-8 (cont.)

Parameters (Type)	Volume Required(mL) 1 Sample[a]	Containers (Material)	Preservation Method	Maximum Holding Period
Drinking water suitability As, Ba, Cd, Cr, Pb, Hg, Se, Ag	Same as above for water quality cations (Fe, Mn, etc.)[f]	Same as above	Same as above	6 months
F⁻	Same as chloride above	Same as above	Same as above	7 days
Remaining organic parameters	As for TOX/TOC, except where analytical method calls for acidification of sample			24 hr

[a] It is assumed that at each site, for each sampling date, replicates, a field blank, and standards must be taken at equal volume to those of the samples.

[b] Temperature correction must be made for reliable reporting. Variations greater than ± 10% may result from a longer holding period.

[c] In the event that HNO_3 cannot be used because of shipping restrictions, the sample should be refrigerated to 4°C, shipped immediately, and acidified on receipt at the laboratory. Container should be rinsed with 1:1 HNO_3 and included with sample.

[d] 28-day holding time if samples are preserved (acidified).

[e] Longer holding times in EPA (1986).*

[f] Filtration is not recommended for samples intended to indicate the mobile substance lead. See Puls and Barcelona (1989)* for more specific recommendations for filtration procedures involving samples for dissolved species.

Note: T = Teflon; S = stainless steel; P = PVC, polypropylene, polyethylene; G = borosilicate glass.

*See Appendix F for reference citations.

Source: Boulding and Barcelona (1991c), after Scalf et al. (1981) and U.S. EPA (1986).

Sample Collection.[8] Water samples should be collected when the major ion solution chemistry of the ground water being pumped has stabilized, as indicated by the pH, specific conductance, and temperature readings (Section 9.5.3). Samples should be collected as close as possible to the well head. A "tee" fitting placed ahead of an in-line device for measuring well-purging parameters makes this more convenient. During sampling, it is important to minimize the disturbance of fine particles that accumulate in the well. This can be achieved by careful placement of the sampling pump intake at the top of the screened interval, low pumping rates, and the avoidance of bailing techniques, which disturb sediment accumulations at the bottom of the well. Well development may have to be repeated at periodic intervals to minimize the collection of turbid samples.

Aeration of ground-water samples should also be minimized. Wells located upgradient of a site (i.e., presumably uncontaminated) should usually be sampled first to minimize the potential for cross contamination of sampling equipment from wells downgradient of the site.

The samples most sensitive to handling should be collected first. Figure 9-7 in Section 9.5 depicts a priority order for a generalized sample collection effort. In the figure, the samples for organic chemical constituent determinations are taken in order of decreasing sensitivity to handling; the inorganic chemical constituents, which may require filtration, are taken afterwards.

Samples collected for total organic carbon (TOC), total organic halogen (TOX), volatile compounds, field alkalinity, dissolved oxygen, and other analyses sensitive to pH and dissolved gases, should not be filtered or transferred from one container to another, because of the potential for the loss of organic materials by volatilization or to the walls of the containers. To minimize volatilization, a "flow" type sampling device (i.e., positive displacement pump) is generally best. Also, no headspace should exist in the sample containers. If solids content is high and may cause interference during organic determinations, the samples may be allowed to settle prior to analysis. Decanting this type of sample is preferable to filtration. If filtration is necessary to determine specific constituents (i.e., dissolved ferrous iron, other metals) pressure filtration should be performed in the field. Vacuum or gravity filtration of ground-water samples containing any parameters is not recommended.

Samples collected for dissolved inorganic chemical constituents, such as metals, alkalinity, and anionic species, are filtered in the field prior to acidifying for preservation. Samples can be filtered using a 0.45µ glass or membrane filter (U.S. EPA, 1986/T9-10). The preferred filtering arrangement is an in-line filtration module that uses sampling pump pressure for its operation. These modules have tubing connectors on the inlet and outlet parts and range in diameter from 2.5 to 15 cm. Large diameter filter holders, which can be rapidly disassembled for filter pad

---

[8] The following discussion focusses on ground-water sample collection procedures. Appendix A in Boulding (1994/T9-10) is recommended for information on protocols for soil sample collection. Lewis et al. (1991/T9-10) address soil sampling for volatile constituents.

replacement, are the most convenient and efficient designs (Kennedy et al., 1976; Skougstad and Scarbo, 1968).[9] Puls and Barcelona (1989) discuss further special considerations in handling samples for analysis of heavy metals.[10]

Sample Preservation. The time between sampling and sample analysis can range from several hours to several weeks. Immediate sample preservation and storage can minimize changes in the chemical composition during this period, and can extend the time a sample can be held in the laboratory prior to analyses. Preservation methods usually involve pH adjustment, chemical addition, and refrigeration or freezing. Table 5-8, which lists sample volumes, also lists the preservation methods and acceptable holding times for the parameters in a detective monitoring program. Samples for dissolved metals analyses are not preserved until after filtration, which is normally performed in the field or in the laboratory as soon as possible after sampling.

Sample Transport. Samples should be chilled in the field immediately with ice and transported to the laboratory as soon as possible. The laboratory staff should be informed of the approximate time of arrival so that they can make all analytical determinations within storage periods. Sample packaging and method of transport must ensure that samples arrive at the laboratory on time and are neither lost nor damaged en route. Several commercial suppliers have sampling kits that include packing materials and freezer packs to keep the samples cold once they have been chilled to approximately 4°C (Kent and Payne, 1988).[11] Special labels or distinctive storage vessels for samples preserved with acid in the field may be required to comply with shipping restrictions.

Actual sample storage and treatment for samples that may contain hazardous constituents are documented using *chain of custody* procedures. The chain of custody record includes the date and times of sample collection, chain of possession, and time and date of receipt by the laboratory. After the laboratory receives the sample, the log is typically documented by the laboratory analytical chronology. The sampling team initiates the chain of custody in the field, and the manager in charge of the sampling program receives a copy of the form with the analytical report to verify sample storage and handling.

---

[9] Kennedy, V.C., E.A. Jenne, and J.M. Burchard. 1976. Backflushing Filters for Field Processing of Water Samples Prior to Trace-Element Analysis. U.S. Geological Survey Water Resources Investigations OFR 76-126.

Skougstad, M.W. and G.F. Scarbo, Jr. 1968. Water Sample Filtration Unit. Environ. Sci. Technol. 2(4):298-301.

[10] Puls, R.W. and M.J. Barcelona. 1989. Ground Water Sampling for Metals Analyses. EPA Superfund Ground Water Issue Paper, EPA/540/4-89/001 (NTIS PB91-133249), 6 pp.

[11] Kent, R.T. and K.E. Payne. 1988. Sampling Groundwater Monitoring Wells: Special Quality Assurance and Quality Control Considerations. In: Principles of Environmental Sampling, L.H. Keith (ed.), American Chemical Society, Washington, DC, pp. 231-260.

5.4 Data Measurement and Reliability

The design of sampling and monitoring plans and interpretation of data resulting from field investigations need to be based on a good understanding of (1) the characteristics of the system being observed (Section 5.4.1), (2) how representative the measurements are (Section 5.4.2), (3) the degree of measurement bias, precision, and accuracy (Section 5.4.3), and (4) sources contributing to error (Section 5.4.4).

5.4.1 Deterministic vs. Random Geochemical Data

Observation or measurement of physical phenomena can be broadly classified as either deterministic or nondeterministic. Deterministic data can be described by an explicit mathematical relationship. Nondeterministic or random data, must be described in terms of probability statements and statistical averages rather than by the use of explicit equations. Figure 5-7 summarizes a classification scheme for deterministic and random data. The classification of physical data as deterministic or nondeterministic is not always clearcut in the real world. In fact, most geochemical data probably fall in a gray area between the two types of data. For example, the total dissolved solids in an aquifer is a function of the chemical composition of the aquifer solids and residence time of the flowing ground water. Consequently, the distribution of sample values over space and time will not be completely random. On the other hand, the factors that determine the precise value of a given sample are sufficiently complex and variable that the distribution often cannot be predicted by an explicit mathematical equation.

The transient, nonperiodic data box in Figure 5-7a is a residual category that includes all data not included in the other boxes. This nonperiodic characteristic of geochemical data allows modeling of the distribution of geochemical species using thermodynamic principles. Essentially all geochemical modeling of the subsurface is done deterministically. The difficulty in accurately modeling the geochemistry of the subsurface can, however, be attributed to large random elements (see Figure 5-7b). Depending on the geochemical parameter, and the time frame of sampling, data may be *stationary*, where characteristics of the population being sampled do not vary over time, or *nonstationary*, where the random process varies with time. Typically, geochemical subsurface data involving contamination are nonstationary, but are not fully random (i.e., the value of one sample may show some correlation with the value of an adjacent sample). This creates special considerations in statistical analysis that are discussed in Section 5.6. Subsurface physical parameters such as hydraulic conductivity, porosity, and soil particle size distribution do not typically change with time, at least not on a time scale of human concern. These parameters, however, are not fully random.

5.4.2 Data Representativeness

In measuring environmental parameters, there is no "true" value, but rather a distribution of values. A representative unit or sample is one selected for measurement from a target population so that it, in combination with other representative samples, will give an accurate picture of the phenomena being studied

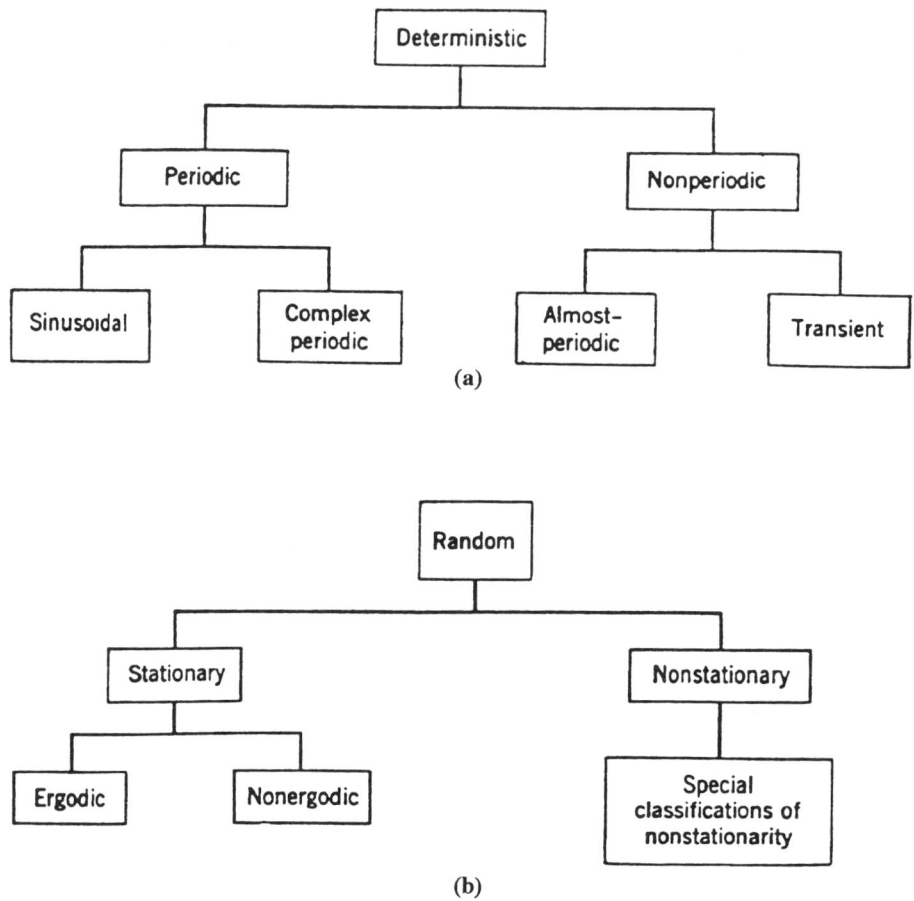

Figure 5-7 Classification of (a) deterministic and (b) random data (Boulding and Barcelona, 1991a, after Bendat and Piersol, 1986).

(Gilbert, 1987/T5-14). Failure to take samples from locations and to use methods that yield samples that are "representative" of a site will result in the collection, at some expense, of analytical data that may be worthless. Representativeness determines whether accurate analysis of the samples will yield results that are close to actual conditions. Quality assurance/quality control systems (QA/QC) in the laboratory or field may be useless if even greater emphasis isn't placed on QA/QC in selecting locations and procedures for sampling (Sections 5.3.3 and 5.3.6).

Thorough site characterization of soils, hydrology, and geology, as described in subsequent chapters, is an essential prerequisite to geochemical sampling. This information provides the basis for developing sampling strategies that will provide some assurance that geochemical samples accurately reflect what is happening in the field. Sample representativeness is essentially knowledge based. For example, sampling locations selected by someone with a rudimentary understanding of sampling theory may yield less accurate results than locations chosen by an individual thoroughly grounded in this theory. At the same time, sampling locations selected without careful site characterization will yield less representative samples than locations selected with thorough site characterization, even with equally sophisticated application of sampling theory.

In contamination investigations, obtaining samples that can be considered representative for assessing one or more particular kinds of environmental exposure is a primary objective. This requires selecting not only the right place (Section 5.3.3), but the right type of sample (Sections 5.3.4 and 5.3.5).

5.4.3 Measurement Bias, Precision, and Accuracy

A measured value that is close to the estimate of the true average value is an unbiased or accurate value. This average or mean can only be estimated by a number of repeat determinations. *Biased* measurements will consistently under- or overestimate the true values in sampled population units. *Precision* is a measure of how closely individual measurements agree and is influenced principally by random measurement uncertainties. Both bias and precision influence *accuracy* as illustrated in Figure 5-8. The center of each target in the figure represents the true value. Both low bias and high precision are required for high accuracy.

Accuracy is largely technologically based. In other words, accuracy can be improved by better drilling and monitoring well installation procedures and better sampling devices and procedures. Pennino (1988/T5-14) has suggested that "there is no such thing as a representative ground water sample" because of geochemical biases inherent in well installation, purging, and sample collection. However, a good understanding of both potential sources of error (see next section) and the way alternative sampling methods may bias results (Section 9.5) minimizes sample disturbance. The final evaluation of the results should be done with full consideration of the unavoidable disturbances involved in subsurface investigations.

5.4.4 Sources of Error

Random error results from slight differences in the execution of the same sampling procedure. Systematic error results from procedures that alter the properties of the sample. Random error is unavoidable, but must be evaluated to determine its effect on accuracy. For example, Figure 5-8b shows data with no systematic bias, but accuracy is low because random error is high. Systematic errors can be minimized by careful selection and consistent application of sampling techniques.

Figure 5-9 illustrates five possible sources of error in ground-water sampling: (1) site selection, (2) sampling, (3) measurement methods, (4) reference samples for calibration, and (5) data handling. Both random and systematic errors may be involved in each stage. Errors at each stage are cumulative, but are not of equal significance or magnitude. Total variance in geochemical data results from the

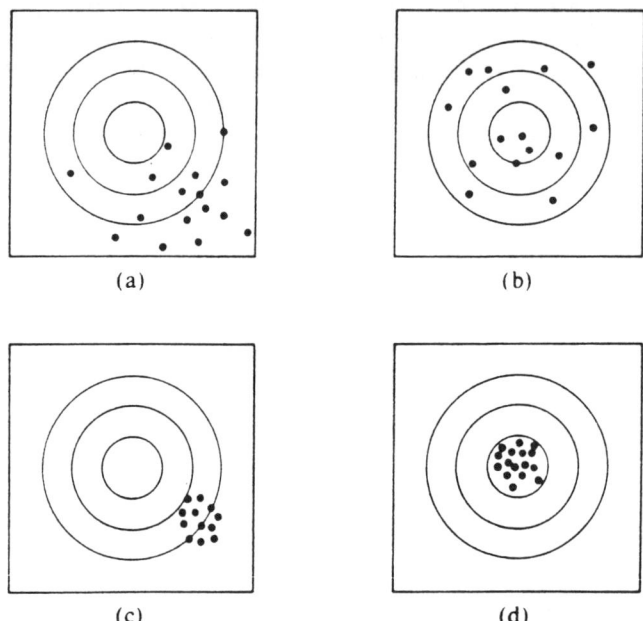

Patterns of shots at a target (after Jessen, 1978, Fig. 1). (a): high bias + low precision = low accuracy; (b): low bias + low precision = low accuracy; (c): high bias + high precision = low accuracy; (d): low bias + high precision = high accuracy.

Figure 5-8 Shots on a target analogy for illustrating influence of bias and precision on accuracy (Boulding and Barcelona, 1991a, after Jessen, 1978).

combination of natural geochemical variability and the cumulative error. The percentage of variance attributable to natural variability may often be greater than either field or laboratory error. Natural variance cannot be reduced; however, variance resulting from field and laboratory error can be reduced so that the actual variance closely approximates the natural variance.

Barcelona et al. (1989)[12] developed estimates of the relative contribution of natural variability, field error, and laboratory error to total variance at three ground-water investigation sites in Illinois. For most chemical constituents, at the three sites, natural variability accounted for more than 90% of the variance. For most inorganic constituents where field and laboratory error could be estimated, field error contributed a larger percentage of total variance. The study also found that organic contaminant indicators (TOC and TOX) showed typically higher percentages of variance due to field and laboratory error than did the inorganic indicators.

---

[12] Barcelona, M.J., D.P. Lettenmaier, and M.R. Schock. 1989. Network Design Factors for Assessing Temporal Variability in Ground-Water Quality. Environmental Monitoring and Assessment 12:149-179.

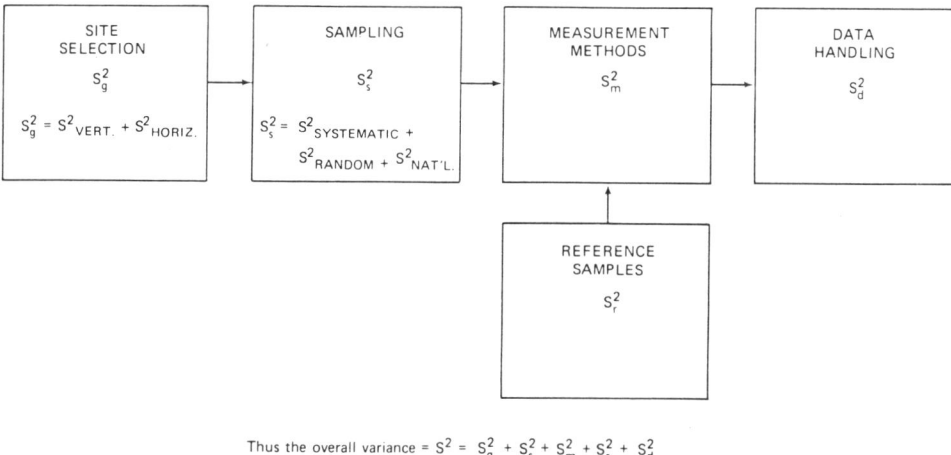

Figure 5-9 Sources of error involved in ground-water monitoring programs contributing to total variance (Barcelona et al., 1983).

Field Error. Figure 5-3 identifies specific possible sources of error at various steps in ground-water sampling. The largest sources of error are unrepresentative sample locations (hence the importance of hydrogeologic site characterization prior to geochemical sampling design) and disturbances caused by drilling and well construction. Sample collection is the next largest source of error. Major sources of systematic sampling error include (1) well construction and screen design preventing representative samples, and (2) improper purging. All of these large sources of systematic error are related to the hydrology of the site over which there is often little QA/QC.

Table 5-9 lists potential contributions of sampling methods and materials to error in ground-water chemical results. This table shows that well purging procedures can result in large variations in pH, TOC, Fe(II), and VOCs. Table 5-9 shows well casing to be the next largest source of error, followed by sampling mechanisms and grouting/sealing. Poorly grouted or cemented wells can greatly alter the pH of water (as much as pH 12). Sampling tubing can result in errors in VOC measurement. Sections 9.5.3 (Purging) and 9.3 (Conventional Monitoring Well Installations) discuss selection criteria for minimizing errors from these sources.

Other possible sources of systematic error in sampling include (1) changing sampling procedures, (2) changing sampling personnel without a strictly defined sampling protocol, and (3) failure to document unavoidable deviations from the sampling protocols, such as no water in the well. Another source of water quality error is mixing from multiple aquifers. Mixing is most common with public water supply wells that penetrate several hydrologically-unconnected aquifers. Improper sealing of ground-water monitor wells also may bias results by mixing water from distinct subsurface formations.

Table 5-9 Potential Contributions of Sampling Methods and Materials to Error[a] in Ground-Water Chemical Results

Sampling Method/ Material	pH	TOC (mg C/L)	Fe(II) (mg/L)	VOC (μg/L)
		Chemical Parameter		
Range of concentration	5-9	0.5-25	0.01-10	0.15-8,000
Drilling muds	---	+, 300%	---	---
Grouts, seals	+, 4 to 5 units (cement)	---	-,[b] 500% cement	---
Well purging	±, 0.1 to 5 units	±, 500%	-,[b] 1,000%	±, 10 to 1,000%[c]
Well casing	---	±, 200%	+, 1,000% iron, galvanized steel	±, 200%[c]
Sampling mechanism	gas lift +, 0.1 to 3 units	bailer +, 150%	gas lift -,[b] 500%	suction -,[b] 1 to 15%[d]
Sampling tubing	---	---	---	-10 to 75%[d]
References	1,5,7	1,4	1,2,5,7	1,3,6

[a] Bias values exceeding >± 100% denoted as gross errors (+ or -); other values expressed as a percent of reported mean.
[b] No data available on the type and extent of error for this parameter.
[c] Concentration range 0.5-15 μg/L (from Barcelona and Helfrich, 1984).
[d] Concentration range 80-8,000 μg/L (from Barcelona et al., 1984; Ho, 1983).

References (see Appendix F for citations):

1 Barcelona and Helfrich (1984)
2 Barcelona et al. (1983)
3 Barcelona et al. (1984)
4 Barcelona et al. (1988)
5 Gibb et al. (1981)
6 Ho (1983)
7 Schuller et al. (1981)

Analytical Error. Figure 5-4 identifies possible sources of error during water sample analysis. Analysis, including measurement methods and reference samples, is typically subject to the most stringent QA/QC procedures, and consequently analytical errors tend to be relatively minor components of total error. Failure to analyze blanks, standards, and samples by exactly the same procedures may result in either a biased blank correction or a biased calibration (Kirchmer, 1983/T5-14). Porter (1986/T5-14) examined in detail the sources of random analytical error for measurement near the limit of detection and how to incorporate this observation error into data analysis procedures.

Einarson and Pei (1988)[13] and Rice et al. (1988),[14] in separate studies of laboratory performance, concluded that the reliability of laboratory analyses should not be taken for granted. Both studies also concluded that the cost of analysis did not necessarily correlate with analytical accuracy. The most expensive of the 10 laboratories evaluated by Einarson and Pei (1988) tied for the worst ranking, while the four least expensive laboratories included the top ranked and other bottom ranked laboratory. Both studies describe criteria and procedures for choosing laboratories that will provide good analytical results. Section 5.5 discusses analytical and QA/QC concepts further.

Data Handling Error. There is probably no large body of scientific records free from human or machine errors. Faulty recording of observations in field or laboratory notebooks or incorrect coding for computer analysis are examples of data handling errors. Misrecorded values that are much larger or smaller than the range of the actual population are called outliers and may distort the results of statistical analysis. Statistical techniques are available for analyzing such data sets (Gilbert, 1987/T5-14), but prevention of data handling error is always better than a cure. Censoring of analytical measurements below the limit of detection (Section 5.5.2) is another serious error introduced by data handling.

Webster (1977/T5-14) suggests some of the following methods to reduce data handling errors: (1) write neatly, forming characters well; (2) distinguish ambiguous digits and letters by a firm convention; (3) restrict the digit 0 to mean zero and use other notations for "missing" or "inapplicable"; (4) eliminate or minimize transcription of field notes; (5) record data on forms designed for the purpose of the investigation with clear headings and ample space; and (6) double-check any transcribed data against the original.

---

[13] Einarson, J.H. and P.C. Pei. 1988. A Comparison of Laboratory Performances. Environ. Sci. Technol. 22:1121-1125.

[14] Rice, G., J. Brinkman, and D. Muller. 1988. Reliability of Chemical Analyses of Water Samples--The Experience of the UMTRA Project. Ground Water Monitoring Review 8(3):71-75.

5.5 Analytical and QA/QC Concepts

Quality assurance and quality control (QA/QC) are accomplished by (1) selecting the best methods for the program purpose, (2) clearly defining protocols or procedures to be followed, and (3) carefully documenting adherence or departures from the protocols. Both field sampling and laboratory analyses require protocols for good QA/QC. Campbell and Mabey (1985) have summarized key elements of data evaluation systems applicable to both field and laboratory measurements.[15] Table 5-13 identifies major reports and papers on QA/QC for environmental investigations.

5.5.1 Instrumentation and Analytical Methods

A bewildering array of methods are available for analyzing geochemical constituents. Most methods used for geochemical analysis involve either emission or adsorption of radiation (Table 9-9, Section 9.6.3). The fine points of instrumentation and analysis are the province of the analytical chemist, but the field scientist can benefit from a general understanding. Section 9.6 provides additional information of field screening and analytical methods. Analytical techniques for specific constituents of geochemical interest may be specified by regulation or, if not so specified, determined by the instrumentation that is most readily available. Table 9-10 identifies major references describing chemical analytical techniques.

5.5.2 Limit of Detection

Ground-water detection monitoring commonly involves measurement of contaminants that are either at or below the detection limit of analytical procedures. The statistical concept of detection limit includes accurately reporting and analyzing data including measurement near or below the detection limit (McNichols and Davis, 1988/T5-14).

Figure 5-10 and Table 5-10 illustrate the definitions of limit of detection and regions of analyte measurement recommended by the Subcommittee on Environmental Analytical Chemistry of the American Chemical Society's (ACS) Committee on Environmental Improvement (1980/T5-14). The zero analyte signal for measuring the limit of detection comes from the field blank (see below). If the actual field blank measurement gives a positive signal, this means that analytical measurements on other samples with a lower signal will be recorded as a negative concentration. For example, a low concentration standard (typically 1 part per billion (ppb) for organic constituents) is made in the laboratory for the contaminant of interest. The standard deviation for analytical measurement of the 1-ppb standard is commonly plus or minus 100%, or 1 μg/L. The detection limit for a contaminated sample is defined as three standard deviations (3 μg/L) above the mean for the standard, or six standard deviations above the zero point defined by the field blank (see Figure 5-10). The limit of detection should be defined every day of analysis. The detection limit is probably

---

[15] Campbell, J.A. and W.R. Mabey. 1985. A Systematic Approach for Evaluating the Quality of Ground Water Monitoring Data. Ground Water Monitoring Review 5(4):58-62.

the most important kind of laboratory quality assurance data and should be reported with the analytical results for each constituent.

Table 5-10 lists the regions of analyte measurement. Following the above example, signals below three standard deviations are considered below the limit of detection. The region of detection is between 3 and 10 standard deviations (5 standard deviations by some rules) and is where the constituent can be said to be present but the precise concentration cannot be stated with certainty. Analyte signals above the limit of quantification (plus 10 standard deviations) can be interpreted quantitatively.

The above-described definition reaffirms the model for limit of detection calculations adopted by the International Union of Pure and Applied Chemistry (IUPAC) in 1975 (IUPAC, 1978/T5-14). However, considerable confusion still surrounds the definition of the limit of detection. This is because (1) acceptance of the above definition by the general analytical community has been slow, and (2) different statistical approaches to calculating limits of detection for constituents can easily vary by an order of magnitude (Long and Winefordner, 1983/T5-14). This is particularly true for chemical constituents at the ppb level.

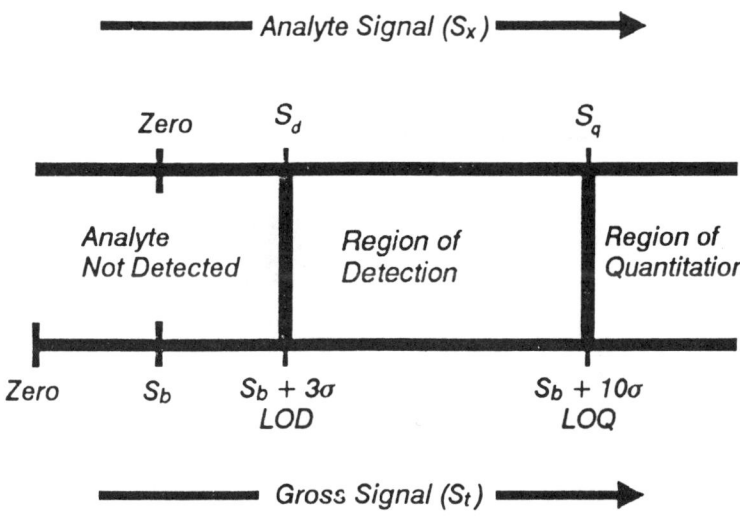

Figure 5-10 The ACS recommended definition of limit of detection (Boulding and Barcelona, 1991a, after ACS Comm. on Environmental Improvements, 1980).

The limit of detection is both a site- (as a result of the field blank) and instrument/operator-specific value. Consequently, the precision and accuracy for low standards must be reported on the analytical report forms. The instrument manufacturer's definition of detection is based normally on carefully controlled conditions (e.g., distilled water solutions) that may not be achievable in routine analyses of complex samples. Consequently, actual limits of detection in contaminated ground water are often higher.

Data Censoring. A major problem with failure to understand the statistical nature of the limit of detection is negative censoring of data. *Negative censoring* involves reporting analyte concentrations that are below the limit of detection as zero, "less than" values, or "not detected." Since 1983 the American Society for Testing and Materials (ASTM) has recommended that data should not be routinely censored by laboratories (ASTM D4210/Table A-14). Nevertheless, censoring of water quality analytical data remains a problem (Porter et al., 1988/T5-14). Section 5.5.1 examines this issue further.

Laboratories should be asked to provide uncensored data on all water samples with measurements near or below the limit of detection. Measurement data should not be discarded unless the lack of statistical control in the measurement process is clearly demonstrated. The general public, and even the uninformed scientist, may find the concept of a negative concentration difficult to understand, so it is prudent to report less than zero values as "trace" when presenting data in tables that present data in the main body of a report. Remediation decisions, however, should be based on concentrations at or above the limit of quantification, not the limit of detection.

Table 5-10 Regions of Analyte Measurement

Analyte Signal (standard deviations in μg/L)	Recommended Inference
<3	Analyte not detected
3 to 10	Region of detection
>10	Region of quantitation

Source: Boulding and Barcelona (1991a), after ACS Committee on Environmental Improvement, 1980).

Analysis of Censored Data. Table 5-11 illustrates the effect of two types of censoring of analytical results near and below the limit of detection. Data reported as less than the limit of detection are heavily censored and yield an average concentration of 3.5 μg/L, since only two values are quantified. In Table 5-11 negative censoring (see definition above) yields an average of 1.2 μg/L. The uncensored data average 0.5 μg/L. The averages of the heavily and negatively censored data would appear to indicate contamination, but the 95% confidence interval for the uncensored data is at best equivocal. Gilliom et al. (1984/T5-14) found that any censoring of trace-level water quality data, even when the censored data were highly unreliable, reduced the ability to detect trends in the data. Unfortunately, censored data continues to be routinely reported by laboratories. Table 5-14 identifies a number of references that discuss statistical techniques for analyzing censored data.

5.5.3 Types of QA/QC Samples

Field scientists tend to consider QA/QC requirements and procedures to be primarily the responsibility of the laboratory. However, QA/QC procedures are equally, if not more important in the field. Field personnel also should be familiar with the different types of samples that may be taken, and their importance for interpreting the analytical results. Field sampling programs typically involve collection of additional sample volumes from a number of wells (typically 10 to 20%) for laboratory QC for matrix spikes and replicate analyses. For specific programs, it may be appropriate to prepare field standards and field spiked samples, and to collect blind control samples to provide an independent check on the performance of the analytical laboratory.

Table 5-11 Effects of Censoring Analyte Signals at and Below the Limit of Detection

Sample	Heavily Censored	Negatively Censored	Uncensored
1	<3	2	2
2	<3	0	-2
3	<3	0	-1
4	4	4	4
5	3	3	3
6	<3	0	-3
7	<3	1	1
8	<3	0	-1
9	<3	0	0
10	<3	2	2
Mean	3.5	1.2	-0.5
95% Conf.	0.14-2.26	1.13	-2.13

Source: ASTM D4210/Table A-14.

Major types of field and QA/QC samples include the following:

A *trip blank* is a sample bottle filled with laboratory analyte-free water that travels to the site with the empty sample bottles and is returned to the laboratory with the samples. The trip blanks are not opened in the field. They provide a method of monitoring bottle preparation, blank water quality, and sample handling. Contaminated trip blanks may indicate poor quality laboratory water or inadequate bottle cleaning.

A *field blank* is a sample of distilled or deionized water taken from the laboratory out into the field, poured into a sampling vial at the site, closed, and returned as if it were a sample. The level of contamination of the field blank is the zero analyte signal for determining the limit of detection.

A *rinse or cleaning blank* is a sample of the final rinse of a sampling mechanism before it is put in a new well. This type of sample is used to evaluate whether a sample may have been contaminated from material taken in the previous sample.

Field samples are those samples that are taken in the field as "representative" of conditions at the site and analyzed in the laboratory for constituents of interest. If sampling points or locations are unrepresentative, or biased sampling procedures are used, no amount of care in QA/QC in subsequent stages will salvage an accurate picture of actual field conditions.

Duplicate samples are collected and not analyzed unless it is later determined that they contain additional useful information. Soil samples are commonly duplicated.

Replicate samples are subsamples of the same sample that are labeled separately to estimate the precision of laboratory analytical results.

Split samples are field samples that are split between two storage vessels or cut in half in the field. One subsample may be analyzed by one laboratory and the other subsample may be archived or given to another laboratory.

Spiked samples are field samples that may be split with one aliquot receiving a spike volume of a reference standard to estimate the recovery of the analyte in the laboratory. Spiked samples allow estimates of accuracy and detect possible matrix interference problems.

Laboratory blanks are similar to field blanks except that the distilled deionized water used in the laboratory at the time each batch of samples is received is analyzed in the same manner as other samples. This type of sampling may detect contamination that occurs in the laboratory.

Standard reference samples have been analyzed previously by outside laboratories. These samples are available from the National Institute of Standards and Technology or the EPA to detect either instrument calibration error or the use of inappropriate laboratory analytical methods (Keith et al., 1983).[16]

---

[16] Keith, S.J., M.T. Frank, G. McCarty, and G. Massman. 1983. Dealing with the Problem of Obtaining Accurate Ground-Water Quality Analytical Results. In: Proc. 3rd Nat. Symp. on Aquifer Restoration and Ground Water Monitoring, National Water Well Association, Dublin, OH, pp. 272-283.

5.6 Statistical Techniques

5.6.1 Statistical Approaches to Geochemical Variability

Virtually all soil sampling and most ground-water sampling that have been done at a high enough level of resolution have shown that chemical constituent concentrations are neither normally distributed nor independent (i.e., noncorrelated). This creates special challenges for statistical analysis of geochemical sampling data because many of the traditional statistical techniques for analyzing sample data, such as linear regression and t-testing, assume that the population sampled has the symmetric, bell-shaped Gaussian ("normal") distribution. Linear regression is probably the most frequently misused statistical technique in this context (Mann, 1987; Kite, 1989).[17]

The first step in analyzing geochemical data is to determine whether they are normally distributed. If they are, traditional techniques described in standard textbooks on statistics can be used. If not, one or more of the following methods must be used: (1) *data transformations* such as logarithmic conversions to create data sets that are normally distributed and hence amenable for analysis by conventional methods--Wilson et al. (1990)[18] discuss how to evaluate bias that may be introduced by this manipulation, (2) *nonparametric* or distribution-free statistical techniques that do not require independent data observations, and (3) *geostatistical* techniques that facilitate differentiation of correlated and noncorrelated data sets and interpolation of values between sample points. The technique of "fuzzy" linear regression may be useful in hydrologic situations where the relationship between variables is imprecise, data are inaccurate, and/or sample sizes are insufficient (Bardossy et al., 1990).[19] Subsurface contamination investigations typically involve measurements of concentration changes in geochemical parameters over time. Consequently, statistical techniques designed specifically for analysis of trends in time-series data are important (Harris et al., 1987; Montgomery et al., 1987/T5-14).

---

[17] Kite, G. 1989. Some Statistical Observations. Water Resources Bulletin 25(3):483-490. See also 1990 discussion by Kirby et al., and reply by Kite in Water Resources Bulletin 26(4)693-698.
    Mann, J.C. 1987. Misuses of Linear Regression in Earth Sciences. In: Use and Abuse of Statistical Methods in the Earth Sciences, W.B. Size (ed.), Oxford University Press, New York, pp. 74-106.

[18] Wilson, B.G., B.J. Adams, and B.W. Karney. 1990. Bias in Log-Transformed Frequency Distributions. J. Hydrology 118:19-37.

[19] Bardossy, A., I. Bogardi, and L. Duckstein. 1990. Fuzzy Regression in Hydrology. J. Hydrology 26(7):1497-1508.

Alhajjar et al. (1990)[20] describe use of the *median-polish* statistical methods of exploratory data analysis developed by Tukey (1977/T5-14) for analyzing highly variable geochemical data collected during a study of chemical pollution from septic systems. This technique is especially well suited for analyzing data in two-way tables (multiple rows and columns) in which each data value is related simultaneously to two factors.

5.6.2 Geostatistics

Geostatistical techniques such as use of correlograms, semivariograms, and kriging have gained increasing popularity in evaluating spatially distributed hydrologic and geochemical data in the last 10 years. Using empirical gold-ore evaluation techniques developed by D.C. Krige in South Africa (hence the term kriging), the French mathematician G. Matheron developed the theory of regionalized variables in the late 1960s (Matheron, 1971/T5-14). This general theory of sampling and estimating spatially dependent (autocorrelated) variables is well suited to analysis of hydrologic and geochemical parameters, which tend to be nonrandom in the classical Gaussian statistical sense.

Geostatistical techniques have three main applications for characterization of subsurface variability: (1) they can assist in reducing spatial sampling intensity, and hence reduce sampling and analytical costs; (2) they can be used to differentiate sample data that are autocorrelated or noncorrelated, elucidating trends for selecting the appropriate statistical analysis of sampling analytical results; and (3) they can be used to interpolate values at locations where measurements have not been made. The last application is done by *kriging*, a weighted moving-averaging technique, that usually provides the most accurate way of contouring data on physical and geochemical parameters. An advantage of kriging is that a standard deviation map can be readily created from kriged contour data, which provides a good indication of the reliability of contours. However, when the variogram (see below) is not well defined and expected values of the mean and variance are not constant over the area of interest, other interpolation techniques, such as inverse distance interpolation methods, may produce superior contouring results compared to kriging.[21]

One of the first steps in geostatistical analysis is to calculate the nonsampling variance (gamma) of samples at different distance spacings. Gamma is a statistical measure of the difference between sample values. For example, if samples were taken from a 50-m grid, gamma would be calculated for the samples spaced at 50 m, 100 m, 150 m, 200 m, and so on. Next, a semivariogram is plotted on an XY plot, where X is distance and Y is the nonsampling variance. Figure 5-11 shows an "ideal" semivariogram. Samples within a certain range of influence, also called the range of correlation (distance *a* in Figure 5-11), show an approximately linear correlation (are

---

[20] Alhajjar, B.J., G. Chesters, and J.M. Harkin. 1990. Indicators of Chemical Pollution from Septic Systems. Ground Water 28(4):559-568.

[21] Weber, D. and E. Englund. 1992. Evaluation and Comparison of Spatial Interpolators. Mathematical Geology 24(4):381-391.

Planning Field Investigations 283

autocorrelated). At some spacing distance, if there is no trend in the data, a sill (c on Figure 5-11) marks a plateau that limits the range of correlation. The nonsampling variance between samples will equal c as long as the distance is greater than a.

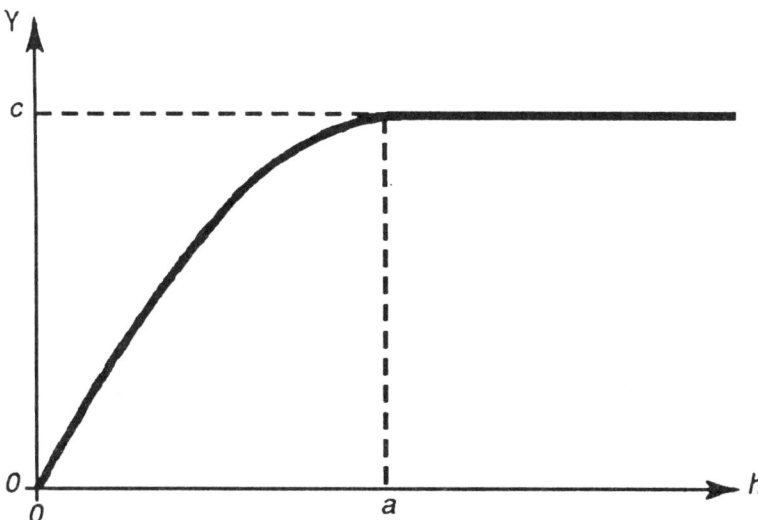

Figure 5-11 The "ideal" shape for a semivariogram-spherical model (Boulding and Barcelona, 1991a, after Clark, 1979).

From a sampling perspective, samples spaced closer than distance a in Figure 5-11 will yield redundant, correlated data, which results in both unnecessary expense and complications in statistical analysis. The minimum distance at which samples are independent (distance a in Figure 5-11) is the optimum sampling distance.

Figure 5-12 shows a semivariogram of lead values in soil sampled by Flatman (1986/T5-14) on a systematic 750-ft grid. The diagram shows that samples for lead that are closer to each other than about 1,200 ft are correlated. In other words, the same information could be obtained by cutting the number of samples almost in half.

Semivariograms may exhibit correlation structures other than the one shown in Figure 5-12, and correct interpretation requires an understanding of the various models that are available for describing semivariogram plots. When data are not normally distributed, such as when a spatial trend is present, estimating the correlation structure is difficult. In these cases, some of the techniques for transforming lognormal data for conventional statistical analysis may be useful.

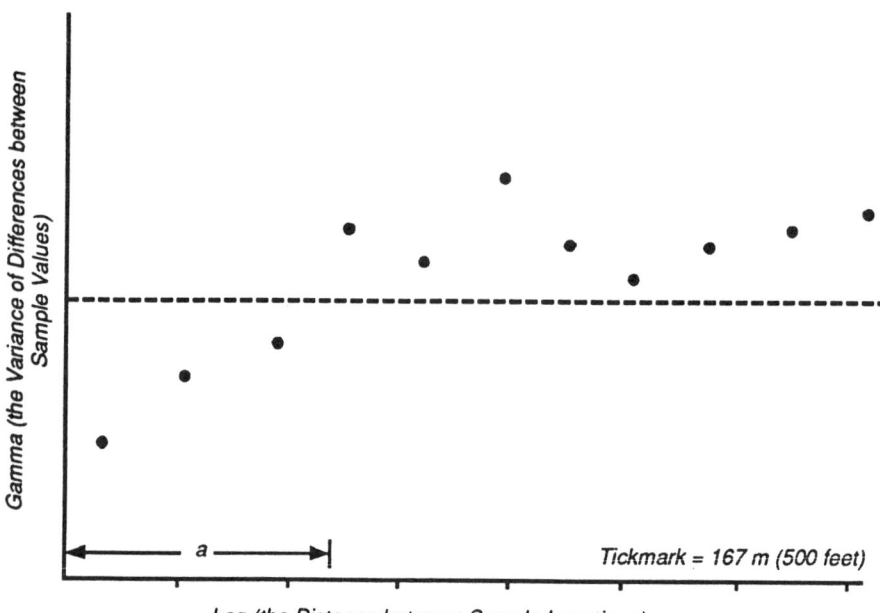

Figure 5-12 A semivariogram of lead samples taken systematically on a 230-m (750-foot) grid (Boulding and Barcelona, 1991a, after Flatman, 1986).

5.7 Guide to Major References

Table 5-12 provides a listing of major conferences and symposia series that address a variety of aspects of soil and ground-water contamination and investigation techniques. NWWA/NGWA conference proceedings are published by the National Ground Water Association (formerly the National Water Well Association) in a series called *Ground Water Management*, costing $240 for 6 coupons, which can be redeemed for whatever proceeding the subscriber wishes to receive. The Hazardous Materials Control Resources Institute (formerly the Hazardous Materials Control Research Institute) also publishes a number of conference series that contain many papers relevant to contaminant investigations. The addresses for these two organizations are:

National Ground Water Association, 6375 Riverside Drive, Dublin, OH 43017 (800/551-7379).

Hazardous Materials Control Resources Institute, Publications Department, 7237-A Hanover Parkway, Greenbelt, MD 20770 (800/397-7161).

Table 5-13 provides an index of major references on existing sources of environmental information, environmental data management, and quality assurance/quality control. Table 5-14 provides an index of major references on basic and environmental statistics and sampling design.

Table 5-12 Conferences and Symposia with Papers Relevant to Subsurface Characterization and Monitoring*

Sponsor	Year	Title
EPA/NWWA	1971	1st National Ground Water Quality Symposium (EPA-16060 GRB, NTIS PB214-614)
	1974	2nd (EPA-68-03-0367, NTIS PB257-312)
	1977	3rd (EPA/600/9-77/014, NTIS PB272-908)
	1979	4th (EPA/600/9-79/029, NTIS PB80-103476)
	1980	5th
	1983	6th (State, County, Regional, and Municipal Jurisdictions of Ground Water Protection)
	1984	7th (Innovative Means of Dealing with Potential Sources of Ground Water Contamination)
	1986	8th (Anatomy of Superfund)
NWWA	1981	1st National Ground Water Quality Monitoring Symposium and Exposition
	1982	2nd National Symposium on Aquifer Restoration and Ground Water Monitoring
	1983	3rd
	1984	4th
	1985	5th
	1986	6th
	1987	1st National Outdoor Action Conference on Aquifer Restoration, Ground Water Monitoring, and Geophysical Methods
	1988	2nd
	1989	3rd
	1990	4th GWM 2
	1991	5th GWM 5
NGWA	1992	6th GWM 11
	1993	7th GWM 15
	1994	8th GWM 18
NWWA/API	1984	[1st] Conference on Petroleum Hydrocarbons and Organic Chemicals in Ground Water--Prevention, Detection, and Restoration
	1985	[2nd]
	1986	[3rd]
	1987	[4th]
	1988	[5th]
	1989	[6th]
	1990	[7th] GWM 4
	1991	[8th] GWM 8
NGWA/API	1992	[9th] GWM 14
	1993	[10th] GWM 17

Modeling

NWWA/IGWMC	1984	1st Conference on Practical Applications of Ground Water Models
	1985	2nd
	1986	3rd Conference on Solving Ground Water Problems with Models
	1988	Conference on Geochemical Modeling of Ground Water Contamination
	1989	4th Conference on Solving Ground Water Problems with Models
NGWA/IGWMC	1992	5th GWM 9

Table 5-12 (cont.)

Sponsor	Year	Title

Geophysics/Vadose Zone/Karst

NWWA/EPA 1984 [1st] Conference on Surface and Borehole Geophysical Methods in Ground Water Investigations
1985 [2nd]
1986 Surface and Borehole Geophysical Methods and Ground Water Instrumentation Conference and Exposition

NWWA/EPA 1983 [1st] Conference on Characterization and Monitoring in the Vadose (Unsaturated) Zone
1985 [2nd]
1986 3rd

NWWA 1986 [1st] Conference on Environmental Problems in Karst Terranes and Their Solutions
1988 2nd
1991 3rd Conference on Hydrogeology, Ecology, Monitoring and Management of Ground Water in Karst Terranes GWM 10

Miscellaneous NWWA Conferences

NWWA/AGWSE 1988 Ground Water Geochemistry Conference
1989 Conference on New Field Techniques for Quantifying Physical and Chemical Properties of Heterogeneous Aquifers
1990 Cluster of Conferences (Agricultural Impacts on Ground Water Quality; Ground Water Geochemistry; Ground Water Management and Wellhead Protection; Environmental Site Assessments: Case Studies and Strategies) GWM 1
1991 Environmental Site Assessments Case Studies and Strategies: The Conference GWM 6
NGWA/AGWSE 1992 [2nd] Environmental Site Assessments Case Studies and Strategies: The Conference GWM 12

NWWA Eastern Regional Conferences

NWWA/AGWSE 1984 [1st] Eastern Regional Ground Water Conference
1985 [2nd]
1986 3rd Annual Eastern Regional Ground Water Conference
1987 4th
1988 [5th] Focus Conference on Eastern Regional Ground Water Issues
1989 [6th]
1990 [7th] GWM 3
1991 [8th] GWM 7
NGWA/AGWSE 1992 [9th] GWM 13
1993 [10th] GWM 16

Other NWWA Regional Conferences

NWWA 1983 Eastern Regional Conference on Ground Water Management
Western Regional Conference on Ground Water Management
1984 Conference on Ground Water Management
1985 Southern Regional Ground Water Conference
Western Regional Ground Water Conference

Planning Field Investigations

Table 5-12 (cont.)

Sponsor	Year	Title

<u>Other NWWA Regional Conferences (cont.)</u>

        1986 Conference on Southwestern Ground Water Issues
              Focus Conference on Southeastern Ground Water Issues
        1987 Focus Conference on Midwestern Ground Water Issues
              Focus Conference on Northwestern Ground Water Issues
        1988 [2nd] Focus Conference on Southwestern Ground Water Issues

<u>Hazardous Materials Control Research Institute Conferences</u>

HMCRI   1980 1st National Conference on Management of Uncontrolled Hazardous Wastes Sites
                  1981 2nd
                  1982 3rd
                  1983 4th
                  1984 5th
                  1985 6th
                  1986 7th
                  1987 8th Superfund '87
                  1988 9th Superfund '88
                  1989 10th Superfund '89
                  1990 11th Superfund '90
                  1991 12th Hazardous Materials Control (HMC-Superfund '91)
                  1992 13th HMC-Superfund '92

<u>Regional Hazardous Materials Control Conferences</u>

HMCRI   1990 HMC—Great Lakes '90
                  1991 HMC—Northeast '91
                  1992 HMC—South '92

HMCRI   1984 1st National Conference on Hazardous Wastes and Environmental Emergencies
                  1985 2nd
                  1986 3rd National Conference on Hazardous Wastes and Hazardous Materials
                  1987 4th
                  1988 5th
                  1989 6th (HWHM '89)

<u>Miscellaneous Conferences</u>

HMCRI   1992 National R&D Conference on the Control of Hazardous Materials
                  1992 Federal Environmental Restoration '92
                  1990 7th (HWHM '90)

[ ]—Indicates that number is not included in the title of the published proceedings.

GRM indicates that proceedings have been published in NWWA/NGWA's Ground Water Management Series.

* See list of acronyms before the Table of Contents for full organization names.

Table 5-13 Index to Major References on Existing Environmental Information and Data Management

Topic	References
<u>Existing Information Sources</u>	
Federal Agencies	<u>EPA</u>: U.S. EPA (1988a, 1991a, 1992, various dates); <u>USGS</u>: Cardin et al. (1986), Dodd et al. (1989), Gilbert and Buchanan (1982), Mercer and Morgan (1982), Rapp et al. (1969), USGS (1979, 1982c, 1982d-ESIS)
Maps/Airphotos	FGDC (1992), Makower (1992), Thompson (1979); <u>Aerial Photography</u>: ASTM D5518 (Table A-14)
Geology	Kaplan (1965), Long (1971), Ward (1972), Wood (1973)
Ground Water Databases	Kaplan et al. (1985), Orr (1984), Way et al. (1984); <u>NAWDEX</u>: Edwards (1978), Edwards et al. (1987), Knecht and Edwards (1980), USGS (1982a, 1982b); <u>STORET</u>: U.S. EPA (1986, 1987c); <u>WATSTORE</u>: Baker and Foulk (1980), Hutchinson (1975)
Other Databases	Olson (1984)
Regional Hydrogeology	See Table 2-5
Climate	Eder et al. (1989), Hatch (1988); <u>Meteorological Tables</u>: Letetsu (1966), List (1966)
<u>Data Handling</u>	
Data Requirements	<u>Ground Water</u>: U.S. EPA (1987a, 1988b, 1990a)
Data Management	Kroenke (1977), Olson and Milleman (1985); <u>Environmental Data</u>: Hopke and Massart (1986); <u>Geology</u>: Frizado (1992); <u>Ground Water</u>: Hix (1983), Hoffman (1986), Kaplan et al. (1985), USGS (1984)
QA/QC*	<u>EPA Reports/Guidance</u>: Barth et al. (1989), Evans et al. (1987), Paulson et al. (1988), Simes (1989), Stanley and Verner (1983), U.S. EPA (1976, 1983, 1987b, 1987c, 1988c, 1989a, 1989b 1990b, 1991b, 1992c/T11-11), van Ee et al. (1990); <u>Other Texts/Reports</u>: NAS (1988), NWQL (1986), Taylor (1987), Taylor and Stanley (1985), U.S. Army Corps of Engineers (1990), <u>Papers</u>: Campbell and Mabey (1985), Evans (1986), Kent and Payne (1988), Kirchmer (1983), Lewis (1988), Mateo et al. (1991), Pennino (1988), Starks and Flatman (1991), van Ee and McMillion (1986); <u>Software</u>: See Table E-2.

* See Table 9-11 for major reference on QA/QC for chemical analytical procedures.

Table 5-13 References (Appendix F contains references for figure and table sources)

Baker, Jr., C.H. and D.G. Foulk. 1980. National Water Data Storage and Retrieval System (WATSTORE): Instruction for Preparation and Submission of Ground Water Data. U.S. Geological Survey Open File Report 75-589-IV. [Revision of 1970 first edition]

Barth, D.S., B.J. Mason, T.H. Starks, and K.W. Brown. 1989. Soil Sampling Quality Assurance User's Guide, 2nd edition. EPA/ 600/8-89/046 (NTIS PB89-189864), U.S. Environmental Protection Agency, Environmental Monitoring Systems Laboratory, Las Vegas, NV, 225+ pp.

Cardin, C.W., J.E. Moore, and J.M. Rubin. 1986. Water Resources Division in the 1980s: A Summary of Activities and Programs of the U.S. Geological Survey's Water Resources Division. U.S. Geological Survey Circular 1005, 79 pp.

Campbell, J.A. and W.R. Mabey. 1985. A Systematic Approach for Evaluating the Quality of Ground Water Monitoring Data. Ground Water Monitoring Review 5(4):58-62.

Dodd, K., H.K. Fuller, and P.F. Clarke (compilers). 1989. Guide to Obtaining USGS Information. U.S. Geological Survey Circular 900, 34 pp. [Earlier edition by Clarke et al. (Circular 777)]

Eder, B.K., L.E. Truppi, and P.L. Finkelstein. 1989. A Climatology of Temperature and Precipitation Variability in the United States. EPA/600/3-89/025 (NTIS PB89-165930).

Edwards, M.D. 1978. NAWDEX: A Key to Finding Water Data. U.S. Geological Survey, National Water Data Exchange, Reston, VA, 15 pp.

Edwards, M.D., A.L. Putnam, and N.E. Hutchinson. 1987. Conceptual Design for the National Water Information System. U.S. Geological Survey Bulletin 1972.

Evans, R.B. 1986. Ground-Water Monitoring Data Quality Objectives for Remedial Site Investigations. In: Quality Control in Remedial Site Investigation: Hazardous and Industrial Solid Waste Testing, 5th Volume, C.L. Perket (ed.), ASTM STP 925, American Society for Testing and Materials, Philadelphia, PA, pp. 21-33.

Evans, R.B., E.N. Koglin, and K.W. Brown. 1987. Ground-Water Monitoring: Quality Assurance for RCRA. EPA/600/X-87/415, U.S. Environmental Protection Agency, Environmental Monitoring Systems Laboratory, Las Vegas, NV.

Federal Geographic Data Committee (FGDC). 1992. Manual of Federal Geographic Data Products. FGDC, Reston, VA.

Frizado, J. 1992. Management of Geological Databases. Computer Methods in the Geosciences 11, Pergamon Press, New York, NY, 250 pp. [Focusses on dBase and PC-File]

Gilbert, B.K. and T.J. Buchanan. 1982. Water-Data Program of the USGS. U.S. Geological Survey Circular 863, 54 pp.

Hatch, W.L. 1988. Selective Guide to Climatic Data Sources. Key to Meteorological Records Documentation No. 4.11. NOAA National Climate Data Center, Asheville, NC.

(Table 5-13 Environmental Data Sources and Management)

Hix, G. 1983. A Data Base Management System Approach to Ground Water Data Management. In: Proc. NWWA Western Regional Conf. on Ground Water Management, National Water Well Association, Dublin, OH, pp. 351-355.

Hoffman, J.L. 1986. A Computer Database for Overview of Ground Water Pollution Investigations. Ground Water Monitoring Review 6(1):76-79.

Hopke, P.K. and D.L. Massart. 1986. Environmental Data Management: Directions for Internationally Compatible Environmental Data. International Council of Scientific Unions, Paris. [Proceedings of CODATA Workshop]

Hutchinson, N.E. (compiler). 1975. WATSTORE—National Water Data Storage and Retrieval System—User's Guide. U.S. Geological Survey Open-File Report 75-426, 791 pp.

Kaplan, S.R. 1965. A Guide to Information Sources in Mining, Minerals, and Geosciences. Interscience Publishers, New York, NY, 599 pp.

Kaplan, E., J. Naidu, M. Hauptman, and A. Meinhold. 1985. Guidebook for the Assembly and Use of Diverse Ground Water Data. BNL-37356, Brookhaven National Laboratory, Department of Applied Science, Brookhaven, NY.

Kent, R.T. and K.E. Payne. 1988. Sampling Groundwater Monitoring Wells: Special Quality Assurance and Quality Control Considerations. In: Principles of Environmental Sampling, L.H. Keith (ed.), ACS Professional Reference Book, American Chemical Society, Washington, DC, pp. 231-260.

Kirchmer, C.J. 1983. Quality Control in Water Analysis. Environ. Sci. Technol. 17(4):174a-181b.

Knecht, W.A. and M.D. Edwards. 1980. Definitions of Components of the Water Data Sources Directory Maintained by the National Water Data Exchange. U.S. Geological Survey Open File Report 79-1541, 106 pp.

Kroenke, D. 1977. Database Processing. Science Research Associates, Chicago, IL.

Letetsu, S. (ed.). 1966. International Meteorological Tables. WMO No. 188 TP 94, World Meteorological Organization, Geneva.

Lewis, D.L. 1988. Assessing and Controlling Sample Contamination. In: Principles of Environmental Sampling, L.H. Keith (ed.), ACS Professional Reference Book, American Chemical Society, Washington, DC, pp. 199-143.

List, R.J. (ed.). 1966. Smithsonian Meteorological Tables, 6th edition. Smithsonian Misc. Collections, Vol. 114, Pub. 4014, Smithsonian Institution, Washington, DC.

Long, H.K. 1971. A Bibliography of Earth Science Bibliographies of the United States. American Geological Institute, Washington, DC.

Makower, J. (ed.). 1992. The Map Catalog, 2nd ed. Random House, New York, NY, 364 pp. [Guide to all types of maps available from government and commercial sources]

(Table 5-13 Environmental Data Sources and Management)

Mateo, J.M., C.M. Andreas, and W. Coakley. 1991. A Quality Assurance Sampling Plan for Emergency Response (QASPAR). In: Proc. Second Int. Symp. Field Screening Methods for Hazardous Waste and Toxic Chemicals, EPA/600/9-91/028 (NTIS PB92-125764), pp. 217-225.

Mercer, J.W. and C.O. Morgan. 1982. Storage and Retrieval of Ground-Water Data at the U.S. Geological Survey. U.S. Geological Survey Circular 856, 9 pp.

National Academy of Science (NAS). 1988. Final Report on Quality Assurance to the EPA. NAS, Washington, DC, 53 pp.

National Water Quality Laboratory (NWQL). 1986. Quality Assurance in the National Water Quality Laboratory. Environment Canada, Canada Centre for Inland Waters, Burlington, Ontario.

Olson, R.J. 1985. Review of Existing Environmental and Natural Resources Databases. ORNL/TM-8928, Oak Ridge National Laboratory, Oak Ridge, TN.

Olson, R.J. and N.T. Millemann (eds.). 1985. Proceedings of the 1983 Integrated Data Users Workshop. ORNL Conf-831117, Oak Ridge National Laboratory, Oak Ridge, TN, 94 pp.

Orr, V.J. 1984. National Ground-Water Information Center. Ground Water 22(2):207-209. [Description of the National Water Well Association's National Ground-Water Information Center and database]

Paulson, S.G., C.L. Chen, K.J. Stetzenbach, and M.J. Miah. 1988. Guide to Application of Quality Assurance Data to Routine Survey Data Analysis. EPA/600/4-88/010 (NTIS PB88-166863).

Pennino, J.D. 1988. There's No Such Things as a Representative Ground Water Sample. Ground Water Monitoring Review 8(3):4-9.

Rapp, J.R., W.W. Doyel, and E.B. Chase. 1969. Geological Survey Water Data Catalog. U.S. Geological Survey.

Simes, G.F. 1989. Preparing Perfect Project Plans: A Pocket Guide for the Preparation of Quality Assurance Project Plans. EPA/600/9-89/087.

Stanley, T.W. and S.S. Verner. 1983. Guidelines and Specification for Preparing Quality Assurance Program Plans. EPA/600/8-83/024 (NTIS PB83-219667), 27 pp.

Starks, T.H. and G.T. Flatman. 1991. RCRA Ground-Water Monitoring Decision Procedures Viewed as Quality Control Schemes. Environmental Monitoring and Assessment 16:19-37.

Taylor, J.K. 1987. Quality Assurance of Chemical Measurements. Lewis Publishers, Chelsea, MI, 335 pp.

Taylor, J.K. and T.W. Stanley (eds.). 1985. Quality Assurance for Environmental Measurements. ASTM STP 867, American Society for Testing and Materials, Philadelphia, PA.

(Table 5-13 Environmental Data Sources and Management)

Thompson, M.M. 1979. Maps of America: Cartographic Products of the U.S. Geological Survey and Others. GPO Stock No. 024 001 03145-1, U.S. Government Printing Office, Washington, DC.

U.S. Army Corps of Engineers. 1990. Chemical Data Quality Management for Hazardous Waste Remedial Activities. Engineering Regulation ER 1110-1-263. [Supersedes ER with the same title and number dated December 30, 1985]

U.S. Environmental Protection Agency (EPA). Various Dates. Data Element Compilations: CERCLIS Data Element Dictionary (March 15, 1988, 300 pp.); Federal Reporting Data System (FRDS-II) Data Element Dictionary (Submitted to Office of Drinking Water by SYSCOM, Inc., October 30, 1987, 239 pp.); RCRIS Data Element Dictionary (Office of Solid Waste, September 20, 1989); UIC Data Element Definitions (Office of Water, Preliminary Draft, September 25, 1989).

U.S. Environmental Protection Agency (EPA). 1976. Minimal Requirements for a Water Quality Assurance Program. EPA/440/9-75-010 (NTIS PB258-807).

U.S. Environmental Protection Agency (EPA). 1983. Guidelines for Assessing and Reporting Data Quality for Environmental Measurements. U.S. EPA Environmental Monitoring and Support Laboratory, Cincinnati, OH.

U.S. Environmental Protection Agency (EPA). 1986. Ground-Water Data Management with STORET. EPA/600/M-86-007 (NTIS PB86-197860).

U.S. Environmental Protection Agency (EPA). 1987a. Ground-Water Data Requirements Analysis. EPA/440/6-87-005 (NTIS PB87-225532).

U.S. Environmental Protection Agency (EPA). 1987b. Data Quality Objectives for Remedial Response Activities; Vol. 1: Development Process; Vol. 2: RI/FS Activities at a Site with Contaminated Soils and Ground Water. Vol. 1 EPA/G-87/003 (NTIS PB88-131370), 156 pp.; Vol. 2 EPA/G-87/004 (NTIS PB88-131388), 287 pp.; both volumes (NTIS PB90-272634).

U.S. Environmental Protection Agency (EPA). 1987c. Ground Water Data Management with STORET. EPA/440/6-87-005.

U.S. Environmental Protection Agency (EPA). 1988a. Compendium of ORD and OSWER Documents Relevant to RCRA Corrective Action. EPA/530/SW-88-010. Available from RCRA Hotline.*

U.S. Environmental Protection Agency (EPA). 1988b. EPA Workshop to Recommend a Minimum Set of Data Elements for Ground Water: Workshop Findings Report. EPA/440/6-88-005 (NTIS PB89-175442).

U.S. EPA. 1988c. Draft Glossary of Quality Assurance Related Terms. Office of Research and Development, September 29, 1988.

U.S. Environmental Protection Agency (EPA). 1989a. Report on Minimum Criteria to Assure Data Quality. EPA/530/SW-90/021, 36 pp. Available from RCRA Hotline.*

U.S. Environmental Protection Agency (EPA). 1989b. Preparation Aids for the Development of RREL Quality Assurance Project Plans. U.S. EPA, Risk Reduction Engineering Laboratory, Cincinnati, OH.

U.S. Environmental Protection Agency (EPA). 1990a. Hydrogeologic Mapping Needs for Ground-Water Protection and Management: Workshop Report 1990. EPA/440/6-90-002. Available from Safe Drinking Water Hotline.*

U.S. Environmental Protection Agency (EPA). 1990b. Quality Assurance/Quality Control Guidance for Removal Activities, Sampling QA/QC Plan and Data Validation Procedures, Interim Final. EPA/540/G-90/004 (NTIS PB90-274481), U.S. EPA, 66 pp.

U.S. Environmental Protection Agency (EPA). 1991a. Compendium of Superfund Program Publications. EPA/540/8-91/014. Available from RCRA Hotline.* [Supersedes 1990 publication titled Catalog of Superfund Program Publications (EPA/540/8-90-015)]

U.S. Environmental Protection Agency (EPA). 1991b. Preparation Aids for the Development of Category III Quality Assurance Project Plans. EPA/600/8-91/005.

U.S. Environmental Protection Agency (EPA). 1992. Catalogue of Hazardous and Solid Waste Publications, 6th ed. EPA/530-B-92-001. Available from RCRA Hotline.* [Supersedes 1991 fifth edition (EPA/530-SW-91-013)]

U.S. Geological Survey (USGS). 1979. Scientific and Technical, Spatial, and Bibliographic Data Bases and Systems of the U.S. Geological Survey, 1979. U.S. Geological Survey Information Circular 817.

U.S. Geological Survey (USGS). 1982a. Definitions of Components of Master Water Data Index Maintained by the National Water Data Exchange. U.S. Geological Survey Open File Report 82-327, 268 pp.

U.S. Geological Survey (USGS). 1982b. Definitions of the Components of the Water Data Source Directory Maintained by the National Water Data Exchange. U.S. Geological Survey Open File Report 82-923, 125 pp.

U.S. Geological Survey (USGS). 1982c. Instructions for Using the U.S. Geological Survey Data Base. U.S. Geological Survey Open File Report 82-568, 189 pp.

U.S. Geological Survey (USGS). 1982d. ESIS User Manual. USGS Information Systems Division, Reston, VA.

U.S. Geological Survey (USGS). 1984. A Data Management System for Use in Ground Water Modeling and Resource Evaluation. U.S. Geological Survey Open File Report 84-4014, 277 pp.

Ward, D.C. 1972. Geological Reference Sources: A Subject and Regional Bibliography of Publications and Maps in the Geological Sciences. Scarecrow Press, Metuchen, NJ, 453 pp.

(Table 5-13 Environmental Data Sources and Management)

Way, S.C., C.R. McKee, and H.K. Wainwright. 1984. A Computerized Ground-Water Monitoring System. Ground Water Monitoring Review 4(1):21-25.

Wood, D.N. (ed.). 1973. Use of Earth Science Literature. Butterworth and Co., London, 459 pp.

* See Preface for information on how to obtain documents from NTIS and Table 5-3 for telephone numbers for the RCRA and Safe Drinking Water Hotlines.

(Table 5-13 Environmental Data Sources and Management)

Table 5-14 Index to Major References on Statistics and Sampling Design

Topic	References
General Statistics	
General	Bendat and Piersoll (1986), Benjamen and Cornell (1970), Bethea et al (1985), Bury (1975), Dixon and Massey (1957), Freund and Wilson (1993), Hoel (1960), Jessen (1978), MacBerthouex et al. (1994), Meyer (1975), Steel and Torrie (1960), Taylor (1994), Wadsworth (1990); Precision and Bias: ASTM (1992)
Sampling Design/ Specific Applications	Environmental Pollution Monitoring: Chapman and El-Sharrawi (1990), Gilbert (1987); Geology: Davis (1973), Till (1974); Soils/Solids: Butler (1980), Gy (1979), Webster (1977), Webster and Oliver (1990); Ground Water: Gillham et al. (1983), Summers et al. (1985), U.S. EPA (1989); Hydrology: Helsel and Hirsch (1992), Riggs (1968); Environmental Biology: Green (1979), EPRI (1985)
Special Statistical Approaches	Exploratory Data Analysis: Hoaglin et al. (1983), Tukey (1977); Nonparametrics: Hollander and Wolfe (1973), Lehmann and D'Abrera (1975), Seigel (1956)
Time Series Data	Texts: Chatfield (1984); Ground-Water: Carosene-Link et al. (1993), Close (1989), Gibbons (1987), Harris et al. (1987), McBean et al. (1988), McNichols and Davis (1988), Montgomery et al. (1987), Nelson and Ward (1981), Pennino (1988), Ross (1993), Rovers and McBean (1981), Sgambat and Stedinger (1981), Schweitzer and Black (1985), Starks (1989), Yevjevich and Harmancioglu (1989)
Spatial Data	Cressie (1991); Geostatistics: See below; Factor Analysis: Lawrence and Upchurch (1976)
Analytical Data	Censored Data: Gilbert (1987), Gilliom and Helsel (1986), Gilliom et al. (1984), Helsel and Gilliom (1986), McBean and Rovers (1984), Porter et al. (1988); Limit of Detection: ACS Committee on Environmental Improvement (1980), ASTM D4210/Table A-14, IUPAC (1978), Long and Winefordner (1983), McNichols and Davis (1988), Porter (1986), Porter et al. (1988)

Table 5-14 (cont.)

Topic	References
Geostatistics	
Texts	Introductory: Clark (1979); Advanced: David (1977), Matheron (1971), Journal and Huijbregts (1978), Isaaks and Srivastava (1989); Hydrologic Applications: ASCE Task Committee (1990), Bárdossy (1992); Glossary: Olea (1991); Software: See Table E-2
Applications	Contaminant Characterization: Flatman (1984, 1986), Flatman and Yfantis (1984), Gilbert and Simpson (1985), Journal (1984); Contour Mapping: Olea (1974, 1975); Soil Characterization: Trangmar et al. (1985), Sinclair (1986), Warrick et al. (1986); Ground Water: Hughes and Lettenmaier (1981), Delhomme (1978, 1979), Sophocleous et al. (1982)

Table 5-14 References (Appendix F contains references for figure and table sources)

ACS Committee on Environmental Improvement. 1980. Guidelines for Data Acquisition and Data Quality Evaluation in Environmental Chemistry. Analytical Chemistry 52:2242-2249.

American Society for Testing and Materials (ASTM). 1992. ASTM Standards on Precision and Bias for Various Applications, 4th ed. ASTM, Philadelphia, PA, 476 pp. [50 standards]

ASCE Task Committee on Geostatistical Techniques in Geohydrology. 1990. Review of Geostatistics in Geohydrology, I. Basic Concepts, II. Applications. ASCE Journal of Hydraulic Engineering 116(5):612-658.

Bárdossy, A. (ed.). 1992. Geostatistical Methods: Recent Developments and Applications in Surface and Subsurface Hydrology. UNESCO, Paris, 161 pp. [14 papers presented at a 1990 international workshop in Karlsruhe, Germany]

Bendat, J.S. and A.G. Piersol. 1986. Random Data, Analysis and Measurement Procedures, 2nd ed. Wiley-Interscience, New York, NY.

Benjamen, J.R. and C.A. Cornell. 1970. Probability Statistics and Decision for Civil Engineers. McGraw-Hill, New York, NY, 684 pp.

Bethea, R.M., B.S. Duran, and T.L. Boullion. 1985. Statistical Methods for Engineers and Scientists. Marcel Dekker, New York, NY, 698 pp.

Bury, K.V. 1975. Statistical Models in Applied Science. John Wiley & Sons, New York, NY.

Butler, B.E. 1980. Soil Classification for Soil Survey, Chapter 2. Oxford University Press, New York, NY.

Carosene-Link, P., H.E. Horsey, J.C. Loftis, and L.D. Rainey. 1993. Ground Water Quality Statistical Analysis: Implementing the New RCRA Regulations. Ground Water Management 15:177-191 [Proc. 7th NOAC]

Chatfield, C. 1984. The Analysis of Time Series: Theory and Practice, 3rd ed. Chapman and Hall, London.

Chapman, D.T. and A.H. El-Sharrawi. 1990. Statistical Methods for the Assessment of Point Source Pollution. Kluwer Academic Publishers, New York. [Proceedings of 1988 workshop held in Burlington, Ontario]

Clark, I. 1979. Practical Geostatistics. Applied Science Publishers, London.

Close, M.E. 1989. Effect of Serial Correction on Ground Water Quality Sampling Frequency. Water Resources Bulletin 25(3):507-515.

Cressie, N. 1991. Statistics for Spatial Data. John Wiley & Sons, New York, NY. [Comprehensive and readable text on the analysis of spatial data through statistical models; unifies a previously disparate subject under a common approach and notation]

(Table 5-14 Statistical Methods References)

David, M. 1977. Geostatistical Ore Reserve Estimation. Elsevier, New York, NY.

Davis, J.C. 1973. Statistics and Data Analysis in Geology. John Wiley & Sons, New York, NY, 550 pp.

Delhomme, J.P. 1978. Kriging in the Hydrosciences. Adv. Water Resources 1:251-266.

Delhomme, J.P. 1979. Spatial Variability and Uncertainty in Groundwater Flow Parameters: A Geostatistical Approach. Water Resources Research 15:269-280.

Dixon, W.J. and F.J. Massey, Jr. 1957. Introduction to Statistical Analysis. McGraw-Hill, New York, NY, 488 pp.

Electric Power Research Institute (EPRI). 1985. Sampling Design for Aquatic Ecologic Monitoring, 5 vols. EPRI EA-4302, EPRI, Palo Alto, CA.

Flatman, G.T. 1984. Using Geostatistics in Assessing Lead Contamination Near Smelters. In: Environmental Sampling for Hazardous Wastes, G.E. Schweitzer and J.A. Santolucito (eds.), ACS Symp. Ser. 267, American Chemical Society, Washington, DC, pp. 43-52.

Flatman, G.T. 1986. Design of Soil Sampling Programs: Statistical Considerations. In: Quality Control in Remedial Site Investigation: Hazardous and Industrial Solid Waste Testing, 5th volume, C.L. Perket (ed.), ASTM STP 925, American Society for Testing and Materials, Philadelphia, PA, pp. 43-56.

Flatman, G.T. and A.A. Yfantis. 1984. Geostatistical Strategy for Soil Sampling: The Survey and the Census. Environmental Monitoring and Assessment 4:335-350.

Freund, R. and W. Wilson. 1993. Statistical Methods. Academic Press, New York, NY, 644 pp.

Gibbons, R.D. 1987. Statistical Prediction Intervals for the Evaluation of Ground-Water Quality. Ground Water 25(4):455-465. [See also 1988 discussion by C.B. Davis and R.J. McNichols in Ground Water 26(1):90-91]

Gilbert, R.O. 1987. Statistical Methods for Environmental Pollution Monitoring. Van Nostrand Reinhold, New York, NY.

Gilbert, R.O. and J.C. Simpson. 1985. Kriging for Estimating Spatial Pattern of Contaminants: Potential and Problems. Environmental Monitoring and Assessment 5:113-135.

Gillham, R.W., M.J.L. Robin, J.F. Barker, and J.A. Cherry. 1983. Groundwater Monitoring and Sample Bias. API Publication 4367. American Petroleum Institute, Washington, DC.

Gilliom, R.J. and D.R. Helsel. 1986. Estimation of Distributional Parameters for Censored Trace Level Water Quality Data: 1. Estimation Techniques. Water Resources Research 22:135-146.

Gilliom, R.J., R.M. Hirsch, and E.J. Gilroy. 1984. Effect of Censoring Trace-Level Water-Quality Data on Trend-Detection Capability. Environ. Sci. Technol. 18:530-536.

(Table 5-14 Statistical Methods References)

Green, R. 1979. Sampling Design and Statistical Methods for Environmental Biologists. John Wiley & Sons, New York, NY, 257 pp.

Gy, P. 1979. Sampling of Particulate Materials: Theory and Practice. Elsevier, New York, NY, 431 pp.

Harris, J., J.C. Loftis, and R.H. Montgomery. 1987. Statistical Methods for Characterizing Ground-Water Quality. Ground Water 25(2):185-193.

Helsel, D.R. and R.J. Gilliom. 1986. Estimation of Distributional Parameters for Censored Trace Level Water Quality Data: 2. Verification and Applications. Water Resources Research 22:146-155.

Helsel, D.R. and R.M. Hirsch. 1992. Statistical Methods in Water Resources. Elsevier, New York, NY, 522 pp.

Hoaglin, D.C., F. Mosteller, and J.W. Tukey. 1983. Understanding Robust and Exploratory Data Analysis. John Wiley & Sons, New York, NY.

Hoel, P.G. 1960. Elementary Statistics. John Wiley & Sons, New York, NY, 261 pp.

Hollander, M. and D.A. Wolfe. 1973. Nonparametric Statistical Methods. John Wiley & Sons, New York, NY.

Hughes, J.P. and D.P. Lettenmaier. 1981. Data Requirements for Kriging: Estimation and Network Design. Water Resources Research 17:1641-1650.

International Union of Pure and Applied Chemistry (IUPAC). 1978. Nomenclature, Symbols, Units and their Usage in Spectrochemical Analysis--II. Spectrochimica Acta B 33B:242.

Isaaks, E.H. and R.M. Srivastava. 1989. Applied Geostatistics. Oxford University Press, New York, NY.

Jessen, R.J. 1978. Statistical Survey Techniques. John Wiley & Sons, New York, NY.

Journal, A.G. 1984. New Ways of Assessing Spatial Distribution of Pollutants. In: Environmental Sampling for Hazardous Wastes, G.E. Schweitzer and J.A. Santolucito (eds.), ACS Symp. Ser. 267, American Chemical Society, Washington, DC, pp. 109-118.

Journal, A.G. and C.J. Huijbregts. 1978. Mining Geostatistics. Academic Press, New York, NY.

Lawrence, F.W. and S.B. Upchurch. 1976. Identification of Geochemical Patterns in Groundwater by Numerical Analysis. In: Advances in Groundwater Hydrology, AWRA Proc. Ser. No. 21. American Water Resources Association, Bethesda, MD. [Factor analysis]

Lehmann, E.L. and H.J.M D'Abrera. 1975. Nonparametrics: Statistical Methods Based on Ranks. McGraw-Hill, New York, NY.

(Table 5-14 Statistical Methods References)

Long, G.L. and J.D. Winefordner. 1983. Limit of Detection, a Closer Look at the IUPAC Definition. Analytical Chemistry 55(7):712A-724A.

MacBerthouex, P. and L.C. Brown (eds.). 1994. Statistics for Environmental Engineers. Lewis Publishers, Boca Raton, FL, 352 pp.

Matheron, G. 1971. The Theory of Regionalized Variables and Its Applications. Cahiers du Centre de Morphologie Mathematique de Fontainebleau. No. 5.

McBean, E.A. and R.A. Rovers. 1984. Alternatives for Handling Detection Limit Data in Impact Assessments. Ground Water Monitoring Review 4(2):42-44.

McBean, E.A., M. Kompter, and F. Rovers. 1988. A Critical Examination of Approximations Implicit in Cochran's Procedure. Ground Water Monitoring Review 8(1):83-87.

McNichols, R.J. and C.B. Davis. 1988. Statistical Issues and Problems in Ground Water Detection Monitoring at Hazardous Waste Facilities. Ground Water Monitoring Review 8(4):135-150.

Meyer, S.C. 1975. Data Analysis for Scientists and Engineers. John Wiley & Sons, New York, NY, 513 pp.

Montgomery, R.H., J.C. Loftis, and J. Harris. 1987. Statistical Characteristics of Ground Water Quality Variability. Ground Water 25(2):176-184.

Nelson, J.D. and R.C. Ward. 1981. Statistical Considerations and Sampling Techniques for Groundwater Quality Monitoring. Ground Water 19(6):617-625.

Olea, R.A. 1974. Optimal Contour Mapping Using Universal Kriging. J. Geophysical Research 79(5):696-702.

Olea, R.A. 1975. Optimum Mapping Techniques Using Regionalized Variable Theory. KGS Series on Spatial Analysis No. 2. Kansas Geological Survey, Lawrence, KS.

Olea, R.A. (ed.) 1991. Geostatistical Glossary and Multilingual Dictionary. Oxford University Press, New York, NY, 192 pp.

Pennino, J.D. 1988. There's No Such Thing as a Representative Ground Water Sample. Ground Water Monitoring Review 8(3):4-9.

Porter, P.S. 1986. A Description of Measurement Error Near Limits of Detection. In: Monitoring to Detect Changes in Water Quality Series, D. Lerner (ed.), Int. Ass. of Hydrological Sciences Pub. No. 157.

Porter, P.S., R.C. Ward, and H.F. Bell. 1988. The Detection Limit. Environ. Sci. Technol. 22:856-861.

Riggs, H.C. 1968. Some Statistical Tools for Hydrology. U.S. Geological Survey Techniques of Water Resource Investigations TWRI 4-A1, 39 pp.

(Table 5-14 Statistical Methods References)

Planning Field Investigations

Ross, D.L. 1993. Multivariate Statistical Analysis of Environmental Monitoring Data. Ground Water Management 17:301-315 [Proc. 10th API/NGWA Hydrocarbon Conf.]

Rovers, F.A. and E.A. McBean. 1981. Significance Testing for Impact Evaluation. Ground Water Monitoring Review 1(2):39-43.

Schweitzer, G.E. and S.C. Black. 1985. Monitoring Statistics. Environ. Sci. Technol. 19(11):1026-1030.

Seigel, S. 1956. Nonparametric Statistics for the Behavioral Sciences. McGraw-Hill, New York, NY.

Sinclair, A.J. 1986. Statistical Interpretation of Soil Geochemical Data. In: Exploration Geochemistry, Design and Interpretation of Soil Surveys. Reviews in Economic Geology 3:97-115.

Sgambat, J.P. and J.R. Stedinger. 1981. Confidence in Ground-Water Monitoring. Ground Water Monitoring Review 1(Spring):62-69.

Sinclair, A.J. 1986. Statistical Interpretation of Soil Geochemical Data. In: Exploration Geochemistry, Design and Interpretation of Soil Surveys. Reviews in Economic Geology 3:97-115.

Sophocleous, M., J.E. Paschetto, and R.A. Olea. 1982. Ground-Water Network Design for Northwest Kansas, Using the Theory of Regionalized Variables. Ground Water 20:48-58.

Starks, T.H. 1989. Evaluation of Control Chart Methodologies for RCRA Waste Sites. EPA/600/4-88/040 (NTIS PB89-138416). [Statistical procedures for ground-water monitoring]

Steel, R.G.D. and J.H. Torrie. 1960. Principles and Procedures of Statistics. McGraw-Hill, New York, NY, 481 pp.

Summers, K.V., G.L. Rupp, G.F. Davis, and S.A. Gherini. 1985. Ground Water Data Analysis at Utility Waste Disposal Sites. EPRI EA-4165. Electric Power Research Institute, Palo Alto, CA.

Taylor, J.K. 1994. Statistical Techniques for Data Analysis. Lewis Publishers, Boca Raton, FL, 300 pp.

Till, R. 1974. Statistical Methods for the Earth Scientist: An Introduction. John Wiley & Sons, New York, NY, 154 pp.

Tukey, J.W. 1977. Exploratory Data Analysis. Addison-Wesley, New York, NY, 506 pp.

Trangmar, B.B., R.S. Yost, and G. Uehara. 1985. Application of Geostatistics to Spatial Studies of Soil Properties. Advances in Agronomy 38:45-93.

(Table 5-14 Statistical Methods References)

U.S. Environmental Protection Agency (EPA). 1989. Guidance Document on Statistical Analysis of Ground-Water Monitoring Data at RCRA Facilities--Interim Final Guidance. Office of Solid Waste Management Division (NTIS PB89-151047). [See also, September 1991 Addendum]

Wadsworth, H. 1990. Handbook of Statistical Methods for Engineers and Scientists. McGraw-Hill, New York, NY, 768 pp.

Warrick, A.W., D.E. Myers, and D.R. Nielsen. 1986. Geostatistical Methods Applied to Soil Science. In: Methods of Soil Analysis, Part I--Physical and Mineralogical Methods, 2nd ed., A. Klute (ed.), ASA Monograph No. 9, American Society of Agronomy, Madison, WI, pp. 53-82.

Webster, R. 1977. Quantitative and Numerical Methods in Soil Classification and Survey. Oxford University Press, New York, NY.

Webster, R. and M.A. Oliver. 1990. Statistical Methods in Soil and Land Resource Survey. Oxford University Press, New York, NY, 328 pp.

Yevjevich, V. and N.B. Harmancioglu. 1989. Description of Periodic Variation in Parameters of Hydrologic Time Series. Water Resources Research 25(3):421-428.

(Table 5-14 Statistical Methods References)

CHAPTER 6

GEOPHYSICAL AND REMOTE SENSING TECHNIQUES

6.1 Overview of Remote Sensing and Geophysical Techniques 304

    6.1.1 Uses of Geophysics in Contamination Studies 305
    6.1.2 General Characteristics of Geophysical Methods 306

6.2 Airborne Remote Sensing 308

    6.2.1 Visible and Near-Infrared Aerial Photography 308
    6.2.2 Other Airborne Remote Sensing Techniques 311

6.3 Surface Geophysical Methods 312

    6.3.1 Electromagnetics 312
    6.3.2 Resistivity and Other Electrical Methods 316
    6.3.3 Seismic Refraction and Shallow Seismic Reflection 318
    6.3.4 Magnetometry 320
    6.3.5 Ground Penetrating Radar 320
    6.3.6 Gravimetrics 321
    6.3.7 Thermal Sensing 321

6.4 Borehole Geophysics 322

    6.4.1 Electrical and Electromagnetic Logging Methods 323
    6.4.2 Nuclear Logging Methods 324
    6.4.3 Acoustic and Seismic Logging Methods 328
    6.4.4 Lithologic and Hydrogeologic Characterization Logs 328
    6.4.5 Downhole Methods in Ground Water
          Contamination Studies 331

6.5 Selection of Geophysical Methods 331

6.6 Guide to Major References* 335

    6.6.1 General Geophysics 336
    6.6.2 Ground Water and Contaminated Sites 336
    6.6.3 Evaluation of Literature References 337

* Appendix F contains citations for table and figure sources.

6.1 Overview of Remote Sensing and Geophysical Techniques

A wide variety of field investigation techniques are available that provide information about a site with little or no disturbance of the surface and subsurface. In the broadest sense most of these techniques involve *remote sensing*, the observation of an object or phenomenon without the sensor being in direct contact with the object being sensed. In this chapter, the term remote sensing is restricted to airborne photographic and geophysical methods, whereas the term geophysics is used loosely to include these remote sensing methods.

Geophysical techniques are used to assess the physical and chemical properties of soils, rock, and ground water based on the response to either (1) various parts of the *electromagnetic* (EM) spectrum, including gamma rays, visible light, radar, microwave, and radio waves (Figure 6-1), (2) *acoustic* and/or *seismic* energy, or (3) other *potential* fields, such as gravity and the earth's magnetic field.

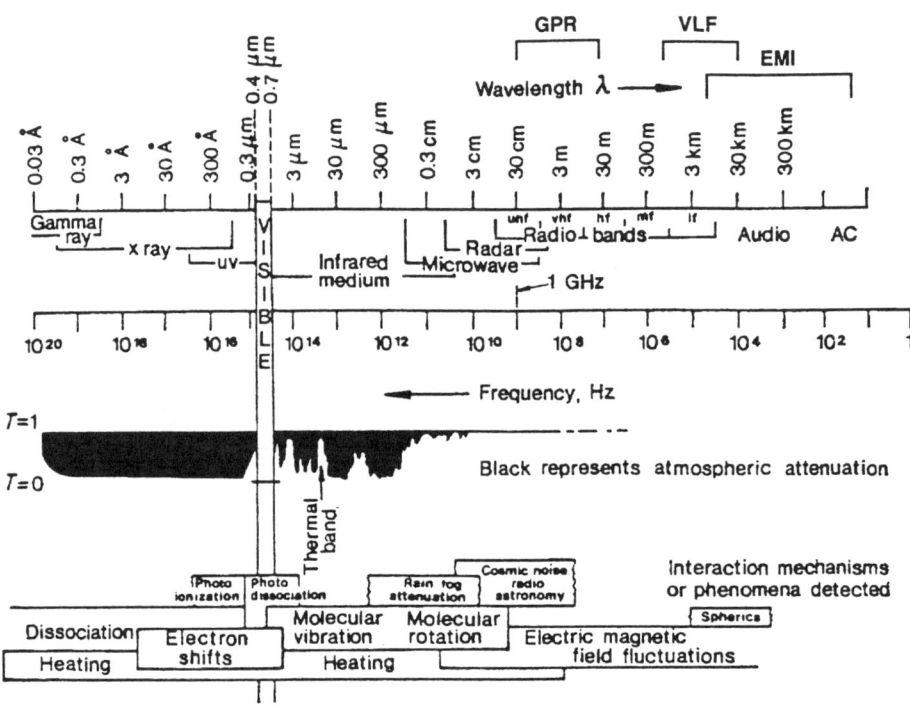

Figure 6-1 The electromagnetic spectrum: the customary divisions and portions used for geophysical measurements (Boulding, 1993a, after Erdélyi and Gálfi, 1988).

Historically, geophysical field methods have been mainly the domain of petroleum and mineral exploration geologists. Early successes in the 1970s using electrical methods (measurement of variations in conductivity and resistivity) to locate contaminant plumes and measure the hydrogeologic properties of aquifers have led to the adaptation of a large number of geophysical methods in ground-water contamination investigations.

In the late 1970s, the availability of microcomputers revolutionized the use of field geophysics by allowing onsite processing of the tremendous amount of data generated by most of these techniques. Use of geophysical methods in hydrogeologic studies became so widespread in the 1980s, that techniques such as electromagnetic induction, electrical resistivity, seismic refraction, and magnetometry are no longer considered innovative but state-of-the-art. Innovations in these and numerous other geophysical methods continue at a rapid rate.

6.1.1 Uses of Geophysics in Contamination Studies

A wide range of subsurface features can be measured or inferred using geophysical methods. These features can be broadly classified into four groups:

1. *Natural Soil/Geologic/Ground Water Conditions.* Preliminary characterization of soil and geologic stratigraphy (i.e., depth and thickness of layering and lateral changes) and subsurface structure (i.e., depth and topography of unconsolidated material-bedrock contact, dip and folding of sediments) are among the primary uses for surface geophysical methods such as ground-penetrating radar, electromagnetic methods, electrical resistivity, and seismic methods. Some other potential uses of these methods for characterizing natural conditions include: (1) measurement of depth to water table and aquifer thickness; (2) mapping of clay layers; and (3) detection of subsurface cavities and sinkholes.

2. *Contaminated Ground Water.* When contaminants in ground water create conductive plumes of leachate (i.e., landfills, injected brines, salt-water intrusion), electromagnetic and resistivity are good methods for preliminary characterization of the extent and direction of flow of a plume. Where contaminant plumes are relatively near the surface, the depth of the top of the plume serves as an indicator of the thickness of the vadose zone.

3. *Buried Wastes.* Ground-penetrating radar, electromagnetic induction, and electrical resistivity can be good methods for preliminary delineation of the location and boundaries of bulk wastes and nonmetallic containers. Electromagnetic induction, metal detectors and magnetometry are primary methods for detection of buried metallic containers (except that magnetometry will not detect nonferrous metals).

4. *Other Anthropogenic Features.* Other subsurface anthropogenic features that can be detected by surface geophysical methods include location of pipes, cables, tanks, and abandoned well casings. Loose fill in trenches associated with buried utilities may form permeable pathways for preferential flow of contaminants in the subsurface. Magnetometry, metal detection, electromagnetic induction and ground penetrating radar are the surface geophysical methods that are primarily used for these applications.

The greatest benefits of geophysical methods come from early use. These methods are typically nondestructive, less risky, cover more area spatially and volumetrically, and require less time and cost than site characterization using monitoring wells. On the other hand, great skill is required in interpreting the data generated by these methods, and their indirect nature creates uncertainties that can only be resolved by direct observation. Consequently, preliminary site characterization by geophysical methods is usually followed by direct observation through the installation of monitoring wells.

Airborne and surface geophysical methods can reduce the number of monitoring wells that must be drilled to adequately characterize a site, which can result in significant cost savings. Figure 6-2 illustrates potential cost savings from using geophysical methods to help site monitoring wells compared to the use of monitoring wells only. When only a few monitoring wells are involved, the costs of the two approaches may be comparable, but when the number of required monitor wells exceeds 10 or so, geophysical methods becomes increasingly more cost effective as the number of wells increases.[1]

6.1.2 General Characteristics of Geophysical Methods

Geophysical investigation techniques can be broadly grouped into three categories: (1) airborne remote sensing, (2) surface, and (3) borehole or downhole methods. Each of these categories has numerous specific techniques, and a specific technique may have a number of variants. Table A-1 (Remote Sensing and Surface Geophysical Methods) and Table A-2 (Characteristics of Borehole Logging Methods) present information on more than 70 specific methods. The rest of this chapter focusses on the more commonly used methods.

Remote sensing and airborne geophysics do not involve direct contact with the object being observed and usually involve airborne instrumentation (see Section 6.2). As the names imply, surface methods involve wave generators and sensors at or near the ground surface (see Section 6.3), and downhole methods involve sensing methods below the surface. Some specific techniques may be used in more than one way. For example, electromagnetic induction (Section 6.3.1) can be used from an airplane, on the ground surface, and in boreholes.

Most of the techniques discussed in this chapter operate in a portion of the electromagnetic spectrum. Electromagnetic radiation can be described in terms of *wavelength*, which is the distance between two crests of the wave, and *frequency*, the number of waves measured passing a certain point in the medium in the course of one second (cycles per second). Airborne remote sensing methods tend to sense higher frequencies (infrared through ultraviolet), whereas surface geophysical methods generally sense the response of subsurface materials to lower frequencies in the EM spectrum (Figure 6-1).

---

[1] Note that this figure applies to installation of permanent monitoring wells. Surveys using push technologies, as discussed in Section 9.1.2, provide many of the benefits of geophysical methods with the added benefit of allowing direct sampling of the subsurface.

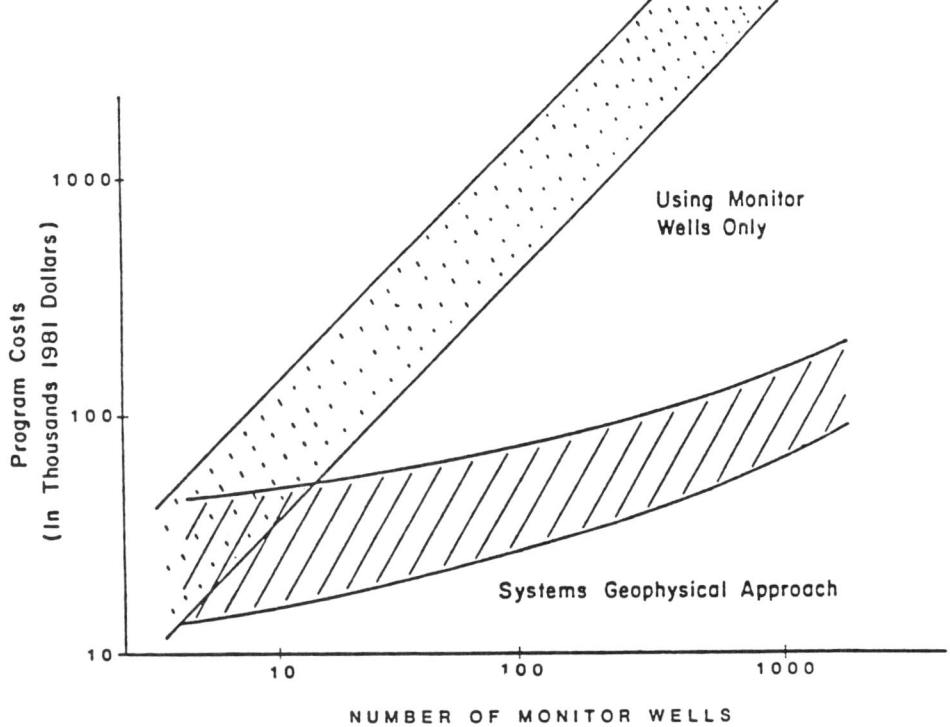

Figure 6-2 Cost comparison for hazardous waste site investigations: geophysics vs. monitoring wells only (Benson et al., 1984).

Geophysical methods tend to measure a larger volume of the subsurface than monitoring wells, thereby increasing the volume sampled for a given measurement. This is usually an advantage, but can be a disadvantage if a feature or anomaly is so small that it may escape detection in a larger sampled volume. Data from these methods may be acquired in the form of (1) *profiles*, which record changes in measured properties in a linear transect along the ground surface, or (2) *soundings*, which measure vertical changes in the measured properties.

Profile measurements can be either *stationary* or *continuous*. Stationary, or station, measurements are taken at discrete intervals, whereas continuous methods measure subsurface parameters continuously along a survey line. Figure 6-3 shows the difference in output from the two types. The figure shows that continuous measurements, where feasible, provide better resolution, but most traditional geophysical techniques involve station measurement. Continuous methods, such as EM induction, are typically limited to a depth of 15 meters or less, but are still preferred when applicable since they can approach 100% site coverage.

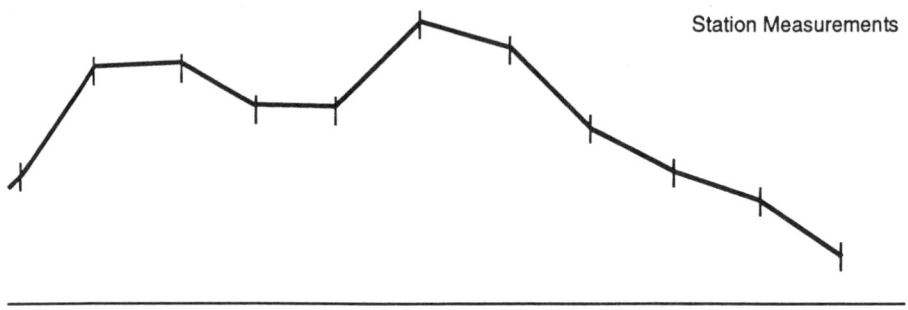

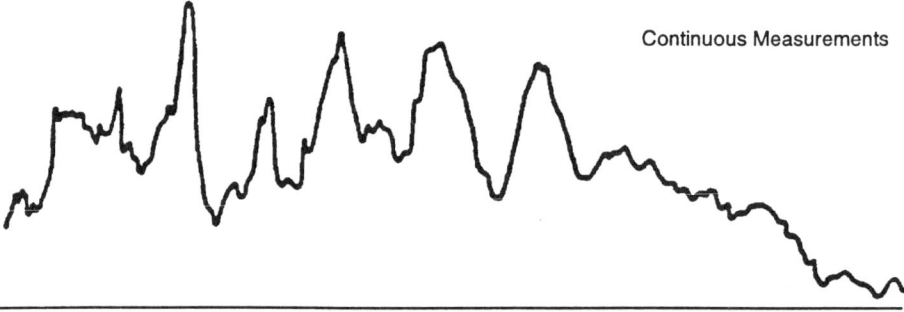

Figure 6-3 Discrete sampling vs. continuous geophysical measurements (Boulding, 1993a, after Benson et al., 1984).

6.2 Airborne Remote Sensing

Although all geophysical methods involve remote sensing in some fashion, the term is used here to apply to airborne instrumentation. Table 6-1 presents a summary description and listing of hydrogeologic applications for six remote sensing techniques. Figure 6-4 identifies features that can be identified in the portions of the spectrum that are sensed using airborne methods.

6.2.1 Visible and Near-Infrared Aerial Photography

Aerial photographs, which record the visible portion of the electromagnetic spectrum, are by far the most common form of remote sensing and are basic to any geologic or hydrogeologic investigation. Much information can be obtained from stereopairs of black-and-white air photos, which provide a three-dimensional image of the surface when viewed with a stereoscope. Patterns of vegetation, variations in grey tones in soil and rock, drainage patterns, and linear features allow preliminary interpretations of geology, soils, and hydrogeology. All air photo interpretations should be field checked and revised where "ground truthing" indicates features that were missed or incorrectly delineated.

Photogrammetric techniques using stereoscopic (overlapping) aerial photographs is often the cheapest way to produce accurate topographic maps (1- or 2-foot contour intervals) for site-specific investigations.

Table 6-1 Use of Airborne Sensing Techniques in Hydrogeologic and Contaminated Site Studies

Method	Description	Applications
Visible and near infrared	Aerial photographs (black and white, color, false color, infrared, multispectral). Imaging limited to surface features.	Air photo interpretation of geologic and surface hydrologic features, fracture trace analysis, soil moisture patterns, and vegetation (infrared).
Photographic ultraviolet	Aerial photographs using special film and filters for sensing reflected ultraviolet radiation.	Mapping of oil spills on surface water bodies; sometimes used for geologic mapping of carbonate formations.
Thermal infrared	Scanners used to detect infrared radiation beyond the range of infrared photography.	Routinely used to detect ground-water discharge into rivers, lakes, and the sea; detects variations in soil moisture content (seepage from leach fields and underground storage tanks), evaporation, and thermal properties.
Side-looking airborne radar (SLAR)	Creates a continuous radar image (reflected radio frequency pulses) of the ground surface.	Similar applications to air photos; can distinguish grain size in alluvium if there is no interference from vegetation; can also be used for fracture trace analysis.
Low frequency airborne electromagnetic methods (AEM)	Uses a low frequency electromagnetic wave transmitter and receiver that responds to changes in the ground electrical conductivity.	Detects variations in soil and rock types; variations in ground-water salinity; location of shallow subsurface aquifers and deeper brine contaminated aquifers.
Aeromagnetic	Measures the earth's total magnetic field.	Primarily used in petroleum and mineral exploration to assist with geological mapping and structural interpretations. Also used to locate abandoned wells with metallic casings.

Source: Boulding (1993a).

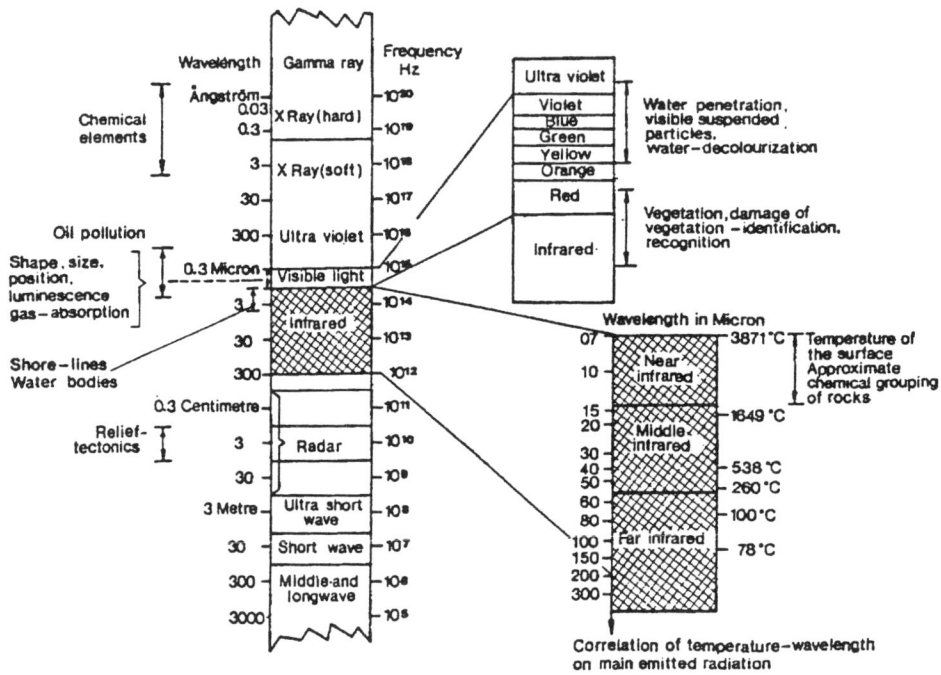

Figure 6-4 Factors and phenomena influencing the radiation of electromagnetic waves (Boulding, 1993a, after Erdélyi and Gálfi, 1988).

Black-and-white (also called panchromatic) air photos are available from various federal agencies for almost any location in the United States and are the cheapest type of air photo to obtain. Black-and-white photographs are reported most frequently as being useful in ground-water contamination studies. Other types of images that can be obtained, usually at greater expense, include:

- *True color* records all colors in the visible spectrum as they appear to the naked eye.

- *Color infrared film* records yellows and reds as green and the near infrared (not visible to the eye) as red. Since vegetation reflects near-infrared radiation, this image is especially useful for observing vegetation patterns. Other types of images that record or display colors differently than they are perceived by the eye (called *false color*) can be created in a similar fashion.

- *Photographic ultraviolet* uses special film and filters to record UV energy. Oil and carbonate minerals are fluorescent in UV bands when photostimulated by sunlight. A disadvantage of UV photography is that UV wavelengths are scattered in the atmosphere and result in a low contrast image, especially when dust or haze is present.

- *Multiband* (also called *multispectral*) images use multiple lenses and filters to record simultaneous exposures of different portions of the visible and near-infrared spectrum of the same area on the ground. Images can also be recorded electronically using a multispectral scanning system.

Air photos often reveal linear features called fracture traces that indicate zones of relatively higher permeability in the subsurface. Fracture-trace analysis using air photos can provide preliminary information on possible preferential movement of contaminants (Parizek, 1976).[2] Aerial photography can also be a valuable tool in documenting preexisting physical conditions and monitoring the progress of cleanup operations at hazardous waste sites (Finkbeiner and O'Toole, 1985).[3] Color infrared photography is most useful where contamination results in vegetation changes such as failed septic tank absorption systems (Farrell, 1985),[4] fertilizers, oil pollution and natural gas leaks (Švoma and Pyšek, 1985).[5]

6.2.2 Other Airborne Remote Sensing Techniques

Table 6-1 describes four other aerial remote sensing techniques that may have applications in hydrogeologic studies. *Thermal infrared* scanning can detect groundwater discharge into surface waters by sensing temperature differences in the ground and surface water. Contaminant plumes often differ in temperature from the surrounding ground water and may also be detected using this method.

Airborne geophysical methods such as *side-looking airborne radar* (SLAR), *airborne electromagnetic methods* (AEM), and *aeromagnetics* have not been widely used much in ground-water contamination studies, although potential exists for their use in regional water quality studies. A special feature of SLAR is its ability to distinguish grain size in alluvium. This technique requires unvegetated surfaces, a condition that is most likely to occur in arid areas.

---

[2] Parizek, R.R. 1976. On the Nature and Significance of Fracture Traces and Lineaments in Carbonate and Other Terranes. In: Karst Hydrology and Water Resources, V. Yevjevich (ed.), Water Resources Publications, Fort Collins, CO, Vol. 1, pp. 3-1 to 3-62.

[3] Finkbeiner, M.A. and M.M. O'Toole. 1985. Application of Aerial Photography in Assessing Environmental Hazards and Monitoring Cleanup Operations at Hazardous Waste Sites. In: Proc. 6th Nat. Conf. on Management of Uncontrolled Hazardous Waste Sites, Hazardous Materials Control Research Institute, Silver Spring, MD, pp. 116-124.

[4] Farrell, S.O. 1985. Evaluation of Color Infrared Aerial Surveys of Wastewater Soil Absorption Systems. EPA/600/2-85/039 (NTIS PB85-189074).

[5] Švoma, J. and A. Pyšek. 1985. Photographic Detection of Groundwater Pollution. In: Hydrological Applications of Remote Sensing and Remote Sensing Data Transmission, B.E. Goodison (ed.), Int. Ass. Hydrological Sciences Publ. No. 145, pp. 561-567.

Surface, rather than airborne, electromagnetic methods are generally better adapted to site-specific ground-water contamination studies, since the spatial resolution of airborne EM methods (on the order of several tens of meters) is usually too coarse for contamination investigations. The U.S. EPA has been supporting research on the use of airborne electromagnetics to locate areas of near-surface brine contamination in the Brookhaven oil field in Mississippi (Smith et al., 1989).[6] Aeromagnetic surveys have been used as a complement to other methods to locate abandoned wells (Frischknecht, 1990).[7]

6.3 Surface Geophysical Methods

Surface geophysical methods have been widely used in ground-water contamination studies for the reasons discussed in Section 6.1.1. Eight major geophysical methods are currently being used at contaminated sites, in the following order of approximate frequency of use: (1) electromagnetic induction (EM), (2) electrical resistivity, (3) magnetometry, (4) ground-penetrating radar, (5) seismic refraction, (6) shallow seismic reflection, (7) gravimetrics, and (8) thermal sensing. Table 6-2 presents a brief summary description of each method and its applications to ground-water contamination studies. The following sections provide additional information on each method. Section 6.5 provides some comparative information on the six most commonly used methods.

6.3.1 Electromagnetics

Electromagnetic methods measure the electrical conductivity of the subsurface. Although most geophysical methods use electromagnetic principles, common usage of the term "EM" in the geophysical industry implies measurement of subsurface conductivities at relatively low frequencies (Figure 6-1). Electrical conductivity is a function of the type of soil and rock, its porosity, permeability, and the fluids that fill the pore space. The conductivity (measured as specific conductance--millimhos per meter) of the pore fluids usually dominates the measurement, especially when dissolved species are present in contaminated water. Consequently, EM is an excellent technique for mapping contaminant plume boundaries, as well as a variety of other subsurface features with contrasting electrical properties.

---

[6] Smith, B., W. Heran, R. Bisdorf, and A.T. Mazzella. 1989. Evaluation of Airborne Geophysical Methods to Map Brine Contamination. EPA/600/4-89/003. U.S. EPA Environmental Monitoring Systems Laboratory, Las Vegas, NV.

[7] Frischknecht, F.C. 1990. Application of Geophysical Methods to the Study of Pollution Associated with Abandoned and Injection Wells. In: Proc. of a U.S. Geological Survey Workshop on Environmental Geochemistry, B.R. Doe (ed.), U.S. Geological Survey Circular 1033, pp. 73-77.

Table 6-2 Major Surface Geophysical Methods for Study of Subsurface Contamination

Method	Description	Hydrogeologic Applications
Electromagnetic induction (EMI), time domain electromagnetics (TDEM), metal detectors (MD) (Section 6.3.1)	**EMI:** Uses a transmitter coil to generate currents that induce a secondary magnetic field in the earth that is measured by a receiver coil. Well suited for areal searches. **TDEM:** Uses a transmitter loop placed on the ground to create a descending eddy current upon termination of current flow and a receiver coil that measures the decaying magnetic field. **MD:** Specialized EMI instruments designed to sense increased conductivity resulting from buried ferrous or nonferrous metals.	Can be used to map a wide variety of subsurface features including natural hydrogeologic conditions, delineation of contaminant plumes, rate of plume movement, buried wastes, and other artificial features (e.g., buried drums, pipelines). **EMI:** Depth of penetration is typically up to 60 meters but depths to 200+ meters are possible. **TDEM:** Penetration of more than 2,000 meters is possible. **MD:** Maximum depth of 3 meters for single drum, and 6 meters for large pile of metallic material.
DC electrical resistivity (Section 6.3.2)	Measures the resistivity of subsurface materials by injecting an electrical current into the ground by a pair of surface electrodes and measuring the resulting potential field (voltage) between a second pair of electrodes.	Similar to electrical conductivity (see above), except not widely used to detect metallic objects, for which magnetic and EMI methods are more effective. Better for depth sounding than frequency domain EMI.
Seismic refraction and reflection (Section 6.3.3)	Uses a seismic source (commonly a sledge hammer), an array of geophones to measure travel time of the refracted/reflected seismic waves, and a seismograph that integrates the data from the geophones.	Can be used to define the thickness and depth to bedrock or water table, thickness of soil and rock layers, and their composition and physical properties; may detect anomalous subsurface features such as pits and trenches.[a]
Magnetometry (Section 6.3.4)	Uses a magnetometer to measure the intensity of the earth's magnetic field.	Used to locate buried metal drums that may be sources of soil and ground-water contamination.

Table 6-2 (cont.)

Method	Description	Hydrogeologic Applications
Ground penetrating radar (GPR) (Section 6.3.5)	Uses a transmitter coil to emit high frequency radio waves that are reflected off subsurface changes in electrical properties (typically density and water content variations) and detected by a receiving antenna.	Can map soil layers, depth of bedrock, buried stream channels, rock fractures, cavities in natural settings, and buried waste materials. Maximum depth of penetration under favorable conditions is around 25 meters. Hundreds of meters penetration may be possible in highly resistive materials (salt or ice).
Gravimetry (Section 6.3.6)	Uses one or more of several types of instruments that measure the intensity of the earth's gravitational field.	Can be used to estimate depth of unconsolidated material over bedrock and boundaries of landfills, which have a different density than natural soil material. Microgravity surveys may be able to detect subsurface cavities and subsidence voids.
Thermal sensing (Section 6.3.7)	Uses temperature sensors anomalies in the soil or surface water.	Can be used to delineate shallow ground-water flow systems, buried valley aquifers, recharge and discharge zones, zones of high permeability, leakage beneath earthen dam embankments, and location of solution channels in karst.

[a] High resolution shallow seismic reflection is increasingly being used as an alternative to seismic refraction. Minimum depth resolution is typically 10 meters but it can be as shallow as 3 meters.

Source: Boulding (1993a).

Electromagnetic induction (EMI) and *metal detectors* are the most commonly used electromagnetic methods used in contamination studies. Figure 6-5 shows the basic principle of operation of an EMI instrument: a transmitter coil generates a sinusoidal electromagnetic field that induces eddy currents in the earth below the instrument. A receiver coil then intercepts both the primary and the secondary electromagnetic fields created by the eddy current loops and produces an output voltage that is

Geophysical and Remote Sensing Techniques

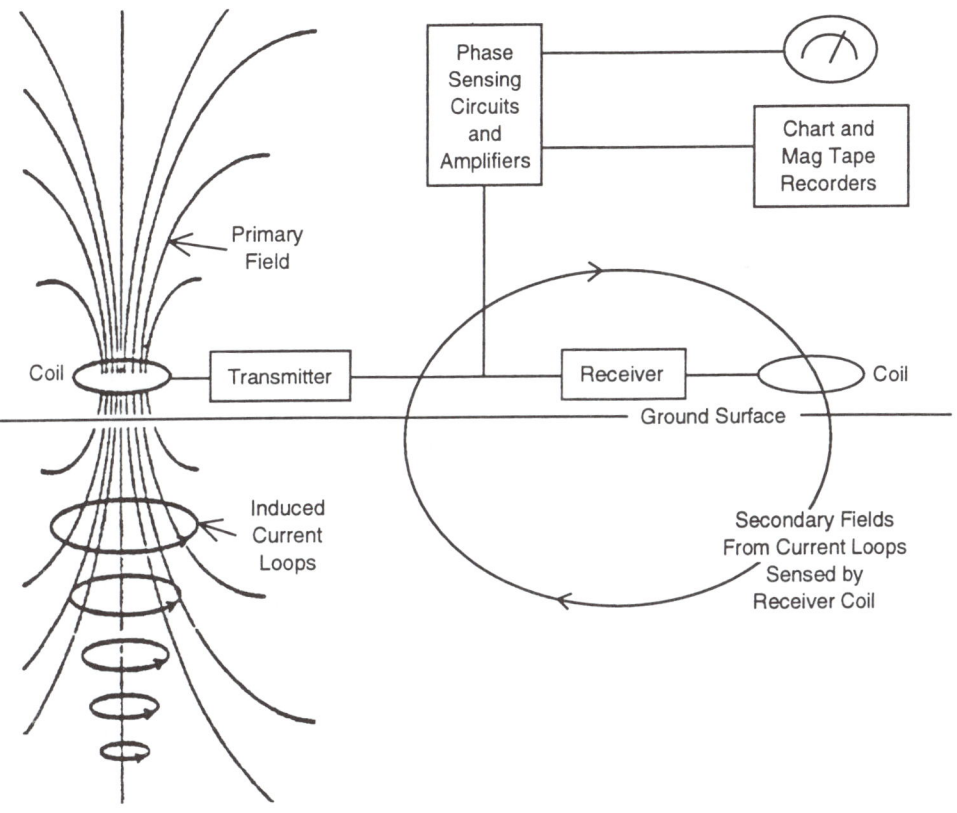

Figure 6-5 Electromagnetic induction principle of operation (Boulding, 1993a, after Benson et al., 1984).

corrected for the primary field and the loop geometry and spacing. This voltage, within limits, is linearly related to subsurface conductivity. The reading represents the weighted cumulative sum of the conductivity variations from the surface to the effective depth of the instrument. The effective depth for EMI is determined by the geometry and spacing of the transmitting and receiving coils, with 60 meters representing a typical maximum depth. Readings to shallow depths can be made continuously since the coils are rigidly connected, whereas greater depth penetration requires stationary measurements.

Metal detectors, EM instruments which have been designed to respond to the electrical conductivity of metals, are commonly used to locate buried metals at uncontrolled hazardous waste sites. They have the advantage of detecting nonferrous metals such as aluminum and copper, which cannot be detected with magnetometers (Section 6.3.4).

Other potentially useful electromagnetic methods include:

- *Time domain electromagnetic* (TDEM), also called *transient* EM instruments, measure the response of the subsurface to a decaying induced current. These instruments can measure to depths exceeding 2,000 meters and are especially useful for detecting fresh-water/salt-water interfaces and brine contamination associated with petroleum production.

- *VLF resistivity* measures the ratio of electric to magnetic fields generated by military communication transmitters. Potential uses are similar to EMI instruments.

- *Magnetotelluric* methods include a variety of techniques more measuring earth response to natural electrical and magnetic currents in the subsurface. These methods, especially the *controlled source audiomagnetotelluric* (CSAMT) method, which uses a remote transmitter to create a signal with known characteristics, have some potential for regional ground-water investigations and detection of deep, conductive contaminant plumes.[8]

6.3.2 Resistivity and Other Electrical Methods

The electrical resistivity method (ER, for electrical resistance) measures the resistance to flow of electricity in subsurface material. Other terms used to describe this method include: direct current (DC) resistivity, galvanic resistivity, and geoelectric resistivity. Resistance is the reciprocal of conductance. Consequently, applications of ER in hydrogeologic studies are similar to those of EM methods, with site-specific conditions determining which may be the preferred method (Section 6.5). In contrast to EM, which does not require direct contact with the ground surface, ER involves the placement of electrodes, called *current electrodes*, on the surface for injection of current into the ground. This current is measured by a volt-meter between two other electrodes, called *potential electrodes*. Apparent resistivity (measured in ohm-meters or ohm-feet) can be calculated from the spacing of the electrodes, the current injected, and the voltage.

DC resistivity methods are identified according to the arrangement of current and potential electrodes, with *Wenner*, *Schlumberger*, and *dipole-dipole* arrays being the most commonly used today. Increasing the spacing between the current and potential electrodes increases the depth of the sounding measurement (in the Wenner array the spacing should be one to two times the depth of interest). *Tri-potential* DC resistivity is a relatively new method that involves taking readings from three electrode arrays (Wenner, dipole-dipole, and bipole-bipole) at each station and can allow resolution of ambiguities from single-array readings. *Azimuthal* resistivity measures the variations

---

[8] Tinlin, R.M., L.J. Hughes, and A.R. Anzzolin. 1988. The Use of Controlled Source Audio Magnetotellurics (CSAMT) to Delineate Zones of Ground-Water Contamination. In: Ground-Water Contamination: Field Methods, A.G. Collins and A.I. Johnson (eds.), ASTM STP 963, American Society for Testing and Materials, Philadelphia, PA, pp. 101-118.

Geophysical and Remote Sensing Techniques 317

in electrical response to changes in the orientation of electrode arrays at a single location and is especially useful to detection of the orientation of subsurface fractures and impermeable caps at waste disposal sites.

Figure 6-6 shows use of resistivity measurements in delineating a leachate plume from a landfill by isopleths of equal resistance measured in ohm-feet. Since landfill leachate contains ions that decrease the resistivity of the ground water, the lower value isopleths in Figure 6-6 delineate the most contaminated areas (140 ohm-feet in the upper map and 180 ohm-feet in the lower map). In the figure, the deep measurements (0 to 45 feet) include an averaging of the resistivity of the shallow measurements and the resistivity of the 15- to 45-foot depth interval.

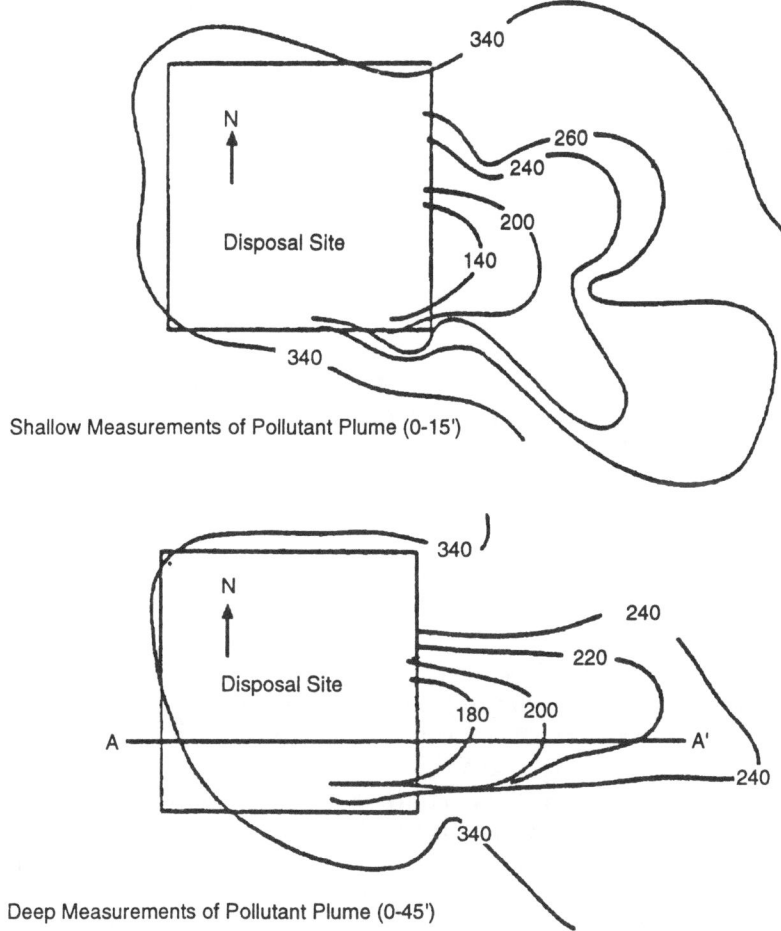

Figure 6-6 Resistivity soundings showing extent of a landfill contaminant plume (Boulding, 1993a, after Benson et al., 1984).

Conductance and resistance are reciprocals, so the output of both EM and ER methods can be expressed in terms of either units (1,000 millimhos/meter = 1 ohm-meter). The published literature on both methods sometimes uses these units interchangeably; thus EM measurements may be reported in terms of resistivity or ER measurements in terms of conductivity. The method used to measure subsurface properties (induction for EM, and current injection by electrodes for ER) will indicate the technique, not necessarily the units in which the measurements are reported. EM and ER methods are by far the most widely used surface geophysical techniques in ground-water contamination studies.

Other, less commonly used electrical methods include:

- *Self potential* uses electrodes to measure natural electrical potentials developed locally in the subsurface. Several types of natural potentials can be measured by this method. *Spontaneous polarization* is a natural voltage difference that occurs as a result of electric currents induced by disequilibria within the earth. *Streaming potential* is an electrokinetic effect related to movement of fluid containing ions through the subsurface. The methods is especially useful for detecting conduit flow in karst limestone (Section 7.5.4 and Figure 7-16) and leaks in impoundments and lined ponds. An advantage of self potential methods is that instrumentation is relatively simple and inexpensive.

- *Induced polarization* (IP) measure the electrochemical response of subsurface material (primarily clays) to an injected current. Equipment and field procedures are similar to that for DC resistivity surveys but are slower and more expensive. *Complex resistivity*, a variant of IP that uses a larger frequency spectrum, has been used with some success experimentally to detect organic contaminant plumes.

6.3.3 Seismic Refraction and Shallow Seismic Reflection

Seismic reflection techniques have been used for many years by the petroleum industry to obtain stratigraphic and structural data on deeply buried sediments. In contrast, seismic *refraction* techniques are designed to obtain data on the near surface (typically to about 30 meters). Seismic refraction provides data on the refraction of seismic waves at the interface between subsurface layers, and their travel time within the layers. Properly interpreted, the refraction data allow estimates of the thickness and depth of unconsolidated materials and bedrock and their properties. Lateral facies changes in aquifer material can also be mapped with this method.

Figure 6-7a shows a field layout for seismic refraction measurements. A seismic source (usually a hammer for near-surface investigations) creates direct compressional waves and refracted waves that are sensed by an array of geophones. The seismograph records the time of arrival of all waves, using the moment the hammer hits the ground as time zero. The processing and interpretation of seismic refraction data require a great deal of skill. Figure 6-7b shows the required steps. First, the seismic signal is recorded on paper or on magnetic tape. A single-channel seismograph plots the

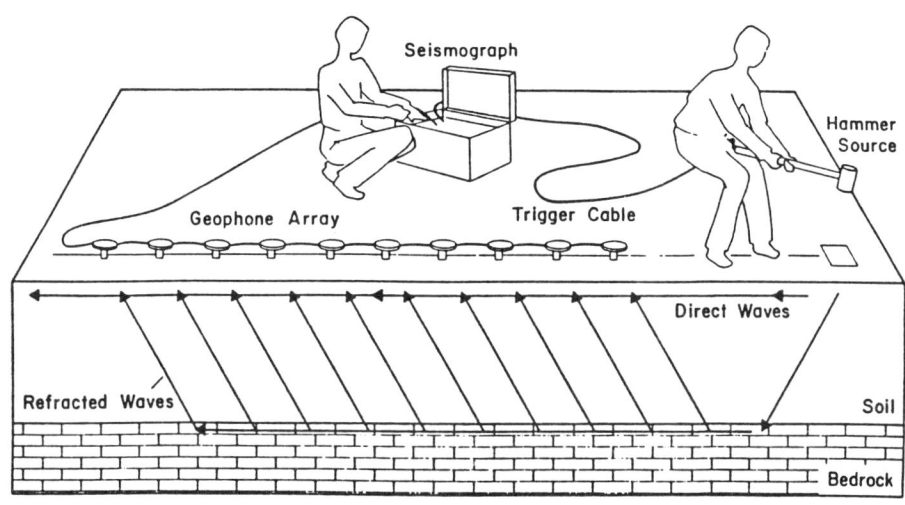

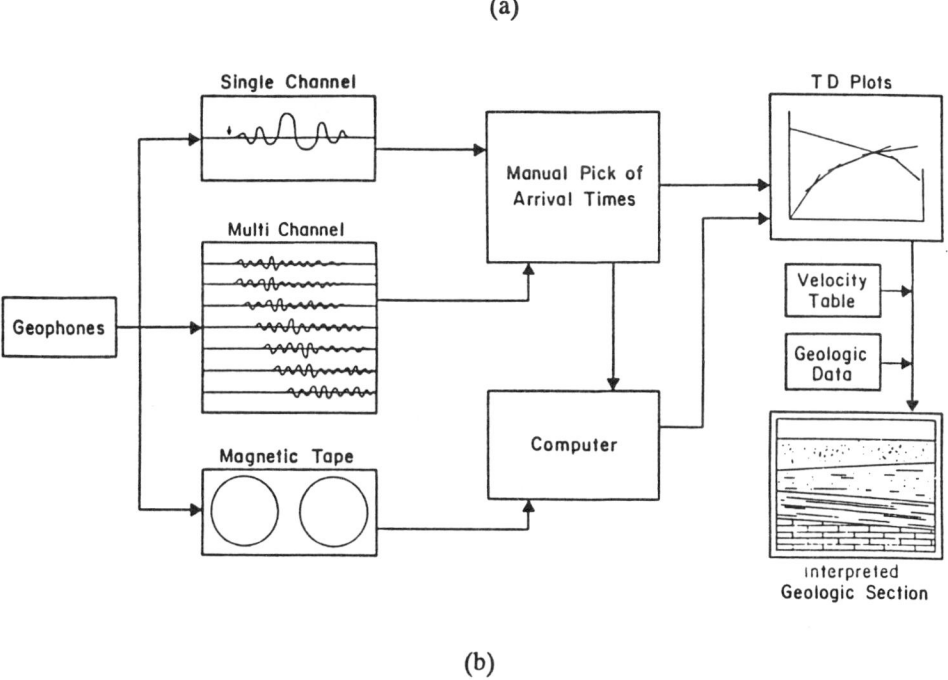

Figure 6-7 Seismic refraction: (a) field layout of 12-channel seismograph; (b) steps in processing and interpretation of seismic refraction data (Boulding, 1993a, after Benson et al., 1984).

waveform against time (milliseconds) from a single geophone, and a multichannel instrument records waveforms from multiple geophones (Figure 6-7b). Then, travel time is plotted against source-to-geophone distance to produce a time/distance (T/D) plot. Finally, line segments, slope, and break points in the T/D can be analyzed to identify the number of layers and depth of each layer.

Relatively recent advances in instrumentation for high resolution seismic *reflection* allows high resolution mapping of depth to bedrock where unconsolidated material is deeper than 10 to 30 meters, and mapping of stratigraphy and rock type at greater depths.

6.3.4 Magnetometry

Magnetic measurements have long been used to map regional geologic structures and in mineral exploration. Their main use in ground-water contamination studies is to locate buried metal drums that may be a source of contamination. A magnetometer locates ferrous metals (iron, steel, and nickel) in drums, buried pipelines, etc. by measuring local perturbations in the strength of the earth's magnetic field. Single 55-gallon drums can be sensed up to a depth of 6 meters, and piles of drums up to 20 meters.

6.3.5 Ground Penetrating Radar

Geophysical methods using the radio- and microwave portion of the electromagnetic spectrum probably have the most confusing terminology of any surface method, although the term *ground penetrating radar* has now gained fairly wide acceptance. Other names that can be found in the literature on this method include: electromagnetic subsurface profiling, electromagnetic pulse radar, pulsed microwave, pulsed radio frequency, ground piercing radar, ground probing radar, and subsurface impulse radar.

In the mid-1960s and early 1970s, the military provided the impetus for development of ground penetrating radar (GPR), primarily for detecting land mines and subsurface tunnels. Since then, GPR has been used increasingly in the mining industry and in soil and contaminant investigations to characterize depth to water table, soil horizon and lithologic contacts, cavities, faults, and bedding joints and planes in rocks, and to detect boundaries of buried trenches and containers.

GPR uses a small microwave transmission antenna to radiate high frequency radio waves into the subsurface and a receiving antenna to record variations in the reflected return signal. Dragging the antennae along the ground surface creates a continuous profile that gives the greatest resolution of all the surface geophysical methods discussed in this book. However, the depth of penetration is generally less than with other methods (1 to 15 meters) and is limited by fluids, soils with high electrical conductivity, and fine-grained materials. GPR has become a popular method for detection of trench boundaries and masses of buried waste at uncontrolled hazardous waste sites where depths of burial have not been too great.

6.3.6 Gravimetrics

Gravimetry involves measurement in variations in the intensity of the earth's gravitational field (expressed as acceleration in centimeters per second squared, or gals). Three principle classes of instruments are used in conventional gravity measurements: torsion balance, pendulum, and gravity meter or gravimeter. All can detect anomalies as small as one-ten-millionth (milligals--10^{-3} gals) of the earth's gravitational field. **Microgravimeters**, measuring in units of microgals (10^{-6} gals), are sufficiently sensitive that they can delineate cavities in the subsurface. This type of instrument usually is used in hydrogeologic and contaminated site investigations.

Station measurements along a transect or on a grid require great care in setting up the instrument, and the elevation of each station must be carefully surveyed. Gravity data obtained in the field must be corrected for elevation, rock density, latitude, earth-tide variations, and the influence of surrounding topographic variations. After corrections, measurements are plotted as Bouger anomaly maps, which look like topographic contour maps, and are interpreted in terms of the size, shape, and position of subsurface structures.

The most common use of gravity measurements for detecting bedrock valleys buried by unconsolidated glacial materials and conducting regional-scale ground-water investigations. Gravity methods are sometimes use for more site-specific investigations of contaminated sites. For example, Roberts et al. (1989)[9] obtained gravity data at a landfill in Tippecanoe County, Indiana, and compared this with gravitational estimates based on prelandfill topographic data to determine density variations within the fill material.

6.3.7 Thermal Methods

Because water has a high specific heat capacity compared to most natural materials, its temperature changes slowly as it migrates through the subsurface. Consequently, shallow-earth temperatures can be related to the occurrence and flow of ground water (Figure 6-8). In recharge areas, ground water tends to be warmer than the normal soil temperature. This effect can be used to detect contaminant plumes. For example, Cartwright and McComas (1968)[10] used soil temperature surveys at several landfills in northeastern Illinois. These surveys indicated the presence of a halo of higher temperatures around the landfills and indicated areas of surface recharge. Shallow geothermal measurements are usually made by measuring subsurface temperatures at a selected depth (up to 40 inches) at numerous stations over a short time span.

---

[9] Roberts, R.G., W.J. Hinze, and D.I. Leap. 1989. A Multi-Technique Geophysical Approach to Landfill Investigations. In: Proc. 3rd Nat. Outdoor Action Conf. on Aquifer Restoration, Ground Water Monitoring and Geophysical Methods, National Water Well Association, Dublin, OH, pp. 797-811.

[10] Cartwright, K. and M.R. McComas. 1968. Geophysical Surveys in the Vicinity of Sanitary Landfills in Northeastern Illinois. Ground Water 6(5):912-918.

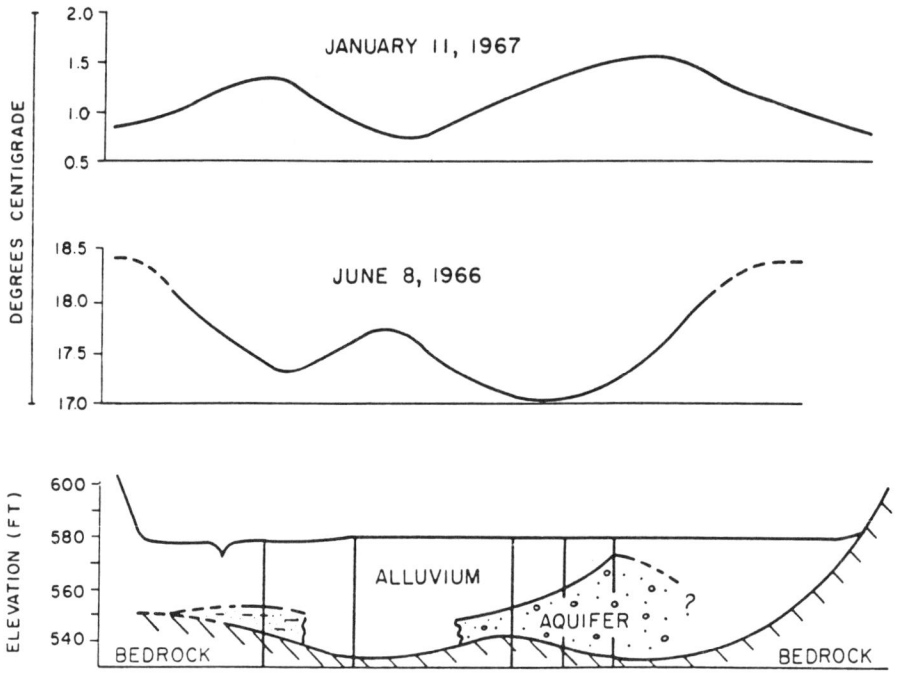

Figure 6-8 Temperature profiles of a discontinuous sand and gravel aquifer within fine-grained alluvium (Boulding, 1993b, after Cartwright, 1968).

6.4 Borehole Geophysics

Borehole geophysics is the science of recording and analyzing continuous or point measurements of physical properties made in wells or test holes. Most specific borehole geophysical techniques have long been in use by the petroleum industry, where holes being logged are usually deep and filled with drilling muds or saline water. Many of these techniques are not suitable or must be adapted for use in fresh-water aquifers, which are the focus of near-surface hydrogeological investigations. Nevertheless, suitable borehole geophysical methods can greatly enhance the geologic and hydrogeologic information obtained from water supply or monitor wells.

Rarely is a single logging method used; many logs require other logs for interpretation. Even when they are not mandatory, multiple logs may interact synergistically to provide more information than individual logs. For example, the minerals gypsum and anhydrite can be distinguished by interpreting gamma and neutron logs together. Figure 6-9 shows typical responses of three electrical logs (spontaneous potential, single-point resistance, and long-normal resistivity--see Section 6.4.1), two nuclear logs (gamma and neutron--see Section 6.4.2), and three other types of logs (acoustic velocity, caliper, and temperature--see Sections 6.4.3 and 6.4.4). In

Geophysical and Remote Sensing Techniques 323

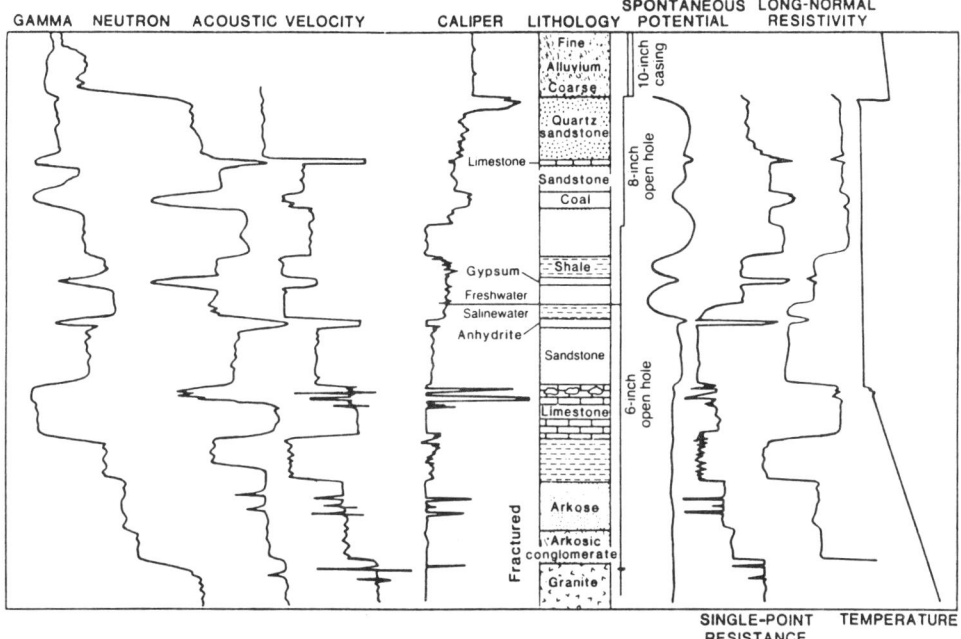

Figure 6-9 Typical response of a suite of hypothetical geophysical well logs to a sequence of sedimentary rocks (Boulding, 1993a, after Keys, 1990).

Figure 6-9, the individual logs do not always show changes with a change in lithology, but for individual strata, one or more logs show changes in measured properties at the top and bottom of the formation. As with surface geophysical methods, most downhole methods require considerable training and skill in recording and interpreting data.

A bewildering number of specific borehole logging methods are available. Table A-2 provides some summary information on more than 40. Equally confusing to the uninitiated is the fact that the same logging technique may be called by several different names. The next three sections describe briefly the major types of logging techniques that have potential for applications of hydrogeologic studies. Section 6.4.5 and Table 6-7 provide some information on specific applications of individual borehole methods.

6.4.1 Electrical and Electromagnetic Logging Methods

Electrical logging measures the flow of electric current in and adjacent to the well. Table 6-3 describes eight major types of electrical logs and their potential for hydrogeologic applications. Spontaneous potential logs, one of the most commonly used electrical logs, simply record the changes in current flow that result from changes in lithology (Figure 6-10).

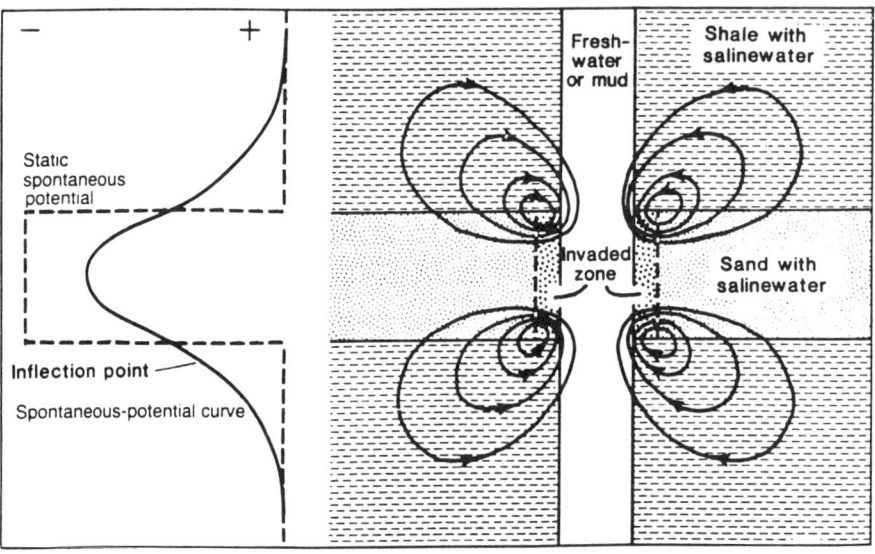

Figure 6-10 The flow of current at typical bed contacts and the resulting spontaneous potential curve (Boulding, 1993b, after Keys, 1990).

Single-point resistance, normal, focused and lateral resistivity logs all measure resistivity using the same principles as surface resistivity measurements. Resistivity logging methods have numerous variants depending on electrode configurations and spacings (see also Section 6.3.2). These logs require conductive drilling mud or ground water with high salinities to work well and, consequently, are not well suited for near-surface investigations in fresh-water aquifers. However, normal resistivity logs are widely used to measure variations in water quality.

Induction logs operate on the same principles as surface EM methods that measure conductivity (see Section 6.3.1). Since direct contact with a conductive medium is not required, they are especially useful for logging the dry portion of boreholes where the water table is deep below the surface.

6.4.2 Nuclear Logging Methods

Nuclear logging includes all methods that either detect the presence of radioisotopes or create such isotopes in the vicinity of a borehole. Table 6-4 describes six types of nuclear logs. All types are potentially useful in hydrogeologic studies of the vadose and/or saturated zones because they do not require conductive media as do most electrical logging methods. Most also allow quantitative interpretation of bulk density, porosity, salinity, and unsaturated moisture content. All of these methods are widely used in the petroleum industry, and neutron logs have been widely used for monitoring of soil moisture content. Gamma and neutron logs are probably the most commonly used in ground-water studies. Gamma spectrometry, gamma-gamma, and neutron activation have been used less frequently and should probably be considered more often.

Geophysical and Remote Sensing Techniques

Table 6-3 Summary of Electrical and EM Borehole Logging Methods in Hydrogeologic Studies

Method	Description	Hydrogeologic Applications
Electric Logs		
Fluid conductivity	A probe that records only the electrical conductivity of the borehole fluids by placing electrodes inside a protective housing.	Provides data related to the salinity (concentration of dissolved solids in the borehole fluid); used to locate sources of salt water leaking into artesian wells; aids in interpretation of electric logs.
Spontaneous potential (SP, self-potential)	Records the potentials or voltages that develop at the contacts between different lithologies.	Widely used in the petroleum industry for determining lithology, bed thickness, and salinity of formation water; generally not applicable for fresh-water aquifers.
Single-point resistance	Measures the resistance in ohms between an electrode in a well and an electrode at the land surface, or between two electrodes in a well.	Excellent for information about changes in lithology; not influenced by bed thickness effects; cannot be used for quantitative interpretation of porosity and salinity.
Normal resistivity (short normal, long normal)	Resistance is measured using four electrodes at various spacings on a single probe that is lowered down the hole.	Widely used in ground-water hydrology, primarily to determine water quality; quantitative interpretations require corrections for bed thickness, borehole diameter, and other factors.
Focused resistivity (guard log, laterolog, dual laterolog)	Uses guard electrodes above and below the current electrode to force the current to flow out into the rocks surrounding the borehole.	Designed to measure the resistivity of thin beds or resistive rocks in wells containing conductive fluids; not generally available to water well loggers.
Lateral resistivity	Similar to normal-resistivity electrode, but electrodes are more widely spaced on the probe.	Designed to measure resistivity of rock farther out from the borehole; suitable only for thick beds (>40 feet); marginal for highly resistive rocks.
Microresistivity (microlog, contact log, microsurvey, microlateral, micronormal)	Numerous variations; all have short electrode spacing and pads or some kind of contact electrode to decrease the effect of borehole fluid.	Designed mainly to determine the presence or absence of mudcake; used primarily by the petroleum industry to determine the resistivity of the 3- to 5-inch zone affected by drilling muds.
Induced polarization (IP)	Probe measures response of formation to an injected current. Requires water-filled hole.	Used to measure clay content and pore fluid chemistry and reactivity.

Table 6-3 (cont.)

Method	Description	Hydrogeologic Applications
Dipmeter	Includes a variety of wall-contact microresistivity probes; electrodes are on pads located 90 or 120 degrees apart and oriented with respect to magnetic north by a magnetometer in the probe.	Probably the best instrument for gathering information on the location and orientation of primary sedimentary structures over a wide variety of hole conditions; provides data on the strike and dip of bedding planes; also on fractures (less precise).
Hole-hole/hole-surface resistivity	Numerous configurations of source and receiver electrodes are possible.	Allows three-dimensional modeling of resistivity data to characterize subsurface inhomogeneities.
Cross-well AC voltage	A low frequency alternating current is introduced into the fracture system of 2 wells and the voltage between the currents and observation wells is measured.	Used to characterize the spatial variation in subsurface fracture systems; uncommon method.

<u>Electromagnetic Logs</u>

Method	Description	Hydrogeologic Applications
Induction (dual induction, slimhole EM probe, borehole conductivity meter)	Probe contains two coils: one for transmitting an alternating current into the surrounding rock, and a second for receiving the return signal; measures conductivity.	Most instruments require a hole with nonconductive oil-based drilling fluids or air; the recently developed EM39 induction logging tool is suitable for use in fresh-water wells.
Microwave sensing (borehole radar, dielectric log)	A variety of methods use microwaves for sensing the subsurface: **single** and **cross-borehole** radar (similar to GPR); **dielectric** log using continuous pulse microwave.	Pulsed microwave systems similar to applications for GPR (Section 6.3.5); dielectric log can be used to measure the thickness of hydrocarbons floating on ground water.
Nuclear magnetic resonance	Similar to proton precession magnetometer, except response of protons in subsurface water is measured.	Measurement of porosity, moisture content, pore-size distribution, available water. Near-surface applications most common.
Surface-borehole CSAMT	Similar to surface CSAMT (Section 6.3.1), except that borehole sensors are used.	Potential for mapping of subsurface conductive zones and three-dimensional characterization of fracture zones in deep boreholes.

Boldface = most commonly used methods.

Source: Adapted from Boulding (1993a).

Table 6-4 Summary of Nuclear Borehole Logging Methods in Hydrogeologic Studies

Method	Description	Hydrogeologic Applications
Gamma (natural gamma)	Records total natural gamma radiation (primarily from K-40, U-238, and Th-232) from a borehole that is within a selected energy range.	The most commonly used nuclear log in ground-water applications; used for identification of lithology (clay and shale particularly) and stratigraphic correlation.
Neutron	Probe contains a source of neutrons and detectors that record neutron interactions in the vicinity of the borehole.	Widely used to measure saturated porosity and moisture content in the unsaturated zone; can also be used for lithology and stratigraphic correlation.
Gamma-gamma (density)	Records the radiation at a detector from a gamma source in the probe after it is attenuated and scattered in the borehole and surrounding rock.	Primarily used to determine bulk density, porosity, and moisture content; distinguishes lithologic units; extensively used in the petroleum industry; less frequently used for ground-water applications.
Gamma spectrometry (spectra(l)-, spectro-, spectronomic-gamma)	Records the amount and energy level of gamma photons either on a continuous basis or at selected depths with a stationary probe. Types and amounts of radioisotopes can be measured.	Allows more precise identification of lithology than gamma log; permits identification of artificial radioisotopes that might be contaminating water supplies; widely used by petroleum industry; should probably be used more frequently in ground-water investigations.
Neutron activation (activation, thermal neutron)	Uses neutrons to "activate" stable isotopes in the borehole and identify the activated element by measuring the amount and energy level of emissions (see gamma spectrometry above).	Permits remote identification of elements present in the ground water and adjacent rocks; relatively new technique with potential for wide application in ground-water hydrology.
Neutron Lifetime (pulsed neutron decay)	Uses a pulsed neutron generator and a synchronously gated neutron detector to measure the rate of decrease of neutron population.	Used to measure salinity and porosity; can provide useful data through casing and cement; used by petroleum industry; to date applications in ground water have been limited.

Boldface = most commonly used methods.

Source: Adapted from Boulding (1993a).

6.4.3 Acoustic and Seismic Logging Methods

Table 6-5 provides information on three types of acoustic logs and various types of borehole seismic methods. Acoustic logging tools incorporate the signal source and the receiver on the same probe and are used in single boreholes. They are especially valuable for characterizing secondary porosity and fractures and for assisting in the interpretation of surface seismic survey results. Borehole seismic methods can use various surface-borehole or borehole-borehole source and geophone/hydrophone configurations. They are used primarily for stratigraphic, fracture, and geotechnical characterization.

6.4.4 Lithologic and Hydrogeologic Characterization Logs

Table 6-6 describes seven types of logs that may be useful for characterizing lithology and hydrogeology. *Caliper* logs have numerous variants but all are intended to measure borehole diameter. They provide essential data for interpreting other types of logs that are affected by variations in borehole diameter and also generate some data on lithology and secondary porosity. Fluid *temperature* can be measured as a gradient (also called thermal resistivity), or changes measured over time at one or more points can be tracked (as when injected water of a different temperature is used as a tracer).

Fluid flow measurements can locate zones of high permeability (fractures and solution porosity) and areas of leakage in artesian wells. The development of thermal and electromagnetic borehole flowmeters that can sense water movement either vertically or horizontally (or both) at very low velocities has greatly enhanced the ability to characterize variations in hydraulic conductivity in boreholes and allows measurement of the direction of ground-water flow in a single well. Borehole *television* cameras have the advantage of allowing visual inspection of a borehole for such things as fracture detection and monitoring well integrity.

Borehole *magnetometers* operate on the same principles as surface magnetometers (Section 6.3.4). Magnetometer probes can be especially useful when drilling is required in areas where the presence of buried ferrous metal wastes is suspected. In such situations, lowering the probe to the bottom of the hole approximately every 5 feet may provide advanced warning of the presence of buried drums that are outside the detection limit of surface instruments. Borehole *gravity* is probably the least commonly used borehole method in contaminated site and hydrogeologic applications, but may have potential value for refining interpretations of data from surface gravity surveys.

Well construction logging is useful for planning cementing operations, installing of casing and screens, performing hydraulic testing, and guiding the interpretation of other logs. The major types of well construction logs are casing logs, for locating cased intervals in wells; cement and gravel pack logs, for locating cement and gravel pack in the annular space outside a casing; and borehole deviation logs, for determining whether a well deviates from the vertical.

Table 6-5 Summary of Acoustic and Seismic Borehole Logging Methods in Hydrogeologic Studies

Method	Description	Hydrogeologic Applications
Acoustic velocity (sonic, transmit time)	Records the travel time of an acoustic wave from one or more transmitters to receivers in the probe.	Useful for providing information on lithology and porosity; limited to consolidated materials in fluid-filled boreholes; beginning to be more widely used in ground-water studies.
Acoustic waveform (variable density, three-dimensional velocity, full waveform sonic)	Received acoustic signals are recorded digitally or photographically using oscilloscope displays; the wave forms are analyzed (e.g., amplitude changes, velocity ratios).	Provides information on lithology and structure; various elastic properties can be determined; vertical compressibility of an aquifer can be estimated; fractures can be characterized. Not yet widely used in hydrogeologic studies.
Acoustic televiewer (seisviewer)	An ATV probe uses a rotating transducer that serves as both transmitter and receiver of high frequency acoustic pulses. An oscilloscope and light-sensitive paper are used to create a 360-degree scan of the borehole wall.	Provides high resolution information on the location and character of secondary porosity, such as fractures and solution openings; can also provide the strike and dip of fractures and bedding planes; not yet used extensively in ground-water studies because of cost and complexity.
Surface-borehole seismic (vertical seismic profiling/ VSP, uphole/ downhole)	Various configurations of surface and borehole geophone and seismic source arrays are possible.	**VSP**: detection of lithologic boundaries, fracture detection, estimation of permeability and hydraulic conductivity; **Uphole/Downhole**: characterization of geotechnical properties.
Crosshole seismic (crosshole shear; crosshole VSP)	Various configurations in which both seismic source and geophones are placed in boreholes.	Stratigraphy, porosity, fracture characterization, cavity detection, and measurement of soil dynamic properties.
Geophysical diffraction tomography	Tomographic imaging principles applied to seismic data. Three configurations are possible for the seismic source: borehole-borehole, surface-borehole, and surface-to boreholes.	High resolution possible; can detect isolated inclusions, lithologic boundaries, and homogeneous areas.

Boldface = most commonly used methods.

Source: Adapted from Boulding (1993a).

Table 6-6 Summary of Miscellaneous Borehole Logging Methods in Hydrogeologic Studies

Method	Description	Hydrogeologic Applications
Caliper	A probe that measures borehole diameter; many types are available: mechanical, electrical, acoustic, and one to four arms.	Provides some information on lithology and secondary porosity; essential to guide the interpretation of other types of logs that are affected by borehole diameter.
Fluid temperature	Temperature probes are used to record temperature or the rate of change in temperature vs. depth.	Widely used in ground-water studies for information on movement of natural or injected water, permeability distribution, and relative hydraulic head.
Flowmeters (mechanical/spinner log, thermal, electromagnetic)	Flow measurement with logging probes most commonly is done mechanically with an impeller flowmeter; thermal and EM flowmeters are relatively recent developments that allow more precise readings.	Used to measure vertical flow in boreholes, locate intervals of leakage in artesian wells, identify fractures producing and accepting water, and locate zones of high permeability; one of the most useful logging methods available for the study of ground water.
Single-borehole tracing	Various methods (injector-detector, injection-withdrawal, borehole dilution) measure direction and speed of water movement using tracers.	Similar to flowmeters (above).
Television/ photography	Borehole television and cameras allow visual inspection of borehole both sideways and downward.	Information on frequency, size, and orientation of fractures; vertical correlation of rock cores where voids are present; inspection of monitoring well integrity.
Magnetic	Probes operating on same principles as surface magnetometers.	Changes in lithology; check for buried ferrous metal containers in boreholes before the next depth increment is drilled.
Gravity	Microgravity instrumentation designed for borehole use.	Complements surface gravity data for structural and stratigraphic interpretation.

Boldface = most commonly used methods.

Source: Adapted from Boulding (1993a).

6.4.5 Downhole Methods in Ground-Water Contamination Studies

Table 6-7 identifies logging techniques of potential value for specific applications in the following major categories: (1) lithology, stratigraphy and formation properties; (2) aquifer properties; (3) ground-water flow and direction; (4) borehole fluid characterization; (5) contaminant characterization; and (6) borehole/casing characterization.

Downhole logging methods that measure changes in lithology are especially valuable when using drilling methods that do not recover discrete cores at specific intervals. Except where split-spoon samples or diamond rotary cores are taken, this is usually the case (Section 9.2). Downhole logs that measure variations in bulk density, porosity, and fracturing are excellent for identifying heterogeneities in the near-surface ground-water system. Geophysical logs are also useful for identifying zones of high permeability and fracturing for deep-well waste injection. Collier and Alger (1988) and Stegner and Becker (1988) provide recent overviews of the use of borehole geophysical methods in hydrogeology.[11]

6.5 Selection of Geophysical Methods

Surface geophysical techniques are most commonly used early in site investigations for preliminary characterization of the geologic and hydrogeologic setting and contaminant plumes. This information serves as a valuable guide for placement of permanent monitoring wells for ground-water sampling and monitoring. The first four major surface geophysical methods identified above are likely candidates for almost any site (ground penetrating radar will not work where conductivity is high near the surface); metal detection and magnetometry are used whenever the presence of buried drums is suspected and to avoid buried pipelines or tanks when drilling. U.S. EPA's Geophysical Advisor Expert System (Olhoeft, 1992)[12] might be useful in determining which of these techniques (plus gravity and radiometric methods) are best suited for specific site and contaminant conditions. Appendix B identifies manufacturers and distributors of surface and borehole geophysical equipment.

---

[11] Collier, H.A. and R.P. Alger. 1988. Recommendations for Obtaining Valid Data from Borehole Geophysical Logs. In: Proc. 2nd Nat. Outdoor Action Conf. on Aquifer Restoration, Ground Water Monitoring and Geophysical Methods, National Water Well Association, Dublin, OH, pp. 897-924.

Stegner, R. and R. Becker. 1988. Borehole Geophysical Methodology: Analysis and Comparison of New Technologies for Ground Water Investigation. In: Proc. 2nd Nat. Outdoor Action Conf. on Aquifer Restoration, Ground Water Monitoring and Geophysical Methods, National Water Well Association, Dublin, OH, pp. 987-1014.

[12] Olhoeft, G.R. 1992. Geophysics Advisor Expert System, Version 2.0. U.S. Geological Survey Open File Report 92-526, 21 pp. plus floppy disk. (Also available from U.S. Environmental Protection Agency Environmental Monitoring Systems Laboratory, P.O. Box 93478, Las Vegas, NV, 89193-3478; Replaces Version 1.0 [EPA/600/4-89/023], released in 1989.) [ER, EMI, complex resistivity, SRR, SRL, GPR, GR, radiometric, soil gas]

Table 6-7 Summary of Borehole Log Applications*

Required Information	Logging Techniques Which Might Be Used
Lithology, Stratigraphy, Formation Properties	
General lithology and stratigraphic correlation	Electric (SP, single-point resistance, normal and focused resistivity, dipmeter, IP, cross-well AC voltage); EM (induction, dielectric); all nuclear (open or cased holes); caliper logs made in open holes, borehole television.
Bed thickness	Single-point resistance, focused resistivity (thin beds), gamma, gamma-gamma, neutron, acoustic velocity.
Cavity detection	Caliper, acoustic televiewer, crosshole radar, crosshole seismic.
Sedimentary structure orientation	Dipmeter, borehole television, acoustic televiewer.
Large geologic structures	Gravity, surface-borehole/crosshole seismic, crosshole radar.
Total porosity/bulk density	Calibrated dielectric, sonic logs in open holes; crosshole radar; calibrated neutron, neutron lifetime, gamma-gamma logs, computer assisted tomography (CAT) in open or cased holes; nuclear magnetic resonance, induced polarization, crosshole seismic.
Effective porosity	Calibrated long-normal and focused resistivity or induction logs.
Clay or shale content	Gamma log, induction log, IP log.
Relative sand-shale content	Gamma, SP log.
Grain size/pore-size distribution	**Grain size**: Possible relation to formation factor derived from electric, induction or gamma logs; **Pore-size distribution**: Nuclear magnetic resonance; **Soil macroporosity**: Computerized axial tomography (CAT).
Compressibility/stress-strain properties	Acoustic waveform, uphole/downhole seismic, crosshole seismic
Geochemistry	Neutron activation log, spectral-gamma log.
Aquifer Properties	
Location of water level or saturated zones	Electric, induction, acoustic velocity, temperature or fluid conductivity in open hole or inside casing. Neutron or gamma-gamma logs in open hole or outside casing.
Moisture content	Calibrated neutron logs, gamma-gamma logs, nuclear magnetic resonance, computerized axial tomography (CAT).

Table 6-7 (cont.)

Required Information	Logging Techniques Which Might Be Used
Lithology, Stratigraphy, Formation Properties (cont.)	
Permeability/hydraulic conductivity	No direct measurement by logging. May be related to porosity, single-borehole tracers methods (injectivity), 2-wave sonic amplitude, temperature, nuclear magnetic resonance. Estimation may be possible using vertical seismic profiling.
Secondary permeability--fractures, solution openings	Caliper, temperature, flowmeters (mechanical, thermal, EM), sonic, acoustic waveform/televiewer, borehole television logs, SP resistance, induction logs, cross-well AC voltage, surface-borehole CSAMT, vertical seismic profiling, crosshole seismic.
Specific yield of unconfined aquifers	Calibrated neutron logs during pumping.
Ground-Water Flow and Direction	
Infiltration	Temperature logs, time-interval neutron logs under special circumstances or radioactive tracers.
Direction, velocity, and path of ground-water flow	Thermal flowmeter; single-well tracer techniques--point dilution and single-well pulse; multiwell tracer techniques.
Source and movement of water in a well	Injectivity profile; mechanical, thermal, EM flowmeters; tracer logging during pumping or injection; temperature logs.
Borehole Fluid Characterization	
Water quality/salinity	Calibrated fluid conductivity and temperature; SP log, Single-point resistance, normal/multielectrode resistivity; neutron lifetime.
Water chemistry	Dissolved oxygen, Eh, pH probes; specific ion electrodes.
Pore-fluid chemistry	Induced polarization log, neutron activation (if matrix effects can be accounted for).
Mudcake detection	Microresistivity, caliper, acoustic televiewer.
Contaminant Characterization	
Conductive plumes	Induction log, resistivity, surface-borehole CSAMT.
Contaminant chemistry	Specific ion electrodes, fiber optic chemical sensors.
Hydrocarbon detection	Dielectric log, IP log.
Radioactive contaminants	Spectral-gamma log.

Table 6-7 (cont.)

Required Information	Logging Techniques Which Might Be Used
Contaminant Characterization (cont.)	
Dispersion, dilution, and movement of waste	Fluid conductivity and temperature logs, gamma logs for some radioactive wastes, fluid sampler.
Buried object detection	Geophysical diffraction tomography.
Borehole/Casing Characterization	
Well diameter and position of casing, perforations, screens	Gamma-gamma, caliper, collar, and perforation locator, borehole television.
Guide to screen setting	All logs providing data on the lithology, water-bearing characteristics, and correlation and thickness of aquifers.
Borehole deviation	Deviation log, dipmeter, single-shot probe, dolly and cage tests.
Cementing/gravel pack	Caliper, temperature, gamma-gamma; acoustic waveform for cement bond; noise/Sonan log.
Casing corrosion/integrity	Borehole television/photography; under some conditions caliper or collar locator.
Casing detection/logging	Casing collar locator, borehole television/photography; various electric, nuclear and acoustic logs.
Casing leaks and/or plugged screen.	Tracer and flowmeters.
Behind casing flow	Neutron activation and neutron lifetime logs.

* See Table A-2 for additional information on methods that are not specifically identified in the chapter text.

Source: Boulding (1993a).

Remote sensing techniques, other than black-and-white and color photography and, less frequently, color infrared aerial photography, will have limited applications in the study of ground-water contamination, except perhaps where relatively large areas must be investigated. Borehole geophysical methods, however, have potential for use in any studies where subsurface borings are made.

The most basic requirement for successful use of geophysical methods is to select the method that is best at detecting the physical property contrasts of the target (i.e., buried waste, soil bedrock contact, conductive plume, etc.). Greenhouse and Monier-

Williams (1985)[13] identified six other considerations in the selection of geophysical methods at contaminated sites: (1) *depth limits of detection and resolution* (see Table A-1); (2) *susceptibility to noise* (electrical, electromagnetic or vibrations); (3) *corroboration* (confirmation of anomalies by multiple readings or use of more than one method); (4) *ties to borehole sampling* (i.e., confirmation of observations by drilling of monitor wells for direct observation); (5) *simplicity* (especially important if time series measurement are to be taken and there is a possibility of multiple contractors taking the measurements); and (6) *cost effectiveness*. To these considerations might be added: (7) *operator experience* (most geophysical methods require specialized training for use and interpretation of results) and (8) *equipment availability*. For example, many of the less commonly used remote sensing and surface geophysical methods would probably be used more frequently if more contractors knew how to use them and/or the equipment was more readily available.

Chapters 1 (Remote Sensing and Surface Geophysical Methods) and 3 (Geophysical Logging of Boreholes) of EPA's **Subsurface Characterization and Monitoring Techniques: A Desk Reference Guide** (Boulding, 1993)[14] are recommended for systematic information of potential applications, and comparative advantages and disadvantages different geophysical methods. A companion document, **Use of Airborne, Surface, and Borehole Geophysical Techniques at Contaminated Sites** (Boulding, 1993),[15] is recommended as a source for journal and conference paper citations that provide examples of specific applications of one or more geophysical methods of interest.

Most geophysical techniques require highly trained and experienced personnel for data collection and interpretation. When dealing with geophysical contractors, there should be a clear understanding about the services being performed. Many geophysical contractors just provide the raw geophysical data as their standard service and charge extra for interpretation of data.

6.6 Guide to Major References

Table 6-8 provides an index to major references on remote sensing, general geophysics and specific surface geophysical methods, and Table 6-9 provides an index to major references on borehole geophysics.

---

[13] Greenhouse, J.P. and M. Monier-Williams. 1985. Geophysical Monitoring of Ground Water Contamination Around Waste Disposal Sites. Ground Water Monitoring Review 5(4):63-69.

[14] Boulding, J.R. 1993. Subsurface Characterization and Monitoring Techniques: A Desk Reference Guide, Vol. I: Solids and Ground Water. EPA/625/R-93/003a. Available from CERI.

[15] Boulding, J.R. 1993. Use of Airborne, Surface, and Borehole Geophysical Techniques at Contaminated Sites: A Reference Guide. EPA/625/R-92/007. Available from CERI.

6.6.1 General Geophysics

Historically, geophysical field methods have been primarily the domain of petroleum and mineral exploration geologists, and textbooks written from this perspective remain an important source of information on basic theory and application of geophysical methods in the study of contaminated sites. Many of the references indexed in Tables 6-8 include annotations of methods covered by individual texts (abbreviations in these annotations are defined at the end of the table). Older texts can provide useful information on basic principles, and even newer texts can become rapidly outdated with respect to specific methods. Information on the latest developments in geophysical methods is most likely to appear in the exploration-oriented geophysical journals: **Geophysics**, **Geophysical Prospecting**, and **Geoexploration** (renamed **Journal of Applied Geophysics** in 1992). The expanded abstracts of the annual meeting of the Society of Exploration Geophysicists (SEG) is another important source of information on recent developments in geophysical methods.

6.6.2 Ground Water and Contaminated Sites

Zohdy et al. (1974), although a relatively old document, is still the best single report covering applications for surface geophysical methods to ground-water investigations. Haeni (1988) is a basic reference on use of seismic refraction in ground water investigations. Benson et al. (1984) is a good reference on for more detailed information on applications of commonly used surface geophysical methods at contaminated sites. Keys and MacCary (1971) and Keys (1990) are good basic references on hydrogeologic applications of borehole geophysical methods.

Information on the latest developments in application of geophysical methods in the investigation of ground-water and contaminated sites is most likely to appear in the hydrogeologic journals **Ground Water** and **Ground Water Monitoring Review** (renamed **Ground Water Monitoring and Remediation** in 1993). Other important journals include **Water Resources Research** and **Journal of Hydrology**.

The Symposium on the Application of Geophysics to Engineering and Environmental Problems (SAGEEP), sponsored by the Society of Engineering and Mineral Exploration Geophysicists (SEMEG), has been held annually since 1988 and is an exceptional source of information on hydrogeologic and contaminated site applications. Each volume of proceeding includes several applications-oriented review papers and numerous case studies. In 1992, SEMEG became the **Environmental and Engineering Geophysical Society (EEGS)**, which continues to sponsor the SAGEEP.

Another important source of information on recent developments are a number of symposium series sponsored by the National Water Well Association (NWWA) or the affiliated Association of Ground Water Scientists and Engineers (AGWSE), and the Hazardous Materials Control Research Institute (HMCRI). NWWA changed its name to the National Ground Water Association (NGWA) in 1992. Table 5-12 lists the year and title of a number of these conference/symposium series. Proceedings of the NWWA's annual **National Outdoor Action Conference (NOAC) on Aquifer**

Restoration, Ground Water Monitoring and Geophysical Methods (titled **National Symposium on Aquifer Restoration and Ground Water Monitoring** prior to 1987) generally provide the largest number of papers related to geophysical methods. The NWWA regional ground-water issues conferences typically have at least six papers related to use of geophysical methods.

The annual **Conference on Petroleum Hydrocarbons and Organic Chemicals in Ground Water—Prevention, Detection and Restoration,** sponsored jointly by NWWA and the American Petroleum Institute, is an important source for papers on developments in the use of geophysical methods for detection of hydrocarbons. Proceedings from the HMCRI's annual **Hazardous Materials Control Conference** (titled **Superfund** from 1987 to 1990 and the **National Conference on Management of Uncontrolled Hazardous Waste Sites** prior to 1987) and **National Conference on Hazardous Waste and Hazardous Materials** usually include a few papers related to geophysical methods.

6.6.3 Evaluation of Literature References

The field of geophysics in general and specific applications in ground-water and contaminated site investigations is changing so rapidly that great care is required when evaluating the literature, especially when dealing with a method that is outside one's area of expertise. Several factors affect the weight that should be given conclusions or recommendations concerning a particular method: (1) whether it is from a peer-reviewed or nonpeer-reviewed source; (2) where the authors come from; and (3) how recently it has been published.

Greatest weight should be given to the content of papers published in peer-reviewed scientific journals such as **Geophysics, Ground Water,** and **Ground Water Monitoring Review.** Most conference proceedings (ASTM conferences being an exception) are not peer-reviewed, consequently there is more likely to be diversity of opinion concerning conclusions or recommendations in individual papers. When nonpeer-reviewed papers are considered, greater weight can be given to those authored by individuals from academic institutions or research-oriented government agencies (e.g., U.S. Geological Survey, personnel from EPA research laboratories) than to papers authored by consultants who may have an interest in promoting a particular method. Finally, more recently published papers can generally be given greater weight than earlier publications because they are more likely to address recent developments and advances in geophysical techniques. As a general rule, review of multiple references from a variety of sources that deal with a specific method should help determine the method's appropriateness for a specific application or for site-specific conditions. When in doubt, one or more experts should be consulted.

Table 6-8 Index to Major References on Remote Sensing and Surface Geophysical Methods

Topic	Reference
Remote Sensing	
General	Colwell (1983), Cracknell and Hayes (1991), Dury (1990), Foody and Curran (1993), Holz (1973), Johnson and Pettersson (1987), Kondratyev (1969), Rees (1990), Reeves (1968, 1975), Regan (1980), Sabins (1978), Ulaby et al. (1982-microwave), Watson and Regan (1983); Hydrology/Geology/Contamination Applications: Boulding (1993a, 1993b), Burgy and Algaz (1974), Deutsch et al. (1979), Drury (1987), Ellyett and Pratt (1975), Goodison (1985), Lund (1978), Reeves (1968), Scherz (1971), Scherz and Stevens (1970), Sers (1971), Thomson et al. (1973)
Aerial Photography	Avery (1968), Ciciarelli (1991), Denny et al. (1968), Drury (1987), Falkner (1994), Johnson and Gnaedinger (1964-bibliography), Lattman and Ray (1965), Lueder (1959), Miller and Miller (1961), Strandberg (1967), Ray (1960), SCS (1973), Way (1973), Wolfe (1974-photogrammetry), Wright (1982)
Airborne Geophysics	AEM: Palacky (1986), Palacky and West (1991), Smith et al. (1989); Aeromag: Smith et al. (1989)
General Geophysics	
General Texts[a]	Beck (1981), d'Arnaud Gerkins (1989), Dobrin and Savit (1988), Eve and Keys (1954), Garland (1989), Grant and West (1965), Griffiths and King (1981), Hansen et al. (1967), Heiland (1940), Howell (1959), Jakosky (1950), Kearey and Brooks (1991), Milsom (1989), Nettleton (1940), Parasnis (1975, 1979), Robinson and Coruh (1988), Sharma (1986), Sheriff (1989), Telford et al. (1990), Valley (1965), Van Blaricom (1980), Ward (1990a); Glossaries: Sheriff (1968, 1991)
Ground Water	Erdélyi and Gálfi (1988), Morely (1970), NWWA (1984, 1985, 1986), Redwine et al. (1985), Rehm et al. (1985), Taylor (1984), U.S. Geological Survey (1980), Violette (1987), Ward (1990b), Zohdy et al. (1974); Bibliographies: Handman (1983), Johnson and Gnaedinger (1964), Lewis and Haeni (1987), Rehm et al. (1985), van der Leeden (1991)
Contaminated Sites	Aller (1984), API (1991), Benson et al. (1984), Boulding (1993a, 1993b), Costello (1980), EC&T et al. (1990), Frischknecht et al. (1983), HRB-Singer (1971), Lord and Koerner (1987), NWWA (1984, 1985, 1986), O'Brien & Gere (1988), Olhoeft (1992), Pitchford et al. (1988), SEMEG (1988-present), Technos (1992), U.S. EPA (1987), Van Eeckhout and Calef (1992), Waller and Davis (1984), Ward (1990b); Review Papers: Benson (1991), Evans and Schweitzer (1984), Hoekstra and Hoekstra (1990)

Geophysical and Remote Sensing Techniques 339

Table 6-8 (cont.)

Topic	Reference
General Geophysics (cont.)	
Engineering	Paillet and Saunders (1990), SEG (various dates), SEMEG (1988-present), U.S. Army Corps of Engineers (1979), Ward (1990c)
Nondestructive Testing Methods	ASTM (Annual), Lord and Koerner (1987), McGonnagle (1961), Sharp (1970)
Specific Surface Geophysical Methods	
Electrical Resistivity	Texts: Bhattacharya and Patra (1968), Goldman (1990-nonconventional methods), Keller and Frischknecht (1970), Kofoed (1979), Kunetz (1966), Mooney (1980), Patra and Mallick (1980), Roux (1978), Soiltest, (1968); Interpretation: Kalenov (1957), Mooney and Wetzel (1956), Orellana and Mooney (1966, 1972), Van Nostrand and Cook (1966), Verma (1980); Geoelectric Properties: Parkhomenko (1967), Wheatcraft et al. (1984)
Induced Polarization	Baizer and Lund (1983), Bertin and Loeb (1976), Bottcher (1952), Fink et al. (1990), Sumner (1976), Wait (1959, 1982), Wheatcraft et al. (1984)
Electromagnetics	Basic Theory: Jackson (1975), Kong (1975), Nabighian (1988), Stratton (1941), Wait (1985); EM Wave Behavior: Chew (1990), Jordon (1963), Kong (1975), Lorrain and Carson (1970), Schelnukoff (1943), Wait (1970, 1981, 1985), Ward and Morrison (1971)
EM Field Methods	EMI: Hoyt (1974), Kaufman and Keller (1983), Kraus (1984), Nabighian (1988, 1991), Rokityanksi (1982), Verma (1982-three-layer interpretation data), Wait (1971, 1982); TDEM: Felsen (1976), Goldman (1990), Kaufman and Keller (1983), Nabighian and Macnae (1991); Magnetotellurics: Kaufman and Keller (1981), Porstendorfer (1975), Vozoff (1986, 1991); VLF: McNeill and Labson (1991); CSAMT: Zonge and Hughes (1991)
Seismic Refraction	Texts: Badley (1985), Dix (1952-oil prospecting), Haeni (1988-hydrogeology), Mooney (1984), Musgrave (1967), Palmer (1986), Redpath (1973), Waters (1981); Analysis/Interpretation: Berkhout (1985, 1988), Fagin (1991), Palmer (1980), Russell (1988), Slotnick (1959), Tucker (1982), Tucker and Yorsten (1973); Wave Theory Texts: Auld (1990), Berkhkout (1987), Bland (1988), Davis (1988), White (1965); Rock Properties: Carmichael (1982)

Table 6-8 (cont.)

Topic	Reference
Continuous Seismic Profiling	<u>Texts</u>: Burdic (1991), Coates (1989), EG&G Environmental Equipment Division (1977), Hassab (1989-signal processing), Sylwester (1983), Trabant (1984); <u>Interpretation</u>: Badley (1985), Leenhart (1969), Roksandic (1978), Sangree and Widmier (1979), Tufekcic (1978)
Other Seismic	<u>Seismic Reflection</u>: Badley (1985), Danbom and Domenico (1987), Kleyn (1983), Steeples and Miller (1988), Waters (1981); <u>Seismic Shear</u>: Danbom and Domenico (1987), Dohr (1985), Ensley (1987-bibliography); <u>Acoustic Emission Monitoring</u>: U.S. EPA (1979), Waller and Davis (1984); <u>Sonar</u>: Saucier (1970)
Ground-Penetrating Radar	Hänninen and Autio (1992), Lucius et al. (1990), Olhoeft (1988-bibliography), Pilon (1992), Pittman et al. (1984), Rossiter and Bazely (1980), SCS (1988), Ulriksen (1982); <u>Subsurface Dielectric Properties</u>: Akhadov (1980), Daniel (1967), Hasted (1974), Kracchman (1970), Tareev (1975), van Beek (1965), von Hippel (1954a,b)
Magnetometry	Bozorth (1951), Breiner (1973), Chikazumi (1964), Hinze (1988), Lahee (1961), Nettleton (1971, 1976)
Gravity	Hinze (1988), Lahee (1961), Nettleton (1971, 1976), Zohdy et al. (1974)
Geothermal Methods	<u>Texts</u>: Brown et al. (1983), Eve and Keys (1954), Gougel (1976), Howell (1959), Jessup (1990), Rehm et al. (1985), Sharma (1986), Sheriff (1989), Stevens et al. (1975), Summers (1971-bibliography); <u>Soil Thermal Properties</u>: Carlslaw (1986), Farouki (1981), Kersten (1949), Lee (1965), Wechsler et al. (1965)

[a]Most texts on geophysics cover electrical, electromagnetic, seismic, magnetic, and gravity methods. Check annotations for major topics covered by texts identified at the beginning of the table.

Method Abbreviations: AEM = airborne electromagnetic, AFMAG = audiofrequency magnetic, AMT = audiomagnetotelluric, ATV = acoustic televiewer, BH = borehole, CSAMT = controlled source audiomagnetotelluric, CSP = continuous seismic profiling, EM = electromagnetic (used when not enough information available to classify further), EMI = electromagnetic induction, ER = electrical resistivity, GDT = geophysical diffraction tomography, GPR = ground penetrating radar, GR = gravity, GT = geothermal, IP/CP = induced polarization/complex resistivity, IR = infrared, MAG = magnetic, MD = metal detection, MT = Magnetotelluric, S = seismic (used when not enough information available to classify further), SASW = spectral analysis of surface waves, SLAR = side-looking airborne radar, SP = self-potential (surface and borehole), SRR = seismic refraction, SRL = seismic reflection, TC = telluric current, TDEM = time domain electromagnetic, VSP = vertical seismic profiling

Geophysical and Remote Sensing Techniques

Table 6-8 References (Appendix F contains references for figure and table sources)

Akhadov, Y. 1980. Dielectric Properties of Binary Solutions. Pergamon, New York, NY, 475 pp.

Aller, L. 1984. Methods for Determining the Location of Abandoned Wells. EPA/600/2-83/123 (NTIS PB84-141530), 130 pp. Also published in NWWA/EPA Series, National Water Well Association, Dublin, OH. [Air photos, color/thermal IR, ER, EMI, GPR, MD, MAG, combustible gas detectors]

American Petroleum Institute (API). 1991. An Evaluation of Soil Gas and Geophysical Techniques for Detection of Hydrocarbons. API Publication No. 4509. API, Washington, DC, 110 pp. [GPR, EMI, ER, complex resistivity]

American Society for Testing and Materials (ASTM). Annual. Book of ASTM Standards: Metals Test Methods and Analytical Procedures, Vol. 3.03: Nondestructive Testing. ASTM, Philadelphia, PA.

Auld, B.A. (ed.). 1990. Acoustic Fields and Waves in Solids, Vols. I and II, 2nd rev. Robert E. Krieger Publishing, Malabar, FL, (I) 435 pp., (II) 421 pp.

Avery, T.E. 1968. Interpretation of Aerial Photographs, 2nd ed. Burgess Publishing Company, Minneapolis, MN, 234 pp.

Badley, M.E. 1985. Practical Seismic Interpretation. International Human Resources Development Corporation, Boston, MA, 266 pp. [SRL, CSP]

Baizer, M.M. and H. Lund (eds.). 1983. Organic Electrochemistry, 2nd ed. Marcel Dekker, New York, NY, 1,166 pp. [IP]

Beck, A.E. 1981. Physical Principles of Exploration Methods. Macmillan, New York, NY, 234 pp. (Reprinted in 1982 with corrections.) [ER, SP, IP, GR, MAG, EMI, VLF, SRR, SRL, radiometric, BH]

Benson, R.C. 1991. Remote Sensing and Geophysical Methods for Evaluation of Subsurface Conditions. In: Practical Handbook of Ground-Water Monitoring, D.M. Nielsen (ed.), Lewis Publishers, Chelsea, MI, pp. 143-194. [GPR, EMI, TDEM, ER, SRR, SRL, GR, MAG, MD, BH]

Benson, R.C., R.A. Glaccum, and M.R. Noel. 1984. Geophysical Techniques for Sensing Buried Wastes and Waste Migration. EPA/600/7-84/064 (NTIS PB84-198449), 236 pp. Also published in NWWA/EPA series by National Water Well Association, Dublin, OH. [EMI, ER, GPR, MAG, MD, SRR]

Berkhout, A.J. 1985. Seismic Migration: Imaging of Acoustic Energy by Wave Field Extrapolation. A. Theoretical Aspects, 2nd ed.; B. Practical Aspects. Elsevier, New York, NY.

Berkhout, A.J. 1987. Applied Seismic Wave Theory. Elsevier, New York, NY, 377 pp.

(Table 6-8 Surface Geophysics References)

Berkhout, A.J. 1988. Seismic Resolution: A Quantitative Analysis of the Resolving Power of Acoustical Echo Techniques. Pergamon, New York, NY, 228 pp.

Bertin, J. and J. Loeb. 1976. Experimental and Theoretical Aspects of Induced Polarization, 2 Vols. Gebrüder Borntraeger, Berlin.

Bhattacharya, P.K. and H.P. Patra. 1968. Direct Current Geoelectric Sounding--Principles and Interpretation. Elsevier, New York, NY, 135 pp.

Bland, D.R. 1988. Wave Theory and Applications. Oxford University Press, New York, NY, 322 pp. [Seismic]

Bottcher, C.F. 1952. Electric Polarization. Elsevier, New York, NY.

Boulding, J.R. 1993a. Use of Airborne, Surface, and Borehole Geophysical Techniques at Contaminated Sites: A Reference Guide. EPA/625/R-92/007, 295 pp. Available from CERI.*

Boulding, J.R. 1993b. Subsurface Characterization and Monitoring Techniques: A Desk Reference Guide, Vol. 1: Solids and Ground Water. EPA/625/R-93/003a. Available from CERI.* [Chapter 1 covers remote sensing and surface geophysical methods and Chapter 3 covers borehole geophysical methods]

Bozorth, R.M. 1951. Ferromagnetism. Van Nostrand Co., New York, NY, 968 pp.

Breiner, S. 1973. Applications Manual for Portable Magnetometers. Geometrics, Sunnyvale, CA, 58 pp.

Brown, R.H., A.A. Konoplyantsev, J. Ineson, and V.S. Kovalensky. 1983. Ground-Water Studies: An International Guide for Research and Practice. Studies and Reports in Hydrology No. 7. UNESCO, Paris. (Originally published in 1972, with supplements added in 1973, 1975, 1977, and 1983.) [Thermal methods for evaluation of ground-water covered in Section 5.5]

Burdic, W.S. 1991. Underwater Acoustic System Analysis, 2nd ed. Prentice-Hall, Englewood Cliffs, NJ, 445 pp. (1st edition, 1984.)

Burgy, R.H. and V.R. Algaz. 1974. An Assessment of Remote Sensing Applications in Hydrologic Engineering. U.S. Army Corps of Engineers Hydrologic Center, Davis, CA, 55 pp.

Carlslaw, H.S. 1986. Conduction of Heat in Solids, 2nd edition. Oxford University Press, New York, NY, 510 pp. (First edition by Carlslaw and Jaeger, published in 1960.)

Carmichael, R.S. 1982. Handbook of Physical Properties of Rocks, Vol. 2. CRC Press, Boca Raton, FL. [Seismic]

Chew, W.C. 1990. Waves and Fields in Inhomogeneous Media. Van Nostrand Reinhold, New York, NY, 611 pp.

Chikazumi, S. 1964. The Physics of Magnetism. John Wiley & Sons, New York, NY.

(Table 6-8 Surface Geophysics References)

Ciciarelli, J. 1991. A Practical Guide to Aerial Photography. Van Nostrand Reinhold, New York, NY, 176 pp.

Coates, R.F.W. 1989. Underwater Acoustic Systems. John Wiley & Sons, New York, NY, 188 pp.

Colwell, R.N. 1983. Manual of Remote Sensing, 2nd ed. American Society of Photogrammetry, Fall Church, VA. (1st edition by Reeves, published in 1975.)

Costello, R.L. 1980. Identification and Description of Geophysical Techniques. USATHAMA Report DRXTH-TE-CR-80084. U.S. Army Toxic and Hazardous Materials Agency, Aberdeen Proving Ground, MD, 215 pp. (Superseded by EC&T et al., 1990.) [ER, GPR, SRR, BH]

Cracknell, A.P., and L.W.B. Hayes. 1991. Introduction to Remote Sensing. Taylor & Francis, Britol, PA, 300 pp.

Danbom, S.H. and S.N. Domenico (eds.). 1987. Shear-Wave Exploration. Geophysical Developments No. 1. Society of Exploration Geophysicists, Tulsa, OK, 282 pp. (Last chapter contains an annotated bibliography on shear-wave exploration seismology--See Ensley, 1987.)

Daniel, V.V. 1967. Dielectric Relaxation. Academic Press, New York, NY.

d'Arnaud Gerkins, J.C. 1989. Foundations of Exploration Geophysics. Elsevier, New York, NY, 667 pp. [SRR, SRL, GR, MAG, SP, TC, MT, IP, ER, EMI, TDEM, radiation]

Davis, J.L. 1988. Wave Propagation in Solids and Fluids. Springer Verlag, New York, NY, 400 pp.

Denny, C.S., C.R. Warren, D.H. Dow, and W.J. Dale. 1968. A Descriptive Catalog of Selected Aerial Photographs of Geologic Features of the United States. U.S. Geological Survey Professional Paper 590, 135 pp.

Deutsch, M., D.R. Wiosnet, and A. Ranjo (eds.). 1979. Satellite Hydrology (Fifth Annual William T. Pecora Memorial Symp. of Remote Sensing). American Water Resources Association, Minneapolis, MN, 730 pp.

Dix, G.H. 1952. Seismic Prospecting for Oil. Harper and Brothers, New York, NY, 213 p.

Dobrin, M.B. and C.H. Savit. 1988. Introduction to Geophysical Prospecting, 4th ed. McGraw-Hill, New York, NY, 867 pp. (Earlier editions by Dobrin: 1960, 1965, 1976.) [SRR, SRL, CSP, GR, MAG, ER, SP, IP, EMI]

Dohr, G. (ed.). 1985. Seismic Shear Waves, Part A. Theory, Part B. Applications. Handbook of Geophysical Exploration, Vols. 15A and 15B, Geophysical Press, London.

Drury, S.A. 1987. Image Interpretation in Geology. Allen and Unwin, London, UK, 243 pp.

Dury, S.A. 1990. A Guide to Remote Sensing: Interpreting Images of the Earth. Oxford University Press, New York, NY, 208 pp.

(Table 6-8 Surface Geophysics References)

EG&G Environmental Equipment Division. 1977. Fundamentals of High Resolution Seismic Profiling. TR760035. EG&G, 31 pp.

Ellyett, C.D. and D.A. Pratt. 1975. A Review of the Potential Applications of Remote Sensing Techniques to Hydrogeological Studies in Australia. Australian Water Resources Council Technical Paper No. 13, Canberra.

Ensley, R.A. 1987. Classified Bibliography of Shear-Wave Seismology. In: Shear-Wave Exploration, S.H. Danbom and S.N. Domenico (eds.), Society of Exploration Geophysicists, Tulsa, OK, pp. 255-275.

Environmental Consulting & Technology (EC&T), Inc., Technos, Inc., and UXB International, Inc. 1990. Construction Site Environmental Survey and Clearance Procedures Manual. U.S. Army Toxic and Hazardous Materials Agency, Aberdeen Proving Ground, MD. [GPR, EMI, MAG, MD, soil gas]

Erdélyi, M. and J. Gálfi. 1988. Surface and Subsurface Mapping in Hydrogeology. Wiley-Interscience, New York, NY, 384 pp. [Chapter 5 covers remote sensing and Chapter 6 geophysical methods: GPR, ER, IP, EMI, SRR, SRL, GR, MAG, geothermal]

Evans, R.B. and G.E. Schweitzer. 1984. Assessing Hazardous Waste Problems. Environ. Sci. Technol. 18(11):330A-339A. [EMI, ER, GPR, MAG, MD, SRR]

Eve, A.S. and D.A. Keys. 1954. Applied Geophysics in the Search for Minerals, 4th ed. Cambridge University Press, New York, NY, 382 pp. (Earlier editions, 1929, 1931, 1938.) [MAG, ER, EM, GR, SRL, geothermal, radiometric]

Fagin, S.W. (ed.). 1991. Seismic Modeling of Geologic Structures. Geophysical Developments No. 2. Society of Exploration Geophysicists, Tulsa, OK, 288 pp.

Falkner, E. 1994. Aerial Mapping. Lewis Publishers, Boca Raton, FL, 336 pp.

Farouki, O.T. 1981. Thermal Properties of Soils. U.S. Army Corps of Engineers Cold Regions Research and Engineering Laboratory Monograph 81-1, Hanover, NH, 151 pp.

Felsen, L.B. (ed.). 1976. Transient Electromagnetic Fields. Springer-Verlag, New York, NY, 274 pp.

Fink, J.B., et al. (eds.). 1990. Induced Polarization. Society of Exploration Geophysicists, Tulsa, OK, 424 pp.

Foody, G. and P. Curran. 1993. Environmental Remote Sensing from Regional to Global Scales. CRC Press, Boca Raton, FL, 224 pp.

Frischknecht, F.C., L. Muth, R. Grette, T. Buckley, and B. Kornegay. 1983. Geophysical Methods for Locating Abandoned Wells. U.S. Geological Survey Open-File Report 83-702, 211 pp. Also published as EPA/600/4-84/065 (NTIS PB84-212711).

(Table 6-8 Surface Geophysics References)

Garland, G.D. (ed.). 1989. Proceedings of Exploration 87. Special Volume 3, Ontario Geological Survey, Toronto, Canada, 914 pp. [77 papers covering surface, borehole, and airborne EM, IP, remote sensing, radiometric, and seismic methods]

Goldman, M.M. 1990. Non-Conventional Methods in Geoelectrical Prospecting. Prentice-Hall, New York, NY, 150 pp. [EM, TDEM, ER]

Goodison, B.E. (ed.). 1985. Hydrological Applications of Remote Sensing and Remote Data Transmission. Int. Ass. Hydrological Sciences Pub. No. 145.

Gougel, J. 1976. Geothermics. McGraw-Hill, New York, NY, 2,090 pp.

Grant, F.S. and G.F. West. 1965. Interpretation Theory in Applied Geophysics. McGraw-Hill, New York, NY, 583 pp. [ER, EM, SRL, SRR, GR, MAG, EMI]

Griffiths, D.H. and R.F. King. 1981. Applied Geophysics for Engineers and Geologists: The Elements of Geophysical Prospecting, 2nd ed. Pergamon Press, New York, NY, 230 pp. (1st edition 1965.) [ER, EM, SRR, SRL, GR, MAG]

Haeni, F.P. 1988. Application of Seismic Refraction Techniques to Hydrogeologic Studies. U.S. Geological Survey Techniques of Water-Resources Investigations TWRI 2-D2, 86 pp.

Handman, E.H. 1983. Hydrologic and Geologic Aspects of Waste Management and Disposal: A Bibliography of Publications by U.S. Geological Survey Authors. U.S. Geological Survey Circular 907, 40 pp. [15 references on geophysics]

Hänninen, P. and S. Autio (eds.). 1992. Fourth International Conference on Ground Penetrating Radar (June 8-13, 1992, Rovaniemi, Finland). Geological Survey of Finland Special Paper 16, 365 pp.

Hansen, D.A., W.E. Heinrichs, Jr., R.C. Holmer, R.E. MacDougall, G.R. Rogers, J.S. Sumner, and S.H. Ward (eds.). 1967. Mining Geophysics, Vol. II, Theory. Society of Exploration Geophysicists, Tulsa, OK, 708 pp. [EMI, ER, IP, MAG, GR]

Hassab, J.C. 1989. Underwater Signal and Data Processing. CRC Press, Boca Raton, FL, 320 pp.

Hasted, T. 1974. Aqueous Dielectrics. Chapman Hall, London.

Heiland, C.A. 1940. Geophysical Exploration. Prentice-Hall, New York, NY, 1,013 pp. (Reprinted under the same title in 1968 by Hafner Publishing, New York, NY.) [S, ER, MAG, GR]

Hinze, W.J. 1988. Gravity and Magnetic Methods Applied to Engineering and Environmental Problems. In: Proc. Symp. on the Application of Geophysics to Eng. and Environmental Problems, Soc. Eng. and Mineral Exploration Geophysicists, Golden, CO, pp. 1-108.

(Table 6-8 Surface Geophysics References)

Hoekstra, B. and P. Hoekstra. 1990. Planning and Executing Geophysical Surveys. In: Proc. Fourth Nat. Outdoor Action Conf. on Aquifer Restoration, Ground Water Monitoring and Geophysical Methods. Ground Water Management 2:1159-1166. [EMI, ER, GPR, GR, MAG, SR]

Holz, R.K. (ed.) 1973. The Surveillant Science: Remote Sensing of the Environment. Houghton Mifflin, Boston, MA, 390 pp.

Howell, B.F. 1959. Introduction to Geophysics. McGraw-Hill, New York, NY, 399 pp. [S, GR, MAG, thermal]

Hoyt, Jr., W.H. 1974. Engineering Electromagnetics, 3rd ed. McGraw-Hill, New York, NY.

HRB-Singer, Inc. 1971. Detection of Abandoned Underground Coal Mines by Geophysical Methods. Project 14010, Report EHN. Prepared for U.S. Environmental Protection Agency and Pennsylvania Dept. of Env. Res. (Cited by Lord and Koerner, 1987.) [VLF, IP, SP]

Jackson, J.D. 1975. Classical Electromagnetics. John Wiley & Sons, New York, NY.

Jakosky, J.J. 1950. Exploration Geophysics. Trija Publishing Co., Los Angeles, CA, 1,195 pp. [S, ER, MAG, GR]

Jessup, A.M. 1990. Thermal Geophysics. Elsevier, New York, NY, 306 pp.

Johnson, A.I. and J.P. Gnaedinger. 1964. Bibliography. In: Symposium on Soil Exploration, ASTM STP 351, American Society for Testing and Materials, Philadelphia, PA, pp. 137-155. [Air photo interpretation (90 references), ER and seismic (60 references), electrical borehole logging (48 references), nuclear borehole logging (40 references), borehole camera (13 references), neutron moisture measurement (50 references)]

Johnson, A.I. and C.B. Pettersson (eds.). 1987. Geotechnical Applications of Remote Sensing and Remote Data Transmission. ASTM STP 967, American Society for Testing and Materials, Philadelphia, PA.

Jordon, E.D. (ed.). 1963. Electromagnetic Wave Theory. Pergamon, New York, NY.

Kalenov, E.N. 1957. Interpretation of Vertical Electrical Sounding Curves. Gostopekhizdat, Moscow, 472 pp.

Kaufman, A.A. and G.V. Keller. 1981. The Magnetotelluric Sounding Method. Elsevier, New York, NY, 686 pp.

Kaufman, A.A. and G.V. Keller. 1983. Frequency and Transient Soundings. Elsevier, New York, NY.

Kearey, P. and M. Brooks. 1991. An Introduction to Geophysical Exploration, 2nd ed. Blackwell Scientific Publications, Boston, MA, 296 pp. (1st edition, 1984.) [SRR, SRL, GR, MAG, ER, SP, IP, EMI, VLF, AFMAG, TC, MT, AEM]

(Table 6-8 Surface Geophysics References)

Keller, G.V. and F.C. Frischknecht. 1970. Electrical Methods in Geophysical Prospecting, 2nd ed. Pergamon Press, New York, NY, 517 pp. (1st edition, 1966.)

Kersten, M.S. 1949. Thermal Properties of Soils. Univ. of Minn. Eng. Exp. Sta. Bulletin 28: 1-225.

Kleyn, A.H. 1983. Seismic Reflection Interpretation. Elsevier, New York, NY.

Kofoed, O. 1979. Geosounding Principles I, Resistivity Sounding Measurements. Elsevier, New York, NY, 276 pp.

Kondratyev, K.Y. 1969. Radiation in the Atmosphere. Academic Press, New York, NY, 912 pp. [Remote sensing]

Kong, J.A. 1975. Theory of Electromagnetic Waves. John Wiley & Sons, New York, NY, 339 pp.

Kracchman, M.B. 1970. Handbook of Electromagnetic Propagation in Conducting Media. U.S. Navy Material Command, NAVMAT P-2302, 128 pp. [Dielectric properties]

Kraus, J.D. 1984. Electromagnetics, 3rd ed. McGraw-Hill, New York, NY, 775 pp.

Kunetz, G. 1966. Principles of Direct Current Resistivity Prospecting. Geoexploration Monograph Series No. 1, Gebrüder Borntraeger, Berlin, 103 pp.

Lahee, F.H. 1961. Field Geology. McGraw-Hill, New York, NY, 926 pp. [GR, MAG, S, ER, EM]

Lattman, L.H. and R.G. Ray. 1965. Aerial Photographs in Field Geology. Holt Rinehart and Winston, New York, NY, 221 pp.

Lee, W.H. (ed.). 1965. Terrestrial Heat Flow. Geophysical Monograph Series, American Geophysical Union, Washington, DC, 276 pp.

Leenhardt, O. 1969. Analysis of Continuous Seismic Profiles. International Hydrology Review 156(1):51-80.

Lewis, M.R. and F.P. Haeni. 1987. The Use of Surface Geophysical Techniques to Detect Fractures in Bedrock--An Annotated Bibliography. U.S. Geological Survey Circular 987. [31 English language and 12 foreign language references]

Lord, Jr., A.E. and R.M. Koerner. 1987. Nondestructive Testing (NDT) Techniques to Detect Contained Subsurface Hazardous Waste. EPA/600/2-87/078 (NTIS PB88-102405), 99 pp. [17 methods; EMI, GPR, MAG, MD best]

Lorraine, P. and D.R. Carson. 1970. Electromagnetic Fields and Waves, 2nd ed. W.H. Freeman, San Francisco, CA.

(Table 6-8 Surface Geophysics References)

Lucius, J.E., G.R. Olhoeft, and S.K. Dukes (eds.). 1990. Third International Conference on Ground Penetrating Radar: Abstracts of the Technical Meeting, Lakewood, CO, May 14-18, 1990. U.S. Geological Survey Open File Report 90-414, 94 pp.

Lueder, D.R. 1959. Aerial Photographic Interpretation: Principles and Applications. McGraw-Hill, New York, NY, 462 pp.

Lund, T. 1978. Surveillance of Environmental Pollution and Resources by Electromagnetic Waves. NATO Advanced Study Institutes Series C, Vol. 45. Reidel Publishing Co., Boston, MA, 402 pp. [9 papers on land/water sensing using microwave and thermal IR]

McGonnagle, W.J. 1961. Nondestructive Testing. Gordon and Breach Publ., New York, NY.

McNeill, J.D. and V. Labson. 1991. Geological Mapping Using VLF Radio Fields. In: Electromagnetic Methods in Applied Geophysics, Vol. 2: Applications, M.N. Nabighian (ed.), Society of Exploration Geophysicists, Tulsa, OK, Part B, pp. 521-640.

Miller, V.C. and C.F. Miller. 1961. Photogeology. McGraw-Hill, New York.

Milsom, J. 1989. Field Geophysics. Halsted Press, New York, NY, 182 pp.

Mooney, H.M. 1980. Handbook of Engineering Geophysics, Vol. 2: Electrical Resistivity. Bison Instruments, Minneapolis, MN, 83 pp.

Mooney, H.M. 1984. Handbook of Engineering Geophysics, Vol. 1: Seismic, 2nd ed. Bison Instruments, Minneapolis, MN, 195 pp. (1st edition published in 1973, reprinted in 1981; 2nd edition has new chapters added, but some planned chapters were not written due to death of the author.) [SRL, SRR]

Mooney, R.M. and W.W. Wetzel. 1956. The Potentials About a Point Electrode and Apparent Resistivity Curves for a Two-, Three- and Four-Layered Earth. University of Minnesota Press, Minneapolis, MN, 145 pp.

Morely, L.W. (ed.) 1970. Mining and Groundwater Geophysics/1967. Economic Geology Report 26. Geological Survey of Canada, Ottawa, Canada. [ER, EM, SRR, BH]

Musgrave, A.W. (ed.). 1967. Seismic Refraction Prospecting. Society of Exploration Geophysicists, Tulsa, OK, 604 pp.

Nabighian, M.N. (ed.). 1988. Electromagnetic Methods in Applied Geophysics, Vol. 1: Theory. Society of Exploration Geophysicists, Tulsa, OK, 528 pp.

Nabighian, M.N. (ed.). 1991. Electromagnetic Methods in Applied Geophysics, Vol. 2, Parts A and B: Applications. Society of Exploration Geophysicists, Tulsa, OK, (A) pp. 1-520, (B) pp. 521-992.

Nabighian, M.N. and J.C. Macnae. 1991. Time Domain Electromagnetic Prospecting Methods. In: Electromagnetic Methods in Applied Geophysics, Vol. 2: Applications, Part A. M.N. Nabighian (ed.), Society of Exploration Geophysicists, Tulsa, OK, pp. 427-520.

(Table 6-8 Surface Geophysics References)

National Water Well Association (NWWA). 1984. NWWA/EPA Conference on Surface and Borehole Geophysical Methods in Ground Water Investigations (San Antonio, TX). NWWA, Dublin, OH.

National Water Well Association (NWWA). 1985. NWWA Conference on Surface and Borehole Geophysical Methods in Ground Water Investigations (Fort Worth, TX). NWWA, Dublin, OH.

National Water Well Association (NWWA). 1986. Surface and Borehole Geophysical Methods and Ground Water Instrumentation Conference and Exposition (Denver, CO). NWWA, Dublin, OH.

Nettleton, L.L. 1940. Geophysical Prospecting for Oil. McGraw-Hill, New York, NY, 444 pp. [GR, MAG, SRR, SRL, ER]

Nettleton, L.L. 1971. Gravity and Magnetics for Geologists and Seismologists. Monograph No. 1. Society of Exploration Geophysicists, Tulsa, OK, 121 pp.

Nettleton, L.L. 1976. Gravity and Magnetics in Oil Prospecting. McGraw-Hill, New York, NY, 464 pp.

O'Brien & Gere Engineering. 1988. Hazardous Waste Site Remediation: The Engineering Perspective. Van Nostrand Reinhold, New York, NY, 400 pp. [SRR, SRL, ER, EM, GPR, MAG]

Olhoeft, G.R. 1988. Selected Bibliography on Ground Penetrating Radar. In: Proc. (1st) Symp. Application of Geophysics to Eng. and Environmental Problems, Soc. Eng. and Mineral Exploration Geophysicists, Golden, CO, pp. 462-520.

Olhoeft, G.R. 1992. Geophysics Advisor Expert System, Version 2.0. U.S. Geological Survey Open File Report 92-526, 21 pp. plus floppy disk. (Also available from U.S. Environmental Protection Agency Environmental Monitoring Systems Laboratory, P.O. Box 93478, Las Vegas, NV, 89193-3478; Replaces Version 1.0 [EPA/600/4-89/023], released in 1989.) [ER, EMI, complex resistivity, SRR, SRL, GPR, GR, radiometric, soil gas]

Orellana, E. and H.M. Mooney. 1966. Master Tables and Curves for Vertical Electrical Sounding Over Layered Structures. Intersciencia, Madrid, 150 pp. (Available from: Technical Information Center, U.S. Army Corps of Engineers Waterways Experiment Station, P.O. Box 631, Vicksburg, MS, 39180.)

Orellana, E. and H.M. Mooney. 1972. Two and Three Layers Master Curves and Auxiliary Point Diagrams for Vertical Electrical Sounding Using Wenner Arrangement. Intersciencia, Madrid, 41 pp.

Paillet, F.L. and W.R. Saunders (eds.). 1990. Geophysical Applications for Geotechnical Investigations. ASTM STP 1101, American Society for Testing and Materials, Philadelphia, PA, 118 pp. [7 peer-reviewed papers on surface and borehole geophysics]

Palacky, G.J. (ed.), 1986. Airborne Resistivity Mapping. Canada Geological Survey Paper 86-22, 195 pp.

(Table 6-8 Surface Geophysics References)

Palacky, G.J. and G.F. West. 1991. Airborne Electromagnetic Methods. In: Electromagnetic Methods in Applied Geophysics, Vol. 2, Part B: Applications. M.N. Nabighian (ed.), Society of Exploration Geophysicists, Tulsa, OK, pp. 811-880.

Palmer, D. 1980. Generalized Reciprocal Method of Seismic Refraction Interpretation. Society of Exploration Geophysicists, Tulsa, OK, 104 pp.

Palmer, D. 1986. Refraction Seismics. Handbook of Geophysical Exploration, Vol. 13. Geophysical Press, London, 269 pp.

Parasnis, D.S. 1975. Mining Geophysics, 2nd ed., revised and updated. Elsevier, New York, NY, 395 pp. (2nd edition, 1973.) [MAG, SP, EMI, TDEM, TC, ER, IP, GR, SRR, SRL, radiometric, BH]

Parasnis, D.S. 1979. Principles of Applied Geophysics, 3rd ed. Chapman and Hall, New York, NY, 269+ pp. (Earlier editions, 1962, 1972.) [MAG, GR, ER, IP, EM, S]

Parkhomenko, E.I. 1967. Electrical Properties of Rocks. Plenum Press, New York, NY, 314 pp.

Patra, H.P. and K. Mallick. 1980. Geosounding Principles, Vol. 2: Time Varying Geoelectric Soundings. Elsevier, New York, NY.

Pilon, J.A. (ed.). 1992. Ground Penetrating Radar. Geological Survey of Canada Paper 90-4, 241 pp.

Pitchford, A.M., A.T. Mazzella, and K.R. Scarbrough. 1988. Soil-Gas and Geophysical Techniques for Detection of Subsurface Organic Contamination. EPA/600/4-88/019 (NTIS PB88-208194). [EMI, ER, complex resistivity, GPR, SRR, SRL, MAG, MD, AEM, radiometric]

Pittman, W.E., Jr., R.H. Church, W.E. Webb, and J.T. McLendon. 1984. Ground-Penetrating Radar: A Review of Its Application in the Mining Industry. U.S. Bureau of Mines Information Circular 8964, 23 pp.

Porstendorfer, G. 1975. Principles of Magneto-Telluric Prospecting. Gebrüder Borntraeger, Stuttgart.

Ray, R.G. 1960. Aerial Photographs in Geologic Interpretation and Mapping. U.S. Geological Survey Professional Paper 373, 320 pp.

Redpath, R. 1973. Seismic Refraction Exploration for Engineering Site Investigations. Technical Report E-73-4. (NTIS AD-768710), U.S. Army Corps of Engineers Explosive Excavation Research Laboratory, Livermore CA, 51 pp.

Redwine, J. et al. 1985. Groundwater Manual for the Electric Utility Industry, Vol. 3: Groundwater Investigations and Mitigation Techniques. EPRI CS-3901. Electric Power Research Institute, Palo Alto, CA, Chapter 3. [SRR, SRL, CSP, sonar, ER, SP, EMI, GPR, GR, BH]

(Table 6-8 Surface Geophysics References)

Rees, W.G. 1990. Physical Principles of Remote Sensing. Cambridge University Press, New York, NY 247 pp.

Reeves, R.G. 1968. Introduction to Electromagnetic Remote Sensing with Emphasis on Applications to Geology and Hydrology. AGI Short Course Lecture Notes. American Geological Institute, Washington, DC.

Reeves, R.G. (ed.). 1975. Manual of Remote Sensing. American Society of Photogrammetry, Falls Church, VA, 2144 pp. (2nd edition, Colwell, 1983.)

Regan, R.D. 1980. Remote Sensing Method. Geophysics 45(11):1685-1689.

Rehm, B.W., T.R. Stolzenburg, and D.G. Nichols. 1985. Field Measurement Methods for Hydrogeologic Investigations: A Critical Review of the Literature. EPRI EA-4301. Electric Power Research Institute, Palo Alto, CA. [Major methods: ER, EMI, SRR, BH; Other: IR, SP, IP/complex resistivity, SRR, GPR, MAG, GR, thermal]

Robinson, E.S. and C. Coruh. 1988. Basic Exploration Geophysics. John Wiley & Sons, New York, NY, 562 pp. [SRR, SRL, GR, MAG, ER, IP, SP, TC, EMI, BH]

Rokityanksi, I.I. 1982. Geoelectromagnetic Investigation of the Earth's Crust and Mantle. Springer-Verlag, New York, NY.

Roksandic, M.M. 1978. Seismic Facies Analysis Concepts. Geophysical Prospecting 26:383-398. [CSP]

Rossiter, J.R. and D.P. Bazeley (eds.). 1980. Proceedings of the International Workshop on the Remote Estimation of Sea Ice Thickness (St. John's, Newfoundland, September 25-26, 1979). C-CORE Publication No. 80-5, Centre for Cold Ocean Resources Engineering, St. John's, Newfoundland. [19 papers]

Roux, P.H. 1978. Electrical Resistivity Evaluations at Solid Waste Disposal Facilities. EPA/SW-729, U.S. Environmental Protection Agency, Office of Solid Waste, Washington, DC, 93 pp.

Russell, B.H. 1988. Introduction to Seismic Inversion Methods. Course Notes No. 2. Society of Exploration Geophysicists, Tulsa, OK, 90 pp.

Sabins, Jr., F.F. 1978. Remote Sensing: Principles and Interpretation. W.H. Freeman, San Francisco, CA, 426 pp.

Sangree, J.B. and J.M. Widmier. 1979. Interpretation of Depositional Facies from Seismic Data. Geophysics 44(2):131-160. [CSP]

Saucier, R.T. 1970. Acoustic Subbottom Profiling Survey--A State-of-the-Art Survey. Miscellaneous Paper S-69-332. U.S. Army Corps of Engineers Waterways Experiment Station, Vicksburg, MS, 71 pp.

Schelkunoff, S.A. 1943. Electromagnetic Waves. Van Nostrand, New York, NY.

(Table 6-8 Surface Geophysics References)

Scherz, J.P. 1971. Monitoring Water Pollution by Means of Remote Sensing Techniques. Remote Sensing Program Report No. 3. University of Wisconsin, Madison, WI, 27 pp.

Scherz, J.P. and A.R. Stevens. 1970. An Introduction to Remote Sensing for Environmental Monitoring. Remote Sensing Program Report No. 1, University of Wisconsin, Madison, WI, 80 pp.

Sers, S.W. 1971. Remote Sensing in Hydrology: A Survey of Applications with Selected Bibliography and Abstracts. Texas A&M University Remote Sensing Center, College Station, TX, 530 pp.

Sharma, P.V. 1986. Geophysical Methods in Geology, 2nd ed. Elsevier, New York, NY, 428 pp. (1st edition, 1976.) [S, GR, MAG, ER, GT]

Sharp, R.S. 1970. Research Techniques in Nondestructive Testing. Academic Press, New York, NY. [Thermal infrared sensing]

Sheriff, R.E. 1968. Glossary of Terms Used in Geophysical Exploration. Geophysics 33(1):181-228.

Sheriff, R.E. 1989. Geophysical Methods. Prentice-Hall, Englewood Cliffs, NJ, 605 pp. [GR, MAG, ER, EM, SRR, geothermal, radiometric, BH]

Sheriff, R.E. 1991. Encyclopedic Dictionary of Exploration Geophysics, 3rd ed. Society of Exploration Geophysicists, Tulsa, OK, 376 pp. (1st edition, 1973; 2nd edition, 1984.)

Slotnick, M.M. 1959. Lessons in Seismic Computing. Society of Exploration Geophysicists, Tulsa, OK, 268 pp. [SRR]

Smith, B., W. Heran, R. Bisdorf, and A.T. Mazzella. 1989. Evaluation of Airborne Geophysical Methods to Map Brine Contamination. EPA/600/4-89/003. U.S. Environmental Protection Agency Environmental Monitoring Systems Laboratory, Las Vegas, NV. [AEM, aeromagnetics]

Society of Engineering and Mineral Exploration Geophysicists (SEMEG). 1988-present. Symposium on the Application of Geophysics to Engineering and Environmental Problems (1st, 1988; 2nd, 1989; 3rd, 1990; 4th, 1991.) SEMEG, Golden, CO.

Society of Exploration Geophysicists (SEG). Various Dates. Annual Meeting Technical Program: Expanded Abstracts and Biographies. SEG, Tulsa, OK. (Publication for the 61st annual meeting in 1991 is a 2 volume set totaling 1707 pages.)

Soil Conservation Service (SCS). 1973. Aerial-Photo Interpretation in Classifying and Mapping Soils. U.S. Department of Agriculture Handbook 294.

Soil Conservation Service (SCS). 1988. Second International Conference on Ground Penetrating Radar (March 6-10, 1988, Gainesville, FL.) U.S. Department of Agriculture, 179 pp.

Soiltest, Inc. 1968. Earth Resistivity Manual. Soiltest, Inc., Evanston, IL, 46 pp.

(Table 6-8 Surface Geophysics References)

Geophysical and Remote Sensing Techniques

Steeples, D.W. and R.D. Miller. 1988. Seismic Reflection Methods Applied to Engineering, Environmental and Ground-Water Problems. In: Proc. (1st) Symp. on the Application of Geophysics to Engineering and Environmental Problems, Soc. Eng. and Mineral Exploration Geophysicists, Golden, CO, pp. 409-461.

Stevens, Jr., H.H., J.F. Ficke, and G.F. Smoot. 1975. Water Temperature--Influential Factors, Field Measurement, and Data Presentation. U.S. Geological Survey Techniques of Water-Resources Investigations TWRI 1-D1, 65 pp.

Strandberg, C.H. 1967. Aerial Discovery Manual. Wiley, New York.

Stratton, J.A. 1941. Electromagnetic Theory. McGraw-Hill, New York, NY.

Summers, W.K. 1971. The Annotated Indexed Bibliography of Geothermal Phenomena. New Mexico Institute of Mining and Technology, Socorro, NW. [More than 14,000 references]

Sumner, J.S. 1976. Principles of Induced Polarization for Geophysical Interpretation. Elsevier, New York, NY, 277 pp.

Sylwester, R.E. 1983. Single-Channel, High-Resolution Seismic Reflection Profiling: A Review of the Fundamentals and Instrumentation. In: Handbook of Geophysical Exploration at Sea, A.A. Geyer (ed.), CRC Press, Boca Raton, FL, pp. 77-122.

Tareev, B. 1975. Physics of Dielectric Materials. Mir, Moscow.

Taylor, R.W. 1984. Evaluation of Geophysical Surface Methods for Measuring Hydrological Variables in Fracture Rock Units. U.S. Bureau of Mines, OFR-17-84 (NTIS PB84-158021), 145 pp.

Technos, Inc. 1992. Application Guide to the Surface Geophysical Methods. Technos, Miami, FL, 19 pp. [GPR, EMI, TDEM, VLF resistivity, ER, SRR, SRL, MAG, MD, GR, thermal, radiation]

Telford, W.M.N., L.P. Geldart, R.E. Sheriff, and D.A. Keys. 1990. Applied Geophysics, 2nd ed. Cambridge University Press, New York, NY, 770 pp. [1st edition, 1976; reprinted, 1982] [GR, MAG, SRR, SRL, ER, IP, SP, MT, EMI, TDEM, AEM, radioactive, BH]

Thomson, K.P.B., R.K. Lane, and S.C. Csallany (eds.). 1973. Remote Sensing and Water Resources Management. AWRA, Proceedings Series No. 17, American Water Resources Association, Urbana, IL, 436 pp.

Trabant, P.K. 1984. Applied High-Resolution Geophysical Methods--Offshore Geoengineering Hazards. International Human Resource Development Corp., Boston, MA, 265 pp. [CSP, GPR]

Tucker, P.M. 1982. Pitfalls Revisited. Monograph No. 3. Society of Exploration Geophysicists, Tulsa, OK, 23 pp. (Expands on Tucker and Yorsten, 1973.)

Tucker, P.M. and H.J. Yorsten. 1973. Pitfalls in Seismic Interpretation. Monograph No. 2. Society of Exploration Geophysicists, Tulsa, OK, 56 pp.

(Table 6-8 Surface Geophysics References)

Tufekcic, D. 1978. A Prediction of Sedimentary Environment from Marine Seismic Data. Geophysical Prospecting 26:329-336. [CSP]

Ulaby, F.T., R.K. Moore, and A.K. Fung. 1982. Microwave Remote Sensing: 3 Vols. Addison-Wesley, Reading, MA.

Ulriksen, P.F. 1982. Application of Impulse Radar to Civil Engineering. Geophysical Survey Systems Inc., Hudson, NH. [GPR]

U.S. Army Corps of Engineers. 1979. Geophysical Exploration. Engineer Manual EM 1110-1-1802, Department of the Army, Washington, DC, 313 pp. [SRR, SRL, SASW, sonar, ER, GR, BH]

U.S. Environmental Protection Agency (EPA). 1979. Acoustic Monitoring to Determine the Integrity of Hazardous Waste Dams. EPA/625/2-79/024 (NTIS PB92-179928).

U.S. Environmental Protection Agency (EPA). 1987. A Compendium of Superfund Field Operations Methods, Part 2. EPA/540/P-87/001 (OSWER Directive 9355.0-14) (NTIS PB88-181557), 644 pp. [Remote sensing, EMI, ER, SRR, SRL, MAG, GPR, BH]

U.S. Geological Survey. 1980. Geophysical Measurements. In: National Handbook of Recommended Methods for Water Data Acquisition, Chapter 2 (Ground Water), Office of Water Data Coordination, Reston, VA, pp. 2-24 to 2-76. [TC, MT, AMT, EMI, ER, IP, SRR, GR, BH]

Valley, S.C. (ed.). 1965. Handbook of Geophysics and Space Environments. McGraw-Hill, New York, NY.

van Beek, L.K.H. 1965. Dielectric Behavior of Heterogeneous Systems. In: Progress in Dielectrics, Vol. 7, J.B. Birks (eds.), CRC Press, Boca Raton, FL, pp. 69-114.

Van Blaricom, R. 1980. Practical Geophysics for the Exploration Geologist. Northwest Mining Association, Spokane, WA, 303 pp.

van der Leeden, F. 1991. Geraghty & Miller's Groundwater Bibliography, 5th ed. Water Information Center, Plainview, NY, 507 pp.

Van Eeckhout, E. and C. Calef (compilers). 1992. Workshop on Noninvasive Geophysical Site Characterization. LA-12311-C, Los Alamos National Laboratories, Los Alamos, NM, 33 pp.

Van Nostrand, R.G. and K.L. Cook. 1966. Interpretation of Resistivity Data. U.S. Geological Survey Professional Paper 499, 310 pp.

Verma, R.K. 1980. Master Tables for Electromagnetic Depth Sounding Interpretation. Plenum, New York, NY.

Verma, R.K. 1982. Electromagnetic Sounding Interpretation Data over Three-Layer Earth, Vols. 1 and 2. IFI/Plenum, New York, NY, (1) 338 pp., (2) 546 pp.

(Table 6-8 Surface Geophysics References)

Violette, P. 1987. Surface Geophysical Techniques for Aquifer and Wellhead Protection Area Delineation. EPA/440/6-87/016 (NTIS PB88-229505). [May also be cited as U.S. EPA (1987)]

Von Hippel, A.R. 1954a. Dielectrics and Waves. MIT Press, Cambridge, MA, 284 pp.

Von Hippel, A.R. (ed.). 1954b. Dielectrics: Materials and Applications. MIT Press, Cambridge MA, 438 pp.

Vozoff, K. (ed.). 1986. Magnetotelluric Methods. Geophysics Reprint Series No. 5, Society of Exploration Geophysicists, Tulsa, OK, 800 pp.

Vozoff, K. 1991. The Magnetotelluric Method. In: Electromagnetic Methods in Applied Geophysics, Vol. 2, Part B: Applications, M.N. Nabighian (ed.), Society of Exploration Geophysicists, Tulsa, OK, pp. 641-712.

Wait, J.R. (ed.). 1959. Overvoltage Research and Geophysical Applications. Pergamon Press, New York, NY. [IP]

Wait, J.R. 1970. Electromagnetic Waves in Stratified Media, 2nd ed. Pergamon Press, New York, NY, 372 pp. (1st edition 1962.)

Wait, J.R. (ed.). 1971. Electromagnetic Probing in Geophysics. The Golem Press, Boulder, CO, 391 pp.

Wait, J.R. 1981. Wave Propagation Theory. Pergamon Press, New York, NY, 349 pp.

Wait, J.R. 1982. Geo-Electromagnetism. Academic Press, New York, NY, 268 pp. [IP, EMI]

Wait, J.R. 1985. Electromagnetic Wave Theory. Harper and Row, New York, NY, 308 pp.

Waller, M.J. and J.L. Davis. 1984. Assessment of Innovative Techniques to Detect Waste Impoundment Liner Failure. EPA/600/2-84/041 (NTIS PB84-157858), 148 pp. [28 methods assessed including ER, seismic, acoustic emission monitoring]

Ward, S.H. (ed.) 1990a. Geotechnical and Environmental Geophysics, Vol. I: Review and Tutorial. Society of Exploration Geophysicists, Tulsa, OK, 397 pp.

Ward, S.H. (ed.) 1990b. Geotechnical and Environmental Geophysics, Vol. II: Environmental and Groundwater. Society of Exploration Geophysicists, Tulsa, OK, 309 pp. [34 papers, including ER, EM multiple methods, thermal, others]

Ward, S.H. (ed.) 1990c. Geotechnical and Environmental Geophysics, Vol. III: Geotechnical. Society of Exploration Geophysicists, Tulsa, OK, 352 pp. [23 papers, including cross-borehole resistivity, seismic shear, radio imaging]

Ward, S.H. and H.F. Morrison (eds.). 1971. Special Issue on Electromagnetic Scattering. Geophysics 36(1):1-183.

Waters, K.H. 1981. Reflection Seismology--A Tool for Energy Resource Exploration, 2nd ed. John Wiley & Sons, New York, NY, 453 pp.

(Table 6-8 Surface Geophysics References)

Watson, K., and R.D. Regan (eds.). 1983. Remote Sensing. Geophysics Reprint Series No. 3. Society of Exploration Geophysicists, Tulsa, OK, 581 pp.

Way, D.S. 1973. Terrain Analysis: A Guide to Site Selection Using Aerial Photographic Interpretation. Dowden, Hutchinson, and Ross, Stroudsburg, PA, 392 pp.

Wechsler, A., P. Glaser, and R. McConnell. 1965. Methods of Laboratory and Field Measurements of Thermal Conductivity of Soils. CRREL Special Rept. 82. U.S. Army Cold Regions Res. and Eng. Lab., Hanover, NH, 31 pp.

Wheatcraft, S.W., K.C. Taylor, and J.G. Haggard. 1984. Investigation of Electrical Properties of Porous Media. EPA/600/4-84/089 (NTIS PB85-137156), 117 pp. [DC and complex resistivity]

White, J.E. 1965. Seismic Waves. McGraw-Hill, New York, NY, 200 pp.

Wolfe, P.R. 1974. Elements of Photogrammetry. McGraw Hill, New York, NY, 562 pp.

Wright, J. 1982. Ground and Air Survey for Field Scientists. Oxford University Press, New York, 340 pp.

Zohdy, A.A., G.P. Eaton, and D.R. Mabey. 1974. Application of Surface Geophysics to Ground-Water Investigations. U.S. Geological Survey Techniques of Water-Resource Investigations, TWRI 2-D1, 116 pp. [ER, GR, MAG, SRR]

Zonge, K.L. and L.J. Hughes. 1991. Controlled Source Audio-Frequency Magnetotellurics. In: Electromagnetic Methods in Applied Geophysics, Vol. 2, Applications, M.N. Nabighian (ed.), Society of Exploration Geophysicists, Tulsa, OK, Part B, pp. 713-810.

* See Preface for information on how to obtain documents from CERI (U.S. EPA Center for Environmental Research Information) and NTIS.

(Table 6-8 Surface Geophysics References)

Geophysical and Remote Sensing Techniques 357

Table 6-9 Index to Major References on Borehole Geophysics

Topic	References
Bibliographies	Prensky (various dates), Rehm et al. (1985), Taylor and Dey (1985), Johnson and Gnaedinger (1964), van der Leeden (1991)
Glossary	Society of Professional Well Log Analysts (1975)
General Texts/Reports	
Log Method Texts	Dresser Atlas (1974, 1982), Ellis (1987), Guyod and Shane (1969), Hallenberg (1983), Hamilton and Myung (1979), Hearst and Nelson (1985), Helander (1983), Kelly (1969), Labo (1987), LeRoy et al. (1987), Lynch (1962), Nelson (1985), Scott and Tibbets (1974), Serra (1984a), Telford et al. (1990), Tittman (1986)
Log Interpretation	Asquith and Gibson (1982), Birdwell Division (1973), Doveton (1986), Dresser Atlas (1975, 1979, 1982), Emerson and Webster (1970), Foster and Beaumont (1990), Hallenberg (1984), Hilchie (1982a,b), Pirson (1963, 1983), Rider (1986), Schlumberger (1972, 1974, 1989a,b, 1991), Serra (1984b), Tearpock and Bischke (1991), Wyllie (1963)
Imaging/Tomography	Borehole Imaging: Lines and Scale (1997), SPWLA (1990-borehole imaging); Tomography: Davis (1989), Desaubies et al. (1990), Stewart (1991), Lines and Scales (1987), Mahannah et al. (1988), Tweeton (1988)
Log Quality Control	Bateman (1985), Theys (1991)
Borehole Logging Symposia	Canadian Well Logging Society (various dates), Killeen (1985), Minerals and Geotechnical Logging Society (1985-91), NWWA (1984, 1985, 1986), SPWLA (1960 to present)
Specific Logging Methods	
Electrical and EM	Electrical: Guyod (1952, 1957, 1958, 1965), Guyod and Pranglin (1959), Hilchie (1979), Keller and Frischknecht (1970), Patten and Bennett (1963), Ross and Ward (1984); Focused Resistivity: Moran and Chemali (1985), Roy (1982); EM Induction: Kaufman and Keller (1989), McNeill (1986), McNeill et al. (1990); Dielectric: Keech (1988); NMR: Abragam (1961), Jackson (1984), Morrison (1983), Schlichter (1963); Dipmeter: Bigelow (1985); Crosshole Resistivity: Daily and Owen (1991), Daniels (1983); Borehole CSAMT: West and Ward (1988)

Table 6-9 (cont.)

Topic	References
Specific Logging Methods (cont.)	
Nuclear Logging	IAEA (1968, 1971), SPWLA (1978a); Neutron: Beck (1981), Bell (1973), Greacen (1981), Institute of Hydrology (1981), IAEA (1970), Johnson (1962), Morrison (1983), Olgaard (1965), SPWLA (1978a), Thompson et al. (1989), Tittle (1961), Weinberg and Wignor (1958); Gamma: Guyod (1965), Killeen (1982), Patton and Bennett (1963), SPWLA (1978a); Gamma-Gamma: Morrison (1983), SPWLA (1978a), Thompson et al. (1989); Gamma Spectrometry: Adams and Gasparini (1970), Schneider (1982); Neutron Activation: Schneider (1982); Protection: Blizard (1958), U.S. Nuclear Regulatory Commission (1985)
Acoustic/Seismic	Acoustic: Guyod and Shane (1979), Paillet and Cheng (1991), SPWLA (1978b); Crosshole: Butler and Curro (1981); VSP: Balch and Lee (1984), CH2M Hill (1991), Gal'perin (1974), Hardage (1985), Toksoz and Stewart (1984); Seismic Tomography: Mahannah et al. (1988), Stewart (1991)
Miscellaneous	Gravity: Hearst and Carlson (1982), Robbins (1986); Flowmeters: Molz et al. (1990), Taylor et al. (1990), Wheatcraft et al. (1986), Young and Pearson (1990); Borehole Television: Morahan and Dorrier (1984); Temperature: Stevens et al. (1975); Fluid Conductivity: Tellam (1992)
Applications	
Ground Water	Bennett and Patten (1960), Emerson and Webster (1970), Hodges and Teasdale (1991), IAEA (1971), Johnson (1968), Jorgenson (1989), Keys (1990), Keys and MacCary (1971), Patten and Bennett (1963), Respold (1989), Taylor and Dey (1985), Welenco (1994); Other Texts with Sections on Borehole Geophysics: Brown et al. (1983), Beesley (1986), Bureau of Reclamation (1981), Campbell and Lehr (1973), Davis and DeWiest (1966), Driscoll (1986), Everett (1985), Redwine et al. (1985), Rehm et al. (1985), U.S. Army Corps of Engineers (1979)
Other Applications	Contaminated Sites: Adams et al. (1983), Benson (1991), Boulding (1993a, 1993b), Stowell (1989), Taylor et al. (1990), Technos (1992), U.S. EPA (1987), Wheatcraft et al. (1986); Fracture/Flow Characterization: Nelson (1985), Robbins and Hayden (1988), Ross and Ward (1984), Serra (1984b), Taylor (1989), Tellam (1992), West and Ward (1988), Williams and Conger (1990), Young and Pearson (1990); Well Construction/Integrity: Gearhart Industries (1982), Nielsen and Aller (1984), Thornhill and Benefield (1990), Yearsley et al. (1991)

Table 6-9 References (Appendix F contains references for figure and table sources)

Abragam, A. 1961. The Principles of Nuclear Magnetism. Clarendon Press, Oxford, England, 599 pp.

Adams, J.A.S. and P. Gasparini. 1970. Methods in Geochemistry and Geophysics: Gamma Ray Spectrometry of Rocks. Elsevier, New York, NY, 280 pp.

Adams, W.M., S.W. Wheatcraft, and J.W. Hess. 1983. Downhole Sensing Equipment for Hazardous Waste Site Investigations. In: Proc. (4th) Nat. Conf. on Management of Uncontrolled Hazardous Waste Sites, Hazardous Materials Control Research Institute, Silver Spring, MD, pp. 108-113.

Asquith, G. and C. Gibson. 1982. Basic Well Log Analysis for Geologists. American Association of Petroleum Geologists, Tulsa, OK, 216 pp.

Balch, A.H. and M.W. Lee (eds.). 1984. Vertical Seismic Profiling: Techniques, Applications, and Case Histories. International Human Resource Development Corporation, Boston, MA, 488 pp.

Bateman, R.M. 1985. Log Quality Control. Boston International Human Resources Development Corporation, Boston, MA, 398 pp.

Beck, A.E. 1981. Physical Principles of Exploration Methods. Macmillan, New York, NY, 234 pp. (Reprinted in 1982 with corrections). [Neutron probe]

Beesley, K. 1986. Downhole Geophysics. In: Ground Water: Occurrence, Development and Protection, T.W. Brandon (ed.), Institute of Water Engineers and Scientists Water Practice Manual 5, London, Chapter 9.

Bell, J.P. 1973. Neutron Probe Practices. Institute of Hydrology Report No. 19, Wallingford, Oxon, U.K.

Bennett, G.D. and E.P. Patten, Jr. 1960. Borehole Geophysical Methods for Analyzing Specific Capacity of Multiaquifer Wells. U.S. Geological Survey Water Supply Paper 1536-A.

Benson, R.C. 1991. Remote Sensing and Geophysical Methods for Evaluation of Subsurface Conditions. In: Practical Handbook of Ground-Water Monitoring, D.M. Nielsen (ed.), Lewis Publishers, Chelsea, MI, pp. 143-194.

Bigelow, E.L. 1985. Making More Intelligent Use of Log Derived Dip Information, Parts I-V. Log Analyst 26(1):41-51; 26(2):25-41; 26(3):18-31; 26(4):21-43; 26(5):25-64.

Birdwell Division. 1973. Geophysical Well Log Interpretation. Birdwell Division, Seismograph Service Corporation, Tulsa, OK. (Birdwell Division is no longer in operation.) [SP, resistivity, gamma, gamma-gamma, neutron, fluid conductivity, temperature, 3-D velocity].

Blizard, E.P. 1958. Nuclear Radiation Shielding. In: Nuclear Engineering, H. Etherington (ed.), McGraw-Hill, New York, NY.

(Table 6-9 Borehole Geophysics References)

Boulding, J.R. 1993a. Use of Airborne, Surface, and Borehole Geophysical Techniques at Contaminated Sites: A Reference Guide. EPA/625/R-92/007, 295 pp. Available from CERI.*

Boulding, J.R. 1993b. Subsurface Characterization and Monitoring Techniques: A Desk Reference Guide, Vol. 1: Solids and Ground Water. EPA/625/R-93/003a, 488 pp. Available from CERI.* [Chapter 1 covers remote sensing and surface geophysical methods and Chapter 3 covers borehole geophysical methods]

Brown, R.H., A.A. Konoplyantsev, J. Ineson, and V.S. Kovalensky. 1983. Ground-Water Studies: An International Guide for Research and Practice. Studies and Reports in Hydrology No. 7. UNESCO, Paris. (Originally published in 1972, with supplements added in 1973, 1975, 1977, and 1983.) [Section 9 covers borehole geophysical techniques]

Bureau of Reclamation. 1981. Ground Water Manual--A Water Resources Technical Publication, 2nd ed. U.S. Department of the Interior, Bureau of Reclamation, Denver, CO, 480 pp. [Chapter 8 covers borehole geophysical methods]

Butler, D.K. and J.R. Curro, Jr. 1981. Crosshole Seismic Testing--Procedures and Pitfalls. Geophysics 46(1):23-29.

Campbell, M.D. and J.H. Lehr. 1973. Water Well Technology. McGraw-Hill, New York, NY, 681 pp. [Annotated bibliography contains over 600 references]

Canadian Well Logging Society. (Various dates). Biannual Formation Evaluation Symposium Series. Canadian Well Logging Society, Calgary. (Published symposia include: 2nd [1968], 6th [1977], 7th [1979], 8th [1981], 9th [1983], 11th [1987], 12th [1989], and 13th [1991].)

CH2M Hill. 1991. Proceedings: NSF/EPRI Workshop on Dynamic Soil Properties and Site Characterization, Vol 1. EPRI NP-7337. Electric Power Research Institute, Palo Alto, CA. [Chapter 3 (Low- and High-Strain Cyclic Material Properties) covers uphole-downhole seismic methods]

Daily, W. and E. Owen. 1991. Cross-Borehole Resistivity Tomography. Geophysics 56(8):1228-1235.

Daniels, J.J. 1983. Hole-to-Surface Resistivity Measurements. Geophysics 48(1):87-97.

Davis, R.W. 1989. Developments in Cross Borehole Tomography. In: Proc. (2nd) Symp. on the Application of Geophysics to Engineering and Environmental Problems, Soc. Eng. and Mineral Exploration Geophysicists, Golden, CO, pp. 262-274.

Davis, S.N. and R.J.M. DeWiest. 1966. Hydrogeology. John Wiley & Sons, New York, NY, 463 pp. [Chapter 8 covers surface and borehole geophysical methods]

Desaubies, Y., A. Tarantola, and J. Zinn-Justin (eds.). 1990. Oceanographic and Geophysical Tomography. Elsevier, New York, NY, 463 pp.

Doveton, S.H. 1986. Log Analysis of Subsurface Geology: Concepts and Computer Models. John Wiley & Sons, New York, NY, 273 pp.

Dresser Atlas. 1974. Log Review 1. Dresser Atlas Division, Dresser Industries, Houston, TX. [Induction, resistivity, acoustic velocity, gamma-gamma, neutron-gamma, diplog, neutron lifetime]

Dresser Atlas. 1975. Log Interpretation Fundamentals. Dresser Atlas Division, Dresser Industries, Houston, TX, 125 pp.

Dresser Atlas. 1979. Log Interpretation Charts. Dresser Atlas Division, Dresser Industries, Houston, TX.

Dresser Atlas. 1982. Well Logging and Interpretation Techniques: The Course for Home Study. Dresser Atlas Division, Dresser Industries, Houston, TX, 350 pp.

Driscoll. F.G. 1986. Groundwater and Wells, 2nd ed. Johnson Filtration Systems Inc., St. Paul, MN, 1089 pp. [Chapter 8 covers borehole geophysical methods: resistivity, SP, gamma, gamma-gamma, neutron, acoustic, temperature, caliper and fluid velocity]

Ellis, D.V. 1987. Well Logging for Earth Scientists. Elsevier, New York, NY, 532 pp. [SP, resistivity, induction, gamma, neutron, acoustic]

Emerson, D.W. and S.S. Webster. 1970. Interpretation of Geophysical Logs in Bores in Unconsolidated Sediments. Australian Water Resources Council Research Project 68/7-Phase I, 212 pp.

Everett, L.G. 1985. Groundwater Monitoring Handbook for Coal and Oil Shale Development. Elsevier, New York, NY. [Section 8 covers borehole geophysical methods: temperature, caliper, gamma, flow, radioactive tracer, 3-D velocity (acoustic waveform), acoustic, gamma-gamma, electric, acoustic-televiewer]

Foster, N.H. and E.A. Beaumont (eds.). 1990. Formation Evaluation I: Log Evaluation; II: Log Interpretation. Reprint Series Nos. 16 and 17, American Association of Petroleum Geologists, Tulsa, OK, (I) 742 pp., (II) 600 pp. [Resistivity, SP, gamma, porosity, dip meter, other logs]

Gal'perin, E.I. 1974. Vertical Seismic Profiling. Society of Exploration Geophysicists, Tulsa, OK, 278 pp.

Gearhart Industries. 1982. Basic Cement Bond Log Evaluation. Gearhart Industries, Inc., Charleston, WV, 36 pp.

Greacen, E.L. (ed.). 1981. Soil Water Assessments by the Neutron Method. ISBN 0-643-004 14-9. CSIRO, East Melbourne.

Guyod, H. 1952. Electrical Well Logging Fundamentals. Well Instruments Developing Co., Houston, TX, 164 pp.

Guyod, H. 1957. Resistivity Determination from Electric Logs. (Published by) Hubert Guyod, Houston, TX.

(Table 6-9 Borehole Geophysics References)

Guyod, H. 1958. Electric Analogue for Resistivity Logging. (Published by) Hubert Guyod, Houston, TX.

Guyod, H. 1965. Interpretation of Electric and Gamma Ray Logs in Water Wells. Am. Geophysical Union Technical Paper. Mandrel Industries, Inc. Houston, TX.

Guyod, H. and J.A. Pranglin. 1959. Analysis Charts for the Determination of True Resistivity from Electric Logs. (Published by) Hubert Guyod, Houston TX, 202 pp.

Guyod, H. and L.E. Shane. 1969. Geophysical Well Logging, Vol. I: Introduction to Geophysical Well Logging and Acoustical Logging. (Published by) Hubert Guyod, Houston TX, 256 pp. [Part I covers general well logging equipment principles, Part II covers acoustical logging]

Hallenberg, J.K. 1983. Geophysical Logging for Mineral and Engineering Applications. Penn Well Books, 264 pp.

Hallenberg, J.K. 1984. Formation Evaluation Programs. Penn Well Books, 120 pp.

Hamilton, R.G. and J.I. Myung. 1979. Summary of Geophysical Well Logging. Birdwell Division, Seismograph Service Corporation, Tulsa, OK, 32 pp.

Hardage, B.A. 1985. Vertical Seismic Profiling, Part A: Principles, 2nd enlarged ed. Seismic Exploration Volume 14A, Geophysical Press, London, 450 pp. (1st edition 1982.)

Hearst, J.R. and R.C. Carlson. 1982. Measurement and Analysis of Gravity in Boreholes. Developments in Geophysical Exploration Methods 4:269-303.

Hearst, J.R. and P.H. Nelson. 1985. Well Logging for Physical Properties. McGraw-Hill, New York, NY, 571 pp.

Helander, D.P. 1983. Fundamentals of Formation Evaluation. Oil & Gas Consultants International Publications, Tulsa, OK, 332 pp. [SP, resistivity, acoustic, radiation]

Hilchie, D.W. 1979. Old (Pre-1958) Electrical Log Interpretation. Institutes for Energy Development, Tulsa, OK.

Hilchie, D.W. 1982a. Applied Open Hole Log Interpretation for Geologists and Engineers, 2nd ed. D.W. Hilchie, Inc., Golden, CO, 400+ pp. [SP, induction, acoustic, gamma, gamma-gamma, neutron, dipmeter]

Hilchie, D.W. 1982b. Advanced Well Log Interpretation. D.W. Hilchie, Inc., Golden, CO, 353 pp.

Hodges, R.E. and W.E. Teasdale. 1991. Considerations Related to Drilling Methods in Planning and Performing Borehole-Geophysical Logging for Ground-Water Studies. U.S. Geological Survey Water Resource Investigations Report 91-409 (NTIS PB92-155688), 22 pp. [Caliper, gamma, gamma-spectral, gamma-gamma, neutron, electric, acoustic velocity, acoustic televiewer, temperature, flowmeters]

(Table 6-9 Borehole Geophysics References)

Institute of Hydrology. 1981. User's Handbook for the Institute of Hydrology Neutron Probe System. Institute of Hydrology Report No. 79, Wallingford, Oxon, U.K.

International Atomic Energy Agency (IAEA). 1968. Guidebook of Nuclear Techniques in Hydrology. Technical Report No. 91. IAEA, Vienna.

International Atomic Energy Agency (IAEA). 1970. Neutron Moisture Gauges. Technical Report No. 112. IAEA, Vienna.

International Atomic Energy Agency (IAEA). 1971. Nuclear Well Logging in Hydrology. Technical Report No. 126. IAEA, Vienna, 92 pp.

Jackson, J.A. 1984. Nuclear Magnetic Resonance Well Logging. The Log Analyst 25(5):16-30.

Johnson, A.I. 1962. Methods of Measuring Soil Moisture in the Field. U.S. Geological Survey Water-Supply Paper 1919-U, 25 pp. [Neutron probe]

Johnson, A.I. 1968. An Outline of Geophysical Logging Methods and their Uses in Hydrogeological Studies. U.S. Geological Survey Water-Supply Paper 1892, pp. 158-164.

Johnson, A.I. and J.P. Gnaedinger. 1964. Bibliography. In: Symposium on Soil Exploration, ASTM STP 351, American Society for Testing and Materials, Philadelphia, PA, pp. 137-155. [Electrical borehole logging (48 refs); nuclear borehole logging (40 refs), borehole camera (13 refs); neutron moisture measurement (50 refs)]

Jorgensen, D. 1989. Using Geophysical Logs to Estimate Porosity, Water Resistivity, and Intrinsic Permeability. U.S. Geological Survey Water-Supply Paper 2321, 24 pp.

Kaufman, A.A. and G.V. Keller. 1989. Induction Logging. Elsevier, New York, NY.

Keech, D.A. 1988. Hydrocarbon Thickness on Groundwater by Dielectric Well Logging. In: Proc. (5th) NWWA/API Conf. on Petroleum Hydrocarbons and Organic Chemicals in Ground Water: Prevention, Detection and Restoration, National Water Well Association, Dublin, OH, pp. 275-289.

Keller, G.V. and F.C. Frischknecht. 1970. Electrical Methods in Geophysical Prospecting, 2nd ed. Pergamon Press, New York, NY, 517 pp. (First edition 1966.)

Kelly, D.R. 1969. A Summary of Geophysical Logging Methods. Pennsylvania Geological Survey Bulletin M61, 88 pp.

Keys, W.S. 1990. Borehole Geophysics Applied to Ground-Water Investigations. U.S. Geological Survey Techniques of Water-Resource Investigations TWRI 2-E2, 150 pp. [Supersedes report originally published in 1988 under the same title as U.S. Geological Survey Open-File Report 87-539, 303 pp., which was published in 1989 with the same title by the National Water Well Association, Dublin, OH, 313 pp. Complements Keys and MacCary (1971)]

(Table 6-9 Borehole Geophysics References)

Keys, W.S. and L.M. MacCary. 1971. Application of Borehole Geophysics to Water Resource Investigations. TWRI 2-E1. U.S. Geological Survey Techniques of Water-Resources Investigations, 126 pp. (Reprinted, 1990; see, also Keys, 1990.)

Killeen, P.G. 1982. Gamma-Ray Logging and Interpretation. Developments in Geophysical Exploration Methods 3:95-150.

Killeen, P.G. (ed.). 1985. Borehole Geophysics for Mining and Geotechnical Applications. Geological Survey of Canada Paper 85-27.

Labo, J. 1987. A Practical Introduction to Borehole Geophysics. Society of Exploration Geophysicists, Tulsa, OK, 336 pp. [Gamma-gamma, gravity, acoustic, VSP, dipmeter]

LeRoy, L.W., D.O. LeRoy, S.D. Schwochow, and J.W. Raese (eds.). 1987. Subsurface Geology, 5th ed. Colorado School of Mines, Golden, CO. (1st edition: LeRoy and Cran [1947], 2nd edition: LeRoy [1951], 3rd edition: Huan and LeRoy [1958], and 4th edition [1977].)

Lines, L.R. and J.A. Scales (eds.) 1987. Geophysical Imaging. Symposium of the Geophysical Society of Tulsa, available from Society of Exploration Geophysicists, Tulsa, OK, 225 pp. [Tomography, inversion, migration, computer-related issues]

Lynch, E.J. 1962. Formation Evaluation. Harper and Row, New York, NY, 422 pp.

Mahannah, J.L., A.J. Witten, and W.C. King. 1988. Use of Geophysical Diffraction Tomography for Hazardous Waste Site Characterization. In: Superfund '88, Hazardous Materials Control and Research Institute, Silver Spring, MD, pp. 152-156.

McNeill, J.D. 1986. Geonics EM39 Borehole Conductivity Meter--Theory of Operations. Technical Note 20, Geonics Ltd., Mississauga, Ontario, 11 pp.

McNeill, J.D., M. Bosnar, and F.B. Snelgrove. 1990. Resolution of an Electromagnetic Borehole Conductivity Logger for Geotechnical and Ground Water Applications. Technical Note TN-25. Geonics Limited, Mississauga, Ontario.

Minerals and Geotechnical Logging Society Symposia Series. 1985-1991. Proc. 1st Int. Symp. Borehole Geophysics for Minerals, Geotechnical and Groundwater Applications (Ottawa, 1985); 2nd (Golden, CO, 1987); 3rd (Las Vegas, NV, 1989); 4th (Ontario, 1991). Available from SPWLA.

Molz, F.J., O. Güven, and J.G. Melville. 1990. A New Approach and Methodologies for Characterizing the Hydrogeologic Properties of Aquifers. EPA/600/2-90/002 (NTIS PB90-187063).

Morahan, T. and R. C. Dorrier. 1984. The Application of Television Borehole Logging to Ground Water Monitoring Programs. Ground Water Monitoring Review 4(4):172-175.

Moran, J.H. and R. Chemali. 1985. Focused Resistivity Logs. Developments in Geophysical Exploration Methods 6:225-260.

(Table 6-9 Borehole Geophysics References)

Morrison, R.D. 1983. Groundwater Monitoring Technology. Timco Mfg., Inc., Prairie du Sac, WI, 105 pp. [Gamma-gamma, neutron probe, nuclear magnetic resonance]

National Water Well Association (NWWA). 1984. NWWA/EPA Conference on Surface and Borehole Geophysical Methods in Ground Water Investigations (San Antonio, TX). NWWA, Dublin, OH.

National Water Well Association (NWWA). 1985. NWWA Conference on Surface and Borehole Geophysical Methods in Ground Water Investigations (Fort Worth, TX). NWWA, Dublin, OH.

National Water Well Association (NWWA). 1986. Surface and Borehole Geophysical Methods and Ground Water Instrumentation Conference and Exposition (Denver, CO). NWWA, Dublin, OH.

Nelson, R.A. 1985. Geologic Analysis of Naturally Fractured Reservoirs. Contributions in Petroleum Geology and Engineering, Vol. 1. Gulf Publishing, Houston, TX, 320 pp.

Nielsen, D.M. and L. Aller. 1984. Methods for Determining the Mechanical Integrity of Class II Injection Wells. EPA/600/2-84/121 (NTIS PB84-215755), 263 pp. Also published in NWWA/EPA Series by the National Water Well Association, Dublin, OH. [Temperature, noise log, EM thickness, flowmeter, radioactive tracers, cement bond]

Olgaard, P.L. 1965. On the Theory of the Neutronic Method for Measuring the Water Content of Soil. Riso Rep. 97. Danish Atomic Energy Commission, Roskilde, Denmark, 74 pp.

Paillet, F.L. and C.H. Cheng. 1991. Acoustic Waves in Boreholes. CRC Press, Boca Raton, FL, 264 pp.

Patten, Jr., E.P. and G.D. Bennett. 1963. Application of Electrical and Radioactive Well Logging to Groundwater Hydrology. U.S. Geological Survey Water-Supply Paper 1544-D, 60 pp. [Resistivity, SP, fluid conductivity, gamma]

Pirson, S.J. 1963. Handbook of Well Log Analysis for Oil and Gas Formation Evaluation. Prentice-Hall, Englewood Cliffs, NJ, 326 pp.

Pirson, S.J. 1983. Geologic Well Log Analysis, 3rd ed. Gulf Publishing Co., Houston, TX. (Earlier editions 1970, 1977.) [SP, Eh, dipmeter]

Prensky, S.E. Various Dates. Log Analyst Geologic Applications Bibliographies. Geological Applications of Well Logs--An Introductory Bibliography and Survey of Well Logging Literature through September 1986, Arranged by Subject and First Author (Log Analyst, 1987: Parts A and B 28(1):71-107; Part C 28(2):219-248); Annual Update, October 1986 through September 1987 (Log Analyst, 1987: 28(6):558-575); Bibliographic Update for October 1987 through September 1988 (Log Analyst, 1988: 29(6):426-443); Bibliography of Well Log Applications: October 1988-September 1989 Annual Update (Log Analyst, 1989: 30(6):448-470); October 1989-September 1990 Annual Update (Log Analyst, 1990:31(6):395-424).

(Table 6-9 Borehole Geophysics References)

Redwine, J. et al. 1985. Groundwater Manual for the Electric Utility Industry, Vol. 3: Groundwater Investigations and Mitigation Techniques. EPRI CS-3901. Electric Power Research Institute, Palo Alto, CA. [Section 3 covers surface and borehole geophysical methods]

Rehm, B.W., T.R. Stolzenburg, and D.G. Nichols. 1985. Field Measurement Methods for Hydrogeologic Investigations: A Critical Review of the Literature. EPRI EA-4301. Electric Power Research Institute, Palo Alto, CA. [Section 5 covers electrical, nuclear, acoustic and flow logs]

Respold, H. 1989. Well Logging in Groundwater Development. International Contributions to Hydrogeology, Vol. 9. International Association of Hydrogeologists, Verlag Heinz Heise, Hannover, West Germany, 147 pp.

Rider, M.H. 1986. The Geological Interpretation of Well Logs. Halstead Press, New York, NY, 175 pp. [SP, resistivity, induction, gamma, spectral gamma, gamma-gamma, neutron, acoustic]

Robbins, S.L. 1986. The Use of Borehole Gravimetry in Water Well and Waste Disposal Site Evaluations. In: Proc. Surface and Borehole Geophysical Methods and Ground Water Instrumentation Conf. and Exp., National Water Well Association, Dublin, OH, pp. 474-496.

Robbins, G.A. and J.M. Hayden. 1988. Application of Cross-Well Voltage Measurement for Assessing Fracture Flow Hydrology. In: Proc. of the Focus Conf. on Eastern Regional Ground Water Issues (Stanford, CT), National Water Well Association, Dublin, OH, pp. 28-38.

Ross, H.P. and S.H. Ward. 1984. Borehole Electrical Geophysical Methods: A Review of the State-of-the-Art and Preliminary Evaluation of the Application to Fracture Mapping in Geothermal Systems. Earth Science Laboratory, Univ. of Utah Res. Inst. Rep. 12196-2.

Roy, A. 1982. Focused Resistivity Logs. Developments in Geophysical Exploration Methods 3:61-94.

Schlichter, C. 1963. Principles of Magnetic Resonance. Harper and Row, New York, 397 pp.

Schlumberger Limited. 1972. Log Interpretation. Vol. I: Principles. Schlumberger Limited, New York, NY.

Schlumberger Limited. 1974. Log Interpretation. Vol. II: Applications. Schlumberger Limited, New York, NY.

Schlumberger Limited. 1989a. Log Interpretation Principles/Applications. Schlumberger Educational Services, Houston, TX. (Earlier edition published in 1987.) [SP, resistivity, induction, dielectric, gamma, gamma-gamma, neutron, acoustic-velocity, VSP]

Schlumberger Limited. 1989b. Cased Hole Log Interpretation Principles/Applications. Schlumberger Educational Services, Houston, TX. [Gamma, spectral gamma, neutron, neutron lifetime, acoustic velocity, spinner flowmeter, temperature, various well construction logs]

(Table 6-9 Borehole Geophysics References)

Schlumberger Limited. 1991. Log Interpretation Charts. Schlumberger Limited, New York, NY. (Earlier charts published in 1972, 1976, 1979, 1984.)

Schneider, G.J. 1982. In Situ Neutron Activation Analysis. In: Premining Investigations for Hardrock Mining, U.S. Bureau of Mines Information Circular 8891, pp. 46-54. [Neutron activation, spectral-gamma]

Scott, J.H. and B.L. Tibbets. 1974. Well Log Techniques for Mineral Deposit Evaluation: A Review. U.S. Bureau of Mines Information Circular 3627, 45 pp.

Serra, O. 1984a. Fundamentals of Well-Log Interpretation, 1: The Acquisition of Logging Data. Developments in Petroleum Science, Vol. 15A. Elsevier, New York, NY, 423 pp. [SP, resistivity, gamma, gamma spectrometry, gamma-gamma, neutron, neutron activation/lifetime, acoustic, dielectric, caliper, temperature, dipmeter, acoustic televiewer, VSP, nuclear magnetic resonance]

Serra, O. 1984b. Fundamentals of Well-Log Interpretation, 2: The Interpretation of Logging Data. Developments in Petroleum Science, Vol. 15B. Elsevier, New York, NY, 684 pp. [Chapters focus on log interpretation for specific applications--sedimentary structure, fractures, etc].

Society of Professional Well Log Analysts (SPWLA). 1960 to Present. Annual Logging Symposium Transactions. SPWLA, Houston, TX. (32nd was held in 1991; recent costs have been $75 for two volume set)

Society of Professional Well Log Analysts (SPWLA). 1978a. Gamma Ray, Neutron, and Density Logging. Reprint Volume Series, SPWLA, Houston, TX.

Society of Professional Well Log Analysts (SPWLA). 1978b. Acoustic Logging. Reprint Volume Series, SPWLA, Houston, TX.

Society of Professional Well Log Analysts. 1985. Glossary of Terms and Expressions Used in Well Logging, Revised. SPWLA, Houston, TX. (1st edition 1975.)

Society of Professional Well Log Analysts (SPWLA). 1990. Borehole Imaging. Reprint Volume Series, SPWLA, Houston, TX. [Optical, acoustic, electrical]

Stevens, Jr., H.H., J.F. Ficke, and G.F. Smoot. 1975. Water Temperature--Influential Factors, Field Measurement and Data Presentation. U.S. Geological Survey Techniques of Water-Resource Investigations TWRI 1-D1, 65 pp.

Stewart, R.R. 1991. Seismic Tomography. Course Notes No. 3. Society of Exploration Geophysicists, Tulsa, OK, 190 pp.

Stowell, J.R. 1989. An Overview of Borehole Geophysical Methods for Solving Engineering and Environmental Problems. In: Proc. 3rd Nat. Outdoor Action Conf. on Aquifer Restoration, Ground Water Monitoring and Geophysical Methods, National Water Well Association, Dublin, OH, pp. 871-890.

(Table 6-9 Borehole Geophysics References)

Taylor, K. 1989. Review of Borehole Methods for Characterizing the Heterogeneity of Aquifer Hydraulic Properties. In: Proc. Conf. on New Field Techniques for Quantifying the Physical and Chemical Properties of Heterogeneous Aquifers, National Water Well Association, Dublin, OH, pp. 121-132.

Taylor, T.A. and J.A. Dey. 1985. Bibliography of Borehole Geophysics as Applied to Ground-Water Hydrology. U.S. Geological Survey Circular 926, 62 pp.

Taylor, K., J. Hess, and S. Wheatcraft. 1990. Evaluation of Selected Borehole Geophysical Methods for Hazardous Waste Site Investigations and Monitoring. EPA/600/4-90/029. U.S. EPA Environmental Monitoring Systems Laboratory, Las Vegas, NV, 82 pp. [Acoustic velocity, thermal flowmeter, borehole dilution, induction logs]

Tearpock, D. and R.E. Bischke. 1991. Applied Subsurface Geological Mapping. Prentice-Hall, Englewood Cliffs, NJ, 648 pp. [Focusses on construction of geological maps from various sources, including geophysical measurements]

Technos, Inc. 1992. Application Guide to Borehole Geophysical Logging. Technos, Miami, FL, 15 pp.

Telford, W.M.N., L.P. Geldart, R.E. Sheriff, and D.A. Keys. 1990. Applied Geophysics, 2nd ed. Cambridge University Press, New York, NY, 770 pp. (1st edition, 1976, reprinted 1982.) [Chapter 11 covers borehole geophysics: SP, resistivity, dipmeter, induction, IP, acoustic, nuclear, gravity, magnetic, temperature]

Tellam, J.H. 1992. Reversed Flow Test: A Borehole Logging Method for Estimating Pore Water Quality and Inflow Rates Along an Uncased Borehole Profile. Ground Water Monitoring Review 12(2):146-154. [Fluid conductivity log]

Theys, P.P. 1991. Log Data Acquisition and Quality Control. Editions Technip, Paris, 326 pp.

Thompson, C.M., L.J. Holcombe, D.H. Gancarz, A.E. Behl, J.R. Erikson, I. Star, R.K. Waddell, and J.S. Fruchter. 1989. Techniques to Develop Data for Hydrogeochemical Models. EPRI EN-6637. Electric Power Research Institute, Palo Alto, CA. [Gamma-gamma, neutron probe]

Thornhill, J.T. and B.G. Benefield. 1990. Injection-Well Mechanical Integrity. EPA/625/9-89/007. Available from CERI.*

Tittle, C.S. 1961. Theory of Neutron Logging I. Geophysics 26(1):27-39.

Tittman, J. 1986. Geophysical Well Logging. Academic Press, New York, NY, 192 pp. [Electrical, nuclear, sonic]

Toksoz, M.N. and R.R. Stewart. 1984. Vertical Seismic Profiling, Part B: Advanced Concepts. Seismic Exploration Volume 14b, Geophysical Press, London, 419 pp.

Tweeton, D.R. 1988. A Tomographic Computer Program with Constraints to Improve Reconstruction for Monitoring In Situ Mining Leachate. U.S. Bureau of Mines Report of Investigation 9159.

(Table 6-9 Borehole Geophysics References)

U.S. Army Corps of Engineers. 1979. Geophysical Exploration. Engineer Manual EM 1110-1-1802, Department of the Army, Washington, DC, 313 pp. [Section II of Chapter 3 covers borehole seismic, SP, resistivity, acoustic, gamma, gamma-gamma, neutron, temperature, caliper and fluid resistivity]

U.S. Environmental Protection Agency (EPA). 1987. A Compendium of Superfund Field Operations Methods, Part 2. EPA/540/P-87/001 (OSWER Directive 9355.0-14) (NTIS PB88-181557/AS). [Section 8.3.4 covers borehole methods]

U.S. Nuclear Regulatory Commission. 1985. Rules and Regulations, Title 10, Chap. 1, Code of Federal Regulations, Part 20, Standards for Protection Against Radiation.

van der Leeden, F. 1991. Geraghty & Miller's Groundwater Bibliography, 5th ed. Water Information Center, Plainview, NY, 507 pp. [4th edition 1987]

Weinberg, A.M. and E.P. Wignor. 1958. The Physical Theory of Neutron Chain Reactors. University of Chicago Press, Chicago, IL, 800 pp.

Welenco, Inc. 1994. Water and Environmental Geophysical Well Logs, 6th ed. Welenco, Bakersfield, CA, 149 pp.

West, R.C. and S.H. Ward. 1988. The Borehole Controlled-Source Audiomagnetotelluric Response of a Three-Dimensional Fracture Zone. Geophysics 53(2):215-230.

Wheatcraft, S.W., K.C. Taylor, J.W. Hess, and T.M. Morris. 1986. Borehole Sensing Methods for Ground-Water Investigations at Hazardous Waste Sites. EPA/600/2-86/111 (NTIS PB87-132783). [Brief overview of SP, single point resistance, induction, gamma, gamma-gamma, neutron, acoustic, temperature, televiewer; main focus on thermal flowmeter]

Williams, J.H. and R.W. Conger. 1990. Preliminary Delineation of Contaminated Water-Bearing Fractures Intersected by Open-Hole Bedrock Wells. Ground Water Monitoring Review 10(4):118-126. [Gamma, SP resistance, caliper, fluid-resistivity, temperature, acoustic televiewer, thermal flowmeter]

Wyllie, M.R.J. 1963. The Fundamentals of Well Log Interpretation, 3rd ed. Academic Press, New York, NY, 238 pp. (Earlier editions 1954, 1957.) [SP, resistivity, neutron, gamma-gamma, acoustic velocity, gamma, gamma-spectrometry, nuclear magnetic resonance, cement bond]

Yearsley, E.N., R.E. Crowder, and L.A. Irons. 1991. Monitoring Well Completion Evaluation with Borehole Geophysical Density Logging. Ground Water Monitoring Review 11(1):103-118. [Acoustic cement bond, gamma-gamma]

Young, S.C. and J.S. Pearson. 1990. Characterization of Three-Dimensional Hydraulic Conductivity Field with an Electromagnetic Borehole Flowmeter. In: Proc. Fourth Nat. Outdoor Action Conf. on Aquifer Restoration, Ground Water Monitoring and Geophysical Methods. Ground Water Management 2:83-97.

* See Preface for information on how to obtain documents from CERI (U.S. EPA Center for Environmental Research Information) and NTIS.

(Table 6-9 Borehole Geophysics References)

CHAPTER 7

CHARACTERIZATION OF VADOSE ZONE AND GROUND WATER HYDROLOGY

7.1 Measurement of Vadose Zone Hydrologic Parameters 372

    7.1.1 Matric Potential and Moisture Content 372
    7.1.2 Infiltration and Hydraulic Conductivity 373
    7.1.3 Other Vadose Water Budget Parameters 375
    7.1.4 Other Hydrologic Parameters 377

7.2 Preparation and Use of Potentiometric Maps 377

    7.2.1 Water Level Measurements 378
    7.2.2 Plotting Equipotential Contours 378
    7.2.3 Flow Nets 380
    7.2.4 Common Errors in Contouring 383
    7.2.5 Common Errors in Interpretation 386

7.3 Field and Laboratory Measurement of Aquifer Parameters 388

    7.3.1 Shallow Water Table Tests 388
    7.3.2 Well Tests 389
    7.3.3 Other Field Tests 392
    7.3.4 Laboratory Measurements 393

7.4 Estimation of Subsurface Hydrologic Parameters 394

    7.4.1 Estimation from Soil Survey Data 394
    7.4.2 Estimation from Aquifer Matrix Type 396
    7.4.3 A Simple Well Test for Estimating Hydraulic Conductivity 397

7.5 Special Considerations in Hydrogeologic Mapping 398

    7.5.1 Delineation of Aquifer Boundaries 398
    7.5.2 Characterization of Aquifer Heterogeneity and Anisotropy 399
    7.5.3 Presence and Degree of Confinement 402
    7.5.4 Characterization of Fractured Rock and Karst Aquifers . . 404

7.6 Guide to Major References* 413

* Appendix F contains citations for table and figure sources.

Many methods are available for characterizing vadose zone and ground-water hydrology. A single chapter can only provide an overview of major techniques and identify key considerations when performing hydrogeologic mapping. The U.S. EPA's guide to **Subsurface Characterization and Monitoring Techniques** (Boulding, 1993/T7-5) is recommended for use as a companion to this chapter. This two-volume document is available at no cost from the U.S. EPA's Center for Environmental research information (see Introduction for ordering information). Appendix A in this handbook contains major method summary tables from that guide, which can be used to quickly identify the location of specific methods of interest. Appendix B provides an index and addresses of manufacturers and distributors of field equipment for subsurface hydrologic characterization.

7.1 Measurement of Vadose Zone Hydrologic Parameters

Historically, measurement of vadose zone hydrologic parameters has been the domain of soil physicists and agronomists because of their significance for plant growth, and engineers because water content of soil materials significantly affects engineering behavior. The U.S. Environmental Protection Agency's increasing emphasis on vadose zone monitoring as an early-warning system for ground-water protection means that familiarity with methods for measuring vadose hydrologic characteristics is necessary for environmental professionals dealing with soil and ground-water protection or remediation. This section addresses the following aspects of vadose zone hydrology: (1) measurement of matric potential and moisture content (Section 7.1.1), (2) measurement of infiltration and hydraulic conductivity, (3) measurement of other water budget parameters, and (4) miscellaneous hydrologic parameters and measurement/estimation of vadose zone water flux.

7.1.1 Matric Potential and Moisture Content

Water state in the vadose zone is measured in terms of *positive pressure head* when the soil is saturated and *negative pressure potential* or *suction* when the soil is unsaturated (Section 2.4.1). Table A-3 provides summary information on six major techniques for measuring soil water potential and a dozen methods for measuring soil moisture content. The measurement of soil water potential and moisture content in the vadose zone is intimately connected, and a specific measurement technique can be classified as measuring potential or moisture content, depending on the perspective of the writer in the literature. Either measurement can be used to obtain the other if a *moisture characteristic curve* has been developed which is a plot of the relationship between moisture content and suction. *Porous cup tensiometers* are the most commonly used method for measuring soil water potential in the vadose zone. The *gravimetric* method is most commonly used to measure moisture content from soil samples, and the *neutron probe* and *gamma-gamma* methods are most commonly used for in situ measurement of soil moisture. The relatively recent commercial availability of *dielectric* or capacitance sensors is likely to increase the use of this method, which provides accuracy similar to the neutron probe without some of the disadvantages of nuclear methods (i.e., radioactive sources). Similarly, *time domain reflectometry*, a relatively new method, is becoming more widely used with the advent of commercially available units. All methods for vadose zone measurement of water content or matric potential have limitations with respect to soils contaminated with nonaqueous phase liquids, due to interference effects.

7.1.2 Infiltration and Hydraulic Conductivity

Characterization of water movement in the vadose zone is complicated by the fact that hydraulic conductivity varies as a function of pressure potential and moisture content. Various terms are used to describe hydraulic conductivity in the vadose zone:

1. *Saturated hydraulic conductivity* (K_{sat}) is the hydraulic conductivity at saturation with no entrapped air. This state rarely is achieved in the vadose zone, except, perhaps, in the zone of seasonal fluctuation of an unconfined water table.

2. *Field-saturated hydraulic conductivity* (K_{fs}), also called the *satiated* hydraulic conductivity, is the hydraulic conductivity when entrapped air is present, which can be as much as 50% below the true K_{sat} (Reynolds and Elrick, 1986).[1] Methods for measuring saturated hydraulic conductivity above the water table usually measure K_{fs}. Another term, $K_{(sat)}$, has been proposed by Bouma (1982)[2] for hydraulic conductivity measurements of the soil matrix without macropore flow (Figure 7-1C). $K_{(sat)}$ will be less than K_{sat} or K_{fs} because water flows more rapidly in macropores than in the soil matrix. The term K_{sat} often is loosely used for reporting measurements that should more accurately be termed K_{fs}.

3. *Unsaturated hydraulic conductivity* (K_{unsat}) is the hydraulic conductivity of soil at negative pressure potentials. $K(\phi)$ is the term usually used to describe the hydraulic conductivity-pressure potential function, and $K(\theta)$ to describe the hydraulic conductivity-moisture content function. Complete characterization of K_{unsat} requires measuring hydraulic conductivity at a range of moisture contents to develop a $K(\theta)$ curve or at a range of pressures to develop a $K(\phi)$ curve. $K(\phi)$ curves for two different soil materials are illustrated in Figure 2-9. These functions are subject to *hysteresis* (i.e., K_{unsat} can differ at the same water content or matric potential, depending on whether the soil is wetting or drying.

Infiltration. The infiltration capacity of a soil is a critical element of water budget calculations because it affects how much precipitation that reaches the ground surface enters the soil and how much moves off a site as surface runoff. The infiltration rate generally is the same as the unsaturated and saturated hydraulic conductivity, except that some processes, such as the initial moisture content (see Figure 7-2), crusting, or sediment clogging, might cause different infiltrations at the ground surface compared to the subsurface with all other soil factors being equal.

---

[1] Reynolds, W.D. and D.E. Elrick. 1986. A Method for Simultaneous In Situ Measurement in the Vadose Zone of Field Saturated Hydraulic Conductivity, Sorptivity and the Conductivity-Pressure Head Relationship. Ground Water Monitoring Review 6(4):84-95.

[2] Bouma, J. 1982. Measuring the Hydraulic Conductivity of Soil Horizons with Continuous Macropores. Soil Sci. Soc. Am. J. 46:438-441.

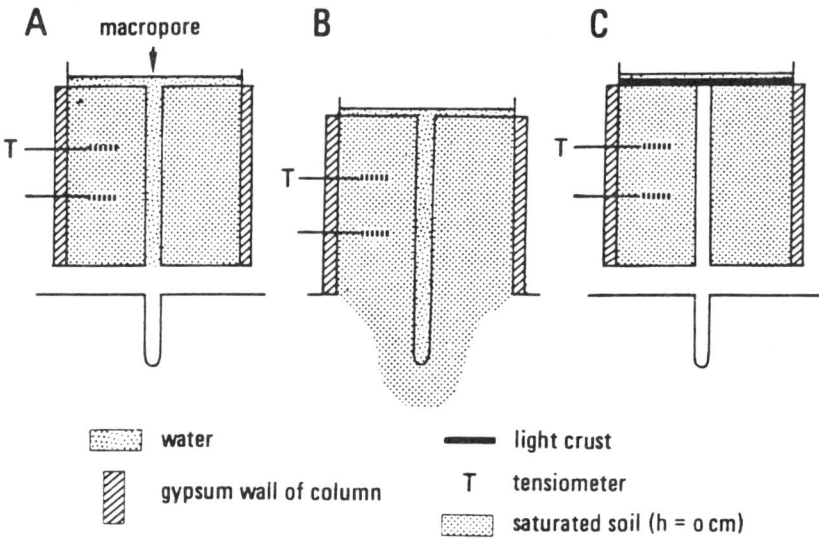

Figure 7-1 Schematic representation of three types of flux measurements using the column-crust method (Boulding, 1993b, after Bouma, 1982).

Table A-4 summarizes information on eight methods for measuring or estimating infiltration. These methods can be grouped into four general categories: (1) *impoundment* methods, where infiltration is below a water surface (seepage meters, instantaneous rate, and impoundment water budget), (2) *land surface* methods (cylinder infiltrometers, infiltration test basins and sprinkler infiltrometers), (3) *watershed* methods for estimating infiltration over larger areas (watershed average and empirical relations), and (4) *infiltration equations*. Often infiltration can be estimated using empirical relations or infiltration equations using other variables which can be obtained with instruments for measuring saturated and unsaturated hydraulic conductivity (see below).

Measurement of Unsaturated Hydraulic Conductivity. Table A-4 summarizes information on nine methods for measuring or estimating unsaturated hydraulic conductivity from field measurements. Most of these methods can be used to develop $K(\phi)$ or $K(\theta)$ relationships, which once established allow subsequent monitoring to focus on either changes in pressure potential or moisture content. The *instantaneous profile method* is the most commonly used method for accurate measurement of unsaturated hydraulic conductivity in the field. Various *draining profile* methods are simpler and less expensive to use if the simplifying assumptions apply to the site of interest. Another common procedure is to collect undisturbed core samples and measure K_{unsat} in the laboratory (Klute and Dirksen, 1986).[3] ASTM D5126 (Table A-

---

[3] Klute, A. and C. Dirksen. 1986. Hydraulic Conductivity and Diffusivity: Laboratory Methods. In: Methods of Soil Analysis, Part 1, 2nd edition, A. Klute (ed.), Agronomy Monograph No. 9, American Society of Agronomy, Madison, WI, pp. 687-734.

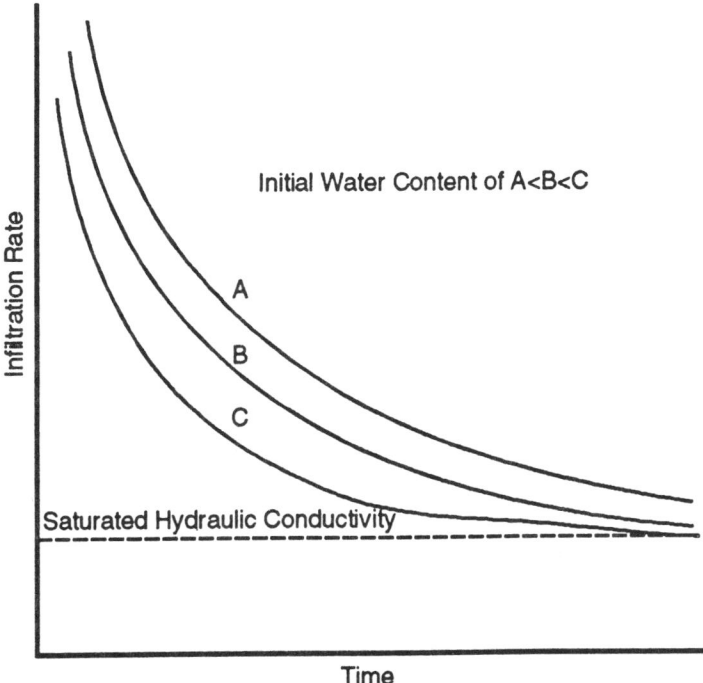

Figure 7-2 The effect of initial water content of soil on infiltration rates (Everett et al., 1983).

14) provides guidance on selecting field methods for measuring unsaturated hydraulic conductivity in the vadose zone.

Measurement of Saturated Hydraulic Conductivity. Table A-4 summarizes information on 10 methods for measuring K_{fs} above a shallow water table and 5 methods for measuring K_{fs} above a deep water table. The *cylinder* or *ring* infiltrometer is a widely used method that measures both infiltration and K_{fs} at the soil surface. Most other shallow methods require a borehole and devices at the surface to control the flow of water into the hole to achieve steady-state infiltration before measurements are taken. The *constant-head borehole infiltration* or *shallow-well pump-in* method and the *Guelph permeameter* probably are the most commonly used methods for measuring K_{fs}. Most of these methods are restricted to a depth of 2 meters or less, but recently developed *compact constant-head permeameter* can be used to depths of 10 meters. Most methods for measuring K_{fs} above a deep water table require drilling or relatively large diameter boreholes (at least 6 inches) and a large supply of water, which can be pumped into the borehole. ASTM D5126 (Table A-14) provides guidance on selecting field methods for measuring saturated hydraulic conductivity in the vadose zone.

7.1.3 Other Vadose Water Budget Parameters

Water movement and transport of contaminants in the vadose zone is determined by the amount of precipitation that enters the ground by infiltration, and the amount of water that is removed from the soil by evaporation from bare soil or by evapotranspiration where vegetation covers the soil. Table A-5 provides information on techniques for measuring or estimating: (1) hydrometeorological parameters and (2) evaporation and evapotranspiration for water budget calculations in the vadose zone and shallow ground-water systems. Methods for measuring and estimating infiltration are covered in the previous section.

Hydrometeorological Data. Table A-5 provides some general summary information on 38 techniques for measuring six major hydrometeorological parameters and identifies sections of this guide where more detailed information can be found. Precipitation is a primary input into water budget calculations, and devices for measuring precipitation fall into two main categories: (1) *manual* gages and (2) *recording* gages. Measurement of *humidity* might be required during field work for protection of health and safety and are required with most micrometeorological methods for measuring evapotranspiration. Other hydrometeorological measurements might be required for monitoring weather conditions, such as *temperature*, *windspeed*, and *wind direction*. Measurement or estimation of these same parameters, as well as *atmospheric pressure* and *insolation* or *solar radiation* measurement, might be required in order to quantify the evapotranspiration component of water budget studies (discussed further below). Although numerous techniques and devices have been developed for hydrometeorological measurements, most of the parameters of interest usually can be estimated for purposes of vadose zone water budget studies by using data from nearby weather stations or interpolations using hydrometeorological tables or maps. Consequently, only those methods relevant to health and safety (temperature, humidity, windspeed, and direction) are likely to be used routinely during site investigations. Table A-5 identifies the specific hydrometeorological techniques or devices that are most commonly used for site investigations.

Evaporation and Evapotranspiration. Water that reaches the earth's surface can return to the atmosphere either by *evaporation* from free water surfaces or bare soil or by *transpiration* by plants. The term *evapotranspiration* (ET) specifically refers to the combined effects of evaporation and transpiration from the land surface, but also might be used loosely to refer to the combined effects of evaporation from water and soil surfaces and transpiration. ET is a critical component of vadose zone water budget calculations and is one of the most difficult of these components to measure accurately. The numerous methods that have been developed for measuring or estimating ET can be broadly classified as *water budget* or *balance* methods and *micrometeorological* methods. Table A-5 summarizes information on 10 water balance methods and 6 micrometeorological methods and identifies specific applications for each method (water evaporation, bare soil evaporation, evapotranspiration, and transpiration). Most of these methods are too complex and time consuming for routine site investigations.

Lysimeters and *soil moisture monitoring* probably are the most commonly used methods for measuring evapotranspiration where site-specific data are required. Most vadose zone hydrologic models use *empirical equations* and use data from nearby weather stations data and published maps. The physically based *Penman equation* (and various methods developed as refinements and adaptations of the Penman equation) probably is the most commonly used method for estimation of evaporation and/or evapotranspiration, where some measurements of meteorological data are feasible but the more complex measurements and instrumentation of other micrometeorological methods are not feasible.

7.1.4 Other Hydrologic Parameters

Miscellaneous Parameters. Other field-measurable hydrologic properties of the vadose zone, which might be of use in evaluating contaminant transport or design of remediation systems, include *sorptivity* (a measure of the capacity of a porous medium to absorb a wetting liquid), *soil diffusivity* (a single parameter of unsaturated soil that relates the hydraulic conductivity and water storage properties), and *available water capacity* (a measure of plant-available water in the soil). Sorptivity and diffusivity are properties that are significant in evaluating infiltration of water into the subsurface. Table A-3 indicates where more information can be obtained about these parameters.

Vadose Zone Water Flux.[4] Various methods are available to measure or estimate the amount of water that passes through the vadose zone and enters the ground-water system. A *water budget* uses a mass balance by measuring inflows, outflows, and storage changes in the area of interest. More often, a simplified water budget approach can be used, in which only changes in *soil moisture* or *matric potential* are measured. A variety of *tracers*, such as chloride and tritium, can be used to estimate the rate of recharge and water flux. Localized water flux can be measured using a *soil-water flux meter*. Finally, a variety of *physical* and *empirical equations* can be used in combination with the methods above or using site-specific data on hydraulic conductivity or soil physical characteristics, such as texture and bulk density. *Tile drains* or *collection lysimeters* (Section 9.3.1 in Boulding, 1993/T7-5) also can be used to measure water flux in the vadose zone, provided the area of vertical infiltration is known and lateral ground-water flow can be excluded or quantified.

7.2 Preparation and Use of Potentiometric Maps

A water table or potentiometric map is one of the most basic and useful tools available for hydrogeologic characterization. A *water table* map usually refers to the hydraulic gradient of an unconfined aquifer (where the top of the saturated zone equal zero pressure head), and a *piezometric* (pressure) surface map usually refers to the pressure potentials of confined aquifers. Either type of map is called a *potentiometric* map. In practice, the terms "water table", "potentiometric", and "piezometric" are often used interchangeably. A potentiometric map is developed by compiling ground- and surface-water level information from all available sources. Section 7.2.1 addresses methods for measuring ground-water levels. Section 7.2.2 describes how

---

[4] Appendix A does not identify these methods, but they are covered in Section 7.5 in Boulding (1993/Table 7-5). Individual subsections in that guide cover italicized terms in this section.

potentiometric contours are developed from water level information. Once a potentiometric map has been developed interpretations of ground-water flow direction and quantity can be developed using flow nets (Section 7.2.3).

7.2.1 Water Level Measurement

Water level measurements in observations wells provide the basic data for a potentiometric map. Table A-6 summarizes information on nine techniques for measuring water levels in open or cased boreholes and three methods for measuring pressure head in flowing (artesian) wells. The *steel-tape* and *electric probe* methods are used most commonly for routine measurement of water levels. *Transducers* are used most commonly in aquifer tests where accurate measurement of changes in multiple wells is required in relatively short time periods. The *air line method* is also useful in pumped wells where water turbulence may preclude using more precise methods. Pressure potential in the saturated zone also can be measured by burying *in situ* piezometers that sense pore pressure.

In addition to providing data for potentiometric maps, the water level in a monitoring well is measured and recorded prior to purging or sample collection. For detailed site investigations, each well requires an accurately surveyed reference point on the casing from which water level measurements are taken. For regional and local investigations (Section 5.1.2) estimating well elevations using a 7.5° topographic map is usually adequate. Documenting the static water levels for all wells at a site will provide historical information on the site's hydraulic conditions. This information may indicate changes in flow paths, documents seasonal changes in water levels. Without careful attention to these measurements, it may be impossible to interpret chemical concentration variability at the site.

7.2.2 Plotting Equipotential Contours

The starting point for a potentiometric map is a base map. The base map identifies well locations and water level elevations in the well and other surface hydrologic features, such as streams, rivers, and water bodies. An accurate potentiometric map requires enough well observations to develop water table contours that do not miss important features of the flow system. Considerable interpretation and judgment may be required in developing contours when well data points do not seem to fit into a coherent pattern. For example, if water level data from wells are drawn from multiple sources, measurements in nearby wells may have been taken at different times of the year and may not be directly comparable. On the other hand, if all the data have been collected so as to minimize effects of short-term or seasonal fluctuations, examination of individual well characteristics may yield explanations for anomalous data points. For example, a single well data point that is far out of line with nearby wells may be tapping a different aquifer. If an anomalous well data point cannot be readily explained as being unrepresentative for any reason, then further field investigation may be required to determine whether any localized hydrogeologic conditions are causing the anomaly.

The contours on a potentiometric map are called *equipotential* lines, indicating that the water has the "potential" to rise to that elevation. In the case of a confined aquifer, however, it cannot reach that elevation unless the confining unit is perforated by a well. Potentiometric surface maps are essential to any ground-water investigation, because they indicate the direction in which ground water is moving and provide an estimate of the gradient, which controls ground-water velocity. As discussed in Section 7.2.5, interpretations of flow directions in aquifers must take into account anisotropy and heterogeneity.

Potentiometric maps provide some information on aquifer homogeneity, provided that well data points are close enough to allow reasonably accurate contouring. A map of a uniform, homogeneous aquifer will have equally spaced equipotential lines and no dramatic changes in hydraulic gradient, because ground water is moving at about the same rate at all points in the aquifer. Irregularly spaced contours and differing hydraulic gradients in different areas of the aquifer indicate lateral changes in aquifer properties.

Preparing a potentiometric map involves plotting water level measurements on a base map and then drawing contours. In isotropic, porous-media aquifers, the direction of ground-water flow is perpendicular to the ground-water contour lines. The next section on flow nets describes in more detail how contour maps can be used to infer the direction of ground-water flow. A minimum of three points is required to determine the general direction of ground-water flow. Figure 7-3 shows a manual graphical depiction of ground-water contours, drawn based on water elevations in three wells. The difference in elevation between each well was calculated and divided into the distance between the wells. This distance was scaled on each line as tick marks that represent a change in elevation of one-tenth of a foot. The lines connecting the points of equal elevation (27.0 and 27.5 feet in Figure 7-3) are potentiometric contours. Ground-water flow direction is on the path line perpendicular to the contours. The method shown in Figure 7-3 and Figure 7-4, described below, are most appropriate for detailed site investigations where the scale is 1" = 200 or larger.

Figure 7-4 illustrates a slightly different approach to determining the direction of ground-water flow from three well points. Steps in this solution involve:

1. Identifying the well that has the intermediate water level.

2. Calculating the position between the well having the highest head and the well having the lowest head at which the head is the same as that in the intermediate well.

3. Drawing a straight line between the intermediate well and the point identified in Step 2. This line represents a segment of the water level contour along which the total head is the same as that in the intermediate well.

4. Drawing a line perpendicular to the water level contour and through the well with the lowest (or highest) head. This indicates the direction of ground-water movement in an isotropic aquifer.

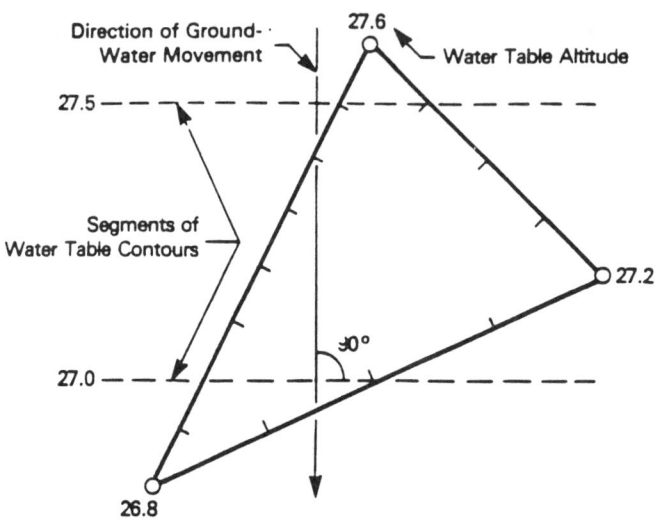

Figure 7-3 The generalized direction of ground-water movement can be determined by means of the water level in three wells of similar depth (U.S. EPA, 1987a, after Heath and Trainer, 1981).

5. Dividing the difference between the head of the well and that of the contour by the distance between the well and the contour. This gives the hydraulic gradient.

A large number of well measurements is needed to develop an accurate potentiometric surface map. Geostatistical methods allow the estimation of water table elevations in unsampled locations where the water table is approximately parallel to the ground surface (Hoeksma et al., 1989).[5]

The most important consideration in preparing a potentiometric map is that the water level measurements should describe a single flow system in an aquifer. Section 7.2.4 describes in detail some common pitfalls in preparing potentiometric maps. Worksheet D-W1 provides a form for compiling well information used to develop an potentiometric map. This information may prove helpful in evaluating individual well elevations that appear to be anomalous.

7.2.3 Flow Nets

A potentiometric surface map can be developed into a flow net by constructing flow lines that intersect the equipotential lines or contour lines at right angles. Flow lines are imaginary paths that trace the flow of water particles through the aquifer. Although there are an infinite number of both equipotential and flow lines, the former

---

[5] Hoeksma, R.J., R.B. Clapp, A.L. Thomas, A.E. Hunley, N.D. Farrow, and K.C. Dearstone. 1989. Cokriging Model for Estimation of Water Table Elevation. Water Resources Research 25(3):429-438.

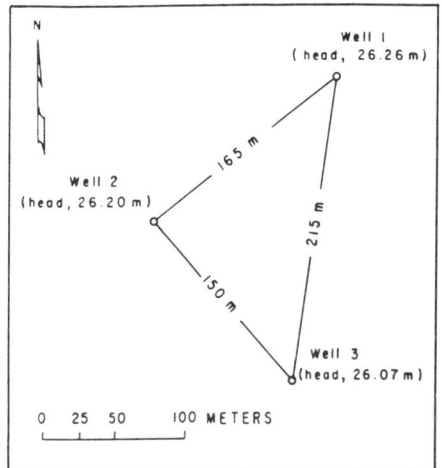 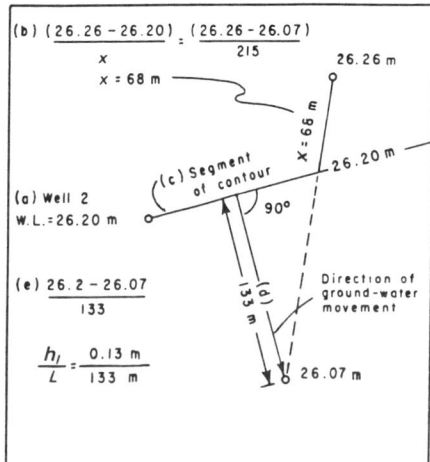

Figure 7-4 Alternative procedure for determination of equipotential contour and direction of ground-water flow in homogeneous, isotropic aquifer (Heath, 1983).

are constructed with uniform differences in elevation between them, while the latter are constructed so that they form, in combination with equipotential lines, a series of squares. A flow net carefully prepared in conjunction with Darcy's Law allows estimation of the quantity of water flowing through an area and of the variability of transmissivity and hydraulic conductivity. Plan view flow nets are a valuable tool in delineating the zone of contribution to a well, as illustrated in Figure 10-3. Scott (1992)[6] provides a good recent review of use of flow net analysis for aquifer identification. U.S. EPA (1986) provides guidance on use of flow net analysis for identifying areas of vulnerable hydrogeology under RCRA.[7]

A standard flow net assumes that the aquifer is isotropic. Figure 7-5 illustrates how anisotropy in a fractured rock aquifer alters the direction of ground-water flow compared to that expected in an isotropic aquifer. When an aquifer is anisotropic, commonly the case in unconsolidated and sedimentary aquifers, the actual direction

---

[6] Scott, D.M. 1992. An Evaluation of Flow Net Analysis for Aquifer Identification. Ground Water 30(5):755-764.

[7] U.S. Environmental Protection Agency (EPA). 1986. Criteria for Identifying Areas of Vulnerable Hydrogeology Under RCRA: A RCRA Interpretive Guidance; Appendix B, Groundwater Flow Net/Flow Line Construction and Analysis. EPA/530/SW-86-022B (NTIS PB86-224979).

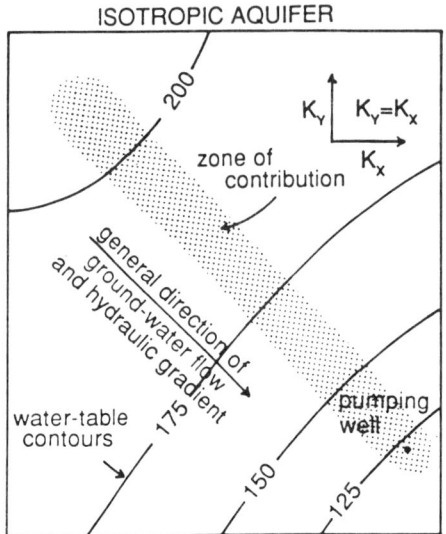 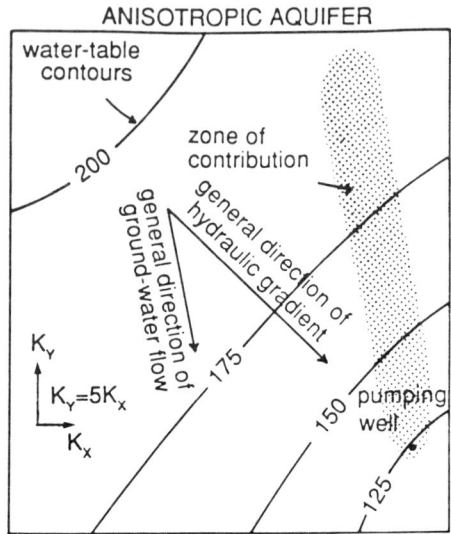

Figure 7-5 Effect of fracture anisotropy on the orientation of the zone of contribution to a pumping well (Bradbury et al., 1991).

of ground-water flow will not be perpendicular to the equipotential contours. Instead, the direction of flow will deviate from the perpendicular at an angle that depends on the ratio of the horizontal to the vertical hydraulic conductivity.[8]

Several methods are available for determining the direction of flow lines where the degree of anisotropy is known. Figure 7-6 illustrates a procedure for transforming a vertical anisotropic flow net to an isotropic section. For potentiometric surface maps, Llakapoulos (1965)[9] developed a graphical technique for determining this deviation. This technique uses a "permeability tensor ellipse", which has semi-axes equal to the inverse square root of the principal permeability values. Figure 7-7 illustrates the five-step sequence for using this method. Fetter (1981)[10] provides some additional guidance on using this technique. Section 7.5.2 provides some guidance on how to determine directional components of hydraulic conductivity in an aquifer.

---

[8] The discussion here assumes that the aquifer is anisotropic in only two directions, with the horizontal conductivity greater than the vertical conductivity, which is typical of horizontally layered sediments. Anisotropy in three directions is possible, but not amenable to simple graphical solutions for determining flow direction. Section 7.5.2 discusses methods for determining anisotropy in three dimensions.

[9] Llakopoulos, A.C. 1965. Variation of the Permeability Tensor Ellipsoid in Homogenous Anisotropic Soils. Water Resources Research 1(1):135-142.

[10] Fetter, Jr., C.W. 1981. Determination of the Direction of Groundwater Flow. Ground Water Monitoring Review 1(3):28-31.

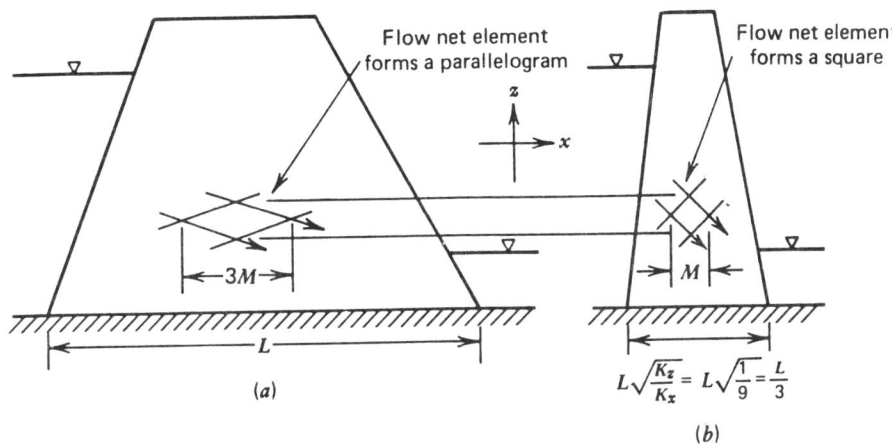

Figure 7-6 Illustration of flow net analysis for anisotropic hydraulic conductivity in an earth dam: (a) true anisotropic section with $K_x = 9K_z$; (b) transformed isotropic section with $K_x = K_z$ (Todd, 1980). © John Wiley & Sons. Reprinted by permission.

When the ratio of a permeability ellipse is around 10:1 (not uncommon in sedimentary formations), the flow line diverges almost 45 degrees from the "expected" direction when the axis of the equipotential line is at a 45-degree angle to the axis of maximum permeability. When interpreting flow direction from potentiometric maps, it is also important to know the angle of the axis of maximum permeability in relation to an equipotential line. For example, flow direction in an anisotropic aquifer will be perpendicular to an equipotential line if the axis of greater permeability in a permeability ellipse and the equipotential line are parallel. Fetter (1981-Footnote 10) provides additional explanation of these relationships.

7.2.4 Common Errors in Contouring

Errors in contouring fall into two general categories: (1) failure to exclude data points that are not representative and (2) failure to take into account subsurface features that change the distribution of potentiometric head as a result of aquifer heterogeneity or boundary conditions. The following are six situations in which contouring errors might occur.

1. **Failure to exclude well measurements from wells cased below the water table surface in recharge and discharge areas.** Figure 7-8a illustrates distortions in contouring that result from this effect, and Figure 7-8b shows the correct interpretation. The effect can perhaps be better understood by referring to the illustration of pressure head distribution in recharge and discharge areas in Chapter 2 (Figure 2-10). In Figure 2-10 only well c gives an accurate reading of the water table surface.

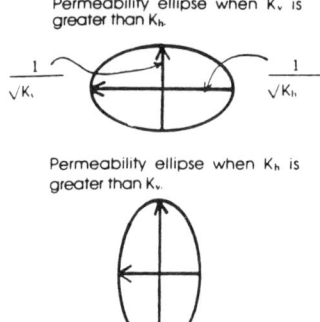

Permeability ellipse when K_v is greater than K_h.

$\dfrac{1}{\sqrt{K_v}}$ $\dfrac{1}{\sqrt{K_h}}$

Permeability ellipse when K_h is greater than K_v.

1. Construct a permeability ellipse.

2. Draw the equipotential line as it is oriented to the permeability axes.

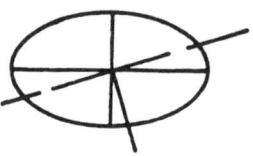

3. Draw the hydraulic gradient vector perpendicular to the equipotential line.

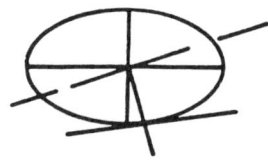

4. Draw a tangent to the ellipse at the point where the hydraulic gradient vector intersects the ellipse.

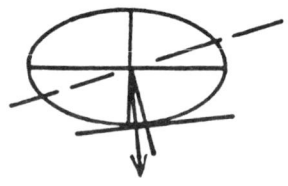

5. Draw the flowline so that it passes through the origin of the ellipse and is perpendicular to the tangent.

Figure 7-7 Steps in the determination of ground-water flow direction in an anisotropic aquifer (Fetter, 1981). See Appendix F for full credit.

2. **Failure to adjust contour lines in areas of topographic depressions occupied by lakes.** Figure 7-9a illustrates the incorrect and correct interpretations in this situation.

3. **Failure to recognize locally steep gradients caused by fault zones.** Figure 7-9b illustrates how conventional contouring methods erroneously portray the ground-water flow systems on the two sides of a fault.

4. **Use of measurements from wells tapping multiple aquifers.** Wells in which the screened interval includes multiple aquifers generally yield inaccurate water level or piezometric measurements because the measured head reflects the interaction between heads of the intersected aquifers. Figure 7-10 illustrates how the failure to differentiate measurements from wells completed in two aquifers, combined with a well that connects the two, results in a apparent depression in the potentiometric surface. An offsetting of nearby ground-water levels with similar elevations to form a "Z" shape is an indicator that a fault may be present.

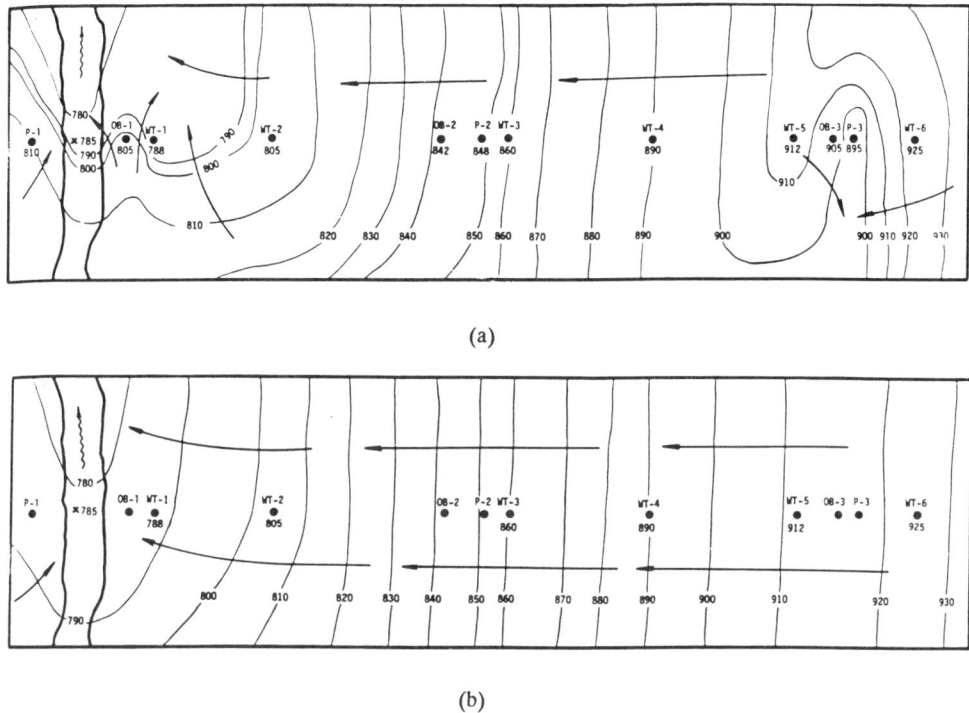

Figure 7-8 Contour errors in recharge and discharge areas: (a) incorrect contours; (b) correct contours after reinterpretation (Saines, 1981). See Appendix F for full credit.

5. **Failure to consider seasonal and other short-term fluctuations in well levels.** If an aquifer experiences seasonal high and low water tables, well measurements are not comparable unless they are taken at the same time of year. Other factors, such as dramatic changes in atmospheric pressure and precipitation events, might reduce the comparability of well measurements even if the measurements are taken at the same time of year.

6. **Failure to consider localized mounding or depression of the potentiometric surface from anthropogenic recharge or pumping.** Pumping wells create a cone of depression around the well (Section 2.6.6) with steepened hydraulic gradients. When the source of recharge is confined to a relatively small area, a localized mound develops with elevations increasing toward the center. Agricultural irrigation, artificial recharge using municipally treated wastewater, and artificial ponds and lagoons usually cause a mounding of water tables. Area-wide recharge will reduce hydraulic gradients compared to natural aquifer conditions. These features are especially significant when they are located near a ground-water divide, because small shifts in the location of a divide may have a major impact on the direction in which contaminants flow.

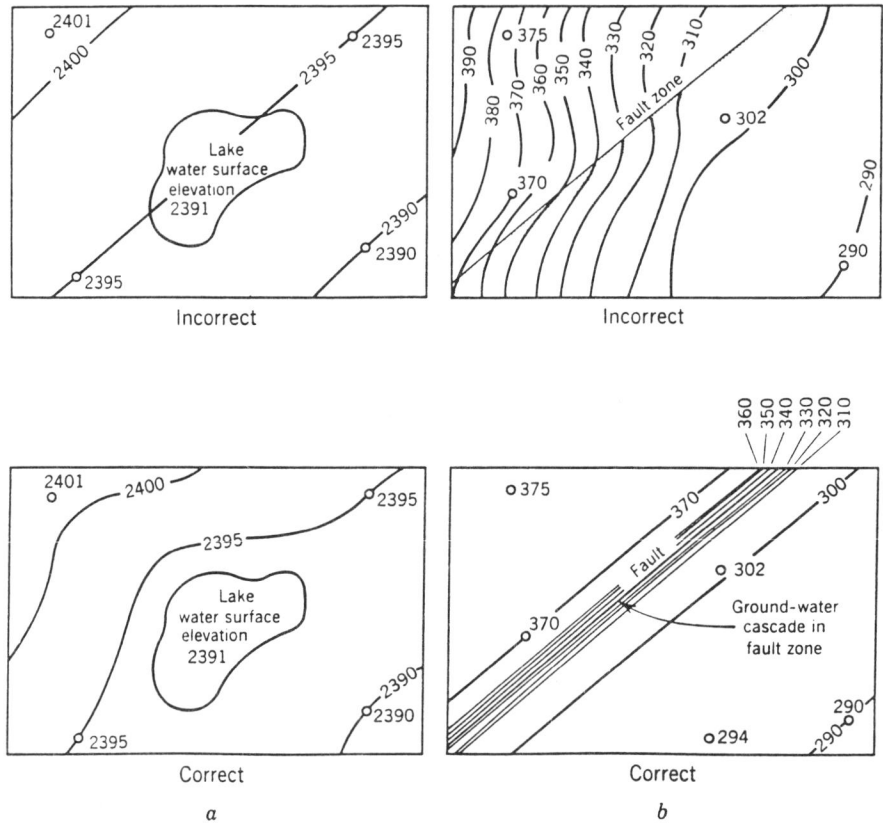

Figure 7-9 Common errors in contouring water table maps: (a) topographic depression occupied by lakes; (b) fault zones (Davis and DeWiest, 1966). © John Wiley & Sons. Reprinted by permission.

7.2.5 Common Errors in Interpretation

As noted earlier, ground-water flow is perpendicular to contours on a potentiometric map if the aquifer is isotropic. However, failure to account for anisotropy and heterogeneities in an aquifer can result in significant errors in the interpretation of ground-water flow direction. Following are three situations in which flow direction will differ from that indicated by conventional flow net construction using an *accurate* potentiometric map:

1. **Homogeneous, anisotropic aquifers.** Figure 7-5, discussed earlier, illustrates how flow direction can diverge from flow in an isotropic aquifer. Section 7.2.3 discussed how to determine the direction of flow in this situation.

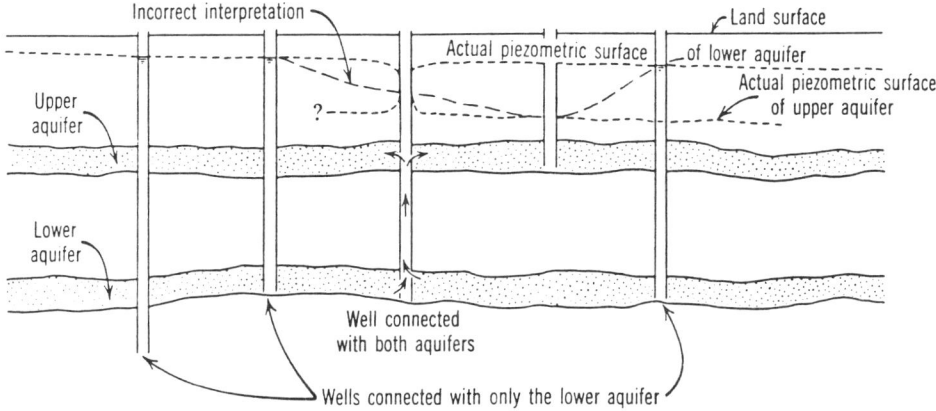

Figure 7-10 Error in mapping potentiometric surface due to mixing of two confined aquifers with different pressures (Davis and DeWiest, 1966). © John Wiley & Sons. Reprinted by permission.

2. **Heterogenous aquifers with contrasting hydraulic conductivity.** Figure 7-11 illustrates an example of divergence of flow from the direction predicted by ground-water contours as a result of a buried channel of higher permeability oriented across the direction of the potentiometric surface. This kind of divergence is difficult to predict accurately. Careful examination of well logs for the areal distribution of materials with contrasting hydraulic conductivity and the use of tracer tests may help modify flow direction interpretations when this situation occurs.

3. **Backwater effects in discharge areas.** Short-term reverses in the direction of ground water occur when streams or rivers are at high stage (Figure 2-6). These effects can extend for hundreds of feet from the stream edge. Wells that may be subject to bank storage can be identified by monitoring changes in water levels in response to stream flood events.

Reverse Flow of Contaminants. Several situations can cause contaminants to flow in a different direction from that indicated by flow net construction using a potentiometric map. Dissolved contaminants follow the direction of ground-water flow. Attention should be paid, however, to the possibility of localized flow patterns that run against the general direction of ground-water flow (mounding of ground water caused by ponds and lagoons and backwater effects in discharge areas). Dense leachates and nonaqueous phase liquids (NAPLs), on the other hand, can flow in an entirely different direction from that of ground-water flow if the slope of the geologic material forming the base of the aquifer does not follow the potentiometric surface. Figure 4-10 illustrates a dense NAPL flowing in the opposite direction of ground-water flow as a result of geologic controls.

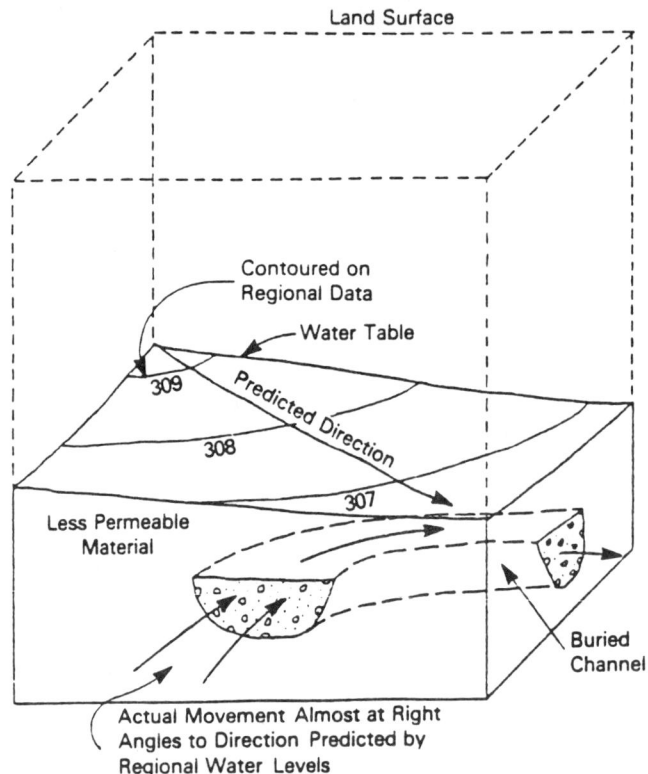

Figure 7-11 Divergence from predicted direction of ground water resulting from aquifer heterogeneity (Davis et al., 1985).

7.3 Field and Laboratory Measurement of Aquifer Parameters

Potentiometric maps provide a useful starting point for developing an understanding of the hydrogeology of a site or area. Site-specific investigations usually require more extensive characterization in the form of field and laboratory tests and measurements of aquifer parameters. These methods can be broadly classified as (1) shallow water table tests (Section 7.3.1), (2) well tests (Section 7.3.2), (3) other field tests (Section 7.3.3), and (4) laboratory measurements (Section 7.3.4).

7.3.1 Shallow Water Table Tests

Table A-7 provides summary information on ten relatively simple techniques that are available for measuring hydraulic conductivity where there is a shallow water table (generally less than 1 m). The *auger hole method* is the most widely used. This method involves boring an open hole below the water table, removing water, and measuring the water level at intervals until water reaches the original level. Other methods may be more appropriate for different site conditions. This type of test measures only hydraulic conductivity of the upper part of the aquifer and is most useful where potential contamination from agricultural chemicals is a concern. Because the tests are relatively fast and inexpensive, they may be useful for measuring spatial variability hydraulic conductivity.

7.3.2 Well Tests

Well tests are the most common and versatile methods for directly measuring aquifer parameters. They fall into three main categories: (1) single-well slug tests, (2) pumping tests (single and multiwell), and (3) packer tests (single and two packer). *Slug* tests involve measuring the rate at which water in a well returns to its initial level after (1) a sudden injection or withdrawal of a known volume of water from a well or (2) instantaneous displacement by a float, weight, or change in pressure. *Aquifer pump* tests involve removing water from a well over a period of time from days to possibly weeks and measuring the changes in water levels in the pumping well (single-well test) and adjacent observation wells (multiple-well test). *Packer* tests are used to measure hydraulic conductivity in isolated sections of a borehole by monitoring the time-pressure response of the aquifer section when water is injected. The data from well tests are plotted and matched against curves calculated using analytical solutions to ground-water flow appropriate for the well construction and aquifer characteristics.

As Table A-7 indicates, all well tests measure hydraulic conductivity, but the types of other aquifer parameters that can be obtained from these tests vary. Slug and packer tests provide information on relatively small portions of an aquifer adjacent to the borehole, but are relatively easy to conduct and consequently are well suited for characterizing aquifer heterogeneity. Aquifer pump tests are more complex and difficult to carry out, but provide information on a larger portion of the aquifer. Pumping tests are the only well test method that information on the aquifer storage properties of an entire aquifer. Most slug tests provide relatively little information on aquifer storage properties.

A key element of aquifer testing is the selection of an appropriate analytical solution, or type curve developed from an analytical solution, to analyze the test data. Characteristics of the aquifer should not violate the assumptions used in developing the analytical solution. Worksheet D-W4 should be used to identify key aquifer characteristics that affect aquifer test results. ASTM D4043 (Table A-14) provides guidance on the selection of aquifer well test methods. Figure 7-12 provides a decision tree for the selection of methods covered in that guide. Table 7-6, at the end of the chapter, provides an index of references that give analytical solutions to aquifer test data according to pump test conditions and type of test. This table includes quite a few references not cited in ASTM D4043 and is most likely to be useful when aquifer conditions depart significantly from assumptions in the most commonly used analytical methods.

Well test methods are best suited for porous media, and most methods tend to give misleading results where fracture or conduit flow is an important component of ground-water flow. Section 7.5.4 discusses how the response of an aquifer to pumping can be used to evaluate whether fracture flow is a significant component of flow in an aquifer.

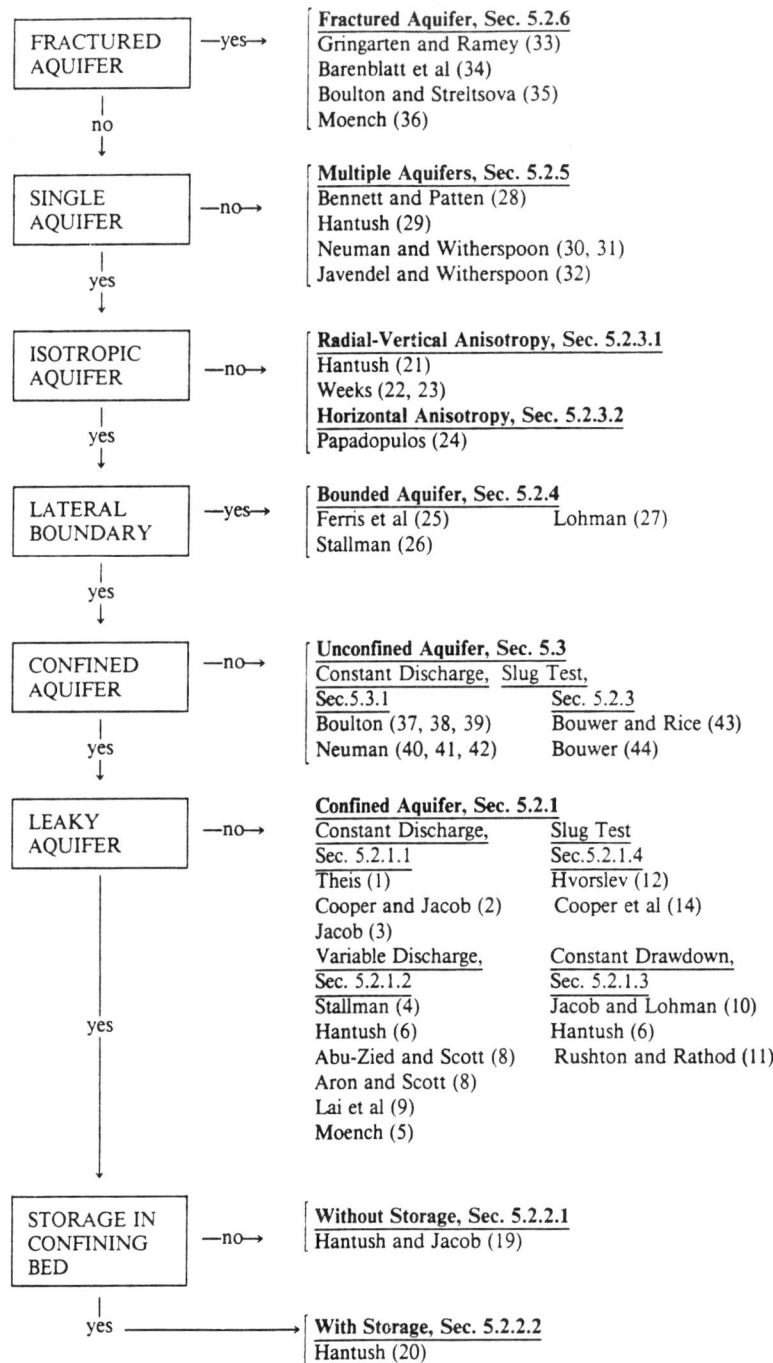

Figure 7-12 Decision tree for selection of aquifer test methods (section numbers refer to sections in ASTM D4043-91/Table A-14). © ASTM. Reprinted with permission.

Single-Well Tests. Hydraulic conductivity values can be estimated by slug or bail tests using a single well or piezometer. These tests record the response of the well or piezometer to a sudden change in the water level and provide in situ values representative of a small volume of the porous material in the immediate vicinity of the piezometer. The water level can be changed by introducing a known quantity of water, slugging the well, or removing a known quantity of water with a bailer. These methods are suitable where changes in water level take place slowly (minutes to hours) and accurate measurements can be made. Table 7-5, at the end of the chapter, identifies major references describing various methods including (1) shallow water table tests, (2) slug tests, and (3) packer tests. Another type of single-well test is the *step drawdown* test, where the well is pumped at successively greater discharge rates for relatively short periods of time and the drawdown at each rate is recorded. The step drawdown test was developed for use in wells that may have turbulent flow (Jacob, 1946b; Bierschenk, 1964/T7-6).

Multiple-Well Aquifer Tests. Multiple-well aquifer tests provide in situ measurements that are averaged over a large aquifer volume. Aquifer test data yield information on both the storativity and transmissivity of the aquifer. The design of a basic aquifer pump test consists of a pumping well that is pumped at a constant rate and at least one observation well. The wells are located far enough from any boundaries so that drawdown trends are not masked by the boundary effects. The interpretation of pump test results requires a good understanding of the hydraulics of ground-water flow. Drawdown data are collected from both the pumping well and the observation well(s); however, measurements in the pumping well may be less reliable due to turbulence caused by pumping. The diameter of an observation well should be just large enough to allow for the accurate measurement of the water levels.

The two most common methods for calculating the aquifer coefficients from the drawdown are data (1) the Theis method (Theis, 1935/T7-6) which involves using the Theis nonequilibrium well equation and curve matching on a log-log plot, and (2) the Jacob method (Cooper and Jacob, 1946/T7-6) which uses a simplified version of the Theis equation and involves plotting the data on a semi-log plot. The Jacob's equation yields an approximation of the Theis analytical solution that slightly reduces the accuracy of predicted head changes at locations far from the well unless pumping is long enough to approach equilibrium or steady-state conditions.

The principal assumptions in the Theis equation are:

- The aquifer is homogenous and isotropic.

- The aquifer is of infinite areal extent, relative to the effects of the well (no boundaries).

- The well is screened over the entire saturated thickness of the aquifer.

- The saturated thickness of the aquifer does not vary as a result of the operation of the well.

- The well has an infinitesimal diameter so that waters in storage in the casing represent an insignificant volume.

- Water is removed from or injected into the aquifer with an instantaneous change in the piezometric head.

Although most aquifers do not conform to all the theoretical conditions assumed by Theis and Jacob, these equations and their graphic relationships are often quite satisfactory as a general indicator of conditions. Numerous analytical equations have been developed for situations where the assumptions in the Theis equation do not apply. These conditions include (1) partially penetrating wells, (2) anisotropic aquifer conditions, and (3) leaky aquifers (confined aquifers where there is significant movement of water upward or downward through the overlying confining layer). Table 7-6 indexes references containing analytical solutions under various combinations of these conditions.

7.3.3 Other Field Tests

Tracer Tests. Table A-7 summarizes the types of hydrogeologic information that can be obtained using ground-water tracers. Chapter 8 discusses use is tracers in some detail.

Other Techniques. Table A-7 also identifies ten miscellaneous techniques for aquifer characterization. Potentiometric maps have been discussed in some detail in Section 7.2. Numerous procedures have been developed for hydrologic analysis based on the *water balance* or *budget* for an area. A simple water balance equation is as follows (Dunne and Leopold, 1978/T2-4):

$$\Delta GWS = P - I - AET - OF - \Delta SM - GWR \qquad (7\text{-}1)$$

where:

ΔGWS = change in ground water-storage
P = precipitation
I = interception
AET = actual evapotranspiration
OF = overland flow
ΔSM = change in soil moisture
GWR = ground-water outflow

Many variants are possible. The usual procedure is to formulate the equation with the parameter of interest on the left-hand side and the other components that define the hydrologic system of an area or aquifer of interest on the right-hand side. Table 7-5 identifies references that are good sources for further information on the water balance approach. In an unconfined aquifer, changes in soil moisture profiles in response to changes in the water table provide an alternative to pumping tests for measurement of *specific yield* (see Section 4.5.2 in Boulding, 1993/T7-5).

The barometric efficiency (Section 2.5.5) of confined aquifers, a measure of the response of a confined aquifer to changes in atmospheric pressure, is being increasingly used to estimate aquifer storage properties and transmissivity (see, for example, Ritzi et al., 1991).[11] Table A-7 also identifies some of the more commonly used borehole geophysical logging methods for measuring aquifer parameters. These methods are used primarily for characterizing aquifer heterogeneity vertically within a single borehole and laterally between boreholes.

7.3.4 Laboratory Measurements

Laboratory methods for measuring hydraulic conductivity provide an important complement to field tests (Petsonk, 1988).[12] Laboratory measurements of the properties of aquifer materials require the collection of undisturbed soil cores using thin-wall samplers for unconsolidated materials or rotating core samplers for rock (Section 9.4.1). Two commonly used methods for laboratory measurement of hydraulic conductivity are the *constant head* and *falling head permeameters* (ASTM D2434 and D5084/Table A-14; Klute and Dirksen, 1986-Footnote 3). In a constant head permeameter, the hydraulic gradient is kept constant, while the discharge is recorded. In a falling head permeameter, the hydraulic gradient and the discharge decreases with time. The constant head system is best suited to samples with hydraulic conductivities grater than 0.01 cm/min, while the falling head system is suited to samples with lower hydraulic conductivity (Klute and Dirksen, 1986/Footnote 13). Each sample is tested several times under different hydraulic gradients, and an average hydraulic conductivity is calculated. Hydraulic conductivity can also be measured using centrifugation (Alemi et al., 1976).[13] Effective porosity is another parameter that is measured in the laboratory (ASTM D4404/Table A-14; Horton et al., 1988).[14] Total porosity is usually calculated by measuring the dry bulk density of a known volume of soil or rock and assuming an average particle density.

A disadvantage of measuring aquifer properties from core samples is that they sample a very small portion of the aquifer. Consequently, values for hydraulic conductivity tend to be low compared to values measured in the field, which include the effects of secondary porosity and aquifer heterogeneities (Bradbury and Muldoon,

---

[11] Ritzi, R.W., S. Sorooshian, and P.A. Hsieh. 1991. The Estimation of Fluid Flow Properties from the Response of Water Levels in Wells to the Combined Atmospheric and Earth Tide Forces. Water Resources Research 27(5):883-893.

[12] Petsonk, A.M. 1988. Hydraulic Conductivity Measurements in Unsaturated Media--Materials vs. Lab Methods. Ground Water Monitoring Review 8(2):50-51.

[13] Alemi, M.H., D.R. Nielsen, and J.S. Biggar. 1976. Determining the Hydraulic Conductivity of Soil Cores by Centrifugation. Soil Sci. Soc. Am. J. 40:212-218.

[14] Horton, R., M.L. Thompson, and J.F. McBride. 1988. Determination of Effective Porosity of Soil Materials. EPA/600/2-88/045 (NTIS PB88-242391).

1990; Bryant and Bodocsi, 1987).[15] Figures 7-19 and 7-20 (Section 7.5.4) illustrate the effect of differences in scale of measurement on observed hydraulic conductivity in karst aquifers. On the other hand, laboratory measurement of multiple samples can provide valuable information on the vertical and lateral variability of aquifer properties. This information is especially important for constructing grids for three-dimensional aquifer modeling (Section 10.2.4).

7.4 Estimation of Subsurface Hydrologic Parameters

The critical aquifer parameters of porosity, specific yield, and hydraulic conductivity are typically not measured for most water wells. Therefore, the initial stages of hydrogeologic investigations often require estimation for one or more of these parameters. Estimation requires some knowledge of the geologic character of the aquifer and data on the ranges or typical values that have been measured in similar settings elsewhere. When used cautiously, such estimates can increase the effectiveness and reduce the cost of any required field measurements and additional data collection.

7.4.1 Estimation from Soil Survey Data

When aquifers are in unconsolidated deposits and the water table is relatively near the surface, soil surveys published by the Soil Conservation Service (SCS) of the U.S. Department of Agriculture are an excellent source of information about the character of subsurface materials and soil hydrologic properties. A two-page soil series description sheet and a two-page soil survey interpretation sheet are available for every established soil series in the United States. Table 7-1 summarizes the information that is available from these records. The table highlights in bold-face type the information that may be useful for geologic and hydrogeologic interpretations.

SCS soils surveys typically do not provide any detailed information deeper than 5 feet below the ground surface, but they do provide a general indication of the type of deeper geologic materials. In the absence of, or in combination with, other geologic data about the area of interest, this information provides a basis for estimating porosity, specific yield, and hydraulic conductivity, as discussed in the next section.

If a published SCS soil survey is available for a site of interest, the information in Table 7-1 will be contained in the report, but scattered in different locations. It is probably useful to obtain the single soil series descriptions and interpretations (usually available from the SCS State Office as a four-page handout) as a convenient consolidated reference for the soil series of interest. However, this sheet should be

---

[15] Bradbury, K.R. and M.A. Muldoon. 1990. Hydraulic Conductivity Determinations in Unlithified Glacial and Fluvial Materials. In: Ground Water and Vadose Zone Monitoring, D.M. Nielsen and A.I. Johnson (eds.), ASTM STP 1053, American Society for Testing and Materials, Philadelphia, PA, pp. 138-151.

Bryant, J. and A. Bodocsi. 1987. Precision and Reliability of Laboratory Permeability Measurements. EPA/600/2-86/097 (NTIS PB87-113791).

checked against data in the published soil survey, since the soil survey often will have additional data specific to the county in question.

Table 7-1 Types of Data Available on SCS Soil Series Description and Interpretation Sheets*

Soil Series Description Sheet

Taxonomic class
Typical soil profile description
Range of characteristics
Competing series
Geographic setting
Geographically associated soils
Drainage and permeability
Use and vegetation
Distribution and extent
Location and year series was established
Remarks
Availability of additional data

Soil Survey Interpretations Sheet

Estimated Soil Properties (major horizons)
    Texture class (USDA, Unified, and AASHTO)
    Particle size distribution
    Liquid limit
    Plasticity index
    Moist bulk density (g/cm^3)
    Permeability (in/hr)
    Available water capacity (in/in)
    Soil reaction (pH)
    Salinity (mmhos/cm)
    Shrink-swell potential
    Sodium absorption ratio
    Cation exchange capacity
    Calcium carbonate (%)
    Gypsum (%)
    Organic matter (%)
    Corrosivity (steel and concrete)
    Erosion factors (K,T)
    Wind erodability group
    Flooding (frequency, duration, months)
    High water table (depth, kind, months)
    Cemented pan (depth, hardness)
    Bedrock (depth, hardness)
    Subsidence (initial, total)
    Hydrologic group
    Potential frost action

* Most of this information is included in published SCS County soil surveys, if available.

Source: Boulding (1993c).

7.4.2 Estimation from Aquifer Matrix Type

Porosity, specific yield, and hydraulic conductivity fall within reasonably well-defined ranges for most aquifer materials, although some rocks, such as basalt, encompass the entire natural range of hydraulic conductivity (see Figure C-3). The following tables and figures in Appendix C provide information compiled from a variety of sources:

Porosity: Table C-1, Figure C-1 and Figure C-3.

Specific Yield: Table C-2 and Figures C-1, C-2, and C-15.

Hydraulic Conductivity: Table C-3 and Figures C-3 through C-15.

Sources may differ somewhat in the ranges given for a specific aquifer material. These differences probably exist because of slight differences in the way the material has been defined or because different sets of data measurements were examined. Below are some guidelines for estimating porosity, specific yield and hydraulic conductivity for a specific aquifer:

1. Define the nature of the aquifer material as thoroughly as possible, using available well logs, soil surveys, geologic maps, and hydrogeologic maps.

2. On the well data worksheet, enter values (or ranges) for porosity, specific yield, and hydraulic conductivity from all sources in the tables and figures identified above that provide data on similar or related aquifer materials.

3. If the sources provide different ranges for the same material, review the tables and/or figures again to see if any subtle distinctions in the way the materials are described might make one more appropriate for the aquifer in question.

4. Select a range of values that seems reasonable based on the information available, and enter the range in the well/aquifer data Worksheet D-W1. For aquifer materials with a wide possible range, the range should be narrowed based on the presence or absence of characteristics that tend to increase or decrease the parameter in question (Table 7-2).

Table 7-2 identifies factors that tend to increase or decrease porosity, specific yield, and hydraulic conductivity. Interactions between factors may mitigate or offset a given tendency. Many of the same factors tend to increase and decrease all three factors, but there are some interesting differences. Porosity tends to decrease as particle size increases, whereas the reverse is true for hydraulic conductivity. This is because clays have a high porosity, but the size of pores is so small that water moves very slowly. Specific yield, on the other hand, is typically highest in sandy materials and generally decreases with larger and smaller particle sizes. This is because as particle size increases to gravels, the pore space available to store water decreases, and as particle size decreases, water drains less readily from the smaller pores.

Table 7-2 Aquifer Characteristics Affecting Porosity, Specific Yield, and Hydraulic Conductivity

Parameter	Tendency to Increase	Tendency to Decrease
Porosity	Well sorted (same size)	Poorly sorted
	Rounded particles	Irregular-shaped particles
	Stratified	Unstratified
	Small particle size	Large particle size
	Unconsolidated	Cemented/lithified
	High secondary porosity	Low secondary porosity
Specific yield	Sand particle size	Gravel, silt, clay
	High secondary porosity	Low secondary porosity
Hydraulic conductivity	Gravel, sand	Clay
	Well sorted (same size)	Poorly sorted
	Stratified	Unstratified
	Unconsolidated	Cemented/lithified
	High secondary porosity	Low secondary porosity

Source: Boulding (1993c).

7.4.3 A Simple Well Test for Estimating Hydraulic Conductivity

The next section describes more complex well tests for measuring aquifer parameters, but a rough estimate of hydraulic conductivity is possible if three easily measured parameters are known: (1) the *static* water level prior to any pumping, (2) the normal well pumping rate, and (3) the level to which water drops after pumping starts and stays when inflow into the well equals the pumping rate. *Drawdown* is the difference between the static level and the level to which the water drops during pumping. The discharge rate of the well divided by the drawdown is the *specific capacity*, not to be confused with *specific yield* (Section 2.6.1). The specific capacity indicates how much water the well will produce per foot of drawdown. It can be calculated by the following equation:

$$S_c = Q/wd \qquad (7\text{-}2)$$

where:

S_c = specific capacity
Q = discharge rate, gpm
wd = well drawdown, feet (elevation of static water surface - elevation when pumped)

If a well produces 100 gpm and the drawdown is 8 feet, the well will produce 12.5 gpm for each foot of available drawdown. Multiplying specific capacity by 2,000 gives a crude estimate of transmissivity (T = 2,000 x specific capacity in units of gpd/ft), which in turn can be used to estimate hydraulic conductivity by rearranging Equation 7-2:

$$K = T/b = 2{,}000 \times S_c/b \qquad (7\text{-}3)$$

However, transmissivity estimates based on specific capacity measurements are commonly low because of well construction details (e.g., screen length is less than the thickness of the aquifer). Worksheet D-W1 contains space for recording information for calculating the specific capacity of a well.

7.5 Special Considerations in Hydrogeologic Mapping

The methods described so far all contribute in some fashion to the process of hydrogeologic mapping. Certain field conditions require special approaches to field investigations when they are present. These include: (1) identification of aquifer boundaries (Section 7.5.1); (2) characterization of aquifer heterogeneity and/or anisotropy (Section 7.5.2); (3) assessing the presence and degree of confinement in aquifers (Section 7.5.3); and (4) characterization of fractured rock and karst aquifers.

7.5.1 Delineation of Aquifer Boundaries

Identification of subsurface aquifer boundaries is an essential part of selecting analytical solutions for aquifer tests and computer modeling of ground-water systems chracterized by porous-media flow (see Section 7.5.4 for approaches to delineating aquifer boundaries in fractured rock and karst aquifers). They must also be known when delineating a well's zone of contribution (ZOC) for wellhead protection areas (Section 11.2.1). Ground-water divides upgradient from a well can be readily identified using a potentiometric surface map (Section 7.2). Section 2.5.6 discusses other major types of aquifer boundaries. Worksheet D-W3 can be used to identify possible aquifer boundaries that may affect a well. Figure 2-13 provides illustrations of most of these types of boundaries. Determining the distance from the boundary to the well will help identify those boundaries that might be most significant for purposes of wellhead protection area delineation.

Additional analysis using simple analytical methods for calculating drawdown (Table 7-6) may be required to determine whether an aquifer boundary actually functions as a boundary to the well's zone of contribution (ZOC). For example, a stream downgradient from a well would represent a potential boundary, but would only serve as an actual boundary if the well's zone of influence (ZOI, see definitions in Table 11-2) extends to the stream.[16] Similarly, an impermeable boundary that lies outside the upgradient ZOC would not represent an aquifer boundary to the ZOC.

---

[16] If the ZOI is within several hundred feet of the stream, some consideration should be given to the possibility of bank storage effects during flooding (Figure 2-6).

7.5.2 Characterization of Aquifer Heterogeneity and Anisotropy

Failure to characterize aquifer heterogeneity and anisotropy can have a number of undesired consequences for contaminant investigations (Section 7.2.5). They are also important considerations in delineation of wellhead protection areas (Section 11.2). Using an average value for hydraulic conductivity in any of the simple wellhead protection area delineation (WHPA) methods summarized in Table 11-1 will underestimate the time of travel or zone of influence based on drawdown, because contaminants will travel faster in fractures or layers of higher permeability, if they are present. Aquifer anisotropy or heterogeneity can result in incorrect delineation of WHPA boundaries based on potentiometric maps and flow net analysis. Figure 7-5 illustrates this effect in an anisotropic aquifer, and Figure 7-11 shows how this can happen in a heterogeneous aquifer. Consequently, any hydrogeologic investigation should assess the presence and degree of variability of hydrologic properties vertically and laterally.

Any method that allows measurement or qualitative observation of the similarities and differences in a particular aquifer characteristic in a vertical or horizontal direction allows assessment of whether an aquifer is homogeneous or heterogeneous. Table 7-3 summarizes a number of field methods that are commonly used or especially well suited for this purpose. Drill logs and geophysical borehole logs allow assessment of vertical changes in lithology, porosity, and permeability. Packer tests allow measurement of variations in hydraulic conductivity at different intervals. Surface geophysical methods, such as seismic refraction, seismic reflection, and electrical resistivity soundings, also allow less precise mapping of vertical changes in lithology.

An accurate potentiometric surface map (7.2.2) is one of the most valuable ways to evaluate aquifer heterogeneity. Hydrochemical maps also provide information that can be specifically related to the hydrogeology of an area. Tracer tests (Chapter 8) may indicate whether fracture flow or zones of high permeability exist. This is indicated when the time of travel of the tracer is faster than the time of travel calculated from estimated aquifer properties or values measured by well tests. Geologic cross-sections, isopach maps, and structural maps, which are generally based on interpolations between borehole logs, allow assessment of lithologic variations. Surface geophysical methods allow relatively rapid measurement of lateral variations in lithology, structure, and water quality where no better subsurface information is available. However, some verification with subsurface borehole data is required.

Geostatistical methods, originally developed for characterizing mineral ore bodies (Section 5.6.2), have been found to be increasingly useful tools for characterizing the variability of aquifer parameters. For example, Poeter and Belcher (1991)[17] recently described a method for characterizing porous medium heterogeneity by "inverse plume analysis", in which the spatial distribution of contaminant concen-

---

[17] Poeter, E.P. and W.R. Belcher. 1991. Assessment of Porous Media Heterogeneity by Inverse Plume Analysis. Ground Water 29(1):56-62.

Table 7-3 Summary of Methods for Characterizing Aquifer Heterogeneity

Method	Properties	Comments
Vertical Variations		
Drill logs	-Changes in lithology. -Aquifer thickness. -Confining bed thickness. -Layers of high/low hydraulic conductivity. -Variations in primary porosity (based on material description).	Basic source for geologic cross sections. Descriptions prepared by geologist preferred over those by well drillers. Continuous core samples provided more accurate descriptions.
Electric logs	-Changes in lithology. -Changes in water quality. -Strike and dip (dipmeter).	Require uncased hole and fluid-filled borehole.
Nuclear logs	-Changes in lithology. -Changes in porosity (gamma-gamma).	Suitable for all borehole conditions (cased, uncased, dry, and fluid-filled).
Acoustic and seismic logs	-Changes in lithology. -Changes in porosity. -Fracture characterization -Strike and dip (acoustic televiewer).	Requires uncased or steel cased hole, and fluid-filled hole.
Other logs	-Secondary porosity (caliper, television/photography). -Variations in permeability (fluid-temperature, flowmeters, single borehole tracing).	Require open, fluid-filled borehole. Relatively inexpensive and easy to use.
Packer tests	-Hydraulic conductivity.	Single-packer tests used during drilling; double-packer tests after hole completed.
Surface geophysics	-Changes in lithology (resistivity, electromagnetic induction, time-domain EM, seismic refraction).	Requires use of vertical sounding methods for electrical and electromagnetic methods.
Lateral Variations		
Potentiometric maps	-Changes in hydraulic conductivity.	Based on interpretation of the shape and spacing of equipotential contours.
Hydrochemical maps	-Changes in water chemistry.	Requires careful sampling, preservation and analysis to make sure samples are representative.

Table 7-3 (cont.)

Method	Properties	Comments
Tracer tests	-Time of travel between points. -Potential distribution of contaminants.	Requires injection point and sufficient downgradient collection points. Essential for investigation of flow in karst.
Geologic maps and cross sections	-Changes in formation thickness. -Structural features, faults.	Result from correlation features observed at the surface and in boreholes.
Isopach maps	-Variations in aquifer and confining layer thickness.	Distinctive strata with large areal extent required.
Geologic structure maps	-Stratigraphic and structural boundary conditions affecting aquifers.	See Table 7-4.
Surface geophysics	-Changes in lithology (seismic). -Structural features (seismic, GPR, gravity). -Changes in water quality/contaminant plume detection (ER, EMI, GPR).	Interpretations require verification using subsurface borehole data.

Source: Boulding (1993c).

trations is used to evaluate variation in aquifer properties. Both of these approaches, however, require a relatively high density of subsurface observations. Special approaches to aquifer characterization are typically required in fractured rock and karst limestone aquifers, as discussed in Section 7.5.4.

Measurement of Anisotropy. Measurement of anisotropy requires determination of the direction of maximum and minimum hydraulic conductivity. In a homogenous, horizontally layered aquifer, the direction of minimum conductivity is usually assumed to be in the vertical direction, and the maximum in the horizontal direction (Section 2.5.3). Fetter (1981-Footnote 10) suggests collecting undisturbed cores for measurement of vertical hydraulic conductivity in the laboratory (Section 7.3.4) and using slug tests (Section 7.3.2), which primarily measure horizontal conductivity, in the test hole. This procedure also requires installation of at least three wells to accurately determine the orientation of equipotential lines.

A number of other methods have been developed for estimating anisotropy in layered aquifers using pumping tests. Most require a minimum of two or three observation wells, in addition to a pumping well, to measure the degree of departure from a circular cone of depression that occurs in an isotropic aquifer. In fractured rock aquifers, anisotropy can occur in three directions with no principle axis aligned in a vertical or horizontal direction. In this situation, various approaches have been developed for measuring anisotropy using packer tests in multiple holes. The *dipole flow test*, recently described by Kabala (1993/T7-5), is a single-hole, multilevel packer test that measures distribution of horizontal and vertical hydraulic conductivity and the specific storativity when applied to different bounded intervals. Table 7-6 provides an index to references where more detailed information on specific methods for measuring anisotropy can be obtained.

7.5.3 Presence and Degree of Confinement

The presence and degree of confinement has a significant impact on the vulnerability of an aquifer to contamination and the size of the WHPA for a given time of travel or drawdown criterion (Section 11.2.1). Figure 7-13 shows the location of major and significant minor confined aquifers in the contiguous United States. Methods for evaluating these aquifer properties can be broadly classified as (1) geologic, (2) hydrologic, and (3) hydrochemical. Table 7-4 identifies 15 indicators of confinement and the characteristics that are associated with highly confined or semiconfined conditions. Kreitler and Senger (1991)[18] provide more detailed discussion of these methods.

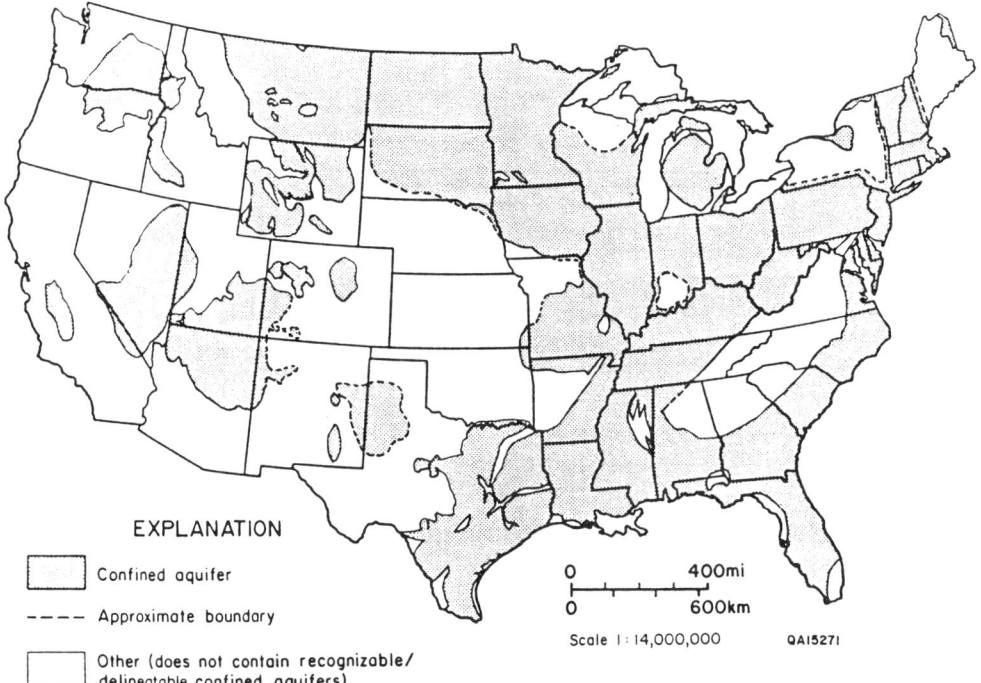

Figure 7-13 Major and significant minor confined aquifers of the United States (Kreitler and Senger, 1991).

---

[18] Kreitler, C.W. and R.K. Senger. 1991. Wellhead Protection Strategies for Confined-Aquifer Settings. EPA/570/9-91-008, 168 pp.

Table 7-4 Indicators of Presence and Degree of Confinement

Information Source	Highly Confined	Semiconfined (Leaky)
Geologic		
Geologic maps and cross-sections	Presence of continuous, unfractured, confining strata (clays, glacial till, shale, siltstone).	Evidence of vertical permeability in confining strata (fracture traces, faults, mineralization or oxidation of fractures observed in cores).
Environmental geologic and hydrogeologic maps	See above.	Presence of artificial penetrations (abandoned or producing oil and gas wells, water wells, exploration boreholes).
Hydrologic		
Water level elevation (single well) of potentiometric surface	Above the top of the aquifer (not diagnostic for differentiation of highly and semiconfined aquifers).	Same.
Hydraulic head differences between aquifers	Large head difference in water levels measured in wells cased in different aquifers (not diagnostic for differentiation of highly and semiconfined aquifers).	Same.
Water level fluctuations (continuous measurement)	Short-lived and diurnal fluctuations in response to changes in barometric pressure, tidal effects, external loading (Table 2-2); no response to recharge events.	Similar to highly confined aquifer, but may also exhibit relatively large and rapid response to recharge events because of leakage through discrete points.
Hydrologic measurements in confining strata	No changes in water levels in response to pumping; diurnal but not seasonal water level fluctuations (see above).	Changes in water levels in response to pumping; seasonal water level fluctuations in response to seasonal variations in precipitation.
Aquifer test for storativity	Storativity less than 0.001.	Between 0.01 and 0.001 (not diagnostic).
Aquifer test for leakage	Pump drawdown vs. time curve matches analytical solution(s) for highly confined aquifer. Estimated or calculated leakage less than 10^{-3} gal/day/ft^2.	Pump drawdown vs. time curve requires use of analytical solution for leaky aquifer. Estimated or calculated leakage 10^{-2} to 10^2 gal/day/ft^2.

Table 7-4 (cont.)

Information Source	Highly Confined	Semiconfined (Leaky)
Numerical modeling	Simulation of potentiometric surface possible without estimates of leakage, or required estimates are low (see above).	Simulation of potentiometric surface requires use of large leakage values.
<u>Hydrochemistry</u>		
General water chemistry	Chemical characteristics indicative of long distance from recharge area (region-specific).	Qualifies as confined using other criteria, but chemical characteristics more similar to ground water in recharge zones.
Anthropogenic atmospheric tracers	No detectable tritium or fluorocarbons in ground water.	Detectable concentrations of tritium or fluorocarbons (less than 40 years old).
Isotope chemistry	Carbon-14 dating of water samples indicates age > 500 years.	See above.
Contaminants	No detectable concentrations of potential contaminants identified by inventory of potential contaminant sources.	Qualifies as confined using other criteria, and contaminants detected in aquifer.
Changes in water chemistry over time	Head declines from long-term pumping have not resulted in changes in water chemistry indicators of vertical leakage.	Head declines from long-term pumping have resulted in changes in water chemistry indicators of vertical leakage (see above).
Time of travel through confining strata	Time of travel calculations based on measured or estimated values of difference in hydraulic head, porosity and hydraulic conductivity exceed 40 years.	Time of travel through confining strata <40 years based on calculations or presence of tritium or fluorocarbons.

Source: Adapted from Kreitler and Senger (1991).

7.5.4 Characterization of Fractured Rock and Karst Aquifers

Where fracture or conduit flow (Section 2.5.4) occurs in an aquifer, special care and techniques are required for delineating wellhead protection areas. Figure 7-14 identifies major areas of the United States and associated territories where unconfined fracture flow is significant, and Figure 1-10 identifies major karst areas of the contiguous United States and other areas where carbonate rocks are at or near the surface. The term "fractured rock" aquifer in this book refers to areas where most of the water supplied to a pumping well comes from fractures with sufficiently narrow apertures that Darcian flow (Section 2.6.3) occurs. Common geologic settings where fractured rock aquifers occur include crystalline intrusive igneous (i.e., granites) and metamorphic rocks, basalts, and some carbonates.

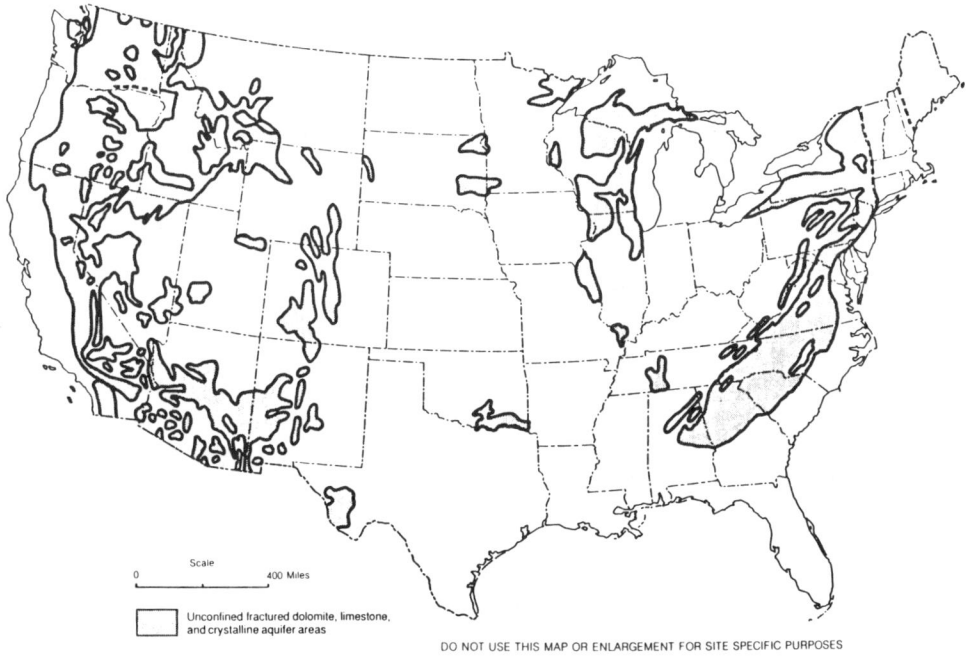

Figure 7-14 Major areas of unconfined fractured rock aquifers in the United States (Bradbury et al., 1991).

The term "karst" aquifer in this book refers to carbonate aquifers where conduit flow is an important component of the ground-water flow system. Generally, where apertures are >5mm in diameter or width and water velocity is >0.001 m/s, ground-water flow is turbulent rather than Darcian. As shown in Figure 1-10, not all carbonate rocks (limestone and dolomite) are karst aquifers. However, whenever carbonate aquifers are present, either fracture or conduit flow should be assumed.

The fundamental objective of hydrogeologic mapping in fractured rock and karst aquifers should be to identify (1) the **boundaries** of the flow system and (2) the **structure** of the flow system. Important elements in defining the structure of a carbonate aquifer include: (1) relative importance of diffuse and conduit flow, (2) the amount of storage in soil above the aquifer, and 3) the extent to which recharge comes from point sources, such as sinkholes, or is dispersed. The rest of this section provides an overview of major methods for characterizing the boundaries and structure of fracture rock and karst systems. Table 1-4 provides an extensive list of major references on karst geology, geomorphology, and hydrology where more detailed information can be obtained.

The primary method for mapping the boundaries of an unconfined fractured rock or karst aquifer is dye tracing (Section 8.4). In karst aquifers this is the only reliable method because conduit flow systems often do not follow surface water drainage systems. For example, Bonacci and Živaljević (1993),[19] using dye tracing and a water budget of a large spring in the Dinaric karst of Montenegro, found the catchment area to be 76 to 79 km^2, while hydrogeologic mapping based on geology and topography indicated a catchment area of 120 to 170 km^2.[20]

Significant differences in flow direction may occur in karst aquifers depending on whether low-flow or high-flow conditions exist. Again, such changes can only be accurately determined using properly designed dye tracer tests. For example, low-flow and high-flow tracer tests were conducted by injecting dye into several wells in the vicinity of Lemon Lane landfill, a Superfund site contaminated with PCBs (McCann and Krothe, 1992).[21] The landfill is located on a topographic divide in a karst area where more than 30 springs have been identified within a mile-and-a-half radius of the landfill. A low-flow tracer test conducted in 1987 found that most water infiltrating in the vicinity of the landfill flowed in a southeasterly direction, but some also flowed to the northeast. A high-flow tracer test, conducted two years later, found that most flow was still in a southeasterly direction, but that some flow occurred in all directions, with dye being detected in springs up to a mile-and-a-half distant in all directions from the landfill.

A variety of methods are available for characterizing the structure of fractured rock and karst flow systems. These can be broadly classified as (1) remote sensing, surface, and borehole geophysical methods; (2) monitoring of natural fluctuations of water levels in wells and their response to pumping; and (3) monitoring of discharge and chemistry of springs.

Remote Sensing and Geophysical Methods. Fracture trace and lineament analysis using air photos is a useful starting point for identifying possible areas of

---

[19] Bonacci, O. and R. Živaljević. 1993. Hydrological Explanation of the Flow in Karst: Example of the Crnojevića Spring. J. Hydrology 146:405-419.

[20] Note that the hydrogeology of karst terranes of the former Yugoslavia are generally very different from karst areas in North America. In the United States, catchments in karst areas typically are larger than would be expected based on an analysis of surface topography.

[21] McCann, M.R. and N.C. Krothe. 1992. Development of a Monitoring Program at a Superfund Site in a Karst Terrane Near Bloomington, Indiana. Ground Water Management 10:349-370 (Proc. 3rd Conf. on Hydrogeology, Ecology, Monitoring and Management of Ground Water in Karst Terranes).

concentration and preferential direction of ground-water flow.[22] Other remote sensing methods, such as near-infrared and thermal infrared scanners, which detect variations in near-surface moisture, may also be useful for mapping the location of sinkholes and fracture trace analysis (LaMoreaux, 1979).[23] Such observations should be supplemented, where possible, with observation and analysis of the character and orientation of rock joint and fracture patterns at surface outcrops (LaPointe and Hudson, 1985).[24]

A number of commonly used surface geophysical methods have potential applications for detection of shallow subsurface cavities in karst areas, including gravity, electrical resistivity, seismic, and ground penetrating radar (Greenfield, 1979).[25] Karous and Mareš (1988)[26] provide detailed treatment of use of geophysical methods for characterizing fractured-rock aquifers, including some methods that are less commonly known. For example, Figure 7-15 illustrates how a conduit feeding a karst spring can be mapped using self-potential measurements. In this example, the current electrode A was grounded at the spring orifice, and potentials measured along transects I through IV. In the U.S., natural potential (another term for self-potential), has been used in karst areas for siting of monitoring wells, delineation of flow paths, mapping of drawdown-recovery patterns associated with pumping, and for minimizing effects of petroleum drilling (Lange and Kilty, 1992).[27]

Figure 7-16 illustrates how repeated seismic velocity measurements at different orientations around a single point provide an indication of the orientation of major fractures. In this example, velocities have been plotted on a polar diagram, with the inferred direction of major fractures based on the higher velocity measurements. Azimuthal resistivity, in which a series of resistivity measurements are taken by shifting the position of the electrodes around a single point, is another possible method for

---

[22] Fracture trace analysis will not necessarily identify major conduits in karst aquifers, however, because these may follow bedding planes with no surface expression.

[23] LaMoreaux, P.E. 1979. Remote-Sensing Techniques and the Detection of Karst. Ass. Eng. Geol. Bull. 16(3):383-392.

[24] LaPointe, P.R. and J.A. Hudson. 1985. Characterization and Interpretation of Rock Mass Joint Patterns. Geological Society of America Special Paper 199, Boulder, CO, 37 pp.

[25] Greenfield, R.J. 1979. Review of Geophysical Approaches to Detection of Karst. Ass. Eng. Geol. Bull. 16(3):398-408.

[26] Karous, M. and S. Mareš. 1988. Geophysical Methods in Studying Fracture Aquifers. Charles University, Prague, 93 pp.

[27] Lange, A.L. and K.T. Kilty. 1992. Natural-Potential Responses of Karst Systems at the Ground Surface. Ground Water Management 10:179-193 (Proc. 3rd Conf. of Hydrogeology, Ecology, Monitoring and Management of Ground Water in Karst Terranes).

detecting fracture orientation (Ritzi and Andolesk, 1992).[28] Azimuthal methods may be useful for identifying fracture zones that contribute to recharge of karst systems, but may not be successful in locating zone of conduit flow, which tends to be along bedding planes, often without any surface expression as fracture traces.

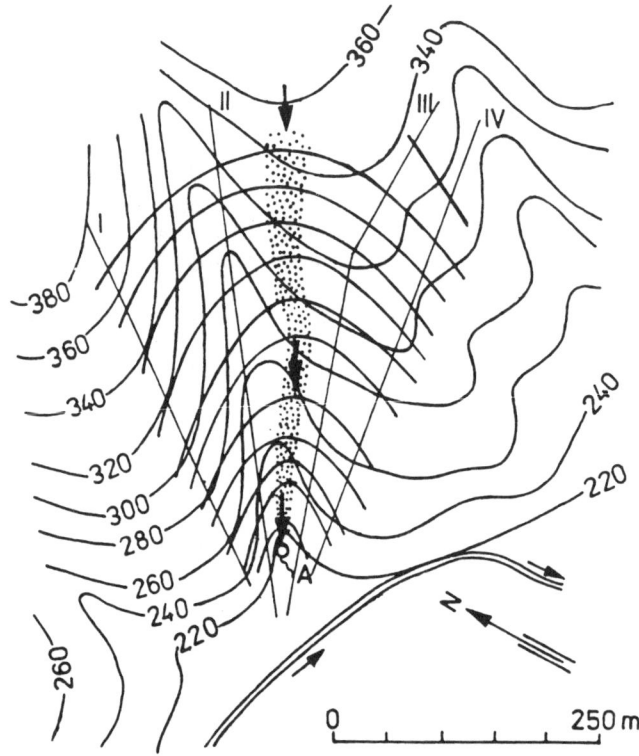

Figure 7-15 Mapping of subsurface conduit using self-potential method (Karous and Mareš, 1988).

Borehole geophysical methods provide a necessary complement to surface geophysical and other characterization techniques (Section 6.4). Acoustic televiewer, borehole television, and dipmeter logs are especially useful for determining the location and orientation of subsurface fractures. Fracture zones can also be detected using borehole flowmeters (mechanical, thermal, and the recently developed electromagnetic flowmeter) with or without pumping. Single-borehole and multiple-well tracer tests ar useful for characterizing the flow at a more local scale. Table 7-6 identifies a number of additional references characterizing fractured rock aquifers.

---

[28] Ritzi, Jr., R.W. and R.H. Andolesk. 1992. Relation Between Anisotropic Transmissivity and Azimuthal Resistivity Surveys in Shallow, Fractured, Carbonate Flow Systems. Ground Water 30(5):774-780.

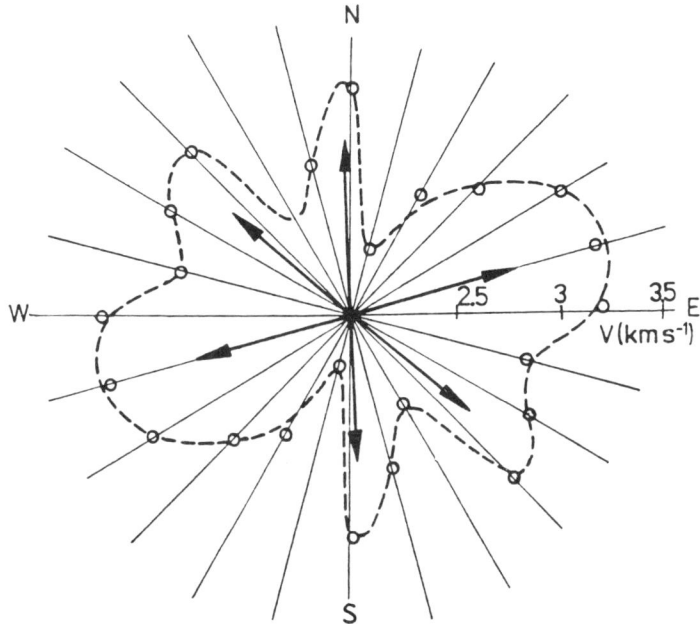

Figure 7-16 Azimuthal seismic survey to characterize direction of subsurface rock fractures (Karous and Mareš, 1988).

Water Level Monitoring. In unconfined fractured rock and karst aquifers, water levels in wells intercepting fractures or conduits commonly show relatively large fluctuations in response to precipitation events (see Figure 1-9). During times of low flow, large differences in water levels in nearby wells serves as an indicator of low matrix permeability (the well with higher water levels) and fracture or conduit flow in the well with the lower water levels.

The response of water levels to pumping provides a basis for judging whether the flow system functions as a "porous medium equivalent" (i.e., the aquifer can be modeled as if it were flowing in a porous medium, even though flow in fractures is occurring).[29] Figure 7-17 illustrates three types of aquifer responses to pumping that indicate a porous medium model should **not** be used for characterizing an aquifer.

---

[29] In the context of wellhead protection, even if a fractured rock or karst aquifer can be modeled using porous medium flow assumptions, results should be interpreted with great caution. Values of hydraulic conductivity calculated from such aquifer tests will reflect average values, whereas actual ground-water flow velocities will be much higher. For example, Quinlan et al. (1992-see Footnote 31) cite a tracer test in the Floridan aquifer using two wells 200 feet apart. The theoretical arrival time of the injected dye, based on geophysical logging and aquifer testing, was about 40 days. Actual breakthrough time was 5 hours.

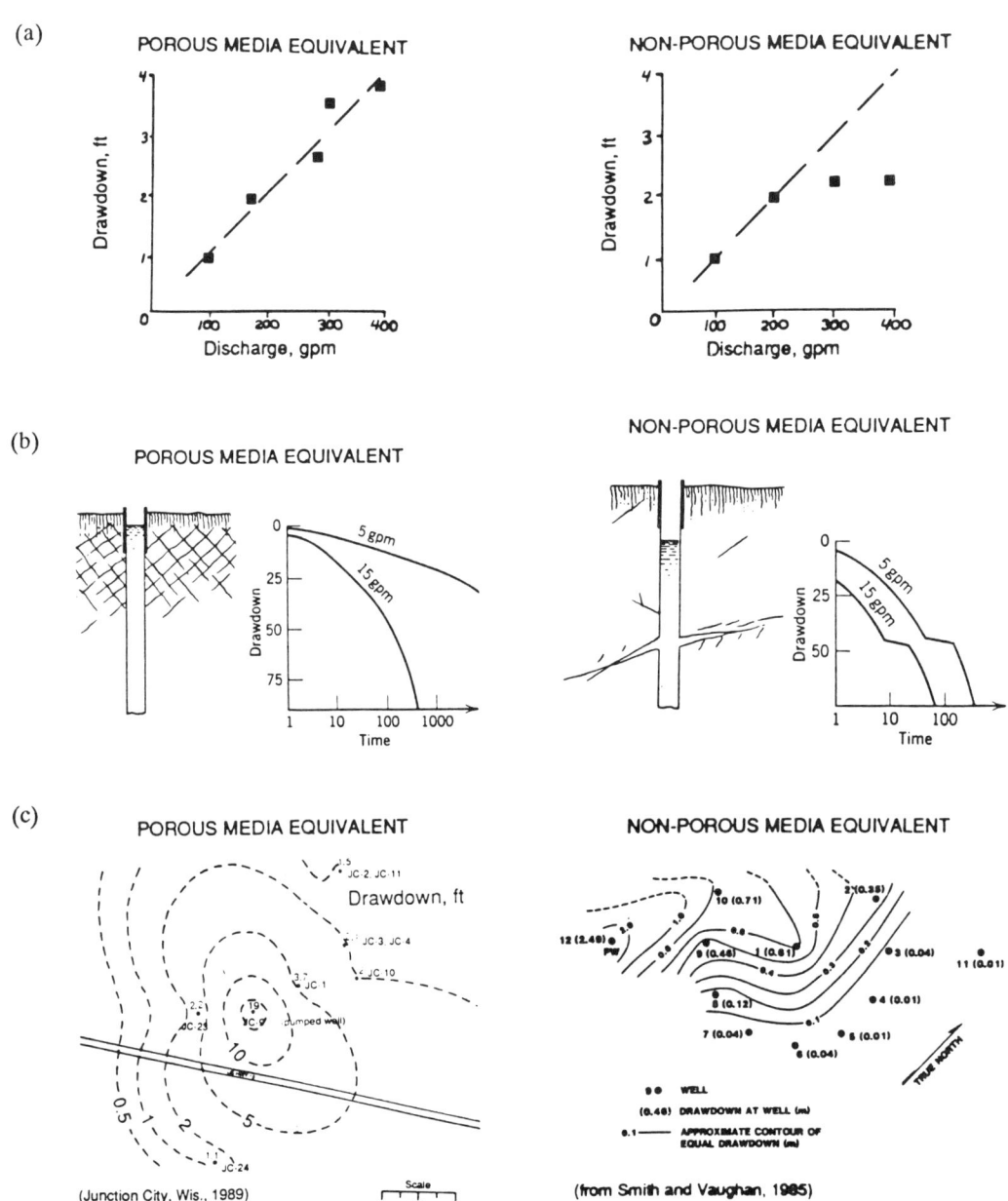

Figure 7-17 Pumping test response indicators of fracture/conduit flow: (a) discharge drawdown plots; (b) time drawdown curves, (c) areal drawdown distribution (Bradbury et al., 1991).

The left-hand side of Figure 7-17 illustrates three types of aquifer response that are indicative of a porous medium equivalent, and the right-hand side indicates aquifers where large fracture or conduit flow is present. Granular aquifers (and fractured rock aquifers where fractures are relatively small and evenly spaced) will generally show a linear relationship between drawdown and pumping rate, whereas aquifers where fracture flow is significant may show a leveling off response in drawdown as pumping rates increase (Figure 7-17a). The presence of large water-bearing fractures is indicated by a temporary leveling off in a drawdown vs. time plot (Figure 7-17b). Finally, if major fractures are feeding a well, the cone of depression may depart significantly from a circular or elliptical shape (Figure 7-17c). Nonporous medium equivalent responses in aquifer tests require use of the appropriate fracture-flow analytical solutions for analyzing pump test data (Table 7-6). All of these responses can also be indicative of conduit flow in carbonate aquifers.

Spring Monitoring. A distinctive characteristic of near-surface karst hydrologic systems is that springs serve as discharge points for subsurface flow. Much useful information about a karst aquifer can be obtained by monitoring the amount and chemistry of flow from a spring. Kresic (1993)[30] provides a review of methods for spring hydrograph analysis and statistical analysis of time series measurements of flow from springs and water level measurements in wells. With antecedent soil moisture conditions being equal, a rapid increase in discharge from a spring in response to a precipitation event indicates that point recharge is a major component of subsurface flow, whereas a relatively small flow response indicates that dispersed recharge contributes most of the flow to a spring. Quantitative interpretations of spring hydrographs require continuous records of both spring discharge and precipitation in the catchment area (which must be delineated by properly designed tracer testing).

Specific conductance, an easily measured ground-water parameter, is widely used for characterizing karst aquifers. Where multiple springs are present in an area, springs with similar specific conductance can be considered to be closely interconnected, while large differences in specific conductance indicate that the flow systems feeding the springs are largely independent. Monitoring of changes in water chemistry with changes in spring discharge is also a useful way to characterize karst aquifers. Specific conductance is the parameter of choice because it is easy to measure and can be monitored continuously (Quinlan et al., 1992).[31] Other parameters such

---

[30] Kresic, N.A. 1993. Review and Selected Bibliography on Quantitative Definition of Karst Hydrogeological Systems. In: Annotated Bibliography of Karst Terranes, Vol. 5 with Three Review Articles, P.E. LaMoreaux, F.A. Assaad, and A. McCarley (eds.), International Contributions to Hydrogeology, Vol. 14, International Association of Hydrogeologists, Verlag Heinz Heise, Hannover, West Germany, pp. 51-87.

[31] Quinlan, J.F., P.L. Smart, G.M. Schindel, E.C. Alexander, Jr., A.J. Edwards, and A.R. Smith. 1992. Recommended Administrative/Regulatory Definition of Karst Aquifer, Principles for Classification and Carbonate Aquifers, Practical Evaluation of Vulnerability of Karst Aquifers, and Determination of Optimum Sampling Frequency at Springs. Ground Water Management 10:573-635 (Proc. 3rd Conf. on Hydrogeology, Ecology, Monitoring and Management of Ground Water in Karst Terranes).

as hardness, degree of saturation with respect to calcite and dolomite, and the Ca/Mg ratio can also be used. A high coefficient of variation of specific conductance (CVC) indicates that point recharge is a major contributor to flow, whereas a low CVC indicates that most recharge comes from dispersed sources. Quinlan et al. (1992-Footnote 31) suggest the following provisional guidelines using CVC as a measure of aquifer vulnerability: moderately sensitive = <5 percent, very sensitive = 5 to 10 percent, hypersensitive = >10 percent.

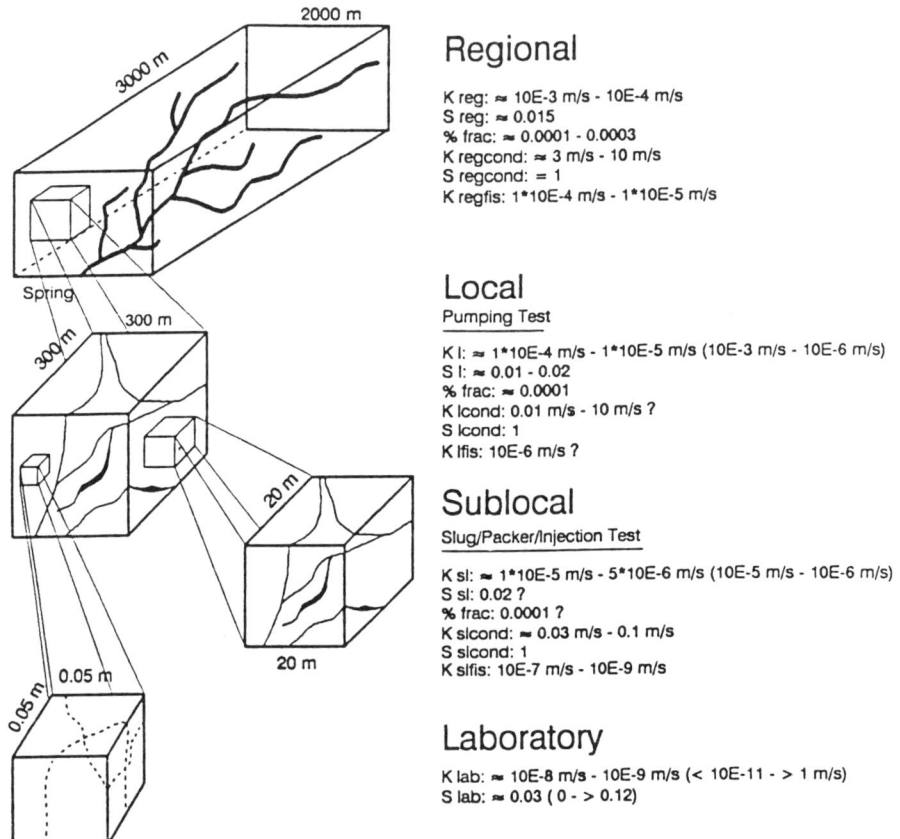

Figure 7-18 Geometrical relationships and hydraulic conductivities at different scales in karst systems (Sauter, 1992). See Appendix F for full credit.

A Cautionary Note: Footnote 28 discussed the possible risk of using porous-medium analytical models for delineating WHPAs in fractured rock or karst areas, even if aquifer test data suggest that flow behavior approximates that in a porous medium. The results of any methods used to quantify storage properties or hydraulic conductivity in fractured rock and karst aquifers described above must be evaluated in the context of the volume of the aquifer that is being measured. As noted in Section 7.3.4, values for hydraulic conductivity tend to increase as larger volumes of an aquifer are measured. This effect is particularly dramatic in karst aquifers. Figure

7-18 shows the effect of scale from laboratory core measurements (centimeters) to regional (thousands of meters) on the storage coefficient (S) and hydraulic conductivity (K) in the Swabian Alps of southwestern Germany. Measurements of K range over six orders of magnitude. Figure 7-19, which summarizes data from many different studies in karst areas, shows an even wider range of eight orders of magnitude for the **predominant** ranges of major methods for estimating average velocity (laboratory core, double packer tests, slug tests, aquifer pump tests, and dye tracer tests). These figures make it clear that time of travel estimates used for delineation of wellhead protection areas in karst aquifers based on any methods other than dye tracer tests are unlikely to provide adequate protection.

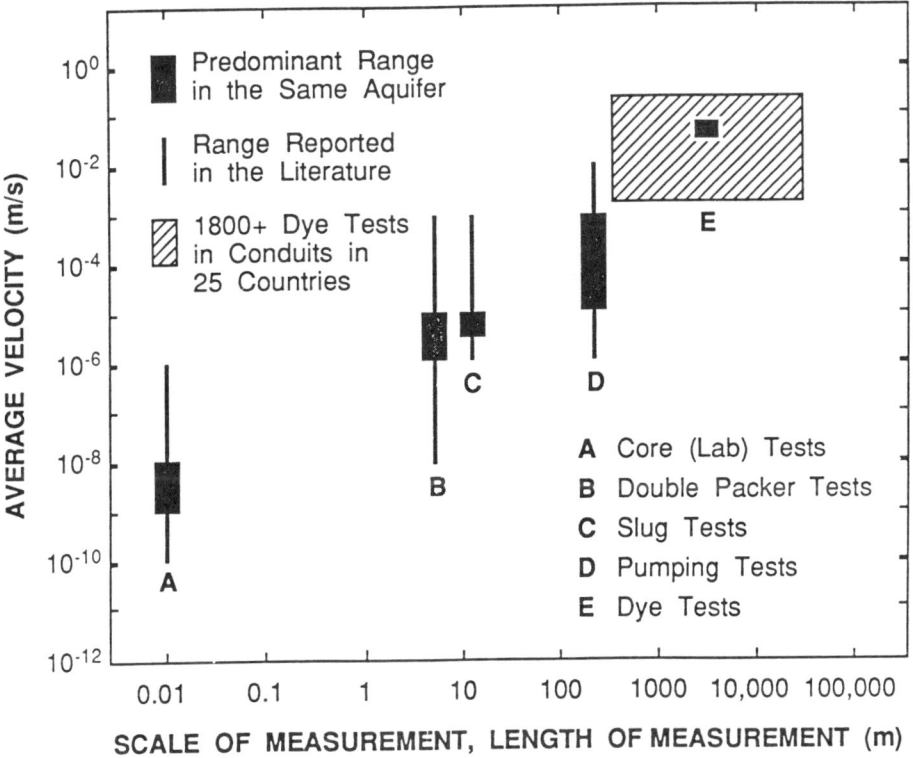

Figure 7-19 Measurement scales and average velocities of different measurement methods in karst systems (modified after Quinlan et al., 1992a and Sauter 1992).

7.6 Guide to Major References

Table 7-5 identifies major references on vadose zone and ground-water hydrologic characterization covering the following major topics: (1) vadose zone hydrology (unsaturated hydraulic conductivity and saturated hydraulic conductivity above a water table), (2) aquifer tests (shallow water table tests, slug tests, packer tests, and multiple-well pump tests), and (3) water balance methods. Dawson and Istok (1991) and Kruseman and de Ridder (1991) are good recent references on design and analysis of aquifer tests. Most of the vadose zone hydrology and hydrogeology texts identified in Table 2-4 also cover field measurement methods. Methods for conducting and analyzing slug and packer tests is a relatively active area of research and development, so Table 7-5 contains a higher portion of journal citations than most of the reference tables in this book.

Table 7-6 provides an index of more than 120 source references covering (1) specific pump test analytical solutions for a variety of aquifer conditions, (2) methods for measuring anisotropy, and (3) methods for characterizing fractured rock. References indexed in this table are almost exclusively from the journal literature, and stands as an exception to the practice in this book of focussing on major texts and references. The first part of this table can be used with Worksheet D-W4 to identify the source literature on analytical solutions where hydrogeologic and well characteristics make use of the Theis and Jacob's solutions inappropriate (see Section 7.3.2 the basic assumptions).

This chapter and elsewhere in this book the importance of considering the effect of anisotropy and fracture flow in ground-water systems has been emphasized. Table 7-6 also provides an index to the literature on these topics, which tend to not be addressed comprehensively in other sources.

Table 7-5 Index to Major References on Hydraulic Conductivity and Water Balance Test Methods

Topic	References
<u>Vadose Zone Hydrology</u> (see also Table 2-4)	
Unsaturated Hydraulic Conductivity	Boulding (1993), Bouma (1977), Bouma et al. (1974), Bouwer and Jackson (1974), Dirksen (1991), Green et al. (1986), Hendrickx (1990), Hillel and Benyamini (1974), U.S. EPA (1986)
Saturated Hydraulic Conductivity (Above Water Table)	Amoozegar and Warrick (1986), Boersma (1965), Boulding (1993), Bouwer and Jackson (1974), Hamilton et al. (1981), Hendrickx (1990), Kessler and Oosterbaan (1974), Lambe (1955), Stephens and Neuman (1980, 1982a, 1982b, 1982c), Sai and Anderson (1991), Stephens et al. (1988), Winger (1960), Youngs (1991); <u>Method Comparisons</u>: Havlena and Stephens (1992), Lee et al. (1985), Reynolds et al. (1983), Roberts (1984), Sai and Anderson (1991), U.S. EPA (1986)
<u>Aquifer Tests</u>	
Shallow Water Table Tests	Amoozegar and Warrick (1986), Boersma (1965), Boulding (1993), Bouma (1979, 1983), Bouwer and Jackson (1974), Johnson and Richter (1967), Kessler and Oosterbaan (1974), Luthin (1957), Schmid (1967), U.S. EPA (1986), Youngs (1991)
Slug Tests*	<u>Texts/Reviews</u>: Bentall (1963a), Campbell et al. (1990), Chapus (1989), Chirlin (1990), Dagan (1978), Dawson and Istok (1991), Herzog and Morse (1990), Kraemer et al. (1990), Leap (1984-method comparisons), Olson and Daniel (1981), Sevee (1991), Wynne (1992); <u>Pressure Displacement</u>: Leap (1984), Levy and Pannell (1991), Prosser (1981), McLane et al. (1990), Orient et al. (1987); <u>Multilevel Tests</u>: Mastrolonardo and Thomsen (1992), Melville et al. (1991), Molz et al. (1990a, b), Widdowson et al. (1989, 1990); <u>Hvorslev Method</u>: Chirlin (1989), Hvorslev (1951); <u>Ferris-Knowles Method</u>: Ferris and Knowles (1954, 1963), Ferris et al. (1962); <u>Cooper-Bredehoeft-Papadopulos Method</u>: Cooper et al. (1967), Papadopulos et al. (1973); <u>Bouwer-Rice Method</u>: Bouwer (1989), Bouwer and Rice (1976); <u>Fissured Aquifers</u>: See Table 7-6; <u>Measurement Volume</u>: Guyonnet et al. (1993); <u>Data Analysis Procedures</u>: Dax (1987), Faust and Mercer (1984), Karasaki and Witherspoon (1988), Keller and van der Kamp (1992), Marschall and Barczewski (1989), Moench and Hsieh (1985), Nguyen and Pinder (1984), Novakowksi (1989), Palmer and Paul (1987), Peres et al. (1989), Sageev (1986), Widdowson et al. (1990)

Table 7-5 (cont.)

Topic	References
Packer Tests	Braester and Thunvik (1984), Brassington and Walthall (1985), Bredeheoft and Papadopoulos (1980), Dagan (1978), Koopman et al. (1962), Sevee (1991), Shuter and Pemberton (1978), Sutcliffe and Joyner (1966); <u>Dipole Flow Test</u>: Kabala (1993); <u>Lugeon Test</u>: Houlsby (1976), Pearson and Money (1977), Roeper et al. (1992); see also, references for multilevel slug tests
Pumping Tests*	Bedinger and Reed (1988), Bentall (1963a,b), Bruin and Hudson (1955), Clarke (1988), Dawson and Istok (1991), Earlougher (1977), Ferris et al. (1962), Johnson and Richter (1966), Kruseman and de Ridder (1990), Osborne (1993), Schicht (1972), Stallman (1971), Streltsova (1989), U.S. Geological Survey (1980), U.S. EPA (1986, 1991), van Haveren (1986), Walton (1962, 1979, 1987), Wenzel (1942); see also Table 7-6 for references on analytical solutions
Waste/Waste Transport	Conrad et al. (1993), Daniel and Trautwein (1994)
<u>Water Balance Methods</u>	
Texts/Reviews	Boulding (1993), Chapman (1964), Downes (1964), Hagan et al. (1967), Meinzer (1932), Phillips (1964, 1969), Rijtema and Wassink (1969), Sokolow and Chapman (1974), Thornthwaite and Mather (1955, 1957). The following references identified in Table 2-4 have sections on water balance methods: ASCE (1952), Brown et al. (1983), Bureau of Reclamation (1981), Childs (1969), Dunne and Leopold (1978), Skeat (1969), Walton (1970)
Aquifer Recharge	Andres (1991), Keefer and Berg (1990), Rehm et al. (1982)
Stormwater Infiltration	Ferguson (1994)

* Most basic hydrogeology and hydraulics texts (Table 2-4) also cover pump test and slug tests. Texts in this table commonly cited in relation to pump test analysis include: Bureau of Reclamation (1981), Freeze and Cherry (1979), Lohman (1972), and Driscoll (1986).

Table 7-5 References (Appendix F contains references for figure and table sources)

Amoozegar, A. and A. W. Warrick. 1986. Hydraulic Conductivity of Saturated Soils: Field Methods. In: Methods of Soil Analysis, Part 1, 2nd edition, A. Klute (ed.), Agronomy Monograph No. 9. American Society of Agronomy, Madison, WI, pp. 735-770.

Andres, A.S. 1991. Methodology for Mapping Ground Water Recharge Areas in Delaware's Coastal Plain. Delaware Geological Survey Open file Report No. 34, 18 pp.

Bedinger, M.S. and J.E. Reed. 1988. Practical Guide to Aquifer Test Analysis. USGS/EPA Interagency Agreement DW 14932431-01-1. U.E. Environmental Protection Agency Environmental Monitoring Systems Laboratory, Las Vegas, NV.

Bentall, R. (ed.). 1963a. Methods of Determining Permeability, Transmissibility, and Drawdown. U.S. Geological Survey Water Supply Paper 1536-I.

Bentall, R. (compiler). 1963b. Shortcuts and Special Problems in Aquifer Tests. U.S. Geological Survey Water-Supply Paper 1545-C. [17 papers]

Boersma, L. 1965. Field Measurement of Hydraulic Conductivity Below a Water Table. In: Method of Soil Analysis, Part 1, 1st edition, ASA Agronomy Monograph 9, American Society of Agronomy, Madison, WI, pp. 222-233.

Boulding, J.R. 1993. Subsurface Field Characterization and Monitoring Techniques: A Desk Reference Guide, Vol. I: Solids and Ground Water, Vol. II: The Vadose Zone, Field Screening and Analytical Methods. EPA/625/R-93/003a&b. Available from CERI.* [Chapter 4 covers aquifer test methods and Chapters 6 and 7 cover vadose zone hydrologic characterization methods]

Bouma, J. 1977. Soil Survey and the Study of Water in Unsaturated Soil. Soil Survey Paper 13, Soil Survey Institute, Wageningen, Netherlands.

Bouma, J. 1979. Field Measurement of Soil Hydraulic Properties Characterizing Water Movement Through Swelling Clay Soils. J. Hydrology 45:149-158.

Bouma, J. 1983. Use of Soil Survey Data to Select Measurement Techniques for Hydraulic Conductivity. Agric. Water Manage. 6:177-190.

Bouma, J., F.G. Baker, and P.L.M. Veneman. 1974. Measurement of Water Movement in Soil Pedons above the Water Table. University of Wisconsin-Extension, Geological and Natural History Survey, Information Circular No. 27.

Bouwer, H. 1989. The Bouwer and Rice Slug Test--An Update. Ground Water 27(3):304-309.

Bouwer, H. and R.D. Jackson. 1974. Determining Soil Properties. In: Drainage for Agriculture, J. van Schilfgaarde (ed.), ASA Agronomy Monograph No. 17, American Society for Agronomy, Madison, WI, pp. 611-672.

(Table 7-5 Hydraulic Conductivity and Water Balance References)

Bouwer, H. and R.C. Rice. 1976. A Slug Test for Determining Hydraulic Conductivity of Unconfined Aquifers with Completely or Partially Penetrating Wells. Water Resources Research 12(3):423-428.

Braester, C. and R. Thunvik. 1984. Determination of Formation Permeability by Double Packer Tests. J. Hydrology 72:375-389.

Brassington, F.C. and S. Walthall. 1985. Field Techniques Using Borehole Packers in Hydrogeological Investigations. Quart. J. Eng. Geol. 18:181-193.

Bredehoeft, J.D. and S.S. Papadopoulos. 1980. A Method for Determining the Hydraulic Properties of Tight Formations. Water Resources Research 16(1):233-238. [Packer]

Bruin, J.K. and H.E. Hudson, Jr. 1955. Selected Methods for Pumping Test Analysis. Illinois State Water Survey Report of Investigation 25, 54 pp.

Campbell, M.D., M.S. Sterrett, S.F. Fowler, and J.J. Klein. 1990. Slug Tests and Hydraulic Conductivity. Ground Water Management 4:85-99 (7th NWWA/API Conf.).

Chapman, T.G. 1964. Effects of Ground-Water Storage and Flow on Water Balance. In: Water Resources Use and Management (1963 Canberra Symp.), Melbourne Univ. Press, Australia, pp. 290-301.

Chapus, R.P. 1989. Shape Factors for Permeability Tests in Boreholes and Piezometers. Ground Water 27(5):647-654). [Slug tests]

Chirlin, G.R. 1989. A Critique of the Hvorslev Method for Slug Test Analysis: The Fully Penetrating Well. Ground Water Monitoring Review 9(2):130-138.

Clarke, D. 1988. Groundwater Discharge Tests: Simulation and Analysis. Dev. in Water Science 37, Elsevier, New York, NY, 375 pp. [Series of analytical programs for analyzing aquifer tests; covers confined, leaky-confined and unconfined aquifers]

Conrad, D.J., S.A. Shumborski, L.Z. Florense, and A.J. Liem. 1993. Parameters Affecting the Measurement of Hydraulic Conductivity for Solidified/Stabilized Wastes. EPA/600/R-93/099 (NTIS PB93-199396). [Laboratory flexible-wall permeameters]

Cooper, Jr., H.H., J.D. Bredehoeft, and I.S. Papadopoulos. 1967. Response of a Finite-Diameter Well to an Instantaneous Charge of Water. Water Resources Research 3(1):263-270.

Dagan, G. 1978. A Note on Packer, Slug and Recovery Tests in Unconfined Aquifers. Water Resources Research 14(5):929-934.

Daniel, D.E. and S.J. Trautwein (eds.). 1994. Hydraulic Conductivity and Waste Contaminant Transport in Soil. ASTM STP 1142, American Society for Testing and Materials, Philadelphia, PA, 606 pp. [28 papers]

Dawson, K.J. and J.D. Istok. 1991. Aquifer Testing: Design and Analysis of Pumping and Slug Tests. Lewis Publishers, Chelsea, MI, 280 pp.

(Table 7-5 Hydraulic Conductivity and Water Balance References)

Dax, A. 1987. A Note on the Analysis of Slug Tests. J. Hydrology 91:153-177.

Dirksen, C. 1991. Unsaturated Hydraulic Conductivity. In: Soil Analysis: Physical Methods, K.A. Smith and C.E. Mullins (eds.), Marcel Dekker, New York, NY, pp. 209-269. [Sprinkler/dripper, crust, instantaneous profile, draining profile methods]

Downes, R.G. 1964. The Water Balance and Land-Use. In: Water Resources Use and Management (1963 Canberra Symp.), Melbourne Univ. Press, Australia, pp. 329-341.

Earlougher, Jr., R.C. 1977. Advances in Well Test Analysis. Monograph No. 5, Society of Petroleum Engineers of AIME, New York, NY, 264 pp.

Faust, C.R. and J.W. Mercer. 1984. Evaluation of Slug Tests in Wells Containing a Finite Thickness Skin. Water Resources Research 20(4):504-506.

Ferguson, B.K. 1994. Stormwater Infiltration. Lewis Publishers, Boca Raton, FL, 288 pp.

Ferris, J.G. and D.B. Knowles. 1954. The Slug Test for Estimating Transmissibility. U.S. Geological Survey Ground Water Note 26.

Ferris, J.G. and D.B. Knowles. 1963. The Slug-Injection Test for Estimating the Coefficient of Transmissibility of an Aquifer. U.S. Geological Survey Water Supply Paper 1536-I.

Ferris, J.G., D.B. Knowles, R.H. Brown, and R.W. Stallman. 1962. Theory of Aquifer Tests. U.S. Geological Survey Water-Supply Paper 1536-E.

Green, R.E., L.R. Ahuja, and S.K. Chong. 1986. Hydraulic Conductivity, Diffusivity, and Sorptivity of Unsaturated Soils: Field Methods. In: Methods of Soil Analysis, Part 1, 2nd edition, A. Klute (ed.), Agronomy Monograph No. 9, American Society of Agronomy, Madison, WI, pp. 771-798.

Guyonnet, D., S. Mishra, and J. McCord. 1993. Evaluating the Volume of Porous Medium Investigated During Slug Tests. Ground Water 31(4):627-633.

Hagan, R.M., H.R. Haise, and T.W. Edminster (eds.). 1967. Irrigation of Agricultural Lands, ASA Monograph 11, American Society of Agronomy, Madison, WI.

Hamilton, J.M., D.E. Daniel, and R.E. Olson. 1981. Measurement of Hydraulic Conductivity of Partially Saturated Soils. In: Permeability and Groundwater Contaminant Transport, T.F. Zimmie and C.O. Riggs (eds.), ASTM STP 746, American Society for Testing and Materials, Philadelphia, PA, pp. 182-196.

Havlena, J.A. and D.B. Stephens. 1992. Vadose Zone Characterization Using Field Permeameters and Instrumentation. In: Current Practices in Ground Water and Vadose Zone Investigations, D.M. Nielsen and M.N. Sara (eds.), ASTM STP 1118, American Society for Testing and Materials, Philadelphia, PA, pp. 93-110. [Air-entry permeameter, air and gas permeameters, constant head borehole, Guelph permeameter, tension infiltrometer, sealed double-ring infiltrometer]

(Table 7-5 Hydraulic Conductivity and Water Balance References)

Hendrickx, J.M.H. 1990. Determination of Hydraulic Soil Properties. In: Process Studies in Hillslope Hydrology, M.G. Anderson and T.P. Burt (eds.), John Wiley & Sons, New York, NY, pp. 43-92. [Saturated: double tube, cylinder permeameter, constant head borehole infiltration, Guelph permeameter, air entry permeameter; unsaturated: crust, sprinkler, instantaneous profile, draining profile, sorptivity (tension infiltrometer), parameter estimation]

Herzog, B.L. and W.J. Morse. 1990. Comparison of Slug Test Methodologies for Determination of Hydraulic Conductivity in Fine-Grained Sediments. In: Ground Water and Vadose Zone Monitoring, D.M. Nielsen and A.I. Johnson (eds.), ASTM STP 1053, American Society for Testing and Materials, Philadelphia, PA, pp. 152-164.

Hillel, D.I., and Y. Benyamini. 1974. Experimental Comparison of Infiltration and Drainage Methods for Determining Unsaturated Hydraulic Conductivity of a Soil Profile In Situ. In: Isotope and Radiation Techniques in Soil Physics and Irrigation Studies 1973, International Atomic Energy Agency, Vienna, pp. 271-275. [Instantaneous profile, sprinkler-imposed flux]

Houlsby, A.C. 1976. Routine Interpretation of the Lugeon Water-Test. Quart. J. Engineering Geology 9:303-313. [Step-pressure packer test]

Hvorslev, M.J. 1951. Time Lag and Soil Permeability in Groundwater Observations. U.S. Army Corps of Engineers Waterways Experiment Station, Bull. 36, Vicksburg, MS.

Johnson, A.I. and R.C. Richter. 1967. Selected Bibliography on Permeability and Capillarity Testing of Rock and Soil Materials. In: Permeability and Capillarity of Soils, ASTM STP 417, American Society for Testing and Materials, Philadelphia, PA, pp. 167-210.

Kabala, Z.J. 1993. The Dipole Flow Test: A New Single-Borehole Test for Aquifer Characterization. Water Resources Research 29(1):99-107. [Packer test]

Karasaki, K., J. Long, and P. Witherspoon. 1988. Analytical Models of Slug Tests. Water Resources Research 24(1):115-126.

Keefer, D.A. and R.C. Berg. 1990. Potential for Aquifer Recharge in Illinois (Appropriate Recharge Areas). Map with Discussion, Illinois State Geological Survey.

Keller, C.K. and G. van der Kamp. 1992. Slug Tests with Storage Due to Entrapped Air. Ground Water 30:3-7.

Kessler, J. and R.J. Oosterbaan. 1974. Determining Hydraulic Conductivity of Soils. In: Drainage Principles and Applications, Vol. III, International Institute for Land Reclamation and Improvement/ILRI, Wageningen, pp. 254-296. [Auger hole, piezometer, pumped borehole methods]

Koopman, F.C., et al. 1962. Use of Inflatable Packers in Multiple-Zone Testing in Water Wells. U.S. Geological Survey Professional Paper 450-B, pp. B108-B109.

Kraemer, C.A., J.B. Hankins, and C.J. Mohrbacher. 1990. Selection of Single-Well Hydraulic Test Methods for Monitoring Wells. In: Ground Water and Vadose Zone Monitoring, D.M. Nielsen and A.I. Johnson (eds.), ASTM STP 1053, American Society for Testing and Materials, Philadelphia, PA, pp. 125-137.

(Table 7-5 Hydraulic Conductivity and Water Balance References)

Kruseman, G.P. and N.A. de Ridder. 1990. Analysis and Evaluation of Pumping Test Data. ILRI Publication No. 47. International Institute for Land Reclamation and Improvement, Wageningen, The Netherlands, 345 pp. [Completely revised edition of the 1979 English version of Bulletin 11; discusses 46 different analytical techniques]

Lambe, T.W. 1955. The Permeability of Fine-Grained Soils. ASTM STP 163, American Society for Testing and Materials, Philadelphia, PA, pp. 56-67.

Leap, D.I. 1984. A Simple Pneumatic Device and Techniques for Performing Rising Water Level Slug Tests. Ground Water Monitoring Review 4(4):141-146.

Lee, D.M., W.D. Reynolds, D.E. Elrick, and B.E. Clothier. 1985. A Comparison of Three Techniques for Measuring Saturated Hydraulic Conductivity. Can. J. Soil Sci. 65(3):563-573. [Guelph permeameter, air entry permeameter, soil core]

Levy, B.S. and L. Pannell. 1991. Evaluation of a Pressure System for Estimating In-Situ Hydraulic Conductivity. Ground Water Management 5:31-45 (5th NOAC).

Lohman, S.W. 1972. Ground-Water Hydraulics. U.S. Geological Survey Professional Paper 708.

Luthin, J.N. 1957. Measurement of Hydraulic Conductivity In Situ. In: Drainage of Agricultural Lands, J.H. Luthin (ed.), ASA Agronomy Monograph 7, American Society of Agronomy, Madison, WI, pp. 420-431.

Marschall, P. and B. Barczewski. 1989. The Analysis of Slug Tests in the Frequency Domain. Water Resources Research 25(11):2388-2396

Mastrolonardo, R.M. and K.O. Thomsen. 1992. Determination of the Vertical Distribution of Horizontal Hydraulic Conductivity-A Simple Method. Ground Water Management 11:205-216 (6th NOAC). [Slug test with packers]

McLane, G.A., D.A. Harrity, and K.O. Thomsen. 1990. A Pneumatic Method for Conducting Rising and Falling Head Tests in High Permeability Aquifers. Ground Water Management 2:1219-1231 (4th NOAC).

Meinzer, O.E. 1932. Outline of Methods for Estimating Ground-Water Supplies. U.S. Geological Survey Water-Supply Paper 638-C.

Melville, J.G., F.J. Molz, O. Güven, and M.A. Widdowson. 1991. Multilevel Slug Tests with Comparisons to Tracer Data. Ground Water 29:897-907.

Moench, A.F. and P.A. Hsieh. 1985. Analysis of Slug Test Data in A Well with Finite Thickness Skin. Int. Congr. Int. Ass. Hydrogeol. 17(1):17-29.

Molz, F.J., O. Güven, and J.G. Melville. 1990a. A New Approach and Methodologies for Characterizing the Hydrogeologic Properties of Aquifers. EPA/600/2-90/002 (NTIS PB90-187063). [Multilevel slug tests, tracers]

(Table 7-5 Hydraulic Conductivity and Water Balance References)

Molz, F.J., O. Güven, and J.G. Melville. 1990b. Measurement of Hydraulic Conductivity Distribution: A Manual of Practice. EPA/600/8-90/046 (NTIS PB91-211938), 71 pp.

Nguyen, V. and G.F. Pinder. 1984. Direct Calculation of Aquifer Parameters in Slug Test Analysis. In: Groundwater Hydraulics, J. Rosenshein and G. Bennett (eds.), American Geophysical Union Water Resources Monograph 9, Washington, DC, pp. 222-239.

Novakowksi, K.S. 1989. Analysis of Pulse Interference Tests. Water Resources Research 25(22):2377-2387.

Olson, R.E. and D.E. Daniel. 1981. Measurement of the Hydraulic Conductivity of Fine-Grained Soils. In: Permeability and Groundwater Contaminant Transport, T.F. Zimmie and C.O. Riggs (eds.), ASTM STP 746, American Society for Testing and Materials, Philadelphia, PA, pp. 18-64. [Slug tests]

Orient, J.P., A. Nazar, and R.C. Rice. 1987. Vacuum and Pressure Test Methods for Estimating Hydraulic Conductivity. Ground Water Monitoring Review 7(1):49-50.

Osborne, P.S. 1993. Suggested Operating Procedures for Aquifer Pumping Tests. Ground Water Issue Paper, EPA/540/S-93/503, 23 pp.

Palmer, C.D. and D.G. Paul. 1987. Problems in the Interpretation of Slug Test Data from Fine-Grained Glacial Tills. In: Proc. Focus Conf. on Northwestern Ground Water Issues, National Water Well Association, Dublin, OH, pp. 99-123.

Papadopulos, S.S., J.D. Bredehoeft, and H.C. Cooper, Jr. 1973. On the Analysis of "Slug Test" Data. Water Resources Research 9(4):1087-1089.

Pearson, R. and M.S. Money. 1977. Improvements in the Lugeon or Packer Permeability Test. Quart. J. Eng. Geol. 10:221-239.

Peres, A.M.M., M. Onur, and A.C. Reynolds. 1989. A New Analysis Procedure for Determining Aquifer Properties from Slug Test Data. Water Resources Research 25(7):1591-1602.

Phillips, J.R. 1964. The Gain, Transfer, and Loss of Soil-Water. In: Water Resources Use and Management (1963 Canberra Symp.), Melbourne Univ. Press, Australia, pp. 257-275.

Phillips, J.R. 1969. Theory of Infiltration. Advances in Hydroscience 5:215-296.

Prosser, D.W. 1981. A Method of Performing Response Tests on Highly Permeable Aquifers. Ground Water 19(6):588-592.

Rehm, B.W., G.H. Groenewold, and W.M. Peterson. 1982. Mechanisms, Distribution, and Frequency of Ground Water Recharge in an Upland Area of Western North Dakota. North Dakota Geological Survey Report of Investigations 75, 72 pp.

Reynolds, W.D., D.E. Elrick, and G.C. Topp. 1983. A Reexamination of the Constant Head Well Permeameter Methods for Measuring Saturated Hydraulic Conductivity Above the Water Table. Soil Science 136:250-268. [Constant head borehole infiltration, Guelph permeameter, air entry permeameter]

(Table 7-5 Hydraulic Conductivity and Water Balance References)

Rijtema, P.E. and H. Wassink (eds.). 1969. Water in the Unsaturated Zone (Proc. Wageningen Symp), 2 Vols. IASH-UNESCO Studies and Reports in Hydrology 2, UNESCO, Paris.

Roberts, D.W. 1984. Soil Properties, Classification, and Hydraulic Conductivity Testing. EPA/SW-925 (NTIS PB87-155784). [K_{sat}: double-ring infiltrometer, cylinder permeameter, modified air-entry permeameter, cube method; K_{unsat}: crust, instantaneous profile]

Roeper, T.R., W.G. Soukup, and R.L. O'Neill. 1992. The Applicability of the Lugeon Method of Packer Test Analysis to Hydrogeologic Investigations. Ground Water Management 13:661-671 ([8th] Focus Conf. Eastern GW Issues). [Not recommended]

Sai, J.O. and D.C. Anderson. 1991. State-of-the-Art Field Hydraulic Conductivity Testing of Compacted Soils. EPA/600/2-91/022 (NTIS PB91-206243), 95 pp. [Air entry permeameters, Guelph permeameter, Boutwell method, surface infiltrometers (ASTM double-ring, modified double-ring, box, single-ring, sealed double-ring), collection lysimeters (in situ monoliths), velocity permeameter, porous plate (tension) infiltrometers]

Sageev, A. 1986. Slug Test Analysis. Water Resources Research 22(8):1323-1333.

Schicht, R.J. 1972. Selected Methods of Aquifer Test Analysis. Water Resources Bulletin 8(1):175-187.

Schmid, W.E. 1967. Field Determination of Permeability by the Infiltration Test. In: Permeability and Capillarity of Soils. ASTM STP 417. American Society for Testing and Materials, Philadelphia, PA, pp. 142-159.

Sevee, J. 1991. Methods and Procedures for Defining Aquifer Parameters. In: Practical Handbook of Ground-Water Monitoring, D.M. Nielsen (ed.), Lewis Publishers, Chelsea, MI, pp. 397-447.

Shuter, E. and R.R. Pemberton. 1978. Inflatable Straddle Packers and Associated Equipment for Hydraulic Fracturing and Hydrologic Tests. U.S. Geological Survey Water Resources-Investigations Report 78-55, 16 pp.

Sokolow, A.A. and T.G. Chapman (eds.). 1974. Methods for Water Balance Computations: An International Guide for Research and Practice. The Unesco Press, Paris.

Stallman, R.W. 1971. Aquifer-Test Design, Observation and Data Analysis. U.S. Geological Survey Techniques of Water Resources Investigations, TWRI 3-B1. [Updated in USGS (1980)]

Stephens, D.B. and S.P. Neuman. 1980. Analysis of Borehole Infiltration Tests above the Water Table. Technical Report No. 35, Department of Hydrology and Water Resources, University of Arizona, Tucson, AZ.

Stephens, D.B. and S.P. Neuman. 1982a. Vadose Zone Permeability Tests: Summary. J. Hydrology Div. ASCE 108(HY5):623-639. [Review of USBR Methods]

Stephens, D.B. and S.P. Neuman. 1982b. Vadose Zone Permeability Tests: Steady State Results. J. Hydrology Div. ASCE 108(HY5):640-659.

(Table 7-5 Hydraulic Conductivity and Water Balance References)

Stephens, D.B. and S.P. Neuman. 1982c. Vadose Zone Permeability Tests: Unsteady Flow. J. Hydrol. Div. ASCE 108(HY5):660-677.

Stephens, D.B., M. Unruh, J. Havlena, R.G. Knowlton, E. Mattson, and W. Cox. 1988. Vadose Zone Characterization of Low-Permeability Sediments Using Field Permeameters. Ground Water Monitoring Review 8(2):59-66. [Air entry, Guelph, constant-head, gas-pressure permeameters]

Streltsova, T.D. 1989. Well Testing in Heterogeneous Formations. John Wiley & Sons, New York, NY. [Focuses on testing of deep oil-bearing formations]

Sutcliffe, Jr., H. and B.F. Joyner. 1966. Packer Testing in Water Wells Near Sarasota, Florida. Ground Water 4(2):23-27.

Thornthwaite, C.W. and J.R. Mather. 1955. The Water Balance. Publications in Climatology, Vol. 8, No. 1, Laboratory of Climatology, Centerton, NJ, 104 pp.

Thornthwaite, C.W. and J.R. Mather. 1957. Instructions and Tables for Computing Potential Evapotranspiration and Water Balance. Publications in Climatology Vol. X, No. 3. Drexel Institute of Technology, Laboratory of Climatology, Centerton, NJ.

U.S. Environmental Protection Agency (EPA). 1986. Criteria for Identifying Areas of Vulnerable Hydrogeology Under RCRA: A RCRA Interpretive Guidance. EPA/530/SW-86/022 (Complete set: NTIS PB86-224946). (Individual Appendices [EPA/530/SW-86/022A to D]: Technical Methods for Evaluating Hydrogeologic Parameters [A, NTIS PB86-224961m 48 pp.]; Groundwater Flow Net/Flow Line Construction and Analysis [B, NTIS PB86-224979]; Technical Methods for Calculating Time of Travel in the Unsaturated Zone [C, NTIS PB86-224987]; Development of Vulnerability Criteria Based on Risk Assessments and Theoretical Modeling [D, NTIS PB86-224995]). [Appendix A covers single and multiple well tests, Appendix C vadose zone tests]

U.S. Environmental Protection Agency (EPA). 1991. Handbook Ground Water, Vol. II: Methodology. EPA/625/6-90/-16b, 141 pp. Available from CERI.* [Chapter 5 covers aquifer test analysis]

U.S. Geological Survey. 1980. Ground Water. In: National Handbook of Recommended Methods for Water Data Acquisition, Office of Water Data Coordination, Reston, VA, Chapter 2. [Section 2.H on aquifer tests contains an update of Stallman (1971)]

Walton, W.C. 1962. Selected Analytical Methods for Well and Aquifer Evaluation. ISWS Bulletin No. 49. Illinois State Water Survey, Champaign, IL.

Walton, W.C. 1979. Progress in Analytical Groundwater Modeling. In: Contemporary Hydrogeology, W. Back and D.A. Stephenson (eds.), Elsevier, New York, NY. [Review paper covering various analytical methods for analyzing pump test data]

Walton, W.C. 1987. Groundwater Pumping Tests: Design and Analysis. Lewis Publishers, Chelsea, MI, 201 pp.

(Table 7-5 Hydraulic Conductivity and Water Balance References)

Widdowson, M.A., F.J. Molz, and J.G. Melville. 1989. Analysis of Multi-Level Slug Test Data to Determine Hydraulic Conductivity Distribution. In: Proc. 4th Int. Conf. on Solving Ground Water Problems with Models (Indianapolis, IN), National Water Well Association, Dublin, OH, pp. 699-714.

Widdowson, M.A., F.J. Molz, and J.G. Melville. 1990. An Analysis Technique for Multilevel and Partially Penetrating Slug Test Data. Ground Water 28:937-945.

Winger, Jr., R.J. 1960. In-Place Permeability Tests and Their Use in Subsurface Drainage. In: Trans. Int. Congr. Comm. Irrig. Drain., (4th, Madrid), pp. 11.417-11.469.

Wenzel, L.K. 1942. Methods for Determining Permeability of Water-Bearing Materials with Special Reference to Discharging Well Methods. U.S. Geological Survey Water Supply Paper 887.

Wynne, D.B. 1992. Specific Capacity and Slug Testing: An Overview and Empirical Comparison of Their Uses in Preliminary Estimating Hydraulic Conductivity. Ground Water Management 11:217-230 (6th NOAC).

Youngs, E.G. 1991. Hydraulic Conductivity of Saturated Soils. In: Soil Analysis: Physical Methods, K.A. Smith and C.E. Mullins (eds.), Marcel Dekker, New York, NY, pp. 161-207. [Laboratory, below water table (auger-hole method, piezometer, multiple-well methods, tile drains); above water table (borehole permeameter, air-entry permeameter, ring infiltrometers)]

* See Preface for information on how to obtain documents from CERI (U.S. EPA Center for Environmental Research Information) and NTIS.

(Table 7-5 Hydraulic Conductivity and Water Balance References)

Table 7-6 Index to Source References on Pump Test Analytical Solutions and Methods for Characterizing Anisotropic and Fractured-Rock Aquifers

Topic	References

Pump Test Solutions: Confined Aquifers[1]

Non-leaky	**Fully Penetrating Wells--**Constant Discharge: Theis (1935), Cooper and Jacob (1946), Jacob (1950); Variable Discharge: Abu-Zied and Scott (1963), Aron and Scott (1965), Hantush (1964), Lai et al. (1973), Moench (1971), Stallman (1962); Constant Drawdown Hantush (1964), Jacob and Lohman (1952), Rushton and Rathod (1980); Unclassified: Boulton and Streltsova (1977a,b)*, Brutsaert and Corapcioglu (1976), Moench and Prickett (1972), Papadopulos (1967), Reed (1980); **Partially penetrating wells--**Hantush (1961a, 1964)
Leaky, fully penetrating wells	No Storage in Confining Bed: Hantush and Jacob (1955); Storage in Confining Bed: Hantush (1960); Multiple Aquifers: Hantush (1967a), Neuman and Witherspoon (1972); Unclassified: Corapcioglu (1976), Hantush (1956, 1959, 1964*, 1967b), Jacob (1946a), Lai and Su (1974), Neuman and Witherspoon (1969b), Reed (1980)

Pump Test Solutions: Unconfined Aquifers[1]

Fully penetrating wells	Constant Discharge: Boulton (1954a, 1954b, 1963), Neuman (1972, 1973); Unclassified: Boulton and Streltsova (1978)*, Cooper and Jacob (1946), Jacob (1944, 1963), Neuman (1975)*, Prickett (1965), Streltsova (1972); Vadose Zone Effects: Guitjens and Luthin (1971), Krozynski and Dagan (1975)
Partially penetrating wells	Hantush (1961a, 1962), Boulton and Streltsova (1976)*, Neuman (1974), Streltsova (1974*, 1976a*)

Tests for Special Aquifer Conditions

Test Analysis	Single Well Step-Drawdown Tests: Bierschenk (1964), Jacob (1946b), Lennox (1966), Rorabaugh (1953); Multiple Aquifers: Aral (1990a, 1990b), Bennett and Patton (1962), Hantush (1967a), Javendal and Witherspoon (1969), Neuman and Witherspoon (1969a-confined; 1972-leaky)
Other Conditions	Aquitard/Aquiclude Hydraulic Conductivity: Neuman and Gardner (1989); Lateral Boundary: Ferris et al. (1962), Lohman (1972), Stallman (1963)

Table 7-6 (cont.)

Topic	References
Tests for Special Aquifer Conditions (cont.)	
Other	Contaminant Transport: Mills et al. (1985); Large Diameter Well: Papadopulos and Cooper (1967); Unclassified: Boulton and Streltsova (1975)
Anisotropy	
General	Bear and Dagan (1965), Fetter (1981), Freeze (1975), Greenkorn (1970), Llakopoulos (1965), Maasland (1957a, 1957b, 1983), Marcus (1962), Scheidegger (1954)
Pump Test Methods[2]	Cited by ASTM D4043: Hantush (1961b), Papadopulos (1965), Neuman (1975), Weeks (1964, 1969); Other Citations: Boulton and Streltsova (1976, 1977a, 1977b, 1978), Butler and Liu (1993), Dagan (1967), Hantush (1964, 1966a, 1966b), Hantush and Thomas (1966), Hsieh and Neuman (1985), Mansur and Dietrich (1965), Neuman (1975), Neuman et al. (1984), Norris and Fidler (1966), Streltsova (1974, 1976a), Way and McKee (1982)
Other Methods	Laboratory Methods: Banton (1993), Rocha and Franciss (1977); Other Field: Loo et al. (1984-surface tiltmeter survey), Maasland (1955-auger hole method), Ritzi and Andolesk (1992-azimuthal resistivity)
Fractured Rock Characterization	
General	Bianchi and Snow (1968), Chen (1989), Duguid and Lee (1977), Gale (1982), Gerke and van Genuchten (1993), Long and Billaux (1987), Long and Witherspoon (1985), Long et al. (1982), Nelson (1985), Neuman and Neretnieks (1990), Sagar and Runchai (1982), Schmelling and Ross (1989), Snow (1969), Streltsova (1976b), Thomas and McGlew (1985), Tsang (1992), Tsang and Tsang (1987), UNESCO (1984), Warren and Root (1973)
Well Test Methods	Cited by ASTM D4043: Barenblatt et al. (1960), Boulton and Streltsova (1977b), Gringarten and Ramey (1974), Moench (1984); Other Citations: Boulton and Streltsova (1977a, 1978), Elkins and Skov (1960), Gale (1982), Gringarten (1982), Gringarten and Witherspoon (1972), Hsieh and Neuman (1985), Hsieh et al. (1983, 1985), Huntley et al. (1992), Jenkins and Prentice (1982), Lewis (1974), Maslia and Randolph (1987), McConnel (1993), Moore (1992), Ramey (1975), Rofail (1967), Sauveplane (1984), Smith and Vaughn (1985), Zekai (1986); Slug Tests: Barker and Black (1983); Tracers: Lewis et al. (1966).

Table 7-6 Footnotes

* Analytical solutions for anisotropic aquifer conditions.

[1] Categories in first column taken from Driscoll (1986); subcategories in the second column taken from ASTM D4043 (full citation in Table A-14). Unclassified references are identified in Driscoll (1986) and other sources, but not ASTM D4043.

[2] See also references for pump test methods in fractured rock, which also characterize anisotropy, when present.

Table 7-6 References (Appendix F contains references for figure and table sources)

Abu-Zied, M. and V.H. Scott. 1963. Nonsteady Flow for Wells with Decreasing Discharge. J. Hydraulic Div. ASCE 89(HY3):119-132.

Aral, M.M. 1990a. Ground Water Modeling in Multilayered Aquifers: Steady Flow. Lewis Publishers, Chelsea, MI, 114 pp. [Includes disks for SLAM — steady layered aquifer model]

Aral, M.M. 1990b. Ground Water Monitoring in Multilayered Aquifers: Unsteady Flow. Lewis Publishers, Chelsea, MI, 143 pp. [includes disks for ULAM — unsteady layered aquifer model]

Aron, G. and V.H. Scott. 1965. Simplified Solutions for Decreasing Flow in Wells. J. Hydraulics Division ASCE 91(HY5):1-12.

Banton, O. 1993. Field- and Laboratory-Determined Hydraulic Conductivities Considering Anisotropy and Core Surface Area. Soil Sci. Soc. Am. J. 47:10-15. [Constant-head permeameter]

Barenblatt, G.I., I.P. Zheltov, and I.N. Kochina. 1960. Basic Concepts in the Theory of Seepage of Homogenous Liquids in Fissured Rocks [Strata]. J. Applied Mathematics and Mechanics 24:1286-1301.

Barker, J.A. and J.H. Black. 1983. Slug Tests in Fissured Aquifers. Water Resources Research 19:1558-1564.

Bear, J. and G. Dagan. 1965. The Relationship Between Solutions of Flow Problems in Isotropic and Anisotropic Soils. J. Hydrology 3:88-96.

Bennett, G.D. and E.P. Patton, Jr. 1962. Constant-Head Pumping Test of a Multiaquifer Well to Determine Characteristics of Individual Aquifers. U.S. Geological Survey Water-Supply Paper 1536-G, 203 pp.

Bianchi, L. and D. Snow. 1968. Permeability of Crystalline Rock Interpreted from Measured Orientations and Apertures of Fractures. Ann. Arid Zone 8(2):231-245.

Bierschenk, W.H. 1964. Determining Well Efficiency by Multiple Step-Drawdown Tests. Int. Ass. Sci. Hydrol. Pub. No. 64, pp. 493-505.

Boulton, N.S. 1954a. Unsteady Radial Flow to a Pumped Well Allowing for Delayed Yield from Storage. Int. Ass. of Hydrological Sciences Publ. No. 37, pp. 472-477.

Boulton, N.S. 1954b. Drawdown of the Water Table Under Non-Steady Conditions Near a Pumped Well in an Unconfined Formation. Proc. Inst. of Civil Engineers (London) 3(Pt3):564-579.

Boulton, N.S. 1963. Analysis of Data from Nonequilibrium Pumping Tests Allowing for Delayed Yield from Storage. Proc. Inst. of Civil Engineers (London) 26:469-482.

(Table 7-6 Pump Test Solutions and Anisotropic/Fractured Rock References)

Boulton, N.S. and T.D. Streltsova. 1975. New Equations for Determining the Formation Constants of an Aquifer from Pumping Test Data. Water Resources Research 11(1):148-153.

Boulton, N.S. and T.D. Streltsova. 1976. The Drawdown Near an Abstraction Well of Large Diameter Under Non-Steady Conditions in an Unconfined Aquifer. J. Hydrology 30:29-46. [Homogenous anisotropic aquifer]

Boulton, N.S. and T.D. Streltsova. 1977a. Unsteady Flow to a Pumped Well in a Two-Layered Water-Bearing Formation. J. Hydrology 35:245-256. [Anisotropic, non-leaky confined fractured rock aquifer]

Boulton, N.S. and T.D. Streltsova. 1977b. Unsteady Flow to a Pumped Well in a Fissured Water-Bearing Formation. J. Hydrology 35:257-269. [Anisotropic, non-leaky confined fractured rock aquifer]

Boulton, N.S. and T.D. Streltsova. 1978. Unsteady Flow to a Pumped Well in a Fissured Aquifer with a Free Surface Level Maintained Constant. Water Resources Research 14(3):527-532. [Anisotropic, unconfined, fractured-rock aquifer]

Brutsaert, W. and M.Y. Corapcioglu. 1976. Pumping of Aquifer with Visco-Elastic Properties. J. Hydraulics Division ASCE 102(HY11):1663-1675.

Butler, Jr., J.J. and W. Liu. 1993. Pumping Test in Nonuniform Aquifers: The Radially Asymmetric Case. Water Resources Research 29(2):259-269.

Chen, Z.-X. 1989. Transient Flow of Slightly Compressible Fluids through Double-Porosity, Double-Permeability Systems--A State-of-the-Art Review. Transport of Porous Media 4:147-184.

Cooper, Jr. H.H. and C.E. Jacob. 1946. A Generalized Graphical Method for Evaluating Formation Constants and Summarizing Well Field History. Trans. Am. Geophysical Union 27(4):526-534.

Corapcioglu, M.Y. 1976. Mathematical Modeling of Leaky Aquifers with Rheological Properties. Int. Ass. of Hydrological Sciences Pub. No. 121, pp. 191-200.

Dagan, G. 1967. A Method of Determining the Permeability and Effective Porosity of Unconfined Anisotropic Aquifers. Water Resources Research 3:1059-1071.

Driscoll, F.G. 1986. Groundwater and Wells, 2nd ed. Johnson Division, UOP Inc., St. Paul, MN, 1089 pp. [First edition by Johnson, UOP (1966); Chapter 9 covers well hydraulics and Chapter 16 discusses collection and analysis of pumping test data]

Duguid, J.O. and P.C.Y. Lee. Flow in Fractured Porous Media. Water Resources Research 13:558-566.

Elkins, L.F. and A.M. Skov. 1960. Determination of Fracture Orientation from Pressure Interference. Trans. Am. Inst. Mining Eng. 219:301-304

(Table 7-6 Pump Test Solutions and Anisotropic/Fractured Rock References)

Ferris, J.G., D.B. Knowles, R.H. Brown, and R.W. Stallman. 1962. Theory of Aquifer Tests. U.S. Geological Survey Water-Supply Paper 1536-E.

Fetter, Jr., C.W. 1981. Determination of the Direction of Groundwater Flow. Ground Water Monitoring Review 1(3):28-31. [Effect of anisotropy]

Freeze, R.A. 1975. A Stochastic-Conceptual Analysis of One-Dimensional Ground Water Flow in Nonuniform Homogeneous Media. Water Resources Research 11:725-741. [See, also, comment by G. Dagan, WRR 12:567 and reply by Freeze, WRR 12:568]

Gale, J.E. 1982. Assessing the Permeability Characteristics of Fractured Rock. In: Recent Trends in Hydrogeology, T.N. Narasimhan (ed.), Geological Society of America Special Paper 189, pp. 163-182.

Gerke, H.H. and M.T. van Genuchten. 1993. A Dual-Porosity Model for Simulating the Preferential Movement of Water and Solutes in Structured Porous Media. Water Resources Research 29(2):305-319.

Greenkorn, R.A. 1970. Dispersion in Heterogeneous Nonuniform Anisotropic Porous Media. EPA/1606 DLL 09/70, 82 pp.

Gringarten, A.C. 1982. Flow Test Evaluation of Fractured Reservoirs. In: Recent Trends in Hydrogeology, Geological Society of American Special paper 189, pp. 237-263.

Gringarten, A.C. and H.J. Ramey, Jr. 1974. Unsteady-State Pressure Distribution Created by a Well with a Single Horizontal Fracture, Partial Penetration, or Restricted Entry. Society of Petroleum Engineers Journal 14(4):413-426.

Gringarten, A.C. and P.A. Witherspoon. 1972. A Method of Analyzing Pump Test Data from Fractured Aquifers. In: Proc. Symp. on Percolation through Fissured Rock (Stuttgart), Int. Soc. Rock Mechanics and Int. Ass. Engineering Geologists, pp. T3-B-1 to T3-B-8.

Guitjens, J.G. and J.N. Luthin. 1971. Effect of Soil Moisture Hysteresis on the Water Table Profile Around a Gravity Well. Water Resources Research 7(2):334-346.

Hantush, M.S. 1956. Analysis of Data from Pumping Tests in Leaky Aquifers. Trans. Am. Geophys. Union 37(6):702-714.

Hantush, M.S. 1959. Non-Steady Flow to Flowing Wells in Leaky Aquifers. J. Geophysical Research 64(8):1043-1052.

Hantush, M.S. 1960. Modification of the Theory of Leaky Aquifers. J. Geophysical Research 65(11):3713-3725.

Hantush, M.S. 1961a. Aquifer Tests on Partially Penetrating Wells. J. Hydraulics Division, ASCE 87(HY5):171-195.

Hantush, M.S. 1961b. Drawdown Around a Partially Penetrating Well. J. Hydraulic Division, ASCE 87(HY4):83-98. [Radial vertical anisotropy]

(Table 7-6 Pump Test Solutions and Anisotropic/Fractured Rock References)

Hantush, M.S. 1964. Hydraulics of Wells. Advances in Hydroscience 1:281-432. [Includes analytical solutions for anisotropic aquifer conditions]

Hantush, M.S. 1966a. Wells in Homogeneous Anisotropic Aquifers. Water Resources Research, 2(2):273-279.

Hantush, M.S. 1966b. Analysis of Data from Pumping Tests in Anisotropic Aquifers. J. Geophys. Res. 71(2):421-426.

Hantush, M.S. 1967a. Flow to Wells in Aquifers Separated by a Semipervious Layer. J. Geophysical Research 72(6):1709-1720.

Hantush, M.S. 1967b. Flow of Groundwater in Relatively Thick Leaky Aquifers. Water Resources Research 3(2):583-590.

Hantush, M.S. and C.E. Jacob. 1955. Non-Steady Radial Flow in an Infinite Leaky Aquifer and Non-Steady Green's Functions for an Infinite Strip of Leaky Aquifer. Trans. Am. Geophysical Union 36(1):95-100.

Hantush, M.S. and R.E. Thomas. 1966. A Method for Analyzing a Drawdown Test in Anisotropic Aquifers. Water Resources Research 2(2):281-285.

Hsieh, P.A. and S.P. Neuman. 1985. Field Determination of the Three-Dimensional Hydraulic Conductivity Tensor of Anisotropic Media, 1. Theory. Water Resources Research 21(11):1655-1665.

Hsieh, P.A., S.P. Neuman, and E.S. Simpson. 1983. Pressure Testing of Fractured Rock--A Methodology Employing Three-Dimensional Cross-Hole Tests. NUREG/CR-3213, U.S. Nuclear Regulatory Commission, Washington DC.

Hsieh, P.A., S.P. Neuman, G.K. Stiles, and E.S. Simpson. 1985. Field Determination of Three-Dimensional Hydraulic Conductivity Tensor of Anisotropic Media. 2. Methodology and Application to Fractured Rocks. Water Resources Research 21(11):1667-1676. [Multi-well, multi-level packer pressure tests]

Huntley, D., R. Nommensen, and D. Steffey. 1992. The Use of Specific Capacity to Assess Transmissivity in Fractured-Rock Aquifers. Ground Water 30(3):396-402.

Jacob, C.E. 1944. Notes on Determining Permeability by Pumping Tests Under Water Table Conditions. U.S. Geological Survey Open File Report.

Jacob, C.E. 1946a. Radial Flow in a Leaky Artesian Aquifer. Trans. Am. Geophysical Union, 27(2):198-205.

Jacob, C.E. 1946b. Drawdown test to determine effective Radius of Artesian Well. Trans. ASCE 112:1047-1070.

Jacob, C.E. 1950. Flow of Ground Water. In: Engineering Hydraulics, H. Rouse (ed.), Wiley and Sons, New York, NY, pp. 321-386.

(Table 7-6 Pump Test Solutions and Anisotropic/Fractured Rock References)

Jacob, C.E. 1963. Determining the Permeability of Water Table Aquifers. In: U.S. Geological Survey Water-Supply Paper 1536-I, pp. 245-271.

Jacob, C.E. and S.W. Lohman. 1952. Nonsteady Flow to a Well of a Constant Drawdown in an Extensive Aquifer. Trans. Am. Geophysical Union 33(4):559-569.

Javendal, I. and P.A. Witherspoon. 1969a. Method of Analyzing Transient Fluid Flow in Multilayered Aquifers. Water Resources Research 5(4):856-869.

Jenkins, D.N. and J.K. Prentice. 1982. Theory for Aquifer Test Analysis in Fractured Rocks Under Linear (Nonradial) Flow Conditions. Ground Water 20:12-21.

Krozynski, V.I. and G. Dagan. 1975. Well Pumping in Unconfined Aquifers: The Influence of the Unsaturated Zone. Water Resources Research 11(3):479-490.

Lai, R.Y.S. and C.-W. Su. 1974. Nonsteady Flow to a Large Well in a Leaky Aquifer. J. Hydrology 22:333-345.

Lai, R.Y., G.M. Karadi, and R.A. Williams. 1973. Drawdown at Time-Dependent Flowrate. Water Resource Bulletin 9(5):854-859.

Lennox, D.H. 1966. Analysis and Application of Step-Drawdown Test. J. Hydraulics Div. ASCE 92:25-48.

Llakopoulos, A.C. 1965. Variation of the Permeability Tensor Ellipsoid in Homogenous Anisotropic Soils. Water Resources Research 1(1):135-142.

Lewis, C. 1974. Introduction a l'Hydraulique des Roches (in French). Bureau de Recherches Geologiques et Miniers, Orleans, France. [Three-dimensional hydraulic conductivity tensor using hydraulic triple probe technique; see Hsieh and Neuman (1985) for description]

Lewis, D.C., G.J. Kriz, and R.H. Burgy. 1966. Tracer Dilution Sampling Technique to Determine Hydraulic Conductivity of Fractured Rock. Water Resources Research 2:533-542. [Fluorescein]

Lohman, S.W. 1972. Ground-Water Hydraulics. U.S. Geological Survey Professional Paper 708.

Long, J.C.S. and D.M. Billaux. 1987. From Field Data to Fracture Network Modeling. J. Hydrology 100:379-409.

Long, J.C.S. and P.A. Witherspoon. 1985. The Relationship of the Degree of Interconnection to Permeability in Fracture Networks. J. Geophys. Res. 90(B4):3087-3098.

Long, J.C.S., J.S. Remer, C.R. Wilson, and P.A. Witherspoon. 1982. Porous Media Equivalents for Networks of Discontinuous Fractures. Water Resources Research 18(3):645-658.

Loo, W.W., K. Frantz, and G.R. Holzhausen. 1984. The Application of Telemetry to Large-Scale Horizontal Anisotropic Permeability Determinations by Surface Tiltmeter Survey. Ground Water Monitoring Review 4(4):124-130.

(Table 7-6 Pump Test Solutions and Anisotropic/Fractured Rock References)

Maasland, M. 1955. Measurement of Hydraulic Conductivity by the Auger Hole Method in Anisotropic Soil. Soil Science 81:379-388.

Maasland, M. 1957a. Theory of Fluid Flow Through Anisotropic Media. In: Drainage of Agricultural Lands, J.H. Luthin (ed.), ASA Monograph 7, American Society of Agronomy, Madison, WI.

Maasland, M. 1957b. Soil Anisotropy and Land Drainage. In: Drainage of Agricultural Lands, J.H. Luthin (ed.), ASA Monograph 7, American Society of Agronomy, Madison, WI, pp. 216-285.

Maasland, M. 1983. Soil Anisotropy and Land Drainage. In: Physical Hydrogeology, R.A. Freeze and W. Back (eds.), Van Nostrand Reinhold, New York, NY, pp. 71-84.

McKay, L.D., J.A. Cherry, and R.W. Gillham. 1993. Field Experiments in Fractured Clay Till 1. Hydraulic Conductivity and Fracture Aperture. Water Resources Research 29(4):1149-1162.

Mansur, C.I. and R.J. Dietrich. 1965. Pumping Test to Determine Permeability Ratio. J. Soil Mech. and Foundation Div. ASCE 91(SM4):151-183.

Marcus, H. 1962. The Permeability of a Sample of an Anisotropic Porous Medium. J. Geophys. Res. 67:5215-5225.

Maslia, M.L. and R.B. Randolph. 1987. Methods and Computer Program Documentation for Determining Anisotropic Transmissivity Tensor Components of Two-Dimensional Ground-Water Flow. U.S. Geological Survey Water-Supply Paper 2308, 46 pp.

McConnell, C.L. 1993. Double Porosity Well Testing in the Fractured Carbonate Rocks of the Ozarks. Ground Water 31:75-83.

Mills, W.B. et al. 1985. Water Quality Assessment: A Screening Procedure for Toxic and Conventional Pollutants (Revised 1985). EPA/600/6-85/002a&b. [Vol. II contains 5 analytical methods for preliminary assessment of contaminant transport]

Moench, A.F. 1971. Ground-Water Fluctuations in Response to Arbitrary Pumpage. Ground Water 9(2):4-8.

Moench, A.F. 1984. Double Porosity Model for a Fissured Groundwater Reservoir with Fracture Skin. Water Resources Research 20(7):831-846.

Moench, A.F. and T.A. Prickett. 1972. Radial Flow in an Infinite Aquifer Undergoing Conversion from Artesian to Water Table Conditions. Water Resources Research 8(2):494-499.

Moore, G.K. 1992. Hydrograph Analysis in a Fractured Rock Terrane. Ground Water 30(3):390-395.

Nelson, R.A. 1985. Geologic Analysis of Naturally Fractured Reservoirs. Contributions in Petroleum Geology and Engineering, Vol. 1, Gulf Publishing Co., Houston, TX, 320 pp.

(Table 7-6 Pump Test Solutions and Anisotropic/Fractured Rock References)

Neuman, S.P. 1972. Theory of Flow in Unconfined Aquifers Considering Delayed Response of the Water Table. Water Resources Research 8(4):1031-1045.

Neuman, S.P. 1973. Supplementary Comments on 'Theory of Flow in Unconfined Aquifers Considering Delayed Response of the Water Table'. Water Resources Research 9(4):1102.

Neuman, S.P. 1974. Effect of Partial Penetration on Flow in Unconfined Aquifers Considering Delayed Gravity Response. Water Resources Research 10(2):303-312.

Neuman, S.P. 1975. Analysis of Pumping Test Data from Anisotropic Unconfined Aquifers Considering Delayed Gravity Response. Water Resources Research 11(2):329-342.

Neuman, S.P. and D.A. Gardner. 1989. Determination of Aquitard/Aquiclude Hydraulic Properties from Arbitrary Water-Level Fluctuations by Deconvolution. Ground Water 27(1):66-76.

Neuman, S.P. and I. Neretnieks (eds.). 1990. Hydrogeology of Low Permeability Environments. Selected Papers Vol. 2, Int. Ass. Hydrogeologists, Verlag Heinz Heise, Hannover, Germany, 268 pp. [Papers from 28th Int. Geol. Congr, Washington, DC, July 13, 1989]

Neuman, S.P. and P.A. Witherspoon. 1969a. Theory of Flow in a Confined Two Aquifer System. Water Resources Research 5(4):803-816.

Neuman, S.P. and P.A. Witherspoon. 1969b. Applicability of Current Theories of Flow in Leaky Aquifers. Water Resources Research 5(4):817-829.

Neuman, S.P. and P.A. Witherspoon. 1972. Field Determination of the Hydraulic Conductivity of Leaky Multiple Aquifer Systems. Water Resources Research 8(5):1284-1298.

Neuman, S.P., G.R. Walter, H.W. Bentley, J.J. Ward, and D.D. Gonzalez. 1984. Determination of Horizontal Aquifer Anisotropy with Three Wells. Ground Water 22(1):66-72.

Norris, S.E. and R.E. Fidler. 1966. Use of Type Curves Developed from Electric Analog Studies of Unconfined Flow to Determine the Vertical Permeability of an Aquifer at Piketon, Ohio. Ground Water 4:43-48.

Papadopulos, I.S. 1965. Nonsteady Flow to a Well in an Infinite Anisotropic Aquifer. In: Proc. Dubrovnik Symp. on Hydrology of Fractured Rocks, Int. Ass. Sci. Hydrol. Publ. No. 23, pp. 21-31. [Horizontal anisotropy]

Papadopulos, I.S. 1967. Drawdown Distribution Around a Large-Diameter Well. In: Ground-Water Hydrology, M. Marion (ed.), AWRA Proc. Series No. 4, American Water Resources Association, Bethesda, MD, pp. 157-167.

Papadopulos, I.S. and H.H. Cooper, Jr. 1967. Drawdown in a Well of Large Diameter. Water Resources Research 3:241-244. [Equations for modifying early part of the Jacob and Theis curves to take casing storage into account]

Prickett, T.A. 1965. Type Curve Solutions to Aquifer Tests Under Water Table Conditions. Ground Water 3(3):5-14.

(Table 7-6 Pump Test Solutions and Anisotropic/Fractured Rock References)

Ramey, Jr., H.J. 1975. Interference Analysis for Anisotropic Formation--A Case Study. J. Petroleum Technology 27(10):1290-1298. [Papadopoulos method]

Reed, J.E. 1980. Type Curves for Selected Problems of Flow to Wells in Confined Aquifers. U.S. Geological Survey Techniques of Water Resources Investigations TWRI 3-B3, 106 pp.

Rocha, M. and F. Franciss. 1977. Determination of Permeability in Anisotropic Rock Masses from Integral Samples. In: Structural and Geotechnical Mechanics, W.J. Hall (ed.), Prentice-Hall, Englewood Cliffs, NJ, pp. 178-202.

Rofail, N. 1967. Analysis of Pumping Test in Fractured Rock. In: Proc. Dubrovnik Symp. on Hydrology of Fractured Rocks, Int. Ass. Sci. Hydrol. Publ. No. 23.

Rorabaugh, M.I. 1953. Graphical and Theoretical Analysis of Step-Drawdown Test of Artesian Well. Proc. Am. Soc. Civil Engineers, Vol. 79.

Ritzi, Jr., R.W. and R.H. Andolsek. 1992. Relation Between Anisotropic Transmissivity and Azimuthal Resistivity Surveys in Shallow, Fractured, Carbonate Flow Systems. Ground Water 30(5):774-780.

Rushton, K.R. and K.S. Rathod. 1980. Overflow Tests Analyzed by Theoretical and Numerical Methods. Ground Water 18(1):61-69.

Sagar, B. and A. Runchai. 1982. Permeability of Fractured Rock: Effect of Fracture Size and Data Uncertainties. Water Resources Research 18(2):266-274.

Sauveplane, C. 1984. Pumping Test Analysis in Fractured Aquifer Formations: State of the Art and Some Perspectives. In: Groundwater Hydraulics, J. Rosenshein and G. Bennett (eds.), American Geophysical Union Water Resources Monograph 9, pp. 171-206.

Scheidegger, A.E. 1954. Directional Permeability of Porous Media to Homogeneous Fluids. Geofis. Pura Appl. 30:17-26.

Schmelling, S.G. and R.R. Ross. 1989. Contaminant Transport in Fracture Media: Models for Decision Makers. EPA/540/4-89/004, 8 pp.

Smith, E.D. and N.D. Vaughan. 1985. Experience with Aquifer Testing and Analysis in Fractured Low-Permeability Sedimentary Rocks Exhibiting Nonradial Pumping Response. In: Hydrogeology of Rocks of Low Permeability, Memoirs, 17th Congress Int. Ass. Hydrogeologists (Tucson, AZ) XVII:137-149.

Snow, D.T. 1969. Anisotropic Permeability in Fractured Media. Water Resources Research 5:1273-1289.

Stallman, R.W. 1962. Variable Discharge without Vertical Leakage (Continuously Varying Discharge). In: Theory of Aquifer Tests, U.S. Geological Survey Water-Supply Paper 1536-E, pp. E118-E122.

Stallman, R.W. 1963. Type Curves for the Solution of Single-Bound Problems. U.S. Geological Survey Water-Supply Paper 1545-C, pp. C45-C47.

(Table 7-6 Pump Test Solutions and Anisotropic/Fractured Rock References)

Streltsova, T.D. 1972. Unsteady Radial Flow in an Unconfined Aquifer. Water Resources Research 8(4):1059-1066.

Streltsova, T.D. 1974. Drawdown in Compressible Unconfined Aquifer. J. Hydraulics Division, ASCE 100(HY11):1601-1616. [Includes solution for anisotropic aquifer conditions]

Streltsova, T.D. 1976a. Analysis of Aquifer-Aquitard Flow. Water Resources Research 12(3):415-422.

Streltsova, T.D. 1976b. Hydrodynamics of Groundwater Flow in a Fractured Formation. Water Resources Research 12(3):4-5-413.

Theis, C.V. 1935. The Relation between the Lowering of the Piezometric Surface and the Rate and Duration of Discharge of a Well Using Ground Water Storage. Trans. Am. Geophysical Union 16(Pt2):519-524.

Thomas, J.E. and P.J. McGlew. 1985. Techniques for Investigation Contaminated Bedrock Aquifers. In: Proc. 6th Nat. Conf. on Management of Uncontrolled Hazardous Waste Sites, Hazardous Material Control Research Institute, Silver Spring, MD, pp. 142-146.

Tsang, Y.W. 1992. Usage of "Equivalent Apertures" for Rock Fractures as Derived From Hydraulic and Tracer Tests. Water Resources Research 28(5):1452-1455.

Tsang, Y.W. and C.F. Tsang. 1987. Channel Model of Flow Through Fractured Media. Water Resources Research 23:467-479.

UNESCO. 1984. Ground Water in Hard Rocks. United Nations Educational, Scientific and Cultural Organization, Paris, France, 227 pp.

Warren, J.E. and P.J. Root. 1963. The Behavior of Naturally Fractured Reservoirs. J. Society Petroleum Engineers 9:245-255.

Way, S.C. and C.R. McKee. 1982. In-Situ Determination of Three-Dimensional Aquifer Permeabilities. Ground Water 20(5):594-603. [Multi-well pump test for homogeneous, anisotropic, leaky aquifer]

Weeks, E.P. 1964. Field Methods for Determining Vertical Permeability and Aquifer Anisotropy. U.S. Geological Survey Professional Paper 501D, pp. D193-D198.

Weeks, E.P. 1969. Determining the Ratio of Horizontal to Vertical Permeability by Aquifer Test Analysis. Water Resources Research 5(1):196-214.

Zekai, S. 1986. Test Analysis in Fractured Rocks with Linear Flow Pattern. Ground Water 24(1):72-78.

(Table 7-6 Pump Test Solutions and Anisotropic/Fractured Rock References)

CHAPTER 8

SOIL AND GROUND WATER TRACERS

8.1 Types and Uses of Tracer Tests 440

    8.1.1 Measurement of Hydrogeologic Parameters 440
    8.1.2 Delineation of Contaminant Sources and Plumes 443
    8.1.3 Characterizing Vadose Zone Hydrology 443

8.2 Tracer Selection 443

    8.2.1 Overview of Types of Tracers 443
    8.2.2 Hydrogeologic Considerations 445
    8.2.3 Tracer Characteristics 447
    8.2.4 Other Considerations 447
    8.2.5 Tracing in Karst vs. Porous Media 448

8.3 Types of Tracers 448

    8.3.1 Ions 448
    8.6.2 Dyes 451
    8.3.3 Gases 456
    8.3.4 Isotopes 458
    8.3.5 Water Temperature 462
    8.3.6 Particulates 463

8.4 Tracer Tests in Karst and Fractured Rock 467

8.5 Tracer Tests in Porous Media 469

    8.5.1 Estimating the Amount of Tracer to Inject 470
    8.5.2 Single-Well Techniques 470
    8.5.3 Two-Well Techniques 472
    8.5.4 Design and Construction of Test Wells 473
    8.5.5 Injection and Sample Collection 474
    8.5.6 Interpretation of Results 475

8.6 Guide to Major References* 478

* Appendix F contains citations for table and figure sources.

In hydrogeology, the term "tracer" refers to distinguishable matter or energy carried by ground water that gives information concerning the ground-water system. A tracer can be entirely natural, such as the heat carried by hot spring waters; it can be accidentally introduced, such as fuel oil from a ruptured storage tank; or it can be introduced intentionally, such as dyes placed in water flowing within limestone caves.

This chapter examines the use of tracers in hydrogeology with special emphasis on the study of contaminant behavior. Section 8.1 discusses the types and uses of tracer tests, Section 8.2 discusses the factors that should be considered in selecting a tracer, and Section 8.3 describes the general characteristics of major types of tracers. The chapter concludes with an examination of methods for tracing in karst systems (Section 8.4) and porous media (Section 8.5).

8.1 Types and Uses of Tracer Tests

The variety of tracer tests is almost infinite, considering the various combinations of tracer types, local hydrologic conditions, injection methods, sampling methods, and geological settings. Tracer tests are used for two main purposes: (1) to measure one or more hydrogeologic parameters of an aquifer and (2) to identify sources, velocity, and direction of movement of contaminants. Tracer tests can also be broadly classified according to whether they rely on natural gradient flow or induced flow by pumping or some other means. Quinlan et al. (1988)[1] discuss how to recognize false negative and false positive tracer results.

8.1.1 Measurement of Hydrogeologic Parameters

Tracers can be used to measure or estimate a wide variety of hydrogeologic parameters, most commonly direction and velocity of flow and dispersion. Depending on the type of test and the hydrogeologic conditions, other parameters such as hydraulic conductivity, porosity, chemical distribution coefficients, source of recharge, and age of ground water can be measured.

Figure 8-1 shows six examples of tracer measurement of hydrogeologic characteristics by natural gradient flow. Figure 8-1a shows the measurement of flow velocity in a cave system, and Figure 8-1b shows the use of a tracer to check subsurface flow patterns in a karst area with sinking and rising streams. The use of a tracer to measure the velocity of movement of dissolved material between two wells is shown in Figure 8-1c. Both velocity and direction of flow can be measured in a single well as shown in Figure 8-1d (see Section 8.5.2) and by use of multiple downgradient sampling wells as shown in Figure 8-1e. Finally, hydrodynamic dispersion can be measured by multiwell, multilevel sampling downgradient (Figure 8-1f).

---

[1] Quinlan, J.F., R.O. Ewars, and M.S. Field. 1988. How to Use Ground-Water Tracing to "Prove" that Leakage of Harmful Materials from a Site in Karst Terrane Will Not Occur. In: Proc. Second Conf. on Environmental Problems in Karst Terranes and Their Solutions (Nashville, TN), National Water Well Association, Dublin, OH, pp. 265-288.

Soil and Ground Water Tracers

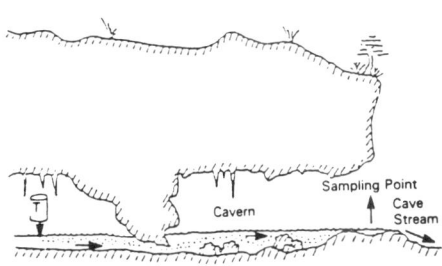
a. To measure velocity of water in cave stream.

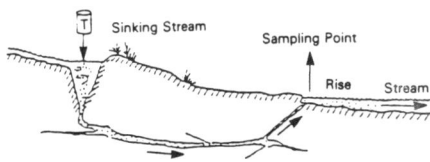
b. To check source of water at rise in stream bed.

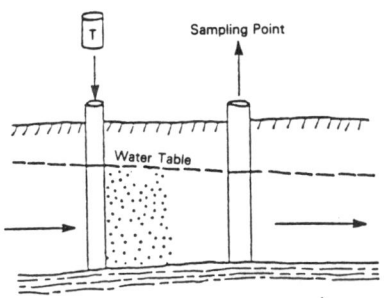
c. To test velocity of movement of dissolved material under natural ground-water gradients.

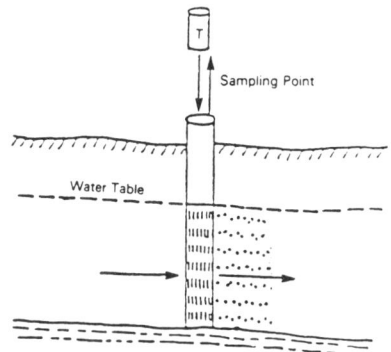
d. To determine velocity and direction of ground-water flow under natural conditions. Injection followed by sampling from same well

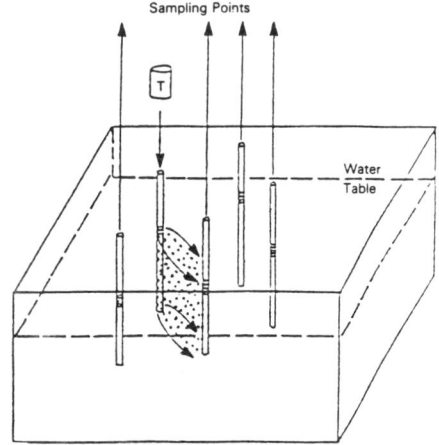
e. To determine the direction and velocity of natural ground-water flow by drilling an array of sampling wells around a tracer injection well.

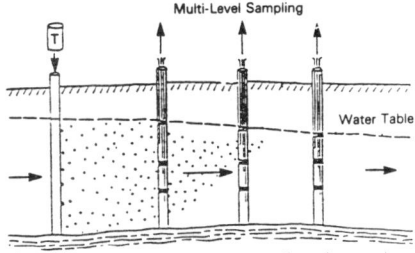
f. To test hydrodynamic dispersion in aquifer under natural ground-water gradients.

Figure 8-1 Common configurations for use of tracers to measure hydrogeologic parameters using natural gradient flow (Boulding, 1991, after Davis et al., 1985).

Figure 8-2 shows four examples of measurement by tracers of hydrogeologic parameters using induced flow. A tracer in surface water combined with pumping from a nearby well can verify a connection, as shown in (Figure 8-2a). Interconnections between fractures can be mapped using tracers and inflatable packers in two uncased wells, as in Figure 8-2b. Figure 8-2c shows the measurement of a number of aquifer parameters using a pair of wells with forced circulation between wells. Figure 8-2d shows the evaluation of geochemical interactions between multiple tracers and aquifer material by alternating injection and pumping.

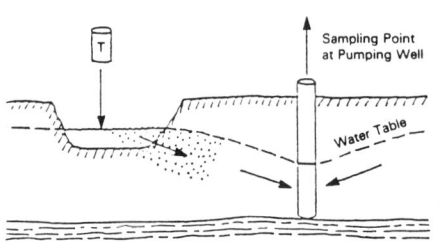

a. To verify connection between surface water and well.

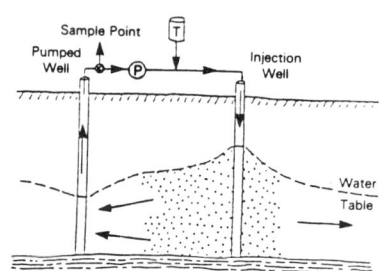

c. To test a number of aquifer parameters using a pair of wells with forced circulation between wells.

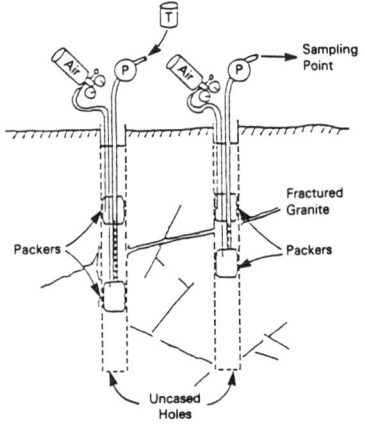

b. To determine the interconnect fractures between two uncased holes. Packers are inflated with air and can be positioned as desired in the holes.

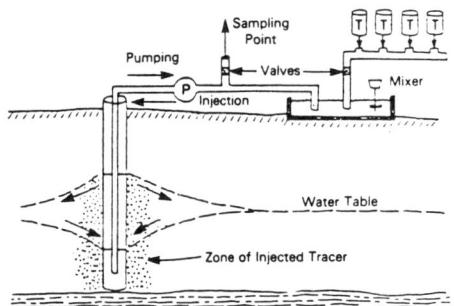

d. To test precipitation of selected constituents on the aquifer material by injecting multiple tracers into aquifer then pumping back the injected water.

Figure 8-2 Common configurations for use of tracers to measure hydrogeologic parameters using induced flow (Boulding, 1991, after Davis et al., 1985).

Other examples of the uses of tracers include determining ground-water recharge using environmental isotopes (Bradbury, 1991; Vogel et al., 1974)[2] and dating of ground water (Davis and Bentley, 1982/T8-5).

8.1.2 Delineation of Contaminant Sources and Plumes

Any contaminant that moves in ground water acts as a tracer; thus the contaminant itself may be mapped, or other tracers may be added to map the velocity and direction of the flow. Contaminant plumes (Section 4.6) are not tracers in the sense used in this chapter and are not discussed further here. However, Figure 8-3 shows three examples of noncontaminant tracers used to identify contaminant sources and flow patterns. Figure 8-3a shows the use of a tracer in a sinkhole to determine if trash at a particular location is contributing to contamination of a spring. Similarly, Figure 8-3b shows that by flushing a dye tracer down a toilet one can determine whether septic seepage is causing contamination of a well or surface water. Use of multiple tracers at multiple sources of potential contamination to pinpoint the actual source is shown in Figure 8-3c.

8.1.3 Characterizing Vadose Zone Hydrology

Tracers are also useful for characterizing water flow in the vadose zone. The travel time of percolating water to the saturated zone can be determined by applying water with tracer such as potassium bromide (KBr) to the soil surface and sampling ground water at intervals until the tracer is detected (Section 8.3.1). Dyes (Section 8.3.2) provide a relatively simple method for assessing the significance of preferential flow in soil. Water with a high concentration of dye is ponded at the surface until a known amount has completely infiltrated into the soil. The soil is then excavated in increments with photographic documentation. If flow of soil water is fairly uniform, the entire soil matrix will be colored, but if there is strong preferential flow, channels and the interfaces of soil structural units will be colored while other parts of the soil matrix will retain its original color.

8.2 Tracer Selection

8.2.1 Overview of Types of Tracers

Ground-water tracers can be broadly classified as *natural* or *environmental* tracers and *injected* tracers. Table 8-1 lists 14 natural tracers and more than 40 injected tracers. The specific characteristics of the different groups shown in Table 8-1 are discussed further in Section 8.3.

---

[2] Vogel, J.C., L. Thilo, and M. Van Dijken. 1974. Determination of Groundwater Recharge with Tritium. J. Hydrology 23:131-140.

Bradbury, K.R. 1991. Tritium as an Indicator of Ground-Water Age in Central Wisconsin. Ground Water 29:398-404.

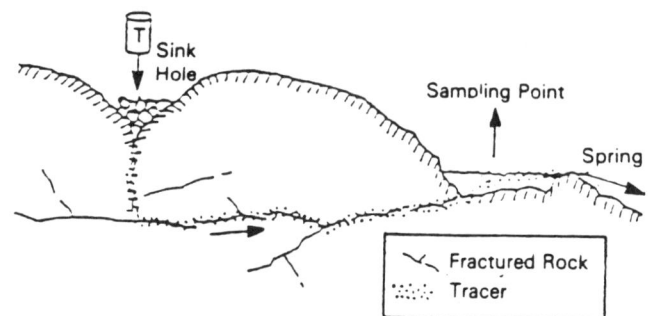

a. To determine if trash in sinkhole contributes to contamination of spring.

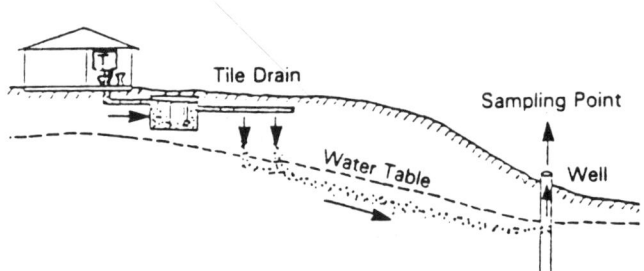

b. To determine if tile drain from septic tank contributes to contamination of well.

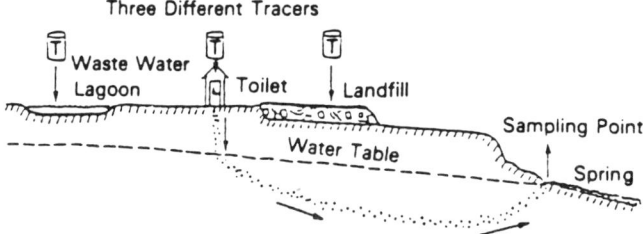

c. To determine source of pollution from three possibilities.

Figure 8-3 Common configurations for use of tracers to identify contaminant sources using natural gradient flow ((Boulding, 1991, after Davis et al., 1985).

Soil and Ground Water Tracers

Table 8-1 List of Major Ground-Water Tracers

Natural Tracers	Injected Tracers		
	Radioactive	Activable	Inactive

Stable Isotopes			Ionic Substances
Deuterium (^{2}H)	Tritium (^{3}H)	Bromine-35	Chloride salts (Na, K, Li,
Oxygen-18	Bromine-32	Indium-39	Ca, NH$_4$)
Carbon-12	Chromium-51	Manganese-25	Bromide salts (K, Li)
Carbon-13	Cobalt-58	Lanthanum-57	Iodide salts (Na)
Nitrogen-14	Cobalt-60	Dysprosium-68	
Nitrogen-15	Gold-198		Nonionic Substances
Strontium-88	Iodine-131		
Sulfur-32	Krypton-85 (gas)		Difluorobenzoates
Sulfur-34	Phosphorus-32		Fluorobenzenes
Sulfur-36	Sodium-24		Sulfur hexafluoride
	Xenon-133 (gas)		
Radioactive Isotopes			Fluorescent Dyes
Tritium (^{3}H)			Fluorescein
Carbon-14			Rhodamine WT
Silicon-32			Eosin (Acid Red 87)
Chlorine-36			Optical brighteners
Argon-37			Tinopal 5Bm6x(FDA 22)
Argon-39			Direct Yellow 96
Krypton-81			Acid Yellow 7
Krypton-85			Amidorhodamine 6 (Acid
Bromine-32			Red 50)
Radon-222			
			Gases
Gases			Helium
			Argon
Fluorocarbons			Neon
			Krypton
			Xenon
			Carbon Monoxide
			Nitrous Oxide
			Physical Characteristics
			Water tmperature
			Flood pulse
			Particulates
			Lycopodium spores
			Bacteria, viruses, fungi
			Sawdust

Source: Adapted from Boulding (1993b).

The potential chemical and physical behavior of the tracer in ground water is the most important selection criterion. For most purposes, a tracer should travel with the same velocity and direction as the water and not interact with solid material. This is known as a *conservative* tracer. *Nonconservative* tracers, which tend to be slowed by interactions with the solid matrix, would be used to measure distribution coefficients (Section 4.5.2) and zones of preferential flow in the vadose zone. For most uses, a tracer should be nontoxic, inexpensive, and easily detected to a low concentration with widely available and simple technology. If the tracer occurs naturally in ground water, it should be present in concentrations well above background concentrations. Finally, the tracer itself should not modify the hydraulic conductivity or other properties of the medium being studied.

No one ideal tracer has been found. Because natural systems are so complex and the requirements for the tracers themselves so numerous, the selection and use of tracers is almost as much an art as a science. The following sections discuss further the factors that should be considered in selecting a tracer.

8.2.2 Hydrogeologic Considerations

The initial step in determining the physical feasibility of a tracer test is to collect as much hydrogeologic information as possible concerning the field area. The logs of the wells at the site to be tested, or logs of the wells closest to the proposed site, will give some idea of the homogeneity of the aquifer, layers present, fracture patterns, porosity, and boundaries of the flow system. Local or regional piezometric maps, or any published reports on the hydrology of the area (including results of aquifer tests), are valuable, as they may give an indication of the hydraulic gradient and hydraulic conductivity.

Major hydrogeologic factors that should be considered when selecting a tracer include:

- *Lithology*. Fine-grained materials, particularly clays, have higher sorptive capacities than coarse-grained material. This must be considered when evaluating the potential mobility of a tracer.

- *Flow Regime*. Whether flow is predominantly through porous media (alluvium, sandstone, soil), solution features (karst limestone), or fractures will influence the choice of tracer. For example, fluorescent dyes work well in karst settings, but are less effective as ground-water tracers in porous media due to sorption effects.

- *Direction of Flow*. Knowledge of the general direction of ground-water movement is essential in tracer studies using two or more wells. Section 7.2.5 discusses possible pitfalls in estimating the direction of flow using potentiometric surface maps.

- *Travel Time*. The equation for estimating travel time is discussed in Section 2.6.5. In two-well tracer tests, travel time is required to estimate spacing for wells.

- *Dispersion.* The importance of hydrodynamic dispersion in contaminant movement has been discussed in Section 4.4.2. Tracer tests are often used to measure dispersion (Figure 8-1f). In two-well tests, some preliminary estimates may be required to estimate the quantity of tracer to inject so that concentrations will be high enough to detect.

8.2.3 Tracer Characteristics

Tracers have a wide range of physical, chemical, and biological characteristics. These properties, as they relate to hydrogeologic (Section 8.2.2) and other factors (Section 8.2.4), will determine the most suitable tracer for the purposes desired.

Detectability. Injected tracers should have no, or very low, natural background levels. The lower the detection limit for instruments (ppm, ppb, ppt), the better. The degree of dilution is a function of type of injection, distance, dispersion, porosity, and hydraulic conductivity. Too much dilution may result in failure to observe the tracer when it reaches a sampling point because concentrations are below the detection limit. Possible interferences due to other tracers and natural water chemistry may have the same effect.

Mobility. Conservative tracers used to measure aquifer parameters such as flow direction and velocity should be (1) stable (i.e., not subject to transformation by biodegradation or nonbiological processes during the length of the test and analysis), (2) soluble in water, (3) of a similar density and viscosity (see Section 4.4.3), and (4) not subject to adsorption or precipitation. Nonconservative nontoxic tracers used to simulate transport of contaminants should have adsorptive and other chemical properties similar to the contaminant of concern.

Toxicity. Nontoxic tracers should be used if at all possible. If a tracer may be toxic at certain concentrations, maximum permissible levels as determined by federal, state, or county agencies must be considered in relation to expected dilution and proximity to drinking water sources (see Section 8.2.4). Most agencies have set no limits, partly because the commonly used tracers are nontoxic in concentrations usually employed, and partly because they never considered tracers to be a problem demanding regulation.

8.2.4 Other Considerations

A tracer may have suitable characteristics for the desired purpose and the hydrogeologic setting, yet still not be suitable for reasons of economics, technological availability or sophistication, or public health.

Economics. The tracer or the instrumentation to analyze samples may be expensive. In this situation, another less-expensive tracer with somewhat less favorable characteristics may suffice.

Technology. Some tracers may be difficult to obtain or may require more complicated sampling methods. Gases, for example, will escape easily from poorly sealed containers. Similarly, instrumentation for gas or isotope analyses may not be available. For example, very few laboratories are able to perform analyses of ^{36}Cl (Davis et al. 1985/T8-5). On the other hand, increasing availability of field analytical methods such as gas chromatographs (Section 9.6) means that technology is less of a limiting consideration that it was, say, ten years ago.

Public Health. The artificial introduction of tracers must involve a careful consideration of possible health implications. Health issues involving specific tracers are identified in Section 8.3. Some local or state health agencies insist on review authority prior to use of artificially introduced tracers, but most do not. Local citizens must be informed of the tracer injections, and usually the results should be made available to the public. Under some circumstances, analytical work associated with tracer studies must be performed in appropriately certified laboratories. These are job-specific decisions.

8.2.5 Tracing in Karst vs. Porous Media

Ground-water flow in karst terranes is characterized by conduit flow and diffuse flow through often complex subsurface channel systems. Ground-water contaminants tend to move rapidly in karst and resurge at the surface in locations that cannot be readily predicted from the morphology of surface drainage patterns. In contrast, ground-water flow in porous media is characterized by slow travel times and more generally predictable flow directions. These differences require substantially different approaches to conducting tracer tests as discussed in Section 8.4 (karst) and Section 8.5 (porous media).

8.3 Types of Tracers

Hundreds, and possibly thousands, of substances have been used as tracers in ground water, considering the full range of organic ground-water contaminants. The most commonly used tracers can be conveniently grouped into six categories: (1) ions, (2) dyes, (3) gases, (4) isotopes (including stable and radioactive isotopes), (5) water temperature, and (6) particulates. These categories are not mutually exclusive (i.e., isotopes may take the form of ions or gases). Selected tracers in each category are discussed in the following sections in relation to applicability in different hydrologic settings, field methods, and type of detection used.

8.3.1 Ions

Inorganic ionic compounds such as common salts have been used extensively as ground-water tracers. This category of tracers includes those compounds that undergo ionization in water, resulting in their separation into charged species possessing a positive charge (cations) or a negative charge (anions). The charge on an ion affects its movement through aquifers by numerous mechanisms.

Ionic tracers have been used as tools for a wide range of hydrologic problems dealing with the determination of flow paths and residence time and the measurement of aquifer properties. Slichter (1902, 1905)[3] was probably the first to use ionic tracers for the study of ground water in the United States. Specific characteristics of individual ions or ionic groups may approach those of an ideal tracer, particularly dilute concentrations of certain anions.

In most situations, anions (negatively charged ions) are not affected by the aquifer medium. Mattson (1929),[4] however, has shown that the capacity of clay minerals for holding anions increases with decreasing pH. Under conditions of low pH, anions in the presence of clay, other minerals, or organic detritus may undergo anion exchange. Other possible effects include anion exclusion and precipitation/dissolution reactions. Cations (positively charged ions) react much more frequently with clay minerals through the process of cation exchange, which displaces other cations such as sodium and calcium into solution (Section 3.2.2). Due to their interaction with the aquifer media, little work has been done with cations. However, natural variations in Ca and Mg concentrations have been used to separate baseflow and stormflow components in a karst aquifer (Dreiss, 1989).[5]

One advantage of simple ionic tracers is that they do not decompose and thus are not lost from the system. However, a large number of ions (including Cl^- and NO_3^-) have high natural background concentrations, thus requiring the injection of a tracer of high concentration. More importantly, several hundred pounds of chloride or nitrate may have an adverse effect on water quality and biota, thus becoming a pollutant. This may also result in density separation and gravity segregation during the tracer test (Grisak and Pickens, 1980).[6] Density differences will alter flow patterns, the degree of ion exchange, and secondary chemical precipitation, all of which may change the aquifer permeability. Comparisons of tracer mobilities under laboratory and field conditions by Everts et al. (1989/T8-5) found bromide (BR^-) to be only slightly less mobile than nitrate. The generally low background concentrations of bromide often make it the ion of choice when a conservative tracer is desired.

---

[3] Slichter, C.S. 1902. The Motions of Underground Waters. U.S. Geological Survey Water-Supply and Irrigation Paper 67.

Slichter, C.S. 1905. Field Measurement of the Rate of Movement of Underground Waters. U.S. Geological Survey Water-Supply and Irrigation Paper 140, pp. 9-34.

[4] Mattson, W. 1929. The Laws of Soil Colloidal Behavior I. Soil Science 27-28:71-87.

[5] Dreiss, S.J. 1989. Regional Scale Transport in a Karst Aquifer 1. Component Separation of Spring Flow Hydrographs. Water Resources Research 25(1):117-125.

[6] Grisak, G.E. and J.F. Pickens. 1980. Solute Transport through Fractured Media 2. Column Study of Fractured Till. Water Resources Research 16(4):731-739.

Various applications of ionic tracers have been described in the literature. Murray (1981)[7] used lithium bromide (LiBr) in carbonate terrane to establish hydraulic connection between a landfill and a fresh-water spring, where use of Rhodamine WT dye tracer proved inappropriate. Sodium chloride (NaCl) was used by Mather et al. (1969)[8] to investigate the influence of mining subsidence on the pattern of ground-water flow. Schmotzer et al. (1973)[9] used post sampling neutron activation to detect a Br⁻ tracer. Chloride (Cl⁻) and calcium (Ca^+) were used by Grisak and Pickens (1980-Footnote 6) to study solute transport mechanisms in fractures. Potassium (K^+) was used to determine leachate migration and the extent of dilution by receiving waters located by a waste disposal site (Ellis, 1980).[10] In vadose zone applications, Tennyson and Settergren (1980)[11] used bromide (Br⁻) to evaluate pathways and transit time of recharge through soil at a proposed sewage effluent irrigation site, and Morrison and Lowery (1990)[12] used KBr to assess the sampling radius of a porous cup soil-water sampler.

The use of *nonionic* organic compounds that are not dyes (see below) as injected tracers is a relatively recent development. Ilgenfritz et al. (1988) suggested using fluorobenzene as a field monitoring tracer that would not be likely to occur in normal industrial and commercial activities, and Bowman and Gibbons (1992) identify difluorobenzoates as good nonreactive tracers in soil and ground water.[13] Wilson and Mackay (1993) found that sodium hexafluoride (SF_6), a *nonionic inorganic* compound, was as conservative a bromide as a tracer in saturated sandy media with the advantage

---

[7] Murray, J.P., J.V. Rouse, and A.B. Carpenter. 1981. Groundwater Contamination by Sanitary Landfill Leachate and Domestic Wastewater in Carbonate Terrain: Principle Source Diagnosis, Chemical Transport Characteristics and Design Implications. Water Research 15(6):745-757.

[8] Mather, J.D., D.A. Gray, and D.G. Jenkins. 1969. The Use of Tracers to Investigate the Relationship between Mining Subsidence and Groundwater Occurrence of Aberdare, South Wales. J. Hydrology 9:136-154.

[9] Schmotzer, J.K., W.A. Jester, and R.R. Parizek. 1973. Groundwater Tracing with Post Sampling Activation Analysis. J. Hydrology 20:217-236.

[10] Ellis, J. 1980. A Convenient Parameter for Tracing Leachate from Sanitary Landfills. Water Research 14(9):1283-1287.

[11] Tennyson, L.C. and C.D. Settergren. 1980. Percolate Water and Bromide Movement in the Root Zone of Effluent Irrigation Sites. Water Resources Bulletin 16(3):433-437.

[12] Morrison, R.D. and B. Lowery. 1990. Sampling Radius of a Porous Cup Sampler: Experimental Results. Ground Water 28(2):262-267.

[13] Ilgenfritz, E.M., F.A. Blanchard, R.L. Masselink, and B.K. Panigrahi. 1988. Mobility and Effects in Liner Clay of Fluorobenzene Tracer and Leachate. Ground Water 26(1):22-30.

Bowman, R.S. and J.F. Gibbons. 1992. Difluorobenzoates as Nonreactive Tracers in Soil and Ground Water. Ground Water 30:8-14.

that much lower amounts needed to be injected compared to bromide because detection limits are much lower (5 µg/L compared to 0.5 mg/L).[14]

8.3.2 Dyes

Dyes are relatively inexpensive, simple to use, and effective. Either fluorescent or nonfluorescent dyes may be useful in studies of water movement in soil if the soil material that has absorbed the dye is excavated and visually inspected (Section 8.1.3). Fluorescent dyes are preferable to nonfluorescent varieties in ground-water tracer studies because they are easier to detect. Dole (1906)[15] was the first to recommend use of dyes for the study of ground water in the United States by reporting the results of use of fluorescein and other dyes in France beginning around 1882. Stiles et al (1927)[16] conducted early experiments using uranine (fluorescein) to demonstrate pollution of wells in a sandy aquifer, and Meinzer (1932)[17] described use of fluorescein as a ground-water tracer. However, extensive use of fluorescent dyes for ground-water tracing did not begin until after 1960. Quinlan (1986)[18] provides a concise, but comprehensive, guide to the literature on dye tracing.

The advantages of using fluorescent dyes include their very high detectability, rapid field analysis, and relatively low cost and low toxicity. Smart and Laidlaw (1977/T8-5) classified commonly used fluorescent dyes by color: orange (Rhodamine B, Rhodamine WT, and Sulforhodamine B), green (fluorescein, Lissamine FF, and pyranine), and blue--also called optical brighteners. Dyes can also be classified according to the detector (also called *bug*) used to recover them: dyes recovered on cotton include optical brighteners (such as Tinopal 5BM GX, and Phorwhite BBH) and Direct Yellow 96; and dyes recovered on activated charcoal (fluorescein and Rhodamine WT).

The literature on use of fluorescent dyes is plagued by a lack of consistency in dye nomenclature (Quinlan, 1986-Footnote 18). The standard reference to dyes is the

---

[14] Wilson, R.D. and D.M. Mackay. 1993. The Use of Sulphur Hexafluoride as a Conservative Tracer in Saturated Sandy Media. Ground Water 31(5):719-724.

[15] Dole, R.B. 1906. Use of Fluorescein in the Study of Underground Waters. U.S. Geological Survey Water Supply and Irrigation Paper 160, pp. 73-85.

[16] Stiles, C.W., H.R. Crohurst, and G.E. Thomson. 1927. Uranin Test To Demonstrate Pollution of Wells. In: Experimental Bacterial and Chemical Pollution of Wells via Ground Water and the Factors Involved, U.S. Public Health Service Hygienic Laboratory Bulletin No. 147, pp. 84-87.

[17] Meinzer, O.E. 1932. Outline of Methods for Estimating Ground Water Supplies. U.S. Geological Survey Water-Supply Paper 638-C, pp. 126-131.

[18] Quinlan, J.F. 1986. Discussion of "Ground Water Tracers" by Davis et al. (1985) with Emphasis on Dye Tracing, Especially in Karst Terranes. Ground Water 24(2):253-259 and 24(3):396-397 (References).

Colour Index (SDC & AATC, 1971-1982).[19] Most dyes are classified according to the CI Generic Name (related to method of dyeing) and chemical structure (the CI Constitution Number). Abrahart (1968, pp. 15-43)[20] provides a concise guide to dye nomenclature. Dyes are also classified according to their use in foods, drugs and cosmetics (Marmion, 1984).[21] There are numerous commercial names for most dyes. Consequently, reported results of dye tracing experiments should always specify (1) the CI Generic Name or CI Constitution Number and (2) the manufacturer and the manufacturer's commercial name. The full name of the dye should be mentioned at least once to distinguish it from other dyes with the same or similar names. For example, in 1985, four structurally different kinds of Rhodamine were sold in the U.S. under 11 different names by five manufacturers, and there are more than 180 kinds of Direct Yellow dye (Quinlan, 1986-Footnote 18).

The first part of the commercial name of a dye should not be confused with the dye itself. For example, Tinopal and Phorwhite are trade names used for whole series of chemically unrelated dyes made by a single company and should be capitalized. Seven chemically different Tinopals and 20 different Phorwhites are currently sold in the U.S. as optical brighteners.

A particularly confusing point of dye nomenclature is that there are two fluorescein dyes with the same CI Name and Number, although they do have different D&C (Drug and Cosmetic) designations: fluorescein ($C_{20}H_{12}O_5$)--D&C Yellow 7--and fluorescein sodium ($C_{20}H_{12}O_5Na_2$)--D&C Yellow 8. Only D&C Yellow 8 is soluble in water and hence suitable for ground-water tracing. In the American and British literature this is referred to as *fluorescein*, whereas in the European literature it is called as *uranine*.

Although fluorescent dyes exhibit many of the properties of an ideal tracer, a number of factors interfere with concentration measurement. Fluorescence is used to measure dye concentration, but the amount of fluorescence may vary with suspended sediment load, temperature, pH, $CaCO_3$ content, salinity, etc. Other variables that affect tracer test results are "quenching" (some emitted fluorescent light is reabsorbed by other molecules), adsorption, and photochemical and biological decay. A disadvantage of fluorescent dyes in tropical climates is poor performance resulting from chemical reactions with dissolved carbon dioxide (Smart and Smith, 1976).[22]

---

[19] SDC & AATC (Society of Dyers & Colorists and American Association of Textile Chemists). 1971-1982. Color Index, 3rd ed.

[20] Abrahart, E.N. 1968. Dyes and their Intermediates. Pergamon Press, Oxford.

[21] Marmion, D.M. 1984. Handbook of U.S. Colorants for Foods, Drugs, and Cosmetics, 2nd ed. Wiley-Interscience, New York.

[22] Smart, P.L. and D.I. Smith. 1976. Water Tracing in Tropical Regions; The Use of Fluorometric Techniques in Jamaica. J. Hydrology 30:179-195.

Fluorescence intensity is inversely proportional to temperature. Smart and Laidlaw (1977/T8-5) described the numerical relationship and provided temperature correction curves. Low pH tends to reduce fluorescence. Figure 8-4 shows that the fluorescence of Rhodamine WT decreases rapidly at increasingly acidic pHs below about 6.0. An increase in the suspended sediment concentration also generally causes a decrease in fluorescence.

Dyes travel slower than water due to sorption and are generally not as conservative as radioactive tracers or some of the ionic tracers. Sorption can occur on organic matter, clays (bentonite, kaolinite, etc.), sandstone, limestone, plants, plankton, and even glass sample bottles. Typically fluorescence decreases with an increase in suspended sediment concentration, and this interferences makes it hard to interpret tracer measurements for water flow.

These sorption effects are a strong incentive to choose a dye that is nonsorptive for the type of medium tested. Different dyes vary greatly in amount of sorption on specific materials. For example, Repogle et al. (1966)[23] measured sorption of three orange dyes on bentonite clay with the following results: Rhodamine WT, 28%; Rhodamine B, 65%; and Sulforhodamine B, 96%.

Smart (1984/T8-5) in a review of the toxicity of 12 fluorescent dyes identified only three tracers (Tinopal CBS-X, Fluorescein, and Rhodamine WT) with no demonstrated carcinogenic or mutagenic hazard. He recommended against use of Rhodamine B because it is a known carcinogen. Use of the other dyes was considered acceptable provided normal precautions are observed during dye handling. Aulenbach et al. (1978/T8-5) concluded that Rhodamine B should not be used as a ground-water tracer simply on the basis of sorption losses.

Currently, the U.S. Geological Survey has a policy of limiting the maximum concentration of fluorescent dyes at water-user withdrawal points to 0.01 ppm (Hubbard et al., 1982/T8-5). This is a conservative, nonobligatory limit, and Field et al. (1990/T8-5) recommend that tracer concentrations not exceed 1 ppm for a period in excess of 24 hours in ground water, and careful evaluation of a tracer before use in a sensitive of unique ecosystem. Dyes should probably not be used where water supplies are chlorinated because dye molecules may react with chlorine to form chlorophenols (Smart and Laidlaw, 1977/T8-5).

Fluorescein, also known as uranine, sodium fluorescein, and by other names, has been one of the most widely used green dyes. Like all green dyes, its use is commonly complicated by high natural background fluorescence, which lowers sensitivity of analyses and makes interpretation of results more difficult. However, fluorescein is the most common dye used for tracing ground water in karst.

---

[23] Repogle, J.A., L.E. Myers, and K.J. Brust. 1966. Flow Measurements with Fluorescent Tracers. J. Hydraulics Division ASCE 92:1-15.

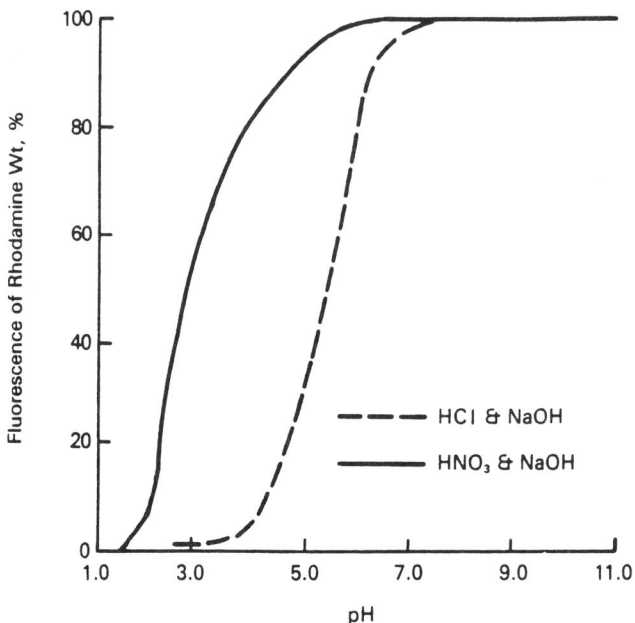

Figure 8-4 The effect of pH on Rhodamine WT (Davis et al., 1985, after Smart and Laidlaw, 1977).

Lewis et al. (1966)[24] used fluorescein in a fractured rock study. Mather et al. (1969-Footnote 8) recorded its use in a mining subsidence investigation in South Wales. Tester et al. (1982)[25] used fluorescein to determine fracture volumes and diagnose flow behavior in a fractured granitic geothermal reservoir. They found no measurable adsorption or decomposition of the dye during the 24-hour exposures to rocks at 392°F. At the other extreme, Rahe et al. (1978/T8-5) did not recover any injected dye in their hillslope studies, even at a distance of 2.5 m downslope from the injection point. The same experiment used bacterial tracers successfully. Sabatini and Austin (1991)[26] used the sorption characteristics of fluorescein (less sorptive) and Rhodamine WT (more sorptive) to delimit the expected appearance times of the

---

[24] Lewis, D.C., G.J. Kriz, and R.H. Burgy. 1966. Tracer Dilution Sampling Technique to Determine Hydraulic Conductivity of Fractured Rock. Water Resources Research 2:533-542.

[25] Tester, J.W., R.L. Bivens, and R.M. Potter. 1982. Interwell Tracer Analysis of a Hydraulically Fractured Granitic Geothermal Reservoir. Soc. Petroleum Engineers J. 8:537-554.

[26] Sabatini, D.A. and T.A. Austin. 1991. Characteristics of Rhodamine WT and Fluorescein as Adsorbing Ground-Water Tracers. Ground Water 29(3);341-349.

pesticides atrazine and arachlor (intermediate sorption characteristics) in a sandy alluvial aquifer.

Another green fluorescent dye, *pyranine*, has a stronger fluorescent signal than does fluorescein, but is much more expensive. It has been used in several soil studies. Reynolds (1966)[27] found pyranine to be the most stable dye used in an acidic, sandy soil. Drew and Smith (1969/T8-5) stated that pyranine is not as easily detectable as fluorescein, but is more resistant to decoloration and adsorption. Pyranine has a very high photochemical decay rate and is strongly affected by pH in the range found in most natural waters (McLaughlin, 1982/T8-5).

Rhodamine WT has been considered one of the most useful tracers for quantitative studies, based on minimum detectability, photochemical and biological decay rates, and adsorption (Knuttson, 1968/T8-5; Smart and Laidlaw, 1977/T8-5; Wilson et al., 1986/T8-5). Rhodamine WT is the most conservative dye available for stream tracing (Hubbard et al.,1982/T8-5) and is also commonly used in karst tracing studies. It is not a suitable tracer, however, for low pH environments (Figure 8-4).

Gann and Harvey (1975)[28] used Rhodamine WT for karst tracing in a limestone and dolomite system in Missouri. Aulenbach et al. (1978/T8-5) compared Rhodamine B, Rhodamine WT, and tritium as tracers in effluent from a sewage treatment plant that was applied to natural delta sand beds. The Rhodamine B was highly adsorbed, while the Rhodamine WT and tritium yielded similar breakthrough curves. Aulenbach and Clesceri (1980)[29] also used Rhodamine WT successfully in a sandy medium. Shiau et al. (1993)[30] found that the two-step breakthrough curve that is often observed when Rhodamine WT is used in porous media is the result of two isomers each having different sorptive characteristics. This effect may lead to an incorrect interpretation of a bicontinuum flow system.

Rhodamine B and Sulforhodamine B are poor tracers for use in ground water and most surface waters. It could be said the "B" stands for "bad". Amidorhodamine G is a significantly better tracer; similarly, it can be said that the "G" stands for "good".[31]

---

[27] Reynolds, E.R.C. 1966. The Percolation of Rainwater through Soil Demonstrated by Fluorescent Dyes. J. Soil Science 17(1):127-132.

[28] Gann, E.E. and E.J. Harvey. 1975. Norman Creek: A Source of Recharge to Maramec Spring, Phelps County, Missouri. J. Res. U. S. Geological Survey 3(1):99-102.

[29] Aulenbach, D.B. and N L. Clesceri. 1980. Monitoring for Land Application of Wastewater. Water, Air, and Soil Pollution 14:81-94.

[30] Shiau, B.-J., D.A. Sabatini, and J.H. Harwell. 1993. Influence of Rhodamine WT Properties on Sorption and Transport in Subsurface Media. Ground Water 31(6):913-920.

[31] Personal communication, James Quinlan, Nashville, TN, July, 1990.

Blue fluorescent dyes, or *optical brighteners*, have been used in increasing amounts in the past decade in textiles, paper, and other materials to enhance their white appearance. Water that has been contaminated by domestic waste entering septic tank soil absorption fields can be used as a "natural" tracer if it contains detectable amounts of the brighteners. Glover (1972)[32] was the first to describe the use of optical brighteners in karst environments. Since then, they have been extensively used in the U.S. The tracer *Amino G acid* is a dye intermediate used in the manufacture of dyes that is sometimes mistakenly classified as an optical brightener (Quinlan, 1986-Footnote 18). Amino G acid is now recognized as being carcinogenic and should not be used in water that might be used for drinking (see Footnote 31). Smart and Laidlaw (1977/T8-5) provide detailed information on the characteristics of the optical brightener Photine CU and Amino G acid.

8.3.3 Gases

Numerous natural and artificially produced gases have been found in ground water. Some of the naturally produced gases can serve as tracers, and gas can also be injected into ground water where it dissolves and serves as a tracer. Only a few examples of gases being used as ground-water tracers are found in the literature, however. Table 8-2 lists gases of potential use in hydrogeologic studies. Gases are useful tracers in the saturated zone. They are less reliable in the unsaturated zone because bleeding into the atmosphere can give false negative results.

Inert Natural Gases. Because of their nonreactive and nontoxic nature, noble gases are potentially useful tracers. Helium is used widely as a tracer in industrial processes. Carter et al. (1959)[33] studied the feasibility of using helium as a tracer in ground water and found that it traveled at a slightly lower velocity than chloride. Advantages of using helium as a tracer are its (1) safety, (2) low cost, (3) relative ease of analysis, (4) low concentrations required, and (5) chemical inertness. Disadvantages identified in this study included (1) relatively large errors in analysis, (2) difficulties in maintaining a constant recharge rate, (3) time required to develop equilibrium in unconfined aquifers, and (4) possible loss to the atmosphere in unconfined aquifers. Relatively recent improvements of field instrumentation for detection of helium as described in tracing experiments conducted in a basalt aquifer in Hawaii now make use of helium a practical and convenient method (Gupta et al., 1994).[34]

---

[32] Glover, R.R. 1972. Optical Brighteners--A New Water Tracing Reagent. Trans. Cave Research Group (Great Britain) 14(2):84-88.

[33] Carter, R.C., W.J. Kaufman, G.T. Orlob, and D.K. Todd. 1959. Helium as a Ground-Water Tracer. J. Geophysical Research 64:2433-2439.

[34] Gupta, S.K., L.S. Lau, and P.S. Moravcik. 1994. Ground-Water Tracing with Injected Helium. Ground Water 32(1):96-102.

Table 8-2 Gases of Potential Use as Tracers

	Approximate Natural Background Assuming Equilibrium with Atmosphere at 20°C (mg gas/L water)	Maximum Amount in Solution Assuming 100% Gas at Pressure of 1 atm at 20°C (mg gas/L water)
Argon	0.57	60.6
Neon	1.7×10^{-4}	9.5
Helium	8.2×10^{-6}	1.5
Krypton	2.7×10^{-4}	234
Xenon	5.7×10^{-5}	658
Carbon monoxide	6.0×10^{-6}	28
Nitrous oxide	3.3×10^{-4}	1,100

Source: Boulding (1991), after Davis et al. (1985).

Neon, krypton, and xenon are other possible candidates for injected tracers because their natural concentrations are very low (Table 8-2). Although the gases do not undergo chemical reactions and do not participate in ion exchange, the heavier noble gases (krypton and xenon) do sorb to some extent on clay and organic material. The solubility of the noble gases decreases with increases in temperature. Therefore, the natural concentrations of these gases in ground water are an indication of surface temperatures at the time of water infiltration. This property has been used to reconstruct palaeoclimatic trends in a sandstone aquifer in England using argon and krypton for age estimates (Andrews and Lee, 1979).[35] Sugisaki (1969)[36] and Mazor (1972)[37] have also used natural inert gases in this way.

---

[35] Andrews, J.H. and D.J. Lee. 1979. Inert Gases in Groundwater from the Bunter Sandstone of England as Indicators of Age and Palaeoclimatic Trends. J. Hydrology 41:233-252.

[36] Sugisaki, R. 1969. Measurement of Effective Flow Velocity of Groundwater by Means of Dissolved Gases. American J. Science 259:144-153.

[37] Mazor, E. 1972. Paleotemperatures and Other Hydrological Parameters Deduced from Noble Gases Dissolved in Ground Waters, Jordan Rift Valley, Israel. Geochimica et Cosmochimica Acta 36:1321-1336.

Anthropogenic Gases. Numerous artificial gases, such as fluorocarbons, have been manufactured during the past decade and several of them have been released in sufficient volumes to produce measurable concentrations in the atmosphere on a worldwide scale. One of the most interesting groups of these gases are the fluorocarbons. These gases generally pose a very low biological hazard, are generally stable for periods measured in years, do not react chemically with other materials, can be detected in very low concentrations, and sorb only slightly on most minerals. They do sorb strongly, however, on organic matter.

Fluorocarbons have two possible applications. First, because large amounts of fluorocarbons were not released into the atmosphere until the late 1940s and early 1950s, the presence of fluorocarbons in ground water indicates that the water was in contact with the atmosphere within the past 30 to 40 years (Thompson and Hayes, 1978).[38] The second possible application of fluorocarbon compounds is as injected tracers (Thompson et al., 1974).[39] Because detection limits are so low, large volumes of water can be labeled with the tracers at a rather modest cost. The problem of sorption on natural material, especially organics matter, and concerns with effect of CFCs on the ozone layer have prevented more widespread use of fluorocarbons as tracers.

8.3.4 Isotopes

An isotope is any of two or more forms of the same element having the same atomic number and nearly the same chemical properties but with different atomic weights and different numbers of neutrons in the nuclei. Isotopes may be *stable* (do not emit radiation) or *radioactive* (emit alpha, beta, and/or gamma rays). There are over 280 isotopic forms of stable elements and 40 or so radioactive isotopes. A wide variety of stable and radioactive isotopes have been used in ground-water tracer studies. There is an extensive literature on the use of isotopes in ground-water investigations (see Table 8-5 at the end of the chapter). Lack of familiarity with techniques for analyzing environmental isotopes has limited their use by practicing field hydrogeologists in ground-water contamination studies. Hendry (1988)[40] recommends the use of hydrogen and oxygen isotopes as a relatively inexpensive way to estimate the age of near-surface ground-water samples.

Stable Isotopes. Stable isotopes are rarely used for artificially injected tracer studies in the field because of (1) the difficulty in detecting small artificial variations of most isotopes against the natural background, (2) the high cost of their analysis, and

---

[38] Thompson, G.M. and J.M. Hayes. 1979. Trichlorofluoromethane in Ground Water, a Possible Tracer and Indicator of Ground-Water Age. Water Resources Research 15(3):546-554.

[39] Thompson, G.M., J.M. Hayes, and S.N. Davis. 1974. Fluorocarbon Tracers in Hydrology. Geophysical Research Letters 1:177-180.

[40] Hendry, M.J. 1988. Do Isotopes Have a Place in Ground-Water Studies? Ground Water 26(4):410-415.

(3) the expense of preparing isotopically enriched tracers. The average stable isotope composition of deuterium (^2H) and ^{18}O in precipitation changes with elevation, latitude, distance from the coast, and temperature. Consequently, measurement of these isotopes in ground water can be used to trace the large-scale movement of ground water and to locate areas of recharge (Gat, 1971).[41]

The two abundant isotopes of nitrogen (^{14}N and ^{15}N) can vary significantly in nature. Ammonia (NH_4) escaping as vapor from decomposing animal wastes, for example, will tend to remove the lighter (^{14}N) nitrogen and will leave behind a residue rich in heavy nitrogen. In contrast, many fertilizers with an ammonia base will be isotopically light. Natural soil nitrate will be somewhat between these two extremes. As a consequence, nitrogen isotopes have been useful in determining the origin of unusually high amounts of nitrate in ground water. Also, the presence of more than about 5 mg/L of nitrate commonly is an indirect indication of contamination from chemical fertilizers and sewage. Aravena et al. (1993)[42] used the stable isotopes ^{18}O and ^{15}N to differentiate a contaminant nitrate plume emanating from a single domestic septic system in an aquifer characterized by high and similar nitrate content outside and inside the plume. Komor and Anderson (1993)[43] used ^{15}N in nitrate as a qualitative indicator of nitrate sources in ground water (animal feedlot, inorganic nitrogen fertilizer, and septic systems).

The stable sulfur isotopes (^{32}S, ^{34}S, and ^{36}S) have been used to distinguish between sulfate originating from natural dissolution of gypsum ($CaSO_4 \cdot 2H_2O$) and sulfate originating from an industrial spill of sulfuric acid (H_2SO_4).

Two stable isotopes of carbon (^{12}C and ^{13}C) and one radioisotope (^{14}C) are used in hydrogeologic studies. Although not as commonly studied as ^{14}C, the ratio of the stable isotopes, $^{13}C/^{12}C$, is potentially useful in sorting out the origins of certain contaminants found in water. For example, methane (CH_4) originating from some deep geologic deposits is isotopically heavier then methane originating from near-surface sources. This contrast forms the basis for identifying aquifers contaminated with methane from pipelines and from subsurface storage tanks.

Isotopes of other elements such as chlorine, strontium, and boron are used more for determining regional directions of ground-water flow than for identifying sources of contamination.

---

[41] Gat, J.R. 1971. Comments on the Stable Isotope Method in Regional Ground Water Investigations. Water Resources Research 7:980-993.

[42] Aravena, R., M.L. Evans, and J.A. Cherry. 1993. Stable Isotopes of Oxygen and Nitrogen in Source Identification of Nitrate from Septic Systems. Ground Water 31(2):180-186.

[43] Komor, S.C. and H.W. Anderson, Jr. 1993. Nitrogen Isotopes as Indicators of Nitrate Sources in Minnesota Sand-Plain Aquifers. Ground Water 31(2):260-270.

Radionuclides. Radioactive isotopes of various elements are collectively referred to as radionuclides. In the early 1950s there was great enthusiasm for using radionuclides both as natural "environmental" tracers and as injected artificial tracers. However, the use of artificially injected radionuclides has all but ceased in many countries, including the United States, because of concerns about possible adverse health effects (Davis et al., 1985/T8-5). Most uses of artificially introduced radioactive tracers are confined to carefully controlled laboratory experiments or to deep petroleum production zones that are devoid of potable water. Table 8-3 lists eight radionuclides commonly used as *injected* tracers, their half-lives, and the chemical form in which they are typically used.

The use of natural environmental tracers has expanded to the point that they are now a major component of many hydrochemical studies. A number of radionuclides are present in the atmosphere from natural and artificial sources, and many of these are carried into the subsurface by rain water (see natural radioisotopes listed in Table 8-1). The most common hydrogeologic use of these radionuclides is in estimating the average length of time ground water has been isolated from the atmosphere. This measurement is complicated by dispersion in the aquifer and mixing in wells that sample several hydrologic zones. Nevertheless, the age of water in an aquifer can usually be established as being older than some given limiting value. For example, detection of atmospheric radionuclides might indicate that ground water was recharged more than 1,000 years ago or that, in another region, all the ground water in a given shallow aquifer is younger than 30 years.

Table 8-3 Commonly Used Radioactive Tracers for Ground-Water Studies

Radionuclide	Half-Life y = year d = day h = hour	Chemical Compound
^3H	12.3y	H_2O
^{32}P	14.3d	Ha_2HPO_4
^{51}Cr	27.8d	EDTA-Cr and $CrCl_3$
^{60}Co	5.25y	EDTA-Co and $K_3Co(CN_6)$
^{32}Br	33.4h	NH_4Br, NaBr, LiBr
^{85}Kr	10.7y	Kr (gas)
^{131}I	8.1d	I and KI
^{198}Au	2.7d	$AuCl_3$

Source: Boulding (1991), after Davis et al. (1985).

Since the 1950s, atmospheric tritium, the radioactive isotope of hydrogen (^{3}H) with a half-life of 12.3 years, has been dominated by tritium from the detonation of thermonuclear devices. Thermonuclear explosions had increased the concentration of tritium in local rainfall to more than 1,000 tritium units (TU) in the Northern Hemisphere by the early 1960s. As a result, ground water in the Northern Hemisphere with more than about 5 TU is generally less than 30 years old. Very small amounts of tritium, 0.05 to 0.5 TU, can be produced by natural subsurface processes, so the presence of these low levels does not necessarily indicate a recent age.

Nuclear bomb testing also released significant amounts of the radioactive isotope of chlorine ^{36}Cl into the atmosphere (which also contains low concentrations of cosmogenic ^{36}Cl), and this isotope occurs as a contaminant in low level nuclear waste from nuclear fuel reprocessing. Beasley et al. (1993)[44] found this isotope to be the best indicator for delineation of contaminant plumes resulting from nuclear fuel reprocessing facilities.

The radioactive isotope of carbon, ^{14}C (with a half-life of 5,730 years), is also widely studied in ground water. In practice, the use of ^{14}C is rarely simple. Sources of old carbon, primarily from limestone and dolomite, will dilute the sample, and a number of processes, such as the formation of CH_4 gas or the precipitation of carbonate minerals, will fractionate the isotopes and alter the apparent age. Interpreting ^{14}C "ages" of water is so complex that it should be attempted only by hydrochemists specializing in isotope hydrology. Despite the complicated nature of ^{14}C studies, they are highly useful in determining the approximate residence time of old water (500 to 30,000 years) in aquifers. In certain circumstances, this information cannot be obtained in any other way. Figure 8-5 illustrates how carbon isotope percentages allow differentiation of bedrock-derived methane leaking from a pipeline or tank from natural methane generated in shallow aquifers of glacial drift.

Inert Radioactive Gases. Chemically inert but radioactive ^{133}Xe and ^{85}Kr appear to be suitable for many injected tracer applications (Robertson, 1969; Wagner, 1977),[45] provided legal restrictions can be overcome. ^{222}Rn, one of the daughter products from the spontaneous fission of ^{238}U, is the most abundant of the natural inert radioactive gases. Radon is present in the subsurface, but owing to the short half-life (3.8 days) of ^{222}Rn, and the absence of parent uranium nuclides in the atmosphere, radon is virtually absent in surface water that has reached equilibrium

---

[44] Beasley, T.M. et al. 1993. Chlorine-36 in the Snake River Plain Aquifer at the Idaho National Engineering Laboratory: Origin and Implications. Ground Water 31(2):302-310.

[45] Robertson, J.B. 1969. Behavior of Xenon-133 Gas after Injection Underground. U.S. Geological Survey Open File Report ID022051.

Wagner, O.R. 1977. The Use of Tracers in Diagnosing Interwell Reservoir Heterogeneities. J. Petroleum Technology 11:1410-1416.

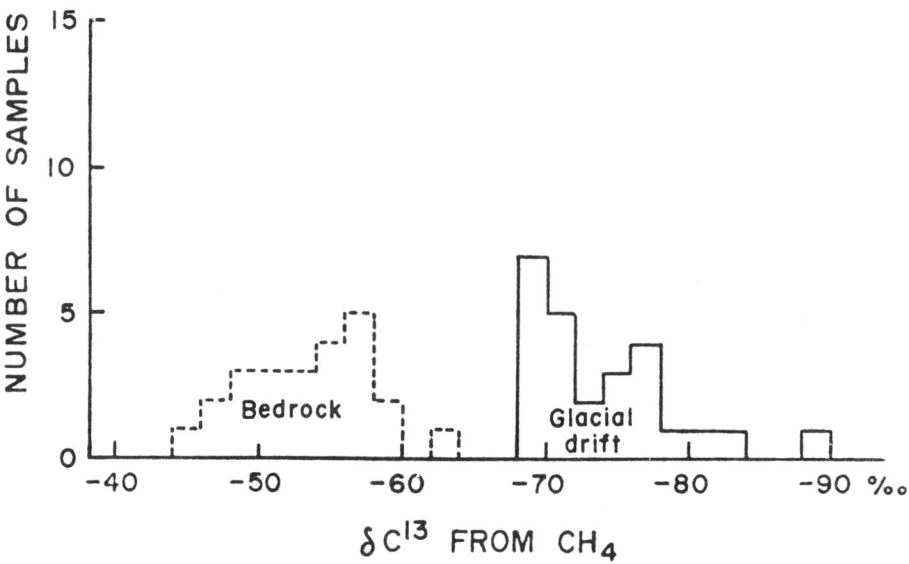

Figure 8-5 Differentiation of methane leak (bedrock) from natural shallow methane source (Davis et al., 1985, after Coleman et al., 1977).

with the atmosphere. Surveys of radon in surface streams and lakes have, therefore, been useful in detecting locations where ground water enters surface waters (Rogers, 1958).[46] Hoehn and von Gunten (1989)[47] measured dilution of radon on ground water to assess infiltration from surface waters to aquifer.

8.3.5 Water Temperature

The temperature of water changes slowly as it migrates through the subsurface, because water has a high specific heat capacity compared to most natural materials. For example, temperature anomalies associated with the spreading of warm wastewater in the Hanford Reservation in south central Washington have been detected more than 8 km (5 mi) from the source.

Water temperature is a potentially useful tracer, although it has not been used frequently. The method should be applicable in granular media, fractured rock, or

---

[46] Rogers, A.S. 1958. Physical Behavior and Geologic Control of Radon in Mountain Streams. U.S. Geological Survey Bulletin 1052E:187-211.

[47] Hoehn, E. and H.J.R. von Gunten. 1989. Radon in Groundwater: A Tool to Assess Infiltration from Surface Waters to Aquifers. Water Resources Research 25(8):1795-1803.

karst regions. Keys and Brown (1978)[48] traced thermal pulses resulting from the artificial recharge of playa lake water into the Ogallala formation in Texas. They described the use of temperature logs (temperature measurements at intervals in cased holes) to detect hydraulic conductivity differences in an aquifer. Temperature logs are a relatively simple and useful borehole logging method (Table 6-6).

Changes in water temperature are accompanied by changes in water density and viscosity, which in turn alter the velocity and direction of flow. For example, injected ground water with a temperature of 40°C will travel more than twice as fast in the same aquifer under the same hydraulic gradient as water at 5°C. Because the warm water has a slightly lower density than cold water, buoyant forces give rise to flow that "floats" on top of the cold water. To minimize problems of temperature-induced convection, very accurately measured small temperature differences should be used if hot or cold water is in the introduced tracer.

Figure 8-6 illustrates use of temperature as a tracer for small-scale field tests, using shallow drive-point wells 2 feet apart in an alluvial aquifer. The transit time of the peak temperature was about 107 minutes, while the resistivity data indicated a travel time of about 120 minutes. The injected water had a temperature of 38°C, while the ground-water temperature was 20°C; the peak temperature obtained in the observation well was 27°C.

In these tests, temperature indicated breakthrough of the chemical tracers, aiding in the timing of sampling. It was also useful as a simple, inexpensive tracer for determining the correct placement of sampling wells.

Another application of water-temperature tracing is the detection of river recharge in an aquifer. Most rivers have large seasonal water temperature fluctuations. If the river is recharging an aquifer, the seasonal fluctuations can be detected in the ground water adjacent to the river (Rorabaugh, 1956).[49]

8.3.6 Particulates

Solid material in suspension, such as spores, can be a useful tracer in areas where water flows in large conduits such as in some basalt, limestone, or dolomite aquifers. Seismic methods at the surface have been used to detect the location of time-delayed explosives floating through a cave system (Arandjelovic, 1969).[50] Small particulate tracers, such as bacteria, can travel through any porous media such as soils

---

[48] Keys, W.S. and R.F. Brown. 1978. The Use of Temperature Logs to Trace the Movement of Injected Water. Ground Water 16(1):32-48.

[49] Rorabaugh, M.I. 1956. Ground Water in Northeastern Louisville, Kentucky. U.S. Geological Survey Water-Supply Paper 1360-B:101-169.

[50] Arandjelovic, D. 1969. A Possible Way of Tracing Groundwater Flows in Karst. Geophysical Prospecting 17(4):404-418.

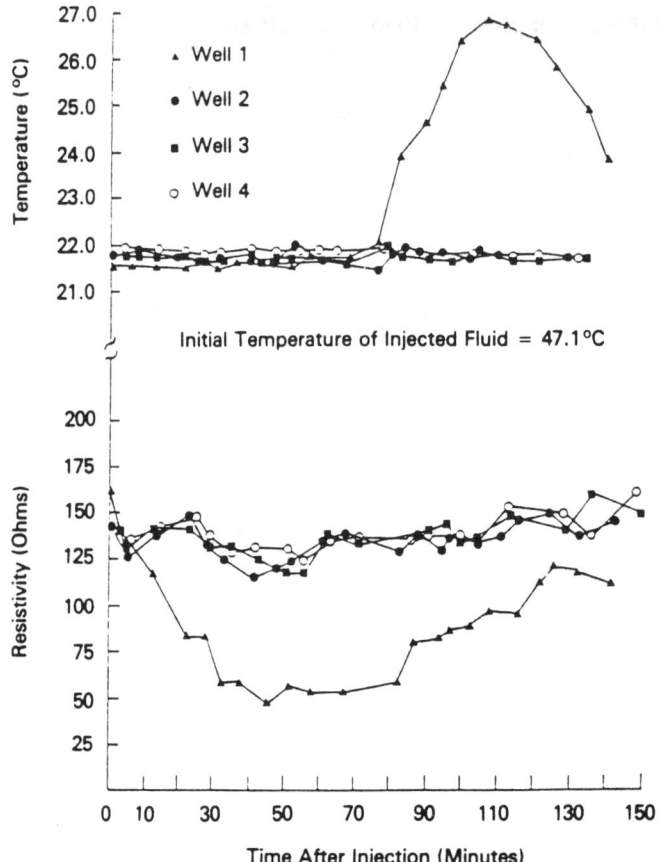

Figure 8-6 Results of field test using a hot water tracer (Boulding, 1991, after Davis et al., 1985).

and fractured bedrock where the pore size is larger than the size of the microorganism. Microorganisms are probably the most commonly used particulate tracers. Table 8-4 compares characteristics of microbial tracers.

Yeast. The use of baker's yeast (*Saccharomyces cerevisiae*) as a ground-water tracer in a sand and gravel aquifer has been reported by Wood and Ehrlich (1978/T8-5). Yeast is a single-celled fungus that is ovoid in shape. The diameter of a yeast cell is 2 to 3 µm, which closely approximates the size of pathogenic bacterial cells. This tracer is probably most applicable in providing information concerning the potential movement of bacteria.

Wood and Ehrlich (1978/T8-5) found that the yeast penetrated more than 7 meters into a sand and gravel aquifer in less than 48 hours after injection. This tracer is very inexpensive, as is analysis. The lack of environmental concerns related to this tracer is another advantage.

Soil and Ground Water Tracers

Table 8-4 Comparison of Microbial Tracers

Tracer	Size (μm)	Time Required for Assay (days)	Essential Equipment Required
Bacteria	1-10	1-2	Incubator*
Spores	25-33	1/2	Microscope Plankton nets
Yeast	2-3	1-2	Incubator*
Viruses:			
Animal (enteric)	0.2-0.8	3-5	Incubator Tissue culture Laboratory
Bacterial	0.2-1.0	1/2-1	Incubator*

* Many may be assayed at room temperature.

Source: Boulding (1991), after Keswick et al. (1982).

Bacteria. Bacteria are the most commonly used microbial tracers, due to their ease of growth and simple detection. Keswick et al. (1982/T8-5) reviewed over 20 case studies of bacteria used as tracers. Some bacteria that have been used successfully are *Escherichia coliform* (*E. coli*), *Streptococcus faecalis*, *Bacillus stearothermophilus*, *Serratia marcescens*, and *Serratia indica*. They range in size from 1 to 10 μm and have been used in a variety of applications.

A fecal coliform, *E. coli*, has been used to indicate fecal pollution at pit latrines, septic fields, and sewage disposal sites. A "marker" such as antibiotic resistance or H_2S production is necessary to distinguish the tracer from background organisms.

The greatest health concern in using these tracers is that the bacteria must be nonpathogenic to humans. Even *E. coli* has strains that can be pathogenic. Davis et al. (1970) and Wilkowske et al. (1970) have reported that *Serratia marcescens* may be life-threatening to patients who are hospitalized with other illnesses.[51] Antibiotic-

---

[51] Davis, J.T., E. Flotz, and W.S. Blakemore. 1970. *Serratia Marcescens*, a Pathogen of Increasing Clinical Importance. J. American Medical Association 214(12):2190-2192.

Wilkowske, C.J., J.A. Washington II, W.J. Martin, and R.E. Ritts, Jr. 1970. *Serratia marcescens*: Biochemical Characteristics, Antibiotic Susceptibility and Clinical Significance. J. American Medical Association 214(12):2157-2162.

resistant strains are another concern, as the antibiotic resistance can be transferred to potential human pathogens. This can be avoided by using bacteria that cannot transfer this genetic information. As is true with most other injected tracers, permission to use bacterial tracers should be obtained from the proper federal, state, and local health authorities.

Viruses. Animal, plant, and bacterial viruses have been used as ground-water tracers. Viruses are generally much smaller than bacteria, ranging from 0.2 to 1.0 µm (see Table 8-4). In general, human enteric viruses cannot be used due to disease potential, although certain vaccine strains, such as a type of polio virus, have been used but are considered risky. Most animal enteric viruses are considered safer as they are not known to infect humans (Keswick et al., 1982/T8-5). However, neither human nor most animal viruses are generally considered suitable tracers for field work because of their potential to infect humans.

Spores. *Lycopodium* spores have been widely used as tracers in karst hydrogeologic systems in Europe since the early 1950s and have been less frequently used in the United States since the 1970s. Much of the literature on the use of spores, however, is in obscure European and American speleological journals. More readily accessible references on the use of spores include Atkinson et al. (1973) and Gardner and Gray (1976).[52]

Lycopodium is a clubmoss that has spores nearly spherical in shape, with a mean diameter of 33 µm. It is composed of cellulose and is slightly denser than water, so that some turbulence is required to keep the material in suspension. Some advantages of using *lycopodium* spores as a tracer are:

- The spores are relatively small.

- They are not affected by water chemistry or adsorbed by clay or silt.

- They travel at approximately the velocity of the surrounding water.

- The injection concentration can be very high (e.g., 8×10^6 spores per cm^3).

- They pose no health threat.

- The spores are easily detectable under the microscope.

- At least five dye colors may be used, allowing five tracings to be conducted simultaneously in a karst system.

---

[52] Atkinson, T.C., D.I. Smith, J.J. Lavis, and R.J. Whitaker. 1973. Experiments in Tracing Underground Waters in Limestones. J. Hydrology 19:323-349.

Gardner, G.D. and R.E. Gray. 1976. Tracing Subsurface Flow in Karst Regions Using Artificially Colored Spores. Association of Engineering Geologists Bulletin 13:177-197.

Some disadvantages associated with the use of *lycopodium* spores include the large amount of time required for their preparation and analysis and the problem of spores being filtered by sand or gravel if flow is not sufficiently turbulent.

The basic procedure involves adding a few kilograms of dyed spores to a cave or sinking stream. The movement of the tracer is monitored by sampling downstream in the cave or at a spring with plankton nets installed in the stream bed. The sediment caught in the net is concentrated and treated to remove organic matter. The spores are then examined under the microscope.

Tracing by *lycopodium* spores is most useful in open joints or solution channels (karst terrane) where there is minimal suspended sediment. It is not useful in wells or boreholes unless the water is pumped continuously to the surface and filtered. The spores survive well in polluted water, but do not perform well in slow flow or in water with a high sediment concentration. A velocity of a few miles per hour has been found sufficient to keep the spores in suspension. According to Smart and Smith (1976),[53] *lycopodium* is preferable to dyes for use in large-scale water resource reconnaissance studies in karst areas. This is true if skilled personnel are available to sample and analyze the spores and a relatively small number of sampling sites are used.

8.4 Tracer Tests in Karst and Fractured Rock

Probably no hydrogeologic system has been more extensively studied by a more diverse group of people with such a plethora of tracing techniques as karst limestone terranes. Geese, tagged eels, computer punch-card confetti, and time bombs are among the more exotic tracers that have been used in karst.

There is an extensive international literature on karst tracing. Table 8-5 identifies major sources of information on this topic. There is a substantial English-language literature in American caving journals, such as **Cave Notes/Caves and Karst** (which ceased publication in 1973), **Missouri Speleology**, and the **National Speleological Society Bulletin**, and similar British periodicals, such as **Transactions of the Cave Research Group** (now **Cave Science**) and the **Proceedings of the University of Bristol Speleological Society**. The international symposia on underground water tracing (SUWT--see Table 8-5) provide the best systematic compilations of international research on this topic. Probably the easiest way to monitor the international literature on dye tracing in karst terranes and other karst and speleological literature is the annual **Speleological Abstracts** published by the Union Internationale de Spèléologie in Switzerland.

Fluorescent dyes are usually the tracer of choice in karst studies because

---

[53] Smart, P.L. and D.I. Smith. 1976. Water Tracing in Tropical Regions; The Use of Fluorometric Techniques in Jamaica. J. Hydrology 30:179-195.

adsorption is usually not a problem in karst hydrogeologic systems. Smart (1985)[54] lists four applications of fluorescent dye tracers in evaluating existing or potential contamination in carbonate rocks: (1) confirmation of leachate contamination, (2) determination of onsite hydrology, (3) determination of hydraulic properties of landfill materials, and (4) prediction of leachate contamination and dilution.

Fluorescein, Rhodamine WT, optical brighteners (Tinopal 5BM GX), and Direct Yellow 96 are the most commonly used dyes. The amount of dye injected depends on whether qualitative or quantitative analysis is planned. **Qualitative** tests involve simple visual detection of dye in flowing water or captured by a detector (see discussion below). **Semiquantitative** results can be obtained by using a fluorometer or spectrofluorometer to detect amounts of dye captured by detectors such as activated charcoal that may not be discernable to the eye. Interpretation of values from such measurements is limited due to lack of precise information on the variation in groundwater flow and dye concentration between collection of detectors. **Quantitative** tests involve precise measurement of dye concentrations in grab sample of water. If the exact amount of injected dye is known, and flow measurements are taken along with each sample, a mass-balance analysis allows estimation of how much dye has been distributed through different parts of the subsurface flow system.

In qualitative tests enough dye must be injected for visual detection; quantitative tests using a fluorometer or spectrofluorometer generally require one-tenth to one-hundredth as much dye. Determination of the correct quantity to inject is as much an art as a science and this should be determined by, or with the assistance of, someone with experience in karst tracer tests.

Dye is recovered with detectors called *bugs* (cotton or activated charcoal, depending on the tracer--see Section 8.3.2), that are typically suspended in streams and springs on hydrodynamically stable stands called a *gumdrop*. Detectors are placed at springs or in streams where flow from the point of injection is suspected of reaching the surface. At chosen time intervals related to the distance from the source of injection, detectors are collected and replaced with fresh detectors. Detectors are usually collected frequently during the first few days after injection to pinpoint the most rapid dye arrival time, and then typically on a daily basis for several weeks. Background tests must always be run before injection, especially with optical brighteners because sewage effluent from individual septic tank absorption fields may increase background levels substantially.

Qualitative tracer tests in which two dyes are injected into two different locations are readily done by combining a fluorescent dye and an optical brightener, which use different detectors. Quantitative techniques are available (developed originally in Europe) for separating mixtures of fluorescent dyes. A 5-dye tracer test has recently been conducted using these techniques (see Footnote 31). Perhaps the most comprehensive karst tracing experiments in a single location were carried out in

---

[54] Smart, P.L. 1985. Applications of Fluorescent Dye Tracers in the Planning and Hydrological Appraisal of Sanitary Landfills. Q. J. Eng. Geol. (London) 18:275-286.

Slovenia, Yugoslavia in the early 1970s where 5 dyes, *lycopodium* spores, lithium chloride, potassium chloride, chromium-51, and detergents were all used (Gospodaric and Habic, 1976/T8-5).

Reports prepared for U.S. EPA by Mull et al. (1988/T8-5) and Quinlan (1989/T8-5) are the most comprehensive references currently available on procedures for dye tracing in karst terranes. Smoot et al. (1987/T8-5) and Smart (1988a/T8-5) describe quantitative dye-tracing techniques in karst, and Smart (1988b/T8-5) describes an approach to the structural interpretation of ground-water traces in karst terrane. Table 8-5 also identifies a number of references on use of tracers for fracture characterization in limestone and other rock types.

8.5 Tracer Tests in Porous Media

Tracer tests in porous media are used primarily to characterize aquifer parameters such as regional velocity (Leap, 1985),[55] hydraulic conductivity distributions (Molz et al., 1988),[56] anisotropy (Kenoyer, 1988),[57] dispersivity (Bumb et al., 1985),[58] and distribution coefficient or retardation (Pickens et al., 1981; Rainwater et al., 1987).[59] Smart et al. (1988/T8-5) have prepared an annotated bibliography on ground-water tracing that focuses on use of tracers in porous media.

The purpose and practical constraints of a potential tracer test must be clearly understood prior to actual planning. Following are a few of the questions that need to be addressed:

- Is only the direction of water flow to be determined?

---

[55] Leap, D.I. 1985. A Simple, Two-Pulse Tracer Methods for Estimating Steady-State Ground Water Parameters. Hydrological Science and Technology: Short Papers 1(1):37-43.

[56] Molz, F.J., O. Güven, J.G. Melville, J.S. Nohrstedt, and J.K. Overholtzer. 1988. Forced-Gradient Tracer Tests and Inferred Hydraulic Conductivity Distributions at the Mobile Site. Ground Water 26(5):570-579.

[57] Kenoyer, G.J. 1988. Tracer Test Analysis of Anisotropy in Hydraulic Conductivity of Granular Aquifers. Ground Water Monitoring Review 8(3):67-70.

[58] Bumb, A.C., J.I. Drever, and C.R. McKee. 1985. In Situ Determination of Dispersion Coefficients and Adsorption Parameters for Contaminations Using a Pull-Push Test. In: Proc. 2nd Int. Conf. on Ground Water Quality Research (Oklahoma), N.N. Durham and A.E. Redelfs (eds.), Oklahoma State University Printing, pp. 186-190.

[59] Pickens, J.F., R.E. Jackson, K.J. Inch, and W.F. Merritt. 1981. Measurement of Distribution Coefficients Using Radial Injection Dual-Tracer Tests. Water Resource Research 17:529-544.
Rainwater, K.A., W.R. Wise, and R.J. Charbeneau. 1987. Parameter Estimation through Groundwater Tracer Tests. Water Resources Research 23:1901-1910.

- Are other parameters such as travel time, porosity, and hydraulic conductivity of interest?

- How much time is available for the test?

- How much money is available for the test?

If results must be obtained within a few weeks, then certain kinds of tracer tests would normally be out of the question. Those using only the natural hydraulic gradient between two wells that are more than about 20 meters apart typically require long time periods for the tracer to flow between the wells. Another primary consideration is budget. Costs for tests that involve drilling of several deep holes, setting of packers to control sampling or injection, and analysis of hundreds of samples in an EPA-certified laboratory could easily exceed $1 million. In contrast, some short-term tracer tests may be possible at costs of less than $1,000.

Choice of a tracer will depend partially on which analytical techniques are easily available and which background constituents might interfere with these analyses. The chemist or technician who will analyze the samples can advise whether background constituents might interfere with the analytical techniques to be used. Bacteria, isotopes and ions are the most frequently used types of tracers in porous media. Fluorescent dyes are less commonly used as tracers because of their tendency to adsorb. A more common use of dyes in porous media is to locate zones of preferential flow in the vadose zone. In this application, adsorption on soil particles is desirable because it allows visual inspection of flow patterns when the soil is excavated.

8.5.1 Estimating the Amount of Tracer to Inject

The amount of tracer to inject is based on the natural background concentrations, the detection limit for the tracer, the dilution expected, and experience. Adsorption, ion exchange, and dispersion will decrease the amount of tracer arriving at the observation well, but recovery of the injected mass is usually not less than 20% for two-hole tests using a forced recirculation system and conservative tracers. The concentration should not be increased so much that density effects become a problem. Lenda and Zuber (1970)[60] presented graphs that can be used to estimate the approximate quantity of tracer needed. These values are based on estimates of the porosity and dispersion coefficient of the aquifer.

8.5.2 Single-Well Techniques

Two techniques, *injection/withdrawal* and *borehole dilution*, produce parameter values from a single well that are valid at a local scale. Advantages of single-well techniques are:

---

[60] Lenda, A. and A. Zuber. 1970. Tracer Dispersion in Groundwater Experiments. In: Proc. IAEA Symp. Isotope Hydrology, International Atomic Energy Agency, Vienna, pp. 619-641.

- Less tracer is required than for two-well tests

- The assumption of radial flow is generally valid, so natural aquifer velocity can be ignored, making solutions easier

- Knowledge of the exact direction of flow is not necessary

Molz et al. (1985)[61] describe design and performance of single-well tracer tests conducted at the Mobile site.

Injection/Withdrawal. The single-well injection/withdrawal (or pulse) technique results in a pore velocity value and a longitudinal dispersion coefficient. The method assumes that porosity is known or can be estimated with reasonable accuracy. In this procedure, a given quantity of tracer is instantaneously added to the borehole, the tracer is mixed, and then two to three borehole volumes of fresh water are pumped in to force the tracer to penetrate the aquifer. Only a small quantity is injected so as not to disturb natural flow.

After a certain time, the borehole is pumped out at a constant rate large enough to overcome the natural ground-water flow. Tracer concentration is measured with time or pumped volume. If the concentration is measured at various depths with point samplers, the relative permeability of layers can be determined. The dispersion coefficient is obtained by matching experimental breakthrough curves with theoretical curves based on the general dispersion equation. A finite difference method is used to simulate the theoretical curves (Fried, 1975).[62]

Fried concluded that the method is useful for local information (2 to 4 meter radius) and for detecting the most permeable strata. A possible advantage of this test is that nearly all of the tracer is removed from the aquifer at the end of the test.

Borehole Dilution. This technique, also called *point dilution*, can be used to measure the magnitude and direction of horizontal tracer velocity and vertical flow (Klotz et al., 1978).[63]

The procedure is to introduce a known quantity of tracer instantaneously into the borehole, mix it well, and then measure the concentration decrease with time. The tracer is generally introduced into an isolated volume of the borehole using packers.

---

[61] Molz, F.J., J.G. Melville, O. Güven, R.D. Crocker, and K.T. Matteson. 1985. Design and Performance of Single-Well Tracer Tests at the Mobile Site. Water Resources Research 21:1497-1502.

[62] Fried, J.J. 1975. Groundwater Pollution: Theory, Methodology Modeling, and Practical Rules. Elsevier, New York.

[63] Klotz, D., H. Moser, and P. Trimborn. 1978. Single-Borehole Techniques, Present Status and Examples of Recent Applications. In: Proc. IAEA Symp. Isotope Hydrology, Part 1, International Atomic Energy Agency, Vienna, pp. 159-179.

Radioactive tracers have been most commonly used for borehole dilution tests, but other tracers can be used.

Factors to consider when conducting a point dilution test include the homogeneity of the aquifer, effects of drilling (mudcake, etc.), homogeneity of the mixture of tracer and well water, degree of tracer diffusion, and density effects.

Ideally, the test should be conducted using a borehole with no screen or gravel pack. If a screen is used, it should be next to the borehole because dead space alters the results. Samples should be very small in volume so that flow is not disturbed by their removal.

A variant of the point dilution method allows measurement of the direction of ground-water flow. In this procedure, a section of the borehole is usually isolated by packers, and a tracer (often radioactive) is introduced slowly and without mixing. Then, after some time, a compartmental sampler (four to eight compartments) within the borehole is opened. The direction of minimum concentration corresponds to the flow direction. A similar method is to introduce a radioactive tracer and subsequently measure its adsorption on the borehole or well screen walls by means of a counting device in the hole. Gaspar and Oncescu (1972/T8-5) describe the method in more detail.

Another common strategy is to inject and subsequently remove the water containing a conservative tracer from a single well. If injection is rapid and pumping to remove the tracer follows immediately, then almost all of the injected conservative tracer can be recovered. If the pumping is delayed, the injected tracer will drift downgradient with the general flow of the ground water and the percentage of tracer recovery will decrease with time. Successive tests with increasingly longer delay times between injection and pumping can be used to estimate ground-water velocities in permeable aquifers with moderately large hydraulic gradients.

8.5.3 Two-Well Techniques

There are two basic approaches to using tracers with multiple wells: one measures tracer movement in uniform (natural) flow and the other measures movement by radial (induced) flow. The parameters measured (dispersion coefficient and porosity) are assumed to be the same for both types of flow.

Uniform Flow. This approach involves placing a tracer in one well without disturbing the flow field, and sampling periodically to detect the tracer in observation wells. This test can be used at a local (2 to 5 meter) or intermediate (5 to 100 meter) scale, but it requires much more time than radial tests. If the direction and magnitude of the velocity are not known, a large number of observation wells are needed. Furthermore, local flow directions may diverge widely from directions predicted on the basis of widely spaced water wells. Failure to intercept a tracer in a well just a few meters away from the injection well is not uncommon under natural-gradient flow conditions.

The quantity of tracer needed to cover a large distance can be expensive. On a regional scale, environmental tracers, including seawater intrusion, radionuclides, or stable isotopes of hydrogen and oxygen, are used. Manmade pollution has also been used. For regional problems, a mathematical model is calibrated with concentration vs. time curves from field data and is used to predict future concentration distributions.

Local- or intermediate-scale uniform flow problems can be solved analytically, semianalytically, or by curve matching. Layers of different permeability can cause distorted breakthrough curves, which can usually be analyzed using one- or two-dimensional models (Gaspar and Oncescu, 1972/T8-5). Fried (1975-Footnote 62) and Lenda and Zuber (1970-Footnote 60) present analytical solutions.

Radial Flow. Radial flow techniques work by altering the flow field of an aquifer by pumping. Solutions are generally easier if radial flow velocity greatly exceeds uniform flow. This method yields values for porosity and the dispersion coefficient, but not natural ground-water velocity. Types of radial flow tests include diverging, converging, and recirculating tests.

A *diverging* test involves constant injection of water into an aquifer. The tracer is introduced into the injected water as a slug or continuous flow, and the tracer is detected at an observation well that is not pumping. Point or integrated samples of small volume are carefully taken at the observation well so that flow is not disturbed. Packers can be used in the injection well to isolate an interval.

In a *converging* test, the tracer is introduced at an observation well, while another well is pumped. Concentrations are monitored at the pumped well. The tracer is often injected between two packers or below one packer; then two to three well-bore volumes are injected to push the tracer out into the aquifer. At the pumping well, intervals of interest are isolated (particularly in fractured rock), or an integrated sample is obtained.

A *recirculating* test is similar to a converging test, but the pumped water is injected back into the injection well. This tests a significantly greater part of the formation because the wells inject to and pump from 360 degrees. The flow lines are longer, however, partially canceling out the advantage of a higher gradient. Sauty (1980)[64] provides theoretical curves for recirculating tests.

8.5.4 Design and Construction of Test Wells

In many tracer tests, construction of the test wells is the single greatest expense. Procedures for the proper design and construction of monitoring wells for sampling ground-water quality (Section 9.3) apply equally to wells used for tracer tests.

---

[64] Sauty, J.P. 1980. An Analysis of Hydrodispersive Transfer in Aquifers. Water Resources Research 16(1):145-158.

Special considerations in designing and constructing test wells for tracer tests include the following:

- Drilling muds and mud additives tend to have a high capacity for the sorption of most types of tracers and therefore should be avoided.

- Drilling methods that alter the hydrologic characteristics of the aquifer being tested (such as clogging of pores) should be avoided.

- Use of packers to isolate the zones being sampled from the rest of the water in the well (see Figure 8-2b) allows the most precise measurements of vertical variations in hydrologic parameters. This approach tends to be more expensive, take longer, and require more technical training than whole-well tests.

- Well casing material should not be reactive with the tracer used.

- If packers are not used, the diameter of the sampling well should be as small as possible in order to minimize the amount of "dead" water in the well during sampling.

- Well-screen slot size and gravel pack must be selected and installed with special care when using single-well tests with alternating cycles of injection and pumping large volumes of water into and out of loose fine-grained sand. On the other hand, if the aquifer being tested contains a very permeable coarse gravel and the casing diameter is small, then numerous holes drilled in the solid casing may be adequate.

- As with any monitoring well, tracer test wells should be properly developed to remove silt, clay, drilling mud, and other materials that would prevent free movement of water in and out of the well.

8.5.5 Injection and Sample Collection

Choice of injection equipment depends on the depth of the borehole and the funds available. In very shallow holes, the tracer can be lowered through a tube, placed in an ampule that is lowered into the hole and broken, or just poured in. Mixing of the tracer with the aquifer water is desirable and important for most types of tests and is simple for very shallow holes. For example, a plunger can be surged up and down in the hole or the tracer can be released through a pipe with many perforations. Flanges on the outer part of the pipe will mix the tracer as the pipe is raised and lowered. For deeper holes, tracers must be injected under pressure and equipment can be quite sophisticated.

Sample collection also can be simple or sophisticated. For tracing thermal pulses, only a thermistor needs to be lowered into the ground water. For chemical tracers, a variety of sampling methods may be used (Section 9.5). Some special sampling considerations for tracer tests include:

Soil and Ground Water Tracers

- Bailers should not be used if mixing of the tracer in the borehole is to be avoided.

- Where purging is required, removal of more than the minimum required to obtain fresh aquifer water may create a gradient toward the well and distort the natural movement of the tracer (Section 9.5.3).

- Use of existing water wells that tap multiple aquifers should be generally avoided in tracer tests except to establish whether a hydrologic connection with the point of injection exists.

8.5.6 Interpretation of Results

The basic plot of the concentration of a tracer as a function of time or water volume passed through the system is called a *breakthrough curve*. The concentration is either plotted as the actual concentration (Figure 8-7) or, quite commonly, as the ratio of the measured tracer concentration at the sampling point, C, to the input tracer concentration, C_o (Figure 8-8).

The measured quantity that is fundamental for most tracer tests is the first arrival time of the tracer as it goes from an injection point to a sampling point. The first arrival time conveys at least two bits of information. First, it indicates that a connection for ground-water flow actually exists between the two points. For many tracer tests, particularly in karst regions, this is all the information that is desired. Second, if the tracer is conservative, the maximum velocity of ground-water flow between the two points may be estimated.

Interpretations more elaborate than the two mentioned above depend very much on the type of aquifer being tested, the velocity of ground-water flow, the configuration of the tracer injection and sampling systems, and the type of tracer or mixture of tracers used in the test.

The value of greatest interest after the first arrival time is the arrival time of the peak concentration for a slug injection or, for a continuous feed of tracers, the time since injection when the concentration of the tracer changes most rapidly as a function of time (Figure 8-8). In general, if conservative tracers are used, this time is close to the theoretical travel time of an average molecule of ground water traveling between the two points.

If a tracer is being introduced continuously into a ditch penetrating an aquifer, as shown in Figure 8-8, then the ratio C/C_o will approach 1.0 after the tracer starts to pass the sampling point. The ratio of 1.0 is rarely approached in most tracer tests in the field, however, because waters are mixed by dispersion and diffusion in the aquifer and because wells used for sampling will commonly intercept far more ground water than has been tagged by tracers (Figure 8-9). Ratios of C/C_o ranging between 10^{-5} and 2×10^{-1} are often reported from field tests.

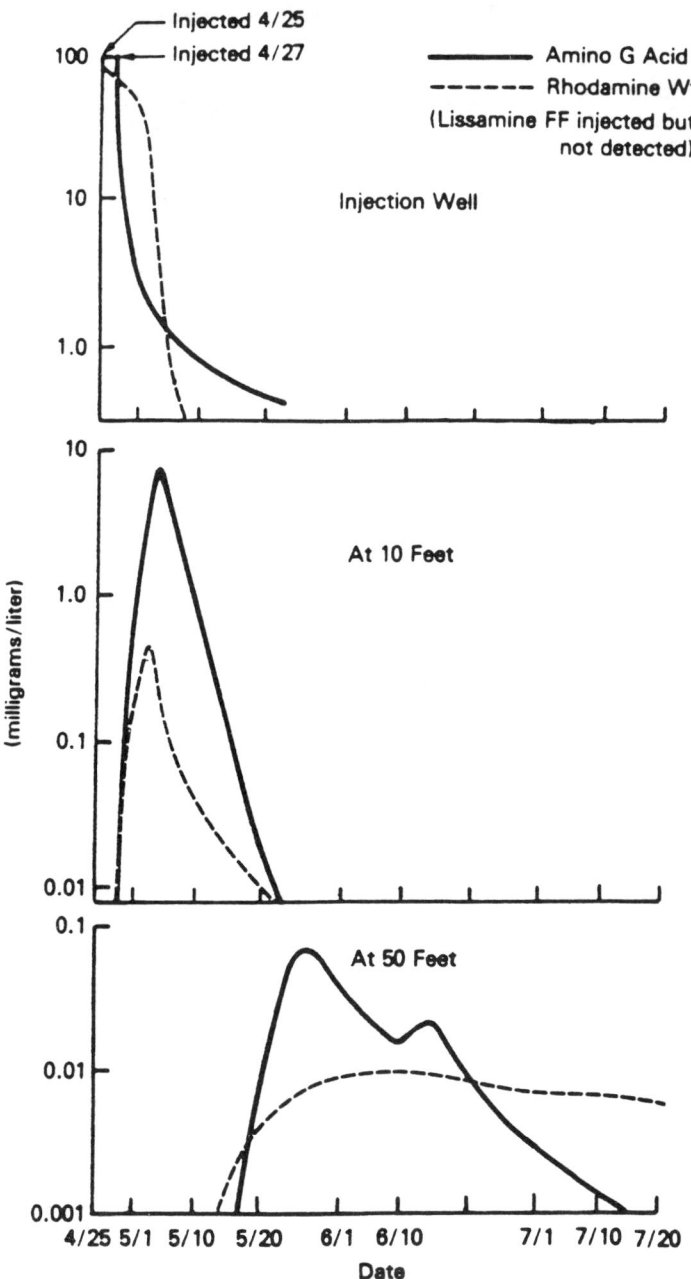

Figure 8-7 Results of tracer tests at the Sand Ridge State Forest, Illinois (Davis et al., 1985, after Naymik and Sievers, 1983).

If a tracer is introduced passively into an aquifer but is recovered by pumping a separate sampling well, then various mixtures of the tracer and the native ground water will be recovered depending on the amount of water pumped, the transmissivity

Soil and Ground Water Tracers

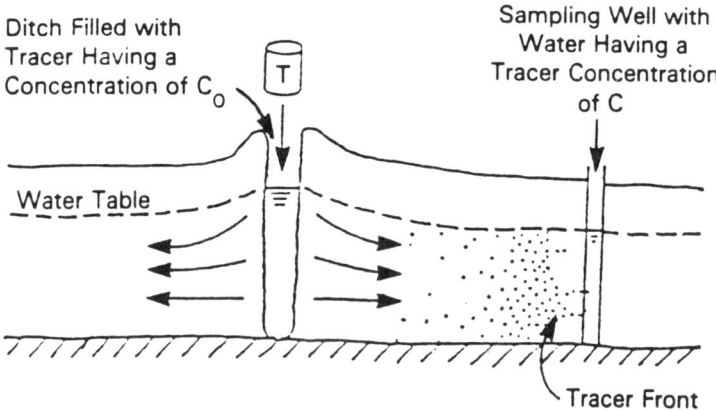

A. Tracer movement from injection ditch to sampling well.

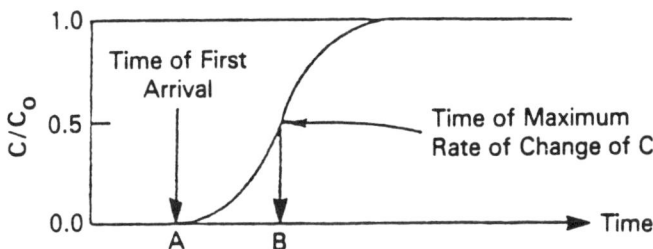

B. Breakthrough Curve.

Figure 8-8 Tracer concentration reported as a ratio (Boulding, 1991, after Davis et al., 1985).

of the aquifer, the slope of the water table, and the shape of the tracer plume. Keely (1984)[65] has presented this problem graphically with regard to the removal of contaminated water from an aquifer.

With the introduction of a mixture of tracers, possible interactions between the tracers and the solid part of the aquifer may be studied. If interactions take place, they can be detected by comparing breakthrough curves of a conservative tracer with the curves of the other tracers being tested (Figure 8-10). Quantitative analyses of tracer breakthrough curves are generally by curve matching of computer-generated type curves or by analytical methods.

---

[65] Keely, J.F. 1984. Optimizing Pumping Strategies for Contaminant Studies and Remedial Actions. Ground Water Monitoring Review 4(3):63-74.

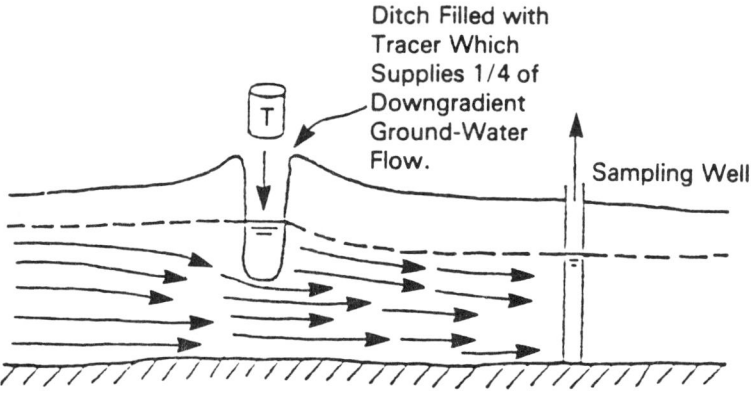

A. Tracer does not fully saturate aquifer.

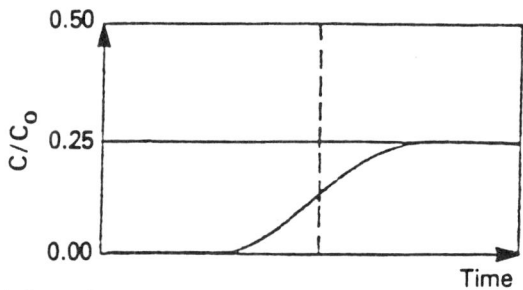

B. Breakthrough curve.

Figure 8-9 Incomplete tracer recovery due to partial penetration of aquifer (Boulding, 1991, after Davis et al., 1985).

8.6 Guide to Major References

Table 8-5 Identifies major references on tracer methods in the following areas: (1) general reviews and bibliographies, (2) specific tracers, (3) karst tracing, and (4) interpretations.

Three good general references on the use of fluorescent dyes are three U.S. Geological Survey publications (Hubbard et al., 1982; Kilpatrick and Cobb, 1985; Wilson et al., 1986). Section 8.4 provides an overview of major sources of information on tracing in karst systems. Two reports prepared for U.S. EPA (Mull, 1988; Quinlan, 1989) are the best recent sources on practical aspects of dye tracing in karst areas.

Because microbial contamination is a widespread and significant contaminant, Table 8-5 contains a number journal references involving site-specific cases in addition to general reviews.

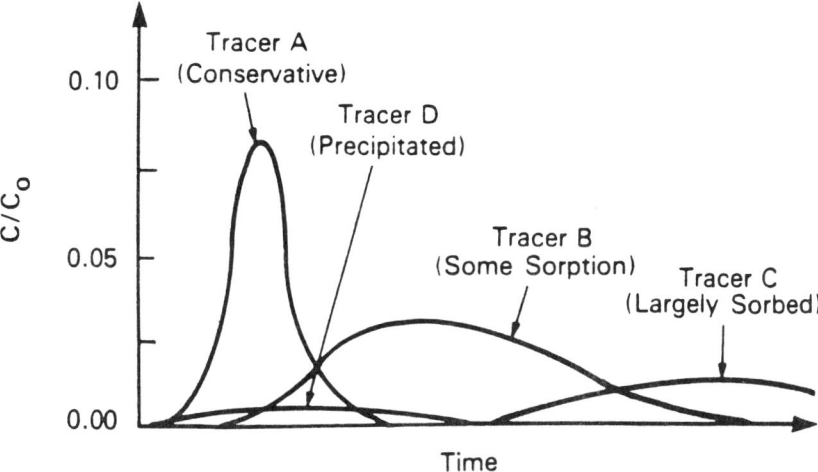

Figure 8-10 Breakthrough curves for conservative and nonconservative tracers (Boulding, 1991, after Davis et al., 1985).

Table 8-5 Index to Major References on Soil and Ground-Water Tracer Methods

Topic	References
General Reviews	AGWSE (1989), Atkinson and Smart (1981), Aulenbach et al. (1978), Davis et al. (1980, 1985), Drew and Smith (1969), Everts et al. (1979), Gaspar (1987), Grisak et al. (1983), Knuttson (1968), Molz et al. (1986, 1987), U.S. EPA (1991-Chapter 4)
Bibliographies	Edwards and Smart (1988a,b), LaMoreaux et al. (1984, 1989), Smart et al. (1988), Taylor and Dey (1985), van der Leeden (1991)
Specific Tracers	
Dyes	Drew and Smith (1969), Hubbard et al. (1982), Feuerstein and Selleck (1963), Kilpatrick and Cobb (1985), McLaughlin (1982), Mull et al. (1988), Quinlin (1989), Smart and Laidlaw (1977), Thrailkill et al. (1983), Wilson et al. (1986); <u>Toxicity</u>: Field et al. (1990), Smart (1984); <u>Quantitative Interpretations</u>: Brown and Ford (1971), Smart (1980a, 1980b), Smoot (1991). See, also, Karst Tracing.
Ionic Tracers	Bowman (1984), Bowman and Gibbons (1992), Kaufman and Orlob (1956a, 1956b), Kaufman and Todd (1955); <u>Single-Well Tests</u>: Hall (1993), Hall et al. (1991), Klotz et al. (1978), Leap (1985), Leap and Kaplan (1988), Molz et al. (1985), Pickens et al. (1981)
Microorganisms	<u>Reviews</u>: Crane and Moore (1984), Gerba (1983, 1985, 1987), Gerba and Bitton (1984), Keswick and Gerba (1980), Keswick et al. (1982), Matthess and Pekdeger (1985), Romero (1970), Sobsey and Shields (1987), Vaughn and Landry (1983), Wood and Ehrlich (1978); <u>Bacteria</u>: Allen and Morrison (1973), Caldwell and Morita (1988), Gerba and Goyal (1985), Gerba et al. (1975), Hagedorn (1984), Hagedorn et al. (1978, 1981), Hendricks et al. (1979), Peterson and Ward (1989), Rahe et al. (1978), Robeck (1969), Van Donsel (1967); <u>Viruses</u>: Drewery and Eliassen (1968), Shaffer (1977), Tyler (1985), Vaughn et al. (1981, 1983)
Stable Isotopes	Back and Cherry (1976), Back and Zoetl (1975), Bowen (1980--Chapter 3), Coleman et al. (1977), Davis and Bentley (1982), Ehleringer et al. (1993), Ferronsky and Polyakov (1982), Fritz and Fontes (1980, 1986), Halevy et al. (1967), Hoefs (1980), IAEA (1967a, 1967c, 1970, 1974a, 1974b, 1978a), Lamoreaux et al. (1984), Moser and Rauert (1985), Payne (1972, 1988), Schimel (1993), Stout (1967), Toran (1982); <u>Nitrogen Isotopes</u>: Knowles and Blackburn (1992)

Table 8-5 (cont.)

Topic	References
Specific Tracers (cont.)	
Radioactive Isotopes	Csallany (1966), Gaspar and Oncescu (1972), Jäeger and Hunziker (1979), Kaufman and Orlob (1956a, 1956b), Kaufman and Todd (1955), IAEA (1963, 1967b, 1967c, 1968, 1974b, 1978b), Ivanonitch and Harmon (1992), Mather (1968), Thornhill and Benefield (1990), Wiebenga et al. (1967); Carbon Isotopes: Coleman and Fry ((1991), Ehleringer et al. (1993); Case Studies: Bradbury (1991-tritium).
Applications	
Karst Tracing	Aley and Fletcher (1976), Bögli (1980), Brown (1972), Ford and Williams (1989), Gospodaric and Habic (1976), LaMoreaux (1984, 1989), Milanovic (1981), Mull et al. (1988), Quinlan (1989), Sweeting (1973), SUWT (1966, 1970, 1976, 1981, 1986), Thrailkill et al. (1983); Review Papers: Back and Zoetl (1975), Brown and Ford (1971), Dunn (1957), Gunn (1982), Jones (1984), Smart (1976)
Fracture Characterization	Kerfoot (1992), Lewis et al. (1966), Morin et al. (1988), Tsang (1992)
Interpretation	Bullivant and O'Sullivan (1989), Fried (1975), Grisak and Pickens (1980a, 1980b), Güven et al. (1985, 1986), Halevy and Nir (1962), Molz et al. (1986, 1987), Sauty (1978), Theis (1963)

Table 8-5 References (Appendix F contains references for figure and table sources)

Aley, T. and M.W. Fletcher. 1976. The Water Tracer's Cookbook. Missouri Speleology 16(3):1-32.

Allen, M.J. and S.M. Morrison. 1973. Bacterial Movement Through Fractured Bedrock. Ground Water 11(2):6-10.

Association of Ground Water Scientists and Engineers (AGWSE). 1989. Tracers in Hydrogeology: Principles, Problems and Practical Applications. Ground Water 27(5):718-728. [Abstracts of 39 papers presented at NWWA meetings in Houston, TX, October 30-November 1, 1989]

Atkinson, T.C. and P.L. Smart. 1981. Artificial Tracers in Hydrology. In: A Survey of British Hydrology, The Royal Society, London, pp.173-190.

Aulenbach, D.B., J.H. Bull, and B.C. Middlesworth. 1978. Use of Tracers to Confirm Ground-Water Flow. Ground Water 61(3):149-157.

Back, W. and J. Zoetl. 1975. Application of Geochemical Principles, Isotopic Methodology, and Artificial Tracers to Karst Hydrology. In: Hydrogeology of Karstic Terrains, A. Burger and L. Dubertret (eds.), Int. Ass. Hydrogeologists, Paris, pp. 105-121.

Back, W. and J.A. Cherry. 1976. Chemical Aspects of Present and Future Hydrogeologic Problems. Advances in Groundwater Hydrology September:153-172.

Bögli, A. 1980. Karst Hydrology and Physical Speleology. Springer-Verlag, New York, NY. [Pages 138-143 review use of tracers]

Bowen, R. 1980. Ground Water. Applied Science Publishers, London. [Chapter 3 covers isotope tracing techniques].

Bowman, R.S. 1984. Evaluation of Some New Tracers for Soil Water Studies. Soil Sci. Soc. Am. J. 48:987--993. [Ionic tracers]

Bowman, R.S. and J.F. Gibbons. 1992. Difluorobenzoates as Nonreactive Tracers in Soil and Ground Water. Ground Water 30:8-14.

Bradbury, K.R. 1991. Tritium as an Indicator of Ground-Water Age in Central Wisconsin. Ground Water 29:398-404.

Brown, M.C. 1972. Karst Hydrology of the Lower Maligne Basin, Jasper, Alberta. Cave Studies No. 13. Cave Research Associates, Castro Valley, CA. [Chapter III reviews tracer methods]

Brown, M.C. and D.C. Ford. 1971. Quantitative Tracer Methods for Investigation of Karst Hydrology Systems, with Reference to the Maligne Basin Area. Cave Research Group (Great Britain) 13(1):37-51.

Bullivant, D.P. and M.J. O'Sullivan. 1989. Matching Field Tracer Test with Some Simple Models. Water Resources Research 25(8):1879-1891.

Caldwell, B.A. and R.Y. Morita. 1988. Sampling Regimes and Bacteriological Tests for Coliform Detection in Groundwater. EPA/600/2-87/083 (NTIS PB88-107230).

Coleman, D.C. and B. Fry. 1991. Carbon Isotope Techniques. Academic Press, New York, NY, 274 pp.

Coleman, D.D., W.F. Meents, C.L. Liu, and R.A. Keogh. 1977. Isotopic Identification of Leakage Gas from Underground Storage Reservoirs--a Progress Report. Petroleum Report No. 111. Illinois State Geological Survey, Urbana, IL.

Crane, S.R. and J.A. Moore. 1984. Bacterial Pollution of Groundwater: A Review. Water, Air, and Soil Pollution 22:67-83.

Csallany, S.C. 1966. Application of Radioisotopes in Water Resources Research. In: Proc. 2nd Annual American Water Resource Conference, American Water Resource Association, Champaign, IL, pp. 365-373.

Davis, S.N. and H.W. Bentley. 1982. Dating Groundwater, a Short Review. In: Nuclear and Chemical Dating Techniques, L. Currie (ed.), ACS Symposium Series 176, pp. 187-222.

Davis, S.N., G.M. Thompson, H.W. Bentley, and G. Stiles. 1980. Ground Water Tracers--A Short Review. Ground Water 18:14-23.

Davis, S.N., D.J. Campbell, H.W. Bentley, and T.J. Flynn. 1985. Introduction to Ground-Water Tracers. EPA 600/2-85/022 (NTIS PB86-100591). Also published under the title Ground Water Tracers in NWWA/EPA Series, National Water Well Association, Dublin, OH, 200 pp. [See also 1986 discussion by J. Quinlan in Ground Water 24(2):253-259 and 24(3):396-397 (references) and reply by Davis in Ground Water 24(3):398-399]

Drew, D.P. and D.I. Smith. 1969. Techniques for the Tracing of Subterranean Drainage. British Geomorphological Research Group Tech. Bulletin 2:1-36. [Focuses on fluorescent dyes and lycopodium sores, but also contains an annotated bibliography on other tracers]

Drewery, W.A. and R. Eliassen. 1968. Virus Movement in Ground Water. J. Water Pollut. Cont. Fed. 40(8):257-271.

Dunn, J.R. 1957. Stream Tracing. National Speleological Society Mid-Appalachian Region 2:1-7.

Edwards, A.J. and P.L. Smart. 1988a. Solute Interaction Processes: An Annotated Bibliography. Turner Designs, Sunnyvale, CA. [61 references]

Edwards, A.J. and P.L. Smart. 1988b. Contaminant Transport Modeling: An Annotated Bibliography. Turner Designs, Sunnyvale, CA. [58 references]

Ehleringer, J.R., A.E. Hall, and G.D. Farquhar (eds.). 1993. Stable Isotopes and Plant Carbon-Water Relationships. Academic Press, New York, NY, 556 pp.

(Table 8-5 Soil/Ground-Water Tracer References)

Everts, C.J., R.S. Kanwar, E.C. Alexander, Jr., and S.C. Alexander. 1989. Comparison of Tracer Mobilities under Laboratory and Field Conditions. J. Environ. Quality 18:491-498.

Ferronsky, V.I. and V.A. Polyakov. 1982. Environmental Isotopes in the Hydrosphere. Wiley-Interscience, New York, NY.

Feuerstein, D.L. and R.E. Selleck. 1963. Fluorescent Tracers for Dispersion Measurements. J. ASCE August:1-21.

Field, M.S., R.G. Wilhelm, and J.F. Quinlan. 1990. Use and Toxicity of Dyes for Tracing Ground Water (Abstract). Ground Water 28(1):154-155. [The complete paper is available from Malcolm S. Field, Exposure Research Group, Office of Research and Development, U.S. Environmental Protection Agency (RO-689), Washington, DC 20460].

Ford, D.C. and P.W. Williams. 1989. Karst Geomorphology and Hydrology. Unwin Hyman, Boston, MA, 601 pp.

Fried, J.J. 1975. Groundwater Pollution: Theory, Methodology Modeling, and Practical Rules. Elsevier, New York, NY.

Fritz, P. and J.C. Fontes (eds.). 1980. Handbook of Environmental Isotope Geochemistry, Vol. 1, The Terrestrial Environment. Elsevier, New York, NY, 545 pp.

Fritz, P. and J.C. Fontes (eds.). 1986. Handbook of Environmental Isotope Geochemistry, Vol. 2, The Terrestrial Environment, Part B. Elsevier, New York, NY.

Gaspar, E. (ed.). 1987. Modern Trends in Tracer Hydrology. CRC Press, Boca Raton, FL, Vol. I, 145 pp.; Vol. II, 137 pp.

Gaspar, E. and M. Oncescu. 1972. Radioactive Tracers in Hydrology. Elsevier, New York, NY, 352 pp. [14 chapters]

Gerba, C.P. 1983. Virus Survival and Transport in Groundwater. Dev. Ind. Microbiol. 24:247-251.

Gerba, C.P. 1985. Microbial Contamination of the Subsurface. In: Ground Water Quality, C.H. Ward, W. Giger, and P.L. McCarty (eds.), Wiley and Sons, New York, NY, pp. 53-67.

Gerba, C.P. 1987. Transport and Fate of Viruses in Soils: Field Studies. In: Human Viruses in Sediments, Sludges, and Soil, V.C. Rao and J.L. Melnick (eds.), CRC Press, Boca Raton, FL.

Gerba, C.P. and G. Bitton. 1984. Microbial Pollutants: Their Survival and Transport Pattern to Groundwater. In: Groundwater Pollution Microbiology, G. Bitton, and C.P. Gerba (eds.), Wiley-Interscience, New York, NY, pp. 66-88.

Gerba, C.P. and S.M. Goyal. 1985. Pathogen Removal from Wastewater During Groundwater Recharge. In: Artificial Recharge of Groundwater, T. Asano (ed.), Butterworth Publishers, Boston, pp. 283-317.

(Table 8-5 Soil/Ground-Water Tracer References)

Gerba, C.P., C. Wallis, and J.L. Melmik. 1975. Fate of Waste-Water Bacteria and Viruses in Soil. J. Irr. Drain. Div. (ASCE) 101(IR3):157-174.

Gospordaric, R. and P. Habic (eds.). 1976. Underground Water Tracing: Investigations in Slovenia 1972-1975. Institute Karst Research, Ljubljana, Jugoslavia.

Grisak, G.E. and J.F. Pickens. 1980a. Solute Transport through Fractured Media 1. The Effect of Matrix Diffusion. Water Resources Research 16(4): 719-730.

Grisak, G.E. and J.F. Pickens. 1980b. Solute Transport through Fractured Media 2. Column Study of Fractured Till. Water Resources Research 16(4):731-739.

Grisak, G.E., J.F. Pickens, F.J. Pearson, and J.M. Bahr. 1983. Evaluation of Ground Water Tracers for Nuclear Fuel Waste Management Studies. Report Prepared by Geologic Testing Consultants, Ltd. for Atomic Energy of Canada, Ltd., Whiteshell Nuclear Research Establishment, Pinawa, Manitoba.

Gunn, J. 1982. Water Tracing in Ireland: A Review with Special References to the Cuillcagh Karst. Irish Geography 15:94-106.

Güven, O., R.W. Falta, F.J. Molz, and J.G. Melville. 1985. Analysis and Interpretation of Single-Well Tracer Tests in Stratified Aquifers. Water Resources Research 21:676-684.

Güven, O., R.W. Falta, F.J. Molz, and J.G. Melville. 1986. A Simplified Analysis of Two-Well Tracer Tests in Stratified Aquifers. Ground Water 24:63-71.

Hagedorn, C. 1984. Microbiological Aspects of Groundwater Pollution Due to Septic Tanks. In: Groundwater Pollution Microbiology, G. Bitton, and C.P. Gerba (eds.), Wiley-Interscience, New York, NY, pp. 181-195.

Hagedorn, C., D.T. Hansen, and G.H. Simonson. 1978. Survival and Movement of Fecal Indicator Bacteria in Soil Under Conditions of Saturated Flow. J. Environ. Qual. 7(1):55-59.

Hagedorn, C., E.L. McCoy, and T.M. Rahe. 1981. Potential for Ground Water Contamination from Septic Tanks Effluents. J. Environ. Qual. 10(1):1-8.

Halevy, E. and A. Nir. 1962. The Determination of Aquifer Parameters with the Aid of Radioactive Tracers. J. Geophysical Research 61:2403-2409.

Halevy, E., H. Moser, O. Zellhoffer, and A. Zuber. 1967. Borehole Dilution Techniques: A Critical Review. In: Isotopes in Hydrology, IAEA, Vienna, pp. 531-564.

Hall, S.H. 1993. Single Well Tracer Tests in Aquifer Characterization. Ground Water Monitoring and Remediation 13(2):118-124. [Bromide Tracer]

Hall, S.H., S.P. Luttrell, and W.E. Cronin. 1991. A Method for Estimating Effective Porosity and Ground-Water Velocity. Ground Water 29(2):171-174. [Bromide tracer]

Hendricks, D.W., F.J. Post, and D.R. Khairnar. 1979. Adsorption of Bacteria on Soils. Water, Air and Soil Pollution 12(2):219-232.

(Table 8-5 Soil/Ground-Water Tracer References)

Hoefs, J. 1980. Stable Isotope Geochemistry, 2nd ed. Springer-Verlag, New York, NY, 208 pp.

Hubbard, E.F., F.A. Kilpatrick, L.A. Martens, and J.F. Wilson, Jr. 1982. Measurement of Time of Travel and Dispersion in Streams by Dye Tracing. U.S. Geological Survey Techniques of Water-Resources Investigations TWRI 3-A9.

International Atomic Energy Agency (IAEA). 1963. Radioisotopes in Hydrology. International Atomic Energy Agency, Vienna, 449 pp. [Proceedings of symposium held in Tokyo]

International Atomic Energy Agency (IAEA). 1967a. Isotopes in Hydrology. International Atomic Energy Agency, Vienna. [21 papers; proceedings of symposium held in Vienna]

International Atomic Energy Agency (IAEA). 1967b. Radioisotopes in Industry and Geophysics--A Symposium (Prague). International Atomic Energy Agency, Vienna.

International Atomic Energy Agency (IAEA). 1967c. Isotopes and Radiation Techniques. (Proc. of Symp. Techniques in Soil Physics and Irrigation Studies, Istanbul), IAEA, Vienna.

International Atomic Energy Agency (IAEA). 1968. Guidebook on Nuclear Techniques in Hydrology. Tech. Rept. Ser. No. 91, IAEA, Vienna, 214 pp.

International Atomic Energy Agency (IAEA). 1970. Isotope Hydrology (1970). Proceedings of a Symposium on Use of Isotopes in Hydrology (Vienna). International Atomic Energy Agency, Vienna. [25 papers]

International Atomic Energy Agency (IAEA). 1974a. Isotope Techniques in Groundwater Hydrology (1970), 2 Vols. Proceedings Vienna Symposium, International Atomic Energy Agency, Vienna. [51 papers]

International Atomic Energy Agency (IAEA). 1974b. Isotope Techniques in Soil Physics and Irrigation Studies. IAEA, Vienna.

International Atomic Energy Agency (IAEA). 1978a. Isotope Hydrology (1978), 2 Vols. Proceedings of an International Symposium on Isotope Hydrology (Neuherberg). International Atomic Energy Agency, Vienna. [41 papers on subsurface hydrology]

International Atomic Energy Agency (IAEA). 1978b. Nuclear Techniques in Groundwater Pollution Research (Proc. 1976 Cracow Symposium). Doc. ISP518 UNIPUB, New York, NY, 285 pp.

Ivanovich, M. and R.S. Harmon. 1992. Uranium-Series Disequilibrium: Applications to Earth, Marine and Environmental Sciences, 2nd ed. Oxford University Press, New York, NY, 944 pp.

Jäeger, E. and J.C. Hunziker (ed.). 1979. Lectures in Isotope Geology. Springer-Verlag, Berlin, 312 pp.

Jones, W.K. 1984. Dye Tracers in Karst Areas. National Speleological Society Bulletin 36:3-9.

(Table 8-5 Soil/Ground-Water Tracer References)

Kaufman, W.J. and G.T. Orlob. 1956a. Measuring Ground-Water Movements with Radioactive and Chemical Tracers. Am. Water Works Ass. J. 48:559-572.

Kaufman, W.J. and G.T. Orlob. 1956b. An Evaluation of Ground-Water Tracers. Trans. Am. Geophys. Union 37:297-306.

Kaufman, W.J. and D.K. Todd. 1955. Methods of Detecting and Tracing the Movement of Ground Water. Inst. Egn. Research Rept. 93-1, University of California, Berkeley, 130 pp. [Radioactive and chemical tracers]

Kerfoot, W.B. 1992. The Use of Borehole Flowmeters and Slow-Release Dyes to Determine Bedrock Flow for Wellhead Protection. Ground Water Management 11:755-763 (Proc. of the 6th NOAC).

Keswick, B.H. and C.P. Gerba. 1980. Viruses in Groundwater. Environ. Sci. Technol. 14:1290-1297.

Keswick, B.H., D. Wang, and C.P. Gerba. 1982. The Use of Microorganisms as Ground-Water Tracers: A Review. Ground Water 20(2):142-149.

Kilpatrick, F.A. and E.D. Cobb. 1985. Measurement of Discharge Using Tracers. U.S. Geological Survey TWRI 3-A16.

Klotz, D., H. Moser, and P. Trimborn. 1978. Single-Borehole Techniques, Present Status and Examples of Recent Applications. In: Proc. IAEA Symp. Isotope Hydrology, Part 1, International Atomic Energy Agency, Vienna, pp. 159-179.

Knowles, R. and T.H. Blackburn (eds.). 1992. Nitrogen Isotope Techniques. Academic Press, New York, NY, 311 pp.

Knuttson, G. 1968. Tracers for Ground-Water Investigations. In: Ground Water Problems, E. Eriksson, Y. Gustafsson, and K. Nilsson (eds.), Pergamon Press, London.

LaMoreaux, P.E., B.M. Wilson, and B.A. Mermon (eds.). 1984. Guide to the Hydrology of Carbonate Rocks. UNESCO Studies and Reports in Hydrology, No. 41. [Pages 196-210 focus on isotope techniques for water tracing]

LaMoreaux, P.E., E. Prohic, J. Zoetl, J.M. Tanner, and B.N. Roche (eds.). 1989. Hydrology of Limestone Terranes: Annotated Bibliography of Carbonate Rocks, Vol. 4. International Association of Hydrogeologists Int. Cont. to Hydrogeology Vol. 10. Verlag Heinz Heise GmbH., Hannover, West Germany. [Focuses on pollution assessment]

Leap, D.I. 1985. A Simple, Two-Pulse Tracer Method for Estimating Steady-State Ground Water Parameters. Hydrological Science and Technology: Short Papers 1(1):37-43.

Leap, D.I. and P.G. Kaplan. 1988. A Single-Well Tracing Methods for Estimating Regional Advective Velocity in a Confined Aquifer: Theory and Preliminary Laboratory Verification. Water Resources Research 23(7):993-998.

(Table 8-5 Soil/Ground-Water Tracer References)

Lewis, D.C., G.J. Kriz, and R.H. Burgy. 1966. Tracer Dilution Sampling Technique to Determine Hydraulic Conductivity of Fractured Rock. Water Resources Research 2:533-542. [Fluorescein]

Mather, J.D. 1968. A Literature Survey of the Use of Radioisotopes in Ground-Water Studies. Tech. Communication No. 1, Great Britain Institute of Geological Sciences, London.

Matthess, G. and A. Pekdeger. 1985. Survival and Transport of Pathogenic Bacteria and Viruses in Ground Water. In: Ground Water Quality, C.H. Ward, W. Giger, and P.L. McCarty (eds.), Wiley and Sons, New York, NY, pp. 474-482.

McLaughlin, M.J. 1982. A Review of the Use of Dyes as Soil Water Tracers. Water S.A., Water Research Commission, Pretoria, South Africa 8(4):196-201.

Milanović, P.T. 1981. Karst Hydrogeology. Water Resource Publications, Littleton, CO. [Pages 263-309 cover karst water tracing]

Molz, F.J., O. Güven, J.G. Melville, and J.F. Keely. 1986. Performance and Analysis of Aquifer Tracer Tests with Implications for Contaminant Transport Modeling. EPA 600/2-86/062 (NTIS PB86-219086).

Molz, F.J., J.G. Melville, O. Güven, R.D. Crocker, and K.T. Matteson. 1985. Design and Performance of Single-Well Tracer Tests at the Mobile Site. Water Resources Research 21:1497-1502.

Molz, F.J., O. Güven, J.G. Melville, and J.F. Keely. 1986. Performance and Analysis of Aquifer Tracer Tests with Implications for Contaminant Transport Modeling. EPA 600/2-86/062 (NTIS PB86-219086).

Molz, F.J., O. Güven, J.G. Melville, and J.F. Keely. 1987. Performance and Analysis of Aquifer Tracer Tests with Implications for Contaminant Transport Modeling--A Project Summary. Ground Water 25:337-341.

Morin, R.H., A.E. Hess, and F.L. Paillet. 1988. Determining the Distribution of Hydraulic Conductivity in a Fractured Limestone Aquifer by Simultaneous Injection and Geophysical Logging. Ground Water 26(5):587-595.

Moser, H. and W. Rauert. 1985. Determination of Groundwater Movement by Means of Environmental Isotopes: State of the Art. In: Relation of Groundwater Quantity and Quality, F.X. Dunin, G. Matthess, and R.A. Gras (eds.), Int. Ass. Hydrological Sciences Pub. No. 146, pp. 241-257.

Mull, D.S., T.D. Lieberman, J.L. Smoot, and L.H. Woosely, Jr. 1988. Application of Dye-Tracing Techniques for Determining Solute-Transport Characteristics of Ground Water in Karst Terranes. EPA 904/6-88-001, Region 4, Atlanta, GA.

Payne, B.R. 1972. Isotope Hydrology. Advances in Hydroscience 8:95-138.

Payne, B.R. 1988. The Status of Stable Isotope Hydrology Today. J. Hydrology 100(1/3):207-237.

(Table 8-5 Soil/Ground-Water Tracer References)

Peterson, T.C. and R.C. Ward. 1989. Bacterial Retention in Soils: New Perspectives, New Recommendations. J. Environ. Health 51(4):196-200.

Pickens, J.F., R.E. Jackson, K.J. Inch, and W.F. Merritt. 1981. Measurement of Distribution Coefficients Using Radial Injection Dual-Tracer Tests. Water Resource Research 17:529-544.

Quinlan, J.F. 1989. Ground-Water Monitoring in Karst Terranes: Recommended Protocols and Implicit Assumptions. EPA 600/X-89/050. U.S. EPA Environmental Monitoring Systems Laboratory, Las Vegas, NV.

Rahe, T.M., C. Hagedorn, E.L. McCoy, and G.F. Kling. 1978. Transport of Antibiotic-Resistant *Escherichia coli* Through Western Oregon Hillslope Soils Under Conditions of Saturated Flow. J. Environ. Qual. 7(4):487-494.

Robeck, G.G. 1969. Microbial Problems in Ground Water. Ground Water 7(3):33-35.

Romero, J.C. 1970. The Movement of Bacteria and Viruses Through Porous Media. Ground Water 8(2):37-48.

Sauty, J.P. 1978. Identification of Hydrodispersive Mass Transfer Parameters in Aquifers by Interpretation of Tracer Experiments in Radial Converging or Diverging Flow (in French). J. Hydrology 39:69-103.

Schimel, D.S. 1993. Theory and Application of Tracers. Academic Press, New York, NY, 119 pp. [Isotope tracers]

Shaffer, P.T.B. 1977. Virus Detection Methods--Comparison and Evaluation. In: Viruses and Trace Contaminants in Water and Wastewater, J.A. Borchardt, J.K. Cleland, W.J. Redman, and G. Olivier (eds.), Ann Arbor Science Publishers, Ann Arbor, MI, pp. 21-32.

Smart, P.L. 1976. Catchment Delimitation in Karst Areas by the Use of Qualitative Tracer Methods. In: Proc. 3rd Int. Symp. of Underground Water Tracing, Bled, Yugoslavia, pp. 291-298.

Smart, P.L. 1984. A Review of the Toxicity of Twelve Fluorescent Dyes Used for Water Tracing. National Speleological Society Bulletin 46:21-33.

Smart, C.C. 1988a. Quantitative Tracing of the Maligne Karst System, Alberta, Canada. J. Hydrology 98:185-204.

Smart, C.C. 1988b. Artificial Tracer Techniques for the Determination of the Structure of Conduit Aquifers. Ground Water 26(4):445-453.

Smart, P.L. and I.M.S. Laidlaw. 1977. An Evaluation of Some Fluorescent Dyes for Water Tracing. Water Resources Research 13(1):15-33.

Smart, P.L., F. Whitaker, and J.F. Quinlan. 1988. Ground Water Tracing: An Annotated Bibliography. Turner Designs, Sunnyvale, CA. [57 annotations]

(Table 8-5 Soil/Ground-Water Tracer References)

Smoot, J.L., D.S. Mull, and T.D. Liebermann. 1987. Quantitative Dye Tracing Techniques for Describing the Solute Transport Characteristics of Ground-Water Flow in Karst Terrane. In: 2nd Multidisciplinary Conf. Sinkholes and the Environmental Impacts of Karst (Orlando), Beck, B.F. and W.L. Wilson (eds.), Balkema, Accord, MA, pp. 29-35.

Sobsey, M.D. and P.A. Shields. 1987. Survival and Transport of Viruses in Soils: Model Studies. In: Human Viruses in Sediments, Sludges, and Soil, V.C. Rao and J.L. Melnick (eds.), CRC Press, Boca Raton, FL.

Stout, G.E. (ed.). 1967. Stable Isotope Techniques in the Hydrologic Cycle. Geophysical Monograph Ser. No. 11, American Geophysical Union, 199 pp.

Sweeting, M.M. 1973. Karst Landforms. Columbia University Press, New York, NY. [Page 218-251 focus on karst hydrology and tracing]

Symposium on Underground Water Tracing (SUWT). 1966. 1st SUWT (Graz, Austria). Published in: Steirisches Beitraege zur Hydrogeologie Jg. 1966/67.

Symposium on Underground Water Tracing (SUWT). 1970. 2nd SUWT (Freiburg/Br., West Germany). Published in: Steirisches Beitraege zur Hydrogeologie 22(1970):5-165, and Geologisches Jahrbuch, Reihe C. 2(1972):1-382.

Symposium on Underground Water Tracing (SUWT). 1976. 3rd SUWT (Ljubljana-Bled, Yugoslavia). Published by Ljubljana Institute for Karst Research: Vol. 1 (1976), 213 pp., Vol. 2 (1977) 182 pp. See also Gospodaric and Habic (1976).

Symposium on Underground Water Tracing (SUWT). 1981. 4th SUWT (Bern, Switzerland). Published in: Steirisches Beitraege zur Hydrogeologie 32(1980):5-100; 33(1981):1-264; and Beitraege zur Geologie der Schweiz--Hydrologie 28 pt.1(1982):1-236; 28 pt.2(1982):1-213.

Symposium on Underground Water Tracing (SUWT). 1986. 5th SUWT (Athens, Greece). Published by Institute of Geology and Mineral Exploration, Athens.

Taylor, T.A. and J.A. Dey. 1985. Bibliography of Borehole Geophysics as Applied to Ground-Water Hydrology. U.S. Geological Survey Circular 926. [42 references on tracers]

Theis, C.V. 1963. Hydrologic Phenomena Affecting the Use of Tracers in Timing Ground-Water Flow. In: Proc. IAEA Tokyo Symp. Radioisotopes in Hydrology, International Atomic Energy Agency, Vienna, Austria. [As cited by Davis et al., 1985]

Thrailkill, J., et al. 1983. Studies in Dye-Tracing Techniques and Karst Hydrogeology. Univ. of Kentucky, Water Resources Research Center Research Report No. 140.

Toran, L. 1982. Isotopes in Ground-Water Investigations. Ground Water 20(6):740-745.

Tsang, Y.W. 1992. Usage of "Equivalent Apertures" for Rock Fractures as Derived From Hydraulic and Tracer Tests. Water Resources Research 28(5):1452-1455.

Tyler, J. 1985. Occurrence in Water of Viruses of Public Health Significance. J. Appl. Bacteriology Symp. Supp. 59:37S-46S.

(Table 8-5 Soil/Ground-Water Tracer References)

U.S. Environmental Protection Agency (EPA). 1991. Handbook Ground Water, Vol. II: Methodology. EPA/625/6-90/-16b, 141 pp. Available from CERI.* [Chapter 4 covers ground water tracers; text of this chapter is largely the same but additional references have been added]

van der Leeden, F. 1991. Geraghty & Miller's Groundwater Bibliography, 5th ed. Water Information Center, Plainview, New York, NY, 507 pp. [90 references on tracers and ground water dating]

Van Donsel, D.J., E.E. Geldreich, and N.A. Clarke. 1967. Seasonal Variations in Survival of Indicator Bacteria in Soil and Their Contribution to Storm-Water Pollution. Appl. Microbiol. 15(6):1362-1370.

Vaughn, J.M. and E.F. Landry. 1983. Viruses in Soils and Groundwaters. In: Viral Pollution of the Environment, G. Berg (ed.), CRC Press, Boca Raton, FL, pp. 163-210.

Vaughn, J.M., E.F. Landry, C.A. Beckwith, and M.Z. Thomas. 1981. Virus Movement in Soil During Saturated Flow. Appl. Environ. Microbiol. 47:335-337.

Vaughn, J.M., E.F. Landry, and M.Z. Thomas. 1983. Entrainment of Viruses from Septic Tank Leach Fields Through a Shallow, Sandy Soil Aquifer. Appl. Environ. Microbiol. 45(5):1474-1480.

Wiebenga, W.A., W.R. Ellis, B.W. Seatonberry, and J.T.G. Andrew. 1967. Radioisotopes as Ground-Water Tracers. J. Geophysical Research 72:4081-4091.

Wilson, Jr. J.F., E.D. Cobb, and F.A. Kilpatrick. 1986. Fluorometric Procedures for Dye Tracing (Revised). U.S. Geological Survey Techniques of Water Resources Investigations TWRI 3-A12. [updates report with the same title by J.F. Wilson, Jr. published in 1968].

Wood, W.W. and G.G. Ehrlich. 1978. Use of Baker's Yeast to Trace Microbial Movement in Ground Water. Ground Water 16(6):398-403.

* See Preface for information on how to obtain documents from CERI (U.S. EPA Center for Environmental Research Information) and NTIS.

(Table 8-5 Soil/Ground-Water Tracer References)

CHAPTER 9

FIELD SAMPLING AND MONITORING OF CONTAMINANTS

9.1 Types of Monitoring Installations 494

    9.1.1 Vadose Zone Monitoring Installations 494
    9.1.2 Ground Water Monitoring Installations 496

9.2 Drilling Methods for Sampling and Well Installation 499

    9.2.1 Selection Criteria 499
    9.2.2 Auger and Rotary Methods 500
    9.2.3 Other Drilling Methods 500

9.3 Conventional Monitoring Well Installations 504

    9.3.1 Well Casing and Screens 504
    9.3.2 Filter Pack, Grouts and Seals 507
    9.3.3 Well Development 509
    9.3.4 Well Maintenance, Rehabilitation, and Abandonment 512
    9.3.5 Common Design Flaws and Installation Problems 514

9.4 Sampling Subsurface Solids and the Vadose Zone 514

    9.4.1 Subsurface Solids Sampling 515
    9.4.2 Soil Gas Sampling 516
    9.4.3 Soil Solute Sampling 516
    9.4.4 Microbiological Sampling and Other Sensitive Constituents .. 519

9.5 Sampling Ground Water 520

    9.5.1 Portable Well Samplers 520
    9.5.2 Portable/Permanent In Situ Samplers and Sensors 522
    9.5.3 Purging 524

9.6 Field Screening and Analytical Methods 525

    9.6.1 Field Analysis vs. CLP Analytical Laboratory 525
    9.6.2 Overview of Specific Techniques 526
    9.6.3 Types of Analytical Instrumentation 526

9.7 Guide to Major References* 528

* Appendix F contains citations for table and figure sources.

This chapter focuses on field procedures and devices for sampling and monitoring of contaminants in the vadose zone and ground water. No field sampling or other procedures should be undertaken until a sampling and monitoring plan has been developed for a site (Section 5.3). The information in this chapter may be useful for selection of sampling devices and types of monitoring installations for a sampling and monitoring plan. Field sampling for *site characterization* involves collection of one-time soil-gas, solid or ground-water samples to define subsurface geology and hydrogeology and the distribution of contaminants. Field sampling for *monitoring* involves collection of a series of samples at specified locations to evaluate changes in a parameter of interest over time. Monitoring usually involves sampling for chemical analysis, but some physical parameters such as moisture content and matric potential (Section 7.1.1) and ground-water levels (Section 7.2.1) are sufficiently time varying that time-series measurements are useful.

As in Chapter 7 (Characterization of Vadose Zone and Ground Water Hydrology), EPA's guide to **Subsurface Characterization and Monitoring Techniques** (Boulding, 1993/T9-10) is recommended for use as a companion to this chapter for illustrations and more detailed information about specific techniques. Appendix A in this book contains major method summary tables from that guide, which can be used to quickly identify the location of specific methods of interest. Appendix B provides an index and addresses of manufacturers and distributors of well drilling, sampling and chemical field screening and analytical equipment.

9.1 Types of Monitoring Installations

Subsurface sampling for the purpose of monitoring typically involves placement of access tubes for portable sampling devices or burial of sensors or samplers in situ. Installations above the saturated zone monitor are used to monitor soil gas, soil moisture or soil solutes. Installations below the saturated zone are used to monitor water levels and collect ground-water samples.

9.1.1 Vadose Zone Monitoring Installations

Specific types of installations that can be used for monitoring the potential for contaminant movement in the vadose zone include: (1) soil-moisture monitoring (Section 7.1.1), (2) soil-gas monitoring (Section 9.4.2), and (3) indirect and direct measurements of solute movement (Section 9.4.3). Figure 9-1 illustrates a number of different types of vadose zone monitoring installations for a hazardous waste landfill. Vadose zone monitoring is appropriate for many potential pollution sources including sanitary landfills, land treatment facilities, spoil piles, septic tank areas, pits, ponds, lagoons, dry channels used for effluent disposal, and irrigated fields (Wilson, 1981/T6-10).

The U.S. Environmental Protection Agency is placing increasing emphasis on vadose zone soil-solute sampling as an early warning system to detect movement of contaminants before they reach the saturated zone (Cullen et al., 1992; Durant et al.,

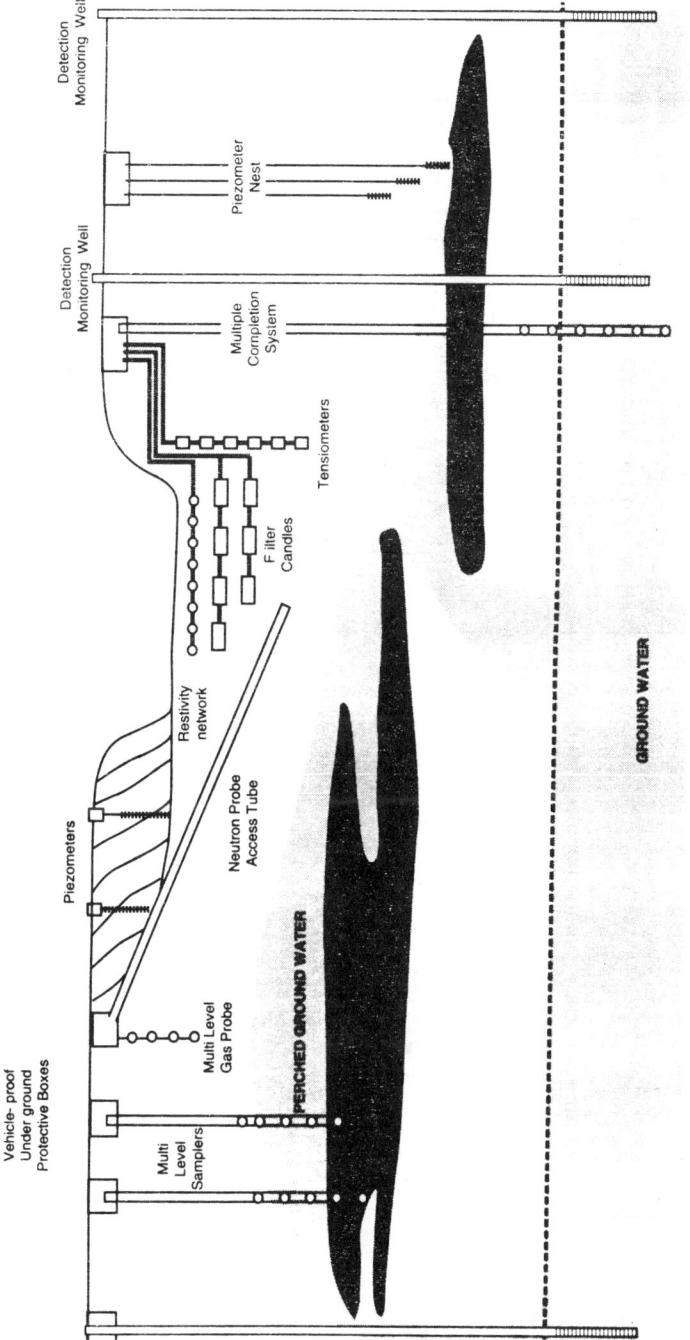

Figure 9-1 Generic monitoring design for existing hazardous waste landfill (Sara, 1994, after Everett et al., 1983). Reprinted with permission.

1993).[1]   Where contaminant sources are located above the saturated zone, and a vadose zone monitoring program indicates no vertical or lateral movement of contaminants, it may be possible to reduce the scope of a ground-water monitoring program. Also, early cleanup of contaminants in the vadose zone may reduce the need for intensive ground-water monitoring and remediation program.

9.1.2 Ground Water Monitoring Installations

Table 9-1 summarizes advantages and disadvantages of six major types of monitoring well installations: (1) *single riser with limited screened interval* (commonly five to ten feet), (2) *single riser/long screen* (in which the full thickness of an aquifer is screened), (3) *nested wells in a single borehole* with screens placed at different depths, (4) *nested wells in multiple boreholes* with screens placed a different depths, (5) *multilevel capsule samplers/sensors* placed in a single borehole with tubing rather than casing running to the surface for sample collection, and (6) *multiple port casings* that allow collection of samples at different levels from a single casing. Various hybrid combinations are also possible. The first type (single riser/limited screened interval) alone or as well nests in multiple boreholes (Type 4 above) are generally required at sites where monitoring is required under a federal or state regulatory program such as Superfund (uncontrolled hazardous waste sites) and RCRA (controlled hazardous and nonhazardous disposal sites). These "conventional" installations are discussed in more detail in Section 9.3.

The installation types described above typically require drilling of a borehole placement of casing or in situ samplers and filling of the remaining space with filter pack or grout. Another general type of monitoring well installation involves driving casing with a well point or other type of tip that allows sample collection manually or hydraulically to the desired sampling depth. *Driven wellpoints* have long been used for shallow well installations. *Push technologies*, such as cone penetration rigs and other truck-mounted hydraulic push and vibratory systems that use small diameter casing (generally 1 inch or less), are being increasingly used for sampling and installation of permanent monitoring wells in environmental studies. Advantages of such installations include: (1) lower cost and easy installation, by hand if necessary, (2) water samples can be collected as driving proceeds, and (3) depending on overburden, a good seal between casing and formation can be achieved. Disadvantages include the following: (1) limited to fairly soft materials--hard to penetrate compact, gravelly materials, (2) hard to develop wells, (3) screen may become clogged if thick clays are penetrated, (4) requirement for metal casing means that the method is not suitable where contaminant or geochemical conditions require plastic casing, and (5) sampling device options for depths below capacity of suction lift sampling devices (around 25 feet) are more limited than for conventional installation because of small casing diameter.

---

[1] Cullen, S.J., J.H. Kramer, L.G. Everett, and L.A. Eccles. 1992. Is Our Ground Water Monitoring Strategy Illogical? Ground Water Monitoring Review 12(3):103-107.

Durant, N.D., V.B. Myers, and L.A. Eccles. 1993. EPA's Approach to Vadose Zone Monitoring at RCRA Facilities. Ground Water Monitoring & Remediation 13(1):151-158.

Table 9-1 Advantage and Disadvantages of Types of Monitoring Well Installations

Type	Advantages	Disadvantages
Single-riser/ limited screened interval	-Simple and suitable for any type of formation. -Easier to install, pack, and seal than multilevel installations. -No potential for vertical cross-contamination between sampling points due to leaky seals. -Maximum flexibility in selection of well diameter (up to diameter of borehole). -Most common well diameters (2 to 4 inches) do not restrict the choice of sample collection methods.	-Provide no information on the vertical distribution of contaminants. -High cost per sampling point compared to multilevel installations, especially at great depth. -Contaminant plume might bypass wells with short screened intervals.
Single riser/long screen	-Simple and suitable for any type of formation. -Easier to install, pack, and seal than multilevel installations. -Maximum flexibility in selection of well diameter (up to diameter of borehole). -Most common well diameters (2 to 4 inches) do not restrict the choice of sample collection methods. -Where flow-through assumptions apply, there is no need to purge the well before sampling and the number of vertical sampling points is not limited by the diameter of the well.	-Do not generally give accurate measurement of maximum concentrations because concentration and hydraulic-head values tend to be averaged over the length of the screen. -Can cause cross-contamination in an aquifer by connecting contaminated zones to uncontaminated zones; consequently they can confirm the presence, but not the absence of a contaminant. -The underlying assumption for flow-through wells that the well screen will not alter the flow of ground water cannot be supported for most natural systems.
Nested wells/ single borehole	-Allow sampling for vertical distribution of ground-water constituents. -Lower cost per sampling point than separate single-riser wells. -Generally smaller diameters of individual wells in a nest compared to single-riser installations means that smaller volumes of water must be removed for purging.	-Installation, packing, and sealing is more difficult than for single-level installations and increases greatly as the number of wells in the boreholes increases. -Screened intervals must be separated by a grout seal with the possibility that small zones of contaminated water might be missed in heterogeneous materials. -Cross-contamination of sampling points might occur as a result of leaky seals (this can be checked using tracer tests). -Number of sampling points per borehole is restricted by the diameter of the borehole and the diameter of the individual piezometers. -Bundle piezometers are suitable only where cohesionless sands will collapse around the tips. -Small diameter of individual piezometers can restrict choice of sampling methods. -In fine-grained material with low hydraulic conductivity, the small storage volume of individual piezometers might make it difficult to collect samples of sufficient volume.

Table 9-1 (cont.)

Type	Advantages	Disadvantages
Nested wells/ multiple boreholes	-Allow sampling for vertical distribution of ground-water constituents. -Somewhat lower cost per sampling point than widely spaced single-riser wells. -Simple design and operation. -Potential for cross-contamination between different levels in the aquifer is eliminated. -Only the drilling method limits well diameter. -If desired, screened intervals can be placed to provide complete vertical coverage of the aquifer.	-More expensive than nested wells in a single borehole. -Small zones of contaminated water might be missed in heterogeneous materials if the screened intervals do not provide complete vertical coverage of the aquifer.
Multilevel capsule samplers/ sensors	-Allow sampling for vertical distribution of ground-water constituents. -Relatively easy to operate and safer than most other installation types where hazardous contaminants are involved. -Minimal purging is required because there is little mixing between incoming water from the formation and stagnant water.	-Proper installation is difficult. -Cost per sampling point is moderately high. -Depending on the type of sampler, number of sampling points might be limited by the diameter of the borehole (commonly three to four sampling points for 6-inch borehole). -Permanent nature of installation means that devices at individual sampling points cannot be retrieved for servicing or repairs, and malfunction means the sampling point is lost. -Cross contamination is a potential concern with multi-level installations requiring grout to isolate sampling points. -The choice of sample collection method is restricted to gas-drive or suction-lift devices (for shallow water table).
Multiple port casings	-Allow sampling for vertical distribution of ground-water constituents. -Cost per sampling point is relatively small (except for Westbay system). -Generally smaller diameters of individual wells in a nest compared to single-riser installations means that smaller volumes of water must be removed for purging. -Seals between sampling points can be obtained using permanent packers or traditional back-filled seals.	-Assembly and placement can be difficult. -Cross-contamination of sampling points possible as a result of leaky seals. -The number of sampling points is limited by the diameter of the casing and the tubing that runs to each sample port (does not apply to Westbay system). -Permanent nature of installation means that devices at individual sampling points cannot be retrieved for servicing or repairs, and malfunction means the sampling point is lost. -The Westbay system is very expensive, but can be cost-effective if a large number of sampling points at great depth is required. -Operation of the Westbay system requires special operator skills and can be time consuming. -The down-hole complexity of the Westbay system might result in mechanical difficulties.

Source: Adapted from Boulding (1993b).

Sampling and Monitoring	499

Figure 9-2 illustrates three types of small-diameter monitoring installations using a hydraulically driven filter-tip probe that can be permanently placed either in the vadose zone or the saturated zone for sampling. Table A-8 (driven wells and cone penetration) provides more information about methods using push techniques.

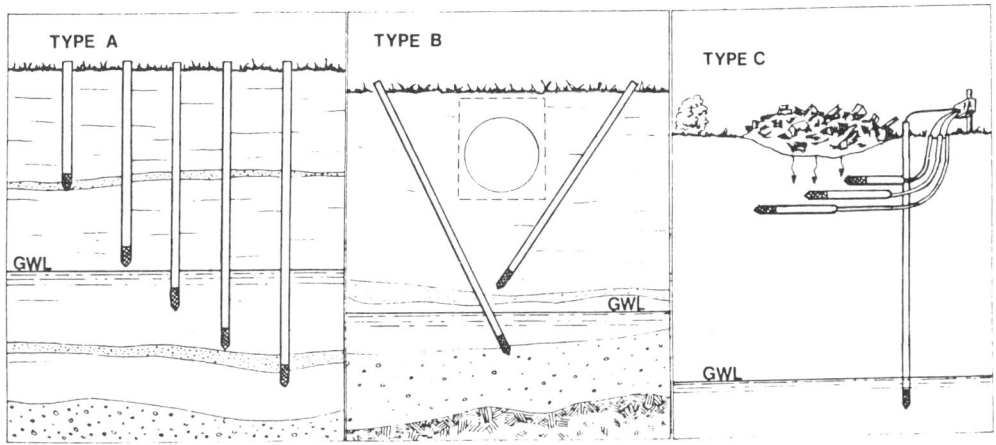

Figure 9-2 Example permanent installations of BAT filter tip probe (Torstensson, 1984). See Appendix F for full credit.

9.2 Drilling Methods for Sampling and Well Installation

Most subsurface investigations require the drilling of boreholes for one or more purposes: (1) collection of solids samples or cores for lithologic logging and laboratory testing, (2) lithologic and hydrogeologic characterization using borehole geophysical logging, and (3) installation of piezometers or monitoring wells.

9.2.1 Selection Criteria

A wide variety of drilling methods have been developed that could be suitable for one or more of the purposes described above. A number of criteria are used when selecting drilling methods including (1) availability and cost, (2) suitability for the type of geologic materials at a site (unconsolidated or consolidated), (3) maximum depth required for sampling or well installation, (4) whether core samples are required, (5) required diameter for well installations, and (6) potential effects on sample integrity (influence by drilling fluids and potential for cross-contamination between aquifers).

Table A-8 provides the following information on 20 drilling methods: (1) whether drilling advances in an open hole or with casing, (2) whether fluids used with the method may affect the chemistry of ground-water samples, (3) whether core samples can be taken, and (4) the sections and tables in Boulding (1993/9-10) where more detailed information about the method can be found.

9.2.2 Auger and Rotary Methods

The most commonly used drilling methods can broadly classified as *augering* and *rotary* methods. Augering methods include *hollow-stem*, *solid-stem*, and *bucket* augers, with the hollow-stem auger being by far the most commonly used method for well installation in unconsolidated deposits. Rotary drilling methods encompass a wide variety of techniques, most of which involve use are an *air rotary* or *mud rotary* rig. Rotary methods can also be classified according to (1) *open hole* or *casing* advancement, (2) normal (air or fluid down the drill stem) or *reverse circulation* (air or fluid up the drill stem), and (3) type of cutting tool (bit, hammer, reamer). Air rotary is probably the most commonly used method for well installation in consolidated formations. Where cross-contamination between aquifers is a concern, some kind of casing advancement method is required, with drill-through methods and dual-wall reverse circulation being the most commonly used. Table 9-2 describes major types of auger and rotary methods and summarizes their advantages and disadvantages.

9.2.3 Other Drilling Methods

Table 9-3 provides brief descriptions and advantages and disadvantage of the following additional drilling methods: (1) sonic drilling, (2) directional drilling, (3) cable tool, (4) jetting and jet percussion, and (5) rotary diamond (which could also be classified as a rotary method). Table A-8 provides additional comparative information on these methods.

Sonic and *directional* drilling have received a lot a recent attention in the context of environmental investigations. Sonic drilling, also called *vibratory* and *rotosonic* drilling, has been experimented with since the 1940s, but recent improvement in equipment design are likely to result in increased use of the method in the future. If further design improvements allow reduction in operating and maintenance costs, sonic drilling could become the method of choice for many settings. Directional drilling, also called *slant rig*, *radial* and *horizontal* drilling, has also been around for a while, but only relatively recently have applications in environmental investigations received much attention. Slant rig and radial drilling systems were originally developed in the petroleum industry and are relatively expensive. The utility industry has long used less sophisticated (and hence less expensive) directional drilling equipment systems for placement of underground cable, and improvements in this technology are likely to result in increased environmental applications.

Cable tool or *percussion* drilling is commonly used for water well installations, but is less common for monitoring well installations. *Jetting* and *jet percussion* have relatively limited capabilities and are not commonly used for monitoring well installations. The term *wash boring* is also used to describe the jetting method in water well applications and the jet percussion method in geotechnical applications, so when this term is encountered the operation of the method needs to be examined to determine which method is involved. *Rotary diamond* drilling is commonly used for mineral exploration in crystalline rock, but less commonly for monitoring well installation.

Table 9-2 Advantage and Disadvantages of Auger and Rotary Drilling Methods

Method	Advantages	Disadvantages
Augers: (Hollow- and Solid-Stem) Successive 5-foot flights of spiral-shaped drill stem are rotated into the ground to create a hole. brought to the surface by the cuttings are turning action of the auger.	-Inexpensive. -Fairly simple operation. -Small rigs can get to difficult-to-reach areas. -Quick set-up time. -Can quickly construct shallow wells in firm, noncavey materials. -No drilling fluid required. -Use of hollow-stem augers greatly facilitates collection of split-spoon samples. -Small diameter wells can be built inside hollow-stem flights when geologic materials are cavey.	-Depth of penetration limited, especially in cavey materials. -Maximum depth around 150 feet. -Can not be used in rock or well-cemented formations. -Difficult to drill in cobbles/boulders. -Log of well is difficult to interpret without collection of split spoons due to the lag time for cuttings to reach ground surface. -Vertical leakage of water through borehole during drilling is likely to occur. -Solid-stem limited to fine-grained, unconsolidated materials that will not collapse when unsupported. -With hollow-stem flights, heaving materials can present a problem. -May need to add water down auger to control heaving or wash materials from auger before completing well.
Rotary Methods **Direct (Mud) Rotary**: Rotating bit breaks formation; cuttings are brought to the surface by a circulating fluid (mud). Mud is forced down the interior of the drill stem, out the bit, and up the annulus between the drill stem and hole wall. Cuttings are removed by settling in a "mud pit" at the ground surface and the mud is circulated back down the drill stem.	-Drilling is fairly quick in all types of geologic materials. -Borehole kept open by formation of a mud wall on sides of borehole by the circulating drilling mud facilitating geophysical logging and well construction. -Geologic cores can be collected. -Virtually unlimited depths possible.	-Expensive, requires experienced driller and fair amount of peripheral equipment. -Completed well may be difficult to develop, especially small-diameter wells, because of mud wall on borehole. -Geologic logging by visual inspection of cuttings is fair due to presence of drilling mud; thin beds of sand, gravel, or clay may be missed. -Presence of drilling mud can contaminate water samples, especially the organic, biodegradable muds. -Circulation of drilling fluid through a contaminated zone can create a hazard at the ground surface with the mud pit and cross-contaminate clean zones during circulation.
Reverse Rotary: Similar to mud rotary method except the drilling fluid is circulated down the borehole outside the drill stem and is pumped up the inside. Water is used as the drilling fluid, rather than a mud, and the hole is kept open by the hydrostatic pressure of the water standing in the borehole.	-Creates a very "clean" hole, not dirtied with drilling mud. -Can be used in all geologic formations. -Very deep penetrations possible. -Split-spoon sampling possible.	-A large water supply is needed to maintain hydrostatic pressure in deep holes and when highly conductive formations are encountered. -Expensive; experienced driller and much peripheral equipment required. -Hole diameters are usually large, commonly 18 inches or greater. -Cross-contamination from circulating water likely. -Geologic samples brought to surface are generally poor, circulating water will "wash" finer materials from sample.

Table 9-2 (cont.)

Method	Advantages	Disadvantages
Air Rotary: Very similar to mud rotary, the main difference being that air is used as primary drilling fluid as opposed to mud or water.	-Can be used in all geologic formations; most successful in highly fractured environments. -Useful at any depth. -Fairly quick. -Drilling mud or water not required.	-Relatively expensive. -Cross-contamination from vertical communication possible. -Air will be mixed with water in the hole and that which is blown from the hole, potentially creating unwanted reactions with contaminants; may affect "representative" samples. -Cuttings and water blown from the hole can pose a hazard to crew and surrounding environment if toxic compounds encountered. -Organic foam additives to aid cuttings removal may contaminate samples.
Air-Percussion Rotary/ Downhole Hammer: Air rotary with a reciprocating hammer connected to the bit to fracture rock.	-Very fast penetrations. -Useful in all geologic formations. -Only small amounts of water needed for dust and bit temperature control.	-Relatively expensive. -As with most hydraulic rotary methods, the rig is fairly heavy, limiting accessibility. -Vertical mixing of water and air creates cross-contamination potential. -Hazard posed to surface environment if toxic compounds encountered. -Organic foam additives for cuttings removal may contaminate samples.
Rotary Casing Advancement Methods: Various methods (rotary drill-through, reverse dual-wall rotary, downhole casing advancers) involving air or mud rotary with a casing.	Varies with specific technique: -General advantage is that advancing casing minimizes of cross-contamination. -Most methods work well in difficult formations, such as unconsolidated materials with boulders, and fractured rock.	Varies with specific technique: -Generally more expensive than conventional air and mud rotary.

Source: Adapted from Boulding (1993b).

Table 9-3 Advantage and Disadvantages of Other Drilling Methods

Method	Advantages	Disadvantages
Sonic Drilling: A high power oscillator with eccentric weights driven by hydraulic motors drive a rotating and vibrating drill bit or core barrel into the ground.	-Collection of continuous, relatively undisturbed unconsolidated and bedrock cores possible. -Higher drilling rates than conventional methods (around twice as fast as air rotary and 8 to 10 times faster than hollow-stem auger and cable tool). -Produces about one-tenth the cuttings of hollow-stem auger and cable tool.	-Higher operation, maintenance, and tooling costs compared to conventional drilling methods. -Present equipment limited to depths of about 300 feet. -Drilling in hard rock generally not recommended. -Driving of material into borehole wall might create problems for borehole logging and aquifer testing. -Limited equipment availability.
Directional Drilling: Various types of rigs capable of drilling slanting or horizontal holes. Common features include a steerable drill stem and devices for detecting drill head location or trajectory.	-Allows borehole access to subsurface areas such as beneath buildings, tanks, landfills, and impoundments where vertical drill rigs cannot go. -Reduces potential for cross-contamination between aquifers -Excellent for remediation techniques that require maximum horizontal access to contaminated zone or contaminant plumes that are not vertically dispersed. -Production from horizontal wells generally is higher than from vertical wells due to greater possible screen length.	-There has been relatively little actual experience using directional drilling methods at contaminated sites, and value for site characterization and monitoring (as opposed to remediation) has yet to be demonstrated. -Drilling costs are high for petroleum industry-related equipment (100 to several hundred dollars a foot). -Utility rigs, although less expensive than petroleum rigs, have more limited depth capabilities (around 20 feet compared to 300 feet for EC slant rig). -Equipment that uses water or other fluids to advance the well bore might affect quality of samples. -Sampling capabilities are currently limited.
Cable-Tool (Percussion): Hole created by dropping a heavy "string" of drill tools into well bore, crushing materials at bottom. Cuttings are removed periodically by bailer. Generally, casing is driven just ahead of the bottom of the hole which is typically greater than 6 inches in diameter.	-Can be used in rock formations as well as unconsolidated formations. -Fairly accurate logs can be prepared from cuttings if collected frequently enough. -Driving a casing ahead of hole minimizes cross-contamination by vertical leakage of formation waters. -Core samples can be obtained easily.	-Requires an experienced driller. -Heavy steel drive pipe used to keep hole open and drilling "tools" can limit accessibility. -Cannot run some geophysical logs due to presence of drive pipe. -Relatively slow drilling method.

Table 9-3 (cont.)

Method	Advantages	Disadvantages
Jetting: Washing action of water forced out of the bottom of the drill rod clears hole to allow penetration. Cuttings brought to surface by water flowing up the outside of the drill rod. **Jet Percussion**: Similar to jetting except a wedge-shaped drill bit is attached to a cable which is alternately raised and dropped.	-Inexpensive. Driller often not needed for shallow holes. -In firm, noncavey deposits where hole will stand open, well construction fairly simple.	-Somewhat slow, especially with increasing depth. -Extremely difficult to use in very coarse materials, i.e., cobbles/boulders. -A water supply is needed that is under enough pressure to penetrate the geologic materials present. -Difficult to interpret sequence of geologic materials from cuttings. -Maximum depth 150 feet, depending on geology and water pressure capabilities.
Diamond Rotary: Rotating bit consists of a tube 10 to 20 feet long with diamond-studded ring fitted to the bottom. Diamond bit cuts through rock, with solid core of rock remaining in tube.	-Takes continuous rock sample for accurate geologic logging. -Can be used with hydraulic or air rotary drill rig.	-Expensive; diamond bits are more expensive than conventional roller bits. -If used with hydraulic rotary, drilling muds may contaminate well. -Slow compared to most other methods.

Source: Adapted from U.S. EPA (1987a) and Boulding (1993b).

9.3 Conventional Monitoring Well Installations

Monitoring well design involves several components including casing and screen material, well casing diameter, screen and gravel pack specifications, screen length and depth of placement, and sealing material. Often, the selection of one component will influence the determination of other components. Elements that are factored into the well design aspect of an overall monitoring program include the geologic setting, the results of previous site investigations, well drilling expertise, well logging and aquifer property determinations, regulatory requirements and specifications, details of the history of the site, and the chemical parameters of concern (ASTM 5092/Table A-14).

9.3.1 Well Casing and Screens

Conventional monitoring wells are typically 2-inches in diameter, but the increasing availability of small-diameter sampling devices has resulted in a trend toward even smaller diameters (1 to 1.5 inches). Smaller-diameter wells reduce cost of materials and purge volumes. Casing materials can be broadly classified as (1) plastic, (2) metallic, and (3) fiberglass reinforced. Table 9-4 summarizes relative advantages and disadvantages of these materials.

Table 9-4 Advantage and Disadvantages of Monitoring Well Casing Materials

Method	Advantages	Disadvantages
Plastic casing	-Lightweight. -PVC is inexpensive. -Generally good to excellent chemical resistance; fluoropolymers have the best chemical resistance, except for fluorinated solvents; PVC has poor resistance to high concentrations of aromatic hydrocarbons (toluene, xylene, trichlorethylene), esters, and ketones.	-Weaker, less rigid, and more temperature sensitive than metallic materials (PTFE/TFE is especially low, PVDF is stronger; ABS has low strength and less heat resistance compared to PVC). -PVC might adsorb some constituents from ground water. -PVC might react with and leach some constituents into ground water and PTFE is prone to sorption of selected organic compounds (proper purging and sampling procedures can minimize these problems). -Fluoropolymers are expensive (PVDF is less expensive than PTFE/TFE). -Some materials are not commonly available (ABS, PVDF). -Tensile strength of wear resistance of PTFE/TFE is low compared to other plastics, and screen slot opening might decrease in size over time. -Antistick properties of fluoropolymer materials make it difficult to achieve an annular seal with neat cement grout, creating potential for alteration of ground-water chemistry by percolating surface water.
Metallic	-Stainless steel has least adsorption of halogenated and aromatic hydrocarbons. -All steel casings have high strength and generally are not temperature sensitive. -Stainless steel has excellent resistance to corrosion and oxidation. -Stainless steel is readily available in all diameters and screen slot sizes. -Mild steel is readily available and less expensive than stainless steel for casing.	-Heavier than plastics. -Stainless steel might corrode and leach some chromium in highly acidic water and might act as a catalyst in some organic reactions -Stainless steel screens are more expensive than plastic screens. -Mild steel might react with and leach some constituents into ground water and is not as chemically resistant as stainless steel. -Under saturated conditions carbon and low carbon steel rust easily, providing highly sorptive surface for many metals, and they deteriorate in corrosive environments. -Zinc might leach from galvanized steel, and if the coating is scratched, will rust, providing a highly sorptive surface for metals.
Fiberglass reinforced	-High strength (almost as strong as stainless steel). -Light (weighs about the same as PVC) -Limited available data indicate that it is relatively inert in most monitoring well environments.	-Some adsorption of volatile organics (can be overcome by proper purging and sampling procedures). -Not readily available and little data available on its performance in the field.

Source: Adapted from Boulding (1993b).

There are two major classes of plastic casing: (1) *thermoplastics*, including polyvinyl chloride (PVC) and acrylonitrile butadiene styrene (ABS), and (2) *fluoropolymers*, including polytetrafluoroethylene/tetrafluoroethylene (PTFE/TFE, Teflon, Halon, Fluon, Hostaflon, Polyflon, Algoflon, Soriflon), fluorinated ethylene propylene (FEP, Neflon, Teflon), perfluoroalkoxy (PFA, Neoflon, Teflon), polyvinylidine fluoride (PVDF, Kynar), and chlorotrifluoroethylene (CTFE, Kel-F, Diaflon). *PVC is by far the most commonly used casing material for monitoring wells.* Although much research has been published about *PTFE* and other fluoropolymers, actual use is uncommon due to expense and low strength.

Metallic casing types include (1) cast iron, (2) mild or soft steel, (3) carbon steel, (4) low carbon steel, (5) galvanized steel, and (6) stainless steel (particularly types 304 and 316). *Stainless steels are the most chemically resistant of the ferrous materials and after PVC are probably the most commonly used material for monitoring wells.* *Fiberglass-reinforced* casings include (1) fiberglass-reinforced epoxy (FRE), and (2) fiberglass-reinforced plastic (FRP). Use of fiberglass-reinforced casings in still largely experimental.

Well *screens* of the appropriate material, length and slot size are attached to solid casing and placed at the depth in the aquifer where sampling is desired. The *slot size* is selected to (1) maximize open area for water to flow through, and (2) minimize entry of fines into the well during pumping. The most commonly used types of well intake screens for monitoring well construction are *factory-slotted pipe* (machine cut to uniform width and length around the pipe) and *continuous-slot screen* (v-shaped continuous wire is wrapped around vertical rods). Figure 9-3 illustrates these two well screen types.

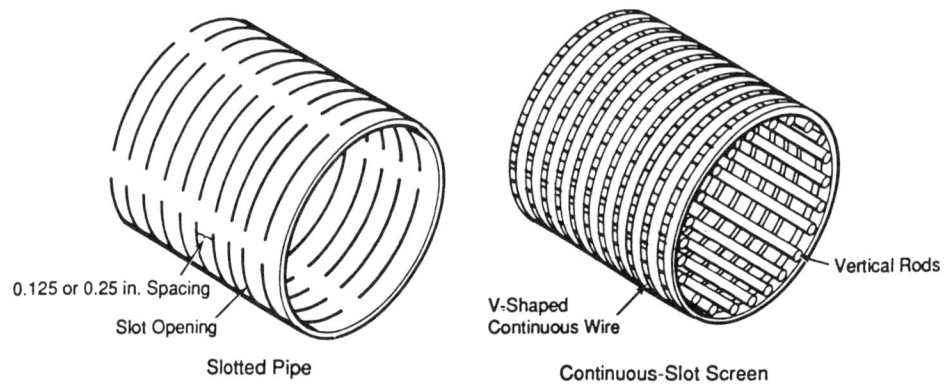

Figure 9-3 Commonly used monitoring well screen types (Boulding 1993b, after Nielsen and Schalla, 1991).

9.3.2 Filter Pack, Grouts and Seals

Filter pack, also called *gravel* and *sand* pack, is usually placed around a well screen to (1) increase hydraulic conductivity around the well screen and (2) keep fine particles from entering the well screen during ground-water sampling. Desirable *artificial* filter pack characteristics include:

- Uniform grain size to minimize loss of material during development and development time.

- Well-rounded grains to increase hydraulic conductivity, porosity, yield, and effectiveness of well development.

- 90 to 95% quartz grains to minimize changes to ground-water chemistry and to eliminate loss of volume by dissolution of minerals.

- Have a uniformity coefficient of 2.5 or less to minimize separation during installation and lower head loss.

Alternatively, when *natural filter pack* is used, well screen slot size is determined based on the particle-size distribution in the aquifer materials and the fines are removed during the development process. Table 9-5 summarizes the relative advantages and disadvantages of artificial and natural filter packs.

In relatively shallow wells, the filter pack can be placed by simply dumping sand down the annulus (provided the annular space is more than 2 inches). More typically, the filter pack is placed by pouring the sand into a *tremie pipe*, a rigid or partially flexible tube of pipe that allows funneling of the material directly to the interval around the well screen. Other methods of emplacement include the *reverse circulation method*, where a sand and water mixture is fed into the annulus around the well screen and the water entering the screen is pumped up to the surface, and *backwashing*, where water is pumped down the well and allowed to rise up around the annular area as filter-pack material filters down through the rising water.

Grouts are used in monitoring well constructions to seal the annular space between the well casing and the formation to prevent contaminants from moving upward or downward to uncontaminated areas. The two major types of grouts are (1) bentonite and (2) neat cement. *Bentonite* can be placed either as unhydrated pellets or chips with water added later or pumped down through a tremie pipe as a slurry. *Neat cement* (a mixture of 5 to 6 gallons of clean water per 1 cubic foot bag of Portland Cement, usually Type I) is mixed manually or with a mechanical mixture and pumped into the annulus. Table 9-5 also summarize the relative advantages and disadvantages of bentonite and cement grouts.

A variety of additives can be mixed with the cement slurry to change the properties of the cement. The more common *cement additives* include:

Table 9-5 Advantage and Disadvantages of Filter Pack, Grouts and Seals

Method	Advantages	Disadvantages
Artificial filter pack	Characteristics of the filter-pack material can be selected for optimum efficiency of well operation.	-Procedure is relatively time consuming and expensive. -Bridging might prevent complete filling around the well screen. -Extension of filter pack above or below the screen area might allow contaminants to move to uncontaminated areas. -Filter pack material might introduce contaminants into the aquifer (a leaching test can be used to determine whether this might be a problem). -Use of reverse circulation and backwashing emplacement methods might alter ground-water chemistry.
Natural filter pack	-Simpler and can be less expensive (depending on time requirements for well development). -Potential for alteration of ground-water chemistry is minimized.	-Well development is more difficult, and success is less assured. -Selection of optimum screen slot size is more difficult.
Bentonite grout	-Readily available. -Inexpensive. -Pellets or slurry can be used.	-Possible constituent interference due to ion exchange. -Complete seal and complete bond to casing not assured. -Pellets can bridge or wet and swell, sticking to the formation or casing before filling the annular space. -Can clog pump if slurry gets too dense.
Cement grout	-Readily available. -Inexpensive -Can use sand and/or gravel filler. -Possible to determine how well the cement has been placed by means of temperature or sonic bond logs.	-Chemical interferences (high pH with attendant change in ground-water chemistry). -Mixer, pump, and tremie lines are required and more cleanup generally is required compared to bentonite. -Possible problems getting the material to set up. -Channeling between the casing and seal can develop because of temperature changes during the curing process, swelling and shrinkage of the grout while the mixture cures, and poor bonding between the grout and the casing surface -Heat from setting can compromise structural integrity of some well casing materials (i.e., thermoplastic).

Source: Adapted from Boulding (1993b).

Sampling and Monitoring

- *Bentonite* improves workability and reduces weight and shrinkage.

- *Calcium chloride* accelerates setting time and creates higher early strength (especially useful in cold climates).

- *Gypsum* creates a quick-setting (but expensive) expanding cement.

- *Aluminum powder* produces a strong, quick-setting cement that expands on setting.

- *Fly ash* increases sulfate resistance and early compressive strength.

- *Hydroxylated carboxylic acid* retards setting time and improves workability without compromising set strength.

- *Diatomaceous earth* reduces slurry density and thickening time, but increases water demand and reduces set strength.

Cement and additives have the potential for affecting ground-water quality (especially pH and iron, see Table 5-9). This is most likely to be a problem if a poor annular seal is achieved (Figure 9-4), but any special additives should be carefully evaluated for the potential effects on ground-water chemistry.

Major *surface sealing* measures include: (1) placement of a sturdy protective outer casing with cover and lock to a depth below the frost line and a drainage hole to prevent moisture buildup between the protective casing and the well casing, and (2) placement of a concrete pad sloping away from the casing to prevent infiltration of surface water and shaped so as to prevent frost heaving. Figure 9-4 illustrates the types of potential pathways for fluid movement when the borehole annulus is incorrectly grouted and sealed.

9.3.3 Well Development

After grouting and sealing are completed, *well development* is essential to remove fines from the filter pack around the well screen in order to improve hydraulic performance and eliminate or reduce collection of sediment in water quality samples. Well development is also necessary to rectify damage done during drilling to borehole wall and adjacent formation, as occurs when drilling muds are used.

A variety of techniques are used to develop wells. In *overpumping* the well is pumped at a rate that substantially exceeds the ability of the formation to deliver water. *Backwashing* often is used in conjunction with overpumping. If the pump does not have a backflow prevention valve, alternately starting and stopping the pump creates a surging effect where water is driven back into the formation during the off cycle. Alternatively, water can be added to the well. In *bailing*, a bailer (Section 9.5.2) is allowed to fall freely through the borehole until it strikes the surface of the water. The impact of the bailer produces an outward surge of water through the well screen and filter pack. As the bailer fills, the flow of water reverses and fines migrate into

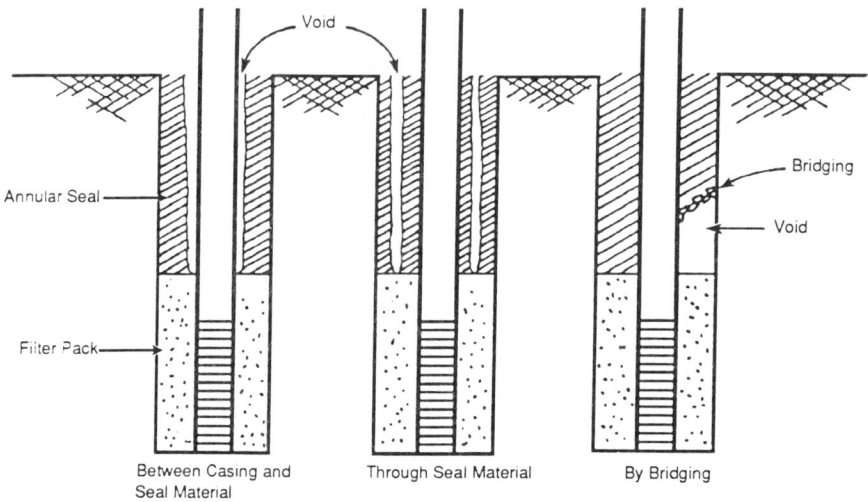

Figure 9-4 Potential pathways for fluid movement in the casing-borehole annulus (Boulding, 1993b, after Aller et al., 1991).

the well and are brought to the surface in the bailer. Sediment in the bottom of the well can be mobilized by short rapid strokes of the bailer near the bottom before retrieving the bailer.

Mechanical surging forces water into and out of the well screen by operating a plunger, called a *surge block*, which is attached to a drill rod or a wireline. The surge block is lowered to the top of the well intake and operated in a pumping action with strokes typically around 3 feet and is gradually worked downward through the screened interval. At regular intervals, the surge block is removed and fines that have entered the well are removed by pumping or with a bailer. *Compressed air* can be used to alternately surge and air-lift pump a well to remove sediment. In *air surging*, injected air lifts the water column until it reaches the top of the casing and the air supply is shut off, causing an outward surging action in the well intake. *Air-lift* pumping uses compressed air to bring water to the surface. *High velocity jetting* uses a single- or multiple-nozzle device, which directs a horizontal stream of water against the well screen opening. The jetting tool is placed near the bottom of the screen and slowly rotated while being pulled upward. Material that enters the screen in the backwash of the jet stream is removed by pumping or bailing. *Jetting/pumping*, which combines jetting with simultaneous pumping, provides for maximum development efficiency. Two development methods that are used for water wells but are not recommended for monitoring well development because they introduce contaminants into the aquifer are (1) *blasting* (used only in solid rock wells), and (2) *acidizing* (used only in limestone aquifers).

Table 9-6 Advantage and Disadvantages of Well Development Methods

Method	Advantages	Disadvantages
Over-pumping	-Is convenient for small wells or poor aquifers. -Minimal time and effort are required. -No new fluids are introduced. -Removes fluids introduced during drilling and fine sediments.	-Not adequate for large wells. -Will not develop maximum efficiency in a well because does not effectively remove fine-grained sediment. -Tends to cause sand to bridge in the formations (can be reduced by alternating pump on and pump off). Requires the use of high capacity pumping equipment. -Can result in a large volume of water to be contained and disposed. -Can leave the lower portion of large screen intervals undeveloped. -Excessive pumping rates can caused well collapse, especially in deep wells. -Equipment for effective overpumping might not fit in small diameter wells.
Back-washing	-Effectively rearranges filter pack. -Effective in breaking down bridging. -No new fluids introduced with on-off overpumping.	-Fine sand, mud, silt, or clay can be washed into the well or filter pack from the formation. -Not fully effective unless combined with surging, bailing, or pumping. -Large quantities of water are required. -Unless combined with pumping or bailing, does not remove drilling fluids. -Backwashing with added water introduces fluid into the well that might alter formation chemistry.
Bailing	-No new fluids are introduced into the aquifer. -Removes fluids introduced during drilling. -Removes fines from well. -Bailers are easily obtained and can double as sampling devices.	-Time-consuming and tiring if done manually. -Not as effective as surge blocks. -Not effective in unproductive wells.
Mechanical surging	-Low cost. -Effectively rearranges filter pack. -Greater suction action and surging than backwashing. -Breaks down bridging in filter pack. -No new fluids are introduced. -Convenient to use for cable-tool rigs.	-Can produce unsatisfactory results when an aquifer contains clay because the casing or screen can collapse if it becomes plugged with fines. -Tends to push fine-grained sediments into the filter pack. -Unless combined with pumping or bailing, does not remove drilling fluids. -Sometimes the well seal can be disturbed when surging. -Excessive sand can result in sand-locking of the surge block.

Table 9-6 (cont.)

Method	Advantages	Disadvantages
Compressed air	-Rapid method.	Not recommended for monitoring wells because: -Air can become entrained in the filter pack and reduce permeability. -Where yield is very weak and drawdown rapid, or submergence is low, other methods will be more satisfactory. -Introduction of air into aquifer can alter chemistry.
Jetting	-Simple to use. -Effectively rearranges and breaks down bridging in filter pack. -Effectively removes mud cake around screen. -Jetting with simultaneous pumping is particularly successful for wells in unconsolidated sands and gravels. -Jetting/pumping removes sediment from the well before it can settle in the screen and jetting waters can be recirculated after sediment has been removed at the surface.	Generally not recommended because: -Foreign water and possible contaminants are introduced to the aquifer. -Air blockage can develop with air jetting. -Air jetting can change water chemistry and biology (iron bacteria) near well. -Unless combined with pumping or bailing, does not remove drilling fluids. -Jetting with simultaneous pumping is not always practicable.

Source: Adapted from Boulding (1993b).

Overpumping and backwashing are probably the most commonly used forms of well development. These methods or bailing combined with mechanical surging will be the most effective methods for most situations. Table 9-6 summarizes relative advantages and disadvantages of the main methods described previously.

9.3.4 Well Maintenance, Rehabilitation and Abandonment

Well *maintenance* involves the routine, ongoing tasks that ensure a well is a representative sampling point. This involves full documentation of design and installation of the well and of all subsequent sampling and other activities involving the well.

Routine well maintenance activities include:

- Periodic bail testing of the well to determine specific capacity (can be done during normal purging for sampling or more frequently if sampling is infrequent).

- Measurement of depth before purging.

- Repair of protective casing, covers, hinges, and any other exposed parts of the well.

- Occasional redevelopment by bailing, surging or bottom pumping.

Rehabilitation involves efforts beyond normal maintenance that are intended to restore the well's original performance or to alter the well to serve other purposes. Common rehabilitation techniques include: (1) deepening because the water table has been lowered, (2) installation of sleeving to repair a physical problem, (3) treatment of screens to reduce plugging or encrustation, and (4) use of aggressive development techniques, such as high-velocity jetting (Section 9.3.3), to improve well performance. In traditional well rehabilitation, three categories of chemicals are used for rehabilitation: (1) *acids*, to dissolve incrustation on the well intake or in the surrounding formation, (2) *biocides*, to kill bacteria in the well or surrounding formation that contribute to clogging, and (3) *surfactants*, to disperse clay and fine materials for easier removal. Chemical treatment of monitoring wells is done only if the integrity of future sample collection can be assured.

Well *abandonment* or *decommissioning* involves the combination of full or partial casing/screen removal and plugging. The two main *casing/screen removal* techniques are: (1) *pulling*, using hydraulic jacks or by pumping the casing with a rig, and (2) *overdrilling*, in which a large-diameter hollow-stem auger is used to drill around the casing. In shallow, sandy aquifers, casing can be removed by *jetting* (Section 9.2.3). *Sandlocking* can be used to remove telescoped well screens, where the diameter is smaller than the casing. A pulling pipe wrapped with burlap strips is lowered to penetrate about 2/3 of the length of the screen. Sand is added to create a locking effect and the screen is pulled to the surface. *Latch-type* tools can be used to remove telescoped well screens that are 2 to 6 inches in diameter. *Partial casing removal* involves cutting the casing off below ground level.

The simplest technique for *plugging* an uncased borehole is to fill the entire hole with grout material, commonly a cement/bentonite mixture (Section 9.3.2), which is chemically compatible with the formation. Where casing is left in place, the interval adjacent to water-bearing zones is ripped or perforated with *casing rippers*, *gun-perforators*, or *jet perforators*, and grouted under pressure to allow penetration outside the casing. *Partial grouting* requires the use of *bridge plugs*, which allow sealing of selected portions of a borehole. A *permanent bridge seal* is the most deeply located plug that forms a bridge upon which fill material can be placed and is used to prevent cross-contamination between lower and upper aquifers. If more than two water-bearing zones are intersected by the wells, *intermediate seals* are placed adjacent to intermediate zones and the remaining permeable zones are filled with clean disinfected sand, gravel, or other material. *Uppermost aquifer seals* keep out surface water and keep artesian aquifers from flowing to the surface. In artesian aquifers, special procedures might be required for plugging, such as (1) Pumping nearby wells to lower hydrostatic head, (2) placing fluids of high specific gravity in the borehole, or (3) elevating the casing high enough to stop the flow. ASTM D5299 (Table A-14) provides more detailed guidance on well decommissioning.

9.3.5 Common Design Flaws and Installation Problems

Nielsen and Schalla (1991/T9-10) have identified six common monitoring well design flaws and installation problems that should be avoided.

1. Use of *well casing* or *well screen materials* that are not compatible with the hydrogeologic environment, known or suspected contaminants, or the requirements of the ground-water sampling program. The result is chemical alteration of samples or failure of the well.

2. Incorrect *screen slot-sizing* practices or use of nonstandard types of well screen, such as field-slotted, drilled, or perforated casing. The result is well sedimentation and turbid samples throughout the monitoring program.

3. Improper *length* and *placement* of *well screens* so that discrete zones of the aquifer are missed or cannot be differentiated. In this situation, water level measurements and water quality samples might provide misleading results.

4. Improper selection and placement of *filter pack* materials. Consequences can include well sedimentation, well screen plugging, ground-water sample alteration, or potential well failure.

5. Improper selection and placement of *annular seal* materials. The results can include alteration of chemistry of water samples, plugging of the filter pack and/or well screen, and cross-contamination between water-bearing units that have not be adequately isolated.

6. Inadequate *surface protection measures*, such as surface seals that are susceptible to frost heave. The results can include surface water entering the well, chemical alteration of water quality samples, and well damage to destruction.

Another common installation problem that can be added to this list occurs after installation has been completed.

7. Use of improper *well development* techniques. The results can include continuing turbidity in water quality samples due to failure to remove fines for the well screen and filter pack, chemical alteration of water quality samples due to the introduction of air or foreign water into the aquifer, and possible damage to the well screen by stresses caused by excessive surging.

9.4 Sampling Subsurface Solids and the Vadose Zone

This sections covers four distinct, but interrelated types of subsurface sampling: (1) soil and geologic materials above and below the saturated zone, (2) soil-gas sampling in the vadose zone, (3) soil-solute sampling in the vadose zone, and (4) microbiological sampling.

9.4.1 Subsurface Solids Sampling

Solids sampling methods can be broadly classified as (1) hand held and (2) power driven. Criteria for selection of *hand-held* equipment includes: (1) whether an undisturbed core is required, (2) soil conditions at the site (cohesion, stones, moisture), (3) the sample size and depth desired, and (4) the number of required operators. Many specific types of hand-held sampling devices are available. Table A-9 provides information on applications and limitations of 20 types in the following categories: (1) grad samplers (spoons, scoops, shovels and picks), (2) augers, and (3) tube samplers. Hand-held soil samplers are usually used for sampling the near surface (2 to 3 meters). *Grab* samplers are commonly used for near-surface sampling for initial screening purposes. *Augers* are typically used for collection of composite near surface samples and in combination with tube samplers to collect undisturbed samples at different depths. *Tube* samplers are used for observation of subsurface soil characteristics and collection of undisturbed cores for laboratory study. Both augers and tube samplers require the use of *liners* when sampling for volatile contaminants in order to minimize contact of samples with the atmosphere.

Power-driven samplers are usually operated in conjunction with drill rigs, cone penetration rigs, or hydraulic devices attached to smaller trucks. Portable power-driven augers that can be used by one or two persons are also available. Table A-10 provides information on applications and limitations on 18 specific power-driven sampling devices in the following general categories: (1) disturbed core samplers and (2) undisturbed core samplers. *Disturbed core* samplers include various types of *barrel* samplers and *rotating core* samplers. *Undisturbed core* samplers might better be called *minimally disturbed* core samplers, since some disturbance is inevitable. These samplers fall into three major categories: (1) *thin-wall open tube* (including the *Shelby tube*, and *Laskey continuous sampler*), (2) *thin-wall piston* samplers which are especially useful for collecting sample on noncohesive materials, and (3) samplers for other specialized situations, such as hard, dense, brittle and cohesive soils.

Collection of soil cores is the preferred method for sampling solids because much more accurate lithologic logging is possible than with cuttings from drill methods that do not obtain cores as part of the drilling process, such as diamond drilling (Section 9.2.3). The most common method for collection of disturbed cores is the *split-spoon* sampler, also called a *split-barrel* sampler. Thin-wall open tube samples are the most common method for collecting undisturbed cores in unconsolidated materials. As noted above, thin-wall piston samplers are usually used where poor cohesion prevents good recovery with conventional thin-wall samplers and specially designed thin-wall samplers are required for gravelly and very stiff or cemented unconsolidated deposits.

ASTM D4700 (Table A-14) provides general guidance selection of soil sampling devices and sample collection in the vadose zone. ASTM guides for specific sampling devices include D1452 (augers), D2113 (diamond core drilling, D3350 (ring-lined barrel), D1586 (split barrel), and D1587 (thin-wall tube).

9.4.2 Soil Gas Sampling

Sampling of soil gas above the zone of saturation and analyzing for the presence of volatile organic compounds (VOCs) is a relatively new technique in contaminant hydrogeology that began in the early 1980s and has gained rapid acceptance. A variety of sampling techniques are used for soil-gas survey. *Passive* sampling uses an in situ adsorbent (usually an activated charcoal rod), which is buried in the soil for a few days to weeks. The adsorbent is then retrieved and analyzed for VOCs in a laboratory by mass spectrometry or gas chromatography. *Dynamic* or *grab* sampling involves the installation of a probe or soil boring in the vadose zone followed by withdrawal of soil gas by a pump. Soil-gas samples taken with this method can be analyzed onsite using portable gas chromatographs (Section 9.6.2). The most common uses of soil-gas data are in planning monitoring well networks and defining the boundaries of contaminant plumes (Marrin and Kerfoot, 1988/T9-10).

Table 9-7 summarizes information on seven major groups of organic contaminants relevant to soil-gas surveying. Contaminants in Group A (halogenated methanes, ethanes, and ethenes) and Group D (lighter petroleum hydrocarbons) are the most likely candidates for this method, followed by Group G (low molecular weight oxygenated compounds) if the leak or spill is in dry soil.

Soil-gas measurements taken areally at the same depth allow creation of an isopleth map that might look something like the resistivity measurements in Figure 6-6. Sampling at different depths at a single point and plotting concentration against depth allows some inferences about subsurface conditions. Figure 9-5 shows VOC concentration profiles with increasing depth in a variety of hydrogeologic settings. Figure 9-5A shows that in homogenous unsaturated porous media VOC concentration is lowest at the surface and increases steadily to a maximum at the contaminant source. Impermeable layers cause different VOC concentration profiles depending on where they are located (see Figure 9-5B and C). Microbiological activity may reduce VOC concentrations near the ground surface (see Figure 9-5D), and a VOC source in the vadose zone creates a profile where concentrations decline with distance in all directions from the source (see Figure 9-5E).

9.4.3 Soil Solute Sampling

Monitoring of soil water in the vadose zone can serve as an early warning system at controlled waste disposal sites that contaminants are entering the subsurface and can allow actions to be taken before contaminants reach the saturated zone. Methods for sampling and monitoring the vadose zone can be broadly categorized as (1) *indirect* (surface geophysical methods and probes that focus on measuring variations in soil salinity), and (2) *direct* (in which soil water is collected directly in the field or extracted from samples of soil solids).

Indirect Soil Salinity Measurements. A variety of methods are available for locating and monitoring areas of high soil salinity. These methods primarily have been developed for agricultural applications to identify saline soils and control irrigation flows where soluble salts can affect crop productivity. Table A-11 summarizes

Sampling and Monitoring 517

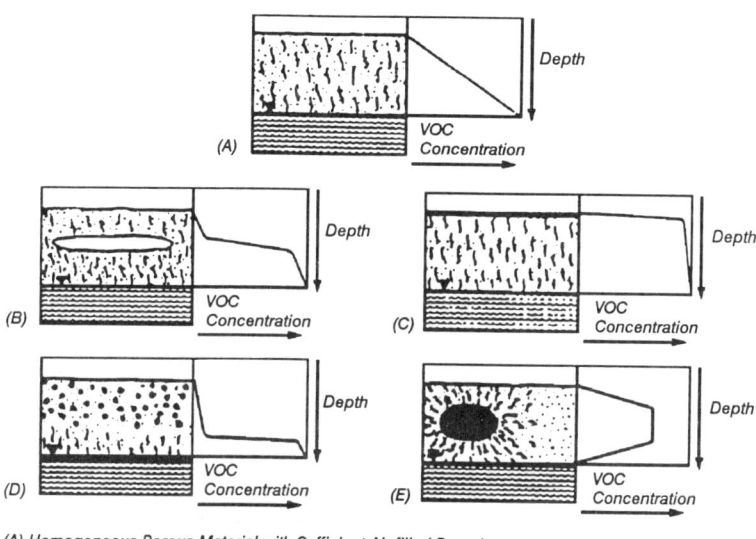

(A) Homogeneous Porous Material with Sufficient Air-filled Porosity
(B) Impermeable Subsurface Layer (e.g., Clay or Perched Water)
(C) Impermeable Surface Layer (e.g., Pavement)
(D) Zone of High Microbiological Activity (Circles and Wavy Lines Indicate Different Compounds)
(E) VOC Source in the Vadose Zone

Figure 9-5 Soil-gas concentrations under a variety of conditions (Boulding and Barcelona, 1991c, after Marrin and Kerfoot, 1988).

information on six indirect methods for monitoring soil salinity. The *four-probe electrical* method is a direct application of the electrical resistivity surface geophysical method (Section 6.3.2), with electrode configurations that measure near-surface resistivity. The *electromagnetic induction sensor* is an instrument that is specifically designed to measure conductivity in the near surface. The other indirect methods involve placement of probes or sensors in the subsurface. The main advantage of indirect methods is that data can be collected quickly. The main disadvantages are: (1) instruments must be calibrated for each soil type by collection of samples where salinity is measured directly to obtain quantitative measurement of soil salinity; and (2) actual chemical constituents that are contributing to soil salinity cannot be determined. The four-probe electrical and porous matrix soil salinity sensors are the most commonly used indirect methods.

Direct Soil-Solute Sampling Methods. Three major types of soil water can be identified in the context of sampling soil water: (1) *macropore* or *gravitational* water, which flows through the soil relatively rapidly in response to gravity (excess of 0.1 to 0.2 bars suction); (2) *soil-pore* or *capillary* water, which is held in the soil at negative pressure potentials from around 0.1 to 31 bars of suction; and (3) *hygroscopic* water

Table 9-7 Characteristics of Contaminants in Relation to Soil Gas Surveying

Group/Contaminants	Applicability of Soil-Gas Survey Techniques
Group A: Halogenated Methanes, Ethanes, and Ethenes	
Chloroform, vinyl chloride, carbon tetrachloride, trichlorofluoromethane, TCA, EDB, TCE	Detectable in soil gas over a wide range of environmental conditions. Dense nonaqueous phase liquid (DNAPL) will sink in aquifer if present as pure liquid.
Group B: Halogenated Propanes, Propenes and Benzenes	
Chlorobenzene, trichlorobenzene, 1,2-dichloropropane	Limited value; detectable by soil-gas techniques only where probes can sample near contaminated soil or ground water. DNAPL.
Group C: Halogenated Polycyclic Aromatics	
Aldrin, DDT, chlordane, heptachlor, PCBs	Do not partition into the gas phase adequately to be detected in soil gas under normal circumstances. DNAPL.
Group D: C_1 - C_8 Petroleum Hydrocarbons	
Benzene, toluene, xylene isomers, methane, ethane, cyclohexane, gasoline, JP-4	Most predictably detected in shallow aquifers or Leaking underground storage tanks where probes can be driven near the source of contamination. Light non-aqueous phase liquids (LNAPLs) float as thin film on the water table. Can act as a solvent for DNAPLs, keeping them nearer the ground surface.
Group E: C_9 - C_{12} Petroleum Hydrocarbons	
Trimethylbenzene, naphthalene, decane, and jet A fuels	Limited value; detectable by soil-gas techniques only where probes can sample near contaminated soil or diesel ground water. DNAPLs.
Group F: Polycylic Aromatic Hydrocarbons	
Anthracene, benzopyrene, fluoranthene, chrysene, motor oils, coal tars	Do not partition adequately into the gas phase to be detected in soil gas under normal circumstances. DNAPLs.
Group G: Low Molecular Weight Oxygenated Compounds	
Acetone, ethanol, formaldehyde, methylethylketone	LNAPLs, but dissolve readily in ground water. May be detected in soil gas if they result from a leak or spill in relatively dry soil.

Source: Adapted from Marrin (1987).

that is held at tensions greater than 31 bars suction. Soil-pore water moves through the vadose zone, but at much slower rates than gravitational water (Section 2.4.1), whereas hygroscopic water, if it moves, does so in primarily in the vapor form. The term *soil-solute* or *solution* sampling has been used loosely in the literature to describe most sampling methods, whereas the term *soil-pore liquid* is typically used in a more restricted sense (and is so used here) to apply to sampling of capillary water. The chemistry of soil-solute sampling methods can differ significantly, depending the method used. Concentrations of inorganic species generally increase as the matric potential increases. In general, ceramic soil suction samplers (which use suctions up to around 0.8 bars) will collect samples that are most representative of the soil solution for the purpose of evaluating contaminant transport.

There are a large number of specific methods by which soil water can be sampled. *Suction* samplers draw water from the soil by applying a vacuum. A variety of *free-drainage* samples collect water percolating through the soil by gravity flow. *Capillary wick* samplers are a relatively new development, which appear to have good potential for collecting more representative samples of soil solutions than either porous cup or free-drainage samplers in the near surface. Other methods include: (1) use of absorbent materials with retrieval and extraction of water in the laboratory, (2) collection of soil solids with extraction of soil water in the laboratory by a variety of methods, and (3) preparation of a soil saturation extract from a solids sample. Table A-11 summarizes some information on six types of suction samplers, seven methods of collecting samples by free drainage, and four miscellaneous methods.

The main advantages of suction samplers is that they are relatively easy to install, and there are essentially no limitations to the depth of sampling when a vacuum-pressure apparatus is used. The main disadvantage of suction samplers is that they might not collect representative samples. Sampling for organic chemicals, microorganisms, volatile chemicals, and metals is especially problematic due to potential sorption/interferences by the porous cup. Vacuum-type and vacuum-pressure type porous cup samplers are by far the most commonly used types of suction samplers. The main advantage of free-drainage samplers is that a relatively large volume of water, which is representative of water that is actually percolating to deeper zones, is obtained. The main disadvantages are that installation procedures are time consuming and complex and limited to relatively shallow depths. *Trench lysimeters* with pan collectors are the most commonly used free-drainage samplers. Figure 9-1 illustrates a variety of configurations for soil-solute sampling.

9.4.4 Microbiological Sampling and Other Sensitive Constituents

Special sampling procedures are required for sampling contaminants that can change in concentration (degassing of volatile compounds) or chemical composition (redox-sensitive chemical species, such as ferrous and ferric iron) when exposed to the air. Similar care is required when sampling for microbiota in the subsurface, especially where oxygen content is low (typically in the zone of saturation). Even where exposure to the air is not a concern for microbiological sampling (typically in the vadose zone), special care is required to make sure that the sample has not been cross-contaminated with soil microorganisms from higher soil horizons. The basic procedure involves collection of subsamples of power-driven sample cores (Section 9.4.1) using smaller-diameter corers.

Liners that fit inside a bucket auger or tube sampler are the best way to collect soil samples suspected of containing volatile organics. Alternatively, samples for volatiles should be quickly transferred to the sample container and sealed with no air headspace in the container. Where cores contained anaerobic bacteria and chemical species of concern that are in a reduced state, samples need to be extracted in an oxygen-free environment. Figure 9-6 shows a plexiglass field glove box for collecting such samples. Sample containers are sterilized and filled with an inert gas such as nitrogen. In the field, the sealed containers are placed in the field glove box before the box is filled with nitrogen. The core sample is pushed into the box through an iris port, and a core paring tool is used to collect subsamples in the oxygen-free environment for placement in sample containers.

9.5 Sampling Ground Water

A wide variety of devices and installations are available for the sampling of ground water. Sampling devices can be broadly classified as: (1) *portable well samplers*, which are used in permanently installed and screened monitoring wells, and (2) *in situ samplers*, which do not require cased monitoring wells. In situ sampling devices can be used for taking one-time samples or permanently installed in the subsurface as discussed in Section 9.1.2.

9.5.1 Portable Well Samplers

Figure 9-7 shows a generalized flow diagram of the following major steps in collecting a sample from a monitoring well using a portable sampler: (1) well inspection, (2) purging (discussed in more detail in Section 9.5.3), (3) sample collection, (4) preservation and QA/QC samples, and (5) storage and transport.

Table A-12 provides the following information on 20 portable sampling devices, which can be used to collect ground-water samples from wells: (1) maximum depth, (2) minimum well diameter, (3) typical ranges of sampling rates, and (4) sections and tables in the book where additional information can be found. Portable well samplers are divided into three main groups: (1) *positive displacement* samplers, (2) *other sampling pumps*, and (3) *grab/depth specific* samplers.

Positive displacement pumps are placed below the static water level of the well and pump the sample to the surface. These pumps include *bladder pumps* (also called *gas-operated squeeze pumps*), *gear-drive pumps*, *helical rotor pumps*, *gas-drive/displacement pumps* (where gas displaces water in the subsurface to force it to surface without mixing with the sample), and *gas-drive piston* and *mechanical piston pumps*.

Other types of portable sampling pumps include *suction-lift* pumps (*peristaltic* pumps being the most common, but surface centrifugal and any other type of surface pump that operates using suction or a vacuum fall in this category), *submersible*

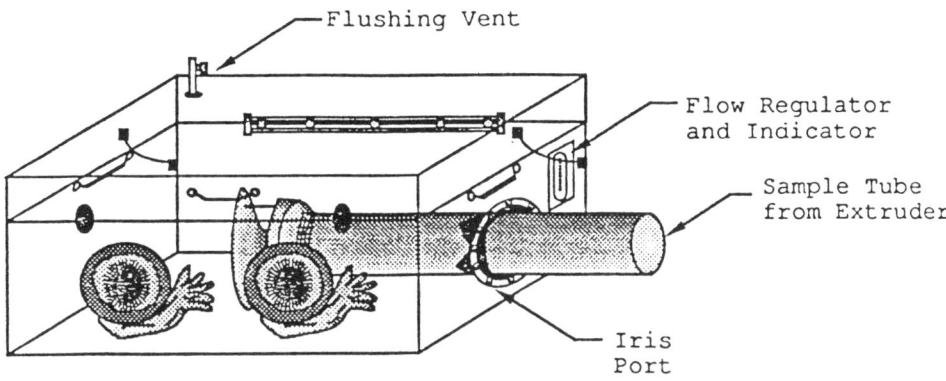

Figure 9-6 Field sampling glove box (Boulding, 1993b, after Leach et al., 1988).

centrifugal pumps (note that *surface* centrifugal pumps are classified as suction-lift pumps), *inertial-lift pumps* (simple mechanisms using foot valves and inertia to bring water to the surface), *gas-lift pumps* (where air or gas mixes with the water to bring ground water to the surface), and *jet* or *venturi* pumps. *Packer* pumps isolate a portion of the well using inflatable packers with the sample being brought to the surface using as suction lift, submersible or other sampling device.

The term *gas lift* (also sometimes called *air lift*) has been used inconsistency in the published literature on ground-water sampling devices. In this book the term refers to methods where gas mixes with water to provide the buoyant force to bring it to the surface, and *gas drive* refers to methods in which gas is used to push water up a tube without the gas becoming mixed with the water that is brought to the surface. For example, Morrison (1983/T9-10) and Scalf et al. (1981/T9-10) have used the term "lift" for samplers that are classified as gas-drive samplers in this guide.

Grab samplers include: *bailers* (open and point-source), *mechanical* or *thief depth-specific* samplers (such as, Kemmerer, Coliwasa, stratified sample thief), and *pneumatic depth specific* samplers, which use vacuum or pressure to activate the sampling mechanism (such as, syringe and Westbay samplers).

Sampling devices vary greatly in their suitability for sampling different chemical constituents. Table 9-8 summarizes the suitability of the 12 most commonly used sampling devices for 14 ground-water parameters. *Bladder* and *helical rotor* pumps are rated as suitable for the largest number of parameters, followed by *point-source bailers*. The *inertial pump*, which is not included in Table 9-8, is a quite new device, which probably would rate favorably for sampling many of the parameters on the table.

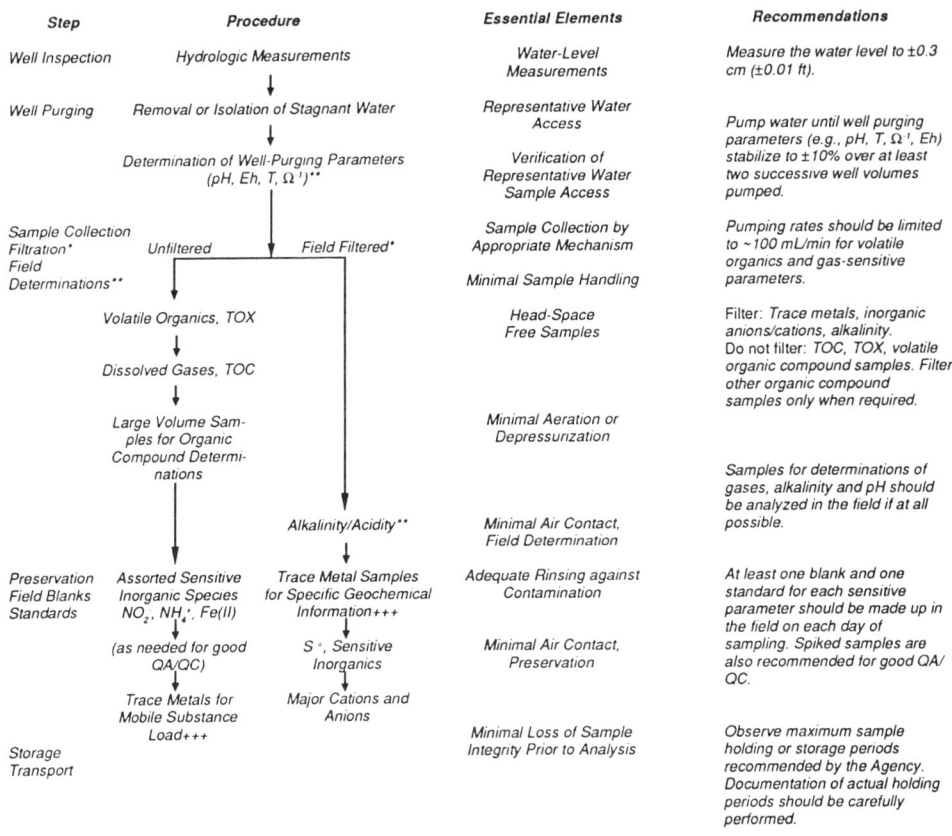

Figure 9-7 General flow diagram of ground-water sampling steps (Boulding and Barcelona, 1991).

9.5.2 Portable/Permanent In Situ Samplers and Sensors

A relatively new development in ground-water sampling technology has been the design of in situ sampling probes, which can be used either for rapid collection of samples without the installation of permanent wells or left in place for continued monitoring. Table A-12 provides summary information on a variety of samplers and sensors of this type. The *Hydropunch*® and *BAT*® systems both operate in conjunction with conventional cone penetrometer rigs. This category also includes a variety of driven probes, which can be retrieved after sampling or left in place as permanent sampling points. Figure 9-2 illustrates several examples of permanent installations using the BAT system. These devices often are best used during the preliminary site characterization stage or where only a shallow water table is to be sampled. Portable in situ samplers can be valuable in deciding the best location of permanent monitoring wells. *Chemical sensors*, such as Eh and pH probes, and ion-selective electrodes usually are used in boreholes. Use of *fiber optic*, *electrochemical*, *piezoelectric*, and other chemical sensors (see also Table A-13) for subsurface chemical characterization is the subject of considerable research and might become more widespread for routine investigations with further refinements in instrumentation. Strictly speaking, the term *in situ* (from Latin, meaning in its original position) only should be applied to sampling devices or chemical sensors that measure ground-water quality in place without bringing the sample to the surface. In common usage, however, the term is applied to methods that allow collection of samples without the installation of a permanent monitoring well, or permanent installations that do not require use of portable sampling equipment.

Table 9-8 Suitability of Major Ground-Water Sampling Devices for Different Ground-Water Parameters

Sampling Device	Inorganic							Organic			Other		
	EC	pH	Redox	Major Ions	Tr. Met.	NO3 F	Diss. Gases	Non Vol.	Vol.	TOC	TOX	Ra-dium	Alpha Beta	Coli-form
Portable Grab/Depth Specific Samplers														
Open Bailer	x			x	x	x		x				x		x
Point-Source Bailer		x	x	x	x	x	x		x	x	x	x	x	x
Syringe Sampler	x	x	x	x	x	x		x				x	x	x
Portable Positive Displacement (Submersible) Pumps														
Bladder	x	x	x	x	x	x	x	x	x	x	x	x	x	x
Centrifugal	x			x		x						x	x	
Helical Rotor	x	x	x	x	x	x	x	x	x	x	x	x	x	
Gas-Drive Piston	x			x	x	x		x				x	x	
Gear-Drive									x					
Other Portable Samplers														
Peristaltic	x			x		x		x				x		x
Gas-Drive/Displacement Gas-Lift			x			x		x		x				x
Portable In Situ Samplers														
Pneumatic	x	x	x	x	x	x		x				x	x	x

TOC = Total Organic Carbon.
TOX = Total Organic Halogen.

Source: Boulding and Barcelona (1991c), after Pohlmann and Hess (1988).

9.5.3 Purging

Purging to remove stagnant water from a well before sample collection is standard practice. The monitoring well is pumped (generally at a rate from 1 to 5 gallons per minute) until a certain number of well volumes have been removed and until water quality indicators, such as pH, conductance, and/ or temperature, have stabilized, indicating that fresh formation water fills the well. Sampling takes place after purging in completed. Recent research (Kearl et al., 1992)[2] has suggested that purging is not desirable because it can mobilize colloidal particles upon which contaminants are sorbed. The alternative to purging is to use a dedicated sampling device set at the level of the well screen capable of low pumping rates (around 100 mL/minute), which will not increase colloid density in the ground-water sample compared to natural colloidal flow through the well screen.

Recommended rules of thumb, such as purging three to five volumes (Fenn et al., 1977/T9-10), should be treated only as a starting point. Accurate estimation of purge volume requires knowing (1) well yield, determined from a slug or pumping test, and (2) the stagnant volumes of both the well casing, and the sand pack. In slowly recovering wells, extra care is required when purging to ensure that water levels do not drop below the level of the well screen because aeration might allow loss of volatile or redox-sensitive contaminants. After stagnant water has been removed or isolated, chemical indicators (pH, conductance, and temperature) should continue to be monitored until they reach a consistent end point (no upward or downward trend). Flow-through cells with multiple probe ports (Garske and Schock, 1986)[3] and multiple parameter probes (Appendix B) are the most effective way to continuously monitor indicator parameters during purging.

Another important consideration in purging is that the pumping rate should not exceed levels that will cause turbulent flow. Turbulent flow in the well can cause pressure changes, which result in loss of carbon dioxide and other volatile gases, subsequently changing pH and dissolved solids content (Meredith and Brice, 1992).[4] The maximum discharge rate during pumping that avoids turbulent flow is a function of hydraulic conductivity, the length of the well screen, width of the screen openings, and the total open area of the screen. Section B.2 in Boulding (1993/T9-10) includes a graph for determining the optimum screen entrance velocity related to the hydraulic conductivity of an aquifer and a table with guidelines for maximum purging rate based on screen type, diameter, slot size, open area, and entrance velocity.

---

[2] Kearl, P.M., N.E. Korte, and T.A. Cronk. 1992. Suggested Modifications to Ground Water Sampling Procedures Based on Observations from the Colloidal Boroscope. Ground Water Monitoring Review 12(2):155-161.

[3] Garske, E.E. and M.R. Schock. 1986. An Inexpensive Flow-Through Cell and Measurement System for Monitoring Selected Chemical Parameters in Ground Water. Ground Water Monitoring Review 6(3):78-84.

[4] Meredith, D.V. and D.A. Brice. 1992. Limitations on the Collection of Representative Samples from Small Diameter Monitoring Wells. Ground Water Management 11:429-439 (6th NOAC).

As noted above, current scientific thinking concerning the need for or the optimum procedures for well purging prior to sampling are in a state of flux. When ground-water sampling is for regulatory purposes, the U.S. EPA or appropriate state regulatory agency should be consulted to determine appropriate procedures.

9.6 Field Screening and Analytical Methods

The term "field screening" has gained widespread use in recent years to describe a wide variety of methods for chemical characterization of contaminated sites. In this book, a distinction is made between field screening and field analytical methods. *Field screening* methods provide an indication of the presence or absence of a particular chemical or chemical class of concern or provide an indication of whether the chemical or chemical class of concern is above or below a predetermined threshold. Screening methods provide relative concentrations for chemical classes, but rarely provide chemical-specific information. This definition is more restrictive than those usually found in the literature. *Field analytical* methods include all chemical analysis methods capable of providing chemical-specific quantitative data in the field or nonlaboratory setting. Field analytical techniques generally are more rapid and less expensive than similar chemical analyses performed in laboratories with fixed facilities. Field screening and analytical techniques can be classified as *portable* (require no external power source, are compact, and are rugged enough to be carried by hand into the field), *fieldable* (require limited external power, are compact, and are rugged enough to be transported in a small van, pick-up, or four-wheel drive), or *mobile* (are small enough to carry in a mobile laboratory, which is feasible for most analytical instruments although power considerations can be a limitation). The standard by which the sensitivity, precision, and accuracy of field-screening techniques are measured are those obtained in fixed-base laboratories in EPA's *contract laboratory program* (CLP). An intermediate option for analysis of samples is the use of a *dedicated* laboratory using CLP procedures but involving more rapid turnaround time (as short as overnight) for sample results.

9.6.1 Field Analysis vs. CLP Analytical Laboratory

Key advantages of field analytical techniques include the following: (1) results can be obtained within hours, compared to the 20 to 40 days required for CLP laboratories, which allows for more rapid definition of the scope of contamination and allows for optimal selection of permanent monitoring wells/locations, (2) lower cost per sample (commonly one-tenth CLP cost) allows for more detailed characterization of contaminant distribution and/or reduced overall costs, and (3) the techniques are best suited for preliminary site characterization, emergency remedial actions, and monitoring of remediation activities. Some general disadvantages of field analytical techniques include the following: (1) application of analytical QA/QC procedures is more difficult in the field, (2) generally, less sophisticated instrumentation and disadvantage #1 results in generally higher detection limits and lower precision and accuracy compared to CLP laboratories, and (3) disadvantages #1 and #2 mean that data are more liable to challenge by litigation.

Cost differences between field analysis and laboratory analysis are strongly dependent on the number of samples from a site that must be analyzed, with the cost advantage tending to shift to field analysis as the number of samples increases. For example, if less than 30 to 50 cumulative samples are required, laboratory gas chromatograph analyses are likely to be less expensive than using portable or mobile GCs. Similarly, around 50 to 80 cumulative samples for field X-ray fluorescence analysis of metals are required to save money over conventional laboratory XRF analyses.

9.6.2 Overview of Specific Techniques

Developments in miniaturization and computer processing of analytical signals and development of innovative analytical techniques mean that almost any instrumental or analytical technique has the potential for being used for field screening. Any attempt to publish a comprehensive compilation of techniques that have been proposed or tested is doomed to be out-of-date before it reaches print. Table A-13 provides summary information on over 80 techniques. This table provides a reasonably comprehensive overview of the state-of-the-art as of early 1993. Techniques are grouped into the following major categories: (1) routine chemical field measurement techniques; (2) major sample extraction procedures; (3) analytical techniques that detect gases or require creation of a gaseous phase during the analytical process if the gaseous phase is not already present; (4) luminescence, spectrophotometric, and other spectroscopic techniques; (5) wet chemistry techniques; and (6) radiological and other miscellaneous techniques.

Photo- and *flame-ionization* detectors are used routinely in contaminant investigations to screen for the presence of volatile compounds. Portable instrumentation for *gas chromatography* has improved to the point where it can be considered a well-established field analytical method for site investigations involving organic chemicals. Portable *X-ray fluorescence* instruments are now commercially available for field screening of heavy metal contamination in soil and waste materials. *Colorimetric wet chemistry* field test kits for analysis of inorganic constituents are being increasingly accepted for use in EPA drinking water and NPDES programs. Colorimetric *enzyme immunoassay* (EIA) test kits are a rapidly developing area of field chemical analysis. EIA kits have been developed for pentachlorophenol, explosives, and pesticides.

9.6.3 Types of Analytical Instrumentation

For the nonchemist, terminology used to describe analytical techniques can be bewildering. A further source of potential confusion is that techniques can be used for different purposes in numerous combinations and configurations. For example, a flame ionization detector (FID) can be used by itself as a total vapor detector or it can be used to detect specific compounds after they have been separated by a gas chromatograph (GC/FID). A gas chromatograph, on the other hand, can be used alone with an FID or other type of detector, or in combination with a mass spectrometer (GC/MS). An understanding of the basic principles of operation of major individual techniques makes it possible to have some idea of how an unfamiliar combination of techniques functions.

A further source of possible confusion is that the different terms can be applied to the same technique. For example, the terms fluorometry, fluorimetry, and spectrofluorometry can be used interchangeably. Furthermore, some terms can be applied to the same technique, but are not necessarily interchangeable. For example, the term luminescence can be applied to any technique involving fluorescence, but the term fluorescence is not applicable to all luminescence techniques (which include phosphorescence). The following discussion might be helpful in developing an understanding of some of the basic principles involved in chemical analysis and in sorting out the relationship between similar techniques. It might also be helpful to think of techniques in terms of the major types of analytical signals as summarized in Table 9-9.

Table 9-9 Major Analytical Signals and Methods

Signal	Analytical Methods Based on Measurement of Signal
Emission of radiation	Emission spectroscopy (X-ray, UV, visible, electron Auger); fluorescence and phosphorescence spectroscopy (X-ray, UV, visible); radiochemistry
Absorption of radiation	**Colorimetry** (visible), **UV-visible/X-ray/IR spectrophotometry**; photoacoustic spectroscopy; **nuclear magnetic resonance** and electron spin resonance spectroscopy
Scattering of radiation	Turbidimetry; nephelometry; **Raman spectroscopy**
Refraction of radiation	Refractometry; interferometry; X-ray diffraction
Rotation of radiation	Polarimetry; optical rotatory dispersion; circular dichroism
Electrical potential	**Potentiometry**; chronopotentiometry
Electrical current	Polarography; amperometry; **coulometry**; voltammetry
Mass-to-charge ratio	**Mass spectrometry**
Rate of reaction	Kinetic methods
Thermal properties	Thermal conductivity and enthalpy methods
Mass	**Gravimetric analysis**
Volume	**Volumetric analysis**

Boldface = Most commonly used in field screening and analytical applications.
Source: Boulding (1993b), after Skoog (1985).

Chromatography refers to processes in which individual components of a mixture migrate through a stationary medium at different rates. In analytical chemistry, chromatography refers to a diverse group of separation methods such as *gas chromatography* and *liquid chromatography* used to separate, isolate, and identify components of mixtures that might otherwise be resolved with great difficulty.

A *spectrum* is the distribution of the phases of a radiated wave cycle or of the intensity of radiation when some property (frequency, mass, or energy) is allowed to vary. *Spectroscopy* encompasses a wide range of techniques involving optical instruments used to form and analyze spectra. *Spectrometry* is a spectroscopic technique in which the instrument measures (1) the deviation of the refracted rays, and (2) wave lengths and angles between two faces of a prism. *Spectrophotometry* involves making comparisons of color intensity between corresponding parts of different spectra or between parts of the same spectrum. *Photometry* involves the measurement of the intensity of light or the relative intensity of different lights. *Luminescence* involves the emission of light at temperatures below that of incandescent bodies and includes *fluorescence* (emission of radiation as a result of absorption of other radiation) and *phosphorescence* (light given off from slow oxidation of phosphorus).

9.7 Guide to Major References

Table 9-10 provides an index of major references dealing with the following general topics: (1) site investigations, (2) ground-water monitoring, (3) microbiological sampling, (4) vadose zone monitoring, (5) soil sampling, and (6) waste sampling. The table also indexes references according to agency or organization. Sara (1994) is recommended as a recent and comprehensive guide to facility assessments for solid and hazardous waste. U.S. EPA (1991a) is another good general reference on site characterization with a focus on soil and ground-water remediation.

U.S. EPA (1986c) provides guidance on unsaturated zone monitoring. ASTM D4696 (Table A-14) provides more detailed guidance on selection and installation of soil-pore liquid sampling devices and ASTM D5314 provides guidance on soil-gas monitoring in the vadose zone. Aller et al. (1991) as updated by Nielsen and Schalla (1992) are good references on design and installation of ground-water monitoring wells. U.S. EPA (1993) provides the most recent detailed EPA guidance on ground-water monitoring methods. Boulding (1991, 1994) are good recent references on description and sampling of contaminated soils.

Table 9-11 provides an index of major references on field and laboratory methods for chemical and physical analysis of water, soil, sediment, chemical contaminants, other solid waste, and microbiology. U.S. EPA (1993) provides in electronic format a compendium of chemical field screening and analytical methods. SW-846 (U.S. EPA, 1986b) is the standard reference for solid waste test methods. Although the latest complete revision was 1986, it is continually being updated and can be obtained on a subscription basis from the U.S. Government Printing Office.

Sampling and Monitoring

Table 9-10 Index to Major Reference Sources on Subsurface Sampling and Monitoring Methods

Topic	References
Site Investigations	
General Methods	Hydrologic Characterization: Alley (1993), Boulding (1993), Brakensiek et al. (1979), Brown et al. (1983), Bureau of Reclamation (1981), Dames & Moore (1974), Driscoll (1986), Kolm (1993), Nielsen and Johnson (1990), Nielsen and Sara (1992), Penn Inc. (1994), Rehm et al. (1985), Sara (1994), Thompson et al. (1989), UNESCO (1983), U.S. EPA (1991a,b), USGS (1977+), Van Haveren (1986), Waste Management of North America (1991), Zimmie and Riggs (1981); Geotechnical Investigations: Dowding (1978), Hanna (1985), Hathaway (1988), Lambe (1951), Corps of Engineers (1984), USNFEC (1982); Ecological Assessment: See Table 11-10
Specific Settings	EPA SOPs: U.S. EPA (1991d, 1991e); Hazardous Waste Sites: Cameron (1991), Cochran and Hodge (1985), Ford and Turina (1985), Lesage and Jackson (1992), Oudjik and Mujica (1989), Perket (1986), Sara (1994), Sisk (1981), U.S. DOE (various dates), U.S. EPA (1987, 1989a, 1991c, 1992b), WPCF/WEF (1988); RCRA Facilities: Sara (1994), U.S. EPA (1986d, 1989b,c); Real Estate Environmental Assessments: AGWSE (1992), Colangelo (1991), Environmental Resource Center (1993), Hess (1993), Jain et al. (1993), Vincoli (1993); USTs/NAPL Sites: Cheremisinoff (1992), Cohen and Mercer (1992), Cole (1994), Texas Water Commission (1993); Low Level Radioactive Wastes: EG&G (1990); Remedial Operations: Byrnes (1994), Cohen et al. (1994), Ross and Keeley (1992), U.S. EPA (1988a, 1991a); Surface Impoundments: Silka and Swearingen (1978); Surface Mining: Barrett et al. (1980)
Ground-Water Monitoring	
General Procedures	Collins and Johnson (1988), Crouch et al. (1976), Devinny et al. (1990), EG&G (1990), Everett (1980), Everett et al. (1976), Fried (1975), Gillham et al. (1983), Keith (1992), Loftis and Ward (1979), Mooij and Rovers (1976), Morrison (1983), Nielsen (1991), Nielsen and Johnson (1990), Nielsen and Sara (1992), Ontario Ministry of the Environment (1989), Ross and Keeley (1992), Todd et al. (1976), U.S. DOE (Various dates), U.S. EPA (1986b, 1990a, 1990b, 1991a, 1991b, 1993), van Duijvenbooden and van Waegeningh (1987), Ward et al. (1990), WEF (1993/T4-4)

Table 9-10 (cont.)

Topic	References
Ground-Water Monitoring (cont.)	
Drilling/Monitoring Wells	Aller et al. (1991), Australian Drilling Association (1992), Barcelona et al. (1983), Campbell and Lehr (1973), Driscoll (1986), Harlan et al. (1989), Howsam (1990), Korte and Kearl (1985), Lehr et al. (1988), Nielsen and Schalla (1991), Roscoe Moss Company (1990), Ruda and Bosscher (1990), Shuter and Teasdale (1989)
Sampling Procedures	API (1987), Barcelona et al. (1983, 1985), Berg (1982), Classen (1982), Holden (1984), Keith (1988), Korte and Kearl (1985), Nash and Leslie (1991), Rainwater and Thatcher (1960), Scalf et al. (1981), Summers and Gherini (1987), Unwin (1982), Wood (1976)
Costs	Crouch et al. (1976), Everett et al. (1976), Loftis and Ward (1979)
Specific Settings	Tinlin (1976); Solid Waste Disposal: Fenn et al. (1977), U.S. EPA (1981a,b, 1986f); RCRA Facilities: U.S. EPA (1983a,b, 1985, 1986a,d,e,f, 1989c); Enhanced Oil Recovery: Beck et al. (1981); Surface Mining: Everett (1979, 1983, 1985), Everett and Hoylman (1980a,b), Williams and Schuman (1987); Oil Shale: Everett (1985), Slawson (1979, 1980a,b); Electric Utilities: GeoTrans (1989), Redwine et al. (1985); Wastewater and Sludge Application: Ho et al. (1978); Waste Spills: Pilie et al. (1975), Yang and Bye (1979); Geothermal: Weiss et al. (1979); WHPAs: Moore (1993/T11-11)
State/Local Guidance Documents*	Connecticut Environmental Protection Agency (1983), Lindorff et al. (1987), NJDEP (1992), Santa Clara County Water District (1985), Stephens (1986)
Microbiological Sampling	Bitton and Gerba (1984), Board and Lovelock (1973), Bordner et al. (1978), Britton and Greeson (1989), Dunlap et al. (1977), Edwards (1993), USGS (1977+); see also Table 9-11
Vadose Zone Monitoring	
General	Everett et al. (1983), Nielsen and Johnson (1990), Nielsen and Sara (1992), Rehm et al. (1985), Rijtema and Wassink (1969), U.S. EPA (1986c), Wilson (1980); Review Papers: Everett et al. (1982, 1984), Wilson (1981, 1982, 1983)
Soil Solute Sampling	Devinny et al. (1990), Morrison (1983), Nash and Leslie (1991), Nielsen (1991), USGS (1977+)

Table 9-10 (cont.)

Topic	References
Vadose Zone Monitoring (cont.)	
Soil Gas	Devitt et al. (1987), Ford et al. (1984), Kerfoot and Barrows (1987), Marrin and Kerfoot (1988), U.S. EPA (1988b)
Soil Sampling	
Field Characterization	Blume et al. (1991), Boulding (1991, 1993, 1994), Brakensiek et al. (1979), Bureau of Reclamation (1974, 1990), Cameron (1991), Hodgson (1978), SCS (1971)
General Sampling	Acker (1974), Barth et al. (1989), Cameron (1966), Corps of Engineers (1972), Keith (1992), Hodgson (1978), Hvorslev (1948, 1949), Mason (1992), McKeague (1978), Mooij and Rovers (1976), SCS (1984), U.S. DOE (Various dates); Sediments: See Table 9-11
Sampling for Soil Contaminants	API (1987, 1992), Boulding (1991, 1994), EG&G (1990), Ford et al. (1984), Goodwin et al. (1982), Keith (1988), Lewis et al. (1991), Neckers and Walker (1952-active sulfides), Scalf et al. (1981), Schweitzer and Santolucito (1984), U.S. EPA (1986b, 1988b, 1989c, 1991a), van Duijvenbooden and van Waegeningh (1987)
Waste Sampling	
Sampling	deVera (1980), Ford et al. (1984), Keith (1988, 1992), Rupp and Jones (1993), Simmons (1991), U.S. DOE (Various dates), U.S. EPA (1986b, 1992a), Wolbach et al. (1984)
Agency/Organization Index	
U.S. EPA	Soils and Ground Water: Aller et al. (1991), Cochran and Hodge (1985), Dunlap et al. (1977), Everett et al. (1976), Fenn et al. (1977), Ford and Turina (1985), Ford et al. (1984), Ross and Keeley (1992), Scalf et al. (1981), Silka and Swearingen (1978), Sisk (1981), U.S. EPA (1986d, 1987, 1989a, 1989b, 1989c, 1990b, 1991a, 1991c, 1991d, 1991e); Vadose Zone: Devitt et al. (1987), Everett et al. (1983), Kerfoot and Barrows (1987), U.S. EPA (1986c), Wilson (1980); General Ground Water: Barcelona et al. (1985), Berg (1982), Crouch et al. (1976), Loftis and Ward (1979), Tinlin (1976), Todd et al. (1976), U.S. EPA (1990a, 1991a), Yang and Bye (1979); Ground-Water Guidance Documents: U.S. EPA (1981a, 1981b, 1983a, 1983b, 1985, 1986a, 1986e, 1986f, 1988a, 1991e, 1992b, 1993); NAPLs: Cohen and Mercer (1993), U.S. EPA (1992b)

Table 9-10 (cont.)

Topic	References
Agency/Organization Index (cont.)	
U.S. EPA (cont.)	Soil and Solid/Liquid Waste: Barth et al. (1989), Boulding (1991), Cameron (1991), deVera (1980), Hatayama et al. (1980), Mason (1992), Pilie et al. (1975), U.S. EPA (1986b, 1991e, 1992a), Yang and Bye (1979); Energy Development Ground-Water Monitoring: Beck et al. (1981), Everett (1979, 1983), Everett and Hoylman (1980a,b); Slawson (1979, 1980a,b), Weiss et al. (1979)
Other Federal	Bureau of Reclamation: Bureau of Reclamation (1974, 1981, 1990); Department of Energy: EG&G (1990), U.S. DOE (Various dates); Fish and Wildlife Service: Brown et al. (1991); Forest Service: Barrett et al. (1980); NASA: Cameron et al.; (1966); Navy: UNSFEC (1982); USATHAMA/Corps of Engineers: Corps of Engineers (1972, 1984), Goodwin et al. (1982), Hvorslev (1949), Plumb (1981); USDA/SCS: Brakensiek et al. (1979), SCS (1971, 1984); U.S. Geological Survey: Classen (1982), Edwards and Glysson (1988), Guy (1969), USGS (1977+), Wood (1976)
Other Government	Canada: McKeague (1978), Mooij and Rovers (1976); States:* Barcelona et al. (1983), Connecticut Environmental Protection Agency (1983), Lindorff et al. (1987), NJDEP (1992), Stephens (1986)
American Society for Testing and Materials (ASTM)	ASTM (Annual, 1992, 1994); Ground-Water and Vadose Zone STPs: Collins and Johnson (1988), Nielsen and Johnson (1990), Nielsen and Sara (1992), Zimmie and Riggs (1980); Hazardous Waste Solid Testing Conference Series: (Papers in this series tend to focus on laboratory methods, but also include papers on field-oriented techniques): 1st (Conway and Mallow, 1981); 2nd (Conway and Gulledge, 1982); 3rd (Jackson et al., 1984); 4th (Petros et al., 1985); 5th (Perket, 1986); 6th (Lorenzen et al., 1986)
Other Organizations	American Chemical Society (ACS): Keith (1988, 1992), Nash and Leslie (1991), Schweitzer and Santolucito (1984); American Petroleum Institute (API): API (1987, 1992), Gillham et al. (1983); Consulting Firms: Dames and Moore (1974), Everett (1980), GeoTrans (1989), Waste Management of North America (1991); Electric Power Research Institute (EPRI): EPRI (1985), Redwine et al. (1985), Rehm et al. (1985), Summers and Gherini (1987), Thompson et al. (1989); UNESCO: Brown et al. (1983), Rijtema and Wassink (1969), UNESCO (1983)

*The appropriate state regulatory agency should be contacted for the most current version of any guidance documents.

Table 9-10 References (Appendix F contains references for figure and table sources)

Acker, W.L. 1974. Basic Procedures for Soils Sampling and Core Drilling. Acker Drill Co., Scranton, PA.

Aller, L., et al. 1991. Handbook of Suggested Practices for the Design and Installation of Ground-Water Monitoring Wells. EPA/600/4-89/034, 221 pp. Available from CERI.* (Also published in 1989 by National Water Well Association, Dublin, OH, in its NWWA/EPA series, 398 pp.) (Nielsen and Schalla [1991] contain a more updated version of the material in this handbook that is related to design and installation of ground-water monitoring wells.)

Alley, W.M. (ed.). 1993. Regional Ground-Water Quality. Van Nostrand Reinhold, New York, NY, 634 pp. [Addresses hydrogeologic, geochemical and statistical principles]

American Petroleum Institute (API). 1987. Manual of Sampling and Analytical Methods for Petroleum Hydrocarbons in Groundwater and Soil. API Publication No. 4449, API, Washington, DC, 230 pp.

American Petroleum Institute (API). 1992. Sampling and Analysis of Gasoline Range Organics in Soil. API Publication 4516, API, Washington, DC, 132 pp.

American Society for Testing and Materials (ASTM). Annual. Books of ASTM Standards: Construction (Soil and Rock), Volume 04.08; Water and Environmental Technology, Volumes 11.01 and 11.02 (Water) and 11.04 (Pesticides, Resource Recovery, Hazardous Substances and Oil Spill Response, Waste Management, Biological Effects). ASTM, Philadelphia, PA. [Volume 4.08 contains most standards related to soil, ground-water, and vadose zone investigations]

ASTM Institute for Standards Research (ASTM-ISR). 1992. Index of ASTM Standards and Special Technical Publications Applicable to Environmental Monitoring and Management. ASTM-ISR, Philadelphia, PA, 64 pp.

American Society for Testing and Materials (ASTM). 1994. ASTM Standards on Ground Water and Vadose Zone Investigations, 2nd ed. ASTM, Philadelphia, PA, 396 pp. [46 ASTM test methods and guides]

Association of Ground Water Scientists and Engineers (AGWSE). 1992. Guidance to Environmental Site Assessments. National Ground Water Association, Dublin, OH, 72 pp.

Australian Drilling Association. 1992. Australian Drillers Manual. Available from S.A. Smith Consulting Service, P.O. Box, Ada, OH 45810, 550+ pp. [Geotechnical, environmental, well drilling]

Barcelona, M.J., J.P. Gibb, and R.A. Miller. 1983. A Guide to the Selection of Materials for Monitoring Well Construction and Ground-Water Sampling. ISWS Contract Report 327, Illinois State Water Survey, Champaign, IL, 78 pp.

(Table 9-10 Sampling and Monitoring References)

Barcelona, M.J., J.P. Gibb, J.A. Helfrich, and E.E. Garske. 1985. Practical Guide for Ground-Water Sampling. EPA/600/2-85/104 (NTIS PB86-137304). (Also published as ISWS Contract Report 374, Illinois State Water Survey, Champaign, IL, 93 pp.) [Covers QA/QC procedures, analyte selection, drilling methods, monitoring well design, well development, sampling, and recommended sampling protocols]

Barrett, J., et al. 1980. Procedures Recommended for Overburden and Hydrologic Studies of Surface Mines. GTR-INT-71, U.S. Forest Service, Intermountain Forest and Experiment Station, Ogden, UT, 106 pp.

Barth, D.S., B.J. Mason, T.H. Starks, and K.W. Brown. 1989. Soil Sampling Quality Assurance User's Guide, 2nd edition. EPA/600/8-89/046 (NTIS PB89-189864), 225+ pp.

Beck, R., B. Aboba, D. Miller, and I. Kaklins. 1981. Monitoring to Detect Groundwater Problems Resulting from Enhanced Oil Recovery. EPA/600/2-81/241 (NTIS PB82-119074), 146 pp.

Berg, E.L. 1982. Handbook for Sampling and Sample Preservation of Water and Wastewater, 2nd edition. EPA/600/4-82/029 (NTIS PB83-124503), 414 pp. (Replaces report with the same title by Huibregste and Mover, EPA/600/4-76/049.) [One chapter covers sampling of ground water and another covers sampling/preservation and storage considerations for trace organics]

Bitton, G. and C.P. Gerba (eds.). 1984. Groundwater Pollution Microbiology. Wiley-Interscience, New York, NY. [Paper by McNabb and Mallard details methods for obtaining uncontaminated subsurface samples for microbiological analysis]

Blume, L.J., et al. 1991. Handbook of Methods for Acid Deposition Studies, Laboratory Analyses for Soil Chemistry. EPA/600/4-90/023 (NTIS PB91-218016). [Nine authors; Includes: sample processing and rock fragment determination, bulk density, field/lab pH, organic matter, particle size, cation exchange capacity, and exchangeable/extractable ions/anions]

Board, R.G. and D.W. Lovelock. 1973. Sampling-Microbiological Monitoring of Environments. Academic Press, New York, NY.

Bordner, R., J. Winters, and P. Scarpino. 1978. Microbiological Methods for Monitoring the Environment: Water and Wastes. EPA/600/8-78/017 (NTIS PB290-329).

Boulding, J.R. 1991. Description and Sampling of Contaminated Soils: A Field Pocket Guide. EPA/625/12-91/002, 122 pp. Available from CERI.*

Boulding, J.R. 1993. Subsurface Field Characterization and Monitoring Techniques: A Desk Reference Guide, Vol. I: Solids and Ground Water, Vol. II: The Vadose Zone, Field Screening and Analytical Methods. EPA/625/R-93/003a&b. Available from CERI.*

Boulding, J.R. 1994. Description and Sampling of Contaminated Soils: A Field Guide, 2nd ed. Lewis Publishers, Chelsea, MI, 220 pp. [Boulding (1991) is recommended for use in the field; the second edition corrects minor errors in the first edition and includes new SCS procedures for field description of redoximorphic soil features]

(Table 9-10 Sampling and Monitoring References)

Brakensiek, D.L., H.B. Osborn, and W.J. Rawls (eds.). 1979. Field Manual for Research in Agricultural Hydrology. Agricultural Handbook No. 224, U.S. Department of Agriculture, Washington, DC. [6 Chapters cover: precipitation, runoff, climate, sedimentation, geology, and soil conditions and watershed characteristics]

Britton, L.J. and P.E. Greeson (eds.). 1989. Methods for Collection and Analysis of Aquatic Biological and Microbiological Samples. U.S. Geological Survey Techniques of Water-Resources Investigations, TWRI 5-A4, 363 pp. (Supersedes Greeson et al., 1977.)

Brown, E., M.W. Skougstad, and M.J. Fishman. 1970. Methods for Collection and Analysis of Water Samples for Dissolved Minerals and Gases. U.S. Geological Survey Techniques of Water-Resource Investigations, TWRI 5-A1. (Superseded by Fishman and Friedman, 1989.)

Brown, R.H., A.A. Konoplyantsev, J. Ineson, and V.S. Kovalensky. 1983. Ground-Water Studies: An International Guide for Research and Practice. Studies and Reports in Hydrology No. 7, UNESCO, Paris. (Originally published in 1972, with supplements added in 1973, 1975, 1977, and 1983.) [Comprehensive guide covering all aspects of ground-water characterization]

Brown, K.W., R.P. Breckinridge, and R.C. Rope. 1991. Soil Sampling Reference Field Methods. U.S. Fish and Wildlife Service Lands Contaminant Monitoring Operations Manual, Appendix J. Prepared by Center for Environmental Monitoring and Assessment, Idaho National Engineering Laboratory, Idaho Falls, ID, 83415. (Final publication pending revisions resulting from field testing of manual.)

Bureau of Reclamation. 1974. Earth Manual, 2nd edition. U.S. Department of the Interior, Bureau of Reclamation, Denver, CO. (First three chapters reprinted in 1990, 326 pp.; remaining chapters superseded by 1990 3rd edition.)

Bureau of Reclamation. 1981. Ground Water Manual--A Water Resources Technical Publication, 2nd edition. U.S. Department of the Interior, Bureau of Reclamation, Denver, CO. [Focuses on test methods for characterizing aquifer properties]

Bureau of Reclamation. 1990. Earth Manual, 3rd edition, Part 2. U.S. Department of the Interior, Bureau of Reclamation, Denver, CO, 1270 pp. (Part 1 consists of a 1990 reprint of the first three chapters of the 1974 2nd edition.)

Byrnes, M.E. 1994. Field Sampling Methods for Remedial Investigations. Lewis Publishers, Boca Raton, FL, 288 pp.

Cameron, R.E. 1991. Guide to Site and Soil Description for Hazardous Waste Sites. EPA/600/4-91/029 (NTIS PB92-146158).

Cameron, R.E., G.B. Blank, and D.R. Gensel. 1966. Sampling and Handling of Desert Soils. Technical Report No. 32-908, Jet Propulsion Laboratory, California Institute of Technology, Pasadena, CA, 37 pp.

Campbell, M.D. and J.H. Lehr. 1973. Water Well Technology. McGraw-Hill, New York, NY, 681 pp. [Annotated bibliography contains 673 references]

(Table 9-10 Sampling and Monitoring References)

Cheremisinoff, P. 1992. A Guide to Underground Storage Tanks: Evaluation, Site Assessment and Remediation. Prentice-Hall, Englewood Cliffs, NJ, 384 pp.

Classen, H.C. 1982. Guidelines and Techniques for Obtaining Water Samples that Accurately Represent the Water Chemistry of the Aquifer. U.S. Geological Survey Open File Report 82-1024, 49 pp.

Cochran, R. and V. Hodge (eds.). 1985. Guidance on Remedial Investigations Under CERCLA. EPA/540/G-85/002 (NTIS PB85-238616), 172 pp.

Cohen, R.M. and J.W. Mercer. 1993. DNAPL Site Evaluation. EPA/600/R-93/002 (NTIS PB93-150217). [Also published by Lewis Publishers as C.K. Smoley edition, Boca Raton, FL, 384 pp]

Cohen, R.M, A.H. Vincent, J.W. Mercer, C.R. Faust, and S.P. Spalding. 1994. Methods for Monitoring Pump-and-Treat Performance. EPA/600/R-94/123, 102 pp.

Colangelo, R.V. 1991. Buyer Be(a)ware: The Fundamentals of Environmental Property Assessments. National Water Well Association, Dublin, OH, 241 pp.

Cole, G.M. 1994. Assessment and Remediation of Petroleum-Contaminated Sites. Lewis Publishers, Boca Raton, FL, 368 pp.

Collins, A.G. and A.I. Johnson (eds.). 1988. Ground-Water Contamination: Field Methods. ASTM STP 963, American Society for Testing and Materials, Philadelphia, PA. [37 papers]

Connecticut Environmental Protection Agency. 1983. Ground-Water Monitoring Guidelines for Hazardous Waste Management Facilities. Hazardous Management Unit of Water Compliance Unit, Connecticut Environmental Protection Agency, Hartford, CT, 20 pp.

Conway, R.A. and W.P. Gulledge (eds.). 1982. Hazardous and Industrial Solid Waste Testing: Second Symposium. ASTM STP 805, American Society for Testing and Materials, Philadelphia, PA.

Conway, R.A. and B.C. Malloy (eds.). 1981. Hazardous Solid Waste Testing: First Conference. ASTM STP 760, American Society for Testing and Materials, Philadelphia, PA.

Corps of Engineers. 1972. Soil Sampling. EM 1110-2-1907, Corps of Engineers, Department of the Army, Washington, DC.

Corps of Engineers. 1984. Engineering and Design--Geotechnical Investigation. Engineer Manual EM 1110-1-1804, U.S. Army Corps of Engineers, Washington, DC.

Crouch, R.L., R.D. Eckert, and D.D. Rugg. 1976. Monitoring Groundwater Quality: Economic Framework and Principles. EPA/600/4-76/045 (NTIS PB260-919), 107 pp.

Dames & Moore. 1974. Manual of Ground-Water Practices. (Cited in Waste Management of North America [1991].)

(Table 9-10 Sampling and Monitoring References)

deVera, E.R. 1980. Samplers and Sampling Procedures for Hazardous Waste Streams. EPA/600/2-80/018 (NTIS PB80-135353).

Devinny, J.S., L.R. Everett, J.C.S. Lu, and R.L. Stollar. 1990. Subsurface Migration of Hazardous Wastes. Van Nostrand Reinhold, New York, NY. [Includes chapters on groundwater monitoring, soil core monitoring, and soil pore-liquid monitoring]

Devitt, D.A., R.B. Evans, W.A. Jury, T.H. Starks, and B. Eklund. 1987. Soil Gas Sensing for Detection and Mapping of Volatile Organics. EPA/600/8-87/036 (NTIS PB87-228516).

Dowding, C.H. (ed.). 1978. Site Characterization Exploration. Proceeding of Specialty Workshop, American Society of Civil Engineers, New York, NY.

Driscoll, F.G. 1986. Groundwater and Wells, 2nd edition. Johnson Filtration Systems Inc., St. Paul, MN, 1089 pp. (First edition Johnson, UOP, 1966.) [Comprehensive text of ground investigation methods, well design and construction]

Dunlap, W.J., J.F. McNabb, M.R. Scalf, and R.L. Cosby. 1977. Sampling for Organic Chemicals and Microorganisms in the Subsurface. EPA/600/2-77/176 (NTIS PB272 679).

Edwards, C. (ed.). 1993. Monitoring Genetically Manipulated Organisms in the Environment. John Wiley & Sons, New York, NY, 198 pp.

EG&G Idaho, Inc. 1990. Environmental Monitoring for Low-Level Waste-Disposal Sites, Revision 2 (2 Vols). Low-Level Radioactive Waste Management Handbook Series, DOE/LLW-13Tg, National Low Level Waste Management Program, Idaho Falls, ID.

Environmental Resource Center. 1993. Environmental Auditing and Compliance Manual. Van Nostrand Reinhold, New York, NY, 224 pp. [1992 edition published by the Environmental Resource Center]

Everett, L.G. (ed.). 1979. Groundwater Quality Monitoring of Western Coal Strip Mining: Identification and Priority Ranking of Potential Pollution Sources. EPA/600/7-79/024 (NTIS PB293-457), 265 pp.

Everett, L.G. 1980. Ground-Water Monitoring. General Electric Company Technology Marketing Operations, Schenectady, NY, 440 pp.

Everett, L.G. 1983. Groundwater Quality Monitoring Recommendations for Western Surface Coal Mines. EPA/600/4-83/057 (NTIS PB84-124619), 154 pp.

Everett, L.G. 1985. Groundwater Monitoring Handbook for Coal and Oil Shale Development. Elsevier, New York, NY.

Everett, L.G. and E.W. Hoylman (eds.). 1980a. Groundwater Quality Monitoring of Western Coal Strip Mining: Preliminary Designs for Reclaimed Mine Sources of Pollution. EPA/600/7-80/109 (NTIS PB80-203193), 50 pp.

(Table 9-10 Sampling and Monitoring References)

Everett, L.G. and E.W. Hoylman (eds.). 1980b. Groundwater Quality Monitoring of Western Coal Strip Mining: Preliminary Designs for Active Mine Sources of Pollution. EPA/600/7-80/110 (NTIS PB80-220502).

Everett, L.G., K.D. Schmidt, R.M. Tinlin, and D.K. Todd. 1976. Monitoring Groundwater Quality: Methods and Costs. EPA/600/4-76/023 (NTIS PB257 133). [Covers ground-water-related measuring techniques applicable to the land surface, topsoil, vadose zone, and zone of saturation; includes cost data on various methods]

Everett, L.G., L.G. Wilson, and L.G. McMillion. 1982. Vadose Zone Monitoring Concepts for Hazardous Waste Sites. Ground Water 20(3):312-324. [Summary information on more than 50 vadose zone monitoring techniques]

Everett, L.G., L.G. Wilson, and E.W. Hoylman. 1983. Vadose Zone Monitoring for Hazardous Waste Sites. EPA/600/X-83/064 (NTIS PB84-212752). (Also published in 1984 with same title by Noyes Data Corporation, Park Ridge, NJ.)

Everett, L.G., E.W. Hoylman, L.G. Wilson, and L.G. McMillion. 1984. Constraints and Categories of Vadose Zone Monitoring Devices. Ground Water Monitoring Review 4(1):26-31.

Fenn, D., E. Cocozza, J. Isbister, O. Braids, B. Yare, and P. Roux. 1977. Procedures Manual for Ground Water Monitoring at Solid Waste Disposal Facilities. EPA/530/SW-611 (NTIS PB84-174820), 283 pp. [Covers monitoring networks, monitoring and well technology, chemical parameters for indicators of leachate, and sampling]

Ford, P.J. and P.J. Turina. 1985. Characterization of Hazardous Waste Sites--A Methods Manual, Vol. I: Site Investigations. EPA/600/4-84/075 (NTIS PB85-215960). [Section 7 covers field investigations. Appendices contains useful forms and checklists]

Ford, P.J., P.J. Turina, and D.E. Seely. 1984. Characterization of Hazardous Waste Sites--A Methods Manual, Vol. II: Available Sampling Methods, 2nd edition. EPA/600/4-84/076 (NTIS PB85-521596). [Covers sampling methods for solids, gases, and liquids]

Fried, J.J. 1975. Groundwater Pollution: Theory, Methodology, Modeling and Practical Rules. Elsevier, New York, NY, 330 pp.

GeoTrans. 1989. Groundwater Monitoring Manual for the Electric Utility Industry. Edison Electric Institute, Washington, DC.

Gillham, R.W., M.J.L. Robin, J.F. Barker, and J.A. Cherry. 1983. Groundwater Monitoring and Sample Bias. API Publication 4367, American Petroleum Institute, Washington, DC. [Examines chemical characteristics of inorganic and organic parameters, sampling installations, sample collection, and methods]

Goodwin, B.E., J.R. Aronson, R.P. O'Neil, M.A. Randel, and E.M. Smith. 1982. Surface Sample Techniques. USATHAMA DRXTH-TE-CR-82179, U.S. Army Toxic and Hazardous Materials Agency, Aberdeen Proving Ground, MD, 91 pp. [Includes annotated bibliography with 47 citations]

(Table 9-10 Sampling and Monitoring References)

Sampling and Monitoring

Hanna, T.H. 1985. Field Instrumentation in Geotechnical Engineering. Trans Tech Publications, Clausthal, Germany, 843 pp.

Harlan, R.L., K.E. Kolm, and E.D. Gutentag. 1989. Water-Well Design and Construction. Elsevier, New York, NY, 205 pp.

Hathaway, A.W. 1988. Manual on Subsurface Investigations. American Association of State Highway and Transportation Officials, Washington, DC.

Hess, K. 1993. Environmental Site Assessment, Phase I: A Basic Guide. Lewis Publishers, Boca Raton, FL.

Ho, L.V., R.D. Morrison, C.J. Schmidt, and J.R. Marsh. 1978. Monitoring of Wastewater and Sludge Application Systems. SCS Engineers, Long Beach, CA, 303 pp.

Hodgson, J.M. 1978. Monograph on Soil Survey, Soil Sampling and Soil Description. Oxford University Press, New York, NY, 241 pp.

Holden, P.W. 1984. Primer on Well Water Sampling for Volatile Organic Compounds. Water Resources Research Center, University of Arizona, Tucson, AZ.

Howsam, P. (ed.). 1990. Proceedings of International Groundwater Engineering Conference on Water Wells: Monitoring, Maintenance, and Rehabilitation. Chapman and Hall, London, 422 pp.

Hvorslev, M.J. 1948. Subsurface Exploration and Sampling of Soils. Engineering Foundation, New York, NY.

Hvorslev, J.J. 1949. Subsurface Exploration and Sampling of Soils for Civil Engineering Purposes. U.S. Army Engineers Waterways Experiment Station, Vicksburg, MS. (Reprinted by the Engineering Foundation in 1962 and 1965.)

Jackson, L.P., A.R. Rohlik, and R.A. Conway (eds.). 1984. Hazardous and Industrial Waste Management and Testing: 3rd Symposium. ASTM STP 851, American Society for Testing and Materials, Philadelphia, PA.

Jain, R.K., L.V. Urban, G.S. Stacey, and H.E. Balbach. 1993. Environmental Assessment. McGraw-Hill, New York, NY.

Johnson, UOP. 1966. Ground Water and Wells. [See Driscoll (1986)]

Keith, L.H. (ed.). 1988. Principles of Environmental Sampling. ACS Professional Reference Book, American Chemical Society, Washington, DC, 480 pp. [30 contributed chapters on principles of environmental sampling covering: (1) general planning and sample design, (2) quality assurance and quality control, (3) sampling waters, (4) sampling air and stacks, (5) sampling biota, and (6) sampling solids, sludges and liquid wastes]

Keith, L.H. 1992. Environmental Sampling and Analysis: A Practical Guide. Lewis Publishers, Chelsea, MI, 143 pp. (In cooperation with ACS Committee on Environmental Improvement.)

(Table 9-10 Sampling and Monitoring References)

Kerfoot, H.B. and L.J. Barrows. 1987. Soil-Gas Measurement for Detection of Subsurface Organic Contamination. EPA/600/2-87/027 (NTIS PB87-174884).

Kolm, K. E. 1993. Conceptualization and Characterization of Hydrologic Systems. GWMI 93-01, International Ground Water Modeling Center, Golden, CO, 58 pp.

Korte, N.W. and P.M. Kearl. 1985. Procedures for the Collection and Preservation of Ground Water and Surface Water Samples and for the Installation of Monitoring Wells, 2nd edition. U.S. Department of Energy Technical Measurements Center Report GJ/TMC-08 (NTIS DE86-006184), 68 pp.

Lambe, W.T. 1951. Soil Testing for Engineers. John Wiley & Sons, New York, NY.

Lehr, J., S. Hurlburt, B. Gallagher, and J. Voyteck. 1988. Design and Construction of Water Wells: A Guide for Engineers. Van Nostrand Reinhold, New York, NY, 224 pp.

Lesage, S. and R.E. Jackson (eds.). 1992. Groundwater Contamination and Analysis at Hazardous Waste Sites. Environmental Science and Pollution Control Series/4, Marcel Dekker, New York, NY, 552 pp.

Lewis, T.E., A.B. Crockett, R.L. Siegrist, and Z. Zarrabi. 1991. Soil Sampling and Analysis for Volatile Organic Compounds. Ground-Water Issue, EPA/540/4-91/001, 24 pp. Available from CERI.*

Lindorff, D.E., J. Feld, and J. Connelly. 1987. Groundwater Sampling Procedures Guidelines. Wisc. Dept. Nat. Resources Rept. PUBL-WR-153, WDNR, Madison, WI.

Loftis, J.C. and R.C. Ward. 1979. Regulatory Water Quality Monitoring Networks-Statistical and Economic Considerations. EPA/600/4-79/055 (NTIS PB80-140882).

Lorenzen, D., R.A. Conway, L.P. Jackson, A. Hamza, C.L. Perket, and W.J. Lacy (eds.). 1986. Hazardous and Industrial Solid Waste Testing and Disposal: 6th Vol. ASTM STP 933, American Society for Testing and Materials, Philadelphia, PA.

Marrin, D.L. and W.B. Kerfoot. 1988. Soil Gas Surveying Techniques. Environ. Sci. Technol. 22(7)-740-745.

Mason, B.J. 1992. Preparation of Soil Sampling Protocols: Sampling Techniques and Strategies. EPA/600/R-92/128 (NTIS PB92-220532). (Supersedes 1983 edition titled, Preparation of Soil Sampling Protocol: Techniques and Strategies, EPA/600/4-03/020 [NTIS PB83-206979], 102 pp.)

McKeague, J.A. (ed.). 1978. Manual on Soil Sampling and Methods of Analysis, 2nd edition. Canadian Society Soil Science, Ottawa, Ontario.

Mooij, H. and F.A. Rovers. 1976. Recommended Groundwater and Soil Sampling Procedures. EPS-4-EC, Environmental Protection Service, Canada.

(Table 9-10 Sampling and Monitoring References)

Morrison, R.D. 1983. Groundwater Monitoring Technology. Timco Mfg., Inc., Prairie du Sac, WI, 105 pp. [Guide to vadose zone and ground-water sampling methods; sections on use of Teflon for suction lysimeters and casing for monitoring wells are out of date; see Sections 9.2.1 and A.1 in Boulding (1993) more current information]

Mueller, W. and D.L. Smith (Compilers). 1991. Compilation of EPA's Sampling and Analysis Methods. Lewis Publishers, Chelsea, MI, 465 pp. (On diskette: EPA's Sampling and Analysis Methods Database, Vol. 1 [Industrial Chemicals], Vol. 2 [Pesticides, Herbicides, Dioxins and PCBs], and Vol. 3 [Elements and Water Quality Parameters].) [Summary information on more than 150 EPA-approved, and a total of 650, sampling and analysis methods for industrial chemicals, pesticides, elements, and water quality parameters]

Nash, R.G. and A.R. Leslie (eds.). 1991. Groundwater Residue Sampling Design. ACS Symposium Series 465, American Chemical Society, Washington, DC, 395 pp. [23 papers on ground-water and vadose zone sampling]

Neckers, J.W. and C.R. Walker. 1952. Field Tests for Active Sulfides in Soil. Soil Science 74:467-470.

Nielsen, D.M. (ed.). 1991. Practical Handbook of Ground Water Monitoring. Lewis Publishers, Chelsea, MI, 717 pp. (Published in cooperation with National Water Well Association, Dublin, OH.) [Covers all aspects of vadose zone and ground-water monitoring]

Nielsen, D.M. and A.I. Johnson (eds.). 1990. Ground Water and Vadose Zone Investigations. ASTM STP 1053, American Society for Testing and Materials, Philadelphia, PA. [22 papers]

Nielsen, D.N. and M.N. Sara (eds.). 1992. Current Practices in Ground Water and Vadose Zone Investigations. ASTM STP 1118, American Society for Testing and Materials, Philadelphia, PA, 431 pp. [28 papers]

Nielsen, D.M. and R. Schalla. 1991. Design and Installation of Ground-Water Monitoring Wells. In: Practical Handbook of Ground-Water Monitoring, D.M. Nielsen (ed.), Lewis Publishers, Chelsea, MI, pp. 239-331.

New Jersey Department of Environmental Protection (NJDEP). 1992. Field Sampling Procedures Manual, 4th ed. Hazardous Waste Program, Division of Hazardous Site Mitigation, NJDEP, Trenton, NJ. [May also be cited with J.R. Schoenlever and P.S. Morton as authors; 3rd ed. 1988]

Ontario Ministry of the Environment. 1989. A Guide to the Collection and Submission of Samples for Laboratory Analysis, 6th ed.

Oudjik, G. and K. Mujica. 1989. Handbook for Identification, Location and Investigation of Pollution Sources Affecting Ground Water. National Water Well Association, Dublin, OH, 185 pp.

(A.W.) Penn, Inc. 1994. The Environmental Field Book. A.W. Penn, Inc., 150 pp.

U.S. Environmental Protection Agency (EPA). 1992b. Estimating the Potential for Occurrence of DNAPL at Superfund Sites. OSWER Directive No. 92355.4.07.

(Table 9-10 Sampling and Monitoring References)

Perket, C.L. (ed.). 1986. Quality Control in Remedial Site Investigation: Hazardous and Industrial Solid Waste Testing, 5th Volume. ASTM STP 925, American Society for Testing and Materials, Philadelphia, PA.

Petros, Jr., J.K., W.J. Lacy, and R.A. Conway (eds.). 1985. Hazardous and Industrial Solid Waste Testing: 4th Symposium. ASTM STP 886, American Society for Testing and Materials, Philadelphia, PA.

Pilie, K., et al. 1975. Methods to Treat, Control, and Monitor Spilled Hazardous Materials. EPA/670/2-75/042 (NTIS PB243-386), 149 pp.

Plumb, Jr., R.H. 1984. Characterization of Hazardous Waste Sites, A Methods Manual: Vol. III, Available Analytical Methods. EPA/600/4-84/038 (NTIS PB84-191048). [Compendium of analytical procedures for characterization of hazardous wastes sites covering water, soil/sediment, biological tissues, and air samples for EPA listed hazardous substances]

Rainwater, F.H. and L.L. Thatcher. 1960. Methods for Collection and Analysis of Water Samples. U.S. Geological Survey Water-Supply Paper 1454. [Describes types of methods, choice of analytical methods for water samples, and specific analytical procedures for over 40 inorganic water parameters]

Redwine, J. et al. 1985. Groundwater Manual for the Electric Utility Industry, Vol. 3: Groundwater Investigations and Mitigation Techniques. EPRI CS-3901, Electric Power Research Institute, Palo Alto, CA.

Rehm, B.W., T.R. Stolzenburg, and D.G. Nichols. 1985. Field Measurement Methods for Hydrogeologic Investigations: A Critical Review of the Literature. EPRI EA-4301, Electric Power Research Institute, Palo Alto, CA. [Comprehensive review of methods for solids, unsaturated zone, and ground-water physical and chemical characterization; bibliography contains over 600 references on these topics]

Rijtema, P.E. and H. Wassink (eds.). 1969. Water in the Unsaturated Zone (Proc. Wageningen Symp.), 2 Vols. IASH-UNESCO Studies and Reports in Hydrology 2, UNESCO, Paris, Vol. 1: pp. 1-156, Vol. 2: pp. 527-944. (Also published as IAHS Publication Nos. 82 and 83.)

Roscoe Moss Company. 1990. Handbook of Ground Water Development. John Wiley & Sons, New York, NY, 493 pp. [Drilling and well installation methods]

Ross, R.R. and J.W. Keeley. 1992. General Methods for Remedial Operations Performance Evaluations. EPA/600/R-92/002, 37 pp.

Ruda, T.C. and P.J. Bosscher (eds.). 1990. Drillers Handbook. National Drilling Contractors Association, Columbia, SC.

Rupp, G.L. and R.R. Jones, Sr. 1993. Heterogeneous Wastes Characterization: Methods and Recommendations. C.K. Smoley/CRC Press, Boca Raton, FL, 144 pp. [Hardback version of U.S. EPA (1992)]

Santa Clara County Water District. 1985. Groundwater Monitoring Guidelines. Santa Clara County, CA, 58 pp.

(Table 9-10 Sampling and Monitoring References)

Sara, M.N. 1994. Standard Handbook of Site Assessment for Solid and Hazardous Waste Facilities. Lewis Publishers, Boca Raton, FL, 976 pp.

Scalf, M.R., J.F. McNabb, W.J. Dunlap, R.L. Cosby, and J. Fryberger. 1981. Manual of Ground-Water Quality Sampling Procedures. EPA/600/2-81/160 (NTIS PB82-103045). (Also published in NWWA/EPA Series, National Water Well Association, Dublin OH.) [Covers drilling methods, collection of ground-water samples, and field tests and preservation, with a short chapter on sampling subsurface solids]

Schoenlever and Morton (1992)-see NJDEP (1992).

Schweitzer, G.E. and J.A. Santolucito (eds.). 1984. Environmental Sampling for Hazardous Wastes. ACS Symp. Series 267, American Chemical Society, Washington, DC, 130 pp. [13 papers focusing on soil contamination]

Shuter, E. and W.E. Teasdale. 1989. Application of Drilling, Coring, and Sampling Techniques to Test Holes and Wells. U.S. Geological Survey Techniques of Water-Resources Investigations TWRI 2-F1, 96 pp.

Silka, L.R. and T.L. Swearingen. 1978. Manual for Evaluating Contamination Potential of Surface Impoundments. EPA/570/9-78/003 (NTIS PB85-211423).

Simmons, M.S. 1991. Hazardous Waste Measurements. Lewis Publishers, Chelsea, MI, 315 pp. [14 contributed chapters on sampling, field techniques and instrumentation, toxicity screening methods for hazardous waste, and quality assurance/quality control]

Sisk, S.W. 1981. NEIC Manual for Groundwater/Subsurface Investigations at Hazardous Waste Sites. EPA/330/9-81/002 (NTIS PB82-103755), 213 pp. [Appendix on information sources is especially useful]

Slawson, G.C. (ed.). 1979. Groundwater Quality Monitoring of Western Oil Shale Development: Identification and Priority Ranking of Potential Pollution Sources. EPA/600/7-79/023 (NTIS PB300-536), 241 pp.

Slawson, G.C. (ed.). 1980a. Groundwater Quality Monitoring of Western Oil Shale Development: Monitoring Program Development. EPA/600/7-80/089 (NTIS PB80-203219), 200 pp.

Slawson, G.C. (ed.). 1980b. Monitoring Groundwater Quality: The Impact of In-Situ Oil Shale Retorting. EPA/600/7-80/132 (NTIS PB81-177453), 300 pp.

Soil Conservation Service (SCS). 1971. Handbook of Soil Survey Investigations Procedures. SCS, Washington, DC, 98 pp.

Soil Conservation Service (SCS). 1984. Procedures for Collecting Soil Samples and Methods of Analysis for Soil Survey. Soil Survey Investigations Report No. 1, U.S. Government Printing Office. (Supersedes 1972 report with same document number titled Soil Survey Laboratory Methods and Procedures for Collecting Soil Samples.) [Laboratory analytical methods for soil physical and chemical properties]

Stephens, E. 1986. Procedures for Conducting a Comprehensive Ground Water Monitoring Evaluation of Hazardous Waste Disposal Facilities. California Department of Health Services, Sacramento, CA, 52 pp.

Summers, K.V. and S.A. Gherini. 1987. Sampling Guidelines for Groundwater Quality. EPRI EA-4952, Electric Power Research Institute, Palo Alto, CA.

Texas Water Commission. 1993. Limited Site Assessment Guidance Document. Petroleum Storage Tank Program, Austin, TX.

Thompson, C.M. et al. 1989. Techniques to Develop Data for Hydrogeochemical Models. EPRI EN-6637, Electric Power Research Institute, Palo Alto, CA, 371 pp. [Overview of hydrologic, physical, and chemical characterization techniques to develop data for hydrogeochemical models; contains summary description of methods for elemental analysis, analysis of anionic species, inorganic and organic carbon, redox sensitive species, and other chemical parameters, along with recommendations for methods best suited for obtaining data for hydrochemical modeling]

Tinlin, R.M. (ed.) 1976. Monitoring Groundwater Quality: Illustrative Examples. EPA/600/4-76/036 (NTIS PB257 936). [Nine case studies illustrating procedures for monitoring various classes of ground-water pollution sources]

Todd, D.K., R.M. Tinlin, K.D. Schmidt, and L.G. Everett. 1976. Monitoring Ground-Water Quality: Monitoring Methodology. EPA/600/4-76/026 (NTIS PB256-068). [15-step procedure]

UNESCO. 1983. Proceedings of the Symposium--Methods and Instrumentation for the Investigation of Groundwater Systems. Committee for Hydrological Research, CHO-TNO, The Hague, The Netherlands. [More than 60 papers]

Unwin, J. 1982. A Guide to Groundwater Sampling. NCPASI Technical Bulletin 362, National Council of the Paper Industry for Air and Stream Improvement, New York, NY.

U.S. Department of Energy (DOE). Various dates. The Environmental Survey Manual. DOE/EH-0053: Vol. 1 (August 1987; Chapter 8, 2nd edition. January 1989—Sampling and Analysis Phase); Vol. 2 (August 1987—Appendices A,B, and C); Vol. 3 (2nd edition. January 1989—Appendix D, Parts 1, 2, and 3; Organic and Inorganic Analysis Methods and Non-Target List Parameters); Vol. 4 (2nd edition. January 1989—Appendix D, Part 4; Radiochemical Analysis Procedures); Vol. 5. (2nd edition. January 1989—Appendices: E, Field Sampling; F, Quality Assurance; G, Decontamination; H, Sample Management; I, Sample Handling, Transport and Documentation; J, Health and Safety; and K, Sampling and Analysis Plan).

U.S. Environmental Protection Agency (EPA). 1981a. Permitting of Land Disposal Facilities: Groundwater and Air Emission Monitoring. EPA/MS-1941.41 (NTIS PB81-246431), 38 pp.

U.S. Environmental Protection Agency (EPA). 1981b. Permitting of Land Disposal Facilities: Groundwater Protection Standard. EPA/MS-1941.40 (NTIS PB81-246423), 39 pp.

(Table 9-10 Sampling and Monitoring References)

Sampling and Monitoring

U.S. Environmental Protection Agency (EPA). 1983a. Ground-Water Monitoring Guidance for Owners and Operators of Interim Status Facilities, Revised Edition. EPA/SW-963-Rev (NTIS DE84-900827; PB83-209445-draft), 190 pp.

U.S. Environmental Protection Agency (EPA). 1983b. Draft RCRA Permit Writer's Manual, Ground-Water Protection, 40 CFR Part 264, Subpart F, 263 pp.

U.S. Environmental Protection Agency (EPA). 1985. RCRA Ground-Water Monitoring Compliance Order Guidance. EPA Office of Solid Waste and Emergency Response (NTIS PB87-193710), 128 pp. [Deals with monitoring violations at interim status land disposal facilities]

U.S. Environmental Protection Agency (EPA). 1986a. RCRA Ground Water Monitoring Technical Enforcement Guidance Document. EPA/530/SW-86/055 (OSWER-9950.1) (NTIS PB87-107751), 332 pp. (Also published in NWWA/EPA Series, National Water Well Association, Dublin, OH. Final OSWER Directive 9950.2 (NTIS PB91-140194). Executive Summary: OSWER 9950.1a (NTIS PB91-140186), 17 pp. See also, U.S. EPA [1986e and 1993].)

U.S. Environmental Protection Agency (EPA). 1986b. Test Methods for Evaluating Solid Waste, 3rd edition. EPA/530/SW-846 (NTIS PB88-239223); First update, 3rd edition. EPA/530/SW-846.3-1 (NTIS PB89-148076). (2nd edition was published in 1982 (NTIS PB87-1200291); current edition and updates available on a subscription basis from U.S. Government Printing Office, Stock #955-001-00000-1. Revised final draft of Chapter 11 (Ground-Water Monitoring System Design, Installation, and Operating Practices contains extensive new guidance.) [Volumes 1A (Metallic Analytes), IB (Organic Analytes), and IC (Miscellaneous Test Methods) cover laboratory methods; Volume II covers field methods (Part IV defines acceptable and unacceptable designs and practice for ground-water monitoring)]

U.S. Environmental Protection Agency (EPA). 1986c. Permit Guidance Manual on Unsaturated Zone Monitoring for Hazardous Waste Land Treatment Units. EPA/530/SW-86/040 (NTIS PB87-215463).

U.S. Environmental Protection Agency (EPA). 1986d. RCRA Facility Assessment Guidance. EPA/530-86/053 (NTIS PB87-107769), 174 pp. [Might also be cited with Rastatter et al. as authors. Chapter 5 covers ground water, Chapter 8 covers subsurface gas, and Chapter 9 covers soils]

U.S. Environmental Protection Agency (EPA). 1986e. Final RCRA Comprehensive Ground-Water Monitoring Evaluation (CME) Guidance Document. Final OSWER Directive 9950.2 (NTIS PB91-140194), 59 pp. [Contains detailed checklist drawing heavily from U.S. EPA (1986a)]

U.S. Environmental Protection Agency (EPA). 1986f. Guidance on Issuing Permits to Facilities Required to Analyze Groundwater for Appendix VIII Constituents. (NTIS PB87-163242).

(Table 9-10 Sampling and Monitoring References)

U.S. Environmental Protection Agency (EPA). 1987. A Compendium of Superfund Field Operations Methods. EPA/540/P-87/001 (NTIS PB88-181557), 644 pp. [Section 7 covers field methods for rapid screening for hazardous material, Section 8 covers methods for geologic characterization, and Section 15 covers field instrumentation]

U.S. Environmental Protection Agency (EPA). 1988a. Guidance on Remedial Actions for Contaminated Ground Water at Superfund Sites. EPA/540/G-88/003 OSWER Directive 9283.1-2 (NTIS PB89-184618), 180 pp. [Section 7.4 covers performance monitoring]

U.S. Environmental Protection Agency (EPA). 1988b. Field Screening Methods Catalog: User's Guide. EPA/540/2-88/005. FSMC System Coordinator, OERR, Analytical Operations Branch (WH-548-A), U.S. EPA, Washington, DC. (Also available as a computerized information retrieval system.) [Guide for selection of instrumental methods for field screening of inorganic and organic contaminants; covers 26 specific field screening methods]

U.S. Environmental Protection Agency (EPA). 1989a. Guidance for Conducting Remedial Investigations and Feasibility Studies Under CERCLA. EPA/540/G-89/004 (NTIS PB89-184626), 195 pp.

U.S. Environmental Protection Agency (EPA). 1989b. RCRA Facility Investigation (RFI) Interim Final Guidance, Vol. I: Development of an RFI Work Plan and General Considerations for RCRA Facility Investigations; Vol. II: Soil, Ground Water and Subsurface Gas Releases; Vol. III: Air and Surface Water Releases; Vol. IV: Case Study Examples. EPA/530/SW-89/031 (NTIS PB89-200299), 1221 pp. [Volume II covers investigation of soil, ground water, and subsurface gas releases]

U.S. Environmental Protection Agency (EPA). 1989c. RCRA Sampling Procedures Handbook. U.S. EPA Region VI, Dallas, TX.

U.S. Environmental Protection Agency (EPA). 1990a. Handbook: Ground Water, Vol. I: Ground Water and Contamination. EPA/625/6-90/16a, 144 pp. Available from CERI.*

U.S. Environmental Protection Agency (EPA). 1990b. Subsurface Contamination Reference Guide. EPA/540/2-90/011. (NTIS PB91-921292), 26 pp. [Three-Part Guide]

U.S. Environmental Protection Agency (EPA). 1991a. Site Characterization for Subsurface Remediation. EPA/625/4-91/026, 259 pp. Available from CERI.* [Part I contains nine chapters on methods for subsurface characterization]

U.S. Environmental Protection Agency (EPA). 1991b. Handbook: Ground Water, Vol. II: Methodology. EPA/625/6-90/16b, 141 pp. Available from CERI.*

U.S. Environmental Protection Agency (EPA). 1991c. Guidance for Performing Preliminary Assessments Under CERCLA. OSWER 9345.0-01A (NTIS PB92-963303), 288 pp.

U.S. Environmental Protection Agency (EPA). 1991d. Environmental Compliance Branch Standard Operating Procedures and Quality Assurance Manual. U.S. EPA Region IV Environmental Services Division, College Station Road, Athens, GA 30613. [Available in Wordperfect 5.1 electronic format; Section 4 covers sampling procedures, Section 6 cover field physical measurements, Appendix E covers design and installation of monitoring wells]

(Table 9-10 Sampling and Monitoring References)

Sampling and Monitoring

U.S. Environmental Protection Agency (EPA). 1991e. Emergency Response Team (ERT) Standard Operating Procedures (SOPs) Compendia: Compendium of ERT Soil Sampling and Surface Geophysics Procedures (EPA/540/P-91/006); Compendium of ERT Groundwater Sampling Procedures (EPA/540/P-91/007); Compendium of ERT Waste Sampling Procedures (EPA/540/P-91/008); Compendium of ERT Toxicity Testing Procedures (EPA/540/P-91/009).

U.S. Environmental Protection Agency (EPA). 1992. Characterizing Heterogeneous Wastes: Methods and Recommendations. EPA/600/R-92/033 (NTIS PB92-216894). [Also published as hardback Smoley edition: Rupp and Jones (1993)]

U.S. Environmental Protection Agency (EPA). 1993. RCRA Ground Water Monitoring: Draft Technical Guidance. EPA/530/R-93/001 (NTIS PB93-139350).

U.S. Geological Survey (USGS). 1977+. National Handbook of Recommended Methods for Water Data Acquisition. USGS Office of Water Data Coordination, Reston, VA. [Individual chapters have come out at different dates; pertinent chapters include: (2) Ground Water (1980); (4) Biological and Microbiological Quality of Water (1983); (5) Chemical Quality (1982); and (6) Soil Water (1982)]

U.S. Naval Facilities Engineering Command (USNFEC). 1982. Soil Mechanics Design Manual, Vol. 7.1. NAVFAC DM-7.1, Department of the Navy. [Includes section on site assessment techniques]

van Duijvenbooden, W. and H.G. van Waegeningh (eds.). 1987. Vulnerability of Soil and Groundwater to Pollutants. Committee for Hydrological Research, CHO-TNO, The Hague, The Netherlands. [Contains a number of papers on soil and ground-water monitoring strategies and vulnerability mapping]

Van Haveren, B.P. 1986. Water Resource Measurements. American Water Works Association, Denver, CO, 132 pp.

Vincoli, J. 1993. Basic Guide to Environmental Compliance. Van Nostrand Reinhold, New York, NY, 256 pp.

Ward, R.C., L.C. Loftis, and G.B. McBride. 1990. Design of Water Quality Monitoring Systems. Van Nostrand Reinhold, New York, NY.

Waste Management of North America. 1991. Site Assessment Manual. Waste Management, Inc., Oak Brook, IL. (Earlier edition published in 1989.)

Water Pollution Control Federation (WPCF/WEF). 1988. Hazardous Waste Site Remediation: Assessment and Characterization. Water Environment Federation, Alexandria, VA, 33 pp.

Weiss, R.B., T.O. Coffee, and T.L. Williams. 1979. Geothermal Environmental Impact Assessment: Ground Water Monitoring Guidelines for Geothermal Development. EPA/600/7-79/218 (NTIS PB80-144801), 232 pp.

(Table 9-10 Sampling and Monitoring References)

Wilson, L.G. 1980. Monitoring in the Vadose Zone: A Review of Technical Elements. EPA/600/7-80-134 (NTIS PB81-125817). [Covers (1) principles of pollutant movement, (2) basic chemical reactions of fluids, and (3) relative advantages and disadvantages of different monitoring techniques]

Wilson, L.G. 1981. Monitoring in the Vadose Zone: Part I. Ground Water Monitoring Review 1(3):32-41. [Review of 9 methods for monitoring water content]

Wilson, L.G. 1982. Monitoring in the Vadose Zone: Part II. Ground Water Monitoring Review 2(1):31-42. [Review of 23 methods for monitoring or estimating flux of wastewater]

Wilson, L.G. 1983. Monitoring in the Vadose Zone: Part III. Ground Water Monitoring Review 3(2):155-166. [Review of 9 methods for monitoring pollutant movement]

Wood, W.W. 1976. Guidelines for Collection and Field Analysis of Ground-Water Samples for Selected Unstable Constituents. U.S. Geological Survey Techniques of Water-Resources Investigations, TWRI 1-D2, 24 pp.

Yang, J.T. and W.E. Bye. 1979. Methods of Preventing, Detecting, and Dealing with Surface Spills of Contaminants Which May Degrade Underground Water Sources for Public Water Systems. EPA/570/9-79/018 (NTIS PB82-204082).

Zimmie, T.F and C.O. Riggs (eds.). 1981. Permeability and Groundwater Contaminant Transport. ASTM STP 746, American Society for Testing and Materials, Philadelphia, PA. [12 papers]

* See Preface for information on how to obtain documents from CERI (U.S. EPA Center for Environmental Research Information) and NTIS.

Sampling and Monitoring

Table 9-11 Index to Major References on Field and Laboratory Analytical Methods

Reference	Description
Instrumentation Principles	Skoog (1985), Willard et al. (1988); QA/QC: AOAC (1975, 1985. 1991)
Field Screening and Analytical Methods	U.S. EPA: Minnich (1993), U.S. EPA (1988, 1993); Other: Clement et al. (1992), Gammage and Berven (1992)
U.S. EPA Analytical Methods	Overviews: Mueller and Smith (1991), Nelson (1988), Smith (1994), Wagner (1994); General: Kopp and McKee (1983); Metals: U.S. EPA (1991a); Organics in Water: Longbottom and Lichtenberg (1982); Solid Waste (SW-846): U.S. EPA (1986b); SW-846 Methods Studies: Edgill (1989), Edgill and Wilburs (1989), Engel et al. (1988); Drinking Water Analysis: Long and Martin (1989), Pfaff (1981), U.S. EPA (1990a, 1990b, 1991d); Pesticides: Watts (1980), U.S. EPA and State Laboratories (1993); Sediment: Guy (1969), Plumb (1981), U.S. EPA (1989); Quality Control: Booth (1979), Provost and Elder (1985), Sharma (1979)
U.S. Geological Survey TWRIs	The Techniques of Water Resource Investigation series includes manuals describing procedures for planning and conducting specialized work in water-resources investigations. Water: Barnett and Mallory (1971-minor elements), Britton and Greeson (1989-aquatic biology and microbiology; supersedes Greeson et al., 1977), Fishman and Friedman (1989-inorganic constituents; supersedes Brown et al., 1970, Skougstad et al., 1979, and Fishman and Bradford, 1982), Friedman and Erdman (1982-quality assurance), Thatcher et al. (1977-radioactive substances), Wershaw et al. (1987-organic substances; supersedes Goerlitz and Brown, 1972), Wood (1976-field analysis of unstable constituents); Fluvial Sediment: Fishman and Friedman (1989—inorganic constituents; supersedes Brown et al., 1970, Skougstad et al., 1979, and Fishman and Bradford, 1982), Friedman and Erdman (1982-quality assurance), Guy (1969-physical analysis), Thatcher et al. (1977-radioactive substances), Wershaw et al. (1987-organic substances; supersedes Goerlitz and Brown, 1972).
<u>Other Major References</u>	
Water	Standard Methods: APHA (1994), ASTM (1966, Annual--Vols. 11.01 and 11.02), Hach (1991), OAOC (1990); Other Major References: Crompton (1993), Fresenius et al. (1988), Rainwater and Thatcher (1960), Thompson et al. (1989), Velthorst (1993); Injected Water/Waste Compatibility: Collins and Crocker (1988), Ostroff (1965), Warner and Lehr (1977), Watkins (1954)

Table 9-11 (cont.)

Reference	Description

<u>Other Major References (cont.)</u>

Soil	Baize (1993), Carter (1993), McKeague (1978); <u>Physical Properties</u>: ASTM (Annual--Vol. 4.08 and 4.09), Carter (1993), Blume et al. (1991), Klute (1986), SCS (1984), Smith and Mullins (1991), Stockham and Fochtman (1979-particle size analysis), Topp et al. (1992); <u>Clay/Other Mineralogy</u>: Carroll (1970), Johnson and Maxwell (1981), Kerr (1959), Moore and Reynolds (1989); see also Table 3-12; <u>Soil Chemistry</u>: Carter (1993), Council on Soil Testing and Plant Analysis (1992), Hoddinott and O'Shay (1993), Jackson (1958, 1979), van Lagen (1993), Walsh and Beaton (1990), Weaver (1994), Westerman (1990); see also Table 3-12
Sediment	<u>Sampling</u>: ASTM D5387/Table A-14, Barth and Starks (1985), Fleischauer and Engelder (1985), Murdoch and MacKnight (1994), Palmer (1985), Plumb (1981), U.S. Interagency Committee (1940); <u>Analysis</u>: Edwards and Glysson (1988), Litchtenberg et al. (1988), Plumb (1981), U.S. EPA (1992-contaminated sediment), U.S. Interagency Committee (1941); see also EPA and USGS TWRI references above.
Contaminants	<u>Method Compilations</u>: Plumb (1984), U.S. DOE (Various dates); <u>Ground water</u>: API (1987-petroleum hydrocarbons), Hach (1991-inorganics), Lesage and Jackson (1992); <u>Hazardous Waste</u>: Francis et al. (1988), Silvestri et al. (1981), Wagner (1990), Wolbach et al. (1984) <u>Soil</u>: API (1987, 1992-petroleum hydrocarbons), Minnich (1993-VOCs), O'Shay and Hoddinott (1994); <u>Environmental Fate</u>: Howard et al. 1975), Mill et al. (1982)
Other Solid Waste	<u>Flue Gas Desulfurization Waste</u>: Noblett and Burke (1990), Radian Corporation (1988); <u>Oil Shale</u>: Wallace et al. (1984); <u>Mine Soils and Overburden</u>: Williams and Schuman (1987)
Subsurface Microbiology	API (1965), Costerton and Colwell (1979), Ghiorse and Balkwill (1985), Pritchard and Bourquin (1984), Rosswall (1973), Weaver (1994), Webster et al. (1985); see also references for Microbiological Sampling in Table 9-10

Sampling and Monitoring

Table 9-11 References (Appendix F contains references for figure and table sources)

American Petroleum Institute (API). 1965. Recommended Practices for Biological Analysis of Subsurface Injection Waters. RP 38. API 1220 L St. NW, Washington, DC, 20005.

American Petroleum Institute (API). 1987. Manual of Sampling and Analytical Methods for Petroleum Hydrocarbons in Groundwater and Soil. API Publication No. 4449, API, Washington, DC, 230 pp.

American Petroleum Institute (API). 1992. Sampling and Analysis of Gasoline Range Organics in Soil. API Publication 4516, API, Washington, DC, 132 pp.

American Public Health Association (APHA). 1994. Standard Methods for the Examination of Water and Wastewater, 19th ed. APHA, Washington, DC, 1644 pp. [Comprehensive compilation of analytical methods for measurement of metals, inorganic nonmetallics, and organic constituents in water samples; published in cooperation with Water Environmental Federation and American Water Works Association]

American Society for Testing and Materials (ASTM). Annual. Books of ASTM Standards: Construction (Soil and Rock), Volume 04.08; Water and Environmental Technology, Volumes 11.01 and 11.02 (Water) and 11.04 (Pesticides, Resource Recovery, Hazardous Substances and Oil Spill Response, Waste Management, Biological Effects). ASTM, Philadelphia, PA.

American Society for Testing and Materials (ASTM). 1966. Manual on Industrial Water and Industrial Wastewater, 2nd edition. ASTM, Philadelphia, PA.

Association of Organic and Analytical Chemists (AOAC). 1975. Statistical Manual of the AOAC. AOAC International, McLean, VA, 96 pp.

Association of Organic and Analytical Chemists (AOAC). 1985. Use of Statistics to Develop and Evaluate Analytical Methods, 2nd ed. AOAC International, McLean, VA, 183 pp.

Association of Organic and Analytical Chemists (AOAC). 1990. Official Methods of Analysis of the AOAC, 15th ed. AOAC International, McLean, VA, 1200 pp.

Association of Organic and Analytical Chemists (AOAC). 1991. Quality Assurance Principles for Analytical Laboratories, 2nd ed. AOAC International, McLean, VA, 192 pp.

Baize, D. 1993. Soil Science Analysis: A Guide to Current Use. John Wiley & Sons, New York, NY, 192 pp.

Barnett, P.R. and E.C. Mallory, Jr. 1971. Determination of Minor Elements in Water by Emission Spectroscopy. U.S. Geological Survey Techniques of Water-Resources Investigations TWRI 5-A2, 31 pp.

Barth, D.S. and T.H. Starks. 1985. Sediment Sampling Quality Assurance User's Guide. EPA/600/4-85/048 (NTIS PB85-233542).

(Table 9-11 Field and Laboratory Analytical References)

Berg, E.L. 1982. Handbook for Sampling and Sample Preservation of Water and Wastewater, 2nd edition. EPA/600/4-82/029 (NTIS PB83-124503), 414 pp. (Replaces report with the same title by Huibregste and Mover, EPA/600/4-76/049.) [One chapter covers sampling of ground water and another covers sampling/preservation and storage considerations for trace organics]

Blume, L.J., et al. 1991. Handbook of Methods for Acid Deposition Studies, Laboratory Analyses for Soil Chemistry. EPA/600/4-90/023 (NTIS PB91-218016). [Nine authors; Includes: sample processing and rock fragment determination, bulk density, field/lab pH, organic matter, particle size, cation exchange capacity, and exchangeable/extractable ions/anions]

Booth, R.L. 1979. Handbook for Analytical Quality Control in Water and Wastewater Laboratories. EPA/600/4-79/019 (NTIS PB297-451), 157 pp.

Britton, L.J. and P.E. Greeson (eds.). 1989. Methods for Collection and Analysis of Aquatic Biological and Microbiological Samples. U.S. Geological Survey Techniques of Water-Resources Investigations TWRI 5-A4, 363 pp. (Supersedes Greeson et al., 1977.)

Brown, E., M.W. Skougstad, and M.J. Fishman. 1970. Methods for Collection and Analysis of Water Samples for Dissolved Minerals and Gases. U.S. Geological Survey Techniques of Water-Resource Investigations TWRI 5-A1. (Superseded by Fishman and Friedman, 1989.)

Carroll, D. 1970. Clay Minerals: A Guide to their X-Ray Identification. GSA Special Paper No. 126. Geological Society of America, Box 9140, Boulder, CO 80301.

Carter, M.R. (ed.). 1993. Soil Sampling and Methods of Analysis. Lewis Publishers, Boca Raton, FL, 823 pp. [75 chapters covering all aspects of physical, chemical and biological analysis of soils; supersedes McKeague (1978)]

Clement, R.E., K.W.M. Siu, and H.H. Hill, Jr. (eds.). 1992. Instrumentation for Trace Organic Monitoring. Lewis Publishers, Chelsea, MI, 319 pp. [15 papers on field/laboratory analytical methods]

Collins, A.G. and M.E. Crocker. 1988. Laboratory Protocol for Determining Fate of Waste Disposed in Deep Wells. EPA-600/8-88-008 (NTIS PB88-166061).

Costerton, J.W. and R.R. Colwell (eds.). 1979. Native Aquatic Bacteria: Enumeration, Activity, and Ecology. ASTM STP 695, Philadelphia, PA. [Papers presented at a Symposium sponsored by ASTM in June, 1977. Contains five papers of methods for direct enumeration of aquatic bacteria, five papers on chemical indices of aquatic bacterial populations, and 6 papers on metabolic potentials of aquatic bacterial populations as indicated by activity measurements]

Council on Soil Testing and Plant Analysis. 1992. Reference Methods for Soil Analysis. Council on Soil Testing and Plant Analysis, Georgia University Station, Athens, GA, 202 pp. [Complete revision of 1980 Handbook on Reference Methods for Soil Testing; oriented towards agricultural applications]

(Table 9-11 Field and Laboratory Analytical References)

Crompton, T.R. 1993. The Analysis of Natural Waters: Vol. 1, Complex-Formation Preconcentration Techniques; Vol. 2: Direct Preconcentration Techniques. Oxford University Press, New York, NY, Vol. 1, 232 pp. and Vol. 2, 264 pp.

Edgill, K.W. 1989. USEPA Method Study 36 SW-846 Methods 8270/3510 GC/MS Method for Semivolatile Organics: Capillary Column Technique; Separatory Funnel Liquid-Liquid Extraction. EPA/600/4-89/010 (NTIS PB89-190581).

Edgill, K.W. and D.M. Wilbers. 1989. USEPA Method Study 38 SW-846 Method 3010 Acid Digestion of Aqueous Samples and Extracts for Total Metals for Analysis by Flame Atomic Absorption Spectroscopy. EPA/600/4-89/011 (NTIS PB89-181945).

Edwards, T.K. and G.D. Glysson. 1988. Methods for Measurement of Fluvial Sediment. U.S. Geological Survey Open File Report 86-531, 118 pp.

Engel, T.M., R.A. Kornfeld, J.S. Warner, and K.D. Andrews. 1988. Screening for Semivolatile Organic Compounds for Extractability and Aqueous Stability by SW-846 Method 3510. EPA/600/4-88/005 (NTIS PB88-161559).

Fishman, M.J. and W.L. Bradford. 1982. A Supplement to Methods for the Determination of Inorganic Substances in Water and Fluvial Sediments. U.S. Geological Survey Open-File Report 82-272. (Superseded by Fishman and Friedman, 1989.)

Fishman, M.J. and L.C. Friedman (eds.). 1989. Methods for Determination of Inorganic Substances in Water and Fluvial Sediments, 3rd edition. U.S. Geological Survey Techniques of Water-Resources Investigations TWRI 5-A1, 545 pp. (Supersedes Brown et al. [1970], Skougstad et al. [1979], and Fishman and Bradford [1982].)

Fleischauer, H.L. and P.R. Engelder. 1985. Procedures for Reconnaissance Stream-Sediment Sampling. GJ/TMC-14 (NTIS DE85-009789). Department of Energy Technical Measurements Center, Grand Junction, CO.

Francis, C.W., M.P. Maskarinec, and D.W. Lee. 1988. Physical and Chemical Methods for the Characterization of Hazardous Wastes. Oak Ridge National Laboratory, Oak Ridge, TN, 27 pp.

Fresenius, W., K.E. Quentin, and W. Schneider (eds.). 1988. Water Analysis: A Practical Guide to Physico-Chemical and Microbiological Water Examination and Quality Assurance. Springer-Verlag, New York, NY.

Friedman, L.C. and D.E. Erdmann. 1982. Quality Assurance Practices for the Chemical and Biological Analyses of Water and Fluvial Sediments. U.S. Geological Survey Techniques of Water-Resources Investigations TWRI 5-A6, 181 pp.

Gammage, R.B. and B.A. Berven (eds.). 1992. Hazardous Waste Site Investigations: Toward Better Decisions. Lewis Publishers, Boca Raton, FL, 288 pp. [Proceedings of 10th ORNL Life Science Symposium, 23 papers mainly focussing on field screening and analytical methods]

(Table 9-11 Field and Laboratory Analytical References)

Ghiorse, W.C. and D.L. Balkwill. 1985. Microbiological Characterization of Subsurface Environments. In: Ground Water Quality, C.H. Ward, W. Giger, and P.L. McCarty (eds.), Wiley Interscience, New York, NY, pp. 386-401.

Goerlitz, D.F. and E. Brown. 1972. Methods for Analysis of Organic Substances in Water. U.S. Geological Survey Techniques of Water-Resources Investigations TWRI 5-A3. [Updated by Wershaw et al. (1987)]

Greeson, P.E., T.A. Ehlke, G.A. Irwin, B.W. Lium, and K.V. Slack (eds.). 1977. Methods for Collection and Analysis of Aquatic Biological and Microbiological Samples. U.S. Geological Survey Techniques of Water-Resources Investigations TWRI 5-A4, 332 pp. [Updated by Britton an Greeson (1989)]

Guy, H.P. 1969. Laboratory Theory and Methods for Sediment Analysis. U.S. Geological Survey Techniques of Water-Resources Investigations TWRI 5-C1, 58 pp.

Hach Company. 1991. Handbook for Waste Analysis, 2nd edition. Hach Company, Loveland, CO, 166 pp. [Methods for use of field kits for 14 elements and 6 solids residue tests]

Hatayama, H.K., J.J. Chen, E.R. de Vera, R.D. Stephens, and D.L. Storm. 1980. A Method for Determining the Compatibility of Hazardous Wastes. EPA/600/2-80/076 (NTIS PB80-221005), 149 pp.

Hoddinott, K.B. and T.A. O'Shay (eds.). 1993. Application of Agricultural Analysis in Environmental Studies. ASTM STP 1162, American Society for Testing and Materials, Philadelphia, PA, 178 pp. [Covers general soil tests, nutrient status, organic constituents, and heavy metal content]

Howard, P.H., J. Saxena, P.R. Durkin, and L.-T. Ou. 1975. Review and Evaluation of Available Techniques for Determining Persistence and Routes of Degradation of Chemical Substances in the Environment. EPA 560/5-75/006.

Jackson, M. 1958. Soil Chemical Analysis. Prentice-Hall, Englewood Cliffs, NJ, 498 pp.

Jackson, M. 1979. Soil Chemical Analysis--Advanced Course, 2nd ed. Madison, WI (published by author), 898 pp.

Johnson, W. and J. Maxwell. 1981. Rock and Mineral Analysis, 2nd ed. John Wiley & Sons, NY, 489 pp.

Kerr, P.F. 1959. Optical Mineralogy, 3rd edition. McGraw-Hill, New York, NY. 442 pp.

Klute, A. (ed.). 1986. Methods of Soil Analysis, Part 1: Physical and Mineralogical Methods, 2nd edition. Agronomy Monograph No. 9, American Society of Agronomy, Madison, WI, 1188 pp. (1965 1st edition was edited by C.A. Black.) [50 chapters covering field and laboratory methods]

Kopp, J.F. and G.D. McKee. 1983. Methods for Chemical Analysis of Water and Wastes, 3rd edition. EPA/600/4-74/020 (NTIS PB84-128677). (Supersedes report with the same title dated 1979.) [chemical analytical procedures used in U.S. EPA laboratories for examining

(Table 9-11 Field and Laboratory Analytical References)

Sampling and Monitoring 555

ground and surface water, domestic and industrial waste effluents, and treatment process samples]

Lichtenberg, J., J. Winter, C. Weber, and L. Fradkin. 1988. Chemical and Biological Characterization of Municipal Sludges, Sediments, Dredge Spoils, and Drilling Muds. ASTM STP 976, American Society for Testing and Materials, Philadelphia, PA, 512 pp.

Long, S.E. and T.D. Martin. 1989. Methods for the Determination of Inorganic Compounds in Drinking Water: Methods 300.0 and 200.8. (NTIS PB90-215021), U.S. Environmental Protection Agency Environmental Monitoring Systems Laboratory, Cincinnati, OH, 158 pp.

Longbottom, J.E. and J.J. Lichtenberg. 1982. Methods for Organic Chemical Analysis of Municipal and Industrial Wastewater. EPA/600/4-82/057 (NTIS PB83-201798). [Describes tests for 15 groups of organic chemicals and includes an appendix defining procedures for determining the detection limit of an analytic methods; the test procedures in this manual are cited in Table IC (organic chemical parameters) and 1D (pesticide parameters) in 40 CFR 136.3(a)]

McKeague, J.A. (ed.). 1978. Manual on Soil Sampling and Methods of Analysis, 2nd edition. Canadian Society Soil Science, Ottawa, Ontario. [Superseded by Carter (1993)]

Mill, T., W.R. Mabey, D.C. Bomberger, T.-W. Chou, D.G. Hendry, and J.H. Smith. 1982. Laboratory Protocols for Evaluating the Fate of Organic Chemicals in Air and Water. EPA-600/3-82-022 (NTIS PB83-150888).

Minnich, M. 1993. Behavior and Determination of Volatile Organic Compounds in Soil: A Literature Review. EPA/600/R-93/140 (NTIS PB94-100153), 104 pp. [Includes chapters on field and laboratory methods to determining soil gas and soil VOCs]

Moore, D.M. and R.C. Reynolds. 1989. X-Ray Diffraction and the Identification and Analysis of Clay Minerals. Oxford University Press, New York, NY, 353 pp.

Mueller, W. and D.L. Smith (Compilers). 1991. Compilation of EPA's Sampling and Analysis Methods. Lewis Publishers, Chelsea, MI, 465 pp. (On diskette: EPA's Sampling and Analysis Methods Database, Vol. 1 [Industrial Chemicals], Vol. 2 [Pesticides, Herbicides, Dioxins and PCBs], and Vol. 3 [Elements and Water Quality Parameters].) [Summary information on more than 150 EPA-approved, and a total of 650, sampling and analysis methods for industrial chemicals, pesticides, elements, and water quality parameters]

Murdoch, A., S.D. MacKnight (eds.). 1994. Handbook of Techniques for Aquatic Sediments Sampling, 2nd ed. Lewis Publishers, Boca Raton, FL, 256 pp.

Nelson, M.R. 1988. Index to EPA (Environmental Protection Agency) Test Methods. EPA/901/3-88/001. U.S. EPA Region 1, Boston, MA. [Index of more than 700 air, water and waste measurement methods]

Noblett, J.G. and J.M. Burke. 1990. FGD Chemistry and Analytical Methods Handbook, 1: Process Chemistry--Sampling, Measurement, Laboratory, and Process Performance Guidelines, Revision 1. EPRI CS-3612, Electric Power Research Institute, Palo Alto, CA. (Originally published in 1984. See Radian Corporation [1988] for Vol. 2.) [Covers sampling,

(Table 9-11 Field and Laboratory Analytical References)

measurement, and laboratory and process performance guidelines]

O'Shay, T.A. and K.B. Hoddinott (eds.). 1994. Analysis of Soils Contaminated with Petroleum Constituents. ASTM STP 1221, American Society for Testing and Materials, Philadelphia, PA, 120 pp. [9 papers]

Ostroff, A.G. 1965. Introduction of Oilfied Water Technology. Prentice-Hall, Englewood Cliffs, NJ, 412 p.

Page, A.L., R.H. Miller, and D.R. Keeney (eds.). 1982. Methods of Soils Analysis, Part 2-- Chemical and Microbiological Properties, 2nd ed. ASA Monograph 9, American Society of Agronomy, Madison, WI. [54 chapters]

Palmer, M. 1985. Methods Manual for Bottom Sediment Sample Collection. EPA/905/4-85/004 (NTIS PB86-107414), 52 pp.

Pfaff, J.D. 1981. Methods for the Determination of Chemical Contaminants in Drinking Water: Instructors Handbook. EPA/430/1-81/023 (NTIS PB81-234312), 179 pp.

Plumb, Jr., R.H. 1981. Procedures for Handling and Chemical Analysis of Sediment and Water Samples. Technical Report EPA/CE-81/1, U.S. Army Engineer Waterways Experiment Station, Vicksburg, MS, 71 pp.

Pritchard, P.H. and A.W. Bourquin. 1984. The Use of Microcosms for Evaluation of Interactions Between Pollutants and Microorganisms. Adv. Microbial Ecol. 7:133-215.

Provost, L.P. and R.S. Elder. 1985. Choosing Cost-Effective QA/QC Programs for Chemical Analysis. EPA/600/4-85/056 (NTIS PB85-241461).

Radian Corporation. 1988. FGD Chemistry and Analytical Methods Handbook, 2: Chemical and Physical Test Methods, Revision 1. EPRI CS-3612, Electric Power Research Institute, Palo Alto, CA. [Originally published in 1984; see Noblett and Burke (1990) for Vol. 1. Presents 54 physical-testing and chemical-analysis methods for FGD reagents, slurries, and solids]

Rainwater, F.H. and L.L. Thatcher. 1960. Methods for Collection and Analysis of Water Samples. U.S. Geological Survey Water-Supply Paper 1454. [Describes types of methods, choice of analytical methods for water samples, and specific analytical procedures for over 40 inorganic water parameters]

Rosswall, T. (ed.). 1973. Modern Methods in the Study of Microbial Ecology. Bulletins from the Ecological Research Committee, Swedish Natural Science Research Council, Stockholm, 17. [Includes about 80 papers and short communications presented at a Symposium held at Uppsala, Sweden in 1972. Major sessions included: (2) Techniques for the observation of microcosms in soil and water; (3) Isolation and characterization of microorganisms; (4) Techniques for the determination of microbial activity in relation to ecological investigations; (5) Estimation of microbial growth rates under natural conditions; (6) Model systems; (7) Mathematical models and systems analysis in microbial ecology. Also includes summary of a panel discussion on problems of assessing the effect of pollutants on microorganisms]

(Table 9-11 Field and Laboratory Analytical References)

Sharma, J. 1979. Manual of Analytical Quality Control for Pesticides and Related Compounds in Humans and Environmental Samples: A Compendium. EPA/600/1-79/008 (NTIS PB298-711).

Silvestri, A., M. Razalis, A. Goodman, P. Vasquez, and A.R. Jones, Jr. 1981. Development of an Identification Kit for Spilled Hazardous Materials. EPA/600/2-8/194 (NTIS PB82-110727).

Skoog, D.A. 1985. Principles of Instrumental Analysis, 3rd edition. Saunders College Publishing, Philadelphia, PA.

Skougstad, M.W., et al. (eds.). 1979. Methods for Determination of Inorganic Substances in Water and Fluvial Sediments, 2nd edition. U.S. Geological Survey Techniques of Water-Resources Investigations TWRI 5-A1, 626 pp. (Superseded by Fishman and Friedman, 1989.)

Smith, K.A. (ed.). 1991. Soil Analysis: Modern Instrumental Methods, 2nd edition. Marcell Dekker, New York, NY. [14 chapters]

Smith, R.-K. 1994. Handbook of Environmental Analysis. Genium Publishing Corp., Schenectady, NY.

Smith, K.A. and C.E. Mullins. 1991. Soil Analysis: Physical Methods. Marcel Dekker, New York, NY, 620 pp.

Soil Conservation Service (SCS). 1984. Procedures for Collecting Soil Samples and Methods of Analysis for Soil Survey. Soil Survey Investigations Report No. 1, U.S. Government Printing Office. (Supersedes 1972 report with same document number titled Soil Survey Laboratory Methods and Procedures for Collecting Soil Samples.) [Laboratory analytical methods for soil physical and chemical properties]

Stockham, J. and E. Fochtman. 1979. Particle Size Analysis. Ann Arbor Science Publishers, Ann Arbor, MI, 140 pp.

Thatcher, L.L., V.J. Janzer, and K.W. Edwards. 1977. Methods for Determinations of Radioactive Substances in Water and Fluvial Sediments. U.S. Geological Survey Techniques of Water-Resources Investigations TWRI 5-A5, 95 pp.

Topp, G.C., W.D. Reynolds, and R.E. Green (eds.). 1992. Advances in Measurement of Soil Physical Properties: Bringing Theory into Practice. SSSA Special Publication 30, Soil Science Society of America, Madison, WI, 304 pp.

U.S. Department of Energy (DOE). Various dates. The Environmental Survey Manual. DOE/EH-0053: Vol. 1 (August 1987; Chapter 8, 2nd edition. January 1989—Sampling and Analysis Phase); Vol. 2 (August 1987—Appendices A,B, and C); Vol. 3 (2nd edition. January 1989—Appendix D, Parts 1, 2, and 3; Organic and Inorganic Analysis Methods and Non-Target List Parameters); Vol. 4 (2nd edition. January 1989—Appendix D, Part 4; Radiochemical Analysis Procedures); Vol. 5. (2nd edition. January 1989—Appendices: E, Field Sampling; F, Quality Assurance; G, Decontamination; H, Sample Management; I, Sample Handling, Transport and Documentation; J, Health and Safety; and K, Sampling and Analysis Plan).

(Table 9-11 Field and Laboratory Analytical References)

U.S. Environmental Protection Agency (EPA). 1986. Test Methods for Evaluating Solid Waste, 3rd edition. EPA/530/SW-846 (NTIS PB88-239223); First update, 3rd edition. EPA/530/SW-846.3-1 (NTIS PB89-148076). (2nd edition was published in 1982 (NTIS PB87-1200291); current edition and updates available on a subscription basis from U.S. Government Printing Office, Stock #955-001-00000-1. [Revised final draft of Chapter 11 (Ground-Water Monitoring System Design, Installation, and Operating Practices) contains extensive new guidance. Vols. 1A (Metallic Analytes), IB (Organic Analytes), and IC (Miscellaneous Test Methods) cover laboratory methods; Vol. II covers field methods (Part IV defines acceptable and unacceptable designs and practice for ground-water monitoring)]

U.S. Environmental Protection Agency (EPA). 1988. Field Screening Methods Catalog: User's Guide. EPA/540/2-88/005. FSMC System Coordinator, OERR, Analytical Operations Branch (WH-548-A), U.S. EPA, Washington, DC. (Also available as a computerized information retrieval system.) [Guide for selection of instrumental methods for field screening of inorganic and organic contaminants; covers 26 specific field screening methods]

U.S. Environmental Protection Agency (EPA). 1992. Sediment Classification Methods Compendium. EPA/823-R-92-006 (NITS PB93-115186). [Compendium of 10 methods for assessing chemically contaminated sediments]

U.S. Environmental Protection Agency. (EPA). 1990a. Manual for Certification of Laboratories Analyzing Drinking Water: Criteria and Procedures Quality Assurance, 3rd edition. EPA/570/9-90/007 (NTIS PB90-220500).

U.S. Environmental Protection Agency (EPA). 1990b. Methods for the Determination of Organic Compounds in Drinking Water, Supplement 1. EPA/600/4-90/020 (NTIS PB91-108266), 225 pp. (Supersedes report with similar title, EPA/600/7-90/XXX [PB90-215039].)

U.S. Environmental Protection Agency (EPA). 1991a. Methods for the Determination of Metals in Environmental Samples. EPA/600/4-91/010 (NTIS PB91-231498), 305 pp. [Covers 13 laboratory analytical methods for 35 metals]

U.S. Environmental Protection Agency (EPA). 1991d. Methods for the Determination of Organic Compounds in Drinking Water, Revise edition. EPA/600/4-88/039 (NTIS PB91-231480), 395 pp. (Supersedes 1988 publication with same title and EPA document number [NTIS PB89-220461].)

U.S. Environmental Protection Agency (EPA). 1993. Field Methods Compendium (FMC) Draft. OERR # 9285.2-11, Analytical Operation Branch, Hazardous Site Evaluation Division, Office of Emergency and Remedial Response. [Available in Wordperfect 5.1 electronic format from OERR, Washington, DC]

U.S. Environmental Protection Agency (EPA) and State Laboratories. 1993. U.S. EPA Manual of Chemical Methods for Pesticides and Devices, 2nd ed. Association of Organic and Analytical Chemists, McLean, VA, 792 pp. [284 methods]

U.S. Interagency Committee. 1940. Field Practice and Equipment Used in Sampling Suspended Sediment. Report No. 1, St. Paul Engineer District Sub-Office, Hydraulic Laboratory, University of Iowa, Iowa City, 175 pp.

(Table 9-11 Field and Laboratory Analytical References)

U.S. Interagency Committee. 1941. Methods of Analyzing Sediment Samples. Report No. 4, St. Paul Engineer District Sub-Office, Hydraulic Laboratory, University of Iowa, Iowa City, 203 pp.

van Lagen, B. (ed.). 1993. Manual for Chemical Soil Analyses. Department of Soil Science and Geology, Agricultural University, Wageningen, The Netherlands, 100 pp. [Emphasis on methods for forms of sulfur]

Velthorst, E.J. 1993. Manual for Chemical Water Analyses. Department of Soil Science and Geology, Agricultural University, Wageningen, The Netherlands, 100 pp. [Emphasis on methods related to acid rain]

Wagner, T.P. 1990. Hazardous Waste Identification and Classification Manual. Van Nostrand Reinhold, New York, NY, 248 pp.

Wagner, R.E. (ed.). 1994. Guide to Environmental Analytical Methods, 2nd ed. Genium Publishing, Schenectady, NY. [Summary/comparisons of methods in: EPA CLP SOW (Inorganics/Organics Analyses), EPA 200 series (water and wastes), EPA 500 series (organic compounds in drinking water), EPA 600 series 40 CFR Part 136, EPA/SW-846 (solid waste), and APHA (1990-17th edition)]

Wallace, J.R., L. Alden, F.S. Bonom, J. Nichols, and E. Sexton. 1984. Method of Chemical Analysis of Oil Shale Wastes. EPA/600/2-84/110 (NTIS PB84-211226), 250 pp.

Walsh, L.M. and J.D. Beaton (eds.). 1973. Soil Testing and Plant Analysis, Rev. Ed. Soil Science Society of America, Madison, WI, 491 pp. [See Westerman (1990) for 3rd edition]

Warner, D.L. and J.H. Lehr. 1977. An Introduction to the Technology of Subsurface Wastewater Injection. EPA 600/2-77-240 (NTIS PB279-207).

Watkins, J.W. 1954. Analytical Methods of Testing Waters to be Injected Into Subsurface Oil-Productive Strata. U.S. Bureau of Mines Report of Investigations 5031, 29 p.

Watts, R.R. (ed.). 1980. Manual of Analytical Methods for the Analysis of Pesticides in Humans and Environmental Samples. EPA/600/8-80/38 (NTIS PB82-208752).

Weaver, R.W. et al. (eds.). 1994. Methods of Soil Analysis: Microbial and Biochemical Processes. SSSA Book Series No. 5, Soil Science Society of America, Madison, WI, 1,121 pp. [Supersedes related sections of Page et al. (1982)]

Webster, J.J. et al. 1985. Determination of Microbial Cell Numbers in Subsurface Samples. Ground Water 23:17-25.

Wershaw, R.L., M.J. Fishman, R.R. Bragge, and L.E. Lowe (eds.). 1987. Methods for the Determination of Organic Substances in Water and Fluvial Sediments. U.S. Geological Survey Techniques of Water-Resources Investigations TWRI 5-A3, 80 pp. (Revision of Goerlitz and Brown, 1972.)

(Table 9-11 Field and Laboratory Analytical References)

Westerman, R.L (ed.). 1990. Soil Testing and Plant Analysis, 3rd edition. Soil Science Society of America, Madison, WI, 812 pp. [Methods for analysis of soil and plants focussing on use for assessing nutritional requirements of crops, efficient fertilizer use, saline-sodic conditions, and toxicity of metals]

Willard, H.H., L.L. Merritt, Jr., J.A. Dean, and F.A. Settle, Jr. 1988. Instrumental Methods of Analysis, 7th edition. Wadsworth Publishing Co., Belmont, CA.

Williams, R.D. and G.E. Schuman (eds.). 1987. Reclaiming Mine Soils and Overburden in the Western United States: Analytic Parameters and Procedures. Soil Conservation Society of America, Ankeny, IA, 336 pp. [Focus on potential toxic elements]

Wolbach, C.D., R.R. Whitney, and U.B. Spannegel. 1984. Design and Development of a Hazardous Waste Reactivity Testing Protocol. EPA/600/2-84/057 (NTIS PB84-158807), 143 pp. [Field test kit for on-site compatibility testing of wastes]

Wood, W.W. 1976. Guidelines for Collection and Field Analysis of Ground-Water Samples for Selected Unstable Constituents. U.S. Geological Survey Techniques of Water-Resources Investigations TWRI 1-D2, 24 pp.

(Table 9-11 Field and Laboratory Analytical References)

CHAPTER 10

USE OF MODELS AND COMPUTERS IN CONTAMINANT INVESTIGATIONS

10.1 Uses of Models and Computers 562

    10.1.1 Government Decision-Making 563
    10.1.2 Site Assessment 564
    10.1.3 Ground-Water Protection and Remediation 564

10.2 Mathematical Approaches to Modeling 565

    10.2.1 Deterministic vs. Stochastic Models 565
    10.2.2 System Spatial Characteristics 569
    10.2.3 Analytical vs. Numerical Models 569
    10.2.4 Grid Design 570

10.3 Classification of Ground Water Computer Codes 573

    10.3.1 Porous Media Flow Codes 576
    10.3.2 Porous Media Solute Transport Codes 577
    10.3.3 Hydrogeochemical Codes 578
    10.3.4 Specialized Codes 579

10.4 General Code Selection Considerations 581

    10.4.1 Ground Water Flow Parameters 582
    10.4.2 Contaminant Transport Parameters 584
    10.4.3 Computer Hardware and Software 585
    10.4.4 Usability and Reliability 586
    10.4.5 Quality Assurance/Quality Control 587
    10.4.6 Potential Pitfalls 589

10.5 Other Geoenvironmental Computer Applications 589

    10.5.1 Parameter Identification 590
    10.5.2 Code Pre- and Postprocessors 590
    10.5.3 Statistical Analysis 591
    10.5.4 Data Plotting and Other Graphics 591
    10.5.5 Geographic Information Systems 594
    10.5.6 Other Data Management and Processing 599

10.6 Guide to Major Information Sources* 599

* Appendix F contains citations for table and figure sources.

10.1 Uses of Models and Computers

Models, in the broadest sense, are simplified descriptions of an existing physical system. Any soil or ground-water investigation that does more than simply collect and tabulate data involves modeling. A preliminary conceptual model, describing the soil and ground-water system, is tested by collecting data. If the data fit the preliminary model, it is accepted; otherwise, the model must be revised.

The meaning of the term *model* varies depending on the context in which it is used. Most models fall into one of the following categories:

- *Qualitative* descriptions of how processes operate in a system. For example, Table 3-10 and Figure 3-6 illustrate two qualitative models for biodegradation of organic contaminants.

- Simplified *physical* representations of the system such as "sand tank" physical aquifer models and laboratory batch experiments to measure adsorption isotherms (Section 4.5.2).

- *Mathematical* representations of a physical or chemical system.

This chapter focuses on models that can be expressed in mathematical form and adapted for use in computer codes. In this context, the American Society for Testing and Materials defines model and computer code as follows (ASTM D978/Table A-14):

- A *model* is an assembly of concepts in the form of a mathematical equation that portrays understanding of a natural phenomenon.

- A *computer code* is the assembly of numerical techniques, bookkeeping, and control languages that represents the model from acceptance of input data and instruction to delivery of output.

The term computer or *digital model* is often used interchangeably with computer code, but may also have a broader meaning to include the conceptual model of a site which forms the basis for entry of spatial and temporal data into a code. Computer codes may take the form of hard-paper documentation in the format of whatever programming language was used, or on an electronic medium (disks or tapes). Vadose zone and ground-water modeling with computers is a specialized field that requires considerable training and experience. In the last few decades, literally hundreds of computer codes for simulating various aspects of soil ground-water systems have been developed. Refinements to existing codes and development of new codes proceed at a rapid pace.

The great advantage of the computer is that large amounts of data can be processed quickly and experimental modifications can be made with minimal effort, so that many possible situations for a given problem can be studied in great detail. The greatest danger in computer modeling is that it is easy to generate outputs that have very little relationship to what is happening in the real world. Section 10.4.6 identifies more specific potential pitfalls associated with computer modeling.

The major uses for computer modeling in the context of soil and ground-water contamination assessment and control can be broadly classified into (1) government decision-making, (2) site assessment, and (3) ground-water protection and remediation. These categories are not mutually exclusive. For example, the results of modeling for site assessment, ground-water protection and remediation all provide information for decision-making.

10.1.1 Government Decision-Making

Computers can assist government decisions concerning ground-water protection in the areas of (1) policy formulation, (2) rulemaking, and (3) regulatory action.

Policy Formulation. Modeling is not explicitly required in any federal water resource legislation, but is often the method of choice to assess the need for new legislation or the requirements of existing legislation (OTA, 1982/T10-6). The Netherlands and Israel have used computer modeling as a cornerstone in the development of ground-water protection policies (van der Heijde, 1985)[1]. The U.S. EPA has used a generic vertical-horizontal spread (VHS) model to determine when solid waste needs to be treated as hazardous waste (National Research Council, 1990/T10-6).

Rulemaking. The U.S. EPA's Underground Injection Program regulations on hazardous waste disposal injection restrictions and requirements for Class I wells exemplify the use of modeling to assist in rulemaking (Proposed Rules: 52 Federal Register 32446-32476, August 27, 1987; Final Rules: 53 Federal Register 28118-28157, July 26, 1988). The 10,000-year no-migration standard in 40 CFR 128.20(a)(1) for injected wastes is based, in part, on numerical modeling of contaminant transport in four major hydrogeologic settings by Ward et al. (1987).[2] Furthermore, worst-case modeling of typical injection sites by EPA formed the basis for the decision not to require routine modeling of dispersion in no-migration petitions.

Regulatory Action. Ground-water flow and possibly solute transport modeling are required to obtain a permit to inject hazardous wastes into Class I wells. Permitting decisions involving activities that may pose a threat to ground-water quality, such as landfills and surface storage of industrial wastes, commonly require ground-water simulations to demonstrate that no hazard exists.

---

[1] van der Heijde, P.K.M. 1985. The Role of Modeling in Development of Ground-Water Protection Policies. Ground Water Modeling Newsletter 4(2).

[2] Ward, D.S., D.R. Buss, and J.W. Mercer. 1987. A Numerical Evaluation of Class I Injection Wells for Waste Confinement Performance, Final Report. Prepared for U.S. EPA by GeoTrans, Herndon, VA.

10.1.2 Site Assessment

Use of modeling and computer codes can be valuable in two phases of site-specific soil and ground-water investigations: (1) site characterization and (2) exposure/risk assessment.

Site Characterization. Relatively simple models (such as analytical equations for ground-water flow that do not require use of computers--see Table 7-6) may be useful at the early stage for roughly defining the possible magnitude of a contaminant problem. Solute transport models that account for dispersion but not retardation (Section 10.3.2) may be useful in providing a worst-case analysis of the situation. They may help in defining the size of the area to be studied and in siting of monitoring wells. If more sophisticated computer modeling is planned, the specific code to be used will, to a certain extent, guide site characterization efforts by the aquifer parameters required as inputs to the model. Site characterization, particularly where water quality samples are tested for possible organic contaminants, can generate large amounts of data. Computers are invaluable in compiling and processing these data sets (Sections 10.5.2 and 10.5.6).

Exposure and Risk Assessment. There is growing use of exposure assessments across the U.S. EPA's regulatory programs (U.S. EPA 1988a and 1088b/T10-6). This applications requires use of solute transport models that estimate the concentration of contaminants in ground water at points of potential exposure. In the case of ground-water contamination, the results of an exposure assessment will often determine whether remediation will be required. Section 11.4 discusses exposure and risk assessment methods in more detail.

10.1.3 Ground-Water Protection and Remediation

Computer models are useful for analyzing site and regional data to develop ground-water protection areas and to design ground-water cleanup systems.

Ground-Water Protection. Ground-water flow models are being increasingly used to assist in the process of delineating wellhead protection areas. About a dozen of ground-water flow codes have been cited in the literature as having been used in actual wellhead delineation investigations. These codes fall into three general categories:

1. *Numerical* codes developed for general ground-water flow modeling (MODFLOW and USGS-2D FLOW) that are used to define zone of influence (ZOI), cone of depression (COD), and/or zone of contribution (ZOC).

2. Simpler analytical and semianalytical *capture zone* codes for defining the zone of influence and/or zone of contribution of one or more pumping wells.

3. Flowpath analysis codes (called *pathline tracing* or *reverse path*), typically analytical or semianalytical, for calculating time-of-travel and/or velocity using the output from numerical modeling or capture zone codes.

Section 10.3 provides additional information on the classification of these codes. Section 11.2 provides additional information on methods for delineating wellhead protection areas.

Remediation. Predictive models can be particularly valuable in estimating the possible effectiveness of alternative approaches to remediating ground-water contamination (Boutwell et al., 1985/T10-6). Chapter 12 discusses approaches to remediation. Table 10-1 summarizes the type of modeling required for the following remediation design features: (1) capping, grading, and revegetation, (2) ground-water pumping, (3) wastewater injection, (4) interceptor trenches, (5) impermeable barriers, (6) subsurface drains, (7) solution mining, and (8) excavation.

10.2 Mathematical Approaches to Modeling

Models and codes are usually described by the number of dimensions simulated and the mathematical approaches used. At the core of any model or computer code are *governing equations* that represent the system being modeled. Many different approaches to formulating and solving the governing equations are possible. The specific numerical technique embodied in a computer code is called an *algorithm*. The following discussion compares and contrasts some of the most important choices that must be made in mathematical modeling.

10.2.1 Deterministic vs. Stochastic Models

A *deterministic* model presumes that a system or process operates such that the occurrence of a given set of events leads to a uniquely definable outcome. The governing equations define precise cause-and-effect or input-response relationships. In contrast, a *stochastic* model presumes that a system or process operates such that factors contributing to an outcome are uncertain. Such models calculate the probability, within a desired level of confidence, of a specific value occurring at any point.

Most available models are deterministic. However, the heterogeneity of hydrogeologic environments, particularly the variability of parameters such as porosity and hydraulic conductivity, plays a key role in influencing the reliability of predictive ground-water modeling (Smith, 1987).[3] Beven (1989)[4] argues that this heterogeneity

---

[3] Smith, L. 1987. The Role of Stochastic Modeling in the Analysis of Groundwater Problems. Ground Water Modeling Newsletter 6(1).

[4] Beven, K. 1989. Changing Ideas in Hydrology--The Case of Physically-Based Models. J. Hydrology 105:157-172.

Table 10-1 Modeling Designed-System Alterations and Corrective Action

Design Feature	Effects on on Ground Water	Type of Model Required	Typical Modeling Problems
Capping, grading, and revegetation	-Reduction of infiltration. -Reduction of successive leachate generation. -Changes in heads, direction of flow, and contaminant migration.	-Unsaturated zone model, vertical layered.	-Parameters related to leaching characteristics of reworked soil
Ground-water pumping (and optional reinjection of treated water)	-Controlled plume removal. -Changes in heads and direction of flow.	-Saturated zone model, two-dimensional areal, axisymmetric or three-dimensional -Well or series of wells assigned to individual node.	-Representing partial penetration.
Wastewater injection	-Plume generation. -Changes in heads, direction of flow, and contaminant migration.	-Saturated zone model, two-dimensional area, axisymmetric or three-dimensional. -Density-dependent flow. -Temperature difference effects.	-Representing density-dependent effects.

Table 10-1 (cont.)

Design Feature	Effects on Ground Water	Type of Model Required	Typical Modeling Problems
Interceptor trenches	-Plume removal.	-Saturated zone model, two-dimensional areal or cross-sectional, or three-dimensional. -Trenches are represented by line of notes with assigned heads.	-Representing partial penetration, resolution near trenches.
Impermeable barrier (optional drainage system to prevent mounding)	-Containment of polluted water. -Routing unpolluted ground water around site. -Changes in heads and direction of flow.	-Saturated zone model, two-dimensional areal or cross-sectional, or three-dimensional. -Possibly two-dimensional cross-sectional unsaturated zone model for liners.	-Representing partial penetration, flow and transport around end of barrier(s). -Conductivity liner or barrier material. -Large changes in conductivity between neighboring elements. -Differences in required grid resolution.
Subsurface drains	-Removal of leachate. -Changes in heads, direction of flow, and contaminant migration.	-Saturated or combined unsaturated-saturated zone model -Two-dimensional cross-sectional or three-dimensional	-Resolution near drain.

Table 10-1 (cont.)

Design Feature	Effects on on Ground Water	Type of Model Required	Typical Modeling Problems
Solution mining	-Removal of contaminants after induced mobilization.	-Saturated or combined unsaturated-saturated zone model. -Two-dimensional areal, cross-sectional or three-dimensional. -Lines of sources (injection) and sinks (removal).	-Parameters related to mobilization (sorption coefficient, retardation coefficient).
Excavation	-Removal of waste material and polluted soil. -Changes in hydraulic characteristics and boundary conditions. -Changes in heads and direction of flow.	-Unsaturated, saturated, or combined unsaturated-saturated zone model. -For unsaturated some models minimal one-dimensional vertical. -For other types minimal two-dimensional, cross-sectional.	-Parameters of backfill material.

Source: Adapted by van der Heijde et al. (1988), from Boutwell et al. (1985).

creates fundamental problems in the application of physically based deterministic models.

The governing equations for both deterministic and stochastic models can be solved either analytically or numerically (van der Heijde et al., 1988/T10-6). Vomvoris and Gelhar (1986/T10-6) provide some simple analytical examples of stochastic prediction of dispersive contaminant transport. Gómez-Hernández and Gorelick (1989)[5] review the literature on approaches to stochastic simulation of ground-water model parameters such as hydraulic conductivity, leakance and recharge and Table 10-6 identifies other major references that address this topic.

10.2.2 System Spatial Characteristics

The spatial characteristics of a system can be modeled in two major ways. *Lumped-parameter* systems are used when the total system is located at a single point. *Distributed-parameter* systems define cause-and-effect relations for specific points or areas. *Input-response* or *black box* models do not explicitly address spatial characteristics, but instead empirically relate observations of different variables, such as response of water levels to recharge.

The distributed-parameter approach is most frequently used in ground-water modeling and the rest of this chapter focuses on models of this type. The mathematical framework for distributed-parameter models includes (1) one or more partial differential equations called field equations; (2) initial and boundary conditions; and (3) solution procedures (Bear, 1979/T2-5). Depending on the solution method used, such models are characterized as analytical, semianalytical, or numerical (next section).

10.2.3 Analytical vs. Numerical Models

A model's governing equation can be solved either analytically or numerically. *Analytical* models use exact closed-form solutions of the appropriate differential equations. The solution is continuous in space and time. In contrast, *numerical* models apply approximate solutions to the same equations. *Semianalytical* models use numerical techniques to approximate complex analytical solutions, allowing a discrete solution in either time or space. Models using a closed-form solution for either the space or time domain and additional numerical approximations for the other domain are also considered semianalytical.

Analytical models provide exact solutions, but employ many simplifying assumptions concerning the ground-water system, its geometry, and external stresses in order to produce tractable solutions (Walton, 1984a/Table E-2). This places a burden on the user to test and justify the underlying assumptions and simplifications (Javendel et al., 1984/T10-6).

---

[5] Gómez-Hernández, J.J. and S.M. Gorelick. 1989. Effective Groundwater Parameter Values: Influence of Spatial Variability of Hydraulic Conductivity, Leakance, and Recharge. Water Resources Research 24(3):405-419.

Semianalytical models can provide streamline and travel-time information through numerical or analytical expression in space or time. This information is especially useful for delineation of wellhead protection areas. *Analytic element models* are a relatively recent development in semianalytical modeling of regional groundwater flow. These use approximate analytic solutions by superposing various exact or approximate analytic functions, each representing a particular feature of the aquifer (Haitjema, 1985;[6] Strack, 1987/T10-6). A major advantage of these models is greater flexibility in incorporating varying hydrogeology and stresses compared to analytic models without significantly increasing the need for data (van der Heijde and Beljin, 1988/T10-6).

Numerical models are much less burdened by the simplifying assumptions used in analytical models and are therefore inherently capable of addressing more complicated problems, but they require significantly more input and their solutions are inexact (numerical approximations). For example, the assumptions of homogeneity and isotropicity are unnecessary because the model can assign point (nodal) values of transmissivity and storage. Likewise, the capacity to incorporate complex boundary conditions provides greater flexibility. The user, however, faces difficult choices regarding time steps, spatial grid designs, and ways to avoid truncation errors and numerical oscillations (Remson et al., 1971; Javendel et al., 1984/T10-6). Improper choices may result in errors unlikely to occur with analytical approaches (e.g., mass imbalances, incorrect velocity distributions, and grid-orientation effects). Table 10-2 summarizes the advantages and disadvantages of analytical and numerical models.

10.2.4 Grid Design

A fundamental requirement of the numerical approach is the creation of a grid that represents the aquifer being simulated (see Figure 10-1). This grid of interconnected nodes, at which process input parameters must be specified, forms the basis for a matrix of equations to be solved. A new grid must be designed for each site-specific simulation based on data collected during site characterization and the conceptual model that is developed for the physical system. Grid design is one of the most critical elements in the accuracy of computational results (van der Heijde et al., 1988/T10-6).

The grid design is influenced by the choice of numerical solution technique. Numerical solution techniques include (1) finite-difference methods (FD), (2) integral finite-difference methods (IFDM), (3) Galerkin and variational finite-element methods (FE), (4) collocation methods, (5) boundary (integral) element methods (BIEM or BEM), (6) particle mass tracking methods, such as random walk (RW), and (7) the method of characteristics (MOC) (Huyakorn and Pinder, 1983/T10-6; Kinzelbach, 1986/T10-6). Figure 10-2 illustrates grid designs involving FD and FE methods for the same well field.

---

[6] Haitjema, H.M. 1985. Modeling Three-Dimensional Flow in Confined Aquifers by Superposition of Both Two- and Three-Dimensional Analytic Functions. Water Resources Research 21(10):1557-1566.

Table 10-2 Advantages and Disadvantages of Analytical and Numerical Methods

Advantages	Disadvantages
Analytical Methods	
1. Efficient when data on the system are sparse or uncertain.	1. Limited to certain idealized conditions with simple geometry; may not be applicable to field problems with complex boundary conditions.
2. Economical.	
3. Good for initial estimation of magnitude of contamination.	2. Most cannot handle spatial or temporal variations in system.
4. Rough estimates often possible from existing data sources.	
5. Input data for computer codes usually simple.	
Numerical Methods	
1. Easily handle spatial and temporal variations of system.	1. Achieving familiarity with complex numerical programs can be time consuming and expensive.
2. Easily handle complex boundary conditions.	2. Errors due to numerical dispersion (artifacts of the computation process) may be substantial for transport models.
3. Three-dimensional transient problems can be treated without much difficulty.	3. More data input is usually required.
	4. Preparation of input data is usually time consuming.[a]

[a] Model-specific preprocessors (Section 10.5.2) and GIS/GSIS (Section 10.5.5) can greatly facilitate data input.

Source: Adapted from Boulding (1991b).

Finite-difference and finite-element methods are the most frequently used numerical solution techniques. The finite-difference method approximates the solution of partial differential equations by using finite-difference equivalents, whereas the finite-element method approximates differential equations by an integral approach.

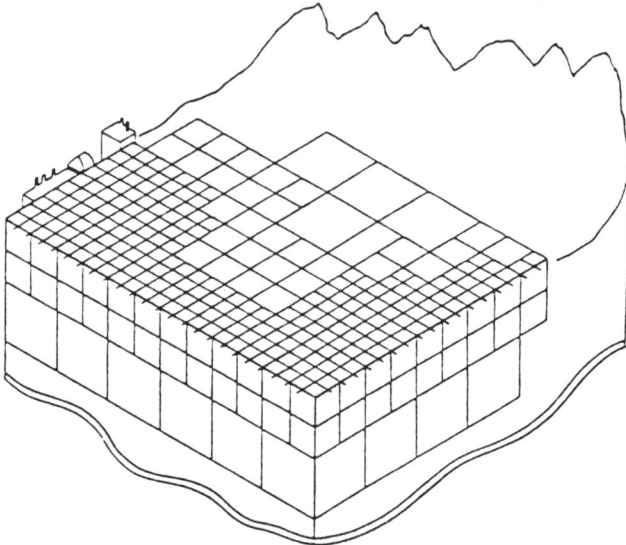

(a) Values for natural process parameters would be specified at each node of the grid in performing simulations. The grid density is greatest at the source and at potential impact locations.

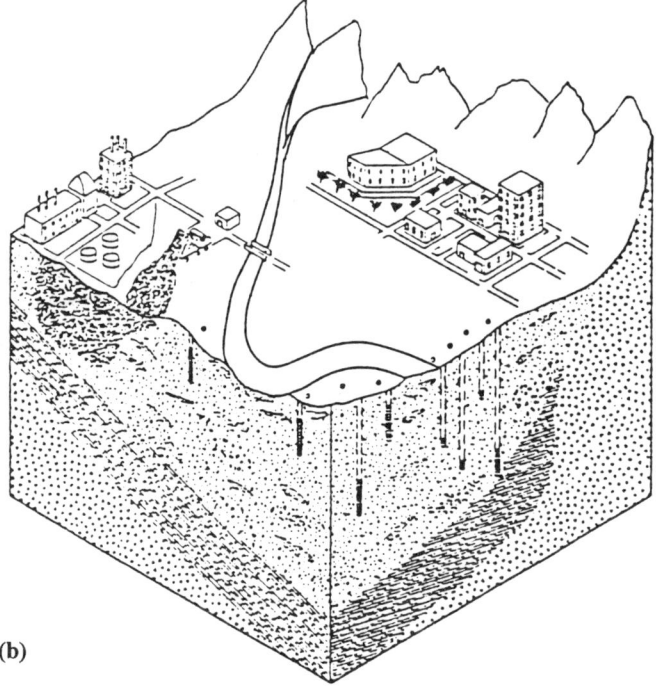

(b)

Figure 10-1 Three-dimensional grid (a) representing a complex geologic setting (b) with pumping wells downgradient from contaminant source (Keely, 1987).

Use of Models and Computers 573

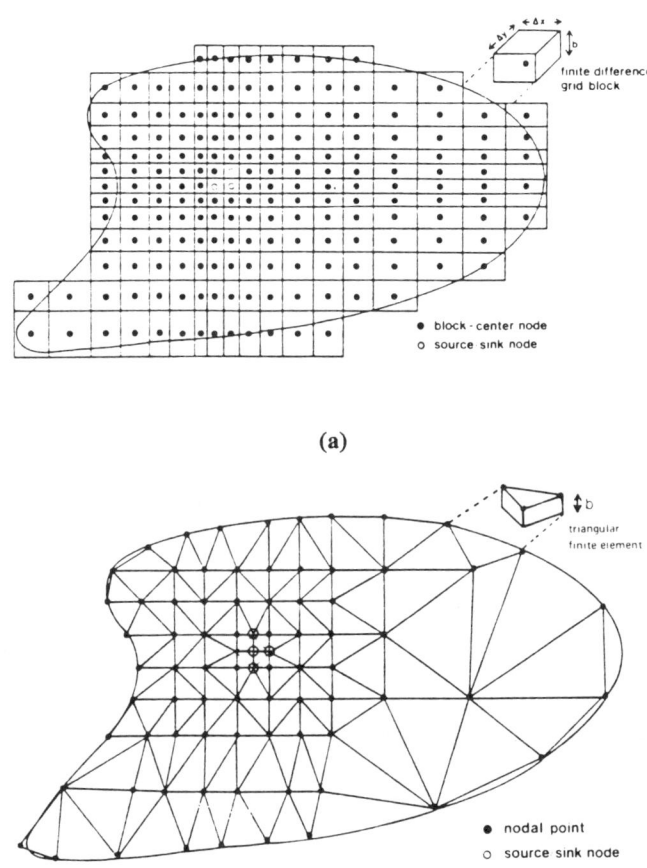

Figure 10-2 Comparison of (a) finite-difference and (b) finite-element grid configurations for modeling the same well field (Mercer and Faust, 1981). See Appendix F for full credit.

Table 10-3 compares the relative advantages and disadvantages of the two methods. Most available numerical computer codes use finite-difference grids. Only in cases of highly irregular geometries do finite-element solutions have a clear advantage.

10.3 Classification of Ground-Water Computer Codes

Terminology for classifying vadose zone and ground-water computer codes according to the kind of system they simulate is sometimes inconsistent and confusing. There are so many different ways that such models can be classified (i.e., porous vs. fractured rock flow, saturated vs. unsaturated flow, mass flow vs. chemical transport, single phase vs. multiphase, isothermal vs. variable temperature) that a systematic classification cannot be developed that would not require placement of single codes in multiple categories.

Table 10-3 Advantages and Disadvantages of Finite-Difference and Finite-Element Numerical Methods

Advantages	Disadvantages
Finite-Difference Method	
Intuitive basis	Low accuracy for some problems
Easy data entry	(mainly solute transport)
Efficient matrix techniques	Rectangular grids required
Programming changes easy	
Finite-Element Method	
Flexible grid geometry	Complex mathematical basis
High accuracy possible	More complex programming
Evaluates cross-product terms better	

Source: Adapted from Boulding (1991b).

The most comprehensive system for describing and classifying ground-water codes has been developed by the International Ground Water Modeling Center (van der Heijde and Einawawy, 1993). This system uses four major categories for describing codes in IGWMC's database:

- **Model objectives**, such as applicability to certain types of ground-water management problems, and code development objectives (research, education, general use). Section 10.1 discusses model objectives or uses further.

- **Processes modeled**, such as saturated or unsaturated flow, solute transport and fate, and hydrogeochemical speciation.

- **Physical system characteristics modeled**, such as porous medium or fractured rock, single or multiple aquifers, confined or unconfined, etc.

- **Mathematical approaches**, such as general nature of equations (deterministic or stochastic) and solution method (analytical or numerical). Section 10.2 discusses this aspect further.

Table 10-4 present a classification system for vadose zone and ground-water computer codes that is useful for evaluating models in relation to contaminant investigations. It combines elements of the process- and physical-system-characteristics categories described above. The table has four major categories of codes and eleven major subdivisions that are discussed below. This classification scheme differs from others (see, for example, Mangold and Tsang, 1987; van der Heijde et al., 1988/T10-6), by distinguishing among solute transport models that simulate (1) only dispersion, (2) chemical reactions with a simple retardation or degradation factor, and (3) complex chemical reactions.

The literature on ground-water codes is most confused when it comes to terminology for models that address subsurface chemistry. For example, the term "hydrochemical" has been applied to completely different types of codes. Rice (1986/T10-6) and van der Heijde et al. (1988/T10-6) used the term hydrochemical for codes in the hydrogeochemical category in Table 10-4, whereas Mangold and Tsang (1987/T10-6) used the term geochemical for such models and the term hydrochemical to describe coupled geochemical and flow models (chemical-reaction transport codes in Table 10-4). More recently van der Heijde and Einawawy (1993/T10-6) have used the term *hydrogeochemical* for codes that model aqueous chemical reactions without regard to transport and that term is used here. The major types of models are discussed briefly below.

Table 10-4 Classification of Vadose Zone and Ground Water Flow and Transport Computer Codes

Type of Code	Description/Uses
Flow (Porous Media)	
Saturated	Simulates movement of water in saturated porous media. Used primarily for analyzing ground-water availability.
Variable saturated (vadose zone)	Simulates unsaturated flow of water in the vadose (unsaturated) zone. Used in study of soil-plant relationships, hydrologic cycle budget analysis.
Solute Transport (Porous Media)	
Dispersion	Simulates transport of conservative contaminants (not subject to retardation) by adding a dispersion factor into flow calculations. Used for nonreactive contaminants such as chloride and for worst-case analysis of contaminant flow.
Retardation/ degradation	Simulates transport contaminants that are subject to partitioning or transformation by the addition of relatively simple retardation or degradation factors to create algorithms for advection-dispersion flow. Used where retardation and degradation are linear with respect to time and do not vary with respect to concentration. Vadose zone transport models of this type use variable-saturated flow governing equations with retardation/degradation factors.
Chemical reaction	Combines an advection-dispersion code with a hydrogeochemical code (see below) to simulate chemical speciation and transport. Integrated codes solve all mass momentum, energy-transfer, and chemical reaction equations simultaneously for each time interval. Two-step codes first solve mass momentum and energy balances for each time step and then reequilibrate the chemistry using a distribution-of-species code. Used primarily for modeling behavior of inorganic contaminants.

Table 10-4 (cont.)

Type of Code	Description/Uses
Hydrogeochemical Codes	
Thermodynamic	Processes empirical data so that thermodynamic data at a standard reference state can be obtained for individual species. Used to calculate reference state values for input into hydrogeochemical speciation calculations.
Distribution of species (equilibrium)	Solves a simultaneous set of equations that describe equilibrium reactions and mass balances of the dissolved elements.
Reaction progress (mass transfer)	Calculates both the equilibrium distribution of species (as with equilibrium codes) and the new composition of the water as selected minerals are precipitated or dissolved.
Specialized Codes	
Fractured rock	Simulates flow of water in fractured rock. Available codes cover the spectrum of advective flow, advection-dispersion, heat, and chemical transport.
Heat transport	Simulates flow where density-induced and other flow variations resulting from fluid temperature differences invalidate conventional flow and chemical transport modeling. Used primarily in modeling of radioactive waste and deep-well injection.
Multiphase liquid flow	Simulates movement of immiscible fluids (water and nonaqueous phase liquids) in either the vadose or saturated zones. Used primarily where contamination involves liquid hydrocarbons or solvents.
Gas flow and vapor transport	Simulate liquid/gas phase changes and movement of vapors in the vadose zone.

Source: Adapted from Boulding (1991b).

10.3.1 Porous Media Flow Codes

Modeling of saturated flow in porous media is relatively straightforward; consequently, by far the largest number of codes are available in this category. Modeling variably saturated flow in porous media (most typically soils and unconsolidated geologic material) is more difficult because hydraulic conductivity varies with changes in water content in unsaturated materials. Such codes typically must model processes such as capillarity, evapotranspiration, diffusion, and plant water uptake. van der Heijde and Einawawy (1993/T10-6) provide summary information on porous media flow codes in the following categories (number in parentheses indicates number of codes): (1) saturated flow/analytical (20); (2) saturated flow/2-dimensional numerical (55); (3) saturated flow/3-dimensional numerical (15); (4) saturated flow/analytical inverse (aquifer test) models (29); (5) saturated flow numerical inverse models (3); (6) saturated flow/pathline models (20). Note that a single model may fall in more than one category. The same report includes information on 25 variably saturated (vadose zone) flow numerical models and seven codes for estimating vadose zone flow parameters.

10.3.2 Porous Media Solute Transport Codes

Solute transport codes fall into three major categories (see Table 10-4 for descriptions): (1) dispersion codes, (2) retardation/degradation codes, and (3) chemical-reaction transport codes. *Dispersion* codes differ from saturated flow codes only in having a dispersion factor and may be required if conservative contaminants such as nitrates are of potential concern. *Retardation/degradation* codes are slightly more sophisticated because they add a retardation or degradation factor to the mass transport and diffusion equations. However, as discussed below, such codes must be used with caution. *Chemical-reaction transport* codes are the most complex (but not necessarily most accurate) because they couple geochemical codes with flow codes. Chemical-reaction transport codes may be classified as *integrated* or *two-step* codes (see Table 10-4).

Mechanisms for reducing the concentration of contaminants in an aquifer are generally too complex and difficult to predict for selection as criteria for wellhead protection (U.S. EPA, 1987).[7] Accurate modeling of contaminant transport is limited by fundamental problems, including (1) inability to describe mathematically some processes, (2) complex mechanisms that are beyond the capability of available numerical techniques, and (3) difficulty in obtaining enough data of sufficient quality to calibrate models (van der Heijde and Beljin, 1988/T10-6).

Hydrodynamic dispersion, the process by which contaminants may travel **faster** than would be expected from simple ground-water flow calculations, must be considered when modeling for delineation of ground-water protection areas or for exposure assessment. As noted in Section 4.4.2, dispersion at the microscopic scale is such a minor component of ground-water movement that it can generally be ignored. Although dispersion at this scale results in a faster arrival time, it also reduces concentration levels and, consequently, can be considered an attenuating process. Transport of contaminants by macroscopic dispersion, on the other hand, is best addressed using methods that account for the effect of aquifer heterogeneity on the speed of ground-water flow (Sections 2.5.3 and 7.5.2). For simple methods this would involve using the upper range of estimated or measured hydraulic conductivity in ground-water flow calculations. Numerical computer codes allow design of the grid to account for more highly transmissive layers.

---

[7] U.S. Environmental Protection Agency (EPA). 1987. Guidelines for Delineation of Wellhead Protection Areas. EPA/440/6-87-010 (NTIS PB88-111430). [R. Hoffer may also be cited as author]

Bradbury et al. (1991)[8] provide a good example of the difference a single highly transmissive layer in an aquifer can make in travel times. At the Sevastopol site in Door County, Wisconsin, where the aquifer is in fractured dolomite, time of travel to the upgradient ground-water divide based on calculations using a potentiometric surface map was 100 years (Figure 10-3). Ground-water simulations using PATH3D that accounted for a fracture zone that was observed by hydrogeologic mapping to occur at a depth of 170 feet below the ground surface resulted in a travel time of **one** year from the ground-water divide (Figure 10-4).

Retardation processes (Section 4.5) provide an unstated safety factor to delineations based on advective flow to the extent that they diminish concentration as a contaminant moves through an aquifer. These codes are most commonly used in heavily contaminated settings to help develop remediation strategies. However, such codes may have value for wellhead protection as a means of quantifying the safety factor that is contained in delineations based on other methods or for further evaluation of the possible risks associated with potential contaminant sources within a wellhead protection area (Section 11.4).

In general, modeling of contaminant transport processes is more complex that modeling ground-water flow alone, and consequently the results of any individual model should be viewed with caution. For example, Arnold (1992/T10-6) used 8 numerical models and 4 analytical models to estimate attenuation of BTX (benzene, toluene, xylene) from a gasoline spill 4,000 feet from the Mississippi River and found a two-order of magnitude range in the predicted concentrations.

van der Heijde and Einawawy (1993) provide summary information on solute transport codes in the following categories (number in parentheses indicates number of codes): (1) saturated zone/analytical (42); saturated zone/2-dimensional numerical (38); (3) saturated zone/3-dimensional numerical (20); (4) unsaturated zone (52).

10.3.3 Hydrogeochemical Codes

Hydrogeochemical codes simulate chemical reactions in ground-water systems without considering transport processes. These fall into three major categories (see Table 10-4): (1) thermodynamic codes, (2) distribution-of-species codes, and (3) reaction progress codes. By themselves, hydrogeochemical codes can provide qualitative insights into the behavior of contaminants in the subsurface. Chemical transport modeling of any sophistication requires coupling geochemical codes with flow codes (see previous section). Over 50 geochemical codes have been described in the literature (Nordstrom and Ball, 1984/T10-6); 27 are sufficiently documented to be summarized by van der Heijde and Einawawy (1993/T10-6).

---

[8] Bradbury, K.R., M.A. Muldoon, A. Zaporozec, and J. Levy. 1991. Delineation of Wellhead Protection Areas in Fractured Rocks. EPA/570/9-91-009, 144 pp.

Use of Models and Computers 579

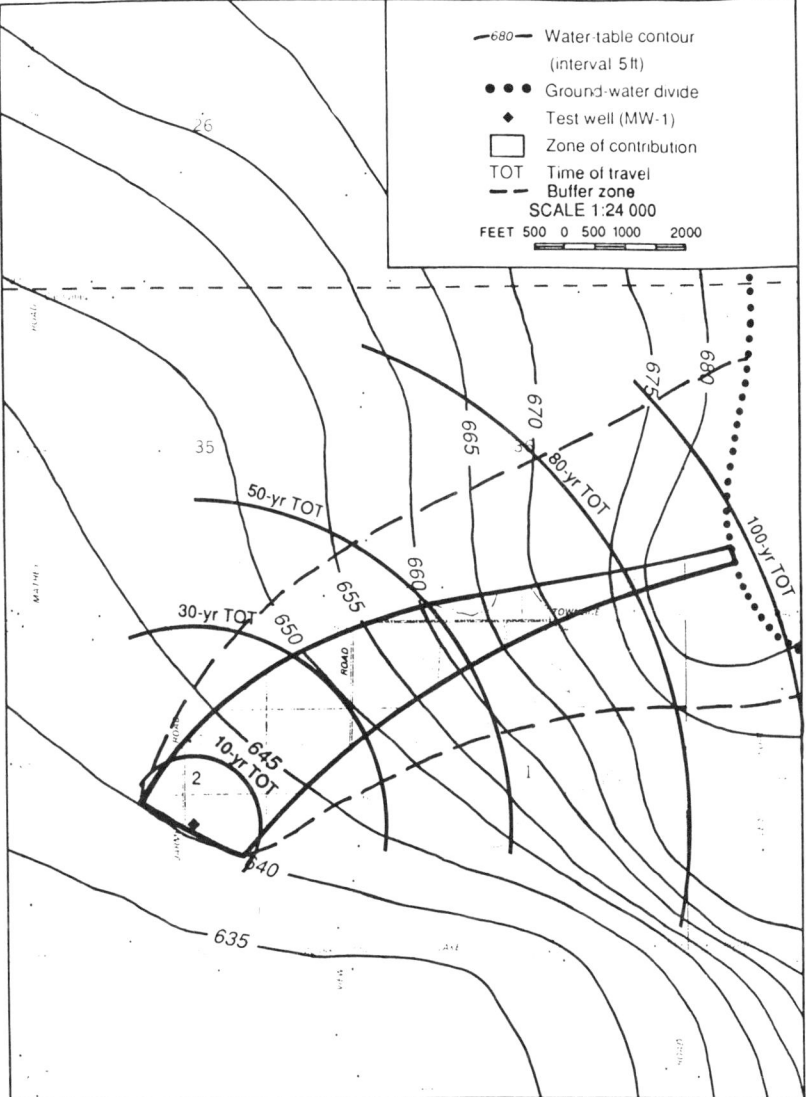

Figure 10-3 Time-of-travel contours in a dolomite aquifer based on potentiometric surface map (Bradbury et al., 1991).

10.3.4 Specialized Codes

This category contains special cases of flow codes and solute transport codes (see Table 10-4), including (1) fractured rock, (2) heat transport, (3) multiphase liquid flow, and (4) gas flow and vapor transport. *Fractured rock* creates special problems in the modeling of contaminant transport for several reasons. First, mathematical representation is more complex due to the possibility of turbulent flow and the need to consider roughness effects. Furthermore, precise field characterization of fracture

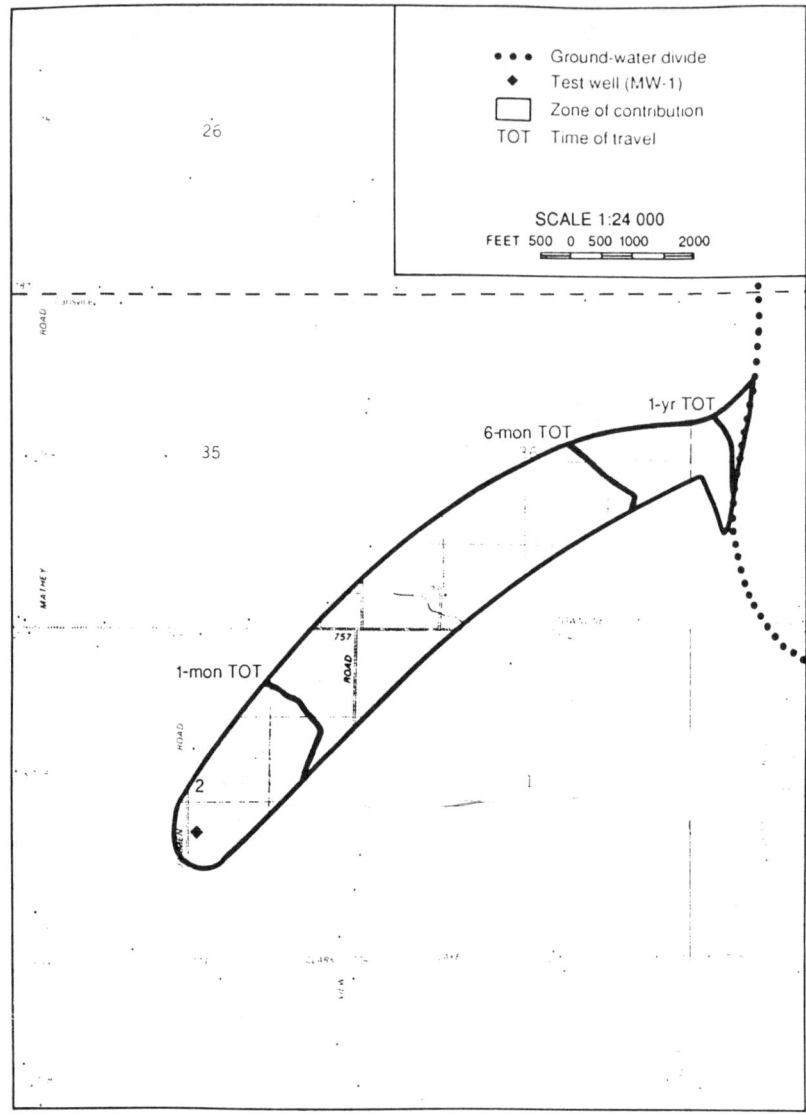

Figure 10-4 Time-of-travel contours in a dolomite aquifer based on numerical modeling of high-conductivity zone (Bradbury et al., 1991).

properties that influence flow, such as orientation, length, and degree of connection between individual fractures, is extremely difficult. In spite of these difficulties, much work is being done in this area (Schmelling and Ross, 1989/T10-6); van der Heijde and Einawawy (1993) provide summary information on 31 fractured rock models, and increase in four models since IGWMC's previous compilation (van der Heijde et al., 1988/T10-6). It is worth noting that no fractured rock models satisfied screening criteria developed by van der Heijde and Beljin (1988/T10-6) for codes potentially suitable for delineation of wellhead protection areas.

Heat transport models have been developed primarily in connection with enhanced oil recovery operations (Kayser and Collins, 1986/T10-6) and programs assessing disposal of radioactive wastes. Van der Heijde and Einawawy (1993/T10-6) present summary information on 51 heat transport models, and increase in 15 since IGWMC's 1988 survey. Early work in *multiphase fluid* flow centered in the petroleum industry focusing on oil-water-gas phases. In the last decade, multiphase behavior of nonaqueous phase liquids in near-surface ground-water systems has received increasing attention. However, the number of codes capable of simulating multiphase liquid flow is still limited, with van der Heijde and Einawawy (1993/T10-6) summarizing 12 codes. It is only relatively recently that the need for modeling *vapor phase transport* of volatile organic contaminants such as gasoline in the vadose zone has been recognized. This is especially important for design of remediation systems involving light nonaqueous phase liquids. This is a rapidly developing area of research, as demonstrated by the fact that 14 of the 18 gas flow and vapor transport listed by van der Heijde and Einawawy (1993) have been released since 1990.

10.4 General Code Selection Considerations

All modeling involves simplifying assumptions concerning parameters of the physical system that is being simulated. Furthermore, these parameters will influence the type and complexity of the equations that are used to represent the model mathematically. Major considerations in selecting a computer code for a particular application include:

- Ground-water flow parameters (Section 10.4.1)

- Contaminant transport parameters (Section 10.4.2)

- Available computer hardware (Section 10.4.3)

- Usability and reliability of the code (Section 10.4.4)

- Model quality assurance and quality control (Section 10.4.5)

Worksheet D-W5 can be used to develop PC-based ground-water computer code specifications and to evaluate code suitability for a specific site. Information on site characteristics should first be entered from Worksheets D-W1 (Water Well/Aquifer Data), D-W3 (Possible Aquifer Boundaries). and D-W4 (Aquifer Characteristics for the Selection of Analytical Solutions to Ground Flow in the Vicinity of Wells). Next fill out the section of Worksheet D-W5 on model system requirements.[9] Make as many copies of the worksheet with the site and system information as codes to be evaluated and fill in the rest of the worksheet using information sources identified in

---

[9] This section of the worksheet can actually be used two ways: (1) determine whether a particular code can run on an existing computer system, and (2) develop minimum specifications for a new computer system for a code or codes that have been identified a suitable for a particular site.

the rest of this section and in Section 10.6. Final code selection will require a qualitative evaluation of the suitability of each code in relation to site characteristics and project objectives.

10.4.1 Ground-Water Flow Parameters

There are six major parameters of ground-water systems that must be considered when selecting a computer code for simulating ground-water flow: (1) type of aquifer (confined, unconfined, leaky), (2) matrix characteristics (porous, fractured) (3) degree of homogeneity and isotropy, (4) phases (density differences, NAPLs), (5) number of aquifers, and (6) flow conditions (steady or transient). Figure 10-5 provides a decision tree for selection of a ground-water flow code based on characteristics of the system.

Type of Aquifer. Confined aquifers with uniform thickness are easier to model than unconfined aquifers because the transmissivity (Section 2.6.2) remains constant.[10] The thickness of unconfined aquifers varies with fluctuations in the water table, thus complicating calculations. Similarly, simulation of variable-thickness confined aquifers is complicated by the fact that velocities will generally increase in response to reductions in the distance between confining beds and will decrease in response to increases in these distances.

Matrix Characteristics. Flow in porous media is much easier to model than in rocks with fractures or solution porosity. This is because (1) equations governing laminar flow are simpler than those for turbulent flow, which may occur in fractures, and (2) porosity and hydraulic conductivity can be more easily estimated for porous media.

Homogeneity and Isotropy. Homogeneous and isotropic aquifers are easiest to model because their properties do not vary in any direction (Section 2.5.3). If hydraulic properties and concentrations are uniform vertically, and in one of two horizontal dimensions, a *one-dimensional* simulation is possible. Horizontal variations in properties combined with uniform vertical characteristics can be modeled in *two dimensions*. Most natural aquifers, however, show variation in all directions and consequently require *three-dimensional* simulation, which also necessitates more extensive site characterization data. The spatial uniformity or variability of aquifer parameters such as recharge, hydraulic conductivity, porosity, transmissivity, and storativity (Section 2.6) will determine the number of dimensions to be modeled.

Phases. Flow of ground water and contaminated ground water in which the dissolved constituents do not create a plume that differs greatly from the unpolluted aquifer in density or viscosity (see Section 4.4.3) are easier to simulate than multiple phases.

---

[10] The statement here and in following discussions about relative complexity of models has more affect on those who develop codes than users. From the user's standpoint more complex models may require more computer memory or be slower in performing calculations compared to models of simpler systems, but are not necessarily more complex to use.

Use of Models and Computers 583

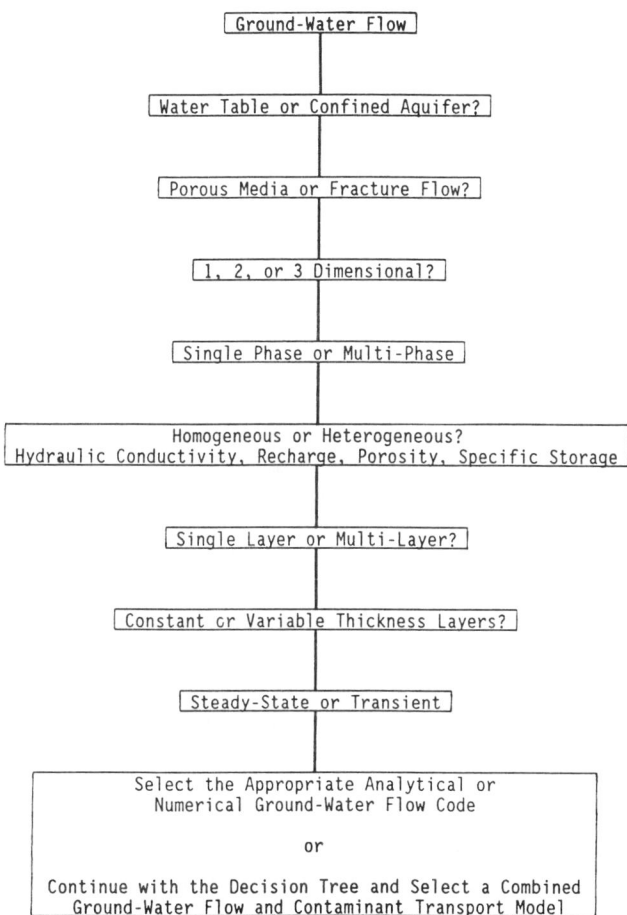

Figure 10-5 Decision tree for selection of ground-water flow code (U.S. EPA, 1988a).

Number of Aquifers. A single aquifer is easier to simulate than multiple aquifers.

Flow Conditions. *Steady-state* flow, where the magnitude and direction of flow velocity are constant with time at any point in the flow field, is much easier to simulate than transient flow. *Transient*, or unsteady flow, occurs when the flow varies in the saturated zone in response to variations in recharge or discharge rates. Both terms can also be applied to *unsaturated* flow in the vadose zone, also called *variably saturated* flow.

10.4.2 Contaminant Transport Parameters

Important contaminant parameters that should be considered when selecting a model for simulating transport or fate of contaminants include (1) concentration, (2) type of source (point, line, or areal), (3) type of source release (slug, continuous, constant, variable), (4) dispersion, (5) sorption, (6) degradation, and (7) density/viscosity effects. Figure 10-6 provides a decision tree for selection of a contaminant transport model based on these factors.

Concentration. The simplest way to model contaminant transport in the subsurface is to specify a starting concentration in the ground water without considering the type of source.

Type of Source. For more sophisticated simulation purposes, sources can be characterized as point, line, area, or volume. A *point* source enters the ground water at a single point such as a pipe outflow or injection well and can be simulated with either a one-, two-, or three-dimensional model. An example of a *line* source would be contaminants leaching from the bottom of a trench. An *area* source enters the ground water through a horizontal or vertical plane. The actual contaminant source may occupy three dimensions outside of the aquifer, but contaminant entry into the aquifer can be represented as a plane for modeling purposes. Leachate from a waste lagoon or an agricultural field are examples of area sources. A *volume* source occupies three dimensions within an aquifer. A DNAPL that has sunk to the bottom of an aquifer (Figure 4-10) would be a *volume* source. Line and area sources may be simulated by either two- or three-dimensional models, whereas a volume source would require a three-dimensional model.

Type of Source Release. Release of an instantaneous pulse, or *slug*, of contaminant is easier to model than a *continuous* release. A continuous release may be either *constant* or *variable*. Figures 4-7b and 4-8b show different contaminant plume configurations resulting from continuous and slug releases, respectively, and Figure 4-15 illustrates some effects of variations in the rate of release on contaminant plume shape.

Dispersion. Accurate contaminant modeling requires incorporation of transport by dispersion (Section 4.4.2). Unfortunately, the conventional convective-dispersion equation often does not accurately predict field-scale dispersion (U.S. EPA, 1988/T10-6) which typically requires numerical modeling of aquifer heterogeneity.

Adsorption. It is easiest to simulate adsorption with a single distribution or partition coefficient (4.5.2). Nonlinear adsorption and temporal and spatial variation in adsorption are more difficult to model.

Degradation. As with adsorption, simulation of degradation is easiest when using a simple first-order degradation coefficient. Second-order degradation coefficients, which result from variations in various parameters such as pH, substrate concentration, and microbial population, are much more difficult to model. Modeling of radioactive decay is complicated but easier to simulate with precision because decay chains are well known.

Use of Models and Computers 585

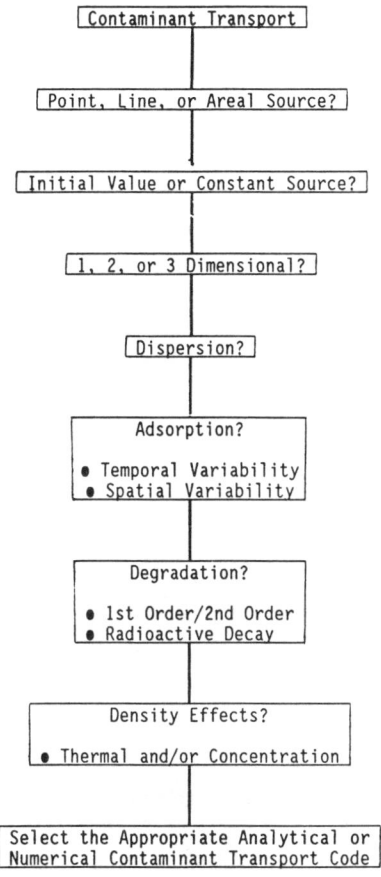

Figure 10-6 Decision tree for selection of a contaminant transport code (U.S. EPA, 1988a).

Density/Viscosity Effects. If temperature or salinity of the contaminant plume is much different than that of the pristine aquifer, simulations must include the effects of density and viscosity variations (see Section 4.4.3).

10.4.3 Computer Hardware and Software

The types of computer hardware (model, random access memory, available use, peripherals for printing code output, etc.) that are available is a primary consideration in selecting a ground-water computer code. Earlier codes depended heavily on mainframe computers (such as CDC, IBM, PRIME, UNIVAC, and VAX models). Rapid advances in microcomputer technology have resulted in increased availability of ground-water modeling software for personal computers.[11] This trend stems from significant improvements in the computing power and quality of printed outputs obtainable from personal computers. It is also due to the improved telecommunications capabilities of personal computers, which are now able to emulate the interactive terminals of large business computers so that vast computational power can be accessed and the results retrieved with no more than a phone call.

Many of the mathematical models and data packages have been "ported" or rewritten from mainframe computers to personal computers. Many more are now being written directly for this market. Table E-2 provides an index to more than 50 pubic domain and commercially available PC-based vadose zone and ground-water flow and contaminant transport codes. A major advantage of PC-based codes is the relatively low cost of both hardware (the necessary computer and peripherals can probably be obtained for less than $3,000) and software. Most of the codes identified in Table E-2 can be obtained for less than $100, and most others for $500 or less.

10.4.4 Usability and Reliability

An ongoing program at the International Ground Water Monitoring Center evaluates codes using performance standards and acceptance criteria (van der Heijde, 1987).[12] The IGWMC rates codes that are in its database using six *usability* and four *reliability* criteria. Favorable ratings for the usability criteria include:

- **Pre- and Postprocessors.** Code incorporates one or more of this type of code.

- **Documentation.** Code has an adequate description of user's instructions and sample problems using example data sets.

- **Hardware Dependency.** Code is designed to function on a variety of hardware configurations.

- **Support.** Code is supported and maintained by the developers or marketers.

Favorable ratings for the reliability criteria include:

- **Review.** Both theory behind the coding and the coding itself are peer reviewed.

---

[11] Most first-generation software for microcomputers was developed for DOC-based IBM PC/AT/XT and compatibles and typically require 640 K (kilobyte) RAM (random access memory). Second-generation DOS-based software typically requires a 386 or 486 CPU (central processing unit) with a math coprocessor and 2 MB (megabyte) RAM. Modeling software is increasingly becoming available for Macintosh computers.

[12] van der Heijde, P.K.M. 1987. Performance Standards and Acceptance Criteria in Groundwater Modeling. Ground Water Modeling Newsletter 6(2).

- **Verification.** Code has been verified (see next section below).

- **Field Testing/Validation.** Code has been extensively field tested for site-specific conditions for which extensive data sets are available.

- **Extent of Use.** Code has been used extensively by other modelers.

Van der Heijde and Einawawy (1993) provide ratings for several hundred models.

10.4.5 Quality Assurance/Quality Control

The increasing use of modeling and computer codes in regulatory settings where decisions may be contested in court requires careful attention to quality assurance and quality control in both model development and application. There are three major aspects to quality control for a site-specific application of a model (ASTM D5447/Table A-14):

1. *Sensitivity* is a measure of the degree to which model results are affected by changes in selected input parameters such as hydraulic properties.

2. *Calibration* is the process of refining input parameters that represent the hydrogeologic system being modeled to achieve a desired degree of correspondence between the simulation and observations of the actual system.

3. *Verification* requires that a code gives results that are in reasonable agreement with analytical solutions or other computer codes using the same data. Verification in a model application requires that a calibrated model successfully simulate a second set of field data measured under similar hydrologic conditions.

These levels of quality control address the soundness and utility of the model alone and do not treat questions of its application to a specific problem. Hence, at least two additional levels of quality control appear justified:

4. Critical review of the problem's conceptualization to ensure that the modeling effort considers all physical, chemical, and biological processes that may affect the problem

5. Evaluation of the specifics of the application, e.g., appropriateness of the boundary conditions, grid design, time steps, etc. Calibration and sensitivity analysis to determine if the model outputs vary greatly with changes in input parameters are important aspects of this process.

The term *validation* has sometimes been used in the ground-water modeling literature to refer to the highest level of QA/QC in which a model has been verified using real-world simulations in multiple locations. However, the this term has been the subject of some recent controversy. Bredehoeft and Konikow (1993)[13] suggested abandoning use of the term validation by the ground-water modeling community because it implies a precision that is not achieved in reality. In response, McCombie and McKinley (1993)[14] argued that the term validation is appropriate for describing the process of ensuring that mathematical models "ensure an acceptable level of predictive accuracy".

When model results are relatively insensitive to changes in values for input parameters, the accuracy of the input values is less of a concern than when a small change in an input parameter causes a large change in the model output. *Sensitivity testing* may be useful in guiding data collection for a site. Less attention need be given to estimating or measuring parameters that do not greatly affect the outcome of the modeling, whereas additional effort may be required to ensure that sensitive input parameters are measured accurately.

Whether the code has been verified or validated is an important criteria for selecting models. Verification is also desirable for site-specific applications, if it is possible to obtain a second set of field data measured under similar hydrologic conditions to the site-calibrated code. The code can be considered verified if it acceptably approximates the second data set. This can be determined by defining an acceptable level of departure between simulated values and the actual data set and calculating the difference between actual and simulated values (residuals). If these residuals fall within the range that was defined as acceptable, the model can be considered verified for application to that particular field situation.

Field validation of a numerical model consists of first calibrating the model using one set of historical records (e.g., pumping rates and water levels from a certain year), and then attempting to predict the next set of historical records. In the calibration phase, the aquifer coefficients and other model parameters are adjusted to achieve the best match between model outputs and known data; in the predictive phase, no adjustments are made (excepting actual changes in pumping rates, etc.).

Presuming that the aquifer coefficients and other parameters were known with sufficient accuracy, a mismatch means that either the model is not correctly formulated or that it does not treat all of the important phenomena affecting the situation being simulated (e.g., does not allow for leakage between two aquifers when this is actually occurring). Field validation is completed by conducting a *postaudit*, in which the predicted changes in responses to changes in the system are confirmed by field measurements.

---

[13] Bredehoeft, J.D. and L.F. Konikow. 1993. Ground-Water Models: Validate or Invalidate. Ground Water 21(2):178-179.

[14] McCombie, C. and I. McKinley. 1993. Validation--Another Perspective. Ground Water 31(4):530-531.

Use of Models and Computers 589

10.4.6 Potential Pitfalls

Computers can easily give a false sense of security or cause unwarranted confidence in the results. The adage "garbage in, garbage out" always applies. Use of the criteria described previously for selection of a code that is appropriate for a particular objective and site conditions should reduce the chances that simulations result in "garbage out". Nevertheless, it is useful to keep in mind pitfalls that can doom a ground-water modeling effort to failure (OTA, 1982; van der Heijde et al., 1985/T10-6):

- Inadequate conceptualization of the physical system, such as flow in fractured bedrock

- Use of insufficient or incorrect data

- Incorrect use of available data

- Use of invalid boundary conditions

- Selection of an inadequate computer code

- Incorrect interpretation of the computational results

- Imprecise or wrongly posed management problems.

Computer modeling requires expertise in both hydrogeology and computer technology. The technology and software may be more readily available than the expertise. When in doubt, experts in government, universities, or consultants with special expertise in computer modeling of ground-water should be consulted.

The Subcommittee D18.21 on Vadose Zone and Ground-Water Investigations of the American Society of Testing and Materials (ASTM) is actively developing guides covering various aspects of subsurface fluid flow modeling. The first to be approved, **Standard Guide for Application of a Ground-Water Flow Model to a Site-Specific Problem** (D5447/Table A-14), provides a good starting point for planning an investigation where computer modeling is going to be used.

10.5 Other Geoenvironmental Computer Applications

In addition to modeling of vadose zone and ground-water flow and contaminant transport, other uses of computers for the assessment of soil and ground-water contamination include: (1) parameter identification codes for estimation of vadose zone and aquifer properties, (2) code pre- and postprocessors to facilitate data entry and simulation outputs, (3) statistical analysis of characterization and monitoring data, (4) simple graphical analysis and presentation of data, (5) geographic and geoscientific information systems (GIS/GSIS), and (6) other data management and processing.

10.5.1 Parameter Identification

Parameter identification codes are most often used to estimate the aquifer parameters determining fluid flow and contaminant transport characteristics. Examples of such codes include annual recharge (Puri, 1984),[15] coefficients of permeability and storage (Shelton, 1982; Khan, 1986a and 1986b),[16] dispersivity (Güven et al., 1984; Strecker and Chu, 1986),[17] transmissivity and leakage factors for leaky confined aquifers (Mukhopadhyay, 1988),[18] and watershed parameters (Rajaram and Georgakakos, 1989).[19] Parameter estimation codes should **not** take the place of critical evaluation and interpretation of site data.

10.5.2 Code Pre- and Postprocessors

Data manipulation codes specifically designed to facilitate ground-water modeling efforts are becoming increasingly popular, because they simplify data entry (*preprocessors*) and facilitate the production of graphic displays of simulation results (*postprocessors*). Preprocessors can be designed for use with a specific code (for example, Moses and Herman, 1986)[20] or for generic use. An early example of a generic preprocessor is PIG (Srinivasan, 1984).[21] Commercial marketing of popular public domain ground-water models, such as USGS's MODFLOW, is typically based on pre- and postprocessors that have been added to the code.

---

[15] Puri, S. 1984. Aquifer Studies Using Flow Simulations. Ground Water 22(5):538-543.

[16] Khan, I.A. 1986a. Inverse Problem in Ground Water: Model Development. Ground Water 24(1):32-38.
   Khan, I.A. 1986b. Inverse Problem in Ground Water: Model Application. Ground Water 24(1):39-48.
   Shelton, M.L. 1982. Ground-Water Management in Basalts. Ground Water 20(1):86-93.

[17] Güven, O., F.J. Molz, and J.G. Melville. 1984. An Analysis of Dispersion in a Stratified Aquifer. Water Resources Research 20(10):1337-1354.
   Strecker, E.W. and W. Chu. 1986. Parameter Identification of a Ground-Water Contaminant Transport Model. Ground Water 24(1):56-62.

[18] Mukhopadhyay, A. 1988. Automated Computation of Parameters for Leaky Confined Aquifers. Ground Water 26(4):500-504.

[19] Rajaram, H. and K.P. Georgakakos. 1989. Recursive Parameter Estimation of Hydrologic Models. Water Resources Research 25(2):281-294.

[20] Moses, C.O. and J.S. Herman. 1986. Computer Notes: WATIN--A Computer Program for Generating Input Files for WATEQF. Ground Water 24(1):83-89.

[21] Srinivasan, P. 1984. PIG--A Graphic Interactive Preprocessor for Ground-Water Models. GWMI 84-15. International Ground Water Modeling Center, Butler University, Indianapolis, IN.

10.5.3 Statistical Analysis

Computers greatly ease statistical analysis by making it possible to perform rapid calculations using multiple statistical tests to a data set. Refer to Section 5.6 for discussion of basic statistical concepts. Major types of statistical analysis in soil and ground-water investigations include (1) trend analysis of time-series ground-water quality monitoring data, (2) multivariate and geostatistical analysis of spatial data, and (3) analysis of chemical analytical data for quality assurance/quality control purposes. Table E-1 identifies about a dozen commercial statistical software packages and Table E-2 identifies EPA-developed statistical software. The rest of this section provides brief descriptions of major EPA software (Refer to Appendix C for full citations).

The U.S. EPA's **GRITS/STAT** (Ground Water Information Tracking System with Statistical Analysis Capability) software functions as a database for ground-water quality monitoring data and performs the statistical tests recommended by EPA for analysis of ground-water monitoring data at RCRA facilities (U.S. EPA, 1992/Table E-2). It is available from EPA's Center for Environmental Research Information (see Introduction for address).

Geo-EAS (Geostatistical Environmental Assessment Software) is a collection of interactive software tools for performing two-dimensional geostatistical analyses of spatially distributed data (Englund and Sparks, 1991/Table E-2). It was developed by EPA's Environmental Monitoring Systems Laboratory in Las Vegas (EMSL/Las Vegas, P.O. Box 93478, Las Vegas, NV 89193-3478) and includes programs for data file management, data transformations, univariate statistics, variogram analysis, cross-validation, kriging, contour mapping, post plots, and line/scatter graphs in a user-friendly format. **GEOPACK** (Yates and Yates, 1990/Table E-2) is a competing geostatistical package developed by EPA's R.S. Kerr Environmental Research Laboratory (see address for CSMoS in Section 10.6). It performs geostatistical tests similar to Geo-EAS and also calculates a number a basic statistical parameters such as the mean, median, variance, standard deviation, etc. Each package is available from the respective EPA laboratory or the International Ground Water Modeling Center (address in Section 10.6).

EPA's **ASSESS** software (U.S. EPA, 1991a/Table E-2), developed for EPA's Environmental Monitoring Systems Laboratory in Las Vegas, performs statistical analysis of regular soil samples and QA/QC samples (duplicates, replicates etc.) to help evaluate whether data quality objectives are being met in sampling at hazardous waste sites. It is available from EPA's EMSL Las Vegas Laboratory (address above).

10.5.4 Data Plotting and Other Graphics

Computers can be used to present spatially related site data in a variety of forms:

- Maps that show the boundaries, surface topography, cultural features, and sampling points at a site.

- Borehole soil or lithologic logs that describe characteristics or changes in soil or geologic materials with depth.

- Monitoring well construction logs that document materials and methods used for constructing ground-water monitoring wells.

- Cross sections and fence diagrams developed from lithologic, geophysical, or other types of subsurface information that provide a two- or three-dimensional view of lateral changes in soils and lithology in the subsurface.

- Isopleth maps that show contours of equal concentration of contaminants or other parameters of interest and maps that represent data as contours (i.e., subsurface maps that show elevation contours of key lithologic units such as bedrock, clay, or sand layers).

- Volumetric data presentations, such as isopach maps, which show variations in the thickness of a horizon or strata, and three-dimensional portrayals of contaminant plumes in ground-water or zones of contamination in soil.

Contour maps, lithologic logs, cross sections, and isopleth maps are essential elements in developing a conceptual model of a site that provides a basis for such undertakings as developing grids and input values for numerical computer models of a site, deciding whether remedial action is required, or selecting the location of permanent ground-water monitoring points for a detection monitoring program. Volumetric data is especially important for design of remediation systems at contaminated sites.

Dozens of commercial software packages of differing sophistication, cost, and complexity are available that facilitate the management, reporting, and analysis of temporal, spatial ground-water, and soils/geologic data (Table E-1). A recent assessment of data needs for development of EPA software for handling soils and geologic data as a complement to the GRITS/STAT software (Section 10.5.3) identified the following major commercial software as having a high potential for developing import/export utilities compatible with GEOS:[22]

> **GEOBASE** is available as a basic package for use in creating site maps, borehole log diagrams, well construction logs, and geological cross sections. The Advanced Lithology Package allows the customizing of logs, use of digitized geophysical logs, and the creation of contour maps and fence diagrams. The data base has some capability to sort records by data. Basic and advanced chemistry packages provide graphical analysis of sample data and can create isopleth maps. Maps can be digitized or DXF files imported. Some programs

---

[22] Boulding, J.R. 1993. Needs Document and Functional Specifications for the Proposed GEOS Module of the GRITS/STAT System. Prepared by Eastern Research Group, Inc. for U.S. EPA Center for Environmental Research Information, Cincinnati, OH, September 30.

allow spreadsheet data entry, and all programs can read ASCII files.

Geotechnical Graphics System (GTGS) is a database that readily imports ASCII files, digitized maps, and widely used database and spreadsheet programs. Graphic outputs include site maps, customized lithologic, geophysical, and well construction logs, cross sections, and fence diagrams. The GTGL script language provides programming flexibility for customizing graphics and preparing tables. GTLog is a downsized version of GTGS for creating lithologic and well construction logs. Graphics created with this version can be exported to most commonly used word processing and graphics programs.

gINT is a database that allows creation of customized lithologic, geophysical, and well construction logs, geologic cross sections, fence diagrams, and contouring. It also can be used to graph data and to print data tables. It imports ASCII files and exports to ASCII, HPGL, or Lotus 1-2-3 WKS formats.

GISKey integrates elements of existing software (e.g., AutoCAD, Fox database, QUICKSURF) for analysis of site characterization and monitoring data. Graphic outputs include boring logs, cross sections, fence diagrams, and contour/isopleth and isopach maps. It handles quality assurance/quality control (QA/QC) requirements for EPA's Contract Laboratory Program (CLP).

ROCKWORKS is an integrated collection of earth science software that is also available separately. Major components include DIGITIZE (a map, log, and general digitizing utility); GRIDZO (a relatively sophisticated gridding and contouring program); LOGGER (develops lithologic logs, cross sections, and fence diagrams); ROCKBASE (for creating base maps); ROCKSOLID (for carving out three-dimensional modeling and volumetric calculations); and ROCKSTAT (for making various statistical calculations). Graphics files can be exported to PCX, ASCII line, DXF, Postscript, HPGL, and PCL formats. The software imports ASCII line and HPGL files.

SPASE is a spatial database manager that runs in a Windows environment and allows the mapping and graphing of data. Contouring is done with mapping programs such as SURFER and GRIDZO (described elsewhere). Image Spase is used to capture or import digitized images. The software can import ASCII files and USGS DLG files (Digital Line Graph) and export files to a variety of programs. Other add-ons to the basic SPASE software include **Petro Spase**, oriented to oil and gas exploration and production, and **Enviro Spase**, which will soon be released for management of site characterization and monitoring data.

StratiFact is a spatial database that creates borehole logs, well construction logs, cross sections, contouring, and three-dimensional fence diagrams. Special features include the ability to plot the elevations of boreholes that deviate from the vertical, and to plot geologic cross sections of surface exposures.

Table E-1 identifies additional commercial software that can be used for more specific graphic presentation of spatial data in the form of (1) borehole logs, (2) cross sections, (3) contouring, (4) three-dimensional presentation, and (5) maps. Addresses for vendors of the software described above and in Table E-1 are located after Table E-1.

10.5.5 Geographic Information Systems

Geographic information systems (GIS) use a common spatial framework for data input, storage, manipulation, analysis, and display of geographic, cultural, political, environmental, and statistical data.[23] Computer processing of spatial data can range from the use of relatively simple graphics software that can plot contours or isopleths from data for which x and y coordinates are known using ASCII or other data files (Section 10.5.4) through to complex systems that can process digitized map data, maintain and manipulate large spatial databases, and generate a wide variety of user-created tables, graphs, and maps (Figure 10-7). This book uses the term *full-scale GIS* to refer to the type of integrated system illustrated in Figure 10-7, and *mini-GIS* to refer to personal computer (PC)-based software that is able to perform most of the functions of full-scale GIS at the scale of a USGS 7.5 minute quadrangle (discussed further below) as an integrated package.[24] The term *desktop GIS* applies to the use independent pieces of PC-based software to achieve the same results that full-scale and mini-GIS systems perform. Table 10-5 identifies major conferences and symposia with papers relevant to geographic information systems and journals focussing on this topic. Table 10-7 provides an index to more than 70 major GIS references.

Full-Scale GIS. The large amount of data that is stored and processed using full-scale GIS requires a workstation or mainframe computer environment with dedicated personnel for data entry and management. The costs of a full-scale geographic information system are substantial, but the greatest cost is the required commitment of personnel for data entry and management.[25] Consequently, the use of full-scale GIS is usually limited to large-scale projects or where GIS coverage of a site has been developed for other purposes. For example, Baker et al. (1993)[26] and

---

[23] The term *geoscientific information system* (GSIS) is also sometimes used.

[24] The geographic area that would exceed the capabilities of a stand-alone PC depends on two main factors: (1) the storage and memory capacity of the computer and (2) the amount and number of layers of data that must be stored and processed. Most stand-alone PCs can readily handle a digitized USGS 7.5 minute quadrangle map, which is adequate for most site-specific investigations. As the computational power of stand-alone PCs continues to increase, the distinction between full-scale and mini-GIS is likely to become more and more blurred.

[25] The cost of most commercial full-scale geographic information systems fall in the range of $10,000 to $100,000 (Rowe and Dulaney, 1991/Table 10-6). The cost of mini-GIS and related PC-based software ranges from hundreds to thousands of dollars.

[26] Baker, C.P., M.D. Bradley, and S.M. Kazco Bobiak. 1993. Wellhead Protection Area Delineation: Linking a Flow Model with GIS. J. Water Resources Planning and Management (ASCE) 119(2):275-287.

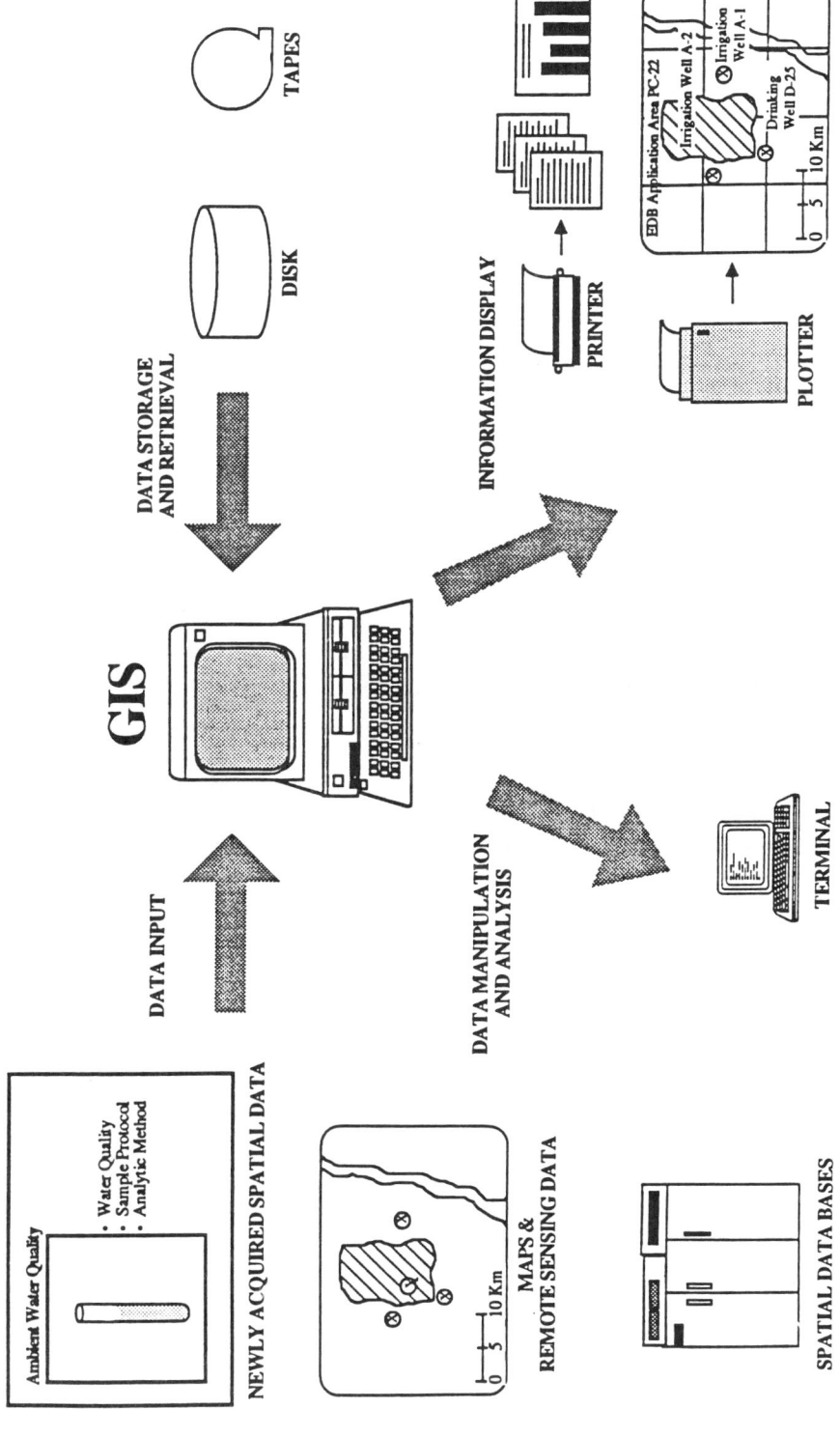

Figure 10-7 Overview of major geographic information system functions (OIRM, 1992).

Table 10-5 Periodicals, Conferences, and Symposia with Papers Relevant to GIS

Sponsor	Year	Title

ACSM/ASPRS Annual Convention Proceedings

	1986	Firm Foundations, New Horizons (Vol. 3, Geographic Information Systems, 286 pp.)
	1987	Technology for the Future, Applications for Today (7 Volumes; Vol. 5, GIS/LIS, 222 pp.)
	1988	The World in Space (6 Volumes; Vol. 5, GIS, 248 pp.)
	1989	Agenda for the Nineties (Vol. 4, GIS/LIS)
	1991	Annual Convention (6 Volumes; Vol. 2 Cartography and GIS/LIS, Vol. 4, GIS)
	1992	Annual Convention (Vol. 1 ASPRS, 592 pp.; Vol. 2 ACSM, 283 pp.)
	1992	Global Change (5 Volumes; Vol. 3, GIS and Cartography)
	1993	Annual Convention (Vol. 1 ACSM, 422 pp.; Vols 2 and 3 ASPRS, 474/458 pp.)

Annual GIS Workshops/Conferences

ASPRS/USFS	1986	Geographic Information Systems Workshop, 220 pp.
ACSM/ASPRS	1987	GIS'87—Into the Hands of the Decision Maker (2 Volumes, 760 pp.; Vol III—post conference proceedings, 234 pp.)
ACSM/ASPRS AAG/URISA	1988	GIS/LIS'88—Accessing the World (2 Volumes, 980 pp.)
	1989	GIS/LIS'89 (2 Volumes, 980 pp.)
	1990	GIS/LIS'90 (2 Volumes, 960 pp.)
	1991	GIS/LIS'91 Proceedings (2 Volumes, 1.045 pp.)
	1992	GIS/LIS'92 Proceedings (2 Volumes, 1,355 pp.)
	1993	GIS/LIS'93 Proceedings (2 Volumes, 835 pp.)

Biannual International Automated Cartography Proceedings

	1987	AutoCarto 8 (775 pp.)
	1989	AutoCarto IX (879 pp.)
	1991	AutoCarto 10 (Vol. 6 of ACSM/ASPRS Annual Convention Proceedings)
	1993	AutoCarto 11 (443 pp.)

Photogrammetric Engineering and Remote Sensing: Special GIS Issues

	1987	October, 184 pp.
	1988	November, 170 pp.
	1989	November, 144 pp.
	1990	_____
	1991	_____
	1992	_____
	1993	_____

Other Conferences/Symposia

| ASTM | 1990 | Geographic Information Systems (GIS) and Mapping: Practices and Standards |
| AWRA | 1993 | Geographic Information Systems and Water Resources |

Table 10-5 (cont.)

Type/Titles

Periodicals/Newsletters

Technical Journals Cartography and Geographic Information Systems (ACMS), GIS/GIMS News (ASPRS*), International Journal of GIS, Photogrammetric Engineering and Remote Sensing (ASPRS)

Vendor Newsletters: ARC News (Environmental Systems Research Institute, Redlands CA*), Grass Clippings (Geographic Resource Analysis Support System, Stennie Space Center, MS*), Monitor (Erdas, Inc., Atlanta, GA*), Remote Sensing and Database Development (James W. Sewall Company, Old Town ME*), TYDAC News (TYDAC Technologies Corporation, Arlington, VA*)

Government Agency Newsletters: Federal Geographic Data Committee (FDC) Newsletter (USGS, Reston, VA*), GIS News layers (Division of Equalization and Assessment, Albany, NY*), GIS Update (Vermont Geographic Information System, Montpelier, VT*), MASS GIS Newsletter (Massachusetts GIS project, Boston, MA*), New Jersey GIS Update (Department of Environmental Protection, Trenton, NJ*), NRGIS News (Minnesota Natural Resources Geographic Information Systems, St. Paul, MN*), RIGIS News (University of Rhode Island, Kingston, RI*)

Other: CAGIS Journal, Environmental Resources Research Institute Newsletter (Pennsylvania State University, University Park, PA*), Geo Info Systems, GIS Review (Greenland, NH*), GIS World (Fort Collins, CO*), Kansas Applied Remote Sensing (KARS) Newsletter (University of Kansas, Lawrence, KS*), The GIS Forum (Spring, TX*), SALIS Journal, URISA News (URISA, Washington, DC*), Wisconsin Land Information Newsletter (Center for Land Information Studies, University of Wisconsin, Madison, WI*)

Abbreviations:

AAG Association of American Geographers
ACMS American Congress on Surveying and Mapping
ASPRS American Society for Photogrammetry and Remote Sensing
ASTM American Society for Testing and Materials
AWRA American Water Resources Association
UIRSA Urban and Regional Information Systems Association

* Addresses listed in: August, P.V. and A. McCann. 1990. Geographic Information Systems (GIS) in Rhode Island. Department of Natural Resources Science Fact Sheet No. 90-23, University of Rhode Island, Kingston, RI, 11 pp.

Rifai et al. (1993)[27] have described use of the semianalytical WHPA code (Blandford and Huyakorn, 1991/Table E-2) in conjunction with full-scale GIS in Rhode Island and Texas, respectively. The Massachusetts Water Resources Authority, which supplies water to 46 communities in Metropolitan Boston, has used GIS to delineate critical recharge areas for local supplies and mapped thousands of point and nonpoint potential sources of contamination (Brandon et al., 1992).[28] Pickus (1992/T10-7) provides detailed guidance on using GIS and ARC/INFO, the full-scale geographic information system used by the U.S. Environmental Protection Agency for hydrogeologic analysis.

Mini- and Desktop GIS. Mini-GIS performs most of the functions of full-scale GIS as an integrated software package that can be used with a stand-alone PC.[29] The specific capabilities of different commercial packages vary, but generally these systems include (1) a spatial database for geologic, hydrologic, and chemical data, (2) the ability to create base maps and special purpose maps using data in the database, and (3) the ability to create geologic cross sections and graphs of time-series data. Often these systems can be used as preprocessors for numerical ground-water models (i.e., create grids and input values into the grid) and as postprocessors for graphic presentation of model output (Section 10.5.2).

PC-based software that performs more specific functions, such as graphic presentation of borehole logs, cross sections, and contour maps, can also facilitate the analysis of geologic and hydrologic data for hydrogeologic mapping.[30] Individual pieces of PC-based software that can handle spatial data can be used in combination to create a desktop GIS. Varljen and Wehrmann (1990)[31] describe using AutoCAD®

---

[27] Rifai, H.S., L.A. Hendricks, K. Kilborn, and P.B. Bedient. 1993. A Geographic Information System (GIS) User Interface for Delineating Wellhead Protection Areas. Ground Water 31(3):480-488.

[28] Brandon, F.O., P.B. Corcoran, and J.L. Yeo. 1992. Protection of Local Water Supplies by a Regional Water Supplier. Ground Water Management 13:525-538 ([8th] Focus Conf. Eastern GW Issues).

[29] Examples of commercially available mini-GIS software packages include GEOBASE, SPASE, GIS\Key, StratiFact, and ROCKWORKS described in Section 10.5.4.

[30] Examples of commercially available software that can create borehole and well construction logs include GTLog, logWRITER, QUICKLOG and LOGGER. Software designed to create cross-sections (also able to construct individual borehole logs) include GTGS, gINT, LOGGCORRELATE, and QUICKCROSS/FENCE. Available contouring software includes CONTUR, CoPlot, GRIDZO, LI-CONTOUR, PS-Plot, QUICKSURF, SURFER, TECKON, and TURBOCON. See Table E-1 for additional information.

[31] Varljen, M.D. and H.A. Wehrmann. 1990. Using AutoCAD® as a Desktop GIS for Hydrogeological Investigations. In: Mapping and Geographic Information Systems, A.I. Johnson, C.B. Pettersson, and J.L. Fulton (eds.), ASTM STP 1126, American Society to Testing and Materials, Philadelphia, PA.

as a desktop GIS for a hydrogeological investigation. The base map contained digital data on terrain elevations, location of transportation and water features, and names of cities, towns, and major landmarks in a CAD (computer-assisted drawing) DXF format [1:24000 scale (7.5 ft. quadrangles)]. Additional layers containing hydrogeologic information were created using SURFER® and exported to AutoCAD® for overlay on the base map.

The advantage of using mini-GIS software compared to using separate software to perform different functions is that import and export of data is minimized, reducing the time required for data processing. The advantage of desktop GIS, especially if one or more of the individual software packages have been purchased and are in use, is possibly lower cost and greater flexibility in processing and presenting data for the particular needs of the user.

Special Considerations in the Handling of Spatial Data. Spatial data is inherent to hydrogeologic mapping. For example, three coordinates are required to accurately locate borehole logging data: x and y coordinates define the position with respect to the surface of the earth, and the z coordinate defines the elevation. U.S. EPA and other federal agencies have adopted latitude and longitude as the standard system for x-y coordinates; new data collection should use that system. U.S. EPA (1992a, 1992b, and 1992c/T10-7) provide guidance for collection of spatial data. Hydrogeologic data compiled from existing sources may be located using a variety of coordinate systems, such as *Township-Range-and-Section*, *state planar coordinates*, or *Universal Transverse Mercator* (UTM). If such data are to be processed electronically, conversion to a standard coordinate system is required. Most mini-GIS software packages include conversion programs. The General Coordinate Transformation Package (GCTP) developed by the U.S. Geological Survey can be used to convert data between any of the commonly used geodetic coordinate systems (see address in Section 10.6). Table E-1 identifies 20 coordinate conversion packages that are available as individual items or as part of mini-GIS systems.

10.5.6 Other Data Management and Processing

The previous sections have identified specific types of computer applications and software for data management and processing in environmental investigations. Generic spreadsheet programs, such as Lotus 1-2-3®, and database management programs, such as dBase®, can also be used for analysis and graphic presentation of data. The advantage of such programs is that they provide high flexibility in handling of data for specific project needs.

10.6 Guide to Major Information Sources

The trend toward development of relatively inexpensive and user-friendly codes for ground-water modeling on personal computers increases the risk that pitfalls identified in the last section will occur because users lack the required breadth of knowledge about hydrogeology and computer modeling. Short courses (usually focusing on a limited number of codes), such as those sponsored by the International Ground Water Modeling Center, the National Ground Water Association, and various universities, are the best way to gain hands on experience with the use of more sophisticated models.

Table 10-6 provides an index to major text references and review papers on principles and applications of vadose zone and ground-water flow and contaminant modeling. Mercer and Faust (1981) provide a good nontechnical introduction to ground-water modeling. Anderson and Woessner (1992) is a good source for advanced coverage of this topic.

The annual **Software Catalog** of the International Ground Water Modeling Center (see address below) contains over 70 PC-based ground-water programs that can be purchased for prices ranging from $50 to $500. Ground-water flow and quality source codes developed by the U.S. Geological Survey can be obtained for IBM-compatible series 360 or 370 computers for $40 per program from: U.S. Geological Survey, WRD, National Water Information System, 437 National Center, 12201 Sunrise Valley Drive, Reston, VA, 22092. Appel and Reilly (1988)[32] provide summary descriptions of these codes. Many commercially developed codes, including enhanced versions of public domain codes, such as MODFLOW, are available. Two good sources of commercially available software are Scientific Software Group and Rockware Scientific Software.[33]

The continuing enhancement existing software and development of new codes makes the task of keeping abreast with new developments a challenge. The following newsletters (available at no cost) are useful for this purpose:

IGWMC Ground Water Modeling Newsletter is published by the International Ground Water Modeling Center, Colorado School of Mines, Golden, CO, 80401-1887 (303/273-3103).

Geraghty & Miller Software Newsletter is a periodic publication of the Geraghty & Miller Modeling Group (10700 Parkridge Boulevard, Suite 600, Reston, VA 22091; 703/758-1200).

GeoTrans Newsletter often contains information on applications and recent developments in ground-water modeling (46050 Manekin Plaza, Suite 100, Sterling, VA 22170; 703/444-7000).

---

[32] Appel, C.A. and T.E. Reilly. 1988. Selected Reports that Include Computer Programs Produced by the U.S. Geological Survey for Simulation of Ground-Water Flow and Quality. Water Resources Investigations Report 87-4271. [Provides summary information on about 40 models; March 6, 1991 update includes information on 16 more references]

[33] Rockware Scientific Software. The 1993 Scientific Software Catalog. Rockware Scientific Software, 4251 Kipling St., Suite 595, Wheat Ridge, CO 80033; 800/775-6745.

Scientific Software Group. Environmental, Engineering and Water Resources Software & Publications, 1993-1994. Scientific Software Group, P.O. Box 23041, Washington, DC, 20026-34041; 703/620-6793.

The scientific journals *Ground Water* and *Water Resources Research* are the best sources of peer-reviewed research on ground-water modeling. Periodic conferences sponsored jointly by the National Water Well Association and IGWMC are excellent sources of information on new developments and practical applications in ground-water modeling (NWWA/IGWMC 1984, 1985, 1987, 1989; NGWA/IGWMC 1992). Table 10-6 identifies other conferences an symposia addressing ground-water modeling.

U.S. EPA's Center for Subsurface Modeling Support (CSMoS) provides ground-water and vadose zone modeling software and services to public agencies and private companies throughout the United States. Its primary aim is to provide direct technical support to EPA and State decision-makers and to coordinate the use of models for risk assessment, site characterization, remedial activities, wellhead protection, and geographic information systems (GIS) applications. Table E-2 identifies PC-based models that are available from CSMoS. The Center's address is:

Center for Subsurface Modeling Support (CSMoS)
U.S. EPA R.S. Kerr Environmental Research Laboratory
P.O. Box 1198
Ada, OK, 74820
(405) 332-8800

Information on EPA exposure assessment modeling can be obtained from:

Center for Exposure Assessment Modeling (CEAM)
U.S. EPA Environmental Research Laboratory
Athens Georgia 30613-0801.
(706) 546-3599; models can be downloaded from CEAM Bulletin Board (706/546-3402).

Table 10-6 Index to Major References on Ground Water and Vadose Zone Flow and Contaminant Transport Modeling

Topic	References
Ground Water Modeling	
General*	Texts: Anderson and Woessner (1992), Bachmat et al. (1980), Bear and Bachmat (1990), Bear and Verruijt (1987), Boonstra and de Ridder (1981), Codell et al. (1982), Dagan (1989), Domenico (1972), Fried (1975), Javendel et al. (1984), Kinzelbach (1986), Mercer and Faust (1981), National Research Council (1990), Pinder and Gray (1977), Remson et al. (1971), van der Heijde et al. (1985), Walton (1988), Wang and Anderson (1982), Zienkiewicz (1977); Computational/Mathematical Methods: Boas (1983), Burden et al. (1981), Celia et al. (1988), Cross and Moscardini (1985), Gerald and Wheatley (1984), Hunt (1983), Huyakorn and Pinder (1983), Istok (1989), James et al. (1977), Ligget and Liu (1983), Press et al. (1986), Rushton and Redshaw (1979), Wood (1993); Review Papers: Anderson (1979), Bear et al. (1992), Faust and Mercer (1980), Faust et al. (1981), Gorelick (1983), Graves (1986), Konikow and Mercer (1988), Prickett (1979); Quality Control: Adrion et al. (1981), Buxton et al. (1989), California Toxic Substance Control Program (1990), Huyakorn et al. (1984), Kovar (1990), Moran and Mezgar (1982), Ross et al. (1982), U.S. EPA (1989), van der Heijde (1989), van der Heijde and Einawawy (1992), van der Heijde et al. (1989)
Conferences/Symposia	Arnold et al. (1982), Beck and Strateen (1983), Buxton et al. (1989), Celia et al. (1988), Custodio et al. (1988), Dickson et al. (1982), Haimes and Bear (1987), Jousma (1989), Kovar (1990), Melli and Zennetti (1992), NWWA/IGWMC (1984, 1985, 1987, 1989), NGWA/IGWMC (1992), Wrobel and Brebbia (1991)
Reviews/Comparisons	Appel and Bredehoeft (1976), Appel and Reilly (1988), Arnold (1992), Bachmat et al. (1978), Beven (1989), Beljin (1988), El-Kadi et al. (1991), IMS/OSWER (1990), Kayser and Collins (1986), Kincaid and Morrey (1984), Kincaid et al. (1984), Mangold and Tsang (1987), Mercer et al. (1982), Morrey et al. (1986), van der Heijde and Beljin (1988), van der Heijde and Einawawy (1993), van der Heijde et al. (1988), Simmons and Cole (1985), Thompson et al. (1989), U.S. EPA (1988), Whelan and Brown (1988)
Applications	Anderson and Woessner (1992), Bachmat et al. (1978), Barnes and Rogers (1988), Boonstra and de Ridder (1981), Boutwell et al. (1985), Bredehoeft et al. (1982), Clark (1987), Haimes and Bear (1987), Holcomb Research Institute (1976), Keely (1987), Moskowitz et al. (1991), National Research Council (1990), OTA (1982), U.S. EPA (1988a, 1988b), van der Heijde and Beljin (1988), van der Heijde et al. (1985), Whelan and Brown (1988)

Table 10-6 (cont.)

Topic	References

<u>Ground Water Modeling (cont.)</u>

Selected Topics	<u>Analytic Element Method</u>: Strack (1987); <u>Stochastic Modeling</u>: Dagan (1989), El-Kadi (1984), Gelher (1993), Vomvoris and Gelhar (1986); <u>Modeling Contaminant Transport/Biodegradation</u>: Beljin (1988), Brusseau et al. (1993), Edwards and Smart (1988-bibliography), Mueller and Crosby (1989-comparison), Pinder (1984), Naymik (1987), Prickett et al. (1986), Travers and Sharp-Hansen (1993); <u>Fracture Flow Modeling</u>: Schmelling and Ross (1989), van der Heijde and El-Kadi (1989); <u>Multiphase Flow Modeling</u>: Abriola (1988), API (1982), Aziz and Settari (1979), El-Kadi et al. (1991), Guarnaccia et al. (1992), Mercer and Cohen (1990), Parker (1989)

<u>Vadose Zone Modeling</u>

Overviews	Breckenridge et al. (1991), Donigian and Rao (1986a), El-Kadi and Beljin (1987), Ghadiri and Rose (1992), Hern and Melancon (1986a), Iskander (1981), Mangold and Tsang (1987), Nielsen et al. (1990), Mackay (1991), Nofziger et al. (1994), Oster (1982), Thompson et al. (1989), U.S. EPA (1984, 1988a, 1988b), van der Heijde (1994), van Genuchten (1987), Weaver et al. (1989), Whelan and Brown (1988), Wierenga (1991); <u>Hydrocarbon Fate</u>: Bonazountas (1991), Bonazountas and Kallidromitou (1993)
Conference/Symposia	Arnold et al. (1982), Beasley et al. (1991), Evans and Nicholson (1987-fractured rock), Wierenga and Bachelet (1988)
Model Comparisons	Addiscott and Wagenet (1985), Donigian and Rao (1986b), Kincaid and Morrey (1984), Kincaid et al. (1984), Oster (1982), van der Heijde et al. (1988)
Field Testing	Hern and Melancon (1986b), Hern et al. (1986), van der Heijde et al. (1989), Weaver et al. (1989)

<u>Hydrogeochemical Modeling</u>

General	Apps (1988), Boulding (1990), Jenne (1979), Jacobs and Whatley (1985), Jenne (1981), Melchior and Bassett (1990), NWWA (1988, 1990), Siegel and Leigh (1985), U.S. EPA (1990), Yeh and Tripathi (1989)
Model Comparisons	Mangold and Tsang (1987), Nordstrom and Ball (1984), Nordstrom et al. (1979), Rice (1986), Schechter et al. (1985), Sposito (1985)

<u>Risk Assessment Modeling</u> See Table 11-10

* Most ground-water and hydraulics texts in Table 2-4 cover governing analytical equations.

Table 10-6 References (Appendix F contains references for figure and table sources)

Abriola, L.M. 1988. Multiphase Flow and Transport Models for Organic Chemicals: A Review and Assessment. EPRI EA-5976. Electric Power Research Institute, Palo Alto, CA.

Addiscott, T.M. and R.J. Wagenet. 1985. Concepts of Solute Leaching in Soils: A Review of Modeling Approaches. J. Soil Science 36:411-424.

Adrion, W.R., M.A. Branstad, and J.C. Cherniasky. 1981. Validation, Verification and Testing of Computer Software. NBS Special Publication 500-75. Institute for Computer Science and Technology, National Bureau of Standards, Washington, DC.

American Petroleum Institute (API). 1982. The Migration of Petroleum Products in Soil and Groundwater--Principles and Countermeasures. Publication 4149, API, Washington, DC.

Anderson, M.P. 1979. Using Models to Simulate the Movement of Contaminants through Groundwater Flow Systems. CRC Critical Reviews on Environmental Control 9(2):97-156. [General review of governing equations and approaches to modeling transport of contaminants]

Anderson, M.P. and W.W. Woessner. 1992. Applied Groundwater Modeling: Simulation of Flow and Advective Transport. Academic Press, New York, NY, 381 pp.

Appel, C.A. and J.D. Bredehoeft. 1976. Status of Groundwater Modeling in the U.S. Geological Survey. U.S. Geological Survey Circular 737. [Summarizes status of development and selected references on 42 ground-water modeling projects supported by the U.S. Geological Survey]

Appel, C.A. and T.E. Reilly. 1988. Selected Reports that Include Computer Programs Produced by the U.S. Geological Survey for Simulation of Ground-Water Flow and Quality. Water Resources Investigations Report 87-4271. [Provides summary information on about 40 models; March 6, 1991 update includes information on 16 more references]

Apps, J.A. 1988. Current Geochemical Models to Predict the Fate of Hazardous Waste in the Injection Zones of Deep Disposal Wells. Lawrence Berkeley Laboratory, Draft Report LBL-26007. (Detailed summary available in Chapter 6 of U.S. EPA [1990b].)

Arnold, F. 1992. A Performance Comparison of Different Analytical and Numerical Saturated Zone Contaminant Transport Models. Ground Water Management 9:21-29 (Proc. 5th Int. Conf. on Solving Ground Water Problems with Models). [Numerical models: AT123D, Bioplume II, Conmig, Hydropal Slug, MOC, Random Walk, SLAEM; 4 analytical models]

Arnold, E.M., G.W. Gee, and R.W. Nelson (eds.). 1982. Proc. of the Symposium on Unsaturated Flow and Transport (Seattle, WA). NUREG/CP-0030, PNL-SA-10325, U.S. Nuclear Regulatory Commission, Washington, DC. [18 papers]

Aziz, K. and A. Settari. 1979. Petroleum Reservoir Simulation. Applied Science Publishers, London, UK.

(Table 10-6 Vadose Zone and GW Modeling References)

Bachmat, Y., B. Andrews, D. Holtz, and S. Sebastian. 1978. Utilization of Numerical Groundwater Models for Water Resource Management. EPA 600/ 8-78/012 (NTIS PB285 782). [Appendix summarizes information on 250 models]

Bachmat et al. (1980)--see van der Heijde et al. (1985).

Barnes, F.J. and J.C. Rogers. 1988. Evaluation of Hydrologic Models in the Design of Stable Landfill Covers. EPA/600/2-88/048 (NTIS PB88-243811).

Bear, J. and Y. Bachmat. 1990. Introduction to Modeling of Transport Phenomena in Porous Media. Kluwer Academic Publishers, Hingham, MA.

Bear, J. and A. Verruijt. 1987. Modeling Groundwater Flow and Pollution. Reidel Publishing Co., Dordrecht, The Netherlands, 414 pp.

Bear, J., M.S. Beljin, and R.R. Ross. 1992. Fundamentals of Ground-Water Modeling. EPA-540/S-92-005, 11 pp.

Beasley, B., W.G. Knisel, and A.P. Rice (eds.). 1991. Proceedings of the CREAMS/GLEAMS Symposium. Publication No. 4. Agr. Eng. Dept., Univ. Georgia, Athens, GA.

Beck, B.F. and G. Van Strateen (eds.). 1983. Uncertainty and Forecasting of Water Quality. Springer-Verlag, New York, NY.

Beljin, M.S. 1988. Testing and Validation of Models for Simulating Solute Transport in Groundwater: Code Intercomparison and Evaluation of Validation Methodology. GWMI 88-11. International Ground Water Modeling Center, Butler University, Indianapolis, IN.* [$10.00]

Boas, M.L. 1983. Mathematical Methods in the Physical Sciences. John Wiley & Sons, New York, NY.

Bonazountas, M. 1991. Fate of Hydrocarbons in Soils: Review of Modeling Practices. In: Hydrocarbon Contaminated Soils and Groundwater, Vol. 1, P.T. Kostecki and E.J. Calabrese (eds.), Lewis Publishers, Chelsea, MI, pp. 167-185. [Baehr/USGS model, POSSM, PRZM, PESTAN, GEOTOX, SESOIL]

Bonazountas, M. and D. Kallidromitou. 1993. Mathematical Hydrocarbon Fate Modeling in Soil Systems. In: Principles and Practices of Petroleum Contaminated Soils, E.J. Calabrese and P.T. Kostecki (eds.), Lewis Publishers, Boca Raton, FL, pp. 131-322 [Summary information on 40 models]

Boonstra, J. and N.A. de Ridder. 1981. Numerical Modelling of Groundwater Basins. International Institute for Land Reclamation and Improvement, Wageningen, The Netherlands. [User-oriented manual]

Boulding, J.R. 1990. Assessing the Geochemical Fate of Deep-Well-Injected Hazardous Waste: A Reference Guide. EPA/625/6-89/025a. Available from CERI.** [Chapter 5 reviews seven geochemical codes; see also U.S. EPA (1990)]

(Table 10-6 Vadose Zone and GW Modeling References)

Boutwell, S.H., S.M. Brown, B.R. Roberts, and D.F. Atwood. 1985. Modeling Remedial Actions at Uncontrolled Hazardous Waste Sites. EPA 540/2-85/001 (NTIS PB85-211357). Also published in 1986 with the same title by Noyes Data Corporation, Park Ridge, NJ. [Covers (1) selection of models, (2) simplified methods for subsurface and waste control action, and (3) numerical modeling of surface, subsurface, and waste control actions]

Breckenridge, R.P., J.R. Williams, and J.F. Keck. 1991. Characterizing Soils for Hazardous Waste Site Assessments. Superfund Ground-Water Issue Paper EPA/600/8-91/008.

Bredehoeft, J.D., P. Betzinski, C. Cruickshank Villanueva, G. de Marsily, A.A. Konoplyntsev, and J.U. Uzoma. 1982. Ground-Water Models, Vol. I: Concepts, Problems, and Methods of Analysis with Examples of Their Applications. UNESCO Studies and Reports in Hydrology No. 34, Paris. [Contains 21 case histories]

Brusseau, M.L., P.S.C. Rao, and C.A. Bellin. 1993. Modeling Coupled Processes in Porous Media: Sorption, Transformation, and Transport of Organic Solutes. In: Interacting Processes in Soil Science, R.J. Wagenet, P. Baveye, and B.A. Stewart (eds.), Lewis Publishers, Boca Raton, FL.

Burden, R.L., J.D. Faires, and A.C. Reynolds. 1981. Numerical Analysis, 2nd ed. Prindle, Weer, and Schmidt, Boston, MA.

Buxton, B.E., S. M. Hogan, L. Copley-Graves, and S.E. Brauning (eds.). 1989. Proceedings of the 1987 DOE/AECL Conference on Geostatistical, Sensitivity, and Uncertainty Methods for Ground-Water Flow and Radionuclide Modeling. Battelle Press, Columbus, OH. [31 papers]

California Toxic Substances Control Program. 1990. Scientific and Technical Standards for Hazardous Waste Sites: Vol. 2, Exposure Assessment. Chapter 4, Draft Standards for Mathematical Modeling of Ground Water Flow and Contaminant Transport at Hazardous Waste Sites.

Celia, M.A., L.A. Ferrand, C.A. Brebbia, W.G. Gray, and G.F. Pinder (eds.). 1988. Computational Methods in Water Resources. Vol. 1 Modeling Surface and Subsurface Flows; Vol. 2 Numerical Methods for Transport and Hydrologic Processes. Elsevier, New York, NY. [7th International conference on computational methods in water resources containing 121 papers, more than half of which are specifically devoted to ground water. Previous conferences were titled "Finite Elements in Water Resources" and were held at Princeton University (1976), Imperial College, UK (1978), University of Mississippi (1980), University of Hanover FRD (1982), University of Vermont (1984) and the Laboratorio Nacional de Engenharia Civil, Portugal (1986)]

Clarke, D. 1987. Microcomputer Programs for Groundwater Studies. Elsevier, New York, NY.

Codell, R.B., K.T. Key, and G. Whelan. 1982. A Collection of Mathematical Models for Dispersion in Surface Water and Groundwater. NUREG-0868. U.S. Nuclear Regulatory Commission, Washington, DC. [Prepared by Battelle Pacific Northwest Laboratory]

Cross, M. and A.O. Moscardini. 1985. Learning the Art of Mathematical Modeling. Ellis Harwood, Ltd., Chichester, UK.

(Table 10-6 Vadose Zone and GW Modeling References)

Custodio, E., A. Garguin, and J.P. Lobo Ferreira (eds.). 1988. Ground Flow and Quality Modeling. NATO ASI Series C Vol. 224. Reidel Publishing Co., Dordrecht, The Netherlands. [Proceedings of workshop on advances in analytical and numerical groundwater flow and quality modeling]

Dagan, G. 1989. Flow and Transport in Porous Formations. Springer-Verlag, New York, NY. [Focuses on stochastic modeling of subsurface flow and transport at different scales]

Dickson, K.L., A.W. Maki, and J. Cairns, Jr. (eds.). 1982. Modeling the Fate of Chemicals in the Aquatic Environment. Ann Arbor Science, Ann Arbor, MI. [21 papers]

Domenico, P.A. 1972. Concepts and Models in Groundwater Hydrology. McGraw-Hill, New York, NY, 405 pp.

Donigian, A.S., Jr. and P.S.C. Rao. 1986a. Overview of Terrestrial Processes and Modeling. In: Guidelines for Field Testing Soil Fate and Transport Models, S.C. Hern and S.M. Melancon, (eds.), EPA/600/4-86/020 (NTIS PB86-209400), pp. 1-32.

Donigian, A.S., Jr. and P.S.C. Rao. 1986b. Overview of Terrestrial Processes and Modeling. In: Vadose Zone Modeling of Organic Pollutants, S.C. Hern and S.M. Melancon (eds.), Lewis Publishers, Chelsea, MI, pp. 3-36. [Summarizes information on 10 codes that simulate transport of contaminants in the vadose zone]

Edwards, A.J. and P.L. Smart. 1988. Contaminant Transport Modeling: An Annotated Bibliography. Turner Designs, Sunnyvale, CA. (58 references).

El-Kadi, A.I. 1984. Modeling Variability in Ground-Water Flow. GWMI 84-10. International Ground-Water Modeling Center, Butler University, Indianapolis, IN.* [$8.50]

El-Kadi, A.I. and M.S. Beljin. 1987. Models for Unsaturated Flow and Solute Transport. GWMI 87-12, International Ground Water Modeling Center, Golden, CO, 27 pp. [Summary information on 62 documented models]

El-Kadi, A.I., O.A. Einawawy, P.K. Kobe, and P.K.M. van der Heijde. 1991. Modeling Multiphase Flow and Transport. GWMI 91-04. International Ground Water Modeling Center, Butler University, Indianapolis, IN.* [$10.00]

Evans, D.D. and T.J. Nicholson (eds.). 1987. Flow and Transport Through Unsaturated Fractured Rock. AGU Geophysical Monograph 42, American Geophysical Union, Washington, DC. [9 out of 22 of the papers of the cover modeling]

Faust, C.R. and J.W. Mercer. 1980. Groundwater Modeling: Recent Developments. Ground Water 18(6):569-577.

Faust, C.R., L.R. Silka, and J.W. Mercer. 1981. Computer Modeling and Ground-Water Protection. Ground Water 19(4):362-365.

Fried, J.J. 1975. Groundwater Pollution: Theory, Methodology Modeling, and Practical Rules. Elsevier, New York, NY, 330 pp.

(Table 10-6 Vadose Zone and GW Modeling References)

Gelher, L.W. 1993. Stochastic Subsurface Hydrology. Prentice-Hall, Englewood Cliffs, NJ, 390 pp.

Gerald, C.F. and P.O. Wheatley. 1984. Applied Numerical Analysis, 3rd ed. Addison-Wesley, Reading, MA.

Ghadiri, H. and C.W. Rose (eds.). 1992. Modeling Chemical Transport in Soils: Natural and Applied Contaminants. Lewis Publishers, Chelsea, MI, 217 pp. [Summary information on more than 70 models for soil erosion, sediment transport and deposition, and subsurface chemical transport]

Gorelick, S. 1983. A Review of Distributed Parameter Groundwater Management Modeling Methods. Water Resources Research 19(2):305-319.

Graves, B. 1986. Ground Water Software--Trimming the Confusion. Ground Water Monitoring Review 6(1):44-53.

Guarnaccia, J.F. et al. 1992. Multiphase Chemical Transport in Porous Media. Environmental Research Brief, EPA/600/S-92/002, 19 pp.

Haimes, Y.Y. and J. Bear (eds.). 1987. Groundwater Contamination: Use of Models in Decisionmaking. Reidel Publishing Co., Dordrecht, The Netherlands.

Hern, S.C. and S.M. Melancon. 1986a. Vadose Zone Modeling of Organic Pollutants. Lewis Publishers, Chelsea, MI. [11 contributed chapters on vadose processes and modeling]

Hern, S.C. and S.M. Melancon. 1986b. Guidelines for Field Testing Soil Fate and Transport Models: Final Report. EPA/600/7-86/020 (NTIS PB86-209400). [PRZM, SESOIL, PESTAN]

Hern, S.C., S.M. Melancon, and J.E. Pollard. 1986. Generic Steps in the Field Validation of Vadose Zone Fate and Transport Models. In: Vadose Zone Modeling of Organic Pollutants, S.C. Hern and S.M. Melancon (eds.), Lewis Publishers, pp. 61-80.

Holcolm Research Institute. 1976. Environmental Modeling and Decision-Making. Praeger Publishers, New York, NY. [Mainly focussed on surface water]

Hunt, B. 1983. Mathematical Analysis of Groundwater Resources. Butterworth, Stoneham, MA, 271 pp.

Huyakorn, P.S. and G.F. Pinder. 1983. Computational Methods in Subsurface Flow. Academic Press, New York, NY, 473 pp. [Paperback edition published in 1986]

Huyakorn, P.S., A.G. Kretschek, R.W. Broome, J.W. Mercer, and B.H. Lester. 1984. Testing and Validation of Models for Simulating Solute Transport: Development, Evaluation, and Comparison of Benchmark Techniques. GWMI 84-13. International Ground Water Modeling Center, Butler University, Indianapolis, IN.*

Information Management Staff, Office of Solid Waste and Emergency Response (IMS/OSWER). 1990. Report of the Usage of Computer Models in Hazardous Waste/Superfund Programs, Phase II Final Report. U.S. Environmental Protection Agency. Washington, DC.

(Table 10-6 Vadose Zone and GW Modeling References)

Iskander, I.K. 1981. Overview of Models Used in Land Treatment of Wastewater. CRREL Special Report 82-1, U.S. Army Corps of Engineers Cold Region Research Engineering Laboratory, Hanover, NH, 27 pp. [Summary information on 29 water and solute flow models]

Istok, J. 1989. Groundwater Modeling by the Finite Element Method. AGU Water Resources Monograph 13, American Geophysical Union, Washington, DC.

Jacobs, G.K. and S.K. Whattley (eds.). 1985. Proceedings of the Conference on the Application of Geochemical Models to High-Level Nuclear Waste Repository Assessment. NUREG/CP-0062, U.S. Nuclear Regulatory Commission, Washington, DC, 126 pp.

James, M.L., G.M. Smith, and J.C. Wolford. 1977. Applied Numerical Methods for Digital Computation with FORTRAN and CSMP, 2nd ed. Harper and Row, New York, NY.

Javendel, I., C. Doughty, and C.F. Tsang. 1984. Groundwater Transport: Handbook of Mathematical Models. AGU Water Resources Monograph No. 10. American Geophysical Union, Washington, DC, 228 pp. [Covers analytical, semianalytical and numerical methods; includes codes for ODAST, TDAST, LTIRD, RESSQ]

Jenne, E.A. (ed.). 1979. Chemical Modeling in Aqueous Systems: Speciation, Sorption, Solubility, and Kinetics. ACS Symp. Series 93. American Chemical Society, Washington, DC.

Jenne, E.A. 1981. Geochemical Modeling: A Review. PNL-3574, Battelle Northwest Laboratory, Richland, WA.

Jousma, G., J. Bear, Y.Y. Haimes, and F. Walter (eds.). 1989. Groundwater Contamination: Use of Models in Decision-Making. Kluwer Academic Publishers, Hingham, MA. [60 papers; proceedings of 1987 International Conference held in Amsterdam]

Kayser, M.B. and A.G. Collins. 1986. Computer Simulation Models Relevant to Ground Water Contamination from EOR or Other Fluids--State-of-the-Art. NIPER-102, National Institute for Petroleum and Energy Research, Bartlesville, OK. [Summarizes recent developments and ongoing work in modeling ground-water contamination from enhanced oil recovery and other fluids]

Keely, J.F. 1987. The Use of Models in Managing Ground-Water Protection Programs. EPA 600/8-87/003 (NTIS PB87-166203), 72 pp.

Kincaid, C.T. and J.R. Morrey. 1984. Geohydrochemical Models for Solute Migration, Vol. 2: Preliminary Evaluation of Selected Computer Codes. EPRI EA-3417-2. Electric Power Research Institute, Palo Alto, CA. [Evaluates 21 codes applicable to the study of leachate migration; vadose zone: SESOIL]

Kincaid, C.T., J.R. Morrey, and J.E. Rogers. 1984. Geohydrochemical Models for Solute Migration, Vol. 1: Process Description and Computer Code Selection. EPRI EA-3417-1. Electric Power Research Institute, Palo Alto, CA. [Summarizes mathematical models and numerical methods for predicting leachate migration and develops criteria for selection of codes; vadose zone: PRZM]

(Table 10-6 Vadose Zone and GW Modeling References)

Kinzelbach, W. 1986. Groundwater Modeling: An Introduction with Simple Programs in BASIC. Elsevier, New York, NY. [Intermediate text]

Konikow, L.F. and J.M. Mercer. 1988. Groundwater Flow and Transport Modeling. J. Hydrology 100(2):379-409.

Kovar, K. (ed.). 1990. Calibration and Reliability in Groundwater Modeling. Int. Ass. Sci. Hydrology Pub. No. 195.

Ligget, J.A. and P.L.-F. Liu. 1983. The Boundary Integral Equation Method for Porous Media Flow. Allen and Unwin, Inc., Winchester, MA.

Mackay, D. 1991. Multimedia Environmental Models: The Fugacity Approach. Lewis Publishers, Chelsea, MI, 257 pp.

Mangold, D.C. and C.-F. Tsang. 1987. Summary of Hydrologic and Hydrochemical Models with Potential Application to Deep Underground Injection Performance. Lawrence Berkeley Laboratory LBL-23497, Berkeley, CA, 54 pp. [Comparative information on 6 flow models, 16 saturated solute transport models, 11 unsaturated solute transport models, 17 chemical-reaction transport models, and 7 geochemical codes]

Melchior, D.C. and R.L. Bassett (eds.). 1990. Chemical Modeling of Aqueous Systems II. ACS Symp. Series 416, American Chemical Society, Washington, DC.

Melli, P. and P. Zannetti (eds.). 1992. Environmental Modeling. Elsevier, New York, NY. [Proceedings of 1990 IBM European Summer Institute Seminar on Environmental Modeling]

Mercer, J.W. and R.M. Cohen. 1990. A Review of Immiscible Fluids in the Subsurface, Properties, Models, Characterization and Remediation. J. Contaminant Hydrology 6:107-163.

Mercer, J.W. and C.R. Faust. 1981. Ground-Water Modeling. National Water Well Association, Dublin, OH, 60 pp. [Introductory text; compilation of 5 papers published in Ground Water]

Mercer, J.W., S.D. Thomas, and B. Ross. 1982. Parameters and Variables Appearing in Repository Siting Models. NUREG/CR-3066, U.S. Nuclear Regulatory Commission, Washington, DC.

Moran, M.S. and L.J. Mezgar. 1982. Evaluation Factors for Verification and Validation of Low-Level Waste Disposal Site Models. DOE/OR/21400-T119, Oak Ridge National Laboratory, Oak Ridge, TN.

Morrey, J.R., C.T. Kincaid, and C.J. Hostetler. 1986. Geohydrochemical Models for Solute Migration, Vol. 3: Evaluation of Selected Computer Codes. EPRI EA-3417-3. Electric Power Research Institute, Palo Alto, CA. [Contains detailed evaluation of five codes identified as best suited for studying leachate migration (EQ3/EQ6, MINTEQ, FEMWATER1/FEMWASTE1, SATURN, and TRANS)]

Moskowitz, P.D., R. Pardi, M.P. DePhillips, and A.F. Meinhold. 1991. Computer Models Used to Support Cleanup Decision-Making at Hazardous Waste Sites. Brookhaven National Laboratory Draft Report. [Cited in Geraghty and Miller Software Newsletter, Spring 1992]

(Table 10-6 Vadose Zone and GW Modeling References)

Mueller, D. and E. Crosby. 1989. Comparison of Microcomputer-Based Groundwater Transport Models. In: Proc. Fourth Int. Conf. on Solving Ground Water Problems with Models (Indianapolis, IN), National Water Well Association, Dublin, OH, pp. 797-820.

National Research Council. 1990. Ground Water Models: Scientific and Regulatory Applications. National Academy Press, Washington, DC, 303 pp.

National Water Well Association (NWWA). 1988. Proceedings of Geochemical Modeling of Ground Water Conference. NWWA, Dublin, OH.

National Water Well Association (NWWA). 1990. Proceedings of the 1990 Cluster of Conferences (Ground Water Geochemistry). Ground Water Management No. 1, NWWA, Dublin, OH. [10 papers on fate and modeling of contaminants]

National Water Well Association/International Ground Water Modeling Center (NWWA/IGWMC). 1984. Proceedings of Conference on Practical Applications of Ground Water Models. NWWA, Dublin, OH. [44 papers]

National Water Well Association/International Ground Water Modeling Center. 1985. Proceedings of Conference on Practical Applications of Ground Water Models. NWWA, Dublin, OH. [27 papers]

National Water Well Association/International Ground Water Modeling Center. 1987. Proceedings of Conference on Solving Ground Water Problems with Models. NWWA, Dublin, OH. [45+ papers]

National Water Well Association/International Ground Water Modeling Center. 1989. Fourth International Conference on Solving Ground Water Problems with Models. NWWA, Dublin, OH. [44+ papers]

National Ground Water Association/International Ground Water Modeling Center. 1992. Fifth International Conference on Solving Ground Water Problems with Models. Ground Water Management No. 9. NGWA, Dublin, OH. [49 papers]

Naymik, T.G. 1987. Mathematical Modeling of Solute Transport in the Subsurface. Critical Reviews in Environmental Control 17(3):229-251.

Nielsen, D.R., D. Shibberu, G.E. Fogg, and D.R. Rolston. 1990. A Review of the State of the Art: Predicting Contaminant Transport in the Vadose Zone. Report 90-17CWP, California Water Resources Control Board, Sacramento, CA, 19 pp.

Nofziger, D.L., J.-S. Chen, and C.T. Haan. 1994. Evaluation of Unsaturated/Vadose Zone Models for Superfund Sites. EPA/600/R-93/184, 188 pp. [RITZ, VIP, CMLS, HYDRUS]

Nordstrom, D.K. and J.W. Ball. 1984. Chemical Models, Computer Programs and Metal Complexation in Natural Waters. In: Complexation of Trace Metals in Natural Waters, C.J.M. Kramer and J.C. Duinker (eds.), Martinus Nijhoff/Dr. W. Junk Publishers, The Hague, pp. 149-164. [Provides information on 57 geochemical codes]

(Table 10-6 Vadose Zone and GW Modeling References)

Nordstrom, D.K., et al. 1979. A Comparison of Computerized Chemical Models for Equilibrium Calculations in Aqueous Systems. In: Chemical Modeling in Aqueous Systems: Speciation, Sorption, Solubility and Kinetics, E.A. Jenne, (ed.), ACS Symp. Series 93, American Chemical Society, Washington, DC, pp. 857-892. [Summarizes information on 14 geochemical codes and comparative results from calculations using a single test case]

Office of Technology Assessment (OTA). 1982. Use of Models for Water Resources Management, Planning, and Policy. OTA, Washington, DC.

Oster, C.A. 1982. Review of Groundwater Flow and Transport Models in the Unsaturated Zone. NUREG/CR-2917, PNL-4427. Pacific Northwest Laboratory, Richland, WA. [Summarizes information on 53 flow and transport codes involving unsaturated flow]

Parker, J.C. 1989. Multiphase Flow and Transport in Porous Media. Reviews in Geophysics 27(3):311-328.

Pinder, G.F. 1984. Groundwater Contaminant Transport Modeling. Environ. Sci. Technol. 18(4):108A-114A.

Pinder, G.F. and W.G. Gray. 1977. Finite Element Simulation in Surface and Subsurface Hydrology. Academic Press, New York, NY, 295 pp.

Press, W.H., B.P. Flannery, S.A. Teukolsky, and W.T. Vetterling. 1986. Numerical Recipes: The Art of Scientific Computing. Cambridge University Press, New York, NY.

Prickett, T.A. 1979. Ground-Water Computer Models--State of the Art. Ground Water 17(2):167-173.

Prickett, T.A., D.L. Warner, and D.D. Runnells. 1986. Application of Flow, Mass Transport, and Chemical Reaction Modeling to Subsurface Liquid Injection. In: Proc. Int. Symp. on Subsurface Injection of Liquid Wastes, National Water Well Association, Dublin, OH, pp. 447-463.

Remson, I., G.M. Hornberger, and F.J. Molz. 1971. Numerical Methods in Subsurface Hydrology. John Wiley & Sons, New York, NY, 389 pp. [Focuses on finite-difference numerical methods applied to transient and steady-state flow problems]

Rice, R. 1986. The Fundamental of Geochemical Equilibrium Models; with a Listing of Hydrochemical Models That Are Documented and Available. GWMI 86-04. International Ground Water Modeling Center, Butler University, Indianapolis, IN, 29 pp.* [$3.50]

Ross, B., J.W. Mercer, S.D. Thomas, and B.H. Lester. 1982. Benchmark Problems for Repository Siting Models. NUREG/CR-3097. U.S. Nuclear Regulatory Commission, Washington, DC.

Rushton, K.R. and S.C. Redshaw. 1979. Seepage and Groundwater Flow: Numerical Analysis by Analog and Digital Methods. John Wiley & Sons, Chichester, UK.

Schechter, R.S., L.W. Lake, and M.P. Walsh. 1985. Development of Environmentally Attractive Leachants, Vol. III. U.S. Bureau of Mines Mining Research Contract Report, Washington, DC. [Evaluates codes with potential for assessment of leachates associated with mining activities]

Schmelling, S.G. and R.R. Ross. 1989. Contaminant Transport in Fractured Media: Models for Decision Makers. Superfund Ground Water Issue Paper. EPA 540/4-89/004.

Siegel, M.D. and C.D. Leigh (eds.). 1985. Progress in Development of a Methodology for Geochemical Sensitivity Analysis for Performance Assessment: Parametric Calculations, Preliminary Databases, and Computer Code Evaluation. NUREG/CR-5085 SAND85-1644. Sandia National Laboratories, Albuquerque, NM, 69+ pp.

Simmons, C.S. and C.R. Cole. 1985. Guidelines for Selecting Codes for Groundwater Transport Modeling of Low-Level Waste Burial Sites, Vol. 1, Guideline Approach. PNL-4980, Vol. 1. Battelle Pacific Northwest Laboratory, Richland, WA.

Sposito, G. 1985. Chemical Models of Inorganic Pollutants in Soils. CRC Critical Reviews in Environmental Control 15(1):1-24. [Compares four geochemical codes for modeling soil chemistry and four models for adsorption]

Strack, O.D.L. 1987. Groundwater Mechanics. Prentice-Hall, Englewood Cliffs, NJ. [Covers analytic element method]

Thompson, C.M., L.J. Holcombe, D.H. Gancarz, A.E. Behl, J.R. Erikson, I. Star, R.K. Waddell, and J.S. Fruchter. 1989. Techniques to Develop Data for Hydrogeochemical Models. EPRI EN-6637. Electric Power Research Institute, Palo Alto, CA. [Summary information on data requirements for 25 saturated and variably saturated flow and transport codes and 5 geochemical codes]

Travers, C.L. and S. Sharp-Hansen. 1993. Leachate Generation and Migration at Subtitle D Facilities: A Summary and Review of Processes and Mathematical Models. EPA/600/R-93/125 (NTIS PB93-217778).

U.S. Environmental Protection Agency (EPA). 1984. Procedures for Modeling Flow Through Clay Liners to Determine the Required Liner Thickness. EPA/530/SW-84/001 (NTIS PB87-191029).

U.S. Environmental Protection Agency (EPA). 1988a. Selection Criteria for Mathematical Models Used in Exposure Assessments: Ground-Water Models. EPA 600/8-88/075 (NTIS PB88-248752). [Contains summary tables and descriptions of 63 analytical solutions and 49 analytical and numerical codes for evaluating ground water contaminant transport]

U.S. Environmental Protection Agency (EPA). 1988b. Superfund Exposure Assessment Manual. EPA/540/1-88/001 (NTIS PB90-135859). [Summary information on 35 vadose zone and ground-water models; include Wilson-Miller nomograph]

U.S. Environmental Protection Agency (EPA). 1989. Resolution on the Use of Mathematical Models by EPA for Regulatory Assessment and Decision-Making. EPA-SAB-EEC-89-012, 7 pp.

(Table 10-6 Vadose Zone and GW Modeling References)

U.S. Environmental Protection Agency (EPA). 1990. Assessing the Geochemical Fate of Deep-Well-Injected Hazardous Waste: Summaries of Recent Research. EPA/625/6-89/025b. Available from CERI.** [Chapter 6 provides a detailed summary of Apps (1988); see also Boulding (1990)]

van der Heijde, P.K.M. 1989. Quality Assurance and Quality Control in Groundwater Modeling. GWMI 89-04. International Ground Water Modeling Center, Butler University, Indianapolis, IN, 26 pp.*

van der Heijde, P.K.M. 1994. Identification and Compilation of Unsaturated/Vadose Zone Models. EPA/600/R-94/028 (NTIS PB94-157773).

van der Heijde, P., and M.S. Beljin. 1988. Model Assessment for Delineating Wellhead Protection Areas. EPA/440/6-88-002 (NTIS PB88-231485 or PB88-238449), 267 pp. [Also available from IGWMC as GWMI 87-21 for $20.00]

van der Heijde, P.K.M. and O.A. Einawawy. 1992. Quality Assurance and Quality Control in the Development and Application of Ground-Water Models. EPA/600/R-93/011 (NTIS PB93-178226), 148 pp.

van der Heijde, P.K.M. and O.A. Einawawy. 1993. Compilation of Ground-Water Models. EPA/600/R-93/118 (NTIS PB93-209401). [Summary information on models for porous media flow and transport, hydrogeochemical models, stochastic models, and fractured rock; also available from IGWMC*]

van der Heijde, P.K.M. and A.I. El-Kadi. 1989. Models for Flow and Transport in Fractured Rocks. GWMI 89-08. International Ground Water Modeling Center, Butler University, Indianapolis, IN, 42 pp.* [$2.00]

van der Heijde, P.K.M., Y. Bachmat, J.D. Bredehoeft, B. Andrews, D. Holz, and S. Sebastian. 1985. Groundwater Management: The Use of Numerical Models. Water Resources Monograph 5, 2nd ed. American Geophysical Union, Washington, DC, 180 pp. [First edition published in 1980 and authored by Bachmat, Bredehoeft, Andrews, Holz, and Sebastian]

van der Heijde, P.K.M., A.I. El-Kadi, and S.A. Williams. 1988. Groundwater Modeling: An Overview and Status Report. EPA/600/2-89/028 (NTIS PB89-224497). Also available from International Ground Water Modeling Center for $15.00 as GWMI 88-10.* [Contains summary listings and usability/reliability ratings for 296 flow and transport codes organized in seven major categories]

van der Heijde, P.K.M., W.I.M. Elderhorst, R.A. Miller, and M.J. Trehan. 1989. The Establishment of A Groundwater Research Data Center for Validation of Subsurface Flow and Transport Models. EPA/600/2-89/040 (NTIS PB89-224455), 238 pp.

van Genuchten, M.Th. 1987. Progress in Unsaturated Flow and Transport Modeling. Reviews of Geophysics 25(2):135-140.

Vomvoris, E.G. and L.W. Gelhar. 1986. Stochastic Prediction of Dispersive Contaminant Transport. EPA/600/2-86/114 (NTIS PB87-141479).

(Table 10-6 Vadose Zone and GW Modeling References)

Walton, W.C. 1988. Practical Aspects of Groundwater Modeling: Analytical and Computer Models for Flow, Mass and Heat Transport, and Subsidence, 3rd ed. National Water Well Association, Dublin, OH. [2nd edition published in 1985. Covers both analytical and numerical methods; includes several tables of field-determined values that can serve as guide for first approximations of unknown aquifer parameters]

Wang, H.F. and M.P. Anderson. 1982. Introduction to Groundwater Modeling: Finite Difference and Finite Element Methods. W.H. Freeman and Company, San Francisco, CA, 237 pp.

Weaver, J., C.G. Enfield, S. Yates, D. Kreamer, and D. White. 1989. Predicting Subsurface Contaminant Transport and Transformation: Considerations for Model Selection and Field Validation. EPA/600/2-89/045 (NTIS PB90-155615).

Whelan, G. and S.M. Brown. 1988. Groundwater Assessment Modeling Under the Resource Conservation and Recovery Act. EPRI EA-5342, Electric Power Research Institute, Palo Alto, CA. [Appendix C presents summary information on RAPCON, PRZM, GRDFLX, AT123D, VTT, and FE3DGW/CFEST codes]

Wierenga, P.J. (ed.). 1991. Validation of Flow and Transport Models for the Unsaturated Zone. J. Contaminant Hydrology 7:1-160. [Special issue with 8 papers]

Wierenga, P.J. and D. Bachelet (eds.). 1988. Proceedings of International Conference and Workshop on the Validation of Flow and Transport Models for the Unsaturated Zone (Ruidoso, NM). New Mexico State University, Las Cruces, NM.

Wood, W.L. 1993. Introduction to Numerical Methods for Water Resources. Oxford University Press, New York, NY, 280 pp.

Wrobel, L.C. and C.A. Brebbia (eds.). 1991. Water Pollution: Modeling, Measuring and Prediction. Computational Mechanics Publications, Billerica, MA. [Proceedings of First International Conference on Water Pollution, held in Southampton, UK]

Yeh, G.T. and V.S. Tripathi. 1989. A Critical Evaluation of Recent Developments in Hydrogeochemical Transport Models of Reactive Multichemical Components. Water Resources Research 25(1):93-108.

Zienkiewicz, O.C. 1977. The Finite Element Method, 3rd ed. McGraw-Hill, London.

* The International Ground Water Modeling Center is now located in Golden, Colorado. Prices subject to change.

** See Preface for information on how to obtain documents from CERI (U.S. EPA Center for Environmental Research Information) and NTIS.

(Table 10-6 Vadose Zone and GW Modeling References)

Table 10-7 Index to Major References on Geographic Information Systems (GIS)

Topic	References
Texts	Green et al. (1993), Luarini and Thompson (1992), Warboys (1994); Introductory: Aronoff (1989), Cadoux-Hudson and Heywood (1992), Pequet and Marble (1990), Ripple (1989), Star and Estes (1990); Cartography: ACSM (1992d), Clarke (1990), Johnson et al. (1992), Tomlin (1990); Technology: ACSM (1992b), Antenucci et al. (1991), Maguire et al. (1992); Land Resource Assessment: Burrough (1986), Gokee and Joyce (1992), Ripple (1986), McCloy (1994), Young and Cousins (1993); Urban Applications: Huxhold (1991); Geoscience/Geotechnical Applications: Johnson et al. (1992), Thomas (1988); Ground-Water and Environmental Applications: Douglas (1994), Goodchild et al. (1993), Johnson et al. (1992), Kovar and Nachtnebel (1993), Pickus (1992); General Applications: Johnson et al. (1992), Maguire et al. (1991), Ripple (1986); Temporal GIS: Langran (1992)
GIS Systems	Arc/Info: ESRI (1990a, 1990b), Pickus (1992); AutoCAD®: Jones and Martin (1988); TIGER: Bureau of Census (1992); Comparison/Evaluation: FICC (1988), Rowe and Dulaney (1991)
Government Use	U.S. EPA: Hewitt et al. (1993), OIRM (1992), U.S. EPA (1992a, 1992c, 1992d); U.S. Geological Survey: USGS (1991a); Soil Conservation Service: SCS (1991); Other Federal: FICC (1990), FGDC (1991a, 1991b, 1993); States: ACSM (1992a), August and McCann (1990), PlanGraphics (1991), Warnecke et al. (1992); Local: ACSM (1992c)
Spatial Data	Analysis: Fotheringham and Rogerson (1994), Goodchild and Gopal (1989), Raper (1989), Samet (1990), Tomlin (1990), Turner (1991); Data Management/Processing: Date (1985, 1990), Ferigno (1986), Fleming and von Halle (1986), Green (1985), Lin and Harbaugh (1984), International Geographical Union Commission on GIS (1992), Michener et al. (1994), Samet (1989, 1990); Standards/Format: Elissal and Caruso (1983), Johnson et al. (1992), National Committee for Cartographic Data Standards (1987), USFWS (1984), USGS (1990a, 1990b, 1991b); Information Exchange: ANSI (1986a, 1986b), Bureau of Census (1992-TIGER), Moellering (1991), Morrison and Wortman (1992), NIST (1992), USGS (1992); Data Coding: NBS (1987, 1988), U.S. EPA (1992d), USGS (1983); Locational Methods/Surveying: Onsrud and Cook (1990), U.S. EPA (1992a, 1992b, 1992c)
Data Sources	Soils: SCS (1991); Topography: Bauer (1989-AutoCad)

Table 10-7 References (Appendix F contains references for figure and table sources)

American Congress on Surveying and Mapping (ACSM). 1992a. State Geographic Information Activities Compendium. ACSM, Bethesda, MD.

American Congress on Surveying and Mapping (ACSM). 1992b. GIS: A Guide to the Technology. ACSM, Bethesda, MD.

American Congress on Surveying and Mapping (ACSM). 1992c. The Local Government Guide to GIS. ACSM, Bethesda, MD.

American Congress on Surveying and Mapping (ACSM). 1992d. GIS: Microcomputer and Modern Cartography. ACSM, Bethesda, MD.

American National Standards Institute (ANSI). 1986a. Specification for a Data Descriptive File for Information Interchange. ANSI/ISO 8211-1985, FIPS PUB 123.

American National Standards Institute (ANSI). 1986b. Computer Graphics Metafile for the Storage and Transfer of Picture Descriptive Information. ANSI X3.122-1986, FIPS PUB 128.

Antenucci, J.C., K. Brown, P.L. Croswell, M.J. Kevany, and H.N. Archer. 1991. Geographic Information Systems: A Guide to the Technology. Van Nostrand Reinhold, New York, NY, 301 pp.

Aronoff, S. 1989. Geographic Information Systems: A Management Perspective. WDL Publications, Ottawa, Canada, 294 pp. [Introduction for users and managers]

August, P.V. and A. McCann. 1990. Geographic Information Systems (GIS) in Rhode Island. Department of Natural Resources Science Fact Sheet No. 90-23, University of Rhode Island, Kingston, RI, 11 pp.

Bauer, M.F. 1989. Digital Map User's Guide. American Digital Cartography, Inc., Appleton, WI. [USGS topographic maps for AutoCad]

Bureau of Census. 1992. TIGER/SDTS™ Prototype Files, 1990 Preliminary Description. Available from Census Bureau, Geography Division, Geographic Base Development Branch, Washington, DC 20233.

Burrough, P.A. 1986. Principles of Geographical Information Systems for Land Resources Assessment. Clarendon/Oxford University Press, New York, NY, 193 pp. [Advanced text]

Cadoux-Hudson, J. and D.I. Heywood (eds.). 1992. Geographic Information 1992/3: Yearbook of the Association for Geographic Information. Taylor & Francis, Bristol, PA, 632 pp.

Clarke, K. 1990. Analytical and Computer Cartography. Prentice-Hall, Englewood Cliffs, NJ.

Date, C.J. 1985. Introduction to Database Systems, Vol II. Addison-Wesley, Reading, MA.

(Table 10-7 Major GIS References)

Date, C.J. 1990. Introduction to Database Systems, Vol I., 5th ed. Addison-Wesley, Reading, MA.

Douglas, W.J. 1994. Environmental GIS: Applications to Industrial Facilities. Lewis Publishers, Boca Raton, FL, 144 pp.

Elissal, A.A. and V.M. Caruso. 1983. Digital Elevation Models. U.S. Geological Survey Circular 895-B.

ESRI, Inc. 1990a. PC Arc/Info User's Manual. Environmental Research Institute, Inc., Redlands, CA.

ESRI, Inc. 1990b. Understanding GIS: The ARC/INFO Method (PC Version). Longman, Publications, White Plains, NY, 480 pp.

Federal Interagency Coordinating Committee on Digital Cartography (FICC). 1988. A Process for Evaluating Geographic Information Systems. Available from U.S. Geological Survey Publications, Reston, VA.

Federal Interagency Coordinating Committee on Digital Cartography (FICC). 1990. A Summary of GIS Use in the Federal Government. Available from U.S. Geological Survey Publications, Reston, VA.

Federal Geographic Data Committee (FGDC). 1991a. A National Geographic Information Resource: The Spatial Foundation of the Information-Based Society. U.S. Government Printing Office, Washington, DC, 10 pp. + 4 Appendices.

Federal Geographic Data Committee (FGDC). 1991b. First Annual Report to the Director, Office of Management and Budget. Available from U.S. Geological Survey Publications, Reston, VA.

Federal Geographic Data Committee (FGDC). 1993 . Manual of Federal Geographic Data Products. Available from U.S. Geological Survey Publications, Reston, VA.

Ferigno, C.F. 1986. A Data-Management System for Detailed Areal Interpretive Data. U.S. Geological Survey Water Resource Investigations Report 86-4091, 103 pp.

Fleming, C. and B. von Halle. 1989. Handbook of Relational Database Design. Addison-Wesley, Reading, MA.

Fotheringham, A.S. and P. Rogerson (eds.). 1994. Spatial Analysis and GIS. Francis & Taylor, Bristol, PA, 254 pp.

Gokee, T.L. and L.A. Joyce. 1992. Analysis of Standards and Guidelines in a Geographic Information System Using Existing Resource Data. Research Paper RM-304, Rocky Mountain Forest and Experiment Station, Fort Collins, CO, 12 pp.

Goodchild, M. and S. Gopal (ed.). 1989. Accuracy of Spatial Databases. Taylor & Francis, Bristol, PA, 308 pp.

(Table 10-7 Major GIS References)

Goodchild, M.F., B.O. Parks, and L.S. Steyaert (eds.). 1993. Environmental Modeling with GIS. Oxford University Press, New York, NY, 512 pp.

Green, W.R. 1985. Computer-Aided Data Analysis, A Practical Guide. John Wiley & Sons, New York, NY, 268 pp.

Green, D.R., D. Rix, and J. Cadoux-Hudson. 1993. Geographic Information 1993: The Sourcebook for GIS. Taylor & Francis, Bristol, PA, 550 pp. [Yearbook of the Association for Geographic Information; 1991 and 1992/93 yearbooks edited by J. Cadoux-Hudson and I. Heywood (429 and 632 pp. respectively)]

Hewitt, M.J., R.L. Webster, and D.E. James. 1993. A Summary of GIS Support to Superfund. EPA/600/X-93/062. U.S. EPA Environmental Monitoring Systems Laboratory, Las Vegas, NV, 54 pp.

Huxhold, W. 1991. Introduction to Urban GIS. Oxford University Press, New York, NY, 337 pp.

International Geographical Union Commission on GIS. 1992. Proceedings: 5th International Symposium on Spatial Data Handling, 2 Vols. [More than 70 papers; held August 3-7, 1992 in Charleston, SC]

Johnson, A.I., C.B. Pettersson, and J.L. Fulton. 1992. Geographic Information Systems (GIS) and Mapping: Practices and Standards. ASTM STP 1126, American Society for Testing and Materials, Philadelphia, PA.

Jones, F.H. and L. Martin. 1988. The AutoCAD® Database Book--Accessing and Managing CAD Drawing Information. Ventana Press, Chapel Hill, NC.

Kovar, K. and H.P. Nachtnebel (eds.). 1993. Application of Geographic Information Systems in Hydrology and Water Resources Management. Int. Ass. Sci. Hydrology Pub. No. 211, 693 pp. [Proc. IAHS/UNESCO conference held in Vienna, Austria, April, 1993; 68 papers]

Langran, G. 1992. Time in Geographic Information Systems. Taylor & Francis, Bristol, PA, 200 pp. [Covers conceptual, logical, and physical design of temporal GISs]

Lin, C. and J.W. Harbaugh. 1984. Graphic Display of Two- and Three-Dimensional Markov Computer Models in Geology. Van Nostrand Reinhold, New York, NY, 180 pp.

Luarini, R. and D. Thompson. 1992. Fundamentals of Spatial Information Systems. Academic Press, New York, NY, 680 pp.

Maguire, D.J., M.F. Goodchild, and D.W. Rhind. 1991. Geographical Information Systems: Principles and Applications. John Wiley & Sons, New York, NY. [2 volume set with 60 papers]

McCloy, K. 1994. Resource Management Information Systems. Francis & Taylor, Bristol, PA, 642 pp.

(Table 10-7 Major GIS References)

Michener, W.K., J.W. Brunt and S.G. Staff. 1994. Environmental Information Management & Analysis. Francis & Taylor, Bristol, PA, 600 pp.

Moellering, H. (ed.). 1991. Spatial DataBase Transfer Standards: Current International Status. Elsevier, New York, NY, 247 pp.

Morrison, J.L. and K. Wortman (eds.). 1992. Implementing the Spatial Data Transfer Standard. Cartography and Geographic Information Systems 19(5):277-334. [Special issue with 12 papers on the federal SDTS]

National Bureau of Standards (NBS). 1987. Codes for the Identification of the State, The District of Columbia and the Outlying Areas of the United States, and Associated Areas. Federal Information processing Standards (FIPS) Publication 5-2, NBS, U.S. Department of Commerce, Washington, DC.

National Bureau of Standards (NBS). 1988. Representation for Calendar Date and Ordinal Date for Information Interchange. Federal Information processing Standards (FIPS) Publication 4-1, NBS, U.S. Department of Commerce, Washington, DC.

National Committee for Digital Cartographic Data Standards. 1987. Issues in Digital Cartographic Data Standards. Report 9.

National Institute of Standards and Technology (NIST). 1992. Spatial Data Transfer Standard. Federal Information Processing Standard Publication 173 (FIPS Pub 173). [Available from NTIS or Internet: isdres.er.usgs.gov(130.11.48.2); user name: anonymous; after connecting: cd usgs.sdts]

Office of Information Resource Management (OIRM). 1992. Geographic Information Systems (GIS) Guidelines Document. OIRM 88-01. U.S. Environmental Protection Agency, Washington, DC.

Onsrud, H.J. and D.W. Cook (eds.). 1990. Geographic and Land Information Systems for Practicing Surveyors: A Compendium. American Congress on Surveying and Mapping, Bethesda, MD, 219 pp. [Collection of 22 articles from the recent GIS/LIS literature]

Pequet, D. and D. Marble (eds.). 1990. Introductory Readings in Geographic Information Systems. Taylor & Francis, Bristol, PA, 387 pp.

Pickus, J. 1992. Data Automation Using GIS and ARC/INFO GIS Support for Hydrogeologic Analysis. Contract No. 68-CO-0050, U.S. EPA Environmental Monitoring Systems Laboratory, Las Vegas, NV, 87 pp.

PlanGraphics. 1991. Summary of State GIS Coordination, Legislation and Funding Sources. PlanGraphics, Frankfort, KY, 9 pp.

Raper, J. (ed.). 1989. Three Dimensional Applications in Geographic Information Systems. Taylor & Francis, Bristol, PA, 189 pp. [Survey of approaches and problems in modeling real geophysical data]

(Table 10-7 Major GIS References)

Ripple, W. (ed.). 1986. Geographic Information Systems for Resource Management: A Compendium. ASPRS, Falls Church, VA/American Congress on Surveying and Mapping, Bethesda, MD, 288 pp. [Papers on land suitability; water, soil, and vegetation resource management; and urban and global GIS applications]

Ripple, W. (ed.). 1989. Fundamentals of Geographic Information Systems: A Compendium. ASPRS, Falls Church, VA/American Congress on Surveying and Mapping, Bethesda, MD, 248 pp.

Rowe, G.W. and S.J. Dulaney. 1991. Building and Using a Groundwater Database. Lewis Publishers, Chelsea, MI, 218 pp. [Appendix includes summary information on more than 80 GIS-related software]

Samet, H. 1989. Applications of Spatial Data Structures. Addison-Wesley, Reading, MA. [Applications in computer graphics, image processing and GIS]

Samet, H. 1990. Design and Analysis of Spatial Data Structures. Addison-Wesley, Reading, MA. [Hierarchical (quad-tree and oc-tree) state structures]

Soil Conservation Service (SCS). 1991. State Soil Geographic Data Base (STATSGO) Data Users Guide. SCS Miscellaneous Publication No. 1492, U.S. Department of Agriculture, Washington, DC, 88 pp.

Star, J. and J. Estes. 1990. Geographic Information Systems: An Introduction. Prentice-Hall, Englewood Cliffs, NJ, 303 pp. [Introductory text for students and professionals]

Thomas, H.F. (ed.). 1988. GIS: Integrating Technology and Geoscience Applications. National Academy of Science, Washington, DC.

Tomlin, D. 1990. Geographic Information Systems and Cartographic Modeling. Prentice-Hall, Englewood Cliffs, NJ.

Turner, A.K. 1991. Three-Dimensional Modeling with Geoscientific Information Systems. Kluwer Academic Publishers, Dordrecth, The Netherlands, 443 pp.

U.S. Environmental Protection Agency (EPA). 1992a. Locational Data Policy Implementation Guidance: Guide to the Policy. EPA/220/B-92-008, Office of Administration and Resources Management, Washington DC.

U.S. Environmental Protection Agency (EPA). 1992b. Locational Data Policy Implementation Guidance: Guide to Selecting Latitude/Longitude Collection Methods. EPA/220-B-92-008, Office of Administration and Resources Management (PMD-211D), Washington DC. [Note that U.S. EPA 1992a and 1992b are separate documents, but have the same document number]

U.S. Environmental Protection Agency (EPA). 1992c. Locational Data Policy Implementation Guidance--Global Positioning System Technology and Its Application In Environmental Programs--GPS Primer. EPA/600/R-92/036.

(Table 10-7 Major GIS References)

U.S. Environmental Protection Agency (EPA). 1992d. Definitions for the Minimum Set of Data Elements for Ground Water Quality. Policy Order 7500.1A; Guidance document EPA/813/B-92/002. Available from CERI.*

U.S. Fish and Wildlife Service (USFWS). 1984. Map Projections for Use with the Geographic Information System. FWS/OBS-84/17, USFWS, Washington, DC.

U.S. Geological Survey (USGS). 1983. Specifications for Representation of Geographic Point Locations for Information Interchange. U.S. Geological Survey Circular 878-B, 23 pp.

U.S. Geological Survey (USGS). 1990a. Digital Elevation Models--Data Users Guide 5. USGS National Mapping Division, Reston, VA, 51 pp.

U.S. Geological Survey (USGS). 1990b. Digital Line Graphs from 1:24,000-Scale Maps--Data Users Guide: National Mapping Program Technical Instructions. USGS National Mapping Division, Reston, VA, 107 pp.

U.S. Geological Survey (USGS). 1991a. National Mapping Program Technical Instructions, FIPS Pub 123 Function Library Software Documentation (Draft). USGS National Mapping Division, Reston, VA.

U.S. Geological Survey (USGS). 1991b. General Cartographic Transformation Package. USGS National Mapping Division, Reston, VA, 87 pp.

U.S. Geological Survey (USGS). 1992. A Prototype SDTS Federal Profile for Geographic Vector Data with Topology (Draft). USGS National Mapping Division, Reston, VA, 17 pp.

Warboys, M.F. 1994. Innovations in GIS. Francis & Taylor, Bristol, PA, 296 pp.

Warnecke, L., J.H. Johnson, K. Marshall, and R.S. Brown. 1992. State Geographic Information Activities Compendium. Council of State Governments, Washington, DC, 603 pp.

Young, R.H. and S. Cousins (eds.). 1993. Landscape Ecology and Geographic Information Systems. Taylor & Francis, Bristol, PA, 300 pp.

* See Preface for information on how to obtain documents from CERI (U.S. EPA Center for Environmental Research Information) and NTIS.

CHAPTER 11

PREVENTION AND MINIMIZATION OF CONTAMINATION

11.1 General Approaches 624

    11.1.1 Facility-Specific Waste Management and Control 624
    11.1.2 Wellhead Protection Area Management 625
    11.1.3 Other Approaches and Programs 625

11.2 Delineation of Wellhead Protection Areas 626

    11.2.1 Criteria for Delineation of Wellhead Protection Areas 626
    11.2.2 Overview of Delineation Methods 628
    11.2.3 Vulnerability Mapping 633
    11.2.4 Selection of Delineation Methods 637

11.3 Identification of Potential Contaminant Sources 641

    11.3.1 Step-By-Step Inventory Procedures 642
    11.3.2 Cross-Cutting Sources: Wells, Storage Tanks
           and Waste Disposal 643
    11.3.3 Nonindustrial Sources 643
    11.3.4 Commercial and Industrial Sources 644

11.4 Assessing The Risk From Potential Contaminants 644

    11.4.1 Risk Ranking Methods 647
    11.4.2 Other Risk Evaluation Methods 652

11.5 Wellhead Protection Area Management 652

    11.5.1 General Regulatory and Nonregulatory Approaches 654
    11.5.2 General Technical Approaches 654
    11.5.3 Specific Regulatory and Technical Approaches 659
    11.5.4 Contingency Planning 659

11.6 Guide to Major References* 661

* Appendix F contains citations for table and figure sources.

11.1 General Approaches

The two major approaches to prevention or minimization of contamination of soil and ground water are waste management and control at specific facilities (Section 11.1.1) and delineation and management of wellhead protection areas (Section 11.1.2). Section 11.1.3 provides a brief overview of other approaches.

11.1.1 Facility-Specific Waste Management and Control

Waste minimization and *pollution prevention* make both economic and environmental sense and have received increasing emphasis in environmental programs in recent years.[1] Techniques for waste minimization fall into two major categories: (1) source reduction, and (2) recycling. The discussion here focuses on waste minimization in the industrial and commercial sector, but the basic approaches can be applied to waste minimization for any activity.

Source Reduction. Source reduction can be achieved by *product changes*, such as product substitution, product conservation, and changes in product composition, and *source control*. Source control can be achieved by various methods:

- *Input material changes*, such as material purification and material substitution.

- *Technology changes*, such as modifications of process, equipment, piping or layout, additional automation, and changes in operational settings.

- *Good operating practices*, including procedural measures for loss prevention, management practices, waste stream segregation, improved material handling and production scheduling.

Recycling. Recycling can take place both onsite and offsite and can be accomplished by *reuse* (either by returning to the original process or use as a raw material substitute for another process) or *reclamation*, where spent materials are processed for resource recovery or processed as a by-product.

Life Cycle Analysis. Waste minimization is most effectively accomplished by life cycle analysis, which provides a complete environmental profile of goods and services. The life cycle consists of each step from the acquisition of raw materials through processing, manufacture, use, and final disposal of all residuals. Source reduction is most easily accomplished using the life cycle design approach for new facilities, but there may also be significant opportunities for source reduction by product changes at existing facilities.

---

[1] The term pollution prevention has a broader connotation that includes waste minimization, and waste treatment and controlled disposal. However, it may also be used interchangeably with waste minimization. For example, EPA's series of guides to pollution prevention (discussed later in this section) actually focus on waste minimization as described in the rest of this section.

A number of good documents are available from U.S. EPA for assessing opportunities for waste minimization and pollution prevention: (1) **Waste Minimization Opportunity Assessment Manual** (U.S. EPA, 1988e/T11-11), (2) **Facility Pollution Prevention Guide** (U.S. EPA, 1992d/T11-11). EPA's **Life Cycle Design Guidance Manual** (U.S. EPA, 1993b/T11-11) is provides up-to-date guidance on use of life cycle analysis to meet environmental objectives.

A number of ongoing publications series by U.S. EPA are useful for identifying opportunities for waste minimization in specific manufacturing, commercial and other activities (all available from EPA's Center for Environmental Research Information, while supplies last). These include the **Guides to Pollution Prevention** (U.S. EPA, 1990-1993/T11-11), **Waste Minimization Assessments** (U.S. EPA, 1990-1992/T11-11), and **Waste Reduction Activities and Opportunities** (U.S. EPA, 1992f/T11-11). There are more than 70 specific activities consolidated in the citations for the three series, and all three should be reviewed if more detailed information about a particular activity is desired. Table 11-11 also provides an index to a number of other major references on waste minimization and pollution prevention in industrial and commercial activities.

11.1.2 Wellhead Protection Area Management

The 1986 Amendments to the Safe Drinking Water Act (SDWA) established the Wellhead Protection Program (WHP), which required states to develop a WHP program to protect drinking water supplies as part of a Comprehensive State Ground-Water Protection Program. The wellhead protection area (WHPA) is the primary unit in this program, and most of the rest of this chapter focusses on the process of delineating WHPAs (Section 11.2), identifying potential contaminant sources (Section 11.3), assessing their risk (Section 11.4) and developing WHPA management strategies to prevent contamination (Section 11.5).

11.1.3 Other Approaches and Programs

Section 11.5 identifies numerous specific approaches to prevention of soil and ground-water contamination. Various federal and state regulatory programs, developed pursuant to legislation such as the Clean Water Act of 1972, Safe Drinking Water Act of 1974, Toxic Substances Control Act of 1976, Resource Conservation and Recovery Act of 1976 (and subsequent amendments to each Act), have as a primary or secondary focus prevention or minimization of soil and ground-water contamination. In addition to wellhead protection program, the Underground Injection Control (UIC) and Sole Source Aquifer (SSA) programs developed pursuant to the Safe Drinking Water Act focus on ground-water protection. The UIC program regulates injection of fluids into 5 classes of wells with the primary focus of ensuring that underground drinking water sources are not contaminated. The SSA program permits EPA to designate aquifers that are the sole or principal drinking water sources for an area and which if contaminated would present a significant hazard to human health. The main protection afforded by sole source aquifer designation is the withholding of federal financial assistance for any project that may contaminate the aquifer or create a significant public health hazard. Table 11-11 provides an index to major references on federal and state ground-water protection programs.

11.2 Delineation of Wellhead Protection Areas

11.2.1 Criteria for Delineation of Wellhead Protection Areas

U.S. EPA (1987a/T11-11) defined five criteria that may be used singly or in combination to define the area around a well in which contamination could represent a threat to drinking water drawn from the well: (1) distance, (2) drawdown, (3) time of travel, (4) flow boundaries, and (5) assimilative capacity. These are described briefly below.

Distance. The distance criterion uses a fixed radius or other dimension from a well to delineate a WHPA. This criterion usually is based on some kind of analysis involving the application of other criteria to generalized hydrogeologic settings. The approach is simple and very inexpensive. It is only suitable as a preliminary step, because the criterion considers ground-water flow or contaminant processes only indirectly. Since the zone of contribution (discussed below) rarely is circular, a fixed radius that provides adequate protection will almost always include areas for which protective actions are not required. Distance is also the *end-product* of the application of other delineation criteria.

Drawdown. Drawdown occurs when water is removed from an aquifer by pumping. The water level declines in the vicinity of the well, creating a gradient that drives water toward the discharge point. The gradient becomes steeper closer to the well, because the flow is converging from all directions and the area through which the water flows gets smaller. This results in a *cone of depression* around the well (Figure 2-17). The cone of depression around a well tapping an unconfined aquifer is relatively small compared to that around a well in a confined system. The former may be a few tens to a few hundred feet in diameter, while the latter may extend outward for miles.

The *zone of influence* (ZOI) is the distance from the well where changes in the ground-water surface can be measured or inferred as a result of pumping (Figure 11-1). In a homogenous, porous aquifer, the ZOI will be circular. In heterogenous porous and fractures aquifers, the ZOI typically has an elliptical or irregular shape. Ground-water velocities increase within the cone of depression of a well, causing contaminants to flow more rapidly toward the well. The drawdown criterion accurately defines areas requiring protection over the aquifer downgradient from the well, but generally does *not* include the zone of contribution upgradient based on flow boundaries (Figure 11-1).

Time of Travel (TOT). The time of travel criterion requires delineation of *isochrones* (contours of equal time) on a map that indicate how long water or a contaminant will take to reach a well from a point within the zone of contribution. The WHPA falls in the portion of the zone of contribution that is downgradient from

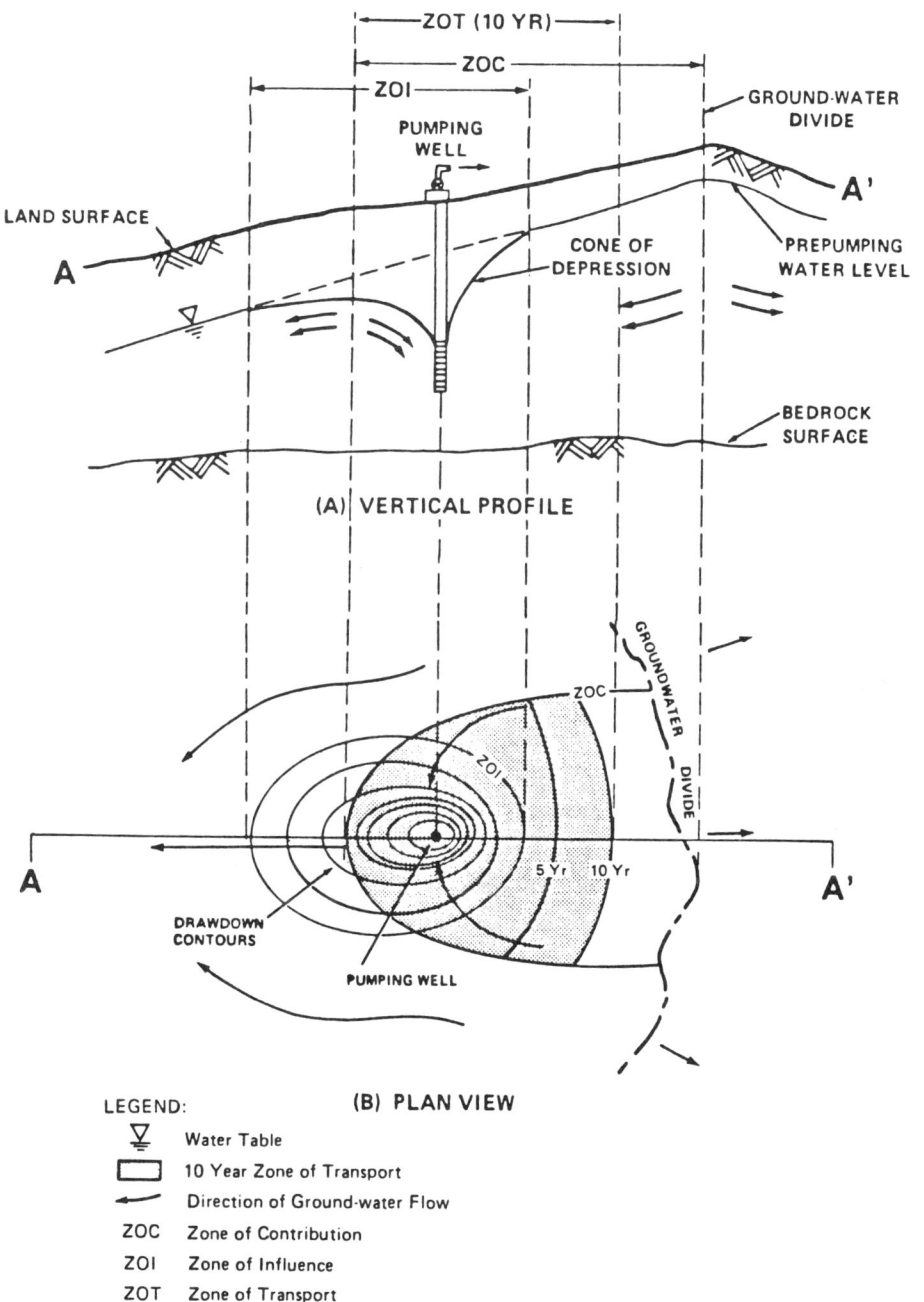

Figure 11-1 Relationship between ZOI, ZOT, and ZOC in an unconfined aquifer with a sloping regional water table (U.S. EPA, 1987b).

the selected isochrone (say 50 years time of travel). This area is called the *zone of transport* (ZOT). When the zone of contribution to a well is large (i.e., ground water from the farthest parts may take hundreds or thousands of years to reach the well), the ZOT will define a smaller area than the ZOC criterion (Figure 11-1). If the ZOC is small, the two will generally overlap.

Flow Boundaries (Zone of Contribution). The flow boundary criterion uses mapping of ground water divides and/or other physical and hydrologic features that control ground-water flow to define the geographic area containing ground water that flows toward a pumping well (Figure 11-1). Designating this *zone of contribution* (ZOC) as the WHPA provides the maximum amount of protection, although there are special cases where the drawdown (zone of influence) and time of travel (zone of transport) criteria will coincide with the ZOC.

Assimilative Capacity. The assimilative capacity criterion allows the reduction of a WHPA if contaminants are immobilized or attenuated while moving through the vadose zone of the aquifer so that concentrations are within acceptable limits by the time they reach a pumping well. This may occur by processes of dilution, dispersion, sorption, chemical precipitation, and biological degradation. A WHPA defined by this criterion would include the *zone of attenuation* (ZOA).

This criterion can be used in several ways. Incorporation of an empirical *retardation factor* for a specific contaminant that represents the combined effects of attenuation processes in the aquifer into time of travel calculations would result in a shift of isochrones closer to the well. A more complex application involves establishing an acceptable concentration of a contaminant at the well and using solute transport models to define the distance required to avoid exceeding of the target concentration.

In practice, this is an unrealistic approach because of the difficulty of characterizing aquifer physical and chemical properties for transport modeling of multiple contaminants. Where only one or two contaminants, such as nitrate loadings from septic tanks or pesticide loadings, are of primary concern, this approach may be very useful.

11.2.2 Overview of Wellhead Protection Delineation Methods

Classification of Delineation Methods. Because the process of wellhead delineation typically involves the use of more than one of the criteria discussed in the previous section, methods for wellhead delineation are not readily classified into distinctive categories. This guide classifies WHPA delineation methods into four major groups of generally increasing complexity:

1. *Geometric* methods involve the use of a pre-determined fixed radius and aquifer geometry without any special consideration of the flow system (such as the cylinder method shown in Figure 11-2), or the use of simplified shapes that have been pre-calculated for a range of pumping and aquifer conditions (Figure 11-3).

Prevention and Minimization of Contamination

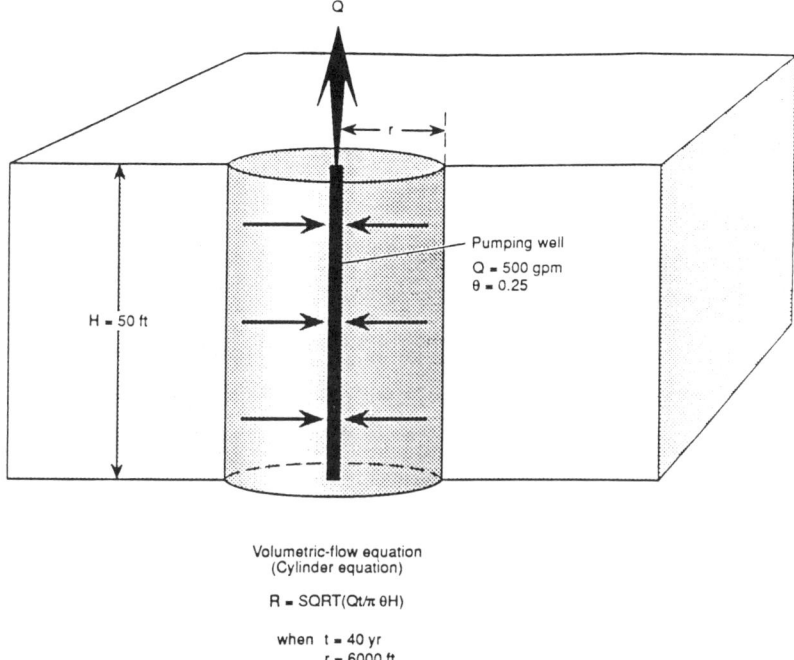

Figure 11-2 WHPA delineation using cylinder method (Kreitler and Senger, 1991).

2. *Simple analytical* methods allow calculation of distances for wellhead protection using equations that can be solved using a hand calculator or microcomputer spreadsheet program. These methods fall into two major groups, which are often used in combination: (1) time of travel calculations (Section 2.6.5), and (2) drawdown calculations, using aquifer test analytical equations that are rearranged to solve for distance to a specific drawdown criterion using measured or estimated values for other aquifer parameters.

3. *Hydrogeologic mapping* involves identification of the zone of contribution (as defined by flow boundaries) based on geomorphic, geologic, hydrologic, and hydrochemical characteristics of an aquifer. This is often used in combination with simple analytical methods and is usually required when using more complex analytical and numerical computer flow and transport models. Section 7.5 addresses special considerations in hydrogeologic mapping.

4. *Computer modeling* methods involve the use of more complex analytical or numerical solutions to ground-water flow and contaminant transport processes. Chapter 10 covers ground-water flow modeling in more detail.

STEP 1 Delineate Standardized Forms for Certain Aquifer Type

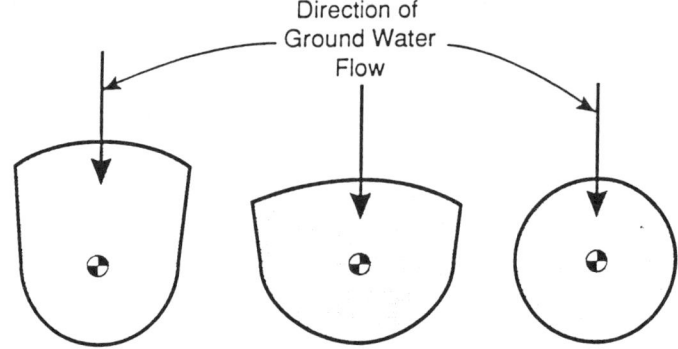

Various standardized forms are generated using analytical equations using sets of representative hydrogeologic parameters. Upgradient extent of WHPA is calculated with Time of Travel equation; downgradient with uniform flow equation.

STEP 2 Apply Standardized Form to Wellhead in Aquifer Type

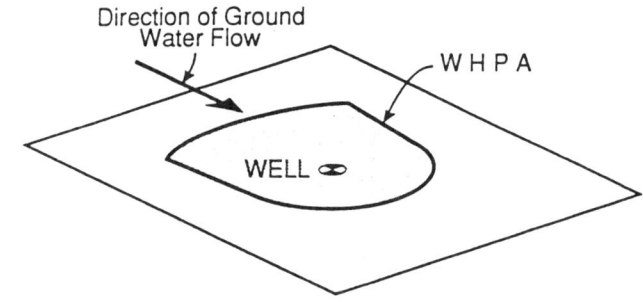

Standardized form is then applied to wells with similar pumping rate and hydrogeologic parameters.

Figure 11-3 WHPA delineation using simplified shapes method (U.S. EPA, 1993).

This classification scheme is generally similar to that used in U.S. EPA (1987a/T11-11) with the following differences: (1) the arbitrary fixed radius, volumetric flow equation (Figure 11-2), and simplified shapes methods (Figure 11-3) are all placed in the geometric category; (2) calculated fixed radius is dropped as a category because the two examples given fall into separate categories (the volumetric equation is geometric, and the Vermont Department of Water Resources method is a simple analytical method using a drawdown criterion); (3) the numerical flow/transport models category includes more complex analytical models that require computer programs for solution.

Table 11-1 summarizes the advantages and disadvantages and identifies the type of threshold criteria used for the three geometric methods and the three other major types of methods for delineating WHPAs (simple analytical methods, hydrogeologic mapping, and computer modeling). Important references focusing on special geologic

Prevention and Minimization of Contamination 631

settings for WHPA delineation include Kreitler and Senger (1991/T11-11) for confined aquifers and Bradbury et al. (1991/T11-11) for fractured rock aquifers. All state submittals to the U.S. Environmental Protection Agency for approval of wellhead protection programs contain a section describing WHPA delineation methods to be used in the state. Often these documents contain state-specific criteria for the application of geometric methods (Section 11.2.4).

Table 11-1 Comparison of Major Methods for Delineating Wellhead Protection Areas

Methods/Criteria	Advantages	Disadvantages
Geometric Methods		
Arbitrary fixed radius (distance)	--Easily implemented. --Inexpensive. --Requires minimal technical expertise.	--Low hydrogeologic precision. --Large threshold radius required to compensate for uncertainty will generally result in overprotection. --Highly vulnerable aquifers may be underprotected. --Highly susceptible to legal challenge.
Cylinder method (calculated fixed radius)	--Easy to use. --Relatively inexpensive. --Requires limited technical expertise. --Based on simple hydrogeologic principles. --Only aquifer parameter required is porosity. --Less susceptible to legal challenge.	--Tends to overprotect downgradient and underprotect upgradient because does not account for ZOC. --Inaccurate in heterogeneous and anisotropic aquifers. --Not appropriate for sloping potentiometric surface or unconfined aquifer.
Simplified variable shapes (TOT, flow boundaries)	--Easily implemented once shapes of standardized forms are calculated. --Limited field data required once standardized forms are developed (pumping rate, aquifer material type and direction of ground-water flow). --Relatively little technical expertise required for actual delineation. --Greater accuracy than calculated fixed radius for only modest added cost.	--Relatively extensive data on aquifer parameters required to develop the standardized forms for a particular area. --Inaccurate in heterogenous and anisotropic aquifers.

Table 11-1 (cont.)

Methods/Criteria	Advantages	Disadvantages
Other Methods		
Simple analytical methods (TOT, drawdown, flow boundaries)	--More accurate than simplified variable shapes because based on site-specific parameters. --Technical expertise required, but equations are generally easily understood by most hydrogeologists and civil engineers. --Various equations have been developed, allowing selection of solution that fits local conditions. --Allows accurate characterization of drawdown in the area closest to a pumping well. --Cost of developing site-specific data can be high.	--Relatively extensive data on aquifer parameters required for input to analytical equations. --Most analytical models do not take into account hydrologic boundaries, aquifer heterogeneities, and local recharge effects.
Hydrogeologic mapping (flow boundaries)	--Well suited for unconfined aquifers in unconsolidated formations and to highly anisotropic aquifers such as fracture bedrock and conduit-flow karst. --Necessary to define aquifer boundary conditions.	--Less suitable for deep, confined aquifers. --Requires special expertise in geomorphic and geologic mapping and judgement in hydrogeologic interpretations. --Moderate to high manpower and data collection costs.
Computer semianalytical and numerical flow/transport models (TOT, drawdown, flow boundaries)	--Most accurate of all methods and can be used for most complex hydrogeologic settings, except where karst conduit flow dominates. --Allows assessment of natural and human-related affects on the ground water system for evaluating management options.	--High degree of hydrogeologic and modeling expertise required. --Less suitable than analytical methods for assessing drawdowns close to pumping wells. --Extensive aquifer-specific data required. --Most expensive methods in terms of manpower and data collection/analysis costs.

Source: Boulding (1994).

Relationship of Protection Areas Based on Different Criteria. Table 11-2 provides summary definitions of types of wellhead areas based on four of the five criteria for wellhead protection: (1) zone of influence (ZOI); (2) zone of travel (ZOT); (3) zone of contribution (ZOC); and (4) zone of attenuation (ZOA). The first criterion, a fixed distance threshold, is based on a qualitative or semiquantitative application of one or more of these criteria. Table 11-2 also defines the hydrogeologic or other conditions required for one zone to be less than, equal to, or greater than another zone, and provides an indication of how commonly the relationship occurs. In general the following relationships occur: ZOA < ZOI < ZOT < ZOC.

11.2.3 Vulnerability Mapping

Ground-water vulnerability mapping involves the delineation of areas of varying susceptibility to ground-water contamination based on the interaction of characteristics that promote or inhibit movement of contaminants in the subsurface. Ground-water vulnerability maps may be developed as specific units within a broader scheme of ground-water classification, or may just delineate highly vulnerable areas without paying special attention to the characteristics of non-vulnerable areas.

Figure 11-4 illustrates WHPAs based on an arbitrary radius and simplified shape marked on a vulnerability map of Door County, Wisconsin. When vulnerability mapping is performed, efforts to inventory potential contaminant sources can be focused on areas where the hazard is greatest. Vulnerability mapping also allows fine-tuning of management approaches within the WHPA. Highly vulnerable areas require stricter management approaches than less vulnerable areas. The rest of this section reviews a number of approaches that have been developed for vulnerability mapping.

DRASTIC. DRASTIC is a widely used method for evaluating the relative vulnerability of mappable hydrogeologic units to ground-water contamination. DRASTIC is an acronym for the seven factors for which numerical ratings are made to develop an index of vulnerability to ground water contamination: Depth to water table, net Recharge, Aquifer media, Soil media, Topography (slope), Impact to vadose zone, and hydraulic Conductivity of the aquifer. Conventional hydrogeologic mapping methods are first used to delineate areas with similar characteristics. A numerical value is given to each of the seven factors, which are multiplied by a weighting factor and added for the DRASTIC index for the map unit. Worksheet D-W6 provides a form for calculating the DRASTIC index.

The DRASTIC index does not have any absolute meaning, but provides a means to assess relative vulnerability. A DRASTIC index of greater than 150 is one means of defining a highly vulnerable aquifer under EPA's ground-water protection strategy (U.S. EPA, 1986a/T11-10). The DRASTIC index has been found to give inconsistent results in karst areas where the water table in relatively deep (Sendlein, 1992/T11-10), and in the arid Tucson basin, Arizona, for reasons that are not entirely clear (Pima Association of Governments, 1992/T11-10). Both of these studies suggest that the relatively high weighting given to depth to water may understate the potential for contamination when preferential pathways allow relatively rapid vertical migration to deep water tables. Another weakness in the DRASTIC index is that is that it does not readily allow differentiation of shallow perched water tables over deeper regional water tables.

Table 11-2 Relationships of WHPAs Based on Zone of Influence, Time of Travel, Zone of Travel, Zone of Contribution and Zone of Attenuation

Terms/Relationship	Description
Zone of Influence	ZOI = area of drawdown or the cone of depression around a well created by pumping.
Zone of Travel[a]	ZOT = area around a well defined by a time of travel (TOT) isochrone and aquifer boundaries. ZOT_{max} = ZOT defined by TOT_{min} isochrone or the edge of the ZOC, whichever is closer to the well.
Zone of Contribution	ZOC = portion of an aquifer in which all recharge and ground water flows toward a pumping well. The boundaries of the ZOC are defined by ground water divides and other aquifer boundaries.
Zone of Attenuation	ZOA = area around an aquifer capable of reducing concentrations of a contaminant entering the area at a specified maximum concentration level to less than a defined acceptable concentration at the well.
ZOI < ZOT	When distance to TOT_{min} isochrone (i.e. ZOT_{max} boundary edge) lies outside the cone of depression. Most common situation for unconfined aquifers.
ZOI = ZOT	When distance to TOT_{min} isochrone = distance to ZOI boundary edge.
ZOI > ZOT	When TOT_{min} isochrone lies within cone of depression for a well. Unlikely to occur in unconfined aquifers, may occur in confined aquifers with very large ZOI.
ZOI < ZOC	When upgradient ground water divide lies outside cone of depression. The case in most hydrogeologic settings.
ZOI = ZOC	Rare. May occur with flat water table, with high recharge from rainfall within ZOI. Also possible when ZOI straddles a ground water divide.
ZOI > ZOC	Cannot occur.
ZOT < ZOC	When distance to TOT_{min} isochrone < distance to ZOC boundary. The most common situation. The difference between the two zone decreases as the TOT threshold criterion increases.
ZOT = ZOC	When distance to TOT_{min} isochrone = distance to ZOC boundary.
ZOT > ZOC	By definition, cannot occur. However, in this situation TOT is less than TOT_{min} indicating that the well is very vulnerable to contamination from sources within the ZOC.
ZOA < ZOT	When assimilative capacity is > 0.
ZOA = ZOT	When contaminant is not attenuated by the aquifer.

[a] Defined by time of travel criterion. TOT = time of travel for ground water or contaminants from a point in an aquifer to a pumping well. TOT_{min} = the minimum acceptable time of travel for purposes of wellhead delineation. TOT isochrone = a line from which TOT is the same at all points to a pumping well.

Source: Boulding (1994).

Prevention and Minimization of Contamination 635

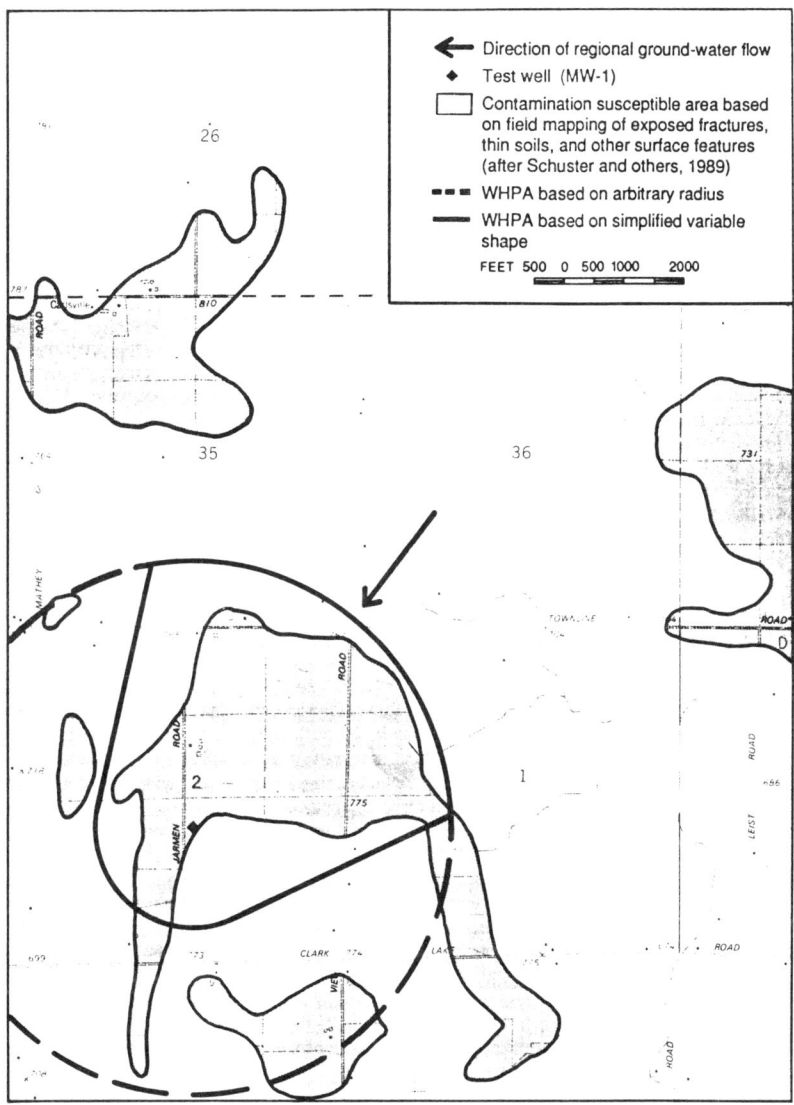

Figure 11-4 WHPAs at Sevastopol site, Door County, Wisconsin, Based on fixed radius, simplified shape and vulnerability mapping (Bradbury et al., 1991).

Other Vulnerability Mapping Methods. Various other methods have been developed for vulnerability mapping. They can be broadly classified as (1) systems using numerical ratings (as with DRASTIC) and (2) non-numerical systems in which map units may be numbered in order of increasing vulnerability, or classified as highly vulnerable and less vulnerable. Table 11-3 describes a number of vulnerability mapping techniques and summarizes the type of criteria used. Knox et al. (1993/T11-10) include tables summarizing criteria for the SAFE, WSSIM, HRS, SRM, and PI methods. Perhaps the simplest application of vulnerability mapping for wellhead protection is to develop criteria based on local conditions for defining highly vulnerable hydrogeologic settings (Figures 11-4 and 7-15). The DRASTIC criteria in Worksheet D-W6 and the information in Table 11-3, and the references indexed in Table 11-10 may be useful for developing locally appropriate vulnerability criteria.

Table 11-3 Summary of Major Ground-Water Vulnerability Mapping Methods

Description	Major Vulnerability Criteria	References
The DRASTIC method can be applied in any hydrogeologic setting. Results in a numerical index based on the sum of weighted ratings for seven criteria. Most widely used method.	See Worksheet D-W6. Highly vulnerable = >150 (U.S. EPA, 1986a).	Aller et al. (1987). Case studies: See Table 11-10.
Illinois ground-water aquifer vulnerability maps and geographic information system. Subsurface geologic data to a depth of 50 feet has been digitized to develop a state-wide stack-unit map.	Has been used for a variety of applications. Uhlman and Smith (1990) defined 8 classes for LUST contamination potential based on depth to uppermost aquifer and presence or absence of major aquifer at depth. Highly vulnerable: aquifer material within 5 feet of land surface, variable underlying materials and major aquifer at depth.	See Table 11-10.
Karst limestone areas are highly vulnerable by definition because conduit flow allows rapid travel of contaminants. Several schemes provide more detailed criteria for assessing relative vulnerability.	Quinlan et al. (1992b): hypersensitive = high point recharge, high conduit flow, low soil storage (Figure 7-15). Schuster et al. (1989); highly vulnerable = shallow or exposed fracture dolomite bedrock; permeable soils; open surface fractures; sinkholes (Figure 11-4).	Quinlan et al. (1992b), Schuster et al. (1989), Sendlein (1992).
Vulnerability to contamination by agricultural chemicals. Various vulnerability indexes have been developed.	DRASTIC pesticide index places greater weight on soil media and topography (Worksheet D-W6). RAVE index (DeLuca and Johnson (1990) uses a numerical index based on depth to ground water, soil texture, percent organic matter, topographic position, distance to surface waster, cropping practice, pesticide application frequency/method, and pesticide leaching index. Scores >60 indicate high concern.	Others include the Pesticide Index (PI)—Rao et al. (1985), U.S. EPA (1986d); SAFE (Soil/Aquifer Field Evaluation)—Roux (1986); See Table 11-10 for additional case study references.

Table 11-3 (cont.)

Description	Major Vulnerability Criteria	References
Numerous schemes have been developed to assess site suitability for solid/hazardous waste land disposal siting or risk from currently contaminated sites. Such suitability ranking systems can also be used to assess ground-water vulnerability.	LSR (landfill site rating) system uses (1) hydraulic conductivity; (2) sorption; (3) aquifer thickness; (4) depth and gradient of water table; (5) topography); (6) distance to wells or streams. High suitability = low vulnerability to ground water contamination. Low suitability = high vulnerability to ground water contamination. Each method has slightly different criteria.	LSR: LeGrand (1964, 1983), LeGrand and Brown (1977); HRS (Hazard Ranking System): Caldwell et al. (1981); SRM (Superfund Site Rating Methodology): Kufs et al. (1980), U.S. EPA (1989, 1991c); SIA (Surface Impoundment Assessment method): Silka and Swearingen (1978), U.S. EPA (1983); WSSIM (Waste-Soil-Site Interaction Matrix): Phillips et al. (1977)
General ground-water classification schemes.	Criteria varies depending on the objective of the classification scheme.	General: U.S. EPA (1985, 1986a); Sole aquifer program: U.S. EPA (1988b)

Source: Boulding (1994); see Table 11-10 references for full citations.

11.2.4 Selection of Delineation Methods

The state wellhead protection coordinator should be contacted to determine if there is any state guidance regarding the methods that can or should be used to delineate WHPAs. For example, Table 11-4 presents proposed guidance from the state of Georgia identifying generic wellhead protection areas: (1) a fixed radius "control zone" in the immediate vicinity of all wells; (2) a fixed radius "inner management zone" based on whether the aquifer is confined, unconfined, or karst; and (3) an "outer management zone" for which different delineation methods are specified, depending on the hydrogeologic setting. Methods used for delineating the outer management zone include: (1) graphical determination of radius based on pumping rate in crystalline rock aquifers (Figure 11-5), (2) hydrogeologic mapping in karst aquifers, and (3) 5-year time of travel or volumetric calculations in unconfined or partially confined porous media aquifers.

The Idaho wellhead protection program, on the other hand, identifies four major zones within a wellhead protection area, with a fixed radius used to Zone IA (Table 11-5). Zones IB and Zone II are delineated based on time of travel using hydrogeologic mapping, semianalytical, analytical, or numerical modeling based on site-specific data. Finally, Zone III includes known recharge areas and flow boundaries based on hydrogeologic mapping.

Table 11-4 Generic Wellhead Protection Areas Proposed for Georgia (Georgia Department of Natural Resources, 1992).

CONTROL ZONE

ALL WELLS

Impervious surface (pavement)	15 feet
Pervious surface (soil)	25 feet

INNER MANAGEMENT ZONE

ALL WELLS

Confined Aquifer Wells	100 feet
Unconfined Aquifer Wells	250 feet
Karst Aquifer Wells	500 feet

OUTER MANAGEMENT ZONE

PIEDMONT AND BLUE RIDGE
(crystalline rocks)

Pumping rate	Radius of Outer Management Zone determined by "Heath Method"

KARSTIC VALLEY AND RIDGE AND COASTAL PLAIN
(unconfined aquifer)

Hydrogeologic Mapping (by EPD)

COASTAL PLAIN
(unconfined or partially confined porous media)

5-year Time of Travel or Volumetric Calculations (by EPD)

COASTAL PLAIN
(completely confined aquifer)

None

Table 11-1 summarizes the relative advantages and disadvantages of the major methods for delineating WHPAs. Figure 11-6 provides a flow chart for delineating a WHPA. Relevant worksheets in Appendix D are identified in this figure, which shows that some form of hydrogeologic mapping is required for any WHPA delineation effort. At a minimum, this would involve collecting and compiling existing data and maps of the area (Worksheet D-W2). Collection of additional data, as needed, is an ongoing process at each step in the process. State wellhead protection programs may specify or provide guidance in selecting criteria (i.e., time of travel isochrones, drawdown limits) for delineating WHPAs using simple analytical methods or computer models.

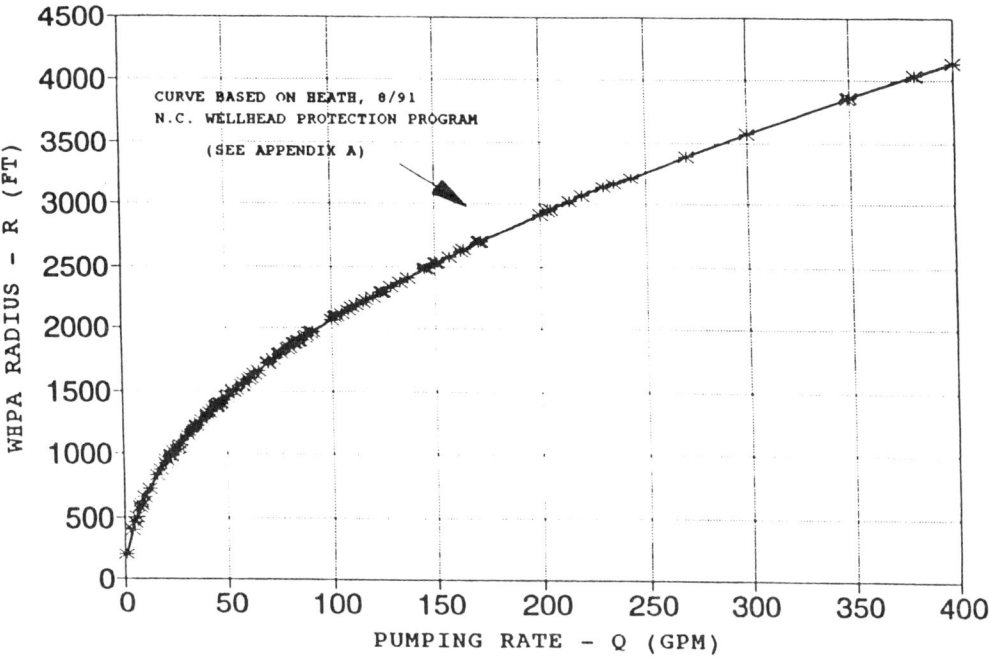

Figure 11-5 Radius of outer management zone based on pumping rates for crystalline aquifers, Piedmont and Blue Ridge (Georgia Department of Natural Resources, 1992).

Table 11-5 Zones for Wellhead Protection Areas in Idaho (Idaho Wellhead Protection Work Group, 1992).

Zone	Criteria and Thresholds	Methods
Zone IA	Minimum distance of 50 feet for wells Minimum distance of 100 feet for springs	Fixed radius
Zone IB	Two-year time of travel	Hydrogeologic mapping, semi-analytical, analytical, or numerical modeling using site specific data
Zone II	Five-year time of travel	Hydrogeologic mapping, semi-analytical, analytical, or numerical modeling using site specific data
Zone III	Known recharge areas and flow boundaries	Hydrogeologic mapping

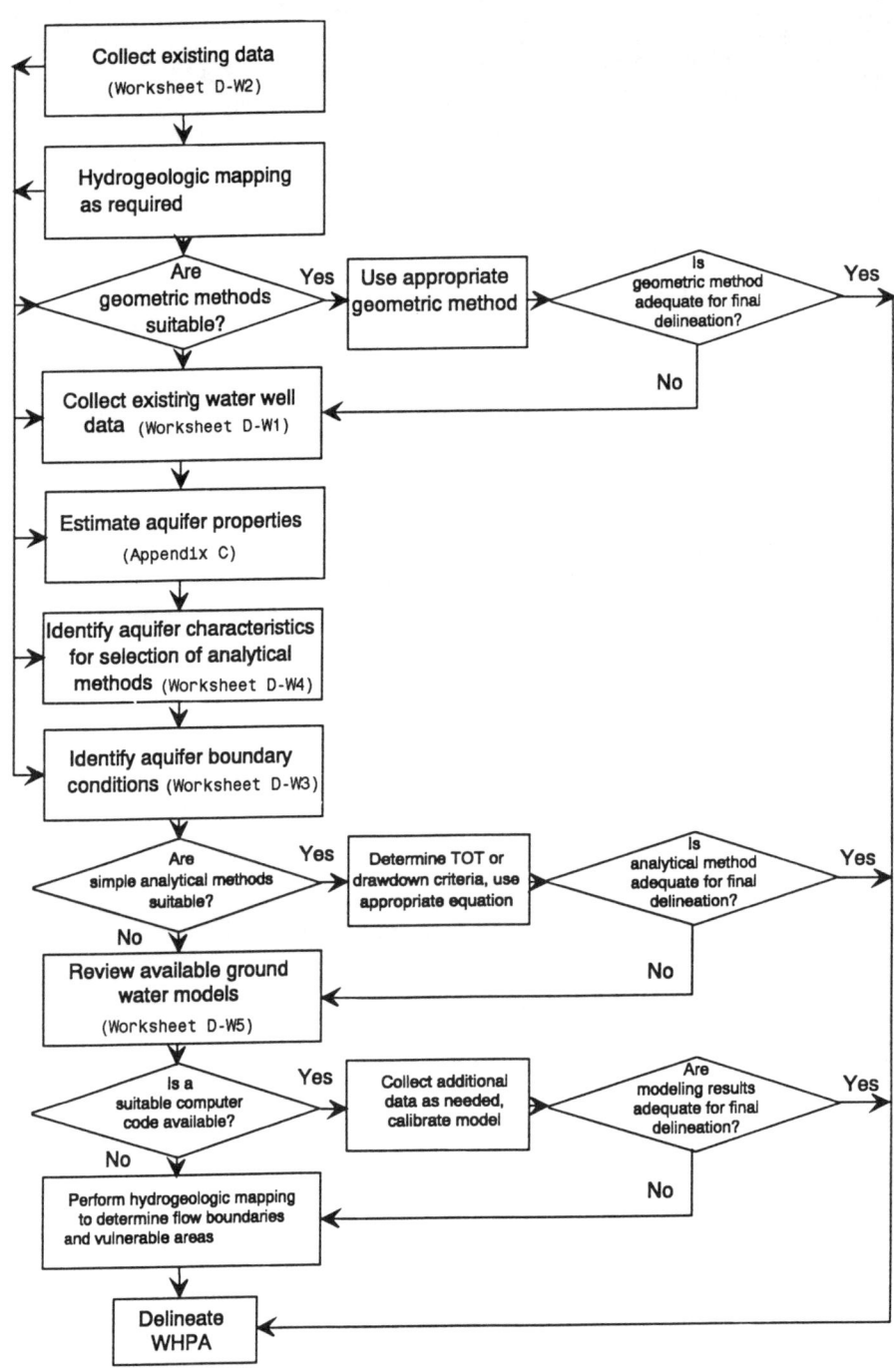

Figure 11-6 Flow chart for selection of wellhead protection delineation methods (Boulding, 1993c).

Use of multiple approaches to delineating a WHPA (i.e. moving as far through the flow chart in Figure 11-6 as time and financial resources allow) increases the likelihood that the area delineated excludes areas that do not actually contribute ground water to the well. Two situations that might require using more sophisticated delineation methods, such as computer modeling, include (1) the presence of a large number of potential sources of contamination, (2) the presence of strong opposition to regulatory controls for wellhead protection. In the first situation, the use of more sophisticated methods may avoid unnecessary effort devoted to inventorying potential contaminant sources outside the zone of contribution. In the second case, opposition may be partly defused by excluding areas from regulatory controls that might otherwise have been included. More sophisticated methods are also easier to defend against legal challenge.

11.3 Identification of Potential Contaminant Sources

The WHPA delineated using one or more methods described in the preceding section provides the focus for efforts to identify potential sources of contamination. The inventory should be comprehensive and should include:

- Potential points sources (underground storage tanks, wells, small commercial and industrial facilities, etc.).

- Potential line sources (sewer lines, gas/petroleum pipelines, highways with traffic that may haul hazardous chemicals, etc.).

- Potential area sources (waste disposal areas, agricultural lands receiving fertilizer and pesticide treatments, etc.).

The inventory should identify the type of source, location, and types of potential contaminants at each source. The next section provides detailed checklists for identifying potential sources. Identification of *active* potential sources is relatively straightforward. Location of *inactive* sources, such as abandoned wells and old waste disposal sites, might require some detective work. All existing maps and sources of information on past human activity in the area should be gathered and reviewed. Interviews with long-time residents in the area could yield valuable information that cannot be obtained in any other way. In areas with a long history of oil and gas exploration and production, or where the exact boundaries of old waste disposal sites are not known, surface geophysical methods and other field investigation techniques might be required to locate and map abandoned features. Table A-1 provides summary information on potential surface geophysical methods. Table 4-4 identifies references that provide more detailed information on methods for locating abandoned wells.

A convenient way to compile the results of the inventory is to assign each source an identification number and plot the identification number on a map of the WHPA. The boundaries of the areal sources should be clearly marked on the map. Repetition of the identifying number along a line source provides a means for distinguishing different types of sources. This map provides the focus for subsequent protective strategy development and land management activities.

Where a large number of commercial and industrial sites with potential contaminants are located within a WHPA, a phased approach may be desirable. The first phase would focus on identifying all potential sources, but would not necessarily involve collection of detailed information of all sites. This information would then be screened to identify sites where contaminants represent a significant potential risk based on the preliminary inventory. In the second phase, these sites would then be revisited to collect more detailed information. The final step in this stage of the well-head protection process would be to evaluate the degree of threat posed by each source (Section 11.4).

11.3.1 Step-By-Step Inventory Procedures

Hundreds of nonindustrial, commercial, and industrial activities that produce or use organic and inorganic substances pose a potential threat to ground-water quality. The number of potential contaminants of concern for a given activity may be restricted to a few or many substances. A single comprehensive list of these activities for inventory purposes would be so large as to be unmanageable. This guide offers a four-step approach to developing an inventory of potential sources of contamination within a WHPA:

1. Checklist D-C1 provides a "short list" of four major categories of potential contamination sources. A "yes" or "uncertain" answer to any of the questions within a major category on this checklist means that use of the detailed checklist for that category should be used (see next step).

2. Checklists D-C2 through D-C5 provide comprehensive lists of activities that may result in ground-water contamination. The first two (cross-cutting sources and nonindustrial sources) will probably be required for most WHPAs. In rural areas, the use of the remaining checklists may not be required. Sections 11.3.2 through 11.3.3 provide additional discussion of these checklists.

3. More detailed information should be compiled for each item that is identified within the WHPA. The following worksheets in Appendix D may provide assistance in gathering information on specific sources: (1) Worksheet D-W7 (Residential Source Inventory); (2) Worksheet D-W8 (Farm Source Inventory); (3) Worksheet D-W9 (Transportation Hazard Inventory); (4) Worksheets D-W10 and D-W11 (Municipal/Commercial/Industrial Source Inventory). Worksheet D-W1 can be used to compile information on active and abandoned wells.

4. A separate inventory worksheet should be filled out for each household or business by contacting the resident, owner or other responsible party. Files maintained by the Local Emergency Planning Committee (LEPC) established under Title III of SARA (the Emergency Planning and

Community Right-to-Know Act—EPCRA) should also be consulted. These files identify locations where hazardous chemicals are stored and used. Table 11-10 identifies references that provide more information on collection and analysis of information collected pursuant to EPCRA.

Many state wellhead protection programs have developed their own checklists, worksheets, and inventory forms for identifying potential contaminant sources. The Checklists and Worksheets in Appendix D represent a synthesis based on a review materials developed by state programs as of late 1993. Any of these, and subsequently state materials, as well as any subsequently developed, can be used as an alternative to or in combination with the materials in Appendix D. This is a complex topic in which improvements are always possible. The best approach is probably to compare the latest materials available for the state's wellhead protection program with the material in Appendix D and select the materials that seem most appropriate for the WHPA of interest. Alternatively, materials should be modified if comparisons show that no single checklist, worksheet, or inventory form addresses all the information needs for the WHPA.

A Few Words About Natural Contamination Sources. The checklists in this chapter do not address contamination sources that result from natural processes. In some areas, particularly in arid and semi-arid areas of the western United States, ground water is of marginal quality, or exceeds drinking water standards for elements such as arsenic, chloride, fluoride, heavy metals, and radionuclides. Little can be done to prevent such contamination, so the options are essentially limited to finding an alternative, higher quality source of drinking water, or treatment to remove contaminants. Human activity may cause degradation of ground water from natural sources. Examples include mobilization of heavy metals and radionuclides by mining activities and salt-water intrusion into fresh-water aquifers by pumping. Such activities are included in the checklists in Appendix D.

11.3.2 Cross-Cutting Sources: Wells, Storage Tanks and Waste Disposal

Checklist D-C2 identifies three major sources of potential contamination: (1) wells and related features, (2) storage tanks, and (3) waste disposal sites. These are called cross-cutting sources because they may be associated with any of the activities identified in the detailed checklists for nonindustrial, commercial, and industrial sources. The high risk of ground water contamination from storage tanks, especially underground storage tanks, and waste disposal sites is another reason for placing them in a separate checklist.

11.3.3 Nonindustrial Sources

Checklist D-C3 identifies five major categories of potential contamination sources that can be broadly classified as nonindustrial: (1) agricultural, (2) residential, (3) other green areas, (4) municipal and other public services, and (5) transportation. The category of "other green areas" includes any nonagricultural and nonresidential area where grass and other vegetation may receive regular applications of agricultural chemicals. In the residential category, each individuals in each residence or living unit should be interviewed, if possible, and a household hazardous waste inventory prepared. Such interviews should increase awareness by individuals and families living within a WHPA of ground-water concerns, and should lay the groundwork for any future public education efforts.

11.3.4 Commercial and Industrial Sources

Checklists D-C4 and D-C5 identify more than 90 commercial and industrial activities that present potential for ground-water contamination. Commercial activities are generally service- and sales-oriented, while industrial activities involve primarily processing and manufacturing. In practice, the dividing line is not always clear, so both checklists should be examined if the classification of an identified source is uncertain. Commercial activities associated with transportation are included in Checklist D-C3.

Checklist D-C4 identifies three major categories of activities: (1) commercial services and sales; (2) activities related to processing and storage of natural products (food, other animal products, and wood); and (3) resource extraction activities. Checklist D-C5 identifies three major categories of industrial activities: (1) chemical processing and manufacturing; (2) metal manufacturing, fabrication, and finishing; and (3) other manufacturing.

A wide array of potential contaminants are associated with commercial and industrial activities. U.S. EPA has developed a series of information sheets, available from the RCRA Hotline (Table 5-3), on 17 business activities that may generate hazardous wastes (U.S. EPA, 1990a/T4-4). Checklists D-C4 and D-C5 indicate activities covered by these summary sheets with the EPA document order number. Table 4-4 provides an index to major references that identify types of contaminants associated with major types of activities. The waste minimization and pollution preventions guides described in Section 11.1.1 are also good sources for identifying types of contaminants associated with specific types of manufacturing and commercial activities. Table 11-6 identifies EPA reference sources where more detailed information can be obtained on potential contaminants for industries for which EPA has developed effluent limitation guidelines.

11.4 Assessing the Risk from Potential Contaminants

Methods for evaluating the risk posed by potential contaminant sources within a WHPA can range from a relatively simple process—classifying sources as high, moderate, and low risk to a comprehensive risk assessment process in which fate and transport of chemicals of concerns are modeled to quantify exposure and risk to people or ecosystems. This section focuses on relatively simple ranking methods for evaluating risk and briefly discusses situations in which more complex methods may be required.

Table 11-6 Index to Development Documents for Effluent Limitations Guidelines for Selected Categories (U.S. EPA, 1987c).

Industrial point source category	Subcategory	EPA publication document No.	NTIS accession No.	GPO stock No.
Aluminum forming	Aluminum forming	EPA 440/1-84/073 Vol. I Vol. II	PB84-244425 PB84-244433	- -
Asbestos manufacturing	Building, construction, and paper	EPA 440/1-74/017a	PB238320/6	5501-00827
	Textile, friction materials, and sealing devices	EPA 440/1-74/035a	PB240860/7	-
Battery manufacturing	Battery manufacturing	EPA 440/1-84/067 Vol. I Vol. II	PB85-121507 PB85-121515	- -
Builders' paper and board mills	Pulp, paper and paperboard, and builders' paper and board mills	EPA 440/1-82/025	PB83-163949	-
Canned and preserved fruits and vegetables	Apple, citrus, and potato processing	EPA 440/1-74/027a	PB238649/8	5501-00790
Canned and preserved seafood processing	Catfish, crab, and shrimp	EPA 440/1-74/020a	PB238614/2	5501-00920
	Fishmeal, salmon, bottom fish, sardine, herring, clam, oyster, scallop, and abalone	EPA 440/1-75/041a	PB256840/0	-
Cement manufacturing	Cement manufacturing	EPA 440/1-74/005a	PB238610/0	5501-00866
Coil coating	Coil coating, Phase I	EPA 440/1-82/071	PB83-205542	-
	Coil coating, Phase II - can-making	EPA 440/1-83/071	PB84-198647	-
Copper forming	Copper	EPA 440/1-84/074	PB84-192459	-
Dairy products processing	Dairy products processing	EPA 440/1-74/021a	PB238835/3	5501-00898
Electroplating and metal finishing	Copper, nickel, chrome, and zinc	EPA 440/1-74/003a	PB238834/AS	5501-00816
	Electroplating - pretreatment	EPA 440/1-79/003	PB80-196488	-
	Metal finishing	EPA 440/1-83/091	PR84-115989	-
Ferroalloy	Smelting and slag processing	EPA 440/1-74/008a	PB238650/AS	5501-00780
Fertilizer manufacturing	Basic fertilizer chemicals	EPA 440/1-74/011a	PR238652/AS	5501-00868
	Formulated fertilizer	EPA 440/1-75/042a	PB240863/AS	5501-01006
Glass manufacturing	Pressed and blown glass	EPA 440/1-75-034a	PB256854/1	5501-01036
	Insulation fiberglass	EPA 440/1-74/001b	PB238078/0	5501-00781
	Flat glass	EPA 440/1-77/001c	PB238-907/0	5501-00814
Grain mills	Grain processing	EPA 440/1-74/028a	PB238316/4	5501-00844
	Animal feed, breakfast cereal, and wheat	EPA 440/1-74/039a	PB240861/5	5501-01007
Inorganic chemicals manufacturing	Inorganic chemicals Phase I	EPA 440/1-82/007	PB82-265612	-
	Inorganic chemicals Phase II	EPA 440/1-84/007	PB85-156446/XAB	-
Iron and steel manufacturing	Iron and steel Volume I Volume II Volume III Volume IV Volume V Volume VI	EPA 440/1-82/024 EPA 440/1-82/024 EPA 440/1-82/024 EPA 440/1-82/024 EPA 440/1-82/024 EPA 440/1-82/024 EPA 440/1-82/024	 PB82-240425a PB82-240433b PB82-240441c PB82-240458d PB82-240466e PB82-240474f	-

Table 11-6 (cont.)

Industrial point source category	Subcategory	EPA publication document No.	NTIS accession No.	GPO stock No.
Leather tanning	Leather tanning	EPA 440/1-82/016	PB83-172593	-
Meat products and rendering	Red meat processing	EPA 440/1-74/012a	PB238836/AS	5501-00843
	Renderer	EPA 440/1-74/031d	PB253572/2	-
Metal finishing	Metal finishing	EPA 440/1-83/091	PB84-115989	-
Metal molding and casting (foundries)	Metal molding and casting	EPA 440/1-85/070	PB86-161452/XAB	-
Nonferrous metals forming	Nonferrous metals forming	EPA 440/1-84/019b Vol. I Vol. II Vol. III	- PB83/228296 PB83/228304 PB83/228312	-
Nonferrous metals manufacturing	Bauxite refining - aluminum segment	EPA 440/1-74/019c	PB238463/4	5501-00116
	Primary aluminum smelting - aluminum segment	EPA 440/1-74/019d	PB240859/9	5501-00817
	Secondary aluminum smelting - aluminum segment	EPA 440/1-74/019e	PB238464/2	5501-00819
Organic chemical manufacturing and plastics and synthetic fibers	Organic chemicals manufacturing and plastics and synthetic fibers	EPA 440/1-87-009	Available from NTIS after publication (1/87)	
Petroleum refining	Petroleum refining	EPA 440/1-82/014	PB83-172569	-
Pharmaceuticals	Pharmaceutical	EPA 440/1-83/084	PB84-180066	-
Phosphate manufacturing	Phosphorus-derived chemicals	EPA 440/1-74/006a	PB241018/1	5503-00078
	Other non-fertilizer chemicals	EPA 440/1-75/043	-	-
Porcelain enameling	Porcelain enameling	EPA 440/1-82/072	-	-
Pulp, paper, and paperboard	Unbleached kraft and semi-chemical pulp	EPA 440/1-74/025a	PB238833/AS	-
	Pulp, paper and paperboard, and builders' paper and board mills	EPA 440/1-82/025	PB83-163949	-
Rubber processing	Tire and synthetic	EPA 440/1-74/013a	PB238609/2	5501-00885
	Fabricated and reclaimed rubber	EPA 440/1-74/030a	PB241916/6	5501-01016
Soaps and detergents	Soaps and detergents	EPA 440/1-74/018a	PB238613/4	5501-00867
Sugar processing	Beet sugar	EPA 440/1-74/002b	PB238462/6	5501-00117
	Cane sugar refining	EPA 440/1-74/002c	PB238147/3	5501-00826
Textile mills manufacturing	Textile mills	EPA 440/1-82/022	PB83-116871	-
Timber products processing	Wood furniture and fixtures	EPA 440/1-74/033a	-	-
	Timber products processing	EPA 440/1-81/023	PB81-227282	-

[a] This list includes only "final" development documents for effluent limitations guidelines. For many industries, these documents are in the draft or proposal stage.

11.4.1 Risk Ranking Methods

Classifying potential contaminant sources into risk categories (high, medium, low) is the simplest way to identify the sources within a WHPA that pose a threat to ground water quality. Figure 11-7 illustrates a matrix developed by the Cape Cod Aquifer Management Project to evaluate pollution potential from 32 land use categories. The top of the matrix contains ratings for 16 groups of chemicals according to (1) overall threat to public health, (2) mobility, (3) and whether they may occur naturally in significant concentrations. The overall threat to public water supply for each land use category in Figure 11-7 is rated as low (L) to high (H) in the right hand column, based on the number of potential contaminants associated with the category and the potential threat posed by each contaminant.

Following the approach in Figure 11-7, once the potential contaminant source inventory has been completed (Section 11.3), each land use category or individual source is placed in a risk category. Figure 11-7 has five categories (low, low-medium, medium, medium-high, and high), but fewer categories (low, medium, and high) can also be used. Figure 11-7 and Checklist D-C6, which identifies high and moderate risk land use activities based on ratings from a variety of sources, can provide some guidance in how to classify potential contaminant sources within a wellhead protection area. Not all sources agree in their classification of specific land use categories, and classification decisions should consider all factors particular to the wellhead protection area in question.

Aquifer vulnerability mapping, as described in Section 11.2.3, is a valuable complement to the risk ranking approach to evaluating potential contaminant sources. For example, any given potential contaminant source represents a less significant threat to a highly confined aquifer than to an unconfined aquifer (see Section 7.5.3).[2] Table 11-10 identifies a number of references that discuss vulnerability mapping in the context of risk assessment.

Whether a land use is classified as high or moderate risk becomes a significant consideration when developing options for managing the WHPA. High-risk land uses are frequently prohibited in high priority wellhead protection areas, and moderate-risk are commonly restricted in such areas. Table 11-7 illustrates how particular high- and moderate-risk land uses have been either prohibited or restricted (i.e., special permit required) in four water resource protection zones on Nantucket Island. High risk activities are either prohibited (P) or a special permit is required (SP). In the most sensitive areas (WR1), all high-risk activities are prohibited, whereas in the least sensitive areas (WR4) all high-risk activities require a special use permit. High-risk activities in water resource protection zones of intermediate sensitivity are prohibited or required a special permit, depending on the activity.

---

[2] An exception to this would be where the source is near an improperly abandoned well that provides a pathway from the surface to the confined aquifer.

Land Use Considerations

Potential Contaminants	Acids	Bases	Chloride	Fluoride	Iron/Manganese (Fe/Mn)	Metals (except Fe & Mn)	Nitrate	Pathogens (Virus/Bacteria)	Pesticides/Herbicides	Petroleum Products	Phenols	Radioactivity	Sodium	Solvents	Sulfate	Surfactants (Detergents)	Overall Threat to Public Water Supply[3]
Overall Threat to Public Health	L-M	L-M	L	L	H	H	M	H	H	H	H	H	H	H	L	L	
Mobility	M	L	H	M	L-H	H	H	L-H	L-H	M	L-H	L-H	H	H	H	H	
Natural Background	■		■	■	■	■						■	■		■		

Land Use Categories

Land Use Category	Overall Threat to Public Water Supply[3]
Agriculture/Golf Courses	M
Airports	M-H
Asphalt Plants	L-M
Beauty Parlors	L(?)
Boat Yards/Builders	L
Car Washes	L
Cemeteries	L
Chemical Manufacture	H
Clandestine Dumping	H
Dry Cleaning	H
Furniture Stripping and Painting	M
Hazardous Materials Storage and Transfer	H
Industrial Lagoons and Pits	H
Jewelry and Metal Plating	M
Junkyards	L
Landfills	H

Figure 11-7 Land use/public supply well pollution potential matrix (Noake, 1988).

Key to Figure 11-7

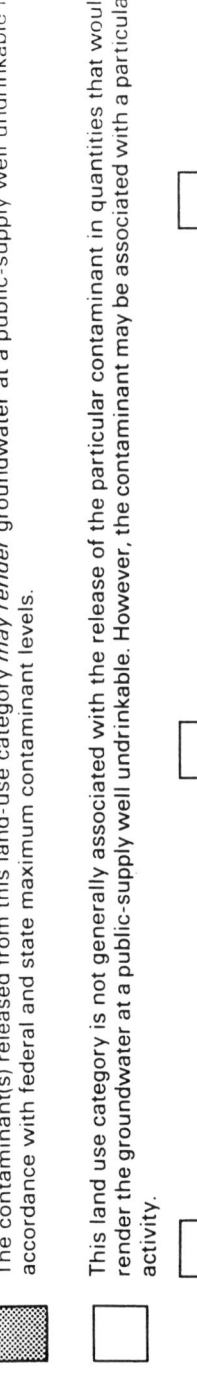

▓ The contaminant(s) released from this land-use category *may render* groundwater at a public-supply well undrinkable in accordance with federal and state maximum contaminant levels.

☐ This land use category is not generally associated with the release of the particular contaminant in quantities that would render the groundwater at a public-supply well undrinkable. However, the contaminant may be associated with a particular activity.

| L | = Low Threat | M | = Medium Threat | H | = High Threat |

This Matrix is based on a literature review and the combined field experience of the Cape Cod Aquifer Management Project (CCAMP). **THIS MATRIX SHOULD BE USED AS A GUIDE AND HANDY REFERENCE.** It is not a substitute for looking at a particular land use in detail. There will always be the potential for a business to use an unusual process using chemicals not normally associated with that business. The land-use categories included in the Matrix and *Guide to Contamination Sources for Wellhead Protection* are those that might be found in the primary recharge area of a public-supply well in Massachusetts. This Matrix may be misleading or erroneous if applied to low-yield private wells.

1. Nitrate has a cumulative impact on groundwater quality. No one category is responsible for the release of nitrate. A variety of land use categories release nitrate. These include animal feedlots, landfills, septic systems, septage lagoons, municipal wastewater and agricultural activities including turf maintenance

2. There are no known instances of beauty parlors contaminating well water in Massachusetts. More research is needed to determine the severity of a threat to groundwater from this land use category.

3. Refer to *Guide to Contamination Sources for Wellhead Protection*, pp. 1-2.

Table 11-7 Regulated Land Uses in Different Water Resource Protection Zones, Nantucket Island, Massachusetts (Horsley, 1990). See Appendix F for full credit.

	WR1	WR2	WR3	WR4
1. Sanitary landfills	P	P	P	SP
2. Junk yards, salvage yards	P	P	P	SP
3. Municipal sewage treatment facilities with on-site disposal of primary or secondary treated effluent	P	P	P	SP
4. Car and truck washes	P	P	SP	SP
5. Road salt stockpiles	P	P	SP	SP
6. Dry cleaning establishments, coin or commercial laundries	P	P	SP	SP
7. Motor vehicle and boat service and repair facilities including body shops	P	P	P	SP
8. Metal plating establishments	P	P	SP	SP
9. Sales, storage or disposal of fuels or hazardous chemicals				
10. Chemical and bacteriological laboratories	P	P	P	SP
11. Trucking or bus terminals	P	P	P	SP
12. Any use which involves as a principal activity the manufacture, storage, use, transportation or disposal of toxic or hazardous materials	P	P	SP	SP
13. Any use which involves the use of toxic and hazardous materials in quantities greater than those associated with normal household use	P	P	SP	SP
14. Residential development at densities exceeding those stated in Section E of this bylaw	P	P	P	SP
15. Golf courses	P	SP	SP	SP

P = Prohibited
SP = Special permit required

11.4.2 Other Risk Evaluation Methods

Risk ranking and aquifer vulnerability mapping methods are probably adequate for many WHPAs. Where many high risk potential contaminant sources exist within a WHPA, more sophisticated risk assessment approaches may be required to help identify the most efficacious and cost-effective options for reducing risk. Factors that need to be considered for a comprehensive risk assessment include (1) chemical toxicity; (2) pathways that can lead to exposure; (3) the characteristics of the population being exposed (density, age, etc.); (4) the probability that health-threatening exposures will actually occur; (5) the cost of options for reducing risk from exposure; and (6) the perception of risk by the exposed population. Figure 11-8 provides and example diagram of an exposure pathway assessment for a landfill.

EPA has developed a relatively sophisticated manual procedure to assess and screen relative threats to ground-water supplies posed by potential contaminant sources (U.S. EPA, 1991a/T11-10). The method involves a series of step-by-step procedures using worksheets that result in an overall risk rating for each contaminant source based on (1) the likelihood of well contamination and (2) the severity of well contamination.

A variety of methods have been developed for evaluating risks addressed by other EPA programs. For example, several methods have been developed to help communities evaluate the risk posed by chemicals that must be reported under EPA's Toxic Release Inventory (TRI) program (FEMA/DOT/EPA, 1989/T11-10; U.S. EPA, 1989b/T11-10). These methods focus more on the risks posed by airborne accidental releases of chemicals. Elements of these methods, however, could be adapted for use in evaluating the risks of ground-water contamination by chemicals reported under the TRI program. Similarly, methods used to assess risk at Superfund sites and for other EPA programs may be useful, under certain circumstances, for evaluating risk in WHPAs. Table 11-10 provides an index to major references on risk assessment in relation to ground-water contamination and other methods for exposure and risk assessment.

11.5 Wellhead Protection Area Management

Management of wellhead protection areas (WHPAs) to prevent ground-water contamination involves several steps:

- Identification of protection options appropriate for the types of potential contaminants present.

- Selection of those that are technically and politically feasible for the area.

- Implementation of the options.

- Monitoring of the effectiveness of management and application of additional management practices, if required.

Prevention and Minimization of Contamination 653

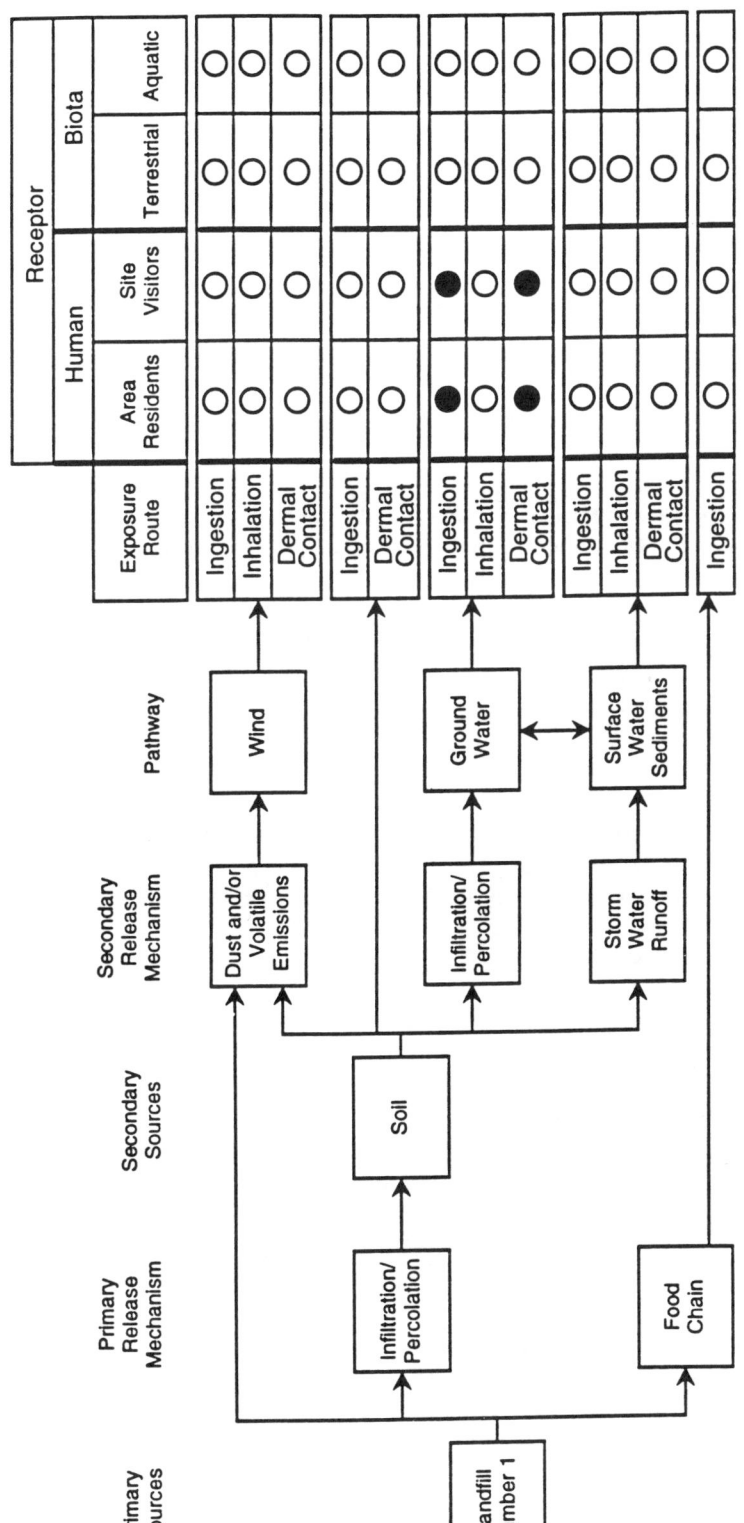

Figure 11-8 Example diagram of exposure pathway assessment for a landfill (adapted from U.S. EPA, 1989a).

- Development of contingency plans to address threats to a water supply as a result of accident or failure of the management practices that have been implemented.

11.5.1 General Regulatory and Nonregulatory Approaches

Wellhead protection management options or tools can be broadly classified as *regulatory* and *nonregulatory*. At the local level, regulatory approaches generally involve the use some form of (1) zoning ordinances, (2) subdivision or individual lot controls, or (3) promulgation of local health and environmental regulations designed to directly or indirectly protect ground water in a WHPA. State-level legislation or regulations may also address wellhead protection. Nonregulatory controls, as the name implies, involve voluntary actions on the part of the public and private sector to enhance ground-water protection.

Wellhead protection management options can also be classified as *technical* and *nontechnical*. Although the dividing line may not always be clear, technical options generally involve controls based on some understanding of the relationship between contaminant characteristics and the hydrogeology of a WHPA. Nontechnical options are generally not directly related to scientific considerations, although indirect relationships exist to the extent that WHPA delineation and contaminant risk assessment processes are scientifically based.

Checklist D-C7 identifies 45 specific wellhead protection tools in three major categories: (1) nontechnical regulatory options, (2) nontechnical nonregulatory options, and (3) technical regulatory and nonregulatory options. Nontechnical options are not discussed further here. However, Checklist D-C7 indicates where Tables 11-8 and 11-9 provide summary information on specific options. The rest of this section focuses on general technical approaches to WHPA management (Section 11.5.2) and specific approaches for different types of land use (Section 11.5.3)

11.5.2 General Technical Approaches

Design Standards and Best Management Practices. *Design standards* define specifications for how a building or onsite wastewater disposal system should be constructed. *Best management practices* (BMPs) define how repeated activities, such as construction and farming, should be carried out so as to minimize adverse environmental impacts. The great advantage of these approaches is their simplicity. They establish an objective standard for monitoring compliance. Design standards usually require inspection for compliance at the time of inspection, although some ongoing monitoring may also be required. BMPs may require ongoing monitoring for compliance. However, design standards and BMPs will only provide adequate protection if the assumptions used in establishing the standard or practice apply within a WHPA. Design standards and BMPs tend to be less flexible than performance standards (next section) because they cannot be readily modified to reflect local conditions.

Table 11-8 Summary of Wellhead Protection Tools (U.S. EPA, 1993)

	Applicability to Wellhead Protection	Land Use Practice	Legal Considerations	Administrative Considerations
Regulatory: Zoning				
Overlay GW Protection Districts	Used to map wellhead protection areas (WHPAs). Provides for identification of sensitive areas for protection. Used in conjunction with other tools that follow.	Community identifies WHPAs on practical base/zoning map.	Well-accepted method of identifying sensitive areas. May face legal challenges if WHPA boundaries are based solely on arbitrary delineation.	Requires staff to develop overlay map. Inherent nature of zoning provides "grandfather" protection to pre-existing uses and structures.
Prohibition of Various Land Uses	Used within mapped WHPAs to prohibit ground-water contaminants and uses that generate contaminants.	Community adopts prohibited uses list within their zoning ordinance.	Well-organized function of zoning. Appropriate techniques to protect natural resources from contamination.	Requires amendment to zoning ordinance. Requires enforcement by both visual inspection and onsite investigations.
Special Permitting	Used to restrict uses within WHPAs that may cause ground water contamination if left unregulated.	Community adopts special permit "thresholds" for various uses and structures within WHPAs. Community grants special permits for "threshold" uses only if ground water quality will not be compromised.	Well-organized method of segregating land uses within critical resource areas such as WHPAs. Requires case-by-case analysis to ensure equal treatment of applicants.	Requires detailed understanding of WHPA sensitivity by local permit granting authority. Requires enforcement of special permit requirements and onsite investigations.
Large-Lot Zoning	Used to reduce impacts of residential development by limiting numbers of units within WHPAs.	Community "down zones" to increase minimum acreage needed for residential development.	Well-recognized prerogative of local government. Requires rational connection between minimum lot size selected and resource protection goals. Arbitrary large lot zones have been struck down without logical connection to Master Plan or WHPA program.	Requires amendment to zoning ordinance.
Transfer of Development Rights	Used to transfer development from WHPAs to locations outside WHPAs.	Community offers transfer option within zoning ordinance. Community identifies areas where development is to be transferred "from" and "to."	Accepted land use planning tool.	Cumbersome administrative requirements. Not well suited for small communities without significant administrative resources.
Cluster/PUD Design	Used to guide residential development outside of WHPAs. Allows for "point source" discharges that are more easily monitored.	Community offers cluster/PUD as development option within zoning ordinance. Community identifies areas where cluster/PUD is allowed (i.e., within WHPAs).	Well-accepted option for residential land development.	Slightly more complicated to administer than traditional "grid" subdivision. Enforcement/inspection requirements are similar to "grid" subdivision.
Growth Controls/ Timing	Used to time the occurrence of development within WHPAs. Allows communities the opportunity to plan for wellhead delineation and protection.	Community imposes growth controls in the form of building caps, subdivision phasing, or other limitation tied to planning concerns.	Well-accepted option for communities facing development pressures within sensitive resource areas. Growth controls may be challenged if they are imposed without a rational connection to the resource being protected.	Generally complicated administrative process. Requires administrative staff to issue permits and enforcement growth control ordinances.
Performance Standards	Used to regulate development within WHPAs by enforcing predetermined standards for water quality. Allows for aggressive protection of WHPAs by limiting development within WHPAs to an accepted level.	Community identifies WHPAs and established "thresholds" for water quality.	Adoption of specific WHPA performance standards requires sound technical support. Performance standards must be enforced on a case-by-case basis.	Complex administrative requirements to evaluate impacts of land development within WHPAs.

Table 11-8 (cont.)

	Applicability to Wellhead Protection	Land Use Practice	Legal Considerations	Administrative Considerations
Regulatory: Subdivision Control				
Drainage Requirements	Used to ensure that subdivision road drainage is directed outside of WHPAs. Used to employ advanced engineering designs of subdivision roads within WHPAs.	Community adopts stringent subdivision rules and regulations to regulate road drainage/runoff in subdivisions within WHPAs.	Well-accepted purpose of subdivision control.	Requires moderate level of inspection and enforcement by administrative staff.
Regulatory: Health Regulations				
Underground Fuel Storage Systems	Used to prohibit underground fuel storage systems (USTs) within WHPAs. Used to regulate USTs within WHPAs.	Community adopts health/zoning ordinance prohibiting USTs within WHPAs. Community adopts special permit or performance standards for use of USTs within WHPAs.	Well-accepted regulatory option for local government.	Prohibition of USTs require little administrative support. Regulating USTs requires moderate amounts of administrative support for inspection followup and enforcement.
Privately Owned Wastewater Treatment Plants (Small Sewage Treatment Plants)	Used to prohibit small sewage treatment plants (SSTP) within WHPAs.	Community adopts health/zoning ordinance within WHPAs. Community adopts special permit or performance standards for use of SSTPs within WHPAs.	Well-accepted regulatory option for local government.	Prohibition of SSTPs require little administrative support. Regulating SSTPs requires moderate amount of administrative support of inspection followup and enforcement.
Septic Cleaner Ban	Used to prohibit the application of certain solvent septic cleaners, a known ground water contaminant, within WHPAs.	Community adopts health/zoning ordinance prohibiting the use of septic cleaners containing 1,1,1-trichloroethane or other solvent compounds within WHPAs.	Well-accepted method of protecting ground water quality.	Difficult to enforce even with sufficient administrative support.
Septic System Upgrades	Used to require periodic inspection and upgrading of septic systems.	Community adopts health/zoning ordinance requiring inspection and, if necessary, upgrading of septic systems on a time basis (e.g., every 2 years) or upon title/property transfer.	Well-accepted purview of government to ensure protection of ground water.	Significant administrative resources required for this option.
Toxic and Hazardous Materials Handling Regulations	Used to ensure proper handling and disposal of toxic materials/waste.	Community adopts health/zoning ordinance requiring registration and inspection of all businesses within WHPA using toxic/hazardous materials above certain quantities.	Well accepted as within purview of government to ensure protection of ground water.	Requires administrative support and onsite inspections.
Private Well Protection	Used to protect private onsite water supply wells.	Community adopts health/zoning ordinance to require permits for new private wells and to ensure appropriate well-to-septic-system setbacks. Also requires pump and water quality testing.	Well accepted as within purview of government to ensure protection of ground water.	Requires administrative support and review of applications.
Non-regulatory: Land Transfer and Voluntary Restrictions				
Sale/Donation	Land acquired by a community with WHPAs, either by purchase or donation. Provides broad protection to the groundwater supply.	As non-regulatory technique, communities generally work in partnership with non-profit land conservation organizations.	There are many legal consequences of accepting land for donation or sale from the private sector, mostly involving liability.	There are few administrative requirements involved in accepting donations or sales of land from the private sector. Administrative requirements for maintenance of land accepted or purchased may be substantial, particularly if the community does not have a program for open space management.

Table 11-8 (cont.)

	Applicability to Wellhead Protection	Land Use Practice	Legal Considerations	Administrative Considerations
Conservation Easements	Can be used to limit development within WHPAs.	Similar to sales/donations, conservation easements are generally obtained with the assistance of non-profit land conservation organization.	Same as above.	Same as above.
Limited Development	As the title implies, this technique limits development to portions of a land parcel outside of WHPAs.	Land developers work with community as part of a cluster/PUD to develop limited portions of a site and restrict other portions, particularly those within WHPAs.	Similar to those noted in cluster/PUD under zoning.	Similar to those noted in cluster/PUD under zoning.
Non-regulatory: Other				
Monitoring	Used to monitor ground water quality within WHPAs.	Communities establish ground water monitoring program within WHPA. Communities require developers within WHPAs to monitor ground water quality downgradient from their development.	Accepted method of ensuring ground water quality.	Requires moderate administrative staffing to ensure routine sampling and response if sampling indicates contamination.
Contingency Plans	Used to ensure appropriate response in cases of contaminant release or other emergencies within WHPA.	Community prepares a contingency plan involving wide range of municipal/county officials.	None.	Requires significant up-front planning to anticipate and be prepared for emergencies.
Hazardous Waste Collection	Used to reduce accumulation of hazardous materials within WHPAs and the community at large.	Communities, in cooperation with the state, regional planning commission, or other entity, sponsor a "hazardous waste collection day" several times per year.	There are several legal issues raised by the collection, transport, and disposal of hazardous waste.	Hazardous waste collection programs are generally sponsored by government agencies, but administered by a private contractor.
Public Education	Used to inform community residents of the connection between land use within WHPAs and drinking water quality.	Communities can employ a variety of public education techniques ranging from brochures detailing their WHPA program, to seminars, to involvement in events such as hazardous waste collection days.	No outstanding legal considerations.	Requires some degree of administrative support for programs such as brochure mailing to more intensive support for seminars and hazardous waste collection days.
Legislative:				
Regional WHPA Districts	Used to protect regional aquifer systems by establishing new legislative districts that often transcend existing corporate boundaries.	Requires state legislative action to create a new legislative authority.	Well-accepted method of protecting regional ground water resources.	Administrative requirements will vary depending on the goal of the regional district. Mapping of the regional WHPAs requires moderate administrative support, while creating land use controls within the WHPA will require significant administrative personnel and support.
Land Banking	Used to acquire and protect land within WHPAs.	Land banks are usually accomplished with a transfer tax established by state government empowering local government to impose a tax on the transfer of land from one party to another.	Land banks can be subject to legal challenge as an unjust tax, but have been accepted as a legitimate method of raising revenue for resource protection.	Land banks require significant administrative support if they are to function effectively.

Table 11-9 Potential Management Tools for Wellhead Protection (Born et al., 1987 and U.S. EPA 1989b)

Regulatory	Nonregulatory
Zoning Ordinances. Zoning ordinances typically are comprehensive land-use requirements designed to direct the development of an area. Many local governments have used zoning to restrict or regulate certain land uses within wellhead protection areas.	**Purchase of Property or Development Rights.** The purchase of property or development rights is a tool used by some localities to ensure complete control of land uses in or surrounding a wellhead area. This tool may be preferable if regulatory restrictions on land use are not politically feasible and the land purchase is affordable.
Subdivision Ordinances. Subdivision ordinances are applied to land that is divided into two or more subunits for sale or development. Local governments use this tool to protect wellhead areas in which ongoing development is causing contamination.	**Public Education.** Public education often consists of brochures, pamphlets, or seminars designed to present wellhead area problems and protection efforts to the public in an understandable fashion. This tool promotes the use of voluntary protection efforts and builds public support for a community protection program.
Site Plan Review. Site plan reviews are regulations requiring developers to submit for approval plans for development occurring within a given area. This tool ensures compliance with regulations or other requirements made within a wellhead protection area.	**Waste Reduction.** Residential hazardous waste management programs can be designed to reduce the quantity of household hazardous waste being disposed of improperly. This program has been used in localities where municipal landfills potentially threaten ground water due to improper household waste disposal in the wellhead area.
Design Standards. Design standards typically are regulations that apply to the design and construction of buildings or structures. This tool can be used to ensure that new buildings or structures placed within a wellhead protection area are designed so as not to pose a threat to the water supply.	**Best Management Practices.** BMPs are voluntary actions that have a long tradition of being used, especially in agriculture. Technical assistance for farmers wishing to apply them is available from local Extension and SCS offices.
Operating Standards. Operating standards are regulations that apply to ongoing land-use activities to promote safety or environmental protection. Such standards can minimize the threat to the wellhead area from ongoing activities such as the application of agricultural chemicals or the storage and use of hazardous substances.	**Training and Demonstration.** These programs can complement many regulations. For example, training underground storage tank inspectors and local emergency response teams or demonstration of agricultural BMPs.
Source Prohibitions. Source prohibitions are regulations that prohibit the presence or use of chemicals or hazardous activities within a given area. Local governments can use restrictions on the storage or handling of large quantities of hazardous materials within a wellhead protection area.	**Ground-Water Monitoring.** Ground-water monitoring generally consists of sinking a series of test wells and developing an ongoing water quality testing program. This tool provides for monitoring the quality of the ground-water supply or the movement of a contaminant plume.
Inspection and Testing. Local governments can use their statutory home rule power to require more stringent control of contamination sources within wellhead protection areas than given in federal or state rules.	**Contingency Planning.** Local governments can develop their own contingency plans for emergency response to spills and for alternative water supply in case of contamination of the existing supply.

Performance and Operating Standards. Performance and operating standards focus on establishing measurable environmental standards that protect human health or the environment. Performance and operating standards alone do not specify how performance should be achieved. Determining compliance for environmental standards, such as minimum acceptable concentrations of a chemical in ground water, is relatively simple, requiring sampling and chemical analysis. However, noncompliance will require additional actions to find the reason for noncompliance and the implementation of methods to bring the system back into compliance. This approach generally provides more flexibility than design standards and BMPs, since almost any method can be used, as long as the performance standard is achieved. To be effective, performance and operation standards must be implemented far enough from the wellhead area that noncompliance can be rectified without posing a threat to the well.

Ground-Water Monitoring. Ground-water monitoring is an essential component of wellhead protection. All WHPA delineation methods involve irreducible uncertainties due to the inherent physical and chemical complexity of hydrogeologic systems. Previous chapters have made suggestions for ways to address uncertainties, but no delineation method or ground-water management practice is fail-safe. For early detection of contamination, monitoring wells should be installed between significant point sources of potential contamination and the wellhead ahead in the most direct ground-water flow path line (Chapter 2). One or more monitoring wells should be installed upgradient of the wellhead along a specified time of travel contour (say 2- to 5-year isochrone) to provide an early warning of the presence of contaminants traveling toward the well.

Installation of ground-water monitoring wells and ground-water sampling require special procedures to ensure that samples are representative. Table 9-10 identifies major EPA and other documents that provide guidance in this area.

11.5.3 Specific Regulatory and Technical Approaches

In addition to Checklist D-C7 and Tables 11-8 and 11-9 discussed earlier, Figure 11-9 provides ratings for the applicability of 10 local regulatory techniques to 34 land use categories.

11.5.4 Contingency Planning

Developing a contingency plan to deal with emergency threats to ground-water quality in the WHPA, such as accidental chemical spills, is an essential part of managing a wellhead protection area. The plan should include information that allows a rapid response to minimize damage from accidental spills or other releases of chemicals, such as during efforts to control a fire at a known chemical storage site. The plan should also include short- and long-term solutions to the temporary or permanent loss of all or a portion of the water system source. A contingency plan should include the following elements:

III: Prevention and Remediation

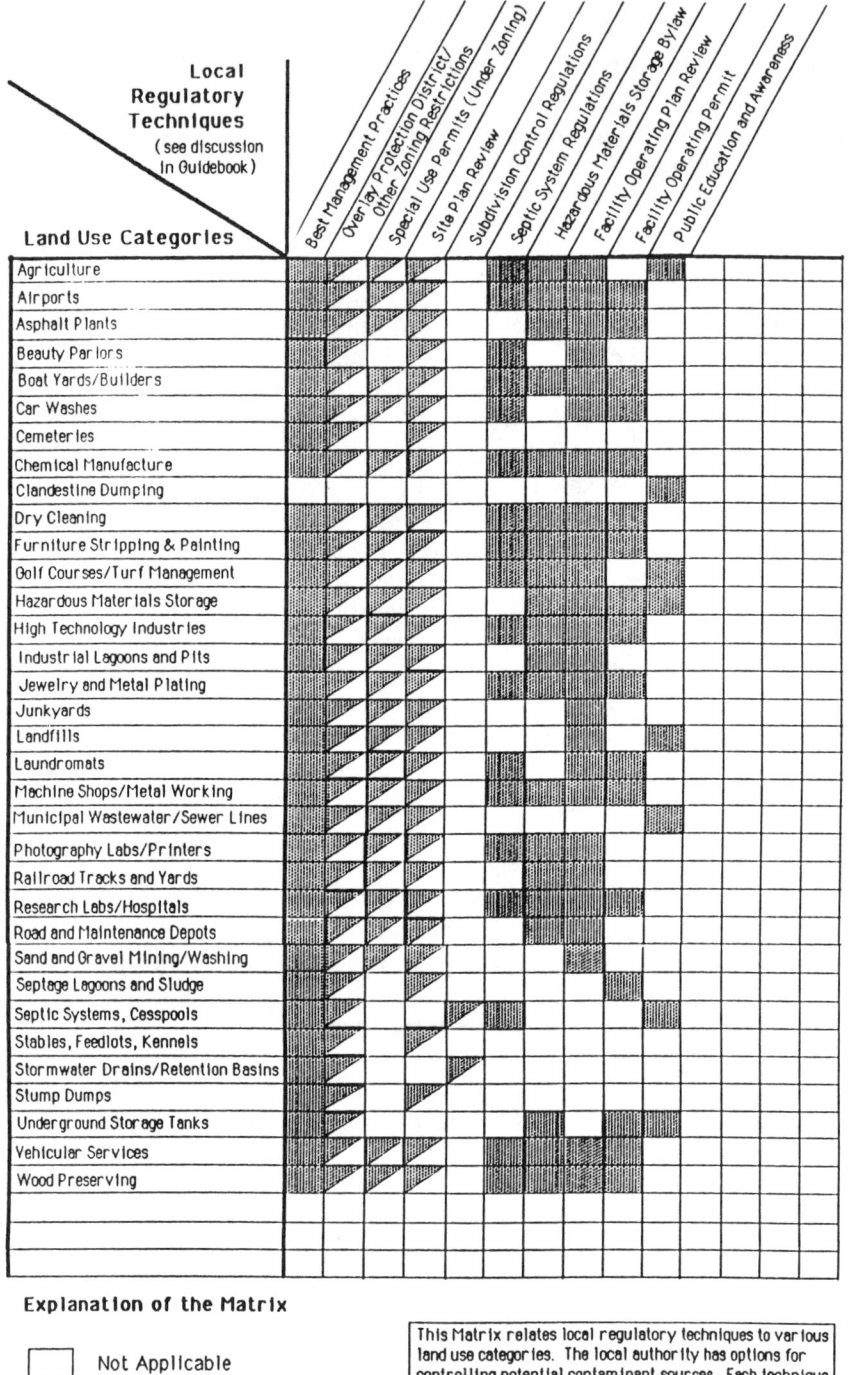

Figure 11-9 Wellhead protection land use/local regulatory techniques matrix (Noake, 1988).

1. Basic information about the water supply system such as population; number of service connections; location of fire hydrants; average daily usage; and the names and telephone numbers of the water system operator, the fire chief, police chief, and other emergency planning officials.

2. A list of potential contaminant sources and their locations (Section 11.3).

3. A map identifying the WHPA boundaries; how they were delineated; and significant aspects of local hydrogeology, geography, and geology that affect movement of contaminants in the subsurface.

4. Fire-fighting plans for specific sites, especially sites within the WHPA that store or handle toxic chemicals. Such plans should be developed in coordination with the Local Emergency Planning Committee (Section 11.3.1).

5. Surface spill emergency response procedures, including the names and phone numbers of agencies and other individuals outside the community who should be informed. These procedures should be developed in coordination with the Local Emergency Planning Committee (Section 11.3.1). Information on the type, location, and amount of spill should be recorded.

6. Short-term emergency water supply options, including a brief description of the type and location of water supply and the names and telephone numbers people who should be contacted in the event that the source must be used.

7. Long-term alternative water supply options.

U.S. EPA (1990c/T11-11) provides general guidance on contingency planning. Many state wellhead protection programs have developed additional guidance. State wellhead protection programs may also have developed guidance documents for development of contingency plans.

11.6 Guide to Major References

Table 11-10 provides an index of major references on ground-water vulnerability mapping and chemical hazard and risk assessment and Table 11-11 provides an index to major references on pollution prevention and soil and ground-water protection management. Section 11.1.1 discusses references for site-specific waste minimization. References on ground-water protection management are grouped into the following main categories: (1) general land use planning, (2) ground-water protection, (3) local planning approaches, (4) community decision-maker and citizen guides (5) federal and state programs, and (6) specific approaches.

Table 11-10 Index to Major References on Ground-Water Vulnerability Mapping and Chemical Hazard and Risk Assessment

Topic	References
Ground-Water Vulnerability Mapping	
Methods/Criteria	General Reviews: Anderson and Gosk (1987), Bachmat and Collin (1987), Barrocu and Biallo (1993), GAO (1992), Hoffer (1986), Kanivetsky et al. (1991), Knox et al. (1993), NRC (1993, 1994); DRASTIC: Aller et al. (1987); Illinois Stack Unit System: Berg and Kempton (1984), Berg et al. (1984), Shafer (1985), Soller and Berg (1992); Indiana Hydrogeologic Terrain System: Fleming (1992), Fleming et al. (1993); Waste Disposal Siting: Caldwell et al. (1981—HRS), Gibb et al. (1983), Halfon (1989), Kufs et al. (1980—SRM), LeGrand (1964, 1983—LSR), LeGrand and Brown (1977—LSR), Phillips et al. (1977—WSSIM), Silka and Swearingen (1978—SIA), U.S. EPA (1983—SIA, 1986b, 1989a—HRS, 1991a—HRS); Other Agricultural Chemical Systems: Curry (1990), DeLuca and Johnson (1990—RAVE), Holman (1986a, 1986b), Rao et al. (1985—PI), Roux et al. (1986—SAFE), Sokol et al. (1993), U.S. EPA (1986c—PI); Karst: Quinlan et al. (1992a, 1992b), Schuster et al. (1989); General Ground Water Classification Schemes: Pettyjohn et al. (1991), U.S. EPA (1985a, 1986a); Sole Source Aquifers: U.S. EPA (1988a)
Applications	Waste Disposal Siting: Gibb et al. (1983); Agricultural Chemicals: Alexander and Liddle (1986), Blanton and Villenueve (1989), Ehtemshemi et al. (1991), Holman (1986a, 1986b), Sokol et al. (1993); Karst: Schuster et al. (1989), Quinlin et al (1992b), Sendlein (1992); Leaking Underground Storage Tanks: Uhlman and Smith (1990)
Case Studies	DRASTIC: Alexander and Liddle (1986), Blanton and Villeneuve (1989), Duda and Johnson (1987), Ehteshami et al. (1991), FDER (undated), LeGrand and Rosen (1992), Pima Association of Governments (1992), Rosen (1994), Sendlein (1992), Soller (1992); Illinois Stack-Unit System: Kempton and Cartwright (1984), Uhlman and Smith (1990)
Chemical Hazard and Risk Assessment	
Chemical Hazards	Conway (1982), FEMA/DOT/EPA (1989), U.S. Department of Agriculture Extension Service (1989), U.S. EPA (1988b, 1988f, 1989b, 1990f, 1992b); Estimating Chemical Releases: See industrial/commercial contaminant source references in Table 4-4 and U.S. EPA (1992a); Chemical Hazard Identification: NIOSH (1990), U.S. DOT (1993)

Table 11-10 (cont.)

Topic	References
Chemical Hazard and Risk Assessment (cont.)	
Risk Communication	Sandman (1986), U.S. EPA (1987-1989, 1988j, 1988k, 1989c, 1989p, 1990a)
SARA Title III*	General: U.S. EPA (1988b, 1988e, 1989f, 1989g, 1989h, 1989i, 1990b, 1992b); Emergency Planning: U.S. EPA (1987b, 1988g, 1988h, 1988i, 1990d)
Exposure Assessment	
General	California Department of Health Services (1990), Neely (1994), U.S. EPA (1986-1988, 1988c, 1990i)
Other	Biomarkers: Saleh et al. (1994); Dermal Exposure: U.S. EPA (1992i); Exposure Factors: AIHC (1994), Schaum (1990), U.S. EPA (1985b, 1989j, 1991d, 1991e); Food Contamination Pathways: U.S. EPA (1986f); Models/Methods: Bird et al. (1991—TEEAM)
Risk Assessment	
General	Asante-Duah (1993), California Department of Health Services (1990), California Department of Toxic Substances Control (1992, 1994), CRAM (1993), Gorsuch et al. (1993), Gots (1992), Mertz et al. 1994), National Research Council (1983), Neely (1994), Renzoni et al. (1994), U.S. EPA (1986-1988, 1987a, 1993a-bibliography); Information Sources: U.S. EPA (1986e); Biological Values: U.S. EPA (1988d); Data Useability: U.S. EPA (1990g, 1992c); Case Studies: Paustenbach (1989); Superfund Risk Assessment: Hoddinott (1993), U.S. EPA (1986d, 1991b, 1992c)
Ground Water	Texts/Reports: McTernan and Kaplan (1990), Reichard et al. (1990), Trojan and Perry (1989), U.S. EPA (1991); Papers: Flanagan et al. (1991), Pfannkuch (1991)
Ecological Assessment	Bartell et al. (1992), Cairns et al. (1992), Calabrese and Baldwin (1993), ICF (1988), Eastern Research Group (1991), Maughn (1993), Norton et al. (1988), Renzoni et al. (1994), Suter (1993), USDI (1987b, 1987c), U.S. EPA (1989e, 1989k, 1989l, 1989m, 1989n, 1990h, 1991f, 1992d, 1992e, 1992g, 1993b); Bioassessment Field/Laboratory Methods: EPRI (1985), U.S. EPA (1989o, 1990k, 1992h, 1993c), USFWS (1989), Warren-Hicks et al. (1989); Ecological Indicators/Functional Values: U.S. EPA (1990j, 1992f), USFWS (1984)

Table 11-10 (cont.)

Topic	References
Risk Assessment (cont.)	
Environmental Toxicology/ Toxocity Testin	Cairns and Niderlehner (1994), Cockerham (1994), Landis and Yu (1994)
Models/Methods	<u>Reviews</u>: Calabrese and Kostecki (1992), U.S. EPA (1990e, 1990f); <u>Specific Methods/Models</u>: Claff (1993), Erbas-White and San Juan (1993); <u>Soil Damage</u>: USDI (1987a)
Other	<u>Drinking Water</u>: Lowrence (1992), U.S. EPA (1985c, 1990e); <u>Public Health</u>: U.S. EPA (1986-1988, 1986d, 1989d, 1990c, 1990i, 1991d, 1991e)

* Commonly referred to as the Emergency Planning and Community Right-To-Know Act (EPCRA).

Table 11-10 References (Appendix F contains references for figure and table sources)

Alexander, W.J. and S.K. Liddle. 1986. Ground Water Vulnerability Assessment in Support of the First Stage of the National Pesticide Survey. In: Proc. Conf. on Agricultural Impacts on Ground Water, National Water Well Association, Dublin, OH, pp. 77-87. [DRASTIC]

Aller, L, T. Bennett, J.H. Lehr, R.J. Petty, and G. Hackett. 1987. DRASTIC: A Standardized System for Evaluating Ground Water Pollution Potential Using Hydrogeologic Settings. EPA/600/2-87/035 (NTIS PB87-213914). [Also published in NWWA/EPA series, National Water Well Association, Dublin, OH. An earlier version dated 1985 with the same title (EPA/600/2-85/018) does not have the chapter on application of DRASTIC to maps or the 10 case studies contained in the later report]

American Industrial Health Council (AIHC). 1994. Exposure Factors Sourcebook. AIHC, Washington, DC.

Anderson, L.J. and E. Gosk. 1987. Applicability of Vulnerability Maps. In: Vulnerability of Soil and Groundwater to Pollutants, W. van Duijvenbooden and H.G. van Waegeningen (eds.), Nat. Inst. of Public Health and Environmental Hygiene, Noordwijk aan Zee, The Netherlands, Vol. 38, pp. 321-332.

Asante-Duah, K. 1993. Hazardous Waste Risk Assessment. Lewis Publishers, Boca Raton, FL, 480 pp.

Bachmat, Y. and M. Collin. 1987. Mapping to Assess Groundwater Vulnerability to Pollution. In: Vulnerability of Soil and Groundwater to Pollutants, W. van Duijvenbooden and H.G. van Waegeningen (eds.), Nat. Inst. of Public Health and Environmental Hygiene, Noordwijk aan Zee, the Netherlands, Vol. 38, pp. 297-307.

Barrocu, G. and G. Biallo. 1993. Application of GIS for Aquifer Vulnerability Evaluation. In: Application of Geographic Information Systems in Hydrology and Water Resources Management, K. Kovar and H.P. Nachtnebel (eds.), Int. Ass. Sci. Hydrol. Pub. No. 211, pp. 571-580.

Bartell, S.M., R.H. Gardner, and R.V. O'Neill. 1992. Ecological Risk Estimation. Lewis Publishers, Boca Raton, FL, 272 pp.

Berg, R.C. and J.P. Kempton. 1984. Potential for Contamination of Shallow Aquifers from Land Burial of Municipal Wastes. 1:500,000 Map. Illinois State Geological Survey, Champaign, IL.

Berg, R.C., J.P. Kempton, and K. Cartwright. 1984. Potential for Contamination of Shallow Aquifers in Illinois. Circular 532. Illinois State Geological Survey, Champaign, IL.

Bird, S.L., J.M. Cheplick, and D.S. Brown. 1991. Preliminary Testing, Evaluation and Sensitivity Analysis for the Terrestrial Ecosystem Exposure Assessment Model (TEEAM). EPA/600/3-91/019 (NTIS PB91-161711).

Blanton, O. and J. Villeneuve. 1989. Evaluation of Groundwater Vulnerability to Pesticides: A Comparison Between the pesticide DRASTIC Index and the PRZM leaching quantities. J. Contaminant Hydrology 4:285-296.

Cairns, Jr., J. and B.R. Niderlehner (eds.). 1994. Ecological Toxicity Testing: Scale, Complexity and Relevance. Lewis Publishers, Boca Raton, FL, 256 pp.

Cairns, Jr., J., B.R. Niederlehner, and D.R. Orvos. 1992. Predicting Ecosystem Risk. Princeton Scientific Publishing Co., Princeton, NJ.

Calabrese, E.J. and L.A. Baldwin. 1993. Performing Ecological Risk Assessments. Lewis Publishers, Boca Raton, FL, 288 pp.

Calabrese, E.J. and P.T. Kostecki. 1992. Risk Assessment and Environmental Fate Methodologies. Lewis Publishers, Boca Raton, FL, 150 pp. [Description and critical review of existing software (AERIS, GEOTOX, LUFT, MYGRT, PCGEMS/SESOIL, POSSM, PPLV, PRZM, RAFT, Risk Assistant, SESOIL], and other methods developed at the state level (California, New Jersey, and Massachusetts)]

Caldwell, S., K.W. Barrett, and S.S. Change. 1981. Ranking System for Releases of Hazardous Substances. In: Proc. of the Nat. Conf. on Management of Uncontrolled Hazardous Waste Sites, Hazardous Materials Control Research Institute, Silver Spring, MD, pp. 14-20. [HRS]

California Department of Health Services. 1990. Draft Scientific and Technical Standards for Hazardous Waste Sites, Book II, Volume 2: Exposure Assessment; Volume 3: Toxicity Assessment and Risk Characterization; Volume 4: Soil Remediation Levels. [Issued in draft form, none had been finalized as of mid-1994]

California Department of Toxic Substances Control. 1992. Supplemental Guidance for Human Health Multimedia Risk Assessments of Hazardous Waste Sites and Permitted Facilities. California Environmental Protection Agency, Sacramento, CA.

(Table 11-10 GW Vulnerability/Risk Assessment References)

California Department of Toxic Substances Control. 1994. Preliminary Endangerment Assessment Guidance Manual. California Environmental Protection Agency, Sacramento, CA. [Guidance manual for evaluating hazardous substance release sites]

Claff, R.E. 1993. The American Petroleum Institute's Decision Support System for Risk and Exposure Assessment. Ground Water Management 17:74-84 (Proc. API/NWGA Hydrocarbon Conf.). [DSS fate and transport modules include: soil (SESOIL, Jury); groundwater (AT123D, SESOIL/AT123D linked, Jury/AT123D linked); air emissions (Jury, Thibodeaux-Hwang, Farmer, SESOIL), air dispersion (box, Gaussian); air particulates (Cowherd)]

Cockerham, L.G. (ed.). 1994. Basic Environmental Toxicology. Lewis Publishers, Boca Raton, FL, 640 pp.

Committee on Risk Assessment Methodology (CRAM). 1993. Issues in Risk Assessment. National Academy Press, Washington, DC, 356 pp.

Conway, R.E. (ed.). 1982. Environmental Risk Analysis for Chemicals. Van Nostrand Reinhold, New York, NY.

Curry, D.S. 1990. Assessment of Empirical Methodologies for Predicting Ground Water Pollution from Agricultural Chemicals. In: Ground Water Quality and Agricultural Practices, D.M. Fairchild (ed.), Lewis Publishers, Chelsea, MI, pp. 227-245.

DeLuca, T. and P. Johnson. 1990. RAVE: Relative Aquifer Vulnerability Evaluation. MDA Technical Bulletin 90-01, Montana Department of Agriculture, Helena, MT, 4 pp. [Pesticide contamination]

Duda, A.M. and R.J. Johnson. 1987. Targeting to Protect Groundwater Quality. J. Soil and Water Conservation 42(4):325-330. [DRASTIC, Tennessee Valley region]

Eastern Research Group, Inc. 1991. Summary Report on Issues in Ecological Risk Assessment: Proceedings of a Colloquium Series March - July, 1990. Prepared for Risk Assessment Forum, U.S. Environmental Protection Agency, Washington, DC.

Ehteshami, M, R.C. Peralta, H. Eisele, H. Deer, and T. Tindall. 1991. Assessing Pesticide Contamination to Ground Water: A Rapid Approach. Ground Water 29(6):862-868. [DRASTIC and CMLS Model]

Electric Power Research Institute (EPRI). 1985. Sampling Design for Aquatic Ecologic Monitoring, 5 Vols. EPRI EA-4302, EPRI, Palo Alto, CA.

Erbas-White, I. and C. San Juan. 1993. Development of a Matrix Approach to Estimate Soil Clean-Up Levels for BTEX Compounds. Ground Water Management 17:74-84 (Proc. API/NWGA Hydrocarbon Conf.). [Uses MULTIMED and VLEACH]

Federal Energy Management Agency, U.S. Department of Transportation and U.S. Environmental Protection Agency (FEMA/DOT/EPA). 1989. Handbook of Chemical Hazard Analysis Procedures. Available from Federal Emergency Management Agency, Publications Department, 500 C St., SW, Washington, DC 20472.

(Table 11-10 GW Vulnerability/Risk Assessment References)

Flanagan, E.K., J.E. Hansen, and N. Dee. 1991. Managing Ground-Water Contamination Sources in Wellhead Protection Areas: A Priority Setting Approach. Ground Water Management 7:415-418 (Proc. Focus Conf. on Eastern Regional Ground-Water Issues).

Fleming, A.H. 1992. The Hydrogeologic Framework of Allen County, Indiana: Hydrostratigraphic Atlas Emphasizing Subsurface Sequence Stratigraphy and Ground-Water Contamination Potential. Indiana Geological Survey Open File Report 92-14, 55 pp. plus 1:24,000 maps. [Scheduled for publication in 1994 as IGS Special Report 57].

Fleming, A.H., S.E. Browning, and V.R. Ferguson. 1993. The Hydrogeologic Framework of Marion County, Indiana: Atlas Emphasizing Hydrogeologic Terrain and Sequence. Indiana Geological Survey Open File Report 93-5, 67 pp. plus 1:24,000 maps.

Florida Department of Environmental Regulation (FDER). Undated. Florida's Ground Water Quality Monitoring Network. FDER, Tallahassee, FL, 20 pp. [DRASTIC]

General Accounting Office (GAO). 1992. Groundwater Protection: Validity and Feasibility of EPA's Differential Protection Strategy. GAO/PEMD-93-6, Washington DC, 104 pp. [Includes discussion of limitation of DRASTIC and other vulnerability mapping methods].

Gibb, J.P., M.J. Barcelona, S.C. Schock, and M.W. Hampton. 1983. Hazardous Waste in Ogle and Winnebago Counties, Potential Risk via Ground Water Due to Past and Present Activities. ISWS Contract Report 336. Illinois State Water Survey, Champaign, IL.

Gorsuch, J.W., F.J. Dwyer, C.G. Ingersoll, and T.W. La Point (eds.). 1993. Environmental Toxicology and Risk Assessment, 2nd Volume. ASTM STP 1216, American Society for Testing and Materials, Philadelphia, PA, 744 pp.

Gots, R.E. 1992. Toxic Risks: Science, Regulations and Perceptions. Lewis Publishers, Chelsea, MI, 288 pp.

Halfon, E. 1989. Comparison of an Index Function and a Vectorial Approach Method for Ranking Waste Disposal Site. Environ. Sci. Technol. 23(5):600-609.

Hoddinott, K.B. (ed.). 1993. Superfund Risk Assessment in Soil Contamination Studies. ASTM STP 1158, American Society for Testing and Materials, Philadelphia, PA, 340 pp. [22 papers]

Hoffer, R.N. 1986. Techniques of Mapping and Protection for Water-Table Aquifers in the United States. In: Proc. 19th Cong. of Int. Ass. of Hydrogeologists (Karlovy Vary, Czechoslovakia), pp. 207-212.

Holman, D. 1986a. A Ground-Water Pollution Potential Risk Index System. In: Proc. Nat. Symp. on Local Government Options for Ground Water Pollution Control, University of Oklahoma, Norman, OK.

Holman, D. 1986b. Groundwater Protection Alternatives for Pesticides and Fertilizers Based on Local Information and Comparable Potential Risk Index for Rock County, Wisconsin. Rock County Health Department, Environmental Health Division, Janesville, WI, 23 pp.

(Table 11-10 GW Vulnerability/Risk Assessment References)

ICF, Inc. 1988. Review of Ecological Risk Assessment Methods. EPA/230/10-88/041.

Kanivetsky, R., B.M. Olsen, and E. Porcher. 1991. An Approach to Ground-Water Protection Based on Hydrogeologic Sensitivity. In: Proc. First USA/USSR Joint Conf. on Environmental Hydrology and Hydrogeology, J.E. Moore et al. (eds.), American Institute of Hydrology, Minneapolis, MN, pp. 146-157.

Knox, R.C., D.A. Sabatini, and L.W. Canter. 1993. Subsurface Transport and Fate Processes. Lewis Publishers, Chelsea, MI, 430 pp. [DRASTIC, SAFE, HRS, SRM, WSSIM, PI]

Kufs, C. et al. 1980. Rating of the Hazard Potential of Waste Disposal Facilities. In: Proc. of the Nat. Conf. on Management of Uncontrolled Hazardous Waste Sites, Hazardous Materials Control Research Institute, Silver Spring, MD, pp. 30-41. [SRM]

Landis, W.G. and M.-H. Yu. 1994. Introduction to Environmental Toxicology: Impacts of Chemicals Upon Ecological Systems. Lewis Publishers, Boca Raton, FL, 384 pp.

LeGrand, H.E. 1964. System for Evaluating the Contamination Potential of Some Waste Sites. Am. Water Works Ass. J. 56(8):959-974.

LeGrand, H.E. 1983. A Standardized System for Evaluating Waste-Disposal Sites, 2nd ed. National Water Well Association, Dublin, OH. [1st edition published 1980]

LeGrand, H.E. and H.S. Brown. 1977. Evaluation of Ground Water Contamination Potential from Waste Disposal Sources. Office of Water and Hazardous Materials, U.S. Environmental Protection Agency, Washington, DC. [LSR]

LeGrand, H.E. and L. Rosen. 1992. Common Sense in Ground-Water Protection and Management in the United States. Ground Water 30(6):867-872. [DRASTIC]

Maughan, J.T. 1993. Ecological Assessment of Hazardous Waste Sites. Van Nostrand Reinhold, New York, NY.

McTernan, W.F. and E. Kaplan (eds.). 1990. Risk Assessment for Groundwater Pollution Control. American Society of Civil Engineers, New York, NY, 368 pp.

Mertz, W., C. Abernathy, and S. Olin. 1994. Risk Assessment of Essential Elements. ISLI Press, Washington, DC. [Proc. International Life Sciences Institute's Risk Science Institute Workshop, March 10-12, 1992].

National Institute for Occupational Safety and Health (NIOSH). 1990. NIOSH Pocket Guide to Chemical Hazards. DHHS (NIOSH) Publication No. 909-117, U.S. Government Printing Office, Washington, DC, 245 pp.

National Research Council (NRC). 1983. Risk Assessment in the Federal Government: Managing the Process. National Academy Press, Washington, DC.

National Research Council (NRC). 1993. Ground Water Vulnerability Assessments: Predicting Relative Contaminant Potential Under Conditions of Uncertainty. National Academy Press, Washington, DC, 204 pp.

(Table 11-10 GW Vulnerability/Risk Assessment References)

National Research Council (NRC). 1994. Ranking Hazardous Waste Sites for Remedial Action. National Academy Press, Washington, DC. [Evaluation of EPA, DOE and DOD methods]

Neely, W.B. 1994. Introduction to Chemical Exposure and Risk Assessment. Lewis Publishers, Boca Raton, FL, 192 pp.

Norton, S., M. McVey, J. Colt, J. Durda, and R. Hegner. 1988. Review of Ecological Risk Assessment Methods. EPA/230/10-88-041 (NTIS PB89-134357), 181 pp. [Review of 16 methodologies]

Paustenbach, D.J. (ed.). 1989. The Risk Assessment of Environmental Hazards: A Textbook of Case Studies. John Wiley & Sons, New York, NY, 1,155 pp.

Pettyjohn, W., M. Savoca, and D. Self. 1991. Regional Assessment of Aquifer Vulnerability and Sensitivity in the Conterminous United States. EPA-600/2-91-043 (NTIS PB92-113141).

Pfannkuch, H.O. 1991. Application of Risk Assessment to Evaluate Groundwater Vulnerability to Non-Point and Point Contamination Sources. In: Proc. First USA/USSR Joint Conf. on Environmental Hydrology and Hydrogeology, J.E. Moore et al. (eds.), American Institute of Hydrology, Minneapolis, MN, pp. 158-168.

Phillips, C.R., J.D. Nathwani, and H. Mooij. 1977. Development of a Soil-Waste Interaction Matrix for Assessing Land Disposal of Industrial Wastes. Water Research 11:859-868. [WSSIM]

Pima Association of Governments. 1992. Application of Historic Well Closure Information for Protection of Existing Wells, Final Technical Report. Prepared for U.S. Environmental Protection Agency. [DRASTIC]

Quinlan, J.F., G.J. Davies, and S.R.H. Worthington. 1992a. Rationale for the Design of Cost-Effective Groundwater Monitoring Systems in Limestone and Dolomite Terranes: Cost Effective as Conceived is Not Cost Effective as Built if the System Design and Sampling Frequency Inadequately Consider Site Hydrogeology. In: Proc. Annual Waste Testing and Water Quality Assurance Symposium, pp. 552-570.

Quinlan, J.F., P.L. Smart, G.M. Schindel, E.C. Alexander, Jr., A.J. Edwards, and A. R. Smith. 1992b. Recommended Administrative/Regulatory Definition of Karst Aquifer, Principles for Classification and Carbonate Aquifers, Practical Evaluation of Vulnerability of Karst Aquifers, and Determination of Optimum Sampling Frequency at Springs. Ground Water Management 10:573-635 (Proc. 3rd Conf. on Hydrogeology, Ecology, Monitoring and Management of Ground Water in Karst Terranes).

Rao, P.S., A.G. Hornsby, and R.E. Jessup. 1985. Indices for Ranking the Potential for Pesticide Contamination of Groundwater. Proc. Soil Crop Sci. Soc. Fla. 44:1-8.

Reichard, E., C.. Cranor, R. Raucher, and G. Zapponi. 1990. Groundwater Contamination Risk Assessment: A Guide to Understanding and Managing Uncertainties. Int. Ass. Hydrological Sciences Publication No. 196.

(Table 11-10 GW Vulnerability/Risk Assessment References)

Renzoni, A., M.C. Fossi, L. Lari, and N. Mattei. 1994. Contaminants in the Environment: A Multidisciplinary Assessment of Risks to Man and Other Organisms. Lewis Publishers, Boca Raton, FL, 312 pp.

Rosen, L. 1994. A Study of the DRASTIC Methodology with Emphasis on Swedish Conditions. Ground Water 32(2):278-285.

Roux, P., J. DeMartinis, and G. Dickson. 1986. Sensitivity Analysis for Pesticide Application on a Regional Scale. In: Proc. Conf. on Agricultural Impacts on Ground Water, National Water Well Association, Dublin, OH, pp. 145-158. [SAFE]

Saleh, M.A., J.N. Blancato, and C.H. Nauman (eds.). 1994. Biomarkers of Human Exposure of Pesticides. ACS Symp. Series No. 542, American Chemical Society, Washington, DC, 328 pp.

Sandman, P.M. 1986. Explaining Environmental Risk. U.S. EPA Office of Toxic Substances, 27 pp. Available from EPCRA Hotline.*

Schaum, J. 1990. Exposure Factors Handbook. 1990. EPA/600/8-89/043 (NTIS PB90-106774).

Schuster, W.E., J.A. Bachhuberrt, and R.D. Steiglitz. 1989. Groundwater Pollution Potential and Pollution Attenuation Potential in Door County, Wisconsin. Door County Soil and Water Conservation Department, Sturgeon Bay, WI. [5 maps, scale 1 inch = 2640 ft]

Sendlein, L.V.A. 1992. Analysis of DRASTIC and Wellhead Protection Methods Applied to a Karst Setting. Ground Water Management 10:669-683 (Proc. 3rd Conf. on Hydrogeology, Ecology, Monitoring and Management of Ground Water in Karst Terranes). [Fayette County, KY]

Shafer, J.M. 1985. An Assessment of Ground-Water Quality and Hazardous Substance Activities in Illinois with Recommendations for a Statewide Monitoring Strategy. Illinois Department of Energy and Natural Resources, Champaign, IL, pp. 79-90.

Silka, L.R. and T.L. Swearingen. 1978. Manual for Evaluating Contamination Potential of Surface Impoundments. EPA-570/9-78-003 (NTIS PB85-211433). [SIA method]

Sokol, G., Ch. Leibundgut, K.P. Schulz, and W. Weinzierl. 1993. Mapping Procedures for Assessing Groundwater Vulnerability to Nitrate and Pesticides. In: Application of Geographic Information Systems in Hydrology and Water Resources Management, K. Kovar and H.P. Nachtnebel (eds.), Int. Ass. Sci. Hydrol. Pub. No. 211, pp. 631-639.

Soller, D.R. 1992. Applying the DRASTIC Model--A Review of County Scale Maps. U.S. Geological Survey Open File Report 92-297, 36 pp.

Soller, D.R. and R.C. Berg. 1992. A Model for the Assessment of Aquifer Contamination Potential Based on Regional Geologic Framework. Environ. Geology and Water Sciences, 19(3):205-213.

Suter, II, G.W. 1993. Ecological Risk Assessment. Lewis Publishers, Chelsea, MI, 538 pp.

(Table 11-10 GW Vulnerability/Risk Assessment References)

Trojan, M.J. and J.A. Perry. 1989. Assessing Hydrogeologic Risk Over Large Geographical Areas. Bull. 585-1988 (Item No. AD-S53-3421), Minn. Ag. Extension Station, University of Minn., St. Paul. [HI—Hazard Index]

Uhlman, K, and L.R. Smith. 1990. LUST Busting: Inventory and Ranking of Leaking Underground Storage Tank Incidents. Ground Water Management 1:565-577 (Proc. of the 1990 Cluster of Conferences: Ground Water Management and Wellhead Protection). [Aquifer vulnerability ranking system]

U.S. Department of Agriculture Extension Service. 1989. *Risk Management for Small Communities Series*. Risk Management Manual: A Reference Tool for Small Local Governments, 220 pp.; Risk Management Workbook: A Guide to Implementation of Risk Management Programs For Small Local Governments, 117 pp.; Risk Reduction Techniques: Methods to Promote Safety and Efficiency for Small Local Governments; Risk Management Instructor's Guide: Techniques for Training Public Officials to Manage Risks. Available from Southern Rural Development Center, PO Box 5446, Mississippi State, MS 39762. [Joint project with Public Risk Management Association and Oklahoma State University Cooperative Extension Service; main focus is on management of liability risks but addresses environmental risks such as emergency response and underground storage tank management]

U.S. Department of the Interior (USDI). 1987a Approaches to the Assessment of Injury to Soil Arising from the Discharge of Hazardous Substances and Oil. Type B Technical Information Document, USDI, Washington DC (NTIS PB88-100144).

U.S. Department of the Interior (USDI). 1987b. Techniques to Measure Damages to Natural Resources. Type B Technical Information Document, USDI, Washington DC (NTIS PB88-100136).

U.S. Department of the Interior (USDI). 1987c. Injury to Fish and Wildlife Species. Type B Technical Information Document, USDI, Washington DC (NTIS PB88-100169).

U.S. Department of Transportation (DOT). 1993. Emergency Response Guidebook. DOT P5800.5, U.S. Department of Transportation, Research and Special Programs Administration, Office of Hazardous Materials Transportation. [Chemical hazard identification; previous edition published in 1990]

U.S. Environmental Protection Agency (EPA). 1983. Surface Impoundment Assessment National Report. EPA 570/9-84/002 (NTIS DE84 901182). [SIA]

U.S. Environmental Protection Agency (EPA). 1985a. Selected State and Territory Ground-Water Classification Systems. EPA/440/6-85-005 (NTIS PB88-111919).

U.S. Environmental Protection Agency (EPA). 1985b. Development of Statistical Distributions or Ranges of Standard Factors Used in Exposure Assessment. EPA/600/8-85/010 (NTIS PB85-242667).

U.S. Environmental Protection Agency. (EPA). 1985c. Techniques for the Assessment of the Carcinogenic Risk to the U.S. Population Due to Exposure to Selected Volatile Organic Chemicals in Drinking Water. EPA/570/9-85-001 (NTIS PB84-213941).

(Table 11-10 GW Vulnerability/Risk Assessment References)

U.S. Environmental Protection Agency (EPA). 1986-1988. *Risk Assessment Guidelines.* Guidelines for Carcinogen Risk assessment (51 FR 33992-34003, 9/24/86); Guidelines for Mutagenicity Risk Assessment (51 FR 34006-34012, 9/24/86); Guidelines for Health Risk Assessment of Chemical Mixtures (51 FR 34028-34040, 9/24/86); Guidelines for the Health Assessment of Suspect Developmental Toxicants (51 FR 34028-34025, 9/24/86); Guidelines for Exposure Assessment (51 FR 34042-24054, 9/24/86); Proposed Guidelines for Assessment Male Reproductive Risk and Request for Comments (53 FR 24850-24869, 6/30/88); Proposed Guidelines for Assessing Female Reproductive Risk (53 FR 24834-24847, 6/30/88); Proposed Guidelines for Exposure-Related Measurements and Request for Comments (53 FR 48830-48853, 12/2/88).

U.S. Environmental Protection Agency (EPA). 1986a. Guidelines for Ground-Water Classification Under the EPA Ground-Water Protection Strategy. Office of Ground-Water Protection, EPA, Washington, DC.

U.S. Environmental Protection Agency (EPA). 1986b. Criteria for Identifying Areas of Vulnerable Hydrogeology Under RCRA: A RCRA Interpretive Guidance, Appendix D: Development of Vulnerability Criteria Based on Risk Assessments and Theoretical Modeling. EPA/530/SW-86-022D (PB86-224995).

U.S. Environmental Protection Agency (EPA). 1986c. Pesticides in Ground Water: Background Document. EPA/440/6-86-002 (NTIS PB88-111976). [Pesticide Index]

U.S. Environmental Protection Agency (EPA). 1986d. Superfund Public Health Evaluation Manual. EPA/540/1-86/060.

U.S. Environmental Protection Agency (EPA). 1986e. Superfund Risk Assessment Information Directory. EPA/540/1-86/061 (NTIS PB87-188918), 200 pp.

U.S. Environmental Protection Agency (EPA). 1986f. Methods for Assessing Exposure to Chemical Substances, Vol. 8, Method for Assessing Environmental Pathways of Food Contamination. EPA/560/5-85-008.

U.S. Environmental Protection Agency (EPA). 1987a. The Risk Assessment Guidelines of 1986. EPA/600/8-87-045.

U.S. Environmental Protection Agency (EPA). 1987b. Hazardous Materials Emergency Planning Guide. NRT-1. Available from EPCRA Hotline.*

U.S. Environmental Protection Agency (EPA). 1987c. Technical Guidance for Hazards Analysis. OSWER-88-001. Available from EPCRA Hotline.* [Used in conjunction with NRT-1]

U.S. Environmental Protection Agency (EPA). 1987-1989. Risk Assessment, Management, Communication: A Guide to Selected Resources. Guide (NTIS PB87-185500); 1st Update (PB87-203402); 2nd Update (PB88-100102); 3rd Update (PB88-128178); Volume 2, No. 1 (PB88-210596); Volume 2, No. 2 (PB89-189641).

U.S. Environmental Protection Agency (EPA). 1987-1989. Risk Assessment, Management, Communication: A Guide to Selected Resources. Guide (NTIS PB87-185500); 1st Update

(Table 11-10 GW Vulnerability/Risk Assessment References)

Prevention and Minimization of Contamination 673

(PB87-203402); 2nd Update (PB88-100102); 3rd Update (PB88-128178); Volume 2, No. 1 (PB88-210596); Volume 2, No. 2 (PB89-189641).

U.S. Environmental Protection Agency (EPA). 1988a. Sole Source Aquifer Designation Petitioners Guidance. EPA/440/6-87-003 (NTIS PB88-111992).

U.S. Environmental Protection Agency (EPA). 1988b. Community Right-to-Know and Small Business. OSWER-88-005. Available from EPCRA Hotline.*

U.S. Environmental Protection Agency (EPA). 1988c. Superfund Exposure Assessment Manual. EPA/540/1-88/001 (NTIS PB90-135859).

U.S. Environmental Protection Agency (EPA). 1988d. Recommendations For and Documentation of Biological Values for Use in Risk Assessment. EPA/600/6-87/008 (NTIS PB88-179874).

U.S. Environmental Protection Agency (EPA). 1988e. Chemicals in Your Community: A Citizen's Guide to the Emergency Planning and Community Right-to-Know Act. OSWER-90-002. Available from EPCRA Hotline.*

U.S. Environmental Protection Agency (EPA). 1988f. List of Extremely Hazardous Substances. OSWER-EHS-1. Available from EPCRA Hotline.*

U.S. Environmental Protection Agency (EPA). 1988g. Criteria for Review of Hazardous Materials Emergency Plans. NRT-1A. Available from EPCRA Hotline.*

U.S. Environmental Protection Agency (EPA). 1988h. Guide to Exercises in Chemical Emergency Preparedness Programs. OSWER-88-006. Available from EPCRA Hotline.* [Compilation of 3 Technical Assistance Bulletins: (1) Introduction to Exercises in Chemical Emergency Preparedness Programs; (2) A Guide to Planning and Conducting Table-Top Exercises; (3) A Guide to Planning and Conducting Field Simulation Exercises; U.S. EPA (1990d) replaces this guide and includes this information]

U.S. Environmental Protection Agency (EPA). 1988i. It's Not Over in October: A Guide for Local Emergency Planning Committees: Implementing the Emergency Planning and Community Right-to-Know Act of 1986. OSWER-90-004. Available from EPCRA Hotline.*

U.S. Environmental Protection Agency (EPA). 1988j. Seven Cardinal Rules of Risk Communication. (Brochure). Available from EPCRA Hotline.*

U.S. Environmental Protection Agency (EPA). 1988k. Report of Conference on Risk Communication and Environmental Management. U.S. EPA Technical Assistance Bulletin 4, 7 pp. Available from EPCRA Hotline.*

U.S. Environmental Protection Agency (EPA). 1989a. Field Test of the Proposed Revised Hazard Ranking System. EPA/540/P-90/001 (NTIS PB90-222746), 140 pp. [HRS Fact Sheets: The Revised Hazard Ranking System: An Improve Tool for Screening Superfund Sites, 1990, 6 pp. (NTIS PB91-921307); The Revised Hazard Ranking System: Background Information, 1990, 14 pp. (NTIS PB91-921303); The Revised Hazard Ranking System: Qs and As, 1990, 10 pp. (NTIS PB91-921305)]

(Table 11-10 GW Vulnerability/Risk Assessment References)

U.S. Environmental Protection Agency (EPA). 1989b. Toxic Chemical Release Inventory Risk Screening Guide, 2 Volumes (Version 1.0). EPA/560/2-89-002 (NTIS PB90-122128).

U.S. Environmental Protection Agency (EPA). 1989c. Risk Communication About Chemicals in Your Community: A Manual For Local Officials. EPA 230/09-89-066, EPA/FEMA/DOT/ATSDR, 76 pp. Available from EPCRA Hotline*. [Facilitator's Manual and Guide (EPA/230/09-89-067) also available].

U.S. Environmental Protection Agency (EPA). 1989d. Risk Assessment Guidance for Superfund, Volume 1: Human Health Evaluation Manual, Part A, Interim Final. EPA/540/1-89/002 (NTIS PB90-155581), 290 pp. [1990 9-page Fact Sheet with same title: NTIS PB90-273830; 1991 Human Health Evaluation Manual, Supplemental Guidance: Standard Default Exposure Factors: NTIS PB91-921314, 28 pp.]

U.S. Environmental Protection Agency (EPA). 1989e. Risk Assessment Guidance for Superfund, Volume 2: Environmental Evaluation Manual, Interim Final. EPA/540/1-89/001 (NTIS PB90-155599), 64 pp.

U.S. Environmental Protection Agency (EPA). 1989f. Emergency Planning and Community Right-to-Know Act of 1986: Questions and Answers. Available from EPCRA Hotline.*

U.S. Environmental Protection Agency (EPA). 1989g. Toxic and Hazardous Chemicals, Title III and Communities: An Outreach Manual for Community Groups. EPA/560/-1-89-002. Available from EPCRA Hotline.*

U.S. Environmental Protection Agency (EPA). 1989h. Information Resources Directory. EPA/OPA 003-89. Available from EPCRA Hotline.*

U.S. Environmental Protection Agency (EPA). 1989i. When All Else Fails! Enforcement of the Emergency Planning and Community Right-to-Know Act. OSWER 89-010, 12 pp. Available from EPCRA Hotline.*

U.S. Environmental Protection Agency (EPA). 1989j. Interim Final Guidance for Soil Ingestion Rates. OSWER Direction 9850.4.

U.S. Environmental Protection Agency (EPA). 1989k. The Nature and Extent of Ecological Risks at Superfund Sites and RCRA Facilities. EPA-230-03-89-043. U.S. EPA Corvalis, OR.

U.S. Environmental Protection Agency (EPA). 1989l. Summary of Ecological Risks, Assessment Methods, and Risk Management Decisions in Superfund Sites and RCRA. EPA/230-03-89-046. U.S. EPA Corvalis, OR.

U.S. Environmental Protection Agency (EPA). 1989m. Ecological Risk Assessment Methods: A Review and Evaluation of Past Practices in the Superfund and RCRA Programs. EPA-230-03-89-044 (NTIS PB90-137324).

U.S. Environmental Protection Agency (EPA). 1989n. Ecological Risk Management in the Superfund and RCRA Programs. EPA-230-03-89-045 (NTIS PB90-137332).

(Table 11-10 GW Vulnerability/Risk Assessment References)

U.S. Environmental Protection Agency (EPA). 1989o. Rapid Bioassessment Protocols for Use in Streams and Rivers: Benthic Macroinvertebrates and Fish. EPA/440/4-89/001.

U.S. Environmental Protection Agency (EPA). 1989p. Chemical Releases and Chemical Risks: A Citizen's Guide to Risk Screening (Pamphlet). EPA/560/2-89-003, 8 pp. Available from EPCRA Hotline*.

U.S. Environmental Protection Agency (EPA). 1990a. Public Knowledge and Perceptions of Chemical Risks in Six Communities: Analysis of a Baseline Survey. EPA/230/01-90-074 (NTIS PB90-217316). Conducted by Georgetown University Medical Center.

U.S. Environmental Protection Agency (EPA). 1990b. Emergency Planning and Community Right-to-Know (Title III) Factsheet. Available from EPCRA Hotline.*

U.S. Environmental Protection Agency (EPA). 1990c. Hazardous Substances in Our Environment: A Citizens' Guide to Understanding Health Risks and Reducing Exposure. EPA/230/09-90-081. Available from U.S. EPA Public Information Center, PM-211-B, 401 M St., SW, Washington, DC 20460. [Brochure titled Understanding Environmental Health Risks and Reducing Exposure: Highlights of a Citizens' Guide (EPA/230/09-90-082) is also available from the same source]

U.S. Environmental Protection Agency (EPA). 1990d. Developing a Hazardous Materials Exercise Program: A Handbook for State and Local Officials. NRT-2. Available from EPCRA Hotline.* [Replaces U.S. EPA (1988h)]

U.S. Environmental Protection Agency. (EPA). 1990e. Risk Assessment Methodologies: Comparing State and EPA Approaches. EPA/570/9-90-012. Available from ODW*.

U.S. Environmental Protection Agency (EPA). 1990f. Computerized System for Performing Risk Assessments for Chemical Constituents of Hazardous Waste. EPA/600/D-90/044 (NTIS PB90-222001), 22 pp. [System combines database, exposure and risk values in an IBM-PC format]

U.S. Environmental Protection Agency (EPA). 1990g. Guidance for Data Useability in Risk Assessment. EPA/540/G-90/008 (NTIS PB91-921208), 272 pp. [Superseded by U.S. EPA (1992c)]

U.S. Environmental Protection Agency (EPA). 1990h. Quantifying Effects in Ecological Site Assessments: Biological and Statistical Considerations. EPA/600/D-90/152 (NTIS PB91-129189), 31 pp.

U.S. Environmental Protection Agency (EPA). 1990i. Statistical Methods for Estimating Risk for Exposure Above the Reference Dose. EPA/600/8-90/065 (NTIS PB90-261504).

U.S. Environmental Protection Agency (EPA). 1990j. Ecological Indicators. EPA/600/3-90/060.

U.S. Environmental Protection Agency (EPA). 1990k. Macroinvertebrate Field and Laboratory Methods for Evaluating the Biological Integrity of Surface Waters. EPA/600/4-90/030.

(Table 11-10 GW Vulnerability/Risk Assessment References)

U.S. Environmental Protection Agency (EPA). 1991a. PA-Score Software; User's Manual and Tutorial: Version 1.0. Manual only: NTIS PB92-963302, 76 pp.; manual and diskette: NTIS PB92-500032. [HRS]

U.S. Environmental Protection Agency (EPA). 1991b. Risk Assessment Guidance for Superfund: Volume I, Human Health Evaluation Manual: (Part B) Development of Risk-Based Preliminary Remediation Goals (OSWER Directive 9285.7-01B); (Part C) Risk Evaluation of Remedial Alternatives (OSWER Directive 9285.7-01C).

U.S. Environmental Protection Agency (EPA). 1991c. Supplemental Guidance on Performing Risk Assessment in Remedial Investigation/Feasibility Studies (RI/FSs) Conducted by Potentially Responsible Parties (PRPs). OSWER Directives 9835.15a and 9835.15b.

U.S. Environmental Protection Agency (EPA). 1991d. Health Effects Assessment Documents: Project Summary. EPA/600/S8-91/041. [Summary table with interim acceptable exposure levels for 51 chemicals and list with NTIS acquisition numbers for obtaining the complete individual reports, which were prepared in 1987-1988]. Available from CERI.*

U.S. Environmental Protection Agency (EPA). 1991e. Updated Health Effects Assessment Documents: Project Summary. EPA/600/S8-91/042. [Summary table with interim acceptable exposure levels for 15 chemicals and list with NTIS acquisition numbers for obtaining the complete individual reports, which were prepared in 1988-1989]. Available from CERI.*

U.S. Environmental Protection Agency (EPA). 1991f. Summary Report on Issues in Ecological Risk Assessment. EPA/625/3-91/018.

U.S. Environmental Protection Agency (EPA). 1992a. Publications Office of Science and Technology: Catalog. EPA-820-B-92-002. Available from U.S. EPA Office of Water Resource Center (WH-556) 401 M Street, SW, Washington DC 20460; 202/260-7786. [List of titles for over 200 EPA documents used to develop industrial effluent limitations and guidelines along with information on how documents can be obtained].

U.S. Environmental Protection Agency (EPA). 1992b. Title III List of Lists: Consolidated List of Chemical Subject to Reporting Under the Emergency Planning and Community Right-To-Know Act. EPA 560/4-92-011/500-B-92-002. Available from EPCRA Hotline.*

U.S. Environmental Protection Agency (EPA). 1992c. Guidance for Data Useability in Risk Assessment (Parts A and B), Final. Part A: OSWER Directive 9285.7-09A (NTIS PB92-963356), 270 pp.; Part B: OSWER Directive 9285.7-09B (NTIS PB92-963362), 85 pp. [Supersedes U.S. EPA 1990g]

U.S. Environmental Protection Agency (EPA). 1992d. Framework for Ecological Risk Assessment. EPA/630/R-92/001.

U.S. Environmental Protection Agency (EPA). 1992e. Peer Review Workshop Report on a Framework for Ecological Risk Assessment. EPA/625/3-93/022.

U.S. Environmental Protection Agency (EPA). 1992f. Evaluation of Terrestrial Indicators for Use in Ecological Assessments at Hazardous Waste Sites. EPA/600/R-92/183.

(Table 11-10 GW Vulnerability/Risk Assessment References)

U.S. Environmental Protection Agency (EPA). 1992g. Report on the Ecological Risk Assessment Guidelines Strategic Planning Workshop. EPA/630/R-92/002.

U.S. Environmental Protection Agency (EPA). 1992h. Sediment Classification Methods Compendium. EPA/823-R-92-006 (NTIS PB93-115186).

U.S. Environmental Protection Agency (EPA). 1992i. Dermal Exposure Assessment: Principles and Applications. EPA/600/8-91/001B.

U.S. Environmental Protection Agency (EPA). 1993a. Selected Publications on Risk Assessment. U.S. EPA Environmental Criteria and Assessment Office, Cincinnati, OH

U.S. Environmental Protection Agency (EPA). 1993b. A Review of Ecological Assessment Case Studies from a Risk Assessment Perspective. EPA/630/R-92/005.

U.S. Environmental Protection Agency (EPA). 1993c. Interim Report on Data and Methods for Assessment of 2,3,7,8-Tetrachlorodibenzo-p-dioxin Risks to Aquatic Wildlife and Associated Wildlife. EPA/600/R-93/055.

U.S. Fish and Wildlife Service (USFWS). 1984. An Overview of Major Wetland Functions and Values. FWS/OBS-84/18.

U.S. Fish and Wildlife Service (USFWS). 1989. Field Assessment of the Effects of Contaminants on Fishes. USFWS, Washington, DC (NTIS PB89-112999).

Warren-Hicks, W., B.R. Parkhurst, and S.S. Baker, Jr. 1989. Ecological Assessment of Hazardous Waste Sites: A Field and Laboratory Reference. EPA/600/3-89/013 (NTIS PB89-205967). [Covers toxicity tests, biomarkers, and ecological field assessments]

* See Introduction for information on how to obtain documents from CERI (U.S. EPA Center for Environmental Research Information and NTIS. See Table 5-3 for number of EPCRA Hotline.

(Table 11-10 GW Vulnerability/Risk Assessment References)

Table 11-11 Index to Major References on Pollution Prevention and Soil and Ground Water Protection Management

Topic	References
Waste Minimization/Pollution Prevention	
Industrial/Commercial	API (1988), Inglese (1992), Licis et al. (1991), NJDEPE (1992), Sell (1992), Theodore and McGuinn (192), U.S. EPA (1991c, 1991d, 1992d, 1992e, 1993b), Ward et al. (1990), see also Table 11-6; <u>Hazardous Waste</u>: Environmental Resource Center Staff (1992), Higgins (1989)
EPA Series	<u>Guides to Pollution Prevention</u>: U.S. EPA (1990-1993); <u>Waste Minimization Assessments</u>: U.S. EPA (1990-1992); <u>Waste Reduction Activities and Opportunities</u>: U.S. EPA (1992f)
Ground Water Management and Protection	
General Land Use Planning	Ellickson and Tarlock (1981), Freund and Goodman (1968), Getzels and Thurow (1979), Global Cities Project (1993), Hendler (1977), Miller and Wood (1983), Mossa (1987), Robinson (1988), Rusmone (1982), Wilson et al. (1979)
Ground Water Protection	<u>Texts/Reports</u>: Cantor et al. (1987), Conservation Foundation (1987), Greeley-Polhemus Group (1985), Kerns (1977), LeGrand and Rosen (1992), Matthess et al. (1985), Montana Environmental Quality Council (1990), National Research Council (1986), Page (1987), Pojacek (1977), Southern Water Authority (1985), Stroman (1987), U.S. EPA (1984, 1984b, 1985a, 1987b, 1987g, 1989e, 1991a, 1991b, 1992c), U.S. OTA (1984), WEF (1993/T4-4), Western Michigan University (1988), Zaporozec (1991); <u>Papers</u>: Amsden and Mullen (1990), Flanagan et al. (1991), Henderson (1987), Hodge and Brown (1990), Holmes (1979), LeGrand and Rosen (1992), Lehr (1987), Milde et al. (1983), Pisanelli and Dutram (1990), Tolman et al. (1991), Tripp and Jaffe (1979), Yanggen and Amrhein (1989)
Local Planning Approaches	<u>Texts/Reports</u>: APA (1975), Born et al. (1988), Cross (1991), DiNovo and Jaffe (1984a), Group for the South Fork (1982), Jaffe and DiNovo (1987), MDEP (1991), Michigan Departments of Natural Resources and Public Health (1993), National Research Council (1986), National Rural Water Association (1991), New Hampshire Office of State Planning (1991), Potter (1984), Redlich (1988), Rusmone (1982), University of Oklahoma (1986), U.S. EPA (1989c, 1989d, 1990c), Yanggen and Weberdorfer (1991); <u>Papers</u>: Allee (1986), Blatt (1986), Boody (1990), Dean (1988), DiNovo and Jaffe (1984b), Jaffe (1987), Massey (1984), Tripp and Jaffe (1979), Yanggen and Amrhein (1989), Oates et al. (1990); <u>Ordinances</u>: Minnesota Project (1984), Trefry (1990);

Table 11-11 (cont.)

Topic	References
Community Decision-Maker/ Citizen Guides	Baize and Gilkerson (1992), Born et al. (1987), Central Connecticut Regional Planning Agency (1981), Clark and Cherry (1992), Community Resource Group (1992), Concern (1989), Dean and Wyckoff (1991), Gordon (1984), Hall Associates and Dight (1986), Harrison and Dickinson (1984), Hrezo and Nickinson (1986), Madarchik (1992), Massachusetts Audubon Society (1984-1987), Massachusetts Department of Environmental Quality Engineering (1985), Mullikin (1984), Murphy (undated), North Dakota State Department of Health (1993), Pierce (1992), Raymond (1986), U.S. EPA (1987i, 1990a, 1992a, 1993)
Federal/State Programs	EPA Program Analyses: U.S. EPA (1985b, 1990b, 1992c); State Programs: Booth and Bronson (1983-New York), Born et al. (1988-Wisconsin), Environmental Law Institute (1990); Henderson et al. (1985), Leavall (1990-Ohio), Meccozi (1989-Wisconsin), National Research Council (1986), NHDES (1991-NH), Pisanelli and Dutram (1990-Maine), Raymond (1981), Roy (1988), Stroman (1987-MA), U.S. EPA (1985c, 1987b, 1987f, 1988a, 1988b, 1989a, 1992b), Walden (1988), Weatherington-Rice and Hottman (1990-Ohio); Financing: Allee (1986), Shafer et al. (1988), U.S. EPA (1987h, 1987f, 1989a, 1989b, 1992b); WHPA Delineation Methods: Bradbury et al. (1991), Kreitler and Senger (1991), Moore (1993), U.S. EPA (1987a)
Specific Approaches	Best Management Practices: Noake (1988), Inglese (1992), U.S. EPA (1993c); Emergency Planning: New York State Department of Health (1984), U.S. EPA (1985d); Nonpoint Source Pollution Control: Bingham et al. (1993-urban runoff), Brown et al. (1993-forest areas), Holmes (1979), ICPRB (1981), Novotny and Chesters (1981), Novotny and Olem (1993); Erosion/Sediment Control: APA (1984), Association of Bay Area Governments (1981), Goldman et al. (1986); Agriculture: Baker (1990-pesticides), Freshwater Foundation (1988-1990), Kemp and Erickson (1989), Lal (1994), Logan et al. (1987), Massey (1984), Stewart (1976), U.S. EPA (1987e, 1988d); Road Salt: Curtis et al. (1986), Greeley-Polhemus Group (1985), NJDEPE (1992); Septic Systems: Lukin (1992), NJDEPE (1992), U.S. EPA (1986a, 1986b, 1987c); Industrial Source Control: See references for waste minimization/pollution prevention above and Table 4-4; Karst: Davis and Quinlin (1991), Fischer et al. (1991), Quinlin et al. (1991), Rubin (1991); Accidental Spills: Yang and Bye (1979a, 1979b); Sole Source Aquifers: U.S. EPA (1987d, 1988c)

Table 11-11 References (Appendix F contains references for figure and table sources)

Allee, D.J. 1986. Local Finance and Policy for Ground Water Protection. The Environmental Professional 8(3):210-218.

American Petroleum Institute (API). 1988. Literature Survey: Subsurface and Groundwater Protection Related to Petroleum Refinery Operations. API Publication 800. API, Washington, DC. [$54.00]

American Planning Association (APA). 1975. Performance Controls for Sensitive Lands: A Practical Guide for Local Administrators. Planning Advisory Service Report #307 and #308, APA, Chicago, IL, 156 pp.

American Planning Association (APA). 1984. State and Local Regulations for Reducing Agricultural Erosion. Planning Advisory Service Report #386, APA, Chicago, IL, 42 pp.

Amsden, T.L. and W.A. Mullen. 1990. Ground Water and Pollution Prevention. Ground Water Management 1:357-363 (Proc. of the 1990 Cluster of Conferences: Ground Water Management and Wellhead Protection).

Association of Bay Area Governments. 1981. Manual of Standards for Erosion and Sediment Control Measures. Association of Bay Area Governments, Oakland, CA, 275 pp.

Baize, D.G. and H.H. Gilkerson. 1992. Wellhead Protection Technical Guidance Document for South Carolina Local Ground-Water Protection. Ground-Water Protection Division, South Carolina Department of Health and Environmental Control, Columbia, SC, 74 pp.

Baker, B. 1990. Groundwater Protection from Pesticides. Garland Publishing, New York, NY, 151 pp.

Bingham, D., W. Boucher, and P. Boucher. 1993. Handbook: Urban Runoff Pollution Prevention and Control Planning. EPA/625/R-93/004, 175 pp. Available from CERI.*

Blatt, D.J.L. 1986. From the Ground Water Up: Local Land Use Planning and Aquifer Protection. J. of Land Use and Environmental Law 2(2):119-148.

Boody, G. 1990. Creating Special Protection Areas for Groundwater and Sustainable Agriculture: A Preliminary Strategy for Local Community Action. Ground Water Management 1:1-15 (Proc. of the 1990 Cluster of Conferences: Agricultural Impacts on Ground Water Quality).

Booth, R.S. and A. Bronson. 1983. Major Institutional Arrangement Affecting Groundwater in New York State. Cornell University Center for Environmental Research, Ithaca, NY.

Born, S.M., D.A. Yanggen, and A. Zaporozec. 1987. A Guide to Groundwater Quality Planning and Management for Local Governments. Special Report 9. Wisconsin Geological and Natural History Survey, Madison, WI, 92 pp.

Born, S.M., D.A. Yanggen, A.R. Czecholinksi, R.J. Tierney, and R.G. Henning. 1988. Wellhead Protection Districts in Wisconsin: An Analysis and Test Applications. Special Report 10. Wisconsin Geological And Natural History Survey, Madison, WI, 75 pp.

Bradbury, K.R., M.A. Muldoon, A. Zaporozec, and J. Levy. 1991. Delineation of Wellhead Protection Areas in Fractured Rocks. EPA/570/9-91-009, 144 pp. Available from ODW*. [May also be cited with Wisconsin Geological and Natural History Survey as author]

Brown, T.C., D. Brown, and D. Binkley. 1993. Laws and Programs for Controlling Nonpoint Source Pollution from Forest Areas. Water Resources Bulletin 29(1):1-13.

Cantor, L.W., R.C. Knox, and D.M. Fairchild. 1987. Ground Water Quality Protection. Lewis Publishers, Chelsea, MI.

Central Connecticut Regional Planning Agency. 1981. Guide to Groundwater and Aquifer Protection. Central Connecticut Regional Planning Agency, Bristol, CT.

Clark, II, E.H. and P.J. Cherry. 1992. Groundwater: Managing the Unseen Resource. World Wildlife Fund Publications, Baltimore, MD, 34 pp.

Community Resource Group, Inc. 1992. The Local Decision-Makers' Guide to Groundwater and Wellhead Protection. 16 pp. Available from RCAP offices. [Cover pages may vary slightly]

Concern, Inc. 1989. Groundwater: A Community Action Guide. Washington, DC, 22 pp.

Conservation Foundation. 1987. Groundwater Protection. Washington, DC, 240 pp.

Cross, B.L. 1991. A Guide to Local Ground Water Protection. Texas Water Commission, Austin, TX.

Curtis, C., C. Walsh, and M. Przybyla. 1986. The Road Salt Management Handbook: Introducing a Reliable Strategy to Safeguard People & Water Resources. Pioneer Valley Planning Commission, West Springfield, MA.

Davis, G.A., and J.F. Quinlan. 1991. Legal Tools for the Protection of Ground Water In Karst Terranes. Ground Water Management 10:637-649 (Proc. 3rd Conf. on Hydrogeology, Ecology, Monitoring and Management of Ground Water in Karst Terranes).

Dean, L.F. 1988. Local Government Regulations for Groundwater Protection: Michigan Case Examples. In: Policy Planning and Resource Protection: A Groundwater Conference for the Midwest, Institute for Water Sciences, Western Michigan University, Kalamazoo, MI, pp. 143-150.

Dean, L.F. and M. A. Wyckoff. 1991. Community Planning and Zoning for Groundwater Protection in Michigan: A Guidebook for Local Officials. Prepared for Office of Water Resources, Michigan Department of Natural Resources. Available from Michigan Society of Planning Officials, 414 Main St., Suite 202, Rochester, MI 48307.

(Table 11-11 Ground Water Protection References)

DiNovo, F. and M. Jaffe. 1984a. Local Groundwater Protection: Midwest Region. American Planning Association, Chicago, IL, 327 pp. [See also Jaffe and DiNovo (1987)]

DiNova, F. and M. Jaffe. 1984b. Local Regulations for Ground-Water Protection Part I: Sensitive Area Controls. Land Use Law and Zoning Digest 30(5):6-11.

Ellickson, R.C. and A.D. Tarlock. 1981. Land Use Controls: Cases and Materials. Little, Brown, and Company, Boston, MA.

Environmental Law Institute. 1990. Appendix: Survey and Analysis of State Ground-Water Programs; Policies, Authorities and Management Tools. Prepared for the Office of Ground-Water Protection, U.S. EPA, Washington, DC.

Environmental Resource Center Staff. 1992. Hazardous Waste Management Compliance Handbook. Van Nostrand Reinhold, New York, NY, 448 pp.

Fischer, J.A., R.J. Canace, and D.H. Monteverde. 1991. Karst Geology and Ground Water Protection Law. Ground Water Management 10:653-666 (Proc. 3rd Conf. on Hydrogeology, Ecology, Monitoring and Management of Ground Water in Karst Terranes). [Hunterdon County, NJ]

Flanagan, E.K., J.E. Hansen, and N. Dee. 1991. Managing Ground-Water Contamination Sources in Wellhead Protection Areas: A Priority Setting Approach. Ground Water Management 7:415-418 (Proc. Focus Conf. on Eastern Regional Ground-Water Issues).

Freshwater Foundation. 1988-1990. *Agricultural Chemicals and Groundwater Protection Conferences Series*: Agricultural Chemicals and Groundwater Protection: Emerging Management and Policy (1987, 23 papers and panel responses); Agrichemicals and Groundwater Protection: Resources and Strategies for State and Local Management (1988, 43 papers plus panel comments); Groundwater and Agrichemicals: Suggested Policy Directions for 1990 (1989, 17 papers/panel presentations). Freshwater Foundation, Navarre, MN.

Freund, E.C. and W.I. Goodman. 1968. Principles and Practices of Urban Planning. International City Managers Association, Washington, DC.

Getzels, J. and C. Thurow (eds.). 1979. Rural and Small Town Planning. American Planning Association, Washington, DC.

Global Cities Project. 1993. Land Use: Stewardship and the Planning Process. An Environmental Guide for Local Government, Volume 10, Global Cities Project, San Francisco, CA, 228 pp.

Goldman, S. T.A. Bursztynsky, and K. Jackson. 1986. Erosion and Sediment Control Handbook. American Planning Association, Chicago, IL, 480 pp.

Gordon, W. 1984. A Citizen's Handbook for Groundwater Protection. Natural Resources Defense Council, New York, NY.

Greeley-Polhemus Group, Inc. 1985. Handbook of Methods for the Evaluation of Water Conservation of Municipal and Industrial Water Supply. U.S. Army Corps of Engineers, Institute of Water Resources, Fort Belvoir, VA.

Group for the South Fork. 1982. Groundwater Management: A Handbook for the South Fork. Group for the South Fork, Inc., Bridgehampton, NY.

Hall and Associates and R. Dight. 1986. Ground Water Resource Protection: A Handbook for Local Planners and Decision Makers in Washington State. Prepared for King County Resource Planning and Washington Department of Ecology, Olympia, WA.

Harrison, E.Z. and M.A. Dickinson. 1984. Protecting Connecticut's Groundwater: A Handbook for Local Government Officials. Connecticut Department of Environmental Protection, Hartford, CT.

Henderson, T.R., J. Traubman, and T. Gallagher. 1985. Groundwater: Strategies for State Action. The Environmental Law Institute, Washington, DC.

Henderson, T.R. 1987. The Institutional Framework for Protecting Groundwater in the United States. In: Planning for Groundwater Protection, G.W. Page (ed.), Academic Press, Orlando, FL, pp. 29-69.

Hendler, B. 1977. Caring for the Land Environmental Principles for Site Design and Review. Planning Advisory Service Report #328, American Planning Association, Chicago, IL, 94 pp.

Higgins, T. 1989. Hazardous Waste Minimization Handbook. Lewis Publishers, Chelsea, MI, 228.

Hodge, R.A. and A.J. Brown. 1990. Ground Water Protection Policies: Myths and Alternatives. Ground Water 28(4):498-504.

Holmes, B.H. 1979. Institutional Bases for Control of Nonpoint Source Pollution. U.S. Environmental Protection Agency Office of Water and Waste Management.

Hrezo, M. and P. Nickinson. 1986. Protecting Virginia's Groundwater: A Handbook for Local Government Officials. Virginia Water Resources Research Center, Virginia Polytechnic Institute and State University, Blacksburg, VA.

Inglese, Jr., O. 1992. Best Management Practices for the Protection of Ground Water: A Local Official's Guide to Managing Class V UIC Wells. Connecticut Department of Environmental Protection, Hartford, CT, 138 pp.

Interstate Commission on the Potomac River Basin (ICPRB). 1981. Proceedings of Nonpoint Pollution Control Symposium. Rockville, MD.

Jaffe, M. 1987. Data and Organizational Requirements for Local Planning. In: Planning for Groundwater Protection, G.W. Page (ed.), Academic Press, Orlando, FL, pp. 89-124.

Jaffe, M. and F.K. DiNovo. 1987. Local Groundwater Protection. American Planning Association, Washington, DC, 262 pp. [see, also DiNovo and Jaffe (1984)]

(Table 11-11 Ground Water Protection References)

Kemp, L. and J. Erickson. 1989. Protecting Groundwater Through Sustainable Agriculture. The Minnesota Project, Preston, MN, 41 pp.

Kerns, W.R. (ed.). 1977. Proceedings of a National Conference on Public Policy on Ground-Water Quality Protection. Virginia Water Resources Research Center, Virginia Polytechnic Institute and State University, Blacksburg, VA, 163 pp.

Kreitler, C.W. and R.K. Senger. 1991. Wellhead Protection Strategies for Confined-Aquifer Settings. EPA/570/9-91-008, 168 pp. Available from ODW.*

Lal, R. (ed.). 1994. Soil Process and Water Quality. Lewis Publishers, Chelsea, MI, 300 pp. [11 chapters focusing behavior of contaminants in the subsurface and minimization of contamination by agricultural practices]

Leavall, D.N. 1990. The Development of Wellhead Protection In Ohio. Ground Water Management 1:669-683 (Proc. of the 1990 Cluster of Conferences: Ground Water Management and Wellhead Protection).

LeGrand, H.E. and L. Rosen. 1992. Common Sense in Ground-Water Protection and Management in the United States. Ground Water 30:867-872.

Lehr, J.H. 1987. Editorial: Wellhead Protection—The Ounce of Prevention That is Now in Jeopardy. Ground Water 25:514-516.

Licis, I.J., H. Skovronek, and M. Drabkin. 1991. Industrial Pollution Prevention Opportunities for the 1990s. EPA/600/8-91/052 (NTIS PB91-220376). [Identifies approaches to source reduction and waste recycling for 17 industries: textile dyes and dyeing, pulp and paper, printing, chemical manufacture, plastics, pharmaceuticals, paint industry, ink manufacture, petroleum industry, steel industry, non-ferrous metals, electronics/semiconductors, automobile manufacture/assembly, laundries/dry cleaning, and automobile refinishing/repair]

Logan, T.J., J.M. Davidson, J.L. Baker, and M.R. Overcash. 1987. Effects of Conservation Tillage on Ground Water Quality—Nitrate and Pesticides. Lewis Publishers, Chelsea, MI, 292 pp.

Lukin, J. 1992. Understanding Septic Systems. Northeast Rural Water Association, Williston, VT.

Madarchik, L.S. 1992. How-To Manual for Ground Water Protection Projects. Texas Water Commission, Austin, TX, 55 pp.

Massachusetts Audubon Society. 1984-1987. *Ground Water Information Flyer Series*. An Introduction to Groundwater and Aquifers (#1, 1984); Groundwater and Contamination: From Watershed into the Well (#2, 1984); Mapping Aquifers and Recharge Areas (#3, 1985); Local Authority for Groundwater Protection (#4, 1985); Underground Storage Tanks and Groundwater Protection (#5, 1985); Protecting and Maintaining Private Wells (#6, 1985); Pesticides and Groundwater Protection (#7, 1986); Landfills and Groundwater Protection (#8, 1986); Road Salt and Groundwater Protection (#9, 1987). Lincoln, MA.

(Table 11-11 Ground Water Protection References)

Massachusetts Department of Environmental Quality Engineering. 1985. Groundwater Quality and Protection: A Guide for Local Officials. Boston, MA.

Massachusetts Department of Environmental Protection (MDEP). 1991. Guidelines and Policies for Public Water Systems (Revised, October 1991). MDEP, Division of Water Supply, Boston, MA, 182 pp. + appendices.

Massey, D.T. 1984. Land Use Regulatory Powers of Conservation Districts in the Midwestern States for Controlling NonPoint Source Pollution. Drake Law Review 33:36-11.

Matthess, G., S.S. D. Foster, and A.C. Skinner (eds.). 1985. Theoretical Background, Hydrogeology, and Practice of Groundwater Protection Zones. International Contributions to Hydrology, Vol. 63, Verlag Heinz Heise, Hannover, Germany, 204 pp.

Mecozzi, M. 1989. Groundwater: Protecting Wisconsin's Buried Treasure. Wisconsin Department of Natural Resources, Madison, WI.

Michigan Departments of Natural Resources and Public Health. 1993. Effective Wellhead Protection Programs: Lesson Learned from Local Communities. Michigan Departments of Natural Resources and Public Health, Lansing, MI, 32 pp.

Milde, G., K. Milde, P. Frisel, and M. Kiper. 1983. Basis in New Developments of Ground-Water Quality Protection Concepts in Central Europe. In: Proc. of the Int. Conf. on Ground-Water and Man, Vol. II, Australian Government Printing Service, Canberra, pp. 287-295.

Miller, C. and C. Wood. 1983. Planning and Pollution: An Examination of the Role of Land Use Planning in the Protection of Environmental Quality. Clarendon Press, Oxford, UK, 232 pp.

Minnesota Project. 1984. Model Ordinance for Groundwater Protection. The Environmental Professional 6:331-349.

Montana Environmental Quality Council. 1990. SJR 22 Interim Study on Ground Water Quality Protection and Management: Final Report to the 52nd Montana State Legislature. Montana Environmental Quality Council, Helena, MT, 123 pp.

Moore, B.A. 1993. Case Studies in Wellhead Protection Area Delineation and Monitoring. EPA/600/R-93/107 (NTIS PB93-213510).

Mossa, E. (ed.). 1977. Land Use Controls in the United States: A Handbook on the Legal Rights of Citizens. Natural Resources Defense Council/The Dial Press, New York, NY.

Mullikin, E.B. 1984. An Ounce of Prevention: A Ground Water Protection Handbook for Local Officials. Vermont Departments of Water Resources and Environmental Engineering, Health, and Agriculture, Montpelier, VT.

Murphy, J. Undated. Groundwater and Your Town: What Your Town Can Do Right Now. Connecticut Department of Environmental Protection, Hartford, CT.

(Table 11-11 Ground Water Protection References)

National Research Council. 1986. Ground Water Quality Protection: State and Local Strategies. National Academy Press, Washington, DC, 309 pp.

National Rural Water Association. 1991. Training Manual: Ground Water/Wellhead Protection Technical Assistance Program. Duncan, OK.

New Hampshire Department of Environmental Services (NHDES). 1991. A Guide to the New Hampshire Wellhead Protection Program and the Groundwater Protection Act. NHDES, Waster Supply and Pollution Control Division, Concord, NH, 15 pp.

New Hampshire Office of State Planning. 1991. Model Health Ordinances to Implement a Wellhead or Groundwater Protection Program. Prepared for New Hampshire Department of Environmental Services, Water Supply and Pollution Control Division, Concord, NH, 63 pp.

New Jersey Department of Environmental Protection and Energy (NJDEPE). 1992. *Ground Water Protection Practices Series*: Motor Vehicle Services (6 pp.), Roadway Deicing (6 pp.), Unregulated Underground Storage Tanks (10 pp.), Urban/Suburban Landscaping (8 pp.), Septic Systems (8 pp.). NJDEPE, Trenton, NJ.

New York State Department of Health. 1984. Emergency Planning and Response - A Water Supply Guide for the Supplier of Water. New York State Department of Health, Albany, NY.

Noake, K.D. 1988. Guide to Contamination Sources for Wellhead Protection (Draft). Massachusetts Department of Environmental Quality Engineering, Boston, MA.

North Dakota State Department of Health. 1993. North Dakota Wellhead Protection User's Guide. NDSDH, Division of Water Quality, Bismarck, ND.

Novotny, V. and G. Chesters. 1981. Handbook of Nonpoint Source Pollution Sources and Management. Van Nostrand Reinhold, New York, NY.

Novotny, V. and H.O. Olem. 1993. Water Quality: Prevention, Identification, and Management of Diffuse Pollution. Van Nostrand Reinhold, New York, NY, 1,008 pp.

Oates, L.E., W.D. Ward, S.P. Roy, and T.N. Blandford. 1990. Tools for Wellhead Protection Delineation and Contingency Planning. Ground Water Management 1:463-477 (Proc. of the 1990 Cluster of Conferences: Ground Water Management and Wellhead Protection).

Page, G.W. (ed.). 1987. Planning for Groundwater Protection. Academic Press, Orlando, FL.

Pierce, J.W. 1992. Wellhead Protection Manual. Massachusetts Department of Environmental Protection, Division of Water Supply, Boston, MA, 17 pp.

Pisanelli, A.J. and P.W. Dutram. 1990. Institutional Constraints to Implementation of the Maine Ground Water Management Strategy. Ground Water Management 3:69-82 (Proc. Focus Conf. on Eastern Regional Ground Water Issues).

(Table 11-11 Ground Water Protection References)

Pojacek, R.B. (ed.). 1977. Drinking Water Quality Enhancement Through Source Protection. Ann Arbor Science Press, Ann Arbor, MI.

Potter, J. 1984. Local Ground-Water Protection: A Sampler of Approaches Used by Local Governments. Misc. Paper 84-2. Wisconsin Geological and Natural History Survey, Madison, WI, 17 pp.

Quinlan, J.F., P.L. Smart, G.M. Schindel, E.C. Alexander, Jr., A.J. Edwards, and A. R. Smith. 1991. Recommended Administrative/Regulatory Definition of Karst Aquifer, Principles for Classification and Carbonate Aquifers, Practical Evaluation of Vulnerability of Karst Aquifers, and Determination of Optimum Sampling Frequency at Springs. Ground Water Management 10:573-635 (Proc. 3rd Conf. on Hydrogeology, Ecology, Monitoring and Management of Ground Water in Karst Terranes).

Raymond, L.S. (ed.). 1981. Groundwater Management in the Northeastern States: Legal and Institutional Issues. Center for Environmental Research, Ithaca, NY.

Raymond, Jr., L.S. 1986. Chemical Hazards in Our Groundwater, Options for Community Action: A Handbook for Local Officials and Community Groups. Center for Environmental Research, Cornell University, Ithaca, NY.

Redlich, S. 1988. Summary of Municipal Actions for Groundwater Protection in the New England/New York Region. New England Interstate Water Pollution Control Commission, Boston, MA.

Robinson, N.A. 1988. Environmental Regulation of Real Property. Law Journal Seminars-Press, New York, NY.

Roy, S. 1988. Developing a State Wellhead Protection Program: A User's Guide to Assist State Agencies Under the Safe Drinking Water Act. U.S. EPA Office of Ground-Water Protection, (NTIS PB89-173751), 48 pp.

Rubin, P.A. 1991. Land-Use Planning and Watershed Protection in Karst Terranes. Ground Water Management 10:769-793 (Proc. 3rd Conf. on Hydrogeology, Ecology, Monitoring and Management of Ground Water in Karst Terranes).

Rusmone, B. (ed.). 1982. Private Options: Tools and Concepts for Land Conservation. Island Press, Covelo, CA, 296 pp. [30 papers]

Sell, N.J. 1992. Industrial Pollution Control: Issues and Techniques, 2nd ed. Van Nostrand Reinhold, New York, NY, 404 pp.

Shafer, P., C. Gesalmen, M. Elliot, A. Maresco, and P.C. Shinn. 1988. Reference Guide on State Financial Assistance Programs. EPA/430/9-88/004 (NTIS PB88-179304), 50 pp.

Southern Water Authority. 1985. Aquifer Protection Policy. Guildbourne House, Worthing, U.K., 47 pp.

Stewart, B.A. (ed.). 1976. Control of Water Pollution from Cropland. U.S. EPA and USDA.

(Table 11-11 Ground Water Protection References)

Stroman, M. 1987. The Aquifer Land Acquisition Program: An Approach for Protecting Ground Water Resources in Massachusetts.

Theodore, L. and Y.C. McGuinn. 1992. Pollution Prevention. Van Nostrand Reinhold, New York, NY, 366 pp.

Tolman, A.L., K.M. Bither, and R.G. Gerber. 1991. Technical and Political Processes in Wellhead Protection. Ground Water Management 7:401-413 (Proc. Focus Conf. on Eastern Regional Ground- Water Issues). [Central Maine]

Trefry, A. 1990. History and Summary of the Wellfield Protection Ordinance, Palm Beach Country, Florida. Ground Water Management 1:559-563 (Proc. of the 1990 Cluster of Conferences: Ground Water Management and Wellhead Protection).

Tripp, J.B. and A.B. Jaffe. 1979. Preventing Groundwater Pollution: Towards a Coordinating Strategy to Protect Critical Recharge Areas. Harvard Environ. Law Review 3(1):1-47.

University of Oklahoma. 1986. Proceedings of a National Symposium on Local Government Options for Ground Water Pollution Control. Norman, OK.

U.S. Environmental Protection Agency (EPA). 1984. EPA Ground-Water Protection Strategy. EPA/440/6-84-002 (NTIS PB88-112107).

U.S. Environmental Protection Agency (EPA). 1985a. Protection of Public Water Supplies from Ground-Water Contamination. Seminar Publication, EPA/625/4-85/016, 181 pp. Available from CERI*.

U.S. Environmental Protection Agency (EPA). 1985b. Ground-Water Monitoring Strategy, 1985. EPA/440/6-85-008 (NTIS PB88-111886).

U.S. Environmental Protection Agency (EPA). 1985c. Overview of State Ground-Water Program Summaries, Vol. 1. EPA/440/6-85-003 (NTIS PB88-112081).

U.S. Environmental Protection Agency. (EPA). 1985d. Emergency Planning for Potable Water Supplies. EPA/570/9-85-SPD-1. Available from ODW*.

U.S. Environmental Protection Agency (EPA). 1986a. Septic Systems and Groundwater Protection: An Executive's Guide. EPA/440/6-86/005 (NTIS PB88-112131), 13 pp.

U.S. Environmental Protection Agency (EPA). 1986b. Septic Systems and Groundwater Protection: A Program Manager's Guide and Reference Book. EPA/440/6-86/005 (NTIS PB88-112123), 134 pp.

U.S. Environmental Protection Agency (EPA). 1987. Guidelines for Delineation of Wellhead Protection Areas. EPA/440/6-87-010 (NTIS PB88-111430). [R. Hoffer may also be cited as author]

U.S. Environmental Protection Agency (EPA). 1987b. An Annotated Bibliography on Wellhead Protection References. EPA/440/6-87-014 (NTIS PB88-148754). [142 references]

(Table 11-11 Ground Water Protection References)

U.S. Environmental Protection Agency (EPA). 1987c. Septic Tank Siting to Minimize the Contamination of Ground Water by Microorganisms. EPA/440/6-87-007 (NTIS PB88-112115).

U.S. Environmental Protection Agency (EPA). 1987d. Sole Source Aquifer Background Study: Cross Program Analysis. EPA/440/6-87-015 (NTIS PB88-230933).

U.S. Environmental Protection Agency (EPA). 1987e. Cross-Program Summary: Pesticides Under EPA Statutes. EPA/440/6-87/001 (NTIS PB88-111448), 30 pp.

U.S. Environmental Protection Agency (EPA). 1987f. State and Territorial Use of Ground-Water Strategy Grant Funds (Section 106 of the Clean Water Act). EPA/440/6-87-008 (NTIS PB88-231493).

U.S. Environmental Protection Agency (EPA). 1987g. Improved Protection of Water Resources from Long-Term and Cumulative Pollution: Prevention of Ground-Water Contamination in the United States. EPA/440/6-87-013 (NITS PB88-111950). [Prepared for the Organization for Economic Cooperation and Development]

U.S. Environmental Protection Agency (EPA). 1987h. Guidance for Applicants for State Wellhead Protection Program Assistance Funds under the Safe Drinking Water Act. EPA/440/6-87-011 (NTIS PB88-111422), 50 pp. [Later versions published in 1988, 1989?]

U.S. Environmental Protection Agency (EPA). 1987i. Wellhead Protection: A Decision Maker's Guide. EPA/440/06-87/009 (NTIS PB88-111893), 24 pp.

U.S. Environmental Protection Agency (EPA). 1988a. Developing a State Wellhead Protection Program: A User's Guide to Assist State Agencies Under the Safe Drinking Water Act. EPA/440/6-88-003 (NTIS PB89-173751).

U.S. Environmental Protection Agency (EPA). 1988b. Survey of State Ground Water Quality Protection Legislation Enacted from 1985 Through 1987. EPA/440/6-88-007 (NTIS PB88-175475).

U.S. Environmental Protection Agency (EPA). 1988c. Sole Source Aquifer Designation Petitioners Guidance. EPA/440/6-87-003 (NTIS PB88-111992).

U.S. Environmental Protection Agency (EPA). 1988d. Protecting Ground Water: Pesticides and Agricultural Practices. EPA/440/6-88-001 (NTIS PB88-230628), 84 pp.

U.S. Environmental Protection Agency (EPA). 1989a. Funding Ground-Water Protection: A Quick Reference to Grants Available Under the Clean Water Act. EPA/440/6-89-004 (NTIS PB92-190255).

U.S. Environmental Protection Agency (EPA). 1989b. Local Financing for Wellhead Protection. EPA/440/6-89-001 (NTIS PB92-188705).

U.S. Environmental Protection Agency (EPA). 1989c. A Local Planning Process for Groundwater Protection. Office of Drinking Water, Washington, DC.

(Table 11-11 Ground Water Protection References)

U.S. Environmental Protection Agency (EPA). 1989d. Wellhead Protection Programs: Tools for Local Governments. EPA/440/6-89-002, 50 pp. Available from ODW.*

U.S. Environmental Protection Agency (EPA). 1989e. Indicators for Measuring Progress on Ground-Water Protection. EPA/440/6-88-006 (NTIS PB92-114425).

U.S. Environmental Protection Agency (EPA). 1990-1992. *Waste Minimization Assessment Series*. **Electronics:** Manufacturer of Printed Circuit Boards (EPA/600/M-91/022), Manufacturer Producing Printed Circuit Boards (EPA/600/S-92/033), Multilayered Printed Circuit Board Manufacturing (EPA/600/M-91/021), Manufacturer of Silicon-Controlled Rectifiers and Schottky Rectifiers (EPA/600/S-92/036); **Metals:** Manufacturer of Aluminum Cans (EPA/600/M-91/025), Manufacturer of Brazed Aluminum Oil Coolers (EPA/600/M-91/018), Manufacturer of Can Manufacturing Equipment (EPA/600/S-92/014), Manufacturer of Compressed Air Equipment Components (EPA/600/M-91/024), Manufacturer of Cutting and Welding Equipment (EPA/600/S-92/029), Manufacturer of Finished Metal Components (EPA/600/S-92/030), Manufacturer of Heating, Ventilating, and Air Conditioning Equipment (EPA/600/M-91/019), Manufacturer of Machined Parts (EPA/600/S-92/031), Manufacturer of Penny Blank and Zinc Products (EPA/600/S-92/037), Manufacturer of Sheet Metal Components (EPA/600/S-92/035), Metal Parts Coating Plant (EPA/600/M-91/015); **Other Manufacturing:** Manufacturer of Military Furniture (EPA/600/S-92/017), Manufacturer of Outdoor Illuminated Signs (EPA/600/M-91/016); **Plastics:** Manufacturer of Custom Molded Plastic Products (EPA/600/S-92/034), Manufacturer of Injection-Molded Car and Truck Mirrors (EPA/600/S-92/032), Manufacturer of Printed Plastic Bags (EPA/600/M-90/017); **Other Chemicals:** Manufacturer of Industrial Coatings (EPA/600/S-92/028), Paint Manufacturing Plant (EPA/600/M-91/023); **Transportation:** Bumper Refinishing Plant (EPA/600/M-91/020), Manufacturer of Motor Vehicle Exterior Mirrors (EPA/600/S-92/020), Manufacturer of Rebuilt Railway Cars and Components (EPA/600/M-91/016). [Environmental Research Briefs ranging from 3 to 6 pages; authors vary but include D.J. Collins, H.W. Edwards, M. Fleischman, C. Hensen, R.J. Jendrucko, F.W. Kirsch, M.F. Kostrzewa, G.P. Looby, J.C. Magnin, P.S. Miller; available from CERI*]

U.S. Environmental Protection Agency (EPA). 1990-1993. *Guide to Pollution Prevention Series (alphabetical by title)*: The Automotive Refinishing Industry (EPA/625/7-91/016); The Automotive Repair Industry (EPA/625/7-91/013); The Commercial Printing Industry (EPA/625/7-90/008); The Fabricated Metal Products Industry (EPA/625/7-90/006); The Fiberglass-Reinforced and Composite Plastics Industry (EPA/625/7-91/014); The Marine Maintenance and Repair Industry (EPA/625/7-91/015); The Mechanical Equipment Repair Industry (EPA/625/R-92/008); Metal Casting and Heat Treating Industry (EPA/625/R-92/009); The Metal Finishing Industry (EPA/625/R-92-011); Non-Agricultural Pesticide Users (EPA/625/R-93/009); The Paint Manufacturing Industry (EPA/625/7-90/005); The Pesticide Formulating Industry (EPA/625/7-90/004); The Pharmaceutical Industry (EPA/625/7-91/017); The Photoprocessing Industry (EPA/625/7-91/012); The Printed Circuit Board Manufacturing Industry (EPA/625/7-90/007); Municipal Pretreatment Programs (EPA/625/R-93/006); Research and Educational Institutions (EPA/625/7-90/010); Selected Hospital Waste Streams (EPA/625/7-90/009). Available from CERI.*

U.S. Environmental Protection Agency (EPA). 1990a. Citizen's Guide to Ground-Water Protection. EPA/440/6-90-004, 33 pp. Available from ODW.*

U.S. Environmental Protection Agency (EPA). 1990b. Progress in Ground-Water Protection and Restoration. EPA/440/6-90-001 (NTIS PB92-188671).

U.S. Environmental Protection Agency (EPA). 1990c. Guide to Ground-Water Supply Contingency Planning for Local and State Governments. EPA/440/6-90-003 (NTIS PB91-145755).

U.S. Environmental Protection Agency (EPA). 1991a. Protecting the Nation's Ground Water: EPA's Strategy for the 1990s. EPA/21Z-1020, 84 pp. Available from ODW.*

U.S. Environmental Protection Agency (EPA). 1991b. Managing Ground Water Contamination Sources in Wellhead Protection Areas: A Priority Setting Approach (Draft). EPA 570/9-91-023. U.S. EPA Office of Ground Water and Drinking Water.

U.S. Environmental Protection Agency. 1991c. Industrial Pollution Prevent Opportunities for the 1990s. EPA/600/8-91/052. Available from CERI.*

U.S. Environmental Protection Agency. 1991d. Achievements in Source Reduction and Recycling for Ten Industries in the United States. EPA/600/2-91/052. Available from CERI.*

U.S. Environmental Protection Agency (EPA). 1992a. Ground Water Protection: A Citizen's Action Checklist. EPA/810-F-91-002, 2 pp. Available from ODW*.

U.S. Environmental Protection Agency (EPA). 1992b. A Handbook for State Ground Water Managers: Using EPA Ground Water-Related Grants to Support the Development and Implementation of Comprehensive Sate Ground Water Protection Programs. EPA/813-B-92-001. Available from ODW.*

U.S. Environmental Protection Agency (EPA). 1992c. Implementing EPA's Ground Water Protection Strategy for the 1990s: Draft Comprehensive State Ground Water Protection Program Guidance. Office of Ground Water and Drinking Water. Available from ODW*.

U.S. Environmental Protection Agency. 1992d. Facility Pollution Prevention Guide. EPA/600/R-92/088, 143 pp. Available from CERI.*

U.S. Environmental Protection Agency. 1992e. User's Guide: Strategic Waste Minimization Initiative (SWAMI) Version 2.0: A Software Tool to Aid in Process Analysis for Pollution Prevention. EPA/625/11-91/004. Available from CERI.*

U.S. Environmental Protection Agency. 1992f. *Waste Reduction Activities and Options Series*. **Chemicals:** Manufacturer of Artists Supply Paints (EPA/600/S-92/045), Manufacturer of Electroplating Chemical Products (EPA/600/S-92/059), Manufacturer of Fine Chemicals Using Batch Processes (EPA/600/S-92/055), Manufacturer of General Purpose Paints and Painting Supplies (EPA/600/S-92/054), Manufacturer of Paints Primarily for Metal Finishing (EPA/600/S-92/040), Manufacturer of Fire Retardant Plastic Pellets and Hot Melt Adhesives (EPA/600/S-92/052), Manufacturer of Plastic Containers by Injection Molding (EPA/600/S-92/060); **Commercial:** Printing Plate Preparation Section of a Newspaper (EPA/600/S-92/053), Scrap Metal Recovery Facility (EPA/600/S-92/058); **Electronics:** Manufacturer of Systems to Produce Semiconductors (EPA/600/S-92/050); **Metals:** Fabricator and Finisher of Steel

(Table 11-11 Ground Water Protection References)

Computer Cabinets (EPA/600/S-92/044), Manufacturer of Commercial Dry Cleaning Equipment (EPA/600/S-92/062), Manufacturer of Commercial Refrigeration Units (EPA/600/S-92/047), Manufacturer of Electroplated Wire (EPA/600/S-92/049), Manufacturer of Hardened Steel Gears (EPA/600/S-92/057), Manufacturer of Orthopedic Implants (EPA/600/S-92/064), Manufacturer of Room Air Conditioning Units and Humidifiers (EPA/600/S-92/042), Manufacturer of Wire Stock Used for Production of Metal Items (EPA/600/S-92/046); **Other Manufacturing:** Laminator of Paper and Cardboard Packages (EPA/600/S-92/056), Manufacturer of Finished Leather (EPA/600/S-92/039); Manufacturer of Writing Instruments (EPA/600/S-92/041); **Public Services:** Local Board of Education in New Jersey (EPA/600/S-92/027); **Transportation:** Autobody Repair Facility (EPA/600/S-92/043), Remanufacturer of Automobile Radiators (EPA/600/S-92/051), State Department of Transportation Maintenance Facility (EPA/600/S-92/026), Transporter of Bulk Plastic Pellets (EPA/600/S-92/048); **Utilities:** Electrical Utility Transmission System Monitoring and Maintenance Facility (EPA/600/S-92/063), Fossil Fuel Fired Electrical Generating Station (EPA/600/S-92/061), Nuclear Powered Electrical Generating Station (EPA/600/S-92/025). [Environmental Research Briefs ranging from 3 to 5 pages; authors vary but include P. Eyraud, K. Gashlin, H. Saqa, A. Ulbrecht, and D.J. Watts; available from CERI*]

U.S. Environmental Protection Agency (EPA). 1993a. Wellhead Protection: A Guide for Small Communities. Seminar Publication EPA/625/R-93-002. Available from CERI*.

U.S. Environmental Protection Agency. 1993b. Life Cycle Design Guidance Manual: Environmental Requirements and The Product System. EPA/600/R-92/226, 181 pp. Available from CERI.*

U.S. Environmental Protection Agency (EPA). 1993c. Guidance Manual for Developing Best Management Practices (BMP). EPA 833-B-93-004. [NPDES manual]

U.S. Office of Technology Assessment (OTA). 1984. Protecting the Nation's Groundwater from Contamination, 2 Vols. OTA-O-233 and OTA-O-276. Washington, DC.

Walden, R. 1988. Ground Water Protection Efforts in Four New England States. EPA/600/9-89/084 (NTIS PB89-229975), 154 pp.

Walker, D, J. Golden, D. Bingham, and E. Driscoll. 1993. Manual: Combined Sewer Overflow Control. EPA/625/R-93/007, 95 pp. Available from CERI.*

Ward, W.D., L.E. Oates, and K.B. McCormack. 1990. Tools for Wellhead Protection: Control and Identification of Light Industrial Sources. Ground Water Management 1:579-593 (Proc. of the 1990 Cluster of Conferences: Ground Water Management and Wellhead Protection).

Weatherington-Rice, J. and A. Hottman. 1990. Beyond a State Ground-Water Protection Strategy: Where Do We Go From Here? Ground Water Management 1:529-544 (Proc. of the 1990 Cluster of Conferences: Ground Water Management and Wellhead Protection). [Ohio case study]

Western Michigan University. 1988. Policy Planning and Resource Protection: A Groundwater Conference for the Midwest. Institute for Water Sciences, Kalamazoo, MI.

Wilson, J.S., P. Tabas, and M. Henneman. 1979. Comprehensive Planning and the Environment: A Manual for Planners. University Press of America, Lanham, MD, 283 pp.

Yang, J.T. and W.C. Bye. 1979a. A Guidance for Protection of Ground-Water Resources from the Effects of Accidental Spill of Hydrocarbons and Other Hazardous Substances. EPA/570/9-79-017 (NTIS PB82-204900), 166 pp.

Yang, J.T. and W.C. Bye. 1979b. Methods for Preventing, Detecting, and Dealings with Surface Spills of Contaminants Which May Degrade Underground Water Sources for Public Water Systems. EPA/570/9-79-018 (NTIS PB82-204082), 118 pp.

Yanggen, D.A. and L.L. Amrhein. 1989. Groundwater Quality Regulation: Existing Governmental Authority and Recommended Roles. Columbia J. of Environmental Law 14(1):1-109.

Yanggen, D.A. and B. Webendorfer. 1991. Groundwater Protection Through Local Land-Use Controls. Wisconsin Geologic and Natural History Survey Special Report 11, Madison, WI, 48 pp.

Zaporozec, A. 1991. Regional Strategies to Protect Ground-Water Quality. In: Proc. First USA/USSR Joint Conf. on Environmental Hydrology and Hydrogeology, J.E. Moore et al. (eds.), American Institute of Hydrology, Minneapolis, MN, pp. 181-187.

---

* See Introduction for information on how to obtain documents from CERI (U.S. EPA Center for Environmental Research Information). Documents available from ODW (U.S. EPA Office of Drinking Water) can be obtained by calling the Safe Drinking Water Hotline (see Table 5-3).

CHAPTER 12

REMEDIATION PLANNING

12.1 Conceptual Approach to Soil and Ground Water Remediation 696

12.2 Methodology 699

    12.2.1 Site Characterization 699
    12.2.2 Assessment of Problem 702
    12.2.3 Interim Corrective Action 706
    12.2.4 Treatment Approaches 706
    12.2.5 Monitoring Program 709
    12.2.6 Health and Safety Considerations 710

12.3 Selection of Treatment Methods 710

    12.3.1 Utility of Mathematical Models 710
    12.3.2 Treatability Studies 713
    12.3.3 Treatment Trains 714

12.4 Measurement and Interpretation of Treatment Effectiveness 715

12.5 Guide to Major References[1] 716

Subsurface remediation includes: (1) identifying, quantifying, and controlling contaminant source(s); (2) considering cleanup levels required for each medium (air, soil, and ground water) to protect human health and the environment; and (3) selecting treatment technologies based on information obtained concerning source(s) and cleanup levels. The challenge is to effectively relate site characterization activities to selecting the most appropriate remediation technologies for contaminated soils and ground water at hazardous waste sites. Effectively relating these activities with technology selection improves the efficiency, purpose, and results of both site characterization and remediation technique selection. This chapter addresses specific subsurface physical, chemical, and biological processes that have been discussed in Part I within the context of (1) site characterization requirements, (2) evaluation and selection of remediation techniques and treatment trains utilizing several techniques, and (3) design of monitoring programs. Specific soil and aquifer remediation techniques are addressed in more detail in Chapters 13 (Soil) and 14 (Ground-Water).

---

[1] Appendix F contains citations for table and figure sources.

12.1 Conceptual Approach to Soil and Ground Water Remediation

This chapter uses the *chemical mass balance* approach as the conceptual framework for soil and ground-water remediation technique evaluation, selection, and monitoring. The concept of a chemical mass balance is familiar to professionals trained in the physical or life sciences or in engineering. It provides a rational and fundamental basis for asking specific questions and obtaining specific information that is necessary for determining fate and behavior, for evaluating and selecting treatment options, and for monitoring treatment effectiveness at both laboratory-scale and field-scale. A mass balance approach also meets the goal of obtaining quantitative accuracy about the amount of contaminants initially present at an uncontrolled site. While a mass balance, or materials balance, is routinely conducted on aboveground treatment processes (Bailey and Ollis, 1986/T12-7; Benefield et al., 1982/T-14-9; Corbitt, 1989/T12-7), and for ground-water processes (Willis and Yeh, 1987/T14-9; Wilson et al., 1989/T12-7), a mass balance approach has generally not been applied to the soil environment or to the subsurface/surface system to link characterization activities and treatment technology selection. The information needed to construct a mass balance for contamination at a site simultaneously addresses site characterization and remediation evaluation and selection.

The mass balance conceptual approach for the soil and ground-water subsurface environment at a contaminated site is illustrated in Figure 12-1. The contaminated subsurface is a system generally consisting of two phases (solid and fluid) and five compartments (gas, an inorganic mineral solid compartment, an organic matter solid compartment, water, and oil [NAPL]) (Sims et al., 1989/T13-9). Generally NAPLs are subdivided into two classes: those that are lighter than water (LNAPLs), and those with a density greater than water (DNAPLs)--Section 4.4.3. LNAPLs include hydrocarbon fuels, such as gasoline, heating oil, kerosene, jet fuel, and aviation gas. DNAPLs include chlorinated hydrocarbons, such as 1,1,1-trichloroethane, carbon tetrachloride, chlorophenols, chlorobenzenes, tetrachloroethylene, and polychlorinated biphenols (PCBs).

Specific subsurface processes concerning water movement, sampling, sorption and reaction, and degradation are discussed in Chapters 3 and 4. The processes and terminology described in the previous chapters will be used in this chapter for the discussion of the components of a mass balance and the mass balance approach to evaluation and selection of soil remediation techniques.

Interphase transfer potential for waste constituents among oil (waste or NAPL), water, air, and solid (organic and inorganic) phases of a subsurface system is affected by the relative affinity of waste constituents for each phase shown in Figure 12-1, and may be quantified through calculation of distribution coefficients (Loehr, 1989/T13-9; Sims et al., 1988/T13-9; U.S. EPA, 1986a/T12-7). Distribution coefficients are calculated as the ratio of the concentration of a chemical in the soil (or aquifer materials), oil, or gas phases to the concentration of a chemical in the water phase. A waste chemical, depending on its tendency to be associated with each phase, will distribute itself among the phases, and can be quantified in terms of distribution coefficients. Distribution coefficients are available for a variety of chemicals and can

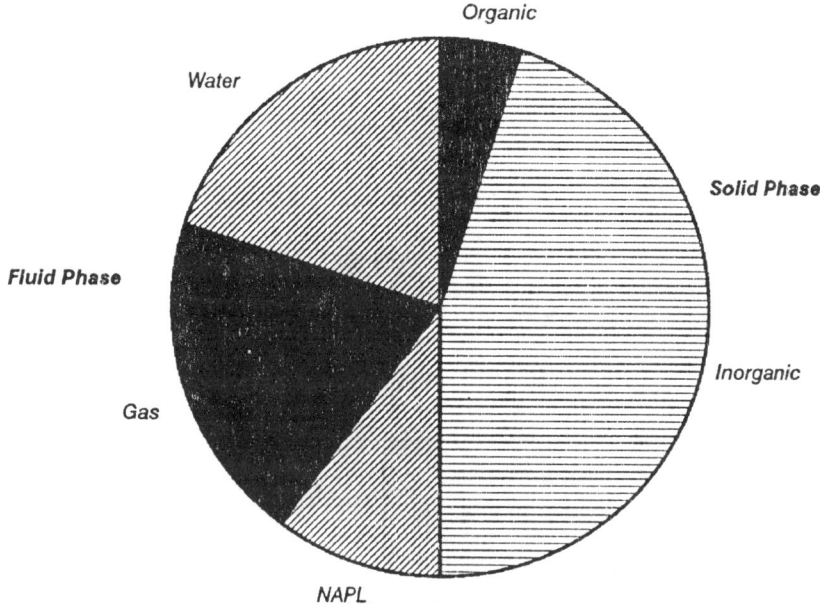

Figure 12-1 Mass balance conceptual framework for the soil and ground water subsurface environment at a contaminated site (Sims and Sims, 1991a).

be expressed as ratios of the concentrations of a chemical between two phases in the subsurface:

K_d = Concentration in solid phase/Concentration in aqueous phase

K_o = Concentration in oil phase/Concentration in aqueous phase

K_h = Concentration in air phase/Concentration in aqueous phase

When distribution coefficients are not available, they can be estimated using structure-activity relationships (SARs) or can be determined in laboratory tests (Sims et al., 1988/T13-9).

Distribution coefficients have been used most successfully with organic chemicals. However, since metals distribute among the phases of the subsurface systems described previously, distribution coefficients also may be used, along with multiphase metal speciation information (Sims et al., 1984/T13-9), to evaluate metal distribution in a contaminated subsurface system.

Knowledge of migration and distribution of chemicals and chemical intermediates among the phases and compartments of a contaminated subsurface system (illustrated in Figure 12-2) provides fundamental information about the fate and behavior of contaminants, which can be used for selecting and evaluating subsurface remedial techniques. Retardation of the downward transport (leaching

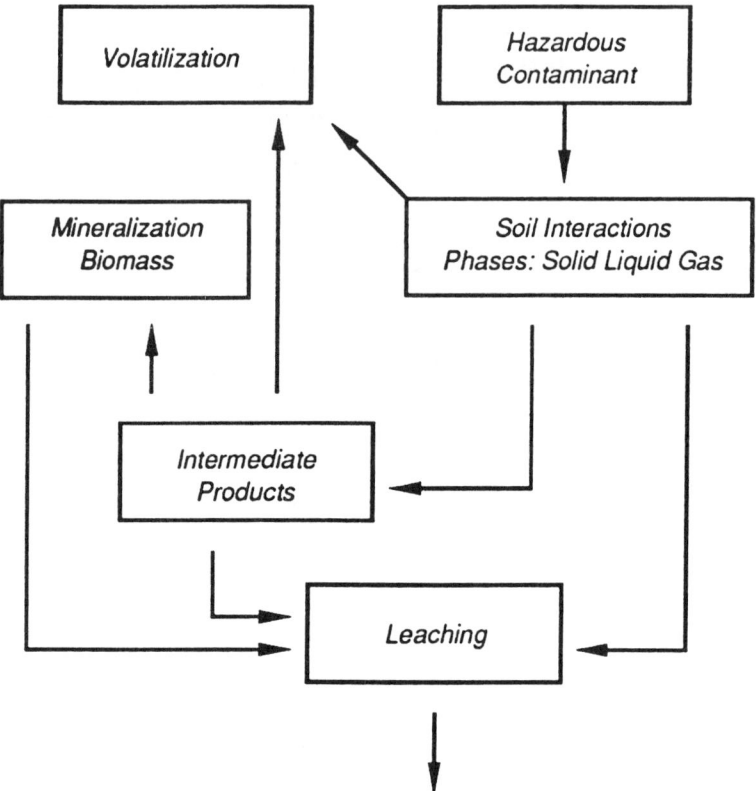

Figure 12-2 Interphase transfer potential of chemicals in the subsurface (Sims and Sims, 1991a).

potential) and upward transport (volatilization potential) is referred to as immobilization of waste constituents, and has been related to the subsurface organic matter content, especially for hydrophobic chemicals (Nkedi-Kizza et al., 1983), soil moisture, and presence and concentration of organic solvents (Mahmood and Sims, 1986; Rao et al., 1985).[2]

In summary, subsurface processes described above, combined with information about the movement of fluids (gases, aqueous phase, and pure product flow) in the

---

[2] Mahmood, R.J. and R.C. Sims. 1986. Mobility of Organics in Land Treatment Systems. Journal of Environmental Engineering (ASCE) 112:236-245.

Nkedi-Kizza, P., P.S.C. Rao, and J.W. Johnson. 1983. Adsorption of Diuron and 2,4,5-T on Soil Particle Separates. Journal of Environmental Quality 12:195-197.

Rao, P.S.C., A.G. Hornsby, D.P. Kilcrease, and P. Nkedi-Kizza. 1985. Sorption and Transport of Hydrophobic Organic Chemicals in Aqueous and Mixed Solvent Systems: Model Development and Preliminary Evaluation. Journal of Environmental Quality 14:376-383.

unsaturated and saturated zones, provide the inputs into the chemical mass balance that can be used for (1) characterizing a site, (2) assessing the problem of mobility, (3) evaluating treatment techniques, and (4) identifying chemicals in specific phases for monitoring treatment effectiveness.

12.2 Methodology

Remediation of contaminated soil and ground water consists of four elements: (1) characterization, (2) assessment of the problem, (3) treatment (train) selection, and (4) monitoring treatment performance (Figure 12-3). The first element involves characterization in the context of waste/subsurface/site interactions to address the question, "Where is the contamination and in what form(s) does it exist?" The second element, assessment of the problem, utilizes subsurface fate and behavior information to address the question "Where is the contamination going under the influence of natural processes?" The problem can be defined in the context of mobility versus degradation for chemicals at a site. Using mathematical models or other tools, the chemicals can be ranked in order of their relative tendencies to leach, to volatilize, to move in a NAPL phase and to remain in-place under site-specific conditions. Containment and/or treatment options then can be selected that are chemical-specific and that address specific escape and attenuation pathways (third element). Therefore, treatment trains can be selected to address specific waste phases at specific times during remediation (volatile, leachate, solid phase, and pure product), with the selection based upon results of a mass balance evaluation through time to identify the fate of each waste phase. Finally monitoring programs can be designed for specific chemicals in specific phases in the subsurface at specific times (fourth element).

The approach for using the methodology described above consists of applying a mass balance for each element of the methodology. This approach assists in the collection of specific information that is transferable among all four elements of the methodology, and also addresses the technical issues of soil remediation within the context of regulatory goals.

12.2.1 Site Characterization

Identifying waste sources by subsurface phases, i.e., identification and amount (if possible) of waste constituents associated with solid and fluid phases (Figure 12-1), allows assessment of the magnitude (mass) and physical form(s) of waste that must be treated. This assessment comprises the first step in the mass balance characterization of waste sources at a site.

Wastewater historically has been characterized and subsequently treated in terms of its interaction and potential impact on the assimilative capacity of surface water receiver systems, generally rivers or lakes (e.g., requiring measurement of characteristics such as oxygen-demanding substances, nutrients, and levels of substances toxic to aquatic organisms). However, a waste characterization program at a contaminated site addresses the vadose zone and ground water, in addition to surface water, as the receiver systems (e.g., requiring measurement of characteristics that reflect individual chemical mobility and destruction in the subsurface environment and

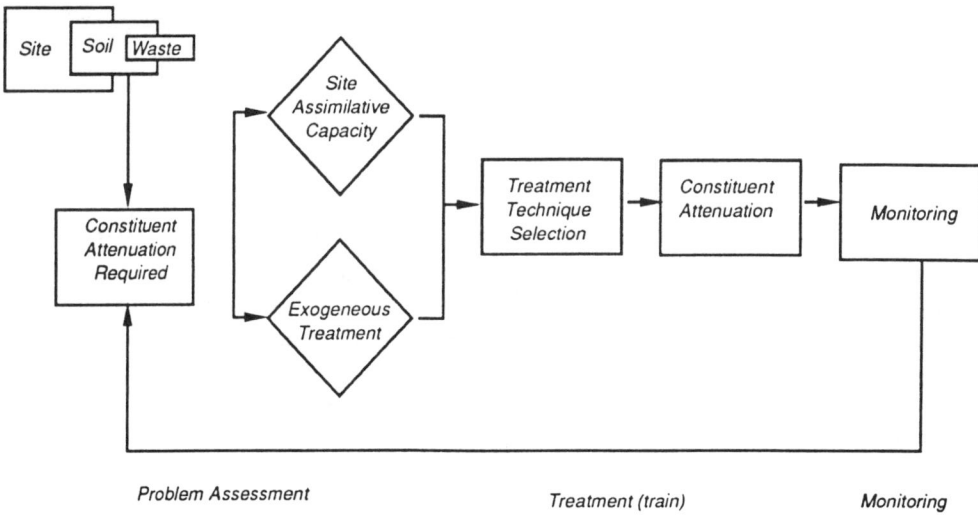

Figure 12-3 Methodology using mass balance approach for integrating data collection activities at a contaminated site (Sims and Sims, 1991a).

those that affect human health as well as characteristics that affect environmental toxicity--see Figure 11-8). Also, it describes the behavioral interaction of waste chemicals in each surface and subsurface phase. Thus hazardous waste is more appropriately characterized in terms of the interaction and potential impact on the subsurface assimilative capacity.

Specific site characteristics important for describing and assessing the environmental behavior and fate of organic constituents in the soil and subsurface are listed in Table 12-1. For each chemical, or chemical class, required information includes (1) characteristics related to potential leaching (e.g., water solubility, octanol/water partition coefficient, solid sorption coefficient), (2) characteristics related to potential volatilization (e.g., vapor pressure, relative volatilization index), (3) characteristics related to potential degradation (e.g., half-life, degradation rate, degradability index), and (4) characteristics related to chemical reactivity (e.g., hydrolysis half-life, soil redox potential) (Sims et al., 1984/T13-9). The information presented in Table 12-1 also is used to assess problem(s) concerning migration potential at a site and to evaluate and select containment- and treatment-management options. Table 12-2 identifies site characterization data needs for major methods used to stabilize and remediate contaminated sites.

If the distribution of waste chemicals among phases that comprise the soil and subsurface at a site are determined, then potential pathways of transport, or escape, from a site can be indicated. Therefore, exposure pathways for human health and the environment may be evaluated, i.e., risk assessment can be made. Through a determination of subsurface flow conditions as part of site characterization activities (aqueous, gas, and pure product flow in the vadose zone and aqueous plume and pure

Remediation Planning

product movement in the saturated zone), the mass of material moving through a site and potential movement off site can be assessed:

Concentration (mass/vol) x Rate of flow (vol/time) = Mass flow at site (mass/time)

This information is combined with additional information, discussed in the next section, that is needed to assess the problem(s) with respect to treatment technique selection.

Table 12-1 Subsurface-Based Waste Characterization

Chemical Class
    Acid
    Base
    Polar neutral
    Nonpolar neutral
    Inorganic
Chemical Properties
    Molecular weight
    Melting point
    Specific gravity
    Structure
    Water solubility
Chemical Reactivity
    Oxidation
    Reduction
    Hydrolysis
    Precipitation
    Polymerization
Soil Sorption Parameters (Section 4.5.2)
    Freundlich sorption constants (K,a)
    Sorption based on organic carbon content (K_{oc})
    Octanol-water partition coefficient (K_{ow})
Soil Degradation Parameters
    Half-life ($t_{1/2}$)
    Rate constant (first order—Section 3.1.1)
    Relative biodegradability (Section 3.5.4)
Soil Volatilization Parameters
    Air-water partition coefficient (K_w)
    Vapor pressure
    Henry's Law constant
    Sorption based on organic carbon content (K_{oc})
    Water solubility
Soil Contamination Parameters
    Concentration in soil
    Depth of contamination
    Date of contamination

Source: Sims and Sims (1991a) after Sims et al. (1984).

Table 12-2 Data Needs for Major Stabilization and Remediation Methods (Bartenfelder et al., 1992)

	Drains and Trenches	Capping	Slurry Wall	Runoff/ Runon Control	Gas Venting	Solidification	In Situ Soil Flushing	Bioremediation	Vacuum Extraction	Pump and Treat
Depth to Water Table	■	❑	■	—	❑	■	❑	❑	■	■
Surface Water/Ground Water Relationship	■	—	❑	—	—	—	❑	—	—	■
Ground-Water Flow Rates and Direction	■	—	■	—	—	—	❑	—	—	■
Seasonal Changes in Ground-Water Elevation	■	❑	—	—	❑	❑	❑	❑	❑	■
Hydraulic Conductivity/Permeability	■	■	■	—	■	—	■	■	■	■
Climate/Precipitation	■	■	—	■	—	■	—	■	■	—
Contaminant Characteristics	■	❑	■	—	■	■	■	■	■	■
Contaminant Concentration	■	—	■	—	■	■	■	■	■	■
Extent of Contamination	■	■	■	—	■	■	■	■	■	■
Types, Thicknesses, and Extents of Saturated and Unsaturated Subsurface Materials	■	—	■	—	❑	❑	■	■	■	■
Soil Characteristics	❑	—	■	■	■	—	■	■	■	■
Soil Water Content	—	—	—	—	■	■	—	■	■	—
Depth of Air Permeable Zone	—	—	—	—	■	—	—	■	■	—
Topography	—	■	■	■	—	—	—	—	—	—
Depth of Aquitard	❑	—	■	—	—	—	❑	—	—	■

■ High Priority; ❑ Medium Priority; — Low Priority

12.2.2 Assessment of Problem

Assessment of the contamination involves organizing the information obtained from site characterization activities to evaluate the transport and degradation behavior of each chemical of concern at a site under consideration. Specifically, the rate of transport can be compared with the rate of degradation to determine if transport is significant relative to degradation. This approach to problem(s) assessment will allow chemicals to be prioritized individually according to (1) magnitude and rate of transport (escape) from a site, (2) persistence, and (3) pathway(s) of migration from a site. Treatment technique evaluation and selection then can be based upon specific combinations of chemical and physical phase-migration pathway(s).

Interfacing subsurface-based behavioral characteristics of specific contaminants (Table 12-1) with specific site and subsurface properties allows an assessment of the problem(s) related to contamination of other media (due to mobility), including the ground water under the contaminated area, the atmosphere over the site or at the site boundaries, surface waters, and/or persistence of chemicals at a site. Pathways of movement and potential mechanisms of removal of contaminants at a specific site are illustrated in Figure 12-2. This element of the methodology functions to identify chemicals that will (1) migrate upward (volatilization), (2) migrate downward (leaching), (3) migrate laterally (aqueous plume and pure product), (4) degrade, and (5) remain at the site as persistent chemicals. By ranking the chemicals in the order in which they migrate or persist, chemicals can be prioritized with regard to urgency for treatment and for monitoring.

Waste characteristics identified in Table 12-1, including potential sorption, degradation, and volatilization at a site, can be determined in laboratory mass balance tests, using waste/soil mixtures from a site. These characteristics can be used to evaluate the fate of the waste at the site, and to generate specific data that can be used to develop treatment approaches. Figure 12-4 illustrates a laboratory flask apparatus that can be used to develop a chemical mass balance by measuring interphase transfer potential of chemicals as well as degradation potential at a site.

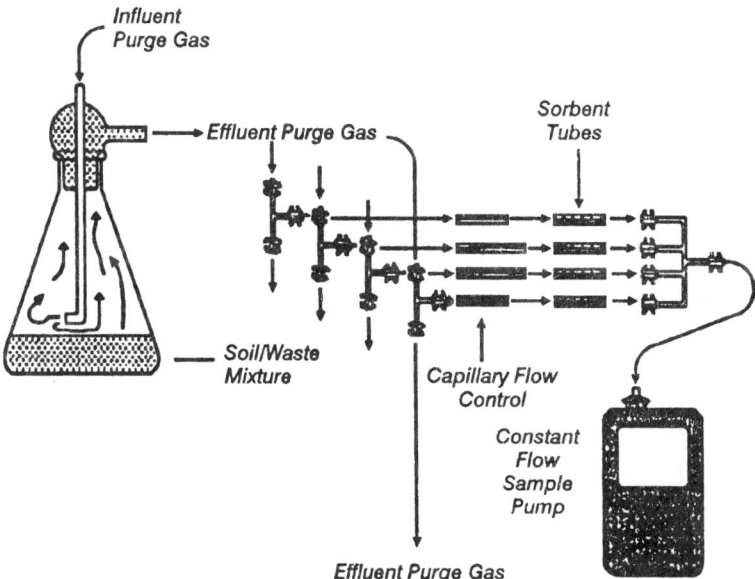

Figure 12-4 Laboratory flask apparatus used for mass balance measurements (Sims and Sims, 1991a, after Park et al., 1990).

The contaminated material is placed in a flask, which is then closed and incubated under controlled conditions for a period of time. During the incubation period, air is drawn through the flask and then through a sorbent material. Volatilized materials are collected by the sorbent and are measured to estimate volatilization loss of the constituents of interest. At the end of the incubation period, a portion of the contaminated soil is treated with an extracting solution to determine the extent of loss of the constituents in the soil matrix. This loss can be attributed to degradation and possible immobilization in the soil materials. It is necessary to select an appropriate extracting solution and procedure to maximize constituent recovery from a soil-waste mixture (Coover et al., 1987).[3] Another portion of the soil is leached with water to determine leaching potential of the remaining constituents. Abiotic and biological processes involved in removal of the parent compound are evaluated by comparing microbially active soil/waste mixtures with mixtures that have been treated with a microbial poison, e.g., mercuric chloride or propylene oxide. Samples generated from the different phases of the system in microcosm mass balance studies identified above can be analyzed for intermediate degradation products and used in bioassay studies to provide information concerning transformation and detoxification processes.

The use of a procedure incorporating features illustrated by the use of this microcosm (Figure 12-4) is crucial to obtain a materials balance of waste constituents in the subsurface system. Examples of such protocols may be found in EPA guidance documents and research reports (Loehr, 1989/T13-9; Sims et al., 1988/T13-9; EPA, 1986a/T12-7; and Park et al., 1990[4]). Contaminated materials also can be spiked with radiolabeled chemicals; tracking the fate of the chemicals as they move through the multiple phases of the soil system also provides a materials mass balance.

The mass balance approach identified above usually represents optimum conditions with respect to mixing, contact of solid materials with waste constituents, and homogeneous conditions throughout the laboratory microcosm; therefore, it does not incorporate site heterogeneity in the evaluation. This aspect must be defined during site characterization activities and evaluated with regard to potential effect on fate and behavior regarding migration and persistence at the site (problem assessment).

In addition to the laboratory tests described, bench-scale reactors, pilot-scale reactors and/or field-scale plots may be used to generate mass balance information for problem assessment. The set of experimental conditions (e.g., temperature, moisture, waste concentration) under which the studies were conducted and experimental results should be presented.

Information from the performance of site characterization and experimental mass balance studies may be integrated with the use of comprehensive mathematical modeling to aid in problem assessment. In general, models are used to analyze the behavior of an environmental system under both current (or past) and anticipated (or

---

[3] Coover, M.P., R.C. Sims, and W.J. Doucette. 1987. Extraction of Polycyclic Aromatic Hydrocarbons from Spiked Soil. Journal of the Association of Official Analytical Chemists 70:1018-1020.

[4] Park, K.S., R.C. Sims, R.R. Dupont, W.J. Doucette, and J.E. Matthews. 1990. Fate of PAH Compounds in Two Soil Types: Influence of Volatilization, Abiotic Loss and Biological Activity. Environ. Toxicol. Chem. 9:187-195.

future) conditions (Donigian and Rao, 1986a/T10-6). A mathematical model provides a tool for (1) integrating degradation and partitioning processes with site-, soil-, and waste-specific characterization; (2) simulating the behavior of waste constituents in a contaminated soil; and (3) predicting the pathways of migration through the contaminated area, and therefore pathways of exposure to humans and to the environment. Diguilio and Suffet (1988)[5] and Weaver et al. (1989/T10-6) have presented guidance on the selection of appropriate subsurface zone models for site-specific applications, focusing on recognition of limitations of process descriptions of models and difficulties in obtaining input parameters required by these process descriptions.

The Regulatory and Investigative Treatment Zone Model (RITZ) is an example of a vadose zone model that has been used to describe the potential fate and behavior of organic constituents in a contaminated soil system (Nofziger and Williams, 1988/Table E-2). The RITZ Model is based on an approach developed by Jury et al. (1983).[6] An expanded version of RITZ, the Vadose Zone Interactive Processes (VIP) model, incorporates predictive capabilities for the dynamic behavior of organic constituents in unsaturated soil systems under conditions of variable precipitation, temperature, and waste concentrations (McLean et al., 1988[7]; Stevens et al. 1993/Table E-2; Symons et al., 1988[8]). Both the RITZ and VIP models simulate vadose zone processes, including volatilization, degradation, sorption/desorption, advection, and dispersion (Grenney et al., 1987).[9]

For example, the VIP model was used to evaluate the relative tendencies for a group of pesticides to volatilize and to leach under specific waste-soil conditions (McLean et al., 1988/Footnote 7). Information input into the model included half-life (measured in laboratory tests), distribution coefficients--K_d, K_h, K_o (calculated), soil texture and moisture (measured), and site-specific climatic data (rainfall and temperature). The ranking of pesticides provided by the model indicated that the

---

[5] Diguilio, D.C. and I.H. Suffet. 1988. Effects of Physical, Chemical, and Biological Variability in Modeling Organic Contaminant Migration through Soil. In: Superfund '88, Hazardous Materials Control Research Institute, Silver Spring, MD, pp. 132-137.

[6] Jury, W.A., W.F. Spencer, and W.J. Farmer. 1983. Behavior Assessment Model for Trace Organics in Soil: Model Description. Journal of Environmental Quality 12:558-564.

[7] McLean, J.E., R.C. Sims, W.J. Doucette, C.L. Caupp, and W.J. Grenney. 1988. Evaluation of Mobility of Pesticides in Soil using U.S. EPA Methodology. Journal of Environmental Engineering (ASCE) 114:689-703.

[8] Symons, B.D., R.C. Sims, and W.J. Grenney. 1988. Fate and Transport of Organics in Soil: Model Predictions and Experimental Results. Journal Water Pollution Control Federation 60:1684-1693.

[9] Grenney, W.J., C.L. Caupp, R.C. Sims, and T.E. Short. 1987. A Mathematical Model for the Fate of Hazardous Substances in Soil: Model Description and Experimental Results. Hazardous Wastes & Hazardous Materials 4:223-239.

tendency of the pesticides to volatilize was not similar to their tendency to leach. Such information can be used to assess which chemicals are likely to volatilize first, which chemicals are likely to leach first, and which chemicals are persistent under site-specific conditions. In addition to assisting in the problem assessment step of the methodology, mathematical models also can be used to design studies for evaluation and selection of treatment options for these chemicals, as well as to design monitoring strategies (i.e., which chemicals to monitor in which media).

Ground-water flow and transport models can also be used as part of the problem assessment (Sections 10.3 and 10.4). Table 10-1 identifies types of modeling that may be useful for design of corrective action measures. Some ground-water models are specifically designed for evaluating contaminated sites. For example, a numerical model, BIOPLUME, was developed to simulate oxygen-limited biodegradation in ground-water environments. BIOPLUME simulates advection, dispersion, and retardation processes as well as the reaction between oxygen and the contaminants under steady, uniform flow (Rifai et al., 1988; Bedient et al., 1989/Table E-2). BIOPLUME has been applied to an aviation gasoline spill site at Traverse City, Michigan. Model predictions for the rates of mass loss closely matched calculated rates from field data. Table E-2 provides an index to PC-based ground water and vadose zone flow and contaminant transport models.

12.2.3 Interim Corrective Action

Depending on the degree of hazard presented by contaminants at a site, interim corrective action measures may be required before site characterization and problem assessment provide enough data for identification of the best permanent treatment approach. Table 12-3 identifies examples of interim measures for a variety of situations where immediate action at a site is justified. The extent and aggressiveness of interim measures will be determined by the degree of risk posed by the contamination (Section 11.4).

12.2.4 Treatment Approaches

Information obtained from an integrated assessment (modeling) of the problem (migration and persistence), based upon a thorough characterization of waste/soil/site interactions, can be used to select treatment approaches for further evaluation with respect to technical and cost-effectiveness factors. Results of characterization and assessment efforts can aid in the identification of constituents that will require treatment in the following phases: (1) air (volatile) phase, (2) leachate phase, and (3) solid (soil) phase. This approach allows evaluation and comparison of different treatment systems identified previously (in situ and prepared bed). Specifically, if treatment is required, the information is used to (1) determine containment requirements to prevent contamination of offsite receiver systems; (2) develop techniques to maximize mass transfer of chemicals affecting a process (e.g., affecting microbial activity through addition of mineral nutrients, oxygen, additional energy sources, pH control products, or removal of toxic products in order to enhance bioremediation); and (3) design a cost-effective and efficient monitoring program to evaluate effectiveness of treatment.

Table 12-3 Examples of Interim Measures for Corrective Action at Contaminated Sites (U.S. EPA, 1988c)

Unit	Possible Interim Measures
Containers	-Overpack/redrum -Construct storage area; move to new storage area -Segregation -Sampling and analysis -Treatment, storage and/or disposal -Temporary cover
Tanks	-Overflow/secondary containment -Leak detection/repair -Partial or complete removal
Surface Impoundments	-Reduce head -Remove free liquids and highly mobile wastes -Stabilize/repair side walls, dikes or liner(s) -Temporary cover -Runoff/runon control (diversion or collection devices) -Sampling and analysis -Interim ground-water measures (see below)
Landfills	-Runoff/runon control (diversion or collection devices) -Reduce head on liner and/or in leachate collection system -Inspect leachate collection/removal system or french drain -Repair leachate collection/removal system or french drain -Temporary cap/cover -Waste removal to more secure location -Interim ground water measures (see below)
Waste Piles	-Runoff/runon control (diversion or collection devices) -Temporary cap/cover -Waste removal to more secure location -Interim ground water measures (see below)
Soils	-Sampling and analysis -Removal to more secure location -Temporary cap/cover
Ground Water	-Delineation/verification of gross contamination -Sampling and analysis -Interceptor trench/sump/subsurface drain -Pump and treat -In situ treatment -Temporary cap/cover

Table 12-3 (cont.)

Unit	Possible Interim Measures
Surface Water Releases (Point and Nonpoint)	-Overflow/underflow dams -Filter fences -Runoff/runon control (diversion or collection devices) -Regrading/revegetation -Sampling and analysis of surface water/sediment/point source
Gas Migration Control	-Barriers -Collection and treatment -Monitoring
Particulate Emissions	-Truck wash (decontamination unit) -Revegetation -Application of dust suppressant
Other Actions	-Fencing to prevent direct contact -Sampling offsite areas -Alternate water supply for contaminated drinking water -Temporary relocation of exposed population -Temporary or permanent injunction -Suspend/revoke authorization to operate under interim status

Source: Adapted from U.S. EPA (1988c).

Containment Requirements. If the major pathway of transport is volatilization, containment and treatment to control volatilization is required. An inflatable plastic dome erected over a contaminated site is a containment method that has been used to control escape of volatile constituents at hazardous waste sites (St. John and Sikes, 1988).[10] Volatiles are drawn from the dome through a conduit and treated in an aboveground treatment system. If leaching has been identified as an important factor, control of soil water movement should be implemented. For example, if contaminated materials are expected to leach downward from the site, run-on and run-off controls can be implemented, or the contaminated materials can be temporarily removed from the site and a plastic or clay liner can be placed under the site (Lynch and Genes,

---

[10] St. John, W.D. and D.J. Sikes. 1988. Complex Industrial Waste Sites. In: Environmental Biotechnology - Reducing Risks from Environmental Chemicals through Biotechnology, G.S. Omenn (ed.), Plenum Press, New York, NY, pp. 237-252.

1989; Ross et al., 1988).[11] When downward as well as upward migration are significant, both volatilization and leaching containment systems can be installed. Some hydrophobic chemicals do not tend to volatilize or to leach but are persistent within the soil solid phase; therefore, containment efforts may not be required. With regard to the saturated zone, containment is generally accomplished by physical barriers (e.g., slurry walls, sheet pilings, grout curtains) or hydraulic barriers (e.g., pumping systems, french drains). Section 14.3 addresses containment of contaminants in more detail.

Maximizing Chemical Mass Transfer. An area of significant research concerns delivery and recovery technologies for maximizing mass transfer of chemicals that affect the rate and/or extent of treatment. Murdoch et al. (1988/T12-7) discussed delivery and recovery technologies, many of which are derived from the petroleum and mining industries. While a liquid phase is usually employed for delivery of chemicals, some technologies utilize vapor and solid phases for delivery. Principal recovery technologies involve hydraulic, thermal, and chemical systems. Delivery and recovery techniques are important in influencing the success of technologies, including bioremediation, vapor extraction, and solidification/stabilization. Specific delivery and recovery systems for in situ treatment systems identified by EPA include hydraulic fracturing, radial well drilling, ultrasonic methods, kerfing, jet-induced slurry methods, carbon dioxide injection, hot brine injection, and cyclic pumping (Chambers et al., 1990/T13-9).

12.2.5 Monitoring Program

A mass balance approach to monitoring, the fourth element in the methodology (Figure 12-3), can be performed at laboratory, pilot, and field scales. Monitoring efforts can be focused on the appropriate environmental phase to evaluate treatment effectiveness for specific chemicals. If a comprehensive and thorough evaluation of a specific contaminated system has been conducted, not all chemicals may need to be monitored in each phase. Specific chemicals will be associated with specific phases; therefore, a monitoring plan can be designed that is chemical/phase specific. This approach also focuses analytical efforts so that methods of development are chemical- and phase-specific.

The level of contamination associated with a particular treatment technology requires monitoring. In addition, the treatment system components, including delivery and recovery systems, maintenance, and structures such as infiltration galleries must be monitored.

---

[11] Lynch, J. and B.R. Genes. 1989. Land Treatment of Hydrocarbon Contaminated Soils. In: Petroleum Contaminated Soils, Vol. 1: Remediation Techniques, Environmental Fate, and Risk Assessment, P.T. Kostecki and E.J. Calabrese (eds.), Lewis Publishers, Chelsea, MI, pp. 163-174.

Ross, D., T.P. Marziarz, and A.L. Bourquin. 1988. Bioremediation of Hazardous Waste Sites in the USA: Case Histories. In: Superfund '88, Hazardous Materials Control Research Institute, Silver Spring, MD, pp. 395-397.

12.2.6 Health and Safety Considerations

A health and safety plan should be part of the sampling plan. The past history of the site and the chemicals used on the site determine the level of personal protection required. The appropriate level of protection should be determined prior to initial entry to the site and changed as more information is obtained. The **Occupational Safety and Health Guidance Manual for Hazardous Waste Site Activities** (NIOSH/OSHA/USCG/EPA, 1985/T12-7) describes the levels of personal protective clothing and equipment and their applications. Table 12-7 identifies other major references addressing health and safety aspects of investigations and remediation of contaminated sites.

12.3 Selection of Treatment Methods

Table 12-4 provides a matrix of general remedial technology categories for eighteen specific site problems, and also identifies sections in U.S. EPA (1985) where more detailed information can be obtained about each category. The number of possible treatment methods for contaminated soil and ground-water borders on being bewildering. Table 12-5 provides a matrix of major types of organic and inorganic contaminants and sixteen major categories of treatment technologies for soils and sludges that can be used as a guide for initial screening of potential technologies for a particular situation. Tables 13-1 through 13-4 and Tables 14-3 through 14-7 provide more detailed information on specific types of soil and ground-water treatment techniques, respectively.

12.3.1 Utility of Mathematical Models

A critical and cost-effective use of modeling in treatment (train) selection and evaluation is for analysis of proposed or alternative future conditions; i.e., the model is used as a management or decision-making tool to help answer "what if" questions (Donigian and Rao, 1986a/T10-6). Models also may be used to approximate and estimate the rates and extent of treatment that may be expected at the field-scale under varying conditions. Attempting to answer such questions through data collection programs would be expensive and practically impossible in many situations. For example, information can be generated to evaluate the effects of using different approaches for enhancing microbial activity and for accelerating biodegradation and detoxification of the contaminated area by altering environmental conditions that affect microbial activity. Therefore, modeling may be used to assist in the design of treatability studies for considering and evaluating the application of different treatment technologies, and therefore to assist in focusing available resources (time and money). Section 12.2.2 (Assessment of Problem) gives some examples of mathematical models for vadose zone and ground-water remediation and Chapter 10 addresses modeling in more detail.

Table 12-4 Matrix of Remedial Technology Categories for Specific Site Problems (U.S. EPA, 1985)

Site Problem	Surface Water Controls (See Section 3)	Air Pollution Controls (See Section 4)	Leachate and Groundwater Controls (See Section 5)	Gas Migration Control (See Section 6)	Waste and Soil Excavation and Removal and Land Disposal (See Section 7)	Contaminated Sediments Removal and Containment (See Section 8)	In-Situ Treatment (See Section 9)	Direct Waste Treatment (See Section 10)	Contaminated Water Supply and Sewer Line Controls (See Section 11)
Volatilization of chemicals into air		•							
Hazardous particulates released to atmosphere		•							
Dust generation by heavy construction or other site activities		•							
Contaminated site run-off	•								
Erosion of surface due to wind or water	•								
Surface seepage of leachate	•								
Flood hazard or contact of surface water body with wastes	•								
Leachate migrating vertically or horizontally			•				•		
High water table which may result in groundwater contamination or interfere with other remedial technologies			•						
Precipitation infiltrating into site to form leachate	•		•						
Evidence of methane or toxic gases migrating laterally underground				•					
On-site waste materials in-non-disposed form: drums, lagooned waste, wastepiles					•		•	•	
Contaminated surface water, groundwater or other aqueous or liquid waste					•		•	•	
Contaminated soils					•		•	•	
Toxic and/or hazardous gases which have been collected								•	
Contaminated stream banks and sediments					•	•		•	
Drinking water distribution system contamination									•
Contaminated sewer lines									•

Table 12-5 Treatment Technology Screening Guide for Different Contaminants in Soils and Sludges (U.S. EPA, 1989c)

Contaminant	A.1-1 Fluidized bed incineration	A.2-1 Rotary kiln incineration	A.3-1 Infrared thermal treatment	A.5-1 Pyrolysis-incineration	A.6-1 Vitrification	B.1-1 Chemical extraction	B.2-1 In situ chemical treatment	B.3-1 Soil washing	B.4-1 In situ soil flushing	B.5-1 Glycolate dechlorination	B.6-1 Low temperature thermal stripping	B.7-1 In situ vacuum/steam extraction	B.8-1 Stabilization/solidification	B.10-1 In situ vitrification	C.1-1 Biodegradation	C.2-1 In situ biodegradation
ORGANIC Halogenated volatiles	◐	●	●	◐	◐	◐	○	◐	◐	◐	◐	◐	●	◐	◐	◐
Halogenated semivolatiles	◐	●	●	◐	◐	◐	○	◐	◐	◐	○	◐	◐	◐	◐	◐
Nonhalogenated volatiles	◐	●	●	◐	◐	◐	○	◐	◐	◐	◐	●	◐	◐	◐	◐
Nonhalogenated semivolatiles	◐	●	●	◐	◐	◐	○	◐	◐	◐	○	◐	◐	◐	◐	◐
PCBs	◐	●	●	◐	◐	◐	○	◐	◐	◐	○	○	◐	◐	◐	◐
Pesticides	◐	●	◐	◐	◐	◐	○	◐	◐	◐	◐	○	◐	◐	◐	◐
Organic cyanides	◐	●	◐	◐	◐	◐	◐	◐	○	○	○	○	◐	◐	◐	◐
Organic corrosives	◐	●	◐	◐	◐	◐	◐	◐	◐	○	○	○	◐	◐	X	X
INORGANIC Volatile metals	X	X	X	○	X	○	○	◐	◐	○	○	○	●	◐	X	X
Nonvolatile metals	○	○	○	○	◐	○	○	◐	◐	○	○	○	●	◐	X	X
Asbestos	○	○	○	◐	●	○	○	○	○	○	○	○	●	◐	○	○
Radioactive materials	○	○	○	○	◐	○	○	○	○	○	○	○	●	●	X	X
Inorganic corrosives	○	○	○	○	◐	○	◐	◐	◐	○	○	○	●	◐	X	X
Inorganic cyanides	◐	◐	◐	◐	◐	○	◐	◐	◐	○	○	○	●	◐	X	X
Oxidizers	◐	●	◐	◐	◐	X	◐	◐	○	○	○	○	◐	◐	X	X
Reducers	◐	●	◐	◐	◐	X	◐	◐	○	○	○	○	◐	◐	X	X

* Do not use this matrix table alone.
Please refer to the cited appendices for guidance

LEGEND
● Demonstrated effectiveness
◐ Potential effectiveness
○ No effectiveness
X Potential adverse impacts to process or environment

12.3.2 Treatability Studies

Treatability studies can be used for evaluating and comparing rate and extent of remediation among several technologies and also to provide specific information about the potential application of treatment technologies at field scale. Treatability studies can be conducted in laboratory microcosms or bench-scale reactors, pilot-scale facilities, or in the field. Laboratory treatability studies are generally screening studies used to (1) establish the validity of a technology, (2) generate data that can be used as indicators of potential to meet performance goals, and (3) identify parameters for investigation during bench- or pilot-scale testing. Laboratory treatability studies are generally not appropriate for generating design or cost data (U.S. EPA, 1989b/T12-7). Pilot-scale testing is conducted to generate information on quantitative performance, cost, and design information. Three categories of treatability testing and associated descriptions are included in Table 12-6.

Table 12-6 General Comparison of Laboratory Screening, Bench-Scale Testing, and Pilot-Scale Testing

	Laboratory Screening	Bench-Scale Testing	Pilot-Scale Testing
Type of data generated	Qualitative	Quantitative	Quantitative
Critical parameters	Several	Few	Few
No. of Replicates	Single/duplicate	Duplicate/triplicate	Triplicate or more
Study size	Jar tests or beaker studies	Bench-top (some larger)	Pilot plant (onsite or offsite)
Usual process type	Batch	Batch or continuous	Batch or continuous
Waste stream volume	Small	Medium	Large
Time required	Hours/day	Days/week	Weeks/month
Cost ($1000)	10-50	50-250	250-1,000

Source: Sims and Sims (1991a) after U.S. EPA (1989d).

Treatability study results are commonly used to provide information on rates and extent of treatment of hazardous organic constituents when mass transfer rates of potential limiting substances are not limiting the treatment. Treatability studies also usually represent optimum conditions with respect to mixing, contact of soil solid materials with waste constituents and with microorganisms, and homogeneous conditions throughout the microcosm. Therefore, treatability studies provide information concerning potential levels of treatment. Rates and extent of remediation in a prepared bed or in situ system are generally limited by accessibility and rate of mass transfer of chemical substances to the contaminated soil and removal of inhibitory microbial degradation products (Symons and Sims, 1988).[12]

Information from mass balance treatability studies, including laboratory screening-, bench- and pilot-scale studies is combined with information about site and waste characteristics to determine applications and limitations of each technology. Information obtained from treatability studies should be focused on identifying ultimate limitations to the use of a remediation technology at a specific site. Limitations are usually related to (1) time required for cleanup, (2) level of cleanup attainable, and (3) cost of cleanup (Sims et al., 1989/T13-9). Table 12-7 identifies other major references on methods for assessing treatability of contaminated materials.

12.3.3 Treatment Trains

The use of treatment trains also is important to consider in an engineering approach for using treatment techniques for subsurface site remediation. For example, vacuum extraction is known to be applicable to unsaturated sites characterized by permeable materials containing volatile chemicals. Vacuum extraction also can be used for the degradation of more semivolatile chemicals. This degradation is accomplished by providing a source of oxygen (air) to the subsurface environment microorganisms where anoxic conditions exist due to relative slow replenishment of oxygen through atmospheric diffusion. This is an example of the use of one technology for the treatment of both volatile and semivolatile chemicals in the subsurface.

Another example of the use of a treatment train for creosote-contaminated soil and ground water involved (1) product removal using a pumping system, (2) flushing with water and surfactants using pump-and-treat technology, and (3) in situ biodegradation of the residual contamination (Kuhn and Piontek, 1989).[13] Each technology was employed in the order of ease of removal of creosote from the

---

[12] Symons, B.D. and R.C. Sims. 1988. Assessing Detoxification of a Complex Hazardous Waste Using the Microtox™ Bioassay. Archives of Environmental Contamination and Toxicology 17:497-505.

[13] Kuhn, R.C. and K.R. Piontek. 1989. A Site-Specific In Situ Treatment Process Development Program for a Wood Preserving Site. Paper presented at EPA Technical Program on Oily Waste Fate, Transport, Site Characterization, and Remediation Seminar, Denver, CO, May 17-18.

subsurface. The treatment train selected was based on a site characterization to identify where the creosote was located and the mass of creosote (including pure product) associated with subsurface phases, i.e., the vadose zone and aquifer materials. The problem assessment identified the following areas of concern: (1) potential offsite migration of pure product; (2) slow leaching of low levels of creosote contaminants sorbed to soil, subsurface, and aquifer materials; and (3) presence of high molecular weight polycyclic aromatic compounds that are toxic to human health, are nonvolatile, and have very low water solubilities. Each technology was evaluated in laboratory-scale treatability tests for treatment effectiveness and for ease of application to contaminated materials obtained from the site. Engineering design and implementation was based on results of site characterization, mass balance determinations at the site, and treatability studies.

Information from treatability studies is used to prepare an approach to the engineering design and implementation of a remediation system at a specific site that combines the treatment techniques evaluated to construct an appropriate treatment train. The formulation of a treatment train for a site generally is based upon information from simulations (e.g., mathematical modeling) generated from mass balance studies, treatability studies, and site/soil characterization data.

12.4 Measurement and Interpretation of Treatment Effectiveness

Typically, subsurface samples are taken from a treatability reactor (in situ or prepared bed) from laboratory-, bench-, or pilot-scale studies, or from a field site. Waste constituents are extracted from the samples with a solvent or are thermally desorbed. Compound concentration is usually measured in the solvent extract or the thermal desorption stream using chemical instrumentation (e.g., gas or liquid chromatographs with appropriate detectors). This information is termed the "apparent loss" of the compound and refers to the observation that the compound only has disappeared from the solvent or extraction phase, but does not necessarily represent a chemical mass balance (Park et al., 1990/Footnote 4). The change in concentration of the compound in the solvent with time often is used to calculate rate and extent of decrease in concentration of the compound in soil. This information is commonly used to interpret treatment effectiveness for different technologies as well as to determine engineering strategies and management approaches, including (1) time required to attain cleanup target concentrations; and (2) effects of environmental factors or experimental variables (chemical, physical, or biological) on treatment effectiveness.

However, additional information is needed to accurately measure and interpret treatment effectiveness. In order to understand treatment mechanisms and to base the selection of treatment technologies on a rational approach, identification and measurement of distribution among the physical phases that comprise a subsurface system is necessary. In addition, the mechanisms by which a compound may be chemically altered in a subsurface system must be identified and differentiated.

Information obtained about the rate of apparent loss of chemicals from a subsurface extract can be enhanced with information about the (1) interphase transfer potential between solid and gas phases of the subsurface, and (2) knowledge of mechanisms of interactions of compounds with subsurface phases. This information then provides the basis for a more rational approach to subsurface remediation. Evaluation of remediation technology effectiveness also can be based upon specific media (solid, air) and upon specific mechanisms, such as recovery of the air phase or enhancement of abiotic destruction or biological degradation, to improve treatment. Evaluation of interphase transfer also allows characterization of routes by which chemicals may migrate from the subsurface to the multimedia environment that then may lead to human exposure. Thus, measuring treatment effectiveness based upon interphase transfer potential (a mass balance approach) is also valuable for determining risk reduction and implementing risk management strategies (Park et al., 1990/Footnote 4). The laboratory flask apparatus used for mass balance determinations (Figure 12-4) also can be used to measure and compare potential effectiveness for different treatment scenarios.

12.5 Guide to Major References

Table 12-7 provides an index of major references that address general aspects of soil and ground-water remediation in the following categories: (1) health and safety, (2) environmental engineering, (3) general references on soil and ground-water remediation, including bioremediation, (4) treatment of hazardous wastes, (5) remedial action at contaminated sites, (6) vapor phase contaminants, (7) underground storage tank and spill remediation, (8) other contaminants (inorganic, radioactive and wood preserving sites), (9) treatability studies, and (10) conferences and symposia. Many of the conference series listed in Table 5-12 also address remediation of contaminated soil and ground water. Tables 13-7 and 14-9 provide an index to major references that focus on treatment and remediation of soil and ground water, respectively.

A major source of information on new developments in technologies for remediating contaminated soil and ground water is the annual **Technology Profiles** published by U.S. EPA's Superfund Innovative Technology Evaluation (SITE) Program, which summarizes results of three SITE programs: (1) technology demonstration, (2) emerging technologies, and (3) monitoring and measurement technologies. Summary information on specific projects is available from a number of series of publications: (1) *Emerging Technology Summaries* and *Bulletins*, (2) *Technology Demonstration Summaries* and *Bulletins*, (3) *Treatability Study Bulletins*, and (4) *Site Technology Capsules*. A **Technology Evaluation** and an **Applications Analysis** report is published for each completed project in the SITE program. The sixth edition of the SITE **Technology Profiles** (U.S. EPA, 1993/T12-7) includes a list of available publications from the SITE Program, which are available at no cost from CERI as long as they are available, at which time they become available from NTIS (see Preface for information on how to obtain documents from CERI and NTIS).

Other useful information sources from U.S. EPA include: (1) *Bioremediation in the Field*, a periodic newsletter which summarizes information on bioremediation projects (available from CERI), (2) *VISITT* (Vendor Information System for Innovative Treatment Technologies) available on diskette from U.S. EPA/NCEPI, P.O. Box 42419, Cincinnati, OH 45242-0419, which is supported by an intermittent newsletter by the same name, and (3) periodic *Engineering Bulletins* and *Superfund Engineering Issue Papers*, prepared by the Risk Reduction Engineering Laboratory in Cincinnati (U.S. EPA, 1991-1992/T12-7). The Engineering Bulletins and Engineering Issue papers are typically 7 to 10 pages and available from CERI.

Table 12-7 Index to Major References on Soil and Ground Water Remediation Planning

Topic	References
Health and Safety	Training: Baldwin (1992); Field: Andrews (1990), National Safety Council (1985), NIOSH/OSHA/USCG/EPA (1985), Streng et al. (1982); Exposure Monitoring/Limits: ACGIH (1992), Ness (1991); Protective Clothing: Forsberg and Mansdorf (1989), Martin and Levine (1993), Schwope et al. (1985); Radiation: Hallenbeck (1994)
Environmental Engineering	Bailey and Ollis (1986), Corbitt (1990); See also references on general design for source controls in Table 14-9
Soil and Ground Water	Ehrenfeld and Bass (1983, 1984), Barkley (1993), Madden and Johnson (1992), Nyer (1993), Russell et al. (1992), U.S. EPA (1980, 1991-1992); Soil: See Table 13-9; Ground Water: See Table 14-9; Bioremediation: Alexander (1994), Flathman and Jerger (1993), Hinchee et al. (1994a, 1994b, 1994c), King et al. (1992), Means and Hinchee (1994), Norris et al. (1993a, 1993b), NRC (1993), Thomas et al. (1987), U.S. EPA (1992c), WEF (1994), Wilson et al. (1989); Cleanup Standards: Oliver et al. (1993)
Hazardous Waste	DeRenzo (1978), Freeman (1988), Grasso (1993), HMCRI (various dates), Madden and Johnson (1992), Martin and Johnson (1987), O'Brien and Gere Engineering (1988), Rich and Cherry (1987), WPCF/WEF (1990), Wentsel et al. (1981); Biotreatment: AWMA/EPA (1989), Levin and Gealt (1993), Lewandowksi et al. (1989), Omenn (1988), Slonim et al. (1985); Drum/Materials Handling: PEI Associates (1993), Wagner et al. (1986b); Land Treatment: Brown et al. (1983), Loehr et al. (1985), U.S. EPA (1986a); Chemical Spills: Andrews (1992), Hosty and Foster (1992), Pilie et al. (1975), Yang and Bye (1979); Case Studies: Allen et al. (1987)
Remedial Action	A.D. Little (1983), Barkley (1993), Bartenfelder et al. (1992), Ehrenfeld and Bass (1983, 1984), U.S. EPA (1985, 1987a-data requirements, 1987b-USTs, 1988-interim measures guidance, 1989a), Wagner et al. (1986a); Contracting: Erickson (1992); Construction Quality Management: Richardson (1992), U.S. EPA (1986c); Costs: Lippitt et al. (1986), Yang et al. (1988-costs); Source Controls: See Table 14-9; State Guidance Documents: California Department of Health Services (1986), California LUFT Task Force (1988), Dixon (1992), Simmons (1989); Case Studies: Cockerin and Furmin (1988), Kingsbury and May (1986), Neely et al. (1981), U.S. EPA (1984, 1989g); Alternative/Innovative Technologies: Murdoch et al. (1988), U.S. EPA (1991a, 1992d, 1993)

Remediation Planning

Table 12-7 (cont.)

Topic	References
Vapor Phase	Landfill Gases/Methane: Campbell et al. (1991), Geyer (1972), Ghassemi et al. (1985), Horz (1986), Pohland and Harper (1986); Hazardous Waste Volatilization: Dupont and Reineman (1986)
Hydrocarbon/Petroleum/ UST/Spill Remediation	API (1980, 1982), Calabrese and Kostecki (1992), California LUFT Task Force (1988), Cole (1994), Dixon (1992), FDER (1992), Heard et al. (1986), Hinchee et al. (1994c), Kostecki and Calabrese (1991), Noonan and Curtis (1989/T14-9), Preslo et al. (1988), Simmons (1989), Thomas et al. (1987), U.S. EPA (1987b), Wilson et al. (1989)
Other Contaminants	Inorganics: Means and Hinchee (1994-metals), SAI Corp (1992-arsenic and mercury); Radioactive: See Table 13-9; Wood Preserving Sites: Barth et al. (1990), Sudell et al. (1992)
Treatability Studies	U.S. EPA (1986a, 1989b, 1991b, 1992a, 1992b); Batch-Tests: Roy et al. (1992)
Conferences/Symposia	Series: HMCRI Bioremediation Series (1988-1989), HMCRI Superfund Series (1980-1992), HMCRI RCRA Series (1984-1990), HMCRI FER Series (1992-1993), NWWA/NGWA Aquifer-NOAC Series (1981-1993), NWWA/API (1984-1992), U.S. EPA Waste Disposal Symposium Series (1975-1993); University of Massachusetts Petroleum/Hydrocarbon Contaminated Soil Series: Calabrese and Kostecki (1989-3rd, 1991-5th), Kostecki and Calabrese (1989-2nd, 1990-4th); West Coast Hydrocarbon Contaminated Soils and Groundwater Series: Calabrese and Kostecki (1992-2nd), Kostecki and Calabrese (1991-1st); Other Conferences: AWMA/EPA (1990-contaminated soils), Kostecki and Calabrese (1992-diesel fuel contaminated soil), National Center for Ground Water Research (1992), McCarthy and Wobber (1993), Schmidt (1985), TNO/BMFT (1985, 1989-contaminated soils)

Table 12-7 References (Appendix F contains references for figure and table sources)

A.D. Little, Inc. 1983. Handbook for Evaluating Remedial Action Plans. EPA/600/2-83/076 (NTIS PB84-118249).

Allen, C., M. Branscome, C. Northheim, K. Leese, and S. Harkins. 1987. Case Studies of Hazardous Waste Treatment to Remove Volatile Organics. EPA/600/2-87/094 (NTIS PB88-125983).

Alexander, M. 1994. Biodegradation and Bioremediation. Academic Press, New York, NY, 436 pp.

American Conference of Governmental Industrial Hygienists (ACGIH). 1992. 1992-1993 Threshold Limit Values for Chemical Substances and Physical Agents and Biological Exposure Indices. ACGIH, Cincinnati, OH, 129 pp.

American Petroleum Institute (API). 1980. Underground Spill Cleanup Manual. API Publication 1628, API, Washington, DC.

American Petroleum Institute (API). 1982. The Migration of Petroleum Products in Soil and Groundwater—Principles and Countermeasures. Publication 4149, API, Washington, DC.

Andrews, L.P. (ed.). 1990. Worker Protection During Hazardous Waste Remediation. Van Nostrand Reinhold, New York, NY, 389 pp.

Andrews, L.P. (ed.). 1992. Emergency Responder Training Manual for the Hazardous Materials Technician. Van Nostrand Reinhold, New York, NY, 505 pp.

AWMA/EPA. 1989. Proceedings of the International Symposium on Hazardous Waste Treatment: Biosystems for Pollution Control. Air and Waste Management Association, Pittsburgh, PA.

AWMA/EPA. 1990. Proceedings of the International Symposium on Hazardous Waste Treatment: Treatment of Contaminated Soils. Air and Waste Management Association, Pittsburgh, PA.

Bailey, J.E. and D.F. Ollis. 1986. Biochemical Engineering Fundamentals, 2nd ed. McGraw-Hill, New York, NY.

Baldwin, D.A. 1992. Safety and Environmental Training: Using Compliance to Improve Your Company. Van Nostrand Reinhold, New York, NY, 256 pp.

Barkley, N.P. 1993. Pilot Study on Demonstration of Remedial Action Technologies for Contaminated Land and Groundwater, Vol. 1 and Vol. 2 (Appendices in two parts). EPA/600/R-93/012a,b&c (Vol. 1, NTIS PB93-218238; Vol. 2 Part 1, PB93-218246, pp. 1 to 662; Vol. 2 Part 2, PB93-218253, pp. 663 to 1389). [Results of 5-year NATO Committee on the Challenges of Modern Society (CCMS) pilot study]

Remediation Planning

Bartenfelder, D., R. Sims, H. Compton, W. Grube, L. Murdoch, and R. Dupont. 1992. RCRA Corrective Action Stabilization Technologies Proceedings. EPA/625/R-92-014, 77 pp. Available from CERI.* [Chapter 4 addresses covers, slurry walls, grouting, and dynamic compaction, Chapter 5 cover recovery of liquids, and Chapter 6 covers vapor extraction and bioventing]

Barth, E., J. Matthews, and R. Wilhelm. 1990. Approaches for Remediation of Uncontrolled Wood Preserving Sites. EPA/625/7-90/001, 21 pp. Available from CERI.*

Brown, K.W. et al. 1983. Hazardous Waste Land Treatment, Rev. ed. EPA 530/SW-874 (NTIS PB89-179014).

Calabrese, E.J. and P.T. Kostecki (eds.). 1989. Petroleum Contaminated Soils, Vol. 2. Lewis Publishers, Chelsea, MI, 515 pp. [36 papers on remediation techniques, environmental fate, and risk assessment]

Calabrese, E.J. and P.T. Kostecki (eds.). 1991. Hydrocarbon Contaminated Soils, Volume I. Lewis Publishers, Chelsea, MI, 747 pp. [46 papers covering remediation, environmental fate, risk assessment, analytical methods, and regulatory considerations]

Calabrese, E.J. and P.T. Kostecki (eds.). 1992. Hydrocarbon Contaminated Soils and Groundwater, Volume 2. Lewis Publishers, Chelsea, MI, 558 pp. [31 papers on sampling and site assessment, environmental fate and modeling, and remediation]

Calabrese, E.J. and P.T. Kostecki (eds.). 1993. Principles and Practices for Petroleum Contaminated Soils. Lewis Publishers, Chelsea, MI, 658 pp. [27 contributed chapters covering analysis and testing, environmental fate and modeling, remediation, and health assessment]

California Department of Health Services. 1986. The California Site Mitigation Decision Tree Manual. Toxic Substance Control Division, Department of Health Services, Sacramento, CA.

California LUFT Task Force. 1988. Leaking Underground Fuel Tank Field Manual: Guidelines for Site Assessment, Cleanup and Underground Storage Tank Closures. State of California, Leaking Underground Fuel Tank Task Force, State Water Resource Board, Sacramento, CA.

Campbell, D., D. Epperson, L. Davis, R. Peer, and W. Gray. 1991. Analysis of Factors Affecting Methane Gas Recovery from Six Landfills. EPA/600/2-91/055 (NTIS PB92-101351).

Cockerin, S.R. and C. Furman. 1988. Case Studies Addendum: 1-8 Remedial Response at Hazardous Waste Sites. EPA/540/2-88/001 (NTIS PB88-204284).

Cole, G.M. 1994. Assessment and Remediation of Petroleum-Contaminated Sites. Lewis Publishers, Boca Raton, FL, 368 pp.

Corbitt, R.A. (ed.). 1990. Standard Handbook of Environmental Engineering. McGraw-Hill, New York, NY, 1152 pp. [Covers air quality control, water supply, wastewater disposal, solid waste management, stormwater and hazardous waste management]

(Table 12-7 Remediation Planning References)

DeRenzo, D. (ed.). 1978. Unit Operations for Treatment of Hazardous Wastes. Noyes Data Corporation. Park Ridge, NJ.

Dixon, R.A. 1992. Site Assessment, Remediation, and Closure Under the Oregon UST Matrix. In: Hydrocarbon Contaminated Soils and Groundwater, Volume 2, E.J. Calabrese and P.T. Kostecki (eds.), Lewis Publishers, Chelsea, MI, pp. 201-210.

Dupont, R.R. and J.A. Reineman. 1986. Evaluation of Volatilization of Hazardous Constituents at Hazardous Waste Land Treatment Sites. EPA/600/2-86/071 (NTIS PB86-233939).

Ehrenfeld, J. and J. Bass. 1983. Handbook for Evaluation Remedial Action Technology Plans. EPA/600/2-83-076.

Ehrenfeld, J. and J. Bass. 1984. Evaluation of Remedial Action Unit Operations of Hazardous Waste Disposal Sites. Pollution Technology Review No. 110. Noyes Publications, Park Ridge, NJ. [Covers control and treatment technologies for ground water/leachate, surface water and contaminated soil and waste materials]

Erickson, R.L. 1992. Environmental Remediation Contracting. John Wiley & Sons, New York, NY, 456 pp.

Flathman, P.E. and D.E. Jerger (eds.). 1993. Bioremediation: Field Experience. Lewis Publishers, Boca Raton, FL, 544 pp.

Florida Department of Environmental Regulation (FDER). 1992. Guidelines for the Assessment and Remediation of Petroleum Contaminated Sites. Division of Waste Management, FDER, 52 pp.

Forsberg, K. and S.Z. Mansdorf. 1989. Quick Selection Guide to Chemical Protective Clothing, 2nd ed. Van Nostrand Reinhold, New York, NY, 60 pp.

Freeman, H.M. (ed.). 1988. Standard Handbook of Hazardous Waste Treatment and Disposal. McGraw-Hill, New York, NY, 992 pp.

Geyer, J.A. 1972. Landfill Decomposition Gases: An Annotated Bibliography. EPA SW-72-1-1 (NTIS PB213-487). [48 articles]

Ghassemi, M, K. Crawford, and M. Haro. 1985. Leachate Collection and Gas Migration and Emission Problems at Landfills and Surface Impoundments. EPA/600/2-86/017 (NTIS PB86-162104).

Grasso, D. 1993. Hazardous Waste Site Remediation: Site Control. Lewis Publishers, Boca Raton, FL, 624 pp.

Hallenbeck, W.H. 1994. Radiation Protection. Lewis Publishers, Boca Raton, FL, 288 pp.

Hazardous Materials Control Research Institute (HMCRI) Monograph Series. Various dates. 26 total volumes as of 1989. HMCRI, Greenbelt, MD. [Pertinent volumes include: Contaminated Groundwater Control (3 Volumes), In Situ Treatment, Site Remediation (3 Volumes)]

(Table 12-7 Remediation Planning References)

Hazardous Materials Control Research Institute (HMCRI) Bioremediation Series. 1988-1989. Proceedings of the National Conference on Bioremediation--The Use of Genetically Engineered or Adapted Microorganisms in the Treatment of Hazardous Waste (1988); Proceedings of the 2nd National Conference on Biotreatment: The Use of Microorganisms in the Treatment of Hazardous Materials and Hazardous Wastes (1989). HMCRI, Greenbelt, MD.

Hazardous Materials Control Research Institute (HMCRI) Superfund Series. 1980-1992. National Conferences on Management of Uncontrolled Hazardous Waste Sites (1st through 7th—1980-1986); Superfund (8th through 11th— '87, '89, '90); Hazardous Materials Control (HMC)-Superfund (12th and 13th—HMC/Superfund '91 and '92). HMCRI, Greenbelt, MD. [Proceedings typically contain more than 100 papers, many of which relate to ground-water remediation]

Hazardous Materials Control Research Institute (HMCRI) RCRA Series. 1984-1990. Proceedings of the National Conference on Hazardous Wastes and Environmental Emergencies (1st and 2nd, 1984-1985); Proceedings of the National Conference on Hazardous Wastes and Hazardous Materials (3rd through 7th, 1986-1990; 6th and 7th may be cited as HWHM '89 and HWHM '90). HMCRI, Greenbelt, MD. [Proceedings contain from 80 to 250 papers, many of which relate to ground-water remediation]

Hazardous Materials Control Research Institute (HMCRI) FER Series. 1992-1993. Proceedings of Conference on Federal Environmental Restoration (1st, FER '92; 2nd FER '93). HMCRI, Greenbelt, MD.

Heard, D.B., L.M. Krasner, and B.G. Vincent. 1986. Documentation and Analysis for Prevention and Control of Hazardous Material Spills. EPA/600/2-86/016 (NTIS PB86-156775).

Hinchee, R.E., D.B. Anderson, F.B. Metting, Jr. and G.D. Sayles (eds.). 1994a. Applied Biotechnology for Site Remediation. Lewis Publishers, Boca Raton, FL, 512 pp.

Hinchee, R.E., A. Leeson, L. Semprini, and S.K. Ong (eds.). 1994b. Bioremediation of Chlorinated and Polycyclic Aromatic Hydrocarbon Compounds. Lewis Publishers, Boca Raton, FL, 560 pp.

Hinchee, R.E., B.C. Alleman, R.E. Hoeppel, and R.N. Miller. 1994c. Hydrocarbon Bioremediation. Lewis Publishers, Boca Raton, FL, 496 pp.

Horz, R.C. 1986. Geotextiles for Drainage, Gas Venting, and Erosion Control at Hazardous Waste Sites. EPA/600/2-86/085 (NITS PB87-129557).

Hosty, J.W. and P. Foster. 1992. A Practical Guide to Chemical Spill Response. Van Nostrand Reinhold, New York, NY, 208 pp.

King, R.B., G.M. Long, and J.K. Sheldon. 1992. Practical Environmental Bioremediation. Lewis Publishers, Chelsea, MI, 300 pp.

Kingsbury, G.L. and R.M. May. 1986. Reclamation and Redevelopment of Contaminated Land: Volume 1. U.S. Case Studies. EPA/600/2-86/066 (NTIS PB87-142121).

(Table 12-7 Remediation Planning References)

Kostecki, P.T. and E.J. Calabrese (eds). 1989. Petroleum Contaminated Soils, Vol. 1. Lewis Publishers, Chelsea, MI, 356. [24 papers on remediation techniques, environmental fate, and risk assessment]

Kostecki, P.T. and E.J. Calabrese (eds). 1990. Petroleum Contaminated Soils, Vol. 3. Lewis Publishers, Chelsea, MI, 423 pp. [28 papers on remediation techniques, environmental fate, and risk assessment]

Kostecki, P.T. and E.J. Calabrese (eds.). 1991. Hydrocarbon Contaminated Soils and Groundwater, Volume 1. Lewis Publishers, Chelsea, MI, 354 pp. [21 papers on analysis, fate, environmental and public health effects, and remediation]

Kostecki, P.T. and E.J. Calabrese (eds.). 1992. Contaminated Soils: Diesel Fuel Contamination. Lewis Publishers, Chelsea, MI, 227. [11 papers focusing on remediation]

Lewandowski, G., P. Armenante, and B. Baltzis (eds.). 1989. Biotechnology Applications in Hazardous Waste Treatment. Engineering Foundation, New York, NY.

Levin, M. and M. Gealt. 1993. Biotreatment of Industrial and Hazardous Wastes. McGraw-Hill, New York, NY, 321 pp.

Lippitt, J., J. Walsh, M. Scott, and A. DiPuccio. 1986. Costs of Remedial Actions at Uncontrolled Hazardous Waste Sites: Worker Health and Safety Considerations. EPA/600/2-86/037 (NTIS PB86-176344).

Loehr, R.C., J.H. Martin, E.F. Neuhauser, R.A. Norton, and M.R. Malecki. 1985. Land Treatment of an Oily Waste--Degradation, Immobilization, and Bioaccumulation. EPA/600/2-85/009 (NTIS PB85-166353).

Madden, M.P. and W.I. Johnson. 1992. Installation Restoration and Hazardous Waste Control Technologies, 1992 Ed. USATHAMA CETHA-TS-CR-92053. U.S. Army Toxic and Hazardous Materials Agency, Aberdeen Proving Ground, MD, 388 pp.

Martin, E.J. and J.H. Johnson, Jr. (eds.). 1987. Hazardous Waste Management Engineering. Van Nostrand Reinhold, New York, NY.

Martin, W.F. and S.P. Levine (eds.). 1993. Protecting Personnel at Hazardous Waste Sites, 2nd ed. Butterworth-Heinemann, Stoneham, MA, 450 pp.

McCarthy, J.F. and F.J. Wobber (eds.). 1993. Manipulation of Groundwater Colloids for Environmental Restoration. Lewis Publishers, Chelsea, MI, 400 pp.

Means, J.L. and R.E. Hinchee (eds.). 1994. Emerging Technology for Bioremediation of Metals. Lewis Publishers, Boca Raton, FL, 160 pp.

Murdoch, L., B. Patterson, G. Losonsky, and W. Harrar. 1988. Innovative Technologies of Delivery or Recovery: A Review of Current Research and a Strategy for Maximizing Future Investigations. EPA/600/2-89/066 (NTIS PB90-156225).

(Table 12-7 Remediation Planning References)

National Center for Ground Water Research. 1992. Proceedings of the Subsurface Restoration Third International Conference of Ground Water Quality Research. Rice University, Houston, TX, 343 pp. [Extended abstracts]

National Research Council (NRC). 1993. In Situ Bioremediation: When Does It Work?. National Academy Press, Washington, DC, 224 pp.

National Safety Council. 1985. Accident Prevention Manual for Industrial Operations.

National Water Well Association (NWWA/NGWA**) Aquifer-NOAC Series. 1981-1993. Proceedings of the National Ground Water Quality Monitoring Symposium and Exposition (1981); Proceedings of the 2nd to 6th National Symposium on Aquifer Restoration and Ground Water Monitoring (1982-1987); Proceedings of the 1st to 7th National Outdoor Action Conference (NOAC) on Aquifer Restoration, Ground Water Monitoring and Geophysical Methods (1987-1993). [Total number of papers usually ranges from 40 to 80, with 20 to 30% related to ground-water remediation]

National Water Well Association/American Petroleum Institute (NWWA**/API). 1984-1992. Proceedings of the NWWA/API Conference on Petroleum Hydrocarbons and Organic Chemicals in Ground Water--Prevention, Detection, and Restoration. NWWA, Dublin, OH. [Typically contain 35 to 50 papers, many of which relate to ground-water remediation]

Neely, D. et al. 1981. Remedial Actions at Hazardous Waste Sites: Survey and Case Studies. EPA/430/9-81-05. [Survey of 169 sites and 9 case studies]

Ness, S.A. 1991. Air Monitoring for Toxic Exposures: An Integrated Approach. Van Nostrand Reinhold, New York, NY, 534 pp.

NIOSH/OSHA/USCG/EPA. 1985. Occupational Safety and Health Guidance Manual for Hazardous Waste Site Activities. DHHS (NIOSH) Publication No. 85-115, U.S. Government Printing Office, Washington, DC. [Prepared by National Institute for Occupational Safety and Health, Occupational Safety and Health Administration, U.S. Coast Guard, and U.S. Environmental Protection Agency]

Norris, R.D. et al. 1993a. Handbook of Bioremediation. Lewis Publishers, Boca Raton, FL, 272 pp.

Norris, R.D. et al. 1993b. In-Situ Bioremediation of Ground Water and Geological Material: A Review of Technologies. EPA/R-93/124 (NTIS PB93-215564). [13 authors]

Nyer, E.K. 1993. Practical Techniques for Groundwater and Soil Remediation. Lewis Publishers, Chelsea, MI, 214 pp.

O'Brien & Gere Engineering. 1988. Hazardous Waste Site Remediation: The Engineering Perspective. Van Nostrand Reinhold, New York, NY.

Oliver, T., P. Kostecki, and E. Calabrese. 1993. State Summary of Soil and Groundwater Cleanup Standards. Soils Magazine, December, various pagings.

(Table 12-7 Remediation Planning References)

Omenn, G.S. (ed.) 1988. Environmental Biotechnology-Reducing Risks from Environmental Chemicals Through Biotechnology. Plenum Press, New York, NY, 505 pp.

PEI Associates. 1991. Survey of Materials-Handling Technologies Used at Hazardous Waste Sites. EPA/540/2-91/010 (NTIS PB91-186924), 225 pp.

Pilie, K. et al. 1975. Methods to Treat, Control, and Monitor Spilled Hazardous Materials. EPA/670/2-75/042 (NTIS PB243-386), 149 pp.

Pohland, F.G. and S.R. Harper. 1986. Critical Review and Summary of Leachate and Gas Production from Landfills. EPA/600/2-86/073 (NTIS PB86-240181)

Preslo, L.M., J.B. Robertson, D. Dworking, E.J. Fleischer, P.T. Kostecki, and E.J. Calabrese. 1988. Remedial Technologies for Leaking Underground Storage Tanks. Lewis Publishers, Chelsea, MI. [Covers technical, environmental and economic aspects of 13 remedial technologies for leaking underground storage tanks]

Rich, G. and K. Cherry. 1987. Hazardous Waste Treatment Technologies. Pudvan Publishing Co., Northbrook, IL. [Covers physical, chemical, biological, thermal, and fixation/encapsulation treatment technologies]

Richardson, G.N. 1992. Technical Guidance Document: Construction Quality Management for Remedial Action and Remedial Design Waste Containment Systems. EPA/540/R-92/073. Available from CERI.*

Roy, W.R., I.G. Krapac, S.F.J. Chou, and R.A. Griffin. 1992. Batch-Type Procedures for Estimating Soil Adsorption of Chemicals. EPA/530/SW-87/006F (NTIS PB92-146190), 100 pp.

Russell, H.H., J.E. Matthews, and G.W. Sewell. 1992. TCE Removal for Contaminated Soil and Ground Water. Ground Water Issue Paper, EPA/540/S-92-002, 10 pp.

Science Applications International (SAI) Corp. (Compiler) 1992. Arsenic and Mercury Workshop on Removal, Recovery, Treatment and Disposal--Abstract Proceedings. EPA/600/R-92/105, 125 pp. Available from CERI.*

Schmidt, K.D. (ed.). 1985. Proceedings of a Symposium on Groundwater Contamination and Reclamation (Tucson). American Water Resources Association, Bethesda, MD. [Contains 5 papers on ground-water reclamation]

Schwope, A.D., R.P. Costas, J.O. Jackson, and D.J. Weitzman. 1985. Guidelines for the Selection of Chemical Protective Clothing, 2nd ed.: Vol. 1, Field Guide; Vol. II, Technical and Reference Manual. American Conference of Governmental Industrial Hygienists, Cincinnati, OH. (513/661-7881).

Simmons, B. (ed.). 1989. Leaking Underground Fuel Tank/Underground Storage Tank Closure. California Leaking Underground Fuel Tank Task Force, State Water Resources Control Board, Sacramento, CA.

Slonim, Z., L.-T. Lien, W.W. Eckenfelder, and J.A. Roth. 1985. Anaerobic-Aerobic Treatment Process for the Removal of Priority Pollutants. EPA/600/2-85/077 (NITS PB85-226900).

Streng, D.R. et al. 1982. Hazardous Waste Sites and Hazardous Substance Emergencies Worker Bulletin. DHHS (NIOSH Publication No. 83-100, U.S. Department of Health and Human Services/National Institute for Occupational Safety and Health, U.S. Government Printing Office, Washington, DC, 22 pp.

Sudell, G., A. Slevakumar, and G. Wolf. 1992. Contaminants and Remedial Options at Wood Preserving Sites. EPA/600/R-92/182.

Thomas, J.M., M.D. Lee, P.B. Bedient, R.C. Borden, L.W. Canter, and C.H. Ward. 1987. Leaking Underground Storage Tanks: Remediation with Emphasis on In-Situ Biorestoration. EPA/600/2-87/008 (NTIS PB87-168084)

TNO/BMFT. 1985. First International Conference on Contaminated Soil. Kluwer Academic Publishers, Hingham, MA.

TNO/BMFT. 1989. Second International Conference on Contaminated Soil. Kluwer Academic Publishers, Hingham, MA.

U.S. Environmental Protection Agency (EPA) Waste Disposal Research Symposium Series. 1975-1993. 1st (1975) Gas and Leachate from Landfill, EPA 600/9-76-004 (PB251 161); 2nd (1976) Residual Management by Land Disposal, EPA 600/9-760-15; 3rd (1977) Management of Gas Leachate from Landfills, EPA 600/9-77-026 (PB272 595); 4th (1978) Land Disposal of Hazardous Wastes, EPA 600/9-78-016 (PB286 956); 5th (1979) Municipal Solid Waste Land Disposal, EPA 600/9-79-023a (PB80-114291; 6th (1980) Disposal of Hazardous Waste, EPA 600/9-80-010 (PB80-175094); 7th (1981) Land Disposal: Municipal/Hazardous Solid Waste EPA 600/9-81-002a,b (PB81-173874 and PB81-173882); 8th (1982) Land Disposal of Hazardous Waste, EPA 600/9-82-002 (PB82-173022); 9th (1983), EPA 600/9-83-018 (PB84-188777); 10th (1984) EPA 600/9-84-007 (PB84-177799); 11th (1985), EPA 600/9-85-013 (PB85-196376); 12th (1986) Land Disposal, Remedial Action, Incineration and Treatment of Hazardous Waste, EPA/600/9-86/022 (PB87-119491); 13th (1987), EPA/600/9-87/015 (PB87-233151); 14th, (1988) EPA 600/9-88/021 (PB89-174403); 15th (1989), EPA 600/9-90/006; 16th (1990), EPA/600/9-90/037 (PB87-233151); 17th (1991), Remedial Action, Treatment and Disposal of Hazardous Waste, EPA/600/9-91/002; 18th (1992), Risk Reduction Engineering Laboratory Research Symposium Abstract Proceedings, EPA/600/R-92/028; 19th (1993), RREL Hazardous Waster Research Symposium Abstract Proceedings, EPA/600/R-93/040.

U.S. Environmental Protection Agency (EPA). 1980. Treatability Manual: 3 Vols. EPA/600/8-80-042a-c. [Volume III covers technologies for control and removal of pollutants]

U.S. Environmental Protection Agency (EPA). 1984. Case Studies No. 1-23: Remedial Response at Hazardous Waste Sites. EPA/540/2-84-002b.

U.S. Environmental Protection Agency (EPA). 1985. Handbook for Remedial Action at Waste Disposal Sites (Revised). EPA 625/6-85-006 (NTIS PB87-201034). Supersedes 1982 report with same title (EPA/625/6-82-006).

(Table 12-7 Remediation Planning References)

U.S. Environmental Protection Agency (EPA). 1986a. Permit Guidance Manual for Hazardous Land Treatment Demonstrations. EPA 530/SW-86/032 (NTIS PB86-229184).

U.S. Environmental Protection Agency (EPA). 1986c. Construction Quality Assurance for Hazardous Waste Land Disposal Facilities; Technical Guidance Document. EPA/530-SW-86-031 (NTIS PB87-132825).

U.S. Environmental Protection Agency (EPA). 1987a. Data Requirements for Selecting Remedial Action Technology. EPA 600/2-87/001, 168 pp. Available from CERI.*

U.S. Environmental Protection Agency (EPA). 1987b. Underground Storage Tank Corrective Action Technologies. EPA 625/6-87/015 (NTIS PB171 278).

U.S. Environmental Protection Agency (EPA). 1988. RCRA Corrective Action Interim Measures Guidance - Interim Final. EPA/530-SW-88-029/OSWER 9902.4. Available from RCRA Hotline (see Table 5-3).

U.S. Environmental Protection Agency (EPA). 1989a. Corrective Action: Technologies and Applications. Seminar Publication EPA/625/4-89/020, 77 pp. Available from CERI.* [Chapter 4 covers containment options and Chapter 5 covers water/leachate treatment options]

U.S. Environmental Protection Agency (EPA). 1989b. Guide for Conducting Treatability Studies Under CERCLA - Interim Final. EPA/540/2-89/058 (NTIS PB90-249772), 118 pp.

U.S. Environmental Protection Agency (EPA). 1991a. Bibliography of Federal Reports and Publications Describing Alternative and Innovative Treatment Technologies for Corrective Action and Site Remediation. EPA/540/8-91/007.

U.S. Environmental Protection Agency (EPA). 1991b. Guide for Conducting Treatability Studies Under CERCLA: Aerobic Biodegradation Remedy Screening Guide. EPA/540/2-91/13a.

U.S. Environmental Protection Agency (EPA). 1991-1992. *Engineering Bulletin Series.* In Situ Soil Vapor Extraction Treatment (EPA/540/2-91/006); Thermal Desorption Treatment (EPA/540/2-91/008); Supercritical Water Oxidation (EPA/540/S-92/006); Design Considerations for Ambient Air Monitoring at Superfund Sites (EPA/540/S-92/012); Air Pathway Analysis (EPA/540/S-92/013); *Superfund Engineering Issue Series.* Treatment of Lead-Contaminated Soils (EPA/540/2-91/009); Considerations for Evaluating the Impact of Metals Partitioning During the Incineration of Contaminated Soils from Superfund Sites (EPA/540/S-92/014). Available from CERI.*

U.S. Environmental Protection Agency (EPA). 1992a. Guide for Conducting Treatability Studies Under CERCLA: Chemical Dehalogenation. EPA/540/R-92/013a (NTIS PB92-169044), 76 pp.

U.S. Environmental Protection Agency (EPA). 1992b. Guide for Conducting Treatability Studies Under CERCLA: Thermal Desorption Remedy Selection--Interim Guidance. EPA/540/R-92/074A, 38 pp. Available from CERI.*

(Table 12-7 Remediation Planning References)

Remediation Planning

U.S. Environmental Protection Agency (EPA). 1992c. Methodologies for Evaluating In-Situ Bioremediation of Chlorinated Solvents. EPA/600/R-92/042 (NTIS PB92-146943), 96 pp.

U.S. Environmental Protection Agency (EPA). 1992d. Federal Programs on Alternative and Innovative Treatment Technologies for Corrective Action and Site Remediation, 2nd ed. EPA/542/B-92/001.

U.S. Environmental Protection Agency (EPA). 1993. The Superfund Innovative Technology Evaluation Program: Technology Profiles, Sixth Edition. EPA/540/R-93/526, 424 pp. Available from CERI.* [Earlier editions include: 2nd (EPA/540/5-89/013), 3rd (EPA/540/5-90/006), 4th (EPA/540/5-91/008), 5th (EPA/540/R-92/077)]

Wagner, K., et al. 1986a. Remedial Action Technology for Waste Disposal Sites, 2nd edition. Noyes Data Corporation, Park Ridge, NJ. [Covers surface water controls, air pollution control, ground-water controls, in-situ and direct waste treatment; 10 authors]

Wagner, K., R. Wetzel, H. Bryson, C. Furman, A. Wickline, and V. Hodge. 1986b. Drum Handling Practices at Hazardous Waste Sites. EPA/600/2-86/013 (NTIS PB86-165362), 177 pp.

Water Environment Federation (WEF). 1994. In-Situ Bioremediation of Contaminated Subsurface Media. WEF, Alexandria, VA, 71 pp.

Water Pollution Control Federation (WPCF/WEF). 1990. Hazardous Waste Site Remediation. Water Environment Federation, Alexandria, VA, 150 pp.

Wentsel, R.S. et al. 1981. Restoring Hazardous Spill-Damage Areas—Technique Identification/Assessment. EPA/600/2-81/208.

Wilson, J.T., L.E. Leach, J. Michalowski, S. Vandegrift, and R. Callaway. 1989. In Situ Bioremediation of Spills from Underground Storage Tanks: New Approaches for Site Characterization, Project Design, and Evaluation of Performance. EPA/600/2-89/042 (NTIS PB89-219976).

Yang, J.T. and W.E. Bye. 1979. Methods of Preventing, Detecting, and Dealing with Surface Spills of Contaminants Which May Degrade Underground Water Sources for Public Water Systems. EPA/570/9-79/018 (NTIS PB82-204082).

Yang, E.C., D. Bauma, L. Schwartz, and J.D. Werner. 1987. Compendium of Costs of Remedial Technologies at Hazardous Waste Sites. EPA/600/2-87/087 (NTIS PB88-113477).

* See Preface for information on how to obtain documents from CERI (U.S. EPA Center for Environmental Research Information) and NTIS.

** In 1992 the National Water Well Association (NWWA) changed its name to the National Ground Water Association (NGWA).

(Table 12-7 Remediation Planning References)

CHAPTER 13

REMEDIATION OF CONTAMINATED SOILS

13.1 General Approaches to Soil Remediation 731

13.2 Soil Vacuum Extraction (SVE) 732

    13.2.1 Significant Chemical Properties 739
    13.2.2 Significant Soil Properties 740
    13.2.3 Design Considerations 742
    13.2.4 Enhanced Biodegradation 742

13.3 Bioremediation 745

    13.3.1 Approaches to In Situ Biological Treatment 746
    13.3.2 Significant Environmental Parameters 746

13.4 Other Treatment Approaches 750

    13.4.1 Sorption, Ion Exchange and Precipitation 751
    13.4.2 Solidification and Stabilization 751
    13.4.3 Soil Flushing 752

13.5 Prepared Bed Reactors 754

13.6 Guide to Major References[1] 756

13.1 General Approaches to Soil Remediation

Soil remediation techniques are applied to the vadose zone ranging from the soil rooting zone to the capillary fringe (Section 2.4.2), and to situations where the saturated zone is engineered to become unsaturated (e.g., when ground water is pumped to create an unsaturated zone). Table 12-5 in Section 12.3 provides information on the suitability of different treatment technologies for contaminated soils and sludges. Remediation techniques for contaminated soils can be broadly classified as follows: (1) physical and chemical treatment, (2) biological treatment, (3) fixation or encapsulation), and (4) thermal destruction. Tables 13-1 through 13-4 provides the following additional summary information on specific techniques in the these five categories:

---

[1] Appendix F contains citations for table and figure sources.

- Ways in which technique can be used (in situ, prepared bed or in-tank). These categories are discussed in Section 13.1).

- The function of the treatment method (detoxification, separation, volume reduction, immobilization).

- Types of wastes that are amenable to treatment using the technique.

- Possible residuals or transformation products.

- Possible limitations to the technique.

In situ treatment consists of treating contaminated soil in place, i.e., the contaminated soil is not moved from the ground. This chapter focuses on *in situ* treatment methods, with special emphasis on soil vacuum extraction (Section 13.2) and bioremediation (Section 13.3).

In a *prepared bed* system, the contaminated soil may be either (1) physically moved from its original site to a newly prepared area, which has been designed to enhance treatment and/or to prevent transport of contaminants from the site; or (2) removed from the site to a storage area while the original location is prepared for use, then returned to the bed, where treatment is accomplished. Preparation of the bed may include placement of a clay or plastic liner to retard transport of contaminants from the site or addition of uncontaminated soil to provide additional treatment medium. Treatment may be enhanced with biological and/or physical/chemical methods, as with in situ systems. Prepared bed treatment approaches are based on modifications of principles developed in the areas of land application of solid and liquid wastes and in land treatment of hazardous wastes. Section 13.5 discusses prepared bed systems in more detail.

In-tank treatment involves removing the contaminated soil and treating it in a vessel or other system designed to optimize treatment efficiency. Many specific treatment techniques can be used in situ, as prepared beds, or in-tank, while others are limited to a specific treatment category. For example, *soil flushing* and *soil washing* (Table 13-1) involve the same principles, but the first term is used for in situ treatment, whereas the latter term applies to in-tank treatment. All of the thermal treatment methods in Table 13-4 are classified as in-tank methods.

13.2 Soil Vacuum Extraction (SVE)

Soil vacuum extraction (SVE), also referred to as *forced air venting*, and *in situ air stripping*, involves extraction of air and contaminants from unsaturated soil. In contrast to a static equilibrium soil system where evaporation of a chemical is equal to the condensation of the chemical, with SVE, clean air is injected or passively flows into the unsaturated zone. Volatile chemicals then partition from soil water into soil air, with relative partitioning based on the air/water partition coefficient (K_h) or Henry's Law constant and the vapor-laden air is removed using vacuum extraction wells.

Table 13-1 Summary of Physical/Chemical Treatment Methods for Contaminated Soil

Method	Characteristics	Limitations
Soil vacuum extraction (SVE)	**Treatment Category**: In situ; prepared bed. **Function**: Separation. **Amenable Wastes**: Volatile organics and toxic metals; may be enhanced by the use of steam. **Possible Residuals/Transformation Products**: Volatile organics and volatile toxic metals.	Soil heterogeneity (e.g., permeability, texture—see Figure 13-4 for most unfavorable textures); not applicable to saturated materials or miscible compounds.
Soil washing	**Treatment Category**: In-tank. **Function**: Separation; volume reduction. **Amenable Wastes**: Organics and inorganics; most suitable for soils contaminated by only a few specific chemicals. **Possible Residuals/Transformation Products**: Extracted materials; water/flushing agent mix.	Unfavorable contaminant separation coefficients; less effective with complex mixtures of waste types and variation in waste composition. **Unfavorable Soil Characteristics**: high humic content, soil/solvent reactions, high silt and clay content, and clay soils containing semivolatiles. **Unfavorable washing fluid characteristics**: difficult recovery of solvent or surfactant, poor treatability of washing fluid, reduction of soil permeability, and high toxicity of washing fluid.
Soil flushing	**Treatment Category**: In situ. **Function**: Separation; volume reduction. **Amenable Wastes**: Organics and inorganics; most suitable for soils contaminated by only a few specific chemicals. **Possible Residuals/Transformation Products**: Extracted materials; water/flushing agent mix.	Similar to soil flushing above. Additional unfavorable soil characteristics include high soil variability. Requires containment of leachate and ground water to prevent offsite ground-water contamination.
Low temperature thermal stripping (including radio frequency heating)	**Treatment Category**: In situ; in-tank. **Function**: Separation. **Amenable Wastes**: Compounds of low water solubility and high volatility. **Possible Residuals/Transformation Products**: Off gas; spent carbon or ash from afterburner; processed soil; hazardous emissions from in situ applications.	Limited to organics with Henry's Law constant greater than 3.0×10^{-3} atm-m^{-3}/mole and boiling points less than 800°; more effective for soils with low contents of organic matter and moisture.

Table 13-1 (cont.)

Method	Characteristics	Limitations
Glycolate dechlorination	**Treatment Category**: In situ; in-tank. **Function**: Detoxification. **Amenable Wastes**: Dehalogenation of aromatic halide compounds. **Possible Residuals/Transformation Products**: Water/reagent mix; reaction products.	Heat and excess reagent required for soils with greater than 20% moisture, contaminant concentrations greater than 5%, and soils containing competing reactive metals such as aluminum.
Neutralization	**Treatment Category**: In situ; prepared bed; in-tank. **Function**: Detoxification; immobilization. **Amenable Wastes**: Waste acids and alkalies to reduce reactivity and corrosiveness. **Possible Residuals/Transformation Products**: Precipitated salts.	Compatibility of waste and treatment chemical to prevent formation of more toxic or hazardous compounds.
Oxidation	**Treatment Category**: In situ; prepared bed; in-tank. **Function**: Detoxification. **Amenable Wastes**: Cyanides and oxidizable compounds. **Possible Residuals/Transformation Products**: Oxidized reaction products.	Possible explosive reactions; production of more toxic of hazardous products; non-selective.
Photolysis	**Treatment Category**: Prepared bed. **Function**: Detoxification. **Amenable Wastes**: Dioxins, nitrated wastes. **Possible Residuals/Transformation Products**: Reaction products.	Inability of light to penetrated the soil.
Precipitation	**Treatment Category**: In situ; prepared bed; in-tank. **Function**: Separation; volume reduction; immobilization. **Amenable Wastes**: Metals; certain anions. **Possible Residuals/Transformation Products**: Precipitated metals.	Unfavorable effects on soil permeability; long-term stability unknown.
Reduction	**Treatment Category**: In situ; prepared bed; in-tank. **Function**: Detoxification. **Amenable Wastes**: Chromium, silver, and mercury. **Possible Residuals/Transformation Products**: Reduced reaction products.	Possible explosive reactions; production of more toxic or hazardous products; non-selective.

Table 13-1 (cont.)

Method	Characteristics	Limitations
Carbon adsorption	**Treatment Category**: In situ; prepared bed. **Function**: Separation; immobilization. **Amenable Wastes**: Organic wastes with high molecular weight and boiling points and low solubility and polarity. **Possible Residuals/Transformation Products**: Processed soil.	Long-term stability unknown.
Ion exchange	**Treatment Category**: In situ; prepared bed. **Function**: Separation; immobilization. **Amenable Wastes**: Metal contaminants. **Possible Residuals/Transformation Products**: Processed soil.	Selectivity/competition limitations; pH requirements.

Source: Adapted from Sims and Sims (1991b).

Table 13-2 Summary of Biological Treatment Methods for Contaminated Soil

Method	Characteristics	Limitations
Aerobic bioremediation	**Treatment Category**: In situ; prepared bed; in-tank. **Function**: Detoxification. **Amenable Wastes**: Biodegradable organic wastes (see Table 3-11). **Possible Residuals/Transformation Products**: Hazardous volatile emissions; incomplete and possibly hazardous degradation products; leachates in soil systems.	Ability to control environmental factors conducive to biodegradation; formation of more toxic or hazardous transformation products. **Prepared Bed**: areal limitation due to cost of bed preparation.
Anaerobic bioremediation	**Treatment Category**: In situ; prepared bed; in-tank. **Function**: Detoxification. **Amenable Wastes**: Certain halogenated organics (Table 3-11). **Possible Residuals/Transformation Products**: Similar to aerobic, plus carbon dioxide, methane and other gases.	May require long treatment periods; incomplete treatment, possibly requiring aerobic conditions to complete degradation process.
Biological seeding	**Treatment Category**: In situ; prepared bed; in-tank. **Function**: Detoxification. **Amenable Wastes**: Many biodegradable organic wastes. **Possible Residuals/Transformation Products**: Similar to aerobic bioremediation.	Survival and activity of organisms in introduced environment (affected by environmental factors and competition with native species).
Composting	**Treatment Category**: Prepared bed; in-tank. **Function**: Detoxification. **Amenable Wastes**: Biodegradable organic wastes. **Possible Residuals/Transformation Products**: Similar to aerobic bioremediation, plus runoff water.	Maintenance of optimum environmental conditions for biological activity; requires large amounts of compost materials mixed with only about 10% wastes.
Enzyme addition	**Treatment Category**: In situ; prepared bed; in-tank. **Function**: Detoxification. **Amenable Wastes**: Certain biodegradable organic wastes. **Possible Residuals/Transformation Products**: Similar to aerobic bioremediation.	Activity and stability of introduced enzymes in natural systems.

Source: Adapted from Sims and Sims (1991b).

Table 13-3 Summary of Fixation/Encapsulation Methods for Contaminated Soil

Method	Characteristics	Limitations
Cement solidification	**Treatment Category**: In-tank; in situ. **Function**: Storage; immobilization. **Amenable Wastes**: Metal cations, latex and solid plastic wastes. **Possible Residuals/Transformation Products**: Leachates; hazardous volatile emissions; solidified waste materials.	Incompatible with large amounts of dissolved sulfate salts or metallic anions such as arsenates or borates; setting time increased by presence of organic matter, lignite, silt or clay; requires complete and uniform mixing of soils and reagents; long-term stability unknown.
Glassification/ vitrification	**Treatment Category**: In-tank; in situ. **Function**: Storage; immobilization. **Amenable Wastes**: Inorganics and some organics in liquids and contaminated soils. **Possible Residuals/Transformation Products**: Leachates; hazardous volatile emissions; glassified or vitrified waste materials; aqueous scrub solution.	Long-term stability unknown; high energy requirements, especially with high soil water contents and low permeability; electrical shorting caused by buried metal drums; possible underground fire from combustible materials; volatile metals near surface may enter air; site may require runoff controls.
Lime solidification (silicate)	**Treatment Category**: In-tank; in situ. **Function**: Storage; immobilization. **Amenable Wastes**: Metals, waste oils, and solvents. **Possible Residuals/Transformation Products**: Leachates; hazardous volatile emissions; solidified waste materials.	Long-term stability unknown; incompatible with borates, sulfates, carbohydrates; requires complete and uniform mixing of soils and reagents.
Thermoplastic microencapsulation	**Treatment Category**: In-tank; in situ. **Function**: Volume reduction; storage; immobilization. **Amenable Wastes**: Complex, difficult to treat hazardous wastes. **Possible Residuals/Transformation Products**: Leachates; hazardous volatile emissions; encapsulated waste materials.	Not suitable for wastes with high water content; strongly oxidizing contaminants, anhydrous inorganic salts, tetraborates, iron and aluminum salts, and organics with low molecular weights and high vapor pressures. Long-term stability unknown; requires complete and uniform mixing of soils and reagents.

Source: Adapted from Sims and Sims (1991b).

Table 13-4 Summary of Thermal Treatment Methods for Contaminated Soil

Method	Characteristics	Limitations
Fluidized bed	**Treatment Category**: In-tank. **Function**: Volume reduction; detoxification. **Amenable Wastes**: Halogenated and non-halogenated organics; inorganic cyanides. **Possible Residuals/Transformation Products**: Off-gases (possibly acidic and with incomplete combustion products); treated materials with residual metals; fly ash; scrubber water.	High maintenance requirements; waste size and homogeneity requirements; applicable to wastes with low sodium and metal contents.
Rotary kiln	**Treatment Category**: In-tank. **Function**: Volume reduction; detoxification. **Amenable Wastes**: Halogenated and non-halogenated organics; inorganic cyanides. **Possible Residuals/Transformation Products**: Similar to fluidized bed.	High particulate emissions; requires small particle size, so may require size reduction equipment.
Infrared	**Treatment Category**: In-tank. **Function**: Volume reduction; detoxification. **Amenable Wastes**: Halogenated and non-halogenated organics; inorganic cyanides. **Possible Residuals/Transformation Products**: Similar to fluidized bed.	Requires small particle size, so may require size reduction equipment.
Pyrolysis	**Treatment Category**: In-tank. **Function**: Volume reduction; detoxification. **Amenable Wastes**: Wastes not conducive to conventional incineration; waste with volatile metals or recoverable residues. **Possible Residuals/Transformation Products**: Nonvolatile char and ash (Metals, salts, and particulates).	Small capacity.

Source: Adapted from Sims and Sims (1991b).

Typically, components of SVE system consist of vacuum extraction wells, air inlet wells, and vapor monitoring wells distributed across a contaminated site, and a blower(s) to control air flow (Figure 13-1). Extraction wells may be placed vertically or horizontally, although vertical alignment is typical for deeper contamination zones and for residues in radial flow patterns (Hutzler, 1990). Schematics of a gas extraction well and a gas monitoring well are presented in Figures 13-2a and 13-2b, respectively.

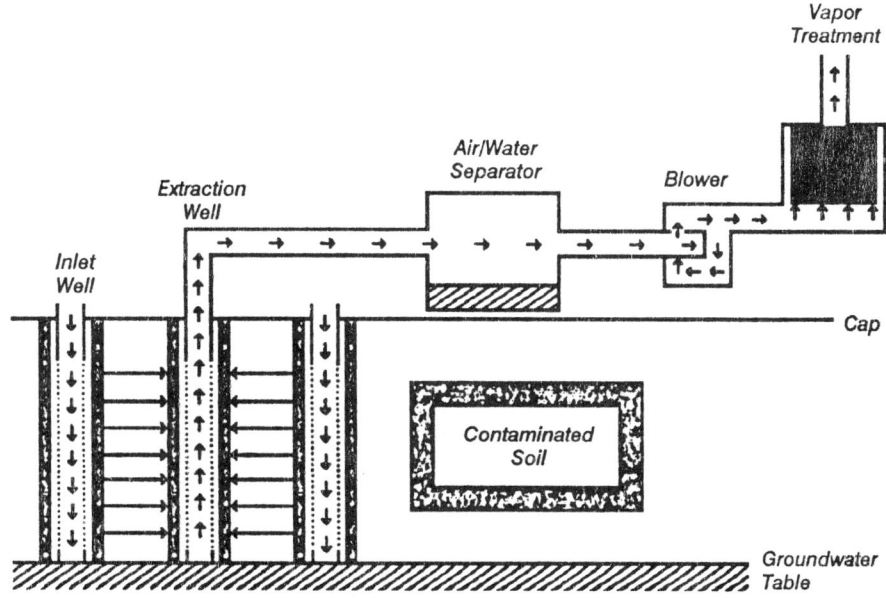

Figure 13-1 Typical components of a soil vacuum extraction system (Sims and Sims, 1991b, after Hutzler et al., 1990).

13.2.1 Significant Chemical Properties

Important system variables that may affect the performance of SVE include properties of the chemical, such as vapor pressure and volatilization, and properties of the site, such as soil moisture content, soil texture, and distribution of contaminants. Vapor pressure is important when a chemical occurs in a pure phase in the subsurface. Vapor pressures above 14 mm Hg at 20°C are desirable for application of SVE. Vapor pressure values for selected subsurface contaminants are given in Table 13-5. When chemicals are distributed in the water phase in the soil, the Henry's Law constant is important, and a dimensionless Henry's constant above 0.01 is desirable for use of SVE.

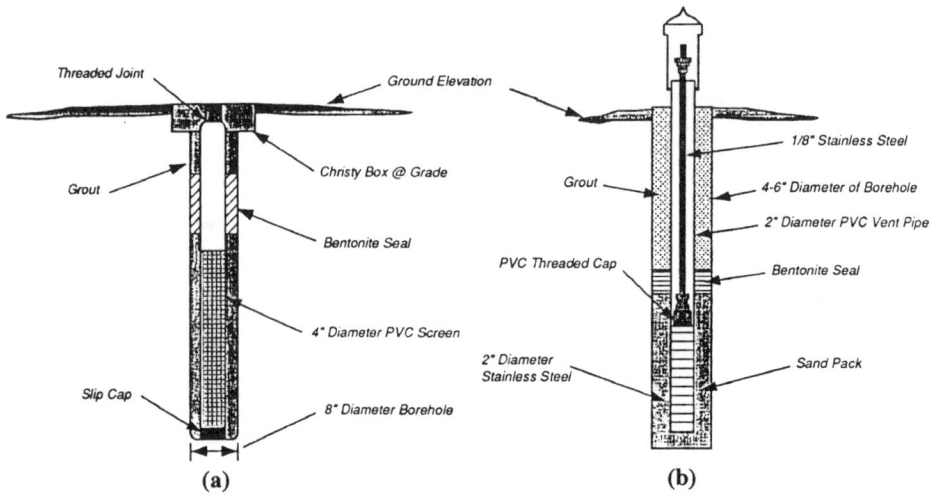

Figure 13-2 Soil vacuum extraction system: (a) gas extraction well; (b) gas monitoring well (Sims and Sims, 1991b).

Table 13-5 Comparative Vapor Pressures and Henry's Constants

Compound	Vapor Pressure (mm Hg)	Henry's Constant (Dimensionless)
Methylene chloride	362	0.13
Acetone	200	miscible
Methyl ethyl ketone (MEK)	100	0.001
1,2-Dichloroethane (EDC)	61	0.037
Bis (chloromethyl) ether	30	0.008
Phenol	0.53	0.00002
Mercury (Hg°)	0.0012	0.48
PCB-1260	4.05×10^{-5}	0.30

Source: Sims and Sims (1991b).

13.2.2 Significant Soil Properties

Since movement of volatile organic chemicals (VOCs) is generally 10,000 times faster in a gas phase than in a water phase, VOC removal is expected to be enhanced by decreasing soil moisture. However, when soil is very dry, which may occur when dry air is drawn through soil, VOCs may adsorb directly onto mineral surfaces, where the magnitude of sorption is increased and consequently volatilization is decreased. Henry's Law constant is not appropriate under these conditions, since partitioning is between air and soil phases only. When moisture is added to soil, the effect is reversible. The moisture content at which a decrease in vapor density becomes apparent is often termed the critical moisture content and generally is equivalent to approximately a monolayer of water molecules coating the soil particles (Spencer et al., 1969, 1973).[2] Johnson and Sterrett (1988) noted that dichloropropane concentrations were correlated with ambient air moisture during the use of SVE at a site in Benson, Arizona.[3]

If contaminated soil contains immiscible fluids in the form of oils, (e.g., petroleum hydrocarbons), the four-compartment system discussed previously is operative (water, air, oil, and soil as discussed in Section 12.1). In this system, chemical volatility will be affected by the chemical vapor pressure and mole fraction within the immiscible oil fluid, and governed by Raoult's Law:

$$P_a = X_a P_a^\circ \qquad (13\text{-}1)$$

where P_a = vapor pressure of solvent over solution (mm Hg), X_a = mole fraction of solvent in solution, and P_a° = vapor pressure of pure solvent (mm Hg).

For contamination by hydrocarbons with multiple components, volatilization will proceed such that lower molecular weight chemicals will volatilize before higher molecular weight compounds. Through this process of weathering of the waste/soil mixture, SVE extraction efficiency is observed to decrease to less than 10 percent when the fraction of gasoline remaining is approximately 40 percent (Figure 13-3). Therefore, measuring general parameters such as total hydrocarbons is not sufficient to indicate the removal efficiency of individual constituents.

Soil texture strongly influences air permeability, and consequently serves as a good indicator of suitability of a site for SVE. In less permeable media, such as glacial till and clayey soils, secondary permeability or porosity (fractures) will dominate air flow. There will be rapid removal of VOCs in fractures and slow removal in the soil matrix. In more permeable media, such as sands, sandy loams, and loamy sands, SVE is appropriate (Figure 13-4). Pneumatic pump tests in the field are recommended for site-specific evaluation of SVE application.

---

[2] Spencer, W.F., M.M. Cliath, and W.J. Farmer. 1969. Vapor Density of Soil-Applied Dieldrin as Related to Soil-Water Content, Temperature and Dieldrin Concentration. Soil Sci. Soc. Am. Proc. 33:509-511.

Spencer, W.F., W.J. Farmer, and M.M. Cliath. 1973. Pesticide Volatilization. Residue Reviews 49:1-47.

[3] Johnson, J.J. and R.J. Sterrett. 1988. Analysis of In Situ Air Stripping Data. In: Proc. 5th National Conference on Hazardous Waste and Hazardous Materials, Hazardous Materials Control Research Institute, Silver Spring MD, pp. 451-455.

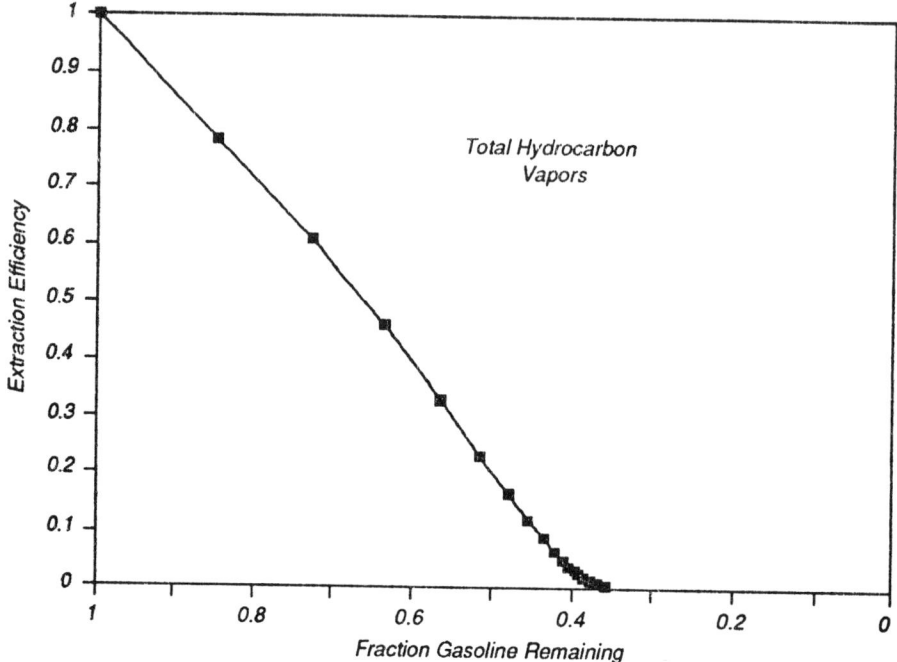

Figure 13-3 Soil vacuum extraction efficiency based on total hydrocarbon vapors (Sims and Sims, 1991b).

Due to release of VOCs from the soil matrix, when extraction wells are temporarily turned off, concentrations of VOC increase in soil air (referred to as "VOC rebound effect"), with an equilibrium concentration that is determined by Henry's Law constant. When blowers are turned on, an increase in the concentration of extracted vapor from the soil will be observed. Diffusive release from subsurface stratigraphy of less permeability will cause the slow continual release of chemicals into the soil-gas phase (Figure 13-5).

13.2.3 Design Considerations

Design considerations that affect SVE include extraction well spacing and extraction well depth. As permeability decreases, well spacing decreases; typical well spacings of 10 m to 30 m are common (Figure 13-6). Also, air circulation generally is not significant below the screened interval for extraction wells. Where contamination is deep and permeability is high throughout the soil profile, the slotted (screened) interval should be extended to the maximum depth possible to maximize treatment, rather than slotted fully vertically. Placement of horizontal extraction wells using directional drilling techniques (described in Table 9-3) in zones with the highest contaminant concentrations can significantly improve vapor recovery.

13.2.4 Enhanced Biodegradation

A promising application of SVE is for enhancement of biodegradation of volatile

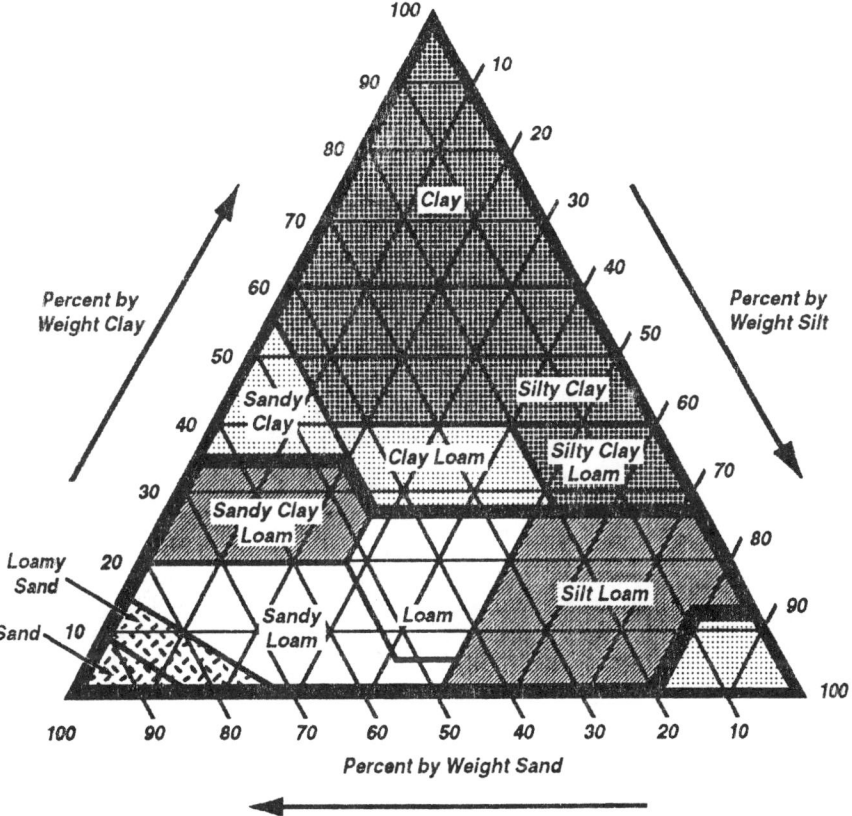

Figure 13-4 Suitability of soil vacuum extraction related to soil texture (Sims and Sims, 1991b).

and semivolatile chemicals in soils. SVE provides air to the vadose zone, and thus carries oxygen that can be used as the terminal electron acceptor (TEA) by soil microorganisms to biodegrade chemicals. Air has a much greater potential than water for delivering oxygen to soil on a weight-to-weight and volume-to-volume basis. The amount of air required to deliver one pound of oxygen is four pounds. In comparison 100,000 pounds of air saturated ground water is required to deliver one pound of oxygen. Oxygen provided by air is more easily delivered since the fluid is less viscous than water; higher oxygen concentrations in air also provide a large driving force for diffusion of oxygen into less permeable areas within a soil formation.

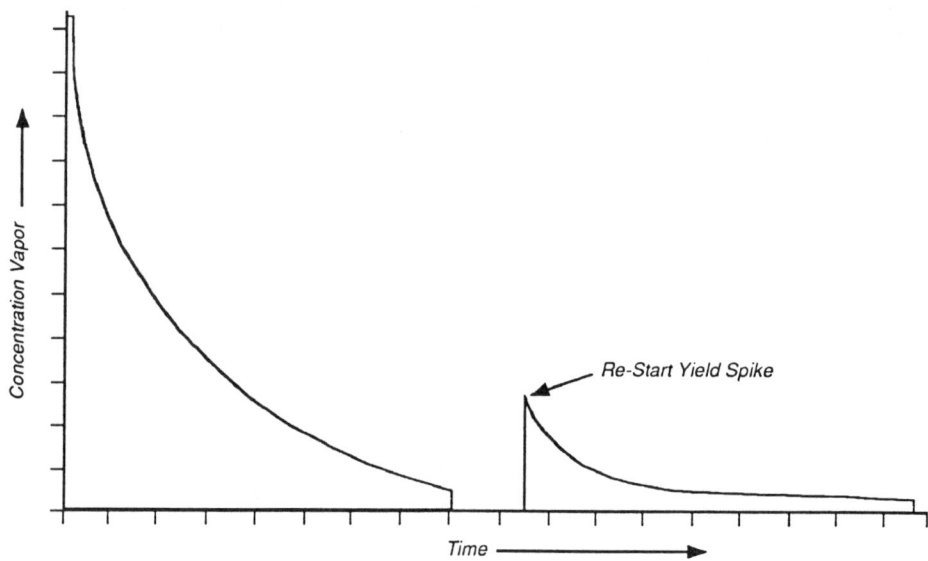

Figure 13-5 Volatile organic compound (VOC) rebound effect with SVE stop and restart (Sims and Sims, 1991b).

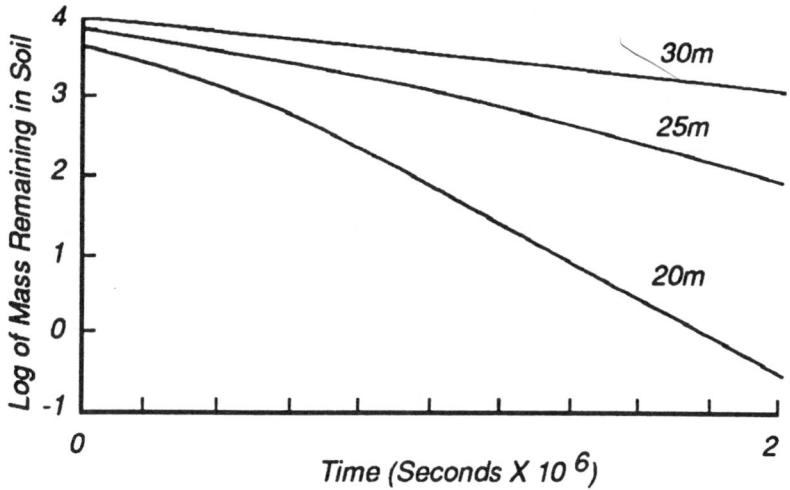

Figure 13-6 Effect of well spacing on total solute mass remaining in soil with vacuum extraction (Sims and Sims, 1991b).

Hinchee and Downey (1990)[4] successfully applied SVE to enhance bio-

---

[4] Hinchee, R., and D. Downey. 1990. In Situ Enhanced Biodegradation of Petroleum Distillates in the Vadose Zone. In: Proceedings of the International Symposium on Hazardous Waste Treatment: Treatment of Contaminated Soils (Cincinnati, OH). Air and Waste Management Association, Pittsburgh, PA.

degradation of petroleum hydrocarbons in JP-4 jet fuel at Hill Air Force Base, Ogden, Utah, by increasing subsurface oxygen concentrations. Soil moisture was found to be a sensitive variable affecting biodegradation, with increased soil moisture (from 20 percent to 75 percent field capacity) related to increased biodegradation (Figure 13-7). Monitoring carbon dioxide and oxygen concentrations, as well as estimating the mass of VOC biodegraded, is recommended for evaluating potential enhancement of biodegradation using SVE.

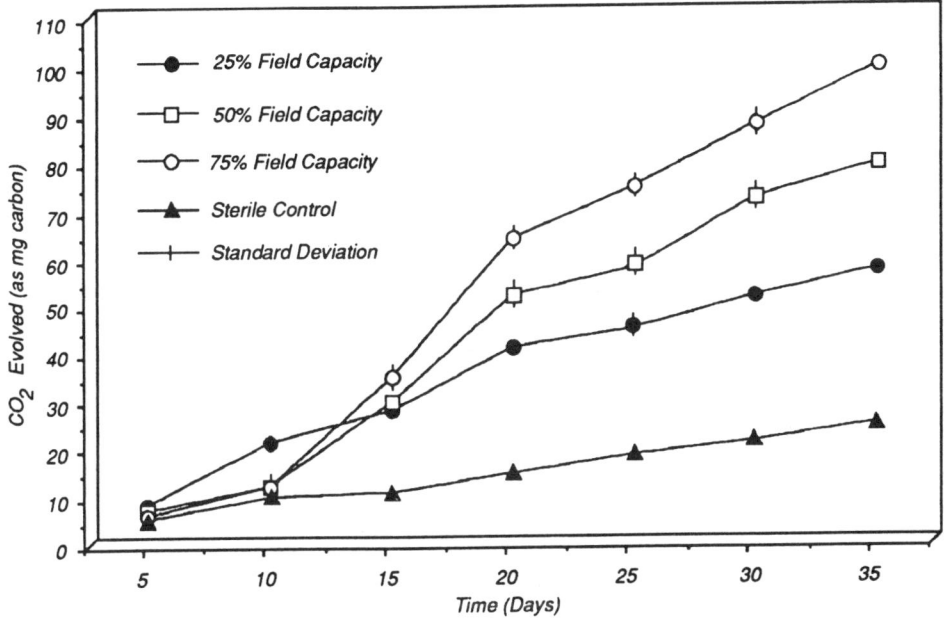

Figure 13-7 Enhanced bioremediation of gasoline-contaminated soil using soil vacuum extraction, nutrient additions and moisture control (Sims and Sims, 1991b).

13.3 Bioremediation

Biotic reactions in the subsurface, including definitions and mechanisms, are addressed in Section 3.5. In situ soil remedial measures using biological processes can reduce or eliminate continuing or potential ground-water contamination, thus reducing the need for extensive ground-water monitoring and treatment requirements (Wilson, 1981, 1982, 1983/T9-10). This section focusses on in situ bioremediation. Table 13-2 provides some summary information on other bioremediation techniques.

In situ bioremediation involves the use of naturally occurring microorganisms (in contrast to genetically engineered microorganisms) to degrade and/or detoxify hazardous constituents in the soil at a contaminated site to protect public health and the environment. The use of bioremediation techniques in conjunction with chemical and physical treatment processes, i.e., the use of a "treatment train," is an effective means for comprehensive site-specific remediation (Ross et al., 1988).[5]

Components of soil bioremediation systems generally include (1) *delivery systems* (such as injection nozzles, plows and irrigation systems) supply water, nutrients, oxygen, organic matter, specialized microorganisms, and/or other amendments, as required and (2) *run-on and run-off controls* for moisture control and waste containment (Chambers et al., 1990/T13-9, Sims et al., 1984/T13-9).

13.3.1 Approaches to In Situ Biological Treatment

Four approaches are generally used for in situ biological treatment: (1) enhancement of biochemical mechanisms for detoxifying or degrading chemicals, (2) augmentation with exogenous acclimated or specialized microorganisms originating from uncontaminated or contaminated environments, (3) application of cell-free enzymes, and (4) vegetative uptake (Chambers et al., 1990/T13-9). Enhancement of biochemical mechanisms may involve (1) control of soil factors such as contaminant concentrations that do not severely inhibit microbial activity, soil moisture, pH, nutrients, and temperature in order to optimize microbial activity; (2) addition of organic amendments to stimulate cooxidation or cometabolism; (3) control of soil oxygen by moisture control to accomplish aerobic or anaerobic biodegradation; and (4) addition of colloidal gas aphrons (microscopic bubbles of gas) to increase the concentration of terminal electron acceptors (oxygen) in the soil and thereby enhance aerobic biodegradation.

13.3.2 Significant Environmental Parameters

The environmental factors presented in Table 13-6, as well as waste and soil/site characteristics identified in Chapter 12 (Table 12-1), interact to affect microbial activity at a specific contaminated site. Computer modeling techniques are useful design and evaluation tools to describe these interactions and their effects on bioremediation treatment techniques for organic constituents in a specific situation.

Soil moisture is a significant factor affecting treatment effectiveness of contaminated soil. For example, Sims (1986)[6] observed more rapid degradation of polynuclear aromatic compounds (expressed as half-life in days) at a soil moisture content of 60 to 80 percent of field capacity than at a soil moisture content of 20 to 40 percent as follows: anthracene (37 days vs 43 days), phenanthrene (54 vs 61 days) and fluoranthene (231 vs 559 days).

---

[5] Ross, D., T.P. Marziarz, and A.L. Bourquin. 1988. Bioremediation of Hazardous Waste Sites in the USA: Case Histories. In: Superfund '88, Hazardous Materials Control Research Institute, Silver Spring, MD, pp. 395-397.

[6] Sims, R.C. 1986. Loading Rates and Frequencies for Land Treatment Systems. In: Land Treatment--A Hazardous Waste Management Alternative, R.C. Loehr, and J.F. Malina (eds.), Water Resources Symposium No. 13, University of Texas Press, Austin, TX, pp. 151-170.

Table 13-6 Critical Environmental Factors for Microbial Activity

Environmental Factor	Optimum Levels
Available soil water	25-85% of water holding capacity; -0.01 MPa.
Oxygen	**Aerobic metabolism:** >0.2 mg/l dissolved O_2; minimum air-filled pore space of 10% by volume. **Anaerobic metabolism:** O_2 concentrations <1% by volume.
Redox potential	**Aerobes and facultative anaerobes:** >50 mV. **Anaerobes:** <50 mV; ph 5.5-8.5.
Nutrients	Sufficient nitrogen, phosphorus, and other nutrients so not limiting to microbial growth (suggested C:N:P ratio of 120:10:1).
Temperature	15-45°C (Mesophiles).

Source: Sims and Sims (1991b), compiled from Huddleston et al. (1986), Paul and Clark (1989), Rochkind et al. (1986), and Sims et al. (1984).

The effect of *temperature* on apparent loss of polycyclic aromatic hydrocarbon (PAHs) compounds in a sandy loam soil is summarized in Table 13-7. Temperature has an important effect on the fate and behavior of PAHs and, therefore, has implications for seasonal effects on the rate of biological remediation of soil contaminated with these chemicals. Microbial ecologists have identified ranges of critical environmental conditions that affect aerobic activity of soil microorganisms (Table 13-6). Many of these conditions are controllable and can be modified to enhance activity.

Oxygen may be consumed faster than it can be replaced by diffusion from the atmosphere, and the soil may become anaerobic. Clay content of soil and the presence of organic matter also may affect oxygen content in soil. Clayey soils tend to retain a higher moisture content, which restricts oxygen diffusion, while organic matter may increase microbial activity and deplete available oxygen. Loss of oxygen as a metabolic electron acceptor induces a change in the activity and composition of the soil microbial population. Obligate anaerobic organisms and facultative anaerobic organisms, which use oxygen when it is present or switch to alternative electron acceptors such as nitrate or sulfate in the absence of oxygen, become the dominant populations. Additional information concerning in situ anaerobic bioremediation can be found in the document, *Handbook on In Situ Treatment of Hazardous Waste-Contaminated Soils* (Chambers et al., 1990/T13-9).

Table 13-7 Effect of Temperature on Degradation of PAHs in a Sandy Loam Soil

Compound	Percent of PAH Remaining			Estimated Half Life (day)[a]			Half lives reported in the literature (day)
	10°C	20°C	30°C	10°C	20°C	30°C	
Acenaphthene	5	0	0	<60	<10	<10	96[b], 45[b], 0.3-4[c]
Fluorene	8	3	2	60 (+11/-10)	47 (+6/-5)	32 (+5/-3)	64[b], 39[b], 2-39[c]
Phenanthrene	36	19	2	200 (+40/-40)	<60	<60	69[b], 23[b], 26[c], 9.7[d], 14[d]
Anthracene	83	51	58	460 (+310/-140)	260 (+160/-70)	200 (+90/-30)	28[b], 17[b], 108-175[c], 17[d], 45[d]
Fluoranthene	94	71	15	f	440 (+560/-160)	140 (+40/-20)	104[b], 29[b], 44-182[c], 39[d], 34[d]
Pyrene	93	89	43	f	1900 (+6200/-800)	210 (+160/-60)	73[b], 27[b], 3-35[c], 58[d], 48[d]
Benz[a]anthracene	82	71	50	680 (+300/-160)	430 (+110/-70)	240 (+40/-40)	52[b], 123[b], 102-252[c], 240[d], 130[d]
Chrysene	85	88	86	980 (+520/-270)	1000 (+900/-250)	730 (+370/-180)	70[b], 42[b], 5.5-10.5[c], 328[d], 224[d]
Benzo[b]fluoranthene	77	75	62	580 (+520/-180)	610 (+590/-200)	360 (+150/-80)	73-130[e], 85[b], 65[b]
Benzo[k]fluoranthene	93	95	89	910 (+690/-270)	1400 (+3300/-560)	910 (+4400/-410)	143[b], 74[b]
Benzo[a]pyrene	73	54	53	530 (+1700/-230)	290 (+570/-120)	220 (+160/-60)	91[b], 69[b], 30-420[c], 347[d], 218[d]
Dibenz[a,h]anthracene	88	87	83	820 (+1100/-300)	750 (+850/-260)	940 (+12000/-450)	74[b], 42[b], 100-190[e]
Benzo[g,h,i]perylene	81	76	75	650 (+650/-230)	600 (+570/-190)	590 (+1800/-250)	179[b], 70[a,b]
Indeno[1,2,3-c,d]pyrene	80	77	70	600 (+310/-150)	730 (+1100/-270)	630 (+2500/-280)	57[b], 42[b], 200-600[e]

[a] $t_{1/2}$ (95 percent confidence interval)
[b] T = 20°C Sims (1986)
[c] T = 15-25°C Sims and Overcash (1983)
[d] T = 20°C PACE (1985)
[e] T = 20°C Sims (1982)
[f] Least squares slope (for calculations of $t_{1/2}$) = zero with 95% confidence

Coover and Sims, 1987

Acclimation of a soil to the presence of a waste is another significant factor affecting bioremediation. Table 13-8 shows this effect for a fossil fuel-contaminated soil. The acclimated soil was exposed to the fossil fuel waste for one year before a repeat application of the waste. Results presented in Table 13-8 indicate that a higher percentage of waste was treated in the acclimated soil. Treatment also occurred more rapidly compared to treatment in unacclimated soil. Management of contaminated soil, therefore, may include the addition of lightly contaminated, preexposed soil to more heavily contaminated and/or newly contaminated soil to increase the rate and extent of treatment.

Table 13-8 Acclimation of Soil to Complex Fossil Fuel Waste

	Unacclimated Soil		Acclimated Soil	
PNA Constituent	Initial Soil Concentration (mg/kg-dry wt)	Reduction in 40 days (%)	Soil Concentration after Reapplication (mg/kg-dry wt)[a]	Reduction in 22 days (%)
Naphthalene	38	90	38	100
Phenanthrene	30	70	30	83
Anthracene	38	58	38	99
Fluoranthene	154	51	159	82
Pyrene	177	47	180	86
Benz(a)anthracene	30	42	40	70
Chyrsene	27	25	33	61
Benz(a)pyrene	10	40	12	50

[a] First reapplication of waste after 168 days incubation at initial level.

Source: Sims and Sims (1991b) after Sims (1986).

The use of *plants* for stimulating microbial activity in soil results in increased biodegradation of target organic chemicals in contrast to the possibility of vegetative accumulation of chemicals for harvesting and removal from a site. This method has been investigated by Aprill and Sims (1990) and Walton and Anderson (1990).[7] In soils with low levels of contamination, plant roots may stimulate the biodegradation

---

[7] Aprill, W. and R.C. Sims. 1990. Evaluation of the Use of Prairie Grasses for Stimulating Polycyclic Aromatic Hydrocarbon Treatment in Soil. Chemosphere 20:253-265.

Walton, B.T. and T.A. Anderson. 1990. Microbial Degradation of Trichloroethylene in the Rhizosphere: Potential Application to Biological Remediation of Waste Sites. Appl. Environ. Microbiol. 56:1012-1016.

of toxic chemicals by providing exudates that serve as carbon and energy substrates for soil microorganisms. Aprill and Sims (1990) reported that for soil with initial concentrations of PAHs of approximately 10 to 50 mg/kg, the presence of vegetation in the soil (prairie grasses) resulted in a statistically significant reduction in PAHs, compared with nonvegetated soil.

Measurement of physical abiotic loss mechanisms and partitioning of organic substances into air and soil phases should be used in degradation studies to ensure that generated information is related to disappearance mechanisms of the constituents in the soil system (Abbott and Sims, 1989; Armstrong and Konrad, 1974).[8] This type of information is needed to more accurately evaluate and select treatment techniques. For example, for organophosphorus pesticides, sorption-catalyzed hydrolysis of ester linkages is known to be an important influence on soil degradation. An understanding of abiotic reactions as influenced by sorption and pH of the system may allow the design of a more effective remediation strategy. If abiotic controls are not used, the disappearance of chemicals may be attributed solely to biological activity, though biological activity may not play the major role in the degradation of the chemical. Therefore, knowledge of the reaction mechanism is directly related to efficiency and effectiveness in remediation strategy design and remediation technique selection.

13.4 Other Treatment Approaches

One way to predict and control the rate of transport of a constituent through a subsurface system is to describe its mobility (or relative immobility) by predicting its retardation (Mahmood and Sims, 1986).[9] Retardation describes the relative velocity of the constituent compared to the rate of movement of water through the subsurface (see Section 4.5 for more information). Retardation in unsaturated soil can be represented as:

$$R = 1 + (\rho K_d/\theta) \tag{13-2}$$

where:

ρ = soil bulk density
K_d = soil/water partition coefficient (Section 4.5.2)
θ = volumetric moisture content

For a saturated system, θ is replaced by the porosity of the system.

---

[8] Abbott, C. and R.C. Sims. 1989. Use of Bioassays to Monitor Polycyclic Aromatic Hydrocarbon Contamination in Soil. In: Superfund '89, Hazardous Materials Control Research Institute, Silver Spring, MD, pp. 23-26.

Armstrong, D.E. and J.G. Konrad. 1974. Nonbiological Degradation of Pesticides. In: Pesticides in Soil and Water, W. D. Guenzi (ed.), Soil Science Society of America, Madison, WI, chapter 7.

[9] Mahmood, R.J. and R.C. Sims. 1986. Mobility of Organics in Land Treatment Systems. Journal of Environmental Engineering (ASCE) 112:236-245.

This information can be used to evaluate treatment techniques for a contaminated soil system (e.g., techniques to modify the soil/water partition coefficient, such as control of soil moisture, changes in bulk density, or addition of amendments to the soil). Constituents can be "captured" or contained within the system by using these techniques, thus allowing time for degradation at the site or for engineering implementation and performance of other remediation treatment techniques, such as soil washing (Sims et al., 1989/T13-9).

13.4.1 Sorption, Ion Exchange and Precipitation

Constituents in in situ and prepared bed treatment systems are generally immobilized through sorption, ion exchange, and/or precipitation reactions. These techniques reduce the rate of contaminant release from the soil environment so that concentrations along exposure pathways are held within acceptable limits. The effects of moisture and distribution coefficient, K_d, on immobilization are illustrated in Figure 13-8. Results indicate that for chemicals with K_d values less than 10, management of soil moisture is important with regard to immobilizing chemicals; for chemicals with K_d values greater than 10, management of soil moisture is less important. Approaches for controlling soil moisture include run-on and run-off controls, temporary capping or covering, and irrigation scheduling.

The cation exchange capacity (CEC) of soil also can be evaluated with regard to organic as well as metal immobilization. Positively charged organic chemicals and metals will generally readily attach to soil materials with negatively charged functional groups and negatively charged clay particles. Addition of clays, synthetic resins, and zeolites will increase the CEC of soils and increase immobilization of chemicals sensitive to CEC characteristics of a soil (Sims et al., 1984/T13-9). For inorganic chemicals that are negatively charged in soil systems and can exist in several oxidation states (e.g., chromium, selenium, and arsenic), immobilization, as well as the toxic form of the chemical, may potentially be controlled by managing the redox and pH of the soil system. Management of redox and pH may be short-term or long-term, depending upon the goals of site management (e.g., temporary immobilization while delivery and recovery systems are designed and implemented, followed by soil flushing with aqueous or surfactant solutions for removal and recovery of the contaminants).

13.4.2 Solidification and Stabilization

Solidification and stabilization are additional immobilization techniques that are applicable to in situ and prepared bed systems. These techniques are designed to accomplish one or more of the following: (1) production of a solid from a liquid or semisolid waste, (2) reduction of contaminant solubility, and/or (3) a decrease in the exposed surface area across which transfer may occur. Solidification may involve encapsulation of fine waste particles (microencapsulation) or large blocks of waste (macroencapsulation). Stabilization refers to the process of reducing the hazardous potential of waste materials by converting contaminants into their least soluble, mobile, or toxic form (Chambers et al., 1990/T13-9). A milestone publication providing

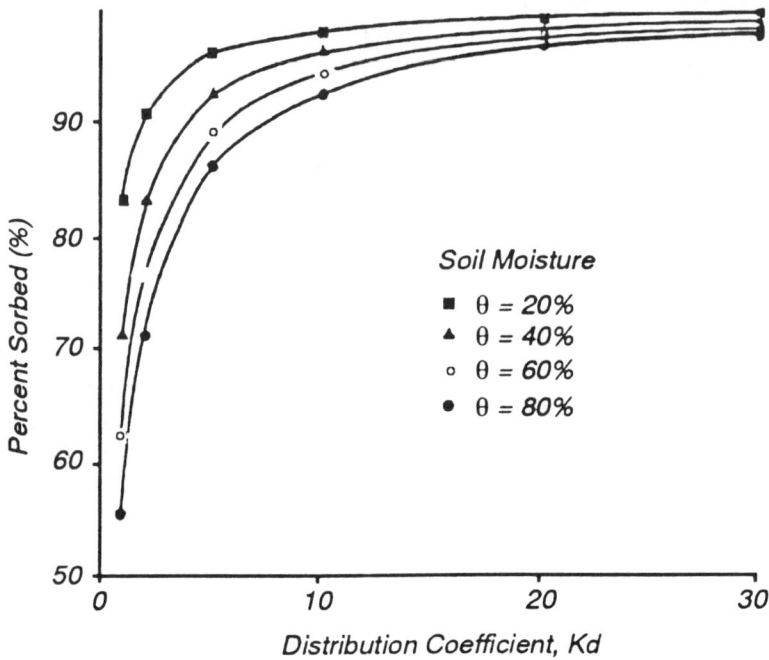

Figure 13-8 Sorption of chemicals to soil as a function of soil moisture content and partition coefficient Kd (Sims and Sims, 1991b, after Sims et al., 1986).

additional detail on this technique is the *Handbook for Stabilization/ Solidification of Hazardous Wastes* (Cullinane et al., 1986/T13-9). Table 13-3 summarizes some information of fixation and capsulization methods for stabilizing contaminated soils.

Systems for delivering reagents to the contaminated area include (1) injection systems; (2) soil surface applicators; and/or (3) delivery and application of electrical energy for melting soils and rocks that contain hazardous materials. Equipment required for preparing, mixing, and applying reagents depends upon the reagent, process, and depth of contamination (Chambers et al., 1990/T13-9).

Important parameters identified by Truett et al. (1983/T13-9) for solidification and stabilization of hazardous wastes include (1) reagent viscosity; (2) permeability of soils; (3) porosity of waste materials and soil; (4) distribution of waste in surrounding material (rocks, soils, etc.); and (5) rate of reaction. The most significant challenge in applying solidification/stabilization treatment in situ is achieving uniform mixing of added chemical agent(s) with the contaminated soils (Chambers et al., 1990/T13-9). Design factors involve delivery and mixing systems to obtain complete and uniform distribution of added reagent throughout the contaminated soil (U.S. EPA, 1990).

13.4.3 Soil Flushing

Mobilization of organic and/or inorganic contaminants from soil may be accomplished using soil flushing and recovery and treatment of the elutriate. Flushing solutions generally include water, acidic and basic solutions, surfactants, and solvents. The solutions partition a contaminant into the liquid phase through the volume of added liquid or by decreasing the distribution coefficient between the soil and the flushing phase (Sims et al., 1984/T13-9; Raghavan et al., 1990/T13-9). A schematic of a soil flushing system is shown in Figure 13-9. Components consist of (1) the flushing solution, and (2) delivery and recovery systems, which may include injection and recovery wells, equipment for surface applications, and holding tanks for storing elutriate for reapplication.

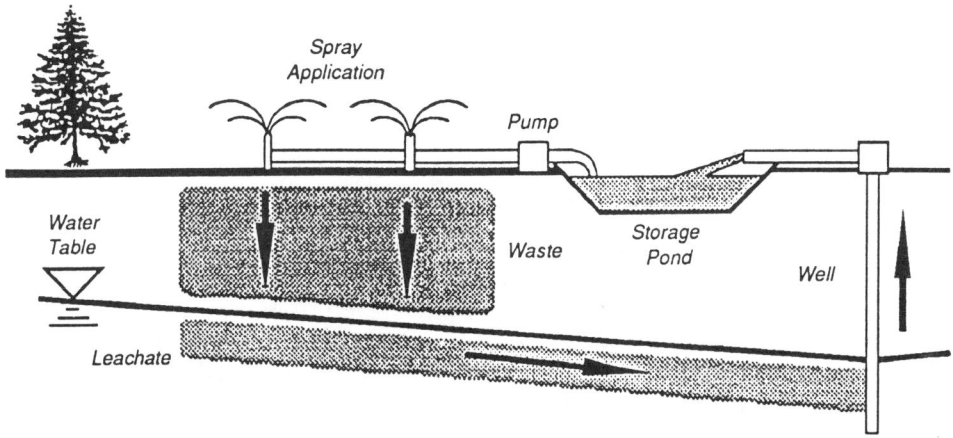

Figure 13-9 Schematic of soil flushing and recycle system (U.S. EPA, 1990).

Variables affecting application of the technique include (1) concentration and volume of contamination; (2) distribution coefficients of waste constituents; (3) interactions of flushing solutions with soil; and (4) suitability of site for installation of wells, drains, etc., for delivery and recovery. Design factors include sizing the delivery and recovery systems to ensure complete recovery of elutriate. Problems with respect to flushing of bulk fluids, or NAPLs, from soil systems are due to the following characteristics of bulk fluids: (1) low water solubility, (2) high interfacial tension, and (3) poor relative permeability. Relative permeability is defined as:

$$M = [K_d/U_d] / [K_o/U_o] \qquad (13\text{-}3)$$

where:

M = mobility ratio
K_d = fluid permeability (water)
K_o = oil permeability
U_d = viscosity of fluid (water)
U_o = viscosity of oil

Strategies for flushing of bulk liquids from soil generally involve control of one or more of the variables affecting the mobility ratio through adding chemicals to decrease mobility of water or increase mobility of oil (e.g., adding surfactants or steam to decrease U_o or adding polymers to increase U_d).

Use of soil flushing in a treatment train with bioremediation has been evaluated by Dworkin et al. (1988) for a wood preserving contaminated site.[10] Flushing using surfactant/polymer combinations was used to remove high concentrations of PAH compounds; residual low concentrations were treated using biological processes. Mahmood and Sims (1985) found that when methanol was used as the solvent in a soil system to flush PAHs from a soil, the resultant concentration of the PAHs in the solution phase was several orders of magnitude higher than the concentration of the PAHs in water.[11]

13.5 Prepared Bed Reactors

The previous sections have focussed on in situ soil treatment methods. In a prepared bed system, the contaminated soil may be either (1) physically moved from its original site to a newly prepared area, which has been designed to enhance remediation and/or to prevent transport of contaminants from the site or (2) removed from the site to a storage area while the original location is prepared for use, then returned to the bed, where the treatment is accomplished. Preparation of the bed may include placement of a clay or plastic liner to retard transport of contaminants from the site or addition of uncontaminated soil to provide additional treatment medium.

Tables 13-1 and 13-2 identify possible prepared bed reactor technologies and provide information on function as well as application and limitations. Treatment of contaminants with a prepared bed may be based on the techniques previously identified and described for in situ treatment.

An example of the use of a prepared bed reactor for soil remediation was described by Lynch and Genes (1989).[12] Prepared bed treatment of creosote-contaminated soils from a shallow, unlined surface impoundment was demonstrated

---

[10] Dworkin, D., D.J. Messinger, and R.M. Shapot. 1988. In Situ Flushing and Bioreclamation Technologies at a Creosote-Based Wood Treatment Plant. In: Proc. 5th National Conference on Hazardous Wastes and Hazardous Materials, Hazardous Materials Control Research Institute, Silver Spring, MD, pp. 67-78.

[11] Mahmood, R.J. and R.C. Sims. 1985. Enhanced Mobility of Polynuclear Aromatic Compounds in Soil Systems. In: Proc. 1985 Environmental Engineering Specialty Annual Conference (Boston, MA), American Society of Civil Engineers, pp. 128-135.

[12] Lynch, J. and B.R. Genes. 1989. Land Treatment of Hydrocarbon Contaminated Soils. In: Petroleum Contaminated Soils, Vol. 1: Remediation Techniques, Environmental Fate, and Risk Assessment, P.T. Kostecki and E.J. Calabrese (eds.), Lewis Publishers, Chelsea, MI, pp. 163-174.

at a disposal facility for a wood-preserving operation in Minnesota. The contaminated soils contained creosote constituents consisting primarily of PAHs at concentrations ranging from 1,000 to 10,000 ppm. Prior to implementation of the full-scale treatment operation, bench-scale and pilot-scale studies simulating proposed full-scale conditions were conducted to define operation and design parameters. Over a 4-month period, 62 to 80 percent removal of total PAHs was achieved in all test plots and laboratory reactors. Two-ring PAH compounds were reduced by 80 to 90 percent, 3-ring PAHs by 82 to 93 percent, and 4+-ring PAHs by 21 to 60 percent.

The full-scale system involved preparation of a treatment area within the confines of the existing impoundment. A lined waste pile for temporary storage of the sludge and contaminated soil from the impoundment was constructed. All standing water from the impoundment was removed, and the sludges were excavated and segregated for subsequent free oil recovery. Three to five feet of "visibly" contaminated soil was excavated and stored in the lined waste pile. The bottom of the impoundment was stabilized as a base for the treatment area. The treatment area was constructed by installation of a polyethylene liner, a leachate collection system, 4 feet of clean backfill, and addition of manure to achieve a carbon:nitrogen ratio of 50:1. A sump for collection of storm water and leachate and a center pivot irrigation system also were installed. The lined treatment area was required because natural soils at the site were highly permeable. A cap also was needed for residual contaminants left in place below the liner. Contaminated soil was periodically applied to the treatment facility and rototilled into the treatment soil. Soil moisture was maintained near field capacity with the irrigation system. During the first year of operation, greater than 95 percent reductions in concentration were obtained for 2- and 3-ring PAHs. Greater than 70 percent of 4- and 5-ring PAH compounds were degraded during the first year. Comparison of half-lives of PNAs in the full-scale facility were in the low end of the range of half-lives reported for the test plot units. Only two PNA compounds were detected in drain tile water samples, at concentrations near analytical detection limits.

Prepared bed treatment of a Texas oilfield site with storage pit backfill soils contaminated with styrene, still bottom tars, and chlorinated hydrocarbon solvents was demonstrated on a pilot scale (St. John and Sikes, 1988).[13] The remediation efforts included biological, chemical, and physical treatment strategies. The pilot-scale, solid-phase biological treatment facility consisted of a plastic film greenhouse enclosure, a lined soil treatment bed with an underdrain, an overhead spray system for distributing water, nutrients, and inocula, an organic vapor control system consisting of activated carbon absorbers, and a fermentation vessel for preparing microbial inoculum or treating contaminated leachate from the backfill soils. Soils were excavated from the contaminated area and transferred to the treatment facility. Average concentrations of volatile organic compounds (VOCs) were reduced by more than 99 percent during the 94-day period of operation of the facility; most of the removal was attributed to air stripping. Biodegradation of semivolatile compounds reduced average concentrations by 89 percent during the treatment period.

---

[13] St. John, W.D., and D.J. Sikes. 1988. Complex Industrial Waste Sites. In: Environmental Biotechnology--Reducing Risks from Environmental Chemicals through Biotechnology, G.S. Omenn (ed.), Plenum Press, New York, NY, pp. 237-252.

13.6 Guide to Major References

Table 13-9 provides an index to major references on soil treatment and remediation including (1) general reviews, (2) soil treatability studies, (3) soil cleanup standards, (4) petroleum/hydrocarbon contaminated soils, (5) contaminated sediments, and (6) radioactive contamination. Major references on the following treatment categories are also indexed in Table 13-9: (1) in-place treatment, (2) bioremediation, (3) stabilization/solidification, and (4) soil vapor extraction and other specific methods.

Currently, many remedial techniques are being used and evaluated for cleanup of contaminated soils. Section 12.5 describes various documents that are available as part of U.S. EPA's Superfund Innovative Technology Evaluation (SITE) program.

Table 13-9 Index to Major References on Soil Treatment and Remediation

Topic	References
General Reviews	Ehrenfeld and Bass (1984/T12-7), Barkley (1993/T12-7), Galer (1988), Madden and Johnson (1992/T12-7), Sims et al. (1984), U.S. EPA (1980/T12-7, 1989, 1990), Wise and Trantolo (1994); See generally the conference and symposium series in identified in Table 12-7; <u>Innovative Technologies</u>: U.S. EPA (1993/T12-7)
Soil Treatability Studies	Loehr (1989), McGinnis et al. (1988—creosote/PCP), Sims et al. (1986, 1988); see also treatability references in Table 12-7
Soil Cleanup Standards	Booz, Allen & Hamilton (1989), Fitchko (1989), Hwang et al. (1987), Lipsky et al. (1989), Oliver et al. (1993/T12-7), WEF (1994)
Petroleum/Hydrocarbon Contaminated Soils	Nash et al. (1992); see also petroleum/hydrocarbon contaminated soil conference series, and UST/Spill Remediation references in Table 12-7.
Contaminated Sediments	Fitchko (1989), Voskuil (1991), Wilson (1988—PCB)
Radioactive Contamination	U.S. EPA (1988, 1990), Voskuil (1992)
<u>Treatment Methods</u>	
In-Place Treatment	Chambers et al. (1990), Sims et al. (1984), U.S. EPA (1986); <u>Heavy Metal Contamination</u>: Czupyma et al. (1989); <u>Soil Flushing/Washing</u>: Nash (1987), Raghavan et al. (1990)
Bioremediation	Sims et al. (1989, 1993); See also references on soil and groundwater bioremediation in Table 12-7
Stabilization/Solidification	Arozarena et al. (1989), Battelle (1993), Cullinane et al. (1986), Means et al. (1994), Truett et al. (1983), U.S. EPA (1979, 1980, 1986), Voskuil (1992-vitrification)
Specific Methods	<u>Vapor Extraction</u>: Bartenfelder et al. (1992/T12-7), DiGuilio (1992); <u>TCE Removal</u>: Russell et al. (1992/T12-7); <u>Wet-Air Oxidation</u>: Unterberg et al. (1987)

Table 13-9 References (Appendix F contains references for figure and table sources)

Arozarena, M.M., et al. 1989. Stabilization/Solidification of CERCLA and RCRA Wastes: Physical Tests, Chemical Testing Procedures, Technology Screening, and Field Activities. EPA/625/6-89/022. Available from CERI.* [Contains summary information on 32 case studies; 8 authors]

Battelle. 1993. Technical Resource Document: Solidification/Stabilization and its Application to Waste Materials. EPA/530/R-93/012.

Booz, Allen & Hamilton, Inc. 1989. Determining Soil Response Action Levels Based on Potential Contaminant Migration to Ground Water: A Compendium of Examples. EPA/540/2-89/057, 144 pp. Available from CERI.*

Chambers, L.D. et al. 1990. Handbook of In Situ Treatment of Hazardous Waste Contaminated Soils. EPA/540/2-90/002 (NTIS PB90-155607), 157 pp. [7 authors]

Cullinane, M.J. L.W. Jones, and P.G. Malone. 1986a. Handbook for Stabilization/Solidification of Hazardous Wastes. EPA/540/2-86/001 (NTIS PB87-116745), 170 pp.

Czupyma, G. et al. 1989. In Situ Immobilization of Heavy-Metal-Contaminated Soils. Noyes Data Corporation, Park Ridge, NJ, 155 pp.

DiGuilio, D.C. 1992. Evaluation of Soil Venting Application. Ground Water Issue Paper, EPA/540/S-92/004, 7 pp.

Fitchko, J. 1989. Criteria for Contaminated Soil/Sediment Cleanup. Pudvan Publishing, Northbrook, IL.

Galer. 1988. Technology Screening Guide for Treatment of CERCLA Soils and Sludges. EPA/540/2-88/004 (NTIS PB89-132674), 136 pp.

Hwang, S.T., J.W. Falco, and C.H. Nauman. 1987. Development of Advisory Levels for Polychlorinated Biphenyls (PCBs Cleanup). EPA/600/6-86/002 (NTIS PB86-232774).

Lipsky, D., W. Tuva, R. Dorrier, B. Johnson, and M. Gardner. 1989. Methods for Evaluating the Attainment of Cleanup Standards, Volume 1: Soils and Solid Media. EPA/230/2-89/042 (NTIS PB89-234959), 264 pp.

Loehr, R. 1989. Treatability Potential for EPA Listed Hazardous Chemicals in Soil. EPA/600/2-89/011 (NTIS PB89-166581/AS).

McGinnis, G.D., H. Borazjani, L.K. McFarland, D.F. Pope, and D.A. Strobel. 1988. Characterization and Laboratory Soil Treatability Studies for Creosote and Pentachlorophenol Sludges and Contaminated Soil. EPA/600/2-88/055 (NTIS PB89-109920).

Means, J.L. et al. 1994. The Application of Solidification/Stabilization to Waste Materials. Lewis Publishers, Boca Raton, FL, 368 pp.

Soil Remediation

Nash, J.H. 1987. Field Studies of In Situ Soil Washing. EPA/600/2-87/110 (NTIS PB88-146808).

Nash, J.H., S. Rosenthal, G. Wolf, and M. Avery. 1992. Potential Reuse of Petroleum-Contaminated Soil: A Directory of Permitted Recycling Facilities. EPA/600/R-92-096 (NTIS PB92-173780), 47 pp.

Raghavan, R., E. Coles, and D. Dietz. 1990. Cleaning Excavated Soil Using Extraction Agents: A State-of-the-Art Review. EPA/600/2-89/034 (NTIS PB89-212757).

Sims, R.C., D.L. Sorensen, J.L. Sims, J.E. McLean, R. Mahmood, and R.R. Dupont. 1984. Review of In-Place Treatment Technologies for Contaminated Surface Soils, 2 Volumes. EPA 540/2-84/003a&b (NTIS PB85-124881 and PB85-124899). [Volume 2 covers background information for in situ treatment]

Sims, R.C., J.L. Sims, D.L. Sorenson, W.J. Doucette, and L.L. Hastings. 1986. Waste-Soil Treatability Studies for Four Complex Industrial Wastes: Methodologies and Results, Volumes 1 and 2. EPA/6-86/002 (NTIS PB87-111738) and EPA/6-86/003 (NTIS PB87-111746).

Sims, R.C., W.J. Doucette, J.E. McLean, W.J. Grenney, and R.R. Dupont. 1988. Treatment Potential for 56 EPA Listed Hazardous Chemicals in Soil. EPA/600/6-88/001 (NTIS PB88-174446), 120 pp.

Sims, J.L., R.C. Sims, and J.E. Matthews. 1989. Bioremediation of Contaminated Soils. EPA/600/9-89/073 (NTIS PB90-164047).

Sims, J.L, R.C. Sims, R.R. Dupont, J.E. Matthews, and H.H. Russell. 1993. In Situ Bioremediation of Contaminated Unsaturated Subsurface Soils. Engineering Issue Paper, EPA/540/S-93/501, 16 pp.

Truett, J.B., R.L. Holberger, and K.W. Barrett. 1983. Feasibility of In Situ Solidification/Stabilization of Landfilled Hazardous Wastes. EPA/600/2-83/088 (NTIS PB83-261099).

Unterberg, W., R.S. Williams, A.M. Balinsky, D.D. Reible, D.M. Wetzel, and D.P. Harrison. 1987. Analysis of Modified Wet-Air Oxidation for Soil Detoxification. EPA/600/2-87/079 (NTIS PB88-102397).

U.S. Environmental Protection Agency (EPA). 1979. Survey of Solidification/Stabilization Technology for Hazardous Industrial Wastes. EPA/600/2-79-056.

U.S. Environmental Protection Agency (EPA). 1980. Guide to the Disposal of Chemically Stabilized and Solidified Wastes. EPA/SW-872 (NTIS PB87-154902).

U.S. Environmental Protection Agency (EPA). 1986. Systems to Accelerate in Situ Stabilization of Waste Deposits. EPA/540/2-86/002 (NTIS PB87-112306).

U.S. Environmental Protection Agency (EPA). 1988. Technological Approaches to the Cleanup of Radiologically Contaminated Superfund Sites. EPA/540/2-88/002, 117 pp.

(Table 13-9 Soil Remediation References)

U.S. Environmental Protection Agency (EPA). 1989. Summary of Treatment Technology Effectiveness for Contaminated Soil. EPA/540/8-89/053.

U.S. Environmental Protection Agency (EPA). 1990. Assessment of Technologies for the Remediation of Radioactivity Contaminated Superfund Sites. EPA/540/2-90/001.

Voskuil, T. 1991. Handbook: Remediation of Contaminated Sediments. EPA/625/6-91/028, 44 pp. Available from CERI.*

Voskuil, T. 1992. Handbook: Vitrification Technologies for Treatment of Hazardous and Radioactive Waste. EPA/625/R-92/002. Available from CERI.*

Water Environment Federation (WEF). 1994. Developing Cleanup Standards for Contaminated Soil, Sediment, & Groundwater: How Clean is Clean?. Specialty Conference Proceedings Series, WEF, Alexandria, VA, 469 pp.

Wilson, D.L. 1988. Report on Decontamination of PCB-Bearing Sediments. EPA/600/2-87/093 (NTIS PB88-113220).

Wise, D.L. and D.J. Trantolo (eds.). 1994. Remediation of Hazardous Waste Contaminated Soils. Marcel Dekker, New York, NY, 952 pp.

* See Preface for information on how to obtain documents from CERI (U.S. EPA Center for Environmental Research Information) and NTIS.

(Table 13-9 Soil Remediation References)

CHAPTER 14

REMEDIATION OF CONTAMINATED GROUND WATER

14.1 Overview 762

    14.1.1 Site Characterization 762
    14.1.2 Treatment Trains 764
    14.1.3 Regulatory Considerations 765

14.2 Source Control 766

    14.2.1 Removal 766
    14.2.2 Surface Water Controls 766

14.3 Containment: Ground-Water Barriers and Flow Control 769

    14.3.1 Slurry Trench Wall 769
    14.3.2 Grouting 770
    14.3.3 Sheet Piling, Membrane and Synthetic Sheet Curtains ... 771
    14.3.4 Interceptor Systems 771
    14.3.5 Hydrodynamic Controls 771

14.4 Ground-Water Collection 772

    14.4.1 Well-Field Pumping 775
    14.4.2 Interceptor Systems 777
    14.4.3 Ground-Water Treatment After Removal 777

14.5 Ground-Water Pump-and-Treat Methods 779

    14.5.1 Physical Treatment 779
    14.5.2 Chemical Treatment 785
    14.5.3 Biological Treatment 787

14.6 In Situ Treatment 790

    14.6.1 In Situ Physical/Chemical Treatment 790
    14.6.2 In Situ Biological Treatment 792

14.7 Guide to Major References* 795

* Appendix F contains citations for table and figure sources.

14.1 Overview

Restoration of contaminated ground water to former background or near background conditions is generally accomplished through one of two overall approaches. One approach involves natural or induced in situ treatment; the other uses engineered systems for the removal of ground water and is usually followed by treatment. For both approaches to ground-water restoration, any source that continues to contaminate ground water in the zone to be restored must be removed, isolated, or concurrently treated. In the former approach, treatment or removal of a still existing contamination source may eventually restore ground-water quality through natural processes. In other situations, contaminated ground water is removed from the aquifer or discharged to a surface water, and natural replacement is relied upon to eventually restore ground-water quality. Typically, however, these processes in the ground-water system take many years, decades, or even centuries. As a result, ground-water restoration typically requires approaches involving ground-water removal and treatment or, if necessary, induced in situ treatment following source control (removal, isolation, treatment). Site-specific conditions, properly defined and understood, provide ground-water engineers with the basic background information for determining an effective approach and for selecting and designing a cost-effective ground-water restoration system appropriate to the site.

This chapter provides an overview of aquifer restoration technologies utilizing techniques derived from interrelated disciplines of hydrology, geochemistry, civil engineering, construction, biology, and agronomy. Many of the technologies have been developed by demonstration and research in conjunction with remedial activities in the Superfund program. Figure 14-1 presents three general response actions and process options for a ground-water remediation program from a management viewpoint. The components of an *active restoration* response include the removal and treatment of contaminated ground water, the discharge of the treated water, and institutional controls. The *containment response* option includes the monitoring and containment of the contaminated ground water as well as institutional controls. The *natural attenuation response* includes institutional controls and monitoring. This chapter, which focuses on the technical aspect of remediation, briefly discusses the extraction and treatment components of active restoration and containment. The major emphasis of the chapter is on ground-water pumping systems and in situ biological treatment for organic contaminants that are found at almost all hazardous waste sites. Table 14-1 lists technologies that are discussed in this chapter.

14.1.1 Site Characterization

Ground-water restoration activities require dedication of sufficient resources to collect and understand background information. Field data provide detailed information about the hydrology and geochemistry of the site as well as the types of contaminants to be removed, their concentrations, and distribution. Monitoring well construction (Section 9.3), ground-water sampling (Section 9.5), and contaminant behavior (Chapter 4) are important elements of site definition that allow the restoration program to be planned and implemented. Understanding the ground-water flow at the site is important in developing the restoration plan. A literature review to

Ground-Water Remediation

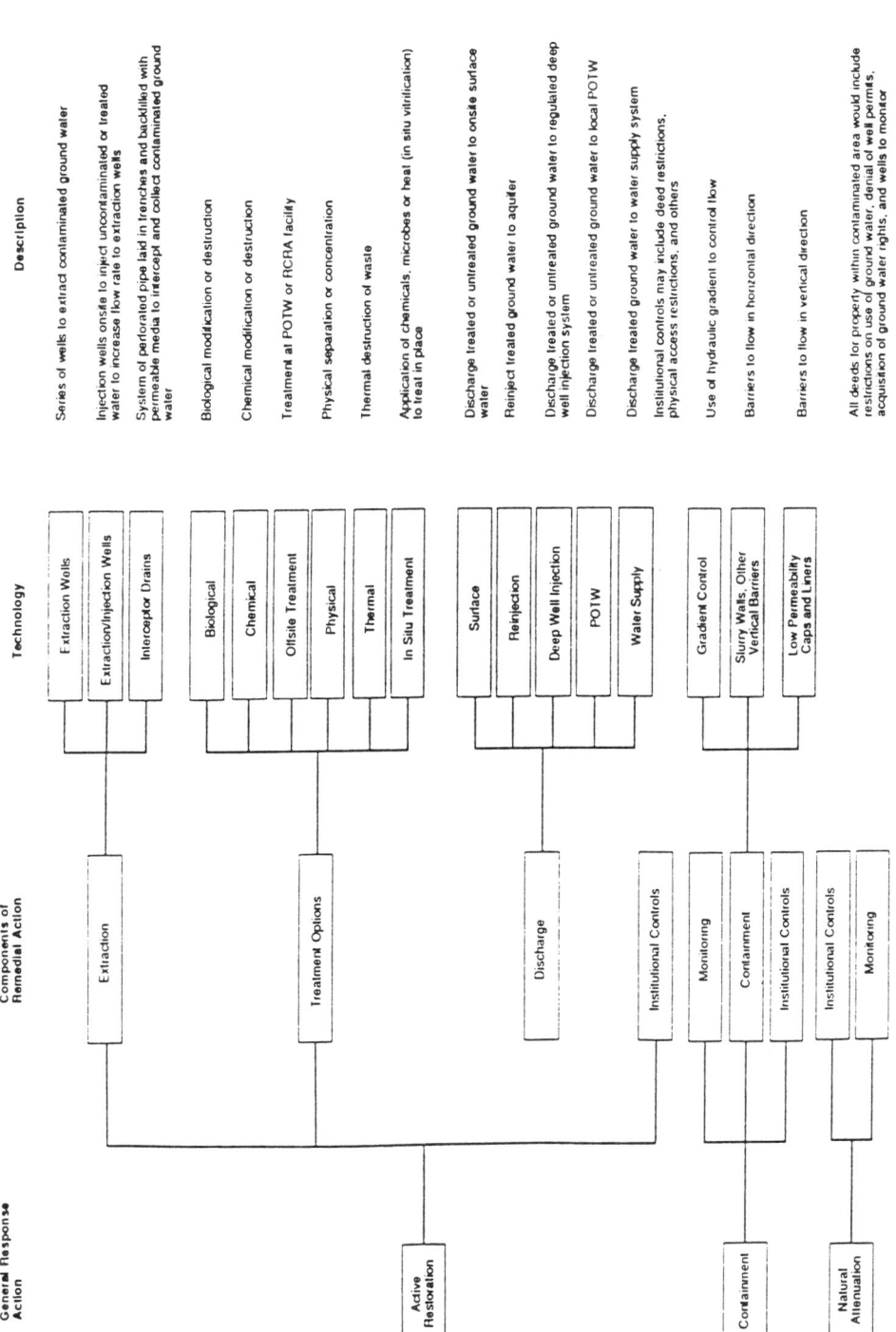

Figure 14-1 General response actions and process options for ground water (U.S. EPA, 1988b).

determine the characteristics of the contaminants is also useful in selecting restoration techniques best suited to the site. Finally, laboratory studies including treatability studies, sorption isotherms, and column and microcosm studies to determine contaminant transport and transformation parameters assist in developing a full understanding of the site conditions and potential alternatives for ground-water remediation.

Table 14-1 Available Technologies for Ground-Water Containment and Restoration

Ground-Water Source Containment and Isolation

Removal (Section 14.2.1)
Surface water control (Section 14.2.2)
Barriers (Sections 14.3.1 to 14.3.3)
Hydrodynamic control (Section 14.3.5)

Ground-Water Collection and Removal

Pumping (Section 14.4.1)
Drains (Section 14.4.2)

Ground-Water Treatment

Physical (Section 14.5.1)
Chemical (Section 14.5.2)
Biological (Section 14.5.3)
In situ treatment (Section 14.6)

14.1.2 Treatment Trains

In most contaminated hydrogeologic systems, the remediation process may be so complex, in terms of contaminant behavior and site characteristics, that no one system or unit is capable of meeting all requirements. Consequently, several unit operations are often combined in series and sometimes in parallel to effectively restore ground-water quality to the required level. Barriers and hydrodynamic controls may serve as temporary plume control measures; however, hydrodynamic processes are integral parts of any withdrawal and treatment or in situ treatment process.

Most remediation projects are typically started by removing the source. The next step may be the installation of pumping systems to remove free product floating on the water surface or the removal of soluble contaminants for treatment at the surface. Barriers might also be constructed to slow an advancing plume or to reduce the amount of water requiring treatment. Enhanced biorestoration techniques may be

Ground-Water Remediation

feasible in some of the more diluted areas of the plume. In some circumstances, a site may reach final restoration goals using natural chemical and biological processes. An adequate monitoring program would be required to establish data on the progress of the restoration program.

Steps in treatment of contaminated ground water include the removal, collection, and delivery of the contaminated water to the treatment units, and in the case of in situ processes, delivery of the treatment materials to the contaminated areas in the aquifer. A thorough knowledge and understanding of the hydrogeologic and geochemical characteristics of the site are required to design a system that will optimize the remediation techniques selected, maximize the predictability of restoration effectiveness, and allow for the development of a cost-effective and lasting remediation program.

14.1.3 Regulatory Considerations

The principal criteria for selecting remediation procedures are the water quality level to which to restore an aquifer and the most economical technology available to reach that level. Institutional limitations, however, sometimes override these criteria in determining if, when, and how remediation will be selected and carried out.

Response to a ground-water contamination problem is likely to require compliance with several local, state, and federal pollution control laws and regulations. If the response involves handling hazardous wastes, discharging substances into the air or surface waters, or injecting wastes underground, federal and state pollution control laws will apply. These laws do not exempt the activities of federal, state, or local officials or other parties attempting to remediate contamination problems. They apply to both generators and responding parties, and it is not unusual for these pollution control laws to conflict. A hazardous waste remediation project must meet RCRA permit requirements governing the transport and disposal of hazardous wastes, which can influence the selection of the remediation plan and the scheduling of cleanup activities.

In situ remediation procedures may be subject to permitting or other requirements under federal or state underground injection control programs. Withdrawal and treatment approaches may be subject to regulation under federal or state air pollution control programs or to pretreatment requirements if contaminated ground water is to be discharged to a surface water or to a municipal wastewater treatment system. A remediation plan involving pumping from an aquifer may be subject to state ground-water regulations on well construction and well spacing, and may need to consider various competing legal rights to extract ground water. U.S. EPA's **CERCLA Compliance with Other Laws Manual: Interim Final** (EPA/540/G-89/006) identifies major laws and regulations that must be considered.

Other factors influencing selection and design of a ground-water remediation program include the availability of alternative sources of water supply, political and judicial constraints, and the availability of funds. Where alternate water supplies are plentiful and economical, there may not be a demand for total remediation; adequate remediation to protect human health and the environment may be sufficient. In the final analysis, responsible agencies can pursue remediation measures to the extent that resources are made available.

14.2 Source Control

The objective of source control strategies is to reduce or eliminate the volume of waste, thereby eliminating or minimizing ongoing contamination of the ground-water environment. Source control techniques include removal of the source, surface water controls, ground-water barriers, interceptors, and hydrodynamic controls. Table 14-2 lists these commonly used source control technologies and advantages and disadvantages of each. A brief discussion of each technique follows.

14.2.1 Removal

Soil and water from a hazardous waste site may be removed for treatment or relocation to a site that is more acceptable from an engineering or environmental viewpoint. While the removal and reburial of contaminated materials at a more controlled site appears to solve the contamination problem, various factors need to be evaluated before excavation commences. These factors include the excavation and handling of bulky partially decomposed or hazardous waste; the distance and means of transportation to the reburial site; the risks to public health and the environment on the route between sites; political, social, and economic factors associated with locating a new site; the disposal of the contaminated ground water at the sites; the control of nuisances and vectors during excavation; reclamation of the excavated site; and the costs involved (Tolman et al., 1978/T14-9). These considerations suggest that excavation and relocation may be a viable alternative only where costs are not significant compared to the importance of the resource protection. In some cases, removal and reburial in an approved facility transfers a problem from one location to another, and possibly creates additional problems.

14.2.2 Surface Water Controls

Surface water control measures are used to minimize the infiltration and percolation of surface water or precipitation into the ground water of a waste site. This can be accomplished by contouring the site, providing a cap or barrier to infiltration, and revegetating the site.

Changing Contours. Several standard engineering techniques can be used to change the contour and runoff or runon characteristics of a particular site. Some of the more common techniques are dikes and berms, ditches, diversion waterways, terraces, benches, chutes, downpipes, levees, seepage basins, sedimentation basins, and surface grading.

Capping. A cover or cap of low permeable material prevents water from entering the site, thus reducing leachate generation. Covers also can control vapors or gases produced in a landfill. Covers or caps may be constructed of native soils, clays, synthetic membranes, soil cement, bituminous concrete, or asphalt.

Revegetation. Revegetation can be a cost-effective method of stabilizing the surface of a waste site especially when preceded by capping and contouring. Vegetation reduces raindrop impact and runoff velocity, and strengthens the soil mass, thereby reducing erosion by wind and water. It also improves the site aesthetically.

Table 14-2 Advantages and Disadvantages of Commonly Used Source Control Technologies

Method	Advantages	Disadvantages
Source Removal	--Source is moved to controlled environment.	--Site accessibility often difficult. --Potential for accidents during transport. --High costs.
Surface Water Controls		
Changing contours and runoff and runon characteristics	--Relatively inexpensive. --Reduces amount of potential infiltration; allows for drainage while controlling erosion.	--Requires maintenance to repair slopes and eliminate depressions.
Capping	--Prevents or minimizes infiltration. --Can be used at almost any site.	--Requires long-term maintenance.
Revegetation (used with capping and contouring)	--Reduces erosion.	--Roots may penetrate cap allowing for infiltration. --Requires maintenance.
Ground-Water Barriers		
Slurry trench wall	--Construction methods are simple. --Adjacent areas are not affected by ground-water drawdown. --Bentonite (mineral) will not deteriorate with age. --Leachate-resistant bentonites are available. --Low maintenance requirements. --Eliminates risks due to strikes, pump breakdowns, or power failures. --Eliminates headers and other aboveground obstructions.	--Cost of shipping bentonite to some areas of the country is high. --Some construction procedures are patented and require a license. --Over-excavation in rocky ground is necessary because of boulders. --Bentonite deteriorates when exposed to high ionic strength leachate --Adequate key to impermeable formation is critical. --Difficult to assess in-place integrity.

Table 14-2 (cont.)

Method	Advantages	Disadvantages
Grouting	--When designed on the basis of thorough preliminary investigations, grouts can be very successful. --Grouts have been used over 100 years in construction and soil stabilization projects. --Many kinds of grout to suit a wide range of soil types are available.	--Grouting is limited to granular types of soils having a pore size large enough to accept grout fluids under pressure yet large enough to prevent significant pollutant migration before implementation. --Grouting in a highly layered soil profile may result in incomplete formation of a grout envelope. --Presence of high water table and rapidly flowing ground water limits groutability through extensive transport of contaminants and rapid dilution of grouts. --Some grouting techniques are proprietary. --Procedure requires careful planning and pretesting. Methods of ensuring that all voids in the wall have been effectively grouted are not readily available. --Grouts may not withstand attack from specific pollutants.
Sheet piling	--Construction is not difficult and no excavation is necessary. --Contractors, equipment, and materials are available throughout the United States. --Construction can be economical. --No maintenance required after construction.	--The steel sheet piling initially is not water-tight. --Driving piles through ground containing boulders is difficult. --Certain chemicals may attack the steel.
Interceptor Systems	--Easy and inexpensive to install. --Useful for intercepting landfill side seepage and runoff. --Useful for collecting leachate in poorly permeable soils. --Large wetted perimeter allows for high rates of flow. --Possible to monitor and recover. --Produces much less fluid to be handled than well-point system (hydrodynamic controls).	--When dissolved constituents are involved it may be necessary to monitor ground water downgradient of the recovery line. --Open systems require safety precautions to prevent fires or explosions. --Interceptor trenches are less efficient well-point systems (hydrodynamic controls). --Operation and maintenance costs are high. --Not useful for deep disposal sites.

Ground-Water Remediation

Table 14-2 (cont.)

Method	Advantages	Disadvantages
Hydrodynamic Controls	--Less costly to construct than physical barriers. --High degree of flexibility in design; additional wells can be added. --Efficient and effective means of assuring ground-water pollution control. --Previously installed monitoring wells can sometimes be employed. --Can sometimes include recharge of aquifer as part of the strategy. --Can be installed readily.	--High operation and maintenance costs. --High cost of monitoring the system. --Withdrawal systems necessarily remove clean (excess) water along with polluted water. --Some systems require the use of sophisticated mathematical models to evaluate effectiveness. --Withdrawal systems will usually require surface treatment prior to discharge. --Applications to fine soils is limited. --System failures due to breakdown or power failures may lead to contaminant movement.

Source: Compiled from U.S. EPA (1985), Knox et al. (1984), Nielsen (1983), and Wagner et al. (1986).

14.3 Containment: Ground-Water Barriers and Flow Control

Subsurface barriers are designed to prevent or control ground-water flow into, through, or from a certain location. Barriers keep fresh ground water from coming into contact with a contaminated aquifer zone or ground water from existing areas of contamination from moving into areas of clean ground water. The types of barriers commonly used are included in Table 14-2 and are described further in the following section.

14.3.1 Slurry Trench Wall

Slurry trench walls are placed either upgradient from a waste site to prevent flow of ground water into the site or around a waste site to prevent the movement of contaminated ground water away from it. A slurry wall may be either keyed in to a low permeability confining layer by extending part-way into the confining layer or it may extend several feet into the water table to act as a barrier to floating contaminants. A slurry trench wall is constructed by excavating a trench at the proper location and to the desired depth, while keeping the trench filled with a clay slurry composed of a 5 to 7% by weight suspension of bentonite in water. The slurry maintains the vertical stability of the trench walls by exerting a hydrostatic pressure against the surrounding ground water and also forms a low permeability filter cake on the walls of the trench. As the slurry trench is excavated, it is simultaneously backfilled with a material that forms the final wall. The three major types of slurry backfill mixtures are soil bentonite, cement bentonite, and concrete.

Slurry trench walls are reported to have a reasonable service life and short construction time, cause minimal environmental impact during construction, and be a highly cost-effective method for enclosing large areas under certain conditions (Nielsen, 1983/T14-9). A concern regarding the use of a slurry wall where contaminated materials are in direct contact with the wall is the long-term integrity of the wall (Wagner et al., 1986a/T12-7). In such cases, the condition of the wall needs to be verified over time by ground-water monitoring.

14.3.2 Grouting

Grouting is the process of injecting stabilizing materials under pressure, into subsurface soils or rocks to fill and, thereby, seal the voids, cracks, fissures, or other openings. *Grout curtains* are fixed, underground physical barriers formed by injecting grout through tubes. The amount of grout needed is a function of the available void space, the density of the grout, and the pressures used in setting the grout; however, two or more rows of grout are normally required to provide a good seal. The grout used may be either particulate (i.e., Portland cement) or chemical (i.e., sodium silicate) depending on the soil type and the contaminant present. Grouting creates a fairly effective barrier to ground-water movement, although the degree of completeness of the grout curtain is difficult to ascertain (Nielsen, 1983/T14-9). Incomplete penetration of the grout into the soil voids would increase the permeability of the curtain.

Semicircular grout curtains provide the most complete containment but require that the grouting take place in contaminated ground water downgradient of the source (Wagner et al., 1986a/T12-7).

A variation of the grout curtain is the *vibrating beam technique* for placing thin (approximately 4 inches) curtains or walls. Although this type of barrier is sometimes called a slurry wall, it is more closely related to a grout curtain since the slurry is injected through a pipe in a manner similar to grouting. A suspended I-beam connected to a vibrating driver-extractor is vibrated through the ground to the desired depth. As the beam is raised at a controlled rate, slurry is injected through a set of nozzles at the base of the beam, filling the void left by the beam's withdrawal. The vibrating beam technique is most efficient in loose, unconsolidated deposits such as sands and gravels.

Another method that uses grouting is *bottom sealing*, where grout is injected through drill holes to form a horizontal or curved barrier below the site to prevent downward migration of contaminants.

Block displacement is a relatively new plume management method, in which a slurry is injected so that it forms a subsurface barrier around and below a specific mass or "block" of material. Continued pressure injection of the slurry produces an uplift force on the bottom of the block, resulting in a vertical displacement proportional to the slurry volume pumped.

14.3.3 Sheet Piling, Membrane and Synthetic Sheet Curtains

Sheet piling cutoff walls can be made of wood, reinforced concrete, or steel, with steel being the most effective material for constructing a ground-water barrier. The construction of a sheet pile cutoff wall involves driving lengths of steel sheets through unconsolidated materials with a pile driver. The individual sheet piles are connected along the edges with various types of interlocking joints, which can provide permeable pathways for ground-water movement if they do not become watertight naturally. It may be desirable to fill these joints with an impermeable material such as grout; however, the success of creating a tight seal in this way has not been fully established.

Membrane and *synthetic sheet curtains* can be used in applications similar to grout curtains and sheet piling. With this method, the membrane is placed in a trench surrounding or upgradient of the plume, thereby enclosing the contaminated source or diverting ground-water flow around the source. Placing a membrane liner in a slurry trench application has also been tried on a limited basis. Attaching the membrane to the impervious layer and forming perfect seals between the sheets is difficult, but necessary in order for membranes and other synthetic sheet curtains to be effective.

14.3.4 Interceptor Systems

Two types of interceptor systems used for source control are the *passive system*, which relies on gravity flow, and the *active system*, which uses pumps. An interceptor system consists of trenches excavated to a depth below the water table with a perforated collection pipe in the bottom. Active interceptor systems have vertical removal wells spaced along the interceptor trench or a horizontal removal pipe in the bottom of the trench. Active systems are usually backfilled with a coarse sand or gravel to maintain the stability of the wall. These interceptor systems can be used as preventive measures or in product recovery from ground water. Interceptor drains are generally used to either lower the water table beneath a contamination source or to collect ground water from an upgradient source to prevent leachate from reaching uncontaminated wells or surface water. Interceptor systems are relatively inexpensive to install and operate; they are not, however, well suited for soils having a low permeability.

14.3.5 Hydrodynamic Controls

Hydrodynamic controls are used to isolate a plume of contamination from the normal ground-water flow regime to prevent the plume from moving into a well field, another aquifer, or the surface water. Hydrodynamic control is achieved through the use of well systems including well point systems, deep-well systems, and pressure ridge systems.

A *well point system* consists of several closely spaced shallow wells connected to a main header pipe, which is then connected to a suction lift pump. Well point systems are used only for shallow aquifers and are designed so that the system drawdown completely intercepts the plume of contamination.

Deep wells are similar to well point systems except they are used at greater depths and are normally pumped individually. These wells are used in consolidated formations where the water table is too deep for the economical use of a suction lift system.

Pressure ridge systems are produced by injecting uncontaminated water into the subsurface through a line of injection wells located upgradient or downgradient of a contamination plume. Upgradient ridges or mounds are used to force upgradient uncontaminated ground water to flow around a contaminant plume. The contaminated ground water is then collected by a line of downgradient pumping wells. The velocity of clean ground water into the plume is increased, thereby increasing flow to the recovery wells, which serves to wash the aquifer. Downgradient pressure ridge systems are normally used in combination with upgradient pumping wells that supply uncontaminated injection water. The injection of clean water produces a mound in the original water table that acts as a barrier by forming a ridge that contains the contaminated plumes and forces flow away from the mound.

A thorough knowledge of the hydrogeological conditions of the site is required for the development of a hydrodynamic control system. The effect of the injection wells on the draw down and the radius of influence of the pumping wells must be analyzed. Monitoring of the system is necessary to ensure that ground water outside of the system is not being contaminated.

14.4 Ground-Water Collection

The cleanup of a contaminated ground-water site involves the collection and/or treatment of the contaminated water. Some of the techniques used for source control are often used as part of a ground-water cleanup program, including interceptor systems, pumping well systems, and some of the techniques used for runoff alteration. In addition, in situ treatment, enhanced desorption, encapsulation, and biodegradation may be part of a ground-water cleanup plan.

A ground-water pumping system combined with a treatment system, also called a *pump-and-treat* system, is often designed for a specific ground-water contamination problem. Figure 14-2 illustrates cross-sectional and plan views of a pump-and-treat system for an aquifer contaminated by an underground storage tank. Ground-water treatment systems are also used in conjunction with pumping systems designed to lower the water table or to contain a contaminant plume. The use of pump-and-treat systems is probably more widespread than that of all other restoration techniques combined. Large expenditures are made each year to prepare for and operate pump-and-treat remediation of ground-water contamination (Keely, 1989/T14-9). The hydrology of the site, the source of the contaminant, and the characteristics of the contaminant must be understood if an efficient and cost-effective pump-and-treat program is to be conducted.

When a significant amount of free product of a nonaqueous phase liquid (NAPL) is present in an aquifer (NAPLs), pump-and-treat systems are designed to maximize recover of the free product. Physical recovery techniques to remove free

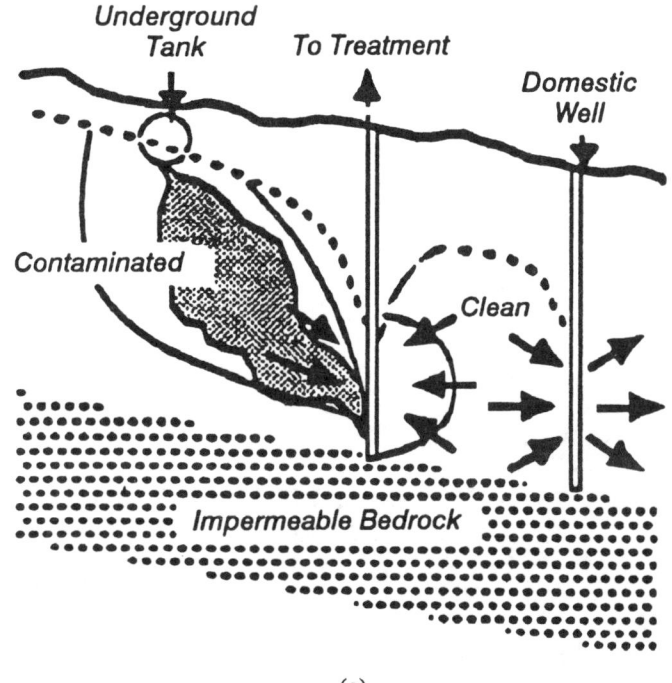

(a)

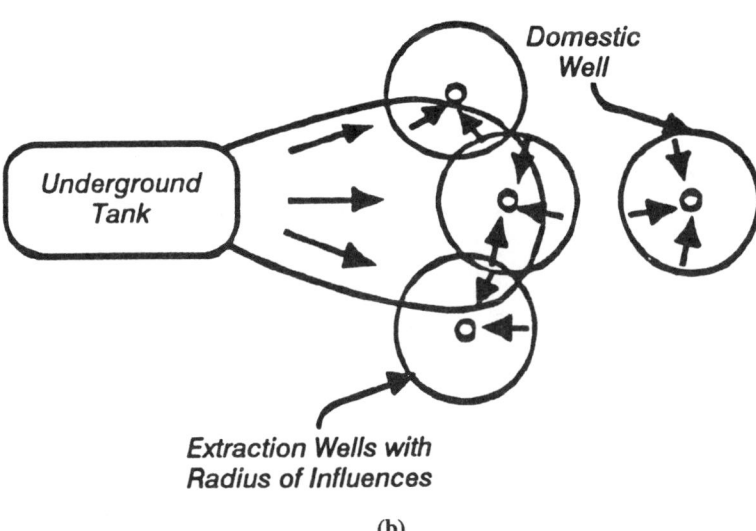

(b)

Figure 14-2 Pump-and-treat system: (a) cross-sectional view; (b) plan view (U.S. EPA, 1985).

product include (1) a single pump system producing a mixture of hydrocarbon and water that must be separated, but requiring minimal equipment and drilling; (2) a two-pump, two-well system utilizing one pump to produce a water table gradient and a second well to recover floating product; or (3) a single well with two pumps in which a lower pump produces a gradient and an upper pump collects free product. Figure 14-3 illustrates a single-well, two-pump system for recovering LNAPLs. An above-ground oil/water separator generally is used to recover product for future use. Vacuum extraction of volatilizing contaminants also may be used to recover floating free product from a perched water table (Section 13.2).

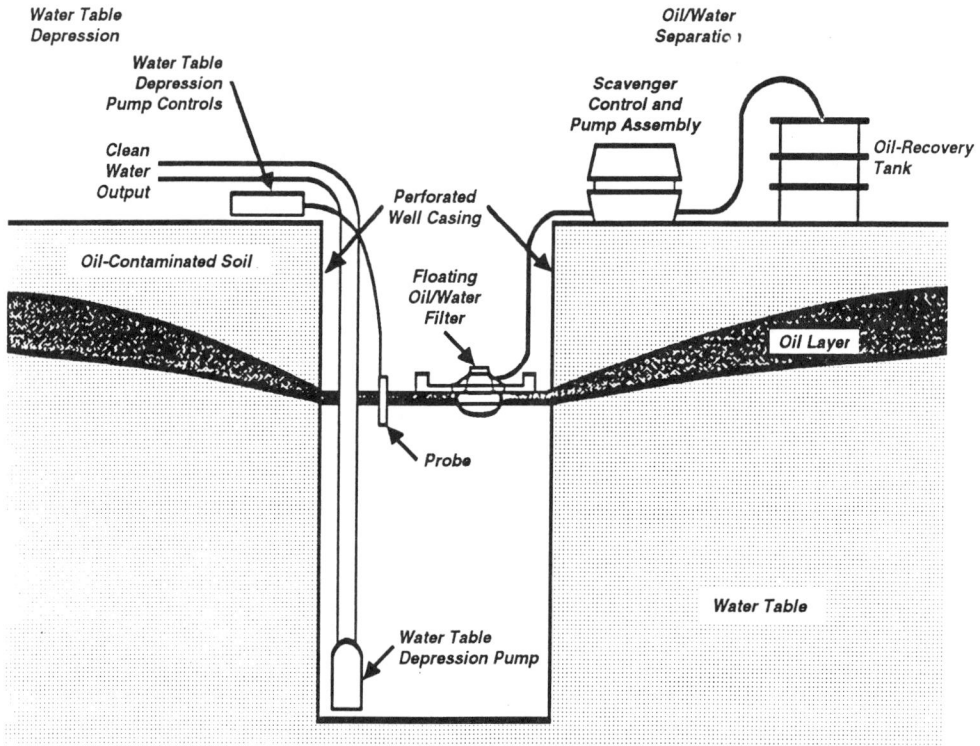

Figure 14-3 Product recovery using a water-table depression pump and a floating oil/water filter (Sims and Sims, 1991c, after Nyer, 1985).

Caution should be exercised during product recovery of LNAPL when an extraction well is used to control local gradients and collect free product in a cone of depression. Due to capillary forces in the subsurface aquifer material, trapped residual will constitute a continuous source of contamination to ground water that will persist after product removal from the water table is completed.

Ground-Water Remediation 775

14.4.1 Well-Field Pumping

The operation of a well field to remove ground water causes the formation of *stagnation zones* downgradient from the extraction wells, which must be considered in the design of a system. For example, if remedial action wells are located within the bounds of a contaminant plume, the portion of the plume lying within the stagnation zones will not be effectively remediated because the contaminants are removed from only the zone of advective ground-water flow. In this case, the only remediation in the stagnation zone will result from the process of chemical diffusion, which is very slow. Proper location of wells based on pumping rates and drawdown mitigate this effect to the extent possible.

The *tailing effect* can affect the removal and renovation of a ground water containing a low solubility contaminant. Tailing is the slow, nearly asymptotic decrease in contaminant concentration in ground water moving through a contaminated geologic material. The contaminants migrate into the finer pore structures of the geologic material and are slowly exchanged with the bulk water present in larger pores that is mobilized during pumping, thus resulting in "tailing" (Figure 14-4).

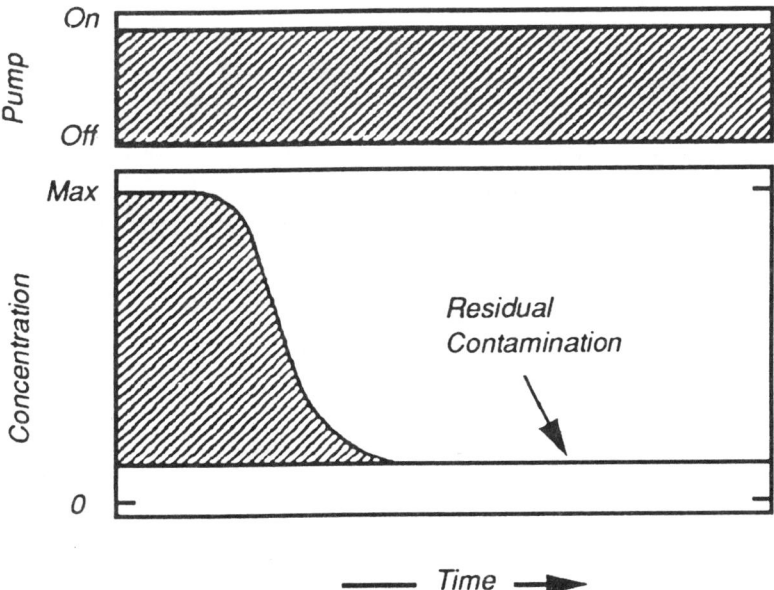

Figure 14-4 Stabilization of aquifer contaminant concentration during pump-and-treat resulting from tailing effect (Keely, 1989).

Many manmade and natural organic compounds found in ground water tend to adsorb to the organic and mineral components of the aquifer material. When water is removed by pumping, the contamination can remain on the aquifer material. The amount of remaining contaminant depends on the geologic materials and characteristics of the contaminants. Once sorbed to the geologic material, contaminants may desorb slowly into the ground water thus requiring extended periods of pumping and treating to attain desired levels of restoration.

The removal of a water-insoluble liquid, such as gasoline, can be difficult since it may become trapped in the pore spaces of the soils and is not easily removed by pumping. Pumping ground water to remove the components of a residual phase initially may reduce the concentration, but this reduction may only be the result of dilution or lowering of the water table below the surface of contamination. A contaminant will not be removed faster than it is released into the ground water, so if the pumping stops for a period of time, water-soluble residual phase components will dissolve into the ground water bringing the concentrations back up to their previous levels.

An innovation in pump-and-treat technology is *pulsed pumping*. This technique involves alternating the periods of pumping, allowing contaminants time to come to equilibrium with the ground water in each cycle (Figure 14-5). Equilibrium is achieved by diffusion from stagnant zones or zones of lower permeability, and by partitioning of sorbed contaminants or those associated with residual contaminant phases. Alternating pumping among wells in an extraction field can also establish active flow paths in the stagnant zones.

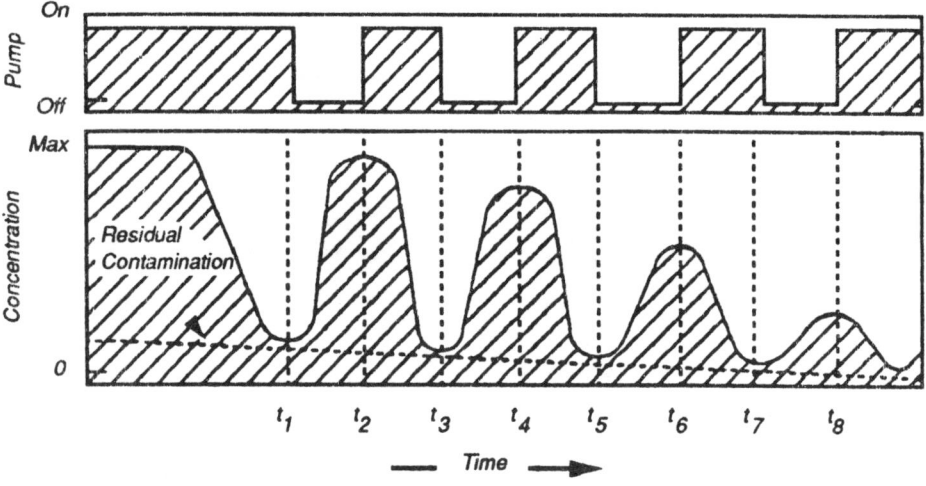

Figure 14-5 Pulsed pumping removal of residual contaminants reduces volume of water pumped and increases contaminant concentration in pumped water (Keely, 1989).

Another innovation is the use of pump-and-treat systems in conjunction with other remediation technologies. Examples are the use of extraction wells with barrier walls to limit plume expansion while reducing the amount of clean water pumped, and the use of surface ponds or flooding to flush contaminants from the vadose zone prior to collection by a pumping system.

14.4.2 Interceptor Systems

Interceptor systems (also discussed in Section 14.3.4) may be an alternative to well-field pumping systems. The subsurface drains used in interceptor systems essentially function as an infinite line of extraction wells, and can perform many of the same functions as the wells. Subsurface drains create a continuous zone of influence in which ground water flows toward the drain (Wagner et al., 1986a/T12-7). Subsurface drains are installed perpendicular to the direction of ground-water flow and collect ground water from an upgradient source for treatment. Interceptor systems prevent leachate or contaminated ground water from moving downgradient toward wells or surface water.

Subsurface drains also have been used for recovery of DNAPLs (Figure 14-6a). When only the oil recovery drainline (ORD) is used (Figure 14-6b), water truncates the flow of product (DNAPL) due to the poor relative permeability of the product as described previously in the discussion of soil flushing. The water table depression drainline (WTDD) is an efficient method (see Figure 14-6c) to drag an oily product across the subsurface by viscous forces and thereby create a hydraulic head of oil above the ORD; however, oil also enters the WTDD, thereby creating the need for aboveground separation of product and water. When both ORD and WTDD are used (Figure 14-6d), subsurface separation of oil and water is achieved, thereby minimizing aboveground separation requirements. This system (Figure 14-6d) is also efficient since the permeability of oil is greatest in the oily contaminated subsurface, and the underground separation maintains water flowing in the water compartment and oil flowing in the oily compartment.

In stratified soils with variable hydraulic conductivities, the drain is normally installed on a layer with a low hydraulic conductivity to minimize leachate leakage under the drain and contamination flowing downgradient. An impermeable liner placed in the bottom of a trench can also be used to control underflow. The design, spacing, and location of drains for various soil and ground water conditions are described further in Wagner et al. (1986a/T12-7).

The difficulty in installing subsurface drains at great depths generally limits the use of interceptor systems to shallow aquifers. Active interceptor systems use pumps to bring the ground water to the surface, but may have an advantage over well fields because fewer pumps are required. Use of passive systems using gravity flow may be possible where the topography allows placement of the surface treatment facility below the outlet of the subsurface drain. Interceptor systems may be preferred over well-field pumping in cases where ground water must be removed over a period of several years, since the operation and maintenance costs are lower than for a well system.

14.4.3 Ground-Water Treatment after Removal

Treatment technologies for pumped or intercepted ground water can be grouped into three broad areas: physical, chemical, and biological. *Physical* treatment methods (Section 14.5.1) include adsorption, density separation, filtration, reverse osmosis, air and steam stripping, and incineration. Precipitation, oxidation/reduction, ion exchange,

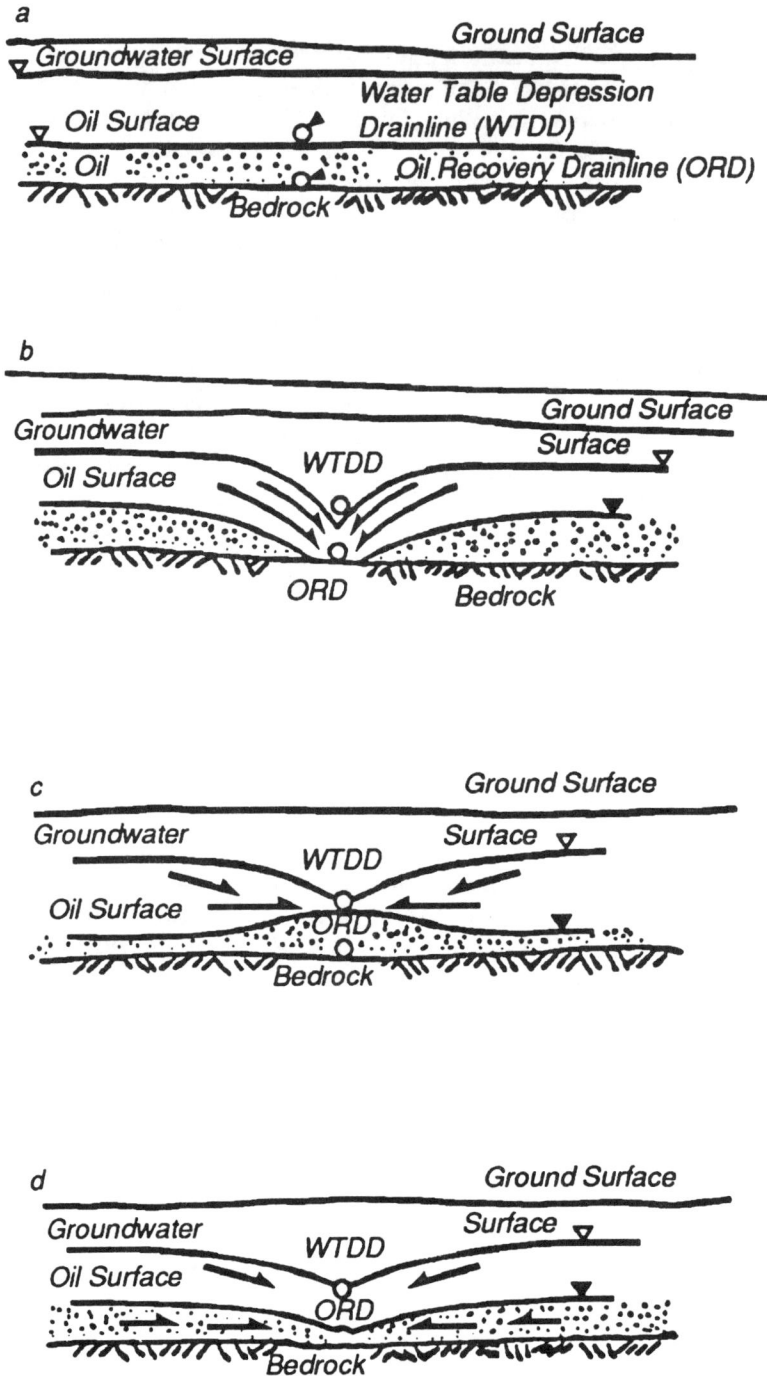

Figure 14-6 Use of drains to separate DNAPL from water where DNAPL rests on shallow bedrock; see text for explanation (Sims and Sims, 1991c, after Sale and Piontek, 1989).

and neutralization are commonly used *chemical* treatment methods (Section 14.5.2). *Biological* treatment methods (Section 14.5.3) include activated sludge, aerated surface impoundments, anaerobic digestion, trickling filters and rotating biological discs. Table 14-3 identifies potential processes for treatment of hazardous waste leachate or contaminated ground water and Table 14-4 summarizes treatment process capabilities for landfill leachate. Figure 14-7 identifies general suitability of different treatment methods for organic and inorganic contaminants. Moving from top to bottom in Figure 14-7, treatment methods are able to handle increasing concentrations of contaminants.

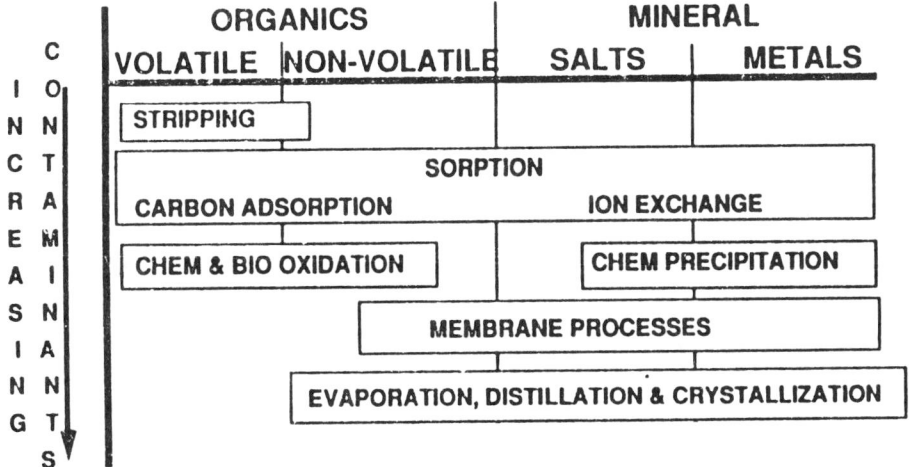

Figure 14-7 Contaminated ground water treatment matrix (U.S. EPA, 1990).

14.5 Ground-Water Pump-and-Treat Methods

14.5.1 Physical Treatment

Table 14-5 summarizes the advantages and disadvantages of seven methods for physically treating contaminated ground water.

Adsorption. *Granular activated carbon* (GAC) is the most widely used material for adsorption of organic contaminants from water. *Synthetic resins* are also used as adsorbents, but they are expensive and their use is currently in the developmental stages. Synthetic resins trap contaminants within the chemical structure of the resin whereas GAC traps contaminants within the physical pore structure of the carbon. The extent of adsorption depends on the strength of the molecular attraction among the adsorbent and the adsorbate, the molecular weight of the contaminants, the type and characteristic of the adsorbent, the electrokinetic charge, the pH, and the surface area (Rich and Cherry, 1987).

Table 14-3 Process Applicability Matrix for Hazardous Waste Leachate Treatment Methods (McArdle et al., 1987)

Technology	Suspended Solids	Oil, Grease, Immiscible Liquids	pH (acidic, basic)	Total Dissolved Solids	Metals	Cyanides	Volatile Organics	Semivolatile Organics	Pesticides, PCB's	Pathogens
Sedimentation	+	+	o	o	o	o	o	o	o	o
Granular-media filtration	+	−	o	o	o	o	o	o	o	o
Oil/water separation	o	+	o	o	o	o	o	o	o	o
Neutralization	o	o	+	o	o	o	o	o	o	o
Precipitation/flocculation sedimentation	+	+	o	+	+	o	o	o	o	o
Oxidation/reduction	−	−	o	o	+	+	o	+	+	+
Carbon adsorption	−	−	o	o	+	+	+	+	+	+
Air Stripping	−	−	o	o	o	o	+	o	o	o
Steam stripping	−	−	−	o	o	o	+	o	o	o
Reverse osmosis	−	−	−	+	+	+	o	o	+	+
Ultrafiltration	−	−	−	+	+	o	o	+	+	+
Ion exchange	−	−	o	+	+	+	o	o	−	o
Wet-air oxidation	o	o	o	o	+	+	+	+	o	o
Activated sludge	−	−	−	o	−	o	+	+	o	o
Sequencing batch reactor	−	−	−	o	−	o	+	+	o	o
Powdered activated carbon treatment (PACT)	+	−	−	o	−	o	+	+	+	o
Rotating biological contactor	o	−	−	o	−	o	+	+	o	o
Trickling filter	−	−	−	o	−	o	+	+	o	o
Chlorination	o	−	−	o	o	+	o	o	o	+

Key: (+) process is applicable for removal of the contaminant; (o) process is not applicable for removal of the contaminant; (−) process is not applicable unless the leachate is pretreated for removal of the contaminant.

Table 14-4 Summary of Landfill Leachate Process Capabilities (Pohland and Harper, 1986).

	BOD$_5$		COD		TKN		Fe		Zn		Ni		Comments
	Rem., %	Effl., mg/l	Rem., %	Effl., mg/l	Rem., %	Effl., mg/l	Rem., %	Effl., mg/l	Rem., %	Effl., mg/l	Rem., %	Effl., mg/l	
Aerobic Biological Processes													
Activated Sludge	95	100	95	500	70-95	10-100	96-99	10-40	96-99	3-10	60	0.25	Θ_c = 6-10 days
Combined Leachate and Sewage	94-99	3-15	92-98	25-60	—	—	—	—	—	—	—	—	ratio <5%
Aerated Lagoon	99	5-60	92-98	300-800	40-70*	40-80	99	0.2	—	—	—	—	Θ_c >10 days
Stabilization Pond	93-99	10-100	99	100-400	70-99	4-100	80-99	1-100	—	—	—	—	τ >40 days
Aerobic Fixed Film*													
Anaerobic Biological Processes													
Attached Growth	85-98	100-900	75-95	200-1000	—	—	80-99	5-25	80-99	0.5-10	10-80	0.1-1	Θ_c >10 days
Suspended Growth	85-98	100-900	75-95	200-1000	—	—	80-99	5-25	80-99	0.5-10	10-80	0.1-1	Θ_c >5 days
Leachate Recycle	NA	<100	NA	<5	NA	20-1000	NA	5-50	NA	0.2-1	NA	—	Θ_c >500 days
Physical/Chemical Processes													
Coagulation	—	—	12	100-10,000	—	—	95-99	2-17	75-98	<1	—	—	Lime, alum, ferric chloride
Oxidation	—	—	10-50		—	—	99	<1	90	<1	—	—	Ozone, chloride permanganate
Reverse Osmosis	—	—	60-90**	1000-8000	—	—	—	—	—	—	—	—	Raw Leachate
			86-94	<10									Pretreated Leachate
Ion Exchange	—	—	40-70	100-300	—	—	40-80	1-10	20-96	<1	14-96	<1	Commercial IX Resins and GG
Adsorption	—	—	75-99	<10	—	—	65-95	2-15	—	—	—	—	GAC and PAC

Rem. = Removal; Effl. = Effluent.
*Insufficient data to make an adequate judgment;
**TOC Basis.

Table 14-5 Advantages and Disadvantages of Physical Methods of Ground-Water Treatment

Method	Advantages	Disadvantages
Absorption	--GAC is effective and reliable for removing low solubility organics. --GAC can be used for treating a wide range of contaminants over a broad range of concentrations. --GAC is not adversely affected by toxics.	--High operation and maintenance costs. --GAC is intolerant of high suspended solids. --Pretreatment is required for oil and grease greater than 10 mg/L. --Synthetic resins are intolerant of strong oxidizing agents and suspended solids.
Density separation	--Minimal operational requirements. --Readily available equipment may be used.	--Incomplete removal of hazardous compounds. --Sludge generated requires disposal.
Filtration	--Reliable and effective means of removing low levels of solids. --Equipment is readily available and is easy to operate and control.	--Filters clog if suspended solids concentration is high. --Backwash water requires further treatment.
Reverse osmosis	--Ability to reduce both inorganic and organic dissolved solids.	--Units are subject to chemical attack, fouling, and plugging. --Pretreatment requirements may be extensive.
Stripping (packed tower)	--Equipment is relatively simple and startup and shutdown can be accomplished quickly. --Modular design is suited for hazardous waste applications.	--Sensitive to pH, temperature, and fluxes in hydraulic load. --May be cost prohibitive at temperatures below freezing. --Often only partially effective and must be followed by another process. --May cause air pollution problems.
Incineration	--Can be used to destroy contaminants in liquid, gaseous, and solid waste.	--Thickening and dewatering may be required prior to incineration. --May pose air pollution problems. --Produces an inorganic ash that may be hazardous. --May require costly fuel or power for operation.
Wet air oxidation	--Effective on concentrated wastes that are not readily biodegradable. --Generates no air pollution.	--Energy intensive.

Source: Compiled from U.S. EPA (1985), Knox et al. (1984), Nielsen (1983), and Wagner et al. (1986).

A typical adsorption system consists of a large vessel filled with adsorbent with an inlet zone for contaminated water and an outlet zone for treated water. Influent water contacts the adsorbent for a specified period of time and then exits for collection, disposal, recharge, or further treatment. GAC systems are often arranged with several units in parallel or in series to allow for the most efficient treatment possible. Once the micropore surfaces are saturated, little or no further adsorption occurs. Some contaminants may be released during treatment, resulting in a gradual increase in the effluent contaminant concentration. When the concentration exceeds the desired limit, the adsorbent must be replaced and the used adsorbent discarded or regenerated.

Adsorption using GAC is an effective and reliable technique for removing low solubility organics and some metals and inorganic species. It can be used for treating a wide range of contaminants over a broad concentration range and is not adversely affected by toxics. Carbon adsorption is often used after preliminary treatment steps to reduce high organic and suspended solids loads to improve the efficiency and extend the life of the GAC or synthetic resin.

Density Separation. Ground water with high concentrations of settleable or floatable organics or excess suspended solids may be treated by density separation. *Clarifiers* separate liquids of differing density. Suspended solids can be removed by *settling chambers* or *sedimentation basins*. Sometimes clay-sized suspended solids are treated with *flocculents* that cause the small particles to aggregate, thus speeding the rate at which they settle.

Filtration. Filtration is a reliable and effective means of removing suspended solids provided the solids concentration is not excessive. In filtration, suspended solids are removed from ground water by forcing the fluid through a porous medium. Filtration systems use a variety of media. Typical high rate gravity or pressure filters consist of a bed of granular particles (typically sand or sand with anthracite coal). The bed is supported by an underdrain system that allows the filtered liquid to be drawn off.

Filtration is often preceded by chemical coagulation and sedimentation or biological treatment, and often precedes activated carbon units to decrease the suspended solids load. The filter must be backwashed at appropriate intervals; the backwash water then requires further treatment.

Reverse Osmosis. Osmosis is the spontaneous flow of a water molecule from a dilute solution, through a semipermeable membrane, to a more concentrated solution by osmotic pressure (see Section 2.4.1). If enough pressure is placed on the concentrated solution to overcome osmotic pressure, the water will flow toward the dilute phase in a process known as reverse osmosis. A reverse osmosis unit is composed of a membrane, a membrane support structure, a containing vessel, and a high-pressure pump. Reverse osmosis allows the concentration of contaminants to build up in a circulating system on one side of the membrane, while relatively pure water passes through the membrane and is discharged. The resulting concentrated solution containing the contaminants and salts requires subsequent special handling.

Reverse osmosis separates ions and small molecules in true solution from water, producing a product low in dissolved solids concentration, both organic and inorganic. This method is used primarily for cleaning low flow waste streams containing highly toxic contaminants. Good removal can be expected for high molecular weight organics and charged univalent ions. However, membrane technology has been developed to the extent that low molecular weight organics such as alcohols, ketones, amines, and aldehydes can also be removed. Reverse osmosis has been demonstrated to be an effective treatment technology, although wastes containing high concentrations of various organics may not be treatable with this method due to excessive membrane dissolution or fouling. Pretreatment such as suspended solids removal, pH adjustment, and removal of oxidizers and oil and grease is also required.

Stripping. Stripping is a mass transfer process in which volatile contaminants are removed from the aqueous phase and transferred to the gaseous phase. This process is accomplished by passing air or steam through the liquid wastes. Air stripping has been applied to ground water for removal of VOCs including trichloroethylene, trihalomethane, chlorobenzene, and vinyl chloride. Biological treatment or carbon adsorption may follow air stripping to remove nonstrippable organics or as a final polishing step.

Air stripping systems include packed tower, cross-flow cooling type tower, coke tray aerator, or stripping lagoons with diffused air. The *packed tower* system works on the principle of countercurrent flow. In this method, the water stream flows down through the packing, while the air flows upward and is exhausted through the top to the atmosphere or to emission control devices (e.g., condensers, carbon adsorption filters). In the *cross-flow tower* method, water flows down through the packing as with the countercurrent packed tower; however, in this method the air is pulled across the water flow path by a fan. The *coke tray aerator* is a simple, low-maintenance process that requires no blower. The water being treated trickles through several layers of trays producing a large surface area for gas transfer and induced draft stripping. *Diffused aeration* stripping uses aeration basins or lagoons similar to those used in wastewater treatment. Water flows through the basin, and the air is dispersed through diffusers at the bottom of the basin. Since the air-to-water ratio is significantly lower in the basins than in either the packed or cross-flow towers, efficiency is typically lower.

Temperature affects the mass transfer coefficient of a compound. This is important when considering stripping for contaminated ground water that contains very soluble compounds (i.e., low Henry's Law constants). High water solubility makes their efficient removal to low levels by ambient temperature air stripping almost impossible. Removal efficiency increases dramatically with increasing temperature and less sharply with the air-to-water ratio. High temperature air stripping or steam stripping offers increased flexibility in such cases.

Incineration. Incineration offers one of the most effective technological methods for complete destruction of organic compounds. It would not, however, normally be applied directly to pumped ground water. Instead, ground water containing high concentrations of miscible organics or organic phases removed from ground water would be potential candidates for incineration. Incineration uses heat under controlled conditions to decompose a substance into products that generally include CO_2, H_2O vapor, SO_2, NO_x, HCl and products of incomplete combustion. Temperature control of the burning and air pollution control equipment are typically required to prevent the release of undesirable combustion products into the atmosphere. Incineration methods can be used to destroy organic contaminants in liquid, gaseous, and solid wastes.

The most common incineration technologies are liquid injection, rotary kiln, fluidized-bed, and multiple hearth. *Rotary kiln* and *multiple hearth* incinerators can be used with most organic wastes including solids, sludges, liquids, and gases. *Liquid injection* incinerators are limited to pumpable slurries and liquids. *Fluidized-bed* incinerators work well for organic liquids, gases, and granular or well-processed solids. Other methods for the thermal destruction of wastes are under development. These methods include molten salt, plasma arch torch, circulating bed, high temperature fluid wall, pyrolysis, supercritical water, advanced electric reactor, and vertical tube reactor. These advanced incineration technologies are briefly described in Wagner et al. (1986).

Wet Air Oxidation. Wet air oxidation is a process in which dissolved or suspended organic substances in the aqueous phase are oxidized at high temperatures and pressures. The extent of oxidation is a function of the nature of the organics and the temperature and detention time in the reactor. As with incineration, its most likely use would be to treat residual liquids with high concentrations of organic and oxidizable inorganic components resulting from treating contaminated ground water. Wet air oxidation is particularly applicable to concentrated streams containing pesticides, herbicides, or other complex organics that are not readily biodegradable (Rich and Cherry, 1987/T12-7).

14.5.2 Chemical Treatment

Table 14-6 summarizes the advantages and disadvantages of five methods for chemically treating contaminated ground water.

Neutralization is a process in which an acid or base is added for pH adjustment. Neutralization is a relatively simple unit process that can be performed using ordinary and commonly available treatment equipment. It is often used prior to other treatment processes such as biological treatment, carbon adsorption, ion exchange, air stripping, or chemical oxidation/reduction processes where the pH of the liquid to be treated is critical.

Chemical precipitation is a physicochemical process whereby a substance in solution is transformed to the solid phase. Precipitation results from the addition of a chemical that will react with the contaminant in solution to form an insoluble compound. Changes in temperature may also reduce contaminant solubility and result in precipitation. The most common application of precipitation in contaminated ground-water treatment is the addition of chemicals to remove metals such as carbonates, hydroxides, or sulfides. Many precipitation reactions (e.g., metal sulfides) do not readily form settleable floc but precipitate as very fine and relatively stable colloidal particles. When this occurs, flocculating agents (e.g., metal salts or polyelectrolytes) are added to cause flocculation of the metal sulfide precipitates. The effectiveness of precipitation/flocculation reactions depends upon the nature and concentration of the contaminants and upon the process design. The process design must consider the optimum chemicals and dosages, suitable chemical addition systems, optimum pH and mixing requirements, temperature, sludge production, and dewatering characteristics.

Table 14-6 Advantages and Disadvantages of Chemical Methods of Ground-Water Treatment

Method	Advantages	Disadvantages
Neutralization	--Uses commonly available equipment.	--Potential for air emissions of acid fumes. --Some neutralization agents are corrosive.
Precipitation	--Useful for most hazardous waste streams.	--Limited by the presence of complexing agents in the waste. --Precipitate may be a hazardous waste.
Oxidation treatment	--Uses simple readily available equipment.	--Costs are generally higher than biological. --Some organics are resistant to most oxidants. --Partial oxidation may generate toxic compounds. --Oxidizing chemicals are potentially hazardous.
Reduction	--Uses simple, readily available equipment and reagents.	--No current applications for reducing organic compounds.
Ion exchange	--Removes a broad range of ionic species. --Columns available commercially. --Units are compact and not energy intensive.	--Must monitor effluent to monitor resin performance. --Effectiveness reduced by high suspended solids and certain organics. --Regeneration solution must be disposed.

Source: Compiled from U.S. EPA (1985), Knox et al. (1984), Nielsen (1983), and Wagner et al. (1986).

Oxidation/reduction reactions raise (oxidation) or lower (reduction) the oxidation state of a substance or substances in order to reduce toxicity or solubility, or to transform the substance to a form that can be handled more easily. Commonly used reducing agents include sulfite salts, sulfur dioxide, and the base metals (i.e., iron, aluminum, and zinc). Chemical *reduction* is used primarily to reduce hexavalent chromium, mercury, and lead. Currently, no practical applications involve reduction of organic compounds. *Oxidation*, however, is used extensively in the treatment of organic wastes. Oxidizers most often used in wastewater treatment include oxygen or air, ozone, ozone with ultraviolet light, chlorine gas, hypochlorites, chlorine dioxide, and hydrogen peroxide.

Ion exchange resins remove toxic ions from the aqueous phase through exchange with relatively harmless ions held by the ion exchange material. Ion exchange systems function well for relatively dilute solutions of variable composition. Ion exchange is used to remove a broad range of ionic species from water including anionic or cationic soluble metallic elements, inorganic anions, organic acids if at an alkaline pH, and organic amines when solution acidity is favorable for the formation of the acid salt (DeRenzo, 1978/T12-7).

The ion exchange effluent is monitored to determine when ion exchange resin bed exhaustion has occurred. The regenerant produced can be an appreciable percentage of the treated flow and will require subsequent disposal or treatment. This aspect of ion exchange can often reduce the cost effectiveness of this technology, particularly if waste concentrations are high.

14.5.3 Biological Treatment

Table 14-7 summarizes the advantages and disadvantages of four methods for biologically treating contaminated ground water.

Activated Sludge. Activated sludge is a suspended growth biological treatment system in which a contaminated solution is mixed in an aeration basin with an active microbial population. The bacteria present are capable of biologically degrading the organic contaminants to cellular material, carbon dioxide, and water. Following the aeration step, the organic matter is removed, the microorganisms are separated from the liquid by gravity settling in clarifiers, and a portion of the settled microorganisms are recycled to the aeration basin. The new cells developed as a result of cell synthesis are removed from the system for further treatment and disposal.

Various modifications of the conventional activated sludge process include pure oxygen aeration, oxygen-enriched aeration, extended aeration, step feed, and contact stabilization. The contaminants present in the liquid to be treated, site considerations, costs, and required effluent limits all influence the type of activated sludge system selected.

Table 14-7 Advantages and Disadvantages of Biological Methods of Ground-Water Treatment

Method	Advantages	Disadvantages
Activated sludge	--Effective for readily treated organics. --Process is reliable in absence of shock loads. --Technology is highly developed. --Can tolerate higher organic loads than most biological treatment processes. --High degree of flexibility.	--High capital costs. --Generates sludge that may be high in metals and refractory organics. --Sensitive to heavy metals, toxic organics at high concentrations and suspended solids. --Fairly energy intensive.
Surface impoundments	--Well demonstrated for stabilization of organics. --Require minimal energy.	--Sensitive to shock loadings and temperature. --Anaerobic lagoons may generate gas. --Must be timed to prevent seepage into ground water.
Trickling filter	--Not highly sensitive to shock loads. --Suitable for removal of suspended or colloidal matter. --Can be used as a roughing filter to even out organic loads.	--Potential for odor problems. --Generates sludge which may be high in refractory organics and sorbed metals. --Sensitive to metals and oil and grease. --Can handle only very low organic loads compared to activated sludge. --Vulnerable to below freezing temperatures.
Rotating biological discs	--Can handle larger flow variations and higher organic shock loads than activated sludge. --Modular construction provides flexibility to meet increased or decreased treatment needs.	--Potential for odor problems. --Vulnerable to climate changes if not covered. --Supplemental aeration may be required. --Can handle only relatively low strength wastes compared to activated sludge. --Excess biological growth of undesirable microorganisms under certain conditions.

Source: Compiled from U.S. EPA (1985), Knox et al. (1984), Nielsen (1983), and Wagner et al. (1986).

Activated sludge has not been used as extensively for ground-water treatment as have activated carbon, filtration, and precipitation followed by flocculation; however, the process is well established for treating a wide variety of organic contaminants (U.S. EPA, 1985/T12-7). Limiting factors in the use of activated sludge for contaminated ground water treatment include its inability to treat low concentrations of contaminants, the relatively long time required for organism acclimation, the long detention times required for some complex contaminant degradation, the sensitivity of the system to shock loads, and the potential for the sludge produced to be hazardous.

Surface Impoundments. Surface impoundments are large shallow ponds or lagoons with suspended microbial populations in much lower concentrations than in activated sludge systems. Detention times on the order of weeks are typical, and natural or applied aeration maintains aerobic conditions. Surface impoundments require large land areas and, because low temperatures adversely affect performance, they are not suitable for climates where the temperature remains below freezing for appreciable lengths of time. Advantages are low operating costs and minimal energy use compared to other biological treatment methods.

Anaerobic lagoons or *anaerobic digesters*, which are totally enclosed, may be used for organic contaminant degradation. The anaerobic digestion process is relatively easy to operate, produces a minimum of sludge, and is energy efficient. Digesters are often maintained at elevated temperatures to increase microbial activity and degradation rate. These systems have applications to high-strength contaminated water and to residuals such as sludges.

Trickling Filters. A trickling filter system is a fixed film biological treatment process in which contaminated water is brought into contact with microorganisms attached to a solid media. The organisms in the slime layer coating the media metabolize the organics in the water, producing biological mass, carbon dioxide, and water. Air moving countercurrent to the waste flow through the filter provides oxygen to the microorganisms.

The *biological tower* is a modification of the trickling filter process in which plastic or wood is stacked in a column that typically reaches 16 to 20 feet in height. The contaminated water is sprayed across the top of the tower and, as this water moves downward, air is pulled upward. A slime layer of microorganisms develops on the media and removes the organic contaminants as the water trickles over it. High void ratio plastic media are used to reduce bed plugging and to improve treatment efficiency.

Trickling filters are more resistant to high organic shock loads and cost less to operate than activated sludge systems. The initial capital cost is higher, however, and final effluent concentrations are typically higher as well.

Rotating Biological Discs. A rotating biological disc system is a fixed film biological treatment process consisting of a series of mechanically rotated discs connected by a shaft and set in a basin or trough. Approximately 40% of each disc's surface area is submerged in the basin, and as the contaminated water passes through the basin, the microorganisms growing on the disc metabolize the organics in the water. As the discs rotate, the microorganisms come in contact with the air from which they obtain oxygen for growth. Rotating biological discs are compact, can handle large flow fluctuations and high organic shock loads, and do not require an aeration system. These systems, however, are subject to problems resulting from continuous excess loading such as excess biological film thickening and snail infestations.

14.6 In Situ Treatment

In situ treatment is an alternative to the removal and subsequent treatment of contaminated ground water. This method requires minimal surface facilities and minimizes exposure to the contaminant. The success of various treatment methods is highly dependent on physical factors including aquifer permeability, the characteristics of the contaminants involved, and the geochemistry of the aquifer material.

In situ treatment technology has not yet been developed to the extent of other currently available technologies for restoring contaminated aquifers. However, some in situ treatment technologies have demonstrated success in actual site remediations (Wagner et al., 1986a/T12-7). Laboratory and pilot-scale testing generally must be performed to evaluate the applicability of a particular technology to a specific site.

In situ treatment technologies may be grouped into two broad categories: physical/chemical and biological. Brief descriptions follow of the available technologies that have potential for success at hazardous waste sites.

14.6.1 In Situ Physical/Chemical Treatment

Organic and inorganic contaminants may be treated chemically to cause immobilization, mobilization for extraction, or detoxification. The application of *oxidation* and *reduction* reactions to in situ treatment is largely conceptual, but potentially may be used to accomplish immobilization by precipitation, mobilization by solubilizing metals or organics, or detoxification of metals and organics (Wagner et al., 1986a/T12-7). The chemicals, however, have the potential to degrade compounds other than the one targeted and to form degradation products that may be more toxic than the one removed.

Precipitation, chelation, and polymerization technologies are used to immobilize contaminants. Mobilization of contaminants is accomplished by soil flushing or vacuum extraction. Neutralization, hydrolysis, and permeable treatment bed technologies may be used for detoxification. In situ physical/chemical treatment processes generally require the installation of a series of injection wells at the head of or within the plume of contaminated ground water.

Immobilization. *Precipitation*, *chelation*, and *polymerization* are three methods used to immobilize a contaminant to prevent its migration out the contaminated area. Precipitation using caustic solutions is effective in immobilizing dissolved metals in ground water. Chelation may also be effective in immobilizing metals, although considerable research is needed (Wagner et al., 1986a/T12-7). Polymerization is effective in immobilizing organic monomers. However, the chemicals added to the contaminants in the ground water may react to form toxic by-products. Precipitation and polymerization will lower the hydraulic conductivities near the injection wells making closely spaced wells necessary for effective treatment. Solidification methods used for treatment of soils can also immobilize contaminants.

Mobilization for Extraction. *Soil flushing* is the process of flooding a contaminated area with water or a solvent to mobilize the contaminant, followed by the collection of the elutriate. This technique is used in combination with previously discussed collection and treatment technologies. The process is based on the solvent solubilizing or chemically reacting with the contaminants and mobilizing them into the solvent phase. Water is used if the contaminant is readily soluble. Acid solutions tend to flush metals and basic organics. Three possible methods for flushing hydrocarbons from soil include *thermal* or *steam flooding* for shallow depths, and *alcohol flooding* on a larger scale. Mobilization of contaminants may also be accomplished by injecting surfactants into the aquifer matrix.

Vacuum extraction, or in situ volatilization, is used to extract volatile organic contaminants from the vadose zone. An extraction system may consist of a series of slotted PVC wells configured to span the area of contamination. Air inlet wells located inside and outside of a contaminated ground-water plume increase the introduction of air from the atmosphere. Vacuum extraction is best suited for areas of high, relatively homogeneous permeability, with no underground structures. Caution must be exercised, however, because extracted vapors may be explosive.

Radio frequency heating has been under development since the mid-1970s and the concept is being applied to in situ decontamination of uncontrolled hazardous waste landfills and sites (Rich and Cherry, 1987/T12-7). In this process, the ground is heated with radio frequency waves that vaporize the hazardous contaminants. The vapors emanating from the soil are then treated.

Detoxification. *Neutralization* of ground water may be accomplished by injecting dilute acids or bases into the aquifer through injection wells to adjust pH to the desired level (see Section 3.2.1). Tolman et al. (1978/T14-9) recommended that neutralization only be applied to ground water at industrial waste disposal sites since municipal landfills, which constantly generate anaerobic decomposition products, would require neutralization over a long period of time.

Hydrolysis may be used for detoxification (see Section 3.3.2); however, the intermediate products formed during hydrolysis of a particular compound must be known since they may be more toxic than the targeted compound. Esters, amides, carbamates, phosphoric and phosphonic acid esters, and pesticides are potentially degradable by hydrolysis (Wagner et al., 1986a/T12-7).

Permeable treatment beds are often used to remediate migrating leachate plumes. In this method, trenches are filled with a reactive permeable medium that reacts with the contaminated ground water to produce a nonhazardous soluble product or a solid precipitate. Limestone, activated carbon, and other ion exchange resins are frequently used in a trench in the aquifer. Permeable treatment beds are applicable in shallow aquifers since the trench must be constructed down to an impermeable layer. The period of effectiveness is limited by the loss of reactive capacity and by plugging with solids.

14.6.2 In Situ Biological Treatment

Biological treatment involves the use of microorganisms to break down hazardous organic compounds into nonhazardous materials. The site hydrology, environmental conditions, and the biodegradability of the contaminants are factors that determine the potential effectiveness of in situ biological treatment. Most compounds are more rapidly degraded aerobically; however, some compounds will only degrade under anaerobic conditions (see Sections 3.5.3 and 3.5.4). Biodegradation in ground water and solids can be a slow process and may take several years for completion depending on the compounds present. In situ biodegradation, however, is a desirable method of treatment because the contaminants are destroyed; thus, removal of ground water for external treatment and residual handling can possibly be avoided.

There are two basic approaches to in situ biodegradation. The first relies on the natural biological activity in the subsurface (Section 3.5). This approach usually involves minimal intervention other than use of containment measures (Section 14.3) to prevent further movement of contaminants. Since organic contaminants tend to create anaerobic conditions, contaminants that are degraded anaerobically are the most likely candidates for using the natural biological activity in an aquifer for treatment. The second approach, called *enhanced biorestoration*, involves the stimulation of the existing microorganisms by adding nutrients and oxygen to promote aerobic degradation of contaminants.

The majority of microbes in the subsurface are firmly attached to soil particles. Nutrients must be brought to the active sites by advection and diffusion of water in the saturated zone or by soil gas in the vadose zone. Most often the compounds to be degraded for microbial energy and cell synthesis are transported in the aqueous phase by infiltrating water, or by advective flow through the ground water. In the vadose zone, volatile organic compounds can move readily as vapors in the soil gas where oxygen is present. Below the water table, aerobic metabolism is limited by the low solubility of oxygen in water. Factors that control the rate of biological activity are the stoichiometry of the metabolic process, the concentration of the required nutrients in the mobile phases, the flow of the mobile phases, the opportunity for colonization in the subsurface by metabolically capable organisms, and the toxicity of the waste.

Much of the development work in the area of ground-water and soil remediation by biodegradation has been performed using petroleum products. The number of gasoline stations, underground tanks, and gasoline pipelines throughout the country and the potential for ground-water contamination have prompted considerable laboratory and field studies on in situ biodegradation of hydrocarbons. Figure 14-8 illustrates a typical system in which microbial nutrients are mixed with ground water

Ground-Water Remediation

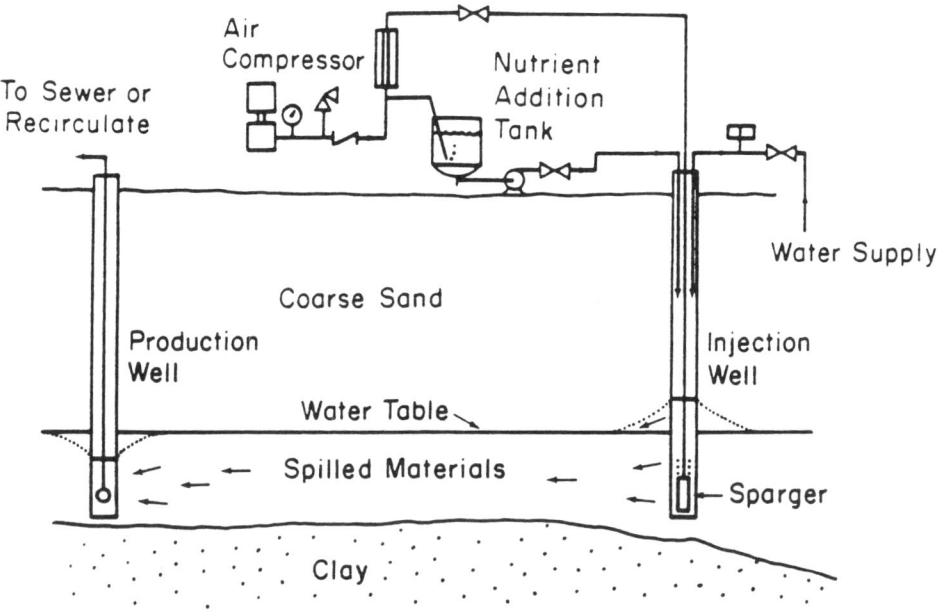

Figure 14-8 Typical schematic for aerobic subsurface bioremediation (Suflita, 1989, after Lee et al., 1988).

and circulated through the contaminated portion of the aquifer through a series of injection and recovery wells. Oxygen is supplied by sparging air into the injection wells. The increased supply of nutrients and oxygen stimulates biodegradation of the hydrocarbons. Wells should be screened to accommodate seasonal fluctuations in the level of the water table. Some operational designs are closed loop in which the water is recycled; thus, unused nutrients can be reinjected, disposal of potentially hazardous ground water is avoided, and the need for make-up water is reduced.

The first step in the process is to use physical methods to recover as much of the gasoline as possible. Then a detailed investigation of the hydrogeology is undertaken to determine the extent of the contamination. Laboratory studies are conducted to determine if the native microbes can degrade the contaminants and to determine the combination of minerals required to promote maximum cell growth at the ambient ground-water temperature and under aerobic conditions. Considerable variations in nutrient requirements among aquifers have been noted.

The field investigations and laboratory studies guide the design and installation of a system of wells for injecting the nutrients and oxygen and controlling subsurface flow. Controlling the ground-water flow is critical to moving oxygen and nutrients to the contaminated zone and optimizing the degradation process.

The technique described above does not provide for treatment above the water table. Soils contaminated by leaking underground storage tanks may be physically removed during the process of removing the tank; however, this may not be practical with deep water tables or large areas of contamination. An alternative to soil removal is the construction of one or more infiltration galleries (Figure 14-9). Infiltration galleries allow movement of the injection solution through the unsaturated zone and the saturated zone, resulting in potential treatment of source materials that may be trapped in the pore spaces of the unsaturated zone. Oxygen may be added to the infiltrated water during an in-line stripping process for volatile organic contaminants or through aeration devices placed in the infiltration galleries.

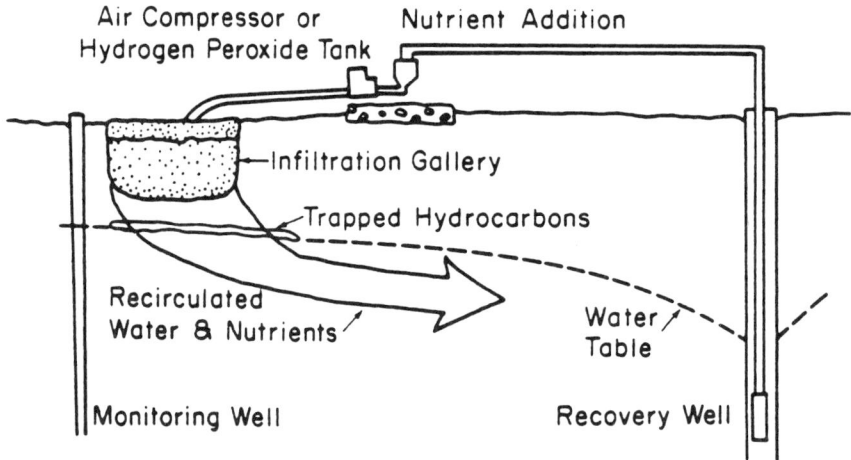

Figure 14-9 Use of infiltration gallery for recirculation of water and nutrients for in situ bioremediation (Suflita, 1989, after Lee et al., 1988).

The rate of bioreclamation of hydrocarbons is effectively the rate of supply of oxygen. Table 14-8 compares the number of times the water in the aquifer, or the air above it, must be replaced to restore subsurface materials of various textures. These values are provided to exemplify the processes involved and would differ at an actual site. The oxygen concentration in the water can be increased by using oxygen rather than air, which would also reduce the volumes of recirculated water required.

Hydrogen peroxide is an alternative source of oxygen in bioreclamation. Iron or an organic catalyst may be used to decompose the hydrogen peroxide to oxygen. The rate at which hydrogen peroxide decomposes to oxygen must be controlled to limit the formation of bubbles that could lead to gas blockage and the loss of permeability. Hydrogen peroxide may mobilize metals such lead and antimony, and, if the water is hard, magnesium and calcium phosphates can precipitate and plug the injection well or infiltration gallery. To determine the microorganism's hydrogen peroxide tolerance level laboratory studies are performed.

Table 14-8 Estimated Volumes of Water or Air Required to Renovate Subsurface Material Containing Hydrocarbons at Residual Concentrations

Texture	Proportion of Total Subsurface Occupied by:			Volumes Required to meet Hydrocarbons Oxygen Demand	
	Hydrocarbons (when drained)[a]	Air (when drained)[a]	Water (when flooded)[a]	Air[b]	Water[c]
Stone to Coarse Gravel	0.005	0.4	0.4	250	5,000
Gravel to Coarse Sand	0.008	0.3	0.4	530	8,000
Coarse to Medium Sand	0.015	0.2	0.4	1,500	15,000
Medium to Fine Sand	0.025	0.2	0.4	2,500	25,000
Fine Sand to Silt	0.040	0.2	0.5	4,000	32,000

[a]Typical values taken from Clapp and Hornberger (1978), see Appendix F for reference.

[b]Oxygen content assumed to be 200 mg/L and the hydrocarbons to be completely metabolized to carbon dioxide.

[c]Water content assumed to be 10 mg/L and the hydrocarbons to be completely metabolized to carbon dioxide.

Source: Adapted from U.S. EPA (1987a).

14.7 Guide to Major References

Table 14-9 provides an index to major references that focus on ground-water remediation in the categories of (1) source controls and (2) ground-water and leachate treatment. The references on cover and liner systems are especially important when designing waste disposal facilities for the prevention of ground-water contamination.

Table 12-7 provides an index to major references that address broader aspects of remediation of contaminated soil and ground water. Many of the conference series identified in Table 5-12 contain useful papers on ground-water treatment and remediation, especially the annual NGWA National Outdoor Action Conference, the annual NGWA/API Petroleum Hydrocarbons and Organic Chemicals Conference, and the various conference series published by the Hazardous Materials Control Resources Institute. Table 12-7 identifies additional conference proceedings that focus on remediation of contaminated ground water. Currently, many remedial techniques are being used and evaluated for cleanup of contaminated soils. Section 12.5 describes various documents that are available as part of U.S. EPA's Superfund Innovative Technology Evaluation (SITE) program.

Table 14-9 Index to Major References on Ground Water and Leachate Treatment and Remediation

Topic	References
General	Ehrenfeld and Bass (1984/T12-7), Barkley (1993/T12-7), HMCRI (various dates/T12-7), U.S. EPA (1978, 1980/T12-7, 1988a, 1989d), Willis and Yeh (1987), Wilson et al. (1976); See generally the conference and symposium series in identified in Table 12-7; <u>Cleanup Standards</u>: Oliver et al. (1993/T12-7); <u>Innovative Technologies</u>: U.S. EPA (1993/T12-7)
<u>Source Controls</u>	
General Design	Atlantic Research Corporation (1980), Bagchi (1990), Chapman (1993), HMCRI (various dates/T12-7), Testa (1993), Tolman et al. (1978), U.S. EPA (1988b, 1989a/T12-7), VanZyl et al. (1987), Wagner et al. (1986b/T12-7)
Cover/Liner Systems	Daniel and Estornell (1991), Fung (1980); <u>Cover Systems</u>: Barnes and Rogers (1988), Dwyer et al. (1986), Landreth et al. (1991), Lutton (1982, 1987), Gilbert and Murphy (1987), Horz (1986), Lutton et al. (1979), McAneny et al. (1985b), U.S. EPA (1987a); <u>Liner Systems</u>: Anderson et al. (1991), ASTM (1994), Brown et al. (1987), Daniel et al. (1991), Matrecon (1988), Mitchell et al. (1989), Telles et al. (1988), U.S. EPA (1983, 1985a, 1985b, 1985c, 1985d, 1986a, 1986b, 1987b, 1989c), Wright et al. (1987); <u>Geosynthetics</u>: Bellen et al. (1987), Goydan et al. (1990), Horz (1986), Koerner (1990), Lord and Koerner (1990), Richardson and Koerner (1987), SAI Corp. (1993), Schwoppe et al. (1985), Telles et al. (1988), U.S. EPA (1989b, 1989c, 1991), Wright et al. (1987); <u>Chemical Effects/Compatibility</u>: Bellen et al. (1987), Brown (1988), Goydan et al. (1990), Schwoppe et al. (1985)
Barriers	Bartenfelder et al. (1992/T12-7), Fung (1980); <u>Grouting</u>: Bodocsi et al. (1988), Bowen (1981), May et al. (1986), Spooner et al. (1984); <u>Slurry Walls</u>: Millet and Perez (1981), Ryan (1980), Xanthakos (1979)
Hydrodynamic Controls	Bartenfelder et al. (1992), Gorelick et al. (1993), Horz (1986); <u>Leachate Plume Management</u>: Repa and Doerr (1985), U.S. EPA (1989c); <u>Surface Impoundments</u>: Johnson and Anderson (1988), Mitchell et al. (1989), Richardson and Koerner (1987)

Table 14-9 (cont.)

Topic	References
Ground Water/Leachate Treatment	
Drinking Water/Wastewater	Fair et al. (1968); <u>Drinking Water</u>: U.S. EPA (1989a); <u>Wastewater</u>: Benefield et al. (1982), Metcalf and Eddy (1979), Patterson (1978), Sundstrom and Klei (1979), U.S. EPA (1992)
Contaminated Ground Water	<u>Texts/Reports</u>: Ahlert and Kosson (1988), Bedient et al. (1994/T4-5), DeRenzo (1978/T12-7), Nyer (1989, 1993/T12-7), Shuckrow et al. (1986), U.S. EPA (1989a/T12-7), Wagner et al. (1986a/T12-7); <u>Review Papers</u>: Nielsen (1983); <u>Cleanup Standards</u>: Oliver et al. (1993/T12-7)
Bioremediation	Lee et al. (1988); See also references on soil and ground-water bioremediation in Table 12-7; <u>In Situ Bioremediation</u>: Semprini et al. (1987), Sims et al. (1992)
Pump and Treat	Cohen et al. (1994/T9-10), Keely (1989), Palmer and Fish (1992), U.S. EPA (1990)
In Situ Treatment	HMCRI (various dates/T12-7); see also references for bioremediation above and UST remediation (Table 12-7)
Leachate Collection/Treatment	<u>Collection</u>: Ghassemi et al. (1985), Kirkham et al. (1986), Pohland and Harper (1986), U.S. EPA (1985c, 1986a, 1989c); <u>Hazardous Waste</u>: Ahlert and Kosson (1988), McArdle et al (1987), Park (1986), Shuckrow et al. (1986)
Leachate Chemistry	<u>Hazardous Waste</u>: Bramlett et al. (1987); <u>Municipal Waste</u>: Kinman et al. (1986), Pohland and Harper (1986, 1987), Williams et al. (1987)
Specific Methods	<u>Air Stripping</u>: Hinchee (1994), Noonan and Curtis (1990); <u>GAC</u>: Cheremisinoff and Ellerbusch (1978), Noonan and Curtis (1990)
Specific Contaminants	<u>Petroleum</u>: Testa and Winegardner (1990), see also Table 12-7; <u>TCE</u>: Russell et al. (1992/T12-7)

Table 14-9 References (Appendix F contains references for figure and table sources)

Ahlert, R.C. and D.S. Kosson. 1988. Treatment of Hazardous Landfill Leachates and Contaminated Groundwater. EPA/600/2-88/064 (NTIS PB89-124648).

American Society for Testing and Materials (ASTM). 1994. ASTM and Other Specifications and Test Methods on the Quality Assurance of Landfill Liner Systems. ASTM, Philadelphia, PA, 500 pp. [79 ASTM standards and 10 other EPA-specified standards]

Anderson, D.C., et al. 1991. Factors Controlling Minimum Soil Liner Thickness. EPA/600/91-008 (NTIS PB91-191346).

Atlantic Research Corporation. 1980. Literature Search on Groundwater. USTHAMA CETHA-TS-CR-91053. U.S. Army Toxic and Hazardous Materials Agency, Aberdeen Proving Ground, MD, 60 pp. [Focusses on containment of contaminated ground water]

Bagchi, A. 1990. Design, Construction, and Monitoring of Sanitary Landfills. John Wiley & Sons, New York, NY, 284 pp.

Barnes, F.J. and J.C. Rogers. 1988. Evaluation of Hydrologic Models in the Design of Stable Landfill Covers. EPA/600/2-88/048 (NTIS PB88-243811).

Bellen, G., R. Corry, and M.L. Thomas. 1987. Development of Chemical Compatibility Criteria for Assessing Flexible Membrane Liners. EPA/600/2-87/067 (NTIS PB87-227310).

Benefield, L.D., J.F. Judkins, and D.L. Weand. 1982. Process Chemistry for Water and Wastewater Treatment. Prentice-Hall, Englewood Cliffs, NJ.

Bodocsi, A., M.T. Bowers, and R. Sherer. 1988. Reactivity of Various Grouts to Hazardous Wastes and Leachates. EPA/600/2-88/021 (NTIS PB88-182936).

Bramlett, J.C. Furman, A. Johnson, W.D. Ellis, H. Nelson, and W.H. Vick. 1987. Composition of Leachates from Actual Hazardous Waste Sites. EPA/600/2-87/043 (NTIS PB87-198743).

Bowen, R. 1981. Grouting in Engineering Practice, 2nd ed. Applied Science Publishers Ltd., Grate Britain.

Brown, K.W. 1988. Review and Evaluation of the Influence of Chemicals on the Conductivity of Soil Clays. EPA/600/2-88/016 (NTIS PB88-170808).

Brown, K.W., J.C. Thomas, R.L. Lytton, P. Jayawickrama, and S.C. Bahrt. 1987. Quantification of Leak Rates Through Holes in Landfill Liners. EPA/600/2-87/062 (NTIS PB87-227666).

Canter, L.W. and R.C. Knox. 1986. Ground Water Pollution Control. Lewis Publishers, Chelsea, MI, 526 pp. [Section 1 covers technologies for ground water pollution control, and Section 2 covers decision-making in aquifer restoration projects. Appendix includes case studies and an annotated bibliography with 225 references]

Chapman, D. 1993. Geotechnical Practice for Waste Disposal. Routledge, Chapman & Hall, 683 pp.

Cheremissinoff, P.N. and F. Ellerbusch (eds.). 1978. Carbon Adsorption Handbook. Ann Arbor Science Publishers, Ann Arbor, MI, 1054 pp.

Daniel, D.E. and P.M. Estornell. 1991. Compilation of Information on Alternative Barriers for Liner and Cover Systems. EPA/600/2-91/002 (NTIS PB91-141846).

Daniel, D.E., C.D. Shackelford, W.P. Liao, and H.M. Liljestrand. 1991. Rate of Flow of Leachate through Clay Soil Liners. EPA/600/2-91/021 (NTIS PB91-196691).

Dwyer, J.R., J.C. Walton, W.E. Grenberg, and R. Clark. 1986. Evaluation of Municipal Solid Waste Landfill Cover Designs. EPA/600/2-86/110 (NTIS PB88-171327).

Fair, G.M., J.C. Geyer, and D.A. Okun. 1968. Water and Wastewater Engineering, Volume 2: Water Purification and Wastewater Treatment and Disposal. John Wiley & Sons, New York, NY.

Fung, R. (ed.). 1980. Protective Barriers for Containment of Toxic Materials. Noyes Data Corporation, Park Ridge, NJ.

Ghassemi, M, K. Crawford, and M. Haro. 1985. Leachate Collection and Gas Migration and Emission Problems at Landfills and Surface Impoundments. EPA/600/2-86/017 (NTIS PB86-162104).

Gilbert, P.A. and W.L. Murphy. 1987. Prediction/Mitigation of Subsidence Damage to Hazardous Waste Landfill Covers. EPA/600/2-87/025 (NTIS PB87-175378).

Gorelick, S.M., R.A. Freeze, D. Donohue, and J.F. Keely. 1993. Groundwater Contamination: Optimal Capture and Containment. Lewis Publishers, Boca Raton, FL, 416 pp.

Goydan, R. A.A. Hawkins, and A.D. Schwope. 1990. Development of a Data Base on Chemical Migration from Polymeric Materials. EPA/600/2-90/029 (NTIS PB90-235102).

Hinchee, R.E. (ed.). 1994. Air Sparging for Site Remediation. Lewis Publishers, Boca Raton, FL, 160 pp.

Horz, R.C. 1986. Geotextiles for Drainage, Gas Venting, and Erosion Control at Hazardous Waste Sites. EPA/600/2-86/085 (NITS PB87-129557).

Johnson, S.H. and D.C. Anderson. 1988. Freeboard Determination and Management in Hazardous Waste Surface Impoundments. EPA/600/2-88/015 (NTIS PB88-243787).

Keely, J.F. 1989. Performance Evaluation of Pump-and-Treat Remediations. Superfund Issue Paper, EPA 540/8-89/005, 14 pp.

Kinman, R.N., J. Rickabaugh, J. Donnelly, D. Nutini, and M. Lambert. 1986. Evaluation and Disposal of Waste Materials Within 19 Test Lysimeters at Center Hill. EPA/600/2-86/035 (NTIS PB86-176336).

Kirkham, R.R., S.W. Tyler, and G.W. Gee. 1986. Estimating Leachate Production from Closed Hazardous Waste Landfills. EPA/600/2-86/057 (NTIS PB86-207503).

(Table 14-9 Ground Water Remediation References)

Knox, R.C., L.W. Canter, D.F. Knicannon, E.L. Stover, and C.H. Ward. 1984. State-of-the Art of Aquifer Restoration. EPA 600/2-84/182a&b (NTIS PB85-181071 and PB85-181089).

Koerner, R.M. 1990. Designing with Geosynthetics, 2nd ed. Prentice-Hall, Englewood Cliffs, NJ. [First edition published 1986]

Landreth, R.E., D.E. Daniel, R.M. Koerner, P.R. Schroeder, and G.N. Richardson. 1991. Design and Construction of RCRA/CERCLA Final Covers. Seminar Publication EPA/625/4-91/025, 190 pp. Available from CERI.*

Lee, M.D., J.M. Thomas, R.C. Borden, P.B. Bedient, C.H. Ward, and J.T. Wilson. 1988. Biorestoration of Aquifers Contaminated with Organic Compounds. CRC Critical Reviews in Environmental Control 18(1):29-87. [Review containing 272 references]

Lord, Jr., A.E. and R.M. Koerner. 1990. Fundamental Approach to Service Life of Flexible Membrane Liners (FMLs). EPA/600/2-90/041 (NTIS PB90-263856).

Lutton, R. 1982. Evaluating Cover Systems for Solid and Hazardous Waste. EPA/SW-867 (NTIS PB87-154894).

Lutton, R.J. 1987. Design, Construction, and Maintenance of Cover Systems for Hazardous Waste: An Engineering Guidance Document. EPA/600/2-87/039 (NTIS PB87-191656).

Lutton, R., G. Regan, and L. Jones. 1979. Design and Construction of Covers for Solid Waste Landfills. EPA/600/12-79-165.

Matrecon, Inc. 1988. Lining of Waste Containment and Other Impoundment Facilities. EPA/600/2-88/052 (NTIS PB89-129670), 1030 pp.

May, J.H., R.J. Larson, P.G. Malone, J.A. Boa, Jr., and D.L. Bean. 1986. Grouting Techniques in Bottom Sealing of Hazardous Waste Sites. EPA/600/2-86/020 (NTIS PB86-158664).

McAneny, C.C., P.G. Tucker, J.M Lorgan, C.R. Lee, and M.F. Kelley. 1985. Covers for Uncontrolled Hazardous Waste Sites. EPA/540/2-85/002 (NTIS PB87-119483), 563 pp.

McArdle, J.L., M.M. Arozarena, and W.E. Gallagher. 1987. A Handbook on Treatment of Hazardous Waste Leachate. EPA/600/S8-87/006 (NTIS PB87-152328).

Metcalf and Eddy, Inc. 1979. Wastewater Engineering: Treatment, Disposal, and Reuse, 2nd ed. McGraw-Hill, New York, NY, 920 pp.

Millet, R.A. and J.Y. Perez. 1981. Current USA Practice: Slurry Wall Specifications. J. Geotch. Eng. Div. ASCE 107(GT8):1041-1056.

Mitchell, D.H, M.A. McLean, and T.E. Gates. 1989. Stability of Lined Slopes at Landfills and Surface Impoundments. EPA/600/2-89/057 (NTIS PB90-251877).

Nielsen, C.M. 1983. Remedial Methods Available in Areas of Ground Water Contamination. In: Proc. 6th Nat. Ground Water Quality Symp., National Water Well Association, Dublin, OH, pp. 219-227.

(Table 14-9 Ground Water Remediation References)

Ground Water Remediation

Noonan, D.C. and J.T. Curtis. 1990. Groundwater Remediation and Petroleum: A Guide for Underground Storage Tanks. Lewis Publishers, Chelsea, MI. [Focuses air stripping and GAC treatment]

Nyer, E.K. 1989. Groundwater Treatment Technology, 2nd ed. Van Nostrand Reinhold, New York, NY, 320 pp. 1st edition 1985. [Focuses on treatment methods for organic and inorganic contaminants in ground water; last chapter contains five examples of field application of design methods]

Palmer, C.D. and W. Fish. 1992. Chemical Enhancements to Pump-and-Treat Remediation. Ground Water Issue Paper, EPA/540/S-92/001, 20 pp.

Park, J.E. 1986. Testing and Evaluation of Permeable Materials for Removing Pollutants from Leachates at Remedial Action Sites. EPA/600/2-86/074 (NTIS PB86-237708).

Patterson, J.W. 1978. Wastewater Treatment Technology, 3rd ed. Ann Arbor Science, Ann Arbor, MI.

Pohland, F.G. and S.R. Harper. 1986. Critical Review and Summary of Leachate and Gas Production from Landfills. EPA/600/2-86/073 (NTIS PB86-240181)

Pohland, F.G. and S.R. Harper. 1987. Retrospective Evaluation of the Effects of Selected Industrial Wastes on Municipal Solid Waste Stabilization in Simulated Landfills. EPA/600/2-87/044 (NTIS PB87-198701).

Repa, E. and D.P. Doerr. 1985. Leachate Plume Management. EPA/540/2-85/004 (NTIS PB86-122330).

Richardson, G.N. and R.M. Koerner. 1987. Geosynthetic Design Guidance for Hazardous Waste Landfill Cell and Surface Impoundments. EPA/600/2-87/097 (NTIS PB88-131263).

Ryan, C.R. 1980. Slurry Cut-Off Walls: Methods and Applications. Geo-Con, Inc., Pittsburgh, PA.

Schwoppe, A.D., P.P. Costas, and W.J. Lyman. 1985. Resistance of Flexible Membrane Liners to Chemicals and Wastes. EPA/600/2-85/127 (NTIS PB86-119955).

Science Applications International (SAI) Corp. (Compiler) 1993. Proceedings of the Workshop of Geomembrane Seaming: Data Acquisition and Control. EPA/600/R-93/112, 64 pp. Available from CERI.*

Semprini, L. P.V. Roberts, G.D. Hopkins, and D.M. Mackay. 1987. A Field Evaluation of In-Situ Biodegradation for Aquifer Restoration. EPA/600/2-87/096 (NTIS PB88-130257).

Shuckrow, A.J., A.P. Pajak, and C.J. Touhill. 1986. Groundwater and Leachate Treatability Studies at Four Superfund Sites. EPA/600/2-86/029 (NTIS PB86-171436).

Sims, J.L., J.M. Suflita, and H.H. Russell. 1992. In Situ Bioremediation of Contaminated Ground Water. Ground Water Issue Paper, EPA/540/S-92/003, 11 pp.

(Table 14-9 Ground Water Remediation References)

Spooner, P.A. et al. 1984. Compatibility of Grouts with Hazardous Wastes. EPA/600/S2-84-015. [Tests using 12 types of grouts and 16 general classes of organic and inorganic compounds]

Sundstrom, K.W. and H.E. Klei. 1979. Wastewater Treatment. Prentice-Hall, Englewood Cliff, NJ.

Telles, R.W., S.L. Unger, H.R. Lubowitz, and H.K. Howard. 1988. Technical Considerations for De Minimis Pollutant Transport Through Polymeric Liners. EPA/600/2-88/042 (NTIS PB88-238332).

Testa, S. 1993. Geological Aspects of Hazardous Waste Management. Lewis Publishers, Boca Raton, FL, 512 pp.

Testa, S.M. and D.L. Winegardner. 1990. Restoration of Petroleum Contaminated Aquifers. Lewis Publishers, Chelsea, MI, 240 pp.

Tolman, A., A. Ballestero, W. Beck, and G. Emrich. 1978. Guidance Manual for Minimizing Pollution from Waste Disposal Sites. EPA 600/2-78/142 (NTIS PB286-905). [Covers surface water control, passive ground water management, active ground water/plume management, chemical immobilization, and excavation/reburial]

U.S. Environmental Protection Agency (EPA). 1978. Surface Impoundments and Their Effects on Ground Water Quality in the U.S.—A Preliminary Survey. EPA-570/9-78-005. [Sections VIII and IX address pollution prevention/remediation methods, including cost data]

U.S. Environmental Protection Agency (EPA). 1983. Lining of Waste Impoundment and Disposal Facilities. EPA/SW-870 (NTIS PB86-192796). [Updated by Matrecon (1988)]

U.S. Environmental Protection Agency (EPA). 1985a. Minimum Technology Guidance on Double Liner Systems for Landfills and Surface Impoundments--Design, Construction, and Operation. EPA/530-SW-85-014 (NTIS PB87-151072).

U.S. Environmental Protection Agency (EPA). 1985b. Minimum Technology Guidance on Double Liner Systems for Landfills and Surface Impoundments; Design, Construction, and Operation (Draft). EPA/530-SW-85-014 (NTIS PB87-151072).

U.S. Environmental Protection Agency (EPA). 1985c. Guidance on Implementation of the Minimum Technological Requirements of HSWA of 1984, Respecting Liners and Leachate Collections Systems. EPA/530-SW-85-012 (NTIS PB87-163-242).

U.S. Environmental Protection Agency (EPA). 1985d. Minimum Technology Guidance on Single Liner Systems for Landfills, Surface Impoundments, and Waste Piles; Design, Construction, and Operation (Draft). EPA/530-SW-85-013 (NTIS PB87-173159).

U.S. Environmental Protection Agency (EPA). 1986a. Supplementary Guidance on Determining Liner/Leachate Collection System Compatibility. OSWER Policy Directive No. 9480.00-13.

(Table 14-9 Ground Water Remediation References)

U.S. Environmental Protection Agency (EPA). 1986b. Design, Construction and Evaluation of Clay Liners for Waste Management Facilities. Draft Technical Guidance Document EPA/530-SW-86-007F (NTIS PB89-181937).

U.S. Environmental Protection Agency (EPA). 1987a. Prediction/Mitigation of Subsidence Damage to Hazardous Waste Landfill Covers. EPA/600/2-97-025 (NTIS PB87-175378).

U.S. Environmental Protection Agency (EPA). 1987b. Background Document on Bottom Liner Performance in Double-Lined Landfills and Surface Impoundments. EPA/530-SW-87-013 (NTIS PB87-182291).

U.S. Environmental Protection Agency (EPA). 1988a. Guidance on Remedial Actions for Contaminated Ground Water at Superfund Sites. EPA/540/G-88/003 (NTIS PB89-184618).

U.S. Environmental Protection Agency (EPA). 1988b. Guide to Technical Resources for the Design of Land Disposal Facilities. EPA/625/6-88/018. Available from CERI.*

U.S. Environmental Protection Agency (EPA). 1989a. Technologies for Upgrading Existing or Designing New Water Treatment Facilities. EPA/625/4-89/023, 209 pp. Available from CERI.*

U.S. Environmental Protection Agency (EPA). 1989b. Technical Guidance Document: The Fabrication of Polyethylene FML Field Seams. EPA/530/SW-89-069 (NTIS PB90-119595).

U.S. Environmental Protection Agency (EPA). 1989c. Requirements for Hazardous Waste Landfill Design, Construction and Closure. Seminar Publication EPA/625/4-89/022, 127 pp. Available from CERI.* [Focuses on liners and leachate collection systems]

U.S. Environmental Protection Agency (EPA). 1989d. Evaluation of Ground-Water Extraction Remedies: Volume 1, Summary Report (EPA/540/2-89/054, NTIS PB90-183583, 66 pp.); Volume 2, Case Studies 1-19 (EPA/540/2-89/054b); and Volume 3, General Site Data Base Reports (EPA/540/2-89/054c).

U.S. Environmental Protection Agency (EPA). 1990. Basics of Pump-and-Treat Ground-Water Remediation Technology. EPA/600/8-90/003.

U.S. Environmental Protection Agency (EPA). 1991. Technical Guidance Document: Inspection Techniques for the Fabrication of Geomembrane Field Seams. EPA/530/SW-91-051 (NTIS PB92-109057).

U.S. Environmental Protection Agency (EPA). 1992. Wastewater Treatment/Disposal for Small Communities. EPA/625/R-92/005, 110 pp. Available from CERI.*

vanZyl, D.J.A., S.R. Abts, J.D. Nelson, and T.A. Shepherd (eds.). 1987. Geotechnical and Geohydrological Aspects of Waste Management. Lewis Publishers, Chelsea, MI.

Williams, N.D., F.G. Pohland, K.C. McGowan, and F.M. Saunders. 1987. Simulation of Leachate Generation from Municipal Solid Waste. EPA/600/2-87/059 (NTIS PB86-227005).

(Table 14-9 Ground Water Remediation References)

Willis, R. and W. W-G. Yeh. 1987. Groundwater Systems Planning and Management. Prentice-Hall, Englewood Cliffs, NJ.

Wilson, J.L., R.L. Lenton, and J. Porras. 1976. Ground Water Pollution: Technology, Economics and Management. TR 208, Massachusetts Institute of Technology, Cambridge, MA.

Wright, T.D., W.M. Held., J.R. Marsh, and L.R. Hovater. 1987. Manual of Procedures and Criteria for Inspecting the Installation of Flexible Membrane Liners in Hazardous Waste Facilities. EPA/600/8-87/056 (NTIS PB88-131313).

Xanthakos, P. 1979. Slurry Walls. McGraw-Hill, New York, NY.

* See Preface for information on how to obtain documents from CERI (U.S. EPA Center for Environmental Research Information) and NTIS.

APPENDIX A

SUMMARY INFORMATION ON MAJOR SUBSURFACE CHARACTERIZATION AND MONITORING TECHNIQUES

Tables A-1 through A-13 in this Appendix are method tables taken from EPA's *Subsurface Characterization and Monitoring Techniques: A Desk Reference Guide* (Boulding, 1993), which contains one- to two-page summary descriptions of more than 280 specific field methods. This two-volume guide can be obtained at no cost from U.S. EPA's Center for Environmental Research Information (see Preface for information on how the obtain documents from CERI). This appendix is intended to serve as a convenient summary of methods covered in that guide. <u>All section numbers in tables and footnotes refer to locations in the EPA guide</u> where more detailed information about specific methods can be found. A list of tables in this appendix and pages on which they are located follows at the bottom of this page.

Table A-14 contains a consolidated list of all standard methods developed by the American Society for Testing and Materials (ASTM) that are mentioned in this handbook. Most ASTM standard relevant to investigation of soil and ground water have been developed by Committee D18 on Soil and Rock. Prior to 1994 all Committee D18 standards were published in Volume 4.08 of ASTM's annual books of standards. Beginning in 1994, Volume 4.08 contains D420 to D4914 and Volume 4.09 contains D4943 to the most recently developed standards. For a specific standard, a volume published prior to the current year can be used, provided its year of issuance is later than the most recent revision to the standard. For example, ASTM D1452-80 could be used from any book of standards published since 1981 (note that there is usually about a year's lagtime from approval of a standard or revision and its publication in the full book of standards). If earlier editions are used, ASTM should be contacted to make sure that there is not a more recent version than indicate in Table A-14.

Table		Page
A-1	Summary Information on Remote Sensing and Surface Geophysical Methods	806
A-2	Characteristics of Borehole Logging Methods	808
A-3	Summary Information on Vadose Zone Water State Measurement and Monitoring Methods	813
A-4	Summary Information on Vadose Hydraulic Conductivity Techniques	814
A-5	Summary Information on Vadose Zone Water Budget Characterization Methods	816
A-6	Summary Information on Ground Water Level/Pressure Measurement	819
A-7	Summary Information on Aquifer Test Methods	820
A-8	Summary Information on Drilling Methods	822
A-9	Summary of Hand-Held Soil Sampling Devices	823
A-10	Summary of Major Types of Power-Driven Samplers	825
A-11	Summary Information on Soil Solute Monitoring and Sampling Methods	827
A-12	Summary Information on Ground-Water Sampling Devices	828
A-13	Summary Information on Sample Processing/Analytical Techniques	829
A-14	ASTM Standard Methods Cited in this Handbook	833

Table A-1 Summary Information on Remote Sensing and Surface Geophysical Methods (all ratings are approximate and for general guidance only)

Technique	Soils/Geology	Leachate	Buried Wastes	NAPLs	Penetration Depth (m)[a]	Cost[b]	Section/Tables
Airborne Remote Sensing and Geophysics							
Visible Photography +	yes	yes[c]	possibly[d]	yes[c]	Surf. only	L	1.1.1/Tb 1.1.1
Infrared Photography +	yes	yes[c]	possibly[d]	yes[c]	Surf. only	L-M	1.1.1/Tb 1.1.1
Multispectral Imaging	yes	yes[c]	no	yes[c]	Surf. only	L	1.1.1/Tb 1.1.1
Ultraviolet Photography	yes	yes[c]	possibly[d]	yes[c]	Surf. only	L	1.1.2/Tb 1.1.1
Thermal Infrared Scanning	yes	yes (T)	no	possibly	0.1-2	M	1.1.3
Active Microwave (Radar) +	yes	possibly	yes	possibly	0-100	M	1.1.4
Airborne Electromagnetics	yes	yes (C)	yes	possibly		M	1.1.5
Aeromagnetics	yes	no	yes	no	10s-100s	M	1.1.6
Surface Electrical and Electromagnetic Methods							
Self Potential	yes	yes (C)	yes	no	S 10s	L	1.2.1
Electrical Resistivity +	yes	yes (C)	yes (M)	possibly	S 60 (km)	L-M	1.2.2, 9.1.1/Tbs 1-2, 1-3, 1.2.1
Induced Polarization	yes	yes (C)	yes	possibly	S km	L-M	1.2.3
Complex Resistivity	yes	yes (C)	yes	yes	S km	M-H	1.2.3
Dielectric Sensors	yes	yes (C)	no	possibly	S 2[e]	L-M	6.2.3/Tb 6-1
Time Domain Reflectometry	yes	yes (C)	no	yes	S 2[e]	M-H	6.2.4/Tb 6-1
Electromagnetic Induction +	yes	yes (C)	yes	possibly	S 60(200)/ C 15(50)	L-M	1.3.1/Tbs 1-2, 1-3, 1.3.1
Transient Electromagnetics	yes	yes (C)	yes	no	S 150 (2000+)	M-H	1.3.2/Tb 1.3.1
Metal Detectors	no	no	yes	no	C/S 0-3	L	1.3.3/Tbs 1-2, 1-3
VLF Resistivity	yes	yes (C)	yes	no	C/S 20-60	M-H	1.3.4
Magnetotellurics	yes	yes (C)	no	no	S 1000+	M-H	1.3.5

Characterization Method Summary Information

Method					
Surface Seismic and Acoustic Methods					
Seismic Refraction +	yes	yes	no	S 1-30(200+)	L-M	1.4.1/Tbs 1-2, 1-3
Shallow Seismic Reflection +	yes	no	no	S 10-30(2000+)	M-H	1.4.2
Continuous Seismic Profiling	yes	no	no	C 1-100	L-M	1.4.3
Seismic Shear/Surface Waves	yes	no	no	S 10s-100s	M-H	1.4.4
Acoustic Emission Monitoring	yes	no	no	S 2[e]	L	1.4.5
Sonar/Fathometer	yes	yes	no	C no limit	L-H	1.4.6
Other Surface Geophysical Methods					
Ground-Penetrating Radar +	yes	yes (C)	yes	C 1-25 (100s)	M	1.5.1/Tbs 1-2, 1-3
Magnetometry +	no	no	yes (F)	C/S 0-20[f]	L-M	1.5.2/Tbs 1-2, 1-3
Gravity	yes	yes	no	S 100s+	H	1.5.3
Radiation Detection	no	no	yes (nuclear)	C/S near surface	L	1.5.4
Near Surface Geothermometry					
Soil Temperature	yes	yes (T)	no	S 1-2[e]	L	1.6.1
Ground-Water Detection	yes	yes (T)	no	S 2[e]	L	1.6.2
Other Thermal Properties	yes	no	no	S 1-2[e]	L-M	1.6.3

Boldface = Most commonly used methods at contaminated sites; + = covered in Superfund Field Operations Manual (U.S. EPA, 1987a); (C) = plume detected when contaminant(s) change conductivity of ground water; (T) = plume detected by temperature rather than conductivity.

[a] S = station measurement; C = continuous measurement. Depths are for typical shallow applications; ( ) = achievable depths
[b] Ratings are very approximate L = low, M = moderate, H = high.
[c] If leachate or NAPLs are on the ground or water surface or indirectly affect surface properties—see Table 1.1.1; field confirmation required.
[d] Disturbed areas which may contain buried waste can often be detected on aerial photographs.
[e] Typical maximum depth, greater depths possible, but sensor placement is more difficult and cable lengths must be increased.
[f] For ferrous metal detection, greater depths require larger masses of metal for detection; 100s of meters depth can be sensed when using magnetometry for mapping geologic structure.

Table A-2 Characteristics of Borehole Logging Methods (information for general guidance only)

Log Type/Section	Casing[a]	Min. Diam.[b]	Borehole Fluid	Radius of Measurement	Required Correction
Electrical Logs					
Spontaneous Potential (3.1.1)	Uncased only	1.5-3.0"	Conductive fluid	Near borehole surface	Drilling fluid resistivity and borehole diameter for quantitative uses
Single-Point Resistance (3.1.2)	Uncased only	1.5-2.0"	Conductive fluid	Near borehole surface	Not quantitative; hole diameter effects significant
Fluid Conductivity (3.1.3)	Uncased or screened	2.0-2.5"	Conductive fluid	Within borehole	Calibration with fluid of known salinity; temperature correction
Resistivity (3.1.4)	Uncased only	2.0-5.5"	Conductive fluid	<1.0-60"	Drilling fluid resistivity, borehole diameter, and temperature log for quantitative uses
Dipmeter (3.1.5)	Uncased only	6.0"	Conductive fluid	Near borehole surface	Orientation; minimum of 6" diam. required for accurate joint/fracture characterization
Induced Polarization (3.1.6)	Uncased only	2.0"	Conductive fluid	2.0-4.0'	Hole diameter
Cross-Well AC Voltage (3.1.6)	Uncased only	?	Wet or dry	10s to 100s of meters	Borehole deviation
Electromagnetic Logs					
Induction (3.2.1)	Uncased or nonmetallic	2.0-4.0"	Wet or dry	30"	Effect of hole diameter and mud negligible

Characterization Method Summary Information

Method	Casing	Vertical Resolution	Wet/Dry	Depth of Investigation	Interferences/Corrections
Borehole Radar (3.2.2)	Uncased or nonmetallic	2.0-6.0"	Wet or dry	meters	Borehole deviation (crosshole)
Dielectric (3.2.3)	Uncased or nonmetallic	5.0"	Wet or dry	30"	Conductive material skin depth, chlorine interference
Nuclear Magnetic Resonance (3.2.4)	Uncased only	7.0"	Required	1.5'	Borehole fluid
Surface-Borehole CSAMT (3.2.4)	Uncased only (?)	?	Wet or dry(?)	?	?
Nuclear Logs					
Natural Gamma (3.3.1)	Uncased or cased	1.0-2.0"	Wet or dry	6.0-12.0"	None for qualitative uses; hole diameter, casing (thickness, composition, and size), and drilling fluid density for quantitative uses
Gamma-Gamma (3.3.2)	Uncased or cased	2.5"	Wet or dry	6.0"	Same as natural gamma with addition of formation fluid and matrix density corrections
Neutron (3.3.3)	Uncased or cased	1.5-4.5"	Wet or dry	6.0-12.0"	Same as natural gamma with addition of temperature, fluid salinity, and matrix composition corrections
Gamma-Spectrometry (3.3.4)	Uncased or cased	2.0-4.0"	Wet or dry	6.0-12.0"	Similar to natural gamma
Neutron-Activation (3.3.5)	Uncased or cased	2.0-4.0"	Wet or dry	< Neutron	?
Neutron-Lifetime (3.3.6)	Uncased or cased	2.0-4.0"	Wet or dry	< Neutron	?

Table A-2 (cont.)

Acoustic and Seismic Logs

Log Type/Section	Casing[a]	Min. Diam.[b]	Borehole Fluid	Radius of Measurement	Required Correction
Acoustic-Velocity/[c] Sonic (3.4.1)	Uncased or bonded metallic	2.0–4.0"	Required	Depends on frequency and rock velocity; several feet	Hole diameter, formation fluid, and matrix velocity corrections for quantitative uses
Acoustic-Waveform[c] (3.4.2)	Uncased or bonded metallic	2.5–3.0"	Required	> sonic	Same as sonic
Acoustic-Televiewer (3.4.3)	Uncased only	3.0" min 16.0" max	Required	Borehole surface	Large number of equipment adjustments required during operation (calibration of magnetometer), borehole diameter response, borehole deviation
Surface-Borehole Seismic (3.4.4)	Uncased or bonded cased	2.5–4.0"	Wet or dry	Depends on geophone configuration	Borehole deviation, correction for geometric spreading of source energy; geophones must be locked in dry holes
Geophysical Diffraction Tomography (3.4.5)	Uncased or nonmetallic	2.5–4.0"	Wet	100'	Borehole deviation
Cross Borehole Seismic (3.4.6)	Cased or uncased	2.0–3.0"	Wet or dry	Depends on borehole spacing	Borehole deviation

Miscellaneous Logging Methods

Characterization Method Summary Information

Method	Casing	Diameter	Fluid	Range/Resolution	Corrections
Caliper (3.5.1)	Uncased or cased	1.5"+	Wet or dry	Arm limit (usually 2.0-3.0')	None
Temperature (3.5.2)	Uncased or cased[d]	2.0"	Required	Within borehole	Calibration to known standard
Mechanical Flowmeter (3.5.3)	e	2.0-4.0"	Required	e	Borehole diameter for velocity and volumetric logging
Thermal Flowmeter (3.5.4)	e	2.0"	Required	e	Borehole diameter for velocity and volumetric logging
EM Flowmeter (3.5.5)	e	2.0"	Required	e	Borehole diameter for velocity and volumetric logging
Single-borehole flow tracing (3.5.6)	e	1.75"+	Required	e	Changes in flow field with time
Colloidal Boroscope (3.5.7)	e	2.0"	Required	e	None
Television/Photography (3.5.7)	Uncased or cased	2.0"+	Wet or dry	Borehole surface	None
Gravity (3.5.8)	Uncased best	6.0"	Wet or dry	10s to 100s of meters	Borehole diameter/inclination; other usual gravity corrections
Magnetic/Magnetic Susceptibility (3.5.8)	Uncased or nonmetallic	?	Wet or dry	1.0-2.0'	Hole diameter correction
Well Construction Logs					
Casing Collar Locator (3.6.1)	Steel Casing	2.0"+	Wet or dry	Casing collar, thickness	None
Cement and Gravel Pack Logs (3.6.2)	Cased	See specific logging methods discussed in Section 3.6.2			e

Table A-2 (cont.)

Log Type/Section	Casing[a]	Min. Diam.[b]	Borehole Fluid	Radius of Measurement	Required Correction
Well Construction Logs (cont.)					
Borehole Deviation (3.6.3)	Uncased	Varies	Wet or dry	Borehole Surface	Magnetic declination
Fluid/Gas Chemical Sensors					
Eh, Ph Probes (5.5.4)	Uncased/screened	1.0"	Required	Within borehole	Calibration to known standards
Ion-Selective Electrodes (5.5.5)	Uncased/screened	1.0"	Required	Within borehole	Calibration to known standards
Fiber Optic Chemical Sensors (5.5.6)	Uncased/screened	<2.0"	Wet or dry	Within borehole	Calibration to known standards
Other Chemical Sensors (10.6.5)	Uncased/screened	<1.0"-2.0"	Wet or dry	Within borehole	Calibration to known standards

Boldface = Most frequently used techniques in ground-water investigations.

[a] Unless otherwise specified, either plastic or steel casing is possible.

[b] Indicates the range of minimum diameters for commercially available probes based on best available information. Various sources were used, with the survey by Adams et al. (1983) being the main source.

[c] Wheatcraft et al. (1986) indicate that acoustic logs are suitable only for uncased boreholes. However, Thornhill and Benefield (1990) report using them for mechanical integrity tests of steel-cased injection wells.

[d] Wheatcraft et al. (1986) indicate that casing is allowable for temperature logs, Benson (1991) indicates that casing should not be used. Uncased holes are required for identification of high permeability zones. Cased hole uses would include measurement of geothermal gradient and cement bond logs (see Section 3.6.2).

[e] Flow measurements are usually made in uncased holes or screened intervals of cased holes. Radius of measurement depends on the permeability and whether natural or induced flow is measured. Natural flow will measure the properties of several well diameters; pumping will measure properties up to 25 to 35 well diameters (Taylor, 1989).

Characterization Method Summary Information

Table A-3 Summary Information on Vadose Zone Water State Measurement and Monitoring Methods

Method	Property Measured	Accuracy/Range	Sections
Vadose Zone Soil Water Potential Measurement[a]			
Porous Cup Tensiometers	Capillary pressure	0 to -85 kPa[b] 0 to -80 kPa[c]	6.1.1
Thermocouple Psychrometers	Relative humidity	-200 to -8,000 kPa[b] -100 to -5,000 kPa[c]	6.1.2
Resistance Sensors	Resistance	-50 to -1,500 kPa[c]	6.1.3
Gypsum Blocks	Resistance	0 to -30 kPa[b]	6.1.3
Fiberglass/Nylon Cells	Resistance	No limits[b]	6.1.3
Electrothermal Methods	Heat transfer	0 to -200 kPa	6.1.4
Osmotic Tensiometers	Osmotic + pressure potential	0 to -1,500 kPa[b]	6.1.5
Filter-Paper Method	Water content	-10 to -100,000 kPa	6.1.6
Electro-Optical Sensors	Optical properties	0 to -2,400 kPa	6.2.6
Water Activity Meter	Relative humidity	0 to -31,600 kPa	6.1.7
Vadose Zone Soil Water Content Measurement[a]			
Gravimetric	Weight	[d]	6.2.1
Gamma-Gamma	Radiation	[d]	6.2.2, 3.3.2
Neutron Moisture Probe	Radiation	[d]	6.2.2, 3.3.3
Dielectric Sensors	Dielectric	[d]	6.2.3
Time Domain Reflectometry	Dielectric	[d]	6.2.4
Nuclear Magnetic Resonance	Magnetic field	[d]	6.2.5, 3.2.4, 10.6.3
Electro-Optical Sensors	Optical properties	[d]	6.2.6
CAT Scan	Radiation	[d]	6.2.7
Thermal Infrared	Remote sensing	[d]	1.1.3
Active Microwave	Remote sensing	[d]	1.1.4
Four-Electrode Method	Resistivity	[d]	9.1.1
Salinity Sensors	Conductivity	[d]	9.1.3
Electromagnetic Induction	Conductivity	[d]	9.1.4
Other Vadose Zone Hydrologic Properties			
Soil Moisture-Potential-Conductivity Relationships			6.3.1
Water Sorptivity/Diffusivity			6.3.2
Available Water Capacity		0.1-1%	6.3.3

Boldface = most commonly used methods.

[a] Moisture content can be determined from measurement of soil water potential and vice versa by the use of a moisture characteristic curve, which relates matric potential to water content (Section 6.3.1). The pascal is the Standard International unit for measuring pressure used by the Soil Science Society of America. The bar is commonly used as a pressure unit in vadose zone investigations: 1 kPa = 1 centibar.
[b] Indicated by Rehm et al. (1985).
[c] Indicated by Bruce and Luxmoore (1986).
[d] Most methods for measuring moisture content are accurate to around 1%. Gravimetric methods and nuclear methods can be accurate to 0.1% or less.

Table A-4 Summary Information on Vadose Hydraulic Conductivity Techniques[a]

Technique	K_{fs} or K_{unsat}	K Direction[b]	Other Parameters Measured	Section	Tables
Infiltration (see also, Sections 7.2.3, 7.2.5, 7.2.6, 7.3.1, 7.3.4)					
Seepage Meters	Saturated	Undefined	I	7.1.1	
Instantaneous Rate	Saturated	Undefined	I	7.1.1	
Impoundment Water Budget	Saturated	Undefined	I	7.1.1	
Sprinkler Infiltrometer	Saturated	Vertical	I	7.1.2	
Infiltration Test Basins	Saturated	Undefined	I	7.1.2	
Watershed Average	Undefined	Undefined	I	7.1.3	
Watershed Empirical Relations	Undefined	Undefined	I	7.1.3	
Infiltration Equations	Both	Vertical	I	7.1.4	7-5
Unsaturated Hydraulic Conductivity					
Instantaneous Profile	Unsaturated	Vertical	D, F, K(ϕ), R	7.2.1	7-3
Draining Profile Methods	Unsaturated	Vertical	D, F, K(ϕ), R, S	7.2.2	7-3
Tension Infiltrometers	Both	Vertical	I, D, F, K(ϕ), R, S	7.2.3	7-3
Crust-Imposed Steady Flux	Unsaturated	Vertical	I, F, K(ϕ)	7.2.4	7-3
Sprinkler/Dripper Methods	Unsaturated	Vertical	I, F, K(ϕ), R, S	7.2.5	7-3
Entrapped Air Method	Unsaturated	Vertical	I, F	7.2.6	7-3
Parameter Identification	Both	Undefined	R	7.2.7	7-3
Empirical Equations	Both	Undefined	Varies	7.2.8	7-5
Column-Crust	Both	Vertical	F, K(ϕ)	7.3.8	7-4
Saturated Hydraulic Conductivity Above Shallow Water Table[c]					
Cylinder Infiltrometers	Saturated	Vertical	I, S	7.3.1	7-4
Constant Head Borehole Infiltration	Saturated	Horizontal	S	7.3.2	7-4
Guelph Permeameter	Both	Vert./Hor.	K(ϕ), S	7.3.3	7-4

Characterization Method Summary Information

Method	Saturation	Direction	Measurement	Section	Table
Air-Entry Permeameter	Both	Vertical	I, K(ϕ), S	7.3.4	7-2, 7-4
Double Tube	Saturated	Vertical	--	7.3.5	7-2, 7-4
Cylinder Permeameter	Saturated	Vertical	--	7.3.6	7-2, 7-4
Infiltration Gradient	Saturated	Vertical[d]	--	7.3.7	7-4
Cube	Saturated	Vert./Hor.	--	7.3.8	7-2, 7-4
Column/Monoliths	Saturated	Vertical	--	7.3.8	7-2, 7-4
Boutwell Method	Saturated	Vert./Hor.	--	7.3.9	
Velocity Permeameter	Saturated	Vertical	--	7.3.10	
Percolation Test	--[e]	--[e]	--	7.3.11	7-4
CP Porous Probe	Saturated	Horizontal	--	2.2.2	
Collection Lysimeter	Saturated	Vertical	F	9.3.1	

Saturated Hydraulic Conductivity Above Deep Water Table[c]

Method	Saturation	Direction	Measurement	Section
USBR Single Well	Saturated	Undefined	--	7.4.1
USBR Multiple-Well	Saturated	Horizontal	--	7.4.2
Stephens-Neuman Single Well	Saturated	Undefined	--	7.4.3
Air Permeability	Saturated	Undefined	--	7.4.4
Packer Tests	Saturated	Vert./Hor.	--	4.3.3

D = diffusivity; F = Flux; I = Infiltration; K(ϕ) = hydraulic conductivity-pressure head relationship; R = Retention (pressure-moisture relationship); S = Sorptivity.

[a] Most methods for measuring or estimating unsaturated hydraulic conductivity also can be used to measure water flux in the vadose zone. Section 7.5 discusses the application of these and other methods for measuring soil water flux.

[b] Directional ratings are qualitative in nature. Different references might give different ratings depending on site conditions and criteria used to define directionality.

[c] These methods measure <u>field-saturated</u> or <u>satiated hydraulic conductivity</u> (K_{fs}), which is lower than <u>saturated</u> hydraulic conductivity, due to the presence of entrapped air.

[d] Differentiation of vertical and horizontal is possible when used with double tube method.

[e] The percolation test does not provide an accurate measure of saturated hydraulic conductivity. See Table 7-4 for sources on information on the relationship between percolation test results and K_{sat}.

Table A-5 Summary Information on Vadose Zone Water Budget Characterization Methods

Technique	Parameters Measured	Manual/ Automatic	S/A/R	Section	Tables
Water-Related Hydrometeorological Measurements			A		
Sacramento Gage	Rain	Manual	± 1 mm[a]	8.1.1	8-2
Storage Gage	Rain	Manual	"	8.1.1	8-2
Automatic Wet/Dry Collectors	Rain/Snow	Either	"	8.1.1	8-2
Weighing Gage	Rain/Snow	Automatic	"	8.1.2	8-2
Tipping Bucket Gage	Rain	Automatic	"	8.1.2	8-2
Float Gage	Rain	Automatic	"	8.1.2	8-2
			S/A/R		
Sling Psychrometer	Humidity	Manual	0.1/0.5/--	8.1.3	8-2
Aspirated Psychrometer	Humidity	Either	0.02/0.1/--	8.1.3	8-2
Thermocouple Psychrometer	Humidity	Either	?	6.1.2	6-1, 6-3
Mechanical Hygrometers	Humidity	Either	1.0/5.0/20 to 100%	8.1.4	8-2
Dew-/Frost-Point Hygrometer	Humidity	Either	.05/0.25/--	8.1.4	8-2
Dew Cell/Probes	Humidity	Automatic	0.5/2.0/10 to 100%	8.1.4	8-2
Electric Hygrometers	Humidity	Either	0.5/2.0/5.0 to 98%	8.1.4	8-2
Diffusion Hygrometers	Humidity	?	?	8.1.4	8-2
Absorption Spectra Hygrometers	Humidity	?	?	8.1.4	8-2
Other Hydrometeorological Measurements		A/R			
Liquid-in-Glass Thermometer	Temperature	Manual	±0.5°C/-40 to +60[b]	8.2.1	
Bi-Metal Thermometer	Temperature	Either	"	8.2.1	
Bourdon Tube Thermometer	Temperature	Either	"	8.2.1	
Thermocouple	Temperature	Either	"	8.2.2	
Metallic Resistance Bulb	Temperature	Either	"	8.2.2	
Thermistor	Temperature	Either	"	8.2.2	

Cup Anemometers	H windspeed	Either	1.0 to 50/±0.5m/s[a]	8.2.3	8-2
Windmill Anemometers	V-H windspeed	Either	"	8.2.3	8-2
Pressure Anemometers	H windspeed	Manual	"	8.2.3	8-2
Hot-Wire Anemometer	V-H windspeed	Automatic	"	8.2.3	8-2
Acoustic Anemometer	V-H windspeed	Automatic	"	8.2.3	8-2
Wind Vanes	Direction	Either	0.5 to 50/±5°	8.2.4	8-2
Wind Cones	Direction	Manual	"	8.2.4	8-2
			S/A		
Mercury Barometer	Air pressure	Manual	?	8.2.5	
Altimeter	Air pressure	Manual	2 hPa/±0.2%	8.2.5	
Precision Aneroid	Air pressure	Either	0.5 hPa/?	8.2.5	
			A		
Thermopile Pyranometers	Global rad.	Automatic	±0.1 to 0.5 mW/cm²	8.2.6	8-2
Bimetallic Pyranometer	Global rad.	Either	±1.0 mW/cm²	8.2.6	8-2
Photovoltaic Pyranometer	Global rad.	Either	?	8.2.6	8-2
Net Radiometers	Net flux	Either	?	8.2.7	8-2
Pyrheliometers	Direct rad.	Either	?	8.2.7	8-2

<u>Evapotranspiration (Water Balance Methods)</u>

			Accuracy		
Lysimeters	WE,SE,ET,T	Either	Moderate to high[c]	8.3.1	8-3
Soil Moisture Monitoring	SE,ET,T	Manual	Moderate to high[c]	8.3.2	8-3
Water Budget Methods	WE,SE,ET,T	Manual	Low to high	8.3.3	8-3
Evaporation Pans	WE	Manual	Moderate	8.3.4	8-3
Evaporimeter	SE	Manual	High[c]	8.3.5	
Atmometers	SE,T	Manual	Moderate	8.3.5	8-3
Chloride Tracer	SE,ET,T	Manual	Moderate	8.3.6	
Ground-Water Fluctuation	SE,ET	Manual	Moderate	8.3.7	8-3
Other Transpiration Methods	T	Manual	Moderate to high[c]	8.3.8	8-3
Thermal Infrared	WE,SE,ET	Either	Low to moderate	1.1.3	1-3

Table A-5 (cont.)

Technique	Parameters Measured	Manual/ Automatic	S/A/R	Section	Tables
Evapotranspiration (Micrometeorological)					
Empirical Equations	WE,SE,ET,T	Manual	Moderate to high	8.4.1	8-3
Physically-Based Equations	WE,SE,ET	Either	Moderate to high	8.4.2	8-3
Mass Transfer Methods	WE,ET,T	Either	Moderate to high	8.4.3	8-3
Energy Budget Methods	WE,SE,ET,T	Either	Moderate to high	8.4.4	8-3
Profile/Gradient Method	WE,SE,ET	Either	Low to moderate	8.4.5	8-3
Eddy Correlation	WE,ET	Either	High	8.4.6	8-3

Boldface = Most commonly used methods.

Abbreviations for hydrometeorological methods: S = Sensitivity = The smallest fraction of a division on a scale on which a reading can be made directly or by estimation; A = Accuracy = The closeness with which an observation approaches the true value; R = Range of relative humidity that can be measured.

Abbreviations for evapotranspiration methods: WE = Water evaporation; SE = Bare soil evaporation; ET = Evapotranspiration; T = transpiration.

[a]Recommended accuracy by World Meteorological Organization. Less precise measurements might be acceptable, depending on the purpose of measurements.
[b]Range and accuracy of specific thermometers can range considerably, value shown is the recommended specification in U.S. EPA (1987b).
[c]For high accuracy, numerous measurements at different locations might be required to adequately characterize the variability of evapotranspiration.

Table A-6 Summary Information on Ground Water Level/Pressure Measurement

Method	Property Measured	Accuracy[a]	Chapter Sections
Monitoring Well Water Level Measurement			
Steel Tape	Water surface	0.01'	4.1.1
Electric Probe	Water surface	0.02-0.1'	4.1.2
Air Line	Pressure head	0.25'	4.1.3
Pressure Transducers	Pressure head	0.01-0.1'	4.1.4
Popper/Acoustic Probe	Water surface	0.1'	4.1.5
Ultrasonic	Water surface	0.02-0.1'	4.1.6
Mechanical Float	Water surface	0.02-0.5'	4.1.7
Potentiometer Float	Water surface	0.01-0.1'[b]	4.1.7
Electromechanical	Water surface	0.02-0.5'	4.1.8
Flowing Well Head Measurement			
Casing Extensions	Water surface	0.1'	4.1.9
Manometer/Pressure Gage	Pressure head	0.1-0.5'	4.1.9
Transducers	Pressure head	0.02'	4.1.9
In Situ Piezometers	Pressure head	0.02-0.5'[c]	4.1.10

[a]Water level measurement accuracy in wells taken from Dalton et al. (1991).
[b]Reported by Rosenberry (1990) as having accuracy similar to pressure transducers.
[c]Lower range for measurements with transducers and upper range for pressure gage.

Table A-7 Summary Information on Aquifer Test Methods

Technique	Confined/ Unconfined	Porous/ Fractured	Aquifer Properties Measured	Chapter Section	Table
Shallow Water Table					
Auger Hole	Unconfined	Porous	K (horizontal)[a]	4.2.1	4-5, 7-2
Pit-Baling	Unconfined	Porous[b]	K (undefined)	4.2.1	4-5
Pumped Borehole	Unconfined	Porous	K (undefined)	4.2.1	4-5
Piezometer	Unconfined	Porous	K (undefined)	4.2.2	4-5, 7-2
Tube	Unconfined	Porous[b]	K (vertical)	4.2.2	4-5
Well Point	Unconfined	Porous	K (undefined)	4.2.2	4-5
Two-Hole	Unconfined	Porous	K (undefined)	4.2.3	4-5
Four-Hole	Unconfined	Porous	K (undefined)	4.2.3	4-5, 7-2
Multiple-Hole	Unconfined	Porous	K (undefined)	4.2.3	4-5
Drainage Outflow	Unconfined	Porous	K (undefined)	4.2.3	4-5
Well Tests					
Slug (Injection/Withdrawal)	Both	Porous	K, H, T	4.3.1	4-5
Slug (Displacement)	Both	Porous	K, H, T	4.3.1	4-5
Single-Well Pump	Both	Porous	K, S, T	4.3.2	4-5
Multiple-Well Pump	Both	Porous	A, K, S, T	4.3.2	4-5
Single-Packer	Both	Both	K, H, T	4.3.3	4-5
Two-Packer[c]	Both	Both	K, H, T	4.3.3	4-5
Tracers					
Ions	Both	Both	D, F, V	4.4.1	4-3
Dyes	Unconfined	Both	D, F, V	4.4.2	4-3, 4-6
Gases	Unconfined	Both	D, F, R, V	4.4.3	4-3
Stable Isotopes	Both	Both	D, F, R, V	4.4.4	4-3, 4-6

Characterization Method Summary Information

Radioactive Isotopes	Both	Both	D, F, R, V, T[d]	4.4.5	4-3, 4-6
Water Temperature	Unconfined	Both	D, F, V	4.4.6	4-3
Particulates/Microorganisms	Unconfined	Both	D, F, V	4.4.7	4-3, 4-6
Other Techniques					
Water Balance	Unconfined	Both	R	4.5.1	4-5
Moisture Profile	Unconfined	Porous	S	4.5.2	
Shallow Geothermal	Unconfined	Porous	F, R	1.6.2	
Fluid Conductivity Log	Both	Both	F	3.1.3	
Neutron Activation	Both	Both	F, H, V	3.3.5	
Differential Temperature Log	Both	Both	F	3.5.2	
Flow Meters	Both	Both	F, H, V	3.5.3-3.5.5	
Single-Well Tracer Methods	Both	Both	F, H, V	3.5.6	
Other borehole methods	Both	Both	H	Section 3	
Piezometric Map	Both	Both	F, H	4.1	

Boldface = most commonly used methods.

A = anisotropy; D = dispersivity; F = flow direction; H = heterogeneity; K = hydraulic conductivity; R = recharge/age; S = specific storage/yield; T = Transmissivity; V = Velocity.

[a]Directional ratings are qualitative in nature. Different references may give different ratings depending on site conditions and criteria used to define directionality. For example, U.S. EPA (1981) and Hendrickx (1990) note that this method often measures primarily horizontal conductivity, whereas Bouma (1983) indicates that the direction is undefined (see Figure 7-2).

[b]Can be used in rocky soils; other methods generally require fine-grained soils.

[c]Can be used to measure saturated hydraulic conductivity both above and below the water table in open holes in consolidated rock.

[d]Actual uses are much more restricted due to health concerns.

Table A-8 Summary Information on Drilling Methods

Drill Method	Casing/ Open Hole	Fluids Affect Chem.?	Core Samples?	Section Number	Tables
Hollow-Stem Auger	Open Hole	Usually No	Possible	2.1.1	2-2, 2.1.1
Open-Hole Rotary Methods					
Direct Air Rotary with Bit	Open Hole	Yes	Possible	2.1.2	2-2, 2.1.2
Direct Air Rotary with Downhole Hammer	Open Hole	Yes	Possible	2.1.2	2-2, 2.1.2
Direct Mud Rotary	Open Hole	Yes	Possible	2.1.3	2-2, 2.1.3
Reverse Rotary (no casing)	Open Hole	Yes	Possible	2.1.3	2-2
Cable Tool	Either	Usually No	Possible	2.1.4	2-2, 2.1.4
Rotary Drill-Through Methods					
Rotary Casing Driver	Casing	Yes	Possible	2.1.5	2-2, 2.1.5
Dual Rotary Advancement	Casing	Yes	Possible	2.1.5	
Reverse Circulation Methods					
Reverse Dual Wall Rotary	Casing	Yes	Possible	2.1.6	2-2, 2.1.6
Reverse Dual Wall Percussion	Casing	Yes	Possible	2.1.6	
Hydraulic Percussion	Casing	Yes	Possible	2.1.6	2-2
Downhole Casing Advancers	Casing	Yes	Possible	2.1.7	
Jet Percussion	Casing	Possible	Possible	2.1.8	2-2, 2.1.8
Jetting	Open Hole	Possible	No	2.1.8	
Solid Stem Auger	Open Hole	No	Possible	2.1.9	2-2, 2.1.9
Bucket Auger	Open Hole	No	Possible	2.1.9	
Rotary Diamond	Open Hole	Possible	Yes	2.1.10	
Directional Drilling	Either[a]	Possible	Possible[b]	2.1.11	
Sonic Drilling	Either	Possible	Yes	2.1.12	
Driven Wells	Either	No	No	2.2.1[c]	2-2
Cone Penetration	Open Hole	No	Possible[d]	2.2.2[c]	

Boldface = Most commonly used methods for monitoring well installation.
[a]EC rig uses casing advancement, other methods may involve open hole advancement.
[b]Sampling with a device resembling a split spoon may be possible with some directional rigs.
[c]Section includes cross references to other sections related to method.
[d]Geoprobe has developed a core sampler for use with a CPT rig.

Characterization Method Summary Information

Table A-9 Summary of Hand-Held Soil Sampling Devices

Sampling Device	Applications	Limitations
Spoons and Scoops (Section 2.3.1)[a]	Surface soil samples or the sides of pits or trenches	Limited to relatively shallow depths; disturbed samples
Shovels and Picks (Section 2.3.1)	A wide variety of soil conditions	Limited to relatively shallow depths
Augers[b] (Section 2.3.2)		
Screw Auger	Cohesive, soft, or hard soils or residue	Will not retain dry, loose, or granular, material
Standard Bucket Auger	General soil or residue	May not retain dry, loose, or granular material
Sand Bucket Auger	Bit designed to retain dry, loose, or granular material (silt, sand, and gravel)	Difficult to advance boring in cohesive soils
Mud Bucket Auger	Bit and bucket designed for wet silt and clay soil or residue	Will not retain dry, loose, or granular material
Dutch Auger	Designed specifically for wet, fibrous, or rooted soils (marshes)	
In-Situ Soil Recovery Auger	Collection of soil samples in reusable liners; closed top reduces contamination from caving sidewalls	Similar to standard bucket auger
Stony Soil Auger	Stony soils and asphalt	
Planer Auger	Clean out and flatten the bottom of predrilled holes	
Post-Hole/Iwan Auger	Cohesive, soft, or hard soils; readily available	Will not retain loose material
Silage Auger	Silage pits and peat bogs	
Spiral Auger	Used to remove rock from auger holes so that borings can continues with other auger-type	
Split core auger[c]	Auger with split core for easier recovery of sample; can be used with liner	

Table A-9 (cont.)

Sampling Device	Applications	Limitations

Tube Samplers[d] (Section 2.3.3)

Soil Probe	Cohesive, soft soils or residue; representative samples in soft to medium cohesive soils and silts	Sampling depth generally limited to less than 1 meter
Thin-Walled Tubes	Cohesive, soft soils or residue; special tips for wet or dry soils available	Similar to Veihmeyer tube
Soil Recovery Probe	Similar to thin-wall tube; cores are collected in reusable liners, minimizing contact with the air	Similar to Veihmeyer tube
Veihmeyer Tube	Cohesive soils or residue to depth of 3 meters (maximum of 4.9 meters)	Difficult to drive into dense or hard material; will not retain dry, loose, or granular material; may be difficult to pull from ground
Geostick[c]	Spot soil sampling and penetrometer tests	
Peat Sampler	Wet, fibrous, organic soils	

[a] Section number in Boulding (1993).
[b] Suitable for soils with limited coarse fragments; only the stony soil auger will work well in very gravelly soil.
[c] Not included in Boulding (1993).
[d] Not suitable for soils with coarse fragments.

Table A-10 Summary of Major Types of Power-Driven Samplers

Tube Type	Applications	Limitations
Disturbed Core Samplers		
Barrel Samplers (Section 2.4.1)[a]		
Solid Barrel Sampler	Sand, silts, clays	Disturbed core, questionable recovery and quality below water table
Split Spoon Sampler	Disturbed samples from cohesive soils	Ineffective in cohesionless sands; not suitable for collection of samples for laboratory tests requiring undisturbed soil
Rotating Core (Section 2.4.2)		
Single Tube	Dense unconsolidated and consolidated formations	
Double-Tube	Friable, erodible, soluble or highly fractured formations	
CP Punch Core[b]	Wireline system with various punch shoes; very effective in mixed formations where deep sampling is needed	
Undisturbed Core Samplers		
Thin-Wall Open Tube Samplers (Section 2.4.3)[a]		
Shelby Tube	Undisturbed samples in cohesive soils, silt, and sand above water table	Ineffective in cohesionless sands or stony soil
Continuous Tube/Laskey Sampler	Same as Shelby tube, except longer barrel designed to operate inside the column of a hollow-stem auger	Same as Shelby tube; no blow counts taken
Thin-Wall Piston Samplers (Section 2.4.4)		
Internal Sleeve Piston Sampler	Collection of sample in heaving sands; used with hollow-stem auger with clamshell bit	Requires use of water or drilling mud for hydrostatic control; only one sample per borehole can be obtained
Fixed-Piston Sampler	Undisturbed samples in cohesive soils, silt, and sand above or below water table	Ineffective in cohesionless sands

Table A-10 (cont.)

Tube Type	Applications	Limitations
Thin-Wall Piston Samplers (cont.)		
Wireline Piston Sampler	Undisturbed samples in cohesive soils and noncohesive sands; used with clam shell device on hollow-stem auger	In heaving sands only one sample per borehole can be collected because clamshell remains open after sampling
Hydraulic Piston Sampler (Osterberg and others)	Similar to fixed-piston sampler	Not possible to limit the length of push or to determine amount of partial sampler penetration during push
Stationary Piston Sampler	Undisturbed samples in stiff, cohesive soils; representative samples in soft to medium cohesive soils, silts, and some sands	
Gus Sampler[b]	Similar to stationary piston sampler, except uses hydraulic action	
Free Piston Sampler	Similar to stationary piston sampler	Not suitable for cohesionless soils
Open Drive Sampler	Similar to stationary piston sampler	Not suitable for cohesionless soils
Specialized Thin-Wall (Section 2.4.5)		
Pitcher Sampler	Undisturbed samples in hard, brittle, cohesive soils and cemented sands; representative samples in soft to medium cohesive soils, silts, and some sands; variable success with cohesionless soils	Frequently ineffective in cohesionless soils; require use of drilling fluid that may affect quality of sample
Denison Sampler	Undisturbed samples in stiff to hard cohesive soils, cemented sands, and soft rocks; variable success with cohesionless materials	Not suitable for undisturbed sampling of loose, cohesionless soils or soft cohesive soils; require use of drilling fluid that may affect quality of sample
Vicksburg Sampler	Similar to Shelby tube, but able to sample denser and coarser material	

[a] Section number in Boulding (1993).
[b] Not included in Boulding (1993).

Characterization Method Summary Information

Table A-11 Summary Information on Soil Solute Monitoring and Sampling Methods

Method	Sampling Method	Depth Limitation	Chapter Sections
Indirect Salinity Measurement Methods			
Four Probe Electrical	Resistivity	Near surface	9.1.1, 1.2.2
Portable EC Probe	Resistivity	1.5 m	9.1.2
In Situ EC Probe	Resistivity	None	9.1.2
Porous Matrix Salinity Sensors	Resistivity	None	9.1.3
Electromagnetic Induction Sensor	Conductivity	2 m	9.1.4, 1.3.1
Dielectric Sensors	Dielectric	2 m[a]	9.1.4, 6.2.3
Time Domain Reflectometry Sensor	Dielectric	Up to 20 m	9.1.4, 6.2.4
Neutron Probe	Nuclear	None	3.3.3, 6.2.2
Direct Soil Solute Sampling Methods			
Vacuum-Type Porous Cup	Suction	2 m	9.2.1
Vacuum-Pressure Porous Cup	Suction	45 ft	9.2.2
Vacuum High-Pressure Porous Cup	Suction	300 ft	9.2.2
Vacuum-Plate Sampler	Suction	2 m[a]	9.2.3
Membrane Filter	Suction	1-4 m[b]	9.2.4
Hollow Fiber	Suction	2 m[a]	9.2.5
Ceramic Tube Sampler	Suction	2 m[a]	9.2.6
Capillary Wick Sampler	Capillary	[d]	9.2.7
BAT Sampler	Suction	45 ft	5.5.2
Trench Lysimeter	Gravity[c]	[d]	9.3.1
Caisson Lysimeter	Gravity	3 m+	9.3.1
Pan Lysimeter	Gravity	[d]	9.3.1
Glass Block Lysimeter	Gravity	[d]	9.3.1
Wicking Type Sampler	Gravity	[d]	9.3.1
Tile Drain Outflow	Gravity	50+ ft	9.3.1
Perched Water Table	Gravity	None	9.3.2
Nylon Sponge	Absorbent	Near surface	9.3.3
Ceramic Rod	Absorbent	Near surface	9.3.3
Solid Soil Water Extraction	[e]	None	9.3.4
Soil Saturation Extract	Slurry	None	9.3.5
SEAMIST	Absorbent	100s ft	9.3.7
Methods for Sampling Sensitive Soil Constituents			
Static Soil-Gas Sampling	Absorbent	Near Surface	9.4.1
Soil-Gas Probes	Suction	[f]	9.4.2
Tank Leak Sensors	Various	Typically <2m	9.4.3
Soil Volatiles/Microorganisms	Core	[f]	9.3.6

Boldface = Most commonly used methods.
[a]With vacuum sampling apparatus; greater depths would be possible using vacuum-pressure sampling system.
[b]Upper limit would require modification of system to use vacuum-pressure sampling apparatus.
[c]Sample is collected by free-drainage in all gravity samplers, but suction can be used to bring sample to the surface.
[d]Depth limited by the depth to which a hole or trench can be safely dug for installation of sampler in the sidewall; typically 2 meters or less.
[e]Various methods can be used to extract soil water from a sample: Squeezing, displacement, displacement/centrifugation, centrifugation, and adsorption.
[f]Depends on density of subsurface material and method of penetration/coring. Soil gas probes used with cone penetration rigs (Sections 2.2.2, 5.5.1, and 5.5.2) can penetrate 100 to 150 feet with favorable soil conditions; greater depths are possible if holes are drilled before insertion of the soil gas probe. Coring depth limits are defined by the type of drilling/coring method used (Sections 2.3 and 2.4).

Table A-12 Summary Information on Ground-Water Sampling Devices (Information is for general guidance only)

Sampling Device	Max. Sample Depth	Min. Well Diameter	Sample Delivery Rate/Vol.[a]	Section	Tables
Portable Positive Displacement Samplers					
Bladder Pumps	1,000'	1.5"	0-3.0 gpm	5.1.1	5-2, 5-3
Gear Pumps	200'	2.0"	0-1.5 gpm	5.1.2	5-2, 5-3
Helical Rotor Pumps	160'	2.0"	0-1.5 gpm	5.1.3	5-2, 5-3
Gas-Drive/Displacement	300'	1.0"	0.1-10 gpm	5.1.4	5-2, 5-3
Gas-Drive Piston Pumps	900'	1.5"	0-1.5 gpm	5.1.5	5-2, 5-3
Mechanical Piston-Pumps	Variable	1.0 to 4.0"	Variable	5.1.6	
Other Portable Ground-Water Sampling Pumps					
Peristaltic Suction Lift	25'	0.5"	0.01-8 gpm	5.2.1	5-2, 5-3
Centrifugal Suction Lift	15'	1.0"	1.0-25 gpm	5.2.1	
Variable-Speed Submersible Centrifugal Pump	290'	1.75"	0.026-8 gpm	5.2.2	5-2, 5-3
Other Submersible Centrifugal Pumps	2,000'	4.0+"	5.0-60 gpm	5.2.2	5-2, 5-3
Inertial-Lift Pump	200'	1.5"	0-2.0 gpm	5.2.3	
Gas-Lift	Variable	1.0"	Variable	5.2.4	5-2
Jet (Venturi) Pump	200'	<1.0"	25-30 gpm	5.2.5	
Packer Pumps[b]	Variable	2.0"	Variable	5.2.6	
Portable Grab/Depth Specific Samplers					
Open Bailer[c]	No limit	0.5"	Variable	5.3.1	5-2, 5-3
Point-Source Bailer[c]	No limit	0.5"	Variable	5.3.1	5-2, 5-3
Syringe Sampler	No limit	1.5"	0.01-0.2 gal	5.3.2	5-2, 5-3
Westbay Sampler	No limit	1.5"	40 mL	5.3.2	
Kemmerer/Van Dorn	No limit	1.0"	Variable	5.3.3	
Coliwasa	5'	2.0"	Variable	5.3.3	
Stratified Sample Thief	No limit	1.5"	Variable	5.3.3	
Swabbing	No limit	6.0"	Variable	5.3.3	
Portable/Permanent In Situ Samplers/Sensors					
Hydropunch	150'[d]	NA	500-1,250 mL	5.5.1	
BAT Sampler	100'[d]	NA	150 mL	5.5.2	
Other CPT Samplers[e]	25'	NA	0.01-0.3 gpm	5.5.2	
Other In Situ Probes[e]	25'	NA	0.01-0.3 gpm	5.5.3	
Eh, pH Probes	No limit	1.0"	NA	5.5.4	
Ion-Selective Electrodes	No limit	1.0"	NA	5.5.5	
Fiber Optic Sensors	No limit	±2.0"	NA	5.5.6	
Other Chemical Sensors	No limit	2.0-6.0"	NA	10.6.5	

Boldface = most commonly used devices.

[a] Sample delivery rates and volumes are averages based on typical field conditions. Actual rates are a function of diameter of monitoring well installation, size and capacity of sampling device, hydrogeologic conditions, and depth to sampling point.

[b] Depends on type of pump used (submersible, gas lift, suction)--see appropriate device for ratings.

[c] Not recommended for use with sensitive chemical constituents (see text discussion).

[d] Unlimited depth if hole is bored to desired depth before using sampler. Otherwise, actual depth of penetration is highly dependent on type of soil material.

[e] Depth and pumping rate depends on type of suction-lift device used. Values shown are for peristaltic pump.

Table A-13 Summary Information on Sample Processing/Analytical Techniques

Technique/Instrumentation	Technology Status[a]	Sample Matrix[b]	Contaminant Type[c]	Detection Limit[d]	Section/Table
Chemical Field Measurement Techniques/Sensors					
ph/Alkalinity/Acidity	I/CP	W,S	--	--	10.1.1, 5.5.4
Eh	I/CP	W,S	--	--	10.1.2, 5.5.4
Dissolved Oxygen	I/CP	W	--	ppm	10.1.2, 5.5.4
Temperature	I/CP	W	--	--	10.1.3
Electrical Conductance	I/CP	W,S	--	--	10.1.3
Filterable Residue	I/CM	W	--	--	10.1.3
Other Specific Ion Electrodes	II/CP	W	EA	ppm	5.5.5
Solid/Porous Fiber Optic	IV	W,S,A	VOC	ppm	5.5.6
Immunochemical Fiber Optic	IV	W,S,A	SVO	ppb-ppm	5.5.6, 10.5.2
Electrochemical Sensors	IV	W,A	VOC,TG	pbb-ppm	10.6.5
SAW Probes	IV	A	VOC,TG	ppm	10.6.5
Piezoelectric Sensors	IV	A	VOC	ppm	10.6.5
Semiconductor Sensors	IV	A,W	VOC	ppm-%	10.6.5
Sample Extraction Procedures					
Headspace Analysis	I	A	VOC	--	10.2.1
Vacuum Extraction	I/CP	A	VOC	--	10.2.1
Purge and Trap	I/CP	W	VOC	--	10.2.2
Solvent Extraction	I/CP	S	SVO,VOC	--	10.2.3
Thermal Digestion	II/CP	W,S	EA,HM	--	10.2.4
Thermal Extraction	II/CP	W,S	SVO	--	10.2.4
Thermal Desorption	III/CP	W,S	VOC,SVO	--	10.2.4
Supercritical Fluid Extract.	III/CP	W,S	VOC,SVO	--	10.2.5
Membrane Extraction	IV	W	VOC	--	10.2.5
Sorbent Extraction	I/CP	A,W	VOC,SVO	--	10.2.5

Table A-13 (cont.)

Technique/Instrumentation	Technology Status[a]	Sample Matrix[b]	Contaminant Type[c]	Detection Limit[d]	Section/Table
Gaseous Phase Analytical Techniques					
Photo-Ionization Detector	I/CP	A	VOC	ppb-ppm	10.3.1
Flame-Ionization Detector	I/CP	A	VOC	ppb-ppm	10.3.1
Argon-Ionization Detector	III/CP	A	VOC	100s ppb-ppm	10.3.1
Explosimeter	I/CP	A	VOC	%	10.3.2
Catalytic Surface Oxidation	I/CP	A	VOC,TG	ppm-%	10.3.2
Detector Tubes	I/CP	A	VOC,TG	high ppm	10.3.2
Gas Chromatography (GC)	II/CP,CM	A,W	VOC,SVO,TR	ppb-ppm	10.3.3/Table 10-3
Gaseous Phase Analytical Techniques (cont.)					
Mass Spectrometry (MS)	II/CF,CM	A	VOC,SVO,TR	ppm	10.3.4/Table 10-3
GC/MS	II/CM	A	VOC,SVO,TR	ppb	10.3.4/Table 10-3
Ion Trap MS	IV	A	VOC,SVO	ppb-ppm	10.3.4
AA Spectrometry	II/CM	A,W	EA,HM	ppb-ppm	10.3.5/Table 10-3
ICP-AES	II	A,W	EA,HM	ppb-ppm	10.3.6/Table 10-3
Ion Mobility Spectrometer	II/CP	A	VOC,SVO,TG	ppt-ppm	10.3.7
Luminescence/Spectroscopic Techniques					
X-Ray Fluorescence	II/CP,CM	S,W	HM	10s-100s ppm	10.4.1
UV Fluorescence	II	S,W	VOC	sub ppm	10.4.2/Tables 10-3, 10.4.2
Room-Temp. Phosphorimetry	III	S,W	VOC,SVO	ppb-ppm	10.4.2/Table 10.4.2
Synchronous Luminescence	III	W	VOC,SVO	ppm	10.4.2
Synchronous Fluorescence	III	W	VOC,SVO	ppm	10.4.2/Table 10.4.2
UV-Visible Spectrophotometry	III	A,W	VOC	ppb-ppm	10.4.3/Tables 10-3, 10.4.3
Infrared Spectroscopy	II	A,W,S	M,VOC,SVO	ppm-1000s ppm	10.4.3/Tables 10-3, 10.4.3

Characterization Method Summary Information

Luminescence/Spectroscopic Techniques (cont.)

Method	Col1	Col2	Col3	Col4	Ref
FTIR Spectroscopy	II/CP,CM	A		VOC	ppb-%	10.4.4/Table 10.4.4
Scattering/Absorption Lidar	IV	A		VOC	ppm	10.4.4
Raman Spectroscopy/SERS	II	W,S		VOC,SVO	ppb-ppm	10.4.4/Table 10.4.4
Near IR Reflectance/Trans. Spect.	IV	S		VOC	100s-1000s ppm	10.4.4/Table 10.4.4

Wet Chemistry

Chemical Colorimetric Kits	II/CP	W	EA,HM,SVO	ppb-100s ppm	10.5.1/Table 10-3
Other Colorimetric Methods	I/CP	W	TR	ppb-100s ppm	10.5.1/Table 10-3
Titrimetry	I/CP	W	EA,HM,TR	ppb-100s ppm	10.5.1
Immunoassay Colorimetric Kits	II/CP	W	SVO	ppb-ppm	10.5.2
Ion Chromatography	II	W	EA	ppm-100s ppm	10.5.3/Table 10-3
High-Pressure Liquid Chromatography	II/CM	W	SVO,TR	ppb-ppm	10.5.3/Table 10-3
Thin-Layer Chromatography	II	W	SVO	ppm	10.5.3
Coulometry	II	W	EA,TR	ppb-ppm	10.5.4
Polarography	II	W	EA	sub-100s ppm	10.5.4
Stripping Voltammetry	II	W	EA	ppt-ppm	10.5.4

Radiological

Neutron Activation/INNA	II	S,W	EA,TR	10s ppm	10.6.1, 3.3.5, 3.3.6
PIXE	II	S,W	EA,HM	10s-100s ppm	10.6.1/Table 10-3
Radiation Detectors	I/CP	A,S,W	R,TR	varies	10.6.1, 3.3.1
X-Ray Diffraction	II	S	M	--	10.6.1
Gamma Spectrometry	I/CP	S	M	--	3.3.2

Other

Gravimetric	I/CP	W,S	P,TDS	%	10.6.2
Volumetric	I/CP	S,W,A	P	--	10.6.2
Nuclear Magnetic Resonance	I/CP	S,W	M,P	--	10.6.3, 3.2.4
Magnetic Susceptibility	II	S	M	--	10.6.3
Electron Spin Resonance	II	S,W,A	M	--	10.6.3

Table A-13 (cont.)

Technique/Instrumentation	Technology Status[a]	Sample Matrix[b]	Contaminant Type[c]	Detection Limit[d]	Section/Table
Other (cont.)					
Optical Microscope	II/CP	S	M,P	--	10.6.4
Scanning Electron Microscope	II	S	M	--	10.6.4
Electron Microprobe	II	S	M	high ppm	10.6.4
Field Bioassessment	II	--	VOC,HM	--	10.6.6
Toxicity Tests	II	W,S,A	VOC,SVO,HM	--	10.6.6
Biomarkers	III	W,S,A	VOC,SVO,HM	--	10.6.6

Boldface = Most commonly used/proven field techniques.

[a] I = Well established and routinely used field technology; II = Well established laboratory technology for which experience in field applications is moderate to limited; III = Relatively well established technology for which there is limited field experience; IV = Developing technology with potentially useful field applications. CP = Commercially available portable instruments; CF = Commercially available fieldable instruments; CM = Commercial/custom mobile laboratories available.

[b] A = Air/gaseous matrix; S = Soil/solid matrix; W = Water/aqueous/liquid matrix. Volatile and semivolatiles in water and solid samples can be extracted for analysis by gaseous phase analytical techniques. Similarly, analytes can be extracted from solids samples for analysis using wet chemistry techniques.

[c] EA = Elemental/ionic analysis; HM = Heavy metals; M = Metals; M = Mineralogy; P = Physical characterization; R = Radioisotopes; SVO = Semivolatile organics; TG = Toxic gases; TDS = Total dissolved solids; TR = Tracer studies; VOC = Volatile organic compounds.

[d] Ranges for specific instruments and analytes might differ from range shown by orders of magnitude. In general, detection limits for soils will be higher than for ground water.

Characterization Method Summary Information

Table A-14 ASTM Standard Methods Cited in this Handbook

Id #/year[a]	Title[b]
D1452-80	Practice for Soil Investigation and Sampling by Auger Borings.
D1586-84	Method for Penetration Test and Split-Barrel Sampling of Soils.
D1587-83	Practice for Thin-Walled Tube Sampling of Soils.
D2113-83	Practice for Diamond Core Drilling for Site Investigation.
D2434-68	Test Method for Permeability of Granular Soils (Constant Head).
D2488-93	Practice for Description and Identification of Soils (Visual-Manual Procedures).
D3550-84	Practice for Ring-Lined Barrel Sampling of Soils.
D4043-91	Guide for Selection of Aquifer-Test Field and Analytical Procedures in Determination of Hydraulic Properties by Well Techniques.
D4210-89	Practice for Interlaboratory Quality Control Procedures and a Discussion on Reporting of Low-Level Data (Vol. 11.01).
D4404-84	Test Method for Determination of Pore Volumes and Pore Volume Distribution of Soil and Rock by Mercury Intrusion Porosimetry.
D4696-92	Guide for Pore-Liquid Sampling From the Vadose Zone.
D4700-91	Guide for Soil Sampling from the Vadose Zone.
D5084-90	Method for Hydraulic Conductivity of Saturated Porous Materials Using a Flexible Wall Permeameter.
D5088-90	Practice for Decontamination of Field Equipment Used at Nonradioactive Waste Sites.
D5092-90	Recommended Practice for Design and Installation of Ground Water Monitoring Wells in Aquifers.
D5126-90	Guide for Comparison of Field Methods for Determining Hydraulic Conductivity in the Vadose Zone.
D5299-92	Guide for the Decommissioning of Ground Water Wells, Vadose Zone Monitoring Devices, Boreholes and Other Devices for Environmental Activities.
D5314-92	Guide for Soil Gas Monitoring in the Vadose Zone.
D5387-93	Guide for Elements of a Complete Data Set for Non-Cohesive Sediments (Vol. 11.02).
D5447-93	Guide for Application of a Ground-Water Flow Model to a Site Specific Problem.
D5518-94	Guide for Acquisition of File Aerial Photography and Imagery for Establishing Historic Site-Use and Surficial Conditions

[a] Last 2 digits indicate the year in which the standard was approved or the last year in which substantive revisions were made to the standard.

[b] In volumes 4.08 (Soil and Rock I: D420 to D4914) and 4.09 (Soil and Rock II: D4943 to latest; Geosynthetics), unless otherwise indicated. Volumes and individual standards can be purchased from ASTM, 1916 Race St., Philadelphia, PA 19103-1187 (215/299-5585).

Appendix A References

Adams, W.M., S.W. Wheatcraft, and J.W. Hess. 1983. Downhole Sensing Equipment for Hazardous Waste Site Investigations. In: Proc. (4th) Nat. Conf. on Management of Uncontrolled Hazardous Waste Sites, Hazardous Materials Control Research Institute, Silver Spring, MD, pp. 108-113.

Benson, R.C. 1991. Remote Sensing and Geophysical Methods for Evaluation of Subsurface Conditions. In: Practical Handbook of Ground-Water Monitoring, D.M. Nielsen (ed.), Lewis Publishers, Chelsea, MI, pp. 143-194.

Boulding, J.R. 1993. Subsurface Field Characterization and Monitoring Techniques: A Desk Reference Guide, Volume I: Solids and Ground Water, Volume II: The Vadose Zone, Field Screening and Analytical Methods. EPA/625/R-93/003a&b. Available from CERI.*

Hendrickx, J.M.H. 1990. Determination of Hydraulic Soil Properties. In: Process Studies in Hillslope Hydrology, M.G. Anderson and T.P. Burt (eds.), John Wiley & Sons, New York, NY, pp. 43-92.

Bouma, J. 1983. Use of Soil Survey Data to Select Measurement Techniques for Hydraulic Conductivity. Agric. Water Manage. 6:177-190.

Bruce, R.R. and R.J. Luxmore. 1986. Water Retention: Field Methods. In: Methods of Soil Analysis, Part 1, 2nd edition, A. Klute (ed.), Agronomy Monograph No. 9, American Society of Agronomy, Madison, WI, pp. 663-686.

Dalton, M.G., B.E. Huntsman, and K. Bradbury. 1991. Acquisition and Interpretation of Water-Level Data. In: Practical Handbook of Ground-Water Monitoring, D.M. Nielsen (ed.), Lewis Publishers, Chelsea, MI, pp. 367-395.

Rehm, B.W., T.R. Stolzenburg, and D.G. Nichols. 1985. Field Measurement Methods for Hydrogeologic Investigations: A Critical Review of the Literature. EPRI EA-4301, Electric Power Research Institute, Palo Alto, CA.

Rosenberry, D.O. 1990. Effect of Sensor Error on Interpretation of Long-Term Water-Level Data. Ground Water 28:927-936. [Pressure transducer, potentiometer float]

Taylor, K. 1989. Review of Borehole Methods for Characterizing the Heterogeneity of Aquifer Hydraulic Properties. In: Proc. Conf. on New Field Techniques for Quantifying the Physical and Chemical Properties of Heterogeneous Aquifers, National Water Well Association, Dublin, OH, pp. 121-132.

Thornhill, J.T. and B.G. Benefield. 1990. Injection-Well Mechanical Integrity. EPA/625/9-89/007. Available from CERI.*

U.S. Environmental Protection Agency (EPA). 1981. Process Design Manual for Land Treatment of Municipal Wastewater. EPA/625/1-81/013. U.S. Army Corps of Engineers, U.S. Department of the Interior, and the U.S. Department of Agriculture also are authors of this report.

U.S. Environmental Protection Agency (EPA). 1987a. A Compendium of Superfund Field Operations Methods, Part 2. EPA/540/P-87/001 (OSWER Directive 9355.0-14) (NTIS PB88-181557), 644 pp.

U.S. Environmental Protection Agency (EPA). 1987b. On-Site Meteorological Program Guidance for Regulatory Modeling Applications. EPA/450/4-87/013 (NTIS PB87-227542), 187 pp.

Wheatcraft, S.W., K.C. Taylor, J.W. Hess, and T.M. Morris. 1986. Borehole Sensing Methods for Ground-Water Investigations at Hazardous Waste Sites. EPA/600/2-86/111 (NTIS PB87-132783).

* See Preface for information on how to obtain documents.

APPENDIX B

MANUFACTURERS AND DISTRIBUTORS OF FIELD CHARACTERIZATION AND MONITORING EQUIPMENT

In general, cost-effective equipment selection decisions require review of equipment specifications and costs from multiple sources. This Appendix includes addresses and telephone numbers of more than 160 manufacturers and distributors of the a wide variety of field equipment used for site characterization and monitoring of contaminated sites. Table B-1 serves as an index for sources of particular kinds of equipment.

Information in this appendix has come from a variety of sources, with principle ones being:

- American Chemical Society's *1993 Environmental Buyers' Guide*.

- 6th Edition of EPA's *SITE Technology Profiles* (U.S. EPA, 1993/T12-7).

- Pollution Equipment News' *1994 Buyers Guide* and the March 1994 edition.

- 1993 Director of Manufacturers in *Ground Water Monitoring and Remediation* and a review of adds in the Winter, 1994 edition.

- Appendix B in Boulding (1994/T1-3).

Where possible, catalogs and product literature have been reviewed in developing the index in Table B-1 and annotations for individual manufacturers and distributors. Every effort has been made to make this appendix comprehensive, but failure to include any manufacturer or distributor of types of equipment indexed in Table B-1 should not be construed as implying that other sources may not be satisfactory. Also, some manufacturers listed in the appendix may make equipment that is not identified in the index in Table B-1 or annotations because it was not possible to completely review product literature from all sources listed.

Table B-1 Index to Manufacturers and Distributors of Field Characterization and Monitoring Equipment

Topic	References
Monitoring Wells	
Well Drilling Equipment	ACI Truespin, Acker Drill Company, Buckeye Drill, Christensen Boyles, Central Mine Equipment, Diedrich Drilling, Drillers Services, GEFCO, Global Drilling Supplies, Gus Pech Mfg., Ingersoll-Rand, Longyear Company, Mobile Drilling, Penndrill Manufacturing Division, Schramm, Southern Iowa Manufacturing; <u>Directional/Horizontal Drilling</u>: American Augers, Eastman Christensen Environmental, GTS, McLaughlin Manufacturing
Direct Push Well Installations	Applied Research Associates, Checkpoint Environmental (CheckWells), Geoprobe, Hogentogler (BAT© system), KVA Analytical, Pine and Swallow Associates (MicroWell© and VibraDrill©), Solinst Canada
Soil and Vadose Zone Sampling Equipment	
Soil (Manual)	Associated Design & Manufacturing, Acker Drill Company, AMS, Ben Meadows, Christensen Boyles, CFE Equipment, Central Mine Equipment, Cole-Parmer Instrument, Concord, Drillers Services, Environmental Instruments, Forestry Suppliers, Geoprobe, Gilson Company, Hansen Machine Works, HAZCO Services, JMC/Clements Associates, LaMotte, Longyear U.S. Products, Nasco, Oakfield Apparatus, Soilmoisture Equipment, Soiltest/ELE, Wheaton Environmental
Soil (Power-Driven)	ACI Truespin, Acker Drill Company, AMS, Christensen Boyles, CFE Equipment, Central Mine Equipment, Concord, Diedrich Drilling, Drillers Services, Forestry Suppliers, GEFCO, Geoprobe, Giddings Machine, Global Drilling Suppliers, KVA Analytical, Little Beaver, Hogentogler, Longyear U.S. Products, Mobile Drilling, Penndrill Manufacturing Division, Solinst Canada, Soiltest/ELE, Southern Iowa Manufacturing
Soil Gas	AMS, Ben Meadows, CFE Equipment, Drillers Services, Eastman Cherrington, Environmental Instruments, Forestry Suppliers, Geoprobe, Gilson Company, HAZCO Services, KVA Analytical, Keck Instruments, Solinst Canada
Soil Water	<u>Suction Samplers/Lysimeters</u>: HAZCO Services, Soilmoisture Equipment, Soil Measurement Systems, Soiltest/ELE, Timco; <u>Absorbent Samplers</u>: Eastman Cherrington

Equipment Manufacturers and Distributors 837

Table B-1 (cont.)

Topic	References
Ground-Water Sampling Equipment	
Grab/Depth-Specific	Bailers: Atlantic Screen, Ben Meadows, Cole-Parmer Instrument, Eco Scientific, Environmental Instruments, Fluoroware, Forestry Suppliers, GeoGuard, Geoprobe, Geotech Environmental, HAZCO Services, Isco Environmental, Longyear U.S. Products, Modern Industrial Plastics, Monoflex, Norton Performance Plastics, PDSCo, QED, Soiltest/ELE, Solinst Canada, Timco, VTI, Wheaton Environmental; Other Grab: AMS, Forestry Suppliers, Slope Indicator, Wheaton Environmental; Direct-Push Samplers: Geoprobe, Hogentogler (BAT System), KVA Analytical, QED (HydroPunch)
Positive Displacement	Bladder Pump: Ben Meadows, GeoGuard, Geotech Environmental, Isco Environmental, Marschalk Corp, Modern Industrial Plastics, QED, Solinst Canada, Timco; Gas-Drive Samplers: Ben Meadows, Cole-Parmer Instrument, GeoGuard, Isco Environmental, Solinst Canada, Timco; Helical Rotor Pump: Forestry Suppliers, Keck Instruments
Other Pumps	Peristaltic: Cole-Parmer Instrument, Geotech Environmental, Isco Environmental; Electric Submersible: Bennett Sample Pumps, Cole-Parmer Instruments, Environmental Instruments, EPG Companies, Fultz Pumps, Geotech Environmental, Grundfos Pumps Corporation, Instrumentation Northwest, KVA Analytical, Longyear U.S. Products; Hand Pump: Ben Meadows, Longyear U.S. Products, Monoflex; Inertial Lift: Solinst Canada, Waterra
Multilevel Samplers	Solinst (Waterloo system), Westbay Instruments (Westbay system)
Packers	Baski, Isco Environmental, Kyle Equipment, QED, Rocktest, RST Instruments, TAM International
Other-Media Sampling Equipment	
Surface Water	Bomb-Type: Ben Meadows, Forestry Suppliers, Wheaton-Environmental; Handle-Grab: AMS, Ben Meadows, Cole-Parmer Instrument, Forestry Suppliers, HAZCO Services, Nasco, Wheaton Environmental; Kemmerer Samplers: Cole-Parmer Instrument, Forestry Suppliers, Wildlife Supply; Other Depth-Specific Grab: Cole-Parmer Instrument, Forestry Suppliers, Flowing Water Grab: Soiltest/ELE

Table B-1 (cont.)

Topic	References

Other-Media Sampling Equipment (cont.)

Sediment	Core: CFE Equipment, Forestry Suppliers, Rossfelder; Dredges: Ben Meadows, Cole-Parmer Instrument, Forestry Suppliers
Waste (liquid/sludge)	Coliwasa: Ben Meadows, Cole-Parmer Instrument, Forestry Suppliers, HAZCO Services, Nasco, Wheaton Environmental; Other Tube Samplers: Ben Meadows, Forestry Suppliers, Gilson Company, HAZCO Services, Nasco, Wheaton Environmental; Dippers: Ben Meadows; Vacuum Liquid: Geotech Environmental, HAZCO Services; Sludge-Grab Sampler: AMS, Nasco, Wheaton Environmental

Ground Water Level Measurements

Water Level	Electric/Conductive Probes: Bedrock Enterprises, Cole-Parmer Instrument, Environmental Instruments, Fisher Research Laboratory, Forestry Suppliers, Geokon, HAZCO Services, Keck Instruments, Longyear U.S. Products, MMC International, QED, Roctest, RST Cable & Tape, RST Instruments, Slope Indicator, Soiltest/ELE, Solinst Canada, Unidata America; Pressure Transducers: Design Analysis Associates, EPG Companies, In-Situ, Keller PSI, Telog Instruments, Unidata America; Acoustic: Bartex; Float Recorders: Leupold & Stevens, Solinst Canada, RST Instruments, Telog Instruments, Unidata America; RF/Admittance Transmitter: Drexelbrook Engineering
Oil-Water Interface	Automation Products, Environmental Instruments, Forestry Supplies, HAZCO Services, Keck Instruments, Longyear U.S. Products, Slope Indicator, Solinst Canada, Veeder-Root; Interface Sampler: Enviro Products; Hydrocarbon Leak Detection Sensors: Gems Sensors, One Plus Corp, PermAlert ESP, Raychem Corp.
Piezometers	Pneumatic: Geokon, Longyear U.S. Products, Roctest, RST Instruments; Electrical/Vibrating Wire: Geokon, Longyear U.S. Products, Roctest, RST Instruments; Small-Diameter Open-Tube: Bartex, Solinst Canada, Soiltest/ELE, Slope Indicator, Timco (see also, listing under Direct Push Well Installations)

Table B-1 (cont.)

Topic	References
Field Soil Characterization	
Slope (Clinometers)	Ben Meadows, Forestry Suppliers
Texture (particle size)	<u>Sieves</u>: CFE Equipment, LaMotte, Soiltest/ELE; <u>Field Test Kit</u>: LaMotte
Munsell soil color charts	Ben Meadows, CFE Equipment, Forestry Suppliers, Soiltest/ELE
Moisture/matric potential	<u>Tensiometers</u>: CFE Equipment, Forestry Suppliers, Irrometer Co., Soilmoisture Equipment, Soil Measurement Systems, Soiltest/ELE; <u>Dielectric Moisture Probe</u>: Troxler Electronics; <u>Neutron Moisture Probe</u>: Campbell-Pacific Nuclear, Longyear U.S. Products, Troxler Electronics; <u>Soil Moisture/Salinity Sensors</u>: CFE Equipment, Delmhorst Instrument, Forestry Suppliers, Gilson Company, Irrometer Co., Soilmoisture Equipment, Soiltest/ELE; <u>Carbide</u>: Gilson Company; <u>Time Domain Reflectometry</u>: Soilmoisture Equipment; <u>Psychrometers</u>: JRD Merrill Specialty Equipment, Wescor
Permeability/Hydraulic Conductivity	<u>Auger-Hole Method</u>: CFE Equipment; <u>Compact Constant Head Permeameter</u>: Ksat, Inc; <u>Double-Ring Infiltrometer</u>: CFE Equipment, Gilson Company, Soiltest/ELE; <u>Guelph Permeameter</u>: Soilmoisture Equipment, Soiltest/ELE; <u>Tension Infiltrometer</u>: Soil Measurement Systems; <u>Laboratory Permeameters</u>: Gilson Company
Penetrometers	<u>Hand-Held</u>: CFE Equipment, Gilson Company, Soiltest/ELE; <u>Power-Driven/CPT</u>: Applied Research Associates, Hogentogler
Bulk Density	Gilson Company (rubber balloon, sand cone)
Soil Temperature	CFE Equipment, Forestry Suppliers
Geophysics/Location	
Global Positioning Systems	Ben Meadows, Forestry Suppliers, Magellan Systems, Navstar Electronics
DC Resistivity	<u>Surface</u>: Advanced Geosciences, Bison Instruments, GISCO, Keck Instruments, SAGA Geophysics, Scintrex, Soiltest/ELE, Terraplus/ABEM; <u>Induced Polarization</u>: Bison Instruments, Geonics; <u>Self Potential</u>: Bison Instruments; <u>EC Probes</u>: CFE Equipment

Table B-1 (cont.)

Topic	References
Geophysics (cont.)	
Electromagnetic	Induction: CFE Equipment (EM38), Geonics, Scintrex; Time Domain EM: Geonics; VLF Resistivity: EDA Instruments, Geonics, Scintrex
Ground Penetrating Radar	Bison Instruments, Geophysical Survey Systems
Metal Detectors	Ben Meadows, Fisher Research Laboratory, Forestry Suppliers, Roctest
Magnetometers	Ben Meadows, Bison Instruments, Forestry Suppliers, EG&G Geometrics, SAGA Geophysics, Scintrex
Surface Seismic	Refraction: Bison Instruments, EG&G Geometrics, GISCO, SAGA Geophysics, Soiltest/ELE; Reflection: Bison Instruments, EG&G Geometrics, GISCO, SAGA Geophysics, Soiltest/ELE
Gravimeters	Bison Instruments, Scintrex
Borehole Logging[1]	Bison Instruments, Geonics, Keck Instruments, KVA Analytical Systems, Mount Sopris Instruments, SAGA Geophysics, Schonstedt Instrument, Scintrex, Terraplus/ABEM; Borehole Video: Laval Underground Surveys, Marks Products; Rental/Logging Services: Colog, Welenco
Field Chemical Screening and Analytical Equipment	
Field Gas Detection	Photoionization Detectors: Environmental Instruments; HNU Systems; Flame Ionization Detectors: Central Instruments, Eagle Monitoring Systems, Environmental Instruments, The Foxboro Co.; Toxic/Explosive Gas Detectors: Biosystems, CEA Instruments, Dynamation, The Foxboro Co., McNeill International, MSA, MTI Analytical, Neotronics, Sensidyne, Viking Instruments
Field Analytical Instruments	Gas Chromatograph (GC):[2] Environmental Instruments Services, Geoprobe, HNU Systems, Microsensor Systems, Microsensor Technology, Photovac International, Sentex Sensing, SRI Instruments; Mass Spectrometer (MS): Bruker Instruments; X-Ray Fluorescence: HNU Systems, Outokumpu Electronics, Spectrace Instruments; Hydrocarbon Analyzer: Eagle Monitoring Systems, The Foxboro Co.; Anodic Stripping Voltammetry: Outokumpu; Fiber-Optic Sensors: FiberChem

Table B-1 (cont.)

Topic	References
Wet Chemistry Test Kits	<u>Enzyme Immunoassay (EIA) Kits</u>: Agri-Diagnostics, Antox/Binax Corporation, EnSys, EM Science, Ohmicron Environmental Diagnostics, Millipore Corporation; <u>Other Colorimetric Kits</u>: Dexsil Corporation, EnSys, EM Science, Hach Company, HNU Systems (Hanby), Sensidyne, Taylor Technologies
GW Downhole Probes[3]	<u>Multiple-Parameter Probes</u>: Campbell Scientific (C/T), Design Analysis Associates (C/T), Geotech Environmental (C/T/pH/Eh/other), Horiba Instruments (C/T/pH/Tb), Hydrolab (C/T/ph/Eh/R/S/TDS/DO), In-Situ (C/T), Martek Instruments (T/C/pH/Eh/DO), Perstorp (C/T/pH/DO), Solomat-Neotronics (pH/DO/T/Tb/Eh/TDS/TSS); <u>Conductivity probes</u>: Solinst Canada, YSI; <u>ph Probes</u>: In-Situ; <u>Ion-Specific Electrodes</u>: ATI/Orion, Hach Company, Innovative Sensors, Solomat-Neotronics, TM Analytic
GW Field Chemistry[3]	<u>Flow-Through Cells</u>: Geotech Environmental, Isco Environmental, QED, YSI; <u>Multiparameter/Specific Meters</u>: Cole-Parmer Instrument, Environmental Instruments, Forestry Suppliers, Gilson Company, HAZCO Services, Horiba Instruments, Soiltest/ELE, Wheaton Environmental, YSI
Soil Field Chemistry	Forestry Suppliers, Gilson Company, HAZCO Services, LaMotte, Soiltest/ELE; see also EIA and other colorimetric test kits under ground-water chemistry
<u>Equipment Rental</u>	Environmental Instruments, HAZCO Services, Keck Instruments

C = conductivity, DO = dissolved oxygen, R = resistivity, S = salinity, TDS = total dissolved solids, T = temperature, Tb = turbidity.

[1] Only companies providing a focus on geophysical logging for environmental applications are listed.

[2] Testing by the U.S. EPA's Superfund Innovative Technology Program (SITE) of some field GC instruments has found that they do not perform well without preconcentration. Manufacturers should be asked to provide any data on performance developed under the SITE program.

[3] These instruments area usually used to monitor ground-water quality parameters during purging and sample collection.

Table B-1 Manufacturer and Distributor Addresses

Advanced Geosciences, Inc., P.O. Box 201087, Austin, TX 78720; 512/335-3338. [DC Resistivity]

ACI Truespin, P.O. Box 164, Hwy 10 West, Laurens, IA 50554; 800/428-4778. [Hollow-stem augers; continuous soil samplers]

Acker Drill Company, P.O. Box 830, Scranton, PA 18501; 800/752-2537. [Well drilling equipment, manual/power-driven/continuous soil samplers; purchased by Christensen Boyles Corporation in 1992]

Agri-Diagnostics Associates, One Executive Drive/Suite 10, Moorestown, NJ 08057; 609/727-4858 [Enzyme immunoassay kits for 5 pesticides]

American Augers, Inc./American Directional Drill, P.O. Box 460, Wooster, OH 44691; 800/324-4930, 216/264-5666. [Directional drilling]

Antox Division, Binax Corporation, 95 Darling Avenue, South Portland, ME 04106; 207/772-3544. [Equate enzyme immunoassay kit for BTX]

Applied Research Associates, Inc., Waterman Rd. RFD 1, South Royalton, VT, 05068; 802/763-8348. [Direct-push ground-water sampler/well installation; cone penetration]

Art's Manufacturing and Supply (AMS), 105 Harrison, American Falls, ID 83211; 800/635-7330. [Manual/power-driven soil samplers (with liners); soil-gas samplers; surface water samplers (handle-grab); waste samplers (sludge grab sampler)]

Associated Design & Manufacturing Co., 814 N. Henry St., Alexandria, VA 22314; 703/549-5999 [Manual/subcore soil samplers]

ATI/Orion, The Schrafft Center, 529 Main St., Boston, MA 02129; 800/225-1480. [pH meters; nitrate and other ion selective electrodes]

Atlantic Screen & Mfg., 118 Broadkill Rd., Milton, DE 19968; 302/684-3197. [Ground-water samplers (bailers)]

Automation Products, Inc., 3030 Max Roy, Houston, TX 77008; 713/869-0361. [Oil/water interface probe]

Bartex, Inc., P.O. Box 3348, Annapolis, MD 21403; 301/261-2224. [Ground-water level measurement (acoustic/sonic); open-tube piezometer]

Baski, Inc., 1586 S. Robb Way, Denver, CO 80232; 303/789-1200. [Inflatable well packers]

Bedrock Enterprises, Inc., P.O. Box 747, Forked River, NJ 08731; 609/693-9434. [Ground-water level probe (electric)]

(Table B-1 Equipment Vendor Addresses)

Equipment Manufacturers and Distributors

Ben Meadows Company, Inc., P.O. Box 80549, Atlanta (Chamblee), GA 30366; 800/241-6401. [Manual soil samplers (with liners); soil gas samplers; ground-water samplers (bailers, bladder pumps, gas drive, hand pump); sediment samplers (dredge); water and waste samplers (bomb-type, coliwasa, dipper, drum-thief, handle-grab); clinometers; Munsell Color Charts; Magellan GPS; metal/magnetic detectors]

Bennett Sample Pumps, Inc., P.O. Box 7644, 6325-B Star Ln., Amarillo, TX 79114; 806/352-0264. [Ground-water samplers (submersible electric pumps)]

Biosystems, Inc., 5 Brookfield Drive, Middle Field, CT 06455; 203/344-1079. [Toxic/combustible gas detectors/sensors]

Bison Instruments, Inc., 5708 W. 36th St., Minneapolis, MN 55416-2595; 612/926-1846. [DC resistivity/IP/SP; ground-penetrating radar; gravimeters; magnetometers; seismic refraction/reflection; vertical/crosshole seismic]

Bruker Instruments, Inc., Manning Park, 19 Fortune Drive, Billerica, MA 01821; 506/667-9580. [Field-portable mass spectrometer]

Buckeye Drill Co., P.O. Box 1190, 999 Zane St., Zanesville, OH 43702. [Well drilling equipment (combination cable tool/rotary drill)]

Campbell-Pacific Nuclear Corp., 130 S. Buchanan Circle, Pacheo, CA 94553; 415/687-6472. [Neutron moisture probe]

Campbell Scientific, Inc., 815 W. 1800 North, Logan, UT 84321; 801/750-9693. [Downhole temperature/conductivity probe]

Canterra-Drill Systems; 6700 9th St. N.E., Calgary, Alberta T2E 8K6 Canada; 403/295-7676. [Auger, mud/air/dual wall reverse rotary]

CEA Instruments, Inc., 16 Chestnut St., Emerson, NJ 07630; 201/967-5660. [Toxic/combustible gas detectors/sensors]

Central Instruments Corp., 25 Law Drive, Fairfield, NJ 07004-3295; 201/575-9114. [Flame ionization detectors]

Central Mine Equipment Co. (CME), 6200 N. Broadway, St. Louis, MO 63147; 800/325-8827. [Well drilling equipment (auger, rotary); power-driven/continuous soil samplers]

C.F.E. Equipment, 9 South Peru Street, Plattsburgh, NY 12901; 800/665-6794. [Manual/power-driven soil (with liners), soil gas, and sediment samplers; sieves; soil moisture (tensiometers, moisture sensors); soil permeability (auger hole-method, double-ring infiltrometer); Munsell soil color charts; geophysical (EM38, EC probes), penetrometers; soil temperature]

Checkpoint Environmental Science and Engineering, Acton, MA 01720; 508/369-8525. [Small-diameter wells installed with vibratory drill rig]

(Table B-1 Equipment Vendor Addresses)

Christensen Boyles Corporation Products Division, 4446 West 1730 South, P.O. Box 30777, Salt Lake City, UT 84130; 800/453-8418, 801/974-5544. [Well drilling equipment (auger, rotary, core); manual/power-driven soil samplers]

Cole-Parmer Instrument Co., 7425 N. Oak Park Ave., Niles, IL 60714-9930; 800/323-4340; 708/647-7600. [Manual soil sampling (with liners); ground-water samplers (bailers, peristaltic pumps, gas-drive, electric submersible); sediment samplers (Ponar, Peterson, Ekman dredges), surface water and waste samplers (coliwasa, other tube samplers, kemmerer, other depth-specific grab, handle-grab); ground-water level (electric) probes; ground-water chemistry]

Colog, Inc., 17301 W. Colfax Ave., Suite 265, Golden, CO 80401; 303/279-0171. [Borehole geophysics (rental)]

Concord, Inc., 2800 7th Ave. N., Fargo, ND 58102; 701/280-1260; [Manual/power-driven soil samplers (with liners)]

Delmhorst Instrument Co., 607 Cedar Street, Boonton, NJ 07005; 201/334-2557. [Gypsum soil moisture sensors]

Dexsil Corporation, One Hamden Park Drive, Hamden, CT 06517; 203/288-3509. [PCB/chloride field test kit]

Diedrich Drilling Equipment, P.O. Box 1670, Laporte, IN 46350; 800/348-8809. [Well drilling equipment; power-driven/continuous soil samplers]

Design Analysis Associates, Inc., 75 W. 100 South, Logan, UT 84321; 801/753-2212. [Water level (pressure transducer)/temperature/conductivity probe]

Drexelbrook Engineering Co., 205 Keith Valley Rd., Horsham, PA 19044; 215/674-1234. [Ground-water level measurement (RF/admittance transmitter)]

Drillers Service, Inc., Environmental Products Division, 1972 Highland Ave. NE, P.O. Drawer 1407, Hickory, NC 28603; 800/334-2308. [Well drilling equipment; manual/power-driven soil samplers (with liners), soil-gas and ground-water samplers]

Dynamation, 3784 Plaza Drive, Ann Arbor, MI 48108; 313/769-0573. [Portable toxic/combustible gas detectors]

Eagle Monitoring Systems, Inc., 23 Mauchly, Suite 109A, Irvine, CA 92718; 714/753-7855. [Flame-ionization detectors; hydrocarbon analyzer]

Eco Scientific, Inc., P.O. Box 1033, Buffalo, NY 14225; 716/892-1017. [Ground-water samplers (bailers)]

EDA Instruments, Inc., 5151 Ward Road, Wheat Ridge, CO 80033; 303/422-9112. [VLF resistivity (OMNI/Plus)]

Eastman Cherrington Environmental, P.O. Box 10129, Santa Fe, NM 87501; 505/983-3199 [SEAMIST system (soil gas and soil-water sampling with absorbent material]

(Table B-1 Equipment Vendor Addresses)

Equipment Manufacturers and Distributors

Eastman Christensen Environmental Systems, P.O. Box 670968, Houston, TX 77267; 713/985-4642. [Directional drilling]

EG&G Geometrics, 395 Java Drive, Sunnyvale, CA 94089; 408/734-4616. [Magnetometer; seismic reflection and refraction]

EnSys, Inc., P.O. Box 14063, Research Triangle Park, NC 27709; 914/941-5509. [Plate and strip immunoassay kits for PCPs (ELISA, developed by Westinghouse Bio-Analytic Systems) and colorimetric kit for TNT]

EM Science, 480 Democrat Rd., P.O. Box 70, Gibbstown, NJ 08027; 800/222-0342, 609/354-9200. [Enzyme immunoassay for PCBs, TNT, RDX; wet chemistry colorimetric test kits for other constituents]

Environmental Instruments Co., 5650 Imhoff Drive, Suite A, Concord, CA 94520-5350; 800/648-9355. [Manual soil samplers (with liners); soil-gas samplers; ground-water samplers (bailers, electric submersible); ground-water level/oil-water interface probes; field screening (photo-/flame-ionization detectors; field portable GC); ground-water chemistry; equipment rental]

EPG Companies, Inc., 9060 Zachary Lane North, Suite 115, Maple Grove, MN 55369; 800/443-7462, 612/424-2613. [Ground-water samplers (electric submersible pump); ground-water level probes (pressure transducers)]

FiberChem, Inc., Las Vegas, NV; 702/361-9873. [Fiber-optic chemicals sensors for air and ground-water monitoring (BTX available; TCE and lead under development)]

Fisher Research Laboratory, 200 W. Willmott Rd., Los Banos, CA 93635-5500; 209/826-3292. [Ground-water level probes (electric); metal detectors]

Fluoroware, Inc., 102 Jonathon Blvd. North, Chaska, MN 55318; 612/448-8181 [Ground-water samplers (bailers)]

Forestry Suppliers, P.O. Box 8397, Jackson, MS 39284-8397; 800/647-5368. [Manual/power-driven soil samplers (with liners); ground-water samplers (depth-specific bomb samplers, helical rotor pump, bailers); soil-gas, sediment samplers (core, LaMotte, ponar, Ekman), water and waste samplers (bomb-type, kemmerer, other depth-specific grab, handle-grab, coliwasa, drum thief); clinometers; Munsell soil color charts; soil moisture (tensiometers, moisture sensors); soil temperature; Pathfinder GPS; field soil and water chemistry; ground-water level probes (electric)/oil-water interface probes; geophysics (magnetometers, metal detectors)]

The Foxboro Co., Environmental Monitoring Operations, 600 N. Bedford St., P.O. Box 500, East Bridgewater, MA; 800/521-0451, 508/378-5556. [FID/PID detectors (OVA); total petroleum hydrocarbon analyzer; gas-specific detectors]

Fultz Pumps, Inc., Box 550 RD 2, Lewistown PA, 17044-0550; 717/248-2300. [Submersible ground water sampler]

GEFCO/George E. Falling Co., P.O. Box 872, Enid, OK 73702; 405/234-4141. [Well drilling equipment (auger, mud/air rotary, core); power-driven soil samplers]

(Table B-1 Equipment Vendor Addresses)

Gems Sensors Division, 203/747-3000. [Hydrocarbon leak detection probes]

GeoGuard, Inc., 536 Orient St., Medina, NY 14103; 716/798-5597. [Ground-water samplers (bailers, bladder and gas-lift pumps); formerly a subsidiary of American Sigma]

Geokon, Inc., 48 Spencer St., Lebanon, NH 03766; 603/448-1562. [Ground-water level probes (electric); pneumatic/vibrating wire piezometers]

Geonics, Ltd., 1745 Meyerside Drive, Unit #8, Mississauga, Ontario, Canada L5T 1C6; 416/670-9580. [Electromagnetic induction (EM-31, EM-34, EM-38); VLF resistivity (EM-16), resistivity (Protem 47); borehole EM (EM-39, Gamma-39)]

Geophysical Survey Systems, Inc., 13 Klein Drive, North Salem, NH 03073-0097; 800/524-3011, 603/893-1109. [Ground penetrating radar]

Geoprobe Systems, 607 Barney St., Salina, KS 67401; 913/825-1842. [Direct-push continuous soil/soil gas/ground-water (bailer) samplers and well installations; field portable gas chromatographs]

Geotech Environmental Equipment, Inc., 1441 W. 46th Ave. #17, Denver, CO 80211; 303/433-7101. [Ground-water samplers (bailers, peristaltic, bladder, and electric submersible pumps); vacuum hazardous waste drum sampler; water chemistry (downhole C/T/pH/Eh probe, flow-through cell)]

Giddings Machine Company, 401 Pine Street, P.O. Box 2024, Fort Collins, CO 80522; 303/482-5586. [Hydraulic soil-core/auger samplers]

Gilson Company, Inc., P.O. Box 677, Worthington, OH 43085-0677; 800/444-1508. [Manual soil samplers (with liners); soil gas samplers; fluid/sludge tube sampler; bulk density (rubber balloon, sand cone); soil moisture (carbide, moisture sensors); sieves; soil permeability (laboratory permeameters, double-ring infiltrometer); soil and ground-water chemistry]

GISCO, 4665 Joliet Street, Denver, CO 80203; 303/371-1940. [DC resistivity; self potential; seismic reflection and refraction]

Global Drilling Suppliers, Inc., 12101 Centron Place, Cincinnati, OH 45246; 800/356-6400. [Small portable auger and drilling unit; power-driven soil samplers]

Grundfos Pumps Corporation, 2555 Clovis Avenue, Clovis, CA 93612; 209/292-8000. [Ground-water samplers (submersible centrifugal pumps)]

GTS, Inc., 1231-B East Main St., Suite 189, Meriden, CT 06450-1019; 203/238-4567. [Directional well drilling equipment]

Gus Pech Mfg. Co., Inc., P.O. Box 96, 1480 Lincoln St. SW, Le Mars, IA 51031-0096; 800/383-7324, 712/546-4145. [Well drilling equipment (auger, rotary, large diameter bucket)]

Hach Company, P.O. Box 389, Loveland, CO 80539; 303/669-3050. [Colorimetric test kits for inorganics]

(Table B-1 Equipment Vendor Addresses)

Equipment Manufacturers and Distributors 847

Hansen Machine Works, 1628 North C Street, Sacramento, CA 95814; 916/443-7755. [Veihmeyer soil probe]

HAZCO Services, Inc., 2006 Springboro West, Dayton, OH 45439; 800/332-0435. [Manual soil samplers (with liners); soil-gas samplers; soil-water samplers (Timco lysimeters), ground-water samplers (bailers); water and waste samplers (bomb-type, coliwasa, other tube-type samplers, handle-grab sampler, vacuum liquid); ground-water level (electric)/oil-water interface probes; soil and ground-water chemistry; equipment rental]

HNU Systems, Inc., 160 Charlemont St., Newton, MA 02161-9987; 800/962-6032, 617/964-6690. [Photoionization detectors; portable gas chromatographs and x-ray fluorescence; Hanby colorimetric field test kits (BTEX, unleaded gasoline, diesel fuel, crude oil, heating oil, PCBs]

Hogentogler & Co., Inc., P.O. Drawer 2219, Columbia, MD 21045; 800/638-8582. [Direct-push soil and ground-water samplers/well installations (BAT System)].

Horiba Instruments, Inc., 17671 Armstrong Ave., Irvine, CA 92714; 714/250-4811. [Ground-water chemistry (downhole C/T/pH/Tb probe, multiparameter/specific meters)]

Hydrolab Corp., P.O. Box 50116, Austin, TX 78763; 800/949-3766, 512/255-8841. [Ground-water chemistry (downhole C/T/ph/Eh/R/S/TDS/DO multiparameter probes)]

Ingersoll-Rand Co., 253 E. Washington Ave., Washington, NJ 07882-9988; 908/850-7837. [Well drilling equipment (auger and rotary)]

Innovative Sensors, Inc., 4745 E. Bryson St., Anaheim, CA 92807; 714/779-8781. [pH, ion-specific electrodes (Br, Cl, F)]

In-Situ, Inc., 210 S. Third Street, P.O. Box 1, 210 S. Thirds St., Laramie, WY 82070; 800/446-7488, 307/742-8213. [Water-level probes,; water chemistry (downhole pH and C/T probes, headspace)]

Instrumentation Northwest, Inc., 14972 NE 31st Circle, Redmond, WA 98052; 800/776-9355. [Ground-water samplers (electric submersible pumps), ground water chemistry]

Irrometer Co., Inc., P.O. Box 2424, Riverside, CA 92516-2424; 909/689-1701. [Tensiometers; soil moisture sensors]

Isco Environmental Division, 531 Westgate Boulevard, Lincoln, NE 68528-1586; 800/228-4373, 402/474-4186. [Ground-water samplers (bailers, peristaltic pumps, gas drive, bladder pumps); inflatable well packers; ground-water chemistry (flow-through cell)]

JMC/Clements Associates, Inc., RR 1 Box 186, Newton, IA 50208-9990; 800/247-6630, 515/742-8285. [Manual soil samplers (with liners)]

JRD Merrill Specialty Equipment Co., RFD Box 140A, Logan, UT 84321; 801/752-8403. [Screen cage psychrometer]

(Table B-1 Equipment Vendor Addresses)

Keck Instruments, Inc., 1099 W. Grand River, Williamston, MI 48895; 800/542-5681, 517/655-5616. [Soil gas samplers; ground-water samplers (helical rotor pump); ground-water level (electric)/oil-water interface probes; geophysics (DC resistivity, borehole: gamma, caliper, resistivity); equipment rental]

Keller PSI, Inc., 503 Vista Bella #11, Oceanside, CA 92057; 619/967-6066. [Ground-water level probes (pressure transducer)]

Ksat, Inc., P.O. Box 30813, Raleigh, NC 27622; 919/676-0025. [Soil permeability (Compact Constant Head Permeameter)]

KVA Analytical Systems, P.O. Box 574, 281 Main St., Falmouth, MA 02541; 508/540-0561. [Direct push soil/soil gas/ground-water samplers/well installations; ground-water samplers (electric submersible pump); borehole flowmeters: Model 40 GeoFlometer (horizontal), Model 80 (vertical); Division of K-V Associates, Inc.]

Kyle Equipment Co., Inc., P.O. Box 658, Sterling, MA 01564; 800/426-6377. [Inflatable well packers]

LaMotte Co., P.O. Box 329, Chesterton, MD 21620; 800/344-3100, 410/778-3100. [Manual soil samplers (3-foot maximum length); sieves; field soil texture kit; soil and water chemistry]

Laval Underground Surveys, 1365 N. Clovis Ave., Fresno, CA 93727; 209/255-1601. [Borehole video (CAM 200, CAM 400); Division of Claude Laval Corp.]

Leupold & Stevens, Inc., P.O. Box 688, Beaverton, OR 97075; 503/646-9171. [Float-type water level recorders]

Little Beaver, Inc., Box 840, Livingston, TX 77351; 409/327-3121. [Truck/hand portable power augers]

Longyear U.S. Products Group, Box 1959, Stone Mountain, GA 30086; 800/241-9468, 404/469-2720. [Hand/power-driven/continuous (GeoBarrel) soil samplers; ground-water samplers (bailers, electric submersible pump, hand pump); neutron moisture probe; ground-water level (electric)/oil-water interface probes; pneumatic/vibrating wire piezometers; subsidiary of Longyear Company]

Longyear Company, 2340 West 1700 South, Salt Lake City, UT 84104; 801/972-6430. [Well drilling equipment (hollow-stem augers)]

Magellan Systems Corporation, 960 Overland Court, San Dima, CA 91773; 714/394-5000. [Global positioning systems]

Marks Products, Inc., RD #1, Box 303, Howard, PA 16841; 800/343-3479. [Borehole video (GeoVISION)]

Marschalk Corp. [Ground-water samplers (bladder pumps)

Martek Instruments, Inc., PO Box 97067, 3216-O Wellington Ct., Raleigh, NC 27624; 919/790-2371. [Ground water chemistry (downhole T/C/pH/Eh/DO probes)

(Table B-1 Equipment Vendor Addresses)

Equipment Manufacturers and Distributors

McLaughlin Manufacturing Co., 2006 Perimeter Road, Greenville, SC 29605; 800/435-9340, 803/277-5870. [Horizontal drilling equipment]

McNeill International, 7041 Hodgson Rd., Mentor, OH 44060; 800/MCNEILL. [Toxic/combustible gas detectors/sensors]

Microsensor Systems, Inc., 62 Corporate Court, Bowling Green, KY 42103; 410/939-1089. [Field-portable gas chromatograph]

Microsensor Technology, Inc., 41762 Christy St., Fremont, CA 94358; 510/490-0900. [Field-portable gas chromatograph]

Millipore Corporation, Analytical Division, 80 Ashby Road, Bedford, MA 01730; 617/275-9200. [EnviroGard PCB enzyme immunoassay kit; also test kits for petroleum, hydrocarbons, and pesticides]

MMC International Corp., 60 Inip Drive, Inwood, NY 11696; 516/239-7339. [Ground-water samplers (bailers); ground-water level probes (electric)]

Mobile Drilling Company, 3807 Madison Ave., Indianapolis, IN 46227; 800/766-3745. [Well drilling equipment; power-driven/continuous soil samplers]

Modern Industrial Plastics, Inc., 365 Carr Drive, Brookville, OH 45309; 513/833-4444. [Ground-water samplers (bailers, bladder pumps)]

Monoflex/Division of Campbell Mfg., Inc., 6450 125th Ave. N., Largo, FL 34643; 800/257-5183. [Ground-water samplers (bailers, hand pump)]

Mount Sopris Instrument Company, 17301 W. Colfax, Suite 255, Golden, CO 80401; 303/279-3211. [Borehole geophysics: Gamma, Resistivity, SP]

MSA, Pittsburgh, PA; 800/MSA-INST, 412/967-3000. [Toxic/combustible gas detectors/sensors]

MTI Analytical Instruments, Inc., Fremont, CA; 510/490-0900. [Toxic/combustible gas detectors/sensors]

Nasco Sampling Products, 901 Janesville Ave., Fort Atkinson, WI 53538-0901; 800/558-9595, 414/563-2446. [Manual soil, Surface water, waste-liquid/sludge samplers]

Navstar Electronics, Inc., 1500 N. Washington Blvd., Sarasota, FL 34236; 813/366-6338. [Global positioning systems]

Neotronics, 2144 Hilton Drive SW, Gainesville, GA 30501-6153; 800/535-0606. [Toxic/combustible gas detectors/sensors]

Norton Performance Plastics Corp. 150 Dey Road, Wayne, NY 07470; 201/696-4700. [Ground-water samplers (bailers)]

Oakfield Apparatus Company, P.O. Box 65, Oakfield, WI 53065; 414/583-4114. [Manual soil samplers]

(Table B-1 Equipment Vendor Addresses)

Ohmicron Environmental Diagnostics, Inc., 375 Pheasant Run, Newton, PA 18940; 800/544-8881. [Quantitative and semi-quantitative enzyme immunoassay kits: 13 pesticides, PCP, PCBs, PAHs, BTEX]

One Plus Corp., 1955 N. Shermer Rd., Northbrook, IL 60062; 708/498-0955. [Hydrocarbon leak detection sensors]

Outokumpu Electronics, Inc., 1900 N.E. Division St., Suite 204, Bend, OR 97701; 800/229-9209, 503/385-6748. [Field-portable X-ray fluorescence; anodic stripping voltammetry]

PDSCo., 105 W. Sharp, El Dorado, AR 71730; 800/243-7544; 501/863-5707. [Ground-water samplers (bailers)]

Penndrill Manufacturing Division, Pennsylvania Drilling Co., 500 Thompsen Ave., McKees Rocks, PA 15136; 800/245-4420, 412/771-2110. [Well drilling equipment; power-driven soil samplers]

PermAlert ESP, 7720 N. Lehigh Ave., Niles, IL 60714-3491; 708/966-2190. [Hydrocarbon leak detection sensors]

Perstorp Analytical Environmental, Box 648, Wilsonville, OR 97070; 800/262-3668. [C/T/pH/DO probes]

Photovac International, Inc., 35B Jefryn Boulevard West, Deer Park, NY 11729; 516/254-4199. [Field-portable gas chromatograph]

Pine & Swallow Associates, 867 Boston Road, Groton, MA 10450; 508/448-9511. [Small-diameter wells installed with vibratory drill rig; affiliated with ProTerra]

QED Groundwater Specialists, 6155 Jackson Rd., P.O. Box 3726, Ann Arbor, MI 48106; 800/624-2547, 313/995-2547 (MI), 415/930-7610 (CA). [Ground-water samplers (bailers, bladder pumps); direct push ground-water samples (HydroPunch); ground-water level probes (electric); inflatable well packers; ground-water chemistry (flow-through cell)]

Raychem Corp., 300 Constitution Dr. MS 204, Menlo Park, CA 94025; 415/361-6209. [Hydrocarbon leak detection sensors]

Roctest Inc., 7 Pond St. Plattsburgh, NY 12901-0118; 518/561-1192. [Metal detectors; ground-water level probes (electric); pneumatic/vibrating wire piezometers]

Rossfelder Corp., 8620-B Production Ave., San Diego, CA 92121-2207; 619/578-3313. [Vibrating core saturated sediment sampler]

RST Cabel & Tape, Inc., 2130 Pond Rd., Ronkonkoma, NY 11779; 516/981-0096. [Ground-water level probes (electric)]

RST Instruments, Inc., 241 Lynch Rd., Yakima, WA 98908; 509/965-1254. [Ground-water level measurement (electric, float-type); pneumatic/vibrating wire piezometers; inflatable well packers]

(Table B-1 Equipment Vendor Addresses)

SAGA Geophysics, Inc., 10710 D-K Range Rd., Bldg. 8, Austin, TX 78759; 512/258-7599. [DC resistivity; seismic refraction; high resolution seismic reflection; magnetometer; borehole geophysics]

Schonstedt Instrument Company, 1775 Wiehle Avenue, Reston, VA 22090-5199; 703/471-1050. [Borehole magnetometer]

Schramm, Inc., 800 E. Virginia Ave., West Chester, PA 19380; 215/696-6950. [Well drilling equipment (auger, mud/air/dual-wall reverse rotary, casing hammer]

Scintrex, Inc., 4600 Witmer Industrial Estates, Unit 4, Niagara Falls, NY 14305; 716/298-1219. [DC resistivity; VLF resistivity; magnetometers; gravimeters; seismic refraction; borehole logging (gamma spectrometry probe); home office in Concord, Ontario]

Sensidyne, 16333 Bay Vista Drive, Clearwater, FL 34620; 800/451-9444. [Toxic/combustible gas detectors/sensors; field chemistry test kits (hazardous chemicals, lead)]

Sentex Systems, Inc., 553 Broad Ave., Ridgefield, NJ 07657; 201/945-3694. [Field-portable gas chromatograph]

Slope Indicator Co., P.O. Box 300316, Seattle, WA 98103-97316; 206/633-3073. [Ground-water samplers (discrete water sampler); ground-water level (electric)/oil-water interface probes; vented piezometers]

Soil Measurement Systems, 7266 N. Oracle Rd., Suite 170, Tucson, AZ 85704; 602/742-4471. [Tensiometers; soil-water suction samplers; permeability (tension infiltrometer)]

Soilmoisture Equipment Corp., P.O. Box 30025, Santa Barbara, CA 93105; 805/964-3525. [Manual soil samplers; soil permeability (Guelph permeameter); soil-water suction samplers; soil moisture (tensiometers, soil moisture/salinity sensors, time domain reflectometry)]

Soiltest Products Division, ELE International, Inc., P.O. Box 8004, Lake Bluff, IL 60044; 800/323-1242. [Manual/power-driven soil samplers (with liners); ground-water samplers (bailers); surface water sampler (watertrap sampler for flowing water); ground-water level probes (electric); geophysics (seismic refraction/reflection, DC resistivity); Munsell soil color charts; penetrometers; piezometers; sieves; soil moisture (tensiometers, soil moisture sensors); soil permeability (Guelph permeameter, double-ring infiltrometer); soil-water suction samplers; soil and water chemistry]

Solinst Canada, Ltd., 515 Main St., Glen Williams, Ontario L7G 3S9; 800/661-2023, 416/873-2255. [Power-driven thin-wall piston soil sampler (with liners); ground-water samplers (bailers, gas-drive, bladder pumps, Waterloo multilevel monitoring system); drive-point piezometers (soil gas sampling); ground-water level (electric, float)/oil-water interface probes; piezometers; ground-water chemistry (conductivity probe)]

Solomat-Neotronics, P.O. Box 370, Gainesville, GA, 30503; 800/765-6628. [Probes/dataloggers (pH/C/DO/T/Tb/Eh/TDS/TSS, ion specific electrodes]

(Table B-1 Equipment Vendor Addresses)

Southern Iowa Manufacturing, Co. (Simco), Drilling Products Division, P.O. Box 448, 802 Furnas Dr., Osceola, IA 50213; 800/338-9925, 515/342-2166. [Well drilling equipment (auger, rotary, core); power-driven soils samplers]

Spectrace Instruments, 345 East Middlefield Road, Mountain View, CA 94043; 414/967-0350. [Field-portable X-ray fluorescence]

SRI Instruments, 3870 Del Amo Boulevard, Suite 506, Torrance, CA 90503; 310/214-5092. [Field-portable gas chromatograph].

TAM International, Inc., 4620 Southerland, Houston, TX 77092; 713/462-7617. [Inflatable well packers]

Taylor Technologies, Inc. 31 Loveton Circle, Sparks, MD 21152; 800/837-8548. [Colorimetric test kits for water and wastewater]

Telog Instruments, P.O. Box 240, West Henrietta, NY 14586; 716/359-1110. [Ground-water level measurement (pressure transducers, float recorders)]

Terraplus USA, Inc./ABEM, 625 W. Valley Rd., Littleton, CO 80124; 303/799-4140. [DC resistivity; electromagnetic induction; magnetometer; high-resolution seismic reflection/refraction; borehole geophysics: resistivity, SP, temperature]

Timco Mfg., Inc. P.O. Box 8, 851 Fifteenth St, Prairie du Sac, WI 53578; 800/236-8534, 608/643-8534. [Ground-water samplers (bailers, gas drive, bladder pumps); soil water suction samplers; piezometers]

TM Analytic, Inc., 1106 N. Parsons Ave., Brandon, FL 33510; 813/684-2660. [Ion-specific electrodes]

Troxler Electronic Laboratory, Inc., P.O. Box 12057, Research Triangle Park, NC 27709; 919/549-8661. [Neutron moisture probe; dielectric moisture probe]

Unidata America, 17408 Boones Ferry Rd., Lake Oswego, OR 97035; 503/697-3570. [Ground-water level measurement (electric/capacitance, pressure transducer, float-type)]

Veeder-Root Co., 125 Powder Forest Dr., Simsbury, CT 06070-2003; 203/651-2700. [Oil-water interface sensor]

Viking Instruments Corp., Reston, VA; 703/758-9339. [Toxic/combustible gas detectors/sensors]

VTI, 4227 Centergate, San Antonio, TX 78217; 800/247-6294. [Ground-water samplers (bailers)]

Waterra Pumps Ltd., 77 Mowat Ave., Suite 101, Toronto, Ontario M6K 3E3 Canada; 416/536-1236. [Ground-water sampler (inertial pump)]

Welenco, 4817 District Blvd., Bakersfield, CA 933313; 800/445-9914, 805/834-8100. [Borehole geophysical logging services for environmental applications]

(Table B-1 Equipment Vendor Addresses)

Wescor, Inc., 459 South Main Street, Logan, UT 84321; 801/752-6011. [Screen cage psychrometer]

Westbay Instruments, Inc., 507 E. Thirds St., North Vancouver, British Columbia, V7L 1G4 Canada; 604/984-4215. [Westbay multilevel ground water sampling system]

Wheaton Environmental Products, 1301 North 10th Street, Millville, NY 08332-9854. 800/225-1437. [Manual soil samplers; ground-water samplers (bailers); surface water and waste samplers (bomb-type, coliwasa, drum thief, handle-grab, sludge-grab sampler); ground-water chemistry]

Wildlife Supply Co., 301 Cass St., Saginaw, MI 48602-2097; 517/799-8100. [Surface and ground-water samplers (Kemmerer)]

YSI, Inc., Box 279, Yellow Springs, OH 45387; 800/765-4974, 513/767-7241. [Water chemistry (ground-water depth/conductivity probe, flow-through cell, multiparameter/specific meters/loggers: DO/T/C/pH/S/ammonia/Eh/Tb, DO, BOD)]

(Table B-1 Equipment Vendor Addresses)

APPENDIX C

TABLES AND FIGURES FOR ESTIMATION OF AQUIFER PARAMETERS

This appendix contains a series of figures and tables that can be used to estimate vadose zone and aquifer hydrologic properties based on texture of unconsolidated materials and rock type in bedrock aquifers. Refer to Section 7.4.2 for a recommended procedure for estimating porosity, specific yield and hydraulic conductivity.

The following table provides information to help identify which figures and tables in this appendix can be used based on the type of information known. Worksheet D-W1 can be used to compile estimates for a particular aquifer location.

Figure /Table	Known Properties				Estimated Parameters		
	Grain-Size* Distribution	USDA si-sa-cl %	ASTM/ USCS Class	Soil/Rock Type	Porosity	Specific Yield	Hydraulic Conductivity
Table C-1				x	x		
Table C-2				x		x	
Table C-3			x	x			x
Figure C-1	x				x	x	
Figure C-2		x				x	
Figure C-3				x	x		x
Figure C-4				x			x
Figure C-5				x			x
Figure C-6				x			x
Figure C-7			x	x			x
Figure C-8	x						x
Figure C-9	x + bulk density						x
Figure C-10	x ($d50, \sigma_I$)			Sands			x
Figure C-11	x (median)			Stratified drift			x
Figure C-12				Glacial till			x
Figure C-13	x + porosity			Sandstone			x
Figure C-14	x	x		Unconsol.			x
Figure C-15				Unconsol.**		x	x

* Full grain size distribution data which allows determination of such characteristics as (1) median grain size, (2) cumulative percent finer than, and (3) standard deviation of grain size. Figure C-14 illustrates a grain-size distribution curve.

** This figure allows estimation of hydraulic conductivity if specific yield is known or vice versa.

Table C-1 Porosity (% of Volume) of Different Aquifer Materials

Soil/Rock Types	(1) P/S*	(2) P/S*	(3)***	(4)	(5)	(6)	(7)****
Unconsolidated Sediments							
Gravel	20/-	30-40/-	23.7-44.1	25-40	25-40		
Coarse						20-35	
Medium						20-35	
Fine						20-40	
Sand and Gravel						20-35	
Sand	25/-		26.0-53.3	25-50	15-48		
Gravelly						20-35	
Coarse		30-40/-				25-45	
Medium						25-45	
Medium to fine		30-35/-					
Fine						25-55	
Dune sand						35-45	
Silt		40-50/yes**	33.9-61.1	35-50	35-50	35-60	
Clay	50/-	45-55/yes**	34.2-56.9	40-70	40-70	35-55	
Sandy						30-60	
Till		45-55/yes**				25-45	
Unstratified drift			22.1-40.6				
Stratified drift			34.6-59.3				
Loess			44.0-57.2			60-80	
Peat						60-80	
Soil	55/-						
Alluvium							10-40(30)
Basin fill							5-30 (20)
Ogallala Formation							15-45(35)
Consolidated Sediments							
Limestone	10/10	1-50/yes**	6.6-55.7	0-20	0-20	5-55	1-20(4)
Karst				5-50	5-50		
Chalk					5-40		
Dolomite		1-50/yes**	19.1-32.7	0-20	0-20		
Sandstone			13.7-49.3	5-30	5-40		1-20(10)
Semiconsolidated	10/1					1-50	
Coarse, medium		<20/yes**					
Fine, argillite		<10/yes**					
Siltstone		-/yes**	21.2-41.0			20-40	
Shale		-/yes**	1.4-9.7	0-10	0-10		
Crystalline Rocks							
Granite (unaltered)	-/0.1				0-2		
Crystalline (fractured)				0-10			
Crystalline (dense)				0-5		0-5	

Table C-1 (cont.)

Soil/Rock Types	(1) P/S*	(2) P/S*	(3)***	(4)	(5)	(6)	(7)****
Crystalline Rocks (cont.)							
Igneous/Metamorphic		-/yes**					
Weathered						40-50	
Unaltered gneiss					0-2		
Quartzite					0-1		
Slates/mica schists					0-10		
Volcanic rocks							
Basalt	10/1	-/yes**					
Fractured				5-50	5-50	5-50	
Volcanic tuff					30-40	10-40	
Acid volcanic rocks							

* P = primary porosity, S = secondary porosity.
** Rarely exceeds 10 percent.
*** Compiled by Barton et al. (1985).
**** Number in parentheses is typical value.

Sources: (1) Heath (1983).
(2) Brown et al. (1983).
(3) Morris and Johnson (1967); compiled by Barton et al. (1985).
(4) Freeze and Cherry (1979).
(5) Sevee (1991).
(6) Devinny et al. (1990).
(7) Wilson (1981).

See also Manger (1963).

Table C-2 Specific Yield (%) for Different Aquifer Materials

Soil/Rock Types	(1)	(2) Mean	(2) Range	(3)	(4)	(5)
Unconsolidated Sediments						
Gravel	19			15-30		
Coarse		21	13-25		10-25	
Medium		24	17-44		15-25	
Fine		18	13-28		15-35	
Sand and Gravel				15-25	15-30	
Sand	22			10-30		
Gravelly					20-35	
Coarse		30	18-43		20-35	
Medium		32	16-46		15-30	
Fine		33	1-46		10-30	
Dune sand		38	32-47		30-40	
Silt		20	1-39		1-30	
Loess		18	14-22		30-50	
Clay	2	6	1-18	1-10	1-20	
Sandy					1-30	
Till					5-20	
Peat					30-50	
Soil	40					
Alluvium					1-25(15)	
Basin fill					1-30 (15)	
Ogallala Formation					1-30(20)	
Consolidated Sediments						
Limestone/Carbonate	18	14	0-36	0.5-5	1-24	1-5(2)
Sandstone				5-15		
Semiconsolidated	6				1-48	0.1-5(1)
Medium		27	12-41			
Fine		21	2-40			
Siltstone		12	1-33		1-35	
Shale				0.5-5		
Volcanic rocks						
Basalt	8					
Fractured					1-30	
Tuff		21	2-47		2-35	
Crystalline Rocks						
Granite	0.09					
Schist		26	22-33			
Crystalline (dense)					0-2	
Igneous/Metamorphic						
Weathered					20-30	

Sources: (1) Heath (1983), (2) Morris and Johnson (1967), as complied by McWhorter and Sunada (1977), (3) Sevee (1991), (4) Devinny et al. (1990), (5) Wilson (1981).

Table C-3 Representative Values for Hydraulic Conductivity of Unconsolidated and Consolidated Sediments

Rock/Soil Type		Hydraulic Conductivity (cm/s)
<u>Unconsolidated Materials</u>*		
Gravel (repacked)		3.1 to 3.4×10^{-2}
Sand		9.0×10^{-2} to 4.7×10^{-6}
Silt		7.1×10^{-3} to 9.4×10^{-9}
Clay		1.4×10^{-6} to 1.4×10^{-9}
Unstratified drift		1.0×10^{-2} to 3.8×10^{-9}
Stratified drift		6.6×10^{-1} to 4.7×10^{-5}
Loess		1.8×10^{-4} to 4.7×10^{-6}
<u>Sedimentary Rocks</u>*		
Sandstone		1.0×10^{-2} to 3.7×10^{-7}
Siltstone		1.4×10^{-6} to 9.4×10^{-10}
Shale		--
Limestone		2.6×10^{-2} to 1.0×10^{-8}
Dolomite		3.3×10^{-6} to 3.8×10^{-9}
<u>Unified Soil Classification (when compacted)</u>**		
GW	Well graded gravels, gravel-sand mixtures, little or no fines.	10^{-2}
GP	Poorly graded gravels, gravel-sand mixtures, little or no fines.	10^{-2}
GM	Silty gravels, gravel-sand-silt mixtures.	10^{-3} to 10^{-6}
GC	Clayey gravels, gravel-sand-clay mixtures.	10^{-6} to 10^{-8}
SW	Well graded sands, gravelly sand, little or no fines.	10^{-3}
SP	Poorly graded sands, gravelly sands, little or no fines.	10^{-3}
SM	Silty sands, sand-silt mixtures.	10^{-3} to 10^{-6}
SC	Clayey sands, sand-clay mixtures.	10^{-6} to 10^{-8}
ML	Inorganic silts and fine sands, silty or clayey fine sands or clayey silts with slight plasticity.	10^{-3} to 10^{-6}
CL	Inorganic clays of low to medium plasticity, gravelly clays, sandy clays, silty clays, lean clays.	10^{-6} to 10^{-8}
OL	Organic silts and organic silty clays of low plasticity.	10^{-4} to 10^{-6}
MH	Inorganic silts, micaceous or diatomaceous fine sandy or silty soils, elastic silts.	10^{-4} to 10^{-6}
CH	Inorganic clays of high plasticity, fat clays.	10^{-6} to 10^{-8}
OH	Organic clays of medium to high plasticity, organic silts.	10^{-6} to 10^{-8}
Pt	Peat and other highly organic soils.	Not classified.

* Compiled from Morris and Johnson (1967) by Barton et al. (1985).
** Compiled by Brown et al. (1991) from SCS (1990).

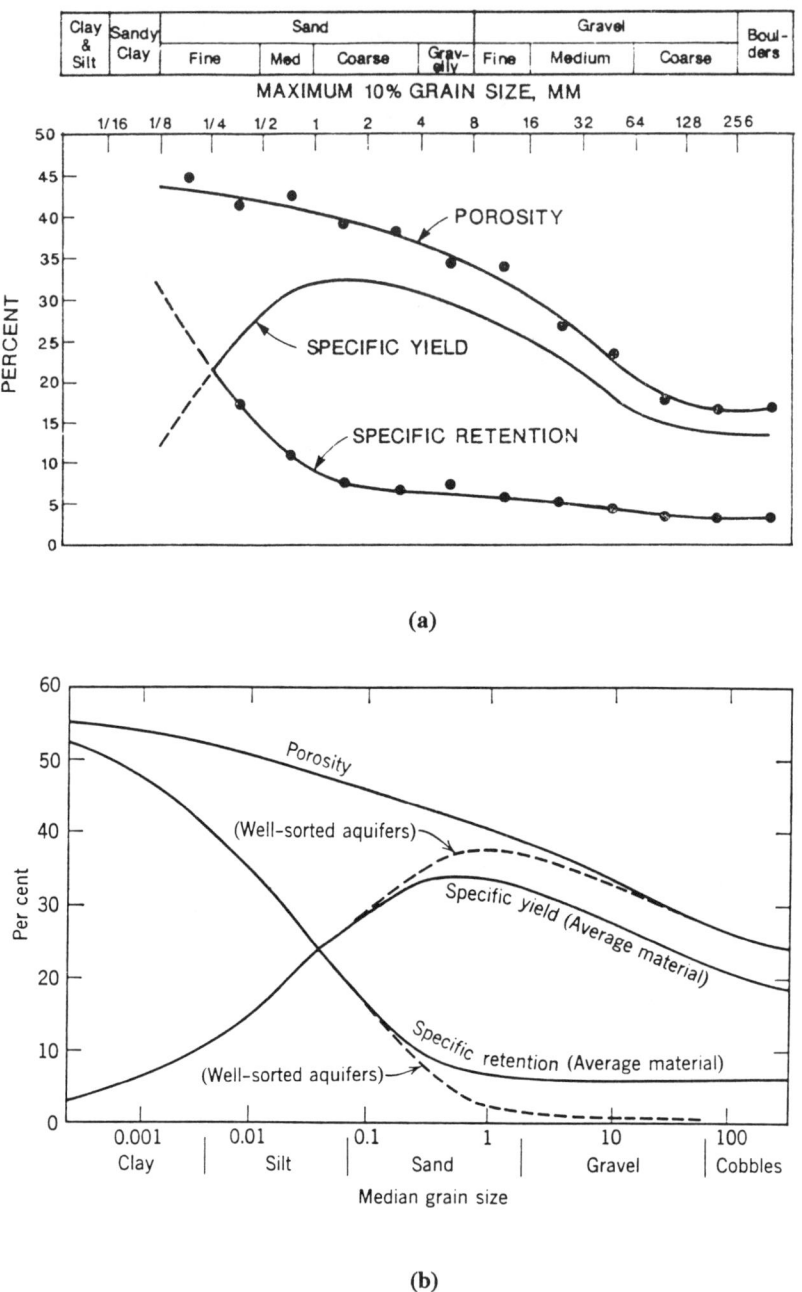

Figure C-1 Porosity, specific yield, and specific retention: (a) mean curves for South Coastal Basin in the Los Angeles area of California (adapted from Todd, 1959 by Devinny et al., 1990); (b) alluvium from large valleys (Davis and DeWiest, 1966 using various sources).

Information for Aquifer Parameter Estimation 861

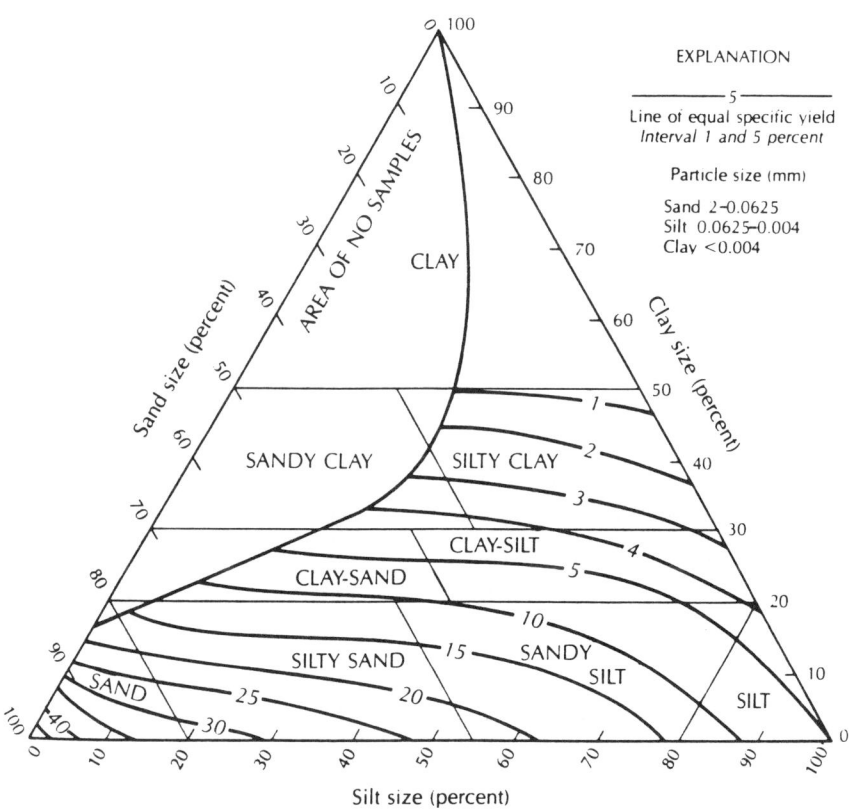

Figure C-2 Textural classification triangle for unconsolidated materials showing the relation between particle size and specific yield (Morris and Johnson, 1967).

Rock types	Porosity		Permeability range (cm/sec)						Well yields			Type of water-bearing unit
	Primary (grain)	Secondary (fracture)[1]	10^2	10^0	10^{-2}	10^{-4}	10^{-6}	10^{-8}	High	Medium	Low	
	%											
Sediments, unconsolidated												
Gravel	30–40								—			Aquifer
Coarse sand	30–40									—		Aquifer
Medium to fine sand	30–35									—		Aquifer
Silt	40–50	Occasional									—	Aquiclude
Clay, till	45–55	Rare (mud cracks)									—	Aquiclude
Sediments, consolidated												
Limestone, dolomite	1–50	Solution joints, planes								—		Aquifer or aquifuge
Coarse, medium sandstone	< 20	Joints and fractures								—		Aquifer or aquiclude
Fine sandstone, argillite	< 10	Joints and fractures									—	Aquifer or aquifuge
Shale, siltstone		Joints and fractures									—	Aquifuge or aquifer
Volcanic rocks												
Basalt		Joints, fractures								—		Aquifer or aquifuge
Acid volcanic rocks											—	Aquifuge or aquifer
Crystalline rocks												
Plutonic and metamorphic		Weathering and fractures decreasing as depth increases									—	Aquifuge or aquifer

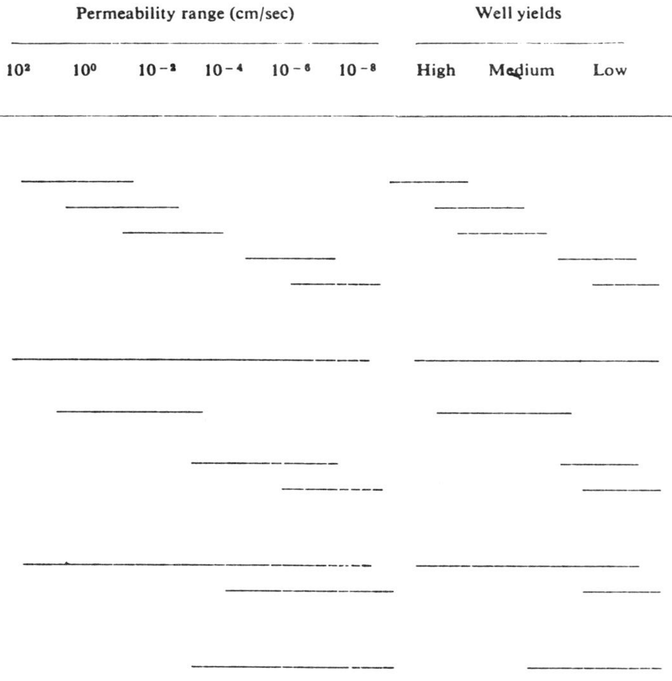

1. Rarely exceeds 10 per cent.

Figure C-3 Porosity, permeability, and well yields of major rock types (Brown et al, 1983).

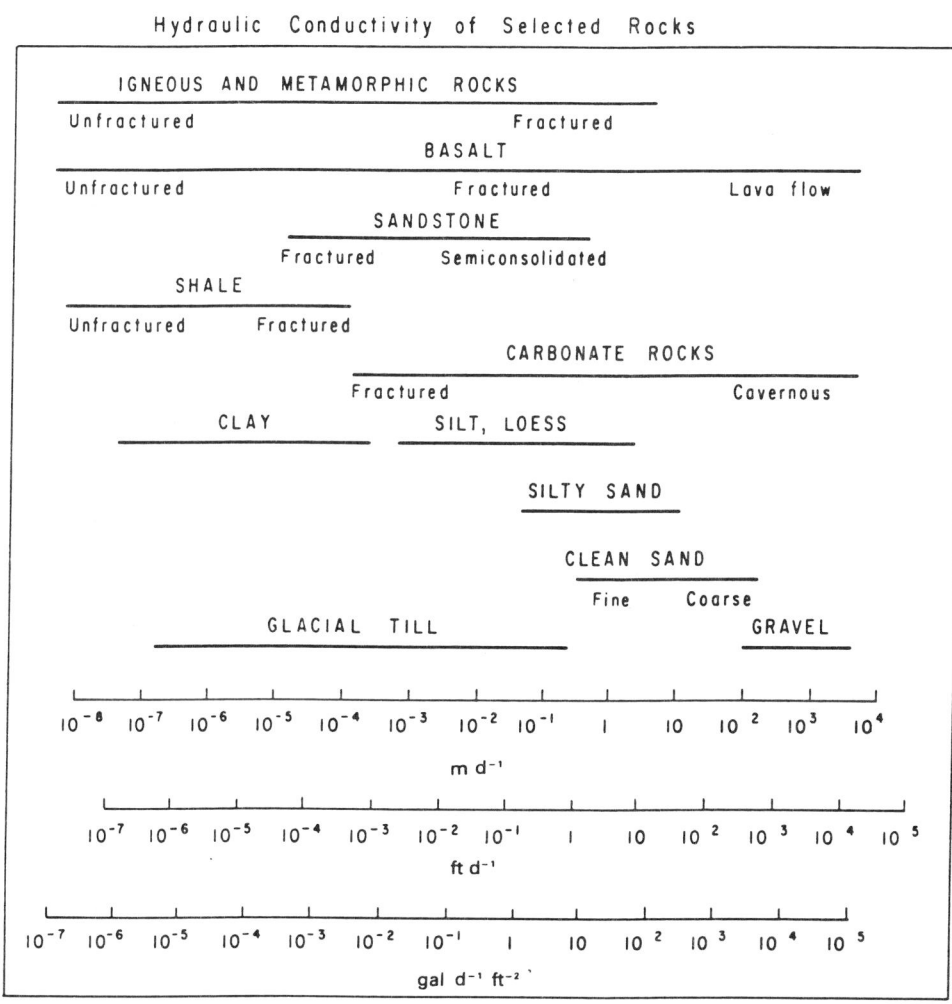

Figure C-4 Hydraulic conductivity of selected rocks (Heath, 1983).

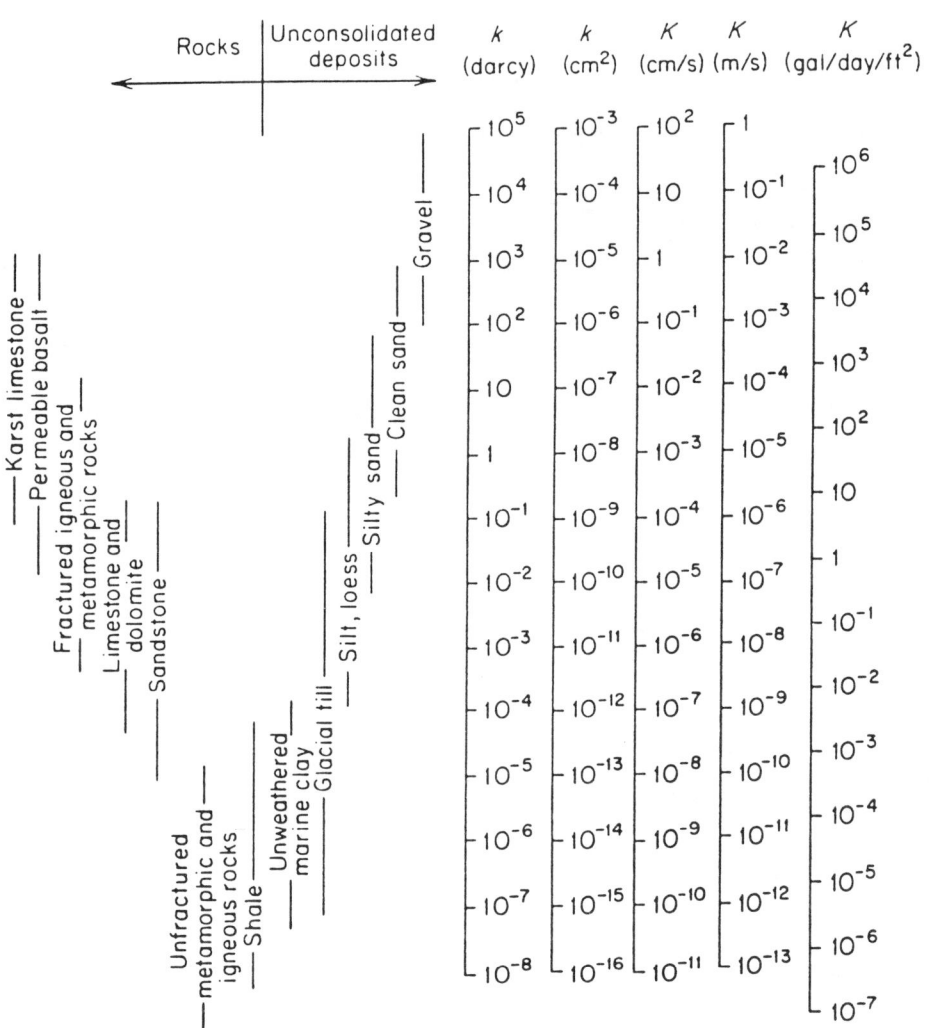

Figure C-5 Range of values of hydraulic conductivity (after Freeze and Cherry, 1979).

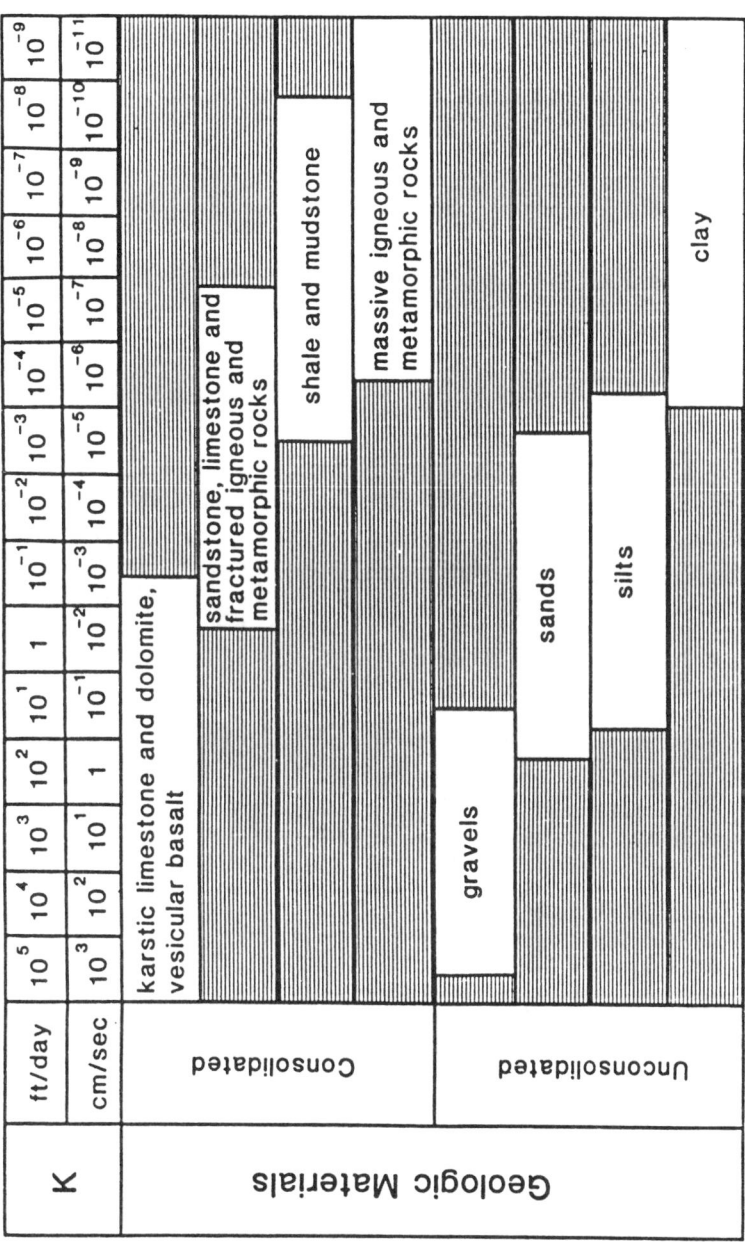

Figure C-6 Representative ranges of saturated hydraulic-conductivity values for geologic materials (adapted from Freeze and Cherry, 1979, by Thompson et al., 1989).

866   Appendix C

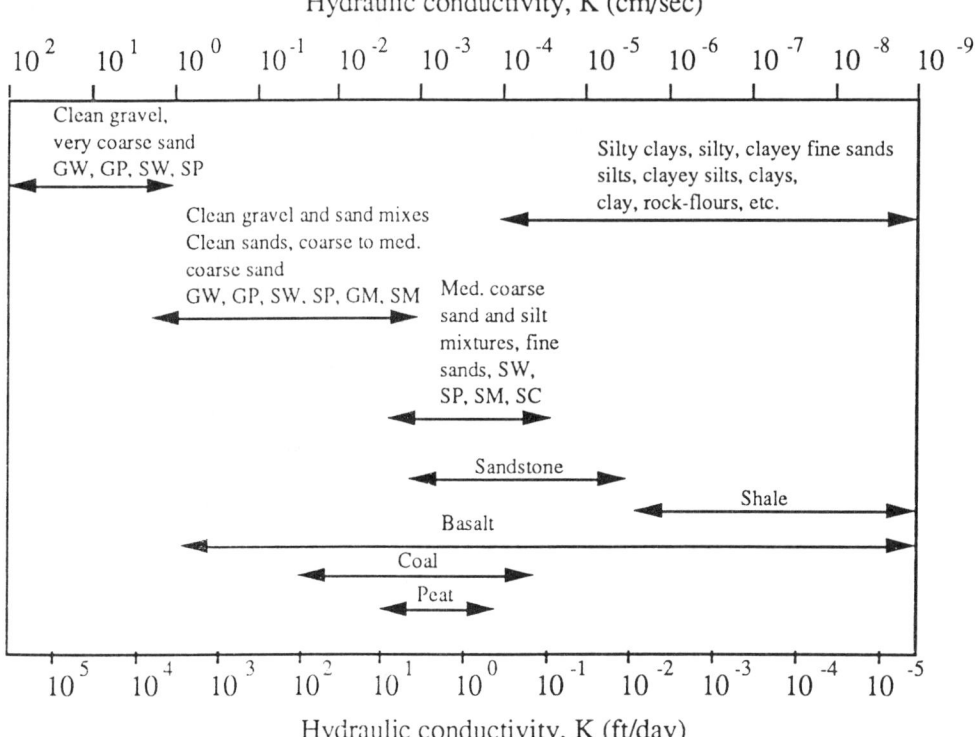

Figure C-7 Typical values of soil and rock hydraulic conductivity (Dawson and Istok, 1991, after Bowles, 1984 and Walton, 1987).

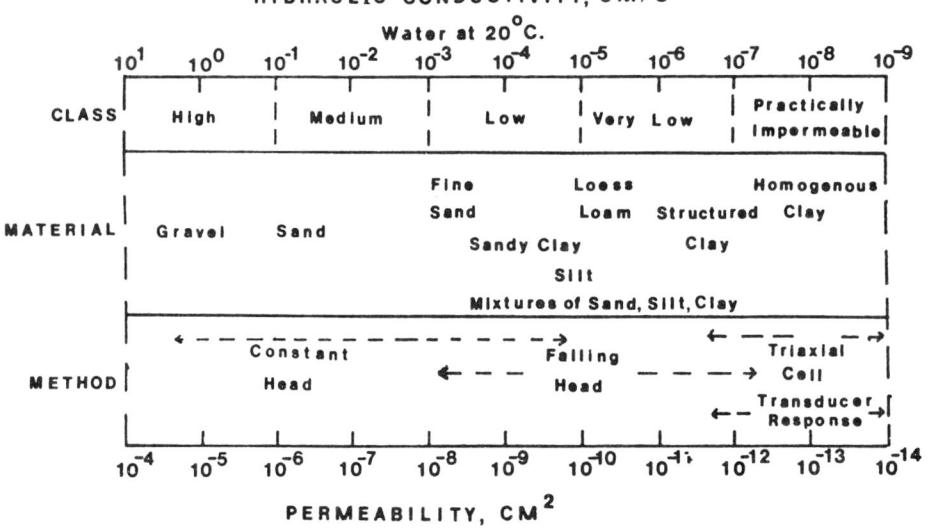

Figure C-8 Saturated hydraulic conductivity of unconsolidated materials (Klute and Dirksen, 1986).

Information for Aquifer Parameter Estimation

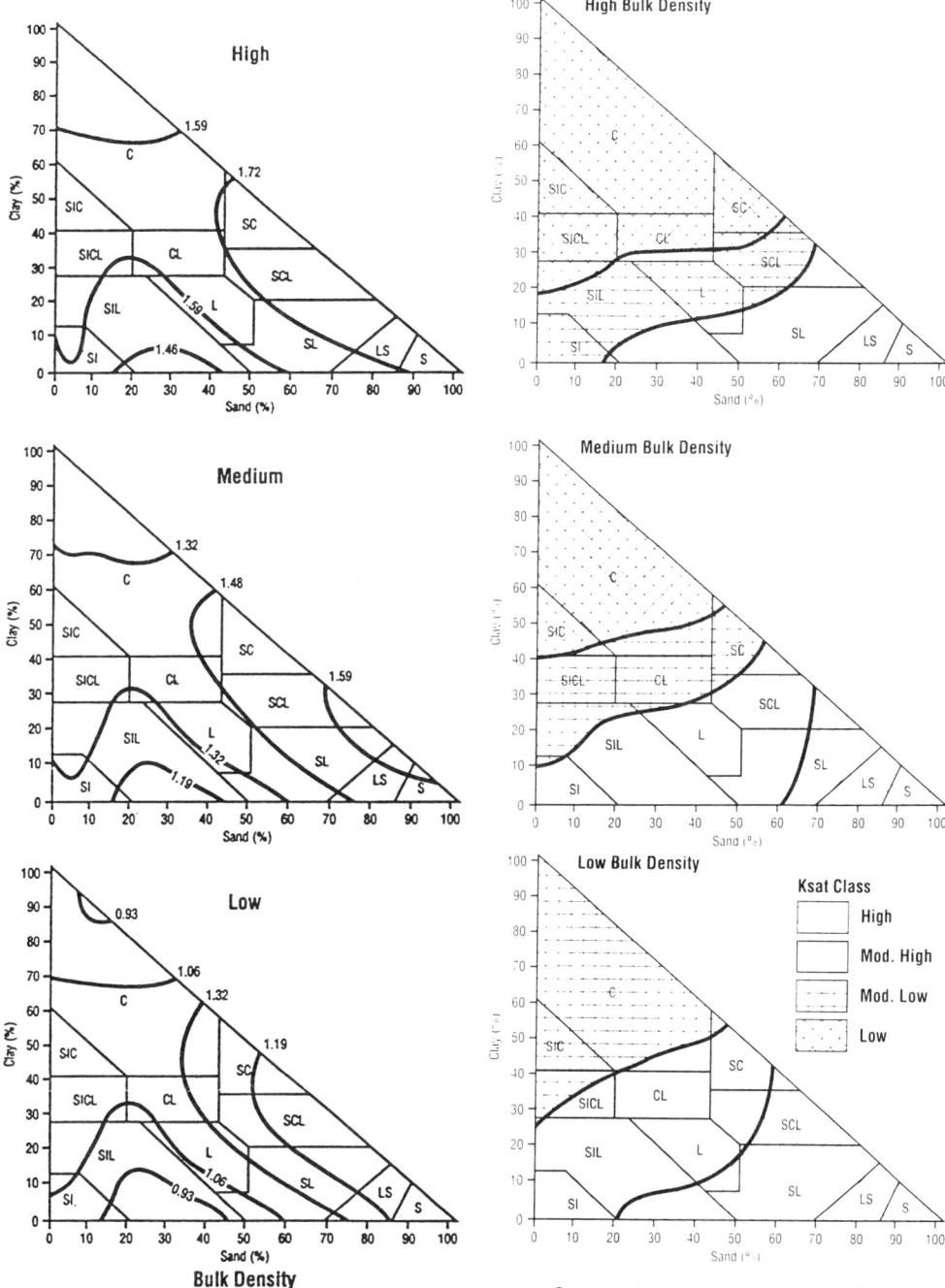

Figure C-9 Estimation of hydraulic conductivity from USDA Texture and bulk density: determine bulk density class and then Ksat class (refer to Table C-4 for class ranges).

Table C-4 SCS Criteria for Hydraulic Conductivity Classes

Class	μ/sec	in./hr
Very Low (VL)	<0.01	<0.001
Low (L)	0.01-0.1	0.001-0.01
Moderately Low (ML)	0.1-1	0.01-0.14
Moderately High (MH)	1-10	0.14-1.4
High (H)	10-100	1.4-14.2
Very High (VH)	>100	>14.2

Source: Boulding (1991)

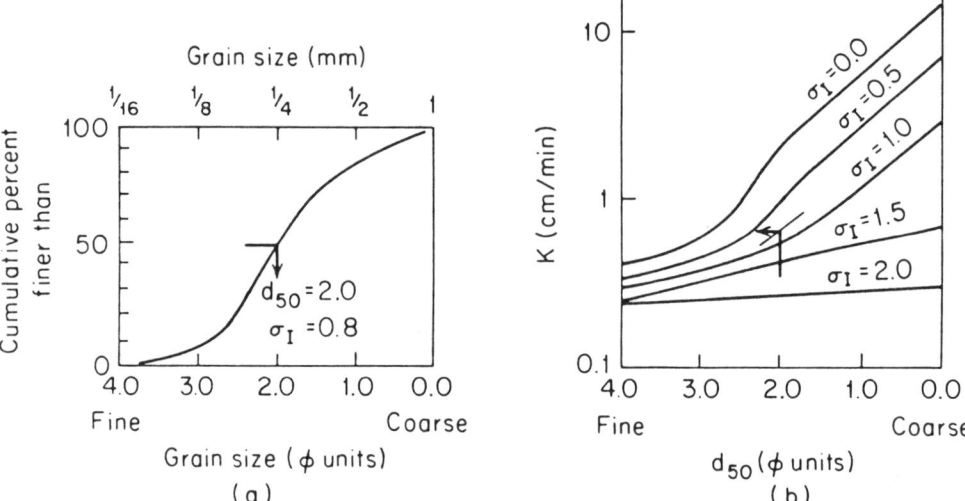

Figure C-10 Determination from grain-size gradation curves for sands (Freeze and Cherry, 1979, after Masch and Denny, 1966).

To use the nomograph above, a particle-size distribution curve, as illustrated in Figure C-10a must be plotted using ρ units, where $\rho = -\log_2 d$, d being the grain size diameter in mm. The *inclusive standard deviation* must also be calculated as follows:

$$\sigma_I = (d_{16} - d_{84})/4 + (d_5 - d_{95})/6.6$$

where the subscripts for d (in ρ units) represent the cumulative percentage finer than that diameter.

Figure C-10 provides an illustrative example. Median grain size d_{50} is first determined from the particle-size curve, C-10a (2.0 in the example). The inclusive standard deviation (calculated from the data used to plot the curve) in the example (0.8) has been interpolated between the curves in the nomograph on the right, C-10b, yielding an approximate K of 0.7 cm/min.

Information for Aquifer Parameter Estimation

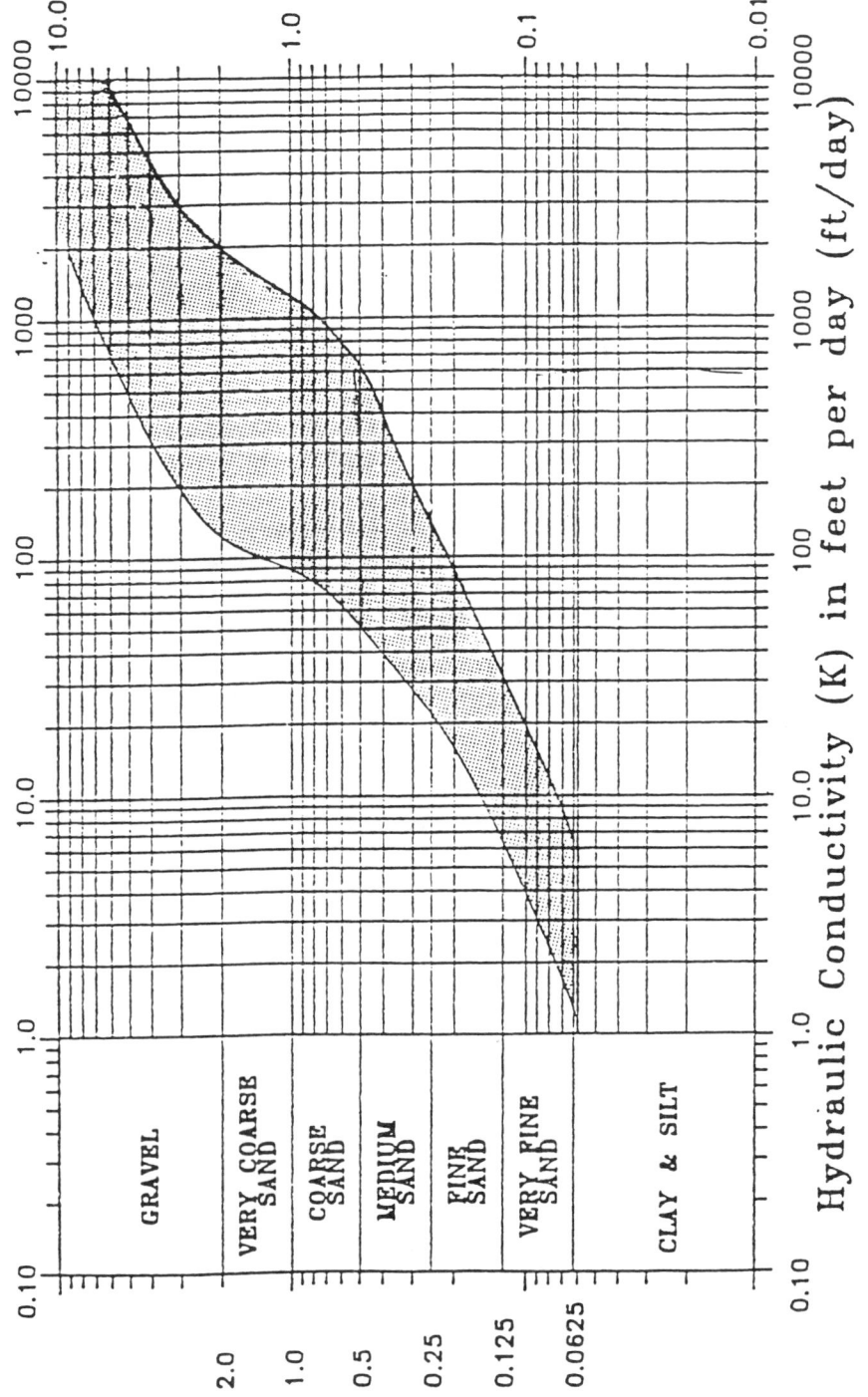

Figure C-11 Relationship between grain size and hydraulic conductivity in stratified drift aquifers (Connecticut Department of Environmental Protection, 1991).

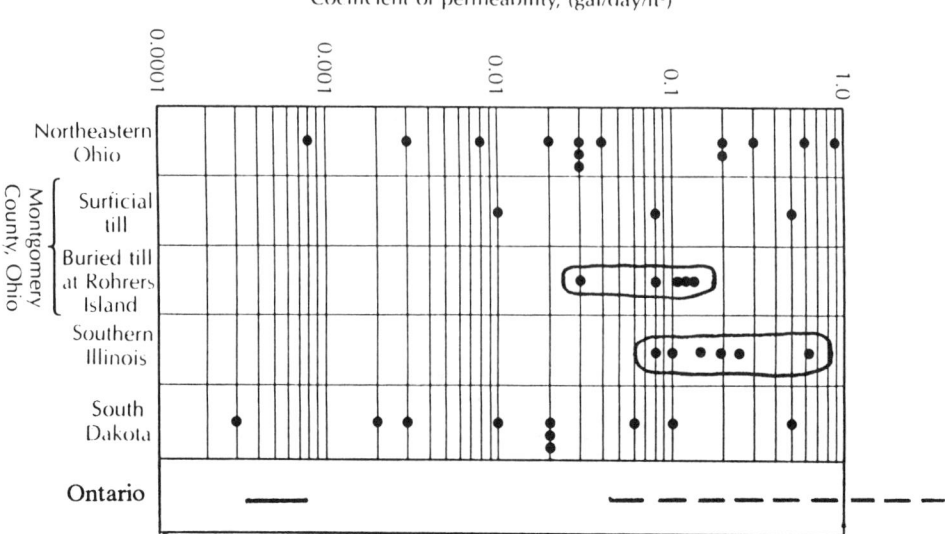

Figure C-12 Range of permeability of glacial tills: • = laboratory measurements (Norris, 1963); circled clusters of dots based on pumping tests (Norris, 1963); Ontario data from McKay et al. (1993) with solid line indicating range of laboratory measurements and dashed line the range of mean values using four different types of piezometer construction for field measurements.

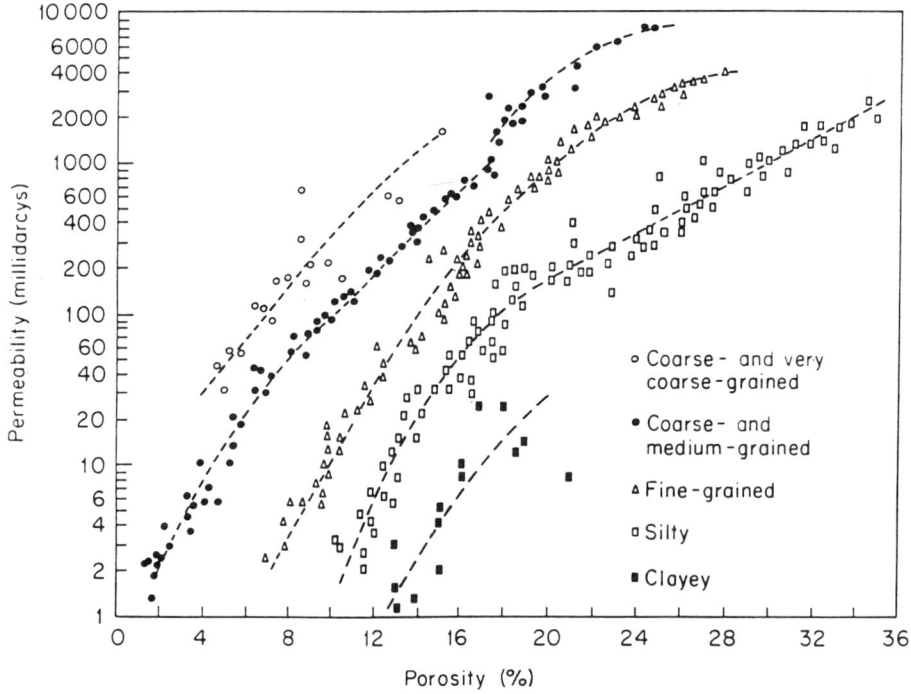

Figure C-13 Relationship between porosity and permeability for sandstone in various grain-size categories (Freeze and Cherry, 1979, after Chilingar, 1963).

Information for Aquifer Parameter Estimation

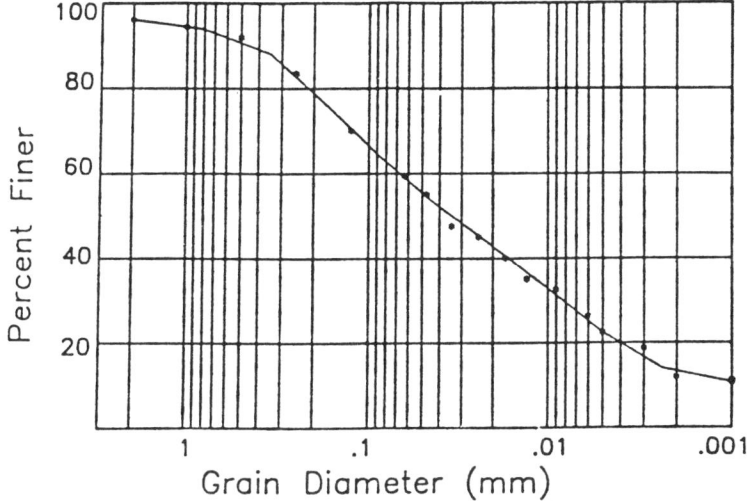

Bedinger: $K_{(gal/day/ft^2)} = 2000 * D50^2$

Hazen: $K_{(cm/sec)} = D10^2$

Krumbein & Monk: $k_{(darcies)} = 760 * Dm^2 * e^{(-1.31 * \sigma_\phi)}$

Cosby et al: $\log K_{(in/hr)} = (.0153 * \%sa) - .884$

Puckett et al: $K_{(m/sec)} = 4.36 \times 10^{-5} * e^{(-.1975 * \%cl)}$

Figure C-14 Sample particle-size distribution curve and five empirical equations used to estimate hydraulic conductivity of unconsolidated materials (Bradbury and Muldoon, 1990).

Definitions of terms in the equations in Figure C-14 include: $D50$ = median diameter, in millimeters; $D10$ = diameter, in millimeters, at which 10% of the sample is finer; Dm = mean diameter, in millimeters; σ_ϕ = phi standard deviation; $\%sa$ = percentage of the sample coarser than 0.05 mm; $\%cl$ = percentage of the total sample finer than 0.002 mm.

Citations for the five empirical equations shown above are: Bedinger (1961), Cosby et al. (1984), Hazen (1893), Krumbein and Monk (1942), and Puckett et al. (1985). Other references that contain empirical equations for estimating hydraulic conductivity include: Alyamani and Sen (1993), Hendry and Paterson (1982), Horn (1971), Uma et al. (1989), Vukovic and Soro (1992), Wiebenga et al. (1970). Empirical equations should only be used when materials being evaluated have similar textural or geologic characteristics to materials from which the equations were derived (see reference annotations).

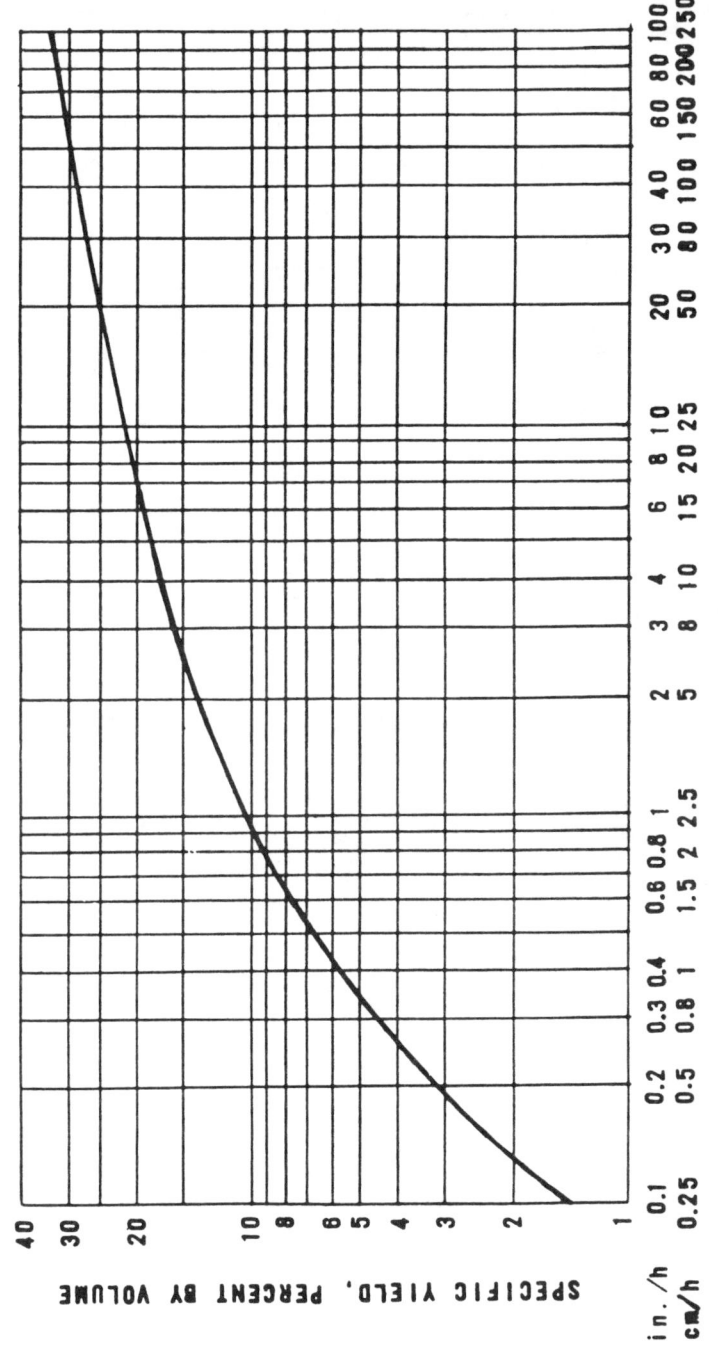

Figure C-15 General relationship between specific yield and hydraulic conductivity (Bureau of Reclamation, 1978).

Appendix C References

Alyamani, M.S. and Z. Sen. 1993. Determination of Hydraulic Conductivity from Complete Grain-Size Distribution Curves. Ground Water 31(4):551-555.

Barton, Jr., A.R. et al. 1985. Groundwater Manual for the Electric Utility Industry, Vol. 1: Geological Formations and Groundwater Aquifers, 1st ed. EPRI CS-3901. Electric Power Research Institute, Palo Alto, CA.

Bedinger, M.S. 1961. Relation Between Median Grain Size and Permeability in the Arkansas River Valley. U.S. Geological Survey Professional Paper 424C, pp. C31-C32. [Empirical equation for K in sandy alluvium]

Boulding, J.R. 1991. Description and Sampling of Contaminated Soils: A Field Pocket Guide. EPA/625/2-91/002, 122 pp. Available from CERI.*

Bowles, J.E. 1984. Physical and Geotechnical Properties of Soils. McGraw-Hill, New York, NY, 816 pp.

Bradbury, K.R. and M.A. Muldoon. 1990. Hydraulic Conductivity Determinations in Unlithified Glacial and Fluvial Materials. In: Ground Water and Vadose Zone Monitoring, D.M. Nielsen and A.I. Johnson (eds.), ASTM STP 1053, American Society for Testing and Materials, Philadelphia, PA, pp. 138-151.

Brown, R.H., A.A. Konoplyantsev, J. Ineson, and V.S. Kovalensky. 1983. Ground-Water Studies: An International Guide for Research and Practice. Studies and Reports in Hydrology No. 7. UNESCO, Paris.

Brown, K.W., R.P. Breckinridge, and R.C. Rope. 1991. Soil Sampling Reference Field Methods. U.S. Fish and Wildlife Service Lands Contaminant Monitoring Operations Manual, Appendix J. Prepared by Center for Environmental Monitoring and Assessment, Idaho National Engineering Laboratory, Idaho Falls, ID, 83415. [Final publication pending revisions resulting from field testing of manual].

Bureau of Reclamation. 1978. Drainage Manual. U.S. Department of The Interior, Denver, CO, 286 pp.

Chilingar, G.V. 1963. Relationship Between Porosity, Permeability, and Grain-Size Distribution of Sands and Sandstones. In: Proc. Int. Sedimentol. Congr., Amsterdam, Antwerp.

Connecticut Department of Environmental Protection (CDEP). 1991b. Guidelines for Mapping Stratified Drift Aquifers to Level B Mapping Standards. CDEP, Hartford, CT, 11 pp.

Cosby, B.J., G.M. Hornberger, R.B. Clapp, and T.R. Ginn. 1984. A Statistical Exploration of the Relationship of Soil Moisture Characteristics to the Physical Properties of Soils. Water Resources Research 20(6):682-690. [Empirical equation for K from soil samples throughout the U.S.]

Dawson, K.J. and J.D. Istok. 1991. Aquifer Testing: Design and Analysis of Pumping and Slug Tests. Lewis Publishers, Chelsea, MI, 280 pp.

Davis, S.N. and R.J.M. DeWiest. 1966. Hydrogeology. John Wiley & Sons, New York, NY, 463 pp.

Devinny, J.S., L.R. Everett, J.C.S. Lu, and R.L. Stollar. 1990. Subsurface Migration of Hazardous Wastes. Van Nostrand Reinhold, New York, NY.

Freeze, R.A. and J.A. Cherry. 1979. Groundwater. Prentice-Hall Publishing Co., Englewood Cliffs, NJ, 604 pp.

Klute, A. and C. Dirksen. 1986. Hydraulic Conductivity and Diffusivity: Laboratory Methods. In: Methods of Soil Analysis, Part 1, 2nd ed., A. Klute (ed.), Agronomy Monograph No. 9. American Society of Agronomy, Madison, WI, pp. 687-734.

Hazen, A. 1893. Some Physical Properties of Sand and Gravels with Special Reference to Their Use in Filtration. 24th Annual Report, Massachusetts State Board of Health, Boston. [Empirical equation for K for clean filter sands]

Heath, R.C. 1983. Basic Ground-Water Hydrology. U.S. Geological Survey Water-Supply Paper 2220. Republished in a 1984 edition by National Water Well Association, Dublin, OH.

Hendry, M.J. and B.A. Paterson. 1982. Relationships Between Saturated Hydraulic Conductivity and Some Physical and Chemical Properties. Ground Water 20(5):604-605. [Texture, depth, electrical conductivity, sodium adsorption ratio, saturation percentage]

Horn, M.E. 1971. Estimating Permeability Rates. J. Irrig. Drainage Div. ASCE 97(IR2):263-274.

Krumbein, W.C. and G.D. Monk. 1942. Permeability as a Function of the Size Parameters of Unconsolidated Sand. Trans. Petroleum Div. AIMME 151:153-163. [Empirical equation for k using laboratory prepared sand samples]

Manger, G.G. 1963. Porosity and Bulk Density of Sedimentary Rocks. U.S. Geological Survey Bulletin 1144E, 55 pp.

Masch, F.D. and K.J. Denny. 1966. Grain-Size Distribution and Its Effect on the Permeability of Unconsolidated Sands. Water Resources Research 2:665-677.

McKay, L.D., J.A. Cherry, and R.W. Gillham. 1993. Field Experiments in Fractured Clay Till 1. Hydraulic Conductivity and Fracture Aperture. Water Resources Research 29(4):1149-1162.

McWhorter, D.B. and D.K. Sunada. 1977. Ground-Water Hydrology and Hydraulics. Water Resources Publications, Littleton, CO, 492 pp. [Later edition published 1981]

Morris, D.A. and A.I. Johnson. 1967. Summary of Hydraulic and Physical Properties of Rock and Soil Materials as Analyzed by the Hydraulic Laboratory of the U.S. Geological Survey. U.S. Geological Survey Water Supply Paper 1839-D, pp. D1-D42.

Norris, S.E. 1963. Permeability of Glacial Till. U.S. Geological Survey Professional Paper 450-E, pp. E150-E151.

Puckett, W.E., J.H. Dane, and B.F. Hajek. 1985. Physical and Mineralogical Data to Determine Soil Hydraulic Properties. Soil Sci. Soc. Am. J. 49:831-836. [Empirical equation for saturated hydraulic conductivity for six Ultisols; soil moisture retention]

Soil Conservation Service (SCS). 1990. Elementary Soil Engineering. In: Engineering Field Manual for Conservation Practices. SCS, Washington, DC, Chapter 4.

Sevee, J. 1991. Methods and Procedures for Defining Aquifer Parameters. In: Practical Handbook of Ground-Water Monitoring, D.M. Nielsen (ed.), Lewis Publishers, Chelsea, MI, pp. 397-447.

Thompson, C.M., L.J. Holcombe, D.H. Gancarz, A.E. Behl, J.R. Erikson, I. Star, R.K. Waddell, and J.S. Fruchter. 1989. Techniques to Develop Data for Hydrogeochemical Models. EPRI EN-6637. Electric Power Research Institute, Palo Alto, CA.

Todd, D.K. 1959. Groundwater Hydrology, 2nd ed. John Wiley & Sons, New York, NY, 535 pp. [2nd edition published 1980]

Uma, K.O., B.C.E. Egboka, and K.M. Onuoha. 1989. New Statistical Grain-Size Method for Evaluating the Hydraulic Conductivity of Sand Aquifers. J. Hydrology 108-434-366. [See also, 1991 comment by Z. Sen in J. Hydrology 130:399-403]

van der Heijde, P.K.M. and O.A. Einawawy. 1993. Compilation of Ground-Water Models. EPA/600/R-93/118 (NTIS PB93-209401). [Also available from IGWMC, see address Section 10.6]

Vukovic, M. and A. Soro. 1992. Determination of Hydraulic Conductivity of Porous Media from Grain-Size Composition. Water Resources Publications, Highlands Ranch, CO, 86 pp.

Walton, W.C. 1987. Groundwater Pumping Tests: Design and Analysis. Lewis Publishers, Chelsea, MI, 201 pp.

Wiebenga, W.A., W.R. Ellis, and L. Kevi. 1970. Empirical Relations in Properties of Unconsolidated Quartz Sand and Silts Pertaining to Water Flow. Water Resources Research 6:1154-1161.

Wilson, L. 1981. Potential for Ground-Water Pollution in New Mexico. In: Environmental Geology of New Mexico, New Mexico Geological Society Special Publication 10, pp. 47-54.

* See Preface for information on how to obtain documents.

APPENDIX D

WORKSHEETS AND CHECKLISTS FOR GROUND-WATER AND WELLHEAD PROTECTION

This appendix contains a series of worksheets and checklists to assist in hydrogeologic characterization with a focus on applications for delineating and managing wellhead protection areas. The first 6 checklists can be used to assist in hydrogeologic mapping, selection of simple analytical equations or computer models for delineation of wellhead protection areas and vulnerability mapping using the DRASTIC system. Worksheets D-W7 through D-W11 and Checklists D-C1 through D-C5 are intended to be used for identification of potential contaminant sources within an area of interest. Section 11.3.1 provides a step-by-step procedure for conducting a potential contaminant source inventory. Checklists D-C6 and D-C7 can be used to help assist the relative risk posed by potential contaminant sources and to identify potential management practices to prevent or minimize ground water contamination. The following provides an easy reference to worksheets and checklists in this appendix:

Worksheets for Hydrogeologic Characterization   Page

D-W1   Water Well/Aquifer Data 878
D-W2   Collection of Existing Data for Wellhead Protection 880
D-W3   Possible Aquifer Boundaries 881
D-W4   Aquifer Characteristics for the Selection of Analytical Solutions to Ground
        Water Flow in the Vicinity of Wells 882
D-W5   Ground Water Computer Code Specifications and Code Suitability for a
        Specific Site 883
D-W6   DRASTIC Evaluation 885

Worksheets for Contaminant Source Inventories

D-W7   Residential Potential Contaminant Source Inventory 886
D-W8   Farm Potential Contaminant Source Inventory 888
D-W9   Transportation Hazard Inventory 891
D-W10   Municipal/Commercial/Industrial Potential Contaminant Source Inventory (1) 892
D-W11   Municipal/Commercial/Industrial Potential Contaminant Source Inventory (2) 893

Checklists for Contaminant Source Inventories

D-C1   Potential Contaminant Source Shortlist for Wellhead Protection 897
D-C2   Cross-cutting Potential Contaminant Sources 899
D-C3   Nonindustrial Potential Contaminant Sources 901
D-C4   Potential Contaminant Sources: Commercial, Natural Products Processing/
        Storage, and Resource Extraction 903
D-C5   Potential Industrial Contaminant Sources 905

Checklists for Risk Assessment and Ground-Water Protection Options

D-C6 Risk Categories of Land Uses and Activities Affecting Ground-Water Quality 906
D-C7 Wellhead Protection Tools 908

Worksheet D-W1 Water Well/Aquifer Data

Well Data (Attach drillers log):

Location: _____

Screen Interval Depth: _____

Water level data:

Date
___ ___ ___ ___ ___ ___ ___ ___ ___ ___

Level (ft.)
___ ___ ___ ___ ___ ___ ___ ___ ___ ___

Pumping Characteristics:
Current non-pumping water level (feet below ground surface) ____
Current pumping rate (gpm) ____
Typical pumping duration (hours/day) ____
Current pumping water level (feet below ground surface) ____
Typical nonpumping duration (hours/day) ____
Estimated annual pumpage (pumping rate x hours/day x 365 x 60) = _____
Specific capacity (pumping rate/(non-pumping water level minus pumping water level) = ____ gpm/ft drawdown*
Estimated transmissivity (specific capacity x 2000) = _____ gpd/ft*
Estimated hydraulic conductivity (transmissivity/aquifer thickness) = _____ gpd/ft^2*

Aquifer Material:		Porosity (%)	Ksat** (_____)	Specific Yield (%)
Unconsolidated Sediments	Low			
___ Gravel				
___ Coarse sand	Average	_____	_____	_____
___ Medium to fine sand				
___ Silt	High	_____	_____	_____
___ Clay, till				
Consolidated Sediments	Sources:			
___ Limestone, Dolomite	Table(s)	_____	_____	_____
___ Coarse, medium sandstone				
___ Fine sandstone	Figure(s)	_____	_____	_____
___ Shale, siltstone				
Volcanic rocks				
___ Basalt				
___ Acid volcanic rocks				
Crystalline Rocks				
___ Granite/gabbro				
___ Metamorphic				

Aquifer Classification:

Unconfined	Confined	Number of Aquifers
___ Perched	___ Semiconfined	___ One
___ Regional	___ Highly confined	___ Two
		___ > Two (#:____)

Worksheet D-W1 (cont.)

Aquifer Boundaries

Recharge Boundaries

___ Interfluv
___ Losing stream
___ Lake, pond
___ Sinkholes (karst)
___ Injection well

___ Ground-Water Divide

Discharge Boundaries

___ Artesian/pumping well
___ Gaining stream
___ Drainage ditch
___ Tile drains
___ Springs
___ Lakes, ponds
___ Semiconfined aquifer leakage

Expected water level fluctuations (see Table 2-3)

Moisture regime

___ High moisture (H)***
___ Moderate moisture (M)
___ Low moisture (L)

Drainage/Slope

___ Well developed/steep (H)***
___ Moderate/upland (M)
___ Poor/flat, bottoms (L)

Zone of Aeration (d)

___ d <0.5 m (H)***
___ d = 0.5 to 4 m (M)
___ d >4 m (L)

Diurnal/Intermittent Fluctuations

___ Evapotranspiration
___ Tidal effects near ocean
___ Atmospheric pressure effects

Seasonal Fluctuations

___ Ground-water recharge area
___ Stream bank storage effects

Long-Term Fluctuations

___ Ground-water pumpage
___ Deep-well injection
___ Artificial recharge
___ Pond, lagoon, landfill leakage
___ Agricultural irrigation
___ Agricultural drainage
___ Geotechnical drainage (open pit mines, slopes, tunnels)

* See Section 7.4.3 for additional discussion of this simple well test for estimating hydraulic conductivity.

** Saturated hydraulic conductivity (specify units).

*** Rating for expected degree of fluctuation: H = high, M = moderate, L = low.

Worksheet D-W2 Collection of Existing Data for Wellhead Protection

<u>Contacts and Phone Numbers</u>

EPA Regional Ground-Water Representative: _____

USGS Water Resources Division State Office: _____

SCS District/State Office: _____

Federal Management Agency Local Office*: _____

State Wellhead Protection Program: _____

State Water Resource Agency**: _____

State Environmental Protection Agency**: _____

State Geological Survey: _____

Local College/University Geology Department: _____

Local College/University Library: _____

<u>Topographic Maps</u>
___ 7 1/2' Topographic
___ 15' Topographic
___ Regional
___ Other

<u>Soils/Vegetation Maps</u>
___ Soil Map
___ Vegetation

<u>Geologic Maps</u>
___ State
___ Regional
___ Local

<u>Aerial Photography</u>
___ Large scale
___ High altitude
___ Satellite

<u>Hydrologic Maps</u>
___ USGS Hydrologic Atlas
___ State-Published Hydrologic Maps
___ Water Table/Potentiometric Surface
___ Watershed
___ Wetlands
___ Flood Plain Maps (FEMA, FIRM)
___ Other

<u>Land Use Maps</u>
___ Ownership/Tax Assessment
___ Subsurface Ownership (if different from surface ownership)
___ Zoning/Planning
___ Utilities
___ Other

* Required only if wellhead protection area includes federal lands (most likely in western U.S.). Possible agencies include: the Bureau of Land Management, U.S. Forest Service, U.S. Fish and Wildlife Service, and U.S. Department of Defense.
** If different from agency responsible for wellhead protection.

Worksheet D-W3 Possible Aquifer Boundaries

Barrier Boundaries	Distance to well	Within ZOC?*
		Yes / No

___ Vertical/Sloping

   ___ Impermeable crystalline rocks

      ___ Fault displacement

___ Horizontal**

Recharge Boundaries

___ Natural ground-water divide
    (unconfined aquifer)

___ Areal recharge from precipitation

___ Loosing stream

___ Lake, other surface water body

    ___ Above water table

    ___ Surface expression of water table

___ Leaky confining layer (downward flow)

___ Injection well

___ Areal artificial recharge

Discharge Boundaries

___ Gaining stream

___ Lake, other surface water body

    ___ Surface expression of water table

    ___ Interior drainage basin

___ Spring(s)

___ Karst conduit flow

___ Leaky confining layer (upward flow)

___ Drainage ditch/tile drain

___ Other pumping wells

* As defined by one or more of the simple methods described in Section 11.2.2.
** Impermeable geologic materials always form the base of an aquifer; see Table 7-4 for criteria for defining the extent to which impermeable confining layers represent boundaries to flow.

Worksheet D-W4 Aquifer Characteristics for the Selection of Analytical Solutions to Ground Water Flow in the Vicinity of Wells

Aquifer Type

___ Water table/unconfined
___ Confined, leaky
___ Confined, non-leaky

Regional Hydraulic Gradient

___ < 0.0005 (nearly flat)
___ 0.0005 to 0.001 (transitional)
___ > 0.001 (sloping)

Number of Aquifers

___ One
___ Two
___ More than two

Well Penetration

___ Fully penetrating well
___ Partially penetrating well

Aquifer Properties

___ Porous media
___ Fracture flow*
___ Karst conduit flow

___ Isotropic
___ Anisotropic

___ Homogeneous hydraulic parameters
___ Heterogeneous hydraulic parameters*

Flow Character/Dimension

___ Steady-state
___ Transient
___ Radial
___ X
___ X-Y
___ X-Y-Z

* Analytical solutions are not able to handle fracture flow or heterogeneous aquifer properties. In this situation, maximum measured or estimated aquifer parameters such as porosity and hydraulic conductivity should be used to account for reduced time of travel resulting from fracture flow and hydrodynamic dispersion caused by localized areas of higher hydraulic conductivity.

Worksheet D-W5 Ground Water Computer Code Specifications and Code Suitability for a Specific Site

Model Name: _____ IGWMC No.: _____

Contact: _____ Available from: ____ IGWMC
Address: _____   ____ Other Location
         _____   _____
         _____   _____
Phone: _____   _____

Site/Model Characteristics	Model System Requirements	Computer Match Requirements? Yes No

Site/Model Characteristics

___ ___ Unconfined (water table)
___ ___ Semiconfined (leaky)
___ ___ Confined
___ ___ Single aquifer
___ ___ Multiple aquifers
___ ___ Isotropic
___ ___ Homogeneous
___ ___ Anisotropic
___ ___ Heterogeneous
___ ___ Radial
___ ___ One-dimensional
___ ___ Two-dimensional
___ ___ Three-dimensional
___ ___ Steady flow
___ ___ Transient flow
___ ___ Variably saturated flow
___ ___ Single-phase flow
___ ___ Multi-phase flow
___ ___ Hydrodynamic dispersion
___ ___ Retardation
___ ___ Decay/degradation

Model System Requirements

____ IBM PC/AT/XT (circle)
____ Other Computer _____
Random Access Memory
____ 640 K
____ 4 MB
____ Other (_____)
Disk Drives
____ Single floppy (HD ___, DD ___)
____ Two floppy (HD ___, DD ___)
____ Hard drive
Disk Operating System
____ DOS 2.1
____ > DOS 2.1 (_____)
Math Coprocessor
____ Required
____ Optional
Graphics
____ CGA
____ EGA
____ VGA

Boundary Conditions See Worksheet D-W3

Site/Model Output

___ ___ Zone of Influence
___ ___ Cone of Depression
___ ___ Time of Travel
___ ___ Velocity
___ ___ Pathways
___ ___ Zone of Contribution
___ ___ Fluxes
___ ___ Concentration

Worksheet D-W5 (cont.)

Usability*

Yes No ?

___ ___ ___ Preprocessor
___ ___ ___ Postprocessor
___ ___ ___ User's instructions
___ ___ ___ Sample problems
___ ___ ___ Hardware dependency
___ ___ ___ Support

Reliability*

Yes No ?

___ ___ ___ Theory peer-reviewed
___ ___ ___ Coding peer-reviewed
___ ___ ___ Verified
___ ___ ___ Field validation

Model Users: ___ many, ___ few, ___ unknown

*Information on usability and reliability for most currently available models can be found in appendix tables in van der Heijde and Einawawy (1993).

Worksheet D-W6 DRASTIC Evaluation (Circle appropriate range and rating).

County: _____ State: _____

General Soil Map Unit Number: _____ **General Description:** _____

1. Depth to Water (ft) **2. Net Recharge (in)** **3. Aquifer Media**

					Rating		
Range	Rating	Range	Rating	Type	Range	Typical	Actual
0-5	10	0-2	1	Massive Shale	1-3	2	_____
5-15	9	2-4	3	Metamorphic/Igneous	2-5	3	_____
15-30	7	4-7	6	Weathered M/I	3-5	4	_____
30-50	5	7-10	8	Glacial Till	4-6	5	_____
50-75	3	10+	9	Bedded SS/LS/Shale	5-9	6	_____
75-100	2			Massive Sandstone	4-9	6	_____
100+	1			Massive Limestone	4-9	6	_____
				Sand and Gravel	4-9	8	_____
				Basalt	2-10	9	_____
				Karst Limestone	9-10	10	_____

4. Soil Media **5. Topography (%)** **6. Vadose Zone Media**

					Rating		
Type	Rating	Range	Rating	Type	Range	Typical	Actual
Thin/		0-2	10	Confining Layer	1	1	_____
Absent	10	2-6	9	Silt/Clay	2-6	3	_____
Gravel	10	6-12	5	Shale	2-5	3	_____
Sand	9	12-18	3	Limestone	2-7	6	_____
Peat	8	18+	1	Sandstone	4-8	6	_____
Structured				Bedded LS/SS/Shale	4-8	6	_____
Clay	7			Sand and Gravel with			
Sandy Loam	6			Sig. Silt and Clay	4-8	6	_____
Loam	5			Metamorphic/Igneous	2-8	4	_____
Silty Loam	4			Sand and Gravel	6-9	8	_____
Clay Loam	3			Basalt	2-10	9	_____
Muck	2			Karst Limestone	8-10	10	_____
Massive							
Clay	1						

DRASTIC Index

7. Hydraulic Conductivity
  (gpd/sq. ft.)

Rating x Weight =   Pesticide Rating x Weight =

Range	Rating
1-100	1
100-300	2
300-700	4
700-1,000	6
1,000-2,000	8
2,000+	10

1. _____ x 5 = _____   1. _____ x 5 = _____
2. _____ x 4 = _____   2. _____ x 4 = _____
3. _____ x 3 = _____   3. _____ x 3 = _____
4. _____ x 2 = _____   4. _____ x 5 = _____
5. _____ x 1 = _____   5. _____ x 3 = _____
6. _____ x 5 = _____   6. _____ x 4 = _____
7. _____ x 3 = _____   7. _____ x 2 = _____

                Total _____ *   Total _____ *

* Aquifers with DRASTIC ratings >150 are considered to be "highly vulnerable" by EPA.

Worksheet DW-7 Residential Potential Contaminant Source Inventory (North Dakota State Department of Health)

DATE: _____
PWS : _____

WELLHEAD PROTECTION AREA SURVEY FORM
<u>RESIDENTIAL</u>

This survey form is designed to inventory activities that may impact groundwater quality within the public water supply wellhead protection area (WHPA).

Name: _____
Address: _____
City: _____
Phone: _____

Please describe all water wells on the property:

First well:
Use/Name: _____
         (e.g., stock, house, irrigation)
Depth: _____ Diameter: _____
Depth to water: _____
Pumping rate (gallons per minute): _____
What year was the well installed? _____
Location: Township _____ Range _____
          Section _____ Quarters _____
  (Please locate on the section/block map provided.)

⟵ 1 mile or 1 block ⟶

SECTION MAP
This map represents an entire section of land. Please take care to plot the location of the source to the nearest 10 acres (see instructions). This map may also be used to represent a one-block area.

Second Well:
Use/Name: _____
         (e.g., stock, house, irrigation)
Depth: _____ Diameter: _____
Depth to water: _____ Pumping rate (gallons per minute): _____
What year was the well installed? _____
Location: Township _____ Range _____ Section _____ Quarters _____
  (Please locate on the section/block map provided.)

Third Well:
Use/Name: _____
         (e.g., stock, house, irrigation)
Depth: _____ Diameter: _____
Depth to water: _____ Pumping rate (gallons per minute): _____
What year was the well installed? _____
Location: Township _____ Range _____ Section _____ Quarters _____
  (Please locate on the section/block map provided.)

Worksheet DW-7 (cont.)

Are there any abandoned wells on the property? _____
If yes, were they plugged and how? _____

If there is a septic tank/drain field on the property, please describe:
Septic tank:
Location: _____
 (township, range, section, quarters, or other description; also locate on map)
Size: _____ Depth: _____ Year: _____ Last pumped out: _____

Drain field size and location: _____

Is there any heating/fuel oil storage on the property? Describe: _____

Are there any livestock on the property? Describe (if farm, please use <u>Farm</u> form): _____

Please describe any chemicals used or stored on the property.
  Storage: _____
  Usage: (fertilizers or pesticides on lawns or gardens? what type? quantity? frequency?) _____
  Disposal: _____

Are there any floor drains in your home or building that do not connect to the city sewer system? _____
If so, what is disposed of there? _____

Other problems or comments: _____
_____

Worksheet DW-8 Farm Potential Contaminant Source Inventory (North Dakota State Department of Health)

DATE: _____
PWS : _____

WELLHEAD PROTECTION AREA SURVEY FORM
FARM

This survey form is designed to inventory activities that may impact groundwater quality within the wellhead protection area (WHPA).

Name: _____
Address: _____
City: _____
Phone: _____

Please describe all water wells on the property:

First well:
Use/Name: _____
          (e.g., stock, house, irrigation)
Depth: _____ Diameter: _____
Depth to water: _____
Pumping rate (gallons per minute): _____
What year was the well installed? _____
Location: Township _____ Range _____
          Section _____ Quarters _____
(Please locate on the section/block map provided.)

Second Well:
Use/Name: _____
          (e.g., stock, house, irrigation)
Depth: _____ Diameter: _____
Depth to water: _____ Pumping rate
(gallons per minute): _____
What year was the well installed? _____
Location: Township _____ Range _____ Section _____ Quarters _____
(Please locate on the section map provided.)

Third Well:
Use/Name: _____
          (e.g., stock, house, irrigation)
Depth: _____ Diameter: _____
Depth to water: _____ Pumping rate (gallons per minute): _____
What year was the well installed? _____
Location: Township _____ Range _____ Section _____ Quarters _____
(Please locate on the section map provided.)

← 1 mile or 1 block →

SECTION MAP
This map represents an entire section of land. Please take care to plot the location of the source to the nearest 10 acres (see instructions). This map may also be used to represent a one-block area.

Worksheet DW-8 (cont.)

Are there any abandoned wells on the property? _____
If yes, were they plugged and how? _____
_____

If there is a septic tank/drain field on the property, please describe:
Septic tank:
Location: _____
  (township, range, section, quarters, or other description; also locate on map)
Size: _____ Depth: _____ Year: _____ Last pumped out: _____
Drain field size and location: _____
_____

Is there any heating/fuel oil storage on the property? Describe: _____
_____

Please list the crops that you typically plant. _____
_____

What is the total acreage that you farm? _____

Please list each crop separately followed by the number of acres that
are generally in that crop or the percentage of the total in that crop.

Crop #1 _____ acres or % _____
Crop #2 _____ acres or % _____
Crop #3 _____ acres or % _____
Crop #4 _____ acres or % _____

Chemicals (pesticides or fertilizers):

Please list the chemicals that you applied to each crop in the last two years.

Crop #	Chemicals applied	# of Years	Volume Kg/hectare/yr
_____	_____	_____	_____
_____	_____	_____	_____
_____	_____	_____	_____
_____	_____	_____	_____

Please describe any chemical storage procedures and the name of the
chemicals which you currently store. _____
_____
_____

Please describe any irrigation or chemigation practices. _____
_____
_____

Please describe any chemical mixing practices. _____
_____
_____

Please describe your container disposal practices. _____
_____
_____

Worksheet DW-8 (cont.)

Are there any livestock on the property? _____

Please list the types of livestock, how many, and their location. _____
_____
_____
_____

Please describe the location, age, and design of any feedlots. _____
_____
_____

Please describe any manure storage on the property. _____
_____
_____

Do you have any underground storage tanks? If so, describe their size, location, and contents. _____
_____
_____

Do you have any above ground storage tanks? If so, describe their size, location, and contents. _____
_____
_____

Other problems or comments: _____
_____
_____

Worksheets for Ground-Water Protection

Worksheet DW-9 Transportation Hazard Inventory (Adapted from Ohio Environmental Protection Agency, 1991)

1. Facility name: _____

2. Facility type: ___ Railroad, ___ Highway, ___ Sewer, ___ Fuel/chemical pipeline, ___ Terminal, ___ Service area, ___ Other: _____

3. Location: _____
   _____
   _____

4. Map No. _____   5. Minimum Distance from Nearest Public Well _____

6. Potential pollution sources (operation and construction information): _____
   _____
   _____
   _____
   _____

7. Describe any past pollution incidents: _____
   _____
   _____
   _____
   _____

8. Date of installation (pipelines): _____

9. Additional information (protection measures, handling practices, etc.):

Worksheet DW-10 Municipal/Commercial/Industrial Potential Contaminant Source Inventory: Short Form (Adapted from Adams et al., 1992)

1a. Facility name: 1b. Facility Map Location Number:

2. Facility address:

3: Owner/operator/other:

4: Type of business/facility:

5: Type of hazard observed:

6. Are storage tanks present? ___ yes ___ no (If no, skip to question 7)
   If yes, are the tanks above ground (AG) ___ , below ground (BG) ___
   If yes, is secondary containment present: ___ yes ___ no

	Age	Size	Tank Material	Material Stored	AG/BG
Tank 1					
Tank 2					
Tank 3					
Tank 4					
Tank 5					
Tank 6					
Tank 7					
Tank 8					
Tank 9					
Tank 10					

Comments:

7. Are solvents presents? ___ yes ___ no (If no, skip to question 8)

	Type	Storage Method	Quantity Stored	Disposal Method	Use
Solv. 1					
Solv. 2					
Solv. 3					
Solv. 4					
Solv. 5					

Comments:

Worksheet DW-10 (cont.)

8. Is the facility sewered? ___ yes ___ no

   If yes, are any floor drains connected to the sewer? ___ yes ___ no

   In no, describe wastewater disposal methods:

   Comments:

9. Is the facility subject to an environmental remediation? ___ yes ___ no (if no, skip to question 10)

   If yes, what type of remediation?

   Is this remediation currently under agency litigation, voluntary cleanup, other?

   Comments:

10. Are there any physical observations which may indicate a potential hazard to the ground water? ___ yes ___ no (If no, skip to question 11)

    If yes, describe:

    Comments:

11. Summarize the results of the findings enumerated above, and indicate the degree of potential hazard this facility may pose to ground water.

Inspector: _____

Date inspected: _____

Worksheet DW-11 Municipal/Commercial/Industrial Potential Contaminant Source Inventory: Long Form (Adapted from Adams et al., 1992)

1. Unique ID Number:

    Distance and Direction from Wellhead:

2. Type of business/facility:

3. Does the facility operate under any permits issued by a state or federal regulatory agency? ___ yes ___ no If yes, provide the following information:

	Issuing Agency	Permit Number	Type of Permit
Permit 1			
Permit 2			
Permit 3			
Permit 4			
Permit 5			

    Comments:

    Does the facility have any history of noncompliance with permit terms? ___ yes ___ no If, yes provide the following information:

	Date	Type of Violation	Disposition
Viol. 1			
Viol. 2			
Viol. 3			
Viol. 4			
Viol. 5			

    Comments:

4. Does the facility have a ground-water monitoring system? ___ yes ___ no If yes, describe:

    Have any ground-water samples exceeded compliance requirements or shown evidence of contamination? ___ yes ___ no If yes, describe:

    Comments:

5. Are storage tanks present? ___ yes ___ no (If no, skip to question 7)
   If yes, are the tanks above ground (AG) ___, below ground (BG) ___
   If yes, is secondary containment present: ___ yes ___ no

	Age	Size	Tank Material	Material Stored	AG/BG
Tank 1					
Tank 2					
Tank 3					
Tank 4					
Tank 5					
Tank 6					
Tank 7					
Tank 8					
Tank 9					
Tank 10					

   Comments:

6. Are solvents presents? ___ yes ___ no (If no, skip to question 8)

	Type	Storage Method	Quantity Stored	Disposal Method	Use
Solv. 1					
Solv. 2					
Solv. 3					
Solv. 4					
Solv. 5					

   Comments:

7. Is the facility sewered? ___ yes ___ no

   If yes, are any floor drains connected to the sewer? ___ yes ___ no

   If no, describe wastewater disposal methods:

8. Has onsite past or present landfilling, land treating, or surface impoundment of waste, other than landscape waste or construction and demolition debris occurred? ___ yes ___ no If yes, describe:

9. Are there currently any onsite piles of special or hazardous waste? ___ yes ___ no
   If yes, describe:

10. Are onsite piles of waste (other than special or hazardous wastes) managed according to Agency guidelines? ___ yes ___ no If no, describe:

11. Have there been any known or suspected releases of hazardous substances or petroleum at the site? ___ yes ___ no If yes, describe:

    Have any of the following actions/events been associated with the release(s) described above?

    ___ Hiring of cleanup contractor to remove obviously contaminated materials including subsoils.

    ___ Replacement or major repair of damaged facilities.

    ___ Assignment of inhouse maintenance staff to remove obviously contaminated materials including subsoils.

    ___ Designation as a "significant" release by a state of federal agency.

    ___ Reordering or other replenishment of inventory due to the amount of substance lost.

    ___ Temporary or more long-term monitoring of ground water at or near the site

    ___ Stop usage of an onsite or nearby water well because of offensive characteristics of the water.

    ___ Coping with fumes from subsurface storm drains or inside basements.

    ___ Signs of substances leaching our of the ground along the base of slopes or at other low points on or adjacent to the site.

12. After considering all of the above information, does this site potentially pose a hazard to ground water? ___ yes ___ no If yes, describe:

Inspector: _____

Date inspected: _____

Checklist D-C1 Potential Contaminant Source Shortlist for Wellhead Protection

If yes, or uncertain is answered to any of the questions below, the appropriate contaminant source worksheets and detailed checklists should be used.

Cross-Cutting Sources (Checklist D-C2)

___ Does the WHPA include natural geologic or hydrogeologic conditions that impair ground-water quality for drinking water? ___ yes ___ no. If yes, evaluate the following options, if this has not already been done:

    ___ Look for alternative, higher quality water supply
    ___ Evaluate effectiveness of existing drinking water treatment system in treating water quality problems
    ___ If there are problems with the existing system evaluate additional or alternative treatment technologies

___ Are any active/abandoned wells or boreholes located within the WHPA? ___ yes ___ no ___ uncertain? If yes, or uncertain, conduct inventory using Checklist D-C2.

___ Are any above- or underground storage tanks in the WHPA? ___ yes ___ no ___ uncertain? If yes, or uncertain, conduct inventory using Checklist D-C2.

___ Are there any areas of controlled or uncontrolled disposal of wastes in the WHPA? ___ yes ___ no ___ uncertain? If yes, or uncertain, conduct inventory using Checklist D-C2.

Nonindustrial Sources (Checklist D-C3)

___ Are there any areas within the WHPA used for agricultural, livestock or forest production? ___ yes ___ no ___ uncertain. If yes, or uncertain, conduct inventory using the Agricultural section of Checklist D-C3.

___ Are there any private homes, apartments or condominiums within the WHPA? ___ yes ___ no ___ uncertain. If yes, or uncertain, conduct inventory using the residential section of Checklist D-C3.

___ Are there any nonagricultural, nonresidential areas within the WHPA that receive treatment with fertilizers or pesticides? ___ yes ___ no ___ uncertain. If yes, or uncertain, conduct inventory using the nonresidential green areas section of Checklist D-C3.

___ Are any areas within the WHPA dedicate for municipal and other public service facilities? ___ yes ___ no ___ uncertain. If yes, or uncertain, conduct inventory using the municipal/public services section of Checklist D-C3.

___ Are any highways, roads, airports, railroads, pipelines, or associated transportation service and support facilities located within the WHPA? ___ yes ___ no ___ uncertain. If yes, or uncertain, conduct inventory using the transportation section of Checklist D-C3.

Checklist D-C1 (cont.)

Sources From Commercial, Natural Products Processing/Storage, and Resource Extraction Activities (Checklist D-C4)

___ Are there nonindustrial commercial activities within the WHPA? ___ yes ___ no ___ uncertain. If yes, or uncertain, conduct inventory using the commercial section of Checklist D-C4.

___ Are there any food, animal, or wood products processing or storage activities located within the WHPA? ___ yes ___ no ___ uncertain. If yes, or uncertain, conduct inventory using the natural products section of Checklist D-C4.

___ Are there any areas within the WHPA affected by current or past mining, oil and gas production or other resource extraction activities? ___ yes ___ no ___ uncertain. If yes, or uncertain, conduct inventory using the resource extraction section of Checklist D-C4.

Industrial Sources (Checklist D-C5)

___ Are there any chemical processing or manufacturing facilities within the WHPA? ___ yes ___ no ___ uncertain. If yes, or uncertain, conduct inventory using the chemical section of Checklist D-C5.

___ Are there any metal manufacturing, fabrication, or finishing facilities within the WHPA? ___ yes ___ no ___ uncertain. If yes, or uncertain, conduct inventory using the metals section of Checklist D-C5.

___ Are there any other manufacturing facilities not included in the two previous categories within the WHPA? ___ yes ___ no ___ uncertain. If yes, or uncertain, conduct inventory using the last section of Checklist D-C5.

Checklist D-C2 Cross-cutting Potential Contaminant Sources

Wells and Related Features

Active Abandoned

___	___	Water supply wells
___	___	Monitoring wells
___	___	Sumps and dry wells for drainage
___	___	Geotechnical boreholes
___	___	Oil and gas production wells
___	___	Mineral, oil and gas exploration boreholes

For each identified feature obtain the following information, if possible:

___ Location
___ Depth
___ Borehole Condition (cased, uncased, sealed, leaky)
___ Depth to ground water
___ Ground-water quality

Storage tanks (see Worksheets D-W8, D-W10, and D-W11)

Above Under
ground ground

___	___	Agricultural
___	___	Residential
___	___	Nonresidential green areas
___	___	Municipal and other public services
___	___	Commercial
___	___	Industrial
___	___	Resource Extraction

For each identified tank obtain the following information, if possible:

___ Location
___ Size
___ Contents
___ Age and condition

Waste Disposal Sites

Residential/Municipal Wastewater Treatment

___ Septic-tank soil absorption systems
___ Cesspools
___ Storage, treatment, and disposal ponds and lagoons
___ Municipal sewage treatment plant
___ Municipal sewer lines/lift stations
___ Wastewater irrigation/artificial ground-water recharge areas
___ Septage/sewage sludge land spreading areas

Checklist D-C2 (cont.)

Controlled Waste Disposal/Handling Sites

___ Municipal solid waste landfill (active)
___ Recycling and waste reduction facility
___ RCRA Hazardous Waste TSD Facility
___ Waste surface impoundments/lagoons
___ Waste injection well
___ Incinerator: ___ municipal waste, ___ medical waste, ___ hazardous waste
___ Demolition/detonation sites
___ Radioactive waste storage sites
___ Fire training facilities
___ Geothermal discharge

Uncontrolled Waste Disposal Sites

___ Accidental spill sites
___ Inactive/abandoned hazardous waste site (Superfund)
___ Other uncontrolled/clandestine waste disposal sites, open dumps
___ Abandoned mine spoils, mine tailings pile/pond
___ Radioactive (uranium mill tailings, laboratory wastes)

For each identified waste disposal obtain the following information, if possible:

___ Location
___ Amount and type of waste
___ Age
___ Inplace or planned measures to control contamination

Checklist D-C3 Nonindustrial Potential Contaminant Sources

Residential (Single-family, apartments and condominiums) — see Worksheet D-W7

___ Common Household products
___ Wall and Furniture treatments
___ Car maintenance
___ Other mechanical repair and maintenance products
___ Lawns and Gardens (EPA/530/SW-90-027i)*
___ Swimming Pools
___ Home-based business (beauty shop, welding, etc.—see Checklist D-C4)

Agricultural (EPA/530/SW-90-027i)* — see Worksheet D-W8

___ Livestock
    ___ Animal feedlots, stables, kennels
    ___ Manure spreading areas and storage pits (line/unlined)
    ___ Livestock waste disposal areas
    ___ Animal burial
___ Chemical storage areas and containers
___ Farm machinery areas
___ Irrigated cropland
___ Irrigation canals
___ Non-irrigated cropland
___ Pasture
___ Orchard/nursery
___ Rangeland
___ Forestland

Other Green Areas (EPA/530/SW-90-027i)*

Building grounds
    ___ Educational/vocational institutions
    ___ Government offices
    ___ Other offices
    ___ Stores
    ___ Processing/manufacturing facilities
___ Camp grounds
___ Cemeteries
___ Country clubs
___ Golf courses
___ Nurseries
___ Parklands
___ Pest-infested areas (specify type of land use)

Municipal and Other Public Services (see also Checklist D-C2, controlled waste sites)
___ Educational/Vocational facilities (EPA/530/SW-90-027l)*
___ Public swimming pools
___ Sewer/stormwater drainage overflows
___ Storm water drains and basins
___ Government service offices
___ Military base/depot

Checklist D-C3 (cont.)

Municipal and Other Public Services (cont.)

Public Utilities
    ___ Electric power and steam generation (coal storage areas, ash/FGD disposal areas)
    ___ Natural gas
    ___ Telephone/communications

Medical/care facilities (EPA/530/SW-90-027m)*
    ___ Doctor/Dentist Offices
    ___ Hospital
    ___ Nursing and rest homes
    ___ Veterinary Services

Transportation — see Worksheet D-W9

Airports
    ___ Active
    ___ Abandoned air fields

Automobile/Truck (EPA/530/SW-90-027a & 027n)*
    ___ Gasoline Service stations
    ___ Truck stops (gasoline plus diesel)
    ___ Dealers without service departments
    ___ Dealers with service departments
    ___ Car rental facilities
    ___ Government vehicle maintenance facilities
    ___ Taxi cab maintenance facilities
    ___ School bus maintenance facilities
    ___ Quick lube shops
    ___ Repair shops
    ___ Muffler repair shops
    ___ Body/paint shops
    ___ Undercoaters/rust proofing
    ___ Car washes

Other point/areal sources
___ Boat yards and marinas
___ Road/highway maintenance depots/road salt storage
___ Passenger transit facilities (local and interurban)
___ Railroad yards (EPA/530/SW-90-027k)*
___ Trucking terminals (EPA/530/SW-90-027k)*

Linear sources
___ Highways and roads
___ Railroad tracks
___ Oil and gas pipelines
___ Other industrial pipelines
___ Powerline corridors

* See U.S. EPA (1990a), Table 4-4 for information on how to obtain reference.

Checklist D-C4 Potential Contaminant Sources: Commercial, Natural Products Processing/Storage, and Resource Extraction (see Worksheets D-W10 or D-W11)

Commercial

    ___ Agricultural chemicals sales/storage (pesticides, herbicides, fertilizers)
    ___ Barber and beauty shops/salons (EPA/530/SW-90-027q)*
    ___ Bowling alleys

Cleaning services (EPA/530/SW-90-027b)*

        ___ dry cleaners
        ___ commercial laundry
        ___ laundromats
        ___ carpet and upholstery cleaners

Construction service/materials (EPA/530/SW-90-027j)*

        ___ plumbing
        ___ heating and air conditioning
        ___ paper hanging/decorating
        ___ drywall and plastering
        ___ carpentry
        ___ carpet flooring
        ___ roofing and sheet metal
        ___ wrecking and demolition
        ___ hardware/lumber/parts stores

    ___ Equipment/appliance repair (EPA/530/SW-90-027d)*
    ___ Florists
    ___ Furniture/wood manufacturing repair and finishing shops (EPA/530/SW-90-027c & 027n)*
    ___ Funeral services and crematories
    ___ Heating oil companies
    ___ Jewelry/metal plating shops (EPA/530/SW-90-027n)*
    ___ Leather/leather products (EPA/530/SW-90-027r)*
    ___ Lawn and garden care services (EPA/530/SW-90-027i)*
    ___ Office buildings and office complexes
    ___ Paint stores (EPA/530/SW-90-027p)*
    ___ Pest extermination services/pesticide application services (EPA/530/SW-90-027i)*
    ___ Pharmacies
    ___ Photography shops, photo processing laboratories
    ___ Printers, publishers and allied industries (EPA/530/SW-90-027g & 027p)*
    ___ Laboratories (research/testing) (EPA/530/SW-90-027m)*
    ___ Scrap, salvage, and junk yards
    ___ Sports and hobby shops
    ___ Taxidermists
    ___ Welders (EPA/530/SW-90-027n)*

Checklist D-C4 (cont.)

Food/Animal/Timber Products Processing and Storage

___ Canned and preserved fruits and vegetables
___ Canned and preserved seafood processing
___ Soft drink bottlers
___ Grain mills (___ grain storage/processing, ___ animal feed, breakfast cereal, and wheat)
___ Sugar processing (___ beet sugar, ___ cane sugar refining)
___ Dairy products processing (creameries and dairies)
___ Leather products (EPA/530/SW-90-027r)*
___ Meat products and rendering (slaughterhouses)
___ Poultry and eggs processing
___ Timber products processing
___ Pulp, paper and paperboard (EPA/530/SW-90-027o)*

    ___ Builders' paper and board mills
    ___ Unbleached kraft and semichemical pulp
    ___ Pulp, paper and paperboard
    ___ Paper coating and glazing

___ Wood preserving facilities (EPA/530/SW-90-027f)*

Resource Extraction

___ Abandoned exploration/production wells
___ Construction materials (sand, gravel)
___ Coal mining (___ active, ___ inactive)
___ Uranium mining (___ active, ___ inactive)
___ Metals mining (___ active, ___ inactive)
___ Phosphate mining (___ active, ___ inactive)
___ Natural gas production
___ Petroleum production/secondary recovery operations
___ Synthetic fuels (coal gasification, oil shale)
___ Waste tailings: ___ heap leaching, ___ non-heap leaching

* See U.S. EPA (1990a), Table 4-4 for information on how to obtain reference.

Checklist D-C5 Potential Industrial Contaminant Sources (see Worksheets D-W10/D-W11)*

Chemical Processing/Manufacturing
___ Explosives (EPA/530/SW-90-027h)*
___ Inorganic chemical manufacturing (EPA/530/SW-90-027h)*
___ Fertilizer manufacturing (___ basic fertilizer chemicals, ___ formulated fertilizer) (EPA/530/SW-90-027p)*
___ Organic chemical manufacturing and plastics and synthetic fibers (EPA/530/SW-90-027h)
___ Paint manufacturing (EPA/530/SW-90-027p)*
___ Pesticide formulation (EPA/530/SW-90-027h & 027p)*
___ Petroleum refining/storage
___ Pharmaceutical manufacturing (EPA/530/SW-90-027p)*
___ Phosphate manufacturing (___ phosphorus-derived chemical, ___ other non-fertilizer chemicals
___ Porcelain enameling
___ Rubber processing (___ tire and synthetic, ___ fabricated and reclaimed rubber) (EPA/530/SW-90-027h)*
___ Soaps and Detergents (EPA/530/SW-90-027q)*

Metals Manufacturing/Fabrication/Finishing
___ Aluminum Manufacturing and forming
   ___ Aluminum forming
   ___ Bauxite refining
   ___ Primary aluminum smelting
   ___ Secondary aluminum smelting
___ Coil coating
___ Copper forming
___ Electroplating (EPA/530/SW-90-027n)*
   ___ Copper, nickel, chrome and zinc
   ___ Electroplating pretreatment
___ Metal manufacturing and fabrication (EPA/530/SW-90-027n)*
   ___ Ferroalloy (smelt and slag processing)
   ___ Iron and steel manufacturing
   ___ Metal molding and casting (foundries)
___ Metal finishing (EPA/530/SW-90-027n)*
___ Machine and metalworking shops (EPA/530/SW-90-027n)*
___ Nonferrous metals forming

Other Manufacturing
___ Asbestos manufacturing
___ Asphalt/tar plants
___ Battery manufacturing (EPA/530/SW-90-027n)*
___ Cement manufacturing
___ Electric/electronic/communications equipment manufacturers (EPA/530/SW-90-027n)*
___ Furniture and fixtures manufacturers (EPA/530/SW-90-027c)*
___ Glass manufacturing
   ___ Pressed and blown glass
   ___ Insulation fiberglass
   ___ Flat glass
___ Stone, and clay manufacturers
___ Textile manufacturing (EPA/530/SW-90-027e)*

* See U.S. EPA (1990a), Table 4-4 for information on how to obtain reference.

Checklist D-C6 Risk Categories of Land Uses and Activities Affecting Ground-Water Quality

<u>High Risk</u> (Frequently Prohibited in High Priority Water Supply Protection Areas)

___ Airport maintenance areas
___ Animal feedlots
___ Appliance/small engine repair shops
___ Asphalt/concrete/coal tar plants
___ Auto repair and body shops*
___ Boat service, repair and washing establishments
___ Beauty parlors/hairdressers
___ Business and industrial uses (excluding agriculture) which involve the onsite disposal of process wastes from operations
___ Car washes
___ Chemical/biological laboratory
___ Chemical manufacturing/industrial areas
___ Cleaning service (dry cleaning, laundromat, commercial laundry)*
___ Disposal of liquid or leachable waste except for properly designed commercial and residential onsite wastewater disposal systems and normal agricultural operations
___ Electroplaters (metal plating and finishing) and metal fabricators*
___ Fuel oil distributors
___ Furniture and wood stripping and refinishing*
___ Gasoline stations
___ Golf courses/parks/nurseries
___ Graveyards
___ Improperly constructed or abandoned wells (perched, confined aquifers)
___ Junkyards and salvage yards*
___ Landfills and dumps
___ Making the surface of more than 10% of any lot impervious
___ Mining operations
___ Medical services (including dental/vet)
___ Military installations
___ Motels/hotels
___ Municipal sewage treatment facilities with onsite disposal of primary or secondary effluent
___ Oil and gas drilling and production
___ Outdoor storage of road salt, or other de-icing materials, the application of road salt and the dumping of salt-laden snow*
___ Outdoor storage of pesticides or herbicides
___ Parking areas of over 50 spaces
___ Pesticide/herbicide stores
___ Petroleum product refining and manufacturing
___ Photo processors/printing establishments
___ RCRA hazardous materials TSDs
___ Sand and gravel extraction

Checklist D-C6 (cont.)

High Risk (cont.)

___ Trucking or bus terminals
___ Underground storage and/or transmission of oil, gasoline or other petroleum products
___ Use of septic system cleaners which contain toxic chemicals (such as methylene chloride, and 1,1,1 trichloroethane)
___ Wood preserving and treating*

Moderate Risk (Frequently restricted in high priority water supply protection areas)

___ Above-ground storage tanks without secondary containment structures
___ Artificial groundwater recharge facilities
___ Excavation for the removal of earth, sand, gravel and other soils
___ Drainage from impermeable surfaces without installation and maintenance of oil, grease and sediment traps
___ Drywells and unlined stormwater drainage channels and impoundments
___ Irrigation in areas with coarse, permeable soils
___ Residential lot size in areas not served by municipal sewers (larger lot sizes reduce the amount of contamination from septic systems and household chemicals)
___ Unlined irrigation canals and tailwater sumps (arid areas)
___ Use of road salt (NaCl)
___ Use of commercial fertilizers, pesticides and herbicides

Sources: Lowrence (1992), Noake (1988), Dean and Wyckoff (1991).

* Highest risk light industrial uses identified in U.S. EPA (1991).

Checklist D-C7 Wellhead Protection Tools

Regulatory Options (Nontechnical)

Zoning Ordinances (Table 11-9)

___ Overlay ground water protection districts (Table 11-8)
___ Land use prohibitions (Table 11-8)
___ Special permitting (Table 11-8)
___ Large-lot zoning (Table 11-8)
___ Transfer of development rights (Table 11-8)
___ Cluster/PUD Design (Table 11-8)
___ Growth controls/timing (Table 11-8)

Subdivision and Individual Lot Controls

___ Subdivision ordinances (Table 11-9, see also Technical Options below)
___ Site plan review (Table 11-9)

Health and Environmental Regulations

___ Prohibit or additional regulation of underground storage tanks (Table 11-8)
___ Other source prohibitions (Table 11-9)
___ Inspection and testing (Table 11-9)
___ Prohibition/regulation of small sewage treatment plants (Table 11-8)
___ Phosphorus buffer zone
___ Septic cleaner ban (Table 11-8)
___ Septic system maintenance/upgrades (Table 11-8)
___ Registration and inspection of businesses using toxic/hazardous materials (Table 11-8)
___ Regulation of household hazardous waste
___ Regulation of agricultural chemicals
___ Regulation of private wells: permits, pump and water quality testing (Table 11-8)

Legislative (State-level)

___ Establishment of regional WHPAs (Table 11-8)
___ Passage of laws authorizing regulation where regulatory powers are limited

Nonregulatory Options (Nontechnical)

___ Land acquisition by purchase or donation (Tables 11-8, 11-9)
___ Purchase of development rights (Table 11-9)
___ Taxation deferments for nondevelopment
___ Conservation easements (Table 11-8)
___ Voluntary limits to development (Table 11-8)
___ Land banking/transfer taxes (Table 11-8)
___ Contingency planning (Tables 11-8, 11-9)
___ Hazardous waste collection program (Table 11-8)
___ Public education (Tables 11-8, 11-9)
___ Training and demonstration (Table 11-9)
___ Waste reduction (Table 11-9)
___ Water conservation

Checklist D-C7 (cont.)

Technical Regulatory and Nonregulatory Options

General

___ Wellhead protection zones
___ Ground water monitoring (Tables 11-8, 11-9)
___ Performance standards (Table 11-8)
___ Operating standards (Table 11-9)
___ Design standards (Table 11-9)
___ Best management practices — BMPs (Table 11-9)
___ Capture zone management

Subdivision Controls

___ Nitrogen/phosphorus loading standards
___ Drainage Requirements (Table 11-8)

Nonpoint Source Pollution Controls

___ Agriculture BMPs
___ Construction Site BMPs

Appendix D References

Adams, S. et al. 1992. Pilot Groundwater Protection Needs Assessment for Illinois American Water Company's Pekin Public Water Supply Facility Number 1795040. Division of Public Water Supplies, Illinois Environmental Protection Agency, Springfield, IL.

Dean, L.F. and M. A. Wyckoff. 1991. Community Planning and Zoning for Groundwater Protection in Michigan: A Guidebook for Local Officials. Prepared for Office of Water Resources, Michigan Department of Natural Resources. Available from Michigan Society of Planning Officials, 414 Main St., Suite 202, Rochester, MI 48307.

Lowrence, J.L. 1992. Vulnerability Assessment Criteria: Public Water Supply Protection (Draft). New Mexico Department of the Environment, Santa Fe, NM. [Criteria for giving waivers for constituents to be monitored by drinking water systems]

Noake, K.D. 1988. Guide to Contamination Sources for Wellhead Protection (Draft). Massachusetts Department of Environmental Quality Engineering, Boston, MA.

North Dakota State Department of Health. 1993. North Dakota Wellhead Protection User's Guide. Division of Water Quality, Bismarck, ND.

Ohio Environmental Protection Agency. 1991. Guidance for Conducting Pollution Source Inventories in Wellhead Protection Areas (Draft). OEPA, Division of Ground Water, Columbus, OH, 17 pp.

U.S. Environmental Protection Agency (EPA). 1991. A Review of Sources of Ground-Water Contamination from Light Industry. EPA/440/6-90-005 (NTIS PB91-145938).

van der Heijde, P.K.M. and O.A. Einawawy. 1993. Compilation of Ground-Water Models. EPA/600/R-93/118 (NTIS PB93-209401). [Also available from IGWMC, see address Section 10.6]

APPENDIX E

PC-BASED GEOENVIRONMENTAL SOFTWARE

This appendix contains information that may be useful for the selection and acquisition of PC-based geoenvironmental software for use in assessment and management of soil and ground-water contamination. Table E-1 provides summary information on more than 80 commercially available software systems for analysis, management and graphic presentation of geoenvironmental data. Software with the following types of capabilities are identified in Table E-1:

- Statistical analysis (conventional and geostatistics-see Table E-2 for statistical software available from U.S. EPA).
- Creation of borehole logs (text and graphic).
- Creation of graphic cross-sections and fence diagrams.
- Data contouring.
- Three-dimensional, volumetric graphic presentation.
- Creation of maps.
- Database entry and management for spatial data.
- Conversion of spatial coordinates from one or more systems to another.

A number of the items identified in Table E-1 can be considered mini- or desktop geographic information systems as discussed in Section 10.5.5. Full-scale geographic information systems are not included in this table. Tables 10-5 and 10-7 provide additional information on reference sources for geographic information systems. The source vendor for a particular item in Table E-1 is identified in parentheses after the name of the software, with vendor names, addresses, and phone numbers provided in an alphabetical listing at the end of the table.

Table E-2 provides an index to user manuals and other key source references for PC-based ground-water and vadose-zone flow and contaminant transport models. References are indexed in four main categories: (1) ground-water flow, (2) solute transport and biodegradation, (3) hydrogeochemical, and (4) vadose zone models. Section 10.3 discusses classification of models in more detail. The name of the model is indicated after the year of the reference source in Table E-2.

Table E-2 also includes an index of public domain software available from U.S. EPA, including ground-water and vadose-zone models, and statistical-QA/QC software. Models available from commercial software vendors are identified at the end of Table E-2, and addresses of the vendors can be found in the alphabetical listing at the end of Table E-1.

Table E-1 PC-Based Geoenvironmental Software for Data Analysis, Management and Graphic Presentation

Name/Source	Stats[1]	Logs	X-Sec Fence	Contour	3-D (vol)	Maps[2]	Spatial Database[3]	Coord. Conv.[4]
AVS/Toolkit (Adv. Vis. Systems)	x		x	x	x	x		
AQUABASE (Tecsoft)							x	
AutoDRILL (Slinn Engineering)*			x		x			x
Auto/GIS (Spatial Utilities)*				x		x	x	
C-Map (Center for Remote Sensing)						x	x	
CoDraw (CoHort Software)						x		
CoPlot (CoHort Software)				x				
CoStat (CoHort Software)	(x)							
CPS/PC (Radian)				x	x	x		
DEMtoDXF/DLGtoDXF (RockWare+)								x
Digital Land Grid (RockWare+)								x
DIGITIZE-PC (RockWare)								x
EarthVision (Dynamic Graphics)			x	x	x	x		
EVS (C Tech Development)			x	x	x	x	x	
GEOBASE—Basic Lithology (Earthware)	x	x						
GEOBASE—Advanced Lithology (")	x	x	x					
Geographic Calculator (RockWare+)								x
GEOKIT (Schreiber Instruments)*				x		x		x
GEOPAL (RockWare)								x
GEOSTAT Toolbox (FSS Int.)	x				x			
GeoUtils (GeoSoft)								x
gINT (GCA)		x	x					
GISKey (GIS\Solutions)		x	x	x	x	x	x	
GMS (Lynx Geosystems)		x	x	x	x	x	x	
GRAPHER (Golden Software)	(x)							
GRIDZO—see ROCKWORKS								
GRIDMGR (WhiteStar Corp)							x	x
GSAS (Int. Dec. Technologies)	(x)							
GSLIB (see Table E-2)	x				x			
GTGS (Geotechnical Graphics)		x	x					
GTLog (Geotechnical Graphics)		x						
K-STAT (SRIE Pty. Ltd.)	x			x		x		
LI-CONTOUR (AB Consulting)				x		x		
LLCALC (WhiteStar Corp.)						x		x
LOGGER—see ROCKWORKS								
LOGGCORRLATE (The Logic Group)				x				
logWRITER (GCA)		x						
MapInfo (see GeoTrans)							x	
MapPlan-DOS/Windows (Wordtech Systems)							x	x
MAPVIEWER (Golden Software)							x	x
MPS (Geosoft)	x			x		x		x
M-STAT (SRIE, Pty. Ltd.)	x			x		x		
The Monitor System (Entech Systems)	(x)						x	
PLANIMETER (The Logic Group)					x			
POLYTREND (Mark Maslyn)	x				x			
PS-Plot (Polysoft)	(x)			x				

Table E-1 (cont.)

Name/Source	Stats[1]	Logs	X-Sec Fence	Contour	3-D (vol)	Maps[2]	Spatial Database[3]	Coord. Conv.[4]
QUICKCROSS/FENCE (MTECH Graphics)			x					
QUICKLOG (MTECH Graphics)		x						
QUICKSURF (Schreiber Instruments)*				x	x			
ROCKSOLID (Rockware)					x			
ROCKWORKS (Rockware)	x	x	x	x	x	x		
GRIDZO (Rockware)	x			x		x		
GEOPAL (Rockware)						x		
LOGGER (Rockware)		x						
ROCKBASE (Rockware)							x	x
ROCKSTAT (Rockware)	(x)							
SiteGIS (GeoTrans)	x			x		x	x	
SiteManager (ConSolve)							x	
SitePlanner (ConSolve)			x	x		x		
SiteView (ConSolve)					x			
SiteWorks (Intergraph)			x	x	x	x		
Spase/Enviro Spase (GCS)						x	x	
SPILLCAD (ES&T)	x			x	x			
Spyglass Plot (Spyglass, Inc.)	(x)							
Spyglass Transform (Spyglass, Inc.)				x				
Spyglass Dicer (Spyglass, Inc.)					x			
S-STAT (SRIE, Pty. Ltd.)	x			x		x		
Statgraphics (STSC, Inc.)	(x)							
StratiFact (GRG Corporation)		x	x	x		x	x	x
SURFER (Golden Software)				x				
TECHBASE (Minesoft)	x		x	x		x	x	
TECGRAF (Tecsoft)	(x)							
TECKON (Tecsoft)				x				
TigerDXF Translator (RockWare+)								x
TOPBASE (Mark Maslyn)							x	
TopoTool (RockWare+)								x
TRALAINE (RockWare+)								x
TURBOCON (Mark Maslyn)				x				
V-CONTOUR (Eng. Desktop Solutions)				x				
WARP (RockWare+)								x
XY CONVERTER/TRANS (GCS)								x
Z/CON-Windows (Rockware)				x				

* Used with AutoCAD.

[1] Statistical software able to graph time series data, spatial data, and or the results of statistical tests; (x) = statistical tests are not distinctive for spatial data (i.e. can be run on spatial data, but spatial location is not a key element of the analysis, as in geostatistical analysis).

[2] Base maps, plotting programs.

[3] Spatial database management capabilities.

[4] Stand-alone packages or modules for converting spatial data from one coordinate system to another or from one digital format to another; direction of conversion and type of coordinate systems handled vary, but most handle lat/long, UTM, Township-Range-Section and State Plane. A RockWare+ notation for stand alone packages indicates that the software is available from RockWare Scientific Software, but was not developed by them.

Appendix E-1 Geoenvironmental Software Vendors

AB Consulting (LI-Contour), 3939 N. 48th St., Lincoln, NE 68504; (402) 464-8021.

Advanced Visualization Systems, Inc. (AVS/Toolkit), 300 Fifth Ave., Waltham, MA 02154; 617/890-4300.

Center for Remote Sensing (C-Map), Michigan State University, 115 Manly Miles Bldg., 1405 S. Harrison Rd., East Lansing, MI 48823-5243; (517) 355-3276.

Consolve (SiteManager, SitePlanner, SiteView), 70 Westview St., Lexington, MA 02173; 617/674-2199.

C Tech, (EVS-Environmental Visualization System), 1908 South O St., Port Angeles, WA 98362; 800/669-4387, 206/452-2275.

(Dr.) Charles R. Fitts, Department of Geosciences, University of Southern Maine, Gorham, ME 04038; (207) 780-5351.*

CoHort Software (CoPlot, CoDraw, CoStat), P.O. Box 1149, Berkeley, CA 94701; (800) 728-9878.

Dynamic Graphics, Inc. (EarthVision), 1015 Atlantic Ave., Alameda, CA 94501; 510/522-0700.

Earthware of California (GEOBASE), 30100 Town Center Drive, Suite 196, Laguna Niguel, CA, 92677; (714) 495-5727.

Engineering Desktop Solutions (V-CONTOUR), 1767 N. State St. # 200, Orem, UT 84057; (801) 225-3133.

Entech Systems, Inc. (The Monitor System), P.O. Box 760, West Meadow Rd., Camden, ME 04843; 207/594-5609.

ES&T/Environmental Systems and Technologies, Inc. (SPILLCAD), 2608 Sheffield Drive, Blacksburg, VA 24060-8270; 800/926-5923.

Field Resource Group, 1 Robert Lane, Unit B, Glen Head, NY 11545; 516/759-7891.

FSS International (GEOSTAT Toolbox), 245 Moonshine Circle, Reno, NV 89523; (702) 345-0448.

Geosoft, Inc. (MPS), 204 Richmond St. W., Toronto, Canada M5V 1V6; (416) 971-7700.

Geotechnical Computer Applications/GCA (gINT, logWRITER), 1200 College Ave., Santa Rosa, CA 95404; (707) 575-8510.

Geotech Computer Systems/GCS (SPASE/Enviro Spase), 7338 South Alton Way, Suite 16F, Englewood, CO 80112; (303) 740-9432.

(Table E-1 Geoenvironmental Software Vendor Addresses)

Geotechnical Graphics (GTGS, GTLog), 930 Dwight Way, Suite 6, Berkeley, CA 94710; (510) 649-4830.

GeoTrans, Inc. (SiteGIS, MapInfo), 46050 Manekin Plaza, Suite 100, Sterling, VA 22170; (703) 444-7000. [SiteGIS is used with MapInfo software developed by MapInfo Corporation, and GeoTrans is an authorized reseller for MapInfo giving a 15% discount on all MapInfo software and data products]

Geraghty & Miller, Inc., Modeling Group, 10700 Parkridge Blvd., Suite 600, Reston, VA 22091; (703) 758-1200.*

Golden Software, Inc. (GRAPHER, MapViewer, SURFER), P.O. Box 281, Golden, CO, 80402; (303) 279-1021.

GIS\Solutions (GIS\Key), 1800 Sutter Street, Suite 830, Concord, CA 94520-2500; (510) 827-5400.

GRG Corporation (StratiFact), 4175 Harlan Street, Wheat Ridge, CO 80033-5150; (303) 423-0221, (800) 783-6250.

Intergraph Corp. (SiteWorks), MS/LR24A2, Huntsville, AL 35894-0001; 205/730-8211. [See also, work station-based InSitu, MGE (Modular GIS Envrironment), and ERMA (Environmental Rsource Management Applications)]

Intelligent Decision Technologies, Ltd. (GSAS), 3308 Fourth St., Boulder, CO 80304; (303) 449-2457. [RCRA Ground-water Statistical Analysis System]

Lynx Geosystems, Inc. (GMS), 400-1199 West Pender St., Vancouver, BC, Canada V6E 2R1; (604) 682-5484.

Mark Maslyn (POLYTREND, TOPBASE, TURBOCON), 1370 S. Logan, Denver, CO 80210; (303) 722-3341.

Minesoft, Inc. (TECHBASE), 165 S. Union Blvd., Suite 510, Lakewood, CO 80228; (303) 980-5300.

MTECH Graphics (QUICKLOG, QUICKCROSS/FENCE), 4950 Eastern Ave., Cincinnati, OH 45208; (513) 321-9964.

Polysoft (PS-Plot), P.O. Box 526368, Salt Lake City, UT 84152; (801) 485-0466.

(S.S.) Papadopoulos & Associates, 7944 Wisconsin Avenue, Bethesda, MD 20814-3620; (301) 718-8900.*

Prickett and Associates, Urbana, IL.*

Radian Corporation (CPS/PC), 8501 Mopac Blvd., Austin, TX 78759; (512) 454-4797.

Rockware Scientific Software (GRIDZO, LOGGER, ROCKWORKS, ROCKSOLID, Z/CON), 4251 Kipling St., Suite 595, Wheat Ridge, CO 80033; (303) 423-5645.

(Table E-1 Geoenvironmental Software Vendor Addresses)

Schreiber Instruments, Inc. (GEOKIT, QUICKSURF), 4800 Happy Canyon Rd #250, Denver CO 80237; (303) 759-1024.

Slinn Engineering Services (AutoDRILL), AA338-808 Peace Portal Dr., Blaine, WA 98230; (604) 688-3692.

Spatial Utilities (Auto/GIS), P.O. Box 466, Temple Hills, MD 20748; (301) 899-2424.

Spyglass, Inc. (Spyglass Plot/Transform/Dicer), 1800 Woodfield Drive, Savoy, IL 61874; (800) 647-2201, (217) 355-6000.

SRIE Pty. Ltd. (K-STAT, M-STAT, S-STAT), P.O. Box 1189, Dee Why N.S.W., Australia 2099; 02-971-9409.

Strack Consulting, North Oaks, MN.*

STSC, Inc. (Statgraphics), 2115 East Jefferson St., Rockville, MD, 20852; (301) 984-5000.

Tecsoft, Inc. (AQUABASE, TECGRAF, TECKON), P.O. Box 888, Fort Collins, CO 80522; (303) 225-2554.

The Logic Group (LOGCORRELATE, PLANIMETER), P.O. Box 50499, Austin, TX 78763; (512) 474-4641.

Waterloo Hydrogeologic Software, 37 Watersdown Crescent, Whitby, Ontario, Canada L1R 1Z1; (416) 404-0991.*

Watershed Research, White Bear Lake, MN.*

Wellware, 3717 Modoc Place, Davis, CA 95616; (916) 758-0290.*

The WhiteStar Corp. (GRIDMGR, LLCALC), 333 West Hampden, Suite 604, Englewood, CO 80110; (303) 781-5182.

Wordtech Systems, Inc. (MapPlan-DOS/Windows), 21 Altarinda Rd., Ordinda, CA 94563; (510) 254-0900.

* See Table E-2 for ground-water/vadose zone models available from this source.

(Table E-1 Geoenvironmental Software Vendor Addresses)

Table E-2 Index to References on PC-Based Ground-Water and Vadose-Zone Flow and Contaminant Transport Models*

Topic	References
Ground Water Flow	Aral (1990a-SLAM, 1990b-ULAM), Bair et al. (1991-CAPZONE), Blandford and Huyakorn ((1991-WHPA), Bonn and Rounds (1990-DREAM), Franz and Guiguer (1990-FLOWPATH), McDonald and Harbaugh (1988-MODFLOW), Pollack (1988-MODPATH), Prickett and Associates (1984-PLASM), Prickett and Lonnquist (1971-PLASM), Shafer (1987, 1990), Strack and Haitjema (in press-WhAEM), Trescott (1975-USGS-3D-FLOW), Trescott et al. (1976-USGS-2D), U.S. EPA (1993-MODFLOW), van der Heijde (1987-THWELLS), Walton (1989a-PLASM, 1989b-WELFLO, 1992-MODFLOW, MODPATH), Wellware (1993-Ressq), Zheng (1992b-PATH3D), Zheng et al. (1992-PATH3D); <u>Analytical Models</u>: Cleary and Ungs (1978), Walton (1962, 1984a, 1984b-WALTON35)
Solute Transport/ Biodegradation	Bedient et al. (1989-BIOPLUMEII), Domenico and Palciauskas (1982-VHS), Freeze et al. (1992), Hostetler et al. (1988-FOWL), Konikow and Bredehoeft (1978-MOC), Newell et al. (1990), Park et al. (1992-VIRALT), Prickett and Associates (1984—Random Walk), Strack (1989-SLAEM), Rifai et al. (1988-BIOPLUMEII), Summers et al. (1989-MYGRT), U.S. EPA (1985-EPA-VHS), Walton (1989a-Random Walk, 1989b-CONMIG, 1992-MOC, SUTRA), Watershed Research (1988-Hydropal Slug), Yeh (1981-AT123D), Zheng (1990, 1992a-MT3D); <u>Analytical Solutions/Nomographs</u>: Dragun (1989-CDT nomograph), Galya (1987), Guswa et al. (1987-Rapid Assessment Nomograph), Kent et al. (1982-Wilson-Miller Nomograph), van Genuchten and Alves (1982)
Hydrogeochemical	Ball and Nordstrom (1991-WATEQ4F), Brown and Allison (1987-MINTEQA1), Parkhurst et al. (1980-PHREEQE)
Vadose Zone	<u>Hydrologic</u>: Perrier and Gibson (1984-HSSWDS), Ross (1990-SWIM), Schroeder et al. (1984a, 1984b-HELP), van Genuchten et al. (1991-RETC), Walton (1992-INTERSAT); <u>Fate and Transport</u>: Bonazountas and Wagner (1984-SESOIL), Brown and Boutwell (1988-POSSM), Carsel et al. (1984-PRZM), CH2M Hill (1990-VLEACH), Davis et al. (1990-GLEAMS), Dean et al. (1989-RUSTIC), General Sciences Corporation (1989-PCGEMS/SESOIL), Knisel (1980-CREAMS), McCone et al. (1987-Geotox), Nofziger and Hornsby (1987-CMLS), Nofziger and Williams (1988-RITZ), Nofziger et al. (1989-CHEMFLO), Pennsylvania Bureau of Waste Management (1989-RAFT), Shields et al. (1987-MCPOSSM), Sims et al. (1991-STF/RITZ/VIP), Stevens et al. (1993-VIP), Varadhan and Johnson (1992-PESTAN), Walton (1992-INTERTRANS)

Table E-2 (cont.)

Topic	References
Vadose Zone (cont.)	<u>Soil Vapor</u>: HyperVentilate (Johnson, 1992; Kruger and Morse, 1992)
Water Budget	<u>Precipitation-Runoff</u>: Leavesly et al. (1983)

<u>EPA Software</u>

Ground-Water Models	See references above for following models: BIOPLUMEII, EPA-VHS, MT3D, OASIS, WhAEM, WHPA; <u>Hydrogeochemical</u>: MINTEQA1
Vadose Zone Models	See references above for following models: CHEMFLO, HELP, HSSWDS, HyperVentilate, PCGEMS/SESOIL, PESTAN, PRZM, RETC, RITZ, RUSTIC, SESOIL, STF, VIP, VLEACH
Statistics	<u>Conventional</u>: Lin (1986, 1993); <u>Geostatistics</u>: England and Sparks (1988, 1991), Yates and Yates (1990); <u>Ground Water</u>: U.S. EPA (1992-GRITS/STAT) <u>Uncertainty Analysis</u>: Klee (1990, 1992); <u>QA/QC</u>: Johnson et al. (1987-PC-QTRAK), Simon et al. (1991-CADRE), U.S. EPA (1991a-ASSESS), van Ee et al. (1990)
Other	Kollig et al. (1991-FATE), U.S. EPA (1991b); <u>MULTIMED</u>: Salhotra et al. (1993), Sharp-Hansen et al. (1992)

<u>Other Statistical Software</u>

Geostatistics	Deutsch and Journal (1993-GSLIB)

<u>Commercial Software Vendors</u>**

See above for references	<u>Charles R. Fitts</u>: TWODAN; <u>GeoTrans</u>: SEFTRAN, SWANFLOW, MOD3D (MODFLOW), BIO1D, FRACFLOW; <u>Geraghty & Miller</u>: AQTESOLV, ModelCad, MODFLOW, MODPATH, MOC, SUTRA, QuickFlow; <u>S.S. Papadopulos & Associates</u>: MT3D, PATH3D; <u>Prickett and Associates</u>: PLASM, Random Walk; <u>Strack Consulting</u>: SLAEM; <u>Waterloo Hydrogeologic Software</u>: AIRFLOW, FLONET, FLOWPATH; <u>Watershed Research</u>: Hydropal; <u>Wellware</u>: AqModel, RessqM

* See IGWMC (1993), Rockware Scientific Software (1993) and Scientific Software Group to identify models available from major software catalogs.

** See addresses after Table E-1.

Appendix E-2 References

Aral, M.M. 1990a. Ground Water Modeling in Multilayered Aquifers: Steady Flow. Lewis Publishers, Chelsea, MI, 114 pp. [Includes disks for SLAM—steady layered aquifer model]

Aral, M.M. 1990b. Ground Water Modeling in Multilayered Aquifers: Unsteady Flow. Lewis Publishers, Chelsea, MI, 143 pp. [Includes disks for ULAM—unsteady layered aquifer model]

Bair, E.S., C.M. Safreed, and B.W. Berdainier. 1991. CAPZONE—An Analytical Flow Model for Simulation Confined, Leaky Confined, or Unconfined Flow to Wells with Superposition of Regional Water Levels, User's Manual. Prepared for OHIO EPA by Dept. of Geological Sciences, Ohio State University, Columbus, OH. [Modification of THWELLS (van der Heijde, 1987)]

Ball, J.W. and D.K. Nordstrom. 1991. User's Manual for WATEQ4F. U.S. Geological Survey Open File Report 91-0183. [Latest version of the WATEQ family of models, available from IGWMC]

Bedient, P.B. et al. 1989. Bioplume II Users Manual. National Center for Ground Water Research, Rice University, Houston, TX.

Blandford, T.N and P.S. Huyakorn. 1991. WHPA: Modular Semi-Analytical Model for the Delineation of Wellhead Protection Areas, Version 2.0. Office of Ground Water Protection; Available from IGWMC. Version 1.0 was released in 1990 [Four modules: MWCAP, RESSQC, GPTRAC, MONTEC; available from IGWMC; most current disk version is 2.1]

Bonazountas, M. and J.M. Wagner. 1984. SESOIL: A Seasonal Soil Compartment Model, Draft Report. EPA Contract No. 68-01-6271. (Prepared by A.D. Little for Office of Toxic Substances, Washington, DC.)

Bonn, B.A. and S.A. Rounds. 1990. DREAM—Analytical Ground Water Flow Programs. Lewis Publishers, Chelsea, MI, 115 pp. [Analytical PC ground water flow program (DREAM) for calculation of drawdown, streamlines, velocities, and water level elevations; includes disk]

Brown, D.S. and J.D. Allison. 1987. MINTEQA1, an Equilibrium Metal Speciation Model: A User's Manual. EPA/600/3-87/012 (NTIS PB88-144167), 103 pp.

Brown, S.B. and S.H. Boutwell. 1988. Chemical Spill Exposure Assessment Methodology. RP 2634-1, Electric Power Research Institute, Palo Alto, CA. [POSSM-PCB Onsite Spill Model]

Carsel, R.F., C.N. Smith, L.A. Mulkey, J.D. Dean, and P. Jowise. 1984. Users Manual for the Pesticide Root Zone Model (PRZM): Release 1. EPA/600/3-84/109 (NTIS PB85-158913).

CH2M Hill. 1990. VLEACH: A One-Dimensional Finite Difference Vadose Zone Leaching Model, Version 1.02. U.S. EPA Region 9.*

Cleary, R.W. and M.J. Ungs. 1978. Analytical Models for Groundwater Pollution and Hydrology. Water Resources Program, Department of Civil Engineering, Princeton, University.

(Table E-2 PC-Based Models References)

Davis, F.M., W.G. Knisel, and R.A. Leonard. 1990. GLEAMS User Manual, Version 1.8.55. Lab. Note SEWRL-030290FMD. ARS-USDA, Tifton, GA.

Dean, J.D., P.S. Huyakorn, A.S. Donigian, Jr., K.A. Voos, R.W. Schanz, and R.F. Carsel. 1989. Risk of Unsaturated/Saturated Transport and Transformation of Chemical Concentrations (RUSTIC), Vol. 1: Theory and Code Verification; Vol. 2: User's Guide. EPA/600/3-89/048a&b. U.S. Environmental Protection Agency Environmental Research Laboratory, Athens, GA. [Includes PRZM, VADOFT, for vadose zone flow/transport, and SAFTMOD for saturated zone flow/transport]

Deutsch, C.V. and A.G. Journel. 1993. GSLIB: Geostatistical Software Library and User's Guide. Oxford University Press, New York, NY, 340 pp. and 2 disks.

Domenico, P.A. and V.V. Palciauskas. 1982. Alternative Boundaries in Solid Waste Management. Ground Water 20(3):303-311. [VHS, available from IGWMC]

Dragun, J. 1989. The Soil Chemistry of Hazardous Materials. Hazardous Material Control Research Institute, Silver Spring, MD. [CDT nomograph]

Englund, E.J. and A.R. Sparks. 1988. Geo-EAS (Geostatistical Environmental Assessment Software) User's Guide. EPA/600/4-88/033a (Guide: NTIS PB89-151252, Software: PB89-151245).

Englund, E. and A. Sparks. 1991. GEO-EAS 1.2.1 User's Guide. EPA/600/8-91/008. Available from U.S. EPA Environmental Monitoring Systems Laboratory, Las Vegas, NV.

Franz, T. and N. Guiguer. 1990. FLOWPATH, Version 4, Steady-State Two-Dimensional Horizontal Aquifer Simulation Model. Waterloo Hydrogeologic Software, Waterloo, Ontario.

Freeze, R.A., J. Massmann, L. Smith, T. Sperling, and B. James. 1992. Hydrogeological Decision Analysis. National Ground Water Association, Dublin, OH, 72 pp. [Coupling of three models: (1) decision model based on a risk-cost-benefit objective function, (2) a simulation model for ground water flow and transport, and (3) an uncertainty model encompassing geological and parameter uncertainty]

Galya, D.P. 1987. A Horizontal Plane Source Model for Ground-Water Transport. Ground Water 25(6):733-739. [HPS analytical chemical transport model]

General Sciences Corporation. 1989. Personal Computer Version of the Graphical Exposure Modeling System-User's Guide. Prepared for USEPA/OTS Contract #68024281. [PCGEMS/SESOIL]

Guswa, J.H., W.J. Lyman, A.S. Donigian, Jr., T.Y.R. Lo, and E.W. Shanahan. 1987. Groundwater Contamination and Emergency Response Guide. Noyes Publications, Park Ridge, NJ. [Rapid Assessment Nomograph; first edition published 1984]

Hostetler, C.J., R.L. Erikson, and D. Rai. 1988. The Fossil Fuel Combustion Waste Leaching (FOWL) Code: Version 1. User's Manual. EPRI EA-5742-CCM, Electric Power Research Institute, Palo Alto, CA.

(Table E-2 PC-Based Models References)

International Ground Water Modeling Center (IGWMC). 1993. IGWMC Software Catalog. Golden, CO, 40 pp. [More than 60 models, most costing less than $100. The following software identified in Table E-2 are available: AT123D, GEOEAS, GEOPACK, MOC, MOCDENSE, MODFLOW, PESTAN, PHREEQE, PLASM, RANDOW WALK, SUTRA, USGS-2D-FLOW, USGS-3D-FLOW, VIRALT, WATEQ4F, WHPA, CAPZONE, EPA-VHS, THWELLS, WALTON35]

Johnson, P.C. 1992. HyperVentilate Users Manual: A Software Guidance System Created for Vapor Extraction Applications. EPA/500-C-B-92-001. Available from Superintendent of Documents, P.O. Box 371954, Pittsburgh, PA 15250-7954.

Johnson, G.L., J.S. Ford, and L.E. Michalec. 1987. PC-QTRAK: An Automated Tracking System for Environmental Quality Assurance Activities. EPA/600/D-87/138, U.S. EPA Air and Energy Engineering Laboratory, Research Triangle Park, NC.

Kent, D.C., W.A. Pettyjohn, F.E. Witz, and T.A. Prickett. 1982. Methods for Prediction of Leachate Plume Migration. In: Proc. 2nd Nat. Symp of Aquifer Restoration and Ground Water Monitoring, National Water Well Association, Dublin, OH, pp. 246-263. [Wilson-Miller Nomograph]

Klee, A.J. 1990. MOUSE (Modular Oriented Uncertainty SystEm): A Computerized Uncertainty Analysis System. EPA/600/8-89/102 (NTIS PB90-172560; diskette PB90-501370).

Klee, A.J. 1992. AutoMOUSE: An Improvement to the MOUSE Computerized Uncertainty Analysis System Operational Manual. EPA/600/R-92/145 (NTIS PB93-500007; diskette PB93-100113).

Knisel, W. (ed.). 1980. CREAMS: A Field-Scale Model for Chemicals, Runoff, and Erosion from Agricultural Management Systems. Conservation Research Report No. 26, U.S. Department of Agriculture, Washington, DC, 643 pp.

Kollig, H.P., K.J. Hamrick, and B.E. Kitchens. 1991. FATE, The Environmental Fate Constants Information System Database. EPA/600/3-91/045 (NTIS PB91-216192). [Online database]

Konikow, L.F. and J.D. Bredehoeft. 1978. Computer Model of Two-Dimensional Solute Transport and Dispersion in Ground Water. U.S. Geological Survey Techniques of Water-Resources Investigation TWRI 7-C2, 90 pp. [MOC]

Kruger, C.A. and J.G. Morse. 1992. Decision-Support Software for Soil Vapor Extraction Technology Application: HyperVentilate. EPA/600/R-93/028 (NTIS PB93-134880, diskettes PB93-5026640).

Leavesley, G.H, R.W. Lichty, B.M. Troutman, and L.G. Saindon. 1983. Precipitation-Runoff Modeling System: User's Manual. U.S. Geological Survey Water-Resources Investigations 83-4238, 207 pp.

Lin, P.C.L. 1986. User-Friendly IBM PC Computer Programs for Solving Sampling and Statistical Problems. EPA/600/4-86/023 (NTIS PB86-203783).

(Table E-2 PC-Based Models References)

Lin, P.C.L. 1993. Updated User-Friendly Computer Programs for Solving Sampling and Statistical Problems. EPA/600/C-93/002 (NTIS PB93-505907).

McDonald, M.G. and A.W. Harbaugh. 1988. A Modular Three-Dimensional Finite-Difference Ground-Water Flow Model. U.S. Geological Survey Techniques of Water Resource Investigations TWRI 6-A1, 575 pp. [MODFLOW, may also be cited with a 1983 or 1984 date as Open File Report 83-875]

McKone, T.E., L.B. Gratt, M.J. Lyon, and B.W. Perry. 1987. Geotox User's Guide and Supplement. Lawrence Livermore National Laboratory/U.S. Army Medical Research and Development Command (Available from NTIS; request Geotox Multimedia Compartment Model User's Guide Project Order Number 83PP3818). [Fate and transport in air, water, soil and sediments]

Newell, C.J. et al. 1990. OASIS: Parameter Estimation System for Aquifer Restoration Models—User's Manual Version 2.0. EPA/600/8-90/039 (NTIS PB90-181314/AS). [Apple Macintosh only]*

Nofziger, D.L. and A.G. Hornsby. 1987. Chemical Movement in Layered Soils: User's Manual. Florida Cooperative Extension Service Circular 78, Institute of Food and Agricultural Science, University of Florida, Gainsville, FL. [CMLS]

Nofziger, D.L. and J.R. Williams. 1988. Interactive Simulation of the Fate of Hazardous Chemicals During Land Treatment of Oily Wastes: RITZ User's Guide. EPA/600/8-88/001 (NTIS PB88-195540)*

Nofziger, D.L., K. Rajender, S.K. Nayudu, and P.Y. Su. 1989. CHEMFLO: A One-Dimensional Water and Chemical Movement in Unsaturated Soils. EPA/600/8-89/076.*

Park, N.-S., T.N. Blandford, and P.S. Huyakorn. 1992. VIRALT 2.0: A Modular Semi-Analytical and Numerical Model to Simulating Viral Transport in Ground Water. Available from IGWMC.

Parkhurst, D.L., D.C. Thorstensen and L.N. Plummer. 1980. PHREEQE - A Computer Program for Geochemical Calculation. U.S. Geological Survey Water Resources Investigation 80-96 (NTIS PB81-167801).

Pennsylvania Bureau of Waste Management. 1989. RAFT-User's Manual for Risk Assessment/Fate and Transport (RAFT) Modeling System. Scientific Services Section, Offices of Special Investigations, Pennsylvania Department of Natural Resources, 18th Floor Fulton Building, Harrisburg, PA 17105. [Includes 7 fate and transport Lotus 1-2-3 spreadsheet models]

Perrier, E.R. and A.C. Gibson. 1982. Hydrologic Simulation of Solid Waste Disposal Sites. EPA/SW-868, Revised Edition. [HSSWDS User's Manual].

Pollack, D.W. 1989. Documentation of Computer Programs to Compute and Display Pathlines Using Results from the U.S. Geological Survey Modular Three-Dimensional Finite-Difference Ground-Water Flow Model. U.S. Geological Survey Open File Report 89-381, 188 pp. [MODPATH]

PC-Based Software

Prickett and Associates, Inc. 1984. Selected Numerical Flow and Mass Transport Groundwater Models for the IBM-PC Micro Computer. Thomas A. Prickett and Associates, Inc., Urbana, IL. [Theis Well Field Model, Analytical Random Walk, PLASM, Discrete Random Walk Mass Transport Model]

Prickett, T.A. and C.E. Lonnquist. 1971. Selected Digital Computer Techniques for Ground-Water Resource Evaluation. Illinois State Water Survey Bulletin 55, Champaign, IL, 66 pp. [PLASM]

Rifai, H.S., P.B. Bedient, R.C. Borden, and J.F. Haasbeek. 1988. BIOPLUME II—Computer Model of Two-Dimensional Contaminant Transport Under the Influence of Oxygen Limited Biodegradation in Ground Water (User's Manual—Version 1.0). EPA/600/8-88/093 (NTIS PB89-151120). [Manual available from CSMoS* is dated 1987]

Rockware Scientific Software. 1993. The 1993 Scientific Software Catalog. Rockware Scientific Software, 4251 Kipling St., Suite 595, Wheat Ridge, CO 80033; 800/775-6745. [**Vadose/Ground Water Flow/Transport:** AqModel, AQTESOLV, FLOWCAD, FLOWPATH, HydroPal, MOC, MOD2DFD, MODFLOW, MODPATH, QUICKFLOW, RessqM, SUTRA, TWODAN; **Vapor/Multiphase Liquid:** AIRFLOW]

Ross, P.J. 1990. SWIM - A Simulation Model for Soil Water Infiltration and Movement. SWIM Project, CSIRO, P.M.M., P.O. Aitkenvale, Qld 4814, Australia.

Salhotra, A.M., P. Mineart, S. Sharp-Hansen, and T.L. Allison. 1993. MULTIMED, the Multimedia Exposure Assessment Model for Evaluating the Land Disposal of Wastes--Model Theory. EPA/600/R-93/081 (NTIS PB93-186252).

Scientific Software Group. Environmental, Engineering and Water Resources Software & Publications, 1993-1994. Scientific Software Group, P.O. Box 23041, Washington, DC, 20026-34041; 703/620-6793. [**Vadose/Ground Water Flow/Transport:** AQUA, AQUIFEM, BIO1D, CTRAN/W, INTERSAT/INTERTRANS, FLOWCAD, FLOWPATH, FLONET, FLOTHRU, FEMSEEP, MOC, MOCDENSE, MODFLOW, MODPATH, SUTRA, SWIFT/486, SWIM; **Vapor/Multiphase Liquid:** AIRFLOW, ARMOS, MOTRANS, VENTING, SPILLVOL, SPILLCAD]

Schroeder, P.R, J.M. Morgan, T.M. Walski, and A.C. Gibson. 1984a. Hydrologic Evaluation of Landfill Performance (HELP) Model: Vol. I. User's Guide for Version 1. EPA/530/SW-84/009 (NTIS PB85-100840).

Schroeder, P.R, A.C. Gibson, and M.D. Smolen. 1984b. Hydrologic Evaluation of Landfill Performance (HELP) Model: Vol. II. Documentation of Version 1. EPA/530/SW-84/010 (NTIS PB85-100832).

Shafer, J.M. 1987. GWPATH: Interactive Ground-Water Flow Path Analysis. Illinois State Water Survey Bulletin 69, 42 pp.

Shafer, J.M. 1990. GWPATH—Version 4.0. John Shafer, 321 Lake Front Drive, Columbia, SC 29212.

(Table E-2 PC-Based Models References)

Sharp-Hansen, S., C. Travers, P. Hummel, and T. Allison. 1992. A Subtitle D Landfill Application Manual for the Multimedia Exposure Assessment Model (MULTIMED). EPA/600/R-93/082 (NTIS PB93-185536).

Shields, W.J., E.W. Strecker, J.D. Dean, and S.M. Brown. 1987. Chemical Spill Uncertainty Analysis. RP 2634-1, Electric Power Research Institute, Palo Alto, CA. [MCPOSSM- Monte Carlo PCB Onsite Spill Model]

Simon, A.W., J.A. Borsak, S.A. Paulson, B.A. Deason, and R.A. Loverio. 1991. Computer-Aided Data Review and Evaluation: CADRE CLP Organic User's Guide. U.S. EPA, EMSL, Las Vegas, NV.

Sims, R.C., J.L. Sims and S.G. Hansen. 1991. STF: Soil Transport and Fate Database 2.0 and Model Management System. U.S. EPA R.S. Kerr Environmental Research Laboratory, Ada, OK.* [Include VIP and RITZ]

Stevens, D.K, W.J. Grenney, and Z. Zan. 1993. VIP: A Model for the Evaluation of Hazardous Substances in Soil, Version 3.0. Civil and Environmental Engineering, Utah State University, Logan, UT 84322-4110.*

Strack, O.D.L. 1989. SLAEM Users Manual. Strack Consulting, North Oaks, MN.

Strack, O.D.L. and H.M. Haitjema. In press. WhAEM Model for Wellhead Protection. [Analytic element method, software currently being beta tested for EPA Ada Laboratory]

Summers, K.V., S.A. Gherini, M.M. Lang, M.J. Ungs, and K.J. Wilkinson. 1989. MYGRT Code Version 2.0: An IBM Code for Simulating Migration of Organic and Inorganic Chemicals in Groundwater. EPRI EN-6531, Electric Power Research Institute, Palo Alto, CA.

Trescott, P.C. 1975. Documentation of Finite difference Model for Simulation of Three-Dimensional Ground Water Flow. U.S. Geological Survey Open-File Report 75-438. [USGS-3D-FLOW]

Trescott, P.C., G.F. Pinder, and S.P. Larson. 1976. Finite-Difference Model for Aquifer Simulation in Two Dimensions with Results of Numerical Experiments. U.S. Geological Survey Techniques of Water Resource Investigations TWRI 7-C1, 116 pp. [USGS-2D]

U.S. Environmental Protection Agency (EPA). 1985. 50 Federal Register 48886, November 27, 1985. [EPA-VHS, available from IGWMC]

U.S. Environmental Protection Agency (EPA). 1989. Statistical Analysis of Ground-Water Monitoring Data at RCRA Facilities, Interim Final Guidance. EPA/530/SW-89/026 (NTIS PB89-151047) plus September 1991 Addendum. [Incorporated into GRITS/STAT]

U.S. Environmental Protection Agency (EPA). 1991a. ASSESS 1.0 User's Guide. EPA/600/8-91/001. Environmental Monitoring Systems Laboratory, Las Vegas, NV.

(Table E-2 PC-Based Models References)

U.S. Environmental Protection Agency (EPA). 1991b. PA-Score Software; User's Manual and Tutorial: Version 1.0. Manual only: NTIS PB92-963302, 76 pp.; manual and diskette: NTIS PB92-500032. [HRS]

U.S. Environmental Protection Agency (EPA). 1992. User Documentation: A Ground Water Information Tracking System with Statistical Analysis Capability GRITS/STAT v4.2 EPA/625/11-91/002. Available from CERI.

U.S. Environmental Protection Agency (EPA). 1993. A Manual of Instructional Problems for the U.S.G.S. MODFLOW Model. EPA/600/R-93/010.

van der Heijde, P.K.M. 1987. THWELLS: A Basic Program to Calculate Head Drawdown or Buildup Caused by Multiple Wells in an Isotropic, Heterogeneous, Nonleaky, Confined Aquifer. IGWMC-PLUTO 6022. International Ground Water Modeling Center, Butler University, Indianapolis, IN, 82 pp.*

van Ee, J.J., L.J. Blume, and T.H. Starks. 1990. A Rationale for the Assessment of Errors in the Sampling of Soils. EPA/600/4-90/013. Environmental Monitoring Systems Laboratory, Las Vegas, NV. [Basis for ASSESS]

van Genuchten, M.Th. and W.J. Alves. 1982. Analytical Solutions of the One-Dimensional Convective-Dispersive Solute Transport Equation. U.S. Department of Agriculture Technical Bulletin 1661, 149 pp.

van Genuchten, M.Th., F.J. Leij, and S.R. Yates. 1991. The RETC Code for Quantifying the Hydraulic Functions of Unsaturated Soils. EPA/600/2-91/065.*

Varadhan, R. and J.A. Johnson. 1992. PESTAN: Pesticide Analytical Model Version 4.0. Center for Subsurface Modeling Support, R.S. Kerr Environmental Research Laboratory, Ada, OK.*

Walton, W.C. 1962. Selected Analytical Methods for Well and Aquifer Evaluation. Illinois State Geological Survey Bulletin 49, 81 pp.

Walton, W.C. 1984a. Handbook of Analytical Ground Water Models. GWMI 84-06. International Ground Water Modeling Center, Holcomb Research Institute, Butler University, Indianapolis, IN.*

Walton, W.C. 1984b. 35 Basic Groundwater Model Programs for Desktop Microcomputers. GWMI 84-06/4. International Ground Water Modeling Center, Butler University, Indianapolis, IN.* [WALTON35; diskette with analytical and simple numerical programs to analyze flow and transport of solutes in confined, leaky, or water table aquifers with simple geometry]

Walton, W.C. 1989a. Numerical Groundwater Modeling: Flow and Contaminant Migration. Lewis Pubishers, Chelsea MI, 272 pp. [Book and disks cover modified version of the Illinois State Water Survey's numerical flow (PLASM) and transport (random walk) models]

Walton, W.C. 1989b. Analytical Groundwater Modeling: Flow and Contaminant Migration. Lewis Publishers, Chelsea MI, 173 pp. [Includes four analytical microcomputer programs on 2 disks: WELFUN, WELFLO, CONMIG, GWGRAF]

Walton, W.C. 1992. Groundwater Modeling Utilities. Lewis Pubishers, Chelsea, MI, 656 pp, 2 5-1/4 diskettes. [MODFLOW, MODPATH/MODPATH-PLOT, MOC, SUTRA, INTERSAT/INTERTRANS]

Watershed Research, Inc. 1988. Hydropal 1 + 2 Iterative Hydrogeologic Applications. White Bear Lake, MN. [Hydropal Slug, Random Walk]

WellWare. 1993. RessqM-DOS. Available from Rockware, Scientific Software, Wheat Ridge, CO [Modified version of RESSQ model (Javendel et al., 1984)]

Yates, S.R. and M.V. Yates. 1990. Geostatistics for Waste Management: A User's Manual for the GEOPACK (Version 1.0) Geostatistical Software System. EPA/600/8-90/004 (NTIS PB90-186420/AS).*

Yeh, G.T. 1981. AT123D: Analytical Transient One-, Two-, and Three-Dimensional Simulation of Waste Transport in the Aquifer System. Environmental Sciences Division Publ. No. 1439, Oak Ridge National Laboratory, Oak Ridge, TN.

Zheng, C. 1990. MT3D: A Modular Three-Dimensional Transport Model for Simulation of Advection, Dispersion and Chemical Reactions of Contaminants in Groundwater Systems. Center for Subsurface Modeling Support, R.S. Kerr Environmental Research Laboratory, Ada, OK.* [Used with MODFLOW]

Zheng, C. 1992a. MT3D: A Modular Three-Dimensional Transport Model, Version 1.5- Documentation and Users Guide, Second Revision 3/15/91. S.S. Papadopulos and Associates, Bethesda, MD.

Zheng, C. 1992b. PATH3D: A Ground Water Path and Trend Simulator, Version 3.2. S.S. Papadopulos and Associates, Bethesda, MD.

Zheng, C., K.R. Bradbury, and M.P. Anderson. 1992. A Computer Model for Calculation of Groundwater Paths and Travel Times in Transient Three-Dimensional Flows. Wisconsin Geological and Natural History Survey Information Circular No. 70. [PATH3D]

* Available from U.S. EPA Center for Subsurface Modeling Support, P.O. Box 1198, Ada Oklahoma, 74820; 505-332-8800.

(Table E-2 PC-Based Models References)

APPENDIX F

MASTER LIST OF FIGURES AND TABLES WITH CREDITS

This appendix provides a master list of figures and tables included in Chapters 1 through 14 in this handbook, and includes the following information:

- The source from which it was obtained. If the figure or table is derived from another source, the original source is also given (i.e., Mercer and Spalding, 1991b, after Portland Cement Association, 1973).

- The page in the handbook where the figure or table is located.

- Full reference citations for figures and tables are given at the end of the list of tables. U.S. EPA publications noted as "Available from CERI" can be obtained from the U.S. EPA Center for Environmental Research Information in Cincinnati, OH (513/569-7562).

Lists of appendix figures and tables are given at the beginning of each appendix, and reference citations at the end of the appendix. Also, the beginning of Appendix D lists and gives the page number of worksheets and checklists that are contained in that appendix.

LIST OF FIGURES   Page

1-1	Particle-size limits of different U.S. textural classification systems (Mercer and Spalding, 1991b, after Portland Cement Association, 1973)	8
1-2	Guide for USDA soil textural classification (SCS, 1971)	9
1-3	Areal extent of glacial deposits in the United States (Heath, 1984)	14
1-4	Block diagram of an anticline and syncline (U.S. EPA, 1987a)	17
1-5	Cross sections of normal and reverse faults and a graben; plan view of a lateral fault (U.S. EPA, 1987a)	19
1-6	Sample fence diagram construction (Mercer and Spalding, 1991b, after Compton, 1962)	20
1-7	Patterns of soil orders and suborders of the United States (Birkeland, 1984)	24
1-8	Geomorphic and hillslope components (Mausbach and Nielsen, 1991, after Ruhe and Walker, 1968)	30
1-9	Diagram of a karst aquifer showing seasonal artesian conditions (Walker, 1956)	31
1-10	Distribution of karst areas in relation to carbonate and sulphate rocks in the United States (Davies and LeGrand, 1972)	33
1-11	Major ground-water regions of the United States (Heath, 1984)	34
1-12	Dissolved solids concentrations in ground water used for drinking in the United States (U.S. EPA, 1987a, after Pettyjohn et al., 1979)	38
2-1	The hydrologic cycle (Muldoon and Payton, 1993)	61
2-2	Mean annual precipitation (Viessman et al., 1972, after USDA Soil Conservation Service)	62
2-3	Hydrograph of Brandywine Creek, Chadd's Ford, Pennsylvania, 1952-1953 (U.S. EPA, 1987a, after Olmsted and Hely, 1962)	67

2-4   Drainage patterns: (a) six basic patterns; (b) drainage density variations (Kolm, 1993, after Way, 1973) 68
2-5   Relation between water table and stream type (U.S. EPA, 1987a) 71
2-6   Movement of water into and out of bank storage along a stream in Indiana (Daniels et al., 1970) 72
2-7   Diagram of the relationship between hydraulic head, H, pressure head, h and gravitational head, Z (Mercer and Spalding, 1991c) 74
2-8   Classification of subsurface water (Sara, 1994) 75
2-9   Generalized relationship between matric suction and hydraulic conductivity for a sandy and a clay soil (adapted from Brady, 1974) 77
2-10 Cross-sectional diagram showing the water level as measured by piezometers located at various depths (Mills et al., 1985) 78
2-11 Confined, unconfined, and perched water in a simple stratigraphic section of sandstone and shale (Davis and DeWiest, 1966) 80
2-12 Heterogeneity and anisotropy: (a) four possible combinations (Physical Properties and Principles Chapter 2 in GROUNDWATER, Freeze/Cherry © 1979, reprinted by permission of Prentice-Hall, Engelewood Cliffs, NJ); (b) three types of aquifer heterogeneity (adapted from Fetter, 1994) 81
2-13 Types of aquifer boundary conditions (Struckmeier et al., 1986, after Castany and Margat, 1977) 87
2-14 Using Darcy's Law to estimate underflow in an aquifer (U.S. EPA, 1987a) 89
2-15 Using Darcy's Law to calculate the quantity of leakage from one aquifer to another (U.S. EPA, 1987a) 91
2-16 Ground water velocity calculations (U.S. EPA, 1987a) 93
2-17 Cones of depression in unconfined and confined aquifers (Heath, 1983) 94
2-18 Geographic index to U.S. Geological Survey Professional Paper 813 ground-water reports 95
2-19 Geographic index to U.S. Geological Survey regional aquifer system studies (Sun and Weeks, 1991) 96

3-1   pe-pH diagram for the $Fe-H_2O$ system (Palmer and Fish, 1991) 127
3-2   Distribution of molecular and ionic species of divalent cadmium, mercury and lead at different pH values (Boulding, 1990, after Hahne and Kroontje, 1973) .. 140
3-3   ph-Eh diagram showing the ranges of various aquatic environments (Johnson et al., 1989) 142
3-4   Horizontal gradients in uncontaminated oxidation-reduction conditions (Boulding and Barcelona, 1991b, after Champ et al., 1979) 143
3-5   Anaerobic food web for aquatic microbial ecosystems (Oremland, 1988) 151
3-6   Geochemical zones with subsurface injection of concentrated toxic wastes (Boulding, 1990, after Leenheer and Malcolm, 1973) 154

4-1   Sources of ground-water contamination (U.S. EPA, 1987a, after Geraghty and Miller, 1985) 174
4-2   Major contaminants at Superfund Sites (Palmer and Fish, 1991) 180
4-3   Plume of leachate migrating from a sanitary landfill on a sandy aquifer using contours of chloride concentration (U.S. EPA, 1987a, after Freeze and Cherry, 1979) 181
4-4   Ground-water contamination from surface water: (a) floodwater entering improperly cased well (Deutsch, 1963); (b) induced flow from pumping (Deutsch, 1965)** 182
4-5   Vertical movement of contaminants along an old, abandoned, or improperly constructed well (U.S. EPA, 1977, after Deutsch, 1961) 183

(List of Figures, continued)

Master Figure and Table Lists

4-6	Movement of a concentration front by advection only: (a) continuous source; (b) slug (U.S. EPA, 1987a)	185
4-7	Effect of dispersion and retardation on movement of a contaminant front from a continuous source: (a) relative concentrations compared to advection only; (b) plan view of plume (U.S. EPA, 1987a)	186
4-8	Effect of dispersion and retardation on movement of a contaminant slug: (a) dispersion over time; (b) plan view of plume from intermittent point source; (c) sorption and biodegradation (U.S. EPA, 1987a)	187
4-9	Movement of LNAPLs into the subsurface (Palmer and Johnson, 1989b)	189
4-10	Movement of DNAPLs into the subsurface (Palmer and Johnson, 1989b)	190
4-11	The three filtration mechanisms that limit particle migration through porous media (Palmer and Johnson, 1989a, after McDowell-Boyer et al., 1986)	193
4-12	Adsorption isotherms: (a) Langmuir; (b) Freundlich (Palmer and Johnson, 1991)	197
4-13	Effect of differences in geology on shapes of contamination plumes (Miller, 1985)	199
4-14	Benzene and chloride appearance in a monitoring well (U.S. EPA, 1987a, after Geraghty and Miller, 1985)	200
4-15	Changes in plumes and factors causing the changes (U.S. EPA, 1987a, after U.S. EPA, 1977 and LeGrand, 1965)	201
4-16	Map view of various types of contaminant plumes (LeGrand, 1965)	202
5-1	Site characterization phases (Mercer and Spalding, 1991a, after Bouwer et al., 1988)	241
5-2	General relationship between site characterization costs and cleanup costs as a function of the characterization approach (Keely, 1987)	244
5-3	Steps in ground-water sampling and sources of error (Barcelona et al., 1985)	254
5-4	Steps in water sample analysis and sources of error (Barcelona et al., 1985)	255
5-5	Some two-dimensional probability sampling designs for sampling over space (Boulding and Barcelona, 1991a, after Gilbert, 1987)	258
5-6	Sampling frequency nomograph (Barcelona et al., 1985)	264
5-7	Classification of (a) deterministic and (b) random data (Boulding and Barcelona, 1991a, after Bendat and Piersol, 1986)	270
5-8	Shots on a target analogy for illustrating influence of bias and precision on accuracy (Boulding and Barcelona, 1991a, after Jessen, 1978)	272
5-9	Sources of error involved in ground-water monitoring programs contributing to total variance (Barcelona et al., 1983)	273
5-10	The ACS recommended definition of limit of detection (Boulding and Barcelona, 1991a, after ACS Comm. on Environmental Improvements, 1980)	277
5-11	The "ideal" shape for a semivariogram-spherical model (Boulding and Barcelona, 1991a, after Clark, 1979)	283
5-12	A semivariogram of lead samples taken systematically on a 230-m (750-foot) grid (Boulding and Barcelona, 1991a, after Flatman, 1986)	284
6-1	The electromagnetic spectrum: the customary divisions and portions used for geophysical measurements (Boulding, 1993a, after Erdélyi and Gálfi, 1988)	304
6-2	Cost comparison curve for hazardous waste site investigations: geophysics vs monitoring wells only (Benson et al. 1984)	307
6-3	Discrete sampling vs. continuous geophysical measurements (Boulding, 1993a, after Benson et al., 1984)	308
6-4	Factors and phenomena influencing the radiation of electromagnetic waves (Boulding, 1993a, after Erdélyi and Gálfi, 1988)	310

(List of Figures, continued)

6-5	Electromagnetic induction principle of operation (Boulding, 1993a, after Benson et al., 1984)	315
6-6	Resistivity soundings showing extent of a landfill contaminant plume (Boulding, 1993a, after Benson et al., 1984)	317
6-7	Seismic refraction: (a) Field layout of a 12-channel seismograph; (b) Steps in processing and interpretation of seismic refraction data (Boulding, 1993a, after Benson et al., 1984)	319
6-8	Temperature profiles of a discontinuous sand and gravel aquifer within fine-grained alluvium (Boulding, 1993b, after Cartwright, 1968)	322
6-9	Typical response of a suite of hypothetical geophysical well logs to a sequence of sedimentary rocks (Boulding, 1993a, after Keys, 1990)	323
6-10	The flow of current at typical bed contacts and the resulting spontaneous potential curve (Boulding, 1993b, after Keys, 1990)	324
7-1	Schematic representation of three types of flux measurements using the column-crust method (Boulding, 1993b, after Bouma, 1982)	374
7-2	The effect of initial water content of soil on infiltration rates (Everett et al., 1983)	375
7-3	The generalized direction of ground-water movement can be determined by means of the water level in three wells of similar depth (U.S. EPA, 1987a, after Heath and Trainer, 1981)	380
7-4	Alternative procedure for determination of equipotential contour and direction of ground-water flow in homogeneous, isotropic aquifer (Heath, 1983)	381
7-5	Effect of fracture anisotropy on the orientation of the zone of contribution to a pumping well (Bradbury et al., 1991)	382
7-6	Illustration of flow net analysis for anisotropic hydraulic conductivity in an earth dam: (a) true anisotropic section with $K_x = 9K_z$; (b) transformed isotropic section with $K_x = K_z$ (Todd, 1980)	383
7-7	Steps in the determination of ground-water flow direction in an anisotropic aquifer (Fetter, 1981)**	384
7-8	Effect of well level measurements in recharge and discharge areas: (a) incorrect contours; (b) correct contours after reinterpretation (Saines, 1981)**	385
7-9	Common errors in contouring water table maps: (a) topographic depression occupied by lakes; (b) fault zones (Davis and DeWiest, 1966)	386
7-10	Error in mapping potentiometric surface due to mixing of two confined aquifers with different pressures (Davis and DeWiest, 1966)	387
7-11	Divergence from predicted direction of ground water resulting from aquifer heterogeneity (Davis et al., 1985)	388
7-12	Decision tree for selection of aquifer test methods (ASTM D4043-91)	389
7-13	Major and significant minor confined aquifer of the United States (Kreitler and Senger, 1991)	402
7-14	Major areas of unconfined fracture rock aquifers in the United States (Bradbury et al., 1991)	405
7-15	Mapping of subsurface conduit using self-potential method (Karous and Mareš, 1988)	408
7-16	Azimuthal seismic survey to characterize direction of subsurface rock fractures (Karous and Mareš, 1988)	409
7-17	Pumping-test response indicators of fracture/conduit flow: (a) discharge drawdown plots; (b) time drawdown curves; (c) areal drawdown distribution (Bradbury et al., 1991)	410
7-18	Geometrical relationships and hydraulic conductivities at different scales in karst systems (Sauter, 1992)***	411

(List of Figures, continued)

Master Figure and Table Lists

7-19 Measurement scales and average velocities of different measurement methods (modified after Quinlan et al., 1992 and Sauter 1992) 413

8-1 Common configurations for use of tracers to measure hydrogeologic parameters using natural gradient flow (Boulding, 1991, after Davis et al., 1985) 441
8-2 Common configurations for use of tracers to measure hydrogeologic parameters using induced flow (Boulding, 1991, after Davis et al., 1985) 442
8-3 Common configurations for use of tracers to identify contaminant sources using natural gradient flow (Boulding, 1991, after Davis et al., 1985) 444
8-4 The effect of pH on Rhodamine WT (Davis et al., 1985, after Smart and Laidlaw, 1977) 454
8-5 Differentiation of methane leak (bedrock) from natural shallow methane source (Davis et al., 1985, after Coleman et al., 1977) 462
8-6 Results of field test using a hot water tracer (Boulding, 1991, after Davis et al., 1985) 464
8-7 Results of tracer tests at the Sand Ridge State Forest, Illinois (Davis et al., 1985, after Naymik and Sievers, 1983) 476
8-8 Tracer concentration reported as a ratio (Boulding, 1991, after Davis et al., 1985) 477
8-9 Incomplete tracer recovery due to partial penetration of aquifer (Boulding, 1991, after Davis et al., 1985) 478
8-10 Breakthrough curves for conservative and nonconservative tracers (Boulding, 1991, after Davis et al., 1985) 479

9-1 Generic monitoring design for existing hazardous waste landfill (Sara, 1994, after Everett et al., 1983) 495
9-2 Example permanent installation of BAT filter tip probe (Torstensson, 1984)** . 499
9-3 Commonly used monitoring well screen types (Boulding, 1993b, after Nielsen and Schalla, 1991) 506
9-4 Potential pathways for fluid movement in the casing-borehole annulus (Boulding, 1993b, after Aller et al., 1991) 510
9-5 Soil-gas concentrations under a variety of hydrogeologic conditions (Boulding and Barcelona, 1991c, after Marrin and Kerfoot, 1988) 517
9-6 Field sampling glove box (Boulding, 1993b, after Leach et al., 1988) 521
9-7 General flow diagram of ground-water sampling steps (Boulding and Barcelona, 1991c) 522

10-1 Three dimensional grid representing (a) representing a complex geologic setting (b) with pumping wells downgradient from contaminant source (Keely, 1987) 572
10-2 Comparison of (a) finite different and (b) finite element grid configurations for modeling the same well field (Mercer and Faust, 1981)** 573
10-3 Time-of-travel contours in a dolomite aquifer based on potentiometric surface map (Bradbury et al., 1991) 579
10-4 Time-of-travel contours in a dolomite aquifer based on numerical modeling of high-conductivity zone (Bradbury et al., 1991) 580
10-5 Decision tree for selection of ground-water flow code (U.S. EPA, 1988a) 583
10-6 Decision tree for selection of contaminant transport code (U.S. EPA, 1988a) .. 585
10-7 Overview of major geographic information system functions (OIRM, 1992) 595

11-1 Relationship between ZOI, ZOT, and ZOC in an unconfined aquifer with a sloping regional water table (U.S. EPA, 1987b) 627
11-2 WHPA delineation using cylinder method (Kreitler and Senger, 1991) 629

(List of Figures, continued)

11-3	WHPA delineation using simplified shapes method (U.S. EPA, 1993)	630
11-4	WHPAs at Sevastopol site, Door County, Wisconsin, Based on fixed radius, simplified shape and vulnerability mapping (Bradbury et al., 1991)	635
11-5	Radius of outer management zone based on pumping rate for crystalline aquifers, Piedmont and Blue Ridge (Georgia Department of Natural Resources, 1992)	639
11-6	Flow chart for selection of wellhead protection delineation methods (Boulding, 1994)	640
11-7	Land use/public-supply well pollution potential matrix (Noake, 1988)	648
11-8	Example diagram of exposure pathway assessment for a landfill (adapted from U.S. EPA, 1989a)	653
11-9	Wellhead protection land use/local regulatory techniques matrix (Noake, 1988)	660
12-1	Mass balance conceptual framework for the soil and ground-water subsurface environment at a contaminated site (Sims and Sims, 1991a)	697
12-2	Interphase transfer potential of chemicals in the subsurface (Sims and Sims, 1991a)	698
12-3	Methodology using mass balance approach for integrating data collection activities at a contaminated site (Sims and Sims, 1991a)	700
12-4	Laboratory flask apparatus used for mass balance measurements (Sims and Sims, 1991a, after Park et al., 1990)	703
13-1	Typical components of a soil vacuum extraction system (Sims and Sims, 1991b, after Hutzler et al., 1990)	739
13-2	Soil vacuum extraction system: (a) gas extraction well; (b) gas monitoring well (Sims and Sims, 1991b)	740
13-3	Soil vacuum extraction efficiency based on total hydrocarbon vapors (Sims and Sims, 1991b)	742
13-4	Suitability of soil vacuum extraction related to soil texture (Sims and Sims, 1991b)	743
13-5	Volatile organic compound (VOC) rebound effect with SVE stop and restart (Sims and Sims, 1991b)	744
13-6	Effect of well spacing on total solute mass remaining in soil with vacuum extraction (Sims and Sims, 1991b)	744
13-7	Enhanced bioremediation of gasoline-contaminated soil using soil vacuum extraction, nutrient additions and moisture control (Sims and Sims, 1991b)	745
13-8	Sorption of chemicals to soil as a function of soil moisture content and partition coefficient Kd (Sims and Sims, 1991b, after Sims et al., 1986)	752
13-9	Schematic of soil flushing and recycle system (U.S. EPA, 1990)	753
14-1	General response actions and process options for ground water (U.S. EPA, 1988b)	763
14-2	Pump-and-treat system: (a) cross-sectional view; (b) plan view (U.S. EPA, 1985)	773
14-3	Product recovery using a water-table depression pump and a floating oil/water filter (Sims and Sims, 1991c, after Nyer, 1985)	774
14-4	Stabilization of aquifer contaminant concentration during pump-and-treat resulting from tailing effect (Keely, 1989)	775
14-5	Pulsed pumping removal of residual contaminants reduces volume of water pumped and increases contaminant concentration in pumped water (Keely, 1989)	776

(List of Figures, continued)

Master Figure and Table Lists

14-6	Use of drains to separate DNAPL from water where DNAPL rests on shallow bedrock; see text for explanation (Sims and Sims, 1991c, after Sale and Piontek, 1989)	778
14-7	Contaminated ground water treatment matrix (U.S. EPA, 1990)	779
14-8	Typical schematic for aerobic subsurface bioremediation (Suflita, 1989, after Lee et al., 1988)	793
14-9	Use of infiltration gallery for recirculation of water and nutrients for in situ bioremediation (Suflita, 1989, after Lee et al., 1988)	794

List of Tables   Page

1-1	Important Characteristics of Silicate Clay Minerals (Boulding, 1990)	6
1-2	Geologic Time Scale (update of U.S. EPA, 1987)	16
1-3	Index to Major References on Geology, Soils and Geomorphology	40
1-4	Index to Major References on Karst Geology, Geomorphology and Hydrology	54
2-1	Drainage Network Patterns as Indicators of Geologic/Hydrogeologic Systems (adapted from Kolm, 1993)	69
2-2	Summary of Mechanisms that Lead to Fluctuations in Ground Water Levels (Freeze and Cherry, 1979)	83
2-3	Factors and Natural Conditions Affecting Natural Ground Water Fluctuations (adapted from Brown et al., 1983)	84
2-4	Index to Major References on Surface and Vadose Zone Hydrology, Hydrogeology and Hydraulics	97
2-5	Index to Major References on Regional Hydrology and Ground Water in the United States	110
3-1	Characteristics of Chemical Processes that May Be Significant in the Subsurface (Boulding, 1990)	120
3-2	Significance of Chemical Processes in the Subsurface (Boulding, 1990)	124
3-3	Major Intermolecular Interactions Involved in Sorption in the Subsurface (Boulding, 1990)	126
3-4	Amenability of Organic Functional Groups to Hydrolysis (Boulding, 1990, after Guswa et al., 1984)	132
3-5	Redox Reactions in a Closed Ground Water System (Boulding, 1990, after Champ et al., 1979)	134
3-6	Relative Oxidation States of Organic Functional Groups (Boulding, 1990, after Valentine, 1986)	135
3-7	Susceptibility of Organic Compounds to Oxidation in Water (Boulding, 1990, after Mill, 1980)	135
3-8	Effects of pH on Subsurface Geochemical Processes and Other Environmental (Boulding, 1990)	138
3-9	Half-Lives of Different Forms of Organic Matter (Boulding, 1990, after Jenkinson and Raynor, 1977)	146
3-10	Redox Zones for Biotransformation of Organic Micropollutants (Boulding, 1990, after Bouwer and McCarty, 1984)	153
3-11	Biodegradable RCRA-Regulated Organic Compounds (U.S. EPA, 1985)	155
3-12	Index to Major References on Soil and Ground Water Geochemistry and Microbiology	157

(List of Figures, continued)

4-1	Sources of Ground Water Contamination (OTA, 1984)	175
4-2	Classification of Types and Sources of Soil Contamination in The Netherlands Based on a Sample of 100 Cases (Zoeteman, 1985)	178
4-3	Explanation of Contaminant Plumes Shown in Figure 4-16 (Adapted from LeGrand, 1965)	203
4-4	Index to Major References on Types and Sources of Contamination in Soil and Ground Water	206
4-5	Index to Major References on Contaminant Chemical Characteristics and Behavior in the Subsurface	219
5-1	Comparison of Approaches to Site Characterization Efforts (Adapted from Keely, 1987)	245
5-2	Sources and Types of Existing Data for Soil and Ground Water Quality Investigations (Adapted from Sisk, 1981)	247
5-3	Environmental Hotlines, Clearinghouses, and Electronic Databases (U.S. EPA, 1991 and other sources)	249
5-4	Summary of Sampling Designs and Conditions for their Use (Boulding and Barcelona 1991a, after Gilbert, 1987)	257
5-5	Chemical Constituents of Interest in Ground Water Monitoring (Boulding and Barcelona, 1991c)	260
5-6	Recommended Analytical Parameters for Detective Monitoring (Boulding and Barcelona, 1991c, after Barcelona et al., 1985)	261
5-7	Estimated Ranges of Sampling Frequency (in Months) to Maintain Information Loss at <10% for Selected Types of Chemical Parameters (Boulding and Barcelona, 1991c)	263
5-8	Recommended Sample Handling and Preservation Procedures for a Detective-Monitoring Program (Boulding and Barcelona, 1991c, after Scalf et al., 1981, and U.S. EPA, 1986)	265
5-9	Potential Contributions of Sampling Methods and Materials to Error in Ground Water Chemical Results (Boulding and Barcelona, 1991c, after Barcelona et al., 1988)	274
5-10	Regions of Analyte Measurement (Boulding and Barcelona 1991a, after ACS Committee on Environmental Improvement, 1980)	278
5-11	Effects of Censoring Analyte Signals at and Below the Limit of Detection (ASTM D4210, see Table A-14)	279
5-12	Conferences and Symposia with Papers Relevant to Subsurface Characterization and Monitoring (Boulding, 1993b with update)	285
5-13	Index to Major References on Existing Environmental Information and Data Management	288
5-14	Index to Major References on Statistics and Sampling Design	295
6-1	Use of Airborne Sensing Techniques in Hydrogeologic and Contaminated Site Studies (Boulding, 1993a)	309
6-2	Major Surface Geophysical Methods for Study of Subsurface Contamination (Boulding, 1993a)	313
6-3	Summary of Electrical and EM Borehole Logging Methods in Hydrogeologic Studies (Adapted from Boulding, 1993a)	325
6-4	Summary of Nuclear Borehole Logging Methods in Hydrogeologic Studies (Adapted from Boulding, 1993a)	327
6-5	Summary of Acoustic and Seismic Borehole Logging Methods in Hydrogeologic Studies (Adapted from Boulding, 1993a)	329
6-6	Summary of Miscellaneous Borehole Logging Methods in Hydrogeologic Studies (Adapted from Boulding, 1993)	330

(List of Tables, continued)

Master Figure and Table Lists

6-7	Summary of Borehole Log Applications (Boulding, 1993a)	332
6-8	Index to Major References on Remote Sensing and Surface Geophysics	338
6-9	Index to Major References on Borehole Geophysics	357
7-1	Types of Data Available on SCS Soil Series Description and Interpretation Sheets (Boulding, 1994)	395
7-2	Aquifer Characteristics Affecting Porosity, Specific Yield, and Hydraulic Conductivity (Boulding, 1994)	397
7-3	Summary of Methods for Characterizing Aquifer Heterogeneity (Boulding, 1994)	400
7-4	Indicators of Presence and Degree of Confinement (Adapted from Kreitler and Senger, 1991)	403
7-5	Index to Major References on Hydraulic Conductivity and Water Balance Methods	415
7-6	Index to Source References on Pump Test Analytical Solutions and Methods for Characterizing Anisotropic and Fracture-Rock Aquifers	426
8-1	List of Major Ground Water Tracers (Boulding, 1993b)	446
8-2	Gases of Potential Use as Tracers (Boulding, 1991, after Davis et al., 1985)	457
8-3	Commonly Used Radioactive Tracers for Ground Water Studies (Boulding, 1991, after Davis et al., 1985)	460
8-4	Comparison of Microbial Tracers (Boulding, 1991, after Keswick et al., 1982)	465
8-5	Index to Major References on Tracer Methods	480
9-1	Advantage and Disadvantages of Types of Monitoring Well Installations (Adapted from Boulding, 1993b)	497
9-2	Advantage and Disadvantages of Auger and Rotary Drilling Methods (Adapted from Boulding, 1993b)	501
9-3	Advantage and Disadvantages of Other Drilling Methods (Adapted from U.S. EPA, 1987a and Boulding, 1993b)	503
9-4	Advantage and Disadvantages of Monitoring Well Casing Materials (Adapted from Boulding, 1993b)	505
9-5	Advantage and Disadvantages of Filter Pack, Grouts and Seals (Adapted from Boulding, 1993b)	508
9-6	Advantage and Disadvantages of Well Development Methods (Adapted from Boulding, 1993b)	511
9-7	Characteristics of Contaminants in Relation to Soil Gas Surveying (Adapted from Marrin, 1987)	518
9-8	Suitability of Major Ground Water Sampling Devices for Different Ground Water Parameters (Boulding and Barcelona, 1991c, after Pohlmann and Hess, 1988)	523
9-9	Major Analytical Signals and Methods (Boulding, 1993b, after Skoog, 1985)	527
9-10	Index to Major Reference Sources on Sampling and Monitoring Methods	529
9-11	Index to Major References on Field and Laboratory Analytical Methods	549
10-1	Modeling Designed System-Alterations and Corrective Action (Adapted by van der Heijde et al., 1988, from Boutwell et al., 1985)	566
10-2	Advantages and Disadvantages of Analytical and Numerical Methods (Adapted from Boulding 1991b)	571
10-3	Advantages and Disadvantages of FDM and FEM Numerical Methods (Adapted from Boulding, 1991b)	574
10-4	Classification of Vadose Zone and Ground Water Flow and Transport Computer Codes (Adapted from Boulding, 1991b)	575

(List of Tables, continued)

10-5	Periodicals, Conferences, and Symposia with Papers Relevant to GIS (Boulding, 1994)	596
10-6	Index to Major References on Ground Water and Vadose Zone Flow and Contaminant Transport Modeling	602
10-7	Index to Major References on Geographic Information Systems (GIS)	616
11-1	Comparison of Major Methods for Delineating Wellhead Protection Areas (Boulding, 1994)	631
11-2	Relationship of WHPAs Based on Zone of Influence, Time of Travel, Zone of Travel, Zone of Contribution and Zone of Attenuation (Boulding, 1994)	634
11-3	Summary of Major Ground Water Vulnerability Mapping Methods (Boulding, 1994)	636
11-4	Generic Wellhead Protection Areas Proposed for Georgia (Georgia Department of Natural Resources (1992)	638
11-5	Zones for Wellhead Protection Areas in Idaho (Idaho Wellhead Protection Work Group, 1992)	639
11-6	Index to Development Documents for Effluent Limitations Guidelines for Selected Categories (U.S. EPA, 1987c)	645
11-7	Regulated Land Uses in Different Water Resource Protection Zones, Nantucket Island, Massachusetts (Horsley, 1990)***	651
11-8	Summary of Wellhead Protection Tools (U.S. EPA, 1993)	655
11-9	Potential Management Tools for Wellhead Protection (Born et al., 1987, U.S. EPA, 1989b)	658
11-10	Index to Major References on Ground Water Vulnerability Mapping and Chemical Hazard and Risk Assessment	661
11-11	Index to Major References on Pollution Prevention and Soil and Ground Water Protection Management	678
12-1	Subsurface-Based Waste Characterization (Sims and Sims, 1991a)	701
12-2	Data Needs for Major Stabilization and Remediation Methods (Bartenfelder et al., 1992)	702
12-3	Examples of Interim Measures for Corrective Action at Contaminated Sites (U.S. EPA, 1988c)	707
12-4	Matrix of Remedial Technology Categories for Specific Site Problems (U.S. EPA, 1985)	711
12-5	Treatment Technology Screening Guide for Different Contaminants (U.S. EPA, 1989c)	712
12-6	General Comparison of Laboratory Screening, Bench-Scale Testing, and Pilot-Scale Testing (Sims and Sims,,1991a, after U.S. EPA, 1989d)	713
12-7	Index to Major References on Soil and Ground Water Remediation Planning	718
13-1	Summary of Physical/Chemical Treatment Methods for Contaminated Soil (Adapted from Sims and Sims, 1991b)	733
13-2	Summary of Biological Treatment Methods for Contaminated Soil (Adapted from Sims and Sims, 1991b)	736
13-3	Summary of Fixation/Encapsulation methods for Contaminated Soil (Adapted from Sims and Sims, 1991b)	737
13-4	Summary of Thermal Treatment Methods for Contaminated Soil (Adapted from Sims and Sims, 1991b)	738
13-5	Comparative Vapor Pressures and Henry's Constants (Sims and Sims, 1991b)	740
13-6	Critical Factors for Microbial Activity (Sims and Sims, 1991b, after Huddleston et al., 1986, Paul and Clark, 1989, Rochkind et al., 1986, and Sims et al., 1984)	747

(List of Tables, continued)

13-7	Effect of Temperature on Degradation of PAHs in a Sandy Loam Soil (Sims and Sims, 1991b, after Coover and Sims, 1987, PACE, 1985, Sims, 1986, and Sims and Overcash, 1983)	748
13-8	Acclimation of Soil to Complex Fossil Fuel Waste (Sims and Sims, 1991b, after Sims, 1986)	749
13-9	Index to Major References on Soil Treatment and Remediation	757
14-1	Available Technologies for Ground Water Containment and Restoration	764
14-2	Advantages and Disadvantages of Commonly Used Source Control Technologies (Compiled from Knox et al., 1984, Nielsen, 1983, and Wagner et al., 1986)	769
14-3	Process Applicability Matrix for Hazardous Waste Leachate Treatment Methods (McArdle et al., 1987)	780
14-4	Summary of Landfill Leachate Treatment Process Capabilities (Pohland and Harper, 1986)	781
14-5	Advantages and Disadvantages of Physical Methods of Ground Water Treatment (Compiled from U.S. EPA, 1985, Rich and Cherry, 1987, and Wagner et al., 1986)	782
14-6	Advantages and Disadvantages of Chemical Methods of Ground Water Treatment (Compiled from U.S. EPA, 1985, Rich and Cherry, 1987, and Wagner et al., 1986)	786
14-7	Advantages and Disadvantages of Biological Methods of Ground Water Treatment (Compiled from U.S. EPA, 1985, Rich and Cherry, 1987, and Wagner et al., 1986)	788
14-8	Estimated Volumes of Water or Air Required to Renovate Subsurface Material Containing Hydrocarbons at Residual Concentrations (Adapted from U.S. EPA, 1987a)	795
14-9	Index to Major References on Ground Water Treatment and Remediation	796

Figure and Table References

ACS Committee on Environmental Improvement. 1980. Guidelines for Data Acquisition and Data Quality Evaluation in Environmental Chemistry. Analytical Chemistry 52:2242-2249.

Aller, L., et al. 1991. Handbook of Suggested Practices for the Design and Installation of Ground-Water Monitoring Wells. EPA/600/4-89/034, 221 pp. [Also published in 1989 by National Water Well Association, Dublin, OH in its NWWA/EPA series, 398 pp.]

Barcelona, M.J. and J.A. Helfrich. 1986. Well Construction and Purging Effects on Ground-Water Samples. Environ. Sci. Technol. 20:1179-1184.

Barcelona, M.J., J.P. Gibb, and R.A. Miller. 1983. A Guide to the Selection of Materials for Monitoring Well Construction and Ground-Water Sampling. ISWS Contract Report 327, Illinois State Water Survey, Champaign, IL, 78 pp.

Barcelona, M.J., J.A. Helfrich, E.E. Garske, and J.P. Gibb. 1984. A Laboratory Evaluation of Ground Water Sampling Mechanisms. Ground Water Monitoring Review 4(2):32-41.

Barcelona, M.J., J.P. Gibb, J.A. Helfrich, and E.E. Garske. 1985. Practical Guide for Ground-Water Sampling. EPA/600/2-85/104 (NTIS PB86-137304). [Also published as ISWS Contract Report 374, Illinois State Water Survey, Champaign, IL, 93 pp.]

(List of Tables, continued)

Barcelona, M.J., J.A. Helfrich, and E.E. Garske. 1988. Verification of Sampling Methods and Selection of Materials for Ground-Water Contamination Studies. In: Ground-Water Contamination: Field Methods, A.G. Collins and A.I. Johnson (eds.), ASTM STP 963, American Society for Testing and Materials, Philadelphia, PA, pp. 221-231.

Bartenfelder, D., R. Sims, H. Compton, W. Grube, L. Murdoch, and R. Dupont. 1992. RCRA Corrective Action Stabilization Technologies Proceedings. EPA/625/R-92-014, 77 pp. Available from CERI.*

Bendat, J.S. and A.G. Piersol. 1986. Random Data, Analysis and Measurement Procedures, 2nd ed. Wiley-Interscience, New York, NY.

Birkeland, P.W. 1984. Soils and Geomorphology. Oxford University Press, New York, NY, 372 pp.

Benson, R.C., R.A. Glaccum, and M.R. Noel. 1984. Geophysical Techniques for Sensing Buried Wastes and Waste Migration. EPA/600/7-84/064 (NTIS PB84-198449), 236 pp. Also published in NWWA/EPA series by National Water Well Association, Dublin, OH.

Born, S.M., D.A. Yanggen, and A. Zaporozec. 1987. A Guide to Groundwater Quality Planning and Management for Local Governments. Special Report 9. Wisconsin Geological and Natural History Survey, Madison, WI, 92 pp.

Boulding, J.R. 1990. Assessing the Geochemical Fate of Deep-Well-Injected Hazardous Waste: A Reference Guide. EPA/625/6-89/025a, 183 pp. Available from CERI.*

Boulding, J.R. 1991. Ground-Water Tracers. In: Handbook: Ground Water; Volume II: Methodology, Chapter 4. Available from CERI.*

Boulding, J.R. 1993a. Use of Airborne, Surface, and Borehole Geophysical Techniques at Contaminated Sites: A Reference Guide. EPA/625/R-92/007, 295 pp. Available from CERI.*

Boulding, J.R. 1993b. Subsurface Characterization and Monitoring Techniques: A Desk Reference Guide, Volume I: Solids and Ground Water (EPA/625/R-93/003a, 488 pp); Volume II: The Vadose Zone, Field Screening and Analytical Methods (EPA/625/R-93/003b, 418 pp). Available from CERI.*

Boulding, J.R. 1994. Technical Manual: Ground-Water and Wellhead Protection. EPA/625/R-93/013. Available from CERI.*

Boulding, J.R. and M.J. Barcelona. 1991a. Geochemical Characterization of the Subsurface: Basic Analytical and Statistical Concepts. In: Site Characterization for Subsurface Remediation, EPA/625/4-91/026, Chapter 7.

Boulding, J.R. and M.J. Barcelona. 1991b. Geochemical Variability of the Natural and Contaminated Subsurface Environment). In: Site Characterization for Subsurface Remediation, EPA/625/4-91/026, Chapter 8.

Boulding, J.R. and M.J. Barcelona. 1991c. Geochemical Sampling of Subsurface Solids and Ground Water. In: Site Characterization for Subsurface Remediation, EPA/625/4-91/026, Chapter 9.

Bouma, J. 1980. Field Measurement of Soil Hydraulic Properties Characterizing Water Movement Through Swelling Clay Soils. J. Hydrology 45:149-158.

(Figure and Table References)

Boutwell, S.H., S.M. Brown, B.R. Roberts, and D.F. Atwood. 1985. Modeling Remedial Actions at Uncontrolled Hazardous Waste Sites. EPA 540/2-85/001 (NTIS PB85-211357). Also published in 1986 with the same title by Noyes Data Corporation, Park Ridge, NJ.

Bouwer, E.J. and P.L. McCarty. 1984. Modeling of Trace Organics Biotransformation in the Subsurface. Ground Water 22:433-440.

Bouwer, E., J.W. Mercer, M. Kavanaugh, and F. DiGiano. 1988. Coping with Groundwater Contamination. J. Water Pollution Control Federation 60(8):1414-1428.

Birkeland, P.W. 1984. Soils and Geomorphology. Oxford University Press, New York, NY, 372 pp.

Bradbury, K.R., M.A. Muldoon, A. Zaporozec, and J. Levy. 1991. Delineation of Wellhead Protection Areas in Fractured Rocks. EPA/570/9-91-009, 144 pp.

Brady, N.C. 1974. The Nature and Properties of Soils, 8th ed. Macmillan Publishing Co., New York, NY, 637 pp.

Brown, R.H., A.A. Konoplyantsev, J. Ineson, and V.S. Kovalensky. 1983. Ground-Water Studies: An International Guide for Research and Practice. Studies and Reports in Hydrology No. 7. UNESCO, Paris.

Cartwright, K. 1968. Thermal Prospecting for Ground Water. Water Resources Research 4(2):395-401.

Castany, G. and J. Margat. 1977. Dictionnaire Français D'Hydrogéolgie. BRGM, Orléans.

Champ, D.R., J. Gulens, and R.E. Jackson. 1979. Oxidation-Reduction Sequences in Ground Water Flow Systems. Can. J. Earth Sci. 16:12-23.

Clapp, R.B. and G.M. Hornberger. 1978. Empirical Equations for Some Soil Hydraulic Properties. Water Resources Research 14:601-604.

Clark, I. 1979. Practical Geostatistics. Applied Science Publishers, London.

Compton, R.R. 1962. Manual of Field Geology. John Wiley & Sons, New York.

Coleman, D.D., W.F. Meents, C.L. Liu, and R.A. Keogh. 1977. Isotopic Identification of Leakage Gas from Underground Storage Reservoirs--a Progress Report. Petroleum Report No. 111. Illinois State Geological Survey, Urbana, IL.

Coover, M.P. and R.C. Sims. 1987. The Effect of Temperature on Polycyclic Aromatic Hydrocarbon Persistence in an Unacclimated Soil. Hazardous Waste and Hazardous Materials 4:69-82.

Daniels, J.F., L.W. Cable, and R.J. Wolf. 1970. Ground Water--Surface Water Relation during Periods of Overland Flow. U.S. Geological Survey Professional Paper 700-B.

Davis, S.N. and R.J.M. DeWiest. 1966. Hydrogeology. John Wiley & Sons, New York, NY, 463 pp.

Davies, W.E. and H.E. LeGrand. 1972. Karst of the United States. In: Karst: Important Karst Regions of the Northern Hemisphere, M. Herak and V.T. Stringfield (eds.), Elsevier, New York, Chapter 15.

(Figure and Table References)

Davis, S.N., D.J. Campbell, H.W. Bentley, and T.J. Flynn. 1985. Introduction to Ground-Water Tracers. EPA 600/2-85/022, NTIS PB86-100591. Also published under the title Ground Water Tracers in EPA/NWWA Series, National Water Well Association, Dublin, OH, 200 pp. [See also 1986 "Discussion of 'Ground Water Tracers' by Davis et al. (1985) with Emphasis on Dye Tracing, Especially in Karst Terranes" in Ground Water 24(2):253-259 and 24(3):396-397, and reply by Davis in Ground Water 24(3):398-399]

Deutsch, M. 1961. Incidents of Chromium Contamination of Ground Water in Michigan. In: Proceedings of the 1961 Symposium, Ground Water Contamination, U.S. Public Health Service Tech. Rept. W61-5, pp. 98-103.

Deutsch, M. 1963. Ground-Water Contamination and Legal Controls in Michigan. U.S. Geological Survey Water-Supply Paper 1691.

Deutsch, M. 1965. Natural Controls Involved in Shallow Aquifer Contamination. Ground Water 3(3):37-40.

Erdélyi, M. and J. Gálfi. 1988. Surface and Subsurface Mapping in Hydrogeology. Wiley-Interscience, New York, NY, 384 pp.

Everett, L.G., L.G. Wilson, and E.W. Hoylman. 1983. Vadose Zone Monitoring for Hazardous Waste Sites. EPA/600/X-83/064 (NTIS PB84-212752). [Also published in 1984 by Noyes Data Corporation, Park Ridge, NJ.]

Fetter, Jr., C.W. 1994. Applied Hydrogeology, 3rd ed. Macmillan, New York, NY, 691 pp.

Fetter, Jr., C.W. 1981. Determination of the Direction of Groundwater Flow. Ground Water Monitoring Review 1(3):28-31.

Flatman, G.T. 1986. Design of Soil Sampling Programs: Statistical Considerations. In: Quality Control in Remedial Site Investigation: Hazardous and Industrial Solid Waste Testing, 5th volume, C.L. Perket (ed.), ASTM STP 925, American Society for Testing and Materials, Philadelphia, PA, pp. 43-56.

Freeze, R.A. and J.A. Cherry. 1979. Groundwater. Prentice-Hall Publishing Co., Englewood Cliffs, NJ, 604 pp.

Georgia Department of Natural Resources. 1992. The Georgia Wellhead Protection Plan (September, 1992). Georgia Department of Natural Resources, Environmental Protection Division, Atlanta, GA.

Geraghty, J.J. and D.W. Miller. 1985. Fundamentals of Ground-Water Contamination, Short Course Notes. Geraghty and Miller, Inc., Syosset, NY.

Gibb, J.P., R.M. Schuller, and R.A. Griffin. 1981. Procedures for the Collection of Representative Water quality Data from Monitoring Wells. ISWS/IGS Cooperative Ground Water Report 7. Illinois State Water Survey, Champaign, IL.

Gilbert, R.O. 1987. Statistical Methods for Environmental Pollution Monitoring. Van Nostrand Reinhold, New York, NY.

Guswa, J.H., W.J. Lyman, A.S. Donigian, Jr., T.Y.R. Lo, and E.W. Shanahan. 1984. Groundwater Contamination and Emergency Response Guide. Noyes Publications, Park Ridge, NJ.

(Figure and Table References)

Hahne, H.C.H. and W. Kroontje. 1973. Significance of pH and Chloride Concentration on Behavior of Heavy Metal Pollutants: Mercury(II), Cadmium (II), Zinc(II), and Lead(II). J. Environ. Qual. 2(4):444-450.

Heath, R.C. 1983. Basic Ground-Water Hydrology. U.S. Geological Survey Water-Supply Paper 2220, 85 pp. Republished in a 1984 edition by National Water Well Association, Dublin, OH.

Heath, R.C. and F.W. Trainer. 1981. Introduction to Ground Water Hydrology, 2nd ed. John Wiley & Sons, New York, 284 pp.

Ho, J.S-Y. 1983. Effect of Sampling Variables on Recovery of Volatile Organics in Water. J. Am. Water Works Ass. 12:583-586.

Horsley, S. 1990. Water Resource Management Plan for Nantucket Island, Massachusetts--A Case Study. Ground Water Management 3:3-20 (Proc. Focus Conf. on Eastern Regional Ground Water Issues).

Huddleston, R.L., C.A. Bleckman, and J.R. Wolfe. 1986. Land Treatment Biological Degradation Processes. In: Land Treatment: A Hazardous Waste Management Alternative, R.C. Loehr and J.F. Malina (eds.), Water Resources Symposium No. 13, University of Texas Press, Austin, TX, pp. 41-61.

Hutzler, N.J., B.E. Murphy, and J.S. Gierke. 1990. State of Technology Review: Soil Vapor Extraction Systems. EPA/600/2-89/024 (NTIS PB89-195184).

Idaho Wellhead Protection Work Group. 1992. Idaho Wellhead Protection Plan (Draft). Division of Environmental Quality, Idaho Department of Health and Welfare, Boise, ID, 86 pp. + appendices.

Jenkinson, D.S. and J.H. Rayner. 1977. The Turnover of Soil Organic Matters in Some of the Rothamsted Classical Experiments. Soil Science 123:298-305.

Jessen, R.J. 1978. Statistical Survey Techniques. John Wiley & Sons, New York, NY.

Johnson, R.L., C.D. Palmer, and W. Fish. 1989. Subsurface Chemical Processes. In: Transport and Fate of Contaminants in the Subsurface, EPA/625/4-89/019, Chapter 5.

Karous, M. and S. Mareš. 1988. Geophysical Methods in Studying Fracture Aquifers. Charles University, Prague, 93 pp.

Keely, J.F. 1987. The Use of Models in Managing Ground-Water Protection Programs. EPA 600/8-87/003 (NTIS PB87-166203), 72 pp.

Keely, J.F. 1989. Performance Evaluations of Pump-and-Treat Remediations. Superfund Groundwater Issue Paper, EPA/540/4-89/005, 19 pp.

Keswick, B.H., D. Wang, and C.P. Gerba. 1982. The Use of Microorganisms as Ground-Water Tracers: A Review. Ground Water 20(2):142-149.

Keys, W.S. 1990. Borehole Geophysics Applied to Ground-Water Investigations. U.S. Geological Survey Techniques of Water-Resource Investigations TWRI 2-E2, 150 pp.

(Figure and Table References)

Knox, R.C., L.W. Canter, D.F. Kincannon, E.L. Stover, and C.H. Ward. 1984. State-of-the Art of Aquifer Restoration. EPA 600/2-84/182a&b (NTIS PB85-181071 and PB85-181089).

Kolm, K. E. 1993. Conceptualization and Characterization of Hydrologic Systems. GWMI 93-01, International Ground Water Modeling Center, Golden, CO, 58 pp.

Kreitler, C.W. and R.K. Senger. 1991. Wellhead Protection Strategies for Confined-Aquifer Settings. EPA/570/9-91-008, 168 pp.

Leach, L.W., F.P. Beck, J.T. Wilson and D.H. Kampbell. 1988. Aseptic Subsurface Sampling Techniques for Hollow-Stem Auger Drilling. In: 2nd Nat. Outdoor Action Conf. on Aquifer Restoration, Ground Water Monitoring and Geophysical Methods, National Water Well Association, Dublin, OH, pp. 31-51.

Lee, M.D., J.M. Thomas, R.C. Borden, P.B. Bedient, C.H. Ward, and J.T. Wilson. 1988. Biorestoration of Aquifers Contaminated with Organic Compounds. CRC Critical Reviews in Environmental Control 18(1):29-87.

Leenheer, J.A. and R.L. Malcolm. 1973. Case History of Subsurface Waste Injection of an Industrial Organic Waste. In: Symposium on Underground Waste Management and Artificial Recharge, J. Braunstein (ed.), Pub. No. 110, Int. Assoc. of Hydrological Sciences, pp. 565-584.

LeGrand, H.E. 1965. Patterns of Contaminated Zones of Water in the Ground. Water Resources Research 1(1):83-95.

Marrin, D.L. 1987. Soil Gas Sampling Strategies: Deep vs. Shallow Aquifers. In: Proc. 1st Nat. Outdoor Action Conf. on Aquifer Restoration, Ground Water Monitoring and Geophysical Methods, National Water Well Association, Dublin, OH, pp. 437-454.

Marrin, D.L. and W.B. Kerfoot. 1988. Soil Gas Surveying Techniques. Environ. Sci. Technol. 22(7)-740-745.

Mausbach, M.J. and R.D. Nielsen. 1991. Some Concepts Concerning Soil Site Assessment for Water Quality. Soil Survey Horizons 31(1):18-25.

McArdle, J.L., M.M. Arozarena, and W.E. Gallagher. 1987. A Handbook on Treatment of Hazardous Waste Leachate. EPA/600/8-87/006 (NTIS PB87-152328).

McDowell-Boyer, L.M., J.R. Hunt, and N. Sitar. 1986. Particle Transport Through Porous Media. Water Resources Research 22:1901-1921.

Mercer, J.W. and C.R. Faust. 1981. Ground-Water Modeling. National Water Well Association, Dublin, OH, 60 pp.

Mercer, J.W. and C.P. Spalding. 1991a. Site Characterization Overview. In: Site Characterization for Subsurface Remediation, EPA/625/4-91/026, Chapter 2.

Mercer, J.W. and C.P. Spalding. 1991b. Geologic Aspects of Site Remediation. In: Site Characterization for Subsurface Remediation, EPA/625/4-91/026, Chapter 3.

Mercer, J.W. and C.P. Spalding. 1991c. Characterization of Water Movement in the Saturated Zone. In: Site Characterization for Subsurface Remediation, EPA/625/4-91/026, Chapter 4.

(Figure and Table References)

Mill, T. 1980. Data Needed to Predict the Environmental Fate of Organic Chemicals. In: Dynamics, Exposure and Hazard Assessment of Toxic Chemicals, R. Haque (ed.), Ann Arbor Science, Ann Arbor, MI, pp. 297-322.

Miller, D.W. 1985. Chemical Contamination of Ground Water. In: Ground Water Quality, C.H. Ward, W. Giger, and P.L. McCarty, (eds.), Wiley Interscience, New York, pp. 39-52.

Mills, W.B. et al. 1985. Water Quality Assessment: A Screening Procedure for Toxic and Conventional Pollutants, Part II. EPA 600/6-85/002b.

Muldoon, M. and J. Payton. 1993. Determining Wellhead Protection Area Boundaries: An Introduction. Publication WR313-92, Wisconsin Department of Natural Resources, Madison, WI, 24 pp.

Naymik, T.G. and M.E. Sievers. 1983. Ground-Water Tracer Experiment (II) at Sand Ridge State Forest, Illinois. ISWS Contract Report 334, Illinois State Water Survey, Champaign, IL.

Nielsen, C.M. 1983. Remedial Methods Available in Areas of Ground Water Contamination. In: Proc. 6th Nat. Ground Water Quality Symp., National Water Well Association, Dublin, OH, pp. 219-227.

Nielsen, D.M. and R. Schalla. 1991. Design and Installation of Ground-Water Monitoring Wells. In: Practical Handbook of Ground-Water Monitoring, D.M. Nielsen (ed.), Lewis Publishers, Chelsea, MI, pp. 239-331.

Noake, K.D. 1988. Guide to Contamination Sources for Wellhead Protection (Draft). Massachusetts Department of Environmental Quality Engineering, Boston, MA.

Nyer, E.K. 1985. Groundwater Treatment Technology. Van Nostrand Reinhold, New York, NY.

Office of Technology Assessment (OTA). 1984. Protecting the Nation's Groundwater from Contamination, Vols. I and II. OTA-0-233 and OTA-0-276. OTA, Washington, DC.

Office of Information Resource Management (OIRM). 1992. Geographic Information Systems (GIS) Guidelines Document. OIRM 88-01. U.S. Environmental Protection Agency, Washington, DC.

Olmsted, F.H. and A.G. Hely. 1962. Relation between Ground Water and Surface Water in Brandywine Creek Basin, Pennsylvania. U.S. Geological Survey Professional Paper 417-A.

Oremland, R.S. 1988. Biogeochemistry of Methanogenic Bacteria. In: Biology of Anaerobic Microorganisms, A.J.B. Zehnder (ed.), John Wiley & Sons, New York, pp. 641-706.

PACE. 1985. The Persistence of Polynuclear Aromatic Hydrocarbons in Soil. PACE Report No. 85-2. Petroleum Association for Conservation of the Canadian Environment, Ottawa, Ontario, Canada.

Paul, E.A. and F.E. Clark. 1989. Soil Microbiology and Biochemistry. Academic Press, San Diego, CA.

Palmer, C.D. and W. Fish. 1991. Physicochemical Processes: Inorganic Contaminants. In: Site Characterization for Subsurface Remediation, EPA/625/4-91/026, Chapter 12.

Palmer, C.D. and R.L. Johnson. 1989a. Physical Processes Controlling the Transport of Contaminants in the Aqueous Phase. In: Transport and Fate of Contaminants in the Subsurface, EPA/625/4-89/019, Chapter 2.

Palmer, C.D. and R.L. Johnson. 1989b. Physical Processes Controlling the Transport of Non-Aqueous Phase Liquids in the Subsurface. In: Transport and Fate of Contaminants in the Subsurface, EPA/625/4-89/019, Chapter 3.

Palmer, C.D. and R.L. Johnson. 1991. Physicochemical Processes: Organic Contaminants. In: Site Characterization for Subsurface Remediation, EPA/625/4-91/026, Chapter 10.

Paul, E.A. and F.E. Clark. 1989. Soil Microbiology and Biochemistry. Academic Press, New York, NY.

Park, K.S., R.C. Sims, R.R. Dupont, W.J. Doucette, and J.E. Matthews. 1990. Fate of PAH Compounds in Two Soil Types: Influence of Volatilization, Abiotic Loss and Biological Activity. Environ. Toxicol. Chem. 9:187-195.

Pettyjohn, W.A., J.R.J. Studlick, R.C. Bain, and J.H. Lehr. 1979. A Ground-Water Quality Atlas of the United States. National Water Well Association, Dublin, OH.

Pohland, F.G. and S.R. Harper. 1986. Critical Review and Summary of Leachate and Gas Production from Landfills. EPA/600/2-86/073 (NTIS PB86-240181).

Pohlmann, K.F. and J.W. Hess. 1988. Generalized Ground Water Sampling Device Matrix. Ground Water Monitoring Review 8(4):82-84.

Portland Cement Association. 1973. PCA Soil Primer. Engineering Bulletin EB007.045, Portland Cement Association, Skokie, IL, 39 pp.

Puls, R.W. and M.J. Barcelona. 1989. Ground Water Sampling for Metals Analyses. Superfund Ground Water Issue Paper, EPA/540/4-89/001, 6 pp. Available from CERI.*

Quinlan, J.F., G.J. Davies, and S.R.H. Worthington. 1992. Rationale for the Design of Cost-Effective Groundwater Monitoring Systems in Limestone and Dolomite Terranes: Cost Effective as Conceived is Not Cost Effective as Built if the System Design and Sampling Frequency Inadequately Consider Site Hydrogeology. In: Proc. Annual Waste Testing and Water Quality Assurance Symposium, pp. 552-570.

Rochkind, M.L., J.W. Blackburn, and G. Sayler. 1986. Microbial Decomposition of Chlorinated Aromatic Compounds. EPA/600/2-86/090 (NTIS PB87-116943).

Ruhe, R.V. and D.H. Walker. 1968. Hillsope Models and Soil Formation: I. Open Systems. Trans. 9th Int. Congress Soil Sci. 4:551-560.

Saines, M. 1981. Errors in Interpretation of Ground Water Level Data. Ground Water Monitoring Review 2(1):56-61.

Sale, T. and K. Piontek. 1989. In Situ Removal of Waste Wood-Treating Oils from Subsurface Materials. Presented at Forum on Remediation of Wood Preserving Sites, Technical Assistance to U.S. EPA Region IX (Edwin Barth, U.S. EPA, Cincinnati, OH, and John Matthews, U.S. EPA, Ada, OK, Technical Coordinators).

(Figure and Table References)

Sara, M.N. 1994. Standard Handbook of Site Assessment for Solid and Hazardous Waste Facilities. Lewis Publishers, Boca Raton, FL, 976 pp.

Sauter, M. 1992. Assessment of Hydraulic Conductivity in a Karst Aquifer at Local and Regional Scale. Ground Water Management 10:38-55 (Proc. 3rd Conf. on Hydrogeology, Ecology, Monitoring and Management of Ground Water in Karst Terranes).

Scalf, M.R., J.F. McNabb, W.J. Dunlap, R.L. Cosby, and J. Fryberger. 1981. Manual of Ground-Water Quality Sampling Procedures. EPA/600/2-81/160 (NTIS PB82-103045). Also published in NWWA/EPA Series, National Water Well Association, Dublin OH.

Schuller, R.M., J.P. Gibb, and R.A. Griffin. 1981. Recommended Sampling Procedures for Monitoring Wells. Ground Water Monitoring Review 1(1):42-46.

Sims, R.C. 1986. Loading Rates and Frequencies for Land Treatment Systems. In: Land Treatment-A Hazardous Waste Management Alternative, R.C. Loehr, and J.F. Malina (eds.), Water Resources Symposium No. 13, University of Texas Press, Austin, TX, pp. 151-170.

Sims, R.C., and M.R. Overcash. 1983. Fate of Polynuclear Aromatic Compounds (PNAs) in Soil-Plant Systems. Residue Reviews 86:1-68.

Sims, R.C. and J.L. Sims. 1991a. Soil and Ground-Water Remediation: Basic Approaches. In: Site Characterization for Subsurface Remediation, EPA/625/4-91/026, Chapter 14.

Sims, R.C. and J.L. Sims. 1991b. Remediation Techniques for Contaminated Soils. In: Site Characterization for Subsurface Remediation, EPA/625/4-91/026, Chapter 15.

Sims, R.C. and J.L. Sims. 1991c. Aquifer Restoration. In: Site Characterization for Subsurface Remediation, EPA/625/4-91/026, Chapter 16.

Sims, R.C., D.L. Sorensen, J.L. Sims, J.E. McLean, R. Mahmood, and R.R. Dupont. 1984. Review of In-Place Treatment Technologies for Contaminated Surface Soils-Volume 2: Background Information for In-Situ Treatment. EPA-540/2-84-003b (NTIS PB85-124899).

Sims, R.C., D. Sorensen, J.L. Sims, J. McLean, R.J Mahmood, R. Dupont, and J. Jurinak. 1986. Contaminated Surface Soils In-Place Treatment Techniques. Pollution Technology Review No. 132. Noyes Publications, Park Ridge, NJ, 536 pp.

Sisk, S.W. 1981. NEIC Manual for Groundwater/Subsurface Investigations at Hazardous Waste Sites. EPA/330/9-81/002 (NTIS PB82-103755), 213 pp.

Skoog, D.A. 1985. Principles of Instrumental Analysis, 3rd ed. Saunders College Publishing, Philadelphia, PA.

Smart, P.L. and I.M.S. Laidlaw. 1977. An Evaluation of Some Fluorescent Dyes for Water Tracing. Water Resources Research 13(1):15-33.

Soil Conservation Service (SCS). 1971. Guide for Interpreting Engineering Uses of Soils. U.S. Government Printing Office Stock No. 0107-0332.

Suflita, J.M.. 1989. Microbial Ecology and Pollutant Biodegradation in Subsurface Ecosystems. In: Transport and Fate of Contaminants in the Subsurface, EPA/625/4-89/019, Chapter 7.

(Figure and Table References)

Sun, R.J. and J.B. Weeks. 1991. Bibliography of Regional Aquifer-System Analysis Program of the U.S. Geological Survey, 1978-1991. U.S. Geological Survey Water-Resources Investigations Report 91-4122, 92 pp.

Struckmeier, W., G.B. Engelen, M.S. Galitzin, and R.K. Shakchnova. 1986. Methods of Representation of Water Data. In: Developments in the Analysis of Groundwater Flow Systems, G.B. Engelen and G.P. Jones (eds.), Int. Ass. of Hydrological Sciences Pub. No. 163, pp. 45-63.

Todd, D.K. 1980. Groundwater Hydrology, 2nd ed. John Wiley & Sons, New York, 535 pp.

Torstensson, B.-A. 1984. A New System for Ground Water Monitoring. Ground Water Monitoring Review 4(4):131-138.

U.S. Environmental Protection Agency (EPA). 1977. The Report to Congress, Waste Disposal Practices and Their Effects on Ground Water. EPA/570/9-77/001 (NTIS PB265-081). [Also published in 1980 by Premier Press with D.W. Miller as editor]

U.S. Environmental Protection Agency (EPA). 1985. Remedial Action at Waste Disposal Sites. EPA/625/6-85/006 (NTIS PB87-201034).

U.S. Environmental Protection Agency (EPA). 1986. Test Methods for Evaluating Solid Waste, 3rd ed., Vol. II Field Manual Physical/Chemical Methods. EPA/530/SW-846 (NTIS PB88-239223); First update, 3rd ed. EPA/530/SW-846.3-1 (NTIS PB89-148076).

U.S. Environmental Protection Agency (EPA). 1987a. Handbook: Ground Water. EPA/625/6-87/016, 212 pp. [Out-of-print; extensively revised second edition available from CERI* in two volumes: I: Ground Water and Contamination (EPA/625/6-90/106a), II: Methodology (EPA/625/6-90/16b)]

U.S. Environmental Protection Agency (EPA). 1987b. Guidelines for Delineation of Wellhead Protection Areas. EPA/440/6-87-010 (NTIS PB88-111430).

U.S. Environmental Protection Agency (EPA). 1987c. Estimating Releases and Waste Treatment Efficiencies for the Toxic Chemical Release Inventory Form. EPA/560/4-88-002. Available from EPCRI Hotline, see Table 5-3.

U.S. Environmental Protection Agency (EPA). 1988a. Selection Criteria for Mathematical Models Used in Exposure Assessments: Ground-Water Models. EPA 600/8-88/075 (NTIS PB88-248752).

U.S. Environmental Protection Agency (EPA). 1988b. Guidance on Remedial Actions for Contaminated Ground Water at Superfund Sites. EPA/540/G-88/003 (NTIS PB89-184618).

U.S. Environmental Protection Agency (EPA). 1988c. RCRA Corrective Action Interim Measures Guidance--Interim Final. EPA/530-SW-88-029/OSWER 9902.4. Available from RCRA Hotline, see Table 5-3.

U.S. Environmental Protection Agency (EPA). 1989a. Guidance for Conducting Remedial Investigations and Feasibility Studies Under CERCLA. EPA/540/G-89/004 (NTIS PB89-184626), 195 pp.

U.S. Environmental Protection Agency (EPA). 1989b. Wellhead Protection Programs: Tools for Local Governments. EPA/440/6-89-002, 50 pp. Available from Drinking Water Hotline, see Table 5-3.

(Figure and Table References)

U.S. Environmental Protection Agency (EPA). 1989c. Technology Screening Guide for Treatment of CERCLA Soils and Sludges.

U.S. Environmental Protection Agency (EPA). 1989d. Guide for Conducting Treatability Studies Under CERCLA--Interim Final. EPA/540/2-89/058 (NTIS PB90-249772), 118 pp.

U.S. Environmental Protection Agency (EPA). 1990. The Superfund Innovative Technology Evaluation Program: Technology Profiles. EPA/540/5-90/006, 171 pp.

U.S. Environmental Protection Agency (EPA). 1991. Access EPA, 1991 Edition. EPA/MSD-91-100. U.S. Government Printing Office S/N 055-000-00378-5, Washington DC, 505 pp.

U.S. Environmental Protection Agency (EPA). 1993. Wellhead Protection: A Guide for Small Communities. Seminar Publication EPA/625/R-93-002. Available from CERI.*

Valentine, R.L. 1986. Nonbiological Transformation. In: Vadose Zone Modeling of Organic Pollutants, S.C. Hern and S.M. Melancon (eds.), Lewis Publishers, Chelsea, MI, pp. 223-243.

van der Heijde, P.K.M., A.I. El-Kadi, and S.A. Williams. 1988. Groundwater Modeling: An Overview and Status Report. EPA/600/2-89/028 (NTIS PB89-224497).

Viessman, Jr., W., T.E. Harbaugh, and J.W. Knapp. 1972. Introduction to Hydrology, 2nd ed. Intext Educational Publishers, New York, NY, 415 pp. [2nd edition published 1977].

Wagner, K., et al. 1986. Remedial Action Technology for Waste Disposal Sites, 2nd edition. Noyes Data Corporation, Park Ridge, NJ. [10 authors]

Walker, E.H. 1956. Ground-Water Resources of the Hopkinsville Quadrangle, Kentucky. U.S. Geological Survey Water-Supply Paper 1328, 98 pp.

Way, D.S. 1973. Terrain Analysis: A Guide to Site Selection Using Aerial Photographic Interpretation. Dowden, Hutchinson, and Ross, Inc., Stroudsburg, PA, 392 pp.

Zoeteman, B.C.J. 1985. Overview of Contaminants in Ground Water. In: Ground Water Quality, C.H. Ward, W. Giger, and P.L. McCarty, (eds.), Wiley-Interscience, New York, pp. 27-37.

* See appendix introduction for phone number for obtaining documents from CERI.

** Reprinted by permission of Ground Water Publishing Company (formerly Water Well Journal Publishing Company), all rights reserved.

*** Reprinted by permission of National Ground Water Association (formerly National Water Well Association), all rights reserved.

(Figure and Table References)

Key Word Index

This index lists the starting page of sections covering major key word topics. A notation of (T) after a page number indicates the starting page of a reference index table covering the topic.

Aquifers (see also Karst), 110(T)
  Anisotropy, 79, 399
  Boundaries, 85, 398
  Confined/Unconfined, 78, 402
  Fractured rock, 80, 404
  Heterogeneity, 79, 399
  Tests, 388

Bioremediation, 718(T)
  Soil, 745, 757(T)
  Ground water, 792, 796(T)

Chemical Analysis, 549(T)
  Field screening, 525
  Methods, 276, 525
  QA/QC, 279

Computer Modeling, 562, 602(T), 912(T), 917(T)
  Ground water, 573, 581
  Other applications, 589, 616(T), 912(T)
  Vadose zone, 602(T)
  Remediation, 710

Contaminant chemistry, 157(T), 219(T)
  Biotransformation, 152
  Nonaqueous phase liquids, 128, 186
  Retardation, 192

Contaminant sources, 172, 206(T), 641,

Contaminant transport, 169 219(T)
  Ground water, 179
  Plume behavior, 198
  Processes, 182

Geochemistry (see also Contaminant chemistry), 117, 157(T)
  Basic concepts, 118
  Distribution processes, 123
  pH, 137, 200
  Eh, 141, 200
  Transformation processes, 129

Field Investigations, 237, 529(T)
  Planning process, 238
  Existing information sources, 246, 288(T)
  Data reliability, 269
  Geophysical methods, 303
  Hydrology, 371
  Maps, 18, 377
  Tracer tests, 439

Geology, 1, 40(T), 199
  Geologic materials, 2
  Geologic processes, 12

Geophysical methods, 303
  Borehole, 322, 357(T)
  Selection, 331
  Surface, 312, 338(T)

Geomorphology, 28, 40(T)

Ground Water (see also Aquifers), 97(T)
  Darcy's law, 88
  Fundamental concepts, 76
  Geologic settings, 32, 110(T)
  Pumping, 92
  Quality, 32
  Remediation, 761
  Tracers, 439, 480(T)
  Storage, 86

Ground water remediation, 761, 718(T), 796(T)
  Collection, 772
  Containment, 769
  In situ, 790
  Pump and treat, 779
  Source control, 766

Hydrologic cycle, 60, 65
  Precipitation, 61
  Infiltration, 63
  Evapotranspiration, 64

Karst, 30, 54(T), 404, 467

Microbial Ecology, 148, 157(T)

Monitoring, 493, 529(T)
  Ground water, 496
  Plans, 253

Monitoring (cont.)
  Remediation, 709
  Wells, 504
  Vadose zone, 372, 494

Pollution prevention, 623, 678(T)
  Containment, 769
  General approaches, 624
  WHPAs, 626, 652

Remediation
  Planning, 696, 678(T)
  Ground water, 761, 796(T)
  Soil, 731, 757(T)

Remote sensing, 304, 308, 338(T)

Risk assessment, 644, 661(T)

Sampling, 493, 529(T)
  Plans, 253
  Ground water, 520
  Microbiology, 519
  Soil, 514

Soil, 40(T)
  Basic concepts, 20
  Sampling, 514
  SVE, 732
  Tracers, 439
  Bioremediation, 745
  Other remediation approaches, 750

Statistics, 281, 295(T)

Surface Water, 60, 97(T)
  Flow characteristics, 65
  Drainage basins, 66
  Stream types, 70

Vadose zone, 97(T), 529(T)
  Basic concepts, 72
  Hydrologic parameters, 372
  Monitoring, 372, 494
  Sampling, 514

Wellhead Protection, 678(T)
  Delineation methods, 626
  WHPA management, 652